SECOND EDITION

Organic Light-Emitting Materials and Devices

SECOND EDITION

Organic
Light-Emitting
Materials
and Devices

EDITED BY

ZHIGANG RICK LI

CRC Press
Taylor & Francis Group
Boca Raton London New York

CRC Press is an imprint of the
Taylor & Francis Group, an **informa** business

CRC Press
Taylor & Francis Group
6000 Broken Sound Parkway NW, Suite 300
Boca Raton, FL 33487-2742

First issued in paperback 2017

© 2015 by Taylor & Francis Group, LLC
CRC Press is an imprint of Taylor & Francis Group, an Informa business

No claim to original U.S. Government works

ISBN-13: 978-1-4398-8223-8 (hbk)
ISBN-13: 978-1-138-74969-6 (pbk)

**Visit the Taylor & Francis Web site at
http://www.taylorandfrancis.com**

**and the CRC Press Web site at
http://www.crcpress.com**

To my mother, wife, and daughter

To my alma mater,

Beijing Institute of Technology, China and LOE du CNRS/University of Paul Sabatier, France

Contents

Preface

In my recent trans-Pacific flight, I had plenty of time to think about the evolution of our everyday display technology. After only eight short years since the first edition of this book was published, display technology has undergone enormous transformation, which is still continuing. Cathode ray tube (CRT) display technology, which dominated for several decades, suddenly disappeared in front of us. Superior organic light-emitting device (OLED) display has moved to center stage. During my trip, I even saw that my 85-year-old mother had started to play with a smartphone, communicating with her sons, grandsons, daughters, and granddaughters in other parts of the world. I could see the enormous joy in her face, brought forth by the new technology. Everywhere I went, I could see people using OLED smartphones. This is thanks to the discovery of organic semiconductors by Nobel laureates A. Heeger, A. MacDiarmid, and H. Shirakawa, and the invention of the first efficient OLED by C. W. Tang and S. VanSlyke. Indeed, OLEDs possess a number of advantages over conventional display devices, such as high brightness and contrast, high luminous efficiency, fast response time, wide viewing angle, low power consumption, and light weight. In addition, the new technologies offer the potential of low manufacturing cost. OLED displays can be fabricated on large-area substrates (including flexible substrates) and offer a virtually unlimited choice of colors. The technological promise of these unique characteristics places OLEDs at the forefront of research efforts of government agencies, industries, and universities. In fact, when in 2012 *Applied Physics Letter* listed the top 50 most cited papers in the journal for the past 50 years, OLED-related publications were ranked as numbers 1, 9, 15, 16, 30 and 35. Many major industrial electronics giants and newcomers have invested heavily in OLED research and development. As a result, a stream of new OLED products has reached the marketplace and a number of large-scale manufacturing facilities are now under construction. Although the field is growing rapidly and its impact is both pervasive and far-reaching, major challenges still remain. Overcoming these drawbacks will require further multidisciplinary studies and breakthroughs.

Until today, several books on related topics have provided the readers with essential information in the field of organic electroluminescence. However, none of these could serve as a comprehensive guide. Our aim is to provide readers with a single source of information covering all aspects of OLEDs, including the systematic investigation of organic light-emitting materials, device physics and engineering, and so on. In this spirit, we titled this book *Organic Light-Emitting Materials and Devices* (second edition)—a compilation of the progress made in recent years and of the challenges facing the future development of OLED technology.

Ten chapters by internationally recognized academic and industrial experts in their respective fields offer a broad perspective of interdisciplinary topics uniting organic materials synthesis with device physics and engineering. Chapter 1 introduces the history, fundamental physics, and potential applications of OLEDs. OLEDs can be divided into two categories: small-molecule and polymer-based light-emitting diodes (SMOLEDs and PLEDs). From the basic structure point of view, both devices employ multilayered architectures with the anode; hole-transporting, emissive, and electron-transporting layers; and the cathode. Developing high-efficiency OLEDs poses a great challenge for material scientists, requiring an understanding of the physics beyond device operation, and of structure–property relationships to allow for new material design. From this perspective,

Chapter 2 through Chapter 6 provide a comprehensive review of the synthesis, properties, and device performance of electroluminescent materials used in OLEDs. Chapter 2 deals with polymer light-emitting materials, subdivided into its most important classes: poly(*p*-phenylenevinylene)s (PPVs), polyfluorenes (PFs), polythiophenes (PTs), and other conjugated and nonconjugated electroluminescent polymers. It describes the progress and the current state of understanding of molecular design in the field, exemplifying >600 light-emitting polymers, and highlighting the most efficient materials and devices. Chapter 3 reviews small molecules–based OLEDs, specifically describing hole- and electron-injection and electron-transport materials, electron- and hole-blocking materials, sensitizers, and fluorescent and phosphorescent light emitters. Solution-processable phosphorescent polymer LEDs are described in Chapter 4, which starts with a brief discussion of the energy transfer processes. Chapter 5 depicts the progress of polarized OLEDs. Chapter 6 is dedicated to anode materials and focuses on novel transparent anode materials with a brief review of other actively investigated anode materials used in transparent OLED devices. Chapter 7 provides readers with well-structured information on vapor deposition manufacturing techniques employed in OLED fabrication. Chapter 8 and part of Chapter 6 focus on flexible display, a unique property of OLED-based display. Chapter 9 describes the backplane circuit technology for organic light-emitting displays. Chapter 10 describes microstructural characterization and performance measurement techniques currently used in the OLED field. The book includes abundant diagrams, device configurations, and molecular structures clearly illustrating the described ideas. Within space limitations, this book provides a comprehensive overview of the field and can serve as a primary reference source to those needing additional information in any particular subarea in organic electroluminescence. Furthermore, the described materials and principles of device physics have broad applications in other areas of organic electronics. A balance between the academic and industrial points of view is presented, enhanced by the diverse backgrounds of the contributing authors. This book should attract the attention of multidisciplinary researchers (e.g., materials scientists, synthetic chemists, solid-state physicists, and electronic device engineers), as well as industrial managers and patent lawyers engaged in OLED-related business areas.

The successful birth of this book is attributed to the hard work of our author teams. I take this opportunity to thank all contributors for their excellent work, from the bottom of my heart. I would like to thank Prof. Y. Cao (South China University and Technology) and Dr. J. Burroughes (CTO, CDT/Sumitomo, UK) for their discussion, first edition coeditor Prof. Meng (Beijing University at Shenzhen) for his contribution, and A. Gasque and J. Jurgensen at CRC Press/Taylor & Francis Group for their valuable support and help during the editing of this book. Finally, my love goes to my wife, daughter, mother, brothers, and sisters for their continuous support and encouragement—now I will have more time to be with you in person or through an OLED display from thousands of kilometers away.

Editors

Dr. Zhigang Rick Li, second edition editor and first edition coeditor, is a researcher at DuPont Central Research and Development, Wilmington, Delaware. He earned a BS degree in optics from Beijing Institute of Technology, Beijing, China, and a PhD degree in applied physics from the Laboratoire d'Optique Electronique du CNRS/Universite de Paul Sabatier, Toulouse, France. Dr. Li is a recipient of the Sino-France Abroad Study Awards.

For more than 20 years, he has worked in the field of optoelectronics, including OLED, magneto-optical recordings, and solar cells. He is the author or coauthor of more than 70 professional publications and a book chapter, and the editor or coeditor of two books. He is the organizer/co-organizer/session chair of many national and international conferences and gave numerous invited talks at the national/international meeting. He and his colleagues proposed the "nano-Ag colloids assisted tunneling" current conduction model of front-side metallization contact of p- and n-type crystalline silicon solar cells.

Dr. Hong Meng, first edition coeditor, was a research chemist at DuPont Central Research and Development, Wilmington, Delaware. His research interests are design and synthesis of conjugated organic materials and their applications in organic electronics, particularly organic thin film transistors and organic light-emitting diodes. He has contributed over 40 peer-reviewed journal articles, 25 conference papers, and 4 book chapters, and has filed several patents.

Dr. Meng received his PhD degree from the University of California, Los Angeles, under the supervision of Prof. Fred Wudl in 2002. Before joining DuPont Company, he pursued internship training at Lucent Technologies, Bell Laboratories, under Prof. Zhenan Bao (now at Stanford University) in the field of organic electronics. He is currently a professor at Beijing University at Shenzhen, China.

Contributors

Peter F. Carcia
University of Delaware
Newark, Delaware

Xu Han
DuPont China R&D
Shanghai, China

Norman Herron
DuPont Company
Wilmington, Delaware

Linfeng Lan
South China University of Technology
Guangzhou, China

Zhigang Rick Li
DuPont CR&D
Wilmington, Delaware

Mang-Mang Ling
Applied Materials
Sunnyvale, California

Hong Meng
Beijing University at Shenzhen
Shenzhen, China

Jeff Meth
DuPont Company
Wilmington, Delaware

Dmitrii F. Perepichka
Centre Energie
Materiaux et Telecommunications
Quebec, Canada

Igor F. Perepichka
Bangor University
Gwynedd, United Kingdom

Daniel Steiger
Ethicon Inc.
Somerville, New Jersey

Shijian Su
South China University of Technology
Guangzhou, China

Jian Wang
South China University of Technology
Guangzhou, China

Lei Wang
South China University of Technology
Guangzhou, China

Qing Wang
South China University of Technology
Guangzhou, China

Michael S. Weaver
Universal Display Corporation
Ewing, New Jersey

Christoph Weder
University of Fribourg
Marly, Switzerland

Weijing Wu
South China University of Technology
Guangzhou, China

Fred Wudl
University of California
Santa Barbara, California

Shidi Xun
DuPont CR&D
Wilmington, Delaware

Gang Yu
CBrite Inc.
Goleta, California

Furong Zhu
Hong Kong Baptist University
Hong Kong, China

1

Organic Light-Emitting Devices and Their Applications for Flat-Panel Displays

Qing Wang, Gang Yu, and Jian Wang

CONTENTS

1.1 Introduction

The electroluminescence (EL) phenomenon was first discovered in a piece of carborundum (SiC) crystal, by H.J. Round in 1907.[1] Commercial research into light-emitting diode (LED) technology started in the early 1960s, when Nick Holonyak, Jr., created the first inorganic LED in 1962.[2,3] Work on gallium arsenide phosphide (GaAsP) led to the introduction of the first mass-produced commercial 655-nm red LEDs in 1968, by Hewlett-Packard and Monsanto. In the 1950s, Bernanose first observed EL in organic material by applying a high-voltage alternating-current field to crystalline thin films of acridine orange and quinacrine.[4,5] The direct current–driven EL cell using single crystals of anthracene was first demonstrated by Pope and his coworkers after the discovery of LEDs made with III–V compound semiconductors.[6] In 1975, the first organic EL devices made with a polymer, polyvinyl carbazole (PVK), were demonstrated.[7]

In early attempts to develop organic EL devices, the driving voltage of such devices was on the order of 100 V or above to achieve a significant light output.[8–10] Vincett et al. achieved an operation voltage of <30 V by using a thermally deposited thin film of anthracene.[11] The research had been mainly in the academic field until Dr. C.W. Tang and his coworkers at Kodak Chemical showed, for the first time, efficient organic light-emitting devices (OLEDs) in multilayer configuration with significant performance improvement.[12] Nowadays, small-molecule OLEDs (SMOLEDs) made by means of a thermal deposition process have been used for commercial display products. Pioneer Corporation has commercialized organic EL (OEL) display panels since 1999 for consumer electronics use, such as in car audio, CD/MP3 player, A/V receiver, etc. Kodak and Sanyo Electric Co. Ltd. demonstrated the first full-color (FC) 2.4-inch (2.4″) active matrix (AM) SMOLED display in 1999. Their joint manufacturing venture, SK Display Corp., produced the world's first AM SMOLED displays for Kodak's EasyShare LS633 camera.[13] The world's first commercially available OLED TV is SONLY XEL-1 with an 11″ size, released by SONY Corporation in 2007.[14] Nowadays, AM SMOLED displays have become ubiquitous in smartphones. In early 2013, LG started selling 55″ OLED TVs worldwide.[15]

Another type of organic semiconductor, conjugated polymer, was discovered in 1977 by Alan J. Heeger, Alan G. MacDiarmid, and Hideki Shirakawa.[16,17] In addition to the focus on its novel physical and chemical properties in heavily doped states, great attention was paid to its intrinsic properties in undoped semiconducting phase, its nonlinear optical properties under photoexcitation,[18,19] and its interfacial behaviors with metal contacts. Schottky diodes made with polyacetylene film were demonstrated in a metal–semiconductor polymer–metal configuration.[20,21] Their optoelectric and electro-optical properties were studied. Although significant photosensitivity was demonstrated, the EL property of this system was intrinsically weak because of its electronic structure. Considerable works in the early and middle 1980s in the field of conjugated polymer were done on searching and developing new materials with solution processability. A popular, well-studied system consisted of polythiophene derivatives, one of which was poly(3-alkyl) thiophene (P3AT) (Figure 1.1). Solution-processed metal/P3AT/metal thin-film devices were demonstrated at the University of California at Santa Barbara in 1987.[22] Following the first demonstration of a light-emitting device with unsubstituted poly(*para*-phenylene vinylene) (PPV) (Figure 1.1) by Burroughes et al. at Cambridge University,[23] a highly efficient polymer LED (PLED) device was made with a solution-processable polymer, poly(2-methoxy-5-(2′-ethyl-hexyloxy)–1,4-phenylene vinylene), MEH–PPV (Figure 1.1), by Alan J. Heeger's group in Santa Barbara, California.[24] As will be discussed in later chapters, the commercialized soluble PPV derivatives nowadays are based on a synthesis approach originally developed by Fred Wudl's group in Santa Barbara in 1988,[25,26] and later modified by UNIAX Corporation (now DuPont Displays) in the middle 1990s and Aventis Research & Technologies GmbH (which later became Covion Organic Semiconductors GmbH, and now is part of Merck) in the late 1990s. Soluble PPV derivatives synthesized following this approach not only have high molecular weight but also show excellent solubility in common organic solvents. Most importantly, these materials have intrinsically low charged impurity, typically $<10^{14}$ cm^{-3}, and high photoluminescent efficiency (typically in the range of 20–60%).[25–28] PLEDs made with such PPV films show high EL efficiency, low operation voltage, and long device lifetime.[29–31] Displays made with PPV emitters were commercialized in 2002 by Philips (Norelco electric razor: Spectra 8894XL).

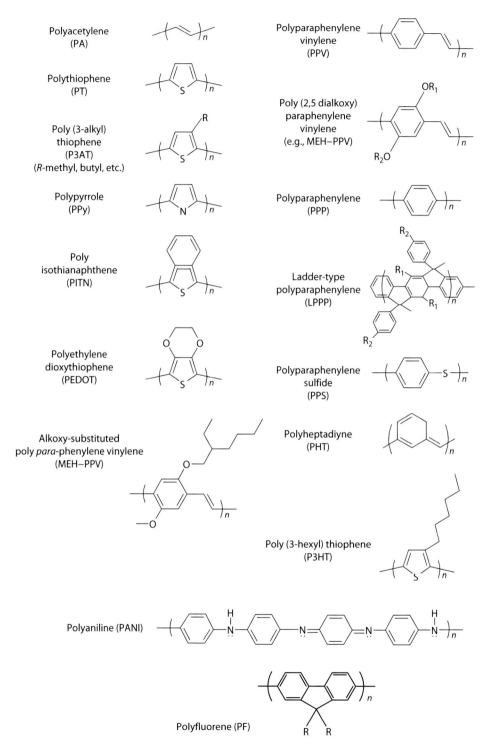

FIGURE 1.1
Chemical structures of popular conjugated polymers.

Although the energy gap in a PPV derivative can be enlarged by reducing the conjugation and planarization between the phenyl group and the vinyl group (as observed in PPVs with phenyl groups attaching at 2- or 5- or both sites),[27,32,33] it is not large enough to produce the saturated blue color needed for FC displays. Conjugated polymers with an optical energy gap of >2.9 eV are needed for PLEDs with blue emission. Significant effort has been made on searching and developing wide-energy-gap polymers (such as poly (*p*-phenyl) and its functional derivatives).[34–46] In addition to being used for making blue emitters, the same building blocks can also be used for making red and green emitters (as the host) by copolymerizing them with a proper emitter group (as the guest).[47–49] The red, green, and blue material sets developed by several companies (including Covion and Dow Chemical) are all soluble in common organic solvents with high optoelectric performance and good film-forming properties.[49,50]

PLED-based displays are attractive owing to their processing advantages in device manufacture. The organic materials used are soluble in common organic solvents or in water. Large-size, uniform, and pinhole-free thin films can be cast from solutions at room temperature by means of spin coating or other coating techniques commonly seen in the printing and painting industries. Because of the characteristic large elongation at rupture of polymers, they are flexible and easily fabricated onto rigid or flexible substrates in flat or curved shapes. Solution processing is also promising for forming patterned color pixels in FC displays. Different EL polymers can be deposited onto predefined locations by means of printing techniques such as ink-jet printing,[51,52] screen printing,[53,54] or by photolithographic patterning.[55] FC PLED displays made with an ink-jet process or with a laser-induced thermal transfer process have been demonstrated to have excellent image qualities.[56,57]

In Section 1.2, a brief review of conjugated polymers in semiconducting phase and metallic phase is given. Section 1.3 discusses device architectures and their corresponding processes. In Section 1.4, we discuss some novel devices and novel functions in thin-film polymer devices. Section 1.5 is devoted to the technical merits of SMOLEDs/PLEDs for use as emitter elements in flat-panel displays.

1.2 Conjugated Polymers in PLEDs

Conjugated polymers are a novel class of semiconductors that combine the optical and electronic properties of semiconductors with the processing advantages and mechanical properties of polymers. The molecular structures of several popular conjugated polymers are shown in Figure 1.1. Before the revolutionary discovery of conjugated polymers, polymer science and technology had focused on "saturated" polymers, i.e., conventionally nonconductive polymers (a term for macromolecules with repeat structure units). In "saturated" polymers, the valence electrons of the carbon atoms in the main chain are hybridized in sp^3 configuration, and each carbon atom is bonded to four other atoms. As a result, the electronic orbitals are fully saturated. Owing to their electronic structures, saturated polymers have wide energy gaps and are electrically insulating.

The fundamental difference between saturated polymers and conjugated polymers is their electronic configuration. Figure 1.2 compares the molecular and electronic structures of saturated (nonconjugated) polyethylene and conjugated polyacetylene. In a conjugated polymer, the carbon orbitals are in the sp^2p_z configuration, which leads to one unpaired

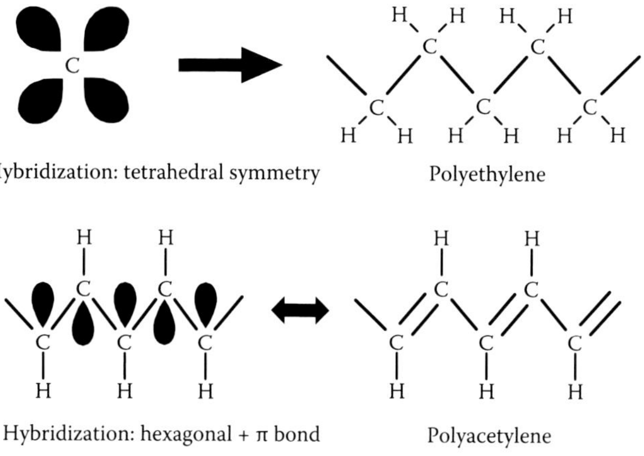

(a) sp^3 Hybridization: tetrahedral symmetry   Polyethylene

(b) sp^2p_z Hybridization: hexagonal + π bond   Polyacetylene

FIGURE 1.2
Electronic and molecular structures of (a) polyethylene and (b) polyacetylene.

electron (the π electron) per carbon atom. Since each carbon atom is covalently bonded to only three other atoms, and p_z orbitals of successive carbon atoms along the backbone overlap, delocalized π bands are therefore formed. As a result, conjugated polymers exhibit semiconducting or metallic properties depending on whether the bands are filled or partially filled. The number of π bands is determined by the number of atoms within the repeat unit. In the case of PPV, since the repeat unit contains eight carbons, the π band is split into eight subbands. Because each subband can hold two electrons per atom, the four π subbands with the lowest energy are filled and the four π* subbands with the highest energy are empty. The energy difference between the highest occupied π subband and the lowest unoccupied π* subband defines the π–π* energy gap, E_g.

One of the advantages of organic semiconductors is that one can modify their mechanical and processing properties while retaining their electric/optical properties. For example, PPV is a semiconductor with E_g ~2.5 eV. It is insoluble in any organic solvent after conversion from its precursor form into its conjugated form.[23,58] However, by attaching alkyl groups to the 2, 5 sites of its benzyl group, alkyl-PPV derivatives are formed. The alkyl-PPV derivatives possess similar energy band gap and luminescent emission profile to those of PPV, whereas they become soluble in most nonpolar organic solvents (such as xylene or toluene) and processable in conjugated form.[59] Another advantage of organic semiconductors is that one can tune the energy band gap of a given system while retaining its processing capability. For example, by replacing the alkyl groups of PPV derivatives with alkoxy groups at the 2- and 5- positions (e.g., poly(2-methoxy-5-(2′-ethyl-hexyloxy)-1,4-phenylene vinylene), MEH–PPV; Figure 1.1), one can reduce the energy band gap from 2.5 to 2.1 eV. Figure 1.3 shows the absorption and EL spectra of a series of PPV derivatives. The energy gaps are in the range of 2.5–1.9 eV, covering an ~0.6 eV range. These engineering flexibilities are especially promising for optoelectric and electro-optic device applications. Along with the change of the energy band gap is the change of luminescent profile and emission color, as shown in Figure 1.3b.

Photonic devices are often classified into three categories: light sources (LEDs, diode lasers, etc.), photodetectors (photoconductors, photodiodes, etc.), and energy conversion devices (photovoltaic devices, solar cells, etc.).[60] Most of the photonic phenomena known in conventional inorganic semiconductors have been observed in these semiconductor

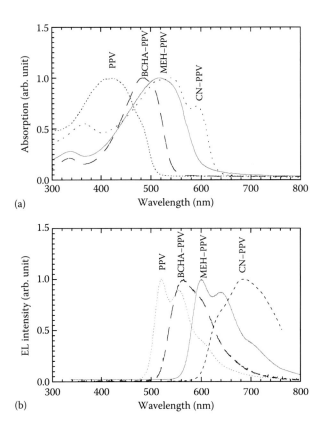

FIGURE 1.3
Absorption (a) and electroluminescence (b) of PPV derivatives. Energy band gap ranges from 2.5 eV (~500 nm) to 1.9 eV (640 nm).

polymers,[29,61] including luminescence and photosensitivity. Photoluminescence describes the phenomenon of light generation under optical radiation. An incoming photon with energy larger than the band gap excites an electron from the filled π band to the unoccupied π^* band to form an electron–hole pair (exciton), which subsequently recombines to emit a photon. In semiconductor polymers used for light emission applications, the photoluminescent quantum efficiency is typically in the 10–60% range. Photoconductivity describes the process of photogeneration of electric current. The electron–hole pairs generated by light illumination can migrate under electric field and be collected at the electrodes. Opposite to the light-emitting process, photoconductivity offers promise for large-area photovoltaic and photosensing applications.[62–64] In the applications of light emission and photoconduction, the carrier mobility of the polymer plays an important role. Depending on the detailed molecular structures, the morphology, and the electric field strength applied, carrier mobility in typical organic semiconductors is in the range of 10^{-7} to 10^{-2} cm²/V·s.

In EL applications, electrons and holes are injected from opposite electrodes into the conjugated polymers to form excitons. Owing to the spin-symmetry, only the antisymmetric excitons known as singlets could induce fluorescent emission. The spin-symmetric excitons known as triplets could not decay radiatively to the ground state in most organic molecules.[65] Spin statistics predict that the maximum internal quantum efficiency for EL cannot exceed 25% of the photoluminescence efficiency, since the ratio of triplets to singlets is 3:1. This was confirmed by the performance data obtained from OLEDs made with fluorescent organic small molecules such as tris(8-hydroxyquinoline) aluminum (Alq_3). In PLEDs made with semiconducting MEH–PPV films, an EL-to-PL efficiency ratio of ~50% was detected by Y. Cao and his coworkers.[66] Since then, this phenomenon has been observed by other groups worldwide and in other polymer systems.[67] It was suggested that, in conjugated polymers, the singlet cross section could be considerably larger than that of triplets by a factor of 3 to 4.[68,69] The finding has triggered considerable interest in further enhancing the singlet recombination cross section and singlet populations in EL polymers. For instance, an EL-to-PL ratio of >75% was reported lately by introducing perturbation of ferromagnetic exchange interaction near the EL polymer chain.[70] Triplet excitons can also emissively recombine, a phenomenon known as phosphorescence. The lifetime of triplet excitons is much longer, typically in the range of 10^{-7} to 10^{-3} s,[71–74] than that of singlet excitons, typically 10^{-10} to 10^{-9} s.[75–78] There have been considerable activities of making SMOLED/PLED devices with the triplet excitons for their potential high quantum efficiency.[72,73,79–83] A challenge in this approach is to prevent the long-life triplet excitons from interacting with impurities in the organic layers. More rigorous requirements on material purity, charge blocking, and device encapsulation are anticipated.

The color of the EL from PLED devices can be selected by modifying the chemical structure of the polymer, either through the main-chain molecular structures or through the side-chain structures, as in the example of PPV derivatives.[23,24,31,34,37,84,85] The EL color can also be tuned by doping the host polymer with luminescent emitters. The emitters could be fluorescent dyes,[86–89] phosphorescent emitters,[79,80,83] or other luminescent polymers.[46,90–93] In such blend systems, the host polymer has a wider energy gap while the dopant has a smaller energy gap. The excitation energy of the host was transferred to the guest molecules through the dipole–dipole interaction (Förster energy transfer), or the direct quantum mechanical electrons transfer (Dexter energy transfer). By selecting appropriate host and guest materials, and adjusting the weight ratio of the guests to the host, white LEDs have also been successfully demonstrated.[87,89,94] To make a PLED-based FC display through the host/guest approach, a stable wide band-gap polymer with high efficiency is a must. Therefore, blue EL materials and devices have received considerable attention and become a focus in the field. In addition to formation of the red and green emitters with host/guest polymer blends, the red and green emitters can also be made by copolymerizing a wider band-gap host molecular unit with one or more guest units with desired emission profiles.[47–50,95] Blue emitters being studied include poly(*para*-phenylene) (PPP),[37–39] ladder-type poly(*para*-phenylene) (LPPP),[40–44] polyfluorene (PF)[34–36,48,95–97], and their stereotype variations.[50] Their molecular structures are provided in Figure 1.1.

Chemical doping and electrochemical doping applied to these semiconducting conjugated polymers lead to a wide variety of interesting and important phenomena. For example, by doping polyaniline (PANI) with phorsulfonic acid (CSA), one can achieve a conducting polymer with a bulk conductivity of 100–300 S/cm.[98] The thin film of the PANI:CSA complex in polyblends with poly(methyl methacrylate) shows optical absorption in the infrared range (due to free-carrier absorption and polaron absorption) and in the UV region (due to interband optical transition), while it is optically transparent in most of visible spectroscopic region.[99] Similar phenomena have also been observed in poly(ethylenedioxythiophene):polystyrene sulfonate (PEDOT:PSS) blends commercialized by Bayer Chemical (Batron-P),[100] and in polypyrrole (PPY).[101] Figure 1.4 shows the optical transmission spectra of PANI:CSA and PEDOT:PSS. The infrared electric conductivities of PANI are shown in Figure 1.5.[102,103] These data can be well described by heavily doped

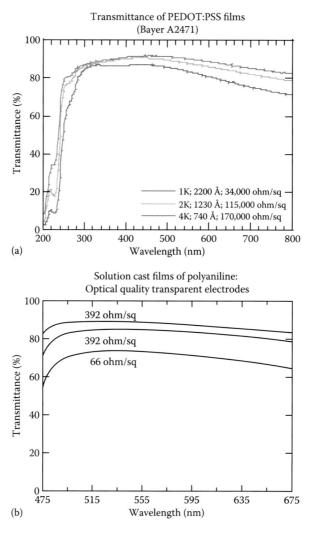

FIGURE 1.4
Optical transmission spectra of (a) PEDOT:PSS and (b) PANI:CSA.

semiconductors in disordered or amorphous systems. Such doped conjugated polymers are a novel class of thin, transparent conducting films that can be cast on rigid or flexible substrates through a solution process. These films have been widely used in PLED devices in single-layer anode form,[99] or as a buffer layer between the indium–tin oxide (ITO) electrode and the EL layer.[104] In addition to optimizing the hole injection, this buffer layer also serves as a planarization layer to eliminate pin holes in the EL layer caused by the rough ITO surface. It also serves as a chemical buffer preventing chemical impurities in the substrate and the transparent ITO electrode from reaching EL polymers, therefore significantly improving the PLED operation lifetime.[30]

The processable organic conductors, semiconductors, and insulators (not discussed in this chapter but well known historically for saturated polymers with sp^3 electronic configuration) form fundamental material sets for device applications. In the following sections, we discuss how to construct a PLED with such material sets.

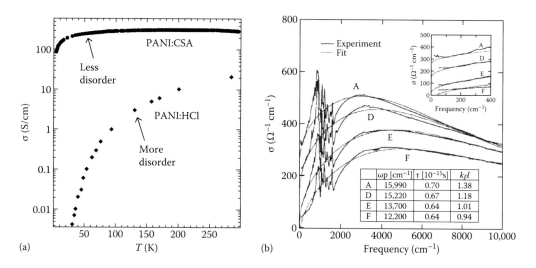

FIGURE 1.5
(a) Temperature dependence of electric conductivity; (b) infrared electrical conductivity of PANI:CSA.

1.3 PLED Structures, Processes, and Performance

The solution-processed PLED is typically prepared with a thin layer of semiconducting polymer film sandwiched between two charge injection contact electrodes, as shown in Figure 1.6. The device is generally made on a glass substrate or a thin plastic film with partially coated transparent electrode (such as ITO). A thin, semiconducting, luminescent polymer film with thickness typically in 50–200 nm range is then coated. Finally, the device is completed by depositing a low-work-function metal (such as calcium)[24] as the cathode electrode. While PLED is a typical single-layer device, the small-molecule OLEDs have a bilayer structure consisting of a hole transport layer and an emissive electron transport layer (ETL), sandwiched between a low-work-function cathode and a transparent anode.[12] To improve the brightness and efficiency of the basic bilayer devices, extra layers are often introduced. A popular multilayer structure used in phosphorescent OLED includes a hole

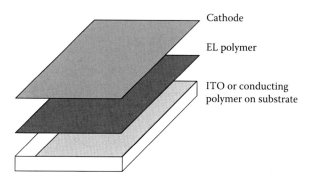

FIGURE 1.6
PLED in sandwich configuration.

injection layer, hole transport layer, electron blocking layer, emission layer, hole blocking layer, electron transport layer, and electron injection layer (Figure 1.7). Those layers help reduce the energy barriers at the electrode–organic interface, facilitate the transportation of the charged carriers, confine the opposite charges inside the emission layer, and therefore ultimately improve the power efficiency and operation lifetime of the device. In contrast, in the PLED field, a popular structure adopted by the whole industry for single-color displays has only one semiconductor polymer layer between the anode and cathode electrode. This layer serves multiple functions, including hole and electron transport and exciton recombination.[61,105]

Owing to the thin profile of the organic layers, the criteria for both the flatness of the transparent contact electrode and the film quality of the EL layer are rigorous. It has been one of the concerns in early days of the SMOLED/PLED development. These requirements are significantly relieved in PLEDs by inserting a conducting polymer buffer layer between the ITO and the EL polymer layer.[30,100,104] Such a conducting polymer buffer layer provides multiple benefits to the PLED device: (i) it serves as a polymeric anode, and it matches the highest occupied molecular orbital (HOMO) of the EL polymer to facilitate hole injection; (ii) it serves as a planization layer smoothing the rough ITO surface and eliminating shorts due to sharp "spikes" on the ITO surface; and (iii) it serves as a chemical barrier preventing inorganic atoms in the ITO layer (such as indium) from diffusing into the EL polymer layer. By applying such buffer layer into PLED devices, the device operation and shelf life has been dramatically improved, from about 10^2 h to longer than 10^4 h.[29,30,61] The so-called single layer (single semiconducting layer between the bilayer anode and the metal cathode) structure has achieved great success in PPV-based PLED devices as well as in PF- and PPP-based red/green emitters. However, it becomes more challenging to use such single-layer device structure for polymers with an energy gap of >2.9 eV. This is similar to that encountered in SMOLED devices, in which the popular host material in the emitter layer (such as Alq_3 or bis-(2-methyl-8-quinolinolate)-4-(phenyl-phenolate)-aluminum, BAlq) has an energy gap of >>2.5 eV. Significant improvements in device efficiency and operation lifetime have been achieved in blue PLEDs with an additional thin cross-linkable interlayer between the buffer layer and the EL layer.[106] After coating the

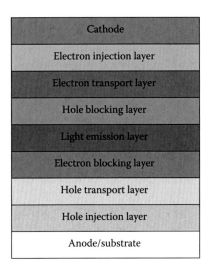

FIGURE 1.7
Multilayer SMOLED structure.

interlayer from solution, it is cross-linked at elevated temperature and becomes insoluble during the process of the following EL layer. A sharp and clear interface could thus be formed. This layer serves as a hole transport from the anode to the EL layer. Since it separates the anode from the recombination zone in EL layer, it eliminates the exciton quenching caused by the contact electrode. With such an interlayer, 10 cd/A was demonstrated in a blue PLED with Commission Internationale de l'Eclairage (CIE) color coordinates (0.15, 0.12) by Cambridge Display Technology. The extrapolated lifetime of such blue devices was reported to be >13,000 h from an initial luminance of 1000 cd/m^2 at room temperature.[107]

For PLEDs with relatively small active area, the anode can be made of a single layer of conducting polymers with relatively high bulk electric conductivity. Gustafsson and coworkers at UNIAX demonstrated a flexible PLED with a conducting PANI:CSA anode, and with MEH–PPV as the light-emitting layer.[108] Conducting polymers with moderate bulk electric conductivity are generally favorable as the buffer layer on top of ITO, especially for applications in segmented or pixelated displays. By proper selections of the bulk conductivity and the thickness of the buffer layer, one can control the lateral conductance between neighbor pixels, so that a continuous, nonpatterned buffer layer can be used in the display matrix, which provides sufficient conducting in the vertical direction and sufficient insulating in the lateral direction. This structure considerably reduces the manufacture process time and cost.

Multilayer OLEDs incorporating electron injection/transport layers (ETL) to improve efficiency is another research focus. The ETLs may serve as a carrier blocking layer, reduce the electron injection barrier, or shift the recombination zone away from the electrode. While it is easy to construct a multilayer device structure in the evaporated SMOLEDs, it is difficult for PLEDs to achieve such a structure, since the solution process of the ETL can easily damage the underlying materials, leading to poorly defined organic–organic interfaces. One solution to such problem is to use orthogonal solvents. Cao and coworkers developed a novel water/alcohol-soluble conjugated polymer, aminoalkyl-substituted polyfluorene copolymer (or its quaternized salt), as an ETL, which significantly improved device performance.[109–114] It is proposed that the improved device performance is due to the reduction of the electron injection barrier as a result of an interfacial dipole at the ETL–cathode interface. Further investigation revealed that the solvent alone could serve as an efficient ETL, which creates an interfacial dipole layer elevating the vacuum level at the cathode side thereby reducing the electron injection barrier.[115]

The main difference in process between SMOLEDs and PLEDs is that SMOLEDs are commonly fabricated by vacuum deposition, whereas PLEDs are made by solution processing, which is a simpler and cheaper process. However, organic molecules with relatively low molecular weight can also be functioned into solution-processable forms. In addition to adding flexible side chains to the luminescent molecule, molecules in oligomer and dendrimer forms have been demonstrated.[116,117] The first PLED made by means of solution processing was demonstrated at University of California at Santa Barbara, right after the discovery of PLED.[23,24] In solution-processing methods, spin coating has been the most popular technique in both the research laboratory and industry for the buffer and EL polymer layer. PLED devices with low operation voltage, high efficiency, and long operation lifetime have been obtained by using the spin-coating method. To pattern fine structures for FC displays, shadow mask is typically used in SMOLED fabrication. The shadow mask process becomes challenging for large display panels. In contrast, FC PLED pixels can be patterned with one of printing techniques such as ink-jet printing,[51,52] screen printing,[53,54] dye diffusion,[53] laser-induced thermal transfer,[118–120] or be patterned with the photoligthographic process.[55] The solution process is typically carried out at low or room temperature, allowing a device to be made onto a flexible, organic substrate.[108] As illustrated in Figure 1.6, the substrate (rigid or flexible) with or without ITO was spin cast with a thin layer of buffer layer. On top of the buffer layer, an EL polymer film with thickness of typically 50–200 nm was spin coated. During these two processing steps, post-baking is generally required to remove the residue solvent in the buffer and polymer layers. The cathode layer(s) is typically made by means of physical vapor deposition (such as thermal deposition) under a vacuum of ≤10^{-6} torr. Calcium, barium, and magnesium are commonly chosen as the cathode owing to their low work functions. Since low-work-function metals are highly reactive, PLED devices must be hermetically sealed for a long operation life. Replacing low-work-function metals with stable high-work-function metals need special engineering at the organic–metal interface. Various surface modification materials, such as conjugated polyelectrolyte (CPE), self-assembled monolayer (SAM), surfactants, or metal compounds, have been successfully developed.[121,122]

PLEDs are two-terminal, dual-carrier thin-film devices. They could be viewed as metal–insulator–metal (M–I–M) or metal–semiconductor–metal (M–S–M) devices with asymmetric metal contacts. The energy structure of a PLED can be approximated by a rigid band model as illustrated in Figure 1.8. This model is justified because the charge carrier concentration in undoped films is so low (≤10^{14} cm^{-3}) that all of the free carriers are swept out by the field that arises from the difference in work functions of the two electrodes. The depletion depth of a high-quality PPV film is beyond microns, which is much larger than the thickness of the EL polymer layer in the PLED device. As a result, there is negligible band bending.[18] Under forward bias, the electrons are injected from the low-work-function cathode into the π* band (conduction band) of the EL polymer, while the holes are injected from a high-work-function electrode into the π band (valence band) of the EL polymer. The oppositely charged carriers in the two different bands encounter each other within the EL polymer film and recombine radiatively to emit light. The light emission process is an intrinsically fast process. The only delay is due to the transport of the holes and electrons from the electrodes to the emission zone. Thus, PLED has a fast response time, typically in the range of 10–10^2 ns, limited by the RC time constant resulting from the geometric factors of the PLED device.[123] For comparison, the fastest response time of LCD displays available in the market is around 10 ms.

Ideally, the work function of the cathode is required to perfectly match to the lowest unoccupied molecular orbital (LUMO) energy level of the EL polymer, and the work function of the anode is required to perfectly match to the HOMO energy level of the EL polymer. In reality, when the electrode's work function lies within ~0.2 eV to the HOMO or LUMO energy level, doping occurs at the contact between the electrode and the EL polymer interface, and *p*- or *n*-type polarons form at the corresponding interfaces. When the

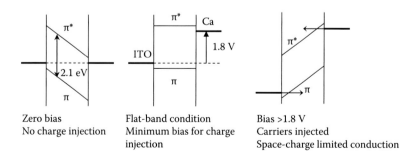

Zero bias   Flat-band condition   Bias >1.8 V
No charge injection   Minimum bias for charge   Carriers injected
                          injection   Space-charge limited conduction

FIGURE 1.8
Energy band structure of a PLED in the configuration of ITO/MEH–PPV/Ca.

work function of the metal electrode is lower than the HOMO or higher than the LUMO of the EL polymer, doping occurs and an inverting layer is formed at the metal–polymer interface. The electrode's work function is "pinned" to the LUMO or HOMO of the EL layer. This effect is similar to that frequently seen in *p–i–n* devices made with inorganic semiconductors.[60] In such a situation, ohmic contact is frequently seen at the polymer–metal interface, which is characterized by exponential behavior in *I–V* curves in forward bias near that corresponding to E_g/e. Figure 1.9 shows a data set taken from an ITO/PEDOT/EL polymer–cathode device. The EL polymer layer is made of Covion® PDY with a thickness of 70–90 nm. Its optical absorption and emission spectra are shown in Figure 1.10, with optical energy gap of ~2.3 eV and peak of emission profile at 560 nm. The anode contact of the device is made of conducting PEDOT layer with work function near ~5.2 eV.

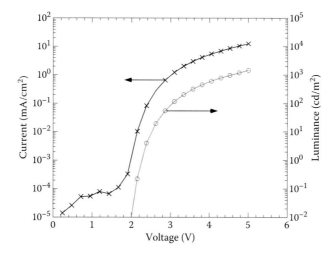

FIGURE 1.9
L–V and *I–V* characteristics of a PLED in the structure of ITO/PEDOT/Covion PDY/Ba/Al.

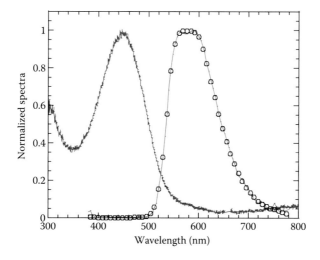

FIGURE 1.10
Optical absorption, photoluminescence, and electroluminescence (circle) of a Covion PDY film.

The cathode contact is made of a thin barium layer with a thick aluminum cap.[50,124] The carrier injection occurs at "flat band" condition (Figure 1.8) at which the forward voltage is close to the optical band gap of the EL polymer minus two polaron binding energies, ~2 V in this example. The exponential *I–V* characteristic is typically seen in the voltage range close to that corresponding to the optical energy gap, similar to that seen in light-emitting devices made with inorganic III–V semiconductor crystals.[60] For current density higher than ~1 mA/cm², the *I–V* deviates from the exponential behavior and becomes bulk limited. This device emits greenish-yellow light with CIE color coordinates at x ~ 0.48 and y ~ 0.52. It reaches 100 cd/m² at 2.4 V (typical luminance of a computer monitor), 1000 cd/m² for $V \leq 4$ V (typical luminance of a fluorescent lamp), and >10,000 cd/m² for $V > 5$ V. These data suggest that conjugated polymer PLEDs are very efficient light emitters with low operating voltages. The external quantum efficiency of such device is >5% ph/el and the luminous efficiency is >10 lm/W at 2–4 V. As the efficiencies indicated, PLED devices are one of the most efficient thin-film light sources. For comparison, the power efficiencies of ZnS-based thin-film EL devices are typically 2–4 lm/W.[125]

When the work functions of the contact electrodes are not well matched to the bands of the EL polymers, energy barriers are formed at the respective interfaces. The height of the barrier for hole injection is determined by the difference between the work function of the anode and the HOMO energy level. The height of the barrier for electron injection is determined by the difference between the work function of the cathode and the LUMO energy level. Carrier tunnels through the barrier can be described primarily by Fowler–Nordheim field emission tunneling: $I \propto V^2 \exp(-1/V)$.[126,127] When the barrier heights are significantly <100 meV (comparable to the thermal energy of room temperature), thermionic emission (thermal-assisted carrier tunneling) becomes the dominating mechanism. Thermionic emission in a PLED device could be well described by Schottky emission, where thermionic emission across the metal–insulator interface or the insulator–semiconductor interface is dominating.[60,127,128] In PPV devices with ohmic contacts and with relatively thick EL polymer layer (>150 nm), the hole current is space-charge limited and the electron current is trap limited.[129,130] The space-charge-limited hole current arises owing to the higher mobility of holes than that of the electrons observed in some popular polymer systems.[129,131,132] The trap-limited electron current results from defects with energy levels just below the conduction band due to disorder, structural imperfectness, or chemical impurities in the polymer. SMOLEDs/PLEDs are electric current–driving devices. Since the light intensity is proportional to the driving current (the number of charges flowing through the device in unit time), when the number of injected holes and injected electrons are balanced, the light emission intensity follows the same dependence as the current on the voltage. This is indeed observed in SMOLEDs/PLEDs with optimized structures and parameters. Figure 1.11 shows the luminous efficiency dependence of a PLED made with Covion PDY on the operating current. The luminous efficiency defined as emitting luminance per unit current is ~15 cd/A in the broad current range without any sign of reduction at the highest testing current.

The operating lifetime, defined as the time for the light emission decaying to half of its initial value under continuous constant current operation, has been a principal concern for the practical application of PLED in the display industry: Can operating lifetime sufficient for commercial display products be achieved using polymers processed from solution? There was skepticism during the early 1990s that solution-processed materials could achieve the level of purity required for semiconductor applications. The operating lifetimes of PLEDs during that time ranged from a few minutes to a few hundred hours at 100 cd/m² initial brightness, significantly less than the operating lifetimes of SMOLEDs for

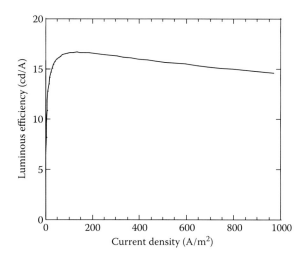

FIGURE 1.11
Luminous efficiency as a function of driving current in a PLED made with Covion PDY.

which >20,000 h operation lifetimes was achieved at a similar brightness level.[133] By means of inserting a conductive PANI layer between the ITO and the MEH–PPV layer, UNIAX demonstrated the first PLEDs with an operating life of >10^4 h.[30,134,135] At ~450 cd/m^2 initial brightness, the luminance of the PLED device drops only ~20% after 2000 h of operation at 35 mA/cm^2. The extrapolated room temperature operation lifetime at 100 cd/m^2 was in excess of 10^4 h, corresponding to a charge density of ~3×10^5 C/cm^2 passing through the device. This number is still a significant number even at the current stage.

Covion PDY and its early versions, including MEH–PPV, represent a class of EL polymers with high quantum efficiencies. Single-layer devices with band-matched electrodes exhibit low operation voltage (as a result of high carrier mobilities), high quantum efficiency, and long operation lifetime. With improved device engineering, PLEDs made with Covion PDY showed an operation lifetime of >700 h under constant current driving with 200 cd/m^2 initial luminance at 80°C. Two data sets taken from two batches of devices (made over 9-month time period) at an accelerated test condition (3 mA/cm^2, 80°C) are shown in Figure 1.12. It is well known that the device operation lifetime at room temperature is 32–35 times longer than that at 80°C.[61,136] The data suggest an ~25,000 h operation lifetime under constant current driving and an ~35,000 h operation lifetime under 200 cd/m^2 constant luminance driving at room temperature. The voltage increase rate is constant for most of the stress period, ~1 mV/h at 80°C, which corresponds to ~30 μV/h at room temperature. It is worth mentioning that under the operating conditions at which device heating is negligible, the operation lifetime of a given SMOLED/PLED is inversely proportional to the operation current density. In other words, the device operation lifetime can be represented by a universal number: the total charge passing through the device during its lifetime. For the device shown in Figure 1.12, the total passing charge at room temperature is approximate 3×10^5 C/cm^2, consistent with the value we found in the MEH–PPV device.[135] Thus, solution-processed PLEDs have met the requirements for commercial applications with operating lifetimes in excess of 10^4 h at display brightness. In 2002, Philips introduced the first commercial PLED product in an electric razor display (Norelco Spectra 8894XL).

Under transient pulsed operation, PLEDs can emit light up to 10^6 cd/m^2. Figure 1.13 provides a data set taken from a PLED made with MEH–PPV.[61] The PLED was made in a

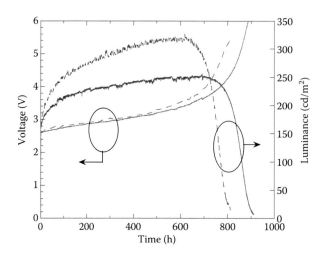

FIGURE 1.12
Operation lifetimes of two PLEDs made with Covion PDY polymer.

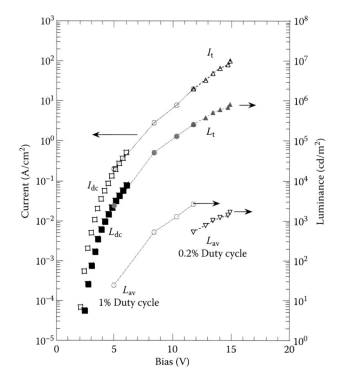

FIGURE 1.13
Light intensity of 10^6 cd/m^2 achieved in PLED.

microcavity configuration with semitransparent Au as the anode electrode and Ca/Al as the cathode electrode. The Au anode not only significantly reduced the contact resistance but also allowed high current density (>100 A/cm^2) to pass through the device. The luminous efficiency was ~2 cd/A for a current density between 1 and 50 mA/cm^2. At 100 A/cm^2, the luminance reached ~10^6 cd/m^2 with an efficiency of ~1 cd/A. This result revealed that the luminous efficiency remained relatively constant for a charge density running over 5 orders of magnitude. It also deserves attention that no substantial emission spectral change was observed even at the highest operation current. Similar results were also obtained in other PPV derivatives in traditional ITO/EL polymer–cathode device configuration.[137] Emission intensity up to 5×10^6 cd/m² was observed in PPV and its copolymer.

Polymer laser diodes have been an attractive topic after the discovery of EL conjugated polymers. Optical lasing in semiconducting luminescent polymer solutions was first demonstrated by D. Moses at the University of California at Santa Barbara by replacing organic molecular dye with MEH–PPV solution.[138] Photopumped solid-state lasing from thin conjugated polymer films was then discovered by several groups in early 1996.[139–141] Since then, considerable efforts have been devoted on the polymer and small-molecule laser diodes.[142] However, no reliable lasing effects have been reported under direct electrical pumping even at the 10^6 cd/m² (current density, 1000 A/cm²) level. The challenges for the electrically pumped laser diodes include (but not limited) the following: (i) good electrode–polymer interfaces that can pass the needed current density; (ii) low optical loss at the metal electrodes; (iii) minimum charge-induced absorption; (iv) incorporating contact electrode with high-Q mirrors; and (v) good thermal dissipation from junction area to the substrate or to a heat sink.

In an FC display, each color pixel contains three subpixels with primary colors in the red, green, and blue emission zones, respectively. A popular approach for fabricating such color pixels is fabricating each color subpixel with EL materials in corresponding emission spectra. Considerable efforts have been given in both the SMOLED and PLED fields to develop red, green, and blue color emitters with performance sufficient for commercial display applications. Figure 1.14 shows a set of luminance–voltage data of PLEDs made with red, green, and blue emitters. Again, light emission occurs at forward bias corresponding to the photon energies of the emitting colors, similar to the fact known in inorganic LEDs. The operation voltages of red, green, and blue emitters at 200 cd/m² are 6.6, 3.6, and 5.4 V, respectively. The respective luminous efficiencies are 2.2, 18, and 6.0 cd/A. These numbers

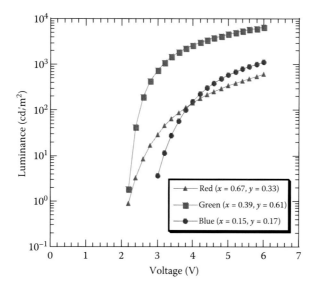

FIGURE 1.14
Luminance–voltage plots of red, green, and blue polymer emitters.

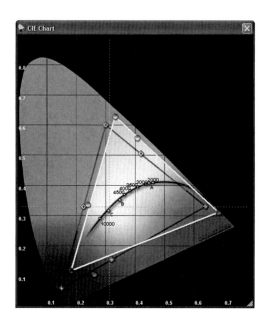

FIGURE 1.15
(See color insert.) CIE coordinates of red, green, and blue polymer emitters.

suggest that they are promising thin-film emitters with high luminous efficiencies and low operation voltages. As shown in Figure 1.15, the CIE color coordinates of these emitters have been improved to a level comparable to those in AMLCD TV screens. The operation lifetimes of these emitters have also been improved dramatically over time, meeting the commercialization's requirement.[143,144]

1.4 Novel Devices and Novel Functions in Thin-Film Polymer Devices

1.4.1 Dual-Function Polymer Device and Display Matrices

The thin-film device illustrated in Figure 1.6, made with a semiconducting polymer as the active layer, can function as both a light-emitting device and a light-detecting device (i.e., a photodetector).[145] The asymmetric electrodes provide a built-in potential equal to the difference between their work functions. Thus, at zero or reverse bias, photogenerated electrons and holes are separated by the internal field, and collected at the respective electrodes. The dual-function utility is illustrated in Figure 1.16 for an ITO/MEH–PPV/Ca device. At forward bias >2 V, light emission becomes visible with bare eyes. The luminance reaches 100 cd/m^2 at ~2.5 V with an EL efficiency of ~2.5 cd/A. The corresponding external quantum efficiency is about 2% ph/el. At –10 V bias, the photosensitivity at 430 nm is around 90 mA/W, corresponding to a quantum yield of ~20% el/ph.[145] The carrier collection efficiency at zero bias was relatively low, in the order of 10^{-3} ph/el. The photosensitivity showed a field dependence with activation energy of 10^{-2} eV.[145] This value is consistent with the trap distribution measured in the PPV-based conjugated polymers.[146,147]

While pure semiconductor polymer films often exhibit low photoconductivities, sensitizers have been used not only to increase the photosensitivity but also to broaden the

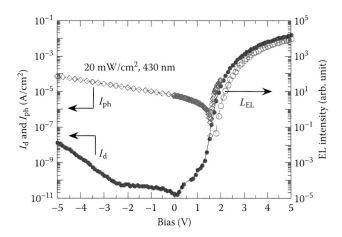

FIGURE 1.16
Dual-function utility of MEH–PPV device: light emission and photodetection.

photoaction spectrum.[148–150] A model system is polymer doped with electron acceptor C_{60} and its functional derivatives. In polymer blends made with MEH–PPV:C_{60} in an ~1:1 weight ratio, the photoluminescence in MEH–PPV is quenched by more than 4 orders of magnitude, while the photoconductivity increases 10–100 times.[151] In a device made in sandwich configuration similar to that shown in Figure 1.6, the carrier collection efficiency at zero bias reaches 30–50% ph/el.[63] The power conversion efficiency reaches 3% at 430 nm. These works in the early 1990s have opened a number of new opportunities for conjugated polymers. Photodetectors, image sensors, solar cells, and photovoltaic cells have since then been successfully demonstrated.[63,64,152,153] Recent works along this direction have resulted in polymer photovoltaic cells with 3–5% power conversion efficiency under AM1.5 solar radiation.[154–157]

The dual-function utility (photon emission in forward bias and photon detection in reverse bias) can be employed to fabricate "smart" display matrices.[158] By presetting the device in the photodetecting mode (zero or reverse bias), each pixel can sense an optical signal and transfer that signal to the memory in the driving circuit, similar to the process in a photodiode array. Subsequently, those pixels that were addressed during the input cycle can be switched to the displaying mode (forward bias), thereby creating an emissive image of the input that was displayed on the same diode array.

1.4.2 Polymer Light-Emitting Electrochemical Cells

Polymer light-emitting electrochemical cell (PLEC) is a new approach to light emission from semiconducting polymers.[159,160] By laminating a blend of a luminescent conjugated polymer and a solid-state electrolyte (such as PEO:Li$^+$) between two air-stable contact electrodes, a dynamic *p–i–n* junction could be created under external bias. The semiconducting polymer is electrochemically doped to *p* type on one side and *n* type on the other side. Light is emitted from the compensated insulating region at the center. Because the doped polymer becomes highly conductive, ohmic contacts are formed at the electrode–polymer interface. As a result, facile electron and hole injections are achieved with stable metals as electrodes, such as Au or Al. The *I–V* characteristic in PLEC is similar to that shown in PLED with optimized interfaces (such as that shown in Figure 1.9). This fact confirms the *p–i–n* structure in PLEDs with optimized interfaces.

The operating mechanism of the PLEC includes the following steps:

1. Electrochemical *p*-type and *n*-type doping in the regions adjacent to the anode and the cathode, respectively, upon application of voltage greater than the π–π^* energy gap of the semiconducting polymer
2. Formation, in situ, of a *p–n* junction within the active layer
3. Radiative recombination of the *p*- and *n*-type carriers within the compensated *p–n* junction

The speed of *p*- and *n*-type doping and the speed of the *p–n* junction formation depend on the ionic conductivity of the solid electrolyte. Because of the generally nonpolar characteristics of luminescent polymers such as PPV and the polar characteristics of solid electrolytes, the two components within the electroactive layer will phase separate. Thus, the speed of the electrochemical doping and the local densities of electrochemically generated *p*- and *n*-type carriers will depend on the diffusion of the counterions from the electrolyte into the luminescent semiconducting polymer. As a result, the response time and the characteristic performance of the LEC device will highly depend on the ionic conductivity of the solid electrolyte and the morphology and microstructure of the composite.

A method for increasing the ionic conductivity within the layer and, simultaneously, for controlling the morphology of the phase-separated microstructure of the EL polymer–solid electrolyte composite has been successfully demonstrated.[161] The idea is to use a bifunctional (surfactant-like) liquid compound with high boiling point as an additive to facilitate the phase separation, to ensure maximum interfacial surface area between these two phases, and ideally, to support the formation of an interpenetrating bicontinuous network in the composite. If the additive compounds have, in addition, a relatively high dielectric constant, the ionic conductivity will also be enhanced. By the said method, the response time of the PLECs has been improved to milliseconds or even submilliseconds. This fast response allows LECs being used as the emitters for *x–y* addressable display arrays in both passive matrix and AM forms, and being operated at a video rate of, e.g., 60 frames per second.

A set of data obtained from a LEC [ITO/MEH–PPV:PEO:Li + OCA/Al] is shown in Figure 1.17. Since the junction is created in situ, the doping profile reverses when the bias polarity is reversed. Consequently, light emission can be observed in both bias directions. By selecting polymers with proper energy gaps, red, green, and blue LECs have been demonstrated with external quantum efficiencies and luminous efficiencies close or even better than those in corresponding PLEDs with optimized parameters.[162,163] Owing to the easiness of controlling the carrier injection and balance, single-layer LEC devices can, in fact, be employed to assess the intrinsic performance of an unknown polymer semiconductor.

On the one hand, the ionic conductor provides a unique feature in creating dynamic junction in LEC. On the other hand, the slow ionic motion and irreversible electrochemical doping under a high biasing field were two of the challenges for the use of PLECs in practical applications. The follow-up works have focused on the following directions:

1. LEC with frozen junction operation[163,164]: With frozen *p–i–n* junction, the LEC device exhibits unipolar (rectifying) behavior, and no hysteresis was observed in fast *I–V* scans. To freeze out ionic mobility, either lower temperature could be applied after the formation of the *p–i–n* junction or an electrolyte with negligible ionic conductivity at room temperature could be used. In the latter method, the *p–i–n* junction was formed at elevated temperatures and the device was operated at room temperature.

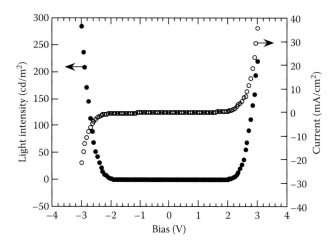

FIGURE 1.17
I–V (open circles) and *L–V* (solid circles) characteristics of a LEC fabricated as ITO/MEH–PPV:PEO:Li + OCA/Al. Forward bias is defined as positive bias with respect to ITO. (From Q. Pei, G. Yu, C. Zhang, Y. Yang, and A. J. Heeger, *Science*, 269, 1086, 1995. With permission.)

2. LEC under pulsed operation with a mean field below the EC doping threshold and pulse width shorter than the response time of ionic charges[165]: This operation scheme allows the induced junction in a LEC to be stabilized, and LEC to be used for emitters in a passive matrix display (see Section 1.5.2).

1.4.3 PLED with Stable Cathode Electrode

To optimize the performance of SMOLEDs/PLEDs, it is important to choose electrodes whose work functions are well matched to the bands of EL organic materials in order to minimize the barriers for charge injections. Although cathodes made of low-work-function metals are necessary to obtain adequate electron injection for high efficiency at low operation voltage, metals such as Ca, Ba, or Mg are air sensitive, highly reactive, and difficult to handle. To achieve long storage life and long stress life, devices fabricated with low-work-function cathodes must be hermetically sealed.

To use air-stable metals with high work function as cathodes, the electron injection barriers have to be reduced. Proposed by Campbell et al., a change in barrier height can be induced by a self-assembled monolayer of oriented dipoles chemically attached to the electrode.[166] Cao and his coworkers have demonstrated that certain surfactant-like additives, such as anionic ether sulfates, $R(OCH_2CH_2)_nOSO_3Li$, dissolved or dispersed in the EL polymer significantly improve electron injection from relatively high-work-function metals such as Al.[167,168] The Al electrode is complexed with the ethyleneoxide group of the surfactant molecule, making the dipoles of the surfactant molecules oriented in the correct direction to reduce the barriers. The Li-NPTEOS–type compounds not only improve the PLEDs' quantum efficiency in the single-layer device, but also enhance the performance of PLEDs and SMOLEDs in bilayer structures with Al as cathodes. With air-stable metals as cathodes, the devices can be stored in air for approximately 30 h without significant degradation.

1.4.4 Highly Efficient White PLEDs

Owing to the potential applications in FC OLED displays,[169,170] backlighting sources for LCDs, and next-generation solid-state lighting sources, white PLEDs (WPLEDs) have generated a lot of interest. Several groups have demonstrated that WPLEDs outperform incandescent light bulbs in efficiency, showing that WPLEDs can find practical application as large-area lighting sources.[171] Various approaches have been developed to realize white light, including the most commonly used method of doping a large band-gap polymer material (host) with different low-band-gap emitters (dopant) to achieve balanced broad emission. Cao's group double-doped a PVK host with a blue emitter, FIrpic, and a yellow-emitting iridium compound in the presence of electron transporting material, OXD-7, to fabricate a highly efficient single-layer WPLED.[172] The devices exhibited a maximum PE of 20.3 lm/W and a maximum LE of 42.9 cd/A with CIE coordinates of (0.395, 0.452).

1.4.5 PLED and PLEC in Surface Cell Configuration

Besides the commonly used sandwich structure, OLEDs could be fabricated in the metal–polymer–metal surface cell configuration, as illustrated in Figure 1.18.[61,173,174] To make such devices, first, two symmetric electrodes are prepared onto a substrate with a gap in between. The metal can be deposited onto the substrate by thermal evaporation, electron beam evaporation, or sputtering. Patterning can be achieved by depositing through a shadow mask, by photolithography, by microcontact printing, or by screen printing. Since the *n*-type doped and *p*-type doped regions in the LEC (under bias) have relatively low resistivity, the LEC in surface cell configuration exhibits similar performance parameters as those in sandwich configuration.[159,161,175] Contrary to LEC devices, the symmetric electrodes in the planar PLED devices make carrier injection extremely difficult. For a planar PLED with a 20 μm gap between two gold electrodes and free-standing PPV as the EL polymer layer, light emission was observed at 77 K and an operation voltage of >500 V was required.[173] To reduce the operation voltage to normal level, a submicron gap of 0.41 μm was made by photolithography, resulting in a turn-on voltage of 35 V at room temperature.[174]

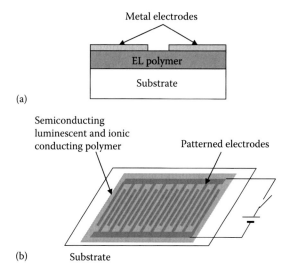

FIGURE 1.18
(a) Planar PLED and (b) planar PLEC in interdigitated cell configuration.

The planar configuration allows the display to be hybridized with integrated circuits on an Si wafer and thereby enables new approaches to fabricating integrated electro-optical devices. The combination of conjugated PLEDs and PLECs with silicon technology relies on the design of structures that allow efficient surface emission. For nontransparent substrates, e.g., Si, a planar device configuration compatible with standard photolithography is a promising approach. With interdigitated electrodes on transparent substrates, such as glass or polymer films, light emission can be viewed from both sides of the device. In addition to these novel features, the planar configuration has manufacturing advantages: (i) the electrodes can be prepared before the active polymer film is cast onto the substrate; (ii) the device performance is insensitive to pinholes or other defects on the polymer films; (iii) the device performance is relatively insensitive to the thickness of the active polymer layer; (iv) the planar emissive devices can be fabricated in a roll-to-roll process at room temperature without vacuum equipment.

1.4.6 Optocouplers Made with Semiconducting Polymers

Optocouplers are a class of devices with input current (I_i) and output current (I_o) coupled optically but isolated electrically. They are used extensively in the automation industry and in laboratory equipment where large common-mode noise/voltage or hazardous electrical shocks are present in circuits between transducers/detectors and controlling equipment. The simplest optocoupler is composed of a LED (input) and a photodiode (output), as shown in Figure 1.19.

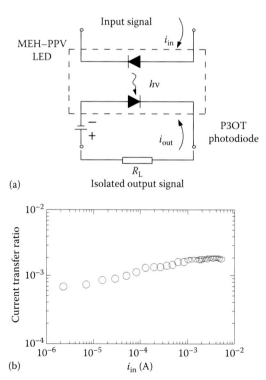

FIGURE 1.19
Circuit diagram (a) and current transfer ratio, i_{out}/i_{in} (b), of the polymer optocoupler. The photodiode was biased at −10 V.

By coupling a PLED with an external EL efficiency (η_{EL}) of >1% ph/el with a polymer photodiode with a quantum yield (η_c) of >20% el/ph, a polymer optocoupler can be constructed with a current transfer ratio, I_o/I_i, of >2 × 10^{-3}. The transfer ratio I_o/I_i is proportional to the product of the LED's external EL efficiency and the photodiode's quantum yield, $A \times \eta_c \times \eta_{EL}$, where A is a coupling constant determined by the spectral and geometrical overlap of the LED and the photodiode. In Figure 1.19, I_o/I_i is plotted vs. the input current for an optocoupler made with an MEH–PPV LED and a P3OT photodiode.[62] The current transfer ratio is 1 × 10^{-3}–2 × 10^{-3} at –10 V bias, comparable to the I_o/I_i of commercial optocouplers made with inorganic semiconductors. By 2004, both the efficiency of the PLED and the efficiency of the photodiode have been improved significantly over those in the devices used to generate the data in Figure 1.19. Using the improved devices currently available, such as SMOLEDs/PLEDs made with fluorescent emitters with η_{EL} of 5–7% or SMOLEDs/PLEDs made with phosphorescent emitters with η_{EL} of 10–15%, and photodiodes made with P3AT:PCBM blend[64] or PFO–BET:PCBM blend with η_c of 50–100%,[156] a current transfer ratio of ~0.1 is expected even when the photodiode is at 0 to –2 V bias. Thus, the polymer optocoupler could be a high-transfer-efficiency electric device for integrated circuits, optical interconnectors, or emitters, receivers, and modulators for telecommunication applications.

1.5 Flat-Panel Displays Made with Solution-Processable Organic Semiconductors

1.5.1 SMOLEDs/PLEDs as Emitter Elements in Flat-Panel Displays

Offering lightweight, thin-panel thickness, wide view angle (Lambertian emitter), high self-electroluminescent efficiencies, less power consumption, fast response time, and high contrast, the SMOLED/PLED technology is being considered as the next-generation flat-panel display technology to replace liquid crystal. Low-temperature processes allow such SMOLED- and PLED-based displays to be made onto plastic substrates in rigid or in flexible forms. Two kinds of processing technologies have been developed for high-information-content FC displays: vacuum deposition, suitable for processing organic molecules with low molecular weight,[12] and solution casting/printing, suitable for processing organic molecules soluble in common solvents.[24] The display development team at DuPont Displays (formerly UNIAX Corporation) has been focusing on pixelated display development with solution-processable EL polymers for more than 10 years. Display panel size, pixel density (number of pixels per inch), and total pixel counts have been improved gradually to the level needed for medium- to large-size high-definition television screens (Figure 1.20). A 9″ AMPLED in WVGA format (800 × RGB × 480), 7″ AMPLED in HVGA format (480 × RGB × 320), and 14.1″ AMPLED in WXGA format (1280 × RGB × 768) with >3 million pixels have been successfully demonstrated.[57,176,177]

1.5.2 PMOLED Displays vs. AMOLED Displays

Display pixels can be connected in the forms of segmental displays and column–row addressable displays. In the form of segmental displays, each display pixel is wired up individually and can be addressed independently. Figure 1.21 shows a drawing of an ITO

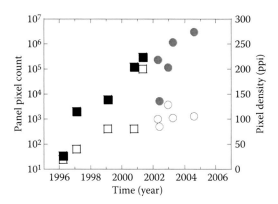

FIGURE 1.20

Development history of monochrome (squares) and FC (circles) displays at DuPont Displays (former UNIAX Corporation). Solid symbols, total pixel counts; open symbols, pixel density.

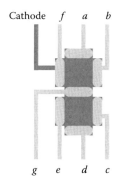

FIGURE 1.21

ITO pattern and cathode pattern of a seven-segment display.

pattern and a cathode pattern for a seven-segment display. Such a display is often operated under direct-current condition (with ~100% duty cycle), or operated in multiplexing mode with low duty cycle. Figure 1.22 shows a four-digit display system operating under one-fourth duty cycle. Since the OLED pixels in segmental displays are operated at low duty cycle, the power efficiency of segmental displays is easy to optimize.

Two kinds of driving schemes have been generally adopted in column–row addressable, pixelated OLED displays: passive matrix (PM) and active matrix (AM). Figure 1.23 shows their respective device structures. Since PLEDs are two-terminal, thin-film devices with rectifying I–V characteristics, a monochrome PMPLED display can be made by laminating an unpatterned EL polymer layer between an array of anode electrodes (transparent ITO has been commonly used) and an array of cathode electrodes. In this way, only the electrode stripes need to be patterned, resulting in a device structure similar to that in twisted nematic or supertwisted nematic LCD displays. The manufacturing process is simple and its cost is relatively low. In AMPLED displays, each display pixel is addressed by the column and row electrodes (bus lines) on an AM backpanel, and a common counter electrode can thus be used on the other side of EL film.

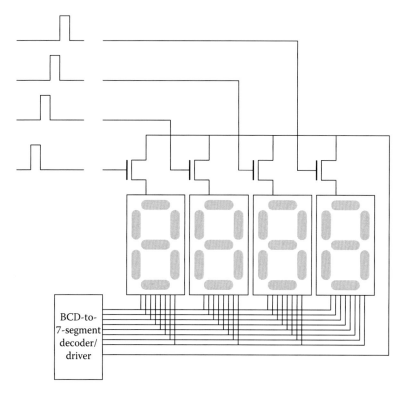

FIGURE 1.22
Multiplexing operation of a four-digit segmental display.

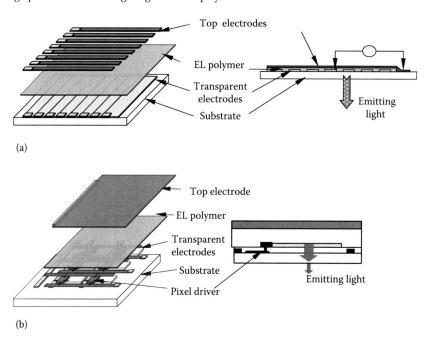

FIGURE 1.23
Cross-section views of the device structures of PMPLED (a) and AMPLED (b).

FIGURE 1.24
(See color insert.) PM displays in 96 × 64 format made at DuPont Display.

OLED pixel elements in PM displays are selectively turned on row by row with transient luminescent intensity N times brighter than the integrated intensity sensed by the human eye. The constraint of the maximum emitting intensity and the maximum operating voltage of the OLED pixels sets the limits on the duty cycle ($1/N$) of the PMOLEDs; 1/32 to 1/120 have been demonstrated in the past. The number of rows in PMOLEDs is equal to the reciprocal of the duty cycle when a single scan is adopted, or twice of the reciprocal of the duty cycle when a dual-scan driving scheme is adopted. Moreover, the resistance of the row and column power lines, the capacitance of the display pixel matrix, the requirement for the frame rate to achieve flicker-free operation, and the dielectric breakdown electric field over the organic EL film set additional limits on the number of rows and/or total number of the display pixels.[178] PMOLEDs are thus only practical for low- or medium-pixel count displays (typically <100,000 pixels). Figure 1.24 shows an engineering prototype for cellphone displays. It is a 96 × 64 PM display with 0.35 mm pitch size, made with Covion SY materials.[179]

By means of a latchable switch made with thin-film transistor technology, the emitting pixels in AMOLED displays are operated continuously during a frame time (with an analog driving scheme). Thus, the transient brightness of the emitting pixel is equal (or close) to the average brightness sensed by human eye. The pixel switch and the AM driving scheme also remove the limitations in PMOLED displays. Displays with high information content can thus be realized with such kind of driving scheme. Monochrome AMOLED in 4″ VGA format (640 × 480 pixels, 200 ppi),[180] and FC AMOLED displays in 2.2″ QCIF+ (176 × RGB × 220, 128 ppi),[181] 3.5″–4″ QVGA (320 × RGB × 240),[182] 9.1″ WVGA (800 × RGB × 480),[176] 13″ (576 × RGB × 324),[183] 13″ WVGA (852 × RGB × 480),[184] 17″ UXGA (1600 × RGB × 1200),[185] 15″–20″ WXGA (1280 × RGB × 768),[186] and 55″ FHD (1920 × RGB × 1080)[15] have been demonstrated with impressive image quality.

1.5.3 Monochrome AMPLEDs Made with Solution-Processable Polymers

Monochrome AMPLED flat-panel displays have been made with performance parameters attractive for battery-powered portable applications. The AM backpanel was made of poly-silicon material with integrated column and row drivers. Table 1.1 shows a data sheet for a 4″ diagonal, monochrome display with 960 × 240 pixels.[187] The resolution in horizontal direction was 300 ppi (85 μm pitch size), while the resolution in vertical direction was 100 ppi

TABLE 1.1

Performance Parameters of a 4″ Diagonal Monochrome AMOLED Panel (without Circular Polarizer)

Active area	3.2″ × 2.4″ (81.6 mm × 61.2 mm)
Pitch size	85 μm × 255 μm
Color coordinates	$x = 0.48, y = 0.53$
Voltage	6.7 V
Panel current	71 mA
Pixel current	0.31 μA
Luminance	205 cd/m^2
Luminous efficiency	14.5 cd/A
Power efficiency	6.8 lm/W
Power consumption	0.47 W (100% pixels on)

(255 μm pitch size). With 100% pixels turned on at 200 cd/m^2, the entire AMPLED panel (including pixel drivers) only consumed electric power less than 500 mW (with 100% pixels on), which was the best number ever reported for AMPLEDs and AMOLEDs. The power efficiency reached 6.8 lm/W (without circular polarizer). These values are three to five times better than those of transflective AMLCDs with fluorescent backlight. Therefore, they are especially suitable for mobile, portable, and handhold applications. The local emission intensity is 1/aperture ratio times the effective areal luminescence. For an AMPLED with aperture ratio of 50%, the display lifetime is approximately half of that measured from a PLED backlight device. This relation has been confirmed in DuPont's monochrome AMPLED panels. Gen-3 to Gen-8.5 panel sizes are currently used for manufacturing AM backpanels.

When an AMOLED display panel is used for graphic or video applications, the average power consumption is only 20–50% of that consumed at 100% pixel light on at full intensity (depending on image content), reducing the power consumption further by two to five times.[187,188] It is worth mentioning that OLED displays are of superb performance at temperatures below room temperature (with fast response time, sustaining EL efficiency and a longer operation lifetime), in contrast to AMLCDs' performances at low temperature.

1.5.4 Full-Color AMPLED Modules

In manufacturing monochromatic OLED displays, the spin-coating process has been widely adopted for solution-based materials. To make a display with FC images, each display pixel consists of three subpixels that each emits one of the three primary colors: red, green, and blue, respectively. A reliable, cost-efficient, high-throughput process for producing color subpixels is essential for commercializing FC AMOLEDs, which has been one of the key focuses in AMOLED developments. For FC AMOLED made of small molecules, thermal deposition with shadow masks has been widely adopted for patterning the EL molecule layer into three-color subpixels. A known challenge for this process is to make large-size FC display panels reliably, efficiently, and uniformly. Another challenge facing thermal deposition is to make FC OLED displays with a large mother glass sheet (Gen-3 to Gen-7 panel sizes are currently used for manufacturing AM backpanels). A popular process used in solution-processed FC OLED to pattern color subpixels is printing, which is a scalable process, as demonstrated in the paper printing industry. Ink-jetting equipment capable of processing Gen-5-sized panels has been demonstrated by Seiko-Epson for manufacturing color filters for AMLCDs. Thus, the FC process by means of ink-jetting is attractive to manufacturing AMPLEDs.

By means of a proprietary nozzle printing process, DuPont Displays has been able to demonstrate FC AMOLEDs at 100 ppi (4″ QVGA and 9.1″ WVGA) and 128 ppi (2.2″ QCIF+). Table 1.2 provides data from a group of 4″ QVGA panels.[187] A photo of such display panel is shown in Figure 1.25a. The total thickness of the display panel is <2 mm, with weight of only 26 g. A panel thickness of <1 mm can also be made with modified packaging schemes. The AM backpanel was made with poly-silicon technology, with integrated column and row drivers. It was driven with a customer-designed display controller in an analog driving scheme. The equivalent white EL efficiency reaches 2.1 cd/A (without circular polarizer). The panel consumes ~260 mA with 100% pixels on at 115 cd/m^2 (white

TABLE 1.2

Performance of a Set of 4″ Diagonal FC AMOLEDs Made with Solution Process

Panel ID	Red (cd/A)	Green (cd/A)	Blue (cd/A)	White (cd/A)
A-DDD-G	1.15	9.2	0.69	1.89
A-DDD-C	0.92	4.7	0.80	1.68
A-DDD-F	0.56	6.6	0.43	1.1
A-DDD-F	0.94	5.07	1.34	2.08

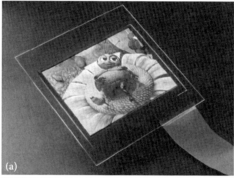

FIGURE 1.25
(See color insert.) Photo of an FC AMOLED in 4″ diagonal QVGA format (a) and 14.1″ diagonal WXGA (1280 × 768) format (b).

TABLE 1.3

Power Consumption of AMOLEDs for Three Sets of OLED Emitters (Portion on Pixel Driver Is Excluded)

Format	Size	Power Consumption (PLED-1)[a]	Power Consumption (PLED-2)[b]	Power Consumption (PLED-3)[c]	Target Applications
QCIF+, 220 × RGB × 176	2.2″, 128 ppi	0.605 W	0.214 W	0.141 W	Mobile phone, digital camera
QVGA, 320 × RGB × 240	4.0″, 100 ppi	1.99 W	0.71 W	0.47 W	Hand PC, PDA, game console, web reader
WQVGA, 480 × RGB × 240	5.4″, 100 ppi	2.97 W	1.04 W	0.70 W	DVD screen, GPS
VGA, 640 × RGB × 480	8.0″, 100 ppi	7.92 W	4.2 W	1.86 W	Portable TV, PC monitors
WVGA, 800 × RGB × 480	9.3″, 100 ppi	9.95 W	3.53 W	2.32 W	Portable TV, PC monitors
WXGA, 1280 × RGB × 768	15″, 100 ppi	25.3 W	8.9 W	5.9 W	Table-top TV, PC monitors
UXGA, 1600 × RGB × 1200	20″, 100 ppi	49.7 W	17.6 W	11.6 W	Table-top and wall-mount TVs
WXGA, 1280 × RGB × 768	40″, 37 ppi	182 W	64 W	42.4 W	Wall-mount TVs

[a] Based on a set of polymer RGB emitters with EL efficiencies of 1 cd/A (R), 5 cd/A (G), and 1 cd/A (B), respectively.

[b] Based on a set of polymer RGB emitters with EL efficiencies of 2 cd/A (R), 10 cd/A (G), and 2 cd/A (B), respectively.

[c] Based on a set of polymer RGB emitters with EL efficiencies of 5 cd/A (R), 17 cd/A (G), and 3 cd/A (B), respectively.

equivalent without a circular polarizer). When running video images, a peak luminescence of 200 cd/m² was achieved. Again, when such a panel is used for a video screen, the average power consumption is only 20–50% of the full power, lower than the power needed for an AMLCD panel with identical size and format.

The performance of AMOLEDs has improved drastically. In contrast to the data shown in Table 1.2 (representing the development stage in 2002), a set of data of a 14.1″ WXGA AMLCD made with solution-processed OLED emitters is shown in Table 1.3.[177,189,190] The color gamut is improved to >60% with respect to NTSC. The luminous and power efficiencies at white point ($x \sim 0.28$, $y \sim 0.31$) are >8 cd/A and >5 lm/W, respectively. The power efficiency surpasses the performance of AMLCDs, plasma displays, and all other known flat-panel displays in commercial market or under development. A photo of the 14.1″ AMOLED display is shown in Figure 1.25b.

1.5.5 Performance Simulation for FC AMOLEDs

A simulation protocol has been developed for designing monochrome and FC AMOLED panels. With input parameters of color coordinates, luminance–voltage, and current–voltage data of the red, green, and blue emitters; geometric dimensions and aperture ratio of the subpixels; and the numbers of rows and columns, one can predict the AMOLED panel performance, including power consumption, at a given brightness and luminous efficiency. In reverse, one can also extrapolate the minimum requirements for

TABLE 1.4

EL Efficiencies of Equivalent White Generated by Color Pixels Made by Three Different Color Reproduction Approaches

FC Production Approach	Luminous Efficiency (cd/A)	Power Efficiency (lm/W)
Color from R, G, B emitters	11.2	11.8
Color from blue emitter with phosphor filters	8.64	7.23
Color from white emitter with transmission filters	0.908	0.759

EL emitters for a given display format, display brightness, and power budget. Table 1.3 provides the power consumption of the PLED portion for a series of display panels of popular formats. The data are obtained at the panel's full brightness of 200 cd/m^2 without circular polarizer when all the pixels are fully turned on. An aperture ratio of 50% was assumed. Power consumptions were calculated on the basis of three groups of polymer emitters available with different maturities. This simulation tool has been used for several AMOLED display panels and proven effective and accurate. Comparing the data in Table 1.3 with the testing results shown in Table 1.2, one can see that the performance of DuPont's 4″ FC AMOLED panels has passed the first performance target, approaching the second one.

The simulation discussed above was based on FC produced by individual red, green, and blue emitters. Other FC reproduction approaches have been proposed for OLED displays, including color from a blue emitter by means of energy down-conversion fluorescent filters[191] and color from white emitters by means of transmission color filter sets similar to that used in the LCD industry.[192,193] Table 1.4 compares the EL efficiency of the equivalent white generated by red, green, and blue subpixels from three different color reproduction approaches.[194] The data were based on EL emitters with equal quantum efficiency (5% ph/el) and identical geometric pixel form factors. In the color-from-blue approach, an energy down-conversion efficiency of 70% was used for the fluorescent filers. The results suggest that the approach of FC reproduced from the red, green, and blue color emitters has the best power efficiency, while the approach of FC reproduced from the white emitter is the least efficient. The display panel operation lifetime (for a given OLED efficiency and lifetime) follows the same trend as that discussed in the foregoing sections, because of the different local brightness needed for achieving a given panel brightness. It is also worth mentioning that the order of manufacturing simplicity is perhaps reversed in the three different color reproduction approaches. The choice of the FC manufacturing process can thus be made differently on the basis of market differentiation, display format and performance requirements, OLED lifetime, process cost, yield, and so on.

1.5.6 AMOLED for Graphic and Motion Picture Applications

An important fact that has not been well addressed in the OLED field is that the power consumption of an OLED display panel changes with the information content. For an AMLCD display panel, the power consumption is almost independent of the information content. For an AMOLED or an AMPLED panel, the power consumption is directly proportional to the number of pixels lighting up. For each display pixel, the power consumption is nearly proportional to the level of brightness (gray level). Thus, AMOLED display only consumes the power necessary, without any waste. This effect is similar to the concept of "pay-per-view" developed in cable and satellite TV industries. Two direct consequences of this effect are as follows:

1. The pay-per-view effect reduces the panel power consumption significantly.
2. The pay-per-view effect extends display operation lifetime substantially.

To investigate the effectiveness of the pay-per-view for graphic and video applications, nine movies were selected randomly and were run with an AMPLED panel. The panel current and power were recorded with respect to the operation time. The total energy consumed for each movie and the respective average power consumption were then derived. An effective power-saving factor (which can also be defined as a lifetime-enhancing factor) was calculated by dividing the full power corresponding to 100% pixels turned on at maximum brightness (provided in Tables 1.1 and 1.3) by the average power consumed for each movie. The results are listed in Table 1.5. The results showed that

1. The power consumption of an OLED display varies over a broad range for videos in different subjects and different categories.
2. The average power of an AMPLED panel for video applications is only 20–50% of the full power.

The average lifetime-enhancing factor for the nine movies is at least 3.3 times longer than that measured with full screen on at full brightness; for example, a panel with 10,000 h operation life under maximum brightness has an effective operation lifetime of ~33,000 h for video applications. This statement is based on the simple charge conservation mechanism discussed in the early sections. It is worth mentioning that the operation lifetime of the colored OLED emitters is inversely proportional to $L^{-\beta}$ (in which $\beta = 1.3–2.2$ was frequently observed in both SMOLED and PLED) at the interesting brightness. Taking this superlinearity into consideration, the operation lifetime in video operation would be larger than the power-saving factor discussed above. Assuming the average intensity for movie operation is one-third of the full intensity, the lifetime-enhancing factor would be 5.3 times that driving at full brightness. The pay-per-view effect makes OLED displays promising for motion image applications.

Another significance of this finding is the effective power saving (with the same factor as lifetime enhancement) for AMOLED displays being used for graphic and video applications. This feature makes OLED displays attractive especially for battery-powered mobile/

TABLE 1.5

Effectiveness in Power Saving and Lifetime Extension

Movie	Operation Time (min)	Average Power (W)	Power-Saving/Lifetime-Enhancing Factor
#1	87.5	0.763	2.13
#2	94.25	0.602	2.70
#3	94.33	0.535	3.03
#4	98.67	0.564	2.88
#5	24.87	0.674	2.41
#6	87.55	0.400	4.06
#7	77.62	0.244	6.65
#8	118.07	0.486	3.34
#9	94.53	0.697	2.33

portable applications. Taking this power-saving factor into account, the AMOLED displays with the performance parameters shown in Tables 1.1 through 1.3 have already won over AMLCD for graphic and video applications. In 2005, DuPont Displays and Samsung demonstrated a 14.1″ diagonal, solution-processed FC AMOLED display in HDTV format at the 2005 Society for Information Display. The current efficiency and power efficiency were improved to >8 cd/A and 5 lm/W.[189,190] AMOLED for video application also minimizes/eliminates the differential aging among emissive display pixels.

All these features above make AMOLED especially suitable for portable DVD players, digital cameras, portable TVs, and game players. The discussion in this section also suggests that the performance of a given OLED display can be maximized by the proper design of display contents.

1.6 Summary and Remarks

The development history of organic light-emitting device and OLED displays provides a great story on how scientific concepts and discoveries can be transformed into application technologies and eventually impact human life in many different ways. Within less than 30 years, the field of organic light emitters has been migrated through concept demonstration, single-device performance improvement, industrial development in matrix formation, color pixel formation, display architecture, and system integration. The performance parameters of OLED displays have been improved to the level better than other existing display technologies and better than that needed for many commercial applications. Products in both PMOLED and AMOLED forms have been in commercial markets for several years. Most recently, LG has commercialized 55″ and larger OLED HDTVs (Figure 1.26).[15] The

FIGURE 1.26
(See color insert.) LG's OLED TV displayed at Society of Information Display 2014 meeting.

market of OLED display industry is expected to reach 35 billion dollars by 2018 up from 4 billion in 2011 estimated by Display Research.[195]

As mentioned in this chapter, AMPLEDs are especially attractive for motion picture applications. The pay-per-view effect in OLED displays reduces power consumption and extends operation lifetime. Motion picture applications also minimize image retention and optimize display homogeneity. AMOLED has been widely viewed as a promising display technology in competing with AMLCD and plasma displays. The dream of using organic semiconductor films for optoelectronic device applications has become reality.

In addition to the light emission effect and its applications, other electric and optoelectronic effects and device applications in organic semiconductors have been well studied, including photovoltaic cells, photodiodes, image sensors, and thin-film transistors. The field of organic electronics has become one of the hottest areas in the past decade. Owing to the excellent optoelectronic and electro-optical conversion efficiencies, and less demand on charge mobilities in thin-film device configuration, significant progress has been achieved in all optoelectric and electro-optical applications. The performance parameters have been improved to the level comparable to or substantially better than their inorganic counterparts, promising a variety of practical applications.

Acknowledgment

The authors would like to acknowledge the many stimulating discussions and the support of our former coworkers in DuPont Displays.

References

1. H. J. Round, *Electron. World* 19, 309 (1907).
2. N. Holonyak Jr., and S. F. Bevacqua, *Appl. Phys. Lett.* 1, 82 (1962).
3. T. S. Perry, *IEEE Spectrum* (June 3, 2003).
4. A. Bernanose, M. Comte, and P. Vouaux, *J. Chim. Phys.* 50, 64 (1953).
5. A. Bernanose, *J. Chim. Phys.* 52, 396 (1955).
6. M. Pope, H. Kallman, and P. Magnante, *J. Chem. Phys.* 38, 2042 (1963).
7. R. H. Partridge, US Patent # 3,995,299 (1976).
8. W. Helfrich, and W. G. Schneider, *Phys. Rev. Lett.* 14, 229 (1965).
9. J. Dresner, *RCA Rev.* 30, 322 (1969).
10. D. F. Williams, and M. Schadt, *J. Chem. Phys.* 53, 3480 (1970).
11. P. S. Vincett, W. A. Barlow, R. A. Hann, and G. G. Roberts, *Thin Solid Films* 94, 171 (1982).
12. C. W. Tang, and S. A. VanSlyke, *Appl. Phys. Lett.* 51, 913 (1987).
13. Press Release. Available at http://www.kodak.com.
14. OLED-Info.com. Available at http://www.oled-info.com/sony-xel-1 (November 17, 2008).
15. OLED-Info.com. Available at http://www.oled-info.com/lgs-55-oled-tv-finally-sale-10000 -korea-will-ship-next-month (January 2, 2013).
16. C. K. Chiang, C. R. Fincher, Y. W. Park, A. J. Heeger, H. Shirakawa, E. J. Louis, S. C. Gau, and A. G. MacDiarmid, *Phys. Rev. Lett.* 39, 1098 (1977).
17. H. Shirakawa, E. J. Louis, A. G. MacDiarmid, C. K. Chiang, and A. J. Heeger, *Chem. Commun.* 16, 578 (1977).

18. M. D. McGehee, E. K. Miller, D. Moses, and A. J. Heeger, in *Advances in Synthetic Metals: Twenty Years of Progress in Science and Technology*, edited by P. Bernier, S. Lefrant, and G. Bidan (Elsevier, Amsterdam, 1999), p. 98.
19. R. Kiebooms, R. Menon, and K. Lee, in *Handbook of Advanced Electronic and Photonic Materials and Devices*, edited by H. S. Nalwa (Academic Press, 2001), p. 1.
20. M. Ozaki, D. L. Peebles, B. R. Weinberg, C. K. Chiang, S. C. Gau, A. J. Heeger, and A. G. MacDiarmid, *Appl. Phys. Lett.* 35, 83 (1979).
21. B. R. Weinberg, S. C. Gau, and Z. Kiss, *Appl. Phys. Lett.* 38, 555 (1981).
22. H. Tomozawa, D. Braun, S. D. Phillips, A. J. Heeger, and H. Kroemer, *Synth. Met.* 22, 63 (1987).
23. J. H. Burroughes, D. D. C. Bradley, A. R. Brown, R. N. Marks, K. Mackay, R. H. Friend, P. L. Burns, and A. B. Holmes, *Nature* 347, 539 (1990).
24. D. Braun, and A. J. Heeger, *Appl. Phys. Lett.* 58, 1982 (1991).
25. F. Wudl, P.-M. Allemand, G. Srdanov, Z. Ni, and D. McBranch, in *Materials for Non-Linear Optics: Chemical Perspectives*; vol. 455, edited by S. R. Marder, J. E. Sohn, and G. D. Stucky (The American Chemical Society, 1991), p. 683.
26. F. Wudl, and G. Srdanov, US Patent # 5,189,136 (1993).
27. H. Spreitzer, H. Becker, E. Kluge, W. Kreuder, H. Schenk, R. Demandt, and H. Schoo, *Adv. Mater.* 10, 1340 (1998).
28. H. Becker, H. Spreitzer, K. Ibrom, and W. Kreuder, *Macromolecules* 32, 4925 (1999).
29. G. Yu, and A. J. Heeger, in *The Physics of Semiconductors*; vol. 1, edited by M. Schleffer, and R. Zimmerman (World Scientific, Singapore, 1996), p. 35.
30. C. Zhang, G. Yu, and Y. Cao, US Patent # 5,798,170 (1998).
31. H. Becker, H. Spreitzer, W. Kreuder, E. Kluge, H. Schenk, I. D. Parker, and Y. Cao, *Adv. Mater.* 12, 42 (2000).
32. D. M. Johansson, G. Srdanov, G. Yu, M. Theander, O. Inganäs, and M. R. Andersson, *Macromolecules* 33, 2525 (2000).
33. D. M. Johansson, X. Wang, T. Johansson, O. Inganäs, G. Yu, G. Srdanov, and M. R. Andersson, *Macromolecules* 35, 4997 (2002).
34. Y. Ohmori, M. Uchida, K. Muro, and K. Yoshino, *Jpn. J. Appl. Phys.* 30, L1941 (1991).
35. M. Uchida, Y. Ohmori, C. Morishima, and K. Yoshino, *Synth. Met.* 55–57, 4168 (1993).
36. Q. Pei, and Y. Yang, *J. Am. Chem. Soc.* 118, 7416 (1996).
37. G. Grem, G. Leditzky, B. Ullrich, and G. Leising, *Adv. Mater.* 4, 36 (1992).
38. B. Grem, G. Leditzky, B. Ullrich, and G. Leising, *Synth. Met.* 51, 383 (1992).
39. Y. Yang, Q. Pei, and A. J. Heeger, *J. Appl. Phys.* 79, 934 (1996).
40. J. Huber, K. Müllen, J. Salbeck, H. Schenk, U. Scherf, T. Stehlin, and R. Stern, *Acta Polym.* 45, 244 (1994).
41. J. Grüner, P. J. Hamer, R. H. Friend, H. J. Huber, U. Scherf, and A. B. Holmes, *Adv. Mater.* 6, 748 (1994).
42. G. Horowitz, P. Delannoy, H. Bouchriha, F. Deloffre, J.-L. Fave, F. Garnier, R. Hajlaoui, M. Heyman, F. Kouki, P. Valat, V. Wintgens, and A. Yassar, *Adv. Mater.* 6, 752 (1994).
43. G. Grem, V. Martin, F. Meghdadi, C. Paar, J. Stampfl, J. Sturm, S. Tasch, and G. Leising, *Synth. Met.* 71, 2193 (1995).
44. S. Tasch, A. Niko, G. Leising, and U. Scherf, *Appl. Phys. Lett.* 68, 1090 (1996).
45. I. Sokolik, Z. Yang, F. E. Karasz, and D. C. Morton, *J. Appl. Phys.* 74, 3584 (1993).
46. C. Zhang, H. von Seggern, K. Pakbaz, B. Kraabel, H.-W. Schmidt, and A. J. Heeger, *Synth. Met.* 62, 35 (1994).
47. A. Kraft, A. C. Grimsdale, and A. B. Holmes, *Angew. Chem. Int. Ed.* 37, 402 (1998).
48. M. Bernius, M. Inbasekaran, E. Woo, W. Wu, and L. Wujkowski, *Thin Solid Films* 363, 55 (2000).
49. M. Bernius, M. Inbasekaran, E. Woo, W. Wu, and L. Wujkowski, *J. Mater. Sci. Mater. Electron.* 11, 111 (2000).
50. H. Becker, S. Heun, K. Treacher, A. Bing, and A. Falcou, *SID 02 Dig.*, 780 (2002).
51. T. R. Hebner, C. C. Wu, D. Marcy, M. H. Lu, and J. C. Sturm, *Appl. Phys. Lett.* 72, 519 (1998).
52. S.-C. Chang, J. Liu, J. Bharathan, Y. Yang, J. Onohara, and J. Kido, *Adv. Mater.* 11, 734 (1999).

53. F. Pschenitzka, and J. C. Sturm, *Appl. Phys. Lett.* 74, 1913 (1999).
54. D. A. Pardo, G. E. Jabbour, and N. Peyghambarian, *Adv. Mater.* 12, 1249 (2000).
55. C. D. Müller, A. Falcou, N. Reckefuss, M. Rojahn, V. Wiederhirn, P. Rudati, H. Frohne, O. Nuyken, H. Becker, and K. Meerholz, *Nature* 421, 829 (2003).
56. SID 2002 Advance Program. Available at http://www.sid.org/conf/sid2002/sid2002.html (2002).
57. Samsung SDI, IMID Exhibition, *Asia Display/IMID'04* (Daegu, Korean, August 23–24, 2004).
58. R. W. Lenz, C.-C. Han, J. Stenger-Smith, and F. E. Karasz, *J. Polym. Sci. Part A* 26, 3241 (1988).
59. M. R. Andersson, G. Yu, and A. J. Heeger, *Synth. Met.* 85, 1275 (1997).
60. S. M. Sze, *Physics of Semiconductor Devices*, 2nd ed. (Wiley, New York, 1981).
61. G. Yu, *Synth. Met.* 80, 143 (1996).
62. G. Yu, K. Pakbaz, and A. J. Heeger, *J. Electron. Mater.* 23, 925 (1994).
63. G. Yu, J. Gao, J. C. Hummelen, F. Wudl, and A. J. Heeger, *Science* 270, 1789 (1995).
64. G. Yu, J. Wang, J. McElvain, and A. J. Heeger, *Adv. Mater.* 10, 1431 (1998).
65. A. R. Brown, K. Pichler, N. C. Greenham, D. D. C. Bradley, R. H. Friend, and A. B. Holmes, *Chem. Phys. Lett.* 210, 61 (1993).
66. Y. Cao, I. D. Parker, G. Yu, C. Zhang, and A. J. Heeger, *Nature* 397, 414 (1999).
67. J.-S. Kim, P. K. H. Ho, N. C. Greenham, and R. H. Friend, *J. Appl. Phys.* 88, 1073 (2000).
68. Z. Shuai, D. Beljonne, R. J. Silbey, and J. L. Brédas, *Phys. Rev. Lett.* 84, 131 (2000).
69. M. Wohlgenannt, K. Tandon, S. Mazumdar, S. Ramasesha, and Z. V. Vardeny, *Nature* 409, 494 (2001).
70. B. Hu, Y. Wu, Z. Zhang, S. Dai, and J. Shen, *Appl. Phys. Lett.* 88, 022114 (2006).
71. V. Cleave, G. Yahioglu, P. Le Barny, R. H. Friend, and N. Tessler, *Adv. Mater.* 11, 285 (1999).
72. M. A. Baldo, M. E. Thomson, and S. R. Forrest, *Nature* 403, 750 (2000).
73. C. Adachi, M. A. Baldo, S. R. Forrest, and M. E. Thompson, *Appl. Phys. Lett.* 77, 904 (2000).
74. A. P. Monkman, H. D. Burrows, L. J. Hartwell, L. E. Horsburgh, I. Hamblett, and S. Navaratnam, *Phys. Rev. Lett.* 86, 1358 (2001).
75. L. Smilowitz, A. Hays, A. J. Heeger, G. Wang, and J. E. Bowers, *J. Chem. Phys.* 98, 6504 (1993).
76. M. Yan, L. J. Rothberg, F. Papadimitrakopoulos, M. E. Galvin, and T. M. Miller, *Phys. Rev. Lett.* 72, 1104 (1994).
77. N. C. Greenham, I. D. W. Samuel, G. R. Hayes, R. T. Phillips, Y. A. R. R. Kessener, S. C. Moratti, A. B. Holmes, and R. H. Friend, *Chem. Phys. Lett.* 241, 89 (1995).
78. S. V. Frolov, M. Liess, P. A. Lane, W. Gellermann, Z. V. Vardeny, M. Ozaki, and K. Yoshino, *Phys. Rev. Lett.* 78, 4285 (1997).
79. J. Kido, K. Nagai, Y. Okamoto, and T. Skotheim, *Chem. Lett.* 20, 1267 (1991).
80. M. D. McGehee, T. Bergstedt, C. Zhang, A. P. Saab, M. B. O'Regan, G. C. Bazan, V. I. Srdanov, and A. J. Heeger, *Adv. Mater.* 11, 1349 (1999).
81. M. A. Baldo, D. F. O'Brien, Y. You, A. Shoustikov, S. Sibley, M. E. Thompson, and S. R. Forrest, *Nature* 395, 151 (1998).
82. W. Zhu, Y. Mo, M. Yuan, W. Yang, and Y. Cao, *Appl. Phys. Lett.* 80, 2045 (2002).
83. X. Gong, J. C. Ostrowski, M. R. Robinson, D. Moses, G. C. Bazan, and A. J. Heeger, *Adv. Mater.* 14, 581 (2002).
84. D. Braun, G. Gustafsson, D. McBranch, and A. J. Heeger, *J. Appl. Phys.* 70, 564 (1992).
85. N. C. Greenham, S. C. Moratti, D. D. C. Bradley, R. H. Friend, and A. B. Holmes, *Nature* 365, 628 (1993).
86. J. Kido, M. Kohda, K. Okuyama, and K. Nagai, *Appl. Phys. Lett.* 61, 761 (1992).
87. J. Kido, K. Hongawa, K. Okuyama, and K. Nagai, *Appl. Phys. Lett.* 64, 815 (1994).
88. C.-C. Wu, J. Chun, P. Burrows, S. Forrest, R. A. Register, J. C. Sturm, and M. E. Thompson, *ACS Polym. Preprints*, in press (1994).
89. J. Kido, H. Shionoya, and K. Nagai, *Appl. Phys. Lett.* 67, 2281 (1995).
90. M. Berggren, O. Inganäs, G. Gustafsson, J. Rasmusson, M. R. Andersson, T. Hjertberg, and O. Wennerström, *Nature* 372, 444 (1994).
91. C. Zhang, S. Hoger, K. Pakbaz, F. Wudl, and A. J. Heeger, *J. Electron. Mater.* 23, 453 (1994).

92. H. Nishino, G. Yu, A. J. Heeger, T.-A. Chen, and R. D. Rieke, *Synth. Met.* 68, 243 (1995).
93. G. Yu, N. Nishino, A. J. Heeger, T.-A. Chen, and R. D. Rieke, *Synth. Met.* 72, 249 (1995).
94. S. Tasch, E. J. W. List, O. Ekström, W. Graupner, G. Leising, P. Schlichting, U. Rohr, Y. Geerts, U. Scherf, and K. Müllen, *Appl. Phys. Lett.* 71, 2883 (1997).
95. M. T. Bernius, M. Inbasekaran, J. O'Brien, and W. Wu, *Adv. Mater.* 12, 1737 (2000).
96. E. P. Woo, W. R. Shiang, M. Inbasekaran, and G. R. Roof, US Patent # 5,708,130 (1998).
97. J. Liu, S. Hu, W. Zhao, Q. Zou, W. Luo, W. Yang, J. Peng, and Y. Cao, *Macromol. Rapid Commun.* 31, 496 (2010).
98. Y. Cao, P. Smith, and A. J. Heeger, *Synth. Met.* 48, 91 (1992).
99. Y. Cao, G. M. Treacy, P. Smith, and A. J. Heeger, *Appl. Phys. Lett.* 60, 2711 (1992).
100. Y. Cao, G. Yu, C. Zhang, R. Menon, and A. J. Heeger, *Synth. Met.* 87, 171 (1997).
101. J. Gao, A. J. Heeger, J. Y. Lee, and C. Y. Kim, *Synth. Met.* 82, 221 (1996).
102. K. Lee, E. K. Miller, A. N. Aleshin, R. Menon, A. J. Heeger, J. H. Kim, C. O. Yoon, and H. Lee, *Adv. Mater.* 10, 456 (1998).
103. R. Menon, C. O. Yoon, D. Moses, and A. J. Heeger, in *Handbook of Conducting Polymers*, 2nd ed., edited by T. A. Skotheim, R. L. Elsenbaumer, and J. R. Reynolds (Marcel Dekker, New York, 1998), p. 27.
104. Y. Yang, and A. J. Heeger, *Appl. Phys. Lett.* 64, 1245 (1994).
105. H. Wang, and G. Yu, Patent Application US03016853 (2003).
106. M. Radler, *IDRC 03* (2003).
107. ICEL2012, Cambridge Display Technology, *Progress in Polymer OLED Efficiency.* Available at http://www.cdtltd.co.uk/pdf/ICEL2012final.pdf.
108. G. Gustafsson, Y. Cao, G. M. Treacy, F. Klavetter, N. Colaneri, and A. J. Heeger, *Nature* 357, 477 (1992).
109. F. Huang, L. Hou, H. Wu, X. Wang, H. Shen, W. Cao, W. Yang, and Y. Cao, *J. Am. Chem. Soc.* 126, 9845 (2004).
110. F. Huang, Y.-H. Niu, Y. Zhang, J.-W. Ka, M. S. Liu, and A. K.-Y. Jen, *Adv. Mater.* 19, 2010 (2007).
111. F. Huang, H. Wu, D. Wang, W. Yang, and Y. Cao, *Chem. Mater.* 16, 708 (2004).
112. L. Wang, B. Liang, F. Huang, J. Peng, and Y. Cao, *Appl. Phys. Lett.* 89, 151115 (2006).
113. H. Wu, F. Huang, Y. Mo, W. Yang, D. Wang, J. Peng, and Y. Cao, *Adv. Mater.* 16, 1826 (2004).
114. H. Wu, F. Huang, J. Peng, and Y. Cao, *Org. Electron.* 6, 118 (2005).
115. Q. Wang, Y. Zhou, H. Zheng, J. Shi, C. Li, C. Q. Su, L. Wang, C. Luo, D. Hu, J. Pei, J. Wang, J. Peng, and Y. Cao, *Org. Electron.* 12, 1858 (2011).
116. J. M. Lupton, I. D. W. Samuel, R. Beavington, P. L. Burn, and H. Bässler, *Adv. Mater.* 13, 258 (2001).
117. S.-C. Lo, E. B. Namdas, P. L. Burn, and I. D. W. Samuel, *Macromolecules* 36, 9721 (2003).
118. S. T. Lee, J. Y. Lee, M. H. Kim, M. C. Suh, T. M. Kang, Y. J. Choi, J. Y. Park, J. H. Kwon, H. K. Chung, J. Baetzold, E. Bellmann, V. Savvateev, M. Wolk, and S. Webster, *SID 02 Dig.*, 784 (2002).
119. M. B. Wolk, P. Γ. Baude, J. M. Florczak, F. B. McCormick, and Y. Hsu, US Patent # 6,114,088 (2000).
120. M. B. Wolk, P. F. Baude, F. B. McCormick, and Y. Hsu, US Patent # 6,194,119 (2001).
121. C. V. Hoven, A. Garcia, G. C. Bazan, and T.-Q. Nguyen, *Adv. Mater.* 20, 3793 (2008).
122. H. Jiang, P. Taranekar, J. R. Reynolds, and K. S. Schanze, *Angew. Chem. Int. Ed.* 48, 4300 (2009).
123. J. Wang, R. G. Sun, G. Yu, and A. J. Heeger, *J. Appl. Phys.* 91, 2417 (2002).
124. B. Werner, J. Posdorfer, B. Webling, H. Becker, S. Heun, H. Vestweber, and T. Hassenkam, *SID 02 Dig.*, 603 (2002).
125. Y. A. Ono, *Electroluminescent Displays* (World Scientific, Singapore, 1995).
126. R. H. Fowler, and L. Nordheim, *Proc. R. Soc. Lond.* 119, 173 (1928).
127. I. D. Parker, *J. Appl. Phys.* 75, 1656 (1994).
128. D. Braun, PhD dissertation Thesis, University of California, Santa Barbara, CA (1991).
129. P. W. M. Blom, M. J. M. de Jong, and J. J. M. Vleggaar, *Appl. Phys. Lett.* 68, 3308 (1996).
130. P. W. M. Blom, and M. C. J. M. Vissenberg, *Phys. Rev. Lett.* 80, 3819 (1998).
131. H. Antoniadis, M. A. Abkowitz, and B. R. Hsieh, *Appl. Phys. Lett.* 65, 2030 (1994).

132. D. J. Pinner, R. H. Friend, and N. Tessler, *J. Appl. Phys.* 86, 5116 (1999).
133. S. A. Van Slyke, C. H. Chen, and C. W. Tang, *Appl. Phys. Lett.* 69, 2160 (1996).
134. G. Yu, C. Zhang, Q. Pei, Y. Cao, Y. Yang, and A. J. Heeger, in *Low-Voltage, High-Brightness Polymer Light-Emitting Diodes with Long Stress Life* (San Francisco, 1996).
135. G. Yu, and A. J. Heeger, *Synth. Met.* 85, 1183 (1997).
136. I. D. Parker, Y. Cao, and C. Y. Yang, *J. Appl. Phys.* 85, 2441 (1999).
137. N. T. Harrison, N. Tessler, C. J. Moss, R. H. Friend, and K. Pichler, *Optic. Mater.* 9, 178 (1998).
138. D. Moses, *Appl. Phys. Lett.* 60, 3215 (1992).
139. N. Tessler, G. J. Denton, and R. H. Friend, *Nature* 382, 695 (1996).
140. F. Hide, M. A. Díaz-García, B. J. Schwartz, M. R. Andersson, Q. Pei, and A. J. Heeger, *Science* 273, 1833 (1996).
141. S. V. Frolov, M. Shkunov, and Z. V. Vardeny, *Phys. Rev. B* 56, R4363 (1997).
142. M. D. McGehee, and A. J. Heeger, *Adv. Mater.* 12, 1655 (2000).
143. DuPont Displays News Release, *SID 03* (2003).
144. M. Leadbeater, N. Patel, B. Tierney, S. O'Connor, I. Grizzi, and C. Towns, *SID 04 Dig.*, 162 (2004).
145. G. Yu, C. Zhang, and A. J. Heeger, *Appl. Phys. Lett.* 64, 1540 (1994).
146. M. Meier, S. Karg, K. Zuleeg, W. Brütting, and M. Schwoerer, *J. Appl. Phys.* 84, 97 (1998).
147. A. Kadashchuk, Y. Skryshevski, Y. Piryatinski, A. Vakhnin, E. V. Emelianova, V. I. Arkhipov, H. Bässler, and J. Shinar, *J. Appl. Phys.* 91, 5016 (2002).
148. Y. Wang, *Nature* 356, 585 (1992).
149. N. S. Sariciftci, L. Smilowitz, A. J. Heeger, and F. Wudl, *Science* 258, 1474 (1992).
150. K. Y. Law, *Chem. Rev.* 93, 449 (1993).
151. C. H. Lee, G. Yu, D. Moses, K. Pakbaz, C. Zhang, N. S. Sariciftci, A. J. Heeger, and F. Wudl, *Phys. Rev. B* 48, 15425 (1993).
152. G. Yu, K. Pakpaz, and A. J. Heeger, *Appl. Phys. Lett.* 64, 3422 (1994).
153. N. S. Sariciftci, D. Braun, C. Zhang, V. I. Srdanov, A. J. Heeger, and F. Wudl, *Appl. Phys. Lett.* 62, 585 (1993).
154. C. J. Brabec, C. Winder, N. S. Sariciftci, J. C. Hummelen, A. Dhanabalan, P. A. van Hal, and R. A. J. Janssen, *Adv. Funct. Mater.* 12, 709 (2002).
155. M. Svensson, F. Zhang, S. C. Veenstra, W. J. H. Verhees, J. C. Hummelen, J. M. Kroon, O. Inganäs, and M. R. Andersson, *Adv. Mater.* 15, 988 (2003).
156. Q. Zhou, Q. Hou, L. Zheng, X. Deng, G. Yu, and Y. Cao, *Appl. Phys. Lett.* 84, 1653 (2004).
157. Siemens Global Website. Available at http://www.siemens.com.
158. A. J. Heeger, and G. Yu, US Patent # 5,504,323 (1996).
159. Q. Pei, G. Yu, C. Zhang, Y. Yang, and A. J. Heeger, *Science* 269, 1086 (1995).
160. Q. Pei, Y. Yang, G. Yu, C. Zhang, and A. J. Heeger, *J. Am. Chem. Soc.* 118, 3922 (1996).
161. Y. Cao, G. Yu, A. J. Heeger, and C. Y. Yang, *Appl. Phys. Lett.* 68, 3218 (1996).
162. Y. Cao, M. R. Andersson, Q. Pei, G. Yu, and A. J. Heeger, *J. Electrochem. Soc.* 144, L317 (1997).
163. G. Yu, Y. Cao, M. Andersson, J. Gao, and A. J. Heeger, *Adv. Mater.* 10, 385 (1998).
164. J. Gao, G. Yu, and A. J. Heeger, *Appl. Phys. Lett.* 71, 1293 (1997).
165. G. Yu, Y. Cao, C. Zhang, Y. Li, J. Gao, and A. J. Heeger, *Appl. Phys. Lett.* 73, 111 (1998).
166. I. H. Campbell, S. Rubin, T. A. Zawodzinski, J. D. Kress, R. L. Martin, D. L. Smith, N. N. Barashkov, and J. P. Ferraris, *Phys. Rev. B* 54, R14321 (1996).
167. Y. Cao, G. Yu, and A. J. Heeger, *Adv. Mater.* 10, 917 (1998).
168. Y. Cao, US Patent # 5,965,281 (1999).
169. Y. Xiong, L. Wang, W. Xu, J. Zou, H. Wu, Y. Xu, J. Peng, J. Wang, Y. Cao, and G. Yu, *Org. Electron.* 10, 857 (2009).
170. Y. Xiong, W. Xu, C. Li, B. Liang, L. Zhao, J. Peng, Y. Cao, and J. Wang, *Org. Electron.* 9, 533 (2008).
171. H. Wu, L. Ying, W. Yang, and Y. Cao, *Chem. Soc. Rev.* 38, 3391 (2009).
172. H. Wu, G. Zhou, J. Zou, C.-L. Ho, W.-Y. Wong, W. Yang, J. Peng, and Y. Cao, *Adv. Mater.* 21, 4181 (2009).
173. U. Lemmer, D. Vacar, D. Moses, and A. J. Heeger, *Appl. Phys. Lett.* 68, 3007 (1996).
174. M. D. Mcgehee, D. Vacar, U. Lemmer, D. Moses, and A. J. Heeger, *Synth. Met.* 85, 1233 (1997).

175. G. Yu, US Patent # 5,677,546 (1997).
176. DuPont Displays, SID Exhibition (Baltimore, May 20–22, 2003).
177. DuPont Displays, SID Exhibition (Boston, May 22–27, 2005).
178. D. Braun, J. Rowe, and G. Yu, *Synth. Met.* 102, 920 (1999).
179. DuPont Displays, SID Exhibition (May 2000); RiTDisplays, SID Exhibition (May 2003).
180. DuPont Displays, SID Exhibition (San Jose, CA, June 5–7, 2001).
181. Toshiba, SID Exhibition (May 2002); Seiko-Epson, SID Exhibition (May 2003); DuPont Displays, SID Exhibition (May 2003); Samsung, SID Exhibition (May 2003); SK Displays, SID Exhibiton (May 2003).
182. DuPont Displays, SID Exhibition (May 2002); Samsung, SID Exhibiton (May 2003).
183. Philips, SID Exhibition (Seattle, WA, May 25–27, 2004).
184. SONY, SID Exhibition (Seattle, WA, May 25–27, 2004).
185. Samsung SDI, SID Exhibition (Seattle, WA, May 25–27, 2004).
186. Toshiba, SID Exhibition (May 2002); Samsung, SID Exhibition (May 2003); IDT/Chi-mei/IBM, SID Exhibiton (May 2003).
187. G. Yu, G. Srdanov, B. Zhang, M. Stevenson, J. Wang, P. Chen, E. Baggao, J. Macias, R. Sun, C. McPherson, P. Sant, J. Innocenzo, M. Stainer, and M. O'Regan, in *Active-Matrix Polymer Displays Made with Electroluminescent Polymers* (SPIE, Orlando, FL, 2003), p. 192.
188. N. C. van de Vaart, E. A. Meulenkamp, N. D. Young, and M. Fleuster, *Asia Display/IMID '04 Dig.*, 337 (2004).
189. G. Yu, in *Solution-Processed OLED Emitters for Active Matrix Displays* (San Diego, CA, 2004).
190. A. K. Saafir, J. Chung, I. Joo, J. Huh, J. Rhee, S. Park, B. Choi, C. Ko, B. Koh, J. Jung, J. Choi, N. Kim, K. Chung, G. Srdanov, C. MacPherson, N. Truong, M. Stevenson, A. Johnson, P. Chen, T. Cardellino, R. Pflanzer, G. Yu, A. Goenaga, M. O'Regan, and D. Keys, *SID 05 Dig.*, 968 (2005).
191. H. Tokailin, C. Hosokawa, and T. Kusomoto, US Patent # 5,126,214 (1992).
192. TDK, SID Exhibition (May 2002 and May 2003); SK Displays, SID Exhibition (May 2003).
193. M. Kashiwabara, K. Hanawa, R. Asaki, I. Kobori, R. Matsuura, H. Yamada, T. Yamamoto, A. Ozawa, Y. Sato, S. Terada, J. Yamada, T. Sasaoka, S. Tamura, and T. Urabe, *SID 04 Dig.*, 1017 (2004).
194. J. Wang, and G. Yu, in *Performance Simulation of Active-Matrix OLED Displays* (Beijing, China, 2004), p. 32.
195. DisplaySearch Q1'12 Quarterly OLED Shipment and Forecast Report.

2

Light-Emitting Polymers

Shidi Xun, Dmitrii F. Perepichka, Igor F. Perepichka, Hong Meng, and Fred Wudl

CONTENTS

2.1 Introduction

The origin of the field of electroluminescent (EL) polymers is connected with the 1990 article by Burroughes and coworkers [1] that describes an EL device based on conjugated poly(*p*-phenylene vinylene) (PPV), although some polymer EL devices based on poly(*N*-vinyl carbazole) (PVK) and doped with luminescent dyes had been reported by Partridge [2,3] long before. In the former paper, a single layer of PPV, placed between indium–tin oxide (ITO) and aluminum (Al) electrodes, emitted green–yellow light under applied direct-current voltage. The device efficiency and relatively low turn-on voltage promised possible technological progress to a state of commercial application. It was clear that such progress would require not only improved device engineering techniques but also sophisticated control of the materials' luminescence efficiency and electron–hole transporting properties, challenging the community of physical organic and polymer chemists.

Since the 1990s until now, light-emitting diodes (LEDs) are probably the most important application, maintaining researchers' interest in conjugated (conducting) polymers; however, in recent years, we witnessed a growing interest in other relevant applications such as sensors and photovoltaics. Hundreds of academic research groups worldwide have contributed to the development of EL polymers. An even more pronounced research activity is held in industries. Several newly born research and development companies, such as Cambridge Display Technologies (CDT, spin-off from Cambridge University), Covion Organic Semiconductors (currently Merck OLED Materials), and UNIAX Corp. (spin-off from University of California of Santa Barbara [UCSB], currently DuPont Displays), are targeted at the development of high-efficiency, long-lifetime EL polymers. A huge commercial potential, connected with the possibility of solution fabrication of EL devices (particularly flat or flexible displays), attracted industrial giants such as Dow Chemical, DuPont, IBM, Kodak, and Philips [4].

Light-emitting polymers (LEPs) have been a subject of many review articles that dealt with various aspects of the design, synthesis, and applications of different classes of LEPs (Table 2.1). Very insightful reviews of a general character have been presented by Holmes and coworkers in 1998, Friend et al. in 1999, and Mitschke and Bäuerle in 2000 (Table 2.1). Among the recent papers, one of the most complete reviews was written by Akcelrud in 2003 (Table 2.1). However, none of the papers mentioned is comprehensive in covering different classes of EL polymers, and cannot be taken as a single source of information on this matter.

This chapter aims to be the most complete collection of references to the existing EL polymers, while discussing the problems of their design, synthesis, physical properties, and the

TABLE 2.1

Reviews Covering the Synthesis and Application of Light-Emitting Polymers

Year	Title	Authors	Publication
1993	Conjugated polymer light-emitting diodes	A.R. Brown, N.C. Greenham, R.W. Gymer, K. Pichler, D.D.C. Bradley, R.H. Friend, P.L. Burn, A. Kraft, and A.B. Holmes	*Intrinsically Conducting Polymers: An Emerging Technology*, NATO ASI Series, Series E: Applied Sciences 246: 87–106
1993	Conjugated polymer electroluminescence	D.D.C. Bradley	*Synth. Met.*, 54: 401–415
1994	Light-emitting diodes fabricated with conjugated polymers—Recent progress	D.R. Baigent, N.C. Greenham, J. Gruener, R.N. Marks, R.H. Friend, S.C. Moratti, and A.B. Holmes	*Synth. Met.*, 67: 3–10
1996	Conjugated polymer electroluminescence	R.H. Friend and N.C. Greenham	*Physical Properties of Polymers Handbook*, J.E. Mark, Ed., AIP Press, New York, pp. 479–487
1997	Polymer electroluminescent devices	Y. Yang	*MRS Bull.*, June: 31–38
1997	Light-emitting polymers: Increasing promise	W.C. Holton	*Solid State Technol.*, 40: 163–169
1997	Single- and heterolayer polymeric light emitting diodes based on poly(*p*-phenylene vinylene) and oxadiazole polymers	W. Rieß	*Organic Electroluminescent Materials and Devices*, S. Miyata and H.S. Nalwa, Eds., Gordon and Breach, Amsterdam, pp. 73–146
1997	Making polymer light emitting diodes with polythiophenes	O. Inganäs	*Organic Electroluminescent Materials and Devices*, S. Miyata and H.S. Nalwa, Eds., Gordon and Breach, Amsterdam, pp. 147–175
1997	Optically detected magnetic resonance (ODMR) studies of π-conjugated polymer-based light emitting diodes (LEDs)	J. Shinar	*Organic Electroluminescent Materials and Devices*, S. Miyata and H.S. Nalwa, Eds., Gordon and Breach, Amsterdam, pp. 177–202
1997	Thin film electroluminescent diodes based on poly(vinyl carbazole)	Z.-L. Zhang, X.-Y. Jiang, S.-H. Xu, and T. Nagatomo	*Organic Electroluminescent Materials and Devices*, S. Miyata and H.S. Nalwa, Eds., Gordon and Breach, Amsterdam, pp. 203–230
1998	Electroluminescent conjugated polymers—Seeing polymers in a new light	A. Kraft, A.C. Grimsdale, and A.W. Holmes	*Angew. Chem. Int. Ed.*, 37: 402–428
1998	The chemistry of electroluminescent organic materials	J.L. Segura	*Acta Polym.*, 49: 319–344
1998	Design and synthesis of polymers for light-emitting diodes	A. Greiner	*Polym. Adv. Technol.*, 9: 371–389
1998	Optical applications	M.G. Harrison and R.H. Friend	*Electronic Materials: The Oligomer Approach*, K. Müllen and G. Wegner, Eds., Wiley-VCH, Weinheim, pp. 515–558

(Continued)

TABLE 2.1 (CONTINUED)

Reviews Covering the Synthesis and Application of Light-Emitting Polymers

Year	Title	Authors	Publication
1998	The chemistry and uses of polyphenylenevinylenes	S.C. Moratti	*Handbook of Conducting Polymers,* T.A. Skotheim, R.L. Elsenbaumer, and J.R. Reynolds, Eds., Marcel Dekker, New York, pp. 343–361
1998	Conjugated ladder-type structures	U. Scherf	*Handbook of Conducting Polymers,* T.A. Skotheim, R.L. Elsenbaumer, and J.R. Reynolds, Eds., Marcel Dekker, New York, pp. 363–379
1998	Electroluminescence in conjugated polymers	R.H. Friend and N.C. Greenham	*Handbook of Conducting Polymers,* T.A. Skotheim, R.L. Elsenbaumer, and J.R. Reynolds, Eds., Marcel Dekker, New York, pp. 823–845
1998	Fundamentals of electroluminescence in *para*-phenylene-type conjugated polymers and oligomers	G. Leising, S. Tasch, and W. Graupner	*Handbook of Conducting Polymers,* T.A. Skotheim, R.L. Elsenbaumer, and J.R. Reynolds, Eds., Marcel Dekker, New York, pp. 847–880
1999	Electro-optical polythiophene devices	M. Granström, M.G. Harrison, and R.H. Friend	*Handbook of Oligo- and Polythiophenes,* D. Fichou, Ed., Wiley-VCH, Weinheim, pp. 405–458
1999	Electroluminescence in conjugated polymers	R.H. Friend, R.W. Gymer, A.B. Holmes, J.H. Burroughes, R.N. Marks, C. Taliani, D.D.C. Bradley, D.A. Dos Santos, J.L. Brédas, M. Lögdlung, and W.R. Salaneck	*Nature,* 397: 121–127
1999	Polarized luminescence from oriented molecular materials	M. Grell and D.D.C. Bradley	*Adv. Mater.,* 11: 895–905
1999	Ladder-type materials	U. Scherf	*J. Mater. Chem.,* 9: 1853–1864
1999	Electroluminescence in organics	J. Kalinowski	*J. Phys. D: Appl. Phys.,* 32: R179–R250
2000	The electroluminescence of organic materials	U. Mitschke and P. Bäuerle	*J. Mater. Chem.,* 10: 1471–1507
2000	Progress in light-emitting polymers	M.T. Bernius, M. Inbasekaran, J. O'Brien, and W. Wu	*Adv. Mater.,* 12: 1737–1750
2000	Synthesis of conjugated polymers for application in light-emitting diodes (PLEDs)	R.E. Martin, F. Geneste, and A.B. Holmes	*C.R. Acad. Sci. Paris,* t. 1, Série IV: 447–470
2000	Blue light emitting polymers	D.Y. Kim, H.N. Cho, and C.Y. Kim	*Prog. Polym. Sci.,* 25: 1089–1139
2000	Fluorene-based polymers— Preparation and applications	M. Bernius, M. Inbasekaran, E. Woo, W. Wu, and L. Wujkowski	*J. Mater. Sci.: Mater. Electronics,* 11: 111–116
2000	Poly(aryleneethynylene)s: Syntheses, properties, structures, and applications	U.H.F. Bunz	*Chem. Rev.,* 100: 1605–1644

(Continued)

TABLE 2.1 (CONTINUED)

Reviews Covering the Synthesis and Application of Light-Emitting Polymers

Year	Title	Authors	Publication
2000	Semiconducting (conjugated) polymers as materials for solid-state lasers	M.D. McGehee and A.J. Heeger	*Adv. Mater.*, 12: 1655–1668
2001	Conjugated polymers. New materials for optoelectronic devices	R.H. Friend	*Pure Appl. Chem.*, 73: 425–430
2001	Conjugated polymers for light-emitting applications	L. Dai, B. Winkler, L. Dong, L. Tong, and A.W.H. Mau	*Adv. Mater.*, 13: 915–925
2001	Polyfluorenes: Twenty years of progress	M. Leclerc	*J. Polym. Sci. Part A: Polym. Chem.*, 39: 2867–2873
2001	Polyfluorene homopolymers: Conjugated liquid-crystalline polymers for bright emission and polarized electroluminescence	D. Neher	*Macromol. Rapid Commun.*, 22: 1365–1385
2002	Recent developments in light-emitting polymers	I.D. Rees, K.L. Robinson, A.B. Holmes, C.R. Towns, and R. O'Dell	*MRS Bull.*, June: 451–455
2002	Semiconducting polyfluorenes—Toward reliable structure–property relationships	U. Scherf and E.J.W. List	*Adv. Mater.*, 14: 477–487
2003	Conjugated polymers as molecular materials: How chain conformation and film morphology influence energy transfer and interchain interactions	B.J. Schwartz	*Annu. Rev. Phys. Chem.*, 54: 141–172
2003	Synthesis of π-conjugated polymers bearing electronic and optical functionalities by organometallic polycondensations. Chemical properties and applications of the π-conjugated polymers	T. Yamamoto	*Synlett*, 425–450
2003	Carbazole-containing polymers: Synthesis, properties and applications	J.V. Grazulevicius, P. Strohriegl, J. Pielichowski, and K. Pielichowski	*Prog. Polym. Sci.*, 29: 1297–1353
2003	Electroluminescent polymers	L. Akcelrud	*Prog. Polym. Sci.*, 28: 875–962
2004	Recent development of polyfluorene-based RGB materials for light emitting diodes	W. Wu, M. Inbasekaran, M. Hudack, D. Welsh, W. Yu, Y. Cheng, C. Wang, S. Kram, M. Tacey, M. Bernius, R. Fletcher, K. Kiszka, S. Munger, and J. O'Brien	*Microelectron. J.*, 35: 343–348
2004	Synthesis of conjugated oligomers and polymers: The organometallic way	F. Babudri, G.M. Farinola, and F. Naso	*J. Mater. Chem.*, 14: 11–34
2004	Electron transport materials for organic light-emitting diodes	A.P. Kulkarni, C.J. Tonzola, A. Babel, and S.A. Jenekhe	*Chem. Mater.*, 16: 4556–4573

(Continued)

TABLE 2.1 (CONTINUED)

Reviews Covering the Synthesis and Application of Light-Emitting Polymers

Year	Title	Authors	Publication
2004	Application of three-coordinate organoboron compounds and polymers in optoelectronics	C.D. Entwistle and T.B. Marder	*Chem. Mater.*, 16: 4574–4585
2005	Electron-transporting materials for organic electroluminescent and electrophosphorescent devices	G. Hughes and M.R. Bryce	*J. Mater. Chem.*, 15: 94–107
2007	Recent development in phosphole-containing oligo- and polythiophene materials	M.G. Hobbs and T. Baumgartner	*Eur. J. Inorg. Chem.*, 23: 3611–3628
2009	Conjugated oligomers and polymers containing dithienosilole units	J. Ohshita	*Macromol. Chem. Phys.*, 210: 1360–1370
2010	Conjugated rod-coil block copolymers and optoelectronic applications	A.d. Cuendias, R.C. Hiorns, E. Cloutet, L. Vignau, and H. Cramail	*Polym. Int.*, 59: 1452–1476
2013	Dithieno[3,2-*b*:2′,3′-*d*] pyrrole-based materials: Synthesis and application to organic electronics	S.C. Rasmussen and S.J. Evenson	*Prog. Polym. Sci.*, 38: 1773–1804

resulting LED performance. In what follows, we describe the main classes of LEPs that have been studied since about 1990 through mid-2014. Although it would not be possible to cover all the related literature in a single chapter (or even a separate book), an attempt has been made to cover all important polymeric EL materials that have been communicated in scientific journals (and, when relevant, in patents). However, considering the enormous number of publications appearing in a broad variety of journals each year, it is possible that some important papers describing a new LEP did not gain our attention. The chapter is written from the viewpoint of an organic materials chemist. It includes description of basic synthetic methods and, through a diversity of discussed structural variations influencing optoelectronic and EL properties, uncovers some general structure–property relationships in the described materials. A short description of LED structure is given along with the data on EL performance, whenever possible. However, the reader should be aware of the limitations of a comparison of the EL data obtained by different groups (even for the same device structure). The conclusions on the practical values of different materials, beyond those given in the chapter, should be made with great care. In Section 2.6, we will list some of the best-performing LEPs and their future perspectives. Finally, in the Appendix (Section 2.7), the interested reader can find some practical synthetic methods for different classes of LEPs.

2.2 Poly(*p*-Phenylene Vinylenes)

Poly(*p*-phenylene vinylene) (PPV) **1** is a highly stable conjugated polymer (Chart 2.1). Its yellow color is due to an absorption band centered at ~400–420 nm (depending on the

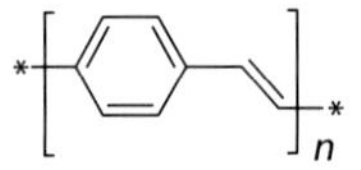

1, PPV

CHART 2.1
Chemical structure of PPV.

method of synthesis) with an onset corresponding to a band gap of ~2.5 eV [5]. The highest occupied molecular orbital (HOMO) and the lowest unoccupied molecular orbital (LUMO) levels in PPV can be accessed through cyclic voltammetry (CV) experiments that, under proper conditions, reveal chemically reversible oxidation and reduction waves (Figure 2.1). The deduced electrochemical gap corresponds reasonably well to the optical band gap. As a relatively good electron donor, PPV and its derivatives can be chemically doped by strong oxidizing agents and strong acids, affording highly conductive p-doped materials (with conductivity up to ~10^4 S/cm [5]). The yellow–green fluorescence of PPV **1** results from a vibronically structured emission band with peak maxima at 520 and 551 nm (Figure 2.1).

The discovery of the EL in PPV in 1990 resulted in a tremendous growth of interest in polymer LEDs (PLEDs) [1]. Since then, numerous derivatives and analogs of PPV with tailored light-emitting properties have been synthesized, and a number of reviews and accounts have described the synthesis and the EL properties of these materials [6–16]. Many new applications of PPV polymers, such as solid-state lasing [17,18], photovoltaics [19],

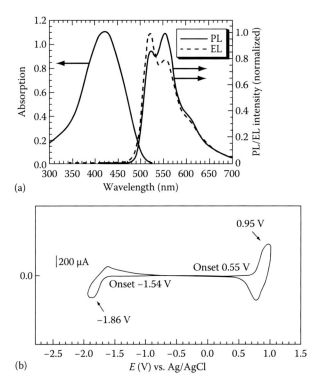

FIGURE 2.1
(a) Typical absorption, photoluminescence, and electroluminescence spectra of PPV (given for dialkyl-PPV **30**). (From Andersson, M.R., Yu, G., and Heeger, A.J., *Synth. Met.*, 85, 1275, 1997. With permission.) (b) Cyclic voltammetry of dialkoxy-PPV **13**. (From Kim, J.H. and Lee, H., *Chem. Mater.*, 14, 2270, 2002. With permission.)

etc., have been explored; however, they are beyond the scope of this book. Below we give a general overview of the basic methods of synthesis of PPV derivatives and the design of PPV materials with controllable properties, as they were widely explored for the last 25 years in order to create high-performance PLEDs.

2.2.1 Synthetic Routes to Poly(*p*-Phenylene Vinylenes)

There are a number of synthetic strategies elaborated for preparation of PPV homopolymers and copolymers [20]:

1. Thermoconversion (Wessling–Zimmerman route)
2. Chemical vapor deposition (CVD)
3. Ring-opening metathesis polymerization (ROMP)
4. Gilch polycondensation
5. Chlorine precursor route (CPR) (Gilch modification)
6. Nonionic route (Gilch modification)
7. Knoevenagel polycondensation
8. Heck-coupling polymerization
9. Wittig(–Horner) condensation
10. Miscellaneous

2.2.1.1 Thermoconversion (Wessling–Zimmerman Route)

Since PPV itself is insoluble and difficult to process, the most widely used method, developed in the early 1960s by Wessling and Zimmerman, is thermoconversion of a processable sulfonium intermediate **2** (Scheme 2.1) [21]. Polymer **2** is soluble in methanol and can be spin coated to give a high-quality thin film, the heating of which results in the formation of PPV **1** via the elimination of hydrogen halide and tetrahydrothiophene. Under proper conditions, the thermoconversion can give pinhole-free thin films of PPV suitable for PLED fabrication. The conversion temperature can be reduced to 100°C by using bromide derivatives instead of chlorides, thus enabling fabrication of PLED on flexible substrates [22,23].

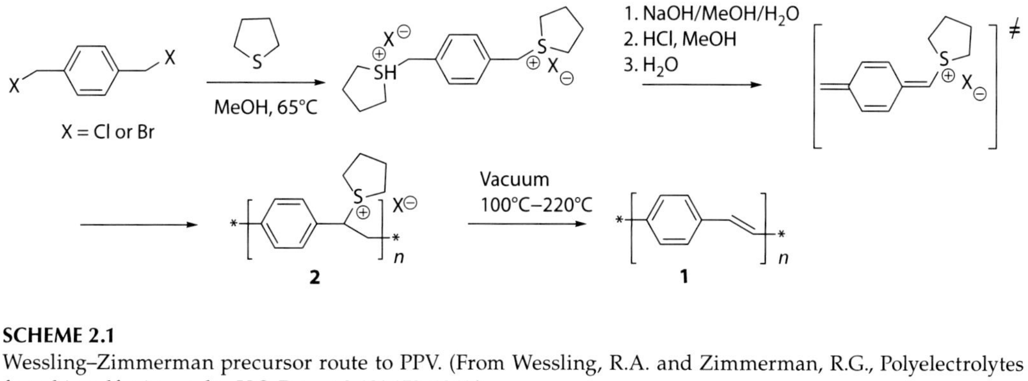

SCHEME 2.1
Wessling–Zimmerman precursor route to PPV. (From Wessling, R.A. and Zimmerman, R.G., Polyelectrolytes from bis sulfonium salts, U.S. Patent 3,401,152, 1968.)

The issues of low stability of the precursor polymer **2** and extremely unpleasant odor of the sulfur-containing by-product can be resolved by the substitution of the sulfonium-leaving groups with a methoxy group (under acid catalysis). The methoxy-substituted precursor polymer requires very harsh conditions for conversion to PPV (HCl gas at 220°C) [24]. On the other hand, the resulting PPV material showed significantly improved photophysical properties (more resolved vibronic structure of the absorbance, higher third-order nonlinearity), which were explained by a higher degree of order of the polymer chains [25]. Some other method modifications, such as employment of a xanthate-leaving group [26,27] or replacing the chloride ion (in precursor **2**) with dodecylbenzenesulfonate ion [28], were reported. For the latter, the long-chain counterion facilitates processing of the precursor polymer in Langmuir–Blodgett films.

2.2.1.2 Chemical Vapor Deposition (CVD)

Another synthetic method, applicable for processing PPV in thin films, is CVD of dichloro-*p*-[2.2]cyclophane [29] or dichloro-*p*-xylene [30] (Scheme 2.2). This method, although it can afford uniform and patterned thin films [31,32], requires heating at 500°C–700°C, which may give rise to by-product impurities. Such harsh conditions and difficulties of complete removal of the halogen (second stage) result in ill-defined material, which affords very-low-performance PLEDs (maximum brightness of 20 cd/m^2) [31,32].

2.2.1.3 Ring-Opening Metathesis Polymerization (ROMP)

The drawback of the CVD method is eliminated in ROMP, which is based on a catalytic (e.g., molybdenum carbene catalyst) reaction, occurring in rather mild conditions (Scheme 2.3). A living ROMP reaction of *p*-cyclophane **3** or bicyclooctadiene **5** results in soluble precursors of PPV, polymers **4** [33] and **6** [34], respectively, with rather low polydispersity. In spite of the all-*cis* (for **4**) and *cis* and *trans* (for **6**) configurations, these polymers can be converted into all-*trans* PPV by moderate heating under acid–base catalysis. However, the film-forming properties of ROMP precursors are usually rather poor, resulting in poor uniformity of the PPV films.

2.2.1.4 Gilch Polycondensation

A general and most widely used method for the synthesis of PPV derivatives was introduced by Gilch and Wheelwright [35]. This method avoids high-temperature conditions and occurs through base-promoted 1,6-elimination of 1,4-*bis*(chloro/bromomethyl)benzenes **7**

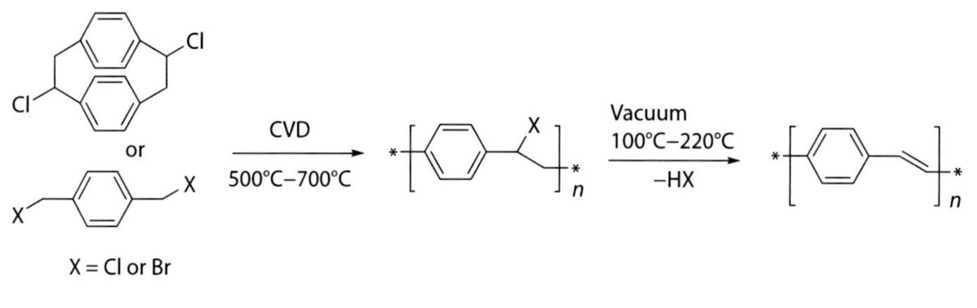

SCHEME 2.2
Chemical vapor deposition route to PPV. (From Iwatsuki, S., Kubo, M., and Kumeuchi, T., *Chem. Lett.*, 20, 1971, 1991.)

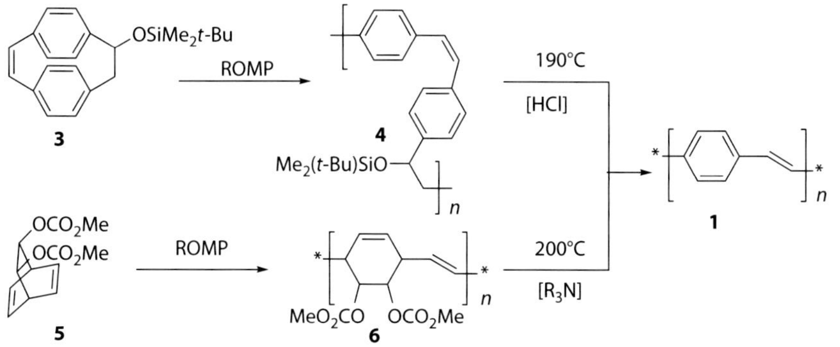

SCHEME 2.3
ROMP route to PPV. (From Miao, Y.-J. and Bazan, G.C., *J. Am. Chem. Soc.*, 116, 9379, 1994; Conticello, V.P., Gin, D.L., and Grubbs, R.H., *J. Am. Chem. Soc.*, 114, 9708, 1992.)

(Scheme 2.4). For device applications, the as-synthesized PPV materials need to be soluble in organic solvents; otherwise, the as-formed polymer powder is completely unprocessable. Alkyl, alkyloxy, and other substituted monomers giving soluble PPVs have been employed in this reaction. The mechanism of the Gilch polymerization is still a subject of some controversy [36–39]. It is well accepted to proceed through a reactive quinodimethane intermediate, followed by either a radical or a living-chain anionic polymerization. A molecular weight decrease upon the addition of chain transfer radical agent (2,2,6,6-tetramethylpiperidyl-1-oxyl) was interpreted as a sign of the radical polymerization mechanism [40], although the same effect imposed by a nucleophilic initiator (4-*tert*-butylbenzyl chloride) was taken as a confirmation of the nucleophilic chain growth mechanism [37]. In the absence of initiators, the latest evidence suggests the radical polymerization mechanism [36]. In both mechanisms, the regularity of the polymer conjugation chain is challenged by the possibility of side

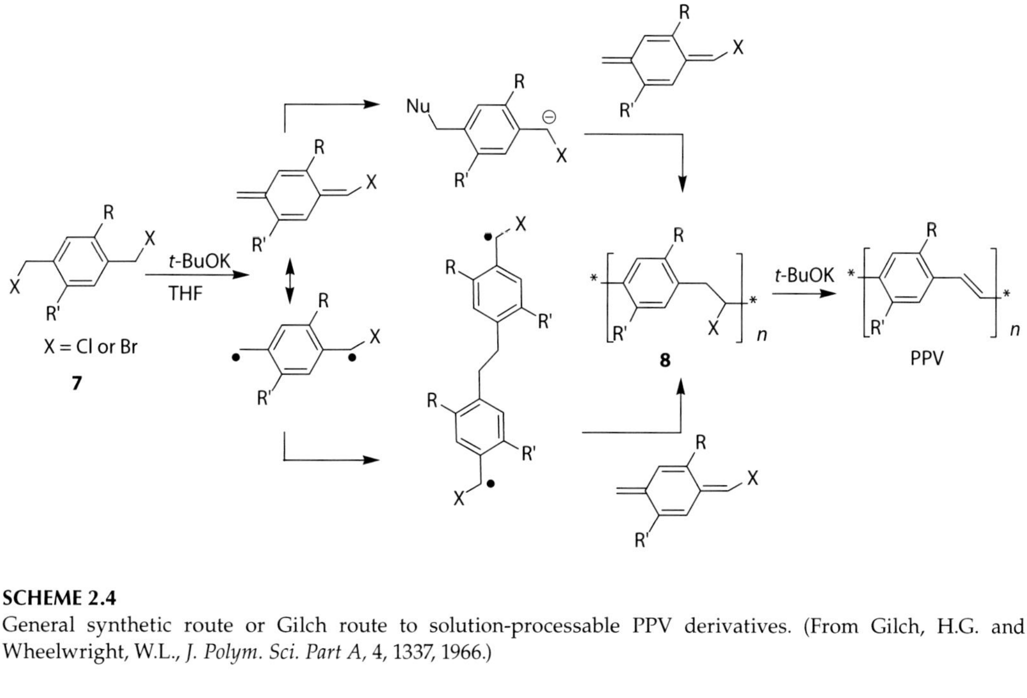

SCHEME 2.4
General synthetic route or Gilch route to solution-processable PPV derivatives. (From Gilch, H.G. and Wheelwright, W.L., *J. Polym. Sci. Part A*, 4, 1337, 1966.)

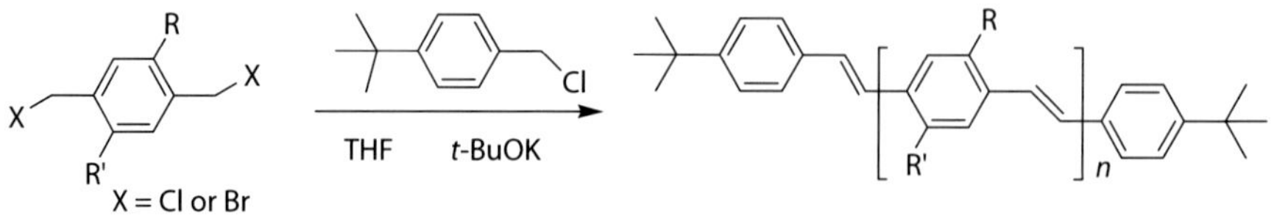

SCHEME 2.5
End-capping modification of the Gilch polymerization. (From Hsieh, B.R., Yu, Y., van Laeken, A.C., and Lee, H., *Macromolecules*, 30, 8094, 1997.)

reactions, which are anomalous "head-to-head" (HH) or "tail-to-tail" (TT) couplings of the dehydrochlorinated intermediate. These reactions lead to the appearance of tolane–bis-benzyl (TBB) defects in the conjugation chain [41]. Although normally the amount of TBB is very low (<1%–2%), certain substitution patterns (such as sterically hindered phenyl-PPV) can greatly enhance the defect formation. Note that (Scheme 2.4) a radical mechanism suggests the formation of at least one TBB defect in the middle of the polymer chain due to sterically preferable HH coupling of two monomer biradicals (although further chain growth should proceed via normal head-to-tail, HT, coupling).

The molecular weight of the polymers can be controlled (from ca. 50,000 to >1,000,000) by changing the reaction temperature and time, the solvent, the concentration of the monomer, and the amount of base [42,43]. High-molecular-weight polymers and a high content of *trans* double bonds are the reasons for the wide usage of Gilch polymerization in the synthesis of PPV homopolymers and copolymers.

Hsieh et al. [37] used 4-*tert*-butylbenzyl chloride as an initiator and end-capping reagent to control the molecular weight of the Gilch synthesis of poly(2-methoxy-5-(2′-ethyl-hexyloxy)–1,4-phenylene vinylene) (MEH–PPV) (Scheme 2.5). Adding different amounts (0.6–60 mol%) of the end capper results progressively in a decrease in the molecular weight of the polymer (M_n = 66,500 for 6% of 4-*tert*-butylbenzyl chloride), suppressing the undesirable gel formation effect, as often observed in Gilch synthesis. However, the polymerization yield under these conditions was found to be rather low (35% for 6% of the initiator and <20% for a higher amount), which can be logically expected, considering possible side reactions of the initiator in the strongly basic media. Admitting the problem of self-coupling reaction of the benzyl chloride initiator, Neef and Ferraris [38,39] attempted to control the molecular weight of the polymer with 4-methoxyphenol as an anionic initiator. The authors claim that relatively small amounts of the phenolic initiator (1%–2%) can decrease the M_n by more than a factor of 2, while keeping the polymerization yield >50% (which is still essentially lower compared with the 76% yield obtained in the absence of an initiator). At the same time, a very low polydispersity, reported by Hsieh et al. [37] (1.04–1.52), is unexpected for this type of reaction, and the reproducibility of the reported results seem to be a problem. Both approaches are based on an arguable hypothesis of anionic living polymerization and are of very limited practical applicability for controlling the molecular weight of PPV, although the use of these and similar reagents in small amounts, as end cappers, may be beneficial for improving the stability of the PPV material.

2.2.1.5 *Chlorine Precursor Route (CPR)*

An important modification of Gilch polymerization (also known as CPR), introduced by Swatos and Gordon in 1990, is based on using one equivalent of the base (instead of four in the classical Gilch method) to stop the polymerization at the stage of formation of precursor polymer **8** (Scheme 2.4) [42,44]. Polymer **8** is very much a soluble material and can be spin coated to afford high-quality films. Thermal annealing at 230°C–280°C results in the elimination of hydrogen chloride and affords the PPV material in high yield.

2.2.1.6 Nonionic Route

Another extension of the Gilch polymerization approach, similar to the CPR method, was developed by Lutsen and coworkers [45]. They have substituted one of the chlorine atoms in monomer **7** with an alkylsulfinyl group (BuS(O)), so that the resulting "precursor polymer" **8** would not contain any chlorine. The alkylsulfinyl group can be removed in the nonionic process by simply heating the polymer at ~110°C for a few hours. The process can be attractive for applications where the low halogen content is a critical issue, although the EL efficiency of PPV prepared by this method is a few times lower than that prepared by traditional Gilch polymerization (presumably due to defects in the polymer structure) [45].

2.2.1.7 Knoevenagel Polycondensation

Knoevenagel condensation, based on the reaction of an aldehyde group with an active methylene component, was one of the first methods used for the synthesis of PPV derivatives [46]. In general, it requires strong electron acceptor substituents (such as a cyano [CN] group) in the methylene component (Scheme 2.6). The method delivers the PPV containing cyano substituents on the vinylene units (**9**), and numerous substituted CN–PPV homopolymers and copolymers have been synthesized [47]. Recently, ruthenium-based catalysis (RuH$_2$(PPh$_3$)$_4$/dppe) was used to replace the strong base (as *t*-BuOK) in the Knoevenagel-type synthesis of CN–PPV, which has the advantage of neutral and mild reaction conditions [48].

2.2.1.8 Heck-Coupling Polymerization

The PPV polymer chain can also be constructed by Heck coupling of aromatic dihalides with a divinylbenzene. In contrast to the above-described methods, the Heck coupling is a Pd-catalyzed reaction occurring in very mild conditions. The polymerization normally leads to all-*trans* geometry with very few side reactions (Scheme 2.7) [49]. Although this method is somewhat complicated for the preparation of PPV homopolymers, the Heck-type synthesis of unsubstituted PPV and its methylated, trifluoromethylated, and phenylated derivatives from divinylbenzene and dibromobenzene, dibromobenzene and ethylene, or bromovinylbenzene (self-coupling) was demonstrated [50]. This method is of great utility in the preparation of alternating copolymers (see Section 2.3).

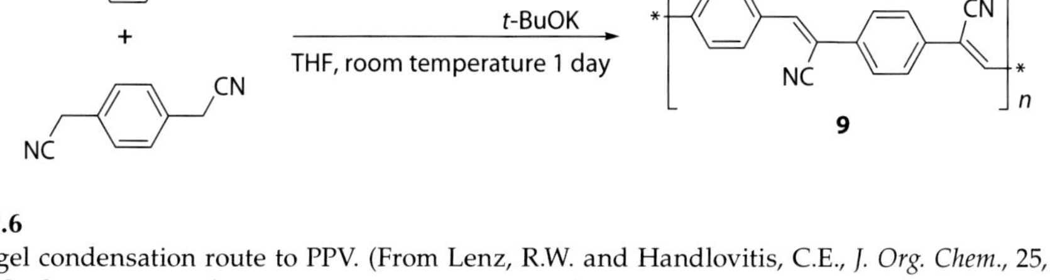

SCHEME 2.6
Knoevenagel condensation route to PPV. (From Lenz, R.W. and Handlovitis, C.E., *J. Org. Chem.*, 25, 813, 1960; Moratti, S.C., Cervini, R., Holmes, A.B., Baigent, D.R., Friend, R.H., Greenham, N.C., Grüner, J., and Hamer, P.J., *Synth. Met.*, 71, 2117, 1995.)

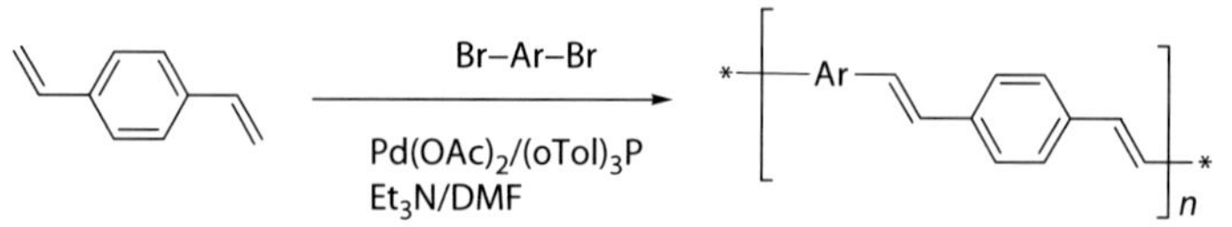

SCHEME 2.7
Heck-coupling route to synthesis of PPV copolymers. (From Hilberer, A., Brouwer, H.-J., van der Scheer, B.-J., Wildeman, J., and Hadziioannou, G., *Macromolecules*, 28, 4525, 1995; Greiner, A. and Heitz, W., *Macromol. Chem. Rapid Commun.*, 9, 581, 1988.)

2.2.1.9 Wittig(–Horner) Condensation

Wittig or Wittig–Horner condensation between substituted terephthalaldehydes and *p*-xylenediylphosphonium salts is also extensively used in the preparation of alternating PPV copolymers, e.g., containing different substituents in adjacent phenylene units (Scheme 2.8) [51].

2.2.1.10 Miscellaneous

A totally different route based on dehydrogenation of a saturated polymer precursor was introduced by Francois et al. [52] (Scheme 2.9). The method is based on anionic copolymerization of cyclohexadiene with styrene, followed by oxidation with chloranil. Owing to possible coupling of two styrene (or two cyclohexadiene) molecules, a block copolymer, containing oligo(phenylene vinylene) units separated by oligo(phenylacetylene) and oligo(phenylene) blocks, is obtained. To the best of our knowledge, it was, thus far, used only in the synthesis of phenyl-substituted PPV **10**.

Aguiar and coworkers [53] reported the preparation of acetoxy-PPV **11** via controlled potential electrolysis of $\alpha,\alpha,\alpha',\alpha'$-tetrabromoxylene precursor on a mercury electrode in Et_4NBr/dimethylformamide (DMF) electrolyte solution (Scheme 2.10). However, the only structural characterization reported was UV–vis and fluorescence spectra.

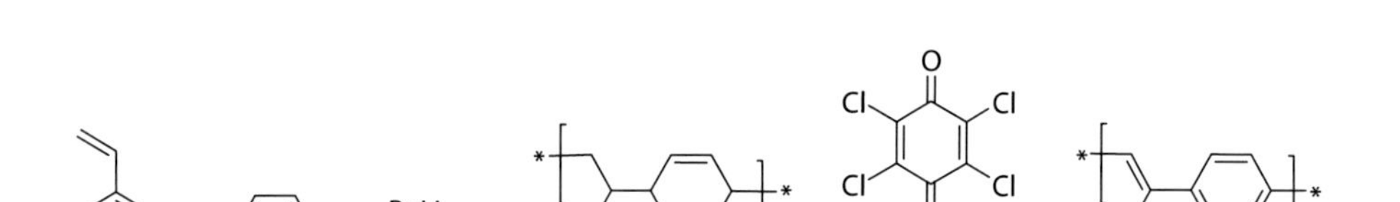

SCHEME 2.8
Wittig condensation route to PPV. (From McDonald, R.N. and Campbell, T.W., *J. Am. Chem. Soc.*, 82, 4669, 1960.)

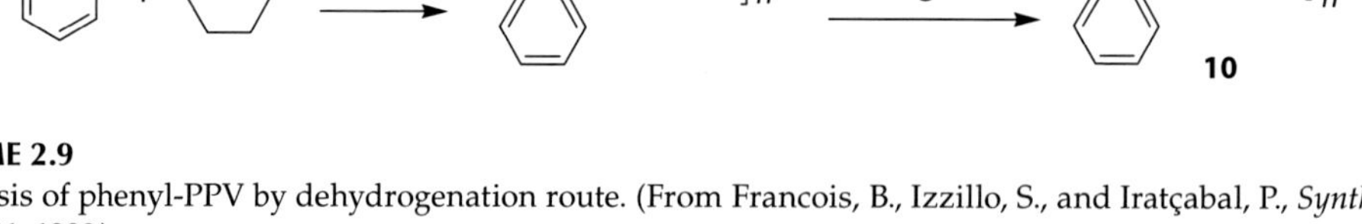

SCHEME 2.9
Synthesis of phenyl-PPV by dehydrogenation route. (From Francois, B., Izzillo, S., and Iratçabal, P., *Synth. Met.*, 102, 1211, 1999.)

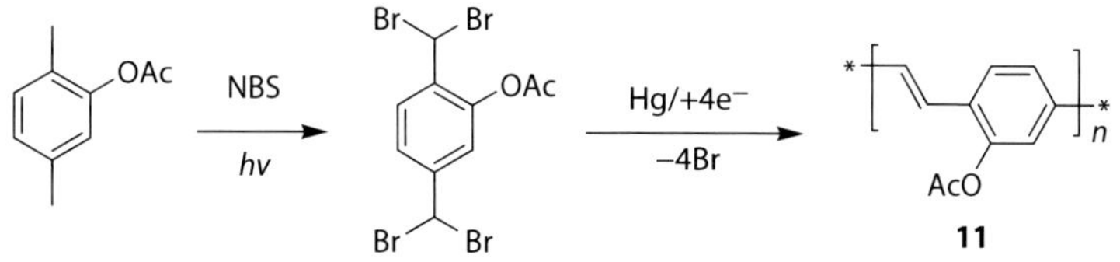

SCHEME 2.10
Synthesis of acetoxy-PPV by electroreduction. (From Aguiar, M., Fugihara, M.C., Hümmelgen, I.A., Péres, L.O., Garcia, J.R., Gruber, J., and Akcelrud, L., *J. Lumin.*, 96, 219, 2002.)

Below we describe the application of the described reactions to the synthesis of PPV derivatives with tailored properties for PLEDs.

2.2.2 Poly(*p*-Phenylene Vinylene) Homopolymers

The first conjugated PLED was prepared by Friend and coworkers [1] by sandwiching unsubstituted PPV **1** (prepared by Wessling–Zimmerman reaction) between a transparent ITO anode and an Al cathode. It showed a maximum Φ_{EL}^{ex} of only 0.01% at room temperature (0.05% at 120 K), and required 14 V to turn on [1]. This low efficiency has several reasons, among which is an imbalance of hole–electron injection barriers. The efficiency could be improved to 0.1% by using a lower-work-function electrode (Ca) [54]. However, Ca is a highly reactive metal, which complicates the fabrication process and reduces the device stability. Alternatively, the EL can be improved by a factor of 30 by placing an electron-transporting layer (ETL) (oxadiazole-based nonconjugated polymer) between PPV **1** and the Al electrode [55]. On the way to electrically pumped lasers, Friend and coworkers [56] reported PLEDs based on PPV **1** having exceptionally high peak brightness. Applying a short-pulsed voltage of ~10–40 V (pulse width of 100–200 ns), a brightness of 5×10^6 cd/m^2 has been achieved for a device, ITO/PEDOT/**1**/Al (PEDOT is poly(3,4-ethylenedioxythiophene; Chart 2.2). Although a relatively high EL efficiency (2.8 cd/A) was found in this device, the authors mention that it is still the device efficiency, and not the brightness, that limits the laser action.

As will be shown throughout the chapter, the parent PPV **1** is extensively used as a hole-transporting layer (HTL) in combination with other EL polymers. Recently, improved photostability of organic–inorganic hybrid EL material, prepared by incorporating PPV **1** into zeolite capsules, was described [57]. However, the material showed only weak EL (at driving voltage of 2.5 V).

Modifications of the chemical structure of PPV provide various opportunities for tuning the optical properties of this material. The most explored modification was introducing the substituents in the benzene ring. These include alkyl, alkoxy, and silyl substituents;

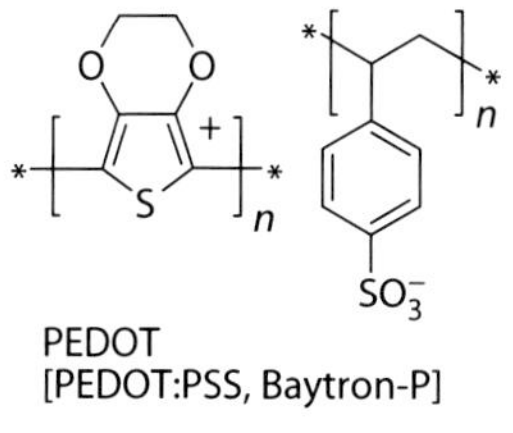

PEDOT
[PEDOT:PSS, Baytron-P]

CHART 2.2
Chemical structure of PEDOT.

aromatic functional side groups; and electron-releasing and -withdrawing groups, as discussed in detail in the following sections.

One should, however, bear in mind that not only the molecular structure of the polymers but also their supramolecular organization defines the performance of a PLED. Thermal annealing of the films and other ordering techniques are widely used to control the properties of the polymers. Particularly, solvents used for casting the film and the casting procedure can substantially change the supramolecular organization of the polymer and, thus, the performance of PLED. Single-molecule fluorescence correlation spectroscopy studies confirmed that the chain collapse and orientation of the single molecules of CN–PPV (**9**) and MEH–PPV (**13**) are highly influenced by the choice of the solvent: the production of oriented species is strongly favored in "poorer" solvents, where the polymer chains have more compact solution-phase structures [58].

2.2.2.1 Alkoxy-Substituted Poly(p-Phenylene Vinylenes)

The insolubility of the PPV **1** and the need for conversion of the precursor polymer on the last stage under rather harsh conditions are obvious drawbacks for a wide application of these materials. The obvious route to increase the solubility of the PPV would be introducing long-chain substituents. Although several groups synthesized and studied dimethoxy- and diethoxy-PPV derivatives starting from 1970, aiming at high-stability conducting polymers [59,60], only a dihexyloxy derivative, DH–PPV **12**, prepared by Askari and coworkers [61] by thermal treatment of the sulfonium salt, appeared to be soluble in common organic solvents (Scheme 2.11).

However, the solubility of this material at room temperature was still not high enough. The simple elongation of the substituents results in a "side-chain crystallization effect" and does not increase the solubility. To solve the problem, Wudl and Srdanov [62] came up with a highly asymmetric substituent pattern (methoxy/2-ethylhexyloxy) for the synthesis of polymer **13** (well-known as MEH–PPV) via the Gilch polymerization route (Chart 2.3). The side-chain disorder brought about by two different substituents as well as nonplanar structure and optical isomers (due to stereogenic 2-ethylhexyloxy substituents) results in a high solubility of this polymer in common organic solvents (toluene, chloroform, tetrahydrofuran [THF], etc.) in spite of extremely high molecular weight (>10^6 Da). Such a high molecular weight can nevertheless result in gelation of the polymer, and several attempts of controlling the degree of polymerization by introducing end-capping reagents were undertaken [37–39]. Also, studies by Burn and coworkers [63] suggest that aggregation of MEH–PPV in solutions might affect the molecular weight determination, and lower M_n values have been obtained when analyzing highly diluted MEH–PPV solutions. A completely insoluble form of MEH–PPV, which can be useful for the preparation of multilayer PLEDs, was prepared by the same group via the CPR (using less than one equivalent of the base) [64].

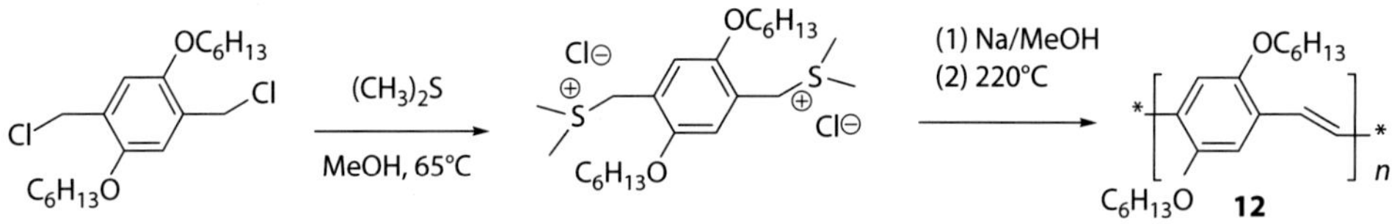

SCHEME 2.11
Synthesis of the first soluble PPV derivative, DH–PPV. (From Askari, S.H., Rughooputh, S.D., and Wudl, F., *Synth. Met.*, 29, E129, 1989.)

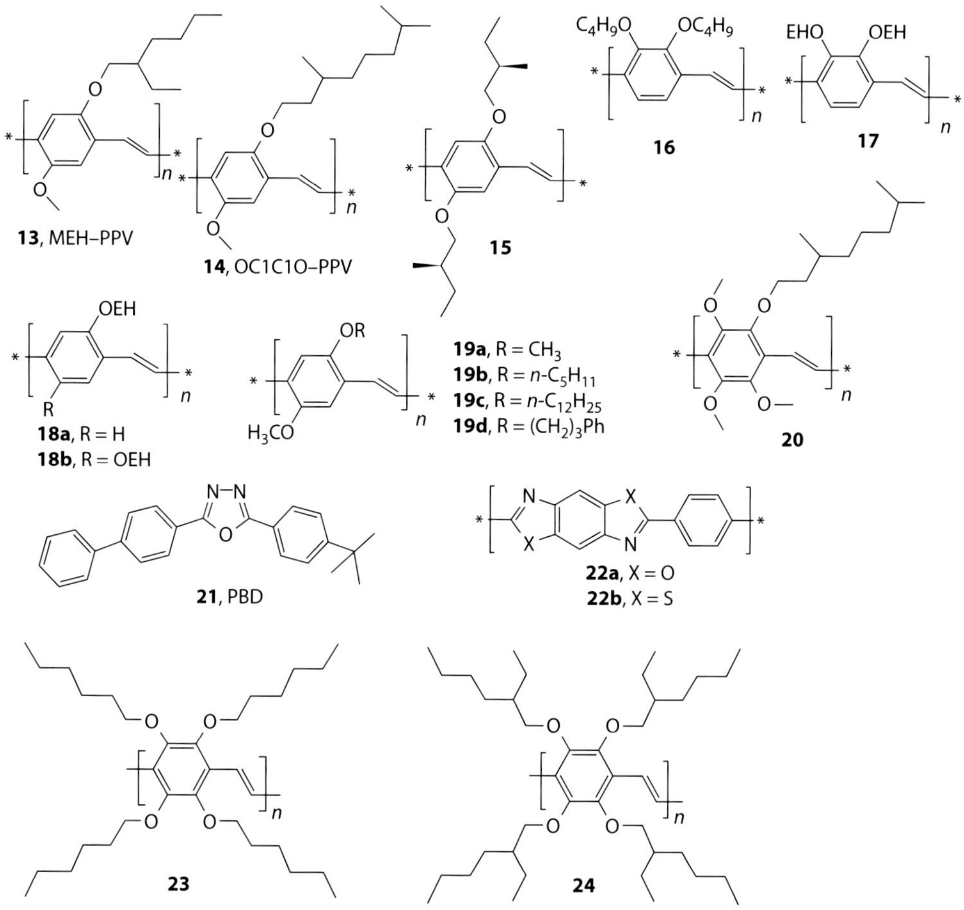

CHART 2.3
Chemical structures of alkoxy-substituted PPVs.

MEH–PPV is a bright-orange material ($\lambda_{max} \sim 490$ nm); upon photoexcitation, it produces a red–orange emission ($\lambda_{PL} \sim 590$ nm, PL is photoluminescence). For the last decade, MEH–PPV has been one of the most studied EL materials [65–76]. It was used as a standard LEP for the demonstration of several innovative concepts in the fabrication of PLEDs, including light-emitting electrochemical cells (LECs) [66], microstructuring the polymer layer for increased light output [72], application of transparent polymer electrodes (doped polyaniline [PANI] or PEDOT films) in place of ITO [68], nanocomposites with inorganic materials [70,72], and others. The first LEDs fabricated with this material were reported to show a Φ_{EL}^{ex} of 0.05% in ITO/**13**/In configuration and ~1% in ITO/**13**/Ca configuration [65]. An external quantum efficiency (QE) of 1% can also be achieved with Al electrode in an LEC device, using a blend of MEH–PPV, poly(ethylene oxide) (PEO), and electrolyte [66], whereas a QE of <0.4% was achieved by the same group in the same device using unsubstituted PPV **1** [67]. Numerous improvements of the EL performance of MEH–PPV by blending this polymer with different organic and inorganic materials were reported. Highly efficient PLEDs (Φ_{EL}^{ex} of 2%, maximum brightness of 10,000 cd/m^2) were fabricated by adding SiO$_2$ nanoparticles to the MEH–PPV layer (between ITO and Ca electrodes) [70].

A Φ_{EL}^{ex} of 1.3% was obtained by blending MEH–PPV with an electron-transport material, 2-(4-biphenyl)-5-(4-*tert*-butylphenyl)-1,3,4-oxadiazole (PBD **21**) [71]. More recently, an efficiency up to 2.5% was reported for multilayer PLEDs with additional polybenzobisazole (**22**) ETL (ITO/PEDOT/**13**/**22b**/Al) [74]. Very high QE values have been also obtained blending MEH–PPV with lithium organophosphonate surfactant (2.3% [68]) or carbazole–thiophene copolymers (3.8% [75]). Recent results from Deng and coworkers [76] show a two orders of magnitude increase of the EL efficiency (up to 2.7 cd/A, maximum brightness up to 5500 cd/m²) of the ITO/MEH–PPV/Al device upon simple dilution of the MEH–PPV with poly(ethylene glycol) (PEG).

The 2-ethylhexyl (EH) substituent became a very popular side-chain group for the synthesis of soluble conjugated polymers of different classes; however, other branched alkyloxy substituents have also been introduced in the PPV backbone. For example, polymer **14** substituted with a 3,7-dimethyloctyl group showed a very similar electronic behavior to that of MEH–PPV; however, an additional branching further improved its solubility and the film-forming properties [45,77]. PLEDs with an EL efficiency of 1.2 cd/A (with maximum brightness of 4000 cd/m²) [45] and even higher, 2.1 cd/A (2.5 lm/W) [77] and ~3 cd/A [78], have been fabricated with polymer **14**. A systematic study of lifetime and degradation effects in PLEDs was reported for this polymer [79]. At low brightness level of ~100 cd/m², a half-lifetime of around 20,000 h was achieved. The device stability strongly depends on the operating temperature (Figure 2.2), and the authors suggested that electron (rather than hole) injection and passage are primarily responsible for the device degradation. Very high Φ_{EL}^{ex} was demonstrated by Cao and coworkers [71] for polymer **14** blended with 20% of

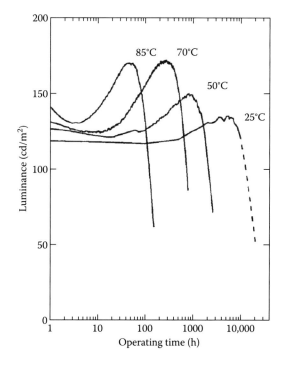

FIGURE 2.2
Operation lifetime of the device ITO/PANI/**14**/Ca/Al. (From Parker, I.D., Cao, Y., and Yang, C.Y., *J. Appl. Phys.,* 85, 2441, 1999. With permission.)

PBD (electron-transport material). Increasing the operating temperature of the device from 25°C to 85°C, the Φ_{EL}^{ex} values increased from 2% to 4%. Comparing these numbers with the PL quantum yield (PLQY) (~8%), measured by exciting the material-incorporated diodes, the authors reached an interesting, although somewhat speculative, conclusion that 50% singlet–triplet ratio is achieved in this material (which exceeds the widely accepted theoretical value of 25%). A related enantiomerically pure (*S*)-2-methylbutoxy-substituted PPV **15** has been used to create circularly polarized PLED, in which the polarization is brought about by the molecular chirality of the polymer and does not require any molecular alignment [80].

Polymer **16** bears two alkoxy substituents at positions 2 and 3 of the benzene ring, which results in a notable blue-shifted absorption and emission bands (λ_{PL} = 519 nm, close to that of unsubstituted PPV) with about twice higher solid-state PL efficiency (40%) compared with 2,5-disubstituted PPVs (15%–20%) [81]. The EL efficiency of 0.03 cd/A (for Al cathode) and 0.07 cd/A (for Ca cathode) at an operating voltage of 10–15 V were reported for a single-layer device, whereas in a double-layer PLED with PPV **1** as an HTL, the high efficiency of 0.68 cd/A was achieved at 7.5 V (and maximum brightness of 4500 cd/m²). Notably, a similar polymer, **17**, showed somewhat lower PL efficiency (28% vs. 40%), and the PLED device fabricated as ITO/PEDOT/**17**/Ca showed an EL efficiency of 0.13 cd/A (and maximum brightness of only 86 cd/A) [82].

Many other monoalkoxy-PPV (e.g., **18a** [83]) as well as symmetric (e.g., polymer **18b** [84] and **19a** [85]) and asymmetric (e.g., polymers **19b–d** [86]) dialkoxy-PPVs have been synthesized. Interestingly, the monoalkoxy-PPV **18a** demonstrated improved PLQY (55 ± 5%) compared with dialkoxy-PPVs, and the authors emphasized a key role of the synthetic conditions determining the photophysical properties of the polymer [83].

EL from tetraalkoxy-substituted PPV **20**, synthesized by Gilch polymerization, was reported [87]. A multilayer device, ITO/PEDOT/**20**/PBD/LiF/Al, with PBD as an ETL emits green–yellow light with a luminescent efficiency of 0.12 lm/W, a maximum brightness of 8200 cd/m², and a turn-on voltage of 5 V. Highly soluble tetra-alkoxy substitute PPV derivatives **23** and **24** were synthesized and characterized [88]. The polymers are readily soluble in a variety of solvents, even in a very apolar solvent of cyclohexane in which most other available PPV derivatives are insoluble. However, the maximum luminance of PLED of polymer **23** is only 100 cd/m, which might be because the conjugation is limited by the out-of-plane distortion decreasing the carrier mobility.

The most branched among the alkoxy-PPVs are cholestanoxy-substituted polymers **25** and **26**, developed by Wudl and coworkers [84,89–91] (Chart 2.4). The authors suggested that a highly amorphous nature, possessed by all known steroids, will result in highly soluble PPVs with very good film-forming properties. Indeed, the solid-state fluorescence efficiency for **25** was reported to be 53%, which is more than twice higher than that of MEH–PPV **13** measured in the same conditions [92].

Oligoethyleneoxy-substituted PPVs **27** [93,94] are also known; however, the PL efficiency of these in the solid state is very low (0.6% for **27a** and 8.8% for **27b** [94]), which is transmitted into a low efficiency of PLED (ITO/**27b**/Al: luminous efficiency of 0.04 lm/W).

A combined theoretical (AM1 and valence-effective Hamiltonian [VEH]) and experimental (ultraviolet photoelectron spectroscopy [UPS]) study of the effect of alkyl and alkoxy substituents on the electronic structure of PPV has been undertaken by Fahlman et al. [95]. The results suggest that the strong influence of the substituents on the HOMO–LUMO levels (and the band gap) is primarily due to change in the torsion angle between the phenylene and vinylene groups. Interestingly, introduction of the alkoxy substituents does not cause significant steric hindrance and weak intramolecular O···H interactions (between

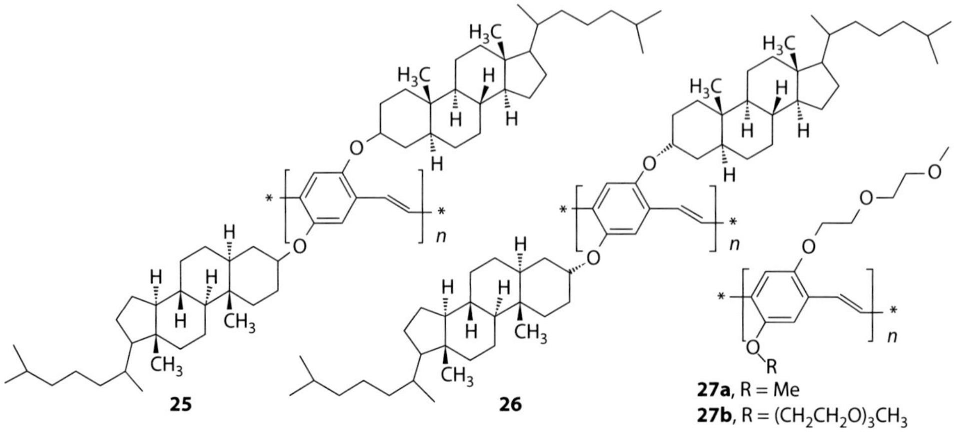

CHART 2.4
Chemical structures of cholestanoxy- and oligoethyleneoxy-substituted PPVs.

the oxygen atom of alkoxy group and a vinylic proton) in dialkoxy-substituted PPV may even favor the planar backbone geometry. In contrast, the dialkyl-substituted PPV is predicted to have a large torsion angle of 34° in the gas phase, and in the solid state the intermolecular packing and planarization will result in a significant change of the band gap.

2.2.2.2 Alkyl-Substituted Poly(p-Phenylene Vinylenes)

A number of alkylated-PPV polymers have been reported, e.g., **28a** [60,85,96], **28b** [97], and **29** [98] (Chart 2.5). The absorption of dialkylated PPVs is very close to that of unsubstituted PPV **1** (**26b**: λ_{max} = 404 nm [solution] [97]; **30**: λ_{max} = 422 nm [film] [92]) and hypsochromically shifted with respect to dialkoxy-PPVs. It was demonstrated that dialkylated PPV **30** can be used as a new type of solid-state laser material [17,99]. As for PLEDs, the first devices ITO/polymer/Ca fabricated with polymers **29** emitted light with λ_{max} = 530–560 nm; however, the reported Φ_{EL}^{ex} = 0.2% was lower [98] compared with alkoxy-PPV (**13**) in the same device configuration [65]. Somewhat later, very high QE PLEDs were fabricated with *t*-butyl/2-ethylhexyloxy-PPV **29** [68]. The device in the configuration ITO/PANI/**31**:surfactant/Al (where surfactant was lithium organophosphonate) emitted green light with Φ_{EL}^{ex} as high as 2.7% (turn-on voltage 10 V), although this high value is mostly due to the surfactant effect because a similar efficiency was obtained for an MEH–PPV **13**-based device in the same configuration.

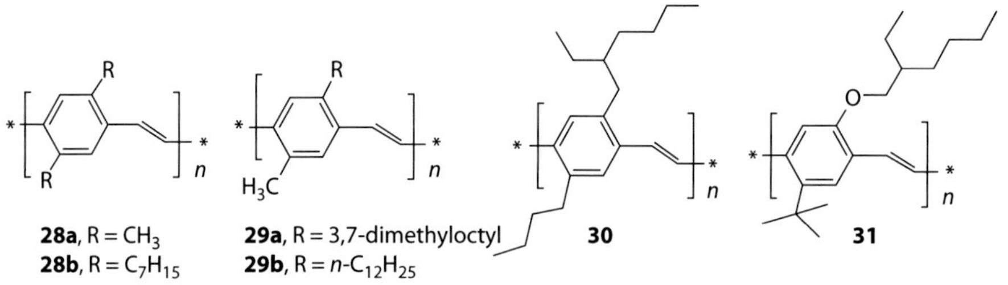

CHART 2.5
Chemical structures of alkyl-substituted PPVs.

2.2.2.3 Silyl-Substituted Poly(p-Phenylene Vinylenes)

Several PPV derivatives possessing trialkylsilyl substituents have been studied with the aim to control the band gap and the emission color of the polymer. The electronic influence of silicon substituents is somewhat difficult to predict. Judging from the variety of reported Hammett σ-constants of silyl substituent, it may act either as a weak electron donor or electron acceptor. Probably, more important in the case of substituted PPV, the bulky trialkylsilyl group increases the torsion angle between the phenylene and vinylene π-systems, thus enlarging the band gap of the polymer. The first Si-substituted PPV **32** was synthesized by Wudl's group [100] (Chart 2.6). Clearly, the trialkylsilyl substituent increases the band gap and shifts the emission to the blue region: thin films of **30** exhibit light-green EL emission with an Φ_{EL}^{ex} up to 0.3% [101].

A simple Si-containing polymer **33a** was reported by Kim and coworkers [102]. The optical band gap of this material in thin films is almost 2.5 eV, and it emits light at λ_{PL} of 515 and 550 nm with a remarkable (as for PPV derivatives) PLQY of 60%. A single-layer PLED device reveals the same emission spectrum and the $\Phi_{EL}^{ex} = 0.05\%$ and 0.08% for ITO/**33**/Al and ITO/**33**/Ca structures, respectively. The QE can be significantly improved (to 0.2% and 0.5%, respectively), by introducing a hole-blocking and electron-injecting layer of PBD (in a blend with polystyrene) between the light-emitting layer and the cathode.

Hwang and coworkers also studied silyl-substituted PPVs **33a** [103] and Chu and coworkers reported electroluminescent properties of sily-bisubstituted PPVs **34a** [104]. Martin and coworkers synthesized PPVs **33a** and **34a** and their copolymers with 2,3-dibutoxy-PPV **16** by using a novel family of 2,3-disubstituted aromatic precursors [82]. The solid-state PL efficiency of **33a** and **34a** is >60%, which is significantly higher than that of PPV **1** (27%) and MEH–PPV **13** (15%). Polymer **31a** reveals bright-green EL with $\Phi_{EL}^{ex} = 0.05\%$ (in ITO/**33a**/Al configuration) and 0.1% (in ITO/**31a**/Ca configuration) [103]. Interestingly, the PLEDs synthesized from **34a** emit light at positive and negative bias; however, the EL efficiency was not reported.

A systematic study of a series of *bis*-silyl-substituted PPVs **34d–h** with different side-chain lengths ranging from C1 to C18 was performed by Chen and coworkers [105]. The long-chain silyl-substituted PPVs show improved processability and film-forming properties, and sharp emission bands, although the thermal stability of the polymer somewhat decreases for the longest-chain substituents. The external QE of the device built with an Al cathode (ITO/**34f**/Al) is modest $\left(\Phi_{EL}^{ex} = 0.05\%\right)$ for this diode structure; however, interestingly, only little improvement of the efficiency (to 0.08%) was observed when replacing the Al cathode with Ca.

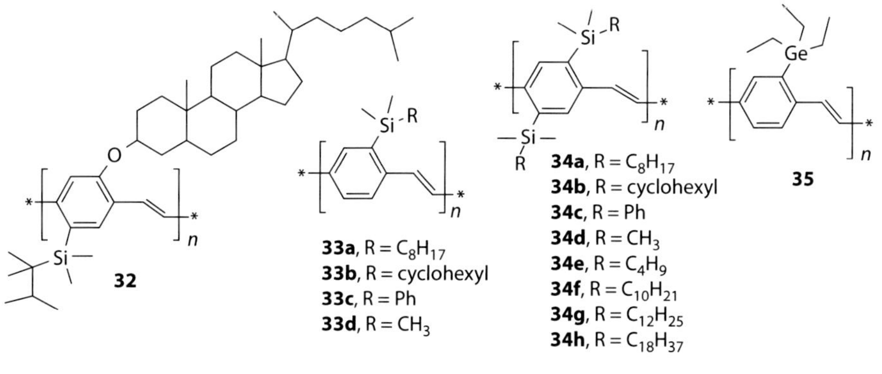

33a, R = C$_8$H$_{17}$
33b, R = cyclohexyl
33c, R = Ph
33d, R = CH$_3$

34a, R = C$_8$H$_{17}$
34b, R = cyclohexyl
34c, R = Ph
34d, R = CH$_3$
34e, R = C$_4$H$_9$
34f, R = C$_{10}$H$_{21}$
34g, R = C$_{12}$H$_{25}$
34h, R = C$_{18}$H$_{37}$

CHART 2.6
Chemical structures of silyl and germyl-substituted PPVs.

To improve the mechanical properties and the thermostability of the silyl-substituted PPVs, Ahn and coworkers [106] replaced the long *n*-alkyl chains in silyl substituents with more sterically demanding cyclohexyl and phenyl groups (polymers **33b**, **34b** and **33c**, **34c**, respectively). The best results (processability and EL efficiency) have been achieved for monosubstituted polymers **33b** and **33c**, which had a high glass-transition temperature (T_g) (~125°C), were thermally stable (5% mass loss at >430°C), and, owing to high molecular weight (M_n ~ 3 × 10⁵), possessed good film-forming properties. The PLED fabricated with these polymers, such as ITO/PVK/**33c**/Al and ITO/PVK/**33b**/Al, emitted green–yellow light (λ_{max} ~ 520 nm) with Φ_{EL}^{ex} = 0.08 and 0.07%, respectively, and the maximum brightness of the device was also rather low (220 cd/m²).

Hwang et al. [107] reported the synthesis of germylated-PPV **35**. Because of the insolubility of the material prepared by Gilch polymerization, the authors employed the thermoconversion route to prepare thin films of **35** from a nonconjugated methoxylated precursor polymer. The rationale beyond this synthesis was to increase steric hindrance (due to bulkier germanium atom) and to prevent the interchain quenching effect; however, no essential improvement vs. silylated analogs has been found. The device fabricated as ITO/**35**/LiF/Al emitted green light (λ_{max} = 514 nm) with an efficiency of 0.015 lm/W and maximum brightness of 600 cd/m² (cf. **33d** [107]: efficiency 0.025 cd/W, brightness 310 cd/m²). The turn-on voltage (13 V) was even somewhat higher than that for silylated-PPV **33** (10.5 V).

Generally, silyl substituents seem to retard the hole-transporting ability of PPV. As a result, devices fabricated from silyl-substituted PPVs suffer from a high turn-on voltage. To improve the EL efficiency of PLEDs fabricated from Si-PPVs, the introduction of additional hole-injection layer or copolymerization with electron-rich comonomers is required.

2.2.2.4 Aryl-Substituted Poly(p-Phenylene Vinylenes)

Numerous studies have been devoted to PPV derivatives possessing pendant aromatic groups. In 1998, Hsieh et al. [108] synthesized a series of soluble diphenyl-PPV derivatives via an ingenious route based on a Diels–Alder reaction of commercially available substituted cyclopentadienone with alkylacetylenes (Scheme 2.12). This is a very versatile method for the preparation of a variety of substituted monomers for PPV. In contrast to

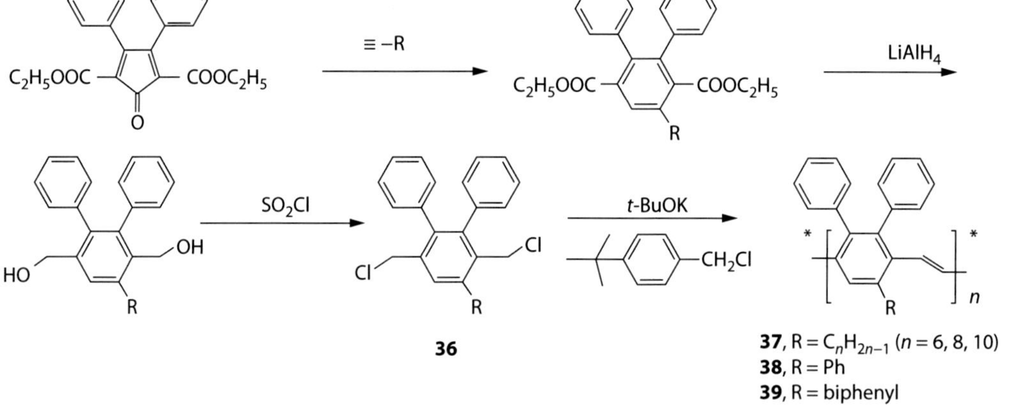

36

37, R = C_nH_{2n-1} (n = 6, 8, 10)
38, R = Ph
39, R = biphenyl

SCHEME 2.12
Synthesis of 2,3-diphenyl-substituted PPVs. (From Hsieh, B.R., Yu, Y., Forsythe, E.W., Schaaf, G.M., and Feld, W.A., *J. Am. Chem. Soc.*, 120, 231, 1998.)

the classical route of chloromethylation of alkyl(alkoxy)-substituted benzenes, the Diels–Alder approach eliminates the problem of isomer formation. The polymerization of the monomer **36** was carried out either through the CPR for insoluble polymers **38** and **39** or via a modified Gilch route involving the end capping with 4-*tert*-butylbenzyl chloride for **37**. However, even with the latter, the extremely high molecular weight of the polymer (M_w exceeds 2×10^6 for 1:0.05 ratio, 4×10^5 for 1:1 ratio) was still an issue, affecting the material processability (which is also the case for MEH–PPV **13**). Owing to significant steric factors of this substitution pattern, the emission of polymer **39** is blue shifted to a λ_{max} of 490 nm, which is very low for fully conjugated PPVs. Furthermore, the solid-state PL efficiency also reached a very high value of 65%, which can be explained by preventing the intermolecular packing of highly distorted polymer chains.

The solubility of the phenyl-PPVs can be greatly improved by the introduction of alkoxy substituents into the pendant phenyl groups. Spreitzer and coworkers [109,110] first reported the alkoxyphenyl-substituted PPVs **40–44** and their numerous copolymers prepared through a modified Gilch route (Chart 2.7). These polymers exhibited high PLQY, and PLEDs fabricated using these alkoxy-substituted phenyl-PPVs showed improved EL performance owing to their good film-forming properties. Thus, the green-emitting PLED ITO/**41**/Ca demonstrates $\Phi_{EL}^{ex} = 3.1\%$ (7.9 cd/A) [110,111].

However, it was later found that the phenyl-substituted PPVs have a significant level of defect TBB moieties built into the polymer chain [110]. The defects have moderate influence on the photophysical properties of polymers but strongly affect the PLED device lifetime. In fact, the amount of TBB defects in phenyl-substituted PPVs is several times higher than in similarly prepared dialkoxy-PPVs (5%–6% in phenyl-PPVs vs. 1.5%–2.2% in dialkoxy-PPVs **13** or **14**). Considering the mechanism of TBB formation (Scheme 2.13), the monophenyl-substituted monomers should favor such defects. Owing to similar acidity of both CH_2Cl groups, two types of quinone intermediate, A and A′, can be formed. At the same time, the steric hindrance brought about by phenyl groups will favor the HH coupling of these monomers rather than a normal HT reaction. The amount of TBB defects can be significantly suppressed (to <0.5%) by introducing an additional methoxy substituent into phenyl-PPV monomer. This is especially important, as it has been shown that the lifetime of the phenyl-substituted PPVs in PLED is increased by >30 times on lowering the TBB content from 6% to 3% [110].

Introducing two alkoxy substituents in positions 2 and 5 of the phenyl side group can efficiently prevent the interchain-fluorescent quenching, as has been demonstrated in the polythiophene (PT) series [112]. On the basis of this observation, Andersson and coworkers [43,113] synthesized phenyl-substituted PPVs **45** and studied their properties and EL performance (Chart 2.8). As in other phenyl-substituted PPVs, the TBB defects were a major

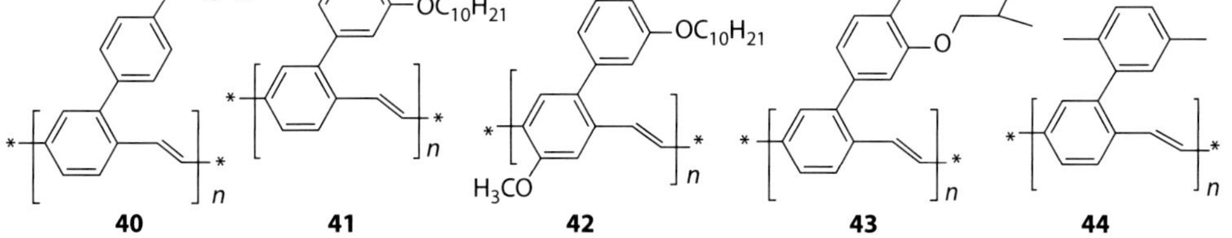

CHART 2.7
Chemical structures of phenyl-substituted PPVs.

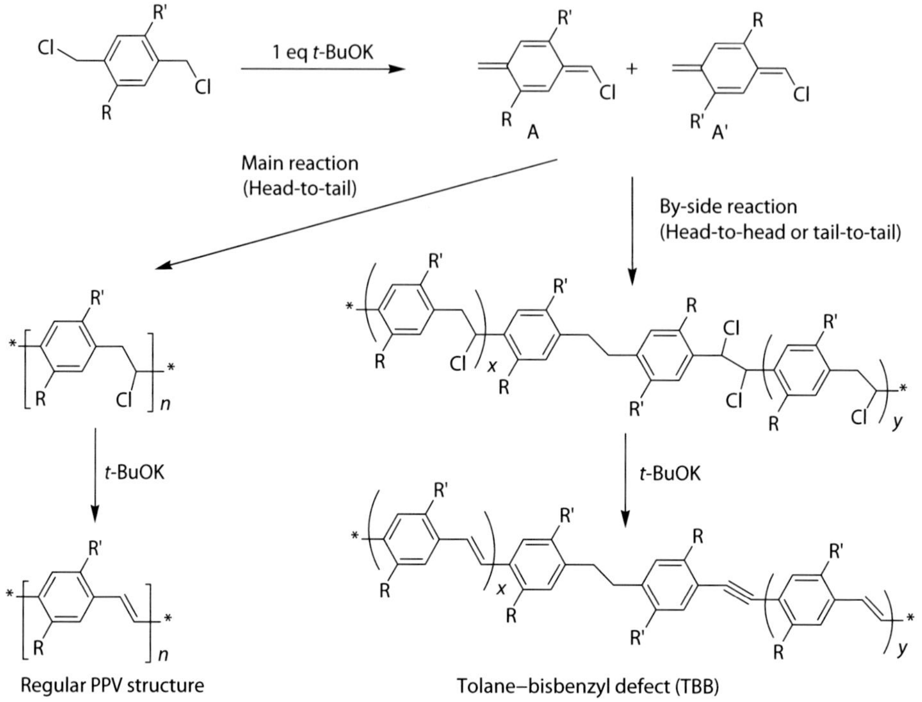

SCHEME 2.13
Mechanism of formation of TBB in the Gilch polymerization. (From Becker, H., Spreitzer, H., Kreuder, W., Kluge, E., Schenk, H., Parker, I., and Cao, Y. *Adv. Mater.*, 12, 42, 2000.)

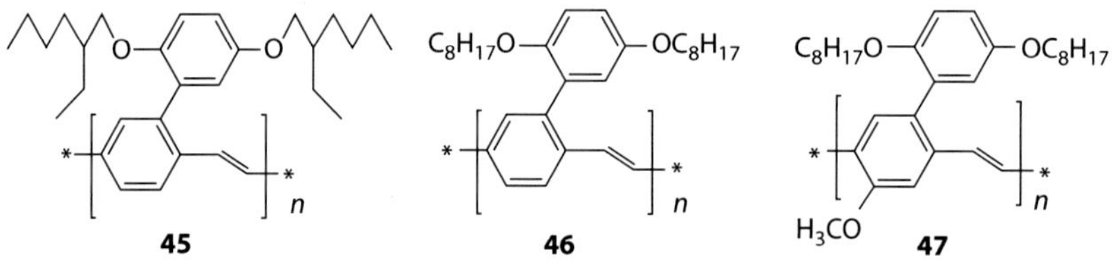

CHART 2.8
Chemical structures of alkoxyphenyl-substituted PPVs.

problem, although the amount of defects could be somewhat decreased by a careful control of the reaction conditions: the TBB content of 9% for the polymer prepared at 144°C reduces to 3% for the polymer prepared at 0°C. The highest $\Phi_{EL}^{ex} = 0.94\%$ was achieved with the polymer prepared at 66°C (in the PLED ITO/PEDOT/**45**/Ca). The same group has also synthesized similar polymers, **46** and **47** [114]. As found earlier, the introduction of the alkoxy group in the phenyl-PPV backbone decreases the amount of TBB content. Polymer **47** showed a few times lower level of TBB defect compared with **46**, and by performing the reaction at −35°C the amount of TBB can be brought to <1%. This difference is clearly reflected in the PLED device performance. The external QE of PLEDs fabricated as ITO/PANI/polymer/Ba/Al is similar for both polymers (1.74% for **47** and 1.34% for **46**); however, the operation lifetime of the low TBB content polymer **47** is prolonged by about 5 times.

Chen et al. [115] synthesized a series of dialkoxybiphenyl-substituted PPV polymers **48–55** and model alkoxyphenyl-PPVs **51** and **52** by the Gilch route (Chart 2.9). Additional

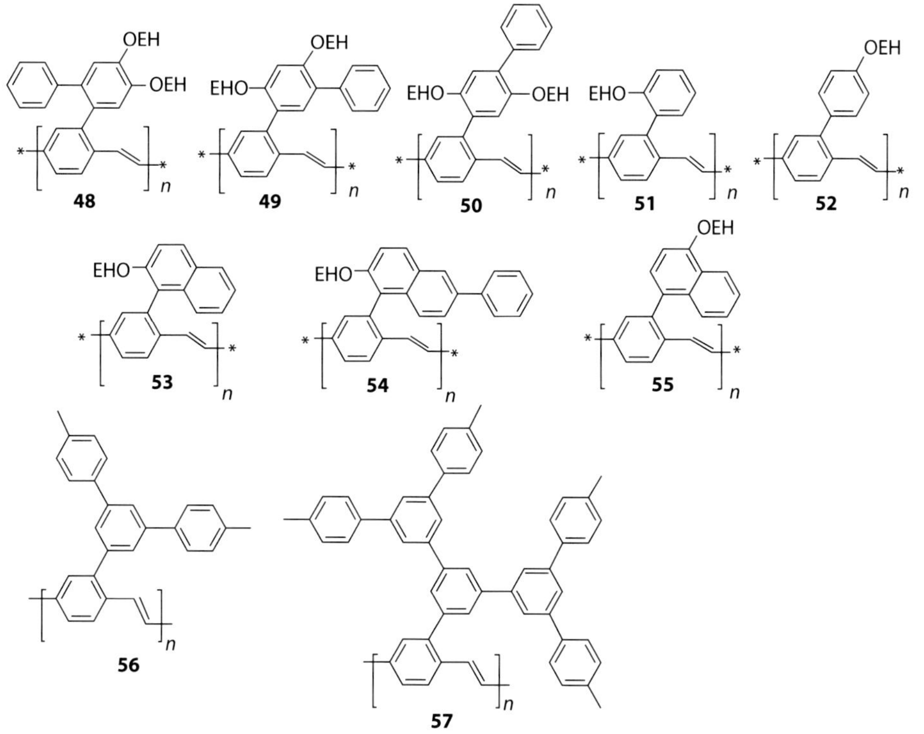

CHART 2.9
Chemical structures of dialkoxybiphenyl-substituted PPVs and naphthyl-substituted PPVs.

phenyl rings were introduced in the side chain to investigate the effect of the steric interaction on the formation of TBB defects as well as to increase the thermal stability of the polymers. The authors described the observed variation of the TBB defect (~0.5% for polymer **48** and ~4% for polymers **49** and **50**) as "expected," although the influence of the structural variations between, e.g., **48** and **49**, is not obvious. The authors also demonstrated that using a more polar solvent, such as THF, during polymerization helps suppress the TBB formation by a factor of 2, compared with *p*-xylene. These results were argued against the previous finding by Andersson and coworkers [43], who attributed the suppression of the TBB formation in THF vs. *p*-xylene solution solely to the lower temperature employed for the former. The green-emitting PLEDs fabricated in configuration ITO/PEDOT/polymer/Ba/Al showed the lowest turn-on voltage for polymer **48**, although the highest QE of 0.66% was achieved for **49** (0.37% for **48** and 0.25% for **50**).

The same group reported that the TBB defects can be brought below the nuclear magnetic resonance (NMR) detection limit by employing similar polymerization conditions (*t*-BuOK in THF at room temperature) in the synthesis of naphthyl-substituted PPVs **53–55** [116]. Although the absorption and PL spectra of all three polymers are similar, the EL can be finely tuned between 486 nm (for **54**) and 542 nm (for **55**). The external QE (studied for ITO/PEDOT/polymer/Ba/Al device) is also sensitive to the substituents pattern in the naphthyl pendant group: 0.08% for **53**, 0.02% for **54**, and 0.54% for **55**.

Kimura et al. [117] synthesized PPVs wrapped with 1,3,5-phenylene-based rigid dendrons by using stepwise Suzuki coupling reaction and polymerization. The fluorescence spectra of the synthesized polymer films demonstrate the effectiveness of side chains in preventing π-stacking among conjugated backbones without causing undesirable shifts in fluorescence spectra. The brightness for **56** and **57** at a driving voltage of 10.0 V were 9500 and 400 cd/m², respectively, with the device configuration of ITO/PEDOT:PSS/**56** or **57**/ Ca/Al. The authors also fabricated an OLED device with MEH–PPV as comparison, and found that the luminescence efficiency decreases in the order **57** < **56** < MEH–PPV, indicating that the attachment of higher-generation dendrons causes a significant reduction in charge mobility.

Jin et al. [118] attached a solubilizing trialkylsilyl substituent in the *meta*-position of the pendant phenyl group (Chart 2.10). The target polymer **58** was purified by membrane dialysis and revealed improved thermal (the decomposition temperature, T_{dec} = 406°C) and color stability. The device fabricated as ITO/PEDOT/polymer/Al:Li emits light at λ_{max} = 525 nm with rather moderate performance: maximum Φ_{EL}^{ex} = 0.08%, maximum brightness of 570 cd/m² (at 43 V), and turn-on voltage as high as 14 V. The authors explained these discouraging results by a high-energy barrier between the HOMO band of **58** (−5.30 eV) and the ITO/PEDOT work function (−5.0 V), although a 0.3 eV barrier can hardly be the only problem with the device. Nevertheless, the device performance was significantly improved by copolymerization with MEH–PPV **13**, which increases the HOMO and also unexpectedly decreases the LUMO level of the copolymer (see Section 2.2.3). Jin et al. [119] investigated the synthesis, characterization, photophysics, and EL of PPVs with various phenyl groups prepared by Gilch polymerization. Compared with polymer **58**, polymers **59** and **60** with disilyl substituents show the maximum luminance of 3098 and 383 cd/m², respectively, with the device configuration of ITO/PEDOT:PSS/polymer/Al. Although the maximum luminance of P6 is much lower than that of P5, the turn-on voltage of the **60** device (3.6 V) is much lower than that of polymer **59** (16 V). The EL spectrum of polymer **59** shows two maximum peaks at 517 and 542 nm. The maximum peak of polymer **60** EL locates at 530 nm.

Polymers **61** and **62** containing 9-phenylanthracene substituents in the PPV backbone have been synthesized via the Gilch polymerization route [120]. The long-wavelength

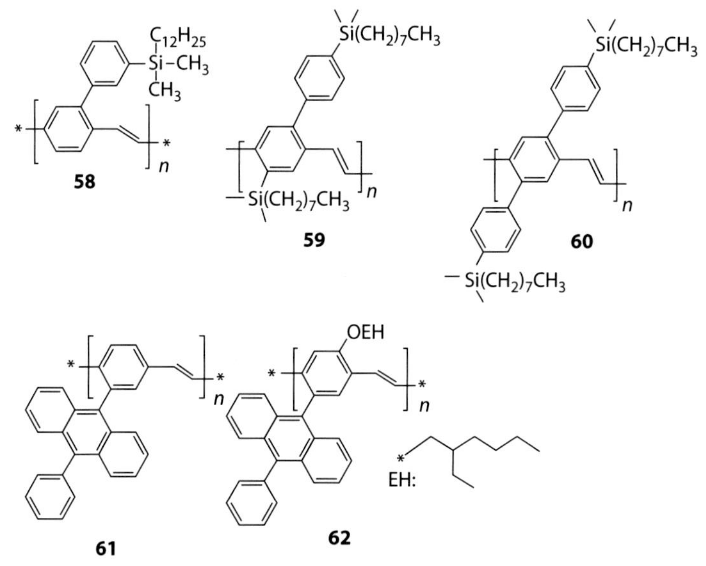

CHART 2.10
Chemical structures of phenyl- and 9-phenylanthracene-substituted PPVs.

absorption of these polymers is blue shifted (compared with PPV and MEH–PPV) owing to a twist in the polymer chain caused by the steric influence of anthracene substituents: band gaps determined from the UV absorption onsets were 2.58 and 2.38 eV for **61** and **62**, respectively. Owing to the presence of two luminophores with different emission properties (alkoxy-PPV and anthracene), **62** showed a very broad emission band (in PL and EL, Figure 2.3). The EL efficiency of PLEDs fabricated in the configuration of ITO/polymer/Al increased in the order PPV **1** < MEH–PPV **13** < **61** < **62**, with an efficiency of the last being almost 10 times higher than that of PPV **1** at the same current [120]. The authors assumed a synergistically enhanced effect of the phenylanthracene and alkoxy substituents. Also, the bulky anthracene group may suppress the interchain interaction, thus increasing the EL efficiency.

An important extension to phenyl-substituted PPVs was first reported by Lee and coworkers [121], who used Gilch polymerization to synthesize fluorenyl-substituted PPVs (**63**, **64**, **65**) and studied their performance in PLEDs (Chart 2.11). Owing to bulky but rigid fluorene substituents, these polymers have excellent solubility and yet are thermally stable up to 320°C and have a T_g of 113°C–148°C. The electron-donating methoxy group or electron-withdrawing cyano group was introduced to adjust the optical and electronic properties of the polymers. However, the influence of the substituents in the fluorene nucleus on the redox and fluorescent properties of polymers **63–65** was found to be very small, indicating that the PPV backbone rather than the pendant fluorene unit determines the optoelectronic properties of the system. At the same time, as was shown later, the substituents in fluorene nucleus retard the hole mobility of the polymer [122]. In fact, the unsubstituted polymer **57** showed rather high hole mobility of 4.5×10^{-4} cm²/(V s) (at electric field of 2.5×10^5 V/cm), which is two orders of magnitude higher than that of MEH–PPV **13**. The *bis*-fluorenyl-substituted PPVs (**67**) was synthesized by Gilch [119]. The T_{dec} was improved to 423°C with *bis*-fluorene substituents compared with mono-fluorenyl-substituted polymer **63**.

All three polymers emit blue–green light with a maximum of around 500 nm and a shoulder of 532 nm with a PLQY of 68%–71% (in solution). The PLEDs were fabricated with

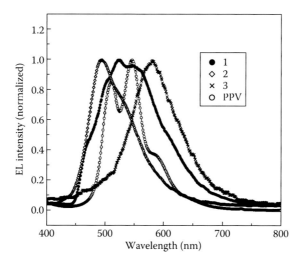

FIGURE 2.3
Electroluminescent spectra of PLEDs ITO/polymer/Al. 1—**62**, 2—**61**, 3—MEH–PPV **13**, and PPV—**1**. (From Chung, S., Jin, J., Lee, C.H., and Lee, C.E., *Adv. Mater.*, 10, 684, 1998. With permission.)

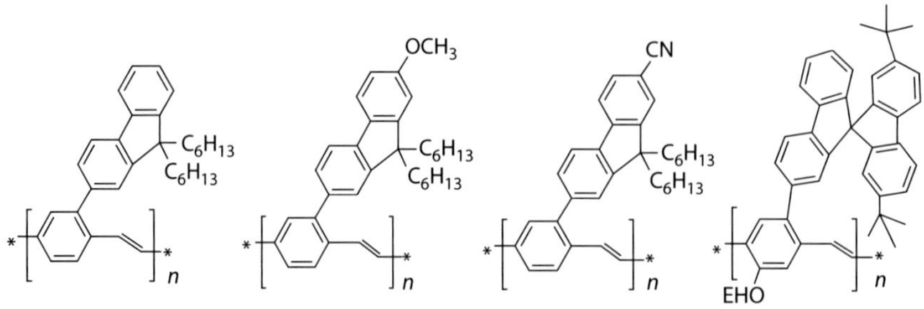

CHART 2.11
Chemical structures of fluorenyl-substituted PPVs.

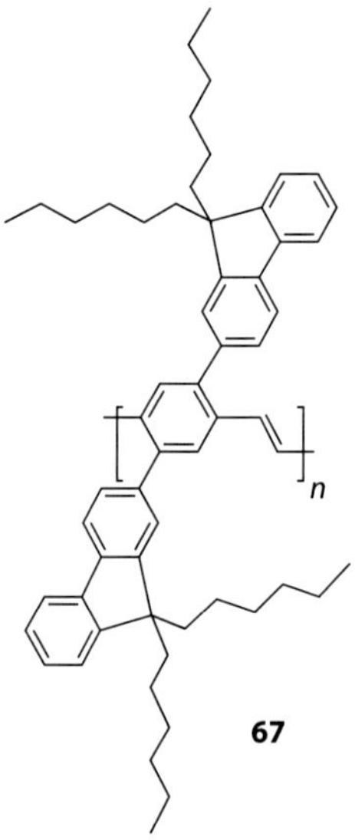

polymers **63**, **64**, and **65** as ITO/PEDOT/polymer/Ca. The EL performance of the device fabricated with **63** had the lowest turn-on voltage (2.8 V) and the highest luminescence efficiency (a maximum luminance of 12,000 cd/m² with a maximum $\Phi_{EL}^{ex} = 0.53\%$). The performance of the other two polymers, **64** and **65**, was similar; however, the QE was 1.6 to 2.8 times lower than that of **63**. The *bis*-fluorenyl-substituted polymer **67** shows lower luminance of 599 cd/m² with the device configuration of ITO/PEDOT:PSS/polymer/Al [119]. Shin and coworkers [123] synthesized even more hindered spiro-bifluorenyl-PPV **66**. Polymer **66** is highly soluble with T_g as high as 200°C, although its solid-state PL efficiency is not very high (26%). The PLED showed a turn-on voltage of 6 V and a maximum brightness of 1150 cd/m² at 12.5 V and a maximum power efficiency of 0.12 lm/W.

2.2.2.5 Poly(p-Phenylene Vinylene) Homopolymers with Electron-Withdrawing and -Donating Substituents

One of the problems of application of conjugated PLEDs is a rather high-lying LUMO energy level, which requires an unstable low-work-function metal electrode (such as Ca) for efficient electron injection. The VEH calculations predict that introduction of an electron-withdrawing group onto either the phenyl ring or the vinyl unit of PPV lowers the HOMO and LUMO energies of the polymer [94]. Significant improvement of the EL efficiency can be achieved by blending dialkoxy-PPV polymers with electron-transporting materials (such as PBD) [71]. In this regard, Chung and coworkers [124] compared the properties of PPV polymers carrying an electron-acceptor 2,5-diphenyl-1,3,4-oxadiazole group (resembling PBD, a widely used electron transporter) and an electron-donor carbazole group (an excellent hole transporter) (Scheme 2.14). In spite of the absence of long-chain substituents, polymers **68** and **69** are very soluble in common organic solvents, probably due to relatively low molecular weight (M_n = 24,000 and 16,000, respectively). Although the absolute efficiencies of the derived PLEDs were rather low $\left(\Phi_{EL}^{ex} < 0.004\% \right)$ owing to unoptimized device structure (ITO/polymer/Al), there is a clear dependence of the EL efficiency on the molecular orbital levels: the lower the barrier between the Al work function (−4.3 eV) and the LUMO of the polymer, the higher the efficiency: **68** > **1** > **69** (Scheme 2.14). This suggests that the electron and hole injections (or transport) in PPV are unbalanced and holes are the dominant charge carriers.

On the other hand, in a later publication, the same group admitted that the charge-injection barrier is not the only consideration playing a role in maximizing the EL efficiency [125]. Surprisingly, the introduction of a donor alkoxy substituent into carbazole–PPV, which further raises the LUMO level (but also the HOMO and thus facilitates the hole injection), increases the EL efficiency. The same-structure PLED, prepared with polymer **70**, possesses a 550-fold increase of external QE (0.01%) compared with polymer **69**. Furthermore, a very-high-performance PLED can be fabricated with a PEDOT-modified anode: the device ITO/PEDOT/**70**/Ca/Al shows an EL efficiency of up to 4.4 cd/A and maximum brightness in excess of 30,000 cd/m² [125]. The device half-lifetime was estimated to be 70 h at a brightness of 1000 cd/m².

In a similar approach, the HOMO level of PPV was controlled by the introduction of a dialkylamino donor group (polymer **71**) [126] (Chart 2.12). The dialkylamino groups render the material with high solubility and good film-forming properties and, similar to alkoxy groups, shift the emission maxima (~560 nm) to red. The PLED ITO/**71**/Mg/Al showed moderate efficiency (0.3% at ~30 cd/m²; 0.2% [0.45 cd/A, 0.08 lm/W] at 300 cd/m²); however, the turn-on voltage (>10 V) was rather high [126].

Balancing the charge transport via introduction of electron-transporting oxadiazole groups was further developed by Huang and coworkers [127,128], who synthesized polymers **72** and **73** via Gilch polymerization. Surprisingly, polymer **72** was completely

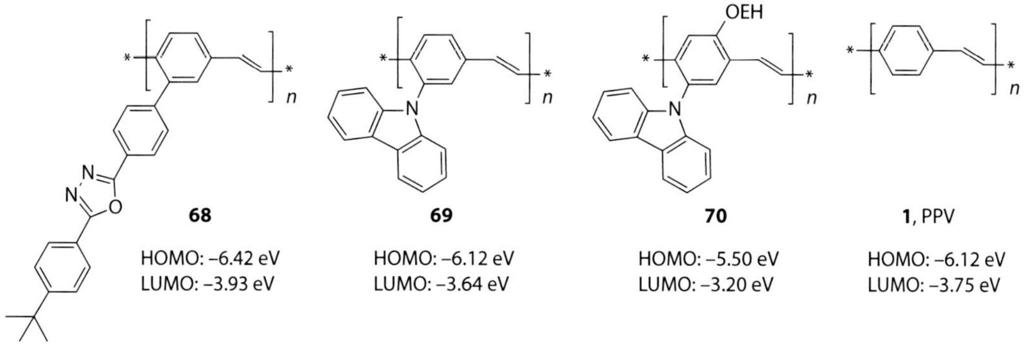

68	**69**	**70**	**1, PPV**
HOMO: −6.42 eV	HOMO: −6.12 eV	HOMO: −5.50 eV	HOMO: −6.12 eV
LUMO: −3.93 eV	LUMO: −3.64 eV	LUMO: −3.20 eV	LUMO: −3.75 eV

SCHEME 2.14
Tuning the energy levels of PPV by introducing pendant charge-transporting units. HOMO defined as the −I_P value (determined from the UPS experiments) and LUMO was deduced by adding the optical gap to the HOMO value.

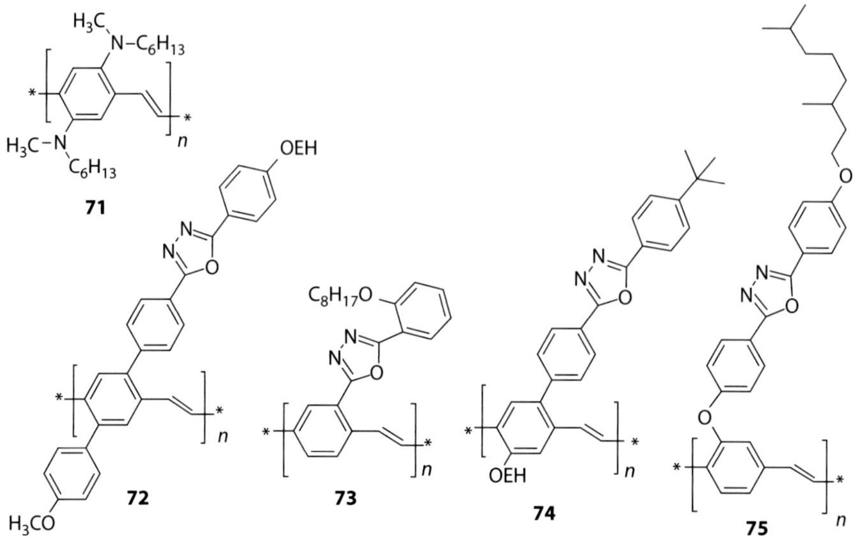

CHART 2.12
Chemical structures of dialkylamino-substituted PPV and PPVs containing oxadiazole units.

insoluble, regardless of the preparation conditions (Gilch polymerization), which was explained by its very rigid structure [127]. This polymer is in drastic contrast with **68**, which does not even have a long-chain 2-ethylhexyloxy substituent and can be due to higher molecular weight. Polymer **73**, where the *o*-alkoxyphenyl-substituted oxadiazole nucleus is directly connected to the PPV backbone, is a highly soluble material with moderate molecular weight of M_n = 20,700 and rather high T_g of 170°C. The PLEDs fabricated with this polymer in simple ITO/polymer/Al configuration showed improved charge-transport properties, as seen from the low turn-on voltage (4.0 V) at which the device starts to emit a yellow–orange light.

Lee and coworkers [129] synthesized PPV **74**, containing an oxadiazole and an alkoxy group. According to a UPS study, the HOMO and LUMO levels in **74** (–6.32 and –3.98 eV) are within the band gap of the parent polymer **68** without alkoxy substituents (Scheme 2.14). The external QE of PLEDs based on polymer **74** is about one order of magnitude higher than that for **68** (0.045% for ITO/**74**/Li:Al), and a maximum brightness of up to 7570 cd/m² was achieved for this material (using a Ca cathode).

PPV **75**, in which the oxadiazole group is separated from the PPV backbone by an oxygen atom, is a very soluble material with optical band gap of 2.36 eV and yellowish–orange emission color (chromaticity coordinates by the Commission Internationale de l'Eclairage [CIE]: x = 0.50, y = 0.47; λ_{max}^{PL} = 591 nm) [130]. An extremely high-performance PLED was claimed for this polymer in a single-layer configuration (ITO/PEDOT/**75**/Al). The device is characterized by a low turn-on voltage of 5 V, achieves the maximum brightness of 19,400 cd/m² at 14 V, and demonstrates a luminance efficiency of 21.1 cd/A (at 5900 cd/m²), which ranks it among the best-performing EL PPV materials.

Using CPR, Burn and coworkers introduced several electron-accepting moieties such as *p*-nitrostyryl (**76** [131]) and methylsulfonyl–phenyl (**77** [132]) groups as substituents in the PPV backbone. However, essentially no difference in EL performance (maximum QE 0.01% for ITO/polymer/Al) was found between polymers **77** and **78**, and the authors concluded

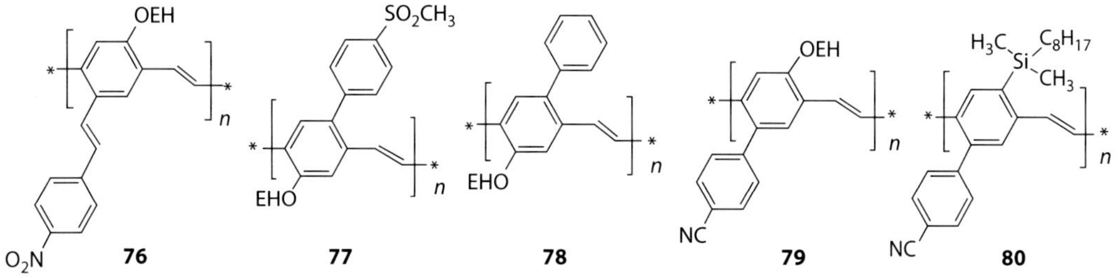

CHART 2.13
Chemical structures of PPVs with electron-accepting substituents.

that the methylsulfonyl group in the pendant phenyl ring does not facilitate electron injection (Chart 2.13).

Similar materials containing electron-withdrawing cyano groups, also attached to pendant phenyl substituents, were recently synthesized by Ko and coworkers [133] via Gilch polymerization. Polymers **79** and **80**, soluble in organic solvents, show good thermal stability (<5% weight loss at 400°C) and very high T_g (192°C and 180°C, respectively). The PL quantum efficiencies of **79** and **80** were 43% and 62%, respectively. PLEDs fabricated in ITO/PEDOT/polymer/LiF/Al configuration emitted greenish–yellow light (CIE: x = 0.455, y = 0.532) for **79** and very pure green light (CIE: x = 0.330, y = 0.599) for **80**, which is very close to the standard green color (CIE: x = 0.30, y = 0.60) used in high-definition television. According to electrochemical measurements, both polymers possess similar LUMO energy values (–2.72 eV for **79** and –2.75 eV for **80**); however, their HOMO energy levels are different (–5.41 eV for **79** and –5.72 eV for **80**), reflecting the difference in electron-donating properties of alkoxy and trialkylsilyl substituents (and also steric, a factor of the trialkylsilyl substituent reducing conjugation). Interestingly, the HOMO energy level of **79** is closer to the work function of PEDOT (–5.0 eV), but its EL performance is lower: the maximum brightness of **80** is 2900 cd/m² at 10 V (maximum QE 0.65%), whereas polymer **79** reaches only 330 cd/m² at 10.5 V (maximum QE 0.025%). Once again, these results indicate a not well-understood yet very beneficial effect of silyl substituents on the EL properties of PPV polymers.

Several groups have investigated the effect of fluorine electron-withdrawing substituents in PPV. The trifluoromethyl electron-withdrawing group (polymer **81**), attached directly to the phenylene units, improves the electron injection rather significantly but probably also acts as a quencher. The PLQY of this polymer (5%–7%) is much lower than that of the parent PPV (27%) [134]. As a result, single-layer LED devices ITO/**81**/Al and ITO/**81**/Ca showed almost the same Φ_{EL}^{ex}, which was one order of magnitude lower than that of ITO/PPV **1**/Ca (Chart 2.14).

Kang and coworkers [135] synthesized poly(2-fluoro-1,4-phenylene vinylene) **82** by the thermal conversion method. This polymer exhibits almost the same absorbance spectra as PPV **1** (λ_{max} ~ 410 nm), but the fluorescence band (λ_{max} = 560 nm) is red shifted by ca. 20 nm. The LUMO level was shifted down by ca. 0.15 eV, facilitating electron injection; however, in contrast to the above polymer **81**, no fluorescence quenching was observed. Consequently, the PLED devices fabricated as ITO/**82**/Al have about 10 times higher EL efficiency than those fabricated with PPV **1** under identical conditions.

Comparative analysis of different fluorine-substituted PPVs **82**–**84** has been performed by Gurge and coworkers [136]. Polymers **82** and **83** exhibit slightly blue-shifted UV absorbance relative to PPV **1**, but, remarkably, have substantially red-shifted PL and EL emission

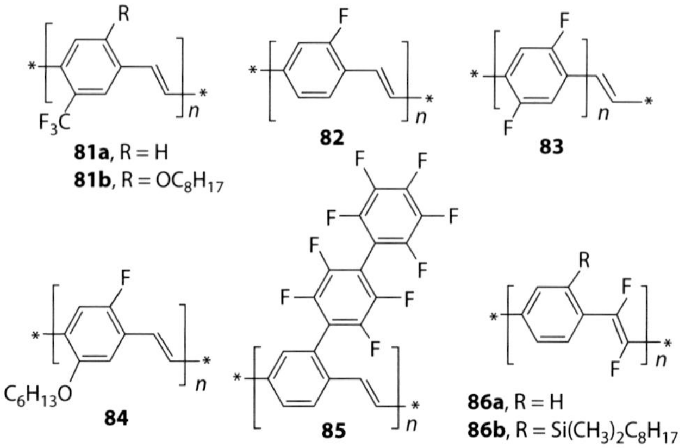

CHART 2.14
Chemical structures of PPVs with fluorine electron-withdrawing substituents.

bands. In the push–pull polymer **84**, both the absorption and emission maxima are red shifted relative to **1**. The LED performance of these materials appeared to be rather low (EL efficiency of ~0.002 cd/A and maximum luminance of ~100 cd/m² were achieved at 30 V), and the turn-on voltage for the push–pull polymer **84** (4 V) was lower than that in the more electron-deficient polymers **82** and **83**.

Jang et al. [137] reported on the high-electron-affinity perfluorobiphenyl-substituted PPV **85**. This polymer was synthesized through the thermoconversion method. The single-layer PLED ITO/**85**/Al showed 64 times higher EL efficiency than that fabricated with unsubstituted PPV **1**. A further (380-fold) increase of efficiency was achieved in a bilayer device ITO/**1**/**85**/Al.

Fluorine has also been introduced into the vinylene fragment of PPV (**86a,b** [138]), resulting in blue shifts in PL (from 580 to 495 nm) and EL (from 565 to 540 nm; yellow to green) spectra. The turn-on voltage of ITO/PEDOT/**86**/Al devices was 3–4 V, and for silyl-substituted **86b** a rather high luminance of 2.7 cd/A (at 6.5 V) and a maximum brightness of 750 cd/m² (at 7.5 V) were achieved.

Chlorine substituents have also been introduced into PPV (in phenylene units); however, no EL or PL properties of these materials have been reported [60].

2.2.3 Poly(*p*-Phenylene Vinylene) Copolymers

2.2.3.1 Poly(p-Phenylene Vinylene) Copolymers with Electron Donor and Aryl Substituents

Owing to a higher variety of possible structures, copolymers allow a better control of the HOMO–LUMO levels necessary to optimize the EL properties of the PPV, compared with homopolymers. Often the optical and electronic properties in copolymers can be finely tuned by simply changing the feed ratio of comonomers (although the structure–property relationship in these systems is even more complex than in homo-PPV polymers). Using different comonomer units, various PPV-based materials with tuned optical and electronic properties have been prepared.

Although MEH–PPV **13** (at the time of discovery) was one of the most efficient soluble polymers for PLEDs application, its performance is not high enough for commercialization as LEP. One of the reasons is unbalanced hole–electron mobility in MEH–PPV (the mobility of holes is 100 times faster than the mobility of electrons) [139]. Copolymerization with other conjugated monomers, to some extent, can improve the electron-transporting properties and increase the EL performance.

The first realization of this approach was reported by the Cambridge group, which synthesized copolymers **87** containing phenylene vinylene and dialkoxy(phenylene vinylene) units by the thermoconversion method [24,140]. A 30-fold improvement in EL efficiency was observed for these copolymers compared with the PPV **1** or MEH–PPV **13** devices fabricated in the same configuration (Chart 2.15).

Since then, most research groups have used the copolymerization approach to tune the properties of EL materials. The synthetic methods include the Wittig–Horner condensation, Gilch polymerization, Heck reaction, and others. A number of PPV copolymers similar to **87** were synthesized (e.g., **88a** [141] and **88b** [142]). Most recently, Huang et al. [143] used a Wittig–Horner reaction to construct polymer **88c** with hydrophilic oligo(ethylene oxide) substituents for LECs. This copolymer is a yellow–green emitter, whose efficiency can be improved from 0.038 lm/W (in classical LED configuration ITO/polymer/Al) to 0.185 lm/W (in LEC ITO/polymer+LiOTf/Al).

The groups of Holmes and Friend reported a series of PPV copolymers containing alkoxy- and trialkylsilyl-substituted phenylene rings in random distribution (**89–90** [93], **91** [82], and **92** [144]) (Chart 2.16). The authors mentioned that introduction of the trialkylsilyl group results in about a five-time increase of the luminous efficiency of copolymers **90a,b**, compared with corresponding dialkoxy-PPV homopolymers **27a,b** (0.2 lm/W for ITO/**90b**/Al). High ion affinity of the oligo(ethylene oxide) pendant group allows to create LECs with an efficiency of 0.5 lm/W (for ITO/**90b**+LiOTf/Al) [93].

EL from related polymers **91a,b**, containing more sterically demanding 2,3-dibutoxyphenylene units, have also been studied. Both copolymers are blue–green emitters ($\lambda_{max} \sim 545$ nm) with moderately high PLQY in the solid state, 35% for **91a** and 28% for **91b**. At the same time, EL from **91a** can only be observed with a Ca (but not Al) electrode, whereas the single-layer device ITO/PEDOT/**91b**/Al shows a current efficiency of 0.72 cd/A and a maximum luminance of 1380 cd/m^2 (turn-on voltage 4 V) [82] (Chart 2.16).

Synthesis of phenyl- or alkoxy-substituted PPV copolymers was first reported by Spreitzer et al. [77], who studied in detail the dependence of the EL on the comonomer ratio in **93** (Scheme 2.15) and other related phenyl-substituted PPV copolymers [108–110]. The polymerization was performed via the Gilch route using different comonomer feed ratios. All copolymers showed a high EL efficiency of >10 cd/A and low-driving voltage (~3.5 V). In addition, very high emission brightness (10,000 cd/m^2) was easily achieved by applying a very reasonable voltage of 6 to 8 V. The emission color of the phenyl-substituted PPV **93a** is green (CIE: $x = 0.35$, $y = 0.61$), whereas increasing the ratio of dialkoxy(phenylene vinylene) unit results in a gradual red shift of the emission, through yellow (for **93e**, CIE: $x = 0.49$, $y = 0.50$) to orange color (for **93f**, CIE: $x = 0.60$, $y = 0.40$). Interestingly, the dialkoxy-PPV

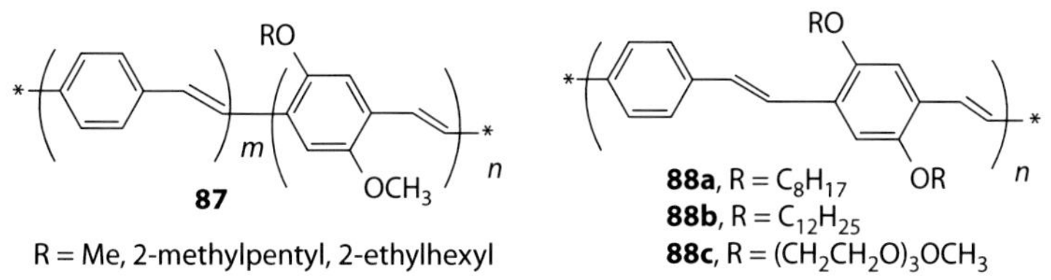

87

R = Me, 2-methylpentyl, 2-ethylhexyl

88a, R = C$_8$H$_{17}$
88b, R = C$_{12}$H$_{25}$
88c, R = (CH$_2$CH$_2$O)$_3$OCH$_3$

CHART 2.15
Chemical structures of alkoxy-substituted PPV copolymers.

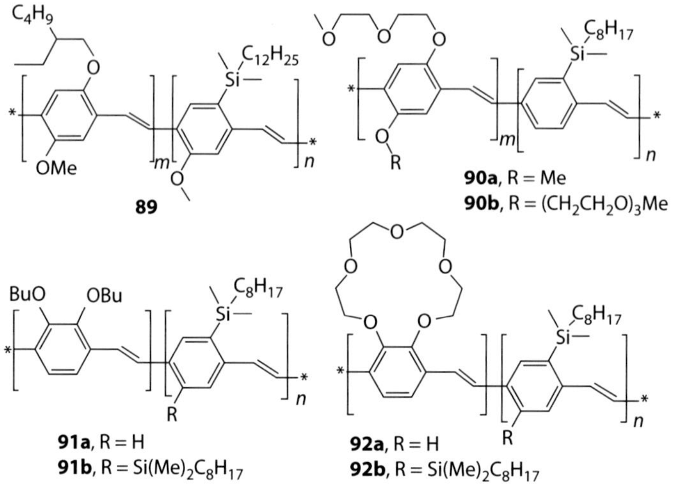

CHART 2.16
Chemical structures of PPV copolymers containing alkoxy- and trialkylsilyl-substituted phenelyenes.

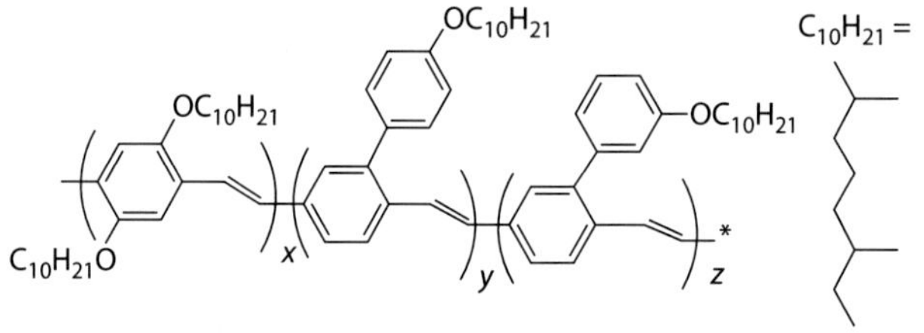

93a, $x{:}y{:}z$ = 0:50:50	EL: 515 nm	11.5 cd/A	10.6 lm/W
93b, $x{:}y{:}z$ = 1:49.5:49.5	EL: 530 nm	12.9 cd/A	11.9 lm/W
93c, $x{:}y{:}z$ = 2:49:49	EL: 540 nm	14.8 cd/A	16.1 lm/W
93d, $x{:}y{:}z$ = 4:48:48	EL: 547 nm	10.5 cd/A	9.8 lm/W
93e, $x{:}y{:}z$ = 10:45:45	EL: 567 nm	9.6 cd/A	8.3 lm/W
93f, $x{:}y{:}z$ = 100:0:0	EL: 583 nm	2.0 cd/A	2.2 lm/W

SCHEME 2.15
Dependence of the electroluminescence of alkoxyphenyl-PPV copolymers on the comonomers ratio (PLED configuration ITO/PEDOT or PANI/polymer/Ca or Yb/Ag).

homopolymer **93f** revealed a significantly lower EL efficiency of 2 cd/A; however, introducing the corresponding unit in the copolymer in small amounts (2%, based on feed ratio) allows increasing the luminous efficiency from 10.6 lm/W (for **93a**) to 16.1 lm/W (for **93c**).

As we mentioned before, phenyl substituents in PPV increase the amount of TBB defects in the Gilch synthesis, affecting the device stability. Spreitzer et al. [109] showed that introducing a second substituent (alkoxy group) can significantly reduce the formation of TBB. The TBB suppression was observed with increasing the feed ratio of alkoxy-phenyl-substituted monomer. Scheme 2.16 shows the chemical structures of two greenish–yellow copolymers and their EL performance along with their TBB defects. These two copolymers have similar optical and electronic properties but their TBB contents are different. Hence, the device fabricated from low-TBB-content copolymer **94** showed 30 times longer lifetime than the device made from high-TBB-content copolymer, although their initial EL performance was nearly identical. Rather high lifetime of copolymers of type **94** made

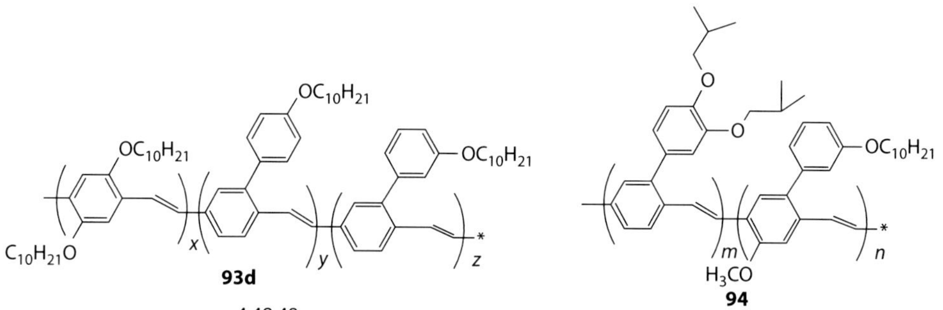

93d

x:y:z = 4:48:48

94

m:n = 50:50

TBB: 6% EL: 547 nm 10.5 cd/A 9.8 lm/W TBB: 2.5% EL: 548 nm 9.3 cd/A 9.2 lm/W

SCHEME 2.16
Concentration of TBB defects and the initial EL performance of two related phenyl-PPV copolymers.

them attractive enough for industrial applications in PLEDs, as pursued by Philips and DuPont [145].

An extremely efficient PLED was fabricated with a similar phenyl-PPV copolymer **95** by the Cambridge group [146]. The PLED ITO/PEDOT/**95**/Ca could be turned on at only 2.2 V and shows Φ_{EL}^{ex} of an amazing 6% (20 cd/A) (estimated internal QE of 15%–20%, close to the theoretical limit of 25%), which is still among the record values in electrofluorescent PPV materials. Although this significant improvement was greatly due to the sophisticated engineering of the anode and EL polymer interface (ITO was modified by multilayer deposition of PEDOT-based materials), the dialkoxy-PPV homopolymer **14**, under the same conditions, showed essentially lower efficiency (1.8%, 2.6 cd/A) (Chart 2.17). Kim et al. synthesized highly alkoxylated PPV copolymers **96a** and **96b** through the Gilch polymerization [147]. EL devices using the copolymers as the emissive layer have been fabricated with a configuration of ITO/PEDOT:PSS/polymers/Ca/Al. The polymer **96b** device resulted in a >4-fold increase in device lifetime (170 h at 1000 cd/m^2) than the homopolymers due to the highly reduced steric hindrance of the polymer backbone by introducing dialkoxy-phenylene vinylene units between the adjacent highly alkoxylated repeating units (tetraalkoxy-phenylene vinylene) in the polymer backbone. Scheme 2.17 illustrates some MEH–PPV random copolymers with trialkylsilylphenyl (**97** [118]) and dihexylfluorene (**98** [148]) units, synthesized by Jin's group using the Gilch polymerization method. The HOMO–LUMO energy levels and the emission color of these copolymers can

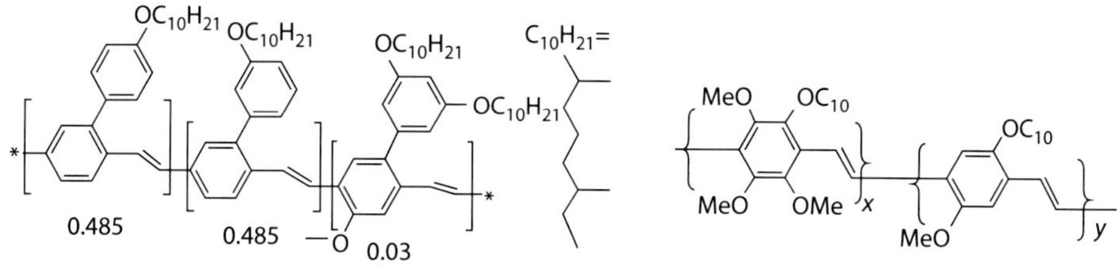

95

96a, x:y = 50.5:45.5
96b, x:y = 23:77

CHART 2.17
Chemical structures of phenyl- and alkoxy-substituted PPV copolymers.

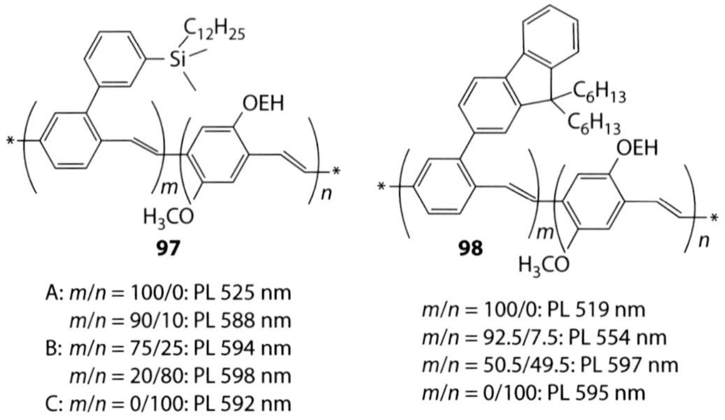

SCHEME 2.17
Tuning the emission maxima in random phenyl-PPV copolymers.

be finely tuned by adjusting the feed ratio of the comonomers (Scheme 2.17, Figure 2.4). The authors found a turn-on voltage dependence on the gap between the HOMO level and the work function of the ITO electrode and suggested that holes are the major carriers in these materials. A high-efficiency red–orange-emitting $\left(\lambda_{max}^{EL} = 590\,nm\right)$ PLED has been fabricated with copolymer **97** using a low-work-function Al:Li alloy electrode [118]. The device ITO/**97**/Al:Li showed a low turn-on voltage of 2.3 V, a high maximum brightness of >19,000 cd/m^2 (at 12 V), a high luminance efficiency of 2.9 lm/W, and a half-life of 120 h at 1000 cd/m^2, which significantly overrides the performance of similar devices prepared from the corresponding homopolymers.

Likewise, copolymers **98** have higher EL efficiency than homopolymers DHF–PPV **63** or MEH–PPV **13**, owing to more balanced charge-injection and -transport properties. Copolymer **98** with 7.5 wt.% loading of dialkoxyphenylene comonomer gave the highest QE. The device with an ITO/PEDOT/**98**/Ca configuration showed remarkably higher efficiency (2.4 cd/A) than devices fabricated with other copolymers in the series (0.65–1.0 cd/A) [148].

One of the problems in the design of organic EL materials is the decrease of the QE of fluorescence in the solid state due to the formation of π-aggregates. Earlier we described one approach to circumvent this problem by introducing bulky aromatic substituents into the PPV backbone to hinder intermolecular π-stacking. However, in homopolymers, very bulky substituents slow down the polymerization reaction, resulting in low-molecular-weight products. The problem can be solved by introducing a second, less sterically demanding comonomer unit. The polymer **99**, synthesized by Peng et al. [149] through a Wittig–Horner reaction, is a highly fluorescent material with a PLQY (in films) of 61%–82% (cf. 10% for dioctyloxy-PPV). Unfortunately, no EL data was reported for these copolymers.

Even more bulky substituents were introduced in copolymers **100** synthesized by Heck coupling [150]. These materials emit blue light with a maximum emission peak at 442 nm, which is among the shortest emission wavelengths of the formally conjugated PPV (although likely, it is more related to the oligophenylene substituents than to the PPV backbone). This high-energy emission was attributed to conjugation interruption caused by oligophenylene substituents, although, as mentioned above, the aromatic substituents can adopt a nearly orthogonal dihedral angle with respect to the PPV chain, minimizing the steric

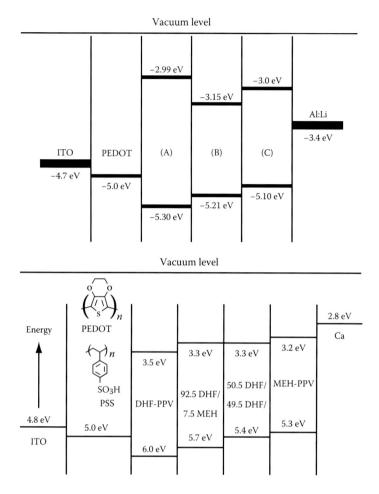

FIGURE 2.4
Energy diagrams of PLEDs based on polymers. Top: **97** (A, B, C). Bottom: **98**, with different feeding ratio (DHF stands for 9,9-dihexylfluorenyl). (From Jin, S., Jang, M., and Suh, H., *Chem. Mater.*, 14, 643, 2002. With permission; Sohn, B., Kim, K., Choi, D.S., Kim, Y.K., Jeoung, S.C., and Jin, J., *Macromolecules*, 35, 2876, 2002. With permission.)

encumbrance. Interestingly, in spite of these bulky substituents, the fluorescence spectra of **100** in films differ from those of solutions by an additional longer wavelength shoulder at 512 nm, ascribed by the authors to aggregate emission (Chart 2.18).

The properties of PPV polymers can also be tailored by introducing additional aromatic units in the PPV chain. In 1995, Hilberer and coworkers [49] reported terphenylene-containing copolymer **101**, synthesized via the Heck reaction (Chart 2.19). A relatively poor conjugation brought about by the alkyl-substituted oligophenylene fragments (see Section 2.5) resulted in blue shift of the emission wavelength (λ_{PL} = 440 nm in films; solution Φ_{PL} = 89%–90%). A nonoptimized single-layer PLED (ITO/**101**/Al) emitted blue light ($\lambda_{EL} \sim$ 450 nm) with Φ_{EL}^{ex} up to 0.03%.

The related copolymer **102**, synthesized in the 1980s by Feast et al. [151], presents a rare example of a PPV-containing phenyl substituents on the vinylene unit. Apparently, the steric hindrance caused by phenyl substituents in **102** is not dramatic, and the optical properties of **102** are similar to those of other PPVs (green emission, $\lambda_{PL} \sim \lambda_{EL} \sim$ 530 nm). An

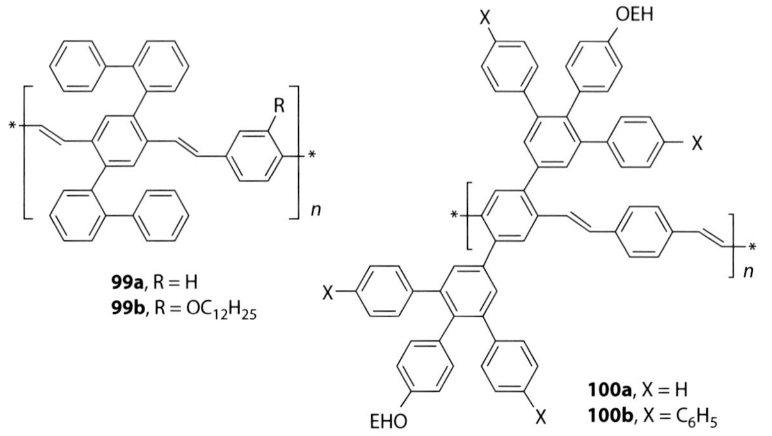

CHART 2.18
Chemical structures of PPV copolymers with bulky substituents.

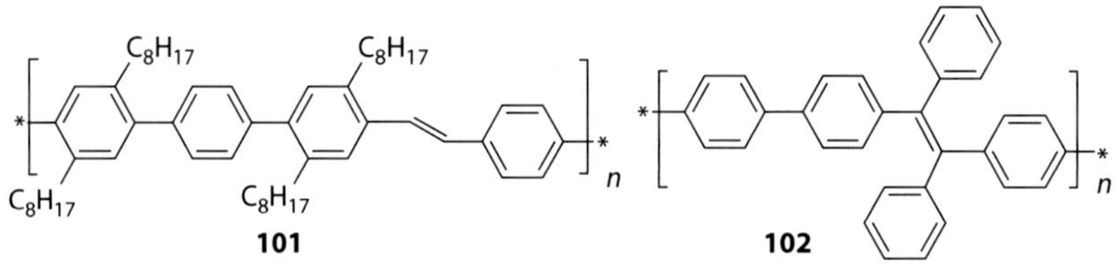

CHART 2.19
Chemical structures of PPV copolymers with additional aromatic units in the PPV chain.

internal QE of up to 1% was reported with multilayer **102**-based PLEDs containing PPV **1** and PVK as HTL [152].

In the effort to make pure blue-emitting materials, Ahn and coworkers [153] synthesized a series of PPV-based copolymers containing carbazole (polymers **103** and **104**) and fluorene (polymers **105** and **106**) units via Wittig polycondensation. The use of trimethylsilyl substituents, instead of alkoxy groups, eliminates the electron donor influence of the latter and leads to chain distortion that bathochromically shifts the emission (λ_{max} = 480 nm for **103** and 495 nm for **105**). In addition, a very high PLQY was found for these polymers in the solid state (64 and 81%, respectively). Single-layer PLEDs fabricated with **103** and **105** (ITO/polymer/Al) showed EL efficiencies of 13 and 32 times higher than MEH–PPV, respectively (see also Ref. [154] for synthesis and PLED studies of polymers **107** and **108**). A PPV derivative containing carbazole as side pendant (**109**) was synthesized by Witting–Horner polymerization [155]. A single-layer PLED with a simple configuration of ITO/**109**(80 nm)/Ca/Al exhibited a bright-yellow emission with a brightness of 1560 cd m² at a bias of 11 V, which is much higher than that of the MEH–PPV device (250 cd/m² at 11V) due to the enhanced hole-injection ability by the introduction of the carbazole functional group (Chart 2.20).

Similar PPV-based copolymers with carbazole and fluorene units in the backbone, **110** and **111** (and also similar copolymers with oxadiazole substituents **112** and **113**), have been synthesized by Kim and coworkers [156]. Much lower PL efficiency in films was found in

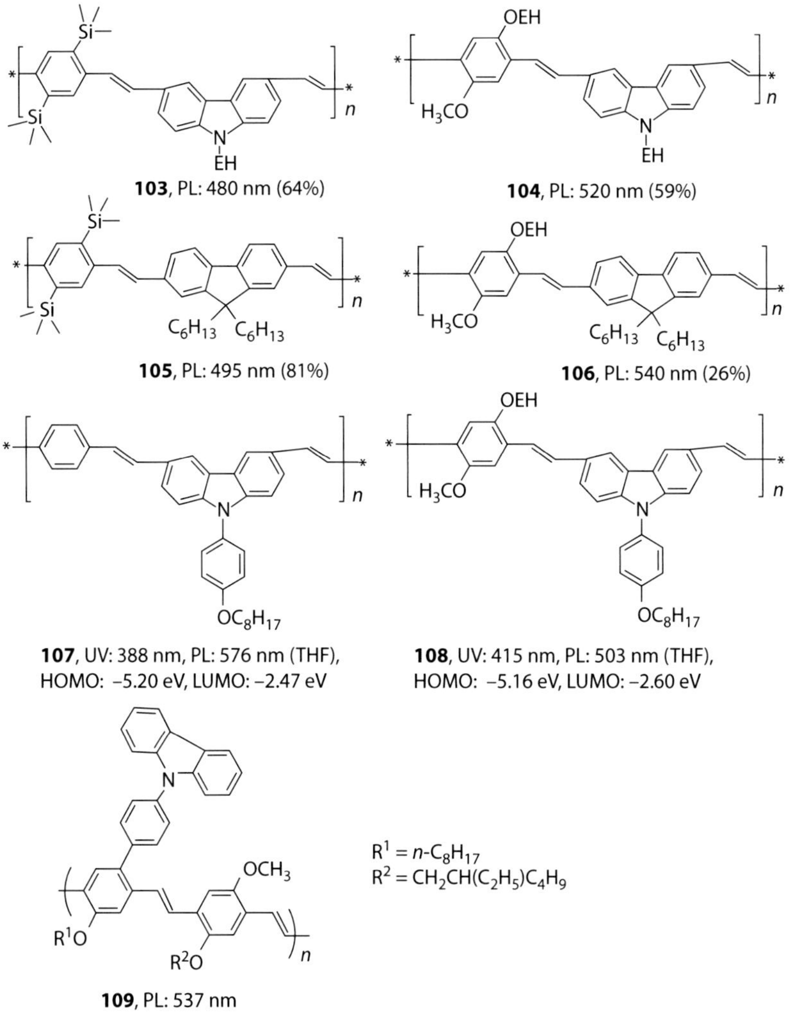

CHART 2.20
Chemical structures of PPV copolymers containing carbazole and fluorene units.

this case for carbazole-containing polymers **111** and **113** (1%–4%), when compared with materials prepared by Shim (**103** and **104**, 59%–64%) (Chart 2.21).

Blends of yellowish–green light–emissive carbazole-containing PPV-based copolymer **114** (λ_{PL} = 490, 520 nm, λ_{EL} = 533 nm) with blue-emissive oxadiazole–poly-*p*-phenylene (PPP) copolymer **115** (λ_{EL} = 426 nm) allowed to tune the emission of PLEDs (ITO/polymer blend/Al) from λ_{EL} = 533 nm to λ_{EL} = 451 nm, although the device turn-on voltage was essentially higher for the blends with increased content of **115** [157] (Chart 2.22).

The diphenylamino-substituted PPV **116**, with solubilizing alkoxy groups, was synthesized by Shi and Zheng [158] via a Wittig–Horner reaction (Chart 2.23). Its PL (555 nm) is very similar to that of diamino-PPV **71** and dialkoxy-PPV **14** homopolymers. The PLQY is rather high in solution (80%); however, it drops to only 8% in films. Consequently, only a moderate EL efficiency of 0.6 cd/A was obtained with this material (device ITO/PEDOT/**116**/Mg/Al) [158]. Almost simultaneously, Pu and coworkers [159] reported a similar

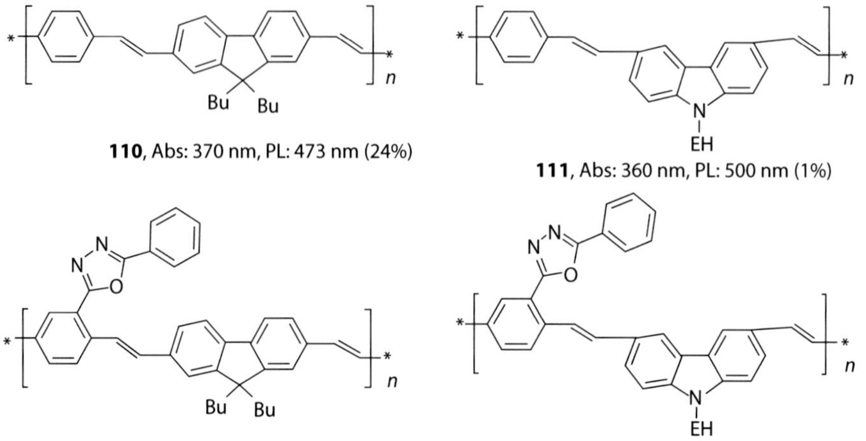

110, Abs: 370 nm, PL: 473 nm (24%)

111, Abs: 360 nm, PL: 500 nm (1%)

112, Abs: 368 nm, PL: 473 nm (14%)

113, Abs: 376 nm, PL: 521 nm (4%)

CHART 2.21
Chemical structures of PPV copolymers containing carbazole, fluorene, and oxadiazole units.

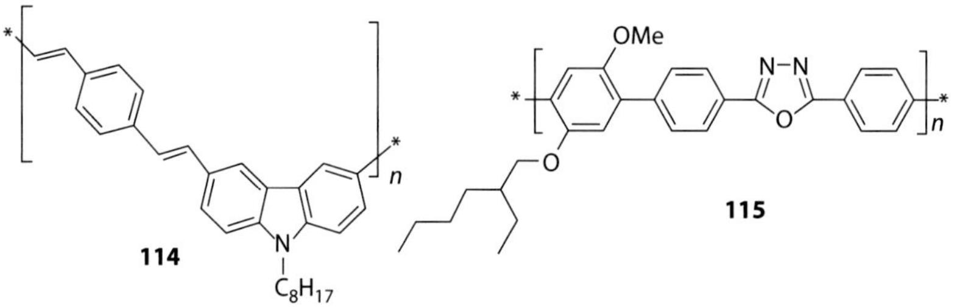

114

115

CHART 2.22
Chemical structures of PPV copolymer containing carbazole and PPP copolymer containing oxadiazole.

diphenylamino-substituted PPV copolymer **117**, which affords very efficient PLEDs. The device ITO/PEDOT/**117**/Ca/Al showed a low turn-on voltage (3 V), a high maximum brightness (29,500 cd/m^2), and a power efficiency of 1.1 lm/W, which can be further improved to 3.0 lm/W, if cesium is used as a cathode.

The amino group has also been introduced in the PPV backbone. Copolymers **118** and **119** (and some derivatives incorporating additional phenylene, naphthalene, or anthracene units in the main chain [160]) have been synthesized by Wittig–Horner reaction as green-emitting materials ($\lambda_{PL} \sim 530$ nm) with moderate PLQY (10%–15% in films) [144]. Preliminary results showed improved hole-transport properties in these materials (manifested as a decrease of the turn-on voltage to 2.4–2.8 V, for ITO/PEDOT/**118b**/Ca(Al)), although the efficiency of the unoptimized device was very low (0.001 cd/A). An even lower turn-on voltage of 1.5 V was reported for PLED ITO/**118a**/Al [160]. Some related imino-substituted PPV copolymers with metal chelating azomethine site have been recently synthesized; however, their applications as EL materials have not yet been explored [161].

Zhang and coworkers [162] synthesized a bipolar luminescent PPV-based polymer **120**, which contained both donor triarylamine and acceptor oxadiazole moieties in the backbone. A device fabricated with this polymer (ITO/PEDOT/**120**/CsF/Al) showed a maximum brightness of 3600 cd/m^2 and a maximum luminescent efficiency of 0.65 cd/A $\left(\Phi_{EL}^{ex} = 0.3\%\right)$, about 15 times brighter and more efficient than the device of the same configuration with

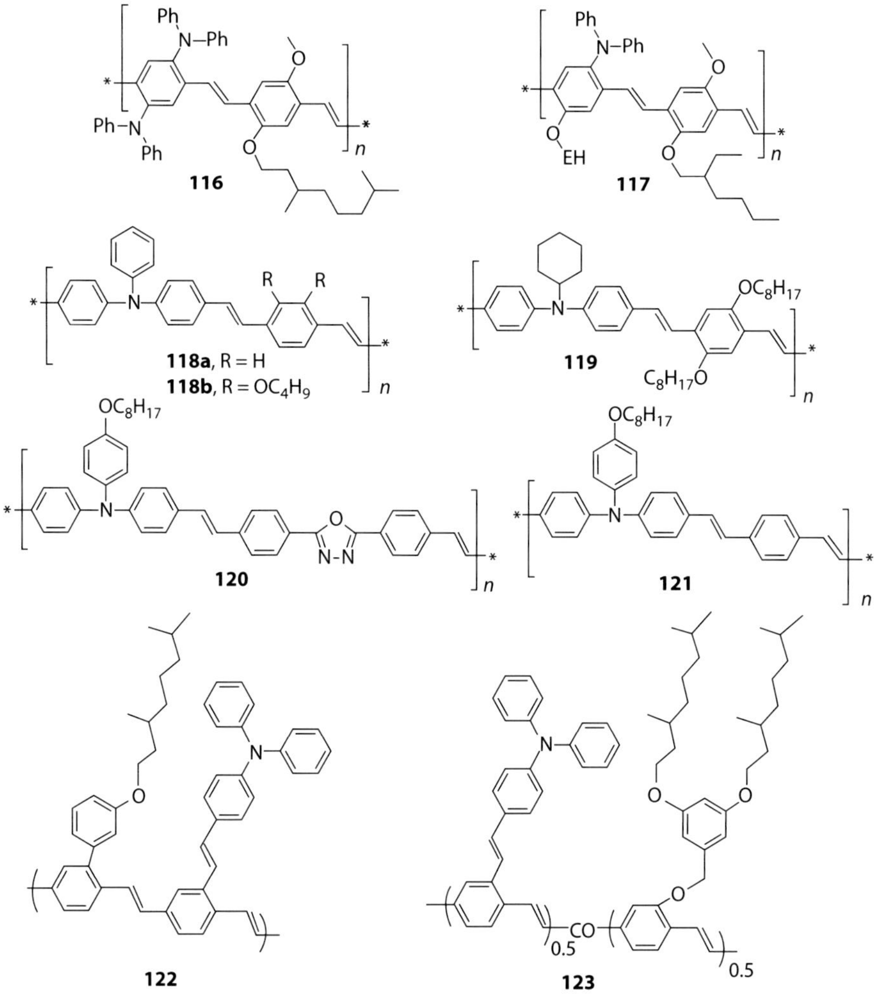

CHART 2.23
Chemical structures of PPV copolymers containing phenylamine derivatives.

a nonoxadiazole polymer **121**. Tan ct al. [163] prepared PPV derivatives bearing triphenyl-amine side chain through a vinylene bridge polymer, **122** and **123**. The single-layer PLED with the configuration of ITO/PEDOT:PSS/P11/Ca/Al emitted yellow–green light with the turn-on voltage of 4.9 V and maximum luminance of 990 cd/m^2 at 15.8 V.

2.2.3.2 Poly(p-Phenylene Vinylene) Copolymers with Electron-Withdrawing Substituents

The first PPV, bearing a cyano group attached to vinylene fragment (**9**), was synthesized as early as 1960 as unprocessable and insoluble material, which could not be used in PLEDs [46]. Electron acceptor cyano substituents lower the HOMO and LUMO levels of the polymer by ca. 0.6 and 0.9 eV, respectively. In 1993, the Cambridge group reported the first soluble phenylene cyanovinylene (CN–PPV) copolymers (**124**, **125**), synthesized via the Knoevenagel condensation polymerization [164]. Actually, the presence of two electron donor alkoxy substituents significantly reduced the electron-accepting effect of the cyano groups. Nevertheless, compared with the most widely used dialkoxy-PPVs, the electron injection (and transport) in **124** and **125** is facilitated, allowing the use of the less-reactive aluminum electrode instead of calcium in PLEDs (both give the same EL QE). On the other hand, the cyano groups reduce the hole-transporting properties of the diode, which had to be adjusted by introducing a second layer of a hole-transporting material (unsubstituted PPV **1**). The double-layer PLEDs ITO/PPV **1**/CN–PPV/Al emitted red (for **124**) or yellow–orange (for **125**) light with rather high (for the time) internal quantum efficiencies of 4% and 2%, respectively. Following this initial report, a series of other CN–PPV derivatives, **126** [165] and **127–130** [47], and thiophene analogs, **131** and **132** [47], have been synthesized via Knoevenagel polycondensation. This reaction appears to be a convenient way to prepare a number of different substituted PPVs with finely tuned band gap and emission wavelength (Scheme 2.18). Thus, PLEDs with blue (**127–130**), red (**124**, **125**), and near-infrared

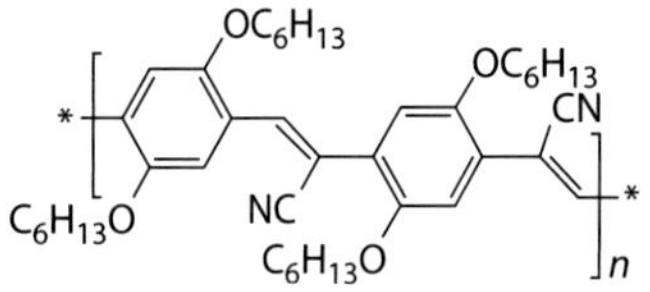

124, PL: 710 nm, E_g = 2.1 eV

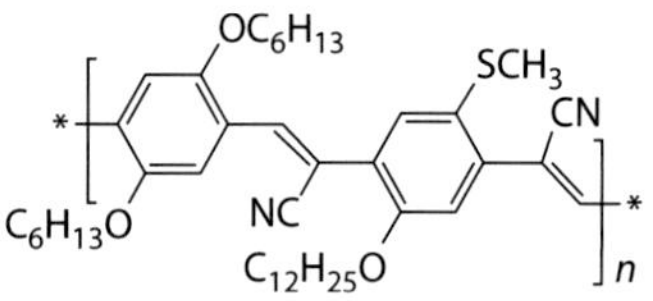

125, PL: 610 nm, E_g = 2.2 eV

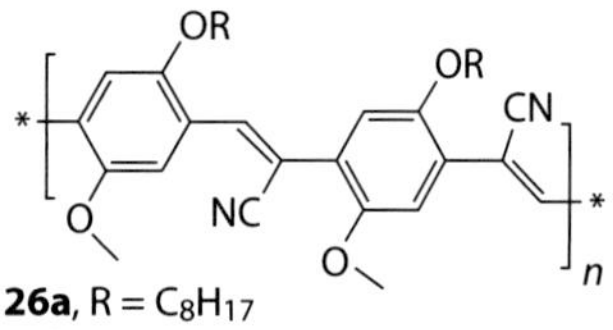

126a, R = C_8H_{17}
126b, R = $C_{10}H_{21}$
126c, R = $C_{12}H_{25}$, $C_{16}H_{33}$, PL: 611 nm
EL: 603–618 nm, E_g = 2.0 eV

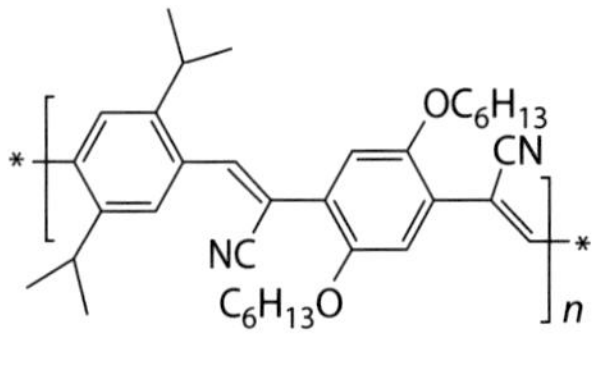

127, PL: 516 nm, EL: 510 nm
E_g = 2.85 eV

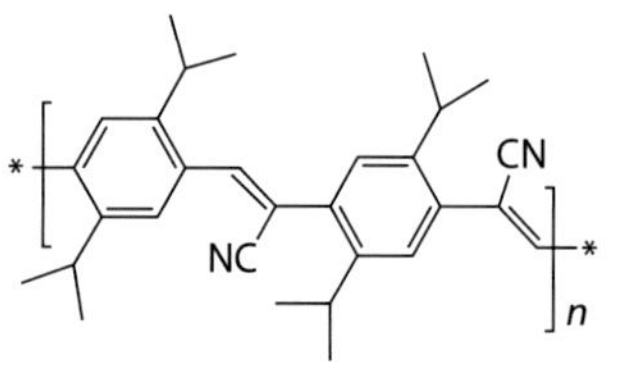

128, E_g = 3.2 eV

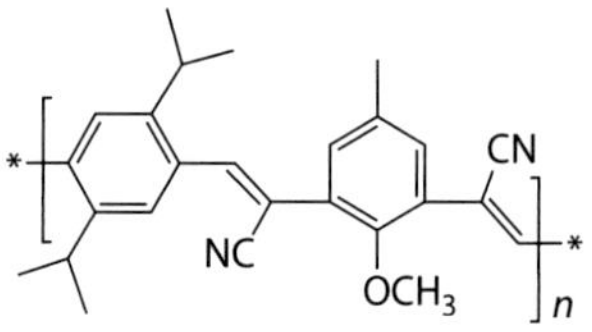

129, E_g = 3.05 eV

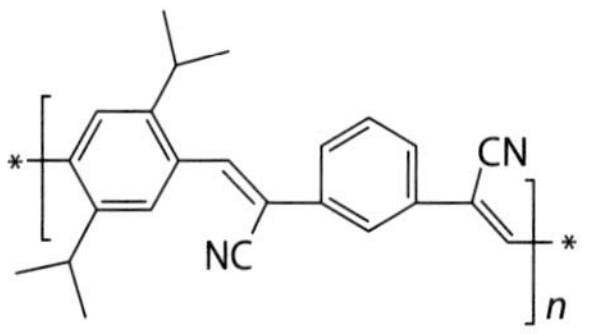

130, E_g = 2.7 eV

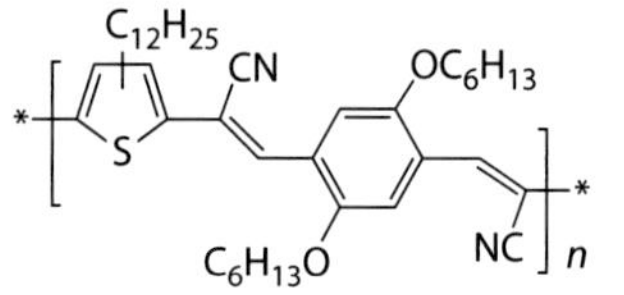

131, PL: 840 nm, EL: 730 nm, E_g = 1.8 eV

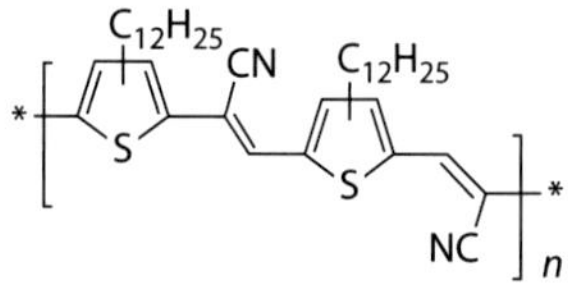

132, PL: 950 nm, E_g = 1.55 eV

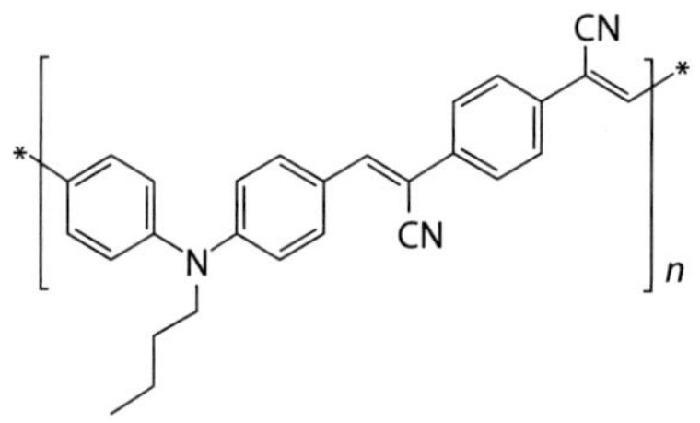

133, EL: 560 nm, E_g = 2.3 eV

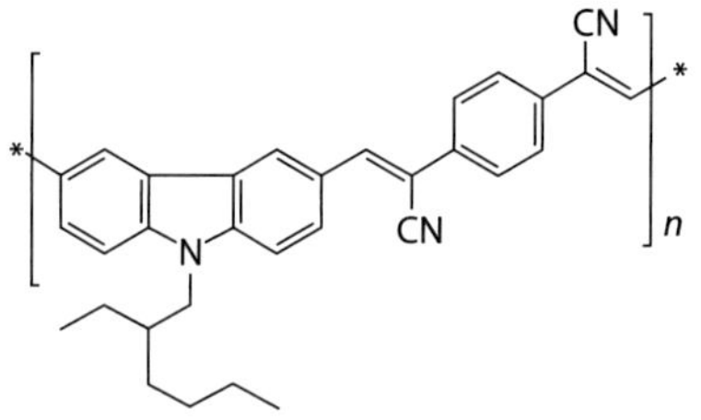

134, PL: 530 nm, EL: 510 nm, E_g = 2.5 eV

SCHEME 2.18
Band-gap and emission tuning in cyano-substituted PPVs.

(NIR) (**131**, **132**) emission have been fabricated with these polymers. Remarkably high (as for PPV) band gaps of >~3 eV were achieved for *bis*(*i*-propyl)-substituted copolymers **127–130**, probably owing to backbone distortion.

Knoevenagel coupling has also been used to synthesize CN–PPV copolymers with a diphenylamino donor unit (**133**, Scheme 2.18) [166]. In spite of a short solubilizing group (butyl), **133** is a very soluble polymer, which is due to its rather low molecular weight (M_n = 5700). A simple PLED ITO/**133**/Al can be turned on at 4.5 V emitting at λ_{max}^{EL} = 560 nm; however, the efficiency of the device was not reported. A related donor–acceptor CN–PPV copolymer **134** was synthesized with essentially higher molecular weight (M_n = 37,000) (Scheme 2.18) [167]. The PLED device thereof was fabricated as ITO/**134**/Al; however, its efficiency and brightness were not reported in the paper.

Heck- and Suzuki-coupling polymerization have been used to synthesize CN–PPV copolymers **135** [168] and **136** [169], respectively. As expected, decreasing the number of cyano groups, compared with CN–PPV **113**, destabilizes the LUMO orbital, increasing the band gap that turned out to be the same for both compounds (**135**: E_g [optical] = 2.37 eV; **136**: E_g [optical] = 2.38 eV, E_g [electrochemical] = 2.37 eV). The solid-state emission maxima are also identical (590 nm). The Φ_{EL}^{ex} = 0.025% demonstrated by **135** in a simple device ITO/ polymer/Al can be increased to 0.062% by applying a second layer of PPV **1** between the ITO and light-emitting layers. For the polymer **136**, a significantly lower EL efficiency was obtained, in spite of a more optimized device structure: the Φ_{EL}^{ex} of ITO/copper phthalocyanine HTL/**136**/Ca/Ag was only 0.011% (and the maximum brightness was 213 cd/m^2) (Chart 2.24).

Hohloch and coworkers [170] reported related cyano-substituted naphthalene vinylene derivatives **137** and **138**. Interestingly, replacing the phenylene unit in CN–PPV **124** with naphthalene in polymers **137** and **138** results in significant blue shift of the emission maxima from 710 to 595 nm (for **137a**) and 500 nm (for **138**). In addition, the efficiency, tested for the double-layer device ITO/**1**/**137**/Mg:Al(3:97), was rather low $\left(\Phi_{EL}^{ex} = 0.017\%\right)$.

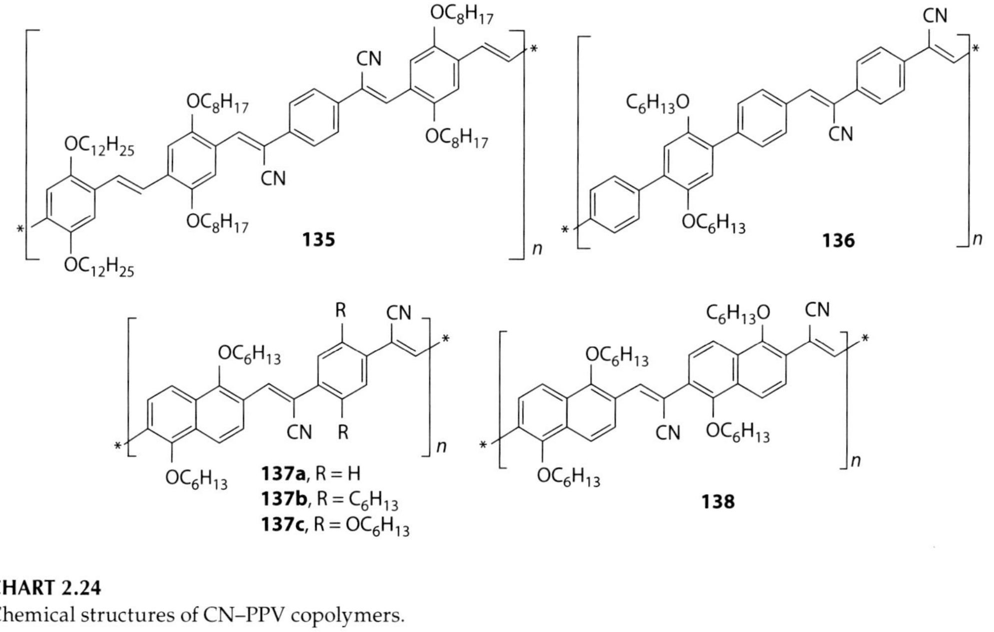

CHART 2.24
Chemical structures of CN–PPV copolymers.

Following the CN–PPV series, another electron-deficient copolymer **139**, in which the cyano groups are attached to the phenylene units, has been synthesized by Huang and coworkers [171,172] (Scheme 2.19). Owing to the more efficient conjugation of two cyano groups within the phenylene unit, this polymer possesses a higher electron affinity than **124–132**. The copolymers **139** with different ratio of dicyanophenylene vinylene and dialkoxyphenylene vinylene units have been synthesized. Changing the feed ratio of comonomers, the HOMO–LUMO energy levels can be finely adjusted, and the electron affinity of the copolymer having a 1:1 ratio of dialkoxy- and dicyano-phosphonium monomers is higher than that of MEH–PPV by >0.8 eV (cf. 0.4 eV for same-ratio CN–PPV copolymer). A single-layer PLED (ITO/**139**($x = y$)/Al) emits pure red light (λ_{PL} = 610 nm); however, no characteristics of the device except a low turn-on voltage (4–6 V) were reported [172].

The 2,5-dicyanophenylene unit has also been used by Liu and coworkers [173] in the synthesis of a series of polyfluorene (PF)–PPV copolymers **140–142** through Suzuki-coupling reactions. An important, although somewhat discouraging, point discovered within this series was an observation of inverse correlation between the electron affinity and the PLQY (17% for **40**, 15% for **141**, and 10% for **142**); that is, strong electron acceptor moieties tend to quench the PL. Nevertheless, polymer **140** showed a rather respectable performance in a double-layer device containing an HTL of bis-(tetraphenyldiamino)biphenyl-perfluorocyclobutane (BTPDPFCB) polymer (**143**): ITO/HTL/**140**/Ca PLED had very low switching voltage (2.6 V), high Φ_{EL}^{ex} = 0.88%, and a brightness of 4730 cd/m² achieved at 1.62 A/cm² (Chart 2.25).

Benjamin and coworkers [174] synthesized tetrafluorinated-PPV copolymer **144** and studied its light-emitting properties. However, this material was rather unsuccessful for LED applications: increasing the amount of fluorinated comonomer resulted in a dramatic decrease of the PLQY and the turn-on voltage of the devices was >30 V (which could only be realized in alternating-current mode because of device shorting). The quenching was less pronounced for an analogous copolymer with MEH–PPV (**145**), which showed an EL efficiency of up to 0.08 cd/A (in ITO/PEDOT/**145**/Ca diode) [175] (Chart 2.26).

As we already mentioned, the electron-transporting properties of PPV polymers can be adjusted by the introduction of an oxadiazole moiety in the polymer structure. A variety of PPV copolymers containing oxadiazole units as pendant groups have been synthesized. Among the first, in 1998, Bao et al. [141] reported copolymers **146**, containing

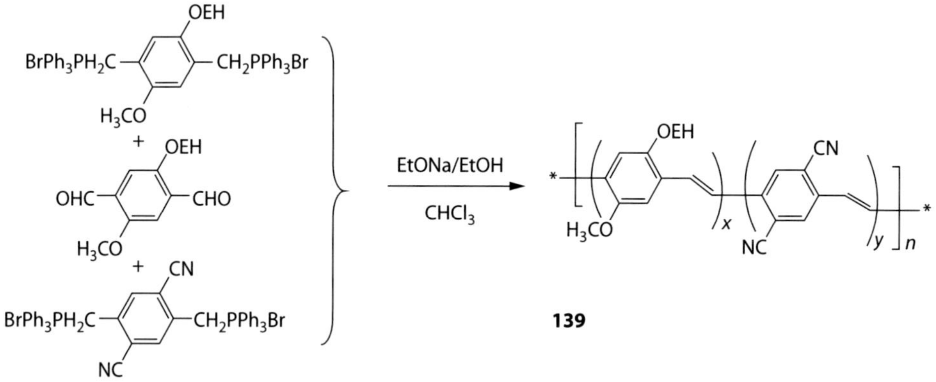

SCHEME 2.19
Synthesis of dicyanophenylene vinylene copolymers by Wittig condensation.

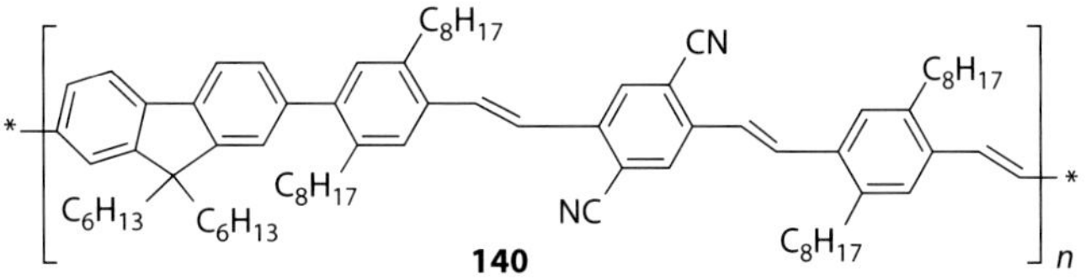

140

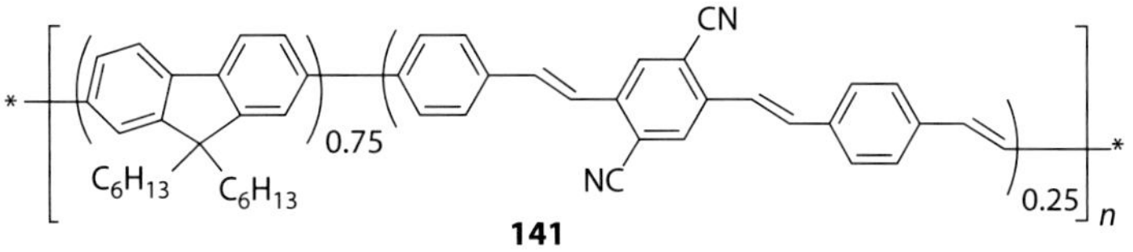

141

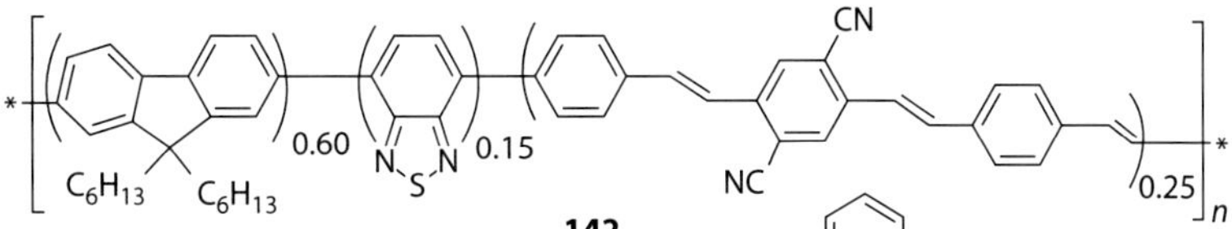

142

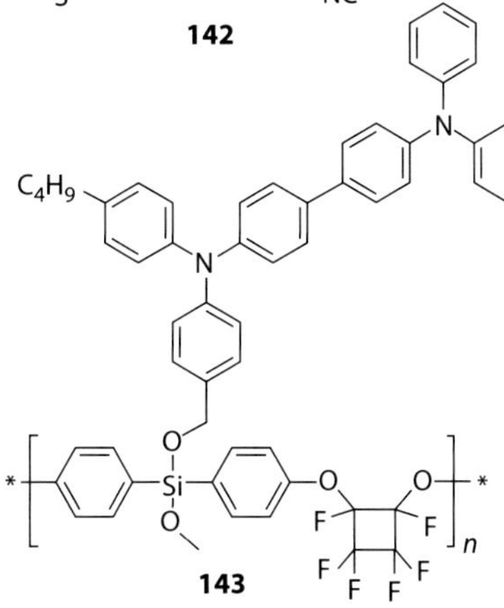

143

CHART 2.25
Chemical structures of PF–CN–PPV copolymers and BTPD–PFCB polymer.

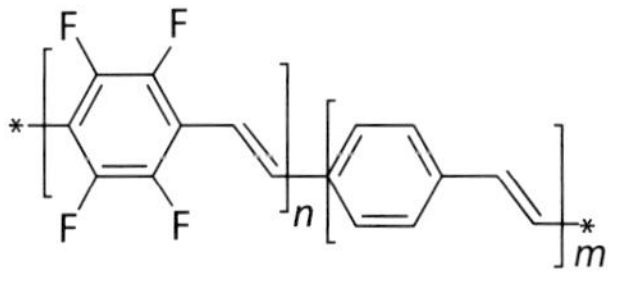

144

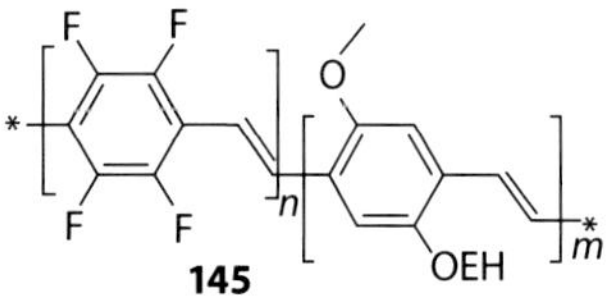

145

CHART 2.26
Chemical structures of tetrafluorinated-PPV copolymers.

phenyl (naphthyl)oxadiazole moieties separated from the PPV backbone by an oxymethylene bridge (Chart 2.27). The PL emission of **146** (λ_{PL} = 580 nm) is almost unperturbed by the presence of the oxadiazole moiety; however, the EL efficiency, measured with Al and Ca cathodes, suggests that the electron transport has been significantly improved in these materials, compared with dialkoxy-PPVs (e.g., Φ_{EL}^{ex} = 0.002% for a related device, ITO/**88c**/

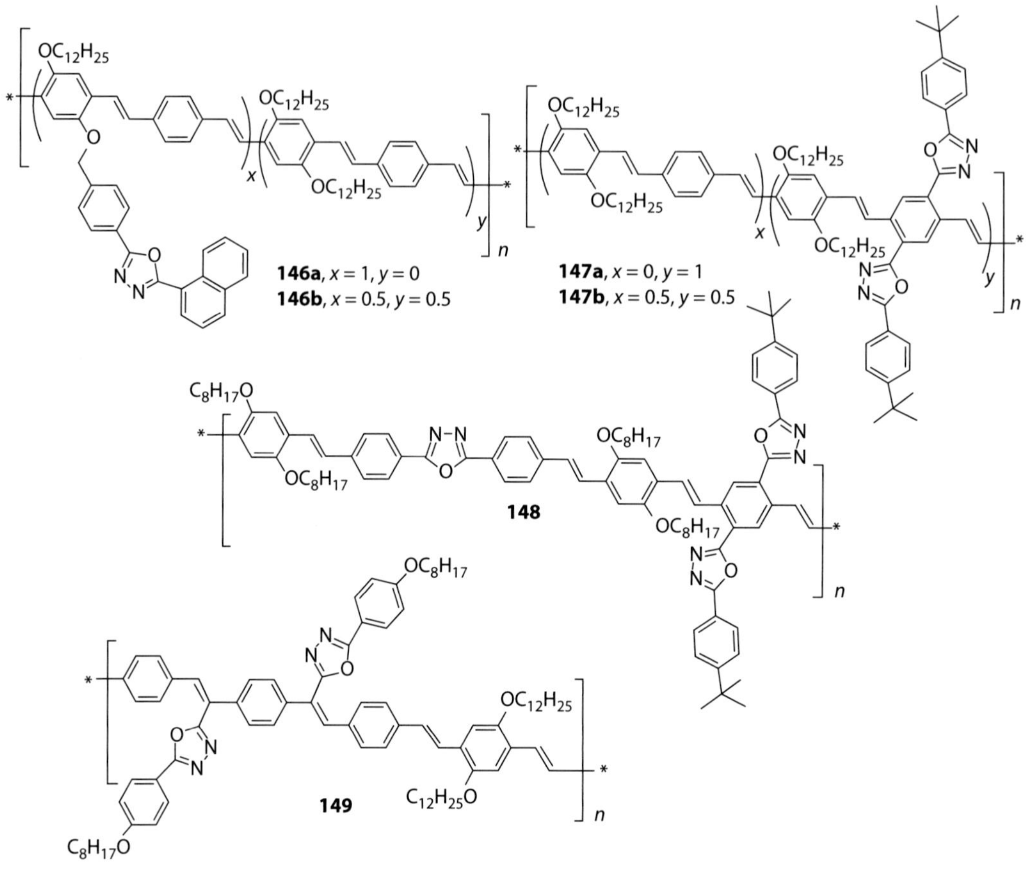

CHART 2.27
Chemical structures of PPV copolymers containing oxadiazole units as pendant groups.

Al) [141]. In fact, Φ_{EL}^{ex} was higher with an Al cathode (0.02%) than with Ca (0.015%) and, in contrast to dialkoxy-PPV, adding PBD as an additional electron-transport material only decreased the device efficiency (0.013%).

A year later, Peng and Zhang [142,176] reported PPV **147**, containing two oxadiazole substituents attached directly to the polymer backbone (to the phenylene unit). Compared with the previous oxadiazole-PPV, Φ_{EL}^{ex} in **147** was further improved to 0.045% (ITO/**147b**/Al) and a maximum brightness reached 1160 cd/m². Even a higher brightness of 3000 cd/m² was achieved with polymer **148** (ITO/**148**/Al device) containing oxadiazole in both pendant groups and in the backbone (Φ_{EL}^{ex} was 0.07% and 0.15% for Al and Ca cathodes, respectively) [176].

Kim and coworkers [177] reported an efficient LEP, containing the oxadiazole groups attached to the vinylene units of PPV. Polymer **149** was synthesized by Heck polymerization of dialkoxy-divinylbenzene with an oxadiazole-containing aromatic dibromide. The PL efficiency of **149** (λ_{max} = 560 nm) in films was 6.5 times higher than that of MEH–PPV **13**, and the energy levels were more favorable for electron transport: **149**, HOMO = −5.30 eV, LUMO = −3.10 eV; cf. MEH–PPV **13**, HOMO = −4.98 eV, LUMO = −2.89 eV (all determined electrochemically). The PLED device ITO/PEDOT/**138**/Al showed relatively high Φ_{EL}^{ex} of 0.34%, with a maximum brightness of 1450 cd/m^2 (at 13 V). Again, changing the Al electrode for Ca resulted in only a small increase of the QE (0.43%), suggesting that the charge-transport properties of **149** are relatively well balanced (Chart 2.27).

Several groups introduced an oxadiazole moiety as a part of the PPV backbone (polymers **150a** [178,179], **150b** [180], **151** [181], **152** [178], and **153** [179]). Not unexpected, the oxadiazole moieties lowered the LUMO energy of these polymers (as demonstrated by CV measurements). The decreased electron-injection barrier is manifested by lowered turn-on voltage (6 V for ITO/**150b**/Al) [180]. However, relatively low efficiency (0.15% for **150b** [180]) was reported for these copolymers (Chart 2.28).

Grice and coworkers [182] synthesized copolymer **154**, containing a similar electron-deficient moiety (triazole) incorporated in the PPV backbone. They have reported an efficient blue emission from this polymer (λ_{PL} = 466 nm [solution], 486 nm [film], Φ_{PL} = 33% [film]), although the efficiency of the PLED fabricated as ITO/PPV/**154**/Al was not very high (Φ_{EL}^{ex} reached 0.08% at a luminance of 250 cd/m^2).

Karastatiris and coworkers [183] synthesized PPV copolymers with quinoxaline as pendants **155** and **156**, as well as a part of the chain (not shown here). These polymers showed reductions with onsets of −1.70 and −1.75 V vs. saturated calomel electrode (SCE), respectively (E_A = 2.70 and 2.65 eV) and greenish–yellow (**155**, λ_{PL} = 563 nm) or blue–green (**156**, λ_{PL} = 470 nm) fluorescence in films. An ITO/PEDOT/**155**/Al diode emitted yellow light (λ_{EL} = 550 nm) with a rather low maximum brightness of 35 cd/m^2 at 12 V (Chart 2.29).

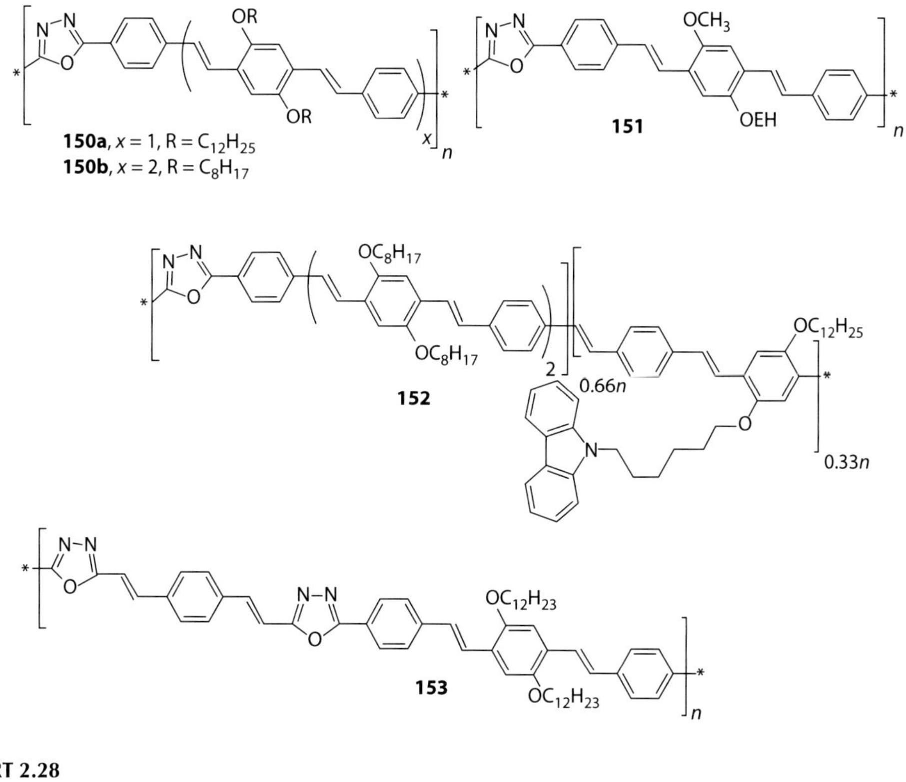

CHART 2.28
Chemical structures of PPV copolymers containing oxadiazole units in backbone chain.

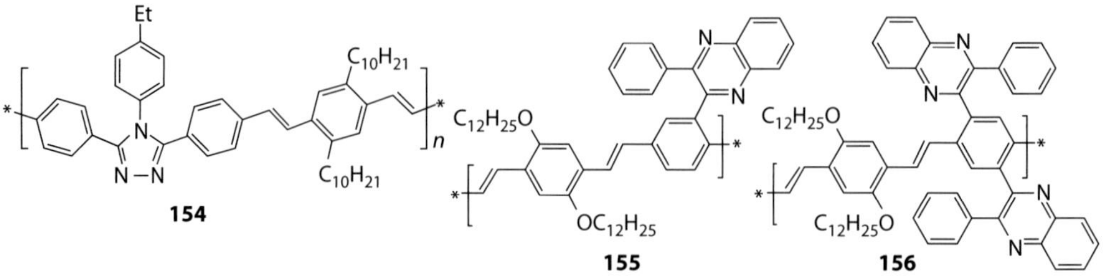

CHART 2.29
Chemical structures of PPV copolymers containing triazole and quinoxaline units.

Yu and Chen [184] studied copolymer **157**, incorporating a triphenyltriazole moiety as a pendant group (Chart 2.30). Increasing the proportion of electron-deficient triazole moieties (*n:m*) improved the electron-transport properties of the material, as demonstrated by an increase of the EL efficiency from 0.2 cd/A (for *n:m* = 0:1, MEH–PPV) to 3.1 cd/A (for *n:m* = 4:1), for the ITO/PEDOT/polymer/Al device configuration. At the same time, for the devices ITO/PEDOT/polymer/Ca, where the electron transport is already improved by using a low-work-function electrode, the device efficiency stayed at the level of 1–2 cd/A for the whole range of polymer compositions (*n:m*). A very high brightness of 17,000–19,000 cd/m² was observed for these devices.

By analogy with Kodak's low-molecular-weight dyes, Kim and Lee [185] introduced an electron acceptor dicyanomethylenepyran moiety into the PPV copolymer chain. The PPV copolymer **158**, synthesized by Heck-coupling polymerization, revealed a strong, pure red emission (λ_{max}: 646 nm; CIE: *x* = 0.67, *y* = 0.33). The downshifted orbital levels of **158** (HOMO: −5.44 eV, LUMO: −3.48 eV) compared with MEH–PPV **13** (HOMO: −4.98 eV, LUMO: −2.89 eV) resulted in more balanced hole–electron injection, and the single-layer PLED fabricated as ITO/**158**/Al showed 8 times higher EL efficiency than the PLED fabricated with MEH–PPV.

Porphyrine chromophore units have also been introduced to the PPV backbone; however, the PLQY of such materials decreased rapidly with increasing ratio of the porphyrin units, and no EL devices have been reported [186,187].

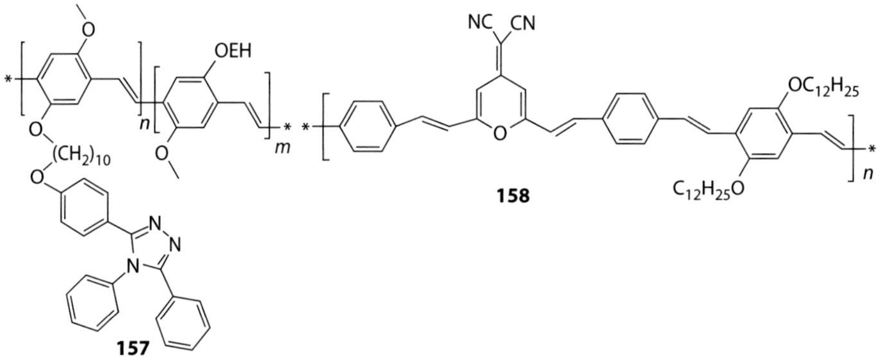

CHART 2.30
Chemical structures of PPV copolymers containing triphenyltriazole and dicyanomethylenepyran units.

2.2.3.3 Poly(p-Phenylene Vinylene) Copolymers with Electron-Withdrawing and Electron-Donating Substituents

For PLEDs, to achieve high-EL emission efficiency, balanced injection and transport of both electrons and holes are essential. From a molecular design perspective, introduction of both electron- and hole-transporting substituents into a PPV polymer backbone or side chains is an efficient way to improve the PLED device performance. In addition, the HOMO and LUMO level of PPVs can be tuned by incorporating these substituents to form a donor–acceptor (D–A) system, and thus the emission color of PLED can be tailored.

Karastatiris et al. [188] reported bipolar PPVs bearing electron-donating triphenylamine or carbazole and electron-accepting quinoxaline moieties (**159** and **160**), which were synthesized via Heck coupling. It is seen that the LEDs with the configuration of ITO/PEDOT/polymer/Al based on bipolar polymers **159** and **160** showed improved performance of turn-on voltage and current density relative to LED based on **161** that does not contain an electron-donating unit. However, the lower luminance was observed for bipolar polymers owing to the reduced PL quantum yields compared with **161**. Huang's group applied the same strategy to improve the quantum efficiency, processability, and stability of the PPVs by introducing pendant 2,4-difluorophenyl and fluorenyl moieties (**162** and **163**) [189]. Fluorine atoms are strongly electron withdrawing and can make electron injection easier. Fluorenyl as an electron donor also possesses high hole-transporting ability and PL efficiency. The devices with the structure of ITOPEDOT:PSS (40 nm)/polymer (80 nm)/Ba (4 nm)/Al (160 nm) showed a low onset voltage (about 4 V) and bright electroluminescence. The maximum luminance of 450 and 4700 cd/m^2 were obtained for **162** and **163** at the current densities of 24 and 15 mA/cm^2 and at voltages of 12 and 7.0 V, respectively (Chart 2.31).

Wang's group synthesized a conjugated poly(*p*-CN-phenylene vinylene) (**164**) containing both electron-donating triphenylamine units and electron-withdrawing cyano groups via Knoevenagel condensation [190] (Chart 2.31). The single-layer ITO/polymer/Mg-Ag device emitted a bright-red light (633 nm); however, the highest brightness of the device is only 150 cd/m^2 at 15 V. The same group also synthesized three series of PPVs containing hole-transporting triphenylamine derivatives and electron-transporting oxadiazole units in the polymer main chain by using improved Wittig polymerization, which was developed by this group. Tributylphosphonium salts were used instead of triphenylphosphonium salts as polymerization monomers to improve the solubility in chloroform and to increase the molecular weights of the polymers [191]. The decomposition temperature of copolymers containing oxadiazole unit was increased to >400°C owing to the rigidity of oxadiazole group. The large solvatochromic shifts of absorption and PL emission were observed in different solvents for bipolar copolymers, which confirms that intramolecular charge transfer (ICT) is obvious for D–A molecules, and ICT effects lead to the red shift of PL and EL. Devices with the configuration of ITO/PEDOT/polymer/Ca/Al were fabricated to investigate the EL properties of the polymers. The device performance based on D–A copolymers was significantly improved. The maximum brightness of the device based on **165a** reaches 4090 cd/m^2 at 21V, which is 29 times brighter than that of the device based on **165c**. The turn-on voltage was dropped to 2.7 V for **165a** from 9.3 V for **165c**. Similar results were observed for other series copolymers. EL device performance improvement is due to the more balanced charge injection and transport by tuning the ratio of electron-withdrawing and electro-donating units in copolymers (Chart 2.32).

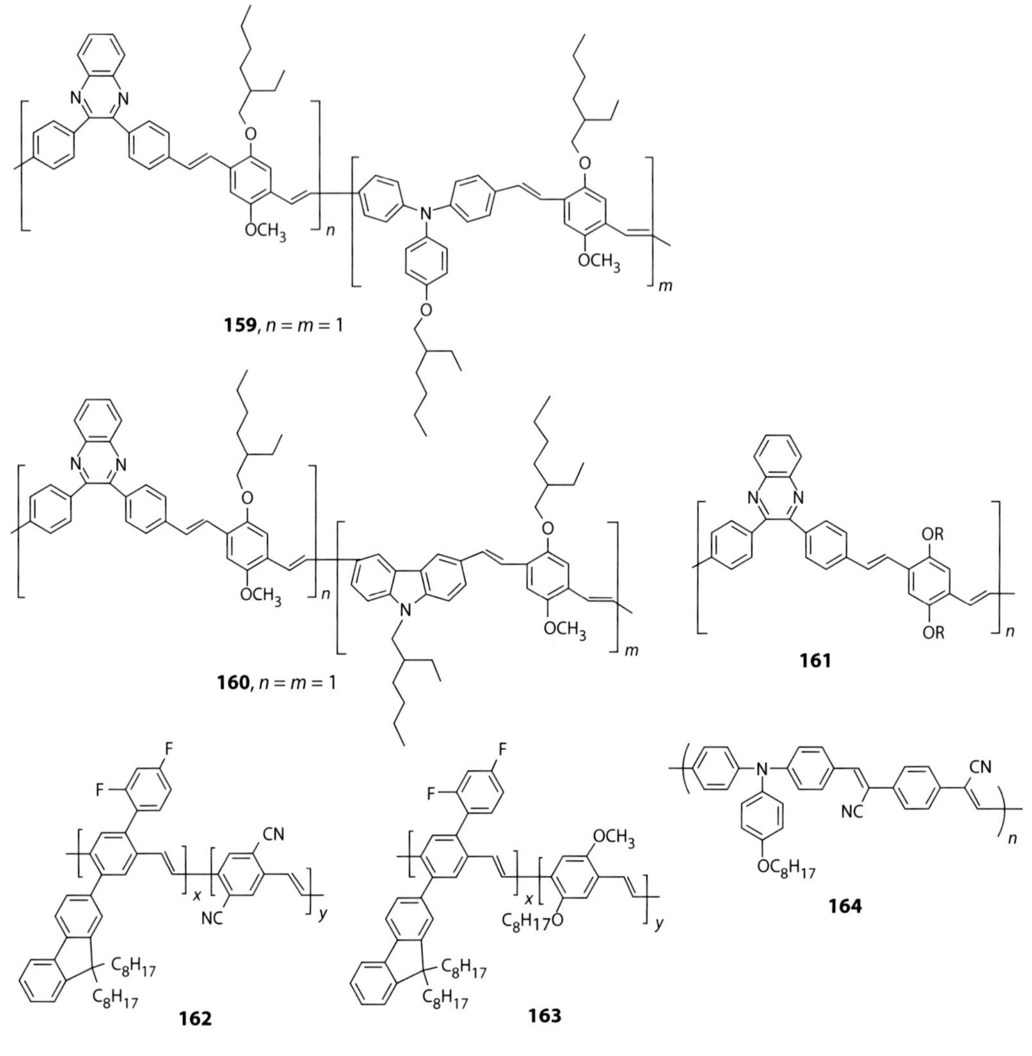

CHART 2.31
Chemical structures of PPV copolymers containing electron-donating and electron-withdrawing substituents.

2.2.4 Poly(*p*-Phenylene Vinylene) Polyelectrolytes

For developing a cost-effective and environment-friendly fabrication process, as well as for possible applications in biological systems, it is desirable to have water-processable LEPs. In this line, Shi and Wudl [192] synthesized water-soluble PPV **166**, containing ionic sulfonate groups, isolated from the PPV backbone by an alkyl chain (Chart 2.33). This material appeared to be highly fluorescent with λ_{PL} in the range of 550–600 nm, depending on the solid-state structure (engineered by self-assembly of the anionic PPV **166** multilayers separated by counterion layers) [193]. A water solution of **166** (M = Na) was used together with a cationic LEP (**508**, Chart 2.118) for a hybrid ink-jet printing of dual-color (blue and red) light-emitting pixels [194]. Taking advantage of the good compatibility of polymer **166** with aqueous media and the known quenching amplification in conjugated polymers,

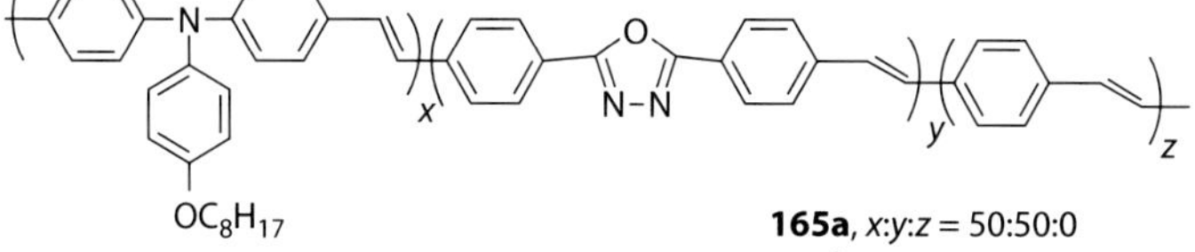

165a, *x*:*y*:*z* = 50:50:0
165b, *x*:*y*:*z* = 50:21:29
165c, *x*:*y*:*z* = 50:0:50

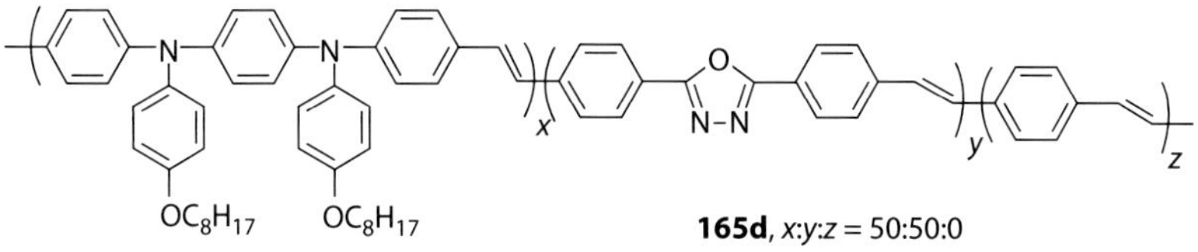

165d, *x*:*y*:*z* = 50:50:0
165e, *x*:*y*:*z* = 50:28:22
165f, *x*:*y*:*z* = 50:0:50

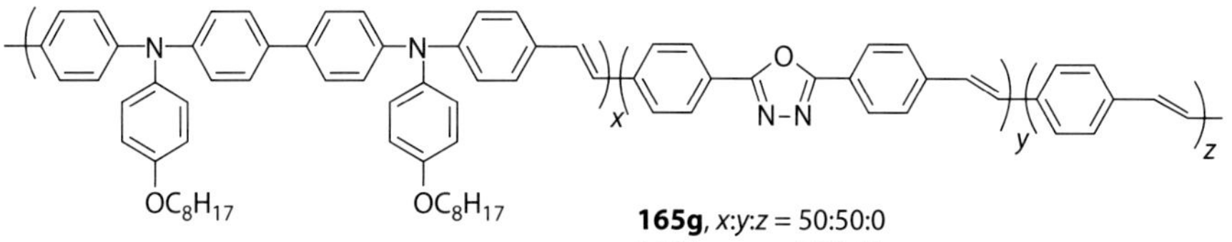

165g, *x*:*y*:*z* = 50:50:0
165h, *x*:*y*:*z* = 50:0:50

CHART 2.32
Chemical structures of PPV copolymers containing triphenylamine derivatives and oxadiazole substituents.

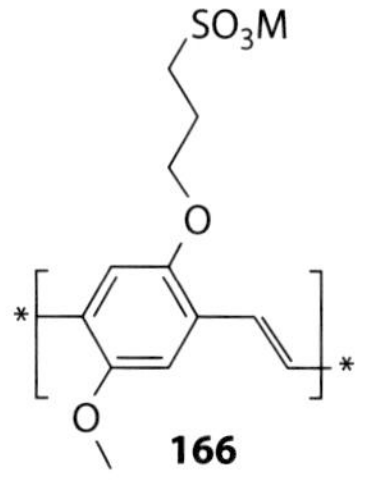

166

M = H, Na, NH_4

CHART 2.33
Chemical structure of a PPV containing sulfonate.

highly efficient fluorescent biological sensors have been designed with this material [195]. Unfortunately, LEC devices with **166** have not been reported.

Anderson and coworkers [196–198] reported water-soluble polyrotaxanes **167** and **168**, containing sulfonated PPV chains surrounded by mechanically bound α- and β-cyclodextrin macrocycles (Chart 2.34). The cyclodextrin rings play the role of a "wire insulator," preventing aggregation and interchain quenching. The effect was demonstrated by atomic force microscopy (AFM), which showed individual polymer chains for cyclodextrin-encapsulated polymer but not for the noncomplexed material. The PL (for

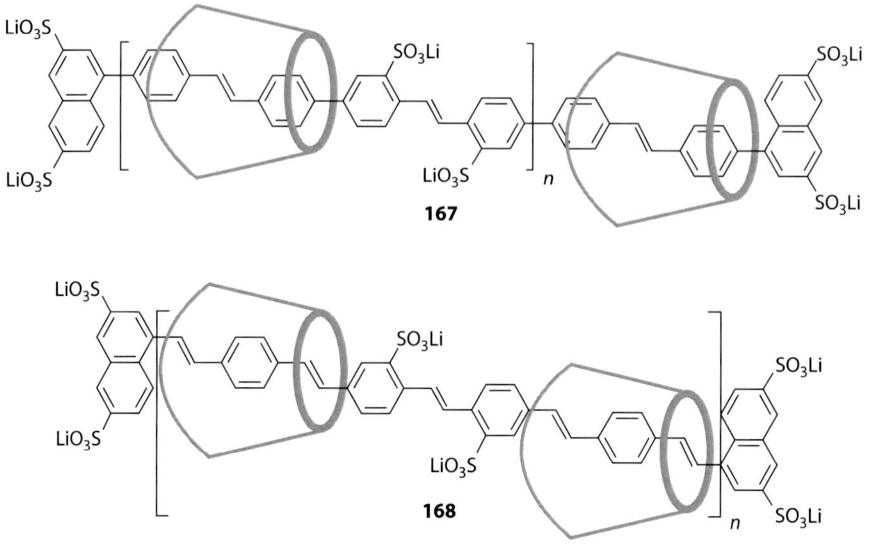

CHART 2.34
Chemical structures of polyrotaxanes containing sulfonated PPV chains.

168 [198]) and EL efficiency (for **167** [196]) of the complexed material was ~3–4 times higher than that for noncomplexed polymer. However, the absolute value of $\Phi_{EL}^{ex} \sim 0.025\%$ was very low for practical application, which was not only due to unoptimized device structure (ITO/**167**/Ca) but also to the generally low PLQY of the polymer containing a sulfonate group directly attached to the backbone.

2.2.5 Controlling the Conjugation in Poly(*p*-Phenylene Vinylene) Polymers

Thus far, we have demonstrated that PPV derivatives are among the most popular materials for PLED, and different color emission can be achieved by substitution; however, with few exceptions [108,150], blue color is not available for fully conjugated PPV. Furthermore, a rigid-rod structure of the highly conjugated chain results in high crystallinity of many PPV materials, which is held responsible for the decrease of the PL efficiency in the solid-state and pinhole defects in thin films. The following two sections present current approaches to the solution of this problem via control of the conjugation length in PPV materials.

2.2.5.1 Formally Conjugated Systems with Twists, Meta-Links, and sp-Hybridized Atoms in the Backbone

Intramolecular π-stacking can be effectively prevented by introducing a twist structure in the backbone of PPV that also reduces the conjugation along the chain and is expected to result in hypsochromic shift of the emission. This kind of twist was achieved by copolymerization with binaphthyl or biphenyl units (**169** [199] and **170** [200]). A twisted binaphthyl copolymer **169**-based PLED (fabricated with ITO and Al electrodes) exhibited blue–green light emission with Φ_{EL}^{ex} of 0.1% and a moderate driving voltage of 6 V. Similar results were found for biphenyl copolymer **170**. The PLEDs fabricated as ITO/PEDOT/**170**/Ca/Al emit green–blue light (λ_{max}^{EL} of 485–510 nm, depending on substituent pattern) with $\Phi_{EL}^{ex} = 0.17\%$.

The authors explained the lower QE of the copolymers to be due to increased nonradiative relaxation resulting from interruption of the conjugation by the twisted units. Importantly, in both cases, because of interrupted conjugation, the emission band undergoes a significant hypsochromic shift (Chart 2.35).

In fact, blue-emitting PPV materials are the subjects of significant research interest, as blue EL is the key for creating either white or full-color EL displays. However, this is generally unavailable for conjugated PPVs owing to relatively low band gap. Consequently, several strategies to decrease the effective conjugation length have been studied in search of blue-emitting PPVs. This can be achieved by introducing either nonconjugated blocks or sp³ "defects" into the PPV chain or changing the attachment mode of the phenylene unit in the chain. The synthesis of substituted poly(m-phenylene vinylene) and poly(o-phenylene vinylene) homopolymers were reported in 1993 by Leung and Chik [96], although with no connection to control the luminescent properties of the polymer. In 1999, Ahn and coworkers [201] synthesized and studied a series of dialkoxy- and *bis*(trimethylsilyl)-substituted PPV copolymers with o-, m-, and p-phenylene linkages (**171–176**) (Chart 2.36). The m-phenylene unit does not allow for direct conjugation, resulting in a hypsochromic shift of both absorption and emission bands. The *ortho*-linking is formally conjugated but because of steric hindrance, the polymer chain has an effectively decreased conjugation length. Combining these structural changes with substituent variations, fine tuning of the EL wavelength was achieved (Figure 2.5).

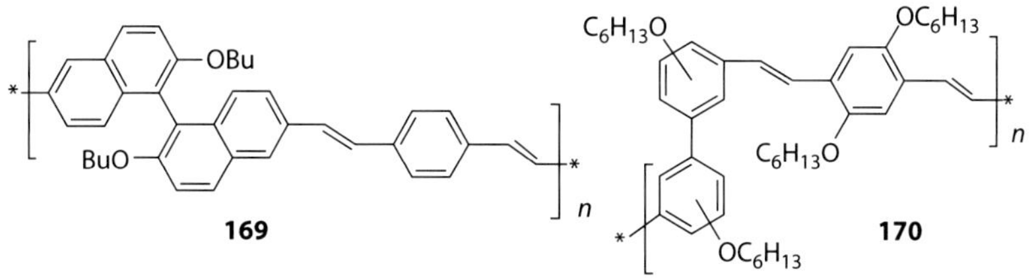

CHART 2.35
Chemical structures of twisted PPV copolymers containing binaphthyl or biphenyl units.

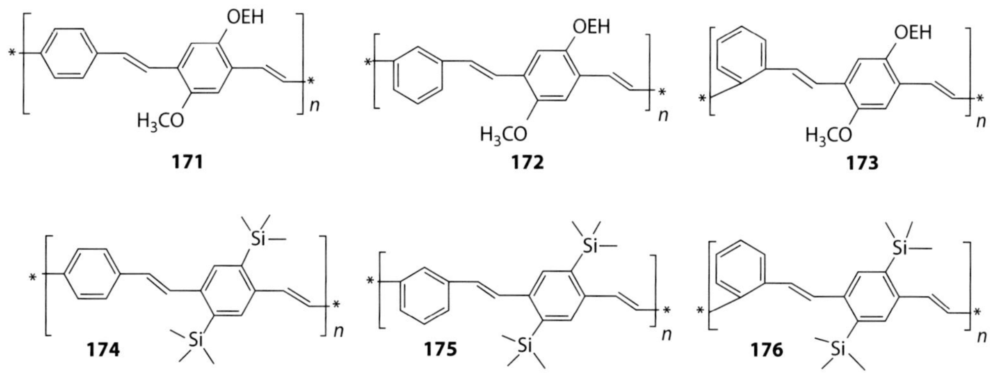

CHART 2.36
Chemical structures of PPV copolymers with o-, m-, and p-phenylene linkages.

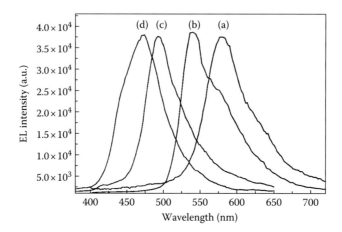

FIGURE 2.5
Tuning the electroluminescence in PPV copolymers through introducing nonconjugated kinks: (a) MEH–PPV **13**, (b) **171**, (c) **174**, and (d) **176**. (From Ahn, T., Jang, M.S., Shim, H.-K., Hwang, D.-H., and Zyung, T., *Macromolecules*, 32, 3279, 1999. With permission.)

Almost simultaneously to the above report, Pang et al. [202] reported another PPV copolymer **177a** containing alternating *p*-phenylene and *m*-phenylene units, also synthesized via Wittig–Horner coupling (Chart 2.37). As expected, **177a** exhibits strong hypsochromic luminescence shift. In solution, the polymer emits blue light with λ_{max}^{PL} of 444 and 475 nm and 60% PLQY; the latter is improved to 82% for the material containing *cis* defects (these are naturally produced in the synthesis but can be converted to *trans* configuration by refluxing in toluene). However, strong aggregation in the solid state resulted in an emission maxima shift to 480 and 530 nm (shoulder). In spite of high PL efficiency, Φ_{EL}^{ex} of a PLED with **177a** was only 0.05% that still was an order of magnitude higher than that of a device with the parent PPV **1**, prepared under the same conditions.

Other structural variations included copolymers **177a,b** [203], **178** [204], **179a,b** [205], and **180** [203]. Low switch-on voltage (4.3 V) and moderately high brightness (1000 cd/m²) were achieved for **179** [205]; however, the purity of blue color was still a problem. Even for the

177a, R = C₆H₁₃
177b, R = C₈H₁₇

178

179a, R = H
179b, R = OC₅H₁₁

180

CHART 2.37
Chemical structures of PPV copolymers containing alternating *p*-phenylene and *m*-phenylene units.

"most blue" copolymer **180**, the CIE coordinates ($x = 0.188$, $y = 0.181$) are still rather far from the pure blue emission ($x = 0.15$, $y = 0.06$) due to a green tail [203].

Several groups have studied the introduction of phenylene ethynylene units into PPV backbones. The first material of this type, copolymer **181,** was reported by Brizius and coworkers [206] (Chart 2.38). The material displayed blue luminescence in solution $\left(\lambda_{max}^{PL} = 460\,nm\right)$; however, owing to the polymer's rigid-rod structure, very strong aggregation in the solid state gave rise to bathochromic shifts and the PLQY in films was only 5%. A series of alkyl- and alkoxy-substituted poly(phenylene vinylene/ethynylene) hybrids **182–184** have been recently reported by Egbe and coworkers [207–209]. Whereas all alkoxy-substituted polymers **182a–g** showed similar PL emission (λ_{PL} from 525 to 554 nm), the performance of the green-emitting PLEDs thereof (ITO/PEDOT/**182**/Ca; λ_{max}^{EL} from 508 to 554 nm) strongly depended on the alkoxy substituents R^1 and R^2 in the copolymers. A large substituent effect was observed in polymers **182a–g** that showed Φ_{EL}^{ex} from 0.02% to 0.047% (0.085–0.20 cd/A; **182a–c,f**) to 0.89%–0.95% (3.5–4.0 cd/A; **182d,e**) [208].

Comparisons between the **183** and **184** series ($\lambda_{abs} \sim 468$–475 nm, $\lambda_{PL} \sim 519$–528 nm) showed that the conjugation pattern has very little effect on photophysical properties of these polymers in solution (blue shift in absorption and PL by only 4–7 nm from **183** to **184**). However, the performance of their PLED (ITO/PEDOT/polymer/Ca) differed drastically; cf. a brightness of 27.9 cd/m² and an external QE of 0.017% for **184b** and a brightness

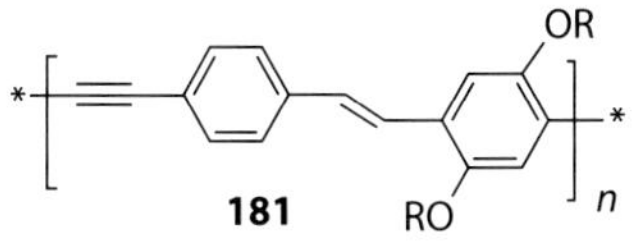

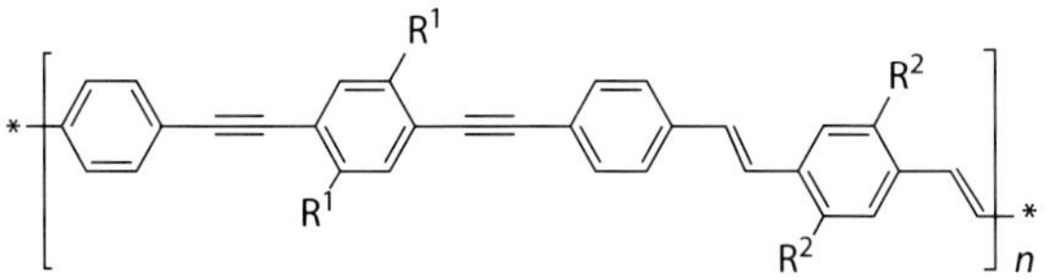

182a, $R^1 = C_{18}H_{37}O$, $R^2 = C_8H_{17}O$
182b, $R^1 = C_8H_{17}O$, $R^2 = C_{18}H_{37}$
182c, $R^1 = C_{18}H_{37}O$, $R^2 = EH$
182d, $R^1 = EH$, $R^2 = C_{18}H_{37}O$
182e, $R^1 = R^2 = C_{18}H_{37}O$
182f, $R^1 = R^2 = C_{12}H_{25}O$
182g, $R^1 = C_{18}H_{37}O$, $R^2 = C_{12}H_{25}O$
182h, $R^1 = EH$, $R^2 = EH$
182i, $R^1 = EH$, $R^2 = C_8H_{17}O$
182j, $R^1 = EH$, $R^2 = C_{12}H_{21}O$
182k, $R^1 = C_{18}H_{37}$, $R^2 = $ methoxy + EH

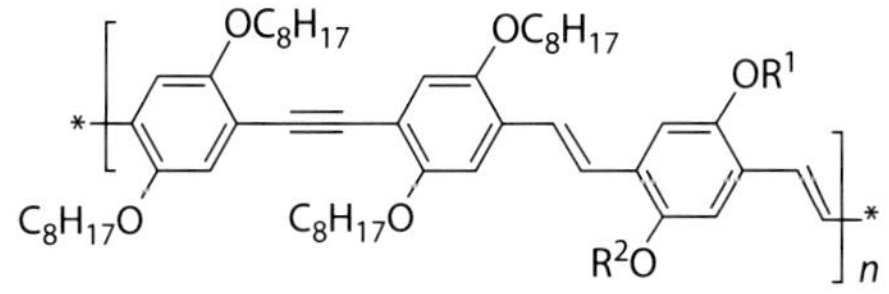

183a, $R^1 = R^2 = C_8H_{17}$
183b, $R^1 = R^2 = C_{18}H_{37}$
183c, $R^1 = R^2 = EH$
183d, $R^1 = CH_3$, $R^2 = EH$

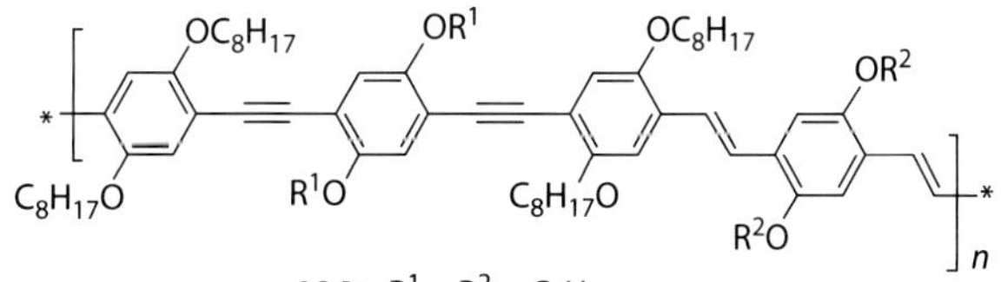

184a, $R^1 = R^2 = C_8H_{17}$
184b, $R^1 = R^2 = C_{18}H_{37}$
184c, $R^1 = R^2 = EH$

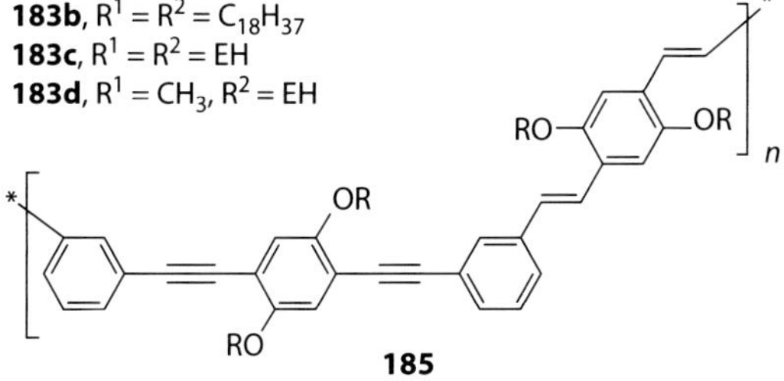

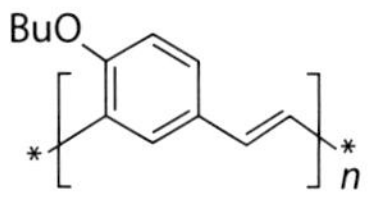

CHART 2.38
Chemical structures of PPV copolymers containing phenylene ethynylene units.

of 595–5760 cd/m² and an external QE of 0.22%–2.15% for **183a–d** (the maximum values are for **183b**) [209]. Chu and coworkers [210] reported that a related polymer **185** possesses *m*-phenylene linking groups, which might reduce the aggregation effect (although this was not investigated). The PLED device ITO/PEDOT/**185**/Ca was reported to emit green light with $\Phi_{EL}^{ex} = 0.32\%$.

Liang et al. [211] reported the first poly(*m*-phenylene vinylene) homopolymer, **186**. Owing to all-*meta*-linking of the phenylene units, the PL maximum (417 nm with a shoulder at 434 nm) of **186** is further shifted in the blue region and the emission band is very narrow, which promises to deliver a pure blue-emitting PLED (although a device fabrication was not yet reported).

2.2.5.2 Conjugated and Nonconjugated Poly(p-Phenylene Vinylene) Block Copolymers

Soon after the first demonstration of the EL of PPV **1** [1], it was shown that introducing saturated (nonconjugated) defects into PPV chains results in an emission blue shift and improves the film quality [24,26]. This could be achieved via modified Wessling–Zimmerman thermoconversion of a precursor polymer containing different (tetrahydrothiophene and methoxy) leaving groups, which can be selectively eliminated to give conjugated and nonconjugated (uneliminated) fragments (**187**) (Chart 2.39) [24,140,212]. A similar effect was obtained by controlled conversion of PPV precursor having ethylxanthate-leaving group (**188**) [26]. For the latter, the ethylxanthate group also favored the formation of *cis*-vinylene defects, preventing the intermolecular stacking effect (Figure 2.6). Owing to very negligible crystallinity and good film-forming properties, the polymers **187** and **188** showed significantly improved EL efficiency ($\Phi_{EL}^{ex} = 0.44\%$ was demonstrated for the device ITO/**188**/PBD/Al) [24,26]. Introduction of nonconjugated fragments (by partial substitution of the tetrahydrothiophenium-leaving group in the PPV precursor **2** with acetoxy group) was also demonstrated to increase the operation lifetime of the PLED (>7000 h without noticeable degradation) [213].

However, owing to random distribution of the conjugation length in these polymers, the emitted light was still essentially green. In 1993, Yang and coworkers [214] developed the idea of preparing a PPV copolymer containing well-defined blocks of rigid conjugated

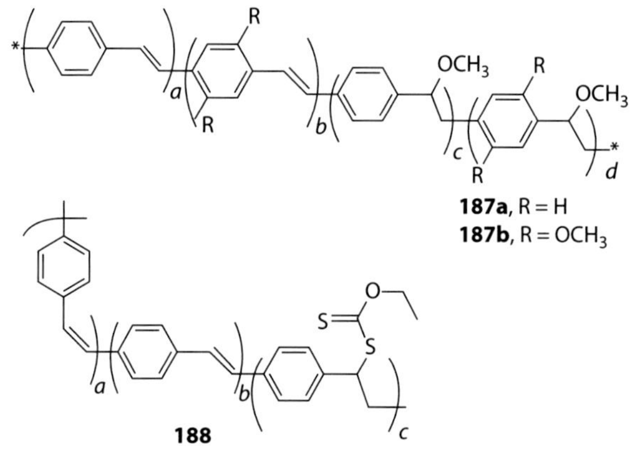

187a, R = H
187b, R = OCH₃

188

CHART 2.39
Chemical structures of PPV block copolymers with nonconjugated units.

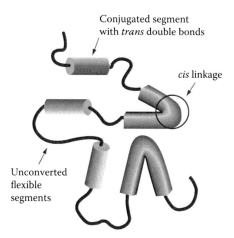

FIGURE 2.6
Schematic structure of rigid and flexible block PPV containing saturated (unconverted) units and *m*-vinylene links. (From Son, S., Dodabalapur, A., Lovinger, A.J., and Galvin, M.E., *Science*, 269, 376, 1995. With permission.)

oligo(phenylene vinylene) and flexible nonconjugated aliphatic units. Copolymer **189a**, synthesized via Wittig–Horner condensation, contained 2.5 phenylene vinylene fragments in each conjugated block and showed PL (and EL) maxima at ~465 nm (Chart 2.40). This low wavelength emission is achieved exclusively owing to very short conjugation length, and increasing the latter by only one more phenylene vinylene unit (**189b** [215] and **189c** [202]) shifts the emission to λ_{max} = 513 nm, so that the PLED ITO/**189b**/Al emits green light with CIE coordinates (x = 0.29, y = 0.47) [215]. Sun et al. [216] have reported a dimethoxy-substituted analog, **190**. The alkoxy substituents in the conjugated block result in a red shift of the emission maxima (vibronic band with peaks at ~500, 540, and 590 nm [shoulder]); however, remarkably, the PLQY in the solid state was as high as 90%. Several other block copolymers of this type (**191a–c**) having shorter nonconjugated block and different substituents in the phenylene vinylene unit have been synthesized [217]. The trimethylsilyl-substituted polymer **191a** showed the most blue PL (and EL) $\left(\lambda_{max}^{PL} = 467, 490 \, nm \right)$, whereas alkoxy substituents result in bathochromic shift of the emission band.

The solubility of such PPV copolymers in both polar and nonpolar media can be dramatically improved when using the oligo(ethylene oxide) flexible block (copolymer **192**), which also allows application in LEC. The first compound of this series copolymer **192c** was synthesized by Benfaremo and coworkers [218] using low-polydispersity PEG block (PEG-900); however, no high-performance PLED could be fabricated with this material. Later, Cacialli and coworkers [219] reported the synthesis of copolymer **192b** as a bluish–green emitter with λ_{max}^{EL} at 490 and 525 nm and moderately high PLQY (34%). The PLED ITO/**192b**/Al can be turned on at 6.5 V and shows a luminescence efficiency of 0.5 cd/A, and the maximum brightness of 2000 cd/m² [219]. Furthermore, the LEC design [66] (blending with LiOTf electrolyte) allows a decrease of the turn-on voltage to 3.8 V. A more pure blue color with a single emission peak at 490 nm has been reported with a similar compound, **192a**, having shorter nonconjugated block. The PLED turn-on voltage achieved with **192a** is lower (4.6 V) due to more complex device structure (ITO/PEDOT/**192a**/Alq3/Ca); however, the device efficiency and brightness were very similar [220]. Alkoxy substituents have been introduced in this polymer structure, but the resulting copolymer **193** showed

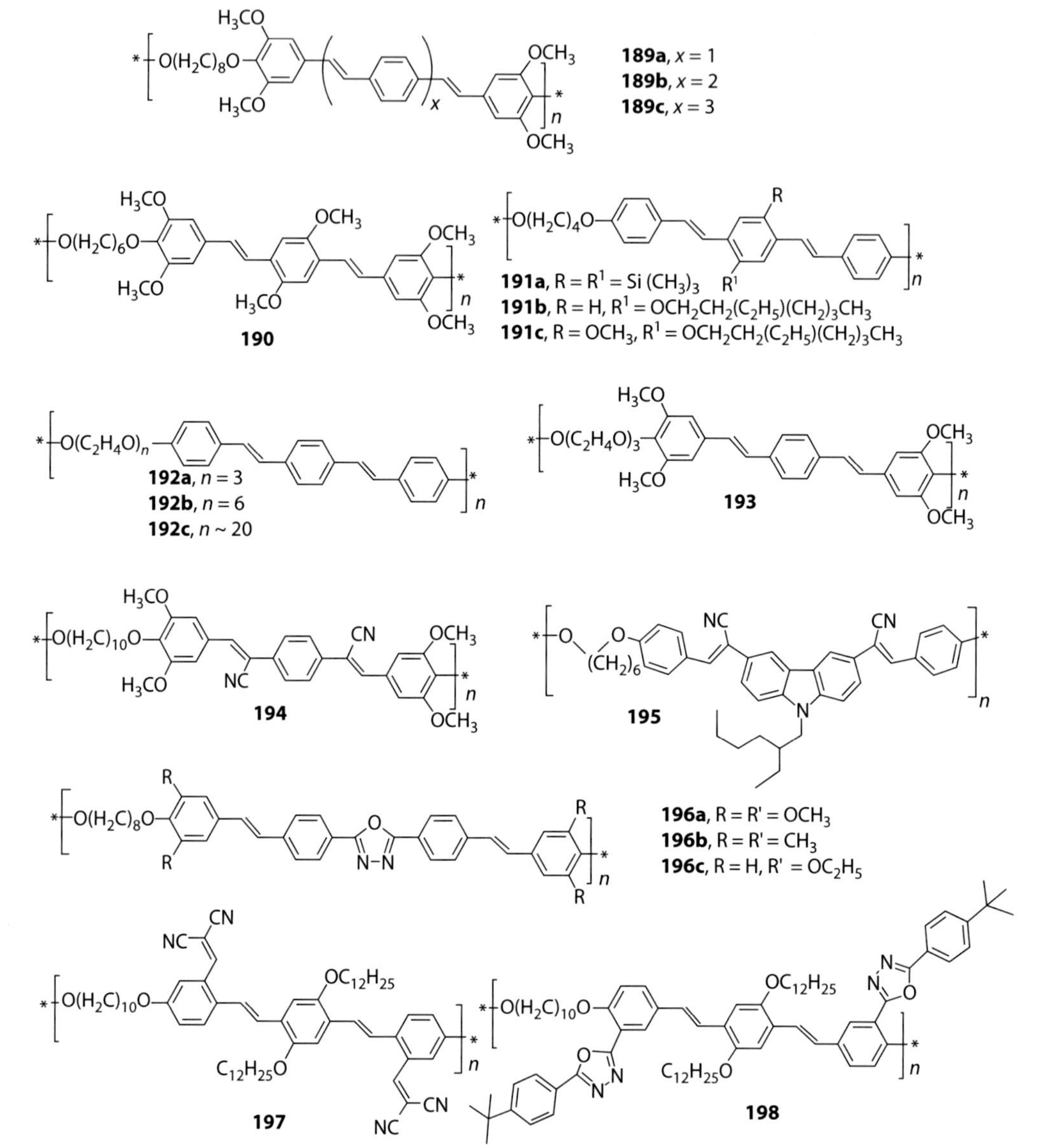

CHART 2.40
Chemical structures of PPV copolymers containing flexible nonconjugated units.

no improvement in the EL properties (λ_{EL} = 475 nm, maximum brightness of 36 cd/m² at 23 V) [221].

Introducing two cyano groups into the conjugated block (copolymer **194** [222] and **195** [167]) not only slightly alters the emission color (λ_{max}^{EL} = 493 nm for **194** and 518 nm for **195**) but also significantly improves the electron-transport properties of the polymer. Nevertheless, even the double-layer (ITO/**1**/**194**/Al) PLEDs showed a modest luminescence efficiency of 0.17 cd/A and a maximum brightness of only 40 cd/m² [167]. Significantly higher brightness (2400 cd/m²) and Φ_{EL}^{ex} = 0.1% were achieved for PPV block copolymer **196** containing an electron acceptor oxadiazole moiety in the backbone, although extending the conjugation through the oxadiazole moiety also resulted in some red shift of the emission maxima ($\lambda_{max}^{EL} = 480 - 509$ nm, depending on R) [223].

Electron-acceptor dicyanovinyl and oxadiazole substituents have been recently introduced into phenylene units of the PPV block copolymers (**197, 198**) [224]. Blue and blue–greenish PL emission was observed for **197** and **198**, respectively; however, the PLQY was relatively low even in solution (13% and 24%), and no EL device has yet been reported.

Sarker and coworkers [225] reported a series of *meta*-linked oligo(phenylene vinylene) block copolymers, **199a–c**. A *meta*-linked phenylene unit imposed an additional hypsochromic shift on the emission of these segmented polymers. The PL maxima were found at 399–416 nm; however, a significant (ca. 70 nm) red shift was observed for EL spectra (ITO/polymer/Ca/Al) (Chart 2.41).

At Kodak, researchers used a rigid adamantane moiety to separate the luminescent oligo(phenylene vinylene) blocks (copolymers **200, 201**) [226]. The EL color can be tuned from blue ($\lambda_{EL} = 470$ nm) to green ($\lambda_{EL} = 516$ nm) by replacing a phenylene unit in **200** for 2,7-naphthylene (**201**). A very low turn-on voltage of 5.5 V (as for this class of materials) was achieved in the device ITO/**200**/Mg:Ag; however, no EL efficiency was reported (Chart 2.42).

Zheng and coworkers [227] have also synthesized a series of block copolymers, **202–205**, having an *m*-xylenedioxy bridge as a flexible unit and studied their optical and electrochemical properties. By changing the substituents in the central ring of the phenylene

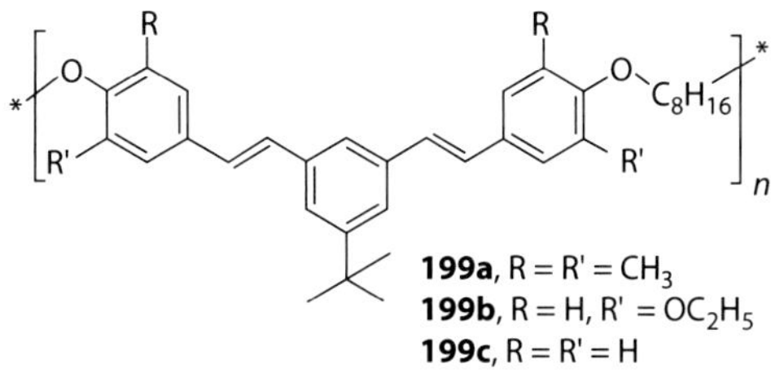

199a, R = R' = CH$_3$
199b, R = H, R' = OC$_2$H$_5$
199c, R = R' = H

CHART 2.41
Chemical structures of *meta*-linked oligo(phenylene vinylene) block copolymers.

200

201

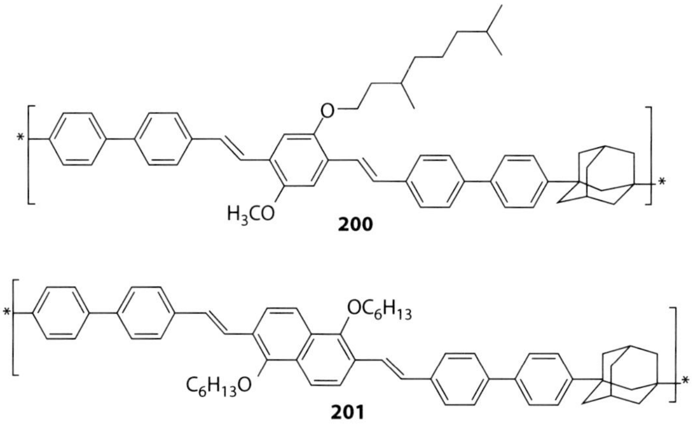

CHART 2.42
Chemical structures of PPV copolymers containing adamantane units.

vinylene block (polymers **202a–e**) or altering the conjugation by changing the aromatic unit in **203–205**, the emission band can be tuned between λ_{max}^{PL} of 413 and 533 nm (Scheme 2.20, Figure 2.7). Breaking the conjugation in the oligo(phenylene vinylene) block by changing the substitution position (1,4-phenylene → 1,2-phenylene → 1,3-phenylene → 9,10-anthracene units) progressively increases the band gap of the polymer, and hypsochromically shifts the emission band (the longest wavelength absorption and emission peaks of **205** are due to isolated anthracene unit; based on the second phenylene vinylene absorption band, its E_g = 3.6 eV). However, the PL efficiencies of compounds **203–205** also decrease very significantly (Scheme 2.20), and no EL devices have been reported for this series.

Kim et al. have introduced silicon atoms in PPV block copolymers to confine the conjugation length and achieve blue EL materials. Copolymers **206–208** [228] and **209** [229] have been synthesized via Wittig–Horner and Knoevenagel condensation, respectively. The emission band in this series can be tuned between 410 and 520 nm, and ITO/polymer/Al PLEDs with turn-on voltages ~7 V have been reported (Chart 2.43).

Monodisperse analogs of such π-electron systems, PPV oligomers (molecular glasses), were studied by Robinson and coworkers [230]. The films prepared from **210** by solution casting showed a completely amorphous structure due to a tetrahedral structure of the molecule and OLEDs ITO/PVK/**210**/Al-emitted green light with an efficiency up to 0.22 cd/A (Chart 2.44).

As we discussed above, nonconjugated blocks in PPV copolymers generally improve the film homogeneity. Furthermore, by changing the properties of the nonconjugated blocks, one can engineer the topology of the films. Introducing highly polar amide

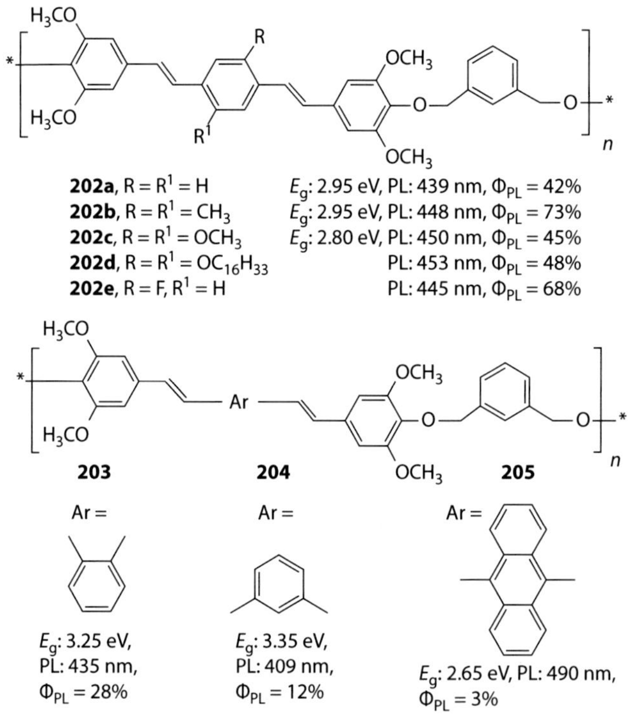

202a, R = R^1 = H	E_g: 2.95 eV, PL: 439 nm, Φ_{PL} = 42%
202b, R = R^1 = CH$_3$	E_g: 2.95 eV, PL: 448 nm, Φ_{PL} = 73%
202c, R = R^1 = OCH$_3$	E_g: 2.80 eV, PL: 450 nm, Φ_{PL} = 45%
202d, R = R^1 = OC$_{16}$H$_{33}$	PL: 453 nm, Φ_{PL} = 48%
202e, R = F, R^1 = H	PL: 445 nm, Φ_{PL} = 68%

203   **204**   **205**

Ar =   Ar =   Ar =

E_g: 3.25 eV,
PL: 435 nm,
Φ_{PL} = 28%

E_g: 3.35 eV,
PL: 409 nm,
Φ_{PL} = 12%

E_g: 2.65 eV, PL: 490 nm,
Φ_{PL} = 3%

SCHEME 2.20
Tuning the band gap and the emission wavelength in PPV block copolymers **202–205** (in chloroform solution).

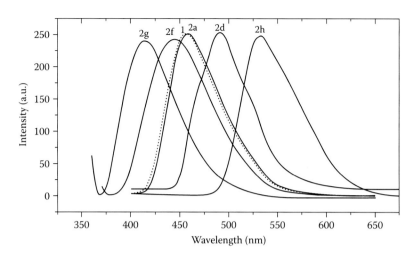

FIGURE 2.7
Tuning the solid-state emission maxima in PPV block copolymers **202–205**: (1) **189**, (2a) **202a**, (2d) **202d**, (2f) **203**, (2g) **204**, and (2h) **205**. (From Zheng, M., Sarker, A.M., Gürel, E.E., Lahti, P.M., and Karasz, F.E., *Macromolecules*, 33, 7426, 2000. With permission.)

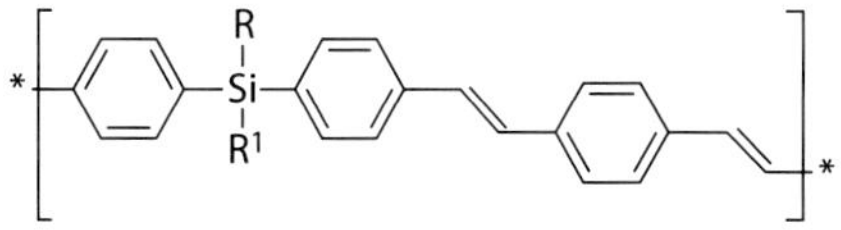

206a, R = R¹ = Ph
206b, R = R¹ = C₄H₉
206c, R = CH₃, R¹ = C₆H₁₃

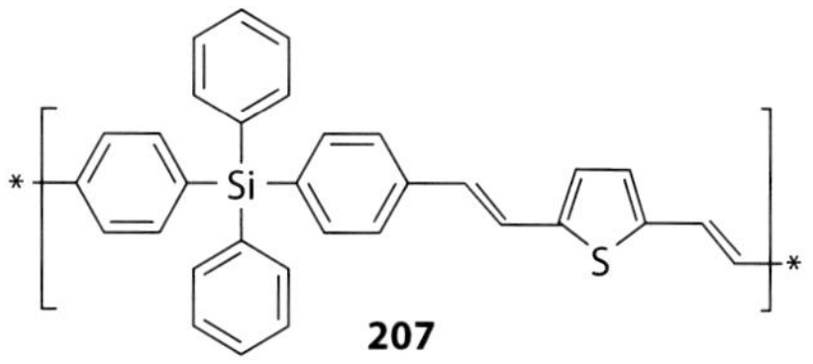

207

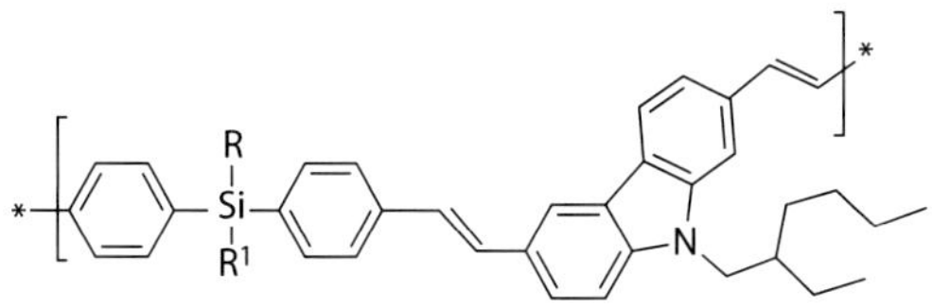

208a, R = R¹ = Ph
208b, R = R¹ = C₄H₉
208c, R = CH₃, R¹ = C₆H₁₃

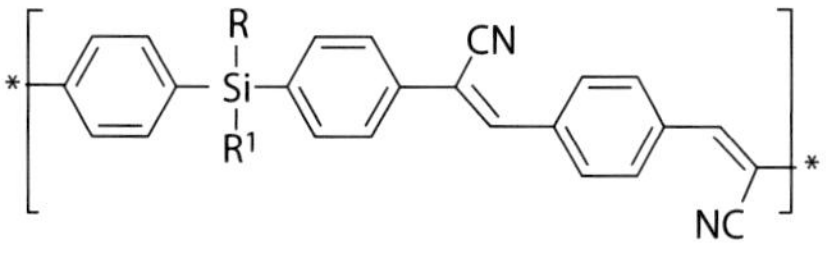

209a, R = R¹ = C₄H₉
209b, R = CH₃, R¹ = C₆H₁₃

CHART 2.43
Chemical structures of PPV copolymers containing silicon atoms.

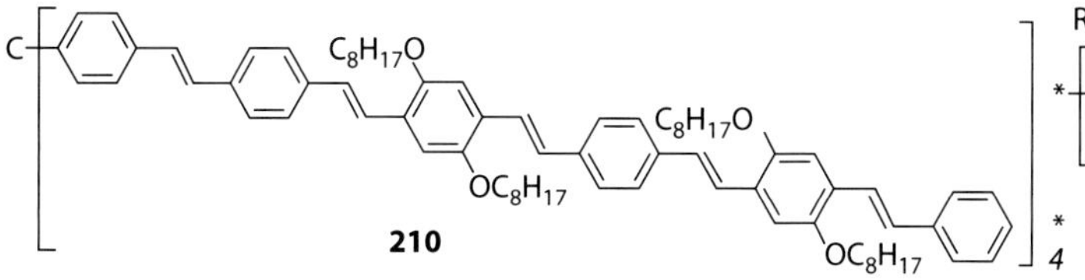

210

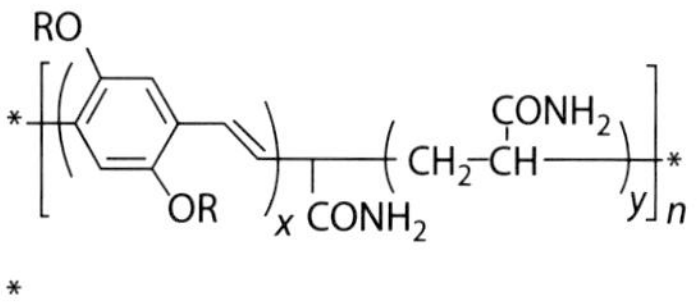

211

CHART 2.44
Chemical structures of PPV oligomers and PPV copolymers containing amide groups.

groups into nonconjugated blocks, Zhang et al. [231] prepared water-soluble copolymer **211** $\left(\lambda_{max}^{PL} = 509 \text{ nm}\right)$, which achieves nanophase separation of conjugated and nonconjugated domains in spin-coated films.

2.2.5.3 Nonconjugated Polymer Containing Oligo(Phenylene Vinylene) Pendant Substituents

The conjugation length and the emission color of PPV-type materials can be also controlled by using short oligo(phenylene vinylene) units as pendant substituents in nonconjugated polymer chain. The advantage of such an approach is the possibility to use well-established polymerization techniques, developed for nonconjugated systems in the last century. Thus, Lee and coworkers [232] reported on high-yield (95%) synthesis of a well-defined polymer, **212** ($n \approx 50$, polydispersity index [PDI] = 1.1), as a blue-emitting material (λ_{PL} [films] = 475 nm). A single-layer PLED ITO/**212**/Ca showed $\Phi_{EL}^{ex} = 0.3\%$ (turn-on voltage of 12 V), which can be improved to 0.55% by blending the polymer with electron-transport material PBD (Chart 2.45).

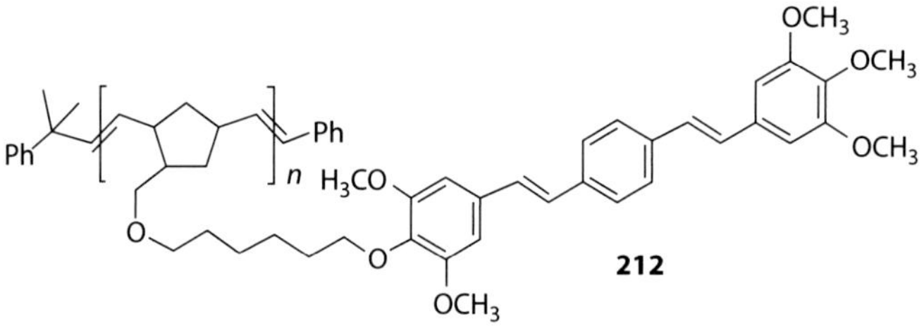

CHART 2.45
Chemical structures of nonconjugated polymer containing PPV oligomer pendant substituents.

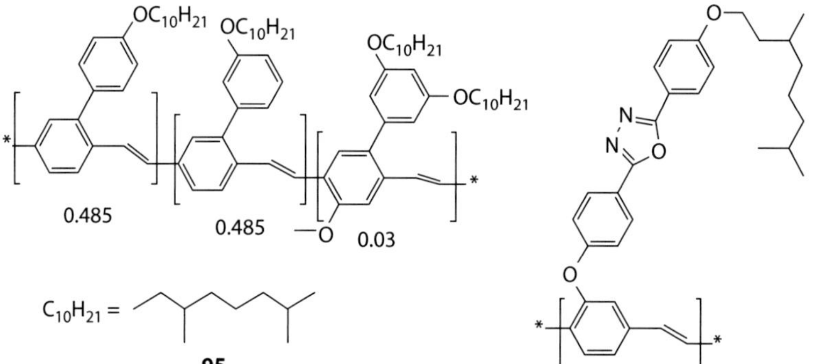

PLED: ITO/PEDOT/polymer/Ca (green)
Turn-on voltage: 2.2 V
Brightness: 1600 cd/m^2 (at 5 V)
Max EL efficiency: 22 cd/A (Φ_{EL}^{ex} 6.5%)

PLED: ITO/PEDOT/polymer/Al
λ_{EL} = 591 nm (CIE = 0.50, 0.47; yellow–orange)
Turn-on voltage: 5.0 V
Max brightness: 19.395 cd/m^2 (at 14 V)
Max EL efficiency: 21.1 cd/A

SCHEME 2.21
Best-performing PPV electroluminescent polymers **95** (From Ho, P.K.H., Kim, J.-S., Burroughes, J.H., Becker, H., Li, S.F.Y., Brown, T.M., Cacialli, F., and Friend, R.H., *Nature*, 404, 481, 2000.) and **75** (From Jin, S., Kim, M., Kim, J.Y., Lee, K., and Gal, Y., *J. Am. Chem. Soc.*, 126, 2474, 2004.).

2.2.6 Summary

As we have seen, 25 years of chemical design in PPV materials, together with remarkable engineering progress, resulted in the creation of PLEDs of remarkable performance. Efficiencies in excess of 20 cd/A were achieved for polymers **95** [146] and **75** [130] (Scheme 2.21), which render them among the best-performing green and orange EL polymers. Low-operating voltages, even with Al cathode, were achieved for PPV-based materials via introduction of electron-accepting fragments. Color tuning, in wide range, covering from blue to the NIR region, was demonstrated in PPV copolymers. The lifetime of the PPV-based PLEDs surpasses 20,000 h (for low brightness of ca. 100 cd/m^2).

2.3 Polyfluorenes

Fluorene (Fl) is a polycyclic aromatic compound that received its name owing to the strong violet fluorescence arising from its highly conjugated planar π-electron system (Chart 2.46).

Positions 2 and 7 in Fl are the most reactive sites toward electrophilic attack, which allows construction of a fully conjugated rigid-rod polymer chain by substitution reactions, whereas the methylene bridge provides an opportunity to modify the processability of the polymer by substituents without perturbing the electronic structure of the backbone. The varieties, excellent optical and electronic properties, and high thermal and chemical stability of PFs make them an attractive class of materials for PLEDs. Different aspects of syntheses, properties, and LED applications of fluorene-based conjugated polymers and copolymers have been highlighted in several recent reviews [233–240]. In fact, PFs are the only class of conjugated polymers that can emit a whole range of visible colors with relatively high QE (Table 2.2).

2.3.1 Synthetic Routes to Polyfluorenes

Parent (unsubstituted) PF was first synthesized electrochemically by anodic oxidation of fluorene in 1985 [241] and electrochemical polymerization of various 9-substituted fluorenes was studied in detail later [233,242]. Cyclic voltammogram of fluorene ($E_{red}^1 = 1.33\,V$, $E_{ox}^2 = 1.75\,V$ vs. Ag/Ag$^+$ in acetonitrile [242]) with repetitive scanning between 0 and 1.35 V showed the growth of electroactive PF film on the electrode with an onset of the p-doping process at ~0.5 V (vs. Ag/Ag$^+$). The unsubstituted PF was an insoluble and infusible material and was only studied as a possible material for modification of electrochemical electrodes. For this reason, it is of little interest for electronic or optical applications, limiting the discussion below to the chemically prepared 9-substituted PFs.

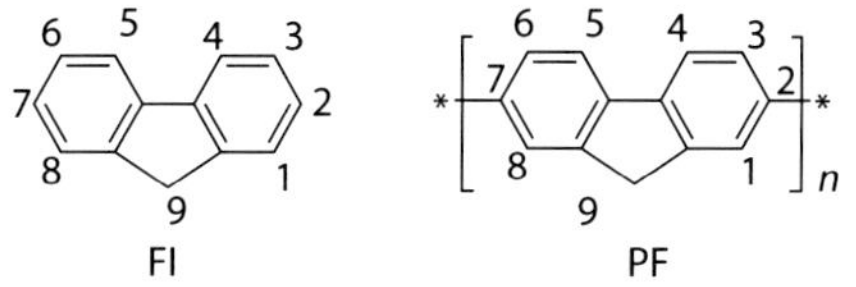

CHART 2.46
Chemical structures of fluorene and polyfluorene.

TABLE 2.2
Physical Properties of PFs

Polymer	M_n (g/mol) (PDI)	T_g (°C)	T_{dec} (°C)	λ_{max}^{abs} (nm) {Solution}	λ_{max}^{abs} (nm) {Film}	λ_{max}^{PL} (nm) {Solution}[a] [Φ_{PL} (%)]	λ_{max}^{PL} (nm) {Film}[a] [Φ_{PL} (%)]	E_g^{opt} (eV)[b]	HOMO–LUMO (eV) [CV Data]	EL Data	Ref.
213	35,700 (2.3)	75	421	385 {p-Xylene}	391	415 [82%]	425	2.91			[243]
213	24,300 (1.4)	103	390	379	385.5	415 (438, 469) [82%]	422 (444, 476) [74%]	2.86	−5.50/−2.37		[244]
213				382 {THF}		417, 436 [79%] {THF}		3.26	−5.87/−2.90		[245]
215	24,000 (2.3)	72	385	386 {CHCl$_3$}	391	422, 443 {CHCl$_3$}	422 [47%]	2.76	−5.77/−2.16	λ_{EL}: 456 nm, Φ_{EL}^{ex}: 0.52%, CIE (0.20, 0.20)	[246,247]
215	36,600 (2.81)			389 {CHCl$_3$}	394	414, 439, 471 [81%] {CHCl$_3$}	424, 448 [55%]	2.93		λ_{EL}: 425, 446 nm, CIE (0.17, 0.09)	[248]
215	41,600 (1.68)	113	418	389 {THF}	383	417, 439 [78%] {THF}	425, 447, 520 [40%]	2.9	−5.8/−2.9		[249]
215	40,000 (2.0)	78	426		380	[80%]	420	2.93	−5.80/−2.87	CIE (0.17, 0.09)	[250]
216					380	[80%]	420	2.91	−5.79/−2.88	λ_{EL}: 419 nm, CIE (0.17, 0.12)	[251,252]
219	5600 (1.6)	130	407	374 {CHCl$_3$}	366	412 (436) [48%]	430		−5.65/−2.38		[253]
220	41,200 (4.3)	128	412	388 {CHCl$_3$}	380	420 (445) [93%]	434		−5.67/−2.30		[253]
223	20,700 (1.75)			389 {THF}			417 (439, 473)				[254]
225	20,300 (1.97)	108			~290	~425	~425, ~452			CIE (0.136, 0.162)	[255]
226	46,000 (3.6)	248	400	388 {CHCl$_3$}	419, 443	427, 448	420				[256]
227				385 {DCM}	413, 437	422, 446				CIE (0.189, 0.137)	[257]
228a	51,400 (2.01)	94	320	388 {THF}	388	414 [78%]	421, 446 [25%]				[258]
228b	9300 (1.66)	50	360	380 {THF}	382	411 [100%]	418, 438 [98%]				[258]
228c	4400 (1.11)	10	320	340 {THF}	345	391 [41%]	419br [15%]				[258]

(Continued)

TABLE 2.2 (CONTINUED)

Physical Properties of PFs

Polymer	M_n (g/mol) (PDI)	T_g (°C)	T_{dec} (°C)	λ_{max}^{abs} (nm) [Solution]	λ_{max}^{abs} (nm) [Film]	λ_{max}^{PL} (nm) [Solution]^a [Φ_{PL} (%)]	λ_{max}^{PL} (nm) [Film]^a [Φ_{PL} (%)]	E_g^{opt} (eV)^b	HOMO–LUMO (eV) [CV Data]	EL Data	Ref.
229a	62,800 (2.22)	73	350	384 [THF]	388	411 [69%]	417, 440 [29%]				[258]
229b	26,200 (2.33)	48	290	383 [THF]	388	411 [92%]	415, 438 [46%]				[258]
229c	68,300 (2.21)	51	320	385 [THF]	388	411 [76%]	416, 449 [35%]				[258]
230a	3900 (1.96)	78	340	372 [THF]	380	411 [53%]	415, 438 [12%]				[258]
230b	16,300 (2.46)	50	370	388 [THF]	388	419 [100%]	419, 440 [73%]				[258]
230c	11,900 (1.65)	49	360	374 [THF]	380	410 [84%]	414, 436 [56%]				[258]
231a	35,000 (1.9)	90	351	392 [THF]	394	419 (443) [92%]	424 (448) [26%]				[259]
231b	18,000 (2.2)	59	371	388 [THF]	392	418 (442) [99%]	424 (448) [58%]				[259]
231c	15,000 (2.5)	56	360	380 [THF]	385	417 (439) [100%]	423 (446) [75%]				[259]
232a	46,700 (2.21)	103		~387 [THF]		420, 444					[260]
232b	131,000 (1.20)	73		~387		420, 444					[260]
232c	143,000 (1.19)	73		~370		415, 440					[260]
236	14,000 (3.3)	–	367	399		425, 446	418	2.94		CIE (0.17, 0.12)	[261]
237	32,000 {M_w}	121	400	~391 [Toluene]		416 [91%] [Toluene]	427				[262]
239	34,000 (2.2)	106	425	385 {p-Xylene}	392	415 [82%]	425	2.91			[243]

(Continued)

TABLE 2.2 (CONTINUED)

Physical Properties of PFs

Polymer	M_n (g/mol) (PDI)	T_g (°C)	T_{dec} (°C)	λ_{max}^{abs} (nm) {Solution}	λ_{max}^{abs} (nm) [Film]	λ_{max}^{PL} (nm) {Solution}[a] [Φ_{PL} (%)]	λ_{max}^{PL} (nm) [Film][a] [Φ_{PL} (%)]	E_g^{opt} (eV)[b]	HOMO– LUMO (eV) [CV Data]	EL Data	Ref.
240	77,000 (2.4)	110	427	386 {p-Xylene}	391	415 [83%]	424	2.91			[243]
241	8860 (1.76)						428, 452 [22%]			CIE (0.184, 0.159)	[263]
242a	7000 (1.8)			382 {THF}		410 (436), 538 {THF}	572				[264]
242b	13,500 (1.8)			382 {THF}		410 (436) {THF}					[264]
253a	80,000	135		380 {THF}		455 [79%] {THF}	[76%]				[265]
253b	89,000	136		380 {THF}		454 [65%] {THF}	[59%]			Φ_{EL}^{ex}: 0.17%	[265]
253c	73,000	139		378 {THF}		453 [61%] {THF}	[53%]			Φ_{EL}^{ex}: 0.09%	[265]
255a	4200 (2.5)						~455			Φ_{EL}^{ex}: 0.08%	[266]
255c	46,200 (1.92)	79	412	384 {THF}	383	417, 438 [74%] {THF}	424, 447 [71%]	2.9	−5.6/−2.7	λ_{EL}: ~455 nm	[249]
255d	27,000 (1.80)		408	374 {THF}	383	417, 440 [81%] {THF}	424, 447 [50%]	3.0	−5.5/−2.5		[249]
256a	28,100 (2.21)	108	422	370 {THF}	365	413, 436 [72%] {THF}	419, 442		−5.37/−2.07		[267]
256b	36,300 (4.02)	101	421	375 {THF}	370	413, 436 [74%] {THF}	419, 443		−5.43/−2.12		[267]
256c	18,500 (2.30)	88	421	380 {THF}	375	414, 438 [74%] {THF}	419, 443		−5.44/−2.13		[267]
257a	26,000 (2.29)	113	422	350 {THF}	346	397 [68%] {THF}	407		−5.33/−2.02		[267]
257b	35,900 (2.51)	108	421	367 {THF}	362	409, 431 [76%] {THF}	414, 438		−5.34/−2.04		[267]

(Continued)

TABLE 2.2 (CONTINUED)

Physical Properties of PFs

Polymer	M_n (g/mol) (PDI)	T_g (°C)	T_{dec} (°C)	λ_{max}^{abs} (nm) {Solution}	λ_{max}^{abs} (nm) {Film}	λ_{max}^{PL} (nm) {Solution}[a] [Φ_{PL} (%)]	λ_{max}^{PL} (nm) {Film}[a] [Φ_{PL} (%)]	E_g^{opt} (eV)[b]	HOMO–LUMO (eV) [CV Data]	EL Data	Ref.
257c	31,000 (2.65)	98	420	372 {THF}	368	421, 435 [76%] {THF}	416, 440		−5.39/−2.17	λ_{EL}: 423 nm	[267]
257d	13,100 (1.97)	86	422	375 {THF}	371	414, 438 [80%] {THF}	419, 443		−5.42/−2.21		[267]
258				381 {CHCl$_3$}	383	417, 440 [78%] {CHCl$_3$}	426, 448				[268]
266	10,000 (1.77)		344	378 {DCM}	378	462 [12%] {Toluene}; 528 [<1%] {NMP}	497 [24%]				[269]
267a						433, 457 [51%] {Toluene}; 466 [50%] {NMP}	443, 465 [51%]				[269]
267b						433, 457 [64%] {Toluene}; 466 [60%] {NMP}	443, 465 [9%]				[269]
268a	14,000 (2.56)	107	394	380 {CHCl$_3$}	406		429	2.88	−5.13/−2.25		[270]
268b	12,500 (3.02)	120	413	385 {CHCl$_3$}	411		433	2.85	−5.10/−2.25		[270]
268c	12,300 (2.37)	65	401		411		433	2.85	−5.10/−2.25		[270]
269a	20,600 (2.0)	70	389	380 {CHCl$_3$}	380	420 {CHCl$_3$}	422 [30%]	2.79	−5.77/−2.19	λ_{EL}: 446 nm, Φ_{EL}^{ex}: 0.16%	[246]
269b	19,000 (1.9)	82	395	385 {CHCl$_3$}	383	420 {CHCl$_3$}	422 [28%]	2.79	−5.75/−2.19	λ_{EL}: 442 nm, Φ_{EL}^{ex}: 0.22%, CIE (0.19, 0.16)	[246]
269c	24,000 (2.6)	80	397	361 {CHCl$_3$}	380	418 {CHCl$_3$}	420 [51%]	2.90	−5.73/−2.22	λ_{EL}: 442 nm, Φ_{EL}^{ex}: 0.40%, CIE (0.17, 0.12)	[246]
269d	12,400 (2.2)	83	406	363 {CHCl$_3$}	370	420 {CHCl$_3$}	420 [60%]	2.90	−5.74/−2.26	λ_{EL}: 438 nm, Φ_{EL}^{ex}: 0.40%, CIE (0.17, 0.11)	[246]

(Continued)

TABLE 2.2 (CONTINUED)

Physical Properties of PFs

Polymer	M_n (g/mol) (PDI)	T_g (°C)	T_{dec} (°C)	λ_{max}^{abs} (nm) {Solution}	λ_{max}^{abs} (nm) {Film}	λ_{max}^{PL} (nm) {Solution}[a] [Φ_{PL} (%)]	λ_{max}^{PL} (nm) [Film][a] [Φ_{PL} (%)]	E_g^{opt} (eV)[b]	HOMO–LUMO (eV) [CV Data]	EL Data	Ref.
269e	22,800 (1.5)			362 {CHCl$_3$}	380	418 [CHCl$_3$]	419 [49%]	2.97	−5.72/−2.27	λ_{EL}: 440 nm, Φ_{EL}^{ex}: 0.45%, CIE (0.15, 0.08)	[246]
269f	15,600 (1.3)	93	398	340 {CHCl$_3$}	350	403 [CHCl$_3$]	413 [15%]	3.15	−5.36/−2.33	λ_{EL}: 424 nm, Φ_{EL}^{ex}: 0.02%, CIE (0.21, 0.20)	[246]
270	13,000 (2.1)	213	403	390 {THF}	390	419 (444) [124%] [THF]	427 (452) [43%]		−5.76/−2.47	λ_{EL}: 428 nm, Φ_{EL}^{ex}: 0.52%	[271]
271	11,000 (2.2)	166	440	389 {THF}		420 (444) [95%] {THF}	426 (450) [42%]		−5.30/−2.54	CIE (0.193, 0.141), Φ_{EL}^{ex}: 1.21%	[272]
272	4900 (1.2)	119	398	321 {CHCl$_3$}	322	445 [15%] {CHCl$_3$}	447	3.14	−5.90/−2.70	λ_{EL}: 452 nm	[273]
273a	6700 (1.7)	109	418	327 {CHCl$_3$}	327	405 [30%] {CHCl$_3$}	411	3.33	−5.97/−2.64	λ_{EL}: 408 nm	[273]
273b	14,800 (1.5)	125	421	329 {CHCl$_3$}	326	402 [39%] {CHCl$_3$}	407	3.32	−5.92/−2.60	λ_{EL}: 409 nm	[273]
273c	26,600 (2.8)	128	414	325 {CHCl$_3$}	323	404 [36%] {CHCl$_3$}	412	3.32	−5.93/−2.61	λ_{EL}: 423 nm	[273]
273d	24,500 (2.7)	85	410	322 {CHCl$_3$}	321	404 [31%] {CHCl$_3$}	409	3.33	−5.96/−2.63	λ_{EL}: 406 nm	[273]
273e	25,200 (2.6)	148	422	323 {CHCl$_3$}	322	430 [17%] {CHCl$_3$}	430	3.32	−5.74/−2.84	λ_{EL}: 431 nm	[273]
273f	12,200 (1.9)	120	407	323 {CHCl$_3$}	324	409 [26%] {CHCl$_3$}	415	3.33	−6.04/−2.71	λ_{EL}: 416 nm	[273]
273g	6500 (1.7)	132	446	328 {CHCl$_3$}	328	410 [38%] {CHCl$_3$}	410	3.34	−5.95/−2.61	λ_{EL}: 417 nm	[273]
275	12,000 (2.28)	150	439	394 (372) {DCM}		402 (426) [73%] {DCM}					[274]

(Continued)

TABLE 2.2 (CONTINUED)

Physical Properties of PFs

Polymer	M_n (g/mol) (PDI)	T_g (°C)	T_{dec} (°C)	λ_{max}^{abs} (nm) {Solution}	λ_{max}^{abs} (nm) {Film}	λ_{max}^{PL} (nm) [Solution][a] [Φ_{PL} (%)]	λ_{max}^{PL} (nm) [Film][a] [Φ_{PL} (%)]	E_g^{opt} (eV)[b]	HOMO–LUMO (eV) [CV Data]	EL Data	Ref.
276	25,100 (2.50)	114	431	396 (380) {DCM}		420 (442) [75%] {DCM}					[274]
277	30,400 (2.70)	98	432	395 (381) {DCM}		422 (443) [72%] {DCM}					[274]
278	21,600 (2.63)	127	430	396 (379) {DCM}		422 (444) [67%] {DCM}					[274]
279a	20,000 (1.71)	143							−6.4/−2.6		[275]
279b	11,000 (1.36)	263							−6.4/−2.6		[275]
279c	33,200 (2.4)	117									[275]
280a	38,600 (1.87)	95							−6.0/−2.8		[275]
280b	36,200 (1.25)	85							−6.0/−2.8		[275]
281a	8700 (2.3)	137	399	366 {CHCl₃}	369	405 {CHCl₃}	414	3.02	−6.24/−3.01		[276]
281b	9100 (2.2)	194	394	349 {CHCl₃}	352	428 {CHCl₃}	422	3.09	−6.38/−3.19		[276]
282a	21,800 (2.7)	119	411	350 (368) {DCE}	354 (373)	379, 396 (415) [41%]	403, 423 (447)	3.22	−6.32/−3.10		[277]
282b	27,600 (3.6)	118	411	352 (371) {DCE}	350 (369)	380, 396 (415) [90%]	399, 421 (447)	3.23	−6.27/−3.04		[277]
282c	11,700 (3.0)	71	352	353 (370) {DCE}	356 (376)	385, 403 (423) [86%]	394, 427 (443)	3.20	−6.08/−2.88		[277]
283	15,600 (1.63)			390 {THF}		404 (425) {THF}					[278]
284a					380		477	2.85	−5.74/−2.89	Φ_{EL}^{ex}: 0.50%	[279]
284b					328, 384		447	2.92	−5.90/−2.98	Φ_{EL}^{ex}: 0.057%	[279]
285					−378, 368		420, 448 [40%]		−5.66/−2.62	λ_{EL}: 420, 448 nm, Φ_{EL}^{ex}: 0.60%	[280,281]
287	11,600 (2.9)	105			379		452	2.88			[282]
288	3800 (3.04)	105	435	374 {THF}	375	413, 436 {THF}	440			Φ_{EL}^{ex}: 0.12%	[283]

(*Continued*)

TABLE 2.2 (CONTINUED)

Physical Properties of PFs

Polymer	M_n (g/mol) (PDI)	T_g (°C)	T_{dec} (°C)	λ_{max}^{abs} (nm) {Solution}	λ_{max}^{abs} (nm) {Film}	λ_{max}^{PL} (nm) {Solution}[a] [Φ_{PL} (%)]	λ_{max}^{PL} (nm) {Film}[a] [Φ_{PL} (%)]	E_g^{opt} (eV)[b]	HOMO–LUMO (eV) [CV Data]	EL Data	Ref.
289	12,800 (7.67)	144	440	376 {THF}	381	414, 438 {THF}	429, 444			λ_{EL}: 423 nm, Φ_{EL}^{ex}: 0.06%, CIE (0.173, 0.098)	[283]
290	4600 (5.25)	93	430	375 {THF}	378	415, 439 {THF}	435				[283]
291	6400 (9.27)	90	445	377 {THF}	381	414, 436 {THF}	427, 440				[283]
292a	13,800 (2.43)	236	449	334 {THF}	345	398 (420) [73%] {THF}	399 (420) [38%]				[284]
292b	6640 (1.15)	237	439	332 {THF}	333	397 [92%] {THF}	402 (419) [65%]				[284]
292c	4530 (1.67)	238	380	355 {THF}	374	407 [99%] {THF}	414 (439) [82%]				[284]
293a	23,500 (1.52)	118	241		341		411 [32%]	2.95	−5.76/−2.85	λ_{EL}: 418 nm, Φ_{EL}^{ex}: 0.43%	[285]
293b	24,200 (1.58)	110	243		341		408 [30%]	2.94	−5.74/−2.80	λ_{EL}: 419 nm, Φ_{EL}^{ex}: 0.60%	[285]
293c	20,700 (1.61)	79	342		344		412 [32%]	2.94	−5.81/−2.91	λ_{EL}: 420 nm, Φ_{EL}^{ex}: 0.41%	[285]
294a	14,100 (4.2)			392, 413 {CHCl₃}	396, 415	427, 447 {CHCl₃}	428				[286]
294b	11,600 (5.8)			392, 412 {CHCl₃}	394, 416	426, 447 {CHCl₃}	429				[286]
294c	9800 (4.4)			290, 410 {CHCl₃}	387, 413	426, 447 {CHCl₃}	429				[285]
294d	17,000 (3.6)			394, 412 {CHCl₃}	392, 416	426, 446 {CHCl₃}	434				[286]
294e	9500 (5.4)			392, 412 {CHCl₃}	396, 424	427, 447 {CHCl₃}	434				[286]
295	19,500 (1.9)				360		428, 445 [44%]	2.97		λ_{EL}: 420, 446 nm, Φ_{EL}^{ex}: 0.82%	[287]

(Continued)

TABLE 2.2 (CONTINUED)

Physical Properties of PFs

Polymer	M_n (g/mol) (PDI)	T_g (°C)	T_{dec} (°C)	λ_{max}^{abs} (nm) {Solution}	λ_{max}^{abs} (nm) {Film}	λ_{max}^{PL} (nm) {Solution}[a] [Φ_{PL} (%)]	λ_{max}^{PL} (nm) {Film}[a] [Φ_{PL} (%)]	E_g^{opt} (eV)[b]	HOMO–LUMO (eV) [CV Data]	EL Data	Ref.
296a	2320 (1.39)			361 {THF}	367	411 (428) [75%] {THF}	417 (440) [42%]				[288,289]
296b	1700 (1.35)					[36%] {THF}					[288]
296c	2780 (1.22)						[18%]				[288]
297a	56,700 (1.56)		330	378 (408) {THF}	386 (410)	471 (427, 448) [22%] {THF}	484 (515)				[289]
297b	48,500 (1.59)		288	384 (411) {THF}	388 (424)	471 (426, 500) [16%] {THF}	492 (518)				[289]
297c	43,300 (1.76)		256	385 (408) {THF}	390 (416)	471 (429, 502) [15%] {THF}	479				[289]
298	35,200 (1.57)			385 {THF}	383	413 (435) {THF}	453, 477 (425)	2.88	E_{ox}: 1.39 V, E_{red}: -1.76 V (vs. SCE)		[290]
299	37,200 (1.49)			424, 448 {THF}	432, 459	463 (492) {THF}	521	2.52	E_{ox}: 1.01 V, E_{red}: -1.60 V (vs. SCE)		[290]
300	59,200 (1.92)			392, 409 {THF}	395, 425	424 (444) {THF}	477, 503 (439)	2.74	E_{ox}: 1.50 V, E_{red}: -1.67 V (vs. SCE)		[290]
302a	36,679 (1.61)		421	364 {CH$_2$Cl$_2$}	361	415 [73%] {CH$_2$Cl$_2$}	411			λ_{EL}: 412 nm, Φ_{EL}^{ex}: 0.086%	[291]
302b	35,633 (1.52)	109	412	370 {CH$_2$Cl$_2$}	370	415 [70%] {CH$_2$Cl$_2$}	420		E_{ox}: 1.17 V, E_{red}: -1.95 V (vs. Ag$^+$/Ag)	λ_{EL}: 422 nm, Φ_{EL}^{ex}: 1.14%	[291]
303				343 {Toluene}	343	395 (405) [68%]	405	3.62	3.8	λ_{EL}: 405 nm, Φ_{EL}^{ex}: 5.02%, CIE (0.16, 0.05)	[292]
304	14,000 (1.5)	185	429	352 {CHCl$_3$}	355	398, 419 {CHCl$_3$}	404, 425 [42%]			CIE (0.165, 0.128)	[293]

(Continued)

TABLE 2.2 (CONTINUED)

Physical Properties of PFs

Polymer	M_n (g/mol) (PDI)	T_g (°C)	T_{dec} (°C)	λ_{max}^{abs} (nm) {Solution}	λ_{max}^{abs} (nm) {Film}	λ_{max}^{PL} (nm) {Solution}[a] [Φ_{PL} (%)]	λ_{max}^{PL} (nm) {Film}[a] [Φ_{PL} (%)]	E_g^{opt} (eV)[b]	HOMO–LUMO (eV) [CV Data]	EL Data	Ref.
305a	9400 (1.5)	50	390	326 {CHCl$_3$}	324	383 (403) [64%] {CHCl$_3$}	404 (389) [88%]	3.32	-5.66/-2.31		[244]
305b	5050 (2.0)	72	377	370.5 {CHCl$_3$}	373.5	414 (436) [63%] {CHCl$_3$}	418 (414) [55%]	2.95	-5.38/-2.41		[244]
305c	22,800 (1.5)	123	410	355 {CHCl$_3$}	370	409.5 (432) [74%] {CHCl$_3$}	423.5 (443) [48%]	2.93	-5.78/-2.81		[244]
305d	48,100 (1.9)	111	376	353 {CHCl$_3$}	350.6	451 [35%] {CHCl$_3$}	443 [20%]	3.04	-5.76/-2.80		[244]
305e	8900 (1.7)	145	390	348 {CHCl$_3$}	343	398 (417) [26%] {CHCl$_3$}	428 (406, 451) [10%]	3.06	-5.28/-2.47		[244]
305f	7300 (1.6)	213	375	349 {CHCl$_3$}	349	418 [58%] {CHCl$_3$}	423.5 [23%]	3.05	-5.64/-2.42		[244]
305g	60,400 (2.0)	122		385 {CHCl$_3$}	395	416 (440, 475) [91%] {CHCl$_3$}	426 (447) [41%]	2.82	-5.69/-2.83		[244]
306	17,500 (2.4), 12,000 (1.8)	55			364	407 [87%] {CHCl$_3$}	425, 443 (cast), 420, 442 {spin coated}; 442, 513		E_{ox}: 1.76 V, E_{red}: -2.44 V (vs. SCE)	λ_{EL}: 419 nm	[294,295][e]
306	11,800 (1.7)		398	369.5 {CHCl$_3$}	371	408 (431.5) [85%] {CHCl$_3$}	422 (441, 470) [78%]	2.92	-5.36/-2.47		[244]
307	15,000 (2.2)	70			365	406 [72%] {CHCl$_3$}	424, 442 (cast), 418, 440 {spin coated}; 443		E_{ox}: 1.50 V, E_{red}: -2.40 V (vs. SCE)	λ_{EL}: 416 nm	[294, 295]
308	2700 (1.4)	60	224	374 {CHCl$_3$}	366	415 (438) [85%] {CHCl$_3$}	424		-5.56/-2.23		[253]
309a	6800 (1.6); 10,000 (2.6)				438	496 [49%] {CHCl$_3$}	490, 515 (cast, 485, 512 {spin coated}; 531		E_{ox}: 1.26 V, E_{red}: -2.40 V (vs. SCE)	λ_{EL}: 480 nm	[294, 295]

(Continued)

TABLE 2.2 (CONTINUED)

Physical Properties of PFs

Polymer	M_n (g/mol) (PDI)	T_g (°C)	T_{dec} (°C)	λ_{max}^{abs} (nm) {Solution}	λ_{max}^{abs} (nm) {Film}	λ_{max}^{PL} (nm) [Solution][a] [Φ_{PL} (%)]	λ_{max}^{PL} (nm) [Film][a] [Φ_{PL} (%)]	E_g^{opt} (eV)[b]	HOMO–LUMO (eV) [CV Data]	EL Data	Ref.
309b	15,000 (2.2)	70			384		469 {cast}, 460 {spin coated}		E_{ox}: 1.35 V (vs. SCE)	λ_{EL}: 468 nm	[294]
310	14,500 (3.0)	73		452, 470 {CHCl$_3$}	427	488 [30%] {CHCl$_3$}	563 {cast}, 548 {spin coated}; 599		E_{ox}: 1.11 V, E_{red}: −2.34 V (vs. SCE)	λ_{EL}: 545 nm	[294, 292]
310	19,000 (2.83)	93	426	452, 470 {CHCl$_3$}	458, 479	495, 530, 571 [20%] {CHCl$_3$}	511, 537, 577 [12%]	2.44	−5.41/−2.48	λ_{EL}: 539 nm, CIE (0.40, 0.58)	[248]
311	2400 (1.3)	65	399	444 {CHCl$_3$}	446	498 (530) [30%] {CHCl$_3$}	510		−5.21/−2.53	λ_{EL}: 588 nm	[253]
312	12,700 (1.4)	129			444		495, 527 {cast}, 494, 526 {spin coated}		E_{ox}: 1.18 V (vs. SCE)	λ_{EL}: 536 nm	[294]
313	6600 (1.9)	105			517		536, 574 {cast}, 532, 574 {spin coated}		E_{ox}: 1.22 V, E_{red}: −2.73 V (vs. SCE)		[294]
315	18,700 (1.4)	203 (T_m)	381	403 {CHCl$_3$}	412	461 (490) [65%] {CHCl$_3$}	492 (477) [41%]	2.50, 2.63	−5.39/−2.56	Φ_{EL}^{ex}: 0.64%	[293, 296,297]
316	18,600 (1.6)	78	374, 410	367 {CHCl$_3$}	375.5	447 (473) [39%] {CHCl$_3$}	458 (475) [25%]	2.76, 2.78, 2.70	−5.63/−2.35; −5.53/−2.65	Φ_{EL}^{ex}: 0.37%	[244,296–298]
317a	15,200 (1.9)	63	394	401 {CHCl$_3$}	403	482 (515) [31%] {CHCl$_3$}	490 (520) [18%]	2.49 2.53, 2.57	−5.38/−2.45	Φ_{EL}^{ex}: 0.25%	[296–298]
317b	22,600 (1.7); 13,000 (1.67)	77	390	398 {CHCl$_3$}	401; 396; 392	483 (520) [28%; 32%] {CHCl$_3$}	493 (520) [21%]; 496	2.58, 2.60, 2.53	−5.40/−2.39; −5.30/−2.85; −5.29/−2.79	λ_{EL}: 493 (515) nm, CIE (0.24, 0.51), Φ_{EL}^{ex}: 0.30%	[296–299]
319	13,600 (1.80)	112	419	439, 457 {CHCl$_3$}	448, 471	478, 510, 554 [32%] {CHCl$_3$}	495, 511, 548 [12%]	2.48	−5.38/−2.40	λ_{EL}: 515 nm, CIE (0.29, 0.63)	[248]
320					390		540 [64%]	2.4	−5.9/−3.2	λ_{EL}: 545 nm, Φ_{EL}^{ex}: 2.5%, CIE (0.394, 0.57)	[300–303]
321	16,800 (2.33)	162	387		339, 395		512 [45%]	2.73	−5.81/−2.98	Φ_{EL}^{ex}: 0.60%	[304]

(Continued)

TABLE 2.2 (CONTINUED)

Physical Properties of PFs

Polymer	M_n (g/mol) (PDI)	T_g (°C)	T_{dec} (°C)	λ_{max}^{abs} (nm) {Solution}	λ_{max}^{abs} (nm) {Film}	λ_{max}^{PL} (nm) {Solution}[a] [Φ_{PL} (%)]	λ_{max}^{PL} (nm) {Film}[a] [Φ_{PL} (%)]	E_g^{opt} (eV)[b]	HOMO–LUMO (eV) [CV Data]	EL Data	Ref.
322a	21,700 (1.53)	111	376		337, 399		508 [56%]	2.71	−5.74/−2.98	Φ_{EL}^{ex}: 1.56%	[304]
322b	25,600 (1.45)	80	380		336, 398		509 [59%]	2.71	−5.71/−2.93	Φ_{EL}^{ex}: 2.53%	[304]
323	10,600 (2.95)		443		415		481, 536		−5.76/−2.79	λ_{EL}: 480, 536 nm	[305]
324a	10,600 (2.1)	173	415	417 {CHCl₃}	419	463 {CHCl₃}	507	2.6	−5.34/−2.75	λ_{EL}: 508 nm	[306,307]
324b	100,000 (1.3)		~400		418		465, 506 (530) [34%]	2.56	−5.73/−3.13	λ_{EL}: 472 nm, CIE (0.23, 0.38)	[308]
324c	39,300 (2.28)	–	~450		420 (370)		505 (480, 550)	2.55	−5.40/−2.85	λ_{EL}: 505	[309]
325	10,700 (3.33)	100	380		375		440, 465, (490), 540	2.85	−5.67/−2.82	λ_{EL}: 540 nm	[310]
326	18,800 (1.91)	149	395	404 {Toluene}		658 {Toluene}	664	2.05		λ_{EL}: 668 nm, CIE (0.682, 0.317)	[158]
330a	33,500 (2.00)	155	422		316, 412		592	2.39	−6.20/−3.79	λ_{EL}: 562 nm	[311]
330b	39,100 (1.83)	122	427		301, 404		584	2.40	−6.18/−3.75	λ_{EL}: 559 nm	[311]
330c	36,000 (1.14)	100	434		304, 406		580	2.42	−6.17/−3.72	λ_{EL}: 557 nm	[311]
331	13,100 (2.9)	114	411	399 {CHCl₃}	403	465 {CHCl₃}	476	2.66	−6.13/−3.08	λ_{EL}: 466 nm, Φ_{EL}^{ex}: 0.2%	[312]
332	13,000 (2.4)	121		500	500	556 (608) [3%]	610	2.2	−5.47/−3.30	λ_{EL}: 582 (610) nm	[313]
333	13,300 (3.08)		388	538 {CHCl₃}	563 (601)	616 [32%] {CHCl₃}	656 (702)	2.00	−5.32/−3.32		[253]
334	7800 (1.54)	–	~450		440		535 (500)	2.38	−5.50/−3.12	λ_{EL}: 535 nm	[309]
335	11,000 (2.5)	106	370	302 {CHCl₃}	372	468 {CHCl₃}	525	2.74	−5.82/−3.08	λ_{EL}: 547 nm, CIE (0.40, 0.53)	[250]
336	18,000 (1.8)	112	391	439 {CHCl₃}	468	513 {CHCl₃}	555	2.28	−5.87/−3.59	λ_{EL}: 554 nm, CIE (0.48, 0.51)	[250]
337	21,000 (2.1)	113	400	516 {CHCl₃}	460	516 {CHCl₃}	602	2.21	−5.65/−3.44	λ_{EL}: 604 nm, CIE (0.63, 0.38)	[250]
338	14,000 (1.9)	118	407	530 {CHCl₃}	537	530 {CHCl₃}	674	1.95	−5.59/−3.64	λ_{EL}: 674 nm, CIE (0.66, 0.33)	[250]

(Continued)

TABLE 2.2 (CONTINUED)

Physical Properties of PFs

Polymer	M_n (g/mol) (PDI)	T_g (°C)	T_{dec} (°C)	λ_{max}^{abs} (nm) {Solution}	λ_{max}^{abs} (nm) {Film}	λ_{max}^{PL} (nm) {Solution}[a] [Φ_{PL} (%)]	λ_{max}^{PL} (nm) {Film}[a] [Φ_{PL} (%)]	E_g^{opt} (eV)[b]	HOMO–LUMO (eV) [CV Data]	EL Data	Ref.
339	4430 (2.1)		367		419		591 [5.7%]			λ_{EL}: 599 (557), CIE (0.48, 0.48)	[314]
340	4620 (2.9)		332		439		602 [14.7%]			λ_{EL}: 596, CIE (0.53, 0.44)	[314]
341	2250 (2.0)		450		417		551 (580) [3.5%]			λ_{EL}: 557, 585, CIE (0.43, 0.45)	[314]
342	23,500 (1.52)	186	407		352, 457		662, 712 [7%]	2.22	−5.78/−3.57	λ_{EL}: 657 (702) nm, Φ_{EL}^{ex}: 0.27%	[315]
343	24,200 (1.58)	119	406		345, 435		641, 704 [5%]	2.32	−5.79/−3.54	λ_{EL}: 636 (694 nm), Φ_{EL}^{ex}: 0.38%	[315]
344	20,700 (1.61)	73	406		347, 438		641, 705 [4%]	2.30	−5.77/−3.53	λ_{EL}: 638 (696) nm, Φ_{EL}^{ex}: 0.21%	[315]
345	11,090 (1.3)				400, 780		1035	1.27		λ_{EL}^{ex}: 970 nm, Φ_{EL}^{ex}: 0.03–0.05%	[316]
350			415	328 [THF]	336	368, 386 [62%] {THF}	376, 384	3.26	−5.89/−2.63	λ_{EL}: 395 nm, Φ_{EL}^{ex}: 0.054%	[245]
351a	23,200 (2.55)	60	415	404				2.79	−5.38/−2.59	λ_{EL}: 445,471 nm, Φ_{EL}^{ex}: 1.15%, CIE (0.16, 0.12)	[317]
351b	17,500 (1.84)	72	421					2.77	−5.44/−2.67	λ_{EL}: 471 nm, Φ_{EL}^{ex}: 1.76%, CIE (0.19, 0.27)	[317]
351c	26,300 (1.67)	89	416					2.80	−5.35/−2.55	λ_{EL}: 445,471 nm, Φ_{EL}^{ex}: 3.00%, CIE (0.16, 0.12)	[317]
351d	20,900 (1.80)	90	423					2.80	−5.44/−2.64	λ_{EL}: 445,471 nm, Φ_{EL}^{ex}: 1.53%, CIE (0.18, 0.19)	[317]

(Continued)

TABLE 2.2 (CONTINUED)

Physical Properties of PFs

Polymer	M_n (g/mol) (PDI)	T_g (°C)	T_{dec} (°C)	λ_{max}^{abs} (nm) {Solution}	λ_{max}^{abs} (nm) {Film}	λ_{max}^{PL} (nm) {Solution}[a] [Φ_{PL} (%)]	λ_{max}^{PL} (nm) {Film}[a] [Φ_{PL} (%)]	E_g^{opt} (eV)[b]	HOMO–LUMO (eV) [CV Data]	EL Data	Ref.
351e	27,800 (3.86)	78	422					2.72	−5.41/−2.69	λ_{EL}: 474 nm, Φ_{EL}^{ex}: 2.80%, CIE (0.18, 0.30)	[317]
354	55,000	90	>400		370, 465		540				[318]
355a	38,000 (3.0)	85	>400		376		466				[318,319]
355b	11,000	95	>400		374		481				[318]
355c	10,000	110	>400		390		510				[318]
357a	20,000 (2.3)		~300		380	[69%]	536		−5.72	λ_{EL}: 532 nm	[251,320]
357b	15,000 (1.5)		~300		380	[57%]	544		−5.73	λ_{EL}: 535 nm	[251,320]
357c	13,000 (2.5)		~300		380	[51%]	583		−5.72	λ_{EL}: 580 nm	[251,320]
357d	22,000 (2.7)		~300		379	[34%]	620		−5.73	λ_{EL}: 630 nm	[251,320]
358a	87,000 (1.4)		~400		418	470, 506 [36%]			−5.53/−2.93	λ_{EL}: 472 nm, CIE (0.23, 0.38)	[308]
358b	57,000 (2.1)		~400		419	470, 506 [36%]			−5.55/−2.95	λ_{EL}: 475 nm, CIE (0.23, 0.38)	
360a	11,900 (2.6)	176	410	415 {CHCl₃}	418	463 {CHCl₃}	576	2.57	−5.34/−2.75	λ_{EL}: 574 nm	[307]
360b	14,100 (1.9)	170	390	418 {CHCl₃}	421	464 {CHCl₃}	580	2.54	−5.36/−2.77	λ_{EL}: 576 nm	[307]
360c	11,400 (2.6)	172	385	418 {CHCl₃}	420	463 {CHCl₃}	589	2.54	−5.16/−2.62	λ_{EL}: 589 nm	[307]
360d	14,400 (3.0)	169	355	417 {CHCl₃}	422	463 {CHCl₃}	593	2.1	−5.09/−2.99	λ_{EL}: 592 nm	[307]
361a	14,000 (1.51)	74	400	383 {CHCl₃}	383	418 {CHCl₃}	422 [29%]		−5.77/−2.16	λ_{EL}: 422 nm, Φ_{EL}^{ex}: 0.36%	[247]
361b	19,000 (1.82)	76	400	375 {CHCl₃}	380	416 {CHCl₃}	421 [25%]		−5.78/−2.17	λ_{EL}: 421 nm, Φ_{EL}^{ex}: 0.34%	[247]
361c	17,000 (1.77)	79	406	369 {CHCl₃}	373	418 {CHCl₃}	420 [18%]		−5.80/−2.12	λ_{EL}: 420 nm, Φ_{EL}^{ex}: 0.22%	[247]
361d	1100 (1.96)	84	401	350 {CHCl₃}	350	415 {CHCl₃}	419 [23%]		−5.81/−2.12	λ_{EL}: 419 nm, Φ_{EL}^{ex}: 0.10%	[247]

(Continued)

TABLE 2.2 (CONTINUED)

Physical Properties of PFs

Polymer	M_n (g/mol) (PDI)	T_g (°C)	T_{dec} (°C)	λ_{max}^{abs} (nm) {Solution}	λ_{max}^{abs} (nm) {Film}	λ_{max}^{PL} (nm) {Solution}[a] [Φ_{PL} (%)]	λ_{max}^{PL} (nm) {Film}[a] [Φ_{PL} (%)]	E_g^{opt} (eV)[b]	HOMO–LUMO (eV) [CV Data]	EL Data	Ref.
361e	1000 (2.18); 5400 (1.9)	125; 130	397	336 {CHCl$_3$}	343	405 {CHCl$_3$}; 381 [37%] {CHCl$_3$}	410 [62%]		−5.89/−2.06	λ_{EL}^{ex}: 417 nm, Φ_{EL}^{ex}: 0.42%	[247,321]
364a	26,000 (2.56)				382		568 [27%]	2.88	−5.68/−2.22	λ_{EL}^{ex}: 574 nm, Φ_{EL}^{ex}: 0.81%, CIE (0.689, 0.314)	[322]
364b	28,000 (2.78)				374, 468		563 [21%]	2.88, 2.30	−5.71/−2.21		[322]
364c	32,000 (3.15)				375, 469		579 [52%]	2.88, 2.32	−5.70/−2.20	λ_{EL}^{ex}: 582 nm, Φ_{EL}^{ex}: 1.0%, CIE (0.698, 0.300)	[322]
364d	16,000 (2.04)				342, 487		595 [16%]	3.18, 2.18	−5.79/−2.14	λ_{EL}^{ex}: 600 nm, Φ_{EL}^{ex}: 0.42%, CIE (0.706, 0.292)	[322]
365a	21,000 (2.23)			391 {CHCl$_3$}			423, 634 [33%]	2.95	−5.78/−3.62	λ_{EL}: 634 nm, Φ_{EL}: 0.56%, CIE (0.37, 0.34)	[323]
365b	60,000 (3.09)			391 {CHCl$_3$}			438, 645 [69%]	2.95	−5.76/−3.62	λ_{EL}: 645 nm, Φ_{EL}: 0.30%, CIE (0.61, 0.35)	[323]
365c	24,000 (2.46)			385, 556 {CHCl$_3$}			423, 647 [77%]	2.95	−5.80/−3.68	λ_{EL}: 657 nm, Φ_{EL}: 3.10%, CIE(0.64, 0.33)	[323]
365d	23,000 (2.32)			390, 559 {CHCl$_3$}			420, 656 [84%]	2.95	−5.76/−3.64	λ_{EL}: 659 nm, Φ_{EL}: 1.14%, CIE (0.67, 0.32)	[323]
365e	18,000 (2.09)			389, 555 {CHCl$_3$}			421, 671 [53%]	2.95, 1.96	−5.74/−3.66	λ_{EL}: 662 nm, Φ_{EL}: 0.35%, CIE (0.68, 0.31)	[323]
365f	8000 (2.61)			353, 563 {CHCl$_3$}			422, 681 [34%]	2.95, 1.95		λ_{EL}: 672 nm, Φ_{EL}^{ex}: 0.22%, CIE (0.69, 0.30)	[323]

(Continued)

TABLE 2.2 (CONTINUED)

Physical Properties of PFs

Polymer	M_n (g/mol) (PDI)	T_g (°C)	T_{dec} (°C)	λ_{max}^{abs} (nm) {Solution}	λ_{max}^{abs} (nm) {Film}	λ_{max}^{PL} (nm) {Solution}[a] [Φ_{PL} (%)]	λ_{max}^{PL} (nm) {Film}[a] [Φ_{PL} (%)]	E_g^{opt} (eV)[b]	HOMO–LUMO (eV) [CV Data]	EL Data	Ref.
366a	23,000 (2.6)				382		635 [11.4%]		−5.72/−2.79	λ_{EL}: 628 nm, Φ_{EL}^{ex}: 0.5%, CIE (0.67, 0.32)	[324]
366b	35,000 (3.1)				382, 535		651 [12.5%]	2.08	−5.65/−2.72	λ_{EL}: 643 nm, Φ_{EL}^{ex}: 0.6%, CIE (0.70, 0.31)	[324]
366c	33,000 (2.7)				384, 536		655 [8.6%]	2.07	−5.61/−3.54	λ_{EL}: 652 nm, Φ_{EL}^{ex}: 0.9%, CIE (0.70, 0.30)	[324]
366d	34,000 (2.6)				382, 538		678 [7.9%]	2.03	−5.55/−3.54	λ_{EL}: 663 nm, Φ_{EL}^{ex}: 1.4%, CIE (0.70, 0.29)	[324]
366e	29,000 (2.4)				384, 542		678 [5.2%]	2.02	−5.50/−3.52	λ_{EL}: 669 nm, Φ_{EL}^{ex}: 0.6%, CIE (0.70, 0.29)	[324]
366f	11,000 (1.7)				388, 551		685 [4%]	2.01	−5.47/−3.46	λ_{EL}: 674 nm, Φ_{EL}^{ex}: 0.5%, CIE (0.70, 0.29)	[324]
367a	16,000 (2.7)				383		629 [88%]	2.91	−6.76/−2.85	λ_{EL}: 613 nm, Φ_{EL}^{ex}: 0.65%, CIE (0.60, 0.38)	[325]
367b	15,000 (2.5)				382, 520		636 [75%]	2.09	−5.68/−3.59	λ_{EL}: 625 nm, Φ_{EL}^{ex}: 1.02%, CIE (0.65, 0.35)	[325]
367c	67,000 (2.9)				381, 520		647 [69%]	2.07	−5.68/−3.61	λ_{EL}: 629 nm, Φ_{EL}^{ex}: 1.10%, CIE (0.66, 0.34)	[325]
367d	18,000 (2.3)				378, 520		642 [70%]	2.06	−5.64/−3.58	λ_{EL}: 634 nm, Φ_{EL}^{ex}: 1.45%, CIE (0.66, 0.34)	[325]

(Continued)

TABLE 2.2 (CONTINUED)

Physical Properties of PFs

Polymer	M_n (g/mol) (PDI)	T_g (°C)	T_{dec} (°C)	λ_{max}^{abs} (nm) {Solution}	λ_{max}^{abs} (nm) {Film}	λ_{max}^{PL} (nm) {Solution}[a] [Φ_{PL} (%)]	λ_{max}^{PL} (nm) {Film}[a] [Φ_{PL} (%)]	E_g^{opt} (eV)[b]	HOMO–LUMO (eV) [CV Data]	EL Data	Ref.
367e	3000 (2.0)				371, 522		670 [50%]	2.06	−5.62/−3.56	λ_{EL}: 672 nm, Φ_{EL}^{ex}: 0.42%, CIE (0.67, 0.33)	[325]
367f	4000 (1.9)				369, 526		678 [44%]	2.03	−5.62/−3.59	λ_{EL}: 671 nm, Φ_{EL}^{ex}: 0.31%, CIE (0.68, 0.32)	[325]
367g	40,000 (2.1)				370, 525		669 [19%]	2.06	−5.60/−3.54	λ_{EL}^{ex}: 669 nm, Φ_{EL}^{ex}: 0.25%, CIE (0.68, 0.32)	[325]
368	2840 (6.45)	167	324	328 [THF]	287	486 [43%] {THF}	490	2.76	−5.1/−2.4	λ_{EL}: ~490 nm, Φ_{EL}: 0.10%	[326]
369a	7650 (1.7)		~350		373	[84%]^d	474	2.86	−5.38	λ_{EL}: 480 nm, CIE (0.16, 0.32)	[252]
369b	13,500 (2.0)		~350		373	[71%]^d	480	2.81	−5.40/−2.59	λ_{EL}: 484, CIE (0.17, 0.37)	[252]
369c	9400 (1.6)		~350		367	[37%]^d	478	2.81	−5.40	λ_{EL}: 480, CIE (0.17, 0.33)	[252]
370	18,500 (1.7)				371		482	2.82	−5.40/−2.58	λ_{EL}: 482	[327]
378	4600 (2.11)	150	396	407, 541 {CHCl$_3$}	412, 563	474, 503, 628	673		−5.90/−4.0	λ_{EL}: 708, CIE (0.55, 0.33) for 6 V	[328]
379a	83,640 (3.75)			382 [THF]	387	421, 443 [THF] [38%]	545	2.95	−5.81/−2.49	λ_{EL}: 530, CIE (0.34, 0.56)	[329]
379b	27,330 (2.55)			380 [THF]	399	421, 445 [THF] [15%]	561	2.48	−5.81/−2.27	λ_{EL}: 540, CIE (0.39, 0.56)	[329]
379c	23.620 (1.70)			374,430 [THF]	412	419, 436, 535 [THF] [12%]	575	2.47	−5.74/−2.30	λ_{EL}: 545, CIE (0.42, 0.55)	[329]

(Continued)

TABLE 2.2 (CONTINUED)
Physical Properties of PFs

Polymer	M_n (g/mol) (PDI)	T_g (°C)	T_{dec} (°C)	λ_{max}^{abs} (nm) {Solution}	λ_{max}^{abs} (nm) {Film}	λ_{max}^{PL} (nm) {Solution}[a] [Φ_{PL} (%)]	λ_{max}^{PL} (nm) {Film}[a] [Φ_{PL} (%)]	E_g^{opt} (eV)[b]	HOMO–LUMO (eV) [CV Data]	EL Data	Ref.
379d	699,600 (1.23)			343, 447 {THF}	426	420, 448, 558 {THF} [8%]	585	2.47	−5.78/−3.10	λ_{EL}: 565, CIE (0.47, 0.52)	[329]
380a	27,760 (1.33)			375 {THF}	372, 485	419, 442 {THF} [66%]	620	2.95	−5.81/−2.24	λ_{EL}: 594, CIE (0.54, 0.43)	[329]
380b	219,200 (2.29)			369, 445 {THF}	379, 489	418, 440, 589 {THF} [30%]	631	2.26	−5.65/−2.57	λ_{EL}: 612, CIE (0.60, 0.39)	[329]
380c	76,210 (1.34)			307, 360, 449 {THF}	380, 494	418, 440, 589 {THF} [12%]	636	2.26	−5.66/−2.61	λ_{EL}: 616, CIE (0.62, 0.38)	[329]
380d	43,680 (1.39)			318, 342, 472 {THF}	364, 475	400, 592, 725 {THF} [3%]	646	2.17	−5.60/−3.22	λ_{EL}: 630, CIE (0.61, 0.38)	[329]
381a	8261 (1.16)			377 {THF}	410	418, 442 {THF} [56%]	439, 465, 500	2.99	−5.72/−2.44	λ_{EL}: 441, CIE (0.22, 0.25)	[329]
381b	47,180 (1.51)			376 {THF}	397	423, 446 {THF} [17%]	420, 440, 465, 498	2.97	−5.73/−2.36	λ_{EL}: 502, CIE (0.22, 0.38)	[329]
381c	6863 (3.60)			368 {THF}	383	420, 447 {THF} [12%]	413, 438, 518	2.97	−5.82/−2.34	λ_{EL}: 556, CIE (0.40, 0.52)	[329]
381d	7694 (2.12)			340 {THF}	364	461 {THF} [9%]	470, 496	2.97	−5.89/−2.83	λ_{EL}: 596, CIE (0.40, 0.46)	[329]

a   Data in parentheses are wavelengths of shoulders and subpeaks.
b E_g stands for the band-gap energy estimated from the onset of the optical absorption.
c   From Marsitzky, D., Murray, J., Scott, J.C., and Carter, K.R., *Chem. Mater.*, 13, 4285, 2001.
d   Quantum yield relative to compound **216**.
e   In some cases, the same group in subsequent publications reported different properties for the given materials, in which case these are marked by italic font.

The first synthesis of poly(9,9-dialkylfluorene) with long-chain solubilizing hexyl groups (**213**) was carried out by Yoshino's group via an oxidative coupling reaction using ferric chloride in chloroform (Scheme 2.22) [330,331]. This resulted in polymers with relatively low molecular weights (M_n up to 5000). The regioregularity of the polymerization process in these conditions was rather poor due to nonregiospecific oxidation reactions resulting in some degree of branching and nonconjugative linkages through positions other than 2 and 7. Some evidences of irregular coupling along the backbone were shown by NMR of soluble low-molecular-weight fractions [332]. The PLED fabricated using this material gave relatively low QE, and in spite of a well-resolved vibronic structure of the PL band, the EL emission showed a very broad band (maximum at 470 nm) [333]. A serious drawback of this synthetic method was also a significant amount of residual metal impurities, which dramatically affected the PLED performance.

The next step toward soluble PF as materials for PLEDs was an application of Yamamoto synthesis to polymerize various 2,9-dihalo-9,9-R^1,R^2-fluorenes. Pei and Yang [334] at UNIAX Corporation first reported a reductive polymerization of 9,9-*bis*(3,6-dioxaheptyl) fluorene in DMF using zinc as a reductant and reactive Ni(0) as a catalyst (generated from NiCl₂ salt), resulting in high-molecular-weight PF **214** (M_n = 94,000 vs. polystyrene standard, PDI = 2.3) (Scheme 2.23). However, a patent application from Dow Chemicals Co., describing polymerization of various 9,9-disubstituted 2,7-dihalofluorenes, Br or Cl as halogen, by the same Yamamoto synthesis in similar conditions was filed almost a year before that in July 1995 [254]. Later on, numerous patents from UNIAX [335,336] and Dow Chemicals [337–340] described the preparation of various PF by Yamamoto synthesis.

Although an Ni-catalyzed reaction allowed improvement of the regiospecificity and minimization of cross-linking and mislinking reactions [332] compared with the FeCl₃ oxidation method, it employs a large amount of metals (Ni, Zn) during the synthesis and the resulting polymer should be carefully purified to remove the metal impurities. In addition, because of the nonpolar hydrophobic nature of poly(9,9-dialkyfluorenes), the polymer chain growth in polar solvents (DMF or *N,N*-dimethylacetamide [DMA]), which are used in Zn/NiCl₂ reductive polymerization, is terminated by polymer precipitation from the

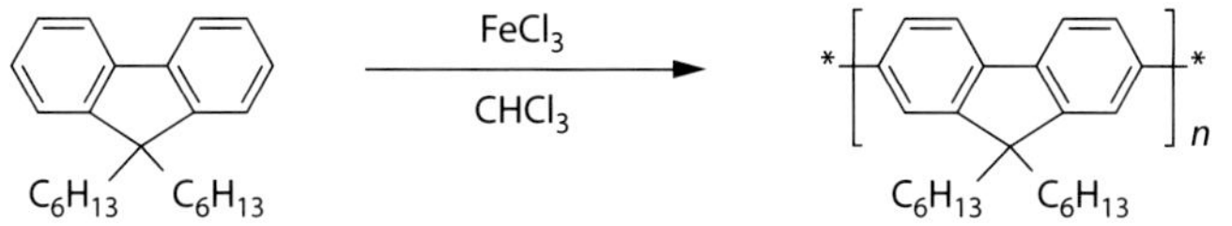

SCHEME 2.22
Oxidative coupling synthesis of poly(9,9-dihexyl)fluorene **213**. (From Fukuda, M., Sawada, K., and Yoshino, K., *Jpn. J. Appl. Phys., Pt. 2 Lett.*, 28, L1433, 1989.)

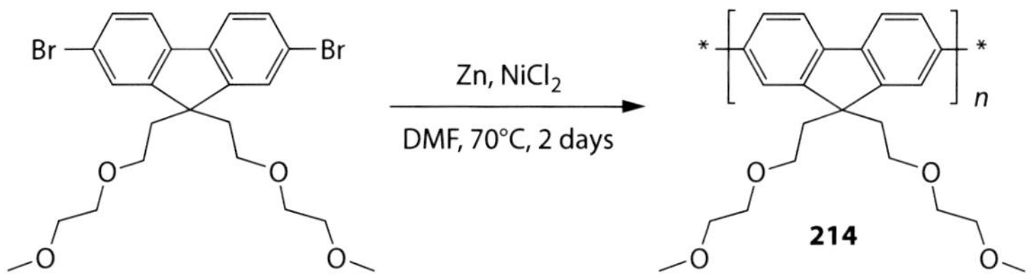

SCHEME 2.23
First Yamamoto synthesis of poly[9,9-*bis*(3,6-dioxaheptyl)fluorene]. (From Pei, Q. and Yang, Y., *J. Am. Chem. Soc.*, 118, 7416, 1996.)

reaction mixture, limiting the molecular weight. Thus, whereas a relatively high molecular weight of $M_n = 94,000$ can be achieved for PF **214**, containing hydrophilic 3,6-dioxaheptyl substituents, in cases of various alkyl substituents, the molecular weights are limited to $M_n \sim 14,000–60,000$ [254,335–337].

Another example of Yamamoto-type polycondensation [341] was demonstrated by Miller and coworkers [342,343], who performed coupling of corresponding 2,7-dibromo-fluorenes using Ni(cod)$_2$/cyclooctadiene/2,2′-bipyridyl in a toluene–DMF solvent mixture (Scheme 2.24). This method allows preparation of PFs with very high molecular weight, M_n up to 250,000 (i.e., up to ~500 fluorene units) [344], and Scherf and List [238] noted that on the laboratory scale, the use of Ni(cod)$_2$ as reductive transition metal-based coupling agent is very convenient.

Suzuki-coupling synthesis of PF, first reported by Leclerc and coworkers [345,346], could minimize the problem of metal impurities by employing catalytic amount of Pd(PPh$_3$)$_4$ (Scheme 2.25), and the use of a phase transfer catalyst gives higher molecular weights ($M_n \sim 50,000$ instead of 15,000) [345–347]. Although the molecular weights of PF achieved by Yamamoto coupling with Ni(cod)$_2$ (up to $M_n \sim 100,000–200,000$) are higher than those obtained by Suzuki coupling (ca. several 10,000 Da), reaching such high molecular weights is controlled not only by the method of the coupling but mainly by careful purification of the monomers and by optimization of the reaction conditions, as well as by the solubility of the polymer in the reaction mixture (determined by substituents on the fluorene nucleus).

Researchers at Dow Chemicals filed a patent, describing the preparation of a wide range of homopolymers and copolymers of a series of dialkylfluorenes by Suzuki- and

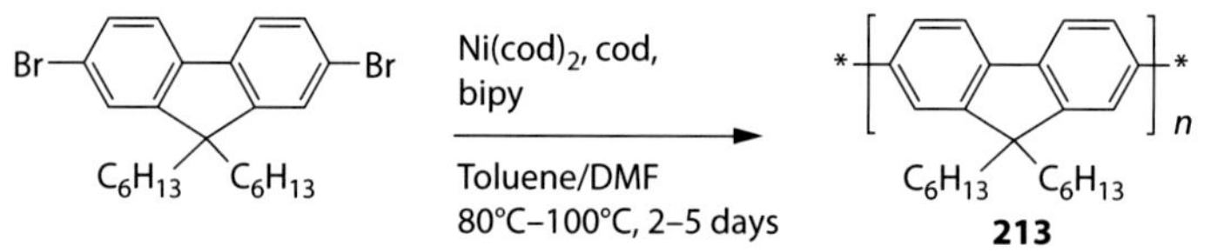

SCHEME 2.24
Synthesis of PF by Yamamoto coupling with Ni(cod)$_2$. (From Klaerner, G. and Miller, R.D., *Macromolecules*, 31, 2007, 1998; Kreyenschmidt, M., Klaerner, G., Fuhrer, T., Ashenhurst, J., Karg, S., Chen, W.D., Lee, V.Y., Scott, J.C., and Miller, R.D., *Macromolecules*, 31, 1099, 1998.)

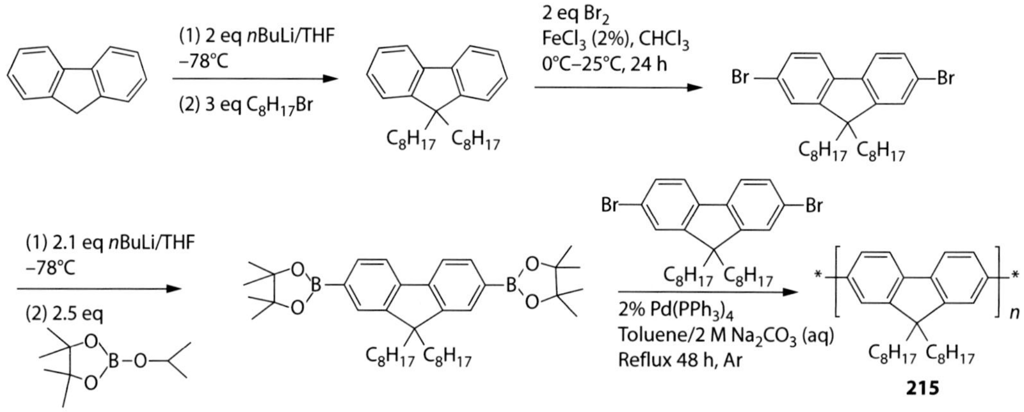

SCHEME 2.25
Synthesis of PF via Suzuki-coupling reaction. (From Ranger, M. and Leclerc, M., *Chem. Commun.*, 1597, 1997.)

Yamamoto-coupling polymerization [338]. Simultaneously and independently, an improved technological procedure for Suzuki coupling polymerization of dialkylfluorenes was also reported by CDT [348].

2.3.2 Physical, Optical, and Electronic Properties

2.3.2.1 Physical Properties

Routine gel permeation chromatography (GPC) (size exclusion chromatography [SEC]) with calibration against polystyrene standard is a common method for the estimation of the molecular weights of PF. The PF homopolymers and copolymers obtained by different synthetic procedures, as will be described below, could substantially differ in molecular weight and polydispersity index, which also depend on the purification procedure. Generally, the M_n ranges from 10,000 to 200,000 with PDI ≈ 1.5–3, using polystyrene as a standard.

In principle, GPC with polystyrene standard overestimate the molecular weight of PFs because of their rigid-rod character [238]. Grell and coworkers [349] determined an overestimation factor of 2.7 for poly(9,9-dioctylfluorene) by comparing the M_n values of coupled GPC and light scattering with those of GPC with polystyrene standard. Dynamic light-scattering experiments on narrow fractions (PDI = 1.22–1.67) of poly (9,9-*bis*(2-ethylhexyl) fluorene-2, 7-diyl), prepared by preparative GPC fractionation, have also displayed reduced absolute M_w values (50%–70%), compared with polystyrene-calibrated SEC results [350]. Nevertheless, use of GPC with the same polystyrene standard throughout most publications on PFs allows comparing more or less adequately the data for different polymers.

Generally, fluorene homopolymers and copolymers show excellent thermal stability: the T_{dec} of many PF exceeds 400°C (according to thermogravimetric analysis under inert atmosphere) [237].

Whereas poly(9,9-dihexylfluorene) (PDHF, **213**) is generally considered as amorphous, PF with longer octyl side chains, PFO **215**, is crystalline material. Many PFs—dioctyl (PFO **215** [349,351,352]) or *bis*(2-ethylhexyl) (**216** [353]) as well as some fluorene copolymers [354]—exhibit liquid crystalline behavior, opening a possibility to fabricate polarized LEDs [237,355,356] (Chart 2.47).

PFO **215** is clearly crystalline with a melting point temperature around 150°C, above which a nematic mesophase exists up to ca. 300°C. The nanoscale crystallinity of PFO **215** was demonstrated by x-ray diffraction (XRD) experiments (Figure 2.8) [357,358]. For the crystalline phase, a periodicity in the plane of the surface of 4.15 Å corresponded to half the fluorene ring repeat distance along the backbone [357,359]. Octyl chains (which are perpendicular to the direction of the PF backbone) of two neighboring polymer backbones are believed to intercalate, allowing a more efficient space filling. This side-chain packing may be responsible for an unusual ability of PFO **215** and related PF to undergo

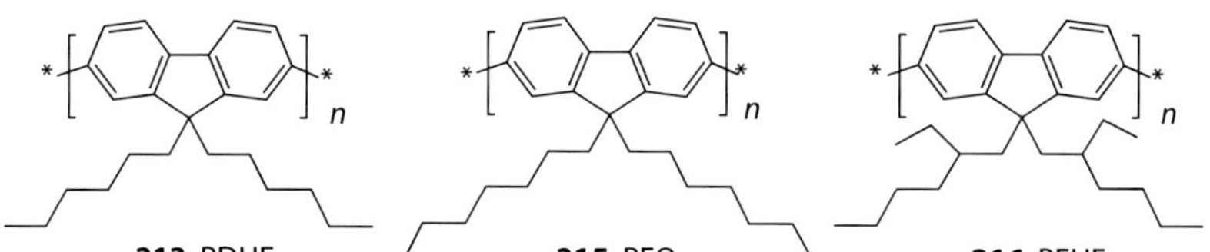

213, PDHF   **215**, PFO   **216**, PEHF

CHART 2.47
Chemical structures of polyfluorenes (PDHF, PFO, and PEHF).

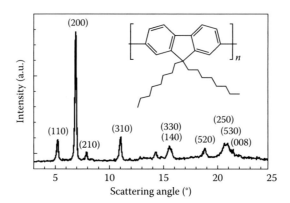

FIGURE 2.8
XRD powder pattern of highly crystalline thick-film PFO **215** specimen after extended heat treatments at elevated temperatures and stepwise cooled to room temperature. (From Chen, S.H., Chou, H.L., Su, A.C., and Chen, S.A., *Macromolecules*, 37, 6833, 2004. With permission.)

thermotropic alignment into monodomain structures [359]. PFO **215** is the most studied PF for its crystallinity and liquid crystallinity having the largest effects and most promising properties for polarized EL, although both liquid crystallinity and crystallinity were also reported for **213** [357,360]. It was also shown that a high degree of alignment in PFO **215** can be achieved by the friction transfer technique with subsequent thermal treatments. Depending on the cooling rate, liquid crystalline or crystalline films are formed [361].

Several reports on PF with optically active side chains imply possible helicity of the fluorene backbone (right- or left-handed 5/2 and 5/1 helical states) [362–365]. Because of the high tendency of PF to form ordered structures, the solvent could have a large effect on the morphology and photophysical properties of PF films and consequently device performance. Grell and coworkers [349] showed dramatic changes in the absorption spectra of PFO **215** resulting from certain treatment protocols (dissolution in moderately poor solvent such as cyclohexane, or exposing spin-coated films to toluene or THF vapor). Banach and coworkers [366] demonstrated that the highest degree of macroscopic order was observed in films that were cast from an aromatic solvent and annealed while still "wet," which was attributed to the plasticizing effect of the solvent during the reorientation process. Ellipsometry was used to study the alignment properties of films cast from six different solvents. Maximum dichroic ratio (D_{max}) was used as a measure of ordering: $D_{max} = 14.6$ (isodurene), 3.7 (tetrachloroethane), 10.2 (*o*-xylene), 9.1 (*p*-xylene), 7.5 (toluene), and 2.8 (chloroform). The ordering did not directly correlate with the boiling point of the solvent (cf. D_{max} for *o*-xylene and tetrachloroethane, which have the same boiling point). These results demonstrate that the choice of the solvent and casting, drying, and annealing techniques are very important for the device performance.

2.3.2.2 Optical and Electronic Properties

The electronic absorption spectra of dilute (typically 5–10 mg/l) solutions of poly(9,9-dialkylfluorenes) show a sharp peak with $\lambda_{max} \sim 385$–390 nm (3.2 eV) of π–π^* electronic transition. Thin solid films (spin coated from 15 to 20 mg/ml solutions) reveal similar absorption with a slightly red-shifted (~10 nm) and relatively broader peak (due to intermolecular interaction) (Figure 2.9) [367].

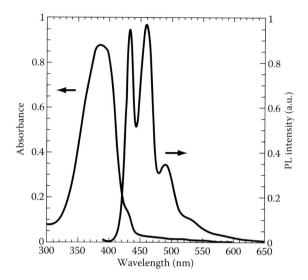

FIGURE 2.9

Typical absorption and emission spectra of polyfluorene in thin films (shown for poly(9,9-dioctylfluorene) **215**). (From Gong, X., Iyer, P.K., Moses, D., Bazan, G.C., Heeger, A.J., and Xiao, S.S., *Adv. Funct. Mater.*, 13, 325, 2003. With permission.)

The PL spectra of the PFs show well-resolved structural features with maxima at 420, 445, and 475 nm assigned to the 0–0, 0–1, and 0–2 intrachain singlet transition, respectively (the 0–0 transition is the most intense) [332]. Owing to the tail emission spectrum of PFs, the thin films emit bright sky-blue light. The QE of the PFs is very high, typically in the range of 40% to 80% and, as shown for PFO **215**, it depends substantially on the morphology of the polymer [368].

The effective conjugation length, estimated by Klaerner and Miller [342] for PDHF **213** from the absorption maxima of a series of monodisperse oligofluorenes (isolated from the mixture of oligomers by high-performance liquid chromatography) is ca. 11.8 fluorene units (Figure 2.10) and similar conjugation length of 9–10 fluorene units was deduced by Koizumi and coworkers [369] from optical band gaps (i.e., red edge of absorbance) in oligo(9,9-dihexylfluorenes) (n = 1–5). Similar estimation of the conjugation length for **216** (λ_{max} = 383 nm) from the linear dependence of $1/\lambda_{max}$ vs. $1/n$ for oligomers with n = 2–7 gave the conjugation length of 14 repeat units [370].

The band gap, determined as the onset of the absorption band in thin films is 2.95 eV (425 nm). Janietz et al. [371] used the onset of the redox waves in CV experiments to estimate the I_P and E_A energies of the dialkyl-PFs (Figure 2.11). The gap between the obtained energy levels (5.8 eV for I_P and 2.12 eV for E_A) $I_P - E_A \sim 3.8$ eV is substantially higher than the optical band gap. Although optical absorption and electrochemistry test two physically different processes (vertical electron excitation and adiabatic ionization) and are not expected to be the same, very good agreement between the methods have been documented, particularly for conjugated polymer systems. On the other hand, the I_P/E_A values derived from the electrochemical measurements in films should be taken with great caution, since they are often obtained under nonthermodynamic conditions (irreversible or quasireversible redox process) and may include a very significant kinetic factor, due to structural rearrangements and counterion diffusion. This is confirmed by comparison of

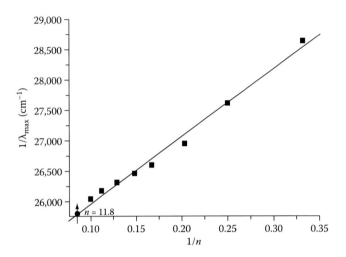

FIGURE 2.10
Plot of $1/\lambda_{max}$ vs. $1/n$ for oligo(9,9-dihexylfluorenes) **213** in THF (n is the number of fluorene units); • is for the polymer with a degree of polymerization of 54, the absorption maximum of which (388 nm) corresponds to effective conjugation length $n = 11.8$. (From Klaerner, G. and Miller, R.D., *Macromolecules*, 31, 2007, 1998. With permission.)

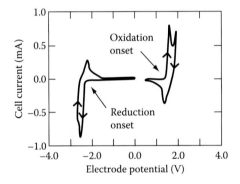

FIGURE 2.11
Cyclic voltammetry of PFO **215** in thin film (potentials vs. Ag/AgCl). (From Janietz, S., Bradley, D.D.C., Grell, M., Giebeler, C., Inbasekaran, M., and Woo, E.P., *Appl. Phys. Lett.*, 73, 2453, 1998. With permission.)

the redox potential of PF in thin films and in solution, where the solution experiments gave significantly lower band gap, similar to the optical band gap [372].

The ultraviolet and x-ray photoelectron spectroscopy (UPS and XPS) measurements are used to calculate the I_p of PFO at -5.6 ± 0.05 eV, and the band gap at 3.1 ± 0.1 eV, which is also much closer to the optical band gap than to the value deduced from the electrochemistry in films [373]. Thus, the HOMO–LUMO levels of PF can be reasonably well matched by work functions of ITO/PEDOT (-5.1 eV) and Ca electrode (ca. -2.9 eV), respectively. However, as shown by Liao and coworkers [374], the energy levels of PF can be shifted significantly in contact with active metals such as Ca, and should be taken with caution. It was also shown that an initial nonohmic PEDOT/PFO **215** contact in ITO/PEDOT/PFO/Al device can be made locally ohmic by electrical conditioning of the device at voltages higher than the EL onset voltage [375,376]. Ohmic injection of holes was also observed from plasma-cleaned ITO electrode or ITO electrode coated with PEDOT (for fluorene–triarylamine copolymer **262**) [377]. The possibility of tuning the HOMO–LUMO energy levels in PF is very important. Besides affecting the emission color, it facilitates the hole–electron injection (and also mobility) by matching the work functions of the electrodes, and thus improves the device performance.

Besides the excellent optical properties and suitable HOMO–LUMO energy levels, the PFs possess great charge-transport properties. Time-of-flight (TOF) measurements of PFO showed nondispersive hole transport with a room temperature mobility of holes of $\mu_p = 4 \times 10^{-4}$ cm^2/(V s) at a field of $E = 5 \times 10^5$ V/cm that is about one order of magnitude higher than that in PPV [378]. The polymer revealed only a weak field dependence of the mobility, from $\mu_p = 3 \times 10^{-4}$ cm^2/(V s) at $E = 4 \times 10^4$ V/cm to $\mu_p = 4.2 \times 10^{-4}$ cm^2/(V s) at $E = 8 \times 10^5$ V/cm.

Because of the great importance of PF as a class of conjugated polymers with excellent optical and electronic properties, several theoretical studies were performed to better understand the electronic structure and the photophysical processes that occur in these materials [379–384].

2.3.3 Problem of Pure Blue Emission in Polyfluorenes

The major problem in the application of PFs in blue PLEDs is color instability. As will be discussed below, the pure blue emission of PFs can be contaminated by the undesired contribution of a green emission band (at ca. 530 nm) upon thermal annealing of the polymer film or during the device operation. The initial hypotheses explained this phenomenon by formation of aggregates [385] or excimers [386–391], which act as energy or charge traps and emit in the long wavelength region. Indeed, the green emission increased during the thermal annealing and was not observed in polymer solution. This hypothesis has borne a tremendous amount of synthetic research activities based on the introduction of bulky substituents in the PF side chain or bulky fragments in the backbone of the polymer to prevent the formation of the excimers. Such modifications resulted, in some cases, in stabilization of blue emission that was, in turn, classified as confirmation of the hypothesis of excimeric nature of green emission in PF.

More recently, it was shown by List et al. [392–395] and later by Gong et al. [367] that the green emission of the PFs is due to fluoren-9-one defects in the polymer chain. This was confirmed by comparison of PL films annealed in an inert atmosphere and in air: a progressive additional band in the green region was observed on annealing in air (Figure 2.12) [367]. A similar increase in the green PL peak was observed on photooxidation of the dialkyl-PF **217** film (Figure 2.13) [392]. Infrared (IR) spectra also indicate an appearance of a fluorenone C=O peak on photooxidation [367,392], and the same peak in the green region appears in the EL spectra during the device operation.

The defects can be either introduced during synthesis or caused by photooxidation during the device preparation and operation. Moreover, the intensity of long wavelength emission is increased in the EL spectrum due to the fact that more electron-deficient fluorenone units can act as electron traps, increasing the probability of electron–hole recombination on the fluorenone defects. It was shown that <1% of the fluorenone defects can almost completely quench the blue fluorescence of the PF, transferring the excitation energy into the long wavelength region [396]. Importantly, as confirmed by theoretical calculations [381], the PF chain planarization and dense intermolecular packing facilitate energy transfer onto the fluorenone defects, which is much less efficient in solutions [384]. This explains the partial success of the strategy of introducing the bulky substituents, which hinders the energy transfer onto fluorenone defect sites.

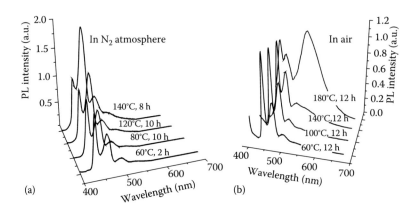

FIGURE 2.12
PL spectra of PFO **215** films after annealing at different temperatures: (a) in a nitrogen atmosphere and (b) in air. (From Gong, X., Iyer, P.K., Moses, D., Bazan, G.C., Heeger, A.J., and Xiao, S.S., *Adv. Funct. Mater.*, 13, 325, 2003. With permission.)

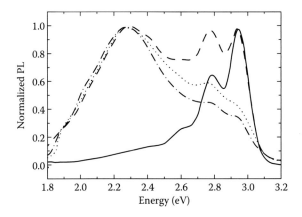

FIGURE 2.13
PL emission spectra of a pristine PF **217** film (solid line), and after photooxidation with a 1000-W xenon lamp under air for 2 min (dashed line), 4 min (dotted line), and 6 min (dashed–dotted line). (From List, E.J.W., Guentner, R., de Freitas, P.S., and Scherf, U., *Adv. Mater.*, 14, 374, 2002. With permission.)

Zhao et al. [397] compared the PL, photoexcitation, GPC, and Fourier transform infrared (FTIR) spectra of silsesquioxane end-capped poly(9,9-dioctylfluorene) in films and solution, before and after annealing. They found that air annealing of PF renders ~80% of the film insoluble, with stronger green emission from the insoluble part of the film. The soluble part of the film, showing negligible green emission in solution, develops the green band after recasting (although with twice lower intensity compared with the insoluble part). Some FTIR features led authors to believe that air annealing also results in β-hydrogen elimination in alkyl chains affording C=C alkene fragments. The latter can be responsible for the observed cross-linking rendering the annealed films insoluble and facilitating the interchain quenching. Another very detailed study of the origin of green emission was performed by Sims and coworkers [398]. They agreed with List et al. on the importance of fluorenone defects for the green emission in PF, which, however, in their opinion, originates from fluorenone-based excimers rather than from monomeric fluorenone π–π* transition. On the basis of comparative studies of matrix-isolated PF chains with nondiluted PF films, the kinetics of PL decay, and the effects of molecular fluorenone additions, they concluded that the green emission band arises from fluorenone-based excimers. Whatever the case, it is clear that solid-state packing plays an important role in the appearance of the green band in PF films.

The polarized emission experiments on partially photooxidized aligned PF films indicate that the emission from the keto defects exhibits a somewhat smaller polarization ratio than the blue emission from the defect-free chains [382]. This observation was explained with the support of quantum mechanical calculations, which showed that the polarization of the fluorenone emission is influenced by local disorder [382].

Although the exact mechanism of the fluorenone formation is not known, it is believed that the monoalkylated fluorene moieties, present as impurities in poly(dialkylfluorenes), are the sites most sensitive to oxidation. The deprotonation of rather acidic C(9)–H protons by residue on Ni(0) catalyst, routinely used in polymerization, or by metal (e.g., calcium) cathode in LED devices form a very reactive anion, which can easily react with oxygen to form peroxides (Scheme 2.26) [392]. The latter are unstable species and can decompose to give the fluorenone moiety. It should also be noted that the interaction of low-work-function metals with films of conjugated polymers in PLED is a more complex phenomenon and the mechanisms of the quenching of PF luminescence by a calcium cathode was studied by Stoessel et al. [399].

Craig and coworkers [400] demonstrated that the purity of the 9,9-dialkylfluorene monomer is of great importance for the stability of the resulting polymer **218**. They performed additional purification of the monomer by treatment with potassium *tert*-butoxide in THF (to deprotonate the monoalkylated by-product) followed by filtration through dried alumina (twice). The material, obtained by polymerization of thus treated monomer, showed significantly less-pronounced green emission, compared with the polymer obtained from unpurified monomer (Figure 2.14). Furthermore, the device operated for 60 h showed no change in the EL spectrum (Chart 2.48).

In this context, a recently proposed procedure for alkylation of 2,7-dibromofluorene by alkylbromides in *t*-BuOK/THF is advantageous over the widely used alkylation in aqueous NaOH or KOH as it could directly result in more pure monomers with good yields (see Section 2.7) [401]. The reaction can be easily monitored by a color change (from yellow to pink) and it is also more convenient than alkylation using BuLi, because of the low cost and ease of handling. In contrast to BuLi, an excess of *t*-BuOK can be used to ensure the complete alkylation.

On the other hand, the stability of the dialkylfluorene moiety to photooxidation and electrooxidation cannot be postulated as well. The n-doping of PF (chemically from Ca cathode, or electrochemically during the device operation) forms radical anion species (see, e.g., theoretical studies of the effect of PF doping with Li atoms [379]), which are not expected to be stable toward oxygen. Thus, not only monomer and polymer purification

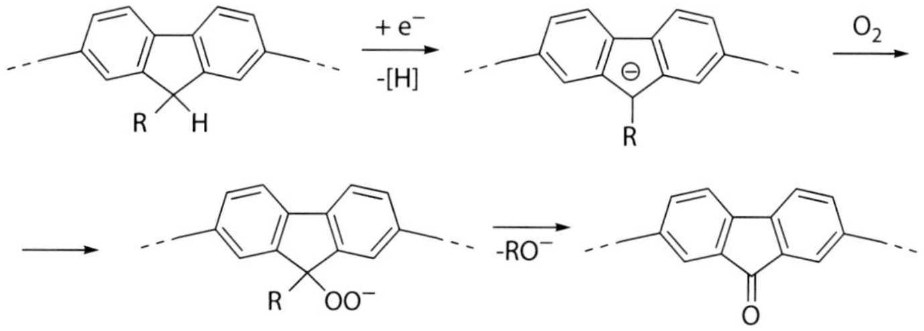

SCHEME 2.26
Mechanism for the generation of keto-defect sites as proposed by List et al. (From List, E.J.W., Guentner, R., de Freitas, P.S., and Scherf, U., *Adv. Mater.*, 14, 374, 2002.)

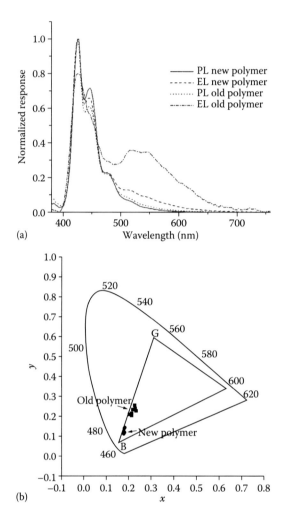

(a)

(b)

FIGURE 2.14
Electro- and photoluminescence (PL: excitation at 265 nm; EL: at 3.6 V [new polymer, i.e., obtained from purified monomer] and 4.0 V [old polymer, i.e., obtained from conventional monomer]) spectra of **218** (a) and CIE (1931) *x–y* color coordinates of old (squares) and new polymers (circles) from 4 to 6 V (b). (From Craig, M.R., de Kok, M.M., Hofstraat, J.W., Schenning, A.P.H.J., and Meijer, E.W., *J. Mater. Chem.*, 13, 2861, 2003. With permission.)

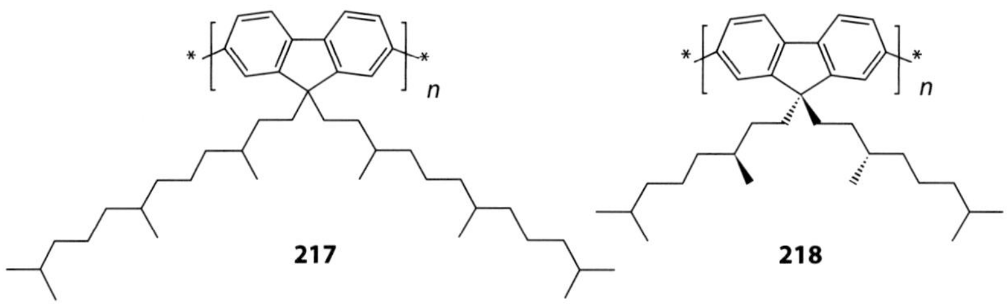

CHART 2.48
Chemical structures of polyfluorenes synthesized with additionally purified monomers.

but also the device structure and operation should be optimized to achieve pure blue emission. In this line, Gong et al. [367] have shown that introducing a buffer layer between the PF and the cathode (Ca) layers block the metal diffusion in the polymer and inhibit fluorenone defect formation, which improves the color stability during the device operation.

The following sections will discuss the modifications of the chemical structure of PF through introduction of side-chain substituents and end-capping units, copolymerization with other conjugated units and polymer blends, aiming to achieve a pure blue LEP. After that, tuning of the emission color via the PF backbone and charge and energy transfer processes will be discussed.

2.3.4 Side-Chain Modifications in Polyfluorenes

Almost all modifications in PF homopolymers consist of variation of substituents at position 9 of the fluorene nucleus. Beaupré and Leclerc [253] reported a new synthetic strategy to polymers **219** and **220** with the aim to modulate the I_P of the PFs (for better injection of holes from the anode in LED) by introducing donor 3,6-dimethoxy substituents into the fluorene moiety (Scheme 2.27). The Φ_{PL} of polymer **219** is relatively low (48%); however, it can be increased to 93% by introducing a dihexylfluorene comonomer unit (copolymer **220**), which is probably due to partial release of the steric hindrances brought by 3,6-substituents.

Many studies on side-chain modifications in PF were initially based on the idea of excimer formation, resulting in the green emission during LED operation or in solid-state PL on annealing PF films. This resulted in several proposed strategies for the design of fluorene side-chain homopolymers, where bulky substituents at position 9 of the fluorene moiety should sterically prevent (hinder) interchain interaction and thus improve the stability of blue emission.

The group of Miller [402] at IBM investigated the intensity of the long wavelength emission in 9,9-dialkyl-PFs with different length alkyl chains. Interestingly, differential scanning calorimetry (DSC) analysis reveals that while dihexyl-PF **213** is amorphous,

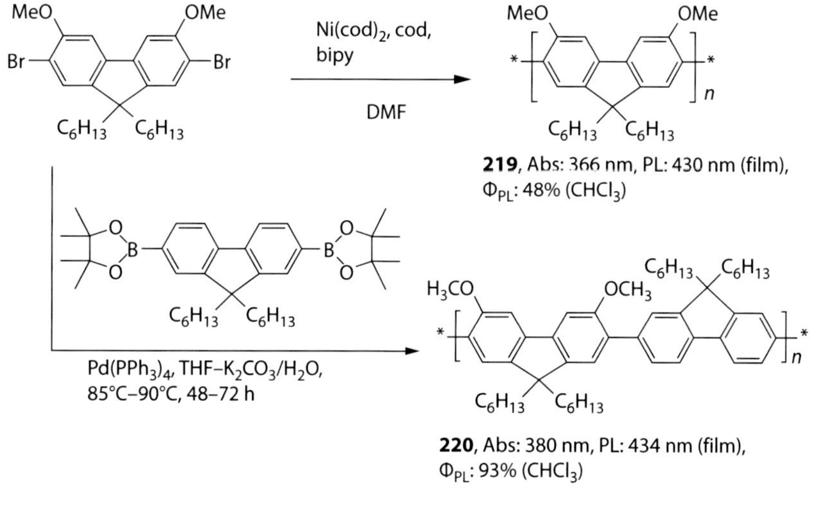
MeO
OMe
Br
Br
C_6H_{13} C_6H_{13}
Ni(cod)$_2$, cod,
bipy
DMF
n
219, Abs: 366 nm, PL: 430 nm (film),
Φ_{PL}: 48% (CHCl$_3$)
Pd(PPh$_3$)$_4$, THF–K$_2$CO$_3$/H$_2$O,
85°C–90°C, 48–72 h
H$_3$CO
OCH$_3$
220, Abs: 380 nm, PL: 434 nm (film),
Φ_{PL}: 93% (CHCl$_3$)

SCHEME 2.27
Synthesis of PFs with 4,5-dimethoxy substituents in the fluorene rings. (From Beaupré, S. and Leclerc, M., *Macromolecules*, 36, 8986, 2003.)

dioctyl-PF **215** and even branched poly(9,9-*bis*(2-ethylhexyl)fluorene) **216** are crystalline. PL spectra of the polymer thin films show the appearance of a green emission band during thermal annealing (appearance of long wavelength emission), regardless of the crystallinity of the films, although the effect was somewhat less pronounced for polymers with larger substituents (**215** and **216**).

In addition to **213–218**, many other alkyl substituents and their derivatives have been introduced at position 9 of the fluorene nucleus in order to create a processable stable blue-emitting PF material, e.g., **221a–h** [335–337,403]. Chiral-substituted PFs **218** and **221g,h** have been synthesized to study their chiroptical properties [404], particularly interesting due to polarized emission in such materials (see Chapter 5 of this book) (Chart 2.49).

Patents of Dow Chemicals first described 9,9-diaryl-substituted PF homopolymers **222** and **223** by Yamamoto polymerization of the corresponding 2,7-dibromo monomers [254], although the methods for monomer preparation were not described. For unsubstituted fluorenone, a convenient method for its conversion into 9,9-(4-hydroxyphenyl)- [405–407] and 9,9-(4-alkoxyphenyl)fluorenes [408] was reported previously, which included condensation of fluorenone with phenol or its ethers in acidic conditions (dry HCl [406,407] or H_2SO_4 [255,405]) in the presence of β-mercaptopropionic or mercaptoacetic acids. Both polymers **222** and **223** showed similar M_n ~ 21,000 with PDI of 1.48 and 1.75, respectively, and spectral data typical for PF (**223**: λ_{abs} = 389 [ε = 50,000 l/(mol cm)]; λ_{PL} = 417, 439, and 473 nm [THF]) (Chart 2.50).

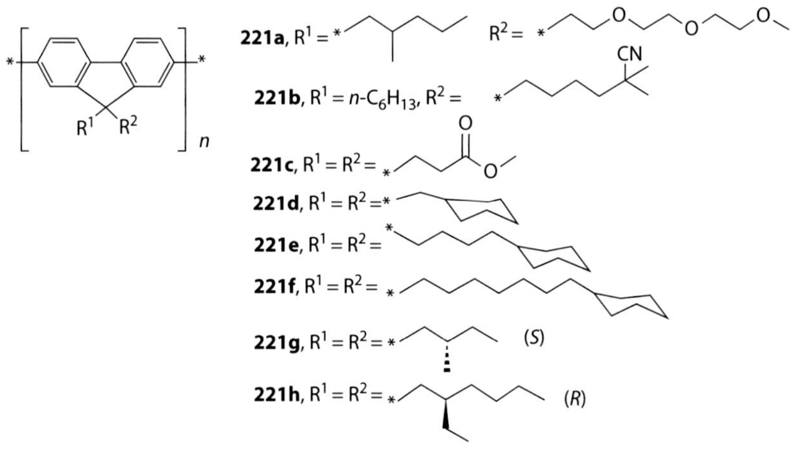

CHART 2.49
Chemical structures of polyfluorenes with alkyl substituents at position 9 of the fluorene.

222, R = Si(Me$_2$)Bu^t
223, R = C(O)CH(C$_2$H$_5$)CH$_2$CH$_2$CH$_2$CH$_3$
Abs: 389 nm (THF), PL: 417, 439, 473 nm (THF)

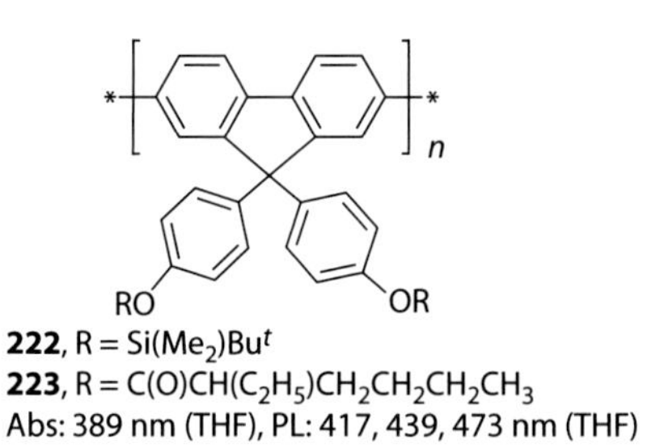

CHART 2.50
Chemical structures of 9,9-diaryl-substituted PF homopolymers.

Polymer **225** obtained by Yamamoto polymerization of monomer **224** (Scheme 2.28) [255] showed bright-blue emission with PL maximum in a film at 430 nm. Its emission was found to be very stable toward thermal annealing. In contrast to poly(9,9-dialkyl)fluorenes [367], this polymer did not show the green component in the PL spectrum after annealing for 2 h at 180°C in air (Figure 2.15), and FTIR spectra also did not show a carbonyl peak after annealing, indicating good resistance of the polymer against oxidation [255]. The device ITO/PEDOT/**225**/Ca/Al showed a turn-on voltage of 3.7 V (~1 cd/m²), with a maximum brightness of 820 cd/m² and an EL efficiency of 0.03 cd/A (CIE: $x = 0.136$, $y = 0.162$). Similar increased stability of pure blue emission toward thermal annealing (compared with dialkyl-PF homopolymers) was also observed in random copolymers of monomer **224** with 2,7-dibromo-9,9-di(2-ethylhexyl)fluorene (absorption and PL spectra of which were very similar to those of homopolymers **225**). However, these showed somewhat lower values of luminous efficiency and maximum brightness, and slightly increased turn-on voltage [409].

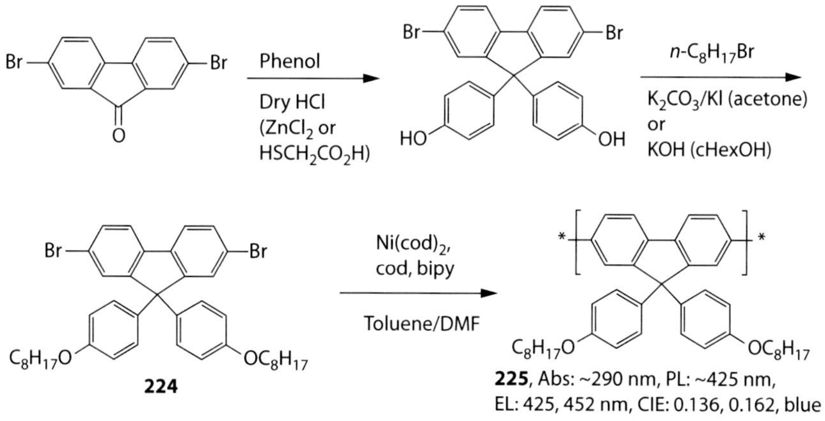

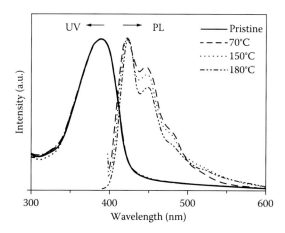

SCHEME 2.28
Synthesis of poly[9,9-di(4-octyloxyphenyl)fluorenes]. (From Lee, J.-H. and Hwang, D.-H., *Chem. Commun.*, 28, 2836, 2003.)

FIGURE 2.15
UV–vis absorption and PL emission spectra of film of **225** after thermal annealing for 2 h at different temperatures in air. (From Lee, J.-H. and Hwang, D.-H., *Chem. Commun.*, 21, 2836, 2003. With permission.)

Müllen's group [256] at Max-Planck Institute was the first to report using dendron sub-stituents to sterically hinder the excimer formation, which delivers pure blue emission in substituted PF material. The dendron-containing monomer was obtained from 2,7-dibro-mofluorene in three elegant steps, including (i) base-catalyzed alkylation of the fluorene methylene group, (ii) Pd(0)-mediated Hagihara–Sonogashira coupling introducing the acetylene functionality, and (iii) Diels–Alder cyclization producing the phenylene dendron (Scheme 2.29). The polymerization was achieved by Yamamoto coupling, and the polymer chain was terminated by arylation with bromobenzene.

The resulting polymer **226** is soluble in toluene, benzene, and chlorinated organic sol-vents and forms high-quality films. The molecular weight was determined by GPC analysis to be M_n = 46,000 g/mol (PDI = 3.6), which corresponds to ~40 repeat units. This poly-mer exhibits the same absorption and emission maxima as alkyl-substituted PFs, which implies that the bulky groups in position 9 do not alter the torsion angle of the conjugated backbone. Thin films of the oligophenylene-substituted PF emit pure blue color without any green emission tail, even after thermal annealing at 100°C for 1 day, confirming the impact of dendron substituents on suppression of the intermolecular aggregation. A PLED device fabricated in the structure of ITO/PEDOT/**226**/Ca/Al gave a pure blue emission, and no green band appeared at driving voltages up to 12 V. However, it was found that the device stability is relatively low due to photooxidation of the benzyl linkage group. This was improved by design of a modified polymer **227** having the phenylene dendron attached directly to the position 9 of PF, for which ITO/PEDOT/**227**/Ca/Al device showed blue emission (CIE: x = 0.189, y = 0.237) with a turn-on voltage of ~6–7 V and a maximum efficiency of 0.06 lm/W at 7.8 V [257] (Chart 2.51).

A different type of a dendron was used by Carter's group at IBM. They demonstrated significant suppression of aggregation by inserting Fréchet-type ether dendrimer

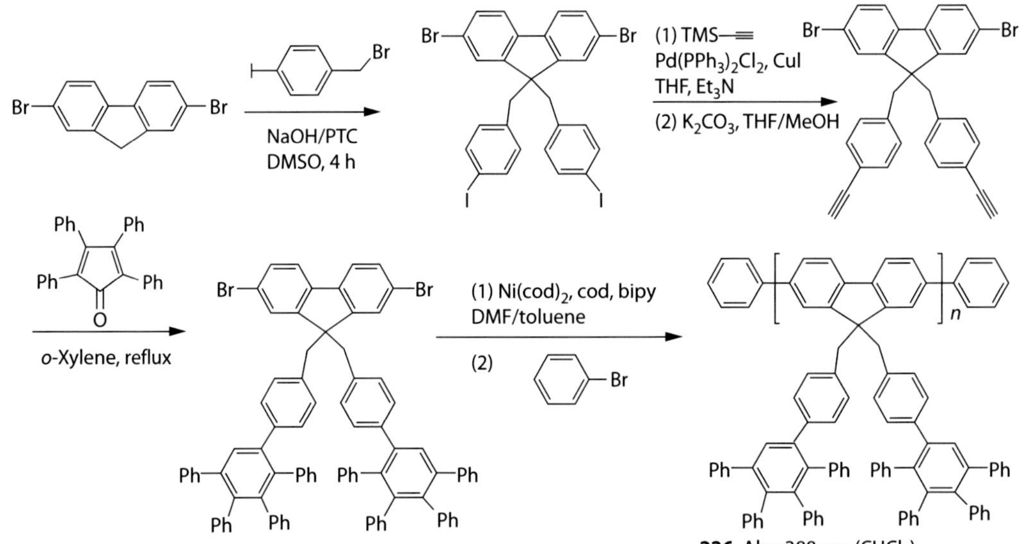

226, Abs: 388 nm (CHCl₃),
PL: 427, 448 nm, EL: 420 nm, blue

SCHEME 2.29
Synthesis of the phenylene–dendron-substituted PFs. (From Setayesh, S., Grimsdale, A.C., Weil, T., Enkelmann, V., Müllen, K., Meghdadi, F., List, E.J.W., and Leising, G., *J. Am. Chem. Soc.*, 123, 946, 2001.)

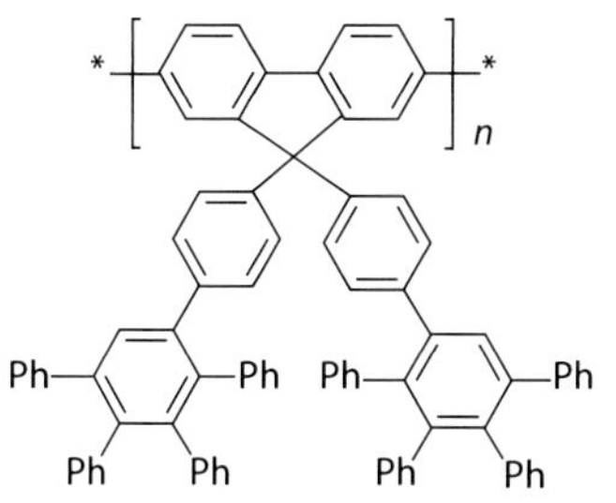

227, Abs: 385 nm (DCM),
PL: 422, 446 nm, CIE (0.189, 0.237), blue

CHART 2.51
Chemical structure of a PF with phenylene dendron substituents.

substituents at position 9 of PFs (Scheme 2.30) [258]. Owing to their large size and flexibility, these substituents can act as encapsulators of the PF chain, hindering the aggregation and increasing the solubility.

These PF dendrimers **228a–c** were synthesized by Yamamoto polymerization, followed by end capping with bromobenzene. A drastically decreased molecular weight (from M_n = 51,400 g/mol for **228a** to 4400 g/mol for **228c**) and glass-transition temperature (from T_g = 94°C for **228a** to 10°C for **228c**) was observed. These polymers show a bright-blue fluorescence in both solution and solid state; the spectral characteristics of **228a,b** are rather similar, whereas some blue shift is observed for **228c**. The polymer **228b** showed the highest QE close to 100% in both solution and in films, whereas **228a,c** displayed much weaker fluorescence (25% and 15%, respectively). Two types of fluorene copolymers, containing dendrimer substituents, have been prepared by Yamamoto (random end-capped polymers **229a–c**) and Suzuki-coupling polymerizations (alternating polymers **230a–c**). Some differences in properties of the two series of polymers were observed. For random polymers **229a–c**, the molecular weights (M_n = 26,200–68,300 g/mol) were generally higher than for alternating polymers **230a–c** (M_n = 3900–16,300 g/mol) and the less steric comonomer with a 2-ethylhexyl side chain allowed a high molecular weight to be achieved with the most bulky dendron **G3**, although the T_g (ranged from 48°C to 78°C) for both series were similar (for given **GX**). Moreover, both absorption (λ_{abs} = 380–388 nm) and PL maxima (λ_{PL} = 414–419 and 436–440 nm) were also rather close. Again the highest PL QE was observed for second-generation dendron, containing polymers **229b** and **230b**. The results indicated that rather bulky dendron substituents (second or third generation) must be used to completely suppress the interchain interaction. Preliminary results show pure blue-light luminescence from PLEDs fabricated in the configuration ITO/PEDOT/**228a–c**/Ca/Al, with driving voltage of 4.5–16 V (depending on the structure). However, nothing was reported on the stability of the devices, which is expected to be low due to possible (photo)oxidation of CH_2 group, as discussed above, and low T_g for the polymers.

A year later, Chou and Shu [259] reported Suzuki coupling of similar alternating fluorene copolymers containing Fréchet-type dendrimers as side chains. In contrast to the above systems (Scheme 2.30), in polymers **231a–c**, the photooxidizable methylene group is separated from the photoactive PF chain by an additional phenylene moiety. The starting

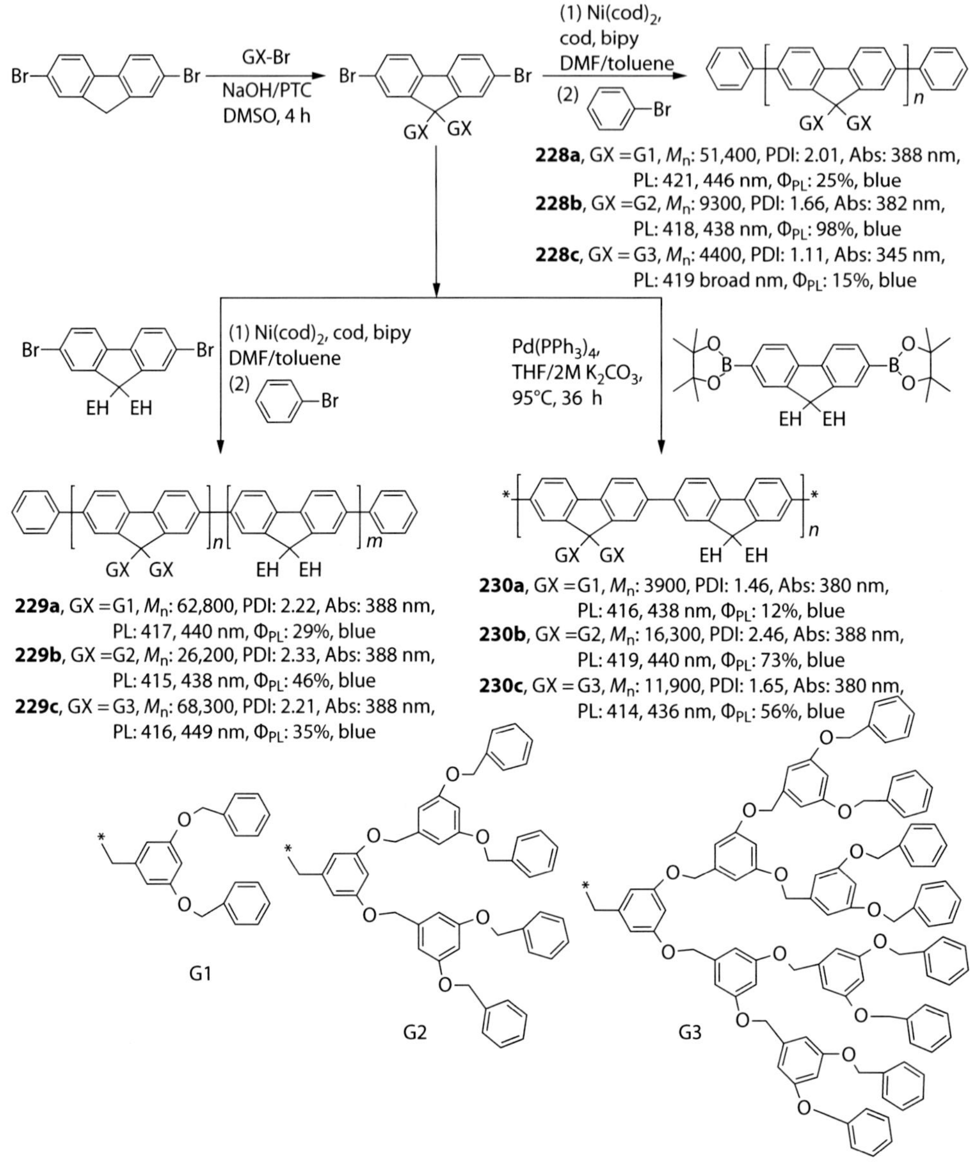

SCHEME 2.30
Synthesis of the Fréchet-type dendrimer-substituted PFs and fluorene random and alternating copolymers. (From Marsitzky, D., Vestberg, R., Blainey, P., Tang, B.T., Hawker, C.J., and Carter, K.R., *J. Am. Chem. Soc.*, 123, 6965, 2001.)

monomer was readily prepared from 2,7-dibromofluorenone by acid-catalyzed arylation with phenol, followed by Williamson alkylation of hydroxy groups (Scheme 2.31).

The highly branched dendronized PF copolymers are readily soluble in common organic solvents. All the three-generation copolymers showed high molecular weights of $M_n \approx 1.5–3.5 \times 10^4$ g/mol with polydispersity in the range of 1.9–2.5 (by GPC analysis). The UV–vis spectra of **231a,b** in solution and in thin films showed the same spectra as dialkyl-PFs; however, a somewhat hypsochromic shift was observed for **231c**. The result is the same as previously

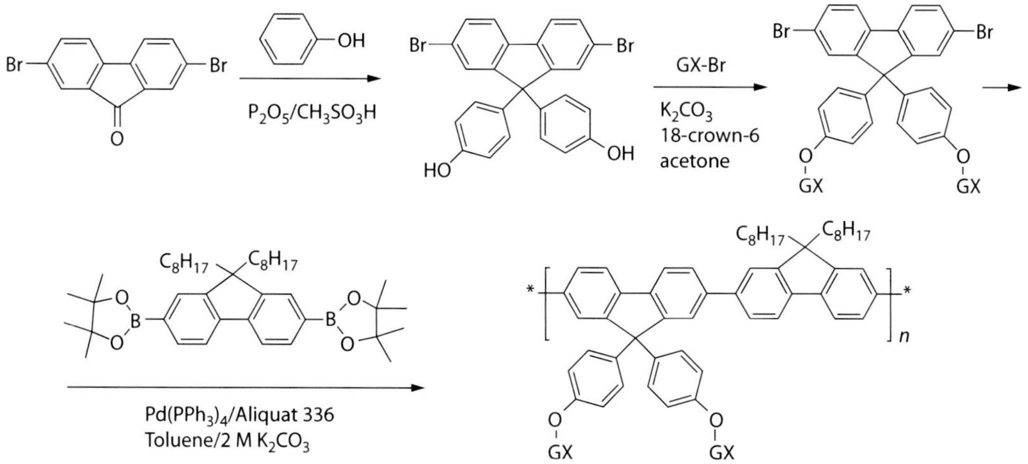

231a, GX = G1, M_n: 35,000, PDI: 1.9, Abs: 394 nm, PL: 424, 448 nm, Φ_{PL}: 26%
231b, GX = G2, M_n: 18,000, PDI: 2.2, Abs: 392 nm, PL: 424, 448 nm, Φ_{PL}: 58%
231c, GX = G3, M_n: 15,000, PDI: 2.5, Abs: 385 nm, PL: 423, 446 nm, Φ_{PL}: 75%

SCHEME 2.31
Dendrimer-PF alternated copolymer synthesized by Suzuki coupling (same functional groups GX as in Scheme 2.30). (From Chou, C.-H. and Shu, C.-F., *Macromolecules*, 35, 9673, 2002.)

observed for the high-generation dendronized PFs with a somewhat different degree of polymerization. The PL spectra of dendritic PFs **231a–c** exhibited the same emission maxima as PFO **215**. The emission spectra of annealed thin films follow the same trend as in homopolymers **228a–c**: the green emission is visible (although suppressed compared with PFO) for **231a**, but completely disappears starting from **231b**. Apart from a strong shielding effect, introduction of dendrimer side groups in PFs may also improve the thermal stability of the material; however, no LED device performance was reported for these materials.

Fu and coworkers [260] reported on dendronized PF with carbazole end groups in the peripheral Fréchet-type dendrons **232a–c**. Polymers emitted blue light with high Φ_{PL} of 86%–96% in solution for all three polymers **232a–c** (**G0–G2**). In the solid state, Φ_{PL} depended on the size of the attached dendron (29%, 55%, and 64% for **G0**, **G1**, and **G2**, respectively) (Chart 2.52).

Tang and coworkers [389] synthesized asymmetrically substituted PFs, bearing a bulky Fréchet-type dendron and a less bulky 3,6-dioxaoctyl group in position 9. The polymers **233–235** showed a pure blue PL emission with rather low green emission band (at 520 nm) for the films annealed at 200°C for 3 h (in vacuum) (Chart 2.53).

Furthermore, the green emission band for the films of asymmetric polymer **235** was much weaker than that of polymer **234** having two straight substituents, and even polymer **233**, having two bulky dendron substituents. The latter fact was attributed to the liquid crystalline properties of relatively well-defined polymer **233**, compared with more disordered asymmetrically substituted **235**. However, it can also be explained by the difficulties in complete dialkylation of fluorene with two bulky substituents (to completely convert the oxidizable C–H bonds). This makes the polymer vulnerable to oxidation and produces fluorenone-originated green emission band. The terminal end capping of polymers **233–235** with 9*H*-fluorene is likely to cause device instability. This is not seen during the annealing *in vacuo* (i.e., in the absence or at least very low concentration of oxygen), but is likely to show up during the operation of the PLED under atmospheric conditions.

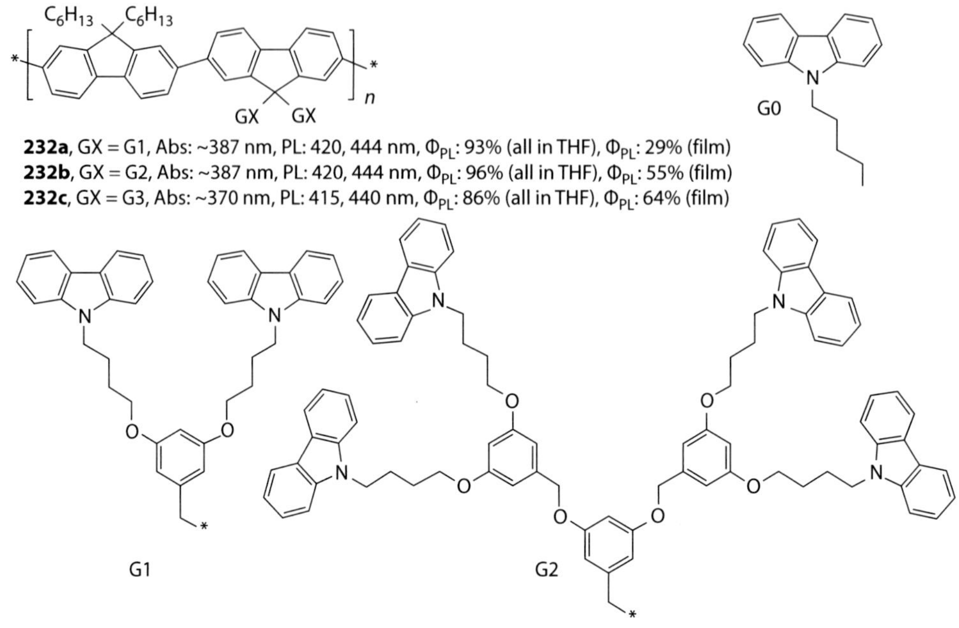
C6H13 C6H13
n
GX GX
G0
232a, GX = G1, Abs: ~387 nm, PL: 420, 444 nm, ΦPL: 93% (all in THF), ΦPL: 29% (film)
232b, GX = G2, Abs: ~387 nm, PL: 420, 444 nm, ΦPL: 96% (all in THF), ΦPL: 55% (film)
232c, GX = G3, Abs: ~370 nm, PL: 415, 440 nm, ΦPL: 86% (all in THF), ΦPL: 64% (film)
N
O
G1
G2

CHART 2.52
Chemical structures of PFs with carbazole end groups in the peripheral Fréchet-type dendrons.

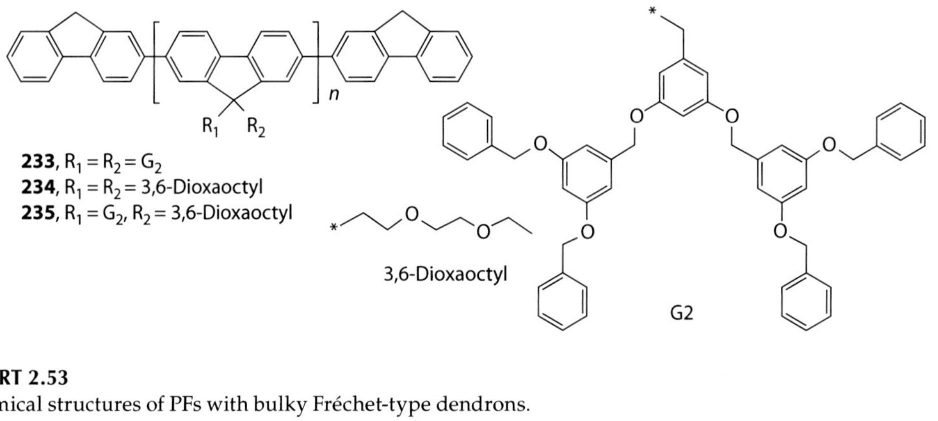
n
R1 R2
233, R1 = R2 = G2
234, R1 = R2 = 3,6-Dioxaoctyl
235, R1 = G2, R2 = 3,6-Dioxaoctyl
O
3,6-Dioxaoctyl
G2

CHART 2.53
Chemical structures of PFs with bulky Fréchet-type dendrons.

Vak and coworkers [261] reported PF derivatives **236** containing spiro-dihydroanthracene units in which the remote C-10 position of the anthracene moiety allows a facile substitution with alkyl groups for improving the solubility (Scheme 2.32). No clear phase transition, including T_g, up to T_{dec} (T_{dec} = 367°C) was observed for this polymer. The polymer showed high spectral stability toward heat treatment, UV irradiation, and current (annealing at 200°C for 15 h did not show any signature of green-band emission usually observed for poly(9,9-dialkylfluorenes)) (Figure 2.16). The ITO/PEDOT/**236**/LiF/Ca/Ag device showed good color coordinates (CIE: x = 0.17, y = 0.12) and a maximum luminance of >1600 cd/m^2.

Another example of spiro-derivatized PF was demonstrated by Wu and coworkers [262], who synthesized soluble spiro-bifluorene-based polymer **237**. This polymer showed stable

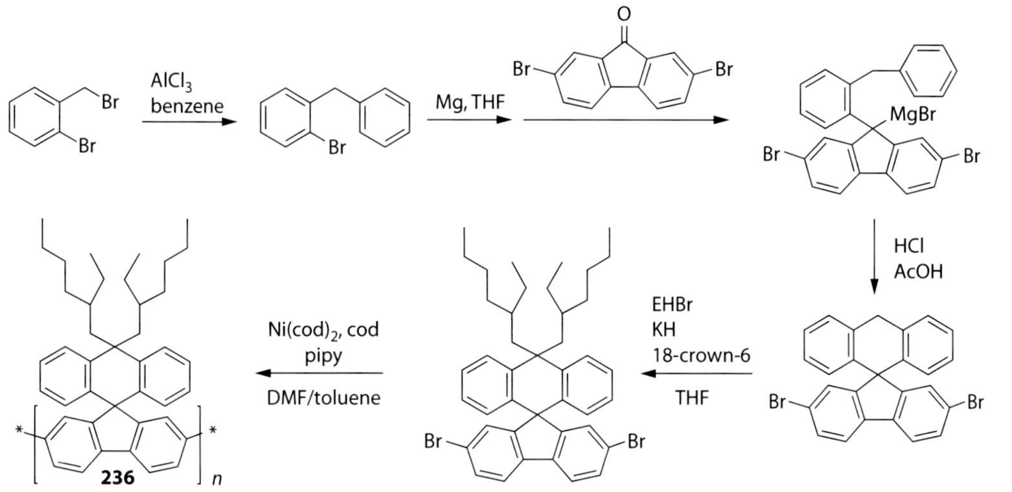

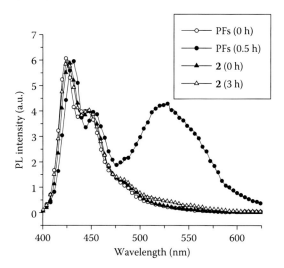

E_g: 2.94 eV, Abs: 399 nm, PL: 425, 446 nm (solution),
PL: 418 nm (film), CIE (0.17, 0.12), deep-blue

SCHEME 2.32
Synthesis of soluble spiro-anthracene–fluorene polymer. (From Vak, D., Chun, C., Lee, C.L., Kim, J.-J., and Kim, D.-Y., *J. Mater. Chem.*, 14, 1342, 2004.)

FIGURE 2.16
PL spectra of polymer **237** and poly(dialkylfluorene) (**2** and PFs in figure, respectively) in films before and after thermal annealing at 200°C for different times. (From Wu, Y., Li, J., Fu, Y., and Bo, Z., *Org. Lett.*, 6, 3485, 2004. With permission.)

bright-blue PL (Φ_{PL} = 91% in toluene), and showed no green emission in the annealed film (although no device preparation has been reported as yet) (Chart 2.54).

Cho and coworkers [243] synthesized cross-linked PF copolymers, containing siloxane bridges **239** and **240**. Ni-mediated copolymerization of 9,9-dihexyl-2,7-dibromofluorene in the presence of 1 or 3 mol% of bridged monomer **238** resulted in copolymers **239** and **240** in which PF backbones are networked by siloxane chains (Scheme 2.33) [243]. Their

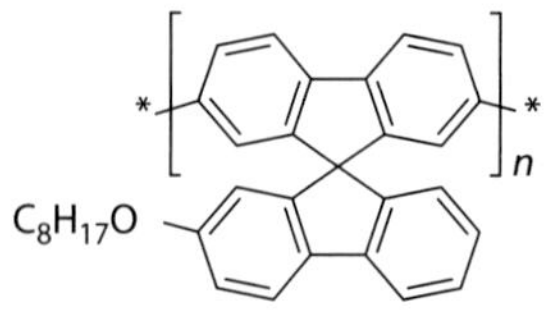

237

Abs: ~391 nm, PL: 416 nm, Φ_{PL}: 91% (toluene),
PL: 427 nm (film), HOMO: −5.8 eV, LUMO: −2.9 eV, blue

CHART 2.54
Chemical structure of a spiro-derivatized PF.

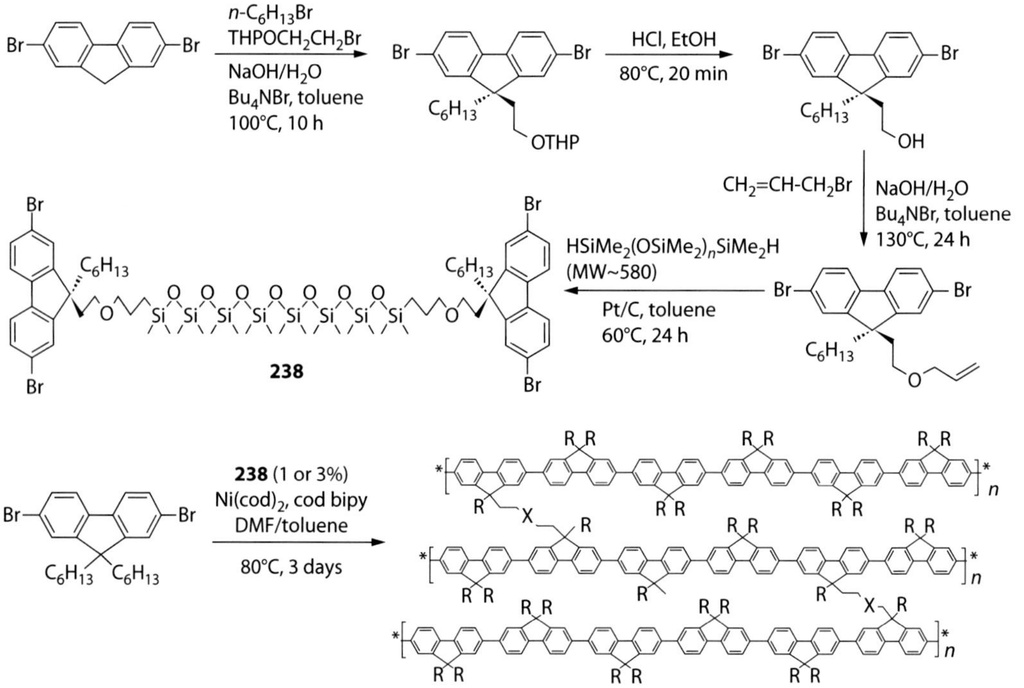

239 (1% of **238**) R = n–C_6H_{13}
240 (3% of **238**) X = –O–$(CH_2)_3$–$(SiMe_2O)_7$–$SiMe_2$–$(CH_2)_3$–O–

SCHEME 2.33
Synthesis of cross-linked polyfluorenes with oligosiloxane bridge. (From Cho, H.-J., Jung, B.-J., Cho, N.S., Lee, J., and Shim, H.-K., *Macromolecules*, 36, 6704, 2003.)

electrochemical and spectral properties were close to that for PDHF **213**. On the other hand, the copolymers showed increased T_g values (106°C and 110°C, respectively), compared with the parent PDHF **213** and almost pure blue emission, which is stable toward annealing (whereas the full width at half-maximum [fwhm] for **213** after annealing of the film at 150°C for 4 h is increased to 85 nm; **239** and **240** showed an fwhm of only 52 and 51 nm, respectively) (Figure 2.17). Authors discussed these results in terms of hindered aggregation and excimer formation in such cross-linked PFs.

The approaches described above only dealt with structural modifications, improving the processability of the polymer and suppressing the undesirable aggregation tendency of PFs.

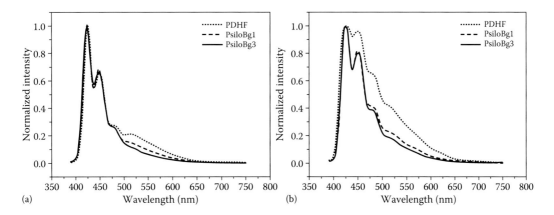

FIGURE 2.17
PL spectra of PDHF **213** (Homo PF), **239** (PsiloBg1), and **240** (PsiloBg3) as spin-coated films (1500 rpm, *p*-xylene) and annealed at 150°C for 30 min (a) and 4 h (b). (From Cho, H.-J., Jung, B.-J., Cho, N.S., Lee, J., and Shim, H.-K., *Macromolecules*, 36, 6704, 2003. With permission.)

For electronic applications, and particularly LEDs, even more important is balancing the charge-transport properties. For the best device performance, the transport of holes and electrons should be equalized without changing the HOMO–LUMO gap of PF (which determines the emission color). To achieve this, Sainova and coworkers [410] blended the light-emitting PF with triarylamine-based hole-transport molecular compounds and demonstrated an increase of the device efficiency (up to 0.87 cd/A) and brightness (up to 800 cd/m^2). Furthermore, a very substantial decrease of the green emission at 520 nm in these blends was observed.

However, a phase separation problem in the above two-component system may affect the device stability. To overcome the problem, Ego and coworkers [263] introduced triphenylamine groups as side chains at the PF backbone. The triphenylamine substituents simultaneously improve the hole-transport properties of PF (facilitating the injection of holes), prevent aggregate formation due to steric shielding effect, and bring high solubility to the material. Commercially available 2,7-dibromofluorenone was arylated with a large excess of triphenylamine in methanesulfonic acid followed by Yamamoto polymerization of the dibromo monomer. The polymer was end capped to give poly[9,9-*bis*(4-diphenylaminophenyl)-fluorene] **241** (Scheme 2.34). The obtained polymer has very good solubility in common organic solvents and has a polymerization degree of ca. 14 units in the chain (M_n = 8860, PDI = 1.76; GPC analysis against a polyphenylene standard). The polymer emits blue light in solution and in the solid state, with a QE of 22% in thin films, a value half of dialkyl-PF (~50% in the solid) [411]; however, the emission spectrum did not change after thermal annealing of the film.

The triarylamine groups in **241** improve the hole-transporting properties and reduce the diode turn-on voltage. A single-layer PLED ITO/PEDOT/**241**/Ba starts to emit blue light (CIE: *x* = 0.184, *y* = 0.159) at as low as 4 V (1 cd/m^2) and shows a maximum current efficiency of 0.67 cd/A (0.36 lm/W). The EL spectrum shows a nearly complete suppression of the green emission. The effect can be rationalized by a charge (hole)-trapping effect of the triarylamine moieties, which compete with electron trapping on the fluorenone defects, and minimize the emission from the defect sites. Introducing an additional HTL (PVK) allows a further improvement of the device performance: a maximum current efficiency of 1.05 cd/A was demonstrated by blue-emitting PLED ITO/PEDOT/PVK/**241**/Ba (CIE: *x* = 0.19, *y* = 0.181), although the maximum brightness demonstrated by both devices was relatively low (up to 200–300 cd/m^2).

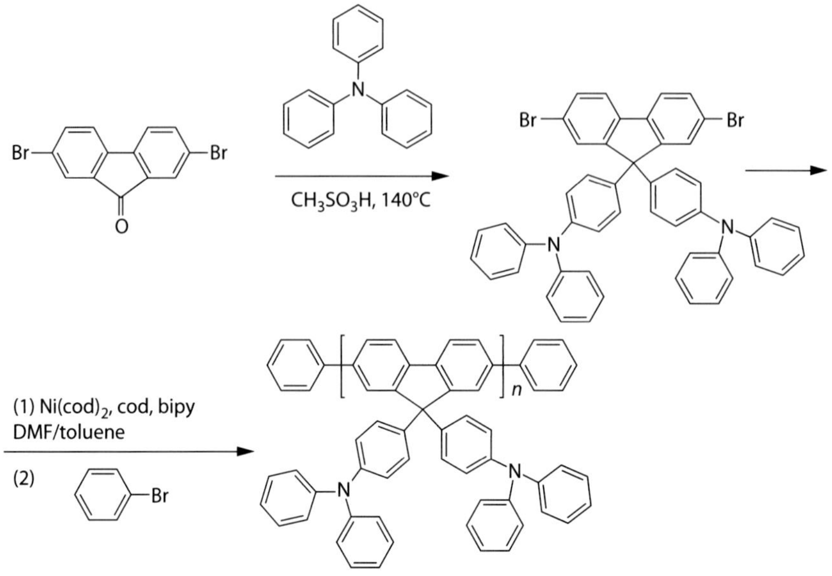

241, PL: 428, 452 nm, Φ_{PL}: 22%, CIE (0.184, 0.159), blue, HOMO: −5.34 eV, LUMO: −2.20 eV

SCHEME 2.34
Synthesis of triphenylamine-substituted PF. (From Ego, C., Grimsdale, A.C., Uckert, F., Yu, G., Srdanov, G., and Müllen, K., *Adv. Mater.*, 14, 809, 2002.)

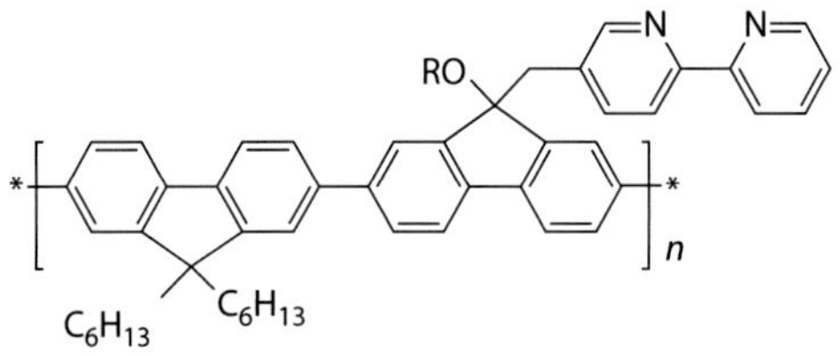

242a, R = H; Abs: 382 nm, PL: 410, 436, 538 nm (THF)
242b, R = n-C_4H_9; Abs: 382 nm, PL: 410, 436 nm (THF)

CHART 2.55
Chemical structures of PFs with 2,2′-bipyridyl side group.

Two PFs, **242a,b**, with 2,2′-bipyridyl side group were reported by Pei et al. [264]. In solution, both polymers show typical PF fluorescence (410 and 436 nm); however, in low-polarity nonhydrogen bonding solvents, the hydroxy polymer **242a** has an additional weak emission at 500–650 nm. In solid films, this low-energy emission becomes dominant, which is not the case for alkylated polymer **242b**. This implies that the hydrogen bonding strengthens the intermolecular interactions in **242a**, changing the emission color from blue (in solutions) to yellow–orange (in films) (Chart 2.55).

2.3.5 End-Capped Polyfluorenes

The very first and obvious reason for introducing the end groups, terminating the polymer chain, was to replace reactive functionalities (halogen, boronic acid, or metal–organic group), which can quench the fluorescence or decrease the stability of the material. Second, varying the feed ratio of the end-capping reagent, one can control the molecular weight of the polymer. Some examples of such utilization of the end-capping approach have been demonstrated in the previous section. The third important reason for end capping is tuning the optoelectronic properties of the polymer by electronically active end cappers. The most important examples of functional end-capped PFs, such as the hole-transport or electron-transport groups, dye moieties, and cross-linkable functionalities are given below.

The initial work introducing end-capping groups in PFs was reported by an IBM group [342]. They also systematically studied the stability of the polymers after end capping [387]. The end-capped polymers **243** and **244** have been synthesized by Yamamoto-coupling polymerization (Scheme 2.35). Comparison of the 9*H*-fluorene and 9,9-dihexylfluorene end-capped polymers unequivocally indicate a higher color stability of the latter, which again confirms the fluorenone-based origin of the green band.

Lee and coworkers [412] reported using an anthracene end capper (which is twisted orthogonally in respect to the neighboring fluorene moiety) in PF **245** to enhance the color stability of PLEDs. However, the results show that the PLED fabricated with the anthracene end-capped polymer still suffer from the color instability (appearance of green emission band), unless the anthracene unit is also introduced as a comonomer (at 15% level) [387] (Chart 2.56).

A high-efficiency PLED with excellent color stability has been fabricated with PFs **246** and **247**, end capped with hole-transporting triarylamine moieties [372]. Rather high HOMO (−5.6 eV for **247** and −5.48 eV **246**) brought by triarylamine moieties facilitates the hole transport, although the authors claim that the current in devices ITO/PEDOT/polymer/Ca is still dominated by electron transport. The best-performing material (maximum luminance of 1600 cd/m^2 at 8.5 V and EL efficiency of 1.1 cd/A) has been achieved at the feed ratio of the end capper (triarylamine) to the fluorene monomer of 4%. Compared with the non-end-capped polymer, the EL efficiency was increased by more than one order of magnitude without disturbing the electronic structure of the backbone. As in

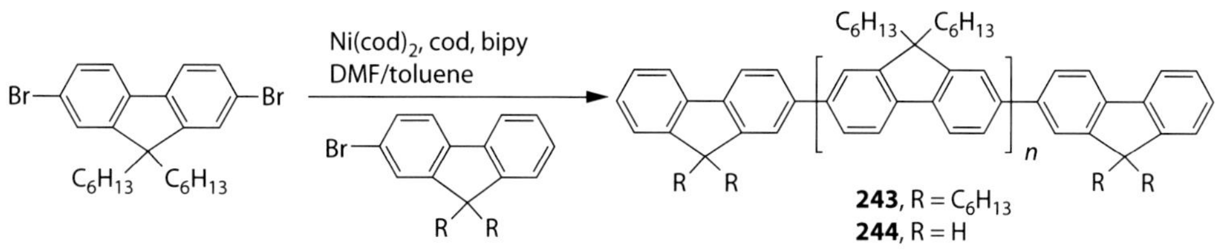

SCHEME 2.35
Synthesis of dihexyl-PF end capped with 2-fluorenyl groups. (From Klaerner, G. and Miller, R.D., *Macromolecules*, 31, 2007, 1998; Lee, J.-I., Klaerner, G., and Miller, R.D., *Chem. Mater.*, 11, 1083, 1999.)

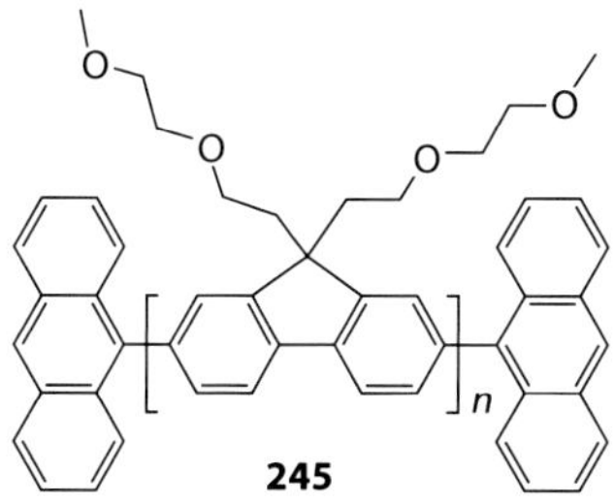

245

CHART 2.56
Chemical structures of PFs end capped with anthracene.

triarylamine-substituted PF **241**, the parasitic green emission in **246** and **247** was completely suppressed, giving a pure blue color (CIE: $x = 0.15$, $y = 0.08$) at voltages above 4 V (Figure 2.18). This effect can be attributed to less effective electron–hole recombination on the green-emitting species (fluorenone defects) due to competing charge trapping on the hole-transporting units. A very high EL efficiency (up to 3 cd/A) can be obtained with polymer **246** in multistructured PLEDs, containing cross-linked HTL made of several triarylamine-based polymer layers with different HOMO levels [413] (Chart 2.57).

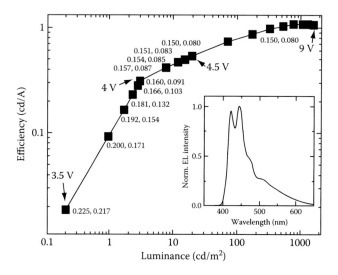

FIGURE 2.18
Efficiency–luminance plot for ITO/PEDOT/**246**/Ca device. Numbers along the curve are the CIE coordinates at the corresponding brightness levels. The inset shows the EL emission spectrum of a diode driven at 3.6 V. (From Miteva, T., Meisel, A., Knoll, W., Nothofer, H.G., Scherf, U., Müller, D.C., Meerholz, K., Yasuda, A., and Neher, D., *Adv. Mater.*, 13, 565, 2001. With permission.)

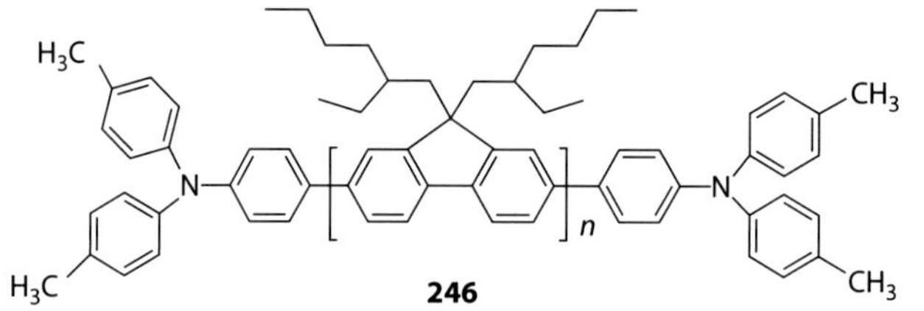

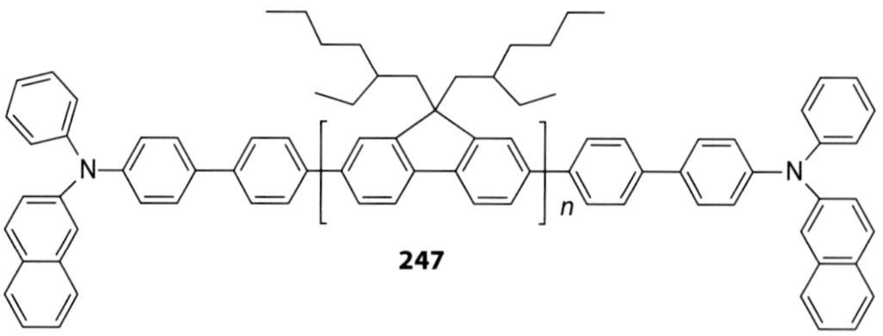

CHART 2.57
Chemical structures of PFs end capped with triarylamine.

Heeger's group [414] reported PF capping with a bulky polyhedral oligomeric silsesquioxane (POSS) group. The hybrid organic–inorganic polymers, containing POSS segments, have several advantages such as increased thermal stability and improved adhesion between the substrate and polymer layer. The molecular weight (M_w) of **248** (GPC vs. polystyrene) was as high as 10^5 g/mol. There is no essential difference in UV–vis and PL between **248** and PDHF **213**; however, the PLED devices fabricated with the former showed increased Φ_{EL}^{ex} and somewhat improved blue color purity (which still suffered from residual green emission). Also, the silsesquioxane end-capped PF show much higher thermal stability (Chart 2.58).

An interesting and important approach of using the end-capping styrene group for cross-linking the PF chains was reported by Miller and coworkers [415,416] (Scheme 2.36). The cross-linking during thermal annealing of the film renders an insoluble material **249**, which can be used as a substrate for spin casting the next layer in multilayer LEDs. In addition, cross-linking drastically increases the T_g temperature, thus suppressing the formation of aggregates and giving pure blue emission.

A styrene-containing polymer is completely soluble in common solvents such as $CHCl_3$, THF, xylene, or chlorobenzene, and can be spin cast as thin films that are easily cross-linked upon heating (as shown by FTIR spectroscopy) to deliver an insoluble material. A relatively high amount of cross-linkable units, however, is required to completely suppress the green emission band. It was achieved by adjusting the fluorene:styrene molar ratio from 85:15 to 67:33. The latter ratio delivered rather short polymer (M_n = 3500 g/mol, ca.

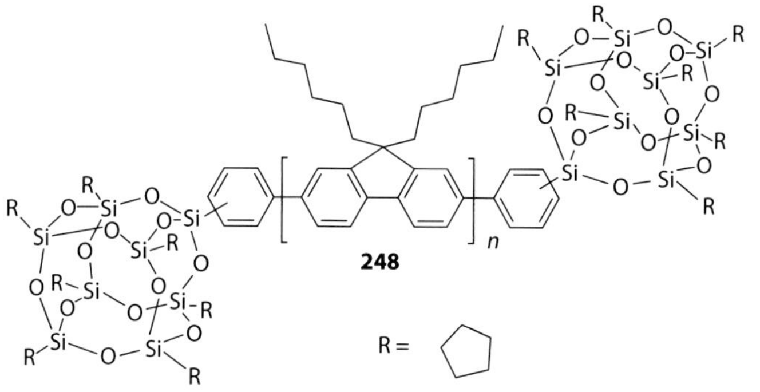

CHART 2.58
Chemical structures of PFs end capped with polyhedral oligomeric silsesquioxane.

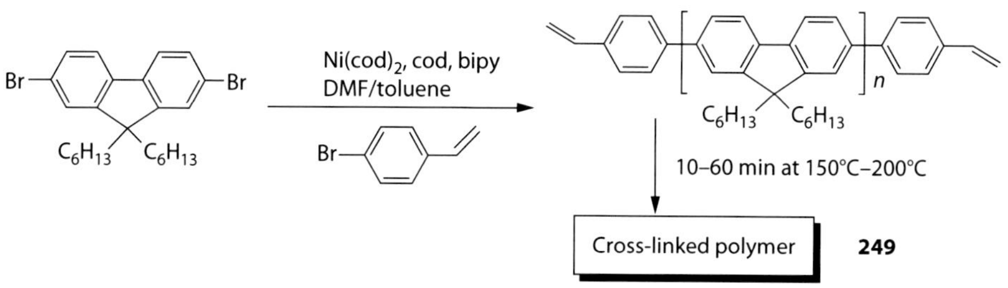

SCHEME 2.36
Synthesis of cross-linked PF **249**. (From Klärner, G., Lee, J.-I., Lee, V.Y., Chan, E., Chen, J.-P., Nelson, A., Markiewicz, D., Siemens, R., Scott, J.C., and Miller, R.D., *Chem. Mater.*, 11, 1800, 1999.)

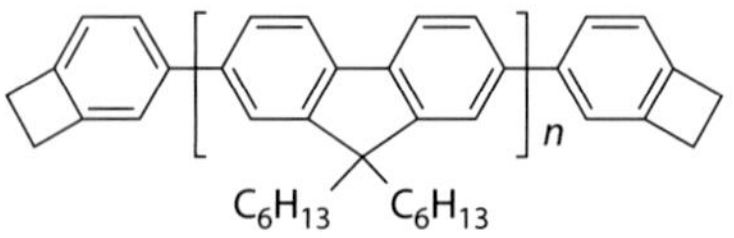

250

CHART 2.59
Chemical structures of PFs end capped with cross-linkable groups.

10 Fl units in the chain), which upon cross-linking (200°C, 10 min), revealed no green band in either the PL or EL spectra. However, incorporating the cross-linkable units in the side chain of fluorene copolymers (in similar ratio) does not lead to complete suppression of green emission, although it does deliver an insoluble cross-linked polymer. The last finding suggests that an improved purity of the end-capped material (due to conversion of reactive chain ends), and not the aggregation suppression due to geometric constraints of cross-linked polymers, is responsible for the pure blue emission.

Some other PFs end capped with cross-linkable groups, such as benzocyclobutene, have been patented by Dow Chemicals [337–339]. For example, the thermal curing of spin-coated polymer **250** gave an insoluble pinhole-free film without alteration of the fluorescent properties (Chart 2.59).

End capping with hole-transporting triarylamine and electron-transporting oxadiazole moieties has been shown to tune the charge injection and transport, without altering the electronic properties of the semiconducting polymer. Comparative studies of polymers **251**, PFO **215**, and **252** showed that the current density increased in the order of **252** < **215** < **251** for "hole-only" devices and in the order of **251** < **215** < **252** for "electron-only" devices (Figure 2.19) [417]. LEDs fabricated with these polymers reach their optimum efficiency whenever hole and electron densities are balanced [78]. Because the hole current measured in the hole-only devices was an order of magnitude larger than the electron current in the electron-only devices (Figure 2.19), improving the electron injection and transport should be more critical for the performance of a real "ambipolar" device. Devices fabricated with polymer **252** instead of PFO **215** demonstrated ca. 20% increase in brightness and luminance efficiency (Chart 2.60).

2.3.6 Blue Light-Emitting Polyfluorene Copolymers

Copolymerization of fluorene with other highly luminescent materials offers a possibility of fine tuning the emitting and charge-transport properties of PF. Thus, Miller and coworkers [265,418] used nickel-mediated copolymerization to synthesize random fluorene–anthracene copolymers **253a–c**, which showed high molecular weights (M_n = 73,000–89,000), good thermal stability ($T_{dec} > 400$°C), high glass-transition temperatures ($T_g = 135$°C–139°C), and a high QE of PL (53%–76% in films) (Scheme 2.37). An increased stability of blue emission (for **253a** even after annealing at 200°C for 3 days) was explained in terms of preventing excimer formation due to incorporation of anthracene units that are orthogonal to the plane of the fluorenes in the backbone, although this can also be an effect of diminishing exciton migration toward fluorenone defects. The device ITO/PANI/**253a**/Ca/Al showed a stable blue EL emission (CIE: $x = 0.17$, $y = 0.25$) with an $\Phi_{EL}^{ex} = 0.17\%$ [415]. Similar fluorene–anthracene copolymers with 3,6-dioxaoctyl substituents on the fluorene moiety and different end-capping groups have also been reported [387,412].

Several groups studied carbazole derivatives as comonomers for blue-emitting PF materials. Carbazole has higher HOMO than fluorene, and many of its oligomers and polymers

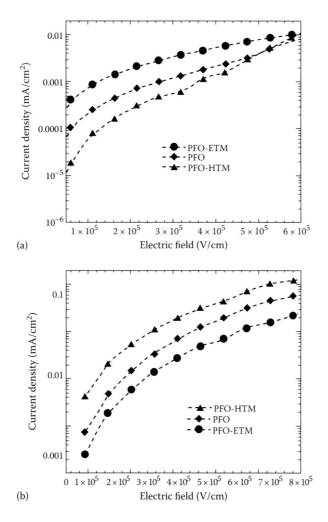

FIGURE 2.19
Current density vs. electric field for PLEDs based on **215** (PFO), **251** (PFO-HTM), and **252** (PFO-ETM). (a) "Electron-only" devices Yb/polymer/Ba/Al; (b) "hole-only" devices ITO/polymer/Au. (From Gong, X., Ma, W., Ostrowski, J.C., Bechgaard, K., Bazan, G.C., Heeger, A.J., Xiao, S., and Moses, D., *Adv. Funct. Mater.*, 14, 393, 2004. With permission.)

(e.g., well-known PVK) are good hole-transporting materials. Therefore, such modification of the fluorene polymers could improve the hole-transporting characteristics of PFs. Kim et al. [419] have reported one of the first alternating fluorene–carbazole polymers **254**, synthesized via the Wittig reaction of carbazole-3,6-dialdehyde with the corresponding fluorene-2,7-*bis*(triphenylmethylenephosphonium) salt. The PLED fabricated by sandwiching a spin-coated copolymer thin film between ITO and Al electrodes is a white emitter with an fwhm of 150 nm. Blending polymer **254** with PVK in a 4:1 ratio narrows the EL emission band (λ_{max} = 460 nm) to give a pure blue color. However, the turn-on voltage for the LED is rather high (13 V) and Φ_{EL}^{ex} is only 0.002% (Chart 2.61).

Later, a random copolymer, **255a,b**, with slightly different substituent pattern was synthesized by Stéphan and Vial [266] by polymerization with Zn/NiCl$_2$ (see Section 2.7). Both

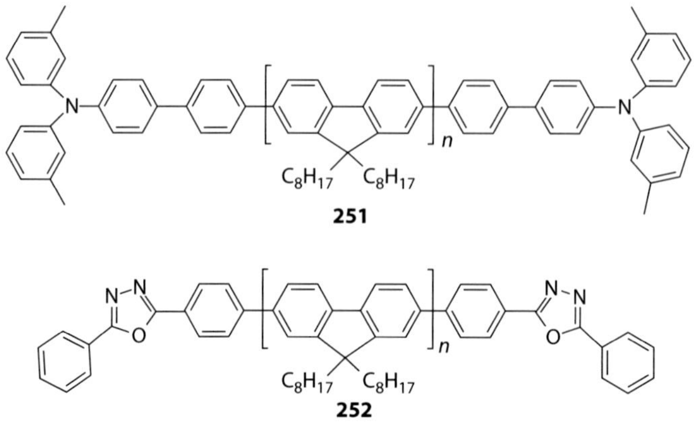

251

252

CHART 2.60
Chemical structures of PFs end capped with hole-transporting and electron-transporting groups.

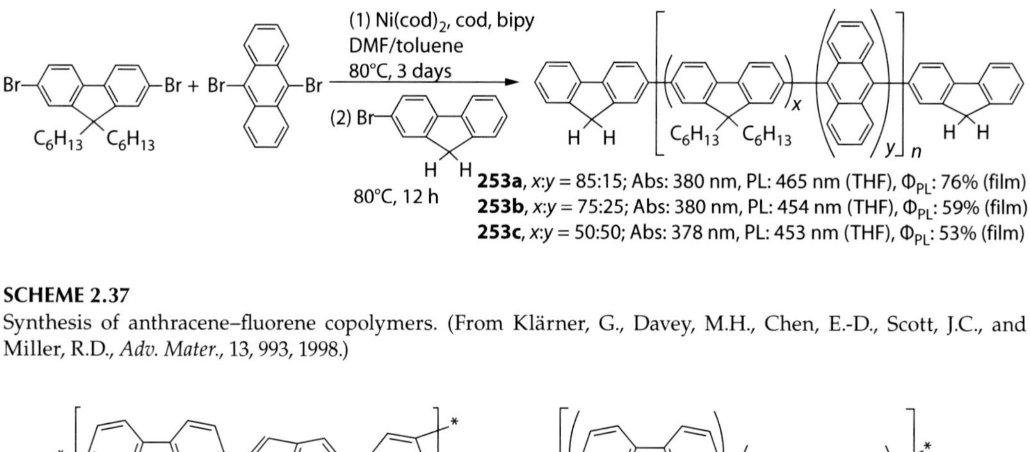

253a, x:y = 85:15; Abs: 380 nm, PL: 465 nm (THF), Φ_{PL}: 76% (film)
253b, x:y = 75:25; Abs: 380 nm, PL: 454 nm (THF), Φ_{PL}: 59% (film)
253c, x:y = 50:50; Abs: 378 nm, PL: 453 nm (THF), Φ_{PL}: 53% (film)

SCHEME 2.37
Synthesis of anthracene–fluorene copolymers. (From Klärner, G., Davey, M.H., Chen, E.-D., Scott, J.C., and Miller, R.D., *Adv. Mater.*, 13, 993, 1998.)

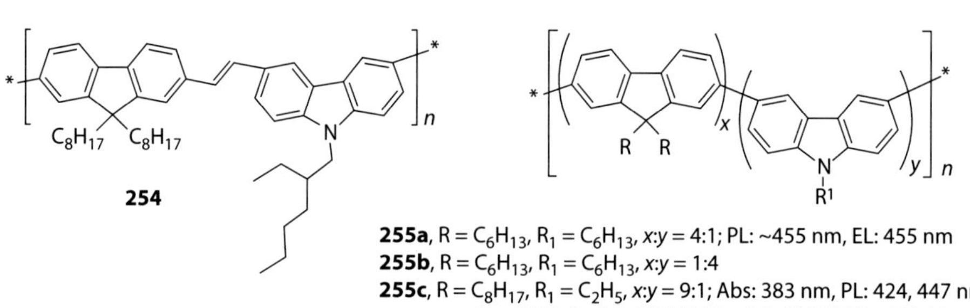

254

255a, R = C_6H_{13}, R_1 = C_6H_{13}, x:y = 4:1; PL: ~455 nm, EL: 455 nm
255b, R = C_6H_{13}, R_1 = C_6H_{13}, x:y = 1:4
255c, R = C_8H_{17}, R_1 = C_2H_5, x:y = 9:1; Abs: 383 nm, PL: 424, 447 nm,
        Φ_{PL}: 71%, E_g: 2.9 eV, HOMO: −5.6 eV, LUMO: −2.7 eV
255d, R = C_8H_{17}, R_1 = C_2H_5, x:y = 7:3; Abs: 370 nm, PL: 424, 447 nm,
        Φ_{PL}: 50%, E_g: 3.0 eV, HOMO: −5.5 eV, LUMO: −2.5 eV

CHART 2.61
Chemical structures of PFs with carbazole derivatives.

copolymers were soluble in organic solvents and had molecular weights of M_n ~ 4000–5000 g/mol (ca. 15 units). The PL spectrum of polymer **255a** is identical to that of the corresponding fluorene homopolymer, and increasing the amount of carbazole units (from 1:4 to 4:1, **255b**) only results in decreasing emission intensity. The authors suggested that only the oligofluorene units are responsible for the emission and assumed the inhomogeneous distribution of comonomers, with relatively long homo-oligomer sections. The low QE of the polymers having high carbazole content can be due to interruption of conjugation, brought about by carbazole-3,6-diyl units (*meta*-substitution effect).

Similar random fluorene–carbazole polymers, **255c,d**, synthesized via Yamamoto polymerization with Ni(cod)$_2$ by Xia and Advincula [249], have much higher molecular weights (M_n = 27,000–46,000). Carbazole units in the backbone of the copolymers do not change the emission of the copolymers in both solution and solid state, which corresponds well to that of PFO homopolymers **215**, but increase the solid state Φ_{PL} (especially for **255c**) and improve the PL color stability toward thermal annealing. This was attributed to a disorder in the polymer chain, brought about by carbazole units (Figure 2.20).

The effect of regularity in fluorene–carbazole copolymers was very recently studied by comparison of random and alternating copolymers **256** and **257** [267]. Both random and alternating copolymers showed progressive blue shifts in absorption with increasing carbazole content. A similar blue shift was observed in PL for copolymers **257**; however, all random copolymers **256** showed almost identical PL spectra, similar to that for PFO **215** (UV: 387 nm, PL: 420, 442 nm [film], E_g^{cv}: 3.25 eV, E_{HOMO}: 5.63 eV, conditions similar to the copolymers in Scheme 2.38). This difference between the alternating and regular copolymers was attributed to the longer fluorene sequences (>5 fluorene units) in random copolymers **256** and migration of the excitons to these segments where the emission occurs. The device ITO/**257c**/F-TBB/Alq3/LiF/Al with 1,3,5-*tris*(4'-fluorobiphenyl-4-yl)benzene (F-TBB) as a hole-blocking layer showed a maximum luminance of 350 cd/m^2 at 27 V and a luminance efficiency of 0.72 cd/A at a practical brightness of 100 cd/m^2, about double that for the device with **215** under the same conditions (160 cd/m^2 and 0.30 cd/A) [267,420]. Pure deep-blue EL with narrow fwhm (39–52 nm) and negligible low-energy emission bands was observed for this device [420].

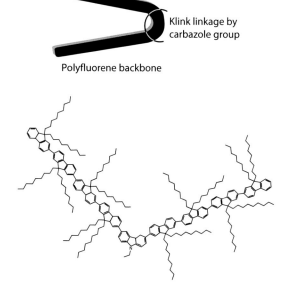

FIGURE 2.20
Schematic illustration of a disorder in polyfluorene introduced by carbazole-3,6-diyl unit in the main chain. (From Xia, C. and Advincula, R.C., *Macromolecules*, 34, 5854, 2001. With permission.)

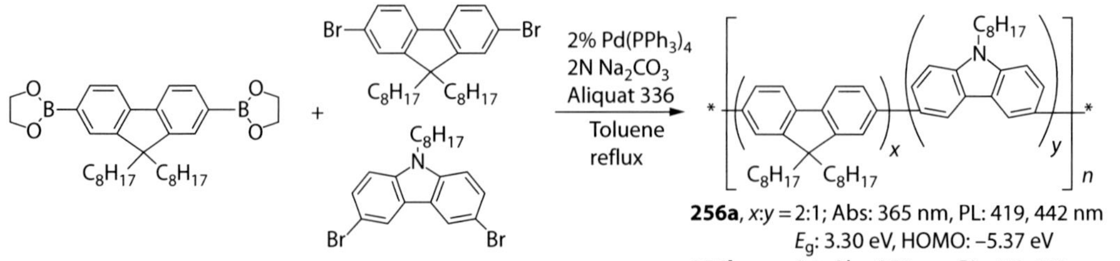

256a, x:y = 2:1; Abs: 365 nm, PL: 419, 442 nm
E_g: 3.30 eV, HOMO: –5.37 eV
256b, x:y = 3:1; Abs: 370 nm, PL: 419, 443 nm
E_g: 3.31 eV, HOMO: –5.43 eV
256c, x:y = 4:1; Abs: 375 nm, PL: 419, 442 nm
E_g: 3.31 eV, HOMO: –5.44 eV

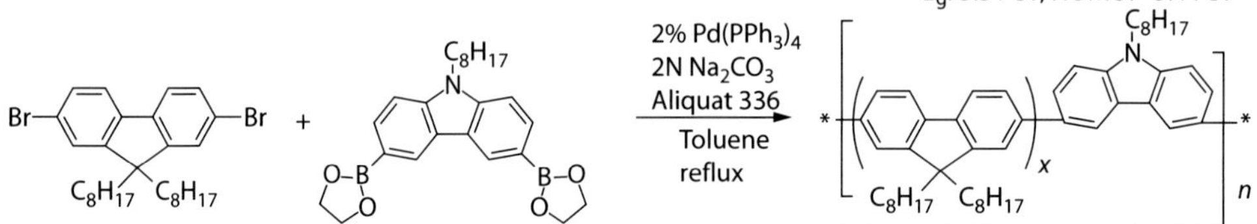

257a, x = 1; Abs: 346 nm, PL: 407 nm,
E_g: 3.31 eV, HOMO: –5.33 eV
257b, x = 2; Abs: 362 nm, PL: 414, 438 nm,
E_g: 3.30 eV, HOMO: –5.34 eV
257c, x = 3; Abs: 368 nm, PL: 416, 440 nm,
E_g: 3.32 eV, HOMO: –5.39 eV
deep-blue
257d, x = 4; Abs: 371 nm, PL: 419, 442 nm,
E_g: 3.21 eV, HOMO: –5.42 eV

SCHEME 2.38
Synthesis of random and alternating fluorene–carbazole copolymers. (From Li, Y., Ding, J., Day, M., Tao, Y., Lu, J., and D'Iorio, M., *Chem. Mater.*, 16, 2165, 2004.)

Leclerc's group [268,421] in Canada first synthesized PF copolymer **258** based on carbazole-2,7-diyl units that, in contrast to the above examples, is a fully conjugated system. Just as in carbazole-3,6-diyl copolymers, polymer **258** showed absorption and PL spectra similar to those of PFO **215**, with almost the same PL QE. However, there was no sign of the green emission band in this copolymer after thermal annealing (Chart 2.62).

Copolymerization of fluorene with triarylamine compounds was shown to increase the hole-transport properties of the polymers. Several copolymers of triarylamine and fluorene (**259–263**) synthesized by Suzuki coupling were reported by Bradley et al. [422,423]. The hole mobility of the copolymers was in the range of 3×10^{-4} cm^2/(V s) to 3×10^{-3} cm^2/(V s), and the I_P was as low as 5.0–5.3 eV (cf. 5.8 eV for PFO **215**), which is well matched with the work function of the ITO/PEDOT electrode. Unfortunately, no PL or EL properties have

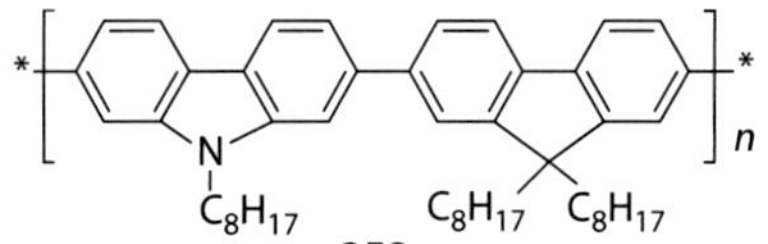

258

Abs: 381 nm, PL: 417, 440 nm, Φ_{PL}: 78% ($CHCl_3$)
Abs: 383 nm, PL: 426, 448 nm (film)

CHART 2.62
Chemical structures of fluorene-carbazole-3,6-diyl copolymers.

been reported in the paper, although PLED devices based on blends of these copolymers with other PF have been patented by Dow Chemicals [424] (Chart 2.63).

The Dow Chemicals group and coworkers [338,425] synthesized similar triarylamine–fluorene copolymers **264** and **265**, possessing carboxylic acid substituents, via hydrolysis of the corresponding ethyl ester polymers, prepared by Suzuki polymerization. Owing to the very polar substituents, the copolymers **264** and **265** are only soluble in polar solvents such as DMF but not in aromatic hydrocarbons as toluene or xylene, which allowed simple fabrication of multilayer PLEDs by solution processes (Chart 2.64).

Fang and Yamamoto [269] reported on postpolymerization functionalization of triarylamine–fluorene copolymer **266**, resulting in copolymers **267a,b** with stilbene pendant groups. Whereas the solid-state absorption and PL maxima of both polymers are essentially the same, PL in solution is strongly influenced by solvent (from 433 nm in toluene to 466 nm in *N*-methylpyrrolidone). Copolymer **267a** showed Φ_{PL} in the solid state of 51%, comparable to that of poly(9,9-dialkylfluorenes) (Chart 2.65).

Molecular triarylamine-based hole-transporting materials are usually synthesized by Ulmann-coupling or Pd-catalyzed amination, although polymerization using these reactions is difficult. Jung and coworkers [270] reported successful Pd-catalyzed copolymerization of dibromofluorene derivatives with anilines and obtained thermally stable, reasonably high molecular-weight (M_n = 12,300–14,000, PDI = 2.4–3.0) copolymers **268a–c** (Scheme 2.39). The HOMO levels of these copolymers (ca. −5.1 eV) matched well with the ITO anode. The LEDs consisting of these polymers as buffer layer demonstrated a lower turn-on voltage, enhanced efficiency, and higher maximum luminance owing to improved hole injection. For comparison, the devices consisting of ITO/**268a**/TPD/Alq$_3$/LiF/Al and ITO/TPD/Alq$_3$/LiF/Al showed, respectively, a turn-on voltage of 2.2 and 3.6 V, and a

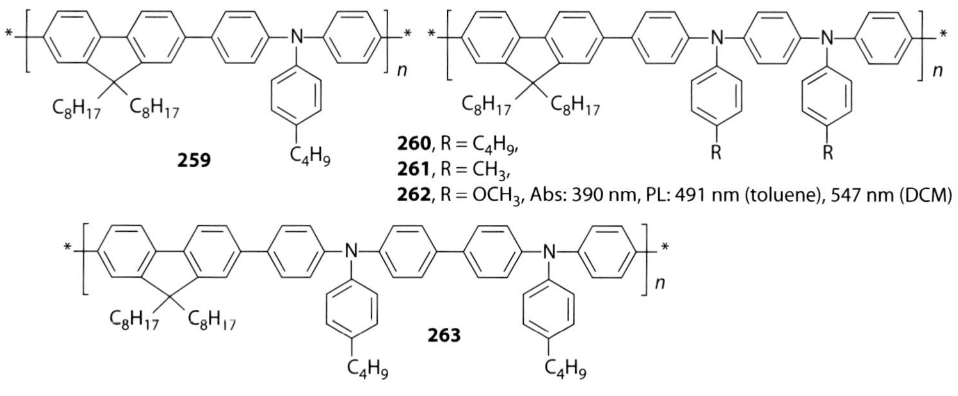

CHART 2.63
Chemical structures of fluorene–triarylamine copolymers.

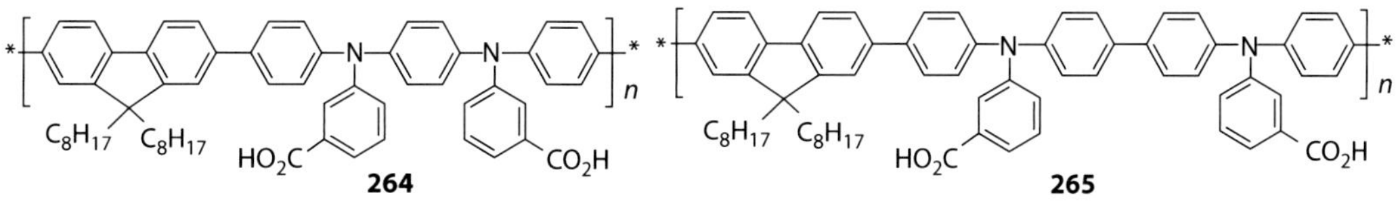

CHART 2.64
Chemical structures of fluorene–triarylamine copolymers with carboxylic acid substituents.

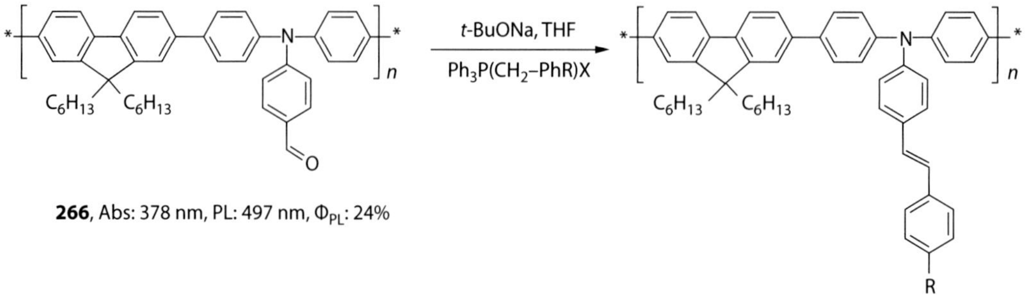

266, Abs: 378 nm, PL: 497 nm, Φ_{PL}: 24%

267a, R = H; Abs: 390 nm, PL: 443, 465 nm, Φ_{PL}: 51%
267b, R = OCH$_3$; Abs: 390 nm, PL: 443, 465 nm, Φ_{PL}: 9%

CHART 2.65
Chemical structures of fluorene–triarylamine copolymers with stilbene pendant groups.

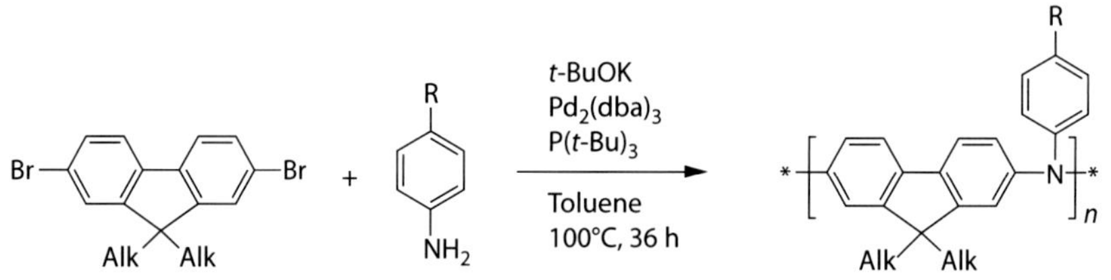

268a, Alk = n-C$_6$H$_{13}$, R = H;
Abs: 406 nm, PL: 429 nm, E_g = 2.88 eV, I_p = 5.18 eV
268b, Alk = n-C$_6$H$_{13}$, R = CH$_3$;
Abs: 411 nm, PL: 433 nm, E_g = 2.85 eV, I_p = 5.08 eV
268c, Alk = n-C$_8$H$_{17}$, R = CH$_3$;
Abs: 411 nm, PL: 433 nm, E_g = 2.85 eV, I_p = 5.07 eV

SCHEME 2.39
Synthesis of fluorene–arylamine electron-rich copolymers. (From Jung, B.-J., Lee, J.-I., Chu, H.Y., Do, L.-M., and Shim, H.-K., *Macromolecules*, 35, 2282, 2002.)

maximum luminance (at the highest current density) of 12,370 cd/m^2 (at 472 mA/cm^2) and 5790 cd/m^2 (at 233 mA/cm^2).

A series of random fluorene–pyridine copolymers (**269a–f**) have been prepared by Suzuki coupling of fluorene monomers with small amounts of 3,5-dibromopyridine (5, 10, 20, 30, 40, 50 mol%) [246]. The *meta*-linkage of the pyridine units is expected to interrupt conjugation; however, no regular spectral dependence was observed for different pyridine loadings. The PL spectra for all copolymers are very close, except for copolymer **269f** (50:50 ratio), for which the PL and EL spectra are shifted by ~10–20 nm, compared with other copolymers. This could be a result of longer fluorene sequences in the random copolymers as discussed above for the fluorene–carbazole and fluorene–phenothiazine copolymers. All materials (except **269f**) showed narrow, pure blue EL emission, and the devices with the configuration ITO/PEDOT/**269a–e**/Ba/Al had a turn-on voltage of 5–6 V as well as moderately high $\Phi_{EL}^{ex} = 0.4\% - 0.5\%$ (Chart 2.66).

The tuning of electron injection and transport in PF has been undertaken by Shu's group [271], who introduced electron-deficient oxadiazole units as pendant groups in fluorene copolymer **270**. The introduction of oxadiazole units into the PF can potentially improve the electron-transport properties of the polymer, while their bulkiness can help suppress aggregation effects (Chart 2.67).

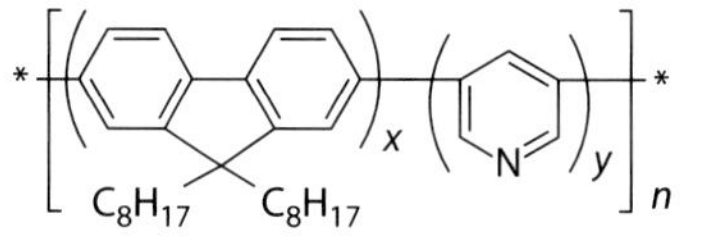

269a, *x:y* = 95:5; Abs: 380 nm, PL: 422 nm, Φ_{PL}: 30%, EL: 446 nm, Φ_{EL}^{ex}: 0.16%

269b, *x:y* = 90:10; Abs: 383 nm, PL: 422 nm, Φ_{PL}: 28%, EL: 442 nm, Φ_{EL}^{ex}: 0.22%

269c, *x:y* = 80:20; Abs: 380 nm, PL: 422 nm, Φ_{PL}: 51%, EL: 442 nm, Φ_{EL}^{ex}: 0.40%

269d, *x:y* = 70:30; Abs: 370 nm, PL: 420 nm, Φ_{PL}: 60%, EL: 438 nm, Φ_{EL}^{ex}: 0.40%

269e, *x:y* = 60:40; Abs: 380 nm, PL: 419 nm, Φ_{PL}: 49%, EL: 440 nm, Φ_{EL}^{ex}: 0.45%

269f, *x:y* = 50:50; Abs: 350 nm, PL: 413 nm, Φ_{PL}: 15%, EL: 424 nm, Φ_{EL}^{ex}: 0.02%

CHART 2.66
Chemical structures of random fluorene–pyridine copolymers.

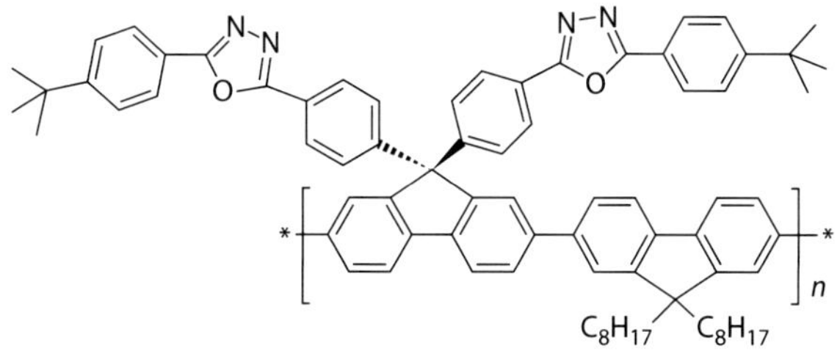

270, Abs: 298 nm, PL: 427 nm, Φ_{PL}: 43%, El: 428 nm,
blue, HOMO: –5.76 eV, LUMO: –2.47 eV

CHART 2.67
Chemical structures of fluorene copolymers with oxadiazole as pendant groups.

Polymer **270** was readily soluble in common organic solvents such as THF, chloroform, chlorobenzene, and xylene. GPC analysis (vs. polystyrene standard) gave a molecular weight, M_n, of 13,000 g/mol, with a PDI of 2.1. The polymer possessed excellent thermal stability with a very high T_g of 213°C. The electrochemically determined HOMO–LUMO levels of **270** were –5.76 and –2.47 eV, respectively, which are similar to those of PFO **215** (–5.8 and –2.6 eV), with only slightly improved electron affinity. The absorption spectrum of **270** in THF solution showed two major peaks at 297 and 390 nm. The former peak is attributed to the aromatic oxadiazole group and the latter to the π–π* transition of the polymer backbone. There is essentially no influence of the attached oxadiazole group on the conjugated backbone: the PL spectrum displays a blue emission with two sharp peaks at 419 and 444 nm, and a small shoulder at 469 nm, very similar to that of PFO **215**. The PL QE in solution was very high (124% vs. 9,10-diphenylanthracene as a standard; for the latter, an absolute fluorescence efficiency of 90% ± 4% was determined [426]), which dropped in thin film to a value of 43%, comparable to PFO (55%). Under thermal annealing of thin films at 150°C for 20 h, the PL spectrum of **270** remains almost intact with only a negligible long wavelength tail, resulting in very pure blue emission in PLED devices (fabricated as ITO/PEDOT/**270**/Ca/Ag). The devices showed a low turn-on voltage of 5.3 V and a luminance of 2770 cd/m^2 at 10.8 V with a current density of 1.12 A/cm^2. The maximum Φ_{EL}^{ex} was 0.52% at 537 cd/m^2 and 7.4 V. The device efficiency is much higher than that of a similar PFO-based device (maximum brightness of 600 cd/m^2 and Φ_{EL}^{ex} = 0.2%).

A highly efficient blue-emitting PF copolymer (**271**) incorporating both triphenylamine and oxadiazole pendant groups was reported [272]. This statistical copolymer was designed to bring together good hole-transporting (due to triphenylamine groups) and electron-transporting properties (due to oxadiazole groups). The copolymer showed high solubility in organic solvents, good stability (T_{dec} = 440°C, T_g = 166°C), and high PL QE (95% in solution and 42% in films) with stable emission color (unchanged after thermal annealing). The electron-rich and electron-deficient substituents endow rather high HOMO (–5.30 eV) and relatively low LUMO (–2.54 eV) energy levels that are expected to facilitate charge injection and transport. Indeed, the PLED device fabricated with the configuration ITO/PEDOT/**271**/Ca/Ag showed a low turn-on voltage of 4.4 V and maximum Φ_{EL}^{ex} = 1.21% (achieved at 7.6-V driving voltage with a brightness of 354 cd/m^2), more than twice higher than that for PF **270**, containing only electron-deficient oxadiazole substituents [271]. In spite of minor additional emission bands at 580 and 660 nm (which increase at higher voltages), this PLED emission falls in the blue region (CIE: x = 0.193, y = 0.141) that, together with a high maximum luminance of 4080 cd/m^2 and an efficiency of 0.63 cd/A (0.19 lm/W), render **271** as a promising LEP (Chart 2.68).

A series of fluorene copolymers (**272** and **273a–g**) with oxadiazole pendant groups was recently synthesized by Sung and Lin [273]. The EL maximum for copolymer **272** (452 nm) is red shifted, compared with **273a–g** (406–431 nm). The devices fabricated with these copolymers in ITO/PEDOT/polymer/Ca/Al configuration showed a relatively high turn-on voltage of 6.5–8.5 V and only moderate maximum brightness (29–462 cd/m^2) (Chart 2.69).

Oxadiazole units were also introduced into the backbone of fluorene-alternating copolymers (Scheme 2.40). The key monomer in the synthesis of a series of fluorene–oxadiazole copolymers was 2,7-*bis*(tetrazolyl)fluorene derivative **274** for the preparation of copolymers **275–278** [274]. This tetrazole route has several advantages over other oxadiazole ring formation reactions: fast and clean reaction, mild reaction conditions, high yields, and high polymer molecular weights. The copolymers showed excellent thermal stability (T_{dec} > 430°C), and their T_g progressively increased from 98°C to 150°C with increasing

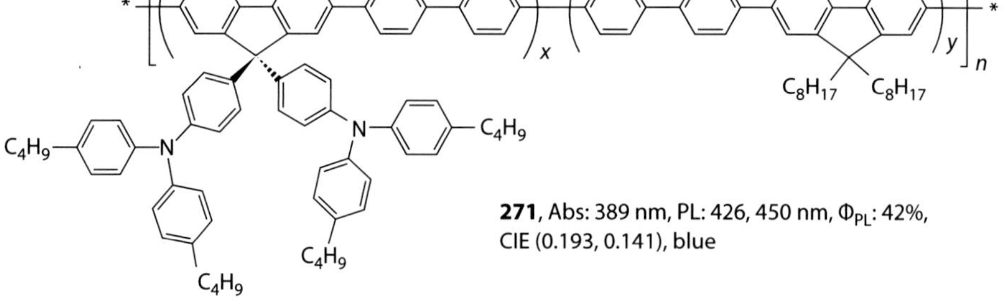

271, Abs: 389 nm, PL: 426, 450 nm, Φ_{PL}: 42%, CIE (0.193, 0.141), blue

CHART 2.68
Chemical structures of fluorene copolymers with triphenylamine and oxadiazole as pendant groups.

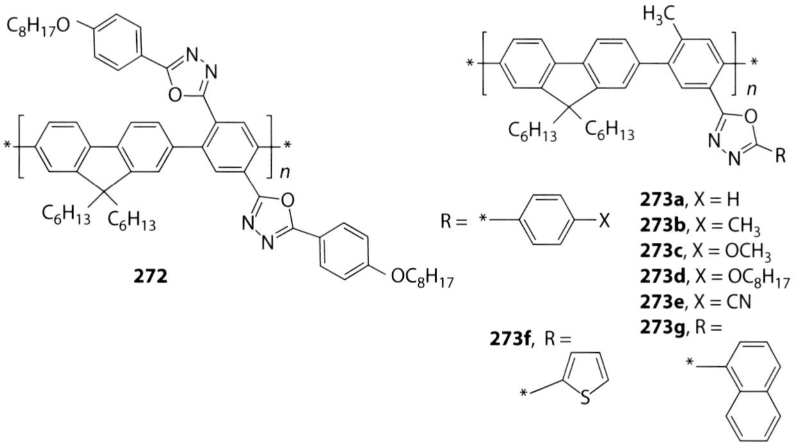
272
R =
273a, X = H
273b, X = CH3
273c, X = OCH3
273d, X = OC8H17
273e, X = CN
273f, R =
273g, R =

CHART 2.69
Chemical structures of fluorene copolymers with oxadiazole as pendant groups.

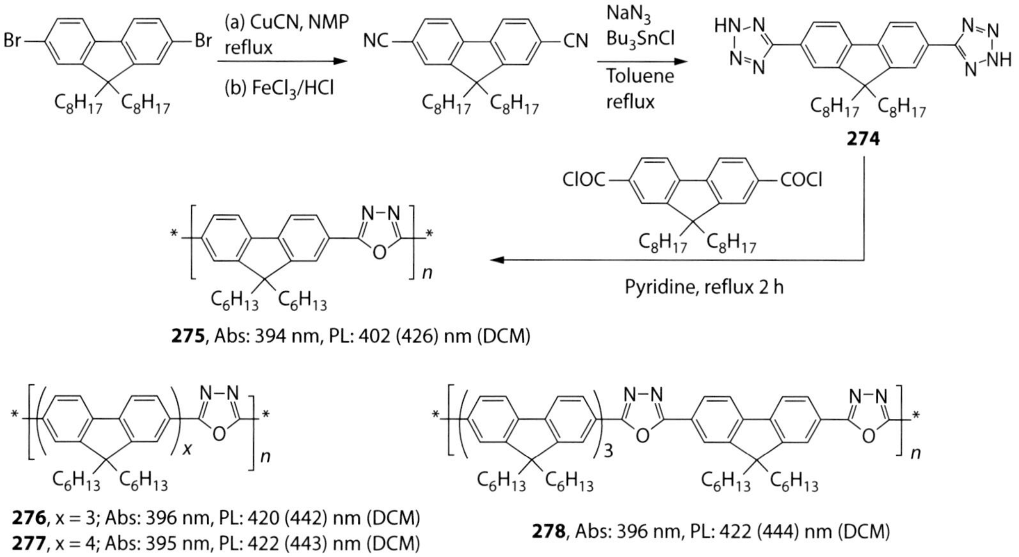
(a) CuCN, NMP reflux
(b) FeCl3/HCl
NaN3 Bu3SnCl
Toluene reflux
274
Pyridine, reflux 2 h
275, Abs: 394 nm, PL: 402 (426) nm (DCM)
276, x = 3; Abs: 396 nm, PL: 420 (442) nm (DCM)
277, x = 4; Abs: 395 nm, PL: 422 (443) nm (DCM)
278, Abs: 396 nm, PL: 422 (444) nm (DCM)

SCHEME 2.40
Synthesis of fluorene–oxadiazole copolymers. (From Ding, J., Day, M., Robertson, G., and Roovers, J., *Macromolecules*, 35, 3474, 2002.)

oxadiazole content in the polymer. Remarkably, the UV and PL spectra of the copolymers were all very similar to those of fluorene homopolymers with only slight red shifts in absorption (10–12 nm) and emission (5–7 nm) spectra. The copolymers also demonstrated high $\Phi_{PL} \approx 70\%$ (in DCM solution), typical for PFs. Several other fluorene–oxadiazole, fluorene–thiadiazole, and fluorene–triazole fully conjugated copolymers **279a–c** and **280a,b** have been prepared in the same way [275] (Chart 2.70).

Other fluorene–oxadiazole copolymers, such as fully conjugated **281a,b** [276] or **282a–c**, with conjugation interrupted by σ-links [277], have been synthesized. For both series of polymers, emission was in the blue region at very similar wavelengths; however, no LED device was reported.

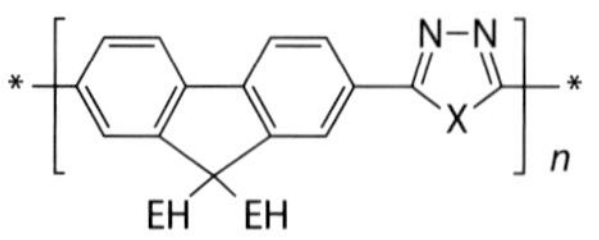

279a, X = O
279b, X = S
279c, X = N-Ph

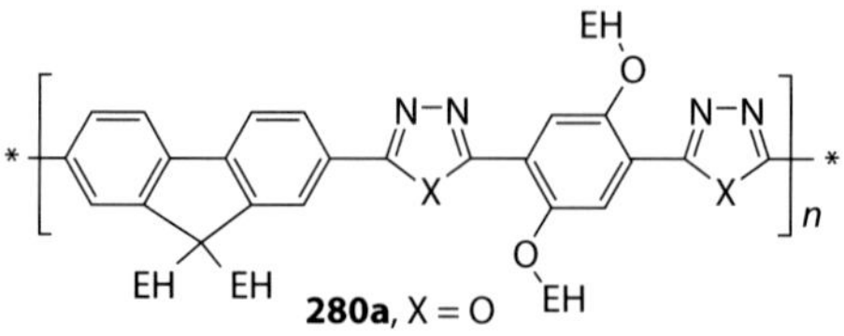

280a, X = O
280b, X = S

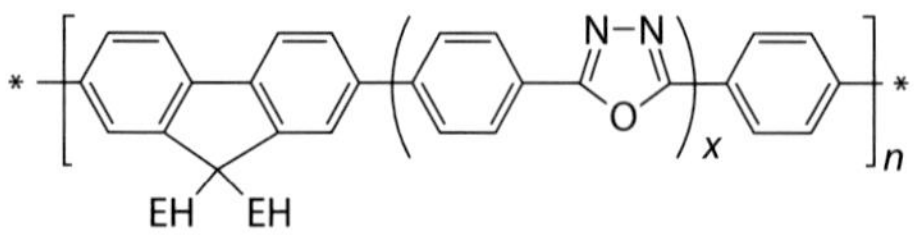

281a, *x* = 1; Abs: 369 nm, PL: 414 nm, fwhm: 76 nm
281b, *x* = 2; Abs: 352 nm, PL: 422 nm, fwhm: 82 nm

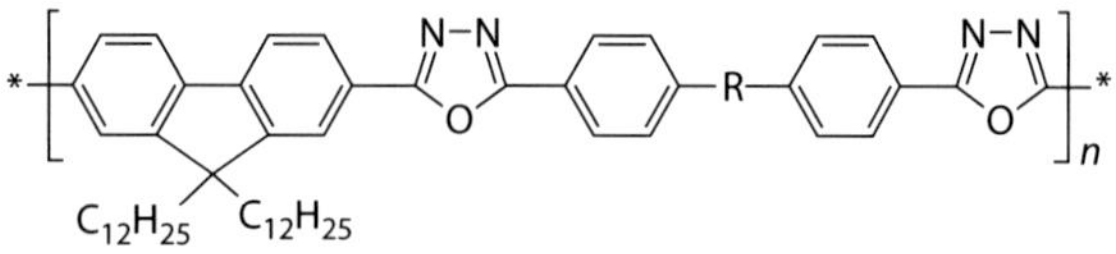

282a, R = $C(CF_3)_2$; Abs: 354 nm, PL: 423 nm, Φ_{PL}: 41% (DCE)
282b, R = O; Abs: 350 nm, PL: 421 nm, Φ_{PL}: 90% (DCE)
282c, R = $O(CH_2)_6$; Abs: 356 nm, PL: 427 nm, Φ_{PL}: 86% (DCE)

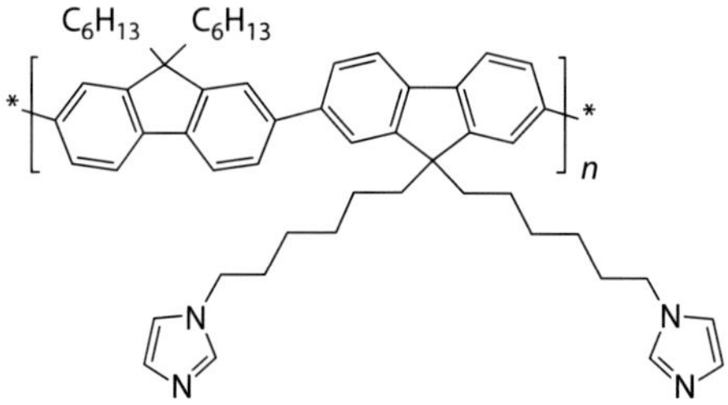

283, Abs: 390 nm, PL: 404, 425 nm (THF)

CHART 2.70
Chemical structures of fluorene–oxadiazole, fluorene–thiadiazole, and fluorene–triazole copolymers.

Zhou and coworkers [278] synthesized fluorene copolymer functionalized with imidazole ligands in the side chains (**83**). The PL emission of **283** was sensitive to the presence of metal cations in solution (particularly efficient quenching was due to Cu^{2+}), which makes it a promising material for fluorescent chemosensing.

Efficient blue emission and good electron affinity and electron-transporting properties were demonstrated for two fluorene copolymers with dicyanobenzene moiety in the main chain, **284a,b** (Scheme 2.41) [279]. Owing to improved electron-transport properties, the device ITO/PEDOT/**284a**/Ca showed a low turn-on voltage (3.4 V), better Φ_{EL}^{ex} (0.5%), and high brightness (5430 cd/m^2), compared with the PDHF **213** device in the same configuration.

Many other PF copolymers that do not contain a particularly electron-active moiety but, nevertheless, can improve the performance of the material in PLED have been synthesized. The Huang group [280,281] at Institute of Materials Research and Engineering (IMRE, Singapore) synthesized the deep-blue copolymer **285** via Suzuki copolymerization of fluorene–diboronic acid with dibromobenzene. The emission band of **285** has a peak at 420 nm and a well-defined vibronic feature at 448 nm with an fwhm of 69 nm, and virtually no green emission, allowing for very pure deep-blue fluorescence, compared with PDHF **213**. The PL QE of thin films was around 40%, similar to that of PDHF. The single-layer PLED device (ITO/**285**/Ca) and the multilayer devices (ITO/PANI/PVK/**285**/Ca and ITO/CuPc/PVK/**285**/Ca) gave identical deep-blue EL emission (λ_{max} = 420 nm, fwhm = 53 nm) [280]. A maximum Φ_{EL}^{ex} = 0.6% and a maximum luminance of 700 cd/m^2 were achieved for a multilayer PLED (Chart 2.71).

A spiro-bifluorene moiety was used to construct alternating copolymers **286** and **287** with dialkylfluorene units. The aim of this modification was to increase the T_g of the material,

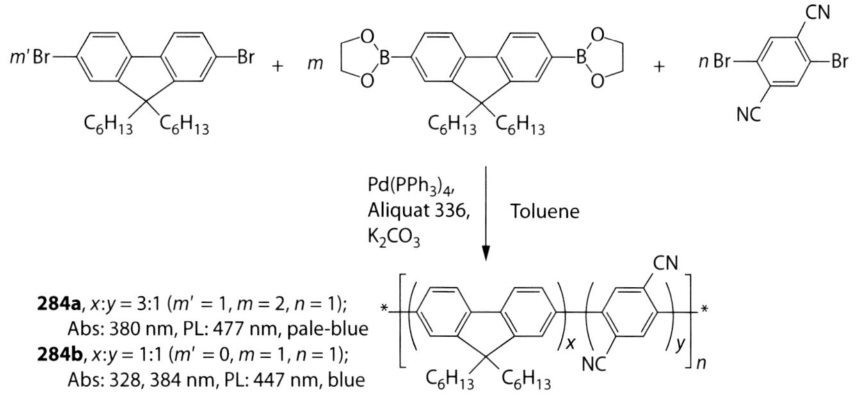

284a, *x*:*y* = 3:1 (*m*′ = 1, *m* = 2, *n* = 1);
Abs: 380 nm, PL: 477 nm, pale-blue
284b, *x*:*y* = 1:1 (*m*′ = 0, *m* = 1, *n* = 1);
Abs: 328, 384 nm, PL: 447 nm, blue

SCHEME 2.41
Synthesis of fluorene–(2,5-dicyanobenzene) copolymers via Suzuki coupling. (From Liu, M.S., Jiang, Z., Herguth, P., and Jen, A.K.-Y., *Chem. Mater.*, 13, 3820, 2001.)

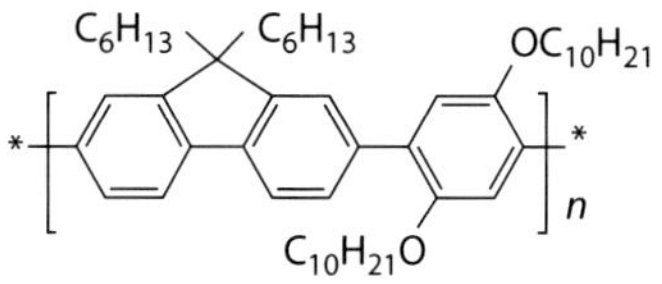

285, Abs: 378 nm, PL: 420, 448 nm, Φ_{PL}: ~40%,
HOMO: −5.66 eV, LUMO: −2.62 eV, deep-blue

CHART 2.71
Chemical structures of PF copolymers without a particularly electron-active moiety.

so that moderate heating during the device operation would not result in aggregate formation [282]. A 90° geometry of spiro-annulated bifluorene units prevents interchain aggregation, whereas the long-alkyl substituents in the second comonomer improve the solubility. The synthesis is given in Scheme 2.42. The Grignard reaction of 2,7-dibromo-9-fluorenone with biphenyl-2-magnesium bromide, followed by acid-promoted cyclization, affords the spiro-annulated monomer. The copolymers were synthesized by Suzuki coupling with

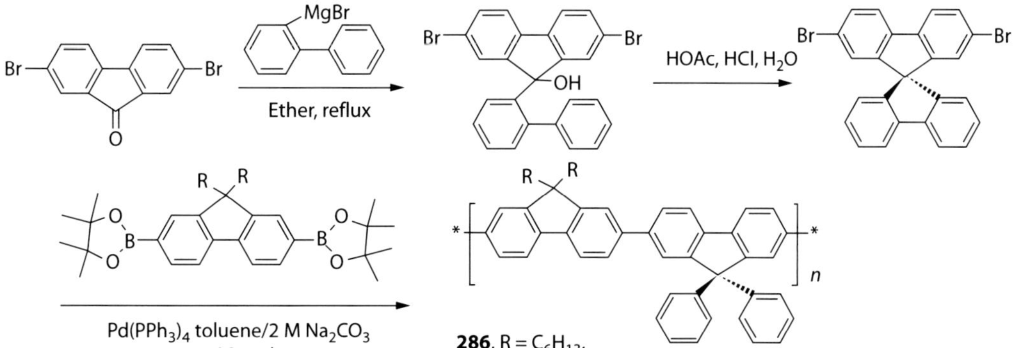

286, R = C_6H_{13},
287, R = C_8H_{17}; Abs: 379 nm, PL: 452 nm, blue

SCHEME 2.42
Synthesis of the PF spiro-copolymers. (From Yu, W.-L., Pei, J., Huang, W., and Heeger, A.J., *Adv. Mater.*, 12, 828, 2000.)

commercially available (from Aldrich) dialkylfluorenediboronic esters. The copolymers **286** and **287** are only partially soluble in solvents such as THF and chloroform, but can be completely dissolved in chlorobenzene (used to spin cast the polymer films). The M_n, determined for the THF-soluble part of **287**, is 11,600 g/mol, PDI = 2.9, although a higher molecular weight is expected for the THF-insoluble part. The DSC-determined T_g of **286** (105°C) is higher than that of PDHF **213** (75°C). Consequently, no long-wavelength emission was found for films annealed at 100°C (in contrast to PDHF), although further annealing above T_g at 150°C does result in the appearance of a strong emission band at 525 nm. Another feature of these copolymers is their narrow emission spectra compared with dialkylfluorene homopolymers, attributed to less dense molecular packing: for **287** and **213**, fwhm = 39 and 62 nm, λ_{onset}^{PL} = 585 and 610 nm, respectively [390]. A PLED device fabricated as ITO/**287**/Ca gave pure blue emission with a maximum Φ_{EL}^{ex} = 0.12%. Adding copper phthalocyanine (CuPc) as an HTL between the ITO and copolymer layers decreased the operating voltage from 16 to 7 V and increased the QE to 0.54% (maximum luminance 24 cd/m²).

Random copolymers **288–291** containing spiro-fluorene moieties were also reported by Carter and coworkers at IBM Almaden Research Center. They used spiro-bifluorene dibrominated at both fluorene moieties. Ni(0)-mediated random copolymerization of 9,9-dihexyl-2,7-dibromofluorene with this spiro-bifluorene monomer resulted in three-dimensional (3D)-branched copolymers **288–291** (Scheme 2.43) [283]. Because of the expected insolubility of such materials at a high degree of polymerization owing to the network formation, the molecular weight was controlled at $M_n \approx 3800$–12,800, by adding 4-bromostyrene as a terminating agent. The latter also serves as a cross-linkable unit that

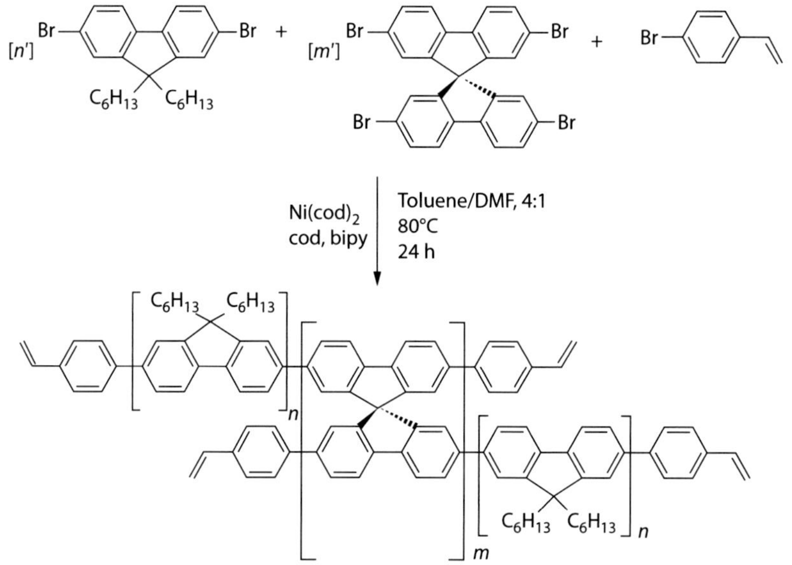

288, n:m′ = 98:2, Abs: 375 nm, PL: 440 nm
289, n:m′ = 96:4, Abs: 381 nm, PL: 429, 444 nm, EL: 424 nm
290, n:m′ = 90:10, Abs: 378 nm, PL: 435 nm
291, n:m′ = 80:20, Abs: 381 nm, PL: 427, 440 nm

SCHEME 2.43
Synthesis of random PFs with spiro-fluorene moieties. (From Marsitzky, D., Murray, J., Scott, J.C., and Carter, K.R., *Chem. Mater.*, 13, 4285, 2001.)

allowed obtaining an insoluble polymer film by heating of the soluble spin-coated polymer. The resulting amorphous polymers (**288–291**) demonstrated excellent thermal stability ($T_{dec} > 430°C$), and their T_g values (105°C, 144°C, 93°C, and 90°C for **288–291**, respectively) were substantially higher than those of PDHF **213** (75°C). A series of PL measurements in films revealed blue emission that remains stable (no green component) after thermal treatment at 120°C for 30 min (in these conditions, the copolymers are cross-linked through the 4-phenylethenyl end groups rendering insoluble films) (Figure 2.21). A bilayer device ITO/PEDOT/x-HTPA/**288**/Ca/Al was fabricated by spin coating the cross-linkable poly(4-hexyl-triphenyl)amine (x-HTPA) as HTL, curing at 100°C for 1 h in a glovebox and repeating the process with **288**. The diode could be switched on at 3 V to emit blue light ($\lambda_{EL} = 424$ nm; CIE: $x = 0.168$, $y = 0.07$) and maximum $\Phi_{EL}^{ex} = 0.08\%$ was achieved at 9 V.

Liu and coworkers [284] synthesized a series of hyperbranched alternating copolymers of tetraphenyl(*p*-biphenyl)-methane and -silane with 9,9-dihexylfluorene by Suzuki coupling (**292a–c**). Remarkably, copolymers **292a–c** were soluble in common solvents, showed high T_g (>200°C), and emitted in the blue region, slightly blue shifted compared with poly(9,9-dialkylfluorenes). Because their 3D structure prevents close packing of fluorene chains, these copolymers are less prone to self-aggregation in the solid state and, accordingly, no green emission was observed. A PLED ITO/PEDOT/polymer/LiF/Ca/Ag showed bright-blue emission peaking at 415 nm with $\Phi_{EL}^{ex} = 0.6\%$ and a turn-on voltage of 6.0 V (Chart 2.72).

Blue light–emitting copolymers **293a–c**, with interrupted conjugation due to *m*-phenylene linkages in the backbone, did not show the typical vibronic structure of PFs in their PL spectra. They possess an increased stability toward the appearance of undesired green emission upon annealing. Fabrication of ITO/PEDOT/**293**/Ba/Al devices demonstrated turn-on voltages of 7.4–8.4 V and Φ_{EL}^{ex} in the range of 0.43%–0.6% [285] (Chart 2.73).

An unusual synthetic approach to PF copolymers was demonstrated by Bunz and coworkers [285], who prepared poly(fluorene ethynylene) **294a–e** by metathesis polymerization reaction (Scheme 2.44) [286]. The aggregation of polymers **294** in concentrated solutions and in solid state is manifested in slight (up to 10–20 nm) red shift of the absorbance and emission peaks, although solutions and films emit pure blue light.

Jiang and coworkers [287] reported a binaphthyl-containing random PF copolymer (**295**) synthesized by Suzuki coupling. The twisted binaphthyl units control the effective

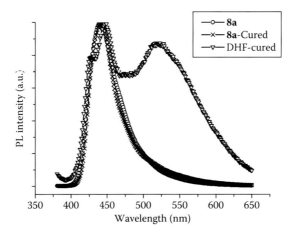

FIGURE 2.21
PL spectra of the uncured and cured amorphous spiro-PF **288** (**8a**) and a cured PDHF **213** (DHF). (From Marsitzky, D., Murray, J., Scott, J.C., and Carter, K.R., *Chem. Mater.*, 13, 4285, 2001. With permission.)

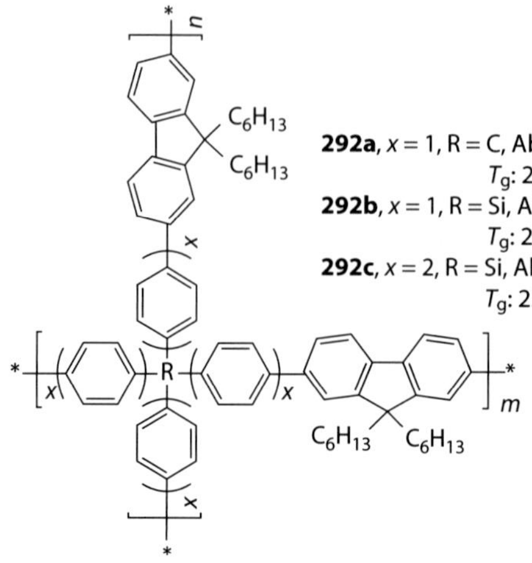

292a, $x = 1$, R = C, Abs: 345 nm, PL: 399, 4520sh nm, H$_{PL}$: 38%, T_g: 236°C, T_{dec}: 449°C

292b, $x = 1$, R = Si, Abs: 333 nm, PL: 402, 419sh nm, H$_{PL}$: 65%, T_g: 237°C, T_{dec}: 439°C

292c, $x = 2$, R = Si, Abs: 374 nm, PL: 414, 439sh nm, H$_{PL}$: 82%, T_g: 238°C, T_{dec}: 380°C

CHART 2.72
Chemical structures of hyperbranched alternating PF copolymers.

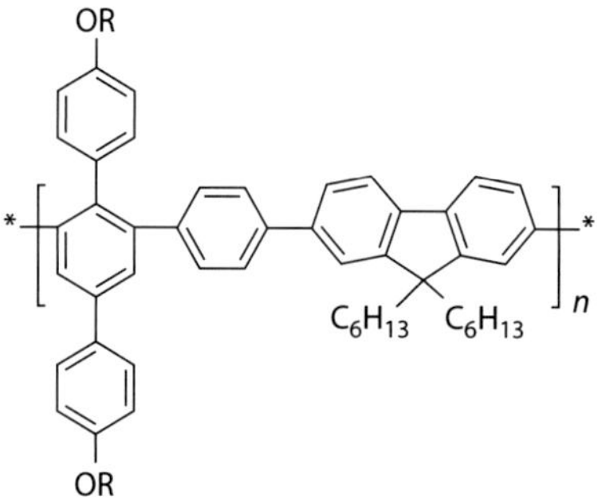

293a, R = C$_6$H$_{13}$; Abs: 341 nm, PL: 411 nm, Φ_{PL}: 32%, EL: 418 nm
293b, R = EH; Abs: 341 nm, PL: 408 nm, Φ_{PL}: 30%, EL: 419 nm
293c, R = C$_{12}$H$_{23}$; Abs: 344 nm, PL: 412 nm, Φ_{PL}: 32%, EL: 420 nm

CHART 2.73
Chemical structures of PF copolymers with *m*-phenylene linkages in the backbone.

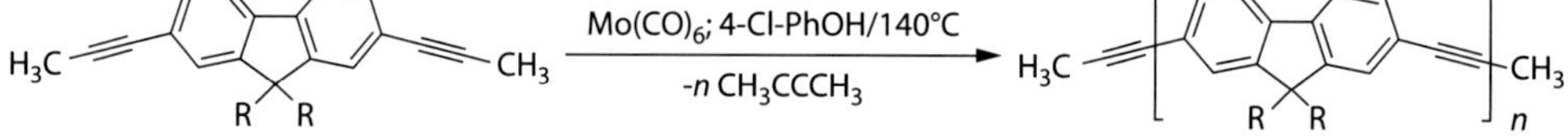

294a, R = C$_6$H$_{13}$; Abs: 396, 415 nm, PL: 428 nm, blue
294b, R = C$_8$H$_{17}$; Abs: 394, 416 nm, PL: 429 nm, blue
294c, R = C$_{12}$H$_{25}$; Abs: 387, 413 nm, PL: 429 nm, blue
294d, R = 2-ethylhexyl; Abs: 392, 416 nm, PL: 434 nm, blue
294e, R = (S)-3,7-dimethyloctyl; Abs: 396, 424 nm, PL: 434 nm, blue

SCHEME 2.44
Metathesis polymerization synthesis of fluorene–ethynyl copolymer **281**. (From Pschirer, N.G. and Bunz, U.H.F., *Macromolecules*, 33, 3961, 2000.)

conjugation length, and prevent fluorescence quenching in the solid state by hindering intermolecular π–π interactions. Accordingly, a pure blue emission (two peaks at 428 and 448 nm, blue shifted compared with PFO **215**) and a relatively high PL efficiency (44%) were observed in solid films of this material. Multilayer PLEDs fabricated with CuPc or BTPD–PFCB (**143**) [427] HTL (ITO/HTL/**295**/Ca configuration) exhibited EL emission peaks at 420 and 446 nm with EL maximum brightness and Φ_{EL}^{ex} up to 3070 cd/m^2 and 0.82%, respectively (although an additional low-intensity emission band was observed for the most efficient devices using BTPD–PFCB as an HTL) [287,427] (Chart 2.74).

Lai and coworkers [288,289] reported fluorene copolymers containing dithia[3.3]cyclophane as a repeating unit. Polymers **296a–c** and **297a–c** were synthesized by Suzuki-coupling fluorene borate derivatives with dibromocyclophane. Polymer **296a** shows pure blue fluorescence with rather high PLQY (75%); however, no LED device performance was reported. Remarkably, the introduction of either electron donor or electron acceptor substituents in the cyclophane moiety results in significant quenching of the fluorescence quantum yield (36% for **296b**, 0% for **296c**), which was explained by a transannular charge-transfer effect [288]. Copolymers **297a–c** also showed lower quantum yields (15%–22%) [289] (Chart 2.75).

Liu and coworkers [290] reported alternating fluorene copolymers, **298**, **299**, and **300**, containing 2,2′-bipyridyl fragments in the main chain using Suzuki, Wittig–Horner, and Heck coupling of 9,9-dioctyl-2,7-dibromofluorene, respectively. All three polymers were responsive to a wide variety of transition metal ions by an absorption spectral red shift (up to 40 nm) and fluorescence quenching (Chart 2.76).

Kong and Jenekhe [428] prepared triblock copolymers **301** by the ring-opening polymerization of γ-benzyl-L-glutamate *N*-carboxyanhydride using benzylamine end-capped PDHF (Scheme 2.45). The polymers retain the emissive properties of PF ($\lambda_{max}^{abs} = 380$ nm,

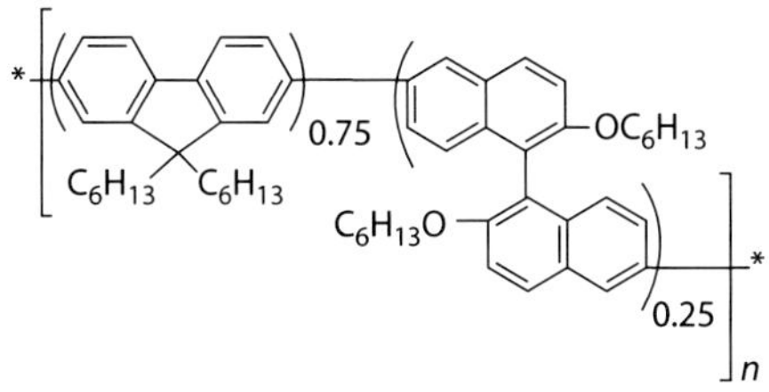

295, Abs: 360 nm, PL: 428, 445 nm, Φ_{PL}: 44%
EL: 420, 446 nm, E_g: 2.97 eV

CHART 2.74
Chemical structure of a binaphthyl-containing random PF copolymer.

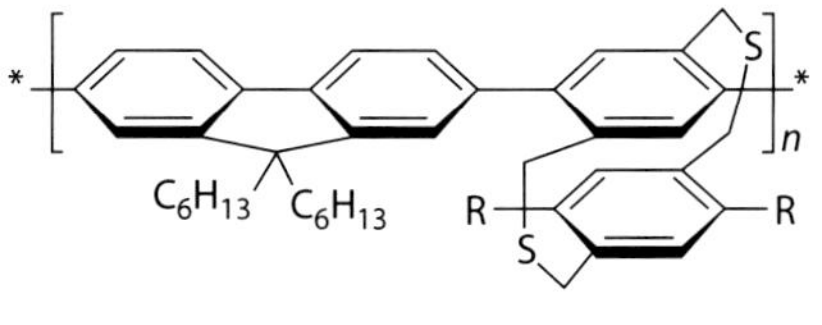

296a, R = H; Abs: 367 nm, PL: 417 nm
296b, R = OCH$_3$
296c, R = CN

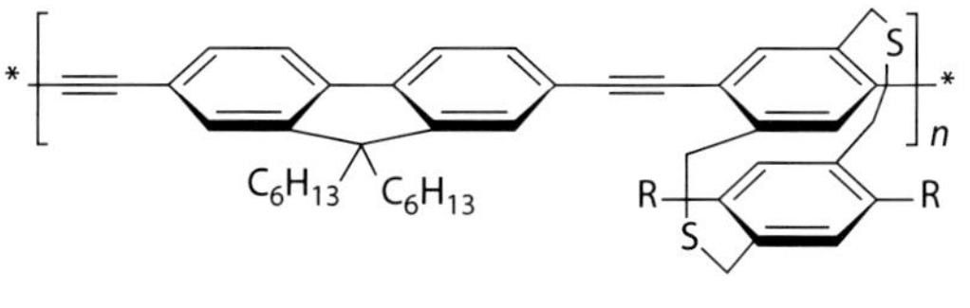

297a, R = H; Abs: 386 nm, PL: 484 nm
297b, R = OCH$_3$; Abs: 388 nm, PL: 492 nm
297c, R = CN; Abs: 390 nm, PL: 479 nm

CHART 2.75
Chemical structures of PF copolymers containing dithia[3,3]cyclophane as a repeating unit.

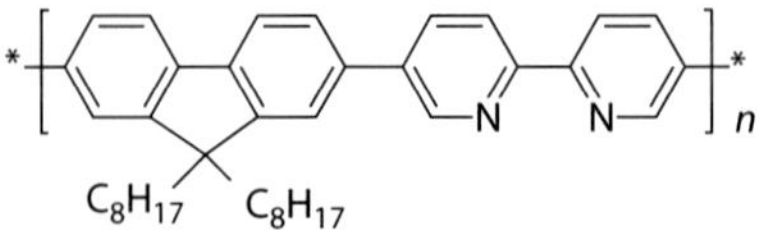

298, Abs: 383 nm, PL: 453, 477 (425) nm

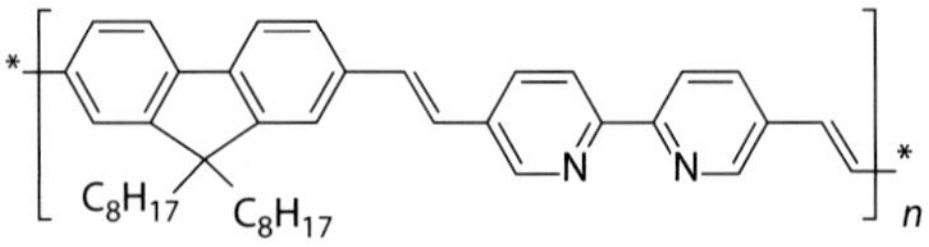

299, Abs: 432 (459) nm, PL: 521 nm

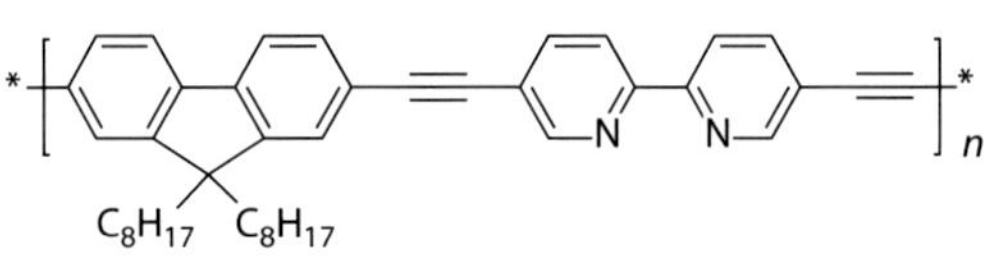

300, Abs: 425 (395) nm, PL: 477, 503 (439) nm

CHART 2.76
Chemical structures of alternating PF copolymers containing 2,2′-bipyridyl units.

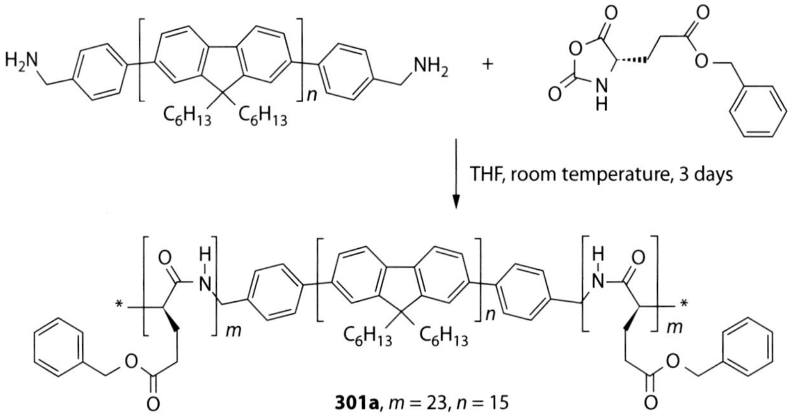

THF, room temperature, 3 days

301a, m = 23, n = 15
301b, m = 16, n = 28

SCHEME 2.45
Synthesis of triblock copolymers **301**. (From Kong, X. and Jenekhe, S.A., *Macromolecules*, 37, 8180, 2004.)

λ_{max}^{PL} = 424 nm in CHCl$_3$) and show EL with a brightness comparable to PF homopolymers. Different nanostructured assemblies of the block copolymers **301** were observed by AFM depending on the copolymer composition and secondary structure (helix or coil) of the polypeptide blocks.

Several groups have demonstrated that pure and stable blue emission of PFs can be achieved by the introduction of the fluorinated groups in backbone (Chart 2.77). Two fluorine-based copolymers containing *p*-difluorophenylene units in backbone, **302a** and **302b**, were reported by Zhang et al. [291]. The introduction of the fluorinated groups increased the fluorescent quantum yield and improved the spectral purity and stability. The green emission band at 520 nm from fluorenone defects was never detected for these copolymers. A maximum external QE of 1.14% (1.14 cd/A) and CIE (0.16, 0.13) were achieved for **302b** device with a configuration of ITO/PEDOT:PSS (40 nm)/polymers (40 nm)/TPBI

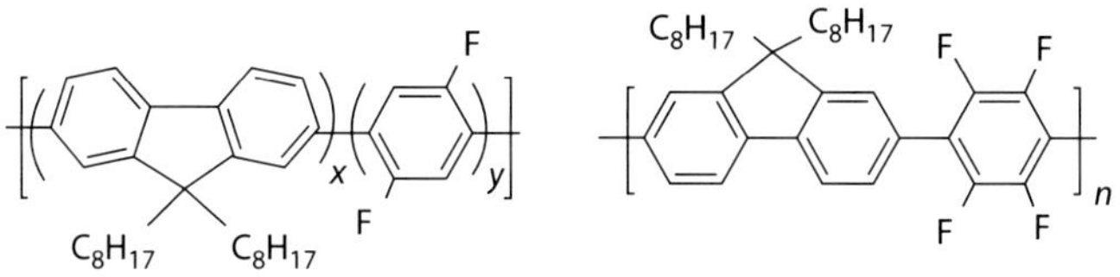

CHART 2.77
Chemical structures of PFs with fluorinated group in backbone.

(20 nm)/LiF (1 nm)/Al (200 nm). An alternating polyfluorene with perfluorinated phenylene unit (**303**) was synthesized by an optimized Suzuki cross-coupling [292]. The PL quantum yields of the solution, as high as 68%, are retained in the solid state. The polymer **303** tested as an undoped active layer in a PLED with a configuration of ITO/PEDOT:PSS/PVK/polymer/TPBI/Ba/Al shows a highly desirable pure deep-blue 405-nm electroluminescence, CIE = (0.17, 0.06), and a remarkable 5.03% external QE.

2.3.7 Color Tuning in Polyfluorene Copolymers

The above strategies were focused on creating highly stable PFs with pure blue emission. However, other emission colors can also be achieved through doping with fluorescent dyes or through a copolymerization strategy in either alternating or random PF copolymers. Through careful selection of suitable comonomers, considering their effect on HOMO–LUMO levels of the polymer, one can achieve a whole range of emission colors from PF-based materials.

2.3.7.1 Doping with Low-Molar-Mass Fluorescent Dyes

The high-energy blue emission of PF polymers can be changed into virtually any color by adding a small amount of low energy–emitting materials. This approach, based on energy transfer, is of prime importance for phosphorescent PLEDs, although nonphosphorescent dyes were used to tune the PF emission color. Thus, efficient Förster energy transfer from blue to red (typical for porphyrins) was demonstrated for tetraphenylporphyrin (TPP)-doped PFO **215**-based LED as a result of a good overlap between the PF emission with the Soret absorption band of the porphyrin [429]. It was shown that for composites with 1%–10% TPP, 95% of photogenerated excitons are transferred from the polymer to TPP. The ITO/**265**/**215**:TPP/Ca LED (**265** here plays the role of HTL) device showed $\Phi_{EL}^{ex} = 0.9\%$ at 33 V, corresponding to 0.18 cd/A and a luminance of 90 cd/m². The EL emission was identical to the PL emission of TPP, with CIE coordinates ($x = 0.65$, $y = 0.29$), representing a rather pure red color.

2.3.7.2 Alternating Copolymers

Wu et al. [293] have reported a new spiro-linked PF (**304**). Unlike the spiro-*co*-PF discussed in the previous section, the conjugation in **304** is completely interrupted by the spiro-bifluorene units. As a result, the copolymer showed significant blue-shifted absorption and emission spectra, compared with PFO **215**. The UV–vis absorption of thin films coated from a toluene solution has a maximum at 355 nm, and the PL emission showed a vibronic fine structure with two sharp peaks at 400 and 422 nm. The QE of the polymer in thin films is 42%, comparable to PFO. This copolymer shows superb color stability: the thermal annealing of thin films has no effect on the emission. While the copolymer itself demonstrates a purple–blue emission at the UV limit of the visible region of the spectra (CIE: $x = 0.165$, $y = 0.128$), it can be used as an energy-transfer material with fluorescent dyes to achieve a desirable color. Thus, pure blue emission is easily achieved with ca. 1% of 2,5,8,11-tetra-*tert*-butylperylene (Chart 2.78).

Using Suzuki polymerization, Liu and coworkers [244] synthesized and studied the optical and electronic properties of a series of fluorene–arylene alternating polymers **305a–g** (Scheme 2.46). These copolymers had a band gap ranging from 2.82 to 3.32 and emitted blue light (λ_{max}^{PL} from 404 to 443 nm), whose PL efficiency varied substantially with the Ar moiety (Φ_{PL} ~10%–88% in films). The polymers showed good thermal stability (T_{dec} ~ 375°C–410°C) with wide variation in T_g (~50°C–213°C). Electron-rich or electron-deficient arylene moieties in the backbone tuned the HOMO–LUMO levels of these materials (see Table 2.2).

Leclerc and coworkers [253,294,295,430] have efficiently tuned the emission from blue to green and yellow by introducing various phenylene and thiophene units in alternating fluorene copolymers **306–311**, although no simple correlation between the polymer structure and the EL color was found. Changing the nature of the comonomer unit from the relatively electron-deficient phenylene to very electron-rich *bis*(ethylenedioxy)thiophene affects the band gap of the polymer (and the emitting color), as well as the charge-transport properties, thus influencing the performance of the PLEDs (see also Ref. [431] on the thermo- and solvatochromic properties of these copolymers). The highest brightness

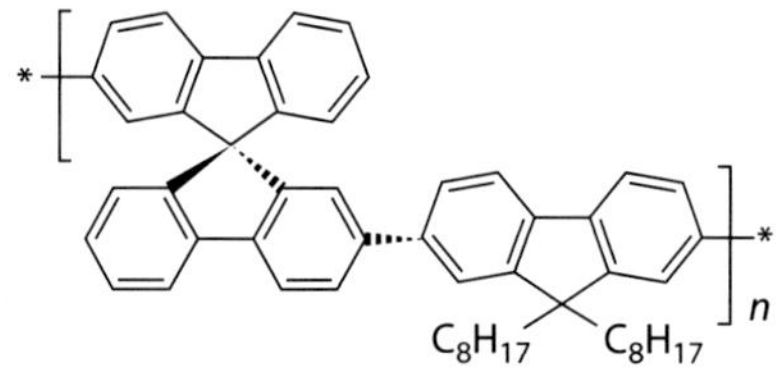

304, Abs: 355 nm, PL: 404, 425 nm, Φ_{FL}: 42%,
Purplish–blue, CIE (0.165, 0.128)

CHART 2.78
Chemical structure of a spiro-linked PF.

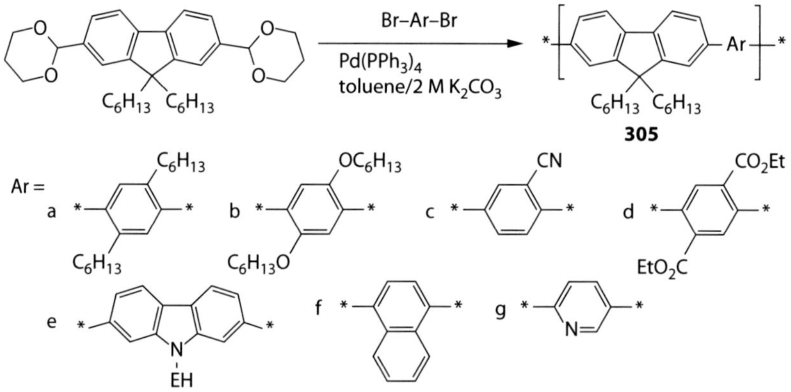

SCHEME 2.46
Suzuki-coupling polymerization route to alternating fluorene–arylene copolymers **305**. (From Liu, B., Yu, W.-L., Lai, Y.-H., and Huang, W., *Chem. Mater.*, 13, 1984, 2001.)

achieved for the bithiophene-based copolymer **310** was of 4.5 cd/m² for a neat copolymer and 18 cd/m² for a blend with 20% of hole-transport material tetraphenylbenzidine, TPD (**314**), values that are far too low for a practical application (Chart 2.79).

Huang's group [296,298,299] has systematically studied the structure–property relationships of fluorene–thiophene-based conjugated polymers **315–317**. In contrast to PT homopolymers (see Section 2.4.2), the regioregularity of the bithiophene fragments in their copolymers show little influence on the optical band gap (**317a,b**: E_g = 2.49 and 2.58 eV [296], or 2.57 and 2.60 eV [297,432], respectively) or the emission maxima; however, the HH copolymer **317b** was significantly more thermally stable (Chart 2.80).

All copolymers showed reversible n- and p-doping in CV experiments, and a schematic diagram showing the HOMO and LUMO energy levels relative to the electrode's work function of EL devices are given in Figure 2.22 [297]. The single-layer ITO/**317b** (100 nm)/ Ca device emitted green light (493, 515 nm) with a high turn-on voltage of 20 V. The introduction of PVK HTL decreased the turn-on voltage to 8 V and increased the maximum Φ_{EL}^{ex} from 0.05% to 0.6% [298,299] (although less pronounced increase was observed for other polymers of the series [297]). Later, this group reported a somewhat lower EL QE of 0.3% for the device ITO/PVK/**317b** (75 nm)/Ca/Al. The highest Φ_{EL}^{ex} in this series was reported for **315** (0.64% for ITO/PVK/**315**/Ca/Al device), and the lowest one was for **317a** (0.07% for ITO/PEDOT/**317a**/Ca/Al device) [297]. Note that this trend has no correlation with the polymer energy levels.

A series of alternating copolymers **318**, containing conjugated thiophene-2,5-diyl (A) and nonconjugated thiophene-3,4-diyl (B) units in different ratios, was recently synthesized by Vamvounis and Holdcroft [433] via Suzuki coupling of fluorene–diboronic ester with

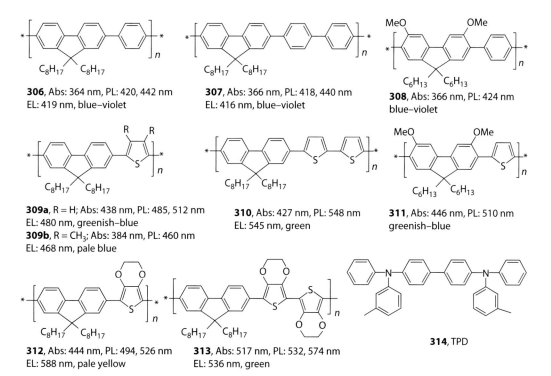

306, Abs: 364 nm, PL: 420, 442 nm
EL: 419 nm, blue–violet

307, Abs: 366 nm, PL: 418, 440 nm
EL: 416 nm, blue–violet

308, Abs: 366 nm, PL: 424 nm
blue–violet

309a, R = H; Abs: 438 nm, PL: 485, 512 nm
EL: 480 nm, greenish–blue
309b, R = CH₃; Abs: 384 nm, PL: 460 nm
EL: 468 nm, pale blue

310, Abs: 427 nm, PL: 548 nm
EL: 545 nm, green

311, Abs: 446 nm, PL: 510 nm
greenish–blue

312, Abs: 444 nm, PL: 494, 526 nm
EL: 588 nm, pale yellow

313, Abs: 517 nm, PL: 532, 574 nm
EL: 536 nm, green

314, TPD

CHART 2.79
Chemical structures of alternating PF copolymers with phenylene and thiophene units.

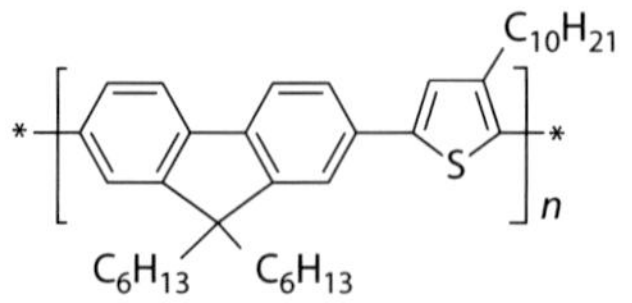

315, Abs: 412 nm, PL: 492, 477 nm
bluish–green

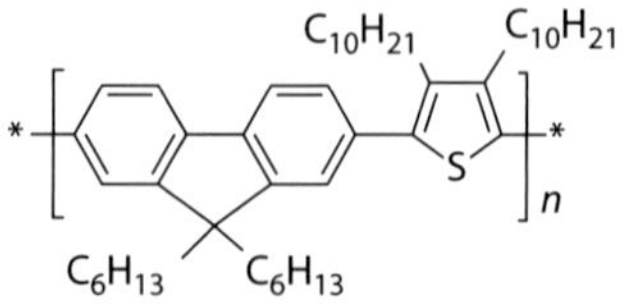

316, Abs: 378 nm, PL: 458, 475 nm
blue

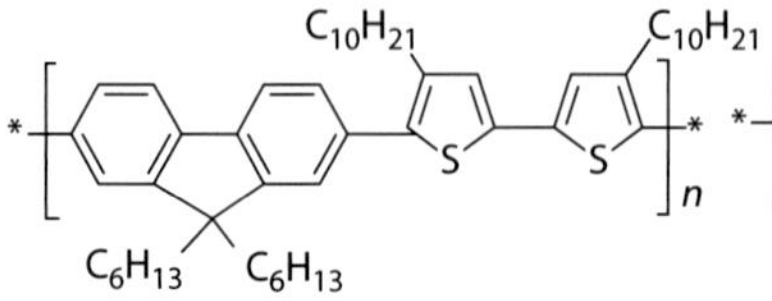

317a, Abs: 403 nm, PL: 490, 520 nm
green

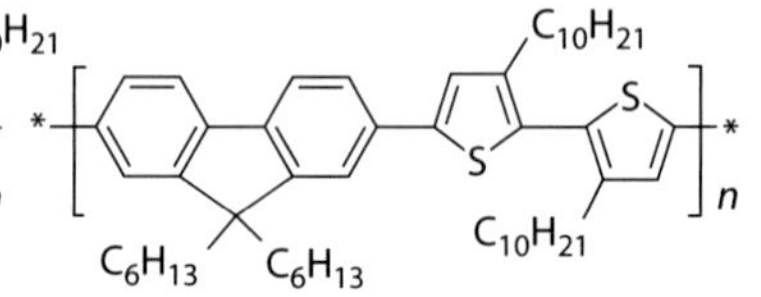

317b, Abs: 401 nm, PL: 493, 520 nm
green

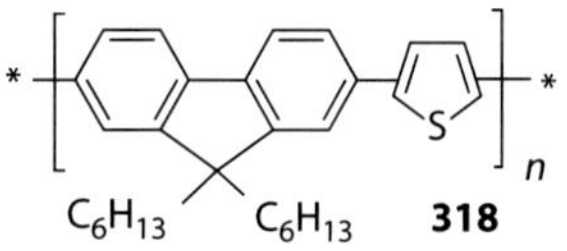

318

CHART 2.80
Chemical structures of fluorene–thiophene copolymers.

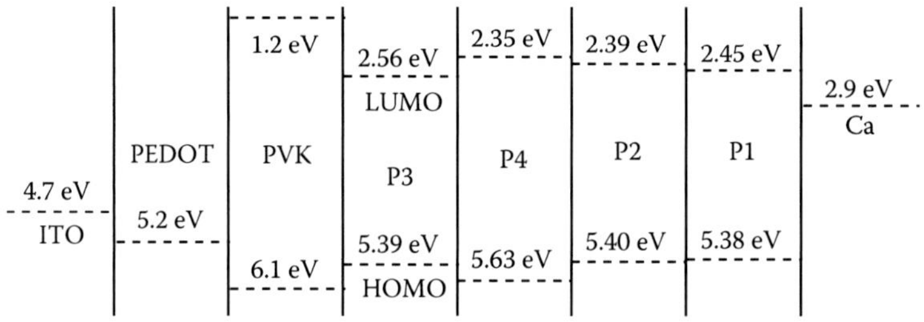

FIGURE 2.22
Schematic energy-level structure for the devices with copolymers **315** (P3), **316** (P4), **317a** (P1), and **317b** (P2). (From Liu, B., Niu, Y.-H., Yu, W.-L., Cao, Y., and Huang, W., *Synth. Met.*, 129, 129, 2002. With permission.)

mixtures of 2,5- and 3,4-dibromothiophenes. There is a significant progressive blue shift of the emission with increased feed ratio of the nonconjugated monomer B (from $\lambda_{PL} = 466$, 482 nm for 100% A-**318** to $\lambda_{PL} = 383$, 482 nm for 100% B-**318**). However, the most interesting finding was that using both of these comonomers allows the suppression of the fluorescence quenching in the solid state of **318**. The solid-state PLQY of a polymer with an A:B ratio of 1:24 was almost as high as that in solution (43% vs. 57%), whereas a nearly 10-fold decrease of the Φ_{PL} in the solid state was observed for copolymers **318** with only A or only B units.

An increase in the PL QE of the fluorene–thiophene copolymers can be achieved by the introduction of *S*-oxidized thiophene units (although no efficient EL from such materials was reported). This aspect and the chemical structures of thiophene-*S,S*-dioxide–fluorene copolymers are discussed in more detail in Section 2.4.

A very efficient green-emitting fluorene copolymer **319** was synthesized by Lim and coworkers [248] via Suzuki coupling of dibromothieno[3,2-*b*]thiophene with dialkylfluorenediboronic acid [248]. The authors compared the EL properties of this copolymer with

PFO homopolymer **215** and PFO-bithiophene copolymer **310**. Both the absorption and emission spectra of **319** are red shifted compared with PFO **215** but slightly blue shifted compared with bithiophene-based copolymer **310**. PLEDs fabricated in the configuration ITO/PEDOT/**319**/LiF/Al showed a pure green emission (CIE: $x = 0.29$, $y = 0.63$) close to the standard NTSC green color (CIE: $x = 0.26$, $y = 0.65$) (NTSC is National Television Systems Committee) with a very low turn-on voltage of 3.3 V. The low turn-on voltage is attributed to the better (compared with PFO) match between the HOMO (–5.38 eV) level and the work function of PEDOT (5.1–5.3 eV). Interestingly, although the PLQY of **319** (12% in films, similar to **310**) is lower than that of PFO (55%), the EL efficiency is much higher, which may reflect an improved balance of electron–hole transport in this copolymer. The maximum current efficiency of **319** is 0.32 cd/A at 0.78 A/cm², which exceeds the performance of similar devices fabricated with **310** or PFO **213** (which showed 0.20 cd/A at 143 mA/cm² and 0.06 cd/A at 25 mA/cm², respectively). The external QE of **319** (0.1%) was twice as high as for the former polymers (Chart 2.81).

A green-emitting fluorene–benzothiadiazole (BT) copolymer **320** was synthesized at Dow Chemicals via Suzuki coupling of fluorene–2,7-diboronic acid with dibromobenzothiadiazole [235,347]. A high-performance green-emission PLED was demonstrated with this copolymer [300,301]. In contrast to other PFs, which demonstrate high-mobility nondispersive hole transport [378], copolymer **320** shows a weak and highly dispersive electron transport [302]. Since this copolymer has an electron-deficient moiety in the polymer backbone, a stable anode electrode such as Al can be used, although a thin (70 Å) HTL (TPD, **214**) should be introduced for optimal EL performance. A double-layer PLED device fabricated in the configuration ITO/**214**/**320**/Al showed green emission with a turn-on voltage of 7.0 V. A maximum Φ_{EL}^{ex} as high as 3.86% (peak efficiency 14.5 cd/A) and a brightness of 5000 cd/m² were achieved at a current of 34 mA/cm². Even higher performance ($\Phi_{EL}^{ex} = 5\%$, efficiency >20 lm/W, brightness 10,000 cd/m²) for this polymer was reported by Millard [303] at CDT, although the exact LED structure was not disclosed. The EL device with **320** exhibited excellent electrical stability even when operated at high current densities (>0.25 A/cm²). When blended with PFO **215**, an efficient Förster energy transfer from excited PFO segments to **320** chain is observed [434].

Green-emitting fluorene copolymers **321** and **322** were also obtained by introducing pyrazoline moieties into the backbone that completely interrupt the conjugation due to sp³ carbons in the ring [304]. These polymers emit green light with a PL efficiency of 49%–59% in films. The PLEDs ITO/PEDOT/polymer/Ba/Ca showed bright-green emission at $\lambda_{EL} = $ 494–500 nm with high $\Phi_{EL}^{ex} = 0.6\% - 2.5\%$, low turn-on voltage (3.7–5.5 V), and a brightness of up to 2400 cd/m² (Chart 2.82).

Introduction of electron-accepting bithieno[3,2-*b*:2′,3′-*e*]pyridine units resulted in copolymer **323** with ca. 0.5 V lower reduction potential compared with the parent homopolymer

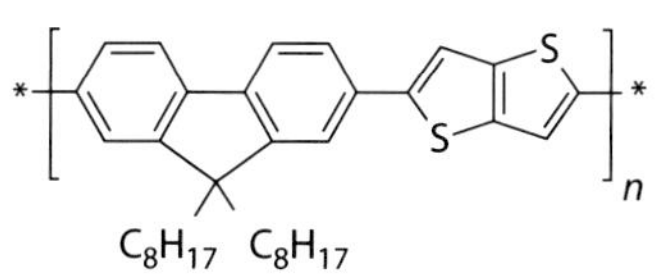

319, Abs: 448, 471 nm, PL: 495, 511, 548 nm
EL: 515 nm, green, CIE (0.29, 0.63)

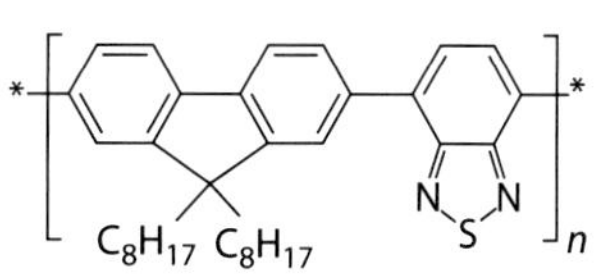

320, Abs: 390 nm, PL: 540 nm
EL: 545 nm, green, CIE (0.394, 0.570)

CHART 2.81
Chemical structure of a PF copolymer synthesized by coupling of dibromothieno[3,2-*b*]thiophene with dialkyl-fluorenediboronic acid.

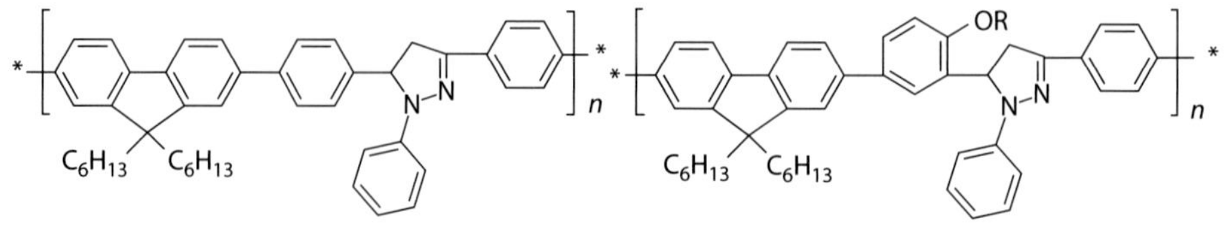

321, Abs: 339, 395 nm, PL: 508, green   **322a**, R = C$_6$H$_{13}$: Abs: 337, 399 nm, PL: 508, green
                                                 322b, R = C$_{12}$H$_{25}$: Abs: 336, 398 nm, PL: 508, green

CHART 2.82
Chemical structures of PF copolymers with pyrazoline units in the backbone.

PFO **213** [305]. Upon excitation at 420 nm $\left(\lambda_{max}^{abs} = 415\,nm\right)$, copolymer **323** exhibited blue–green emission with two peaks at 481 and 536 nm. Preliminary EL studies of an ITO/PEDOT/**323**/Al device showed two peaks positioned as in the PL spectra. The PLED exhibited low turn-on voltage (~4 V); however, at higher voltages of 6–9 V, a slight increase in the green component was observed (Chart 2.83).

Various fluorene–phenylene vinylene alternating copolymers with different emission colors have been synthesized, e.g., **324a** [306,308,435], **324b** [309], **325** [310], **326** [157], **327–329** [436], and **330a–c** [311] (Scheme 2.47). Introducing electron-rich (as in polymers **327** and **329**), electron-deficient (as in compounds **325**, **330**, and **342–344**), or both of these units (as in compounds **326** and **328**) in the PF chain allow for precise tuning of the emission wavelength (Scheme 2.47), the HOMO–LUMO levels, and the charge-injection and -transport properties. However, no high-performance PLEDs based on the above copolymers have yet been reported. One of the best-performing devices built in the configuration ITO/PEDOT/**324a**/Ca/Al showed a maximum brightness of 870 cd/m^2 (at 10 V) and an EL efficiency of only 0.16 cd/A.

Several fluorene-containing arylene–vinylene copolymers with a cyano group attached to the vinylene fragment have been reported. The copolymer **331**, containing cyanovinylene–phenylene segments in the main chain, in contrast to the red-emitting copolymers CN–PPV [162], emitted a narrow band with blue light (fwhm = 71 nm) [312]. The ITO/PVK:**331**/Alq$_3$/Al device reached a brightness of 784 cd/m^2 at a bias voltage of 18 V and Φ_{EL}^{ex} = 0.2% (at 123 mA/cm^2).

The structural analog **332** of copolymer **331**, in which the benzene ring was replaced with thiophene, showed substantial red shifts in both the absorption (97 nm) and emission (134 nm) spectra, and PLEDs based on this copolymer (ITO/LiF/**332**/PBD/LiF/Al/Ag) emitted red–orange light (brightness of 45 cd/m^2 at 10 V; turn-on voltage of 5 V) [313]. A further bathochromic shift in the PL (pure red emission) was observed in copolymer **333**, for which HOMO (–5.32 eV) and LUMO (–3.32 eV) energies were calculated from

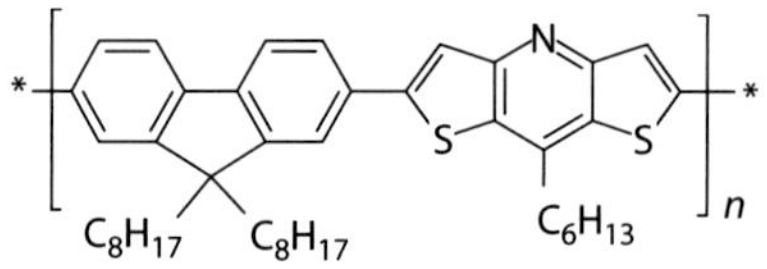

323, Abs: 415 nm, PL: 481, 536 nm
EL: 480, 536 nm, blue–green

CHART 2.83
Chemical structures of PF copolymers with bithieno[3,2-*b*:2′,3′-*e*]pyridine units.

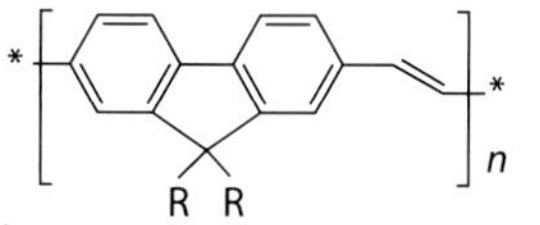

324a, R = C$_8$H$_{17}$; Abs: 418 nm, PL(EL): 465, 500, 530 nm, greenish–blue, CIE (0.23, 0.38), HOMO: –5.73 eV, LUMO: –3.13 eV
324b, R = C$_6$H$_{13}$

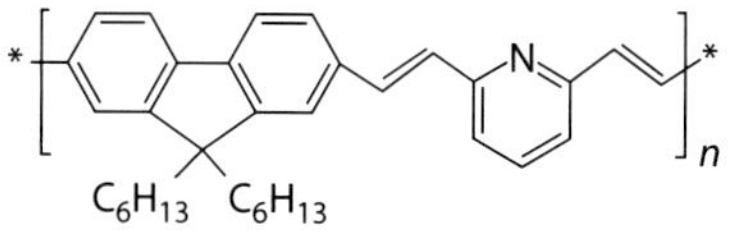

325, Abs: 375 nm, PL: 440, 465, 540 nm, EL: 540 nm, yellowish–green, HOMO: –5.67 eV, LUMO: –2.82 eV

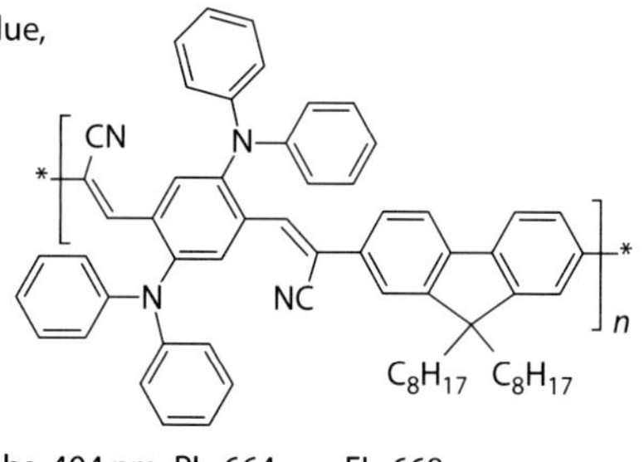

326, Abs: 404 nm, PL: 664 nm, EL: 668 nm, saturated red, CIE: (0.68, 0.32), E_g: 2.05 eV

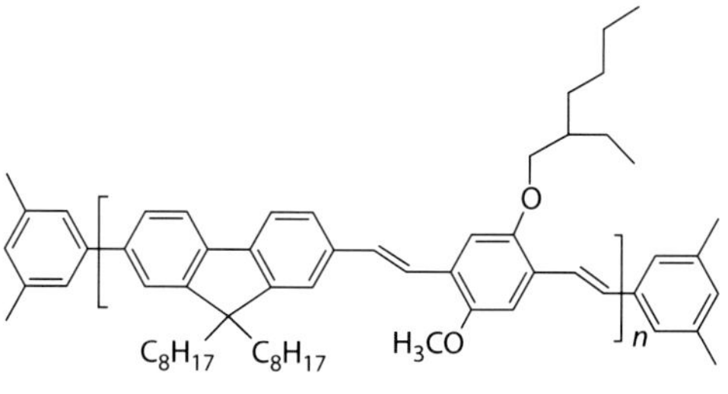

327, PL: 539 nm, green

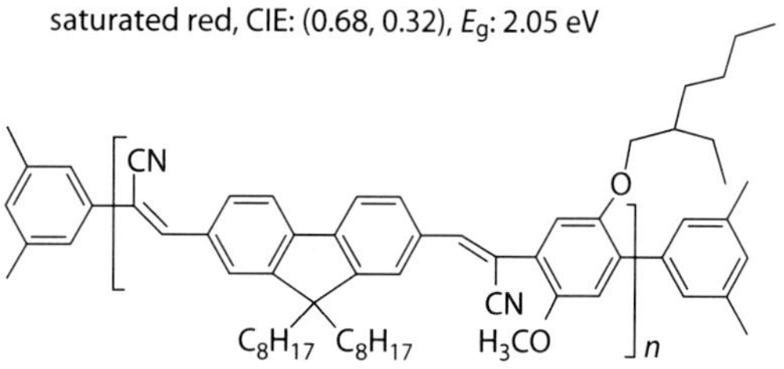

328, PL: 495 nm (THF), green

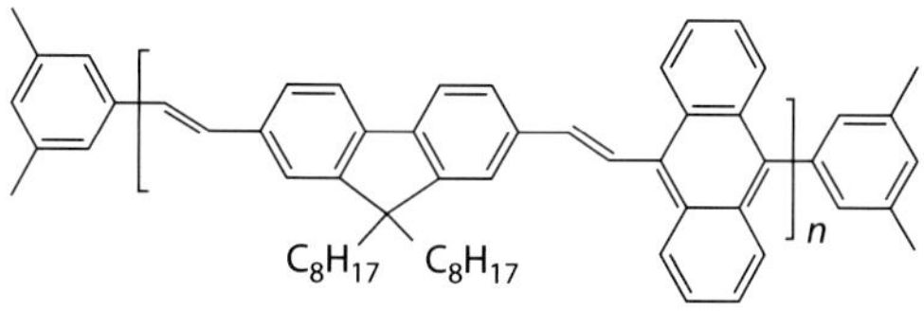

329, PL: 576 nm, orange–red

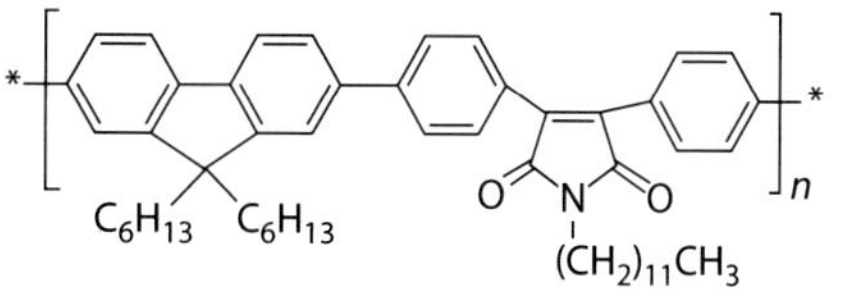

330a, R = C$_4$H$_9$: Abs: 412 nm, PL: 592 nm, EL: 562 nm, yellow, HOMO: –6.20 eV, LUMO: –3.79 eV
330b, R = C$_8$H$_{17}$: Abs: 404 nm, PL: 584 nm, EL: 559 nm, yellow, HOMO: –6.18 eV, LUMO: –3.75 eV
330c, R = C$_{12}$H$_{25}$: Abs: 406 nm, PL: 580 nm, EL: 557 nm, yellow, HOMO: –6.17 eV, LUMO: –3.72 eV

SCHEME 2.47
Fluorene–phenylene vinylene copolymers and their optical and electronic properties in solid state.

electrochemical data (both oxidation and reduction appeared as quasireversible processes) [253]. No devices were fabricated with this copolymer. Green-emitting polymer **334** ITO/ PEDOT/**334**/Al devices showed a turn-on voltage of 4.8 V, a brightness of 600 cd/m^2 (at 5.8 V), and a maximum power efficiency of 0.85 lm/W (at 5.6 V) [309].

A series of four fluorene–phenylene vinylene copolymers **335**–**338** clearly demonstrates the effect of the exact position of CN groups in the vinylene fragment on the emission of the materials (Scheme 2.48) [250]. Substitution of benzene rings in copolymers **335** and **336** by thiophene results in red-shifted PL and EL, where copolymers **337** and **338** exhibit pure red emission with chromaticity values very close to the standard red (CIE: $x = 0.66$, $y = 0.34$), although no PLQY values were reported. The ITO/PEDOT/**337**/Ca/Al device showed a very low turn-on voltage of 2.6 V and a maximum brightness of 3100 cd/m^2 at 6 V. Its maximum Φ_{EL}^{ex} was 0.46% at 4 V, with a brightness of 652 cd/m^2.

Yellow to orange emission was observed in another series of fluorene–phenylene copolymers with CN groups in the vinylene fragment **339**–**341** (Scheme 2.48) [314]. The PLQY of

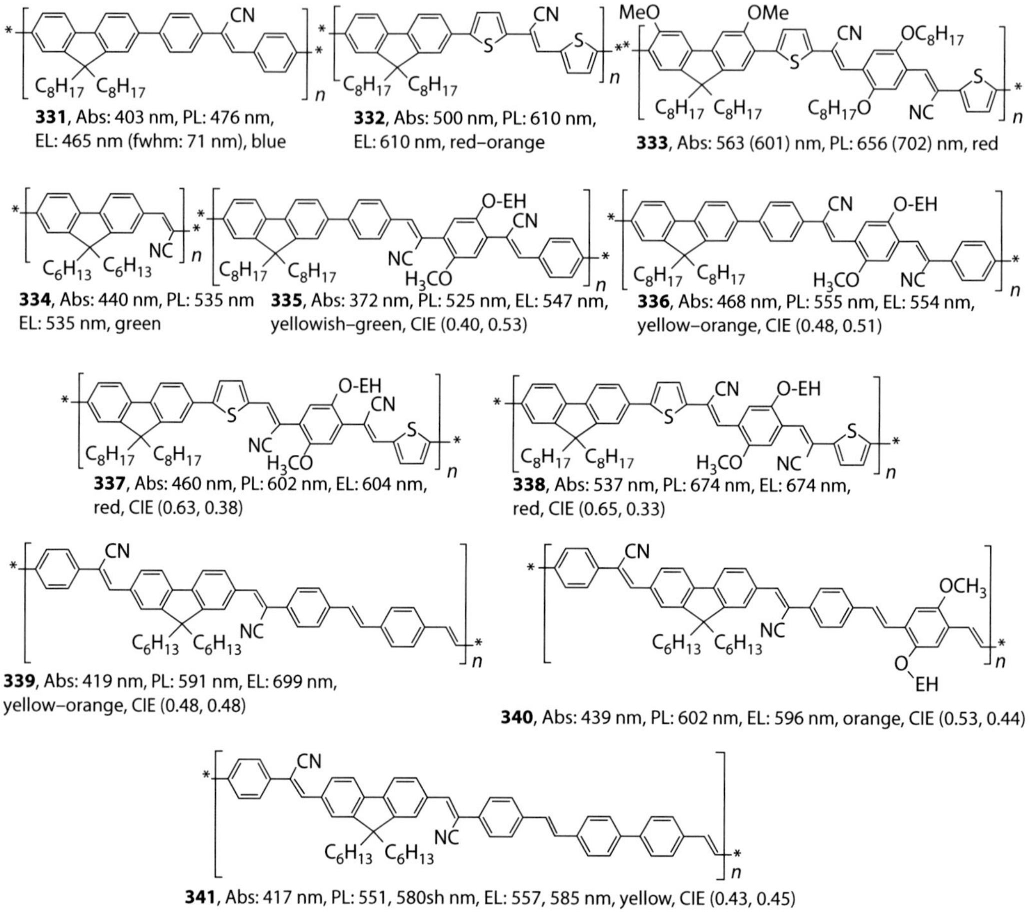

SCHEME 2.48
Fluorene–phenylene vinylene copolymers with CN groups at vinylene fragment, and their optical and electronic properties in solid state.

the copolymers was relatively low (from 3.5% for **341** to 14.7% for **340**), and the best results in PLED testing were achieved for copolymer **340**. The device ITO/PEDOT/**340**/Al showed a turn-on voltage of 5.0 V and a maximum brightness of 7500 cd/m² at 20 V, with a maximum luminance efficiency of 0.21 lm/W at 6.7 V.

Another series of red-emitting fluorene-containing copolymers of arylene–vinylene type was obtained by introducing 4-dicyanomethylenepyrane-2,6-diyl moiety in the main polymeric chain (**342–344**) [315]. The PL and EL spectra revealed a broad band at 600–800 nm and no emission from the fluorene segments (expected at ~450 nm). The PLEDs fabricated as ITO/PEDOT/polymer/Ba/Al emitted red light with maximum $\Phi_{EL}^{ex} = 0.21\%-0.38\%$, rather high turn-on voltages (10.4–11.7 V), and a brightness of ~200–450 cd/m² at a bias voltage of ~16–18 V (Chart 2.84).

PL and EL emissions from a very-low-band-gap copolymer **345** ($E_g = 1.27$ eV) were demonstrated by Swedish researchers [316]. The material has two absorption peaks at 400 and 780 nm, and emits light in the NIR region. The PL spectrum of thin films has one peak at 1035 nm, which is blue shifted by ca. 60 nm on annealing at 200°C for 10 min. The ITO/PEDOT/**345**/Ca/Al diode was positively biased when the Al/Ca electrode was connected

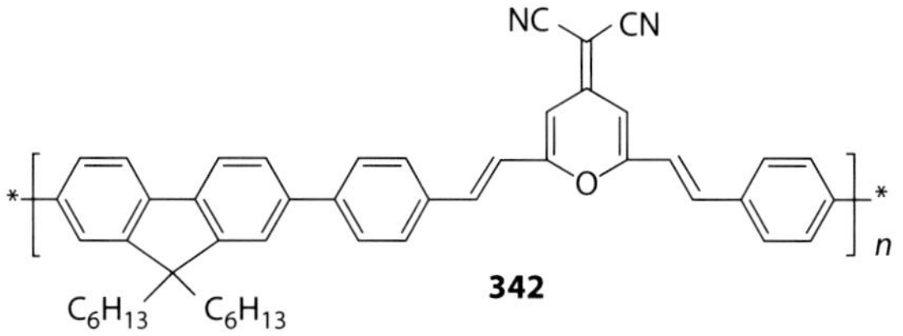

342

Abs: 352, 457 nm; PL: 682, 712 nm, EL: 657, 702sh nm, E_g = 2.22 eV

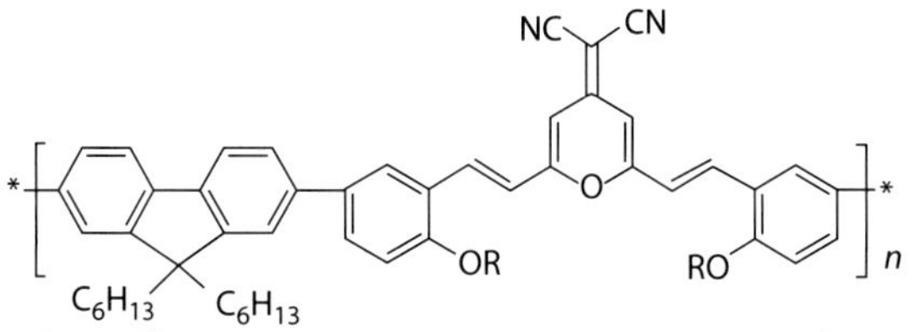

343, R = n-C_6H_{13}; Abs: 345, 435 nm; PL: 641, 704 nm, EL: 636, 694sh nm, E_g = 2.32 eV
344, R = n-$C_{12}H_{25}$; Abs: 347, 438 nm; PL: 641, 705 nm, EL: 638, 696sh nm, E_g = 2.30 eV

CHART 2.84
Chemical structures of arylene–vinylene-type PF copolymers.

to lower potential and the EL emission became observable at 1.1 V (λ_{EL} = 970 nm). The Φ_{EL}^{ex} for a nonoptimized device was rather low (0.03%–0.05%); nevertheless, demonstration of EL from PLED in the NIR can be important for communication and sensor technologies (Chart 2.85).

Pei et al. [437] reported an alternating fluorene copolymer **346** with 2,2′-bipyridyl in a side chain that emitted at 422 nm. Treating this polymer with Eu^{3+} chelates formed the polymeric complexes **347–349**. Their emission was governed by intramolecular Förster energy transfer, whose efficiency depends on the structure of the ligands and the Eu^{3+} content (Scheme 2.49) [437]. The most effective energy transfer manifested itself in a single red emission band at 612 nm for the complex **347** with a maximum intensity achieved at ~25 mol% content of Eu^{3+}.

The color of fluorene copolymers can also be shifted into the UV, as exemplified by copolymer **350** [245]. Its PL spectrum in THF solution is blue shifted by 49 nm compared with PDHF **213**, showing a somewhat lower PLQY (62%). The ITO/PEDOT/**350**/Ba/Al device emits violet light with λ_{max}^{EL} = 395 nm. However, performance is poor (turn-on voltage 8 V, maximum Φ_{EL}^{ex} = 0.054%, brightness 10 cd/m^2). When it was blended with 5% PDHF **213**,

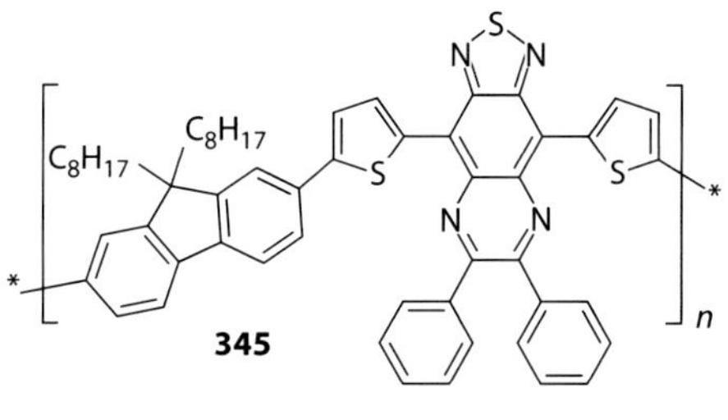

345

Abs: 400, 780 nm; PL: 1035 nm, EL: 970 nm, NIR, E_g = 1.27 eV

CHART 2.85
Chemical structures of low-band-gap PF copolymers.

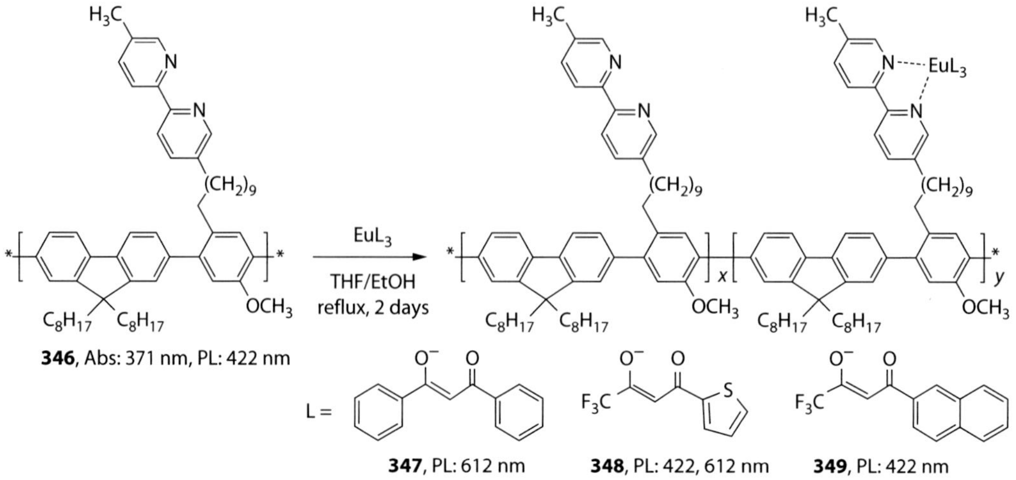

SCHEME 2.49
Synthesis of Eu-containing chelated PFs. (From Pei, J., Liu, X.-L., Yu, W.-L., Lai, Y.-H., Niu, Y.-H., and Cao, Y., *Macromolecules*, 35, 7274, 2002.)

Φ_{EL}^{ex} increased to 0.3% (slightly lower than for pure PDHF **213** measured in the same conditions) and the color purity (blue) was improved (CIE: $x = 0.18$, $y = 0.10$ at potentials 8–20 V), compared with pure PDHF **213** (CIE: $x = 0.21$, $y = 0.26$ at 5 V and $x = 0.26$, $y = 0.37$ at 10 V) (Chart 2.86).

A series of blue light–emitting conjugated polymers with trifluoren-2-yl-amine (TPA) as a building block were synthesized via Suzuki cross-coupling polycondensation [317]. The emission color can be effectively tuned in the region of deep blue with CIE coordinates of

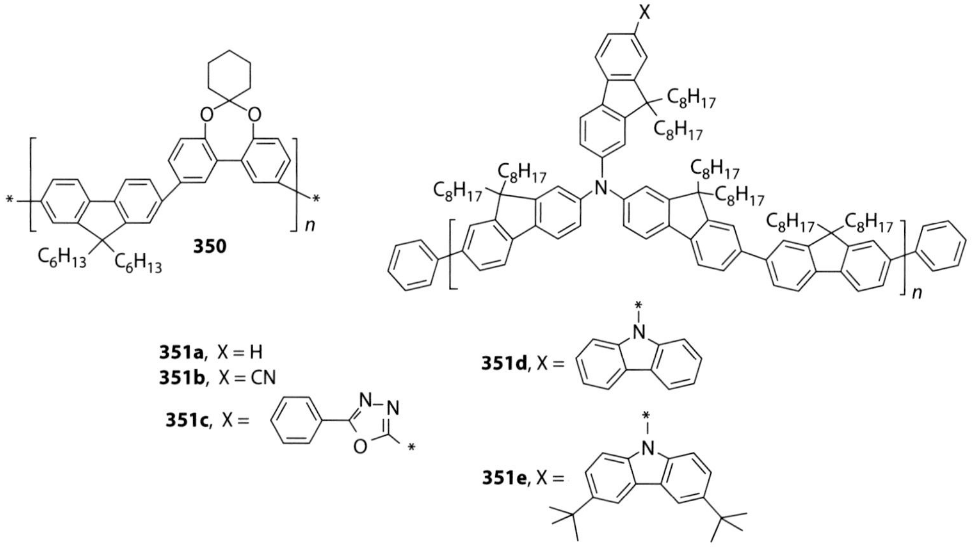

CHART 2.86
Chemical structures of PF copolymers with tunable emission.

(0.16, 0.12) (**351a**) and light blue with CIE coordinates (0.18, 0.30) (**351c**) by introducing various substituents onto the TFA as the pendants (Chart 2.86). Deep-blue light emission was achieved for the device based on polymer **351d** as an emitting layer, giving a maximum current efficiency of 2.44 cd/A, corresponding to an external QE of 3.00%. In addition, the authors also explored **351d** as an HTL for PLEDs due to its good hole-injection and -transport ability. Red and green light–emitting devices were fabricated by using **351d** as an HTL, giving maximum CE values of 7 and 25 cd/A, respectively, which are comparable to the ones based on 4,4'-bis(N-(1-naphthyl)-N-phenylamino)-biphenyl (NPB) as HTL [317].

2.3.7.3 Random (Statistical) Copolymers

Another example of efficient Förster energy transfer in Eu^{3+} complexes of fluorene copolymers (similar to the alternating copolymers described in Scheme 2.49) was demonstrated by Ling and coworkers [438] for random copolymers. They synthesized copolymers **352** with a different ratio between the fluorene and the benzene units in the backbone, and converted them into europium complexes **353** (Scheme 2.50) [438]. The complexes **353** were capable of both blue and red emission under UV excitation. In solution, blue emission was the dominant mode. However, the blue emission was significantly reduced or completely suppressed in the solid state, and nearly monochromatic (fwhm ≈ 4 nm) red emission at 613 nm was observed.

Miller's group [318,319,439] at IBM reported two series of statistical PF copolymers using perylene and cyano-substituted phenylene vinylene chromophoric segments (Scheme 2.51). Copolymers **354** and **355a–c** are soluble in organic solvents, and their molecular weights are in the range of $M_n \sim 10,000$–55,000 g/mol. They are thermally stable and show no color change upon thermal annealing. The emission color is strongly dependent on the feed ratio of the comonomers. For example, in the case of copolymers **355**, the emission color can be turned from blue (**355a**, 466 nm) through blue–green (**355b**, 481 nm) to pure green (**355c**, 510 nm).

After Miller's report, Shim and coworkers [251,320] reported a series of random copolymers synthesized by Yamamoto coupling of 2,7-dibromo-9,9-bis(2-ethylhexyl)fluorene with **356** (Scheme 2.52). Varying the feed ratio of the thiophene comonomer from 1:99 to 15:85, green-, yellow-, and red-emitting copolymers were synthesized. The actual ratio of the comonomer, calculated from nitrogen analysis, is somewhat higher than the feed ratio, reflecting a higher reactivity of **356** monomer in the Yamamoto reaction (y = 1.4, 3.1, 7.0, and 17.5% for **357a–d**, respectively). All the copolymers are soluble in common organic solvents and a molecular weight (M_n) in the range of 13,000–22,000 g/mol (PDI = 1.5–2.7) was

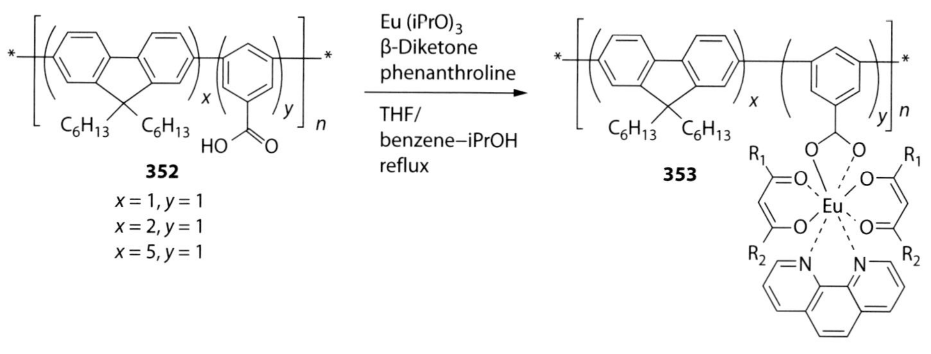

SCHEME 2.50
Synthesis of polymeric Eu complex **353**. β-Diketonate ligands are the same as in complexes **347–349** (Scheme 2.49). (From Ling, Q.D., Kang, E.T., Neoh, K.G., and Huang, W., *Macromolecules*, 36, 6995, 2003.)

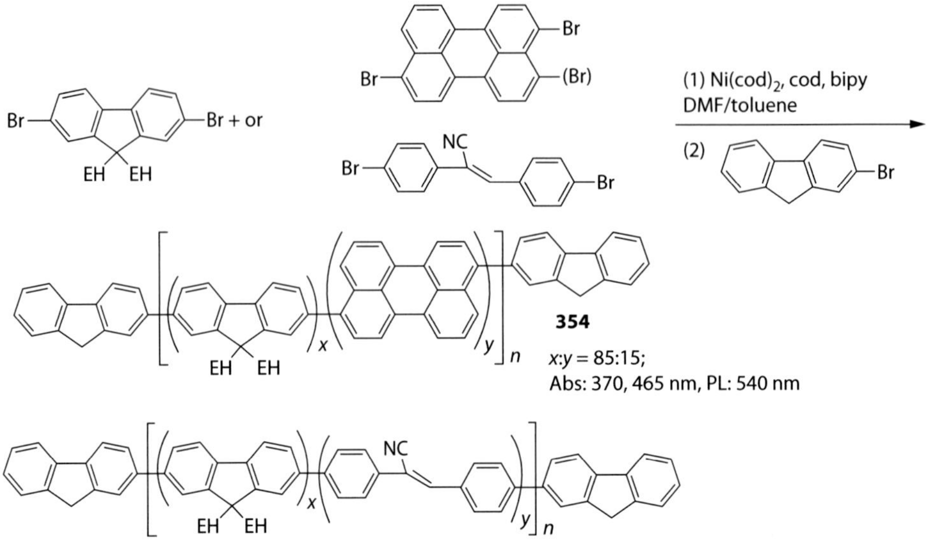

354
x:y = 85:15;
Abs: 370, 465 nm, PL: 540 nm

355a, x:y = 95:5; Abs: 376 nm, PL: 466 nm
355b, x:y = 75:25; Abs: 374 nm, PL: 481 nm
355c, x:y = 50:50; Abs: 390 nm, PL: 510 nm

SCHEME 2.51
Synthesis of perylene–fluorene and α-cyanostilbene–fluorene copolymers via Yamamoto coupling. (From Klärner, G., Lee, J.-I., Davey, M.H., and Miller, R.D., *Adv. Mater.*, 11, 115, 1999.)

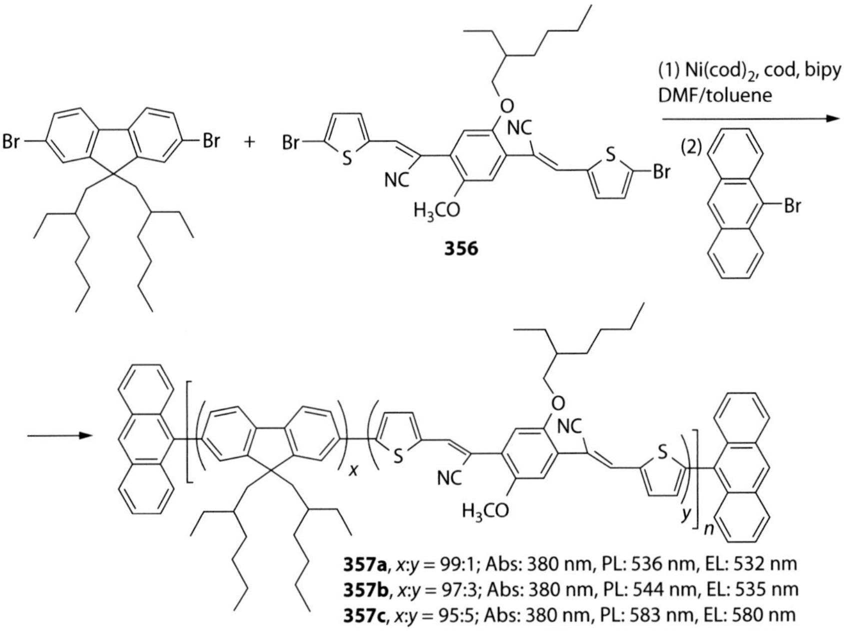

356

357a, x:y = 99:1; Abs: 380 nm, PL: 536 nm, EL: 532 nm
357b, x:y = 97:3; Abs: 380 nm, PL: 544 nm, EL: 535 nm
357c, x:y = 95:5; Abs: 380 nm, PL: 583 nm, EL: 580 nm
357d, x:y = 85:15; Abs: 379 nm, PL: 620 nm, EL: 630 nm

SCHEME 2.52
Synthesis of PF–PPV copolymers **357a–d** with tunable emission color. (From Hwang, D.-H., Cho, N.S., Jung, B.-J., Shim, H.-K., Lee, J.-I., Do, L.-M., and Zyung, T., *Opt. Mater.*, 21, 199, 2002; Cho, N.S., Hwang, D.-H., Lee, J.-I., Jung, B.-J., and Shim, H.-K., *Macromolecules*, 35, 1224, 2002.)

reported. Interestingly, regardless of composition, all the copolymers showed a similar absorption with λ_{max} at ~380 nm, with the exception of a weak shoulder at ~470 nm, which is more intense for polymers with a higher ratio of the **356** unit. In contrast, their emission colors are progressively red shifted with increasing comonomer **356** fraction. As a result, **357a** emits green (540 nm), **357b** emits yellow (585 nm), and **357c** emits bright red (620 nm), all with reasonably high PLQY (34%–69%). The EL spectra obtained from PLEDs fabricated as ITO/PEDOT/polymer/LiF/Al are almost identical to the PL spectra. As the feed ratio of **356** increases, the threshold voltage drops from 13 V to only 5 V, and the emission intensity at low voltages also increases, which suggests improved charge transport and balance in the material. As can be calculated from the current–voltage–luminance (*I–V–L*) plot, a luminous efficiency of ~0.7 cd/A (at 15 V), ~2.4 cd/A (at 12 V), and ~1.5 cd/A (at 9.5 V) is achieved with polymers **357a–c**, respectively.

Hwang et al. [308] reported random copolymers of PF with PPV units in different feed ratios through Gilch polymerization (Scheme 2.53). The molecular weights of the obtained copolymers are in the range of 57,000–87,000 g/mol with PDI of 1.4 to 2.1. Interestingly, the copolymers **358** with different feed ratio *x:y* showed the same optical properties as those of the corresponding fluorene–vinylene homopolymer (*y* = 0, **324a**). The UV–vis spectrum of the thin films showed a peak absorption at 418 nm with onset absorption of 485 nm and PL emission maximum at 470 nm plus well-defined vibronic bands at ~500, 530, and 560 nm. The PLED devices, consisting of ITO/PEDOT/**358**/Al, gave very similar EL emission spectra (CIE: *x* = 0.23, *y* = 0.38) and significantly improved device performance (0.71–1.05 cd/A), compared with the homopolymer **324** (0.13 cd/A). The latter was explained to be due to improved hole injection and transport, owing to the PPV segments. Indeed, the ionization potential of the copolymers is around 5.53–5.55 eV, which is much closer to the HOMO level of PEDOT (5.2 eV), when compared with homopolymer (5.73 eV).

It is interesting that statistical copolymers **359**, containing *m*-phenylene linkages that are supposed to interrupt conjugation, showed a PL maximum of 475 nm, similar to **358**. Owing to efficient energy transfer from the *meta*- to the *para*-linked chromophores, the emission maxima did not depend on the ratio of *m*- and *p*-divinylbenzenes, unless 100% loading of the *meta* units was used [440] (Scheme 2.54).

The emission wavelength of PF–PPV copolymer can be red shifted and the band gap reduced by introducing alkoxy substituents, as demonstrated for compounds **360** [307]. Consequently, the transport of both holes and electrons is facilitated and the PLED built in the configuration ITO/PEDOT/**360**/Al produces an orange–red color (λ_{EL} = 574–592 nm) with a maximum brightness for **360b** of 1350 cd/m^2 and a luminous efficiency of 0.51 cd/A, at a rather low turn-on voltage of 2.5 V (Chart 2.87).

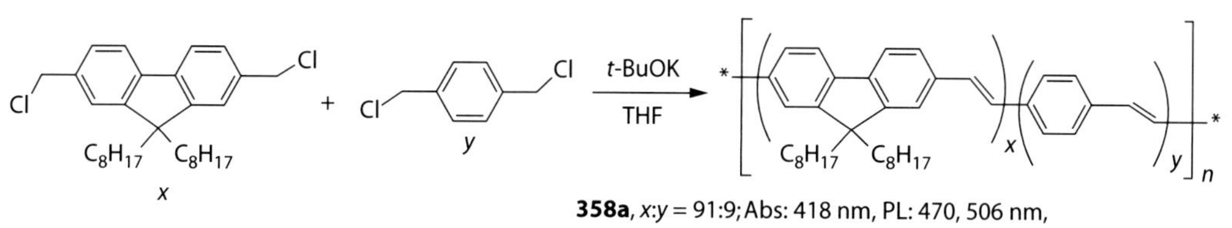

358a, *x:y* = 91:9; Abs: 418 nm, PL: 470, 506 nm,
Φ_{PL}: 36%, EL: 472 nm, CIE (0.23, 0.38)
358b, *x:y* = 59:41; Abs: 419 nm, PL: 470, 506 nm,
Φ_{PL}: 36%, EL: 475 nm, CIE (0.23, 0.38)

SCHEME 2.53
Gilch polymerization to synthesis of random copolymers. (From Hwang, D.-H., Lee, J.-D., Kang, J.-M., Lee, S., Lee, C.-H., and Jin, S.-H., *J. Mater. Chem.*, 13, 1540, 2003.)

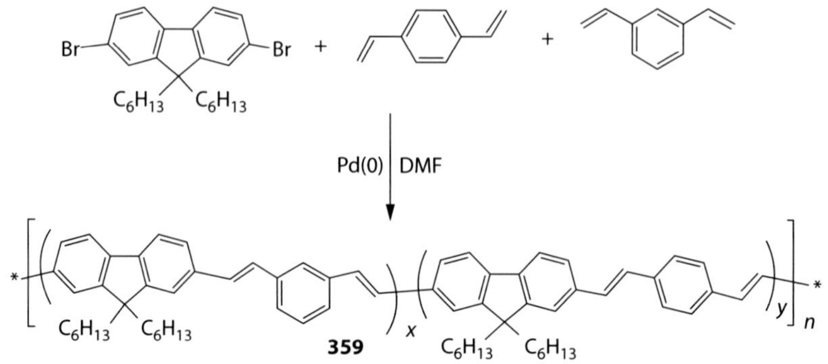

SCHEME 2.54
Random fluorene–divinylbenzenes copolymers by Heck polymerization. (From Cho, H.N., Kim, J.K., Kim, D.Y., Kim, C.Y., Song, N.W., and Kim, D., *Macromolecules*, 32, 1476, 1999.)

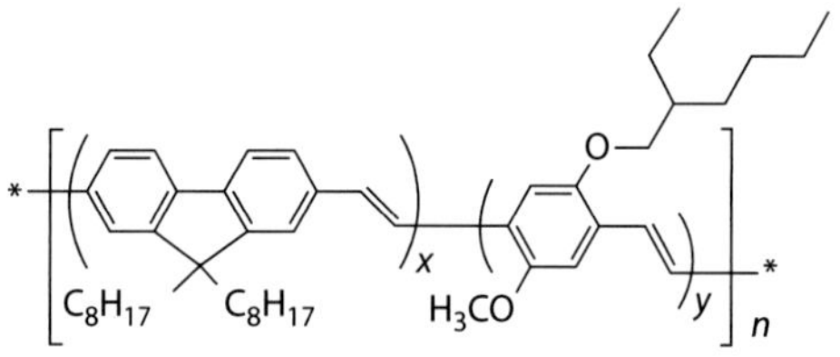

360a, *x:y* = 95:5; Abs: 418 nm, PL: 576 nm, EL: 574 nm, HOMO: −5.36 eV, LUMO: −2.77 eV
360b, *x:y* = 90:10; Abs: 421 nm, PL: 576 nm, EL: 576 nm, HOMO: −5.16 eV, LUMO: −2.62 eV
360c, *x:y* = 80:20; Abs: 420 nm, PL: 589 nm, EL: 589 nm, HOMO: −5.31 eV, LUMO: −2.77 eV
360d, *x:y* = 50:50; Abs: 422 nm, PL: 593 nm, EL: 592 nm, HOMO: −5.09 eV, LUMO: −2.99 eV
324a, *x:y* = 100:0; Abs: 419 nm, PL: 507 nm, EL: 508 nm, HOMO: −5.34 eV, LUMO: −2.75 eV
13, *x:y* = 0:100; Abs: 502 nm, PL: 582 nm, EL: 586 nm, HOMO: −4.90 eV, LUMO: −2.80 eV

CHART 2.87
Chemical structures of PF–PPV copolymers.

Several random fluorene–thiophene copolymers such as **361a−e** [247,321], **362** [441], **363** [440,442] have been investigated. Because of the possible fine tuning of the comonomer ratio, many of these have shown an EL performance far greater than that of the PF homopolymer or the corresponding alternating copolymers. Thus, the PLED ITO/PEDOT/**362**/Ba/Al showed a brightness of ca. 2600 cd/m² (at 8 V) and Φ_{EL}^{ex} = 1.25%. The *"meta"* linkage of the dibenzothiophene in copolymers **361a−e** limits the effective conjugation length and restricts emission to the deep-blue color range (λ_{max} ~ 420 and 440 nm) [247]. Furthermore, hindered conjugation (within the chain) and π-aggregation (between chains) suppress the parasitic green emission at 520 nm in the solid state (Chart 2.88).

Cao and coworkers synthesized three series of copolymers **364** [322], **365** [323], **366** [324], and **367** [325], exploiting random copolymerization of fluorene fragments with dibromo-derivatives of Se,N and S,N heterocycles (Scheme 2.55). Fluorene–benzoselenadiazole alternating copolymer **364** (*x:y* = 50:50) showed substantial (55 nm) red shift in PL, compared with its BT analog **320** (595 nm [251] and 540 nm [300,399], respectively) because of a narrower π−π* gap of the benzoselenadiazole unit. Increasing the content of the fluorene moieties in the copolymers **364** results in a regular PL blue shift of (568 nm for *x:y* = 98:2), although irregular variations in Φ_{PL} (between 16% and 51% in films) was reported for this series of copolymers. Whereas the main PL in solution is observed in the region of

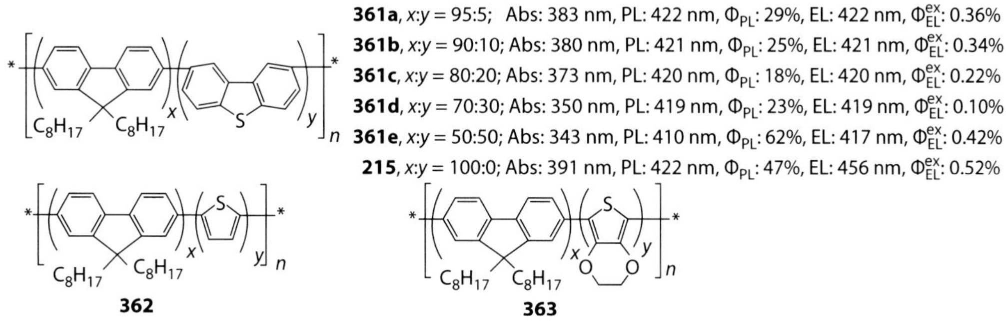
361a, x:y = 95:5; Abs: 383 nm, PL: 422 nm, Φ_{PL}: 29%, EL: 422 nm, Φ_{EL}^{ex}: 0.36%
361b, x:y = 90:10; Abs: 380 nm, PL: 421 nm, Φ_{PL}: 25%, EL: 421 nm, Φ_{EL}^{ex}: 0.34%
361c, x:y = 80:20; Abs: 373 nm, PL: 420 nm, Φ_{PL}: 18%, EL: 420 nm, Φ_{EL}^{ex}: 0.22%
361d, x:y = 70:30; Abs: 350 nm, PL: 419 nm, Φ_{PL}: 23%, EL: 419 nm, Φ_{EL}^{ex}: 0.10%
361e, x:y = 50:50; Abs: 343 nm, PL: 410 nm, Φ_{PL}: 62%, EL: 417 nm, Φ_{EL}^{ex}: 0.42%
215, x:y = 100:0; Abs: 391 nm, PL: 422 nm, Φ_{PL}: 47%, EL: 456 nm, Φ_{EL}^{ex}: 0.52%
362
363

CHART 2.88
Chemical structures of random fluorene–thiophene copolymers.

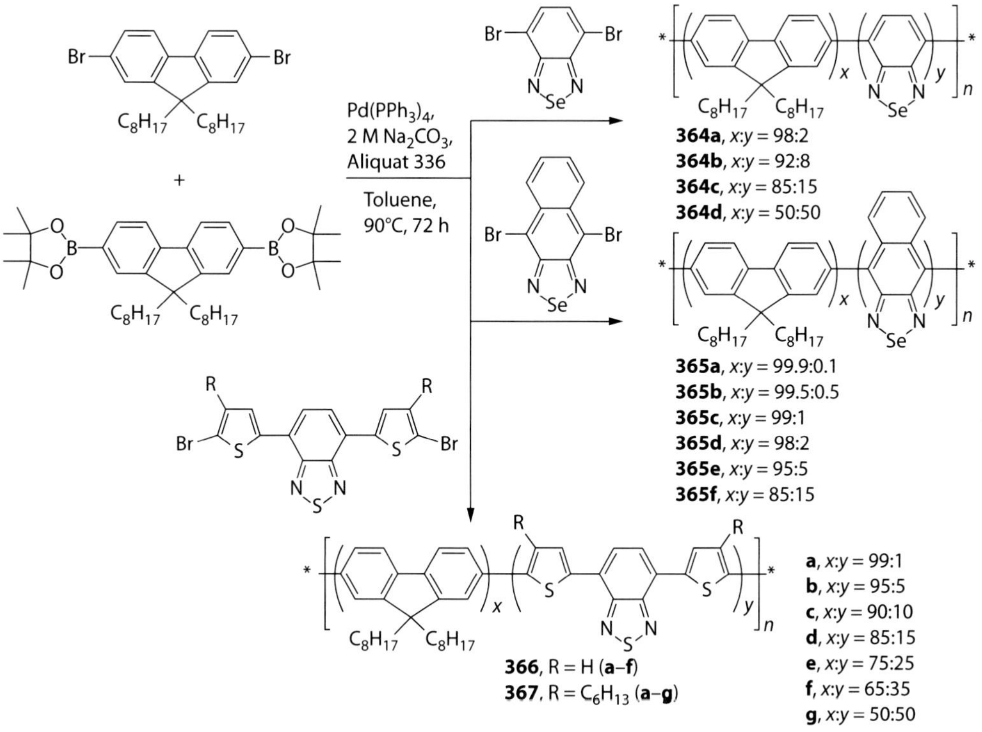
Pd(PPh3)4, 2 M Na2CO3, Aliquat 336
Toluene, 90°C, 72 h
364a, x:y = 98:2
364b, x:y = 92:8
364c, x:y = 85:15
364d, x:y = 50:50
365a, x:y = 99.9:0.1
365b, x:y = 99.5:0.5
365c, x:y = 99:1
365d, x:y = 98:2
365e, x:y = 95:5
365f, x:y = 85:15
366, R = H (**a–f**)
367, R = C6H13 (**a–g**)
a, x:y = 99:1
b, x:y = 95:5
c, x:y = 90:10
d, x:y = 85:15
e, x:y = 75:25
f, x:y = 65:35
g, x:y = 50:50

SCHEME 2.55
Synthesis of benzothiaselanazole–, naphthoselenathiazole–, and (thiophene–benzothiazole–thiophene)–fluorene copolymers via Suzuki coupling. (From Yang, R., Tian, R., Hou, Q., Yang, W., and Cao, Y., *Macromolecules*, 36, 7453, 2003; Yang, J., Jiang, C., Zhang, Y., Yang, R., Yang, W., Hou, Q., and Cao, Y., *Macromolecules*, 37: 1211, 2004; Hou, Q., Xu, Y., Yang, W., Yuan, M., Peng, J., and Cao, Y., *J. Mater. Chem.*, 12, 2887, 2002; Hou, Q., Zhou, Q., Zhang, Y., Yang, W., Yang, R., and Cao, Y., *Macromolecules*, 37, 6299, 2004.)

570–600 nm with a low-intensity short-wavelength band (due to fluorene emission), the latter is completely suppressed in films. PLEDs fabricated as ITO/PEDOT (or PVK)/**364**/Ba/Al showed orange–red emission ($\lambda_{max}^{EL} = 573 - 600$ nm, depending on the x:y ratio) with no blue emission, even at the lowest concentration (2%) of the heterocyclic units. Comparison of devices with two different HTL (PEDOT or PVK) showed that LEDs with a PVK layer

(ITO/PVK/**364** ($x{:}y$ = 85:15)/Ba/Al) demonstrated a better Φ_{EL}^{ex} that reached a value of 1% (λ_{max}^{EL} = 582 nm; CIE: x = 0.698, y = 0.300) [322].

A much more pronounced bathochromic shift in emission was achieved for copolymers **365**, which emit in the red with λ_{PL} = 634–681 nm (Φ_{PL} = 33%–84%). Although blue emission at 423–438 nm from the fluorene fragments was also observed, its intensity decreased with decreasing $x{:}y$ ratio [323]. In EL spectra, the blue emission from the PF segment was completely quenched at very low naphthoselenadiazole content (0.5%), which could be due to efficient exciton and charge trapping on the narrow-band-gap naphthoselenadiazole sites. The highest Φ_{EL}^{ex} = 3.1% was reported for ITO/PEDOT/**365** ($x{:}y$ = 99:1)/Ba/Al diodes (although it corresponds to a relatively low luminous efficiency of 0.91 cd/A very likely due to insensitivity of the human eye to this particular spectral distribution) with a maximum brightness of up to 2100 cd/m^2 (CIE: x = 0.64, y = 0.33).

The last examples of red-emitting fluorene copolymers in this series, the copolymers **366** and **367**, contain both BT acceptor and thiophene donor units in the main chain [324,325]. As expected for donor–acceptor alternation in the main chain of the conjugated polymers, a substantial red shift in PL, compared with copolymer **320** or even **364**, was observed (λ_{PL} = 635–685 nm (**366** [324]) 629–678 nm (**367b** [325])) in films for $x{:}y$ = 99:1–65:35). In dilute solution, additional emission from fluorene segments is completely suppressed by increasing the polymer concentration or the content of heterocyclic fragment in the polymer chain. ITO/PEDOT/**366**/Ba/Al devices showed Φ_{EL}^{ex} in the range of 0.5%–1.4%. The highest efficiency of 1.4% and the luminance of 256 cd/m^2 at a bias of 5.1 V was achieved for the copolymer having an $x{:}y$ = 85:15 ratio. This device reached a maximum luminance of 3780 cd/m^2 at 8.2 V. Even higher Φ_{EL}^{ex} = 1.93%–2.54% was achieved with copolymers **367b** ($x{:}y$ = 95:5–85:15) for the device ITO/PEDOT/PVK/**367b**/Ba/Al [325]. These copolymers showed saturated red emission with λ_{EL} = 634–647 (x = 0.66–0.67, y = 0.33–0.34).

Kong and coworkers [326] synthesized a phenothiazine-containing alternating fluorene copolymer **368** by Suzuki coupling (Chart 2.89). The phenothiazine-3,7-diyl fragment in the polymer backbone interrupts the conjugation, and substantially blue shifts the absorption compared with PDHF homopolymer **213** (328 and 385 nm, respectively). Nevertheless, the ionization potential of **368** (estimated from the electrochemical data as $I_p = E_{ox}^{onset} + 4.4\,V$) is much higher (5.1 eV) than that for PDHF **213** due to the electron-rich phenothiazine unit, suggesting good hole-transport properties. Greenish–blue EL $\left(\lambda_{max}^{EL} = 490\,nm\right)$ with a luminance of up to 320 cd/m^2 and a maximum Φ_{EL}^{ex} = 0.10% at 12.5 V was observed for an ITO/PEDOT/**368**/Al diode. Similar random fluorene–phenothiazine copolymers **369a–c** showed close EL maxima but much better device performance (e.g., for **369b**: maximum

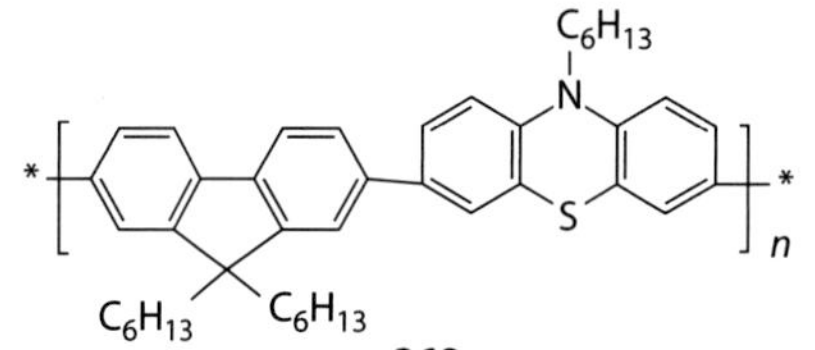

368

Abs: 328 nm, PL: 490 nm, EL: 490 nm

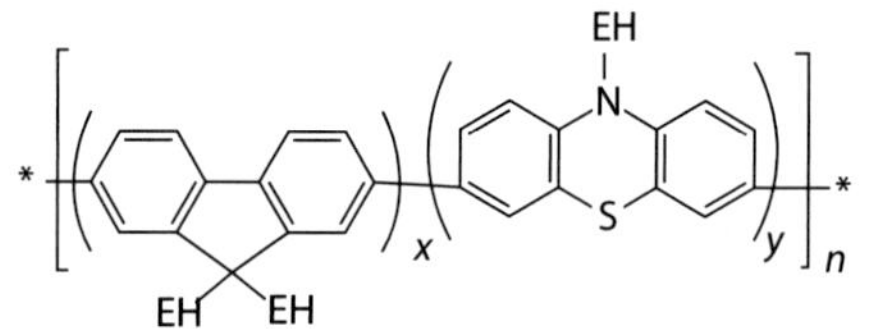

369a, $x{:}y$ = 92:8; Abs: 373 nm, PL: 474 nm, EL: 480 nm
369b, $x{:}y$ = 88:12; Abs: 373 nm, PL: 480 nm, EL: 484 nm
369c, $x{:}y$ = 74:26; Abs: 367 nm, PL: 478 nm, EL: 480 nm

CHART 2.89
Chemical structures of phenothiazine-containing PF copolymers.

brightness 4170 cd/m², power efficiency 2.08 cd/A, turn-on voltage 3.8 V; CIE: $x = 0.17$, $y = 0.37$) [252].

Hwang et al. [327] studied EL from the devices fabricated using blends of similar blue-emissive fluorene–phenothiazine copolymer **370** with MEH–PPV (Chart 2.90). The maximum brightness of the devices ranged from 1580 to 2640 cd/m² with $\Phi_{EL}^{ex} = 0.3\% - 0.4\%$. The inefficient energy transfer between these blue and red LEP enabled the production of white-light emission through control of the blend ratio; with an increasing amount of **370** in the blends, CIE coordinates of EL emission are shifted from $x = 0.19$, $y = 0.45$ (for 1% of **370**) to $x = 0.36$, $y = 0.51$ (for 3% of **370**), although even the most optimal blend (2.5% of **370**: $x = 0.19$, $y = 0.45$) is rather far from the real white point [443].

Ego and coworkers [444] achieved efficient color tuning in PF via excitation energy transfer onto perylene dye fragments introduced as (i) randomly distributed comonomer in the PF chain, (ii) end-capping group, or (iii) side chain. Perylene compounds are known for their high stability, great QE, and large range of emission colors, achieved via introduction of different substituents. The random copolymers **371–377** were synthesized via Yamomoto coupling of dibromodialkylfluorenes with corresponding brominated perylenes (1%–5%) (Chart 2.91). The M_n in the range of 30,000–140,000 (90–400 units in a chain) and very high polydispersity (4.1–7.7) observed in these polymers were explained by relatively low reactivity of bromoperylene comonomers. The PL spectra of the copolymers in solution were essentially identical to those of PF homopolymers, while in the solid state an efficient energy transfer onto the dye fragments occurred, giving rise to an additional red-shifted emission band, whose position depends on the perylene structure. The energy transfer appears to be as efficient with 1% of the dye as with 5%. Generally, the PL efficiencies of the copolymers in solid films varied from 38% to 56%, with the exception of material **375** that showed a PL efficiency of only 7%. The latter contains 3% of both perylene fragments. The low QE was explained by aggregation of the perylene units in the solid state. The devices made from these copolymers in configuration ITO/PEDOT/copolymer/Ca showed stable emission color with $\Phi_{EL}^{ex} = 0.2\% - 0.6\%$ and an EL efficiency of 0.9–1.6 cd/A, which are rather high for nonoptimized devices. In contrast to PL spectra, there was a complete energy transfer in the EL spectra (no fluorene emission), and the emission color can be finely tuned by structural modifications in the perylene units (Table 2.3, Figure 2.23).

A very efficient energy transfer (producing emission at 613 nm) was observed in the PL spectra of the perylene end-capped polymer **377** in solid films. This material had the highest QE (>60%) among the fluorene–perylene polymers, although the performance of its PLED has not yet been reported [444].

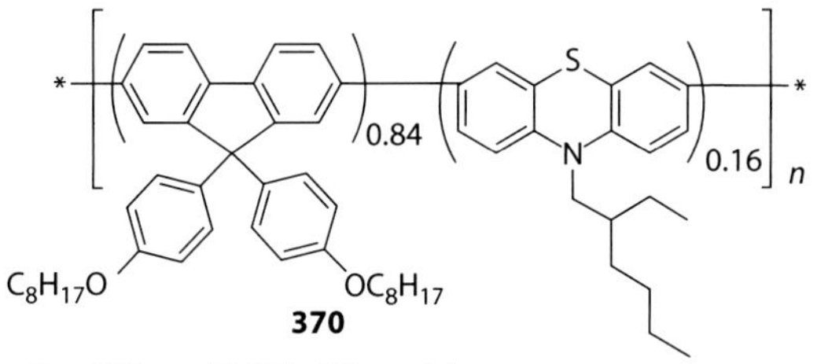

370

Abs: 371 nm, PL (EL): 482 nm, blue,
E_g: 2.82, HOMO: −5.40 eV, LUMO: −2.58 eV

CHART 2.90
Chemical structure of a fluorene–phenothiazine copolymer.

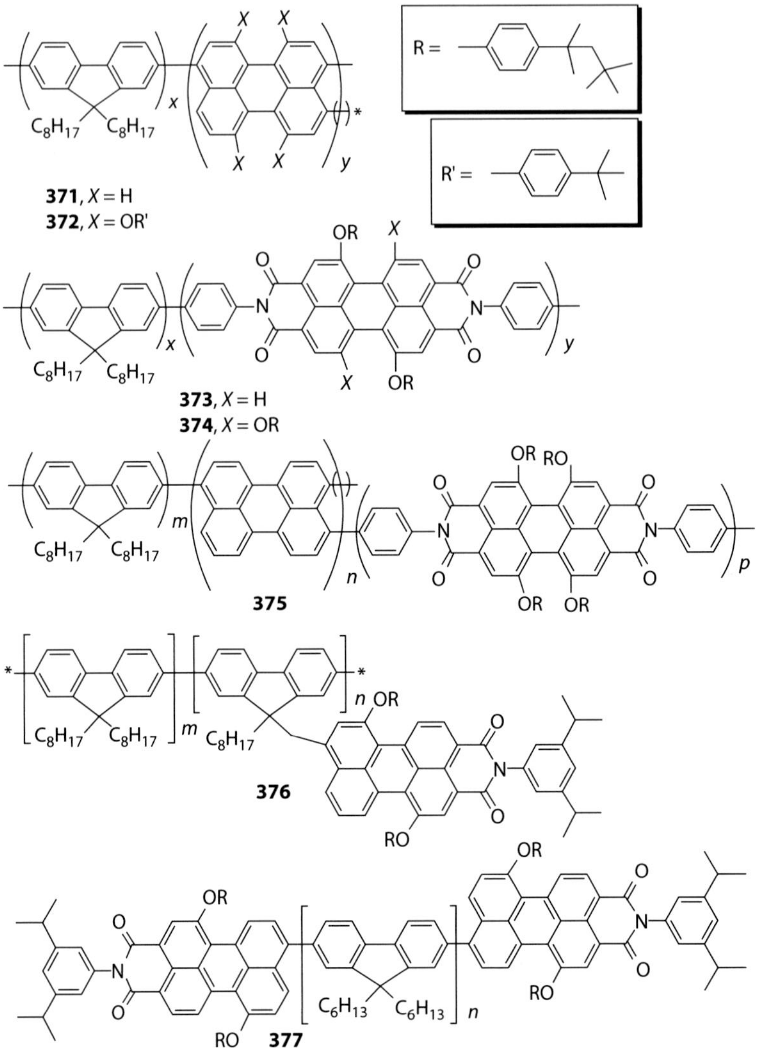

CHART 2.91
Chemical structures of PFs with perylene units.

TABLE 2.3

Optical and Electronic Properties of Fluorene–Perylene Copolymers

Polymers (% Dye), Emission Color	M_n (g/mol) (PDI)	Φ_{PL} (Film)	λ_{EL}^{ex} (nm) $\left[\Phi_{EL}^{ex}\right]$, CIE (x, y)	Turn-On Voltage (V)	Luminous Efficiency (cd/A)
371 (3%), bright green	47,930 (4.1)	51%	520 [0.6%], (0.362, 0.555)	12	0.9
372 (5%), yellow	32,300 (4.9)	40%	558 [0.2%], (0.414, 0.519)	11	0.4
373 (5%), deep red	63,510 (3.6)	33%	675 [0.5%], (0.636, 0.338)	8	1.6
374 (1%), red–orange	142,500 (3.8)	42%	600 [0.3%], (0.590, 0.365)	15	1.4

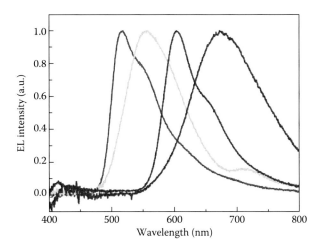

FIGURE 2.23
EL spectra of the copolymers (from left to right) **371**, **372**, **374**, and **373** in devices ITO/PEDOT/polymer/Ca/Al. (From Ego, C., Marsitzky, D., Becker, S., Zhang, J., Grimsdale, A.C., Müllen, K., MacKenzie, J.D., Silva, C., and Friend, R.H., *J. Am. Chem. Soc.*, 125, 437, 2003. With permission.)

Attaching perylene moieties as side groups allows achievement of high concentration without affecting the electronic structure of the polymer backbone. Putting 16% perylene moieties as side chains predictably results in more efficient energy transfer, observed with polymer **376**, both in solution and solid state (emission band at 599 nm). Although no PLED device with **376** has been reported, this material showed excellent performance in solar cells (external photovoltaic QE = 7%, in blend with PPV) [444].

A soluble terpolymer (**378**) containing fluorine, thiophene, and benzothiadiazole units were synthesized via a Stille coupling reaction [318]. A deep red–emitting OLED device with the configuration of ITO/PEDOT:PSS/**378**/TPD (weight ratio being 1:1)/BCP/Alq$_3$/LiF/Al was obtained, and the maximum luminance was measured to be 226 cd/m^2 at the bias voltage of 10 V. The EL peak located at 708 nm, and the spectrum covered both the red and infrared regions (Chart 2.92). Ozturk's group reported three groups of fluorine-dithieno[3,2-*b*;2′,3′-*d*]-thiophene-*S*,*S*-dioxides copolymers (**379–381**) synthesized through a Suzuki coupling method [329]. A spread of colors from light blue (border of white) to red through green and yellow was obtained with OLED applications of the copolymers. The highest quantum yield was obtained with **380** as 0.66 (Chart 2.92).

Alex Jen's group [445] reported a series of highly efficient random conjugated copolymers (**382–386**), containing dialkylfluorene and electron-deficient BT units, synthesized via Suzuki-coupling polymerization (Scheme 2.56). To balance the charge-transport and charge-injection properties, a third, electron-rich unit was introduced in the polymer. All of the copolymers showed very similar absorption spectra (λ_{max} = 380 nm). The luminescence properties of the copolymers are dominated by the BT unit; all the copolymers have similar red-shifted luminescence peaks at λ_{PL} = 540 nm (Φ_{PL} = 23%–55%, depending on the monomer), and the third comonomer does not contribute to the emission spectra. This can be explained by a charge transfer or energy transfer process between the electron-rich segments and electron-deficient BT units. Although the luminescence properties of the copolymers are the same, their charge-transport properties are rather different. The authors studied current–voltage characteristics of metal–polymer–metal junctions, fabricated with

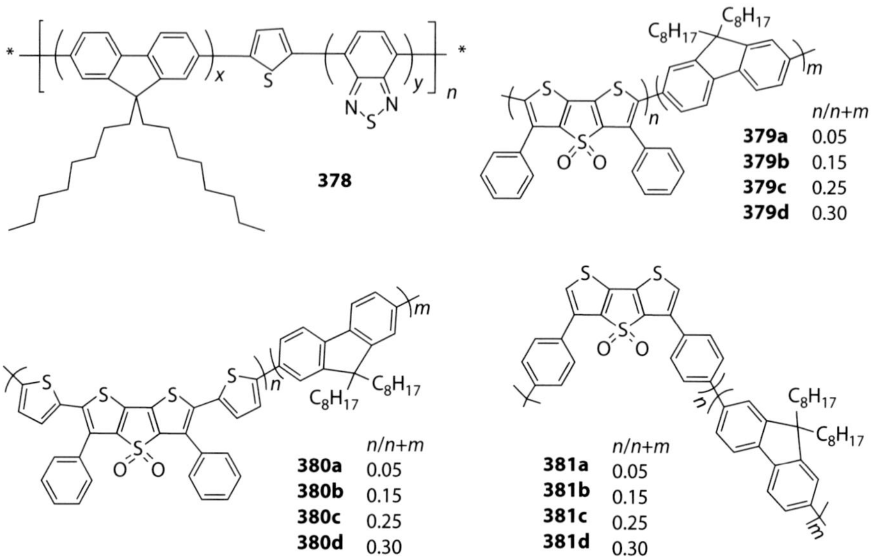

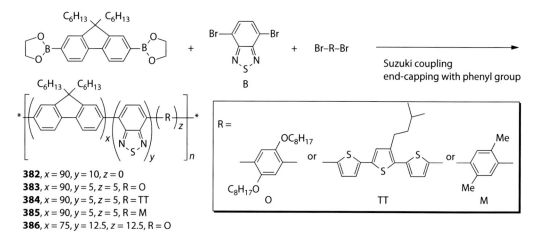

CHART 2.92
Chemical structures of PF terpolymers.

382, $x = 90, y = 10, z = 0$
383, $x = 90, y = 5, z = 5, R = O$
384, $x = 90, y = 5, z = 5, R = TT$
385, $x = 90, y = 5, z = 5, R = M$
386, $x = 75, y = 12.5, z = 12.5, R = O$

SCHEME 2.56
Synthesis of fluorene–benzothiadiazole copolymers. (From Herguth, P., Jiang, X., Liu, M.S., and Jen, A.K.-Y., *Macromolecules*, 35, 6094, 2002.)

metals of high and low work function. As expected, the highest hole conductivity belongs to terthiophene-containing polymer **384** (at the expense of the lowest electron conductivity). The highest electron conductivity was achieved with polymer **386**, having the highest ratio of electron-deficient BT component. Remarkably, the last polymer showed the best performance in LED devices, demonstrating that increasing the electron conductivity balances the charge-transport properties in PF materials. The PLEDs were fabricated in the configuration ITO/HTL/copolymer/Ca (HTL was either PEDOT or BTPD–PFCB, **143**).

The lowest performance was revealed by copolymer **384**, attributed to the oligothiophene units' quenching effect. The best device, fabricated in the configuration ITO/BTPD–PFCB **143**/**386**/Ca/Al, had a remarkable Φ_{EL}^{ex} of 6.0% and a photometric power efficiency of 18.5 cd/A (an even higher efficiency of 28.6 cd/A for an unspecified device structure is reported in Section 2.6 [445]). The highest brightness of 59,400 cd/m² was achieved with this device at 15.2 V. Interestingly, when PEDOT was used in the same device structure, instead of BTPD–PFCB **143**, the PLED performance was lower: $\Phi_{EL}^{ex} = 1.5\%$, an EL current to light efficiency of 4.66 cd/A, and a maximum brightness of 21,000 cd/m². However, as seen in Figure 2.24, the higher performance of HTL **143** appears only at relatively high voltage (>12 V) and thus cannot be taken as a general rule for all PLEDs.

A further development of this approach with multicomponent PF copolymers for tuning the emission color was recently exemplified by fabrication of a red–blue–green (RGB) prototype display, where pure red, green, and blue colors were achieved by simple variation of the feed ratio of several monomers (Scheme 2.57) [446]. The resulting polymers were very soluble in organic solvents, had high molecular weight ($M_n \sim 50,000$), and revealed a respectable

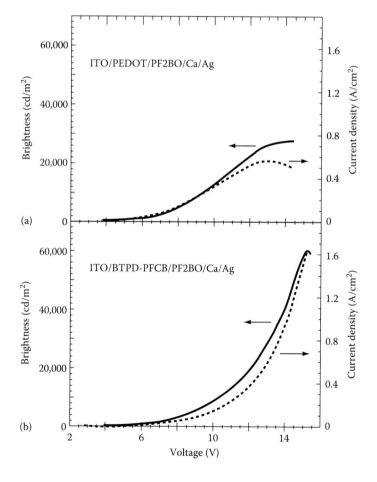

FIGURE 2.24

Plot of brightness (dashed line) and current density (dotted line) vs. applied voltage for PLEDs (a) ITO/PEDOT/**386**/Ca/Ag and (b) ITO/**143**/**386**/Ca/Ag. (From Herguth, P., Jiang, X., Liu, M.S., and Jen, A.K.-Y., *Macromolecules*, 35, 6094, 2002. With permission.)

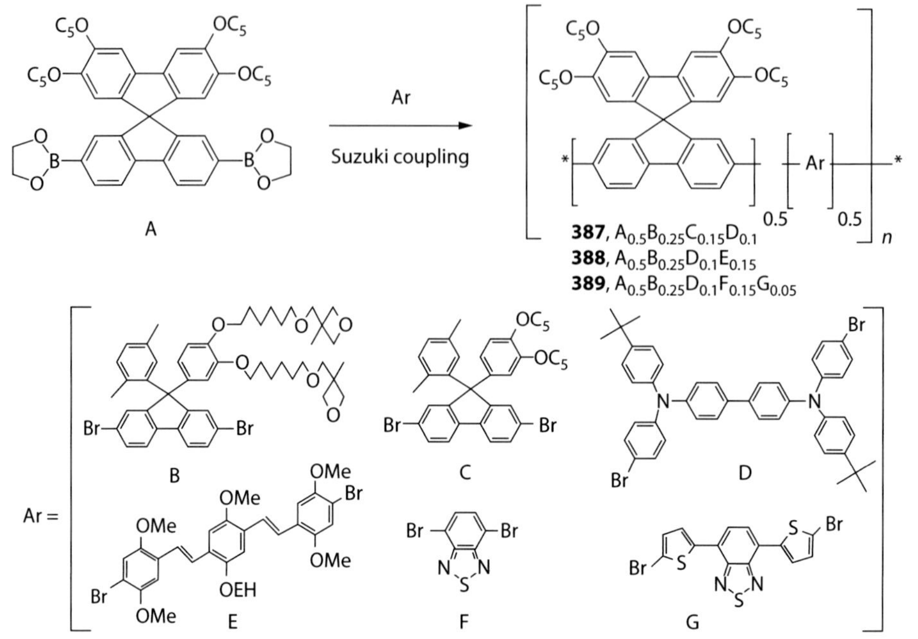

SCHEME 2.57
Suzuki synthesis of multicomponent copolymers for RGB PLED display. (From Müller, C.D., Falcou, A., Reckefuss, N., Rojahn, M., Widerhirn, V., Rudati, P., Frohne, H., Nuyken, O., Becker, H., and Meerholz, K., *Nature*, 421, 829, 2003.)

EL performance. The PLED fabricated as ITO/PEDOT/polymer/Ca/Al showed switch-on voltages of 4.5 V for blue emitter **387**, 3.8 V for green emitter **388**, and 7.5 V for red emitter **389**, with a maximum EL efficiency of 3.0, 6.5, and 1.1 cd/A, respectively. The presence of a photopolymerizable (in the presence of photoacid) oxetane unit in the comonomer B renders insoluble cross-linked polymer upon photolithographic development, allowing solution process fabrication of the PLED display, bearing different emitting materials.

White PLEDs have received particular attention lately for their potential applications in lighting sources to replace the incandescent light bulb and fluorescent lamp, and as backlight in liquid-crystal displays. Wang's group has developed a series of polyfluorene copolymers (**390–396**) chemically doped with other color-emitting components to achieve white-light emission in a single polymer, which shows advantages over the polymer blending system in terms of both efficiency and color stability (Chart 2.93) [447–450].

White light–emitting polyfluorene copolymers with orange fluorescent 1,8-naphthalimide moieties were synthesized via Yamamoto polycondensation reaction. A white polymeric LED with a structure of ITO/PEDOT/**390**/Ca/Al showed a current efficiency of 5.3 cd/A and a power efficiency of 2.8 lm/W at 6 V with the CIE coordinates at (0.25, 0.35) [447] (Chart 2.93). The same group reported a novel strategy to realize white electroluminescence with simultaneous blue, green, and red emission from a single polymer [448]. This is achieved by attaching a small amount of a green-emissive component (4-diphenylamino-1,8-naphthalimide) to the pendant chain and incorporating a small amount to a red-emissive component (4,7-*bis*(5-(4-(*N*-phenyl-*N*-(4-methylphenyl)amino)phenyl)-thienyl-2-)-2,1,3-benzothiadiazole) into the main chain of the macromolecule, which itself has a blue emission. White electroluminescence with simultaneous blue (445 nm), green (515 nm), and red (624 nm) emission is achieved from a single-layer device with a

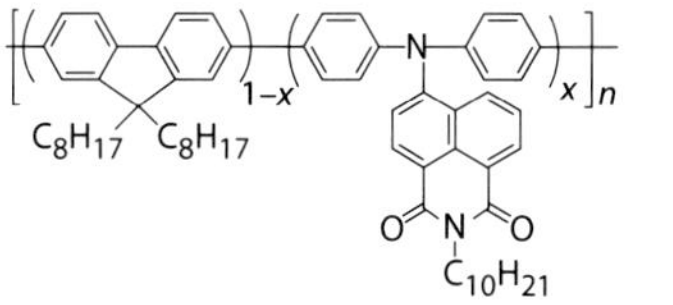

390a, *x* = 0.08
390b, *x* = 0.02
390c, *x* = 0.005
390d, *x* = 0.0005

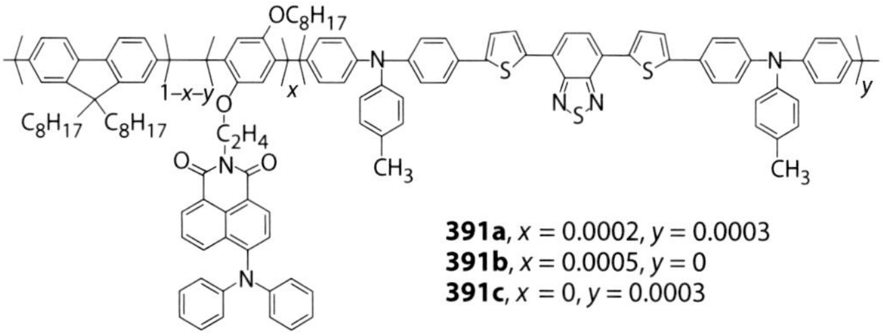

391a, *x* = 0.0002, *y* = 0.0003
391b, *x* = 0.0005, *y* = 0
391c, *x* = 0, *y* = 0.0003

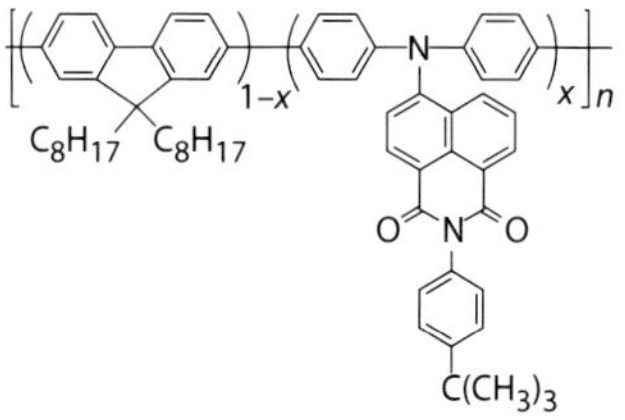

392, *x* = 0.005

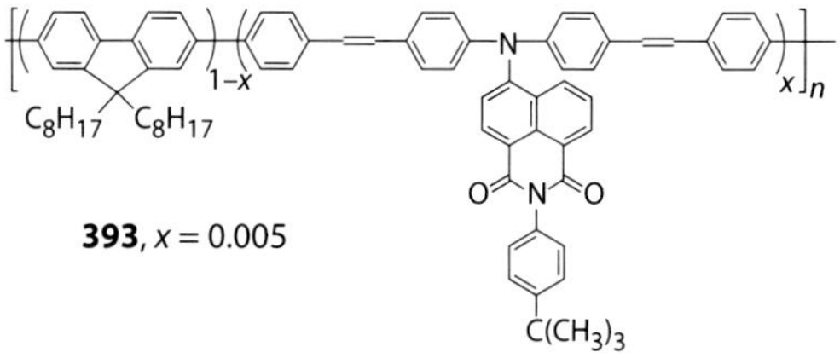

393, *x* = 0.005

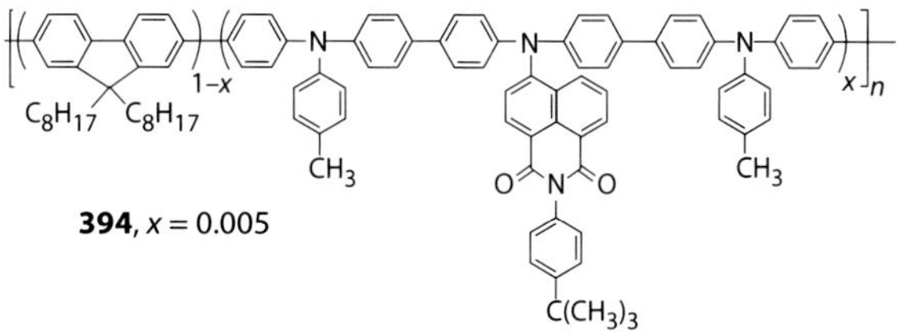

394, *x* = 0.005

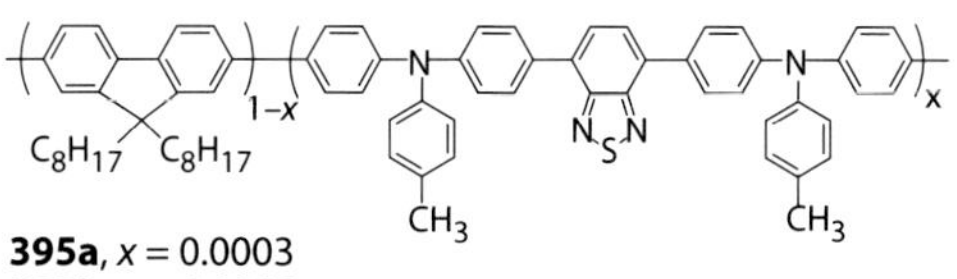

395a, *x* = 0.0003
395b, *x* = 0.0005

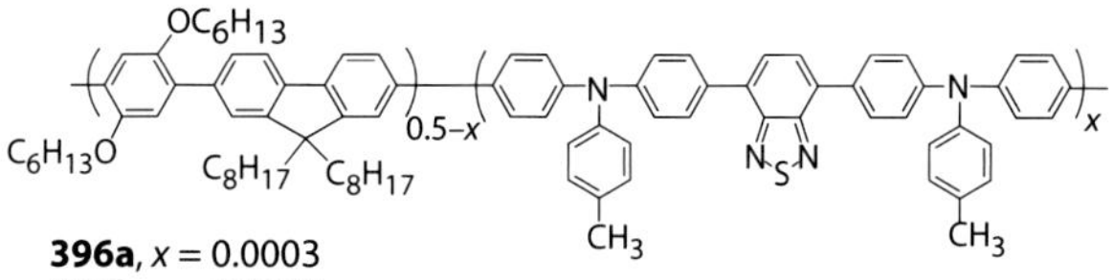

396a, *x* = 0.0003
396b, *x* = 0.0005

CHART 2.93
Chemical structures of white light-emitting fluorene copolymers.

configuration of ITO/PEDOT:PSS (50 nm)/polymer (70 nm)/Ca (10 nm)/Al (100 nm). CIE coordinates of (0.31, 0.34) and a luminance efficiency of 1.59 cd/A have been obtained [448].

Another highly efficient pure white light–emitting polyfluorene with naphthalimide moieties was synthesized [449]. By adjusting the emission wavelength of the 1,8-naphthalimide components and optimizing the relative content of 1,8-naphthalimide derivatives in the resulting polymers, white-light electroluminescence from a single polymer has been obtained in a device with a configuration of ITO/PEDOT (50 nm)/polymer (80 nm)/ Ca (10 nm)/Al (100 nm). The device with polymer **394** exhibits CIT coordinates of (0.32, 0.36), a maximum brightness of 11,900 cd/m², a current efficiency of 3.8 cd/A, a power efficiency of 2.0 lm/W, an external QE of 1.50%, and stable color coordinates at different driving voltages, even at high luminances of >5000 cd/m² [449].

Wang's group further improved white light–emitting device performance by designing a single-polymer system: polyfluorene as the blue host and 2,1,3-benzothiadiazole derivative as the orange dopant on the main chain [450]. For a single-layer device fabricated in ari (ITO/PEDOT:PSS)/**395a**/Ca/Al), pure-white electroluminescence with CIE coordinates of (0.35, 0.32), maximum brightness of 12,300 cd/m², luminance efficiency of 7.30 cd/A, and power efficiency of 3.34 lm/W was achieved (Figure 2.25). Thermal treatment of the single-layer device before cathode deposition leads to the further improvement of the device performance, with CIE coordinates of (0.35, 0.34), turn-on voltage of 3.5 V, luminance efficiency of 8.99 cd/A, power efficiency of 5.75 lm/W, external QE of 3.8%, and maximum brightness of 12,680 cd/m² (Figure 2.25) [450].

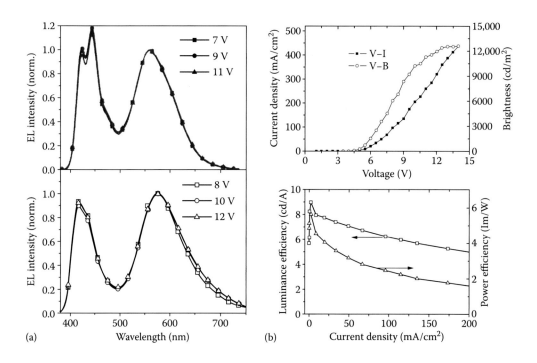

FIGURE 2.25
(a) EL spectra of the devices of **396a** and **395a** under different bias in devices ITO/PEDOT:PSS/polymer/Ca/Al. (b) Voltage–current density–brightness curve of the device of **395a** after thermal annealing and dependence of the luminance efficiency and power efficiency on the current density of the device of **395a** with thermal annealing. (From J. Liu, Q. Zhou, Y. Cheng, Y. Geng, L. Xiang, D. Ma, X. Jing, and F. Wang, *Adv. Funct. Mater.*, 16, 957, 2006. With permission.)

2.3.8 Polyfluorene-Based Polyelectrolytes

Fluorene–[2,5-di(aminoethoxy)benzene] copolymers **397a,b** have been synthesized by Huang and coworkers [451,452] as precursors to the first water-soluble cationic PFs **398** and **399a–c** (Scheme 2.58). Whereas the neutral polymers **397a,b** readily dissolve in common organic solvents such as THF, chloroform, toluene, and xylene (but not in dimethyl sulfoxide [DMSO], methanol, or water), their quaternization produces the material **398**, which is insoluble in chloroform or THF but completely soluble in DMSO, methanol, and water. For copolymers **399a–c**, with partial degree of quaternization, the solubility is intermediate between the neutral and fully quaternized polymers. Whereas neutral polymers **397a,b** showed good thermal stability (T_{dec} = 400°C and 340°C, respectively, in nitrogen), quaternized polymers **398** and **399a–c** begin to decompose at ca. 230°C [452], although a higher value of T_{dec} = 300°C was reported in the preliminary communication [451]. Both neutral and quaternized polymers absorb and emit in the region typical for PFs with only small variations (~10 nm) in PL depending on the structure and the solvent, e.g., **397a**: λ_{abs} = 370.5 nm, λ_{PL} = 414, 428sh nm (THF); **399a–c**: λ_{abs} = 360.5 nm, λ_{PL} = 410 nm (methanol). The Φ_{PL} in solutions is very high for both neutral (**397a**: 87% [CHCl$_3$]; **397b**: 57% [THF]) and quaternized (**398**: 76% [methanol], 25% [H$_2$O]) polymers. The decrease of Φ_{PL} for **398** in

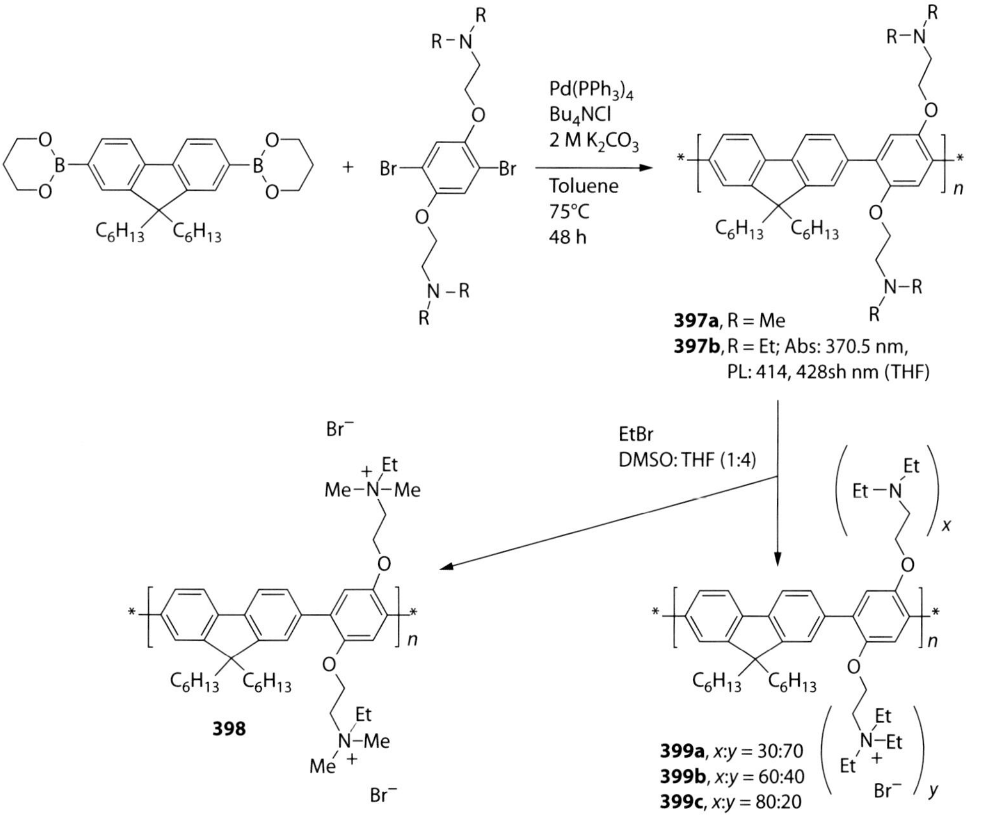

SCHEME 2.58
Synthesis of water-soluble blue-emitting fluorene copolymers. (From Liu, B., Yu, W.-L., Lai, Y.-H., and Huang, W., *Chem. Commun.*, 551, 2000; Liu, B., Yu, W.-L., Lai, Y.-H., and Huang, W., *Macromolecules*, 35, 4975, 2002.)

water was attributed to the aggregation, and a further decrease in Φ_{PL} for this polymer was observed in films (4%). No EL devices with these copolymers were reported.

A series of fluorene copolymers with amino-functionalized side chains **400** and **401** has been prepared by the same group (Chart 2.94). Upon quaternization, they gave copolymers **402** and **403**, which were soluble in polar solvents (methanol, DMF, and DMSO) [453]. Devices from the neutral copolymers **400** and **401** and the quaternized copolymers **402** and **403** showed similar absorption and PL spectra but very different EL spectra. For the neutral polymers, high-energy peaks observed in the PL spectra are replaced with a new broad low-energy peak. At the same time, the main peaks in the EL spectra of quaternized copolymers coincide well with PL emission, with only one additional shoulder around 492–497 nm that the authors assigned to excimer emission. Using **400** as an electron-injection layer in ITO/PEDOT/MEH–PPV/**400**/Al configuration gave devices with a luminance of 3000 cd/m^2 and $\Phi_{EL}^{ex} = 2.3\%$.

When an electron-deficient BT unit was incorporated into the backbone of these polymers, an efficient energy transfer resulted in complete fluorescence quenching from the fluorene sites already at BT concentrations as low as 1% (for both neutral and quaternized copolymers, **404** and **405**) [454] (Chart 2.95). These macromolecules emit green (544–550 nm, **404**) to yellow (555–580 nm, **405**) light, and can be processed from environment-friendly solvents such as alcohols. The PLED fabricated with these polymers showed high Φ_{EL}^{ex} of higher than 3% and 1% for **404** and **405**, respectively (Al cathode).

Bazan and coworkers [455,456] reported similar water-soluble random cationic fluorene copolymers **406** and **407** with quaternary amine side groups attached at the C-9 position of the fluorene moiety (Chart 2.96). There was a progressive blue shift in absorption spectra of copolymers **406** with increasing amount of *meta*-linker in the polymer chain. Increasing the *para* content above a 50:50 ratio, however, did not perturb the emission

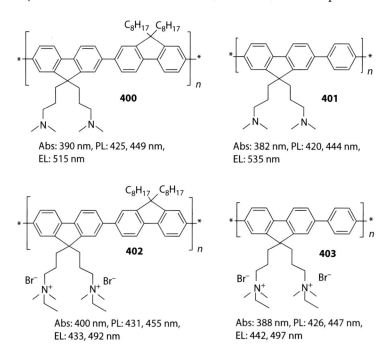

400
Abs: 390 nm, PL: 425, 449 nm,
EL: 515 nm

401
Abs: 382 nm, PL: 420, 444 nm,
EL: 535 nm

402
Abs: 400 nm, PL: 431, 455 nm,
EL: 433, 492 nm

403
Abs: 388 nm, PL: 426, 447 nm,
EL: 442, 497 nm

CHART 2.94
Chemical structures of PFs with amino-functionalized chains.

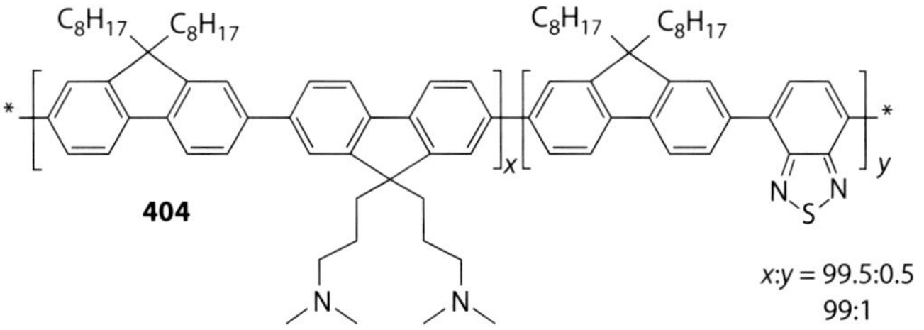

x:y = 99.5:0.5
99:1
95:5
85:15

UV: 364–375 nm, PL: 544–550 nm, Φ_{PL}: 85%–71%

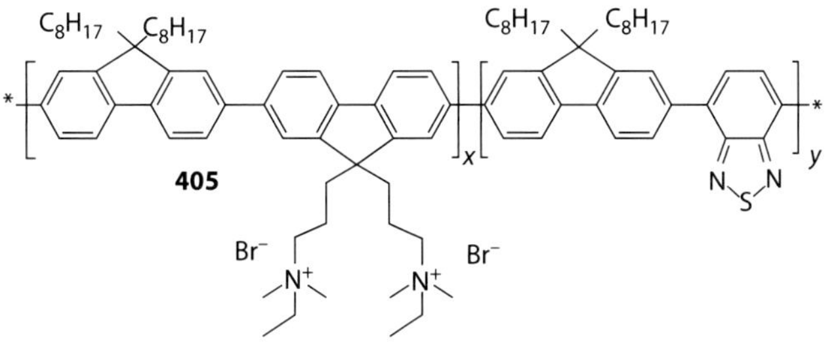

UV: 391–382 nm, PL: 555–580 nm, Φ_{PL}: 23%–32%

CHART 2.95
Chemical structures of PFs with electron-deficient BT units.

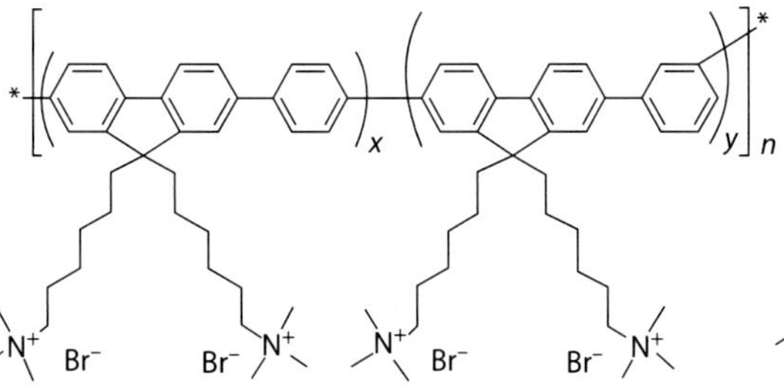

406

x:y = 100:0; Abs: 384 nm, PL: 369 nm, Φ_{PL}: 42%
x:y = 90:10; Abs: 376 nm, PL: 403 nm, Φ_{PL}: 42%
x:y = 70:50; Abs: 361 nm, PL: 410 nm, Φ_{PL}: 44%
x:y = 50:50; Abs: 347 nm, PL: 421 nm, Φ_{PL}: 50%
x:y = 25:75; Abs: 337 nm, PL: 417 nm, Φ_{PL}: 57%
x:y = 0:100; Abs: 335 nm, PL: 417 nm, Φ_{PL}: 51%
(all spectral data are for aqueous solutions)

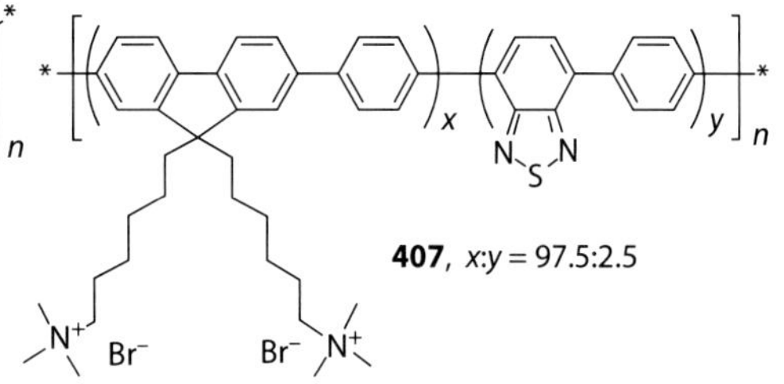

407, x:y = 97.5:2.5

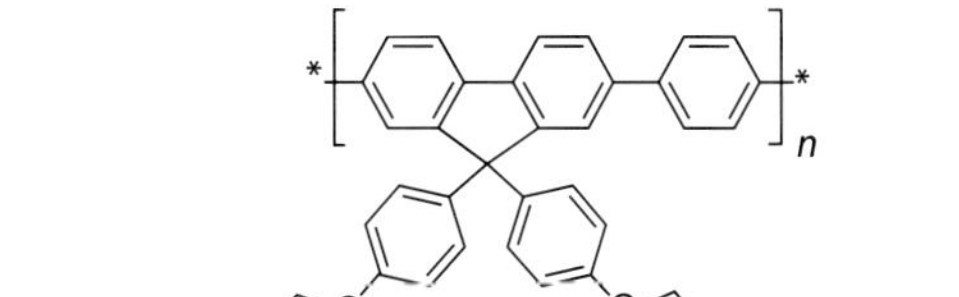

408

CHART 2.96
Chemical structures of water-soluble random cationic PFs.

maxima, interpreted to be due to localization of excitons on the longest conjugated segments [455]. At low concentrations in solution, the absorption and emission spectra of **407** (λ_{max}^{abs} = 380 nm, λ_{PL} ~ 400–500 nm; blue emission) is nearly identical to that of **406**, which lacks BT units. However, at concentrations >10^{-6} M, the emergence of green emission (λ_{PL} ~ 500–650 nm) characteristic of BT sites was observed, resulting from aggregation that leads to enhanced energy transfer to units containing lower-energy BT chromophores [456].

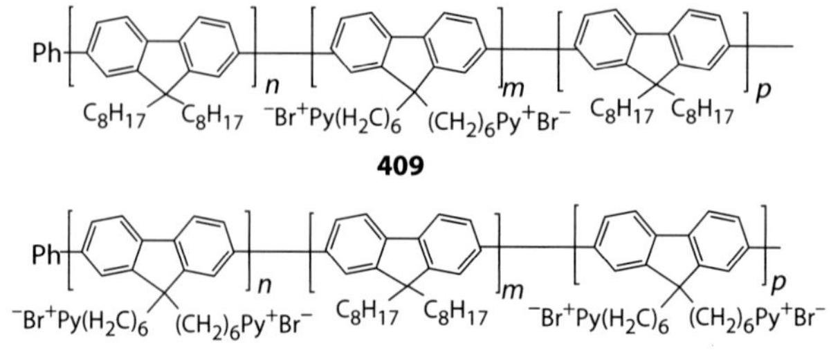

409

410

CHART 2.97
Chemical structures of all-conjugated triblock polyelectrolytes with PF backbone.

Whereas all above water-soluble PFs are tetra-alkylammonium-based salts, Burrows et al. [457] reported on anionic fluorene-based copolymer **408** that showed a blue shift in PL (from 424 to 411 nm) as well as a dramatic increase in the fluorescence quantum yield (from 10%–15% to 60%) when incorporated into *n*-dodecylpentaoxyethylene glycol ether micelles [457].

Bazan's group also synthesized all-conjugated triblock polyelectrolytes with polyfluorene backbone and *n*-octyl side chain [458]. The generation of the ionic component requires incorporation of segments containing alkylbromide side groups, which in a subsequent step are ionized with pyridine. The function of polymer **409** and **410** as the interlayer in PLEDs were examined with the devices with a configuration of ITO/PEDOT:PSS/MEH–PPV/cathode, where cathode can be Al, Ba/Al, **409**/Al, or **409**/Al. Both polymer **409** and **410** work efficiently as electron-injection layers for Al in PLEDs, and the resultant device efficiencies are similar to those achieved using the less stable Ba/Al cathode (Chart 2.97).

2.3.9 Summary

Clearly, at present, PFs are the most promising class of blue-emitting materials. The original problem associated with the undesirable "green emission band" was shown to be a result of exciton trapping on the electron-deficient fluorenone defect sites. The color purity can be reestablished via (i) careful purification of the monomer (complete elimination of monosubstituted units); (ii) inserting a protecting layer between the PF and reactive cathode material; (iii) introducing hole-trapping sites (most commonly, triarylamine units), which would compete with fluorenone defects, minimizing the exciton formation on the latter; and (iv) introducing bulky substituents to the PF backbone that minimize the exciton trapping on fluorenone defects. Furthermore, introduction of different conjugated moieties to the PF backbone allows for efficient color tuning in these materials.

Thus far, the most efficient PLED based on PFs in terms of pure red, green, and blue emission can be summarized with the following examples (Chart 2.98).

Blue emitter **246** [372]: A single-layer device ITO/PEDOT/**246**/Ca can be turned on at 3.5 V, and emits blue light (CIE: $x = 0.150$, $y = 0.080$) with an EL efficiency of 1.1 cd/A (at 8.5 V; power efficiency of 0.40 lm/W) and a maximum brightness of 1600 cd/m². A multilayer device with a structured triarylamine-based HTL results in an EL efficiency of 2.7 cd/A and higher (maximum brightness of 5000 cd/m²).

Blue emitter **411** [446]: A single-layer PLED ITO/PEDOT/**411**/Ca emits blue light (CIE: $x = 0.15$, $y = 0.16$) with a current efficiency of 3.0 cd/A and an operating voltage of 4.6 V (at 100 cd/m²).

Green emitter **386** [445]: The device with the structure ITO/BTPD–PFCB/**386**/Ca/Ag works at the operating voltage of 3.6 V and reaches extremely high $\Phi_{EL}^{ex} = 6\%$ (18.5–28.6 cd/A) and

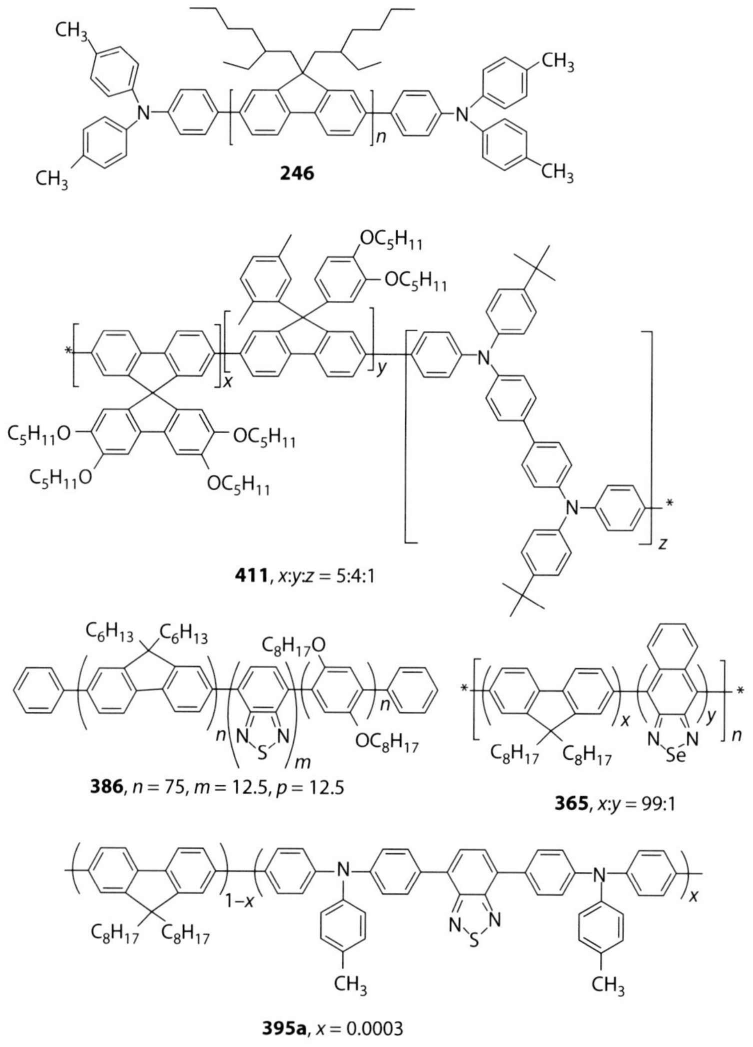

CHART 2.98
Chemical structures of PFs for most efficient PLED.

a maximum brightness of 59,400 cd/m^2. These unusually high values are unique and await to be reproduced and surpassed by researchers in the field.

Red emitter **365** [323]: A single-layer PLED ITO/PEDOT/polymer/Ba operates at 8.9 V, emitting saturated red color (CIE: $x = 0.67$, $y = 0.33$) with very high $\Phi_{EL}^{ex} = 3.1\%$, corresponding to a photometric current efficiency of 0.9 cd/A. A brightness in excess of 2000 cd/m^2 was achieved for this device.

White emitter **395a** [450]: For a single-layer device fabricated in ari (ITO/PEDOT:PSS/**395a**/Ca/Al), pure-white electroluminescence with CIE coordinates of (0.35, 0.32), maximum brightness of 12,300 cd/m^2, luminance efficiency of 7.30 cd/A, and power efficiency of 3.34 lm/W was achieved. Thermal treatment of the single-layer device before cathode deposition leads to the further improvement of the device performance, with CIE coordinates of (0.35, 0.34), turn-on voltage of 3,5 V, luminance efficiency of 8.99 cd/A, power efficiency of 5.75 lm/W, external QE of 3.8%, and maximum brightness of 12,680 cd/m^2.

2.4 Polythiophenes

Polythiophenes (including oligothiophenes) are one of the most studied and important classes of linear conjugated polymers [459,460]. Versatile synthetic approaches to PTs (chemical [461] and electrochemical [462]), easy functionalization and unique, widely tunable electronic properties have been the source of tremendous interest in this class of polymers.

Owing to their electron-rich character, the thiophene rings in PTs can be easily and reversibly oxidized by chemical or electrochemical means to form p-doped, usually highly conducting materials. The first electronic transition of undoped PT that strongly depends on structure lies between 300 and 500 nm (ε ~10,000 l/(mol cm)) [463], and on doping undergoes dramatic bathochromic shifts concomitant with the formation of a so-called "conducting" band that tails from the visible to the deep IR region. In contrast to undoped PTs, known to exhibit reasonably strong luminescence in the visible region of the spectrum, doped PTs are not luminescent, although partially doped PTs have been used in LECs and doped PEDOT is routinely used as an electrode for PLED (mostly as a second layer on ITO-covered glass).

Although in terms of EL materials, PTs have not been studied as widely as PPVs or PFs, they present an important class of LEPs. PTs emit orange-to-red light, consistent with their band gap of ca. 2 eV. Often the luminescence efficiency of PTs in the solid state is relatively low [464,465], much lower than that of PPV and PFs. A possible explanation is that it originates from their solid-state structure and has a tendency of strong interchain interactions (especially for low-molecular-weight oligomers). This feature is an advantage of PTs in some electronic applications as, for example, field-effect transistors (FETs) [466]. However, it becomes one of the most critical drawbacks for application as emissive materials in LED. Whereas in solution the PL efficiency (Φ_{PL}) of poly(3-alkylthiophenes) is ~30%–40%, it drastically decreases to 1%–4% and lower in the solid state due to increased contribution of nonradiative decay via interchain interactions and intersystem crossing caused by the heavy-atom effect of sulfur [467]. Thiophene-based polymers have stronger spin–orbital coupling than phenylene-based polymers due to the internal heavy atom effect of the sulfur heteroatom, and hence triplet-state processes play a greater role in their photophysics [468].

Another feature of PTs is the phenomenon of thermochromism [469], which has been shown for poly(3-alkyl)thiophenes in many publications. It is believed that the thermochromism observed in poly(3-alkyl)thiophene films originates from the thermal movement of the side chains, shifting a predominantly planar structure of chains at low temperatures to a random coil conformation when the temperature is increased, thus forcing the polymer backbone out of planarity. This leads to a decreased orbital overlap and effective conjugation length, resulting in band gap increase and blue-shifted polymer absorbance (from red to purple or purple–blue) [470,471]. The process is completely reversible, and on cooling the initial color is restored. Although thermochromism is of theoretical interest for understanding the effect of structural and electronic features of PTs, it is undesirable for LED applications, as it could lead to changes of the emission wavelength and the QE of the device during the operation.

2.4.1 Synthetic Routes to Polythiophenes

Polymerization of thiophenes can be carried out in different ways, and the most used methods can be generalized into three categories: (i) electropolymerization, (ii) metal-catalyzed-coupling reactions, and (iii) chemical oxidative polymerization. Electropolymerization is a widely used method of preparing insoluble films of PTs, and represents a simple and efficient way to study the optical and electronic properties of PTs [462], although it is rarely used in preparation of EL materials. In 1980, Yamamoto et al. [472] reported the Ni-catalyzed polycondensation of 2,5-dibromothiophene **412**. The latter was allowed to react with Mg in THF, affording 2-magnesiobromo-5-bromothiophene **413** that, in the presence of Ni(bipy)Cl$_2$, produced PT **414** (Scheme 2.59). In the same year, Lin and Dudek [473] described another example of a metal-catalyzed route to unsubstituted PT **414**, exploiting acetylacetonates of Ni, Pd, Co, or Fe as catalysts.

The PT synthesized by these methods is a low-molecular-weight material because even at low molecular weights, the material is insoluble and precipitates from THF; moreover, the elemental analysis indicates the presence of 1%–3% of Mg [461]. Later an Ni(dppp) Cl$_2$ catalyst was exploited for polycondensation polymerization of bromo(iodo)-Grignard reagents of type **413** [474]. Another polycondensation approach to PT, also reported by Yamamoto et al. [475], included heating of **413** with Ni(cod)$_2$ and triphenylphospine in DMF at 60°C–80°C. Owing to very high yield (near quantitative), this reaction has been applied in the syntheses of a wide range of conjugated polymers.

The FeCl$_3$-mediated polymerization of thiophene in chloroform was described 20 years before [476] and currently is the most widely exploited oxidative route to 3(3,4)-substituted PTs (Scheme 2.60). It produces rather high molecular-weight polymers (often $M_w = 20,000$– $400,000$) with polydispersity from 1.3 to 5 [477,478]. Other oxidative agents (e.g., ammonium persulfate) are of limited use, although oxidative coupling with CuCl$_2$ is widely used as a dimerization reaction of lithiothiophenes in the syntheses of oligothiophenes.

As already mentioned, unsubstituted PT is an insoluble and infusible material. Once the polymer is prepared, it is difficult (if not impossible) to further process it as a material for electronic applications. The solubility can be greatly enhanced by the introduction of side chains at position 3 (or at both positions 3 and 4). The most widely studied side chains are *n*-alkyl substituents that can be easily introduced into the thiophene core by reaction of 3-bromothiophene with alkyl-Grignard reagents [479].

While 3-substitution efficiently improves the solubility and the processability of the PTs, polymerization of 3-*R*-thiophenes can result in three different types of coupling of the thiophene rings along the polymer main chain, i.e., HT, HH, and TT (Chart 2.99).

Generally, both oxidative polymerization and metal-catalyzed polycondensation afford all three possible types of isomers, although this process is not completely random and

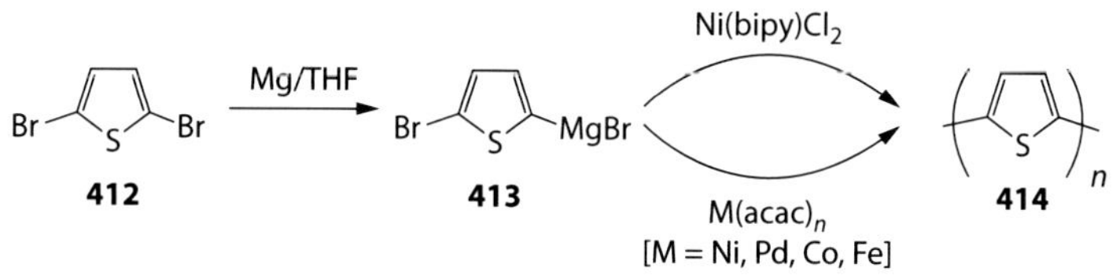

SCHEME 2.59
Synthesis of polythiophene via metal-catalyzed couplings.

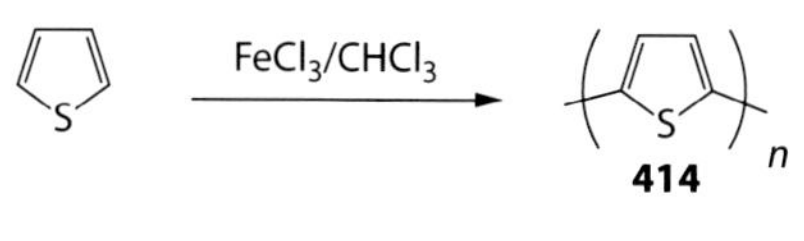

SCHEME 2.60
Synthesis of polythiophene via chemical oxidation polymerization.

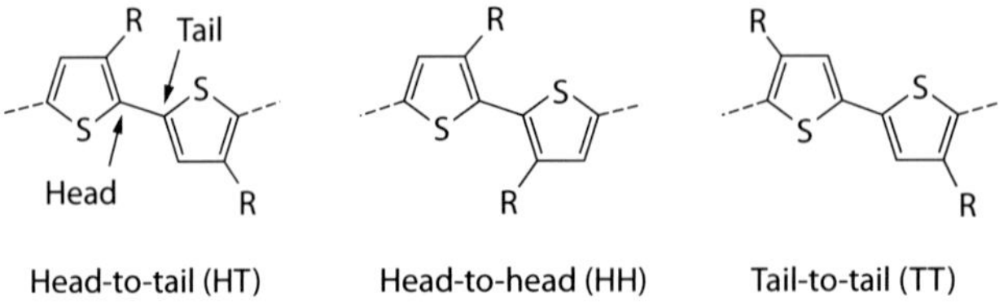

CHART 2.99
Chemical structures of HT, HH, and TT.

electronic and steric factors have an influence on somewhat preferred HT coupling, which could reach more than 80%–94% [480,481]. Amou et al. [480] showed that the regioregularity of FeCl$_3$-synthesized poly(3-hexylthiophene) (P3HT, **424**) depends on the temperature of the reaction and the concentration, and in diluted solutions at –45°C, the regioregularity of P3HT approaches 90% [480].

Several approaches leading to selective formation of the least sterically hindered HT-regioregular PTs have been developed in the last decade. The McCullough method presents the first synthesis of regioregular HT poly(alkyl)thiophenes (HT-PATs, **HT-417**) (yielding almost 100% HT coupling) (Scheme 2.61) [482,483]. It is based on regiospecific metallation of 2-bromo-3-R-thiophene (**415**) with lithium diorganoamide (LDA) at position 5 and its further transformation into Grignard derivative **416**. The latter is polymerized with catalytic amounts of Ni(dppp)Cl$_2$ using the Kumada cross-coupling method. The important modification of this synthetic route replaces the lithiation reaction by treatment of the dibromothiophene derivative with methylmagnesium or vinylmagnesium bromide, which affords the Grignard intermediate in one step [484–486]. Other methods for preparing regioregular PTs, exploiting Stille [487] or Suzuki [488] coupling instead of Grignard reagents, have also been developed. Rieke and coworkers [489,490] have used highly reactive "Rieke zinc" (Zn*) that reacts with 2,5-dibromo-3-R-thiophenes (**418**) at low temperature, affording predominantly the 5-metallated compound (**419b**). Polymerization with the Kumada catalyst Ni(dppe)Cl$_2$ produced HT-regioregular PATs **HT-417**; in contrast, the monodentate Pd(PPh$_3$)$_4$-ligated catalyst yields regiorandom PATs **417** under the same conditions. The detailed aspects of synthesis and characterization of regioregular PTs were reviewed recently by McCullough [461].

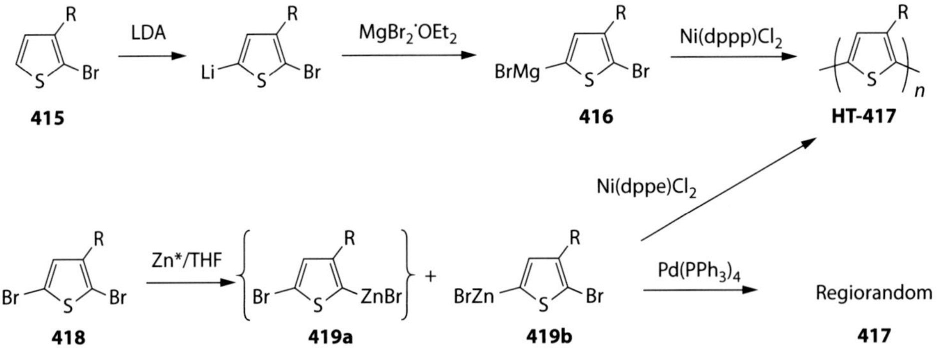

SCHEME 2.61
McCullough and Rieke methods of synthesis of regioregular HT-poly(3-alkylthiophenes) **HT-417**. (From McCullough, R.D. and Lowe, R.D., *J. Chem. Soc., Chem. Commun.*, 70, 1992; McCullough, R.D., Lowe, R.D., Jayaraman, and Anderson, D.L., *J. Org. Chem.*, 58, 904, 1993.)

The presence of HH coupling in irregular PTs causes an increased twisting of thiophene units (due to steric repulsion) with concomitant loss of conjugation. This results in an increased band gap (blue shift in an absorption and luminescence), decreased conductivity, and other undesirable changes in electronic properties. As will be shown below, regioregularity also plays an important role in luminescence properties of PTs and is used as a tool to tune the properties of PT-based LEDs.

2.4.2 Polythiophene Homopolymers

2.4.2.1 *Polythiophenes as Red-Light Emitters*

PT LEDs were first reported by Ohmori et al. [491,492] in 1991. Poly(3-alkylthiophenes) **420–422** (prepared by oxidation of 3-alkylthiophenes with $FeCl_3$ in chloroform [476]) as red–orange-emitting material (peak emission at 640 nm for **420**) in single-layer ITO/**420–422**/Mg:In devices were described (Chart 2.100). It was shown that the luminescence efficiency follows a linear dependence on the length of the alkyl chain, showing about a 4-fold increased EL efficiency for PT with R = $C_{22}H_{45}$, compared with R = $C_{12}H_{25}$ (although no quantum yield values have been given in these reports), possibly owing to the improved confinement of excitons on the main chain with an increase of the side-chain length [493]. The use of a carrier-confining layer (TPD **314**) was shown to markedly increase (by 3–5 times) the device efficiency [492]. Shortly afterward, Heeger's group [494] reported EL in poly(3-octylthiophene) **423** that showed red–orange luminescence with Φ_{EL}^{ex} at room temperature of 0.025% in ITO/**423**/Ca configuration. Greenham et al. [495], while studying the EL from PTs **422** and **424** (also prepared via oxidation of the monomers with $FeCl_3$), achieved a significantly higher emission efficiency of 0.2% for **422**, also using Ca as a cathode (**424** gave 0.05% and use of Al cathode gave 0.01% and 0.003% yield, respectively).

Bolognesi and coworkers [496–498] reported the Ni-initiated polymerization of 2,5-diiodothiophenes yielding poly(3-alkylthiophene)s **426–428**, containing a methoxy group at the end of the alkyl chain (Scheme 2.62). Small shifts in EL of polymer **428** to higher energies compared with polymer **423** (from 1.8 to 1.95 eV) were mentioned [494,496]. The effect could be the result of asymmetry of the rather wide band (comparison with P3DT **425** reveals smaller blue shifts of 0.05 eV [499]). Polymers **426** and **427** showed high (for PTs) PLQYs in solution (38%–45% in THF) that, however, decreased in the films [498]. A general explanation of this decrease as a result of the interchain interactions is supported by temperature-dependent PL experiments. On gradual heating of the sample to 140°C, the PL intensity increased by 2 and 6 times for **426** and **427**, respectively. Φ_{EL}^{ex} for **426** and **427** in ITO/polymer/Al diodes were relatively low (10^{-5}–10^{-4} and 5×10^{-3}%, respectively, at the same voltages).

A larger blue shift in fluorescence was observed for alkoxycarbonyl-substituted PTs **429** and **430**. The polymers were prepared from 2,5-dibromo-substituted monomers by two methods: (i) Ullmann reaction with Cu powder and (ii) Ni(0)-mediated polymerization (Scheme 2.63) [500]. Both polymers have similar molecular weights ($M_n \sim 3000$), although

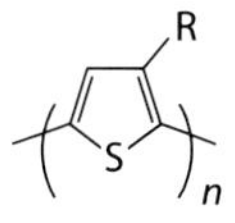

420, R = $C_{22}H_{45}$
421, R = $C_{18}H_{37}$
422, R = $C_{12}H_{25}$
423, R = C_8H_{17}
424, R = C_6H_{13}
425, R = $C_{10}H_{21}$

CHART 2.100
Chemical structures of poly(3-alkylthiophenes).

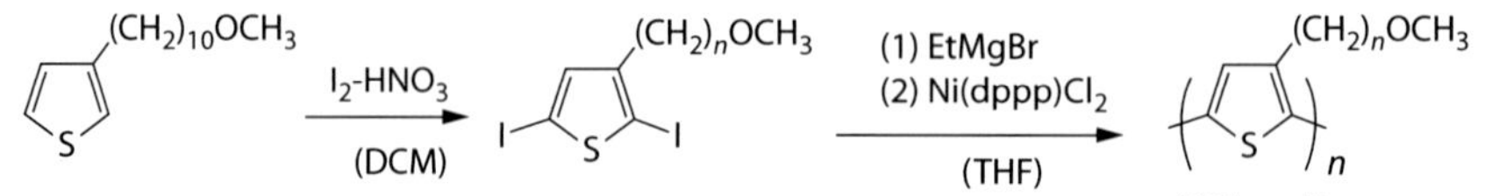

SCHEME 2.62
Synthesis of alkylthiophenes **426–428** through oxidative iodination of the thiophene followed by Ni-catalyzed polymerization. (From Bolognesi, A., Botta, C., Geng, Z., Flores, C., and Denti, L., *Synth. Met.*, 71, 2191, 1995; Bolognesi, A., Porzio, W., Bajo, G., Zannoni, G., and Fanning, L., *Acta Polym.*, 50, 151, 1999.)

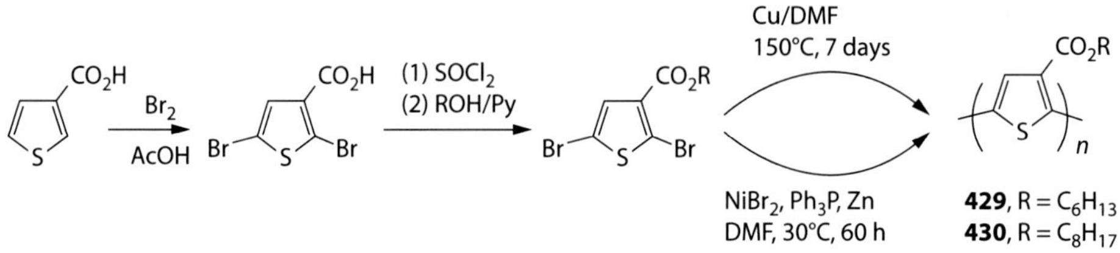

SCHEME 2.63
Synthesis of alkoxycarbonyl-PTs. (From Pomerantz, M., Yang, H., and Cheng, Y., *Macromolecules*, 28, 5706, 1995.)

the Cu-prepared polymers showed higher quality and lower polydispersity. The PL emission maxima for the Cu-prepared polymers **429** and **430** were red shifted, compared with the Ni-prepared polymers (by 13–15 nm [≈0.05–0.06 eV] in solution and 25–30 nm [≈0.08–0.10 eV] in films, Table 2.4). This demonstrates that the properties of the polymer depend on the preparation method and, consequently conclusions from small shifts of 0.05–0.1 eV in PL–EL energies of the materials, prepared by different methods, should be made with care.

Another example of the effect of the polymerization method on the optical properties of the resulting polymer is the synthesis of polymer **431** (Scheme 2.64) [501]. Polymers obtained by oxidative polymerization of the corresponding 3-(methoxyphenyl)thiophene with $FeCl_3$ ($CHCl_3$-soluble fraction), and with Mg/Ni(dppp)Cl_2 or Ni(cod)$_2$ polymerizations of the corresponding dibromothiophene derivatives showed somewhat different maxima in absorption (and PL emission) spectra: 405 (520), 433 (555), and 435 (560) nm, respectively.

Jin et al. [502] synthesized and studied the PL and EL properties of polymers **432** and **433** that differ by the position of the alkoxy substituent in the phenyl ring, expecting different distortion of the polymer main chain (and consequently conjugation length) due to different steric factors for *para*- and *ortho*-substitution (Chart 2.101). The absorption spectrum of the "*ortho*-polymer" **432** showed a substantial blue shift of 40 nm compared with "*para*" **433**, and a decrease in EL turn-on voltage (4.5 and 6.5 V, respectively). Both polymers demonstrated nearly the same PL and EL maxima (Table 2.1).

A series of PT derivatives (**434–436**) with electron-transporting benzotriazole moieties in the side chains was prepared (Scheme 2.65) [503]. Polymer **434** was insoluble in common organic solvents (THF, $CHCl_3$, or DMF). In contrast, polymers **435** and **436**, endowed with longer tethers, possessed improved solubility as well as high molecular weight and thermal stability ($T_{dec} > 300°C$). They showed pronounced blue shifts of 50–70 nm in absorption, PL and EL maxima, compared with P3OT **423** (Table 2.4), somewhat speculatively attributed to steric hindrance [503]. The EL efficiency was not enhanced as dramatically as the authors expected, although they showed 3–7 times higher Φ_{EL}^{ex} values. This was

TABLE 2.4

Properties of Poly(3-R)Thiophenes, Prepared by Ulmann (Cu), Yamamoto (Ni), or FeCl$_3$ Polymerization

Polymer (Method)	M_n (g/mol) (PDI)	λ_{max}^{abs} (nm) (Solution)	λ_{max}^{abs} (nm) (Film)	λ_{max}^{PL} (nm) (Φ_{PL}, %)	λ_{max}^{PL} (nm) (Film) (Φ_{PL}, %)	λ_{max}^{EL} (nm) (Φ_{EL}^{ex}, %)	Ref.
423						1.8 eV	[494]
423 (FeCl$_3$)	16,800 (6.22)		500		655	650 (0.012%)[a]	[503]
424 (50% HT)	4000 (1.7)	413 (CHCl$_3$)	420	567, 600 (9%)	608 (0.8%)		[504]
424 (60% HT)	3000 (1.6)	420 (CHCl$_3$)	432	572, 600 (12%)	608, 643 (0.3%)		[504]
424 (70% HT)			456		650	630 (1.3 × 10^{-5}%)[b]	[465]
424 (80% HT)	40,000 (2.3)	440 (CHCl$_3$)	518	580, 614 (14%)	670, 714 (0.2%)		[504]
HT-424 (98% HT)			510		717	662 (3.85 × 10^{-4}%)[b]	[465]
425			496				
426 (Ni)	5900 (1.4)	451 (Toluene)	550				[497]
427 (Ni)	9900 (1.5)	450 (Toluene)	535				[497]
428 (Ni)	9450 (1.6)	448 (Toluene)	530				[497]
428 (Ni)	8000 (2)		470		660	635	[496]
429 (Cu)	3030 (2.3)	423 (THF)	447	570 (THF)	620		[500]
430 (Cu)	4060 (1.9)	430 (THF)	450	568 (THF)	630		[500]
429 (Ni)	3050 (3.2)	408 (THF)	429	555 (THF)	595		[500]
430 (Ni)	3510 (2.8)	408 (THF)	430	555 (THF)	600		[500]
429	6700 (2.5)	410 (THF)	434		600	600 (0.016%)[a]	[505]
430	9400 (3.2)	439 (THF)	460		610	615 (0.018%)[a]	[505]
HHTT-429	8100 (1.8)	387 (THF)	377		590	590 (0.0085%)[a]	[505]
HHTT-430	8700 (2.0)	389 (THF)	381		600	600 (0.0047%)[a]	[505]
432 (FeCl$_3$)	3100 (2.8)		400	525 (CHCl$_3$)		600[c]	[502]
433 (FeCl$_3$)	3400 (1.6)		440	534 (CHCl$_3$)		607[c]	[502]
435 (FeCl$_3$)	17,000 (2.07)		444		580	580 (0.09%)[a]	[503]
436 (FeCl$_3$)	20,000 (1.89)		446		588	600 (0.04%)[a]	[503]

[a] ITO/polymer/Al.
[b] ITO/polymer/Mg/Al.
[c] ITO/PEDOT/polymer/Al.

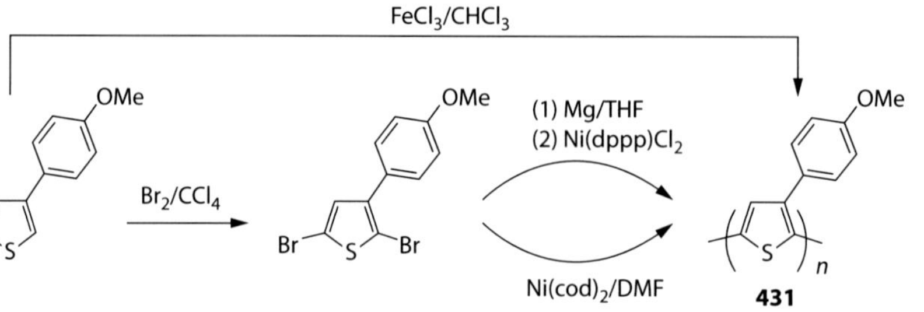

SCHEME 2.64
Preparation of methoxyphenyl-PT **431**. (From Yamamoto, T. and Hayashi, H., *J. Polym. Sci., Part A: Polym. Chem.*, 35, 463, 1997.)

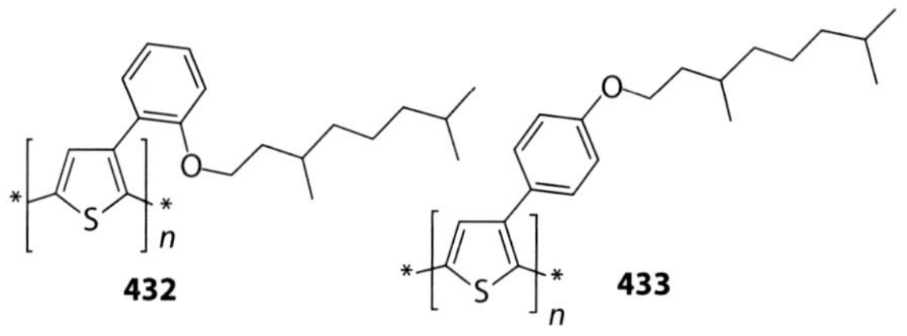

CHART 2.101
Chemical structures of poly(3-alkoxyphenylthioenes).

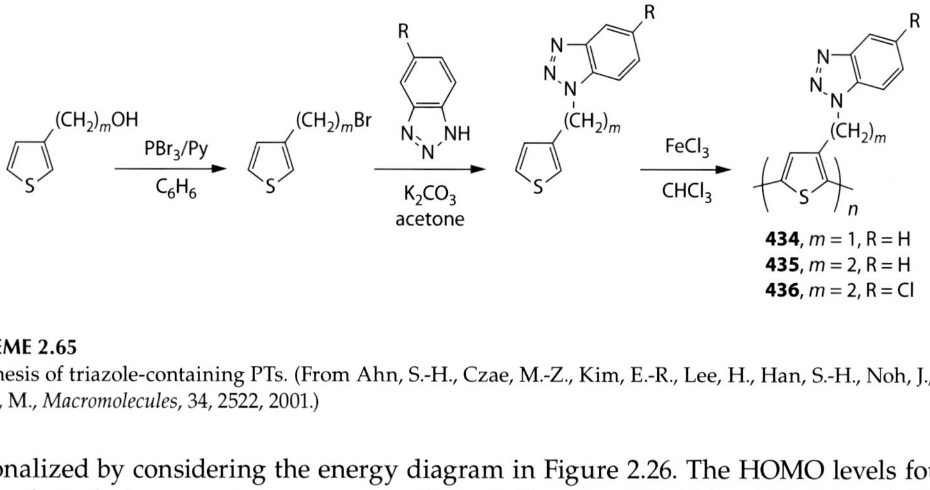

434, m = 1, R = H
435, m = 2, R = H
436, m = 2, R = Cl

SCHEME 2.65
Synthesis of triazole-containing PTs. (From Ahn, S.-H., Czae, M.-Z., Kim, E.-R., Lee, H., Han, S.-H., Noh, J., and Hara, M., *Macromolecules*, 34, 2522, 2001.)

rationalized by considering the energy diagram in Figure 2.26. The HOMO levels found from photoelectron spectroscopy were −5.45, −5.62, and −4.57 eV for **435**, **436**, and P3OT (**423**), respectively. The LUMO energy levels (estimated as E_{HOMO} plus optical band gap from UV–vis spectra) were at −3.31, −3.42, and −2.61 eV. Thus, both HOMO and LUMO levels of **435** and **436** were lowered compared with those of P3OT **423** through the introduction of the electron-withdrawing benzotriazole moieties. This strong effect can certainly be attributed to the replacement of the alkoxy substituents with alkyl groups and not due to the triazole moieties, although the latter might have contributed to the reduced LUMO. The authors note that the total barrier to charge injection was not reduced, although Φ_{EL}^{ex} enhancement was observed. The observed enhancement is not surprising, in our opinion, because the hole- and electron-injection balance (which was improved significantly) rather than total barrier is expected to determine the EL efficiency.

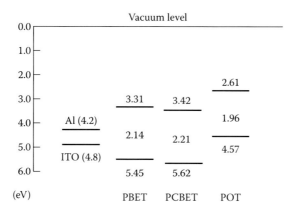

FIGURE 2.26
Energy diagram of substituted PT **435** (PBET), **436** (PCBET), and **423** (POT). (From Ahn, S.-H., Czae, M.-Z., Kim, E.-R., Lee, H., Han, S.-H., Noh, J., and Hara, M., *Macromolecules*, 34, 2522, 2001. With permission.)

2.4.2.2 Color Tuning in Polythiophenes

Although the first publications on EL of poly(3-alkylthiophenes) described materials with red–orange emission (610–640 nm), a large number of PTs with emissions covering the full visible region, i.e., from blue to red and NIR, were later reported. These EL color changes were achieved by structural variations in PT side chains, as well as by controlling the regioregularity.

Regioregularity in PTs plays an important role in their band-gap control. The random polymerization of 3-alkylthiophenes leads to a larger degree of HH coupling, which in turn become the sites showing the largest twist distortion between thiophene units, resulting in a decreased effective conjugation length. On the other hand, HT-regioregular PTs show longer conjugation lengths and a red shift of their absorption and PL.

Xu and Holdcroft [504] studied the effect of regioregularity on luminescent properties of P3HT **424**. They found that increasing the percentage of HT coupling in P3HT from 50% HT to 80% HT results in red shifts of both absorption and emission maxima, as well as fluorescence efficiency in solution from 9% to 14% (Table 2.4). On the other hand, an increased planarization of regioregular HT PT facilitates aggregation, which results in a decrease of the Φ_{PL} emission efficiency in the solid state (from 0.8% for 50% HT to 0.2% for 80% HT). Later, McCullough and coworkers [465] performed comparative studies on EL performance of HT-regioregular and regiorandom P3HT **424**. Compared with nonregioregular material, the HT-regioregular polymer showed a narrower EL spectrum and an approximate doubling in Φ_{EL}^{ex} (1.5×10^{-4}% vs. 7×10^{-5}%, at 6 mA current; Table 2.4). However, very low QE and preferential degradation of LEDs with irregular P3HT might be responsible for the difference.

Regioregular HT-coupled poly(3-decylthiophene) (**HT-425**, Chart 2.102) is an EL polymer exhibiting a well-defined vibronic structure in its absorption spectrum (Figure 2.27a; 0–0 transition at 2.0 eV), a red emission with good color purity, but a rather low QE in the solid state $\left(\Phi_{PL}^{film} = 1 \pm 0.1\%\right)$ [506]. Its regioregular HH–TT-coupled isomer, **HHTT-425** [507], showed large blue shifts (Figure 2.27b) in absorption, fluorescence, and EL spectra. In addition, **HHTT-425** emits green light with one order of magnitude higher PL $\left(\Phi_{EL}^{film} = 11 \pm 0.1\%\right)$ and two orders of magnitude higher EL ($\Phi_{EL}^{int} = 0.001$ and 0.25%–0.30%, respectively), but requires a higher turn-on voltage [506]. The blue shifts and the increased emission efficiency of HHTT-regioregular polymers were explained by pronounced

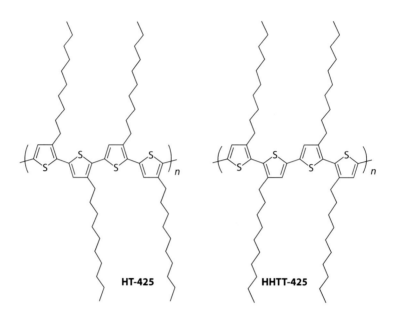

CHART 2.102
Chemical structure of a regioregular HT-coupled poly(3-decylthiophene).

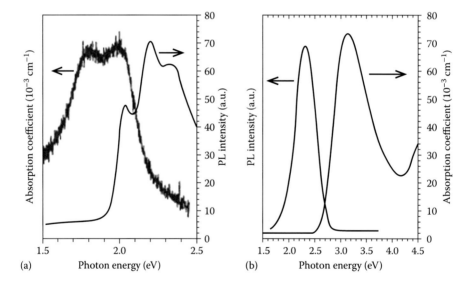

FIGURE 2.27
PL and absorption spectra of thin films of regioregular polythiophenes **HT-425** (a) and **HHTT-425** (b), spin coated on a fused silica substrate. (From Barta, P., Cacialli, F., Friend, R.H., and Zagórska, M., *J. Appl. Phys.*, 84, 6279, 1998. With permission.)

interannular conformational distortion in the HH fragment (~70° in HH and ~0° in HT fragments, as suggested by AM1 semiempirical calculations [508]).

Pomerantz et al. [505] prepared the regioregular polymers **HHTT-429** and **HHTT-430** (Scheme 2.66) and compared them with previously synthesized irregular polymers **429** and **430** (Scheme 2.63) [500]. Regioregular polymers showed blue shifts in absorption

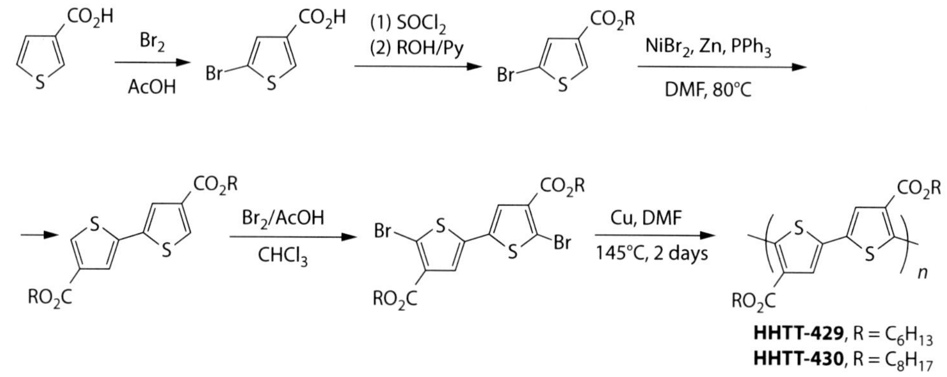

SCHEME 2.66
Synthesis of regioregular HHTT alkoxycarbonyl-PTs. (From Pomerantz, M., Cheng, Y., Kasim, R.K., and Elsenbaumer, R.L., *J. Mater. Chem.*, 9, 2155, 1999.)

for solution and films (23–30 nm and 57–79 nm, respectively; Table 2.4), interpreted in terms of shorter conjugation length. Blue shifts in PL and EL were much less pronounced (10–15 nm), and the PLED showed 2–4 times lower Φ_{EL}^{ex} (Table 2.4) [505].

Gill and coworkers [509] synthesized a number of regioregular-alkylated polymers **HHTT-423, 437–439** (Scheme 2.67) and demonstrated PL and EL color tuning through a

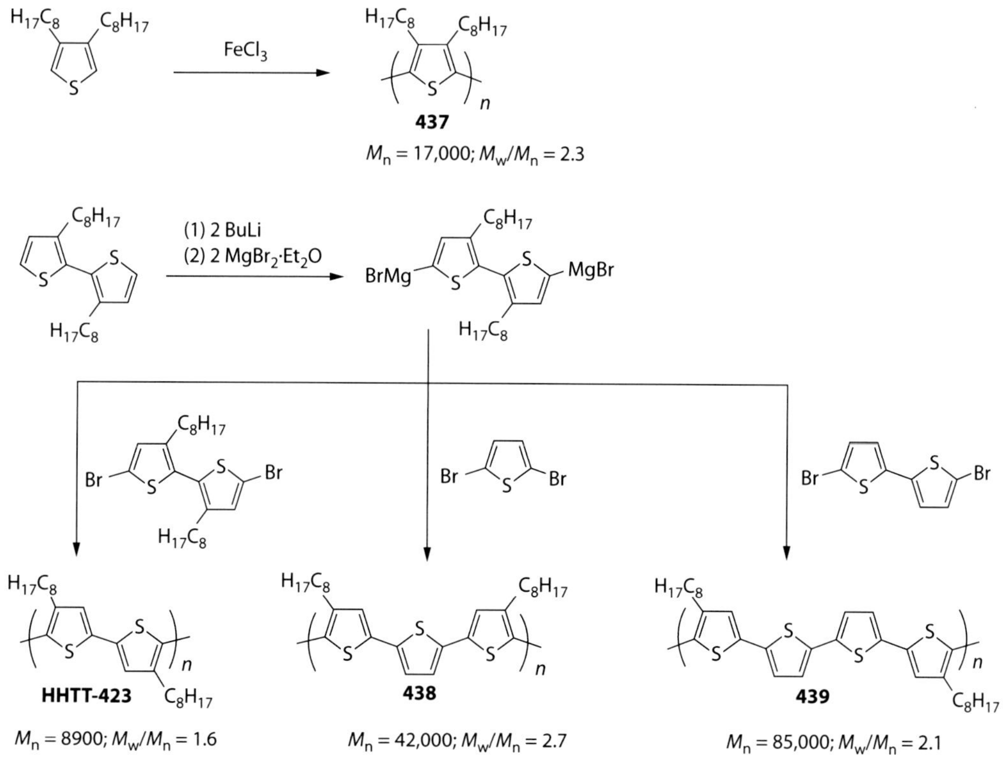

SCHEME 2.67
Synthesis of regioregular HHTT octyl-PTs. (From Gill, R.E., Malliaras, G.G., Wildeman, J., and Hadziioannou, G., *Adv. Mater.*, 6, 132, 1994.)

variation of the length of the coplanar blocks between the HH links. They also found blue shifts of >100 nm in absorption, PL and EL spectra in the sequence **439** → **438** → **HHTT-423** → **437**, i.e., increasing HOMO–LUMO gap with increasing steric hindrance. The energies of absorption and emission maxima were linear functions of the inverse number of thiophene units ($1/n$), between the two consecutive HH links (Figure 2.28), in agreement with theoretical predictions that the band gap scales linearly with the inverse of the chain length [510]. These results clearly demonstrate that the effective conjugation length is limited by HH links.

EL color tuning through introduction of various substituents is widely used in other LEPs but, probably, is not as impressive there. To understand the wide range of colors available from PTs, it is necessary to look at the underlying phenomena. The PT emission color depends directly on the effective conjugation length, determined by the twist angle between the thiophene units. Theory predicts a large change in the band gap of PT, depending on the torsion angle between consecutive thiophene units: the difference in the band gap of fully planar and 90°-twisted PTs is calculated to be 1.7 eV [511].

These observations inspired Inganäs and coworkers [111,512,513] to exploit the principle of steric hindrance to design PTs with emission colors spanning the full visible spectrum. A wide range of 3-substituted and 3,4-disubstituted thiophenes has been synthesized and successfully polymerized by $FeCl_3$ in chloroform, affording polymers **440–453** [481,513–516] (Chart 2.103).

Although the emission of substituted PTs is not very predictable owing to the interplay of several factors (steric effects, regioregularity, electronic effects, side-chain crystallization, etc.), the full visible range of PL and EL emissions from red to blue can be covered by variations of the PT substituents in positions 3(3,4) (Figure 2.29). A shift in band gap can also be seen through the change of electrochemical oxidation potentials. Additional evidence for the modification of the effective conjugation in these PTs was also found from Raman spectroscopy studies (shift of the symmetrical C=C stretching: from 1442 to 1506 cm^{-1})

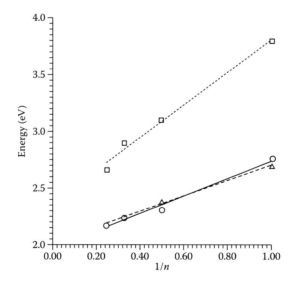

FIGURE 2.28

Absorption (□---), PL (Δ---), and EL (○----) energies of polymers **HHTT-423, 437–439** in thin films vs. an inverse number of thiophene units between the head-to-head links. (From Gill, R.E., Malliaras, G.G., Wildeman, J., and Hadziioannou, G., *Adv. Mater.*, 6, 132, 1994. With permission.)

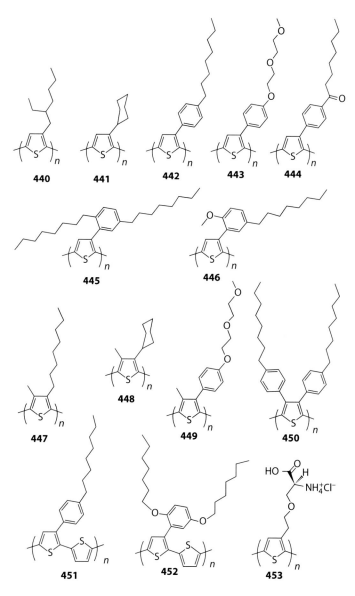

CHART 2.103
Chemical structures of 3-substituted and 3,4-disubstituted PTs.

[513]. However, calculation of the chromatic coordinates for these polymers showed that they cover only part of the chromatic space (Figure 2.30): no deep green colors are available in this family of PTs, mostly due to the broadness of the emission spectra. Some absorption and emission data of these polymers are collected in Table 2.5. The large steric hindrance in **448** allowed a shift of the EL down to 460 nm, with a concomitant drop in QE.

A certain balance should be kept in distortion of the thiophene planarity as a way to prevent the formation of interchain aggregates. Introducing two substituents at positions 3 and 4 of PT allows a shift of the emission through the entire visible range and prevents interchain interactions (resulting in a smaller decrease of the quantum yield in the solid

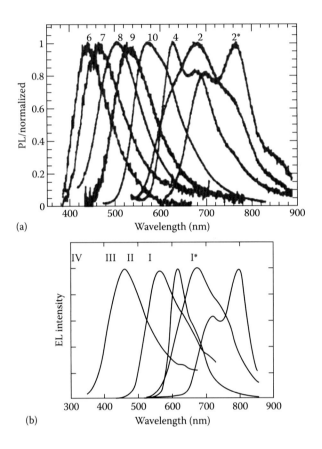

(a)

(b)

FIGURE 2.29
(a) PL spectra from spin-coated films of PTs: 2 = **442**, 4 = **451**, 6 = **448**, 7 = **447**, 8 = **450**, 9 = **449**, 10 = **441**, and 2* = **442** treated with chloroform vapor at room temperature. (From Andersson, M.R., Thomas, O., Mammo, W., Svensson, M., Theander, M., and Inganäs, O., *J. Mater. Chem.*, 9, 1933, 1999. With permission.) (b) EL spectra of ITO/polymer/Ca/Al devices: I = **442**, II = **451**, III = **441**, IV = **447**; I* = **442** upon thermal treatment or by chloroform vapor. (From Andersson, M.R., Berggren, M., Inganäs, O., Gustafsson, G., Gustafsson-Carlberg, J.C., Selse, D., Hjertberg, T., and Wennerström, O., *Macromolecules*, 28, 7525, 1995. With permission.)

state compared with solution). Highly crowded disubstituted PTs **447–450** show very low luminescence efficiency already in solution (Table 2.5) due to substantial distortion of the backbone [111,513].

In this context, polymer **445** represents a well-balanced material showing high PL efficiency in solution and solid state. The steric factor of the dialkylphenyl substituent is similar to those in **442–444**, as follows from the similarity of their absorption and emission energies. Molecular structure simulation shows that two octyloxy groups in the phenyl ring force its rotation to ca. 90° out of the thiophene plain. In this configuration, the bulky side chains prevent interchain interactions yet allow conjugation within the backbone. Spin-cast films of **445** show clear vibronic features in absorption (λ_{max} = 532 nm, ΔE = 0.18 eV) and PL (λ_{max} = 659 nm, ΔE = 0.16 eV) that, together with a very small Stokes shift of only 0.10 eV, suggest a highly ordered film (Table 2.5). Several PLEDs prepared with this polymer with the ITO/**445**/Ca/Al and ITO/**445**/PBD/Ca/Al configurations showed Φ_{EL}^{ex} = 0.1% and 0.7%, respectively [517].

The highly regioregular polymer **442**, obtained by oxidative polymerization with FeCl$_3$ (94 ± 2% HT) required some special attention because it could exist in two different forms.

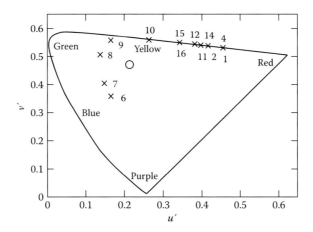

FIGURE 2.30
Chromatic coordinates for PTs: 1 = **423**, 2 = **442**, 4 = **451**, 6 = **448**, 7 = **447**, 8 = **450**, 9 = **449**, 10 = **441**, 11 = **443**, 12 = **444**, 14 = **445**, 15 = **440**, 16 = **446** and the white spot (○). (The u'–v' coordinates are 1976 modification of CIE x, y coordinates; the white point $x = y = 0.33$ corresponds to $u' = 0.211$, $v' = 0.474$). (From Andersson, M.R., Thomas, O., Mammo, W., Svensson, M., Theander, M., and Inganäs, O., *J. Mater. Chem.*, 9, 1933, 1999. With permission.)

In its pristine form, the absorption maximum of spin-cast films is at 493 nm (2.68 eV) [481,518]. On treating the films with chloroform vapor, the maximum was shifted to 602 nm (2.06 eV) and the spectrum showed fine vibronic structure with $\Delta E = 0.19$ eV, typical of a more planar ordered conformation (Figure 2.31a). This conversion could also be achieved by heating the film and, in contrast to the widely observed thermochromism in PTs, is irreversible. The PLEDs prepared with spin-coated **442** show red EL centered at 670 nm (1.85 eV). When the diode is heated, the emission is shifted continuously well into the NIR as a function of heating time and increasing temperature. On prolonged heating, **442** is converted into an ordered film with an EL emission maximum at 800 nm (1.55 eV) (Figure 2.31b) [519]. It is noteworthy that simple exposure to chloroform vapor results in a more pronounced ordering of the polymer than when heating to 200°C.

As an alternative strategy, the high-luminescence efficiency observed in solution can be preserved by incorporating the polymer in an inert-solid matrix. A possible problem of phase segregation can be solved by careful design of the polymer and the matrix. Thus, polymer **453**, when blended with polyacrylic acid (PAA), did not suffer phase segregation probably due to attractive ionic–hydrogen bonding interactions between the PT and the matrix [517,520]. As a result, the PL QE of **453**:PAA (1:100) was 16% (cf. 26% in solution), much higher than in **453** film (4%). Unfortunately, no PLED with the material was reported, and one can suggest that the performance of such a device would be low due to a very low concentration of the emitting (and conducting) component.

Another approach to tuning the optical properties and improving the emission of PT derivatives in the solid state was proposed by Li and coworkers [521] (Scheme 2.68). They used postfunctionalization of P3HT **424** by electrophilic substitution reaction to afford polymers **454** followed by Pd-catalyzed coupling (Suzuki, Stille, or Heck methodologies) of **454a** to afford polymers **455** [522]. In this case, functionalization with bulky aryl substituents (**455**) allowed a substantial increase of Φ_{PL} in the films from 1.6% to 13%–22%.

Saxena and Shirodkar [523] prepared copolymers **456** by oxidative (FeCl$_3$) polymerization of a mixture of 3-hexyl and 3-cyclohexylthiophenes, varying the monomer ratio from 1:9 to 9:1. The copolymers emitted yellow–green to green light (550–580 nm), and the

TABLE 2.5

Tuning the Properties of Electroluminescent PTs via Conjugation Control

Polymer (Regioregularity, % HT)	M_n (g/mol) (PDI)	λ_{abs} (nm) (Film)	λ_{PL} (nm) (Film)	Φ_{PL} (%) (CHCl₃)	Φ_{PL} (%) (Film)	λ_{EL} (nm)[a]	Φ_{EL}^{ex} (%)[a]	Turn-On Voltage (V)	Ref.
P3OT **423** (70)	35,000 (3.14)	506		27	4				[112]
440 (~70)		464	593	26	9				[112]
441 (77)	6000 (9.3)	405 [413[b]]	574	27	9	555	0.01 (7 V)	2.4	[112,513]
442 (94)	8000 (6.25)	482 (555[c])	677 (764[c])	18	9 (3[c])				[112]
	23,000 (2.26)	485	670			670	0.3 (6 V)	1.4	[512]
443 (88)	7800 (3.2)	476 (552[c])	616 (783[c])	20	8				[112]
444 (85)	9400 (2.9)	454 (555[c])	638 (795[c])	14	10				[112]
445 (90)	36,000 (1.94)	494, 532, 577	606, 659, 720	37	24		0.1[a]/0.7[d]		[112,517]
446 (90)	46,000 (3.48)	470	590	29	11				[112]
447	42,000 (2.17)	326	468	4.6	2.2				[112]
448	26,000 (2.77)	303 [305[b]]	442	1.3	0.8	460[d]	0.6 (25 V)[d] <10⁻⁴ (4 V)	7[d]	[112,513]
449	16,000 (3.0)	380	532	3.8	2.8				[112]
450	21,000 (3.0)	346	504	1.1	1.0				[112]
451	9000 (9.1)	513 [518[b]]	627	27	5	610	0.1 (8 V)	1.6	[112,513]
452	24,500 (4.3)	510	598	31	4				[112]
456			550–580				(2–16) × 10⁻³ (2.8–3.6 V)		[523]

[a] ITO/polymer/Ca/Al.

From Andersson, M.R., Berggren, M., Inganäs, O., Gustafsson, G., Gustafsson-Carlberg, J.C., Selse, D., Hjertberg, T., and Wennerström, O., *Macromolecules*, 28, 7525, 995.

[c] Annealed film.

[d] ITO/polymer/PBD/Ca/Al.

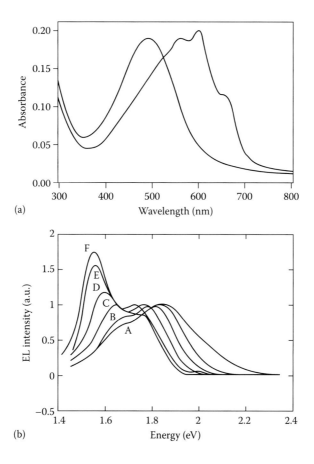

FIGURE 2.31
(a) UV–vis absorption spectra of a spin-coated film of **442** on glass (λ_{max} = 493 nm) and the same film treated with chloroform vapor (λ_{max} = 602 nm, ΔE = 0.19, 0.15, 0.19 eV). (From Andersson, M.R., Selse, D., Berggren, M., Järvinen, H., Hjertberg, T., Inganäs, O., Wennerström, O., Österholm, J.-E., *Macromolecules*, 27, 6503, 1994. With permission.) (b) EL of ITO/**442**/Ca/Al device at different heating temperatures: (A) unheated device; (B) T = 100°C, 6 s; (C) 100°C, +6 s; (D) 200°C, +10 s; (E) 200°C, +20 s; and (F) treated with chloroform before evaporating the contact. (From Berggren, M., Gustafsson, G., Inganäs, O., Anderson, M.R., Wennerström, O., and Hjertberg, T., *Appl. Phys. Lett.*, 65, 1489, 1994. With permission.)

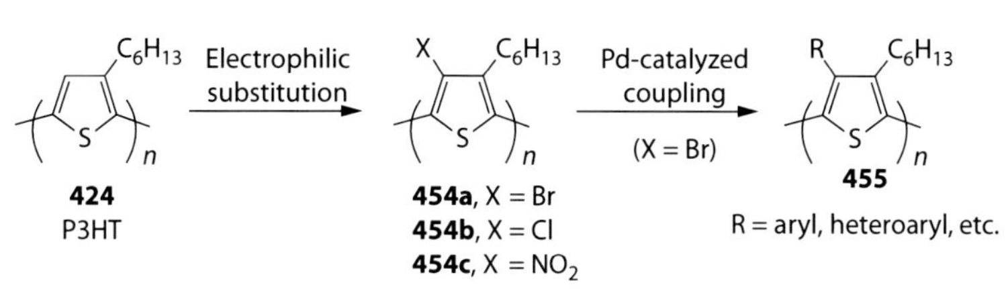

SCHEME 2.68
Postfunctionalization of PT **424**. (From Li, Y., Vamvounis, G., and Holdcroft, S., *Macromolecules*, 35, 6900, 2002; Li, Y., Vamvounis, G., Yu, J., and Holdcroft, S., *Macromolecules*, 34, 3130, 2001.)

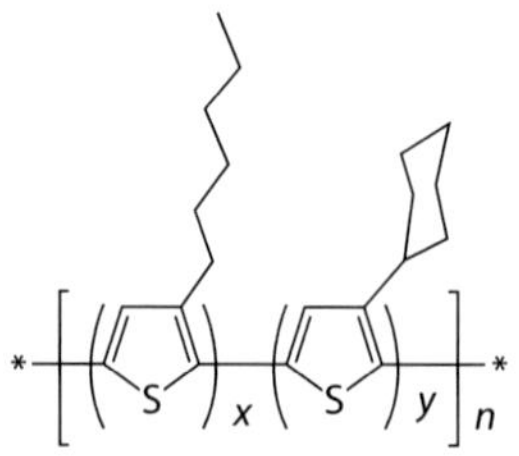

456

CHART 2.104
Chemical structures of polythiophene copolymers.

ITO/**456**/Al devices showed Φ_{EL}^{ex} in the range of 0.002%–0.016% [523]. However, even for the best emissive copolymer (x:y = 2:3), Φ_{EL}^{ex} was only 0.016% with a charge-carrier mobility of 5.6×10^{-4} cm^2/(V s) (Chart 2.104).

2.4.3 Blends of Polythiophenes

It was demonstrated that blends of 3(3,4)-substituted PT derivatives of different band gaps gave rise to a voltage-controlled variable color light source [524]. Because of different turn-on voltages of high and low band-gap polymers, the emission color can be potentially tuned applying different bias. When a relatively low voltage was applied to the polymer blend PLED, the low-band-gap polymer started to emit first, followed by higher band-gap emission as the voltage increased. An effective phase separation, however, is required to minimize the Förster energy transfer onto the low-band-gap species. A blend of PTs **448**:**451** (50:50) at 22 V showed EL at ca. 630 nm, originating from the emission of **451**. When the voltage was increased, an additional emission at ca. 440 nm (corresponding to the EL of **448**) appeared, reaching a comparable intensity at 27 V [524,525].

Other combinations of PTs in a blend, **447**:**441**:**451**:PMMA (PMMA is poly(methyl methacrylate), 10:4:1:1) produced EL emission of the ITO/polymer blend/PBD/Ca/Al device at 20 V, very close to the equienergy white point as defined by the CIE, while providing a relatively high Φ_{EL}^{ex} = 0.4%–0.6% (at 20 V) (Figure 2.32) [526]. PMMA was used in this case to diminish energy transfer from the high-band-gap to the low-band-gap polymers. Several other inert polymer matrices (polystyrene, polycarbonate, polyvinylchloride, and poly(2,6-dimethyl-1, 4-phenyleneoxide)) showed similar effects [527].

Destri et al. [528] reported the electrochemical synthesis of polymer **457**, which produced a mixture of oligomers up to dodecamer. According to the MALDI–TOF mass spectrum, the maximum peak corresponded to the tetramer (12 thiophene units). The ITO/**457**/Ca/Al device emits red light (580–650 nm) comparable to that for regioregular PATs. Blending **457** with PVK and PBD resulted in a white-emitting diode with 0.03% EL efficiency [528]. Blending with PBD also was an effective way to increase Φ_{EL}^{ex} of highly sterically hindered disubstituted PT **448** from <0.0001% to 0.6% with no change in EL maximum [513] (Chart 2.105).

The emission spectrum of some PT and PBD polymer bilayer devices cannot be explained by a linear combination of emissions of the components. Thus, white emission of the PLEDs ITO/**451**/PBD/Al showed Φ_{EL}^{ex} of 0.3% at 7 V, and consisted of blue (410 nm), green (530 nm), and red–orange (620 nm) bands. Whereas the first and the last EL peaks are due to the EL from the PBD and the PT layers, respectively, the green emission probably originates from a transition between electronic states in the PBD layer and hole states in the polymer layer

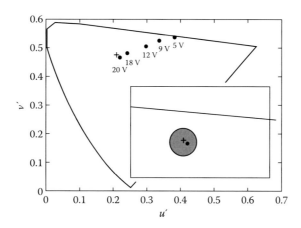

FIGURE 2.32
Chromaticity diagram showing the color of the LED with **447:441:451**:PMMA (10:4:1:1) polymer blend at different voltages. Inset: magnified part of the chromaticity diagram; the shaded circle represents the area, which is defined as white (radius 0.028 and centered at equienergy white point; the u'–v' coordinates are 1976 modification of CIE x, y coordinates, the white point $x = y = 0.33$ corresponds to $u' = 0.211$, $v' = 0.474$). (From Granström, M. and Inganäs, O., *Appl. Phys. Lett.*, 68, 147, 1996. With permission.)

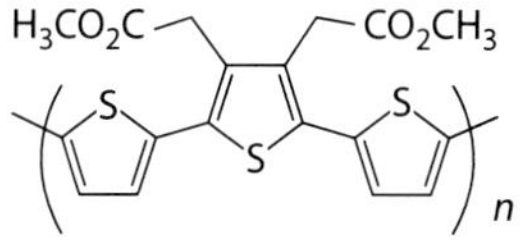

457

CHART 2.105
Chemical structures of thiophene oligomers to dodecamers.

[529]. Similar results (additional green–blue EL at ~495 nm) were demonstrated by PLED ITO/**451**/PBD/Ca/Al [530].

Blending low-band-gap PTs with other EL polymers was employed to increase the EL efficiency of a PLED, and it was demonstrated that only small additions of PTs can improve the device performance. The Φ_{EL}^{ex} of red-emitting ITO/P3HT(**424**):MEH–PPV/Ca diodes initially increased with P3HT content and went through a maximum at 1 wt.% of P3HT with $\Phi_{EL} = 1.7\%$ [531], which is 2–3 times higher than in the neat ITO/MEH–PPV/Ca diode and three orders of magnitude higher than the ITO/P3HT(**424**)/Ca diode. Later, List and coworkers [532,533] reported a similar observation of efficient yellow-light emission from the blend of blue-emitting ladder-PPP (LPPP **543b**, Chart 2.125) with small additions (0.5–2%) of orange-emitting **425**. When the concentration of P3DT **425** was as small as 1%, the external EL efficiency of the Al/polymer blend/ITO device was also significantly higher $\left(\Phi_{EL}^{ex} = 4.2\%\right)$ than in pure LPPP **543b** (2%).

Apart from the tunable color emission covering the full visible range, there are several other aspects supporting the interest in PTs for PLEDs. PTs are examples of classical conjugated polymers with intrinsic one-dimensionality of the polymer chain. Alignment can induce anisotropy in macroscopic properties such as electron-transport or optical properties. Polarized EL with $\Phi_{EL}^{ex} = 0.05\%$ was observed in multilayer LB-film PLED ITO/**428**/Al, with a ratio in EL between the parallel and perpendicular orientations of 1.3 [534]. An even higher ratio of 2.4 was achieved in ITO/**451**/Ca/Al diode made from a stretch-oriented polymer film [535]. For more information on polarized LEDs, see Chapter 5 of this book.

Among other specific applications of PTs as light-emitting materials, it is necessary to mention microcavity LEDs prepared with PTs **451** and **445** [536,537] and nano-LEDs demonstrated for a device with patterned contact structure, and PT **451** blended in a PMMA matrix that emits from phase-separated nanodomains (50–200 nm) [538,539].

2.4.4 Oligothiophenes

High order and crystallinity of oligothiophenes in the solid state determine their unique optical and electrical properties, i.e., high charge-carrier mobility, anisotropy of electrical and optical properties, etc. In particular, oligothiophenes are widely studied in FETs showing a high level of hole and electron mobilities and high on–off ratios. In this context, one would expect poor suitability of such materials as light-emitting layers in LED. Nevertheless, several publications demonstrate EL from some oligothiophene-based LEDs. For a deeper understanding of the effect of the conjugation length on the electrical properties and emission in PTs, Geiger et al. [540] studied a series of end-capped oligothiophenes **458** (Chart 2.106). The ITO/**458**/Al devices prepared by vacuum sublimation of oligomers showed relatively low turn-on voltages of ca. 2.5 V and moderate current densities (e.g., 7 mA/cm^2) with maximum efficiency at ca. 8 V; however, the EL efficiency was rather low (estimated internal efficiency $\Phi_{EL}^{int} \approx 10^{-2}\% - 10^{-3}\%$). LEDs showed a red shift in the EL peak with increasing number of thiophene units ($n = 3 \rightarrow n = 5$) and a linear dependence of the EL band energy on the inverse number of monomer units. Averaging over EL, PL, and absorbance data, the effective conjugation length was estimated at approximately 9–10 thiophene units.

Several other oligothiophenes were studied in this regard. Variations in the main chain length of phenyl end-capped oligothiophenes **460** also showed red shifts of absorbance

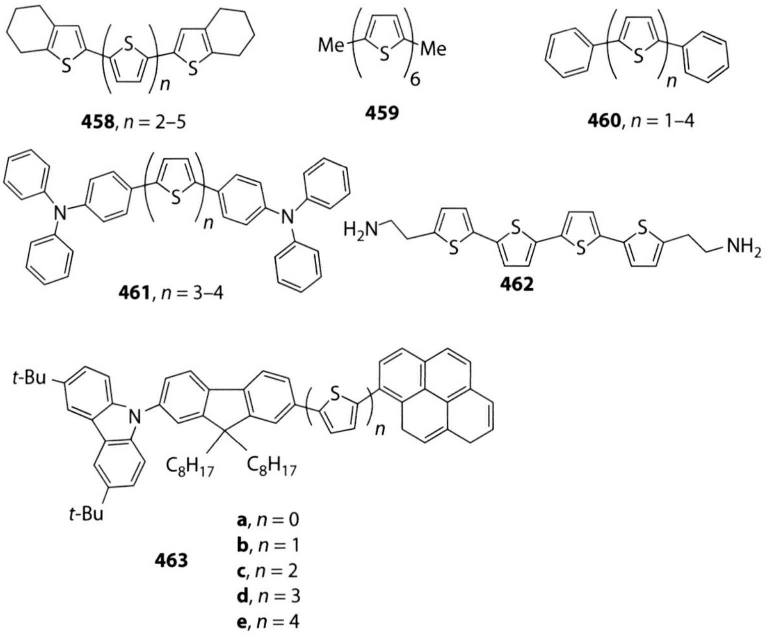

458, *n* = 2–5   **459**   **460**, *n* = 1–4

461, *n* = 3–4   **462**

463

a, *n* = 0
b, *n* = 1
c, *n* = 2
d, *n* = 3
e, *n* = 4

CHART 2.106
Chemical structures of oligofluorene–thiophenes.

(from 375 to 524 nm) and emission (from 470 to 620 nm) with increasing number of thiophene units [541–543]. Sexithiophene end capped with CH_3 groups (**459**) in ITO/**459**/Al configuration (vacuum sublimed) showed red–orange emission with a very low QE of ~10^{-9}% [544]. Low efficiencies of LEDs, based on crystalline oligothiophenes, were somewhat improved by end capping of terthiophene and quaterthiophene with triphenylamino groups (**461**), which led to stable amorphous glasses with luminescence efficiencies of 0.03 and 1.1 lm/W, respectively (at a luminance of 300 cd/m^2) [545,546]. A double-layer device with oligomer **461** ($n = 3$) as emitting layer and Alq$_3$ as ETL showed significantly improved performance, exhibiting a maximum luminance of 13,000 cd/m^2 at a driving voltage of 18 V. Undoubtedly, improving the hole-transport properties in these oligomers by end capping with triphenylamino fragments is also an important factor. Terthiophene end capped with 2-aminoethyl groups was also used in hybrid organic–inorganic perovskite materials. When **462** was incorporated within lead halide perovskite layers in an ITO/**462**:PbCl$_4$/OXD7/Mg/Ag device (OXD7 is 1,3-*bis*[4-(*tert*-butylphenyl)-1,3,4-oxadiazolyl]phenylene), a bright-green emission (530 nm) from the organic layer was found [547].

Promarak's group reported a series of oligofluorene–thiophenes end capped with 3,6-di-*tert*-butylarbazole and pyrene as color tunable emissive materials [548]. The carbazole moiety attached to the end of the molecule is nearly perpendicular to oligofluorene–thiophene–pyrene plane, and π-electrons in the ground state delocalize over the entire molecule. PLED devices of these materials emitted brightly in various colors from deep blue to orange. Deep-blue (CIE coordinates of 0.16, 0.14) device (ITO/PEDOT:PSS/**463a**/BCP/LiF/Al) shows a luminance efficiency of 1.14 cd/A, where 2,9-dimethyl-4,7-diphenyl-1,10-phenanthroline (BCP) was used as a hole-blocking and electron-transporting layer. Green device with a configuration of ITO/PEDOT:PSS/NPB/**463c**/BCP/LiF/Al exhibits a high luminance efficiency of 11.15 cd/A, where *N,N*'-diphenyl-*N,N*'-*bis*(1-naphthyl)-(1,1'-biphenyl)-4,4'-diamine (NPB) was used as electron blocking layer (Chart 2.106).

A quinquethiophene oligomer unit was used as a core in a light-emitting dendrimer with redox-active triarylamine peripherals (**464**) [549] (Chart 2.107). In this material, an excitation of the peripheral amines at 310 nm results in energy transfer to the highly luminescent fluorophore at the core of the dendrimer with subsequent green emission (λ_{PL} = 550 nm) exclusively from the oligothiophene. In an LED containing PBD as the electron-transporting material, hole transport occurs solely through the peripheral triarylamines, whereas the core oligothiophene plays the role of a light emitter. The EL spectrum was essentially identical to PL (λ_{EL} = 560 nm), with no emission from either PBD (390 nm) or peripheral amines (425 nm), and the maximum Φ_{EL}^{ex} was 0.12%.

2.4.5 Thiophene-*S,S*-Dioxides as Emissive and Electron-Transport Moieties

Poly- and oligothiophenes are generally p-type (hole-transporting) semiconducting materials. Recently, Barbarella and coworkers [550,551] reported a novel approach to tailoring the frontier orbitals of thiophene oligomers through a chemical transformation of the thiophene ring into the corresponding thiophene-*S,S*-dioxide (via oxidation with 3-chloroperbenzoic acid). This modification results in "dearomatization" of the thiophene unit and increases the electron deficiency, thus offering an efficient methodology to increase the electron delocalization and the electron-transport and -injection properties of the material. A comparison of two quaterthiophenes, **465** and **466**, indicates that a single thiophene-*S,S*-dioxide moiety leads to only a slight increase of the oxidation potential (from 0.95 to 1.04 V vs. Ag/AgCl), whereas the reduction potential is drastically shifted into positive potentials (from –2.12 to –1.28 V) that result in a band-gap contraction by >0.7 eV [550]

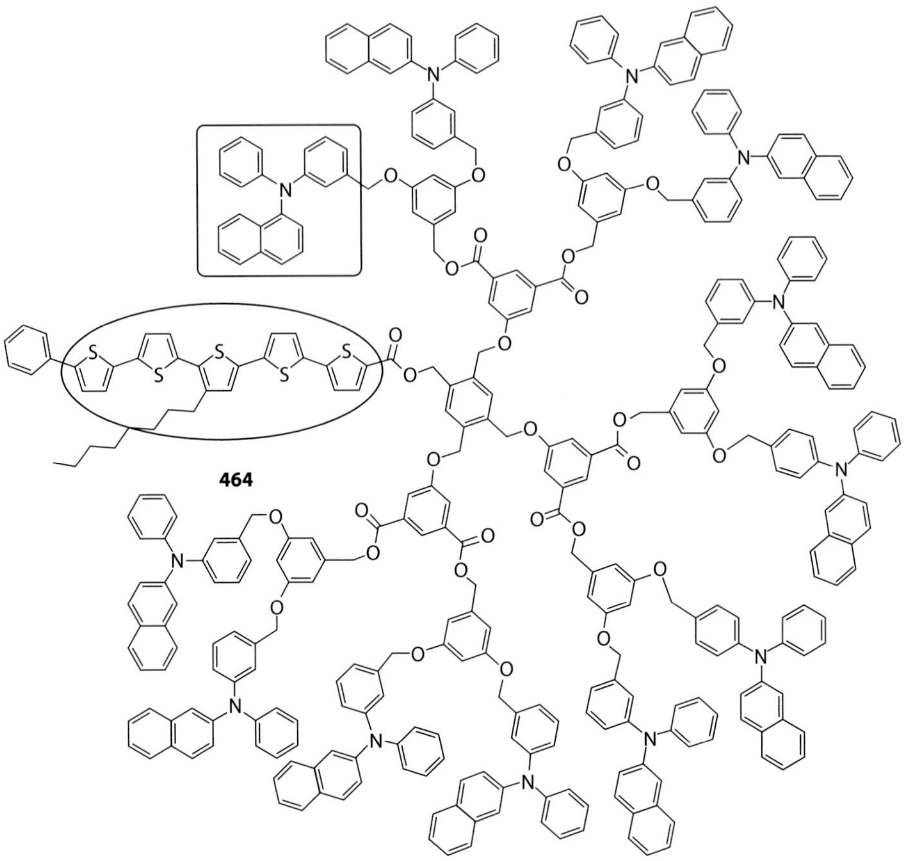

CHART 2.107
Chemical structure of a quinquethiophene oligomer with triarylamine peripherals.

(Chart 2.108). Another feature of this modification is a decreased aggregation tendency, resulting in decreased exciton migration to the nonradiative centers. Consequently, oligomers incorporating thiophene-S,S-dioxide units possess good PL properties in solution and the solid state, as well as high (for PTs) EL efficiency. Particularly interesting in this case are the oligomers with a central location of the thiophene-S,S-dioxide unit for which the solid-state PLQYs were reported to be as high as 37% (**467** [552]) and 45% (**468** [553]) (and up to 70% for a thiophene-S,S-dioxide unit incorporated into oligophenylenes [554]).

Incorporation of thiophene-S,S-dioxide units in oligothiophenes allows to vary both absorption and PL energies in a wide range ($\lambda_{max}^{abs} \sim 400-540\,nm$, $\lambda_{max}^{PL} \sim 525-725\,nm$) [550,554]. Polymers obtained by chemical polymerization of oligomers **467** and **468** with $FeCl_3$ showed PL in the NIR region (801 and 910 nm, respectively), although the quantum yields were not reported for these materials [554]. A nonoptimized LED with **467** as an active layer (ITO/**467**/Ca/Al) showed a luminance of ~100 cd/m² at 7 V and a rather low EL efficiency of 0.03 cd/A at ~180 mA/cm² [555]. However, further studies showed that these parameters can be sufficiently improved by blending **467** with PVK and introducing a PEDOT layer: the PLED built as ITO/PEDOT/**467**:PVK, 85:15/Ca/Al configuration showed a maximum luminance of ~200 cd/m² at 7 V and an EL efficiency of ~0.9 cd/A at 3 mA/cm² [552].

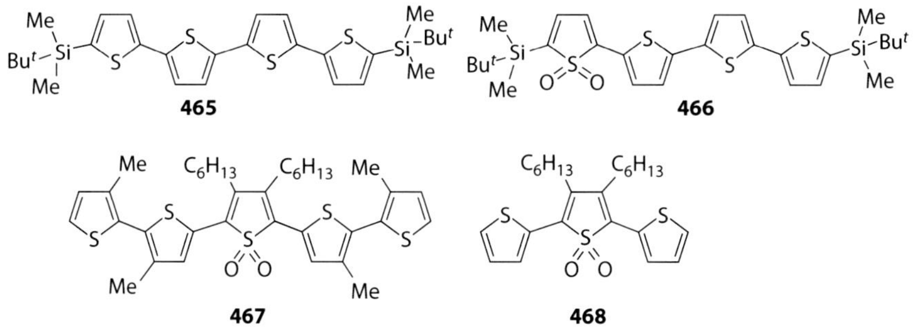

CHART 2.108
Chemical structures of thiophene-*S*,*S*-dioxides.

Other thiophene–thiophene-*S*,*S*-dioxide copolymers were reported by Berlin et al. [556], who synthesized copolymers **469** and **470** with an alternating electron acceptor thiophene-*S*,*S*-dioxide unit and donor ethylenedioxythiophene (EDOT) units (Chart 2.109). The polymers absorbed at 535 nm (E_g = 2.3 eV) in chloroform solution and in films (which is consistent with their electrochemistry: $E_{ox} \approx 0.40$–0.50 V, $E_{red} \approx -1.75$–1.8 V; $\Delta E \approx 2.2$–2.25 V) and emitted at 650 nm (Φ_{PL} [film] ~ 1%). Such a high band gap (which exceeds that in PEDOT homopolymer by ~0.6 eV) strongly suggests a disruption of the conjugation (possibly owing to two alkyl substituents in the thiophenedioxide moiety). The EL emission spectrum was entirely the same as PL emission, and $\Phi_{EL}^{ex} = 0.01\%$ at 100 cd/m² was found for ITO/TPD:**469**:PC (40:40:20)/Ca diode (PC is bisphenol-A-polycarbonate).

These pioneering works stimulated recent research activities in incorporating the thiophene-*S*,*S*-dioxide unit into various copolymers PLEDs built with such copolymers were reported by several groups. Charas and coworkers studied PLEDs based on copolymer **471** obtained by Suzuki coupling of 2,5-dibromothiophene-*S*,*S*-dioxide with diboronic ester of 9,9-*bis*(2-ethylhexyl)fluorene [557] and its blends with PFO **215** [558,559]. The copolymer **471** emitted orange light $\left(\lambda_{PL}^{film} = 615\,\text{nm}\right)$, and there was a strong suppression of PLQY going from solution to the solid state $\left(\Phi_{PL}^{cyclohexane} = 19\%,\ \Phi_{PL}^{film} = 0.5\%\right)$. A single-layer ITO/**471**/Ca PLED exhibited rather low EL efficiency $\left(\Phi_{EL}^{ex} = 2.2 \times 10^{-4}\%\right)$, which was attributed to a combination of low PL efficiency and charge-transport limitations. Upon inserting a hole-injection PEDOT layer, the EL efficiency was increased to $\Phi_{EL}^{ex} = 9 \times 10^{-4}\%$ and the maximum luminance increased from 0.2 cd/m² to about 5.3 cd/m². The optoelectronic characteristics of the devices were improved by blending **471** with PFO, allowing an increase in Φ_{EL}^{ex} up to 0.21% (for ITO/PEDOT/PFO:**471**(95:5)/PBD/Ca architecture) and a

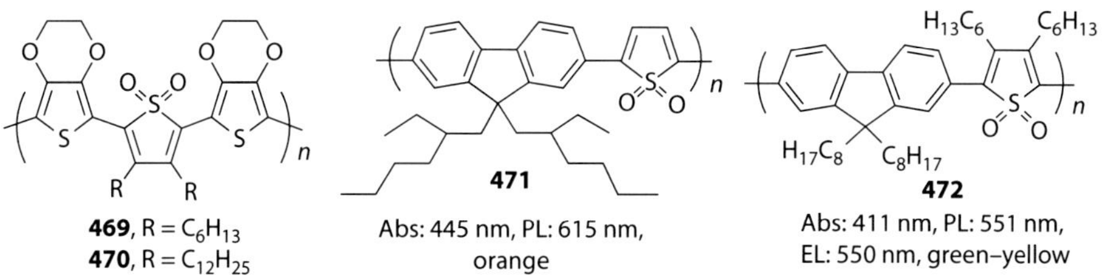

469, R = C₆H₁₃
470, R = C₁₂H₂₅

471
Abs: 445 nm, PL: 615 nm, orange

472
Abs: 411 nm, PL: 551 nm, EL: 550 nm, green–yellow

CHART 2.109
Chemical structures of thiophene–thiophene–*S*,*S*-dioxide copolymers.

decrease in the turn-on voltage from 16 to 5–5.5 V. Remarkably, the EL of the blend in this case was almost exclusively from the copolymer **471**, in spite of the low ratio of the latter.

The same Suzuki methodology was used to synthesize a similar copolymer **472** [560]. The polymer showed a solvent-dependent green–yellow emission (from 545 nm in THF to 565 nm in chloroform) as often observed for polar chromophores. The PL QE also varied with the solvent (from 11% in THF to 21% in decalin); however, in contrast to copolymer **471**, no strong decrease in emission efficiency was observed in the solid state $\left(\Phi_{PL}^{film} = 13\%\right)$ that could be attributed to the effects of substituents at the thiophene ring. LEDs based on **472** showed, for an ITO/PEDOT/**472**/Ca/Al architecture, a turn-on voltage of ca. 10 V with a maximum brightness of 340 cd/m² at 22 V and appreciable $\Phi_{EL}^{ex} = 0.14\%$.

Beaupré and Leclerc [313] reported fluorene–thiophene copolymers, in which fluorene and thiophene-*S,S*-dioxide fragments were separated by one or two thiophene units (**473** and **474**, respectively) (Chart 2.110). The electronic effect of an additional thiophene unit (the system can be viewed as an alternating donor–acceptor polymer) and the planarization factor known for longer oligothiophene units resulted in a pronounced band-gap contraction. These copolymers are p- and n-dopable, as followed from their electrochemistry, with band gaps of 2.0 and 2.2 eV for **473** and **474**, respectively. The PLEDs, fabricated with configurations ITO/LiF/polymer/PBD/LiF/Al/Ag, showed rather low turn-on voltages of 4 V; however, the maximum brightness (120 cd/m² at 7 V and 15 cd/m² at 8 V, for **473** and **474**, respectively) was lower than that for copolymer **472**. Although highly efficient $\left(\Phi_{PL}^{film} \approx 40\% - 70\%\right)$ solid-state PL was demonstrated from some oligothiophenes and oligophenylenes containing thiophene-*S,S*-dioxide units [554], the efficiency of similar fluorene copolymers is remarkably lower.

The combination of thiophene and thiophene-*S,S*-dioxide units in a copolymer allows tuning the emission color from green to pure red [313,561]. However, the PLEDs fabricated with these materials showed a rather low $\Phi_{EL}^{ex} < 0.01\%$, which further decreased with an increasing number of thiophene units. Similar results (significant decrease of the PL QE) were observed for thiophene–thiophene-*S,S*-dioxide copolymers containing 3,6-dimethoxyfluorene (**475** [253]) and carbazole units (**476** [562]) ($\Phi_{PL} = 20\% - 25\%$ in solution).

Also, a danger of potential instability toward n-doping of the thiophene-*S,S*-dioxide containing oligomeric and polymeric EL materials can be foreseen, based on the known

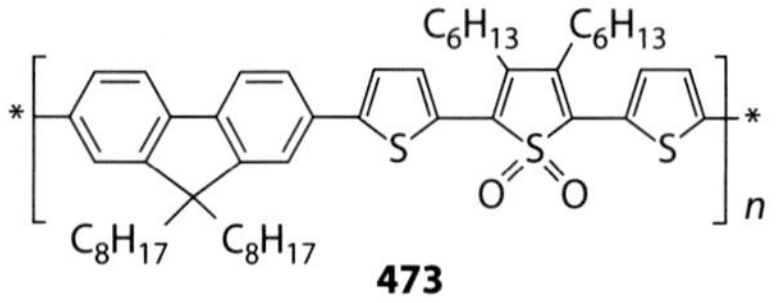

Abs: 509 nm, PL: 610, 660 nm,
EL: 610, 650 nm, red–orange

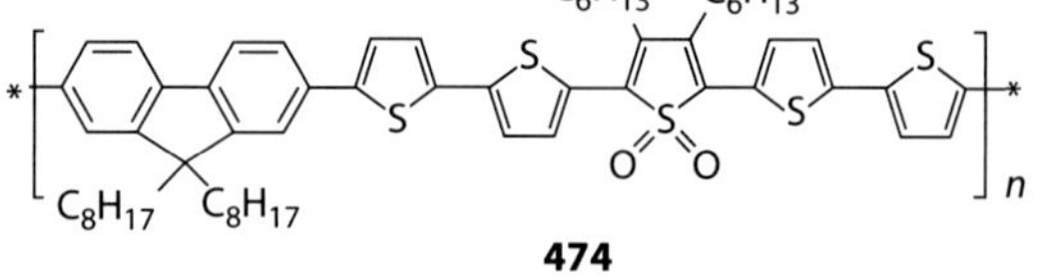

Abs: 547 nm, PL: 666, 708 nm,
EL: 668, 708 nm, red

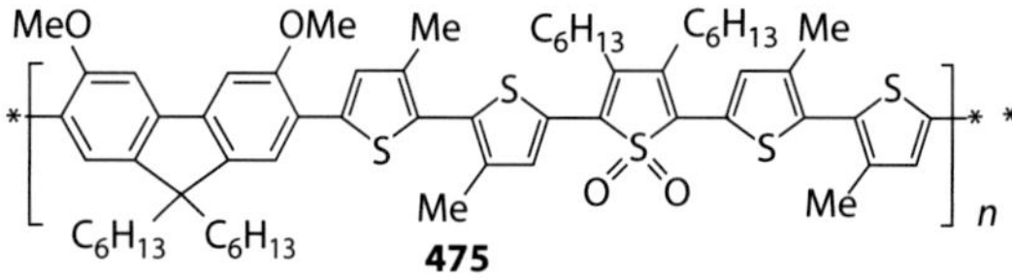

Abs: 392, 478 nm, PL: 662 nm, red

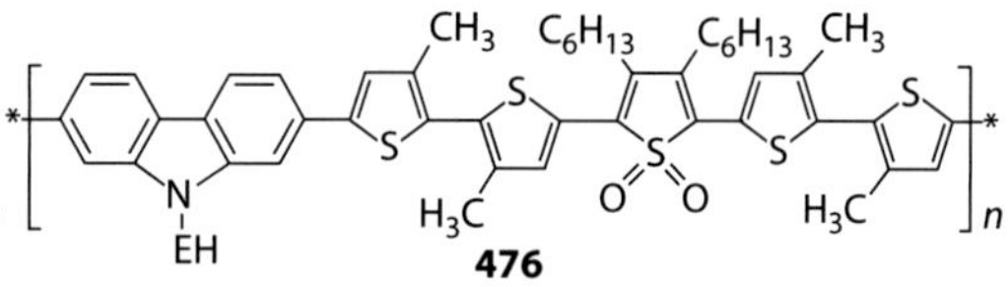

Abs: 512 nm, PL: 671 nm (Φ_{PL}: 25%)

CHART 2.110
Chemical structures of fluorene–thiophene copolymers.

instability of other classes of heterocyclic systems with SO_2 fragment in the cycle toward reductive cleavage. However, it should be mentioned that such a behavior has not, thus far, been reported for the above thiophene-*S*,*S*-dioxide-containing materials.

2.4.6 Polythiophene Copolymers

2.4.6.1 Thiophene Block Copolymers with Conjugation Break

The above approaches used the idea of conjugation length control in PTs by distorting the polymer backbone with bulky substituents as side groups. Hadziioannou and coworkers [563,564] demonstrated PL and EL tuning via exciton confinement with block copolymers **477a–d** and **478a–f**, containing oligothiophene and alkylsilylene units (Chart 2.111). Precise control of the conjugation length of the oligothiophene blocks, interrupted by silylene units, allowed emission tuning from blue to orange–red (Table 2.5). Later, Yoshino et al. [565] reported similar extended block copolymers **477d–h** that showed changes in EL color from green to red with increasing oligothiophene block length.

2.4.6.2 Thiophene Copolymers with Aromatic Moieties

Fahlman and coworkers [566] first reported EL from alternating phenylene–thiophene copolymer **480**. Its band gap, ionization potential, and electron affinity, calculated with the VEH method, are 3.08, 5.29, and 2.22 eV, respectively. These values are between the corresponding values for PPP **500** (3.28, 5.43, and 2.15 eV [567]) and PT **414** (1.6, 5.0, and 3.4 eV [568]). The steric hindrance of heptyl side groups in this polymer results in interannular torsion angles of 50° that are substantially larger than that of PPP (23°); nevertheless, their band gaps are smaller than that of PPP. Phenylene–thiophene copolymers **479** and **480** emit blue light at ca. 450–475 nm, with somewhat different reported Φ_{EL}^{ex} of ~0.2% [569] and 0.03% [467], respectively, for ITO/polymer/Ca configurations. It has also been shown that the efficiency of the device can be substantially improved (up to 2%) by blending of **480** with substituted PPP (1:10) (**503e**, Chart 2.118) [569]. This is among the highest values reported for thiophene-based LEP (Chart 2.112).

An important series of copolymers, **481**, containing thiophene–phenylene–thiophene repeating units, have been reported by Huang and coworkers [299,570,571]. The polymers were obtained via $FeCl_3$ oxidation of corresponding thiophene–phenylene–thiophene trimers that were synthesized by Pd-catalyzed coupling of 1,4-R^2,R^3-2,5-dibromobenzenes with the corresponding 3-R^1-2-thienylzinc chlorides. By changing the substituents R^1 and R^2, the polymer emission can be tuned from greenish–yellow to pure green.

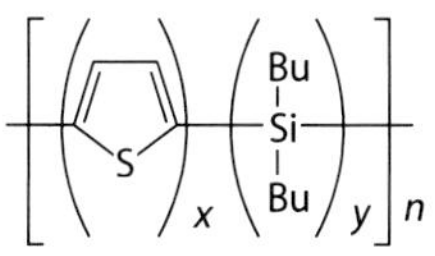

477

a, $x = 2$, $y = 1$	**e**, $x = 4$, $y = 2$
b, $x = 2$, $y = 2$	**f**, $x = 5$, $y = 2$
c, $x = 3$, $y = 1$	**g**, $x = 6$, $y = 2$
d, $x = 3$, $y = 2$	**h**, $x = 7$, $y = 2$

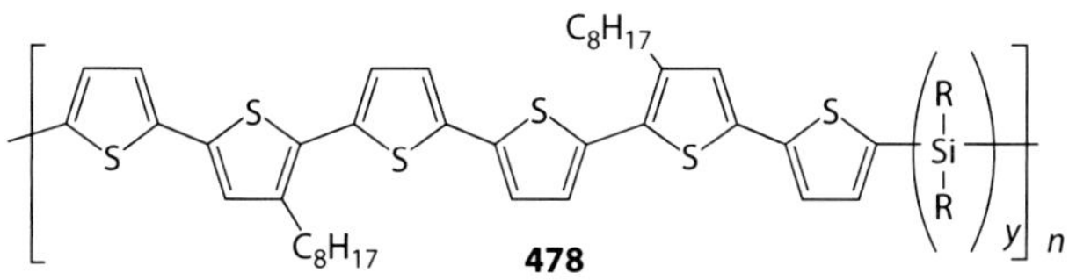

478

a, R = Bu, $y = 1$	**c**, R = Me, $y = 1$
b, R = Bu, $y = 2$	**d**, R = Me, $y = 2$
	e, R = Me, $y = 4$
	f, R = Me, $y = 8$

CHART 2.111
Chemical structures of block copolymers containing oligothiophene and alkylsilylene units.

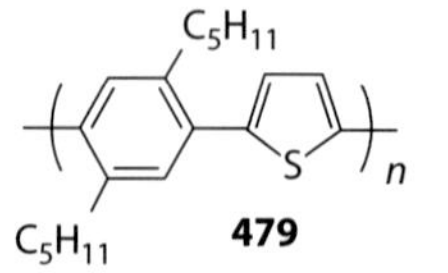

479

Abs: 335 nm, PL: 455 nm,
EL: 455 nm, Φ_{EL}^{ex} = 0.03%,
blue

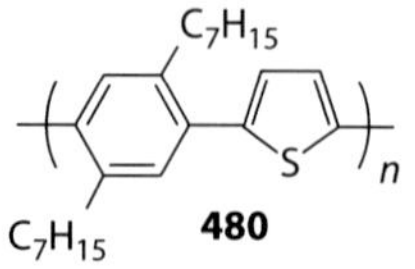

480

Abs: ~330 nm, PL: ~450 nm,
EL: ~475 nm, Φ_{EL}^{ex} = 0.03%,
blue

CHART 2.112
Chemical structures of phenylene–thiophene copolymers.

While retaining much of the substituted PT character (e.g., good hole-transport properties and stability), these materials exhibit significantly improved fluorescence efficiency in the solid state (Φ_{PL} up to 29%) that leads to Φ_{EL}^{ex} of up to 0.1% for ITO/**481**/Ca PLED (Table 2.6). Other widely studied thiophene copolymers with aromatic 9,9-disubstituted fluorene units were described in Section 2.3.

Alternating oligothiophene-containing copolymers **482–486** with 1,1-binaphthyl units, which interrupt the conjugation due to the large torsion angle between the naphthalene rings, were reported. The nonplanar structure could prevent the self-quenching processes in the solid state, and variation in the length of the oligothiophene segment from one to seven thiophene rings tuned the emission color from yellow–green to red (Table 2.7) [572]. A single-layer device ITO/**483**/Al prepared with copolymer **483** emitted orange light (λ_{EL} at 568 nm with a shoulder at 590 nm) with a turn-on voltage of 5.7 V, luminance of 25 cd/m^2 at 8.0 V, and Φ_{EL}^{ex} = 0.005% [573] (Chart 2.113).

Copolymer **487**, prepared by Stille coupling of dibromophenylene with 2,5-*bis*(tributylstannyl)thiophene, represents another example of a phenylene-*alt*-thiophene backbone, where the substituted phenylene unit forms an oligophenylene–vinylene fragment that is not in the main conjugation chain [574]. A PLED fabricated with this polymer (ITO/**487**/Al) emitted green light (520 nm) with a turn-on voltage of ca. 9.5 V, but no other data on luminance or efficiency of the device were reported (Chart 2.114).

2.4.6.3 Thiophene Copolymers with Heteroaromatic Moieties

Several copolymers (**488–494**) containing electron-rich thiophene and electron-deficient 1,3,4-oxadiazole units have been reported by Huang and coworkers [575–578]. Structural variations, particularly different lengths of oligothiophene fragments, allowed the tuning of the band gap and PL energy of these materials (Table 2.8). An ITO/**488c**/Ca single-layer LED emitted blue light with a turn-on voltage of 8 V at forward bias. Although the polymer showed very high Φ_{PL} = 79%, the Φ_{EL}^{ex} of the device was only ~0.0001% due to unoptimized device structure and, possibly, a purity problem (Chart 2.115).

Jenekhe and coworkers [579,580] reported the synthesis of other *n*-type conjugated copolymers with alternating bithiophene and *bis*-quinoline units, **495a–f**, that showed reversible reduction at –1.65 to –1.80 eV vs. SCE (onsets at ca. –1.4 to –1.52 V) and thus expected to exhibit good electron-transport properties (Scheme 2.69). These polymers emit green light $\left(\lambda_{PL}^{sol} = 502–504\,\text{nm},\ \lambda_{PL}^{film} = 517–524\,\text{nm}\right)$ with moderate efficiency in solution (Φ_{PL} = 22%–28%) that, however, decreased to 1%–3% in the solid state. The poor luminescence quantum yield of **495**, compared with other polyquinolines, might be due to charge-transfer quenching in these donor–acceptor copolymers. As an emissive material, **495** exhibits weak green EL (529–538 nm) with Φ_{EL}^{ex} = 0.004% – 0.015% for an ITO/PEDOT/**495**/Al configuration. A red

TABLE 2.6

Tuning the Properties of Substituted Phenylene–Thiophene Copolymers

481

R^1	R^2	R^3	M_n (g/mol)	PDI	λ_{max}^{abs} (nm)	λ_{max}^{PL} (nm)	Φ_{PL} (%)	Φ_{EL}^{ex} (%)	Turn-On Voltage (V)[a]	Ref.
n-C$_6$H$_{13}$	CH$_3$	CH$_3$	20,400	2.92	340	477	10	0.004	17	[299,570]
			12,600	1.48	346	470	15 ± 1	0.02	13	[299,570]
n-C$_6$H$_{13}$	H	H	19,700	2.74	396	524	20			
			31,200	2.31	378	505	22 ± 2	0.1	8	[299,570]
n-C$_6$H$_{13}$	OC$_{10}$H$_{21}$	OC$_{10}$H$_{21}$	14,600	1.92	430	530	29			
			26,100	1.27	405	520	27 ± 3			
c-C$_6$H$_{11}$	CH$_3$	CH$_3$			360	466	6			[570]
c-C$_6$H$_{11}$	H	H			376	495	11			[570]
c-C$_6$H$_{11}$	OC$_{10}$H$_{21}$	OC$_{10}$H$_{21}$			410	505	16	0.05	9	[570]
p-(n-C$_4$H$_9$)Ph	n-C$_{10}$H$_{21}$	n-C$_{10}$H$_{21}$	82,300	2.19	442	532	11	(~0.1)	(~7)	[571]
p-(n-C$_4$H$_9$)Ph	CH$_3$	EH	68,500	1.51	426	538	12	(~0.1)	(~7)	[571]
p-(n-C$_4$H$_9$)Ph	EH	EH	43,600	1.68	429	533	11	0.1	7	[571]

[a] ITO/polymer/Ca.

TABLE 2.7

Properties of Binaphthyl-Linked Thiophene Block Copolymers with Different Conjugation Lengths

Compound	M_n (g/mol)	PDI	λ_{max}^{abs} (nm)	λ_{max}^{PL} (nm) (DCM)	Φ_{PL} (%) (DCM)[a]
482, *n* = 1	13,900	2.6	368	421, 446, 475sh	54
483, *n* = 2	18,100	1.7	406	463, 498	26
484, *n* = 4	5100	1.1	440	515, 549sh	23 (19[b])
485, *m* = 2	17,400	1.6	434	530, 568sh	5.4 (23[b])
486, *m* = 3	2300	2.6	454	545, 583, 631sh	6.5 (7.2[b])

[a] Excitation at 380 nm.
[b] Excitation at the longest wavelength absorption maximum.

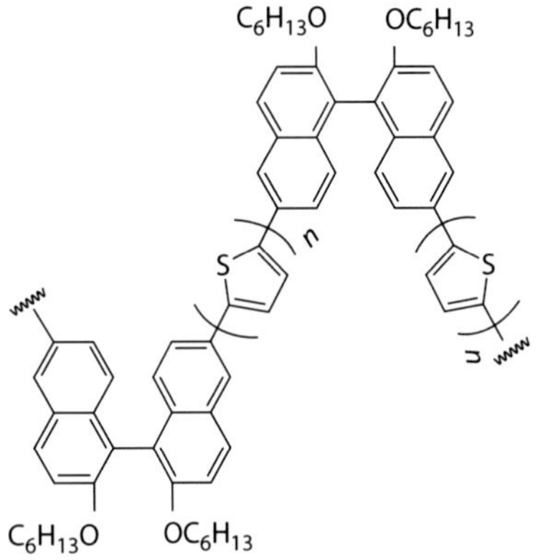

482, *n* = 1; Abs = 368 nm, PL = 421, 446, 475sh nm
483, *n* = 2; Abs = 406 nm, PL = 463, 498 nm
484, *n* = 4; Abs = 440 nm, PL = 515, 549sh nm

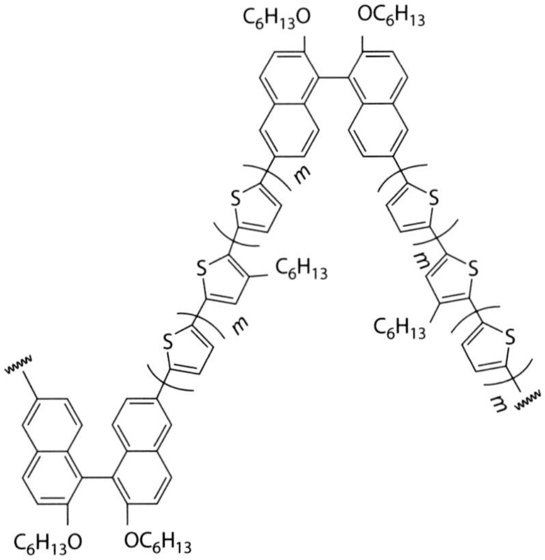

485, *m* = 2; Abs = 434 nm, PL = 530, 568sh nm
486, *m* = 3; Abs = 454 nm, PL = 545, 583, 631sh nm

CHART 2.113
Chemical structures of alternating oligothiophene-containing copolymers.

shift in absorption (37 nm), PL (44 nm), and EL (33 nm) was observed for films of HT polymer **495f**, compared with a HH analog **495b** [580]. This had no effect on Φ_{PL} of materials and performance of the devices. A large improvement in performance of PLEDs was found for bilayer devices ITO/MEH–PPV/**495**/Al, where polymers **495** act as ETL materials. The diodes showed bright-orange–red EL emission of MEH–PPV with turn-on voltages of 8.5–9 V and luminance in the range of 948 cd/m² for **495a** to 2170 cd/m² for **495b** (Φ_{EL}^{ex} was in the range of 0.86% for **495e** to 1.4% for **495b**) [580]. Similarly, polymer **495c** can be used in polymer blend systems, where a single-layer blend PLED ITO/MEH–PPV:**495c** (72:28)/Al showed a turn-on voltage of 6 V, luminance of 1480 cd/m² at 14 V, current density of 279 mA/cm², and Φ_{EL}^{ex} = 0.64% [579].

A low-band-gap ($E_g \sim 1.6$ eV) conjugated thiophene copolymer **496** with pyrrole and BT units was synthesized via Stille coupling [581]. They showed emission in the NIR region ($\lambda_{EL} \sim 800$ nm) with turn-on voltage below 4 V but with very low efficiency (Chart 2.116).

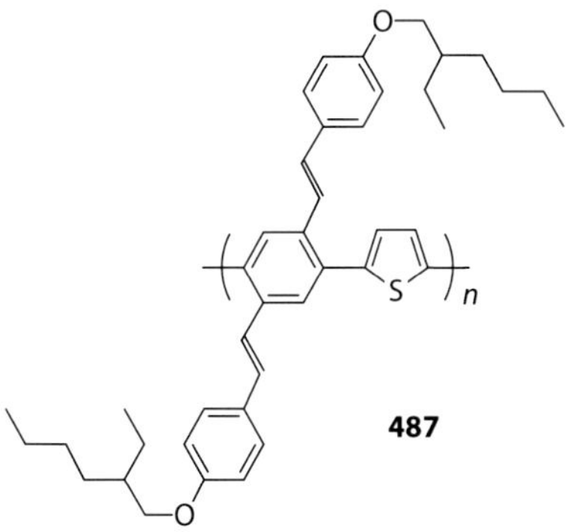

487

Abs: 362 nm, PL: 530 nm, EL: 520 nm, green

CHART 2.114
Chemical structure of a phenylene–thiophene copolymer.

TABLE 2.8

Electrochemical and Optical Properties of Oxadiazole–Thiophene Copolymers

Compound	M_n (g/mol) (PDI)	λ_{max}^{abs} (nm) Film (Solution)	λ_{max}^{PL} (nm) Film (Solution)	Φ_{PL} (%)	E_g^{opt} (eV) (Film)	E_{red} (V) (Ref. Electrode)
488a		(330)	(425)	7.6[a]	(2.87)	
488b		395 (364)	475 (433)	68[a]	2.83 (2.85)	Onset −1.86 (SCE)
488c		396, 440sh (375)	486 (438)[b] 455, 469, 478[c]	79[a]	2.79 (2.84)	Onset −1.88 (SCE); onset −1.54 (Ag/Ag⁺)
490	5419 (1.9)	420, 443	489			pc/pa −1.83/−1.60 (SCE)
489	7574 (2.0)	441, 471	530			pc/pa −1.95/−1.70 (SCE)
491	2814 (1.4)	439sh, 461, 494sh	580			pc/pa −1.79/−1.70 (SCE)
492	3830 (1.8)	358	462		3.04	pc/pa −1.73/−1.53 (SCE); E_g^{CV} = 3.02
493	24,900 (1.41)	376 (342)	498, 526 (444, 462)		2.83	pc/pa −1.86/−1.75 (SCE); E_g^{CV} = 2.84
494	2870 (2.0)	430	568		2.54	pc/pa −1.76/−1.67 (SCE); E_g^{CV} = 2.51

[a] Relative quantum yield vs. quinine sulfate.
[b] From Huang, W., Meng, H., Yu, W.-L., Gao, J., and Heeger, A.J., *Adv. Mater.*, 10, 593, 1998.
[c] From Huang, W., Yu, W.-L., Meng, H., Pei, J., and Li, S.F.Y., *Chem. Mater.*, 10, 3340, 1998.

Blending with dialkoxy-PPV **14** in a device (ITO/PEDOT/polymer blend layer/LiF/Ca) substantially improved the EL efficiency (by about two orders of magnitude). A moderately efficient energy transfer from the higher-band-gap PPV (λ_{EL} = 650 nm) to PT **496** (λ_{EL} = 830 nm) allowed fine tuning of the emission color by changing the component ratio (Figure 2.33) [582].

Oligo-2,5-thienylenevinylenes (OTVs) have been recognized as a potential class of linear conjugated systems for micro- and nanoelectronics [583]. Comparison of optical data

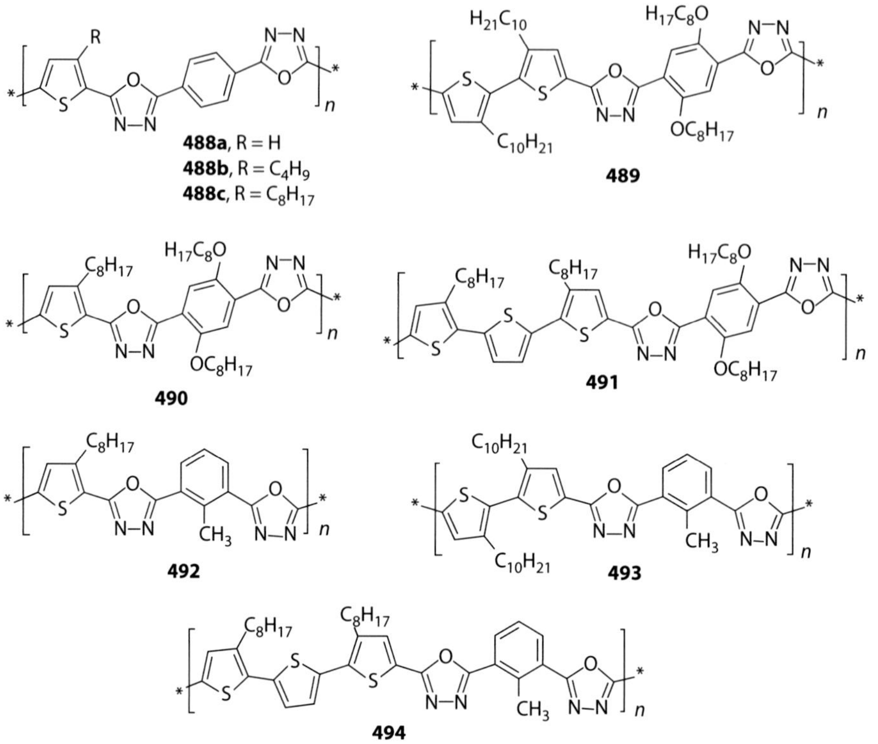

CHART 2.115
Chemical structures of PTs containing 1,3,4-oxadiazole unit.

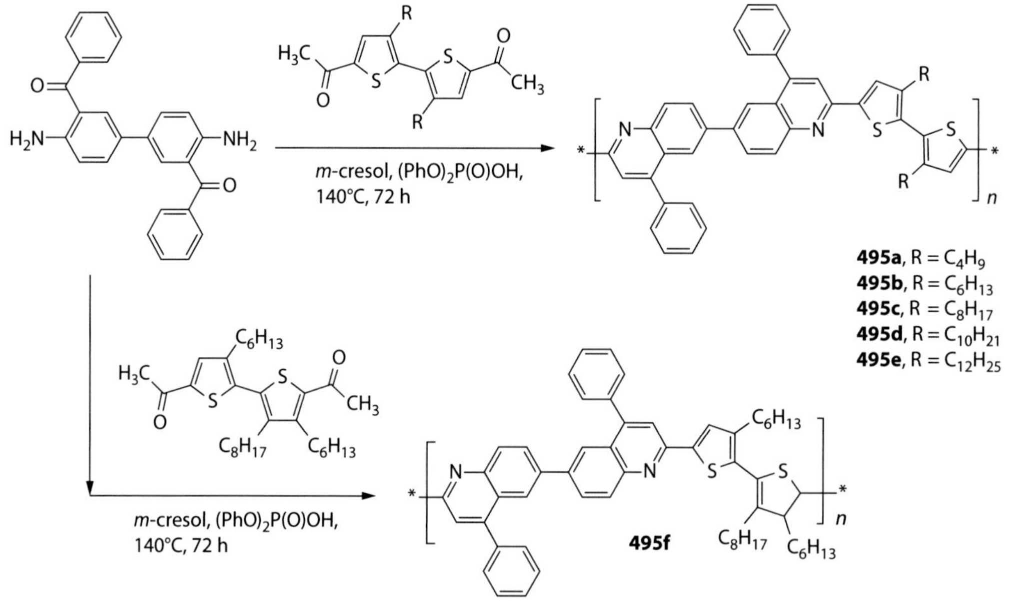

SCHEME 2.69
Synthesis of bisquinoline–thiophene copolymers. (From Tonzola, C.J., Alam, M.M., and Jenekhe, S.A., *Adv. Mater.*, 14, 1086, 2002; Tonzola, C.J., Alam, M.M., Bean, B.A., and Jenekhe, S.A., *Macromolecules*, 37, 3554, 2004.)

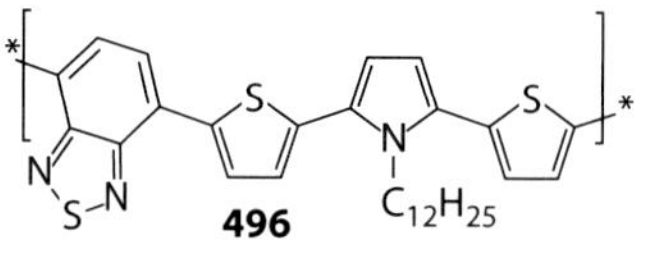

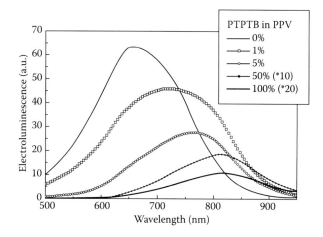

CHART 2.116
Chemical structures of PTs containing pyrrole and BT units.

FIGURE 2.33
Electroluminescence spectra of ITO/PEDOT/active layer/LiF/Ca devices with **496** (PTPTB) and **14** (PPV) as an active layer. (From Brabec, C.J., Winder, C., Sariciftci, N.S., Hummelen, J.C., Dhanabalan, A., van Hal, P.A., and Janssen, R.A.J., *Adv. Funct. Mater.*, 12, 709, 2002. With permission.)

for OTV with other classes of conjugated oligomers (oligothiophenes, oligo-2,5-thiophene ethynylenes, oligo-1,4-phenylene vinylenes, oligo-1,4-phenylene ethynylenes, and oligo-acetylenes) showed that OTVs exhibit the longest effective conjugation length among the known systems and the smallest $\Delta E_{HOMO-LUMO}$ values, promising the lowest band gap for the corresponding polymers. We are not aware of any report of poly-2,5-thienylenevinylene (PTV, **497** [27]), a fluorescent material, in spite of strong NIR PL and EL observed in cyano-substituted PTV (**132** [47]). Furthermore, blending 5%–25% of PTV **497** into PPV **1** completely quenched the luminescence of the latter, and the resulting blend was not emissive (Chart 2.117). A thiophene-benzobisthiazole copolymer (**498**) synthesized via Stille coupling reaction was reported by Kim et al. [584]. The EL result with an architecture of

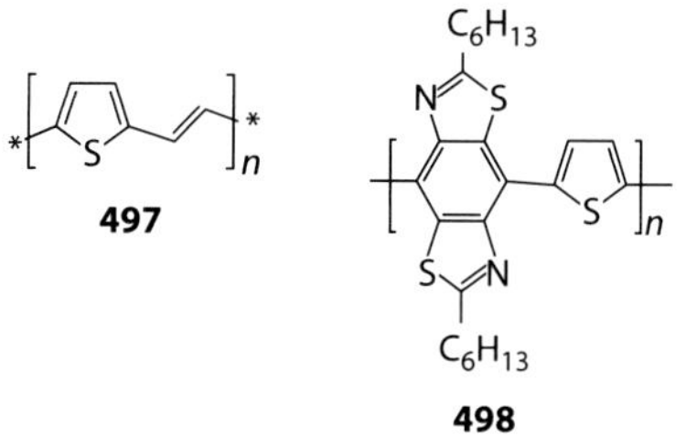

CHART 2.117
Chemical structures of poly(2,5-thienylenevinylene) and thiophene-benzobisthiazole copolymer.

ITO/PEDOT/498/Al demonstrates that the maximum luminescence of the device reaches 55 cd/m² at the driving voltage of 9.3 V with EL emission around 600 nm.

2.4.7 Summary

PTs represent an important class of (generally) low-band-gap conjugated polymers for LED applications. Variation in substituents that changes the torsion angle of the thiophene rings allows tuning of the emission over a wide range, from blue–greenish to deep red and NIR. PTs possess a strong aggregation tendency that decreases the PL and EL emission efficiency but that can be minimized by introducing bulky substituents. Regioregularity in mono-3-substituted PTs offers an additional control over the light-emitting properties of these materials. PATs have higher HOMO energies than PPVs or PFs (e.g., for poly-3-octylthiophene: HOMO = −4.57 eV, E_g = 1.96 eV), thus decreasing the hole-injection barrier from the ITO electrode.

Relatively efficient blue (polymer **448**, $\Phi_{EL}^{ex} = 0.6\%$ [513]), red (polymer **445**, $\Phi_{EL}^{ex} = 0.7\%$ [517]), and white (blend of **441**, **447**, and **451**, $\Phi_{EL}^{ex} = 0.3\%$ [526]) emitters have been reported for thiophene homopolymers, although their performance is far from the champions of other classes of LEP.

On the other hand, very respectable performances were demonstrated by blends of PT with other emitting polymers, as exemplified by a yellow-emitting blend of LPPP **543b** with **425** $\left(\Phi_{EL}^{ex} = 4.2\%\right)$ [532]. Furthermore, easy functionalization of the thiophene nucleus and its electron-rich character make it attractive for the design of various copolymers with other classes of aromatics and heteroaromatic systems, allowing EL color tuning and hole–electron-transport properties of the materials. In fact, many of the best-performing LEPs contained a certain amount of thiophene comonomer units in the structure (e.g., pure red emitter **357** [320]).

2.5 Miscellaneous Classes of Light-Emitting Polymers

In the previous sections, we described three main classes of LEP: PPV, PF, and PTs, although many other conjugated and nonconjugated polymers have also been used as EL materials for LEDs. Without making an attempt to cover all types of polymers ever used as EL materials, we describe below the most important classes and the most prominent examples of EL polymers, not covered in the previous sections. Some more examples of such systems can be found in recent reviews on blue LEPs [240], polycarbazoles [585], and the general topic of organic EL materials [6,10,20].

2.5.1 Poly-*p*-Phenylenes

Material instability is one of the major limitations of PLEDs. In this light, a very high thermal stability of polyphenylenes, combined with high PLQY, renders them attractive as materials for device applications. The first PPP-based PLED was described by Leising and coworkers [586,587] in 1992. Polymer **500** was synthesized by aromatization of a soluble precursor **499** (Scheme 2.70). Owing to the relatively high band gap of PPP **500** (2.7–3.0 eV), the PLED ITO/**500**/Al emitted blue light with λ_{EL} ~460 nm ($\Phi_{EL}^{ex} = 0.01\%$, turn-on voltage

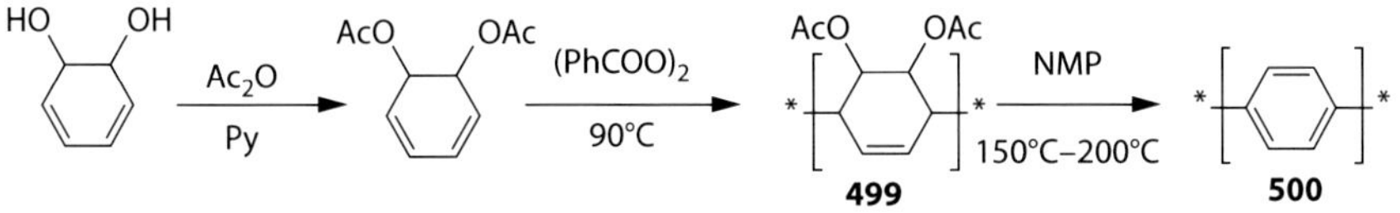

SCHEME 2.70
Synthesis of PPP by soluble precursor way.

of 10 V) [587]. High-efficiency EL was never reported for this particular polymer, which might be due to intrinsic defects in the structure (~15% of *meta*-linkage) associated with the synthetic method.

Considerable interest has been paid to solution-processable PPP materials, prepared by introducing long-solubilizing substituents into the phenylene rings of the backbone (**501–503**) (Chart 2.118). Synthesis of such systems was typically achieved via Ni-catalyzed Yamamoto coupling of 1,4-dihalophenylenes or Pd-catalyzed Suzuki coupling of halides with boronic acids or esters, both of which delivered well-defined polymers with molecular weight of ~10⁴ [588–590]. Numerous substituted PPPs, including dialkyl (**501a,b** [589] and **503** [590]), alkoxy (**501c–e** [591] and **501f–h** [592]), benzoyl (**501i** [593]), and dialkoxy (**501j** [594], **501k** [595], **501l** [596], **501m** [66]) PPPs have been synthesized and studied in PLEDs.

Steric hindrance due to alkyl substituents increases the torsion angle between phenylene units, which results in an additional (unwanted) increase of the band gap. The emission band shifts into the violet region of the spectrum (λ_{PL} ~400 nm) [597]. The hypsochromic shift due to less sterically demanding alkoxy group is less dramatic (λ_{EL} = 420, with much weaker bands at ~500 and ~600 nm). These polymers show rather high solid-state

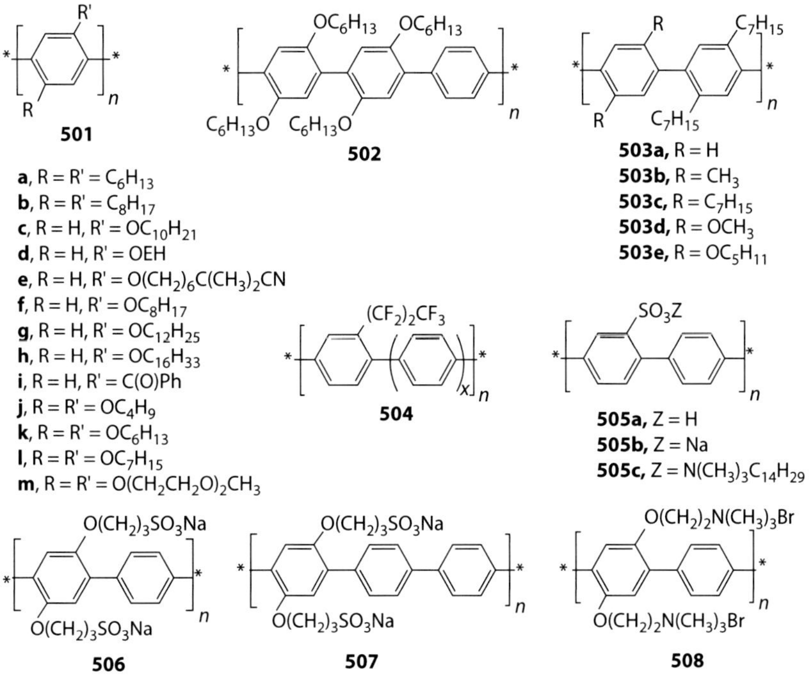

CHART 2.118
Chemical structures of PPPs with long-solubilizing substituents.

PLQY (35%–45%) [591]. Although a relatively poor EL performance ($\Phi_{EL}^{ex} = 0.15\%$, brightness 30 cd/m^2) was reported by Chen and Chao [592] for alkoxy derivatives **501f–h**, Heeger and coworkers [591,598] succeeded in fabrication of the very efficient violet blue-emitting PLEDs with alkoxy-PPPs **501c–e**. The $\Phi_{EL}^{ex} = 1.8\%$ was reported for the single-layer device ITO/**501c**/Ca [598]. It can be further increased to 3% using a second HTL (in devices ITO/PPV **1**/**501c**/Ca [583] or ITO/PVK/**501c**/Ca [591]). About five times lower efficiency was obtained using a high-work-function electrode (Al). Somewhat lower but also very respectable efficiencies were demonstrated by PLEDs with two other alkoxy-PPPs: ITO/PVK/**501d**/Ca (2%) and ITO/PVK/**501e**/Ca (1.4%) [591].

A second alkoxy substituent in PPP may increase the chain twisting and further enlarge the band gap. Thus, a band gap of 3.4 eV and PL–EL maxima at ~400 and ~500 nm have been reported for polymer **501k**, synthesized by oxidative (FeCl$_3$) polymerization of di(heptyloxy) benzene [595]. On the other hand, this could be a result of defective polymer structure (due to the synthetic method), as a PL maximum of 435 nm was reported for dibutoxy-PPP **501j** [594]. By blending the latter with PVK **535** (10%) and PBD **21** (10%) to improve the charge-transport properties of the material, a very efficient $\left(\Phi_{EL}^{ex} = 1.2\%\right)$ PLED was fabricated with a stable Al cathode; however, the reported turn-on voltage was high (20 V). In 1995, Pei and coworkers [66] used the dialkoxy-PPP **501m** to demonstrate the idea of LEC, in which a $\Phi_{EL}^{ex} = 2\%$ was achieved with an Al cathode (in the device configuration ITO/**501l**:PEO:LiOTf/Al).

The band gap in substituted PPPs can be tuned to some extent by "diluting" the substituted phenylene rings with unsubstituted phenylenes. A bathochromic shift in absorption was observed for copolymer **502** ($E_g = 2.95$ eV) compared with the "all-substituted" homopolymer **501k** ($E_g = 3.1$ eV) [595]. Sandwiching this polymer between ITO and Ca electrodes afforded a blue-emitting PLED with $\Phi_{EL}^{ex} = 0.5\%$. Even a higher-efficiency PLED $\left(\Phi_{EL}^{ex} = 2\%\right)$ was fabricated from PPP alkoxy and alkyl copolymer **503a**, when blended with 10% of diheptylphenylene–thienylene copolymer **480** (device ITO/**503**:**480**/Ca) [599]. Several other PPP homopolymers and copolymers **503b–e** based on the diheptylphenylene units have been synthesized [590].

A very different route to soluble PPP derivatives was demonstrated by Hamaguchi and coworkers [600], who introduced perfluorinated alkyl substituents into PPP **500** by reaction with perfluorobutanoyl peroxide. The resulting modified polymer **504** was soluble in common organic solvents and a solution-fabricated PLED ITO/**504**/Mg:In emitted blue to green light (depending on voltage) with band half-width of >200 nm.

Some EL devices based on sulfonated copolymer **505** were reported by Cimrová and coworkers [601]. These polyelectrolyte materials are soluble in DMSO (or in mixed solvents, for **505c**). The solution PL maximum (400 nm) does not depend on the counterion; however, the EL maximum is more sensitive to coulombic interactions and can be shifted from ~480–500 nm (for the sodium salt **505b**) to 454 nm (for the acid **505a**). The Φ_{EL}^{ex} of the devices, fabricated between ITO and Al electrodes, is between 0.5% and 0.8%. Later, Kim and coworkers [602] reported related polymers **506** and **507**, where the ionic sulfonate substituents are separated from the PPP backbone by an alkoxy chain. In contrast to **505**, these polymers were soluble in water. A PLED device fabricated by sandwiching the blend of **506** with polyethyleneimine (as a counterion source) between ITO and Al electrode showed very low $\Phi_{EL}^{ex} \sim 0.01\%$ [602,603]. An interesting approach to fabrication of multilayer PLEDs via sequential adsorption of PPP cationic (**506**) and anionic (**508**) layers from a water solution was demonstrated by Baur et al. [603]. Although the Φ_{EL}^{ex} of the device ITO/35 bilayers (**506**/**508**)/Al was very low (0.002%), the importance of this work is in the demonstration of a novel device fabrication method. Another approach taking advantage of the cationic–anionic LEP interaction was reported by Chang and coworkers [194], who prepared dual-color LED matrix by ink-jetting a solution of an anionic PPV derivative **166** on a spin-coated film of **508**.

In the previous sections, we described the effective tuning of EL properties of other classes of LEP (PPV, PF, PT) by introduction of phenylene units into their backbone (e.g., copolymers **136**, **284**, **285**, **306–308**, and **479–481**). Likewise, a series of very effective EL materials were designed by separating the oligophenylene ($n = 2$–5) blocks by ethylene, vinylene, or ethynylene units (copolymers **509–519**) [604] (Chart 2.119). Changing the length of the phenylene block and separating bridges, the polymer absorption (336–406 nm), and PL maxima (401–480 nm) can be varied over a wide range. A solid-state PLQY of up to 60%–70% was reported for some of these materials. The EL can be tuned over an even wider range (423–650 nm) in multilayer PLEDs, and by varying the polymer composition and the device structure, certain PLEDs with an internal QE of up to 4% were engineered (Table 2.9).

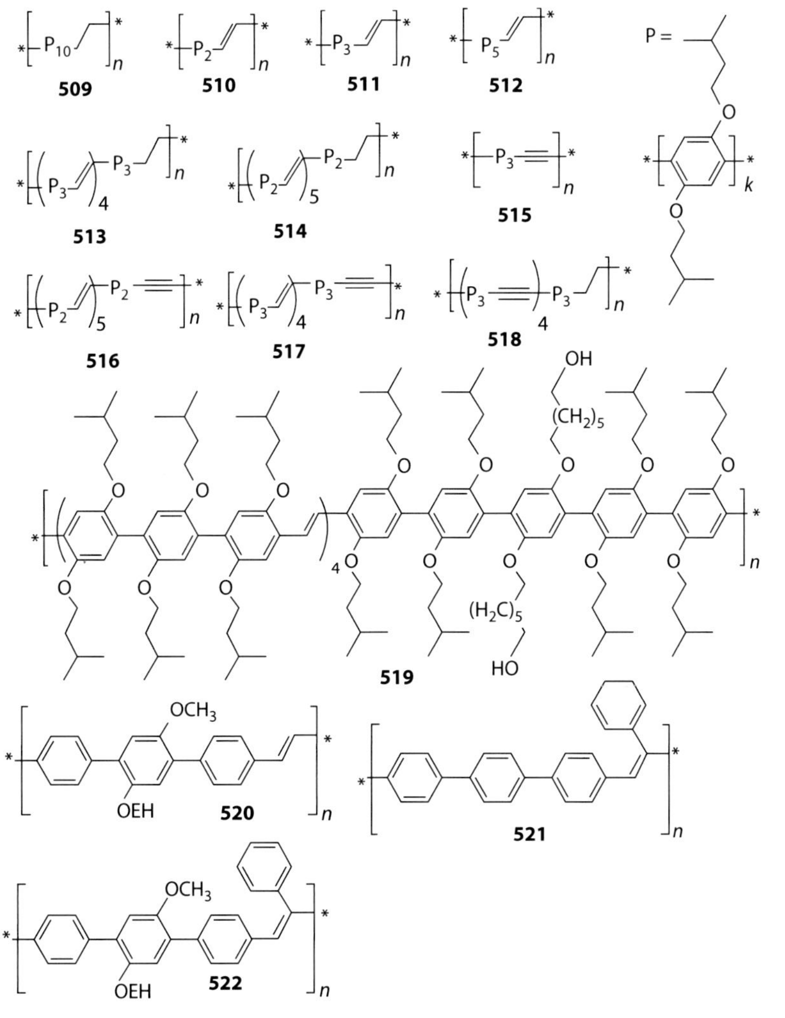

CHART 2.119
Chemical structures of copolymers containing oligophenylene with ethylene, vinylene, or ethynylene units.

TABLE 2.9

Tuning Device Performance (EL Maxima, Turn-On Field, and Internal QE) in PPP Copolymers

Polymer	Monolayer (ITO/Pol/Ca)			Bilayer (ITO/PVK/Pol/Ca)			Triple Layer (ITO/PVK/Pol/PBD[a]/Ca)			Monolayer (ITO/Pol/Al)		
	λ_{max} (nm)	Onset (MV/cm)	Eff (%)	λ_{max} (nm)	Onset (MV/cm)	Eff (%)	λ_{max} (nm)	Onset (MV/cm)	Eff (%)	λ_{max} (nm)	Onset (MV/cm)	Eff (%)
PPP	450	3.2	0.01		1.3	0.16	423	0.44	0.6		1.5	0.01
P3V		0.94	0.04		0.82	0.10	464	0.83	0.4			
P3/5V		1.2	0.24		0.44	0.13	459	0.69	4.0		2.4	0.02
P2V	483	1.3	0.06				486	0.50	2.0[b]	482	1.7	0.02
P3A		1.0	0.3		0.5	0.2	650					
P10E	452 (400)	1.4	0.1	450	1.5	0.09	450	0.70	1.0	450	2.1	0.10
P3VE				464	0.75	0.15	463	0.54	1.4			0.12

Source: Remmers, M., Neher, D., Grüner, J., Friend, R.H., Gelinck, G.H., Warman, M., Quattrocchi, C., dos Santos, D.A., and Brédas, J.-L., *Macromolecules*, 29, 7432, 1996. With permission.

Note: PPP, **500**; P3V, **511**; P3/5V, **519**; P2V, **510**; P3A, **515**; P10E, **509**; and P3VE, **513**.

[a] 25% in PMMA matrix.

[b] No PVK layer.

Aggregation-induced quenching has been a problem for several polymers of this series. Thus, a relatively low solid-state PLQY was observed for rigid polymers with short PPP blocks (9% for **510** and **511**). Quenching can be suppressed by blending the light-emitting PPP **511** with a hole-transporting poly(phenylbiphenylylsilylene). The result was a 5-fold increase of the PL efficiency [605]. An ITO/polymer blend/Al device exhibited a $\Phi_{EL}^{ex} = 0.2\%$ that was up to two orders of magnitude higher than that obtained in a device with a pure polymer **511** layer.

A series of terphenylene–vinylene copolymers **520–522** was recently synthesized by Kim and coworkers [606] via Suzuki coupling. Interestingly, neither alkoxy groups nor pendant phenyl substituents (attached to the vinylene moiety) have any significant effect on the emission properties. The PLEDs fabricated with each of these copolymers (ITO/polymer/Al) emit blue light with λ_{EL} ~450 nm (above a turn-on voltage of 8–10 V).

A Diels–Alder cyclization polymerization was recently used by Park and coworkers [607] for the synthesis of sterically hindered copolymers **523a,b**, in which terphenylene or quinquephenylene blocks are separated by dihexylfluorene units (Scheme 2.71). On the basis of the absence of any metal catalyst or initiator in the synthesis, the authors claim this method to be a way to intrinsically high-purity LEPs. However, similar to fluorene homopolymers, the emission of **523** in the solid state suffers from a defect-dependent (see Section 2.3) "green band" at ~530 nm that became more pronounced in the EL spectra (the main peak is at ~450–470 nm).

Keegstra and coworkers [608] reported copolymers **524a,b** containing unsubstituted PPP blocks separated by long-chain-substituted aliphatic blocks (Chart 2.120). Compound **524a** showed a liquid crystalline behavior. The PL emission maxima can be tuned by changing the length of the oligophenylene block (396 nm for **524a** and 429 nm for **524b**). A low-efficiency $\left(\Phi_{EL}^{ex} = 0.05\%\right)$ PLED was fabricated by blending a very small amount of **524a** into a PVK matrix (ITO/PVK:**524** (0.045%)/Al).

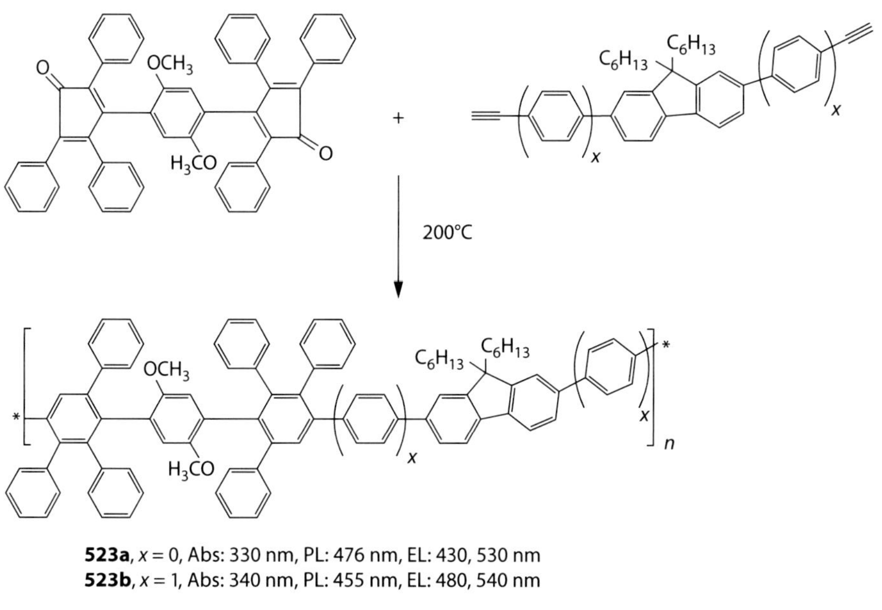

523a, *x* = 0, Abs: 330 nm, PL: 476 nm, EL: 430, 530 nm
523b, *x* = 1, Abs: 340 nm, PL: 455 nm, EL: 480, 540 nm

SCHEME 2.71
Synthesis of oligophenylene–fluorene copolymers via Diels–Alder cyclization. (From Park, S.J., Jung, S.-H., Kim, J.M., and Cho, H.-N., *Mater. Sci. Eng.* C, 24, 99, 2004.)

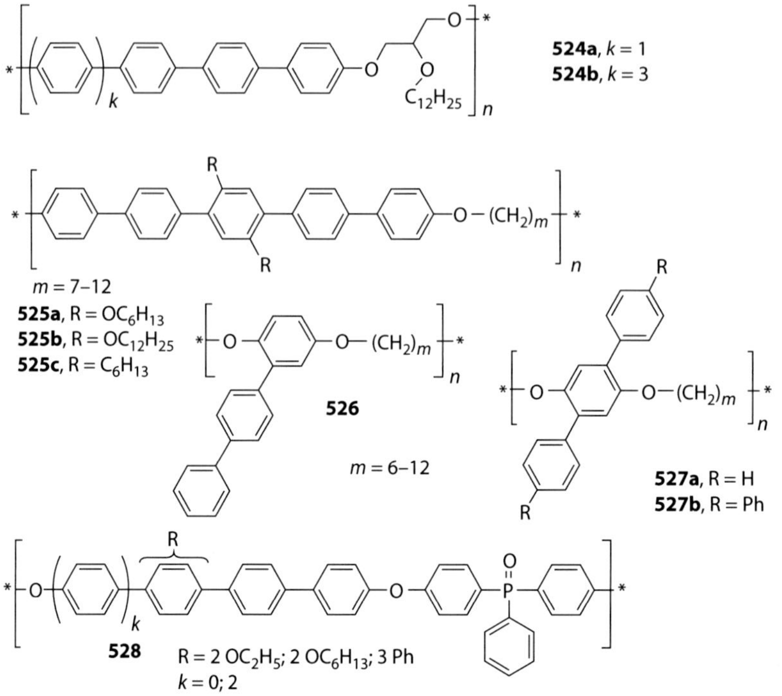

CHART 2.120
Chemical structures of copolymers containing unsubstituted PPP with aliphatic blocks.

Later, Kallitsis and coworkers reported similar PPP copolymers, with oligophenylene blocks separated by nonconjugated aliphatic chains. Copolymers **525** [609], **526**, and **527** [610] contained the oligophenylene blocks in a main chain or as pendant substituents, respectively. Elongation of the oligophenylene block from terphenylene (**526**, **527a**) to quinquephenylene (**527b**) resulted in a small but observable red shift of PL maxima, from ~390 to 407 nm. A somewhat longer wavelength emission was demonstrated by polymers **525a,b**, having quinquephenylene blocks in the polymer chain; however, introduction of alkyl groups to the oligophenylene block (**525c**) results in an expected emission blue shift to 372 nm, due to the above-mentioned steric factor. Some related blue-emitting quinquephenylene block copolymers **528** with triphenylphosphine oxide units were reported to have a very high T_g (270°C); however, their luminescence efficiency was not investigated in detail [611].

Several blue luminescent PPP-type polymers containing naphthalene and anthracene moieties were synthesized by Suzuki polymerization (**529**, **530** [612]) as well as by a soluble precursor route (**531**, **532** [613]) (Chart 2.121). The anthracene-2,6-diyl polymers **531** and **532** showed the longest wavelength emission (green–yellow); however, no data on the EL performance was reported. Later, Jen et al. [614] reported a very efficient PLED based on polynaphthalene **533**. An orthogonal 1,1′-binaphthyl connection confines the conjugation within the 1,1′-binaphthyl-6,6′-diyl unit, enlarging the band gap of the polymer (3.33 eV) and shifting the fluorescence band into the blue. The PL spectrum of **533** is characterized by emission peaks at 390 and 410 nm and a broad shoulder at 500–600 nm. Only the latter is observed in the EL spectrum (λ_{EL} ~ 540 nm) attributed to an excimer emission. Alternatively, the higher wavelength emission could be due to planarization of the normally twisted 1,1′-binaphthyl unit. The device ITO/**533**/ETL/Al (ETL is perfluorinated

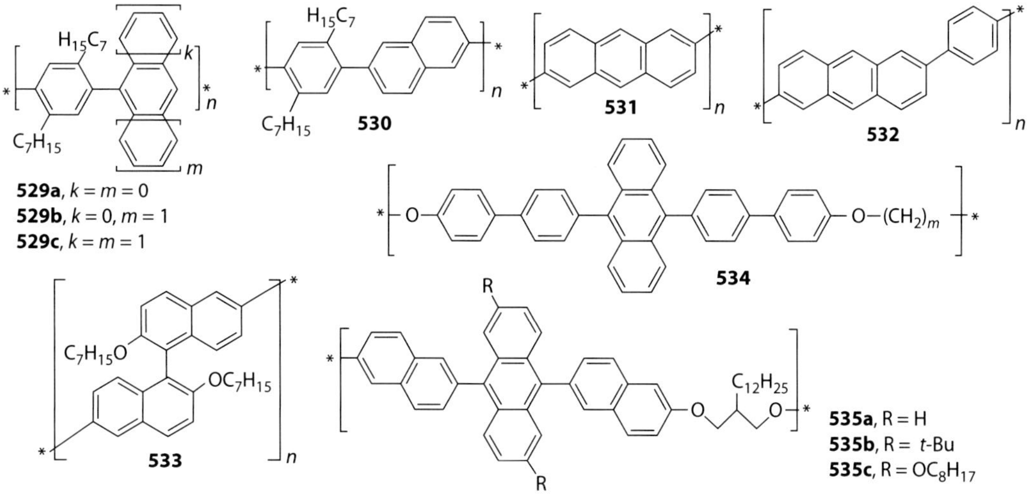

CHART 2.121
Chemical structures of PPPs containing naphthalene and anthracene units.

copper phthalocyanine) showed $\Phi_{EL}^{ex} = 2\%$ (4.9 lm/W) and a maximum brightness of 9400 cd/m², although the turn-on voltage was rather high (15 V for a single-layer device and 20 V for double layer with ETL). Anthracene-containing block copolymers 534 [615] and 535 [616] have been prepared as highly fluorescent blue-emitting materials. The luminous efficiency of 0.4 cd/A was reported for the device ITO/PEDOT/535b/Mg:Ag [616].

Other light-emitting polycyclic aromatic hydrocarbons have also been introduced as pendant groups in nonconjugated polymers, delivering amorphous EL materials. Thus, blue EL was recently reported from perylene-containing poly(methylacrylamide) 536 [617]. The PLED fabricated from 536 emitted blue light with λ_{EL} = 478–491 nm (depending on the polymer ratios) and showed a moderate maximum brightness of 500 cd/m² (at 15 V) but rather low $\Phi_{EL}^{ex} \sim 0.01\%$. A higher-efficiency $\left(\Phi_{EL}^{ex} = 0.34\%\right)$ PLED device was fabricated with nonconjugated polymer 537 containing dialkoxy-substituted anthracene pendant moieties [618] (Chart 2.122).

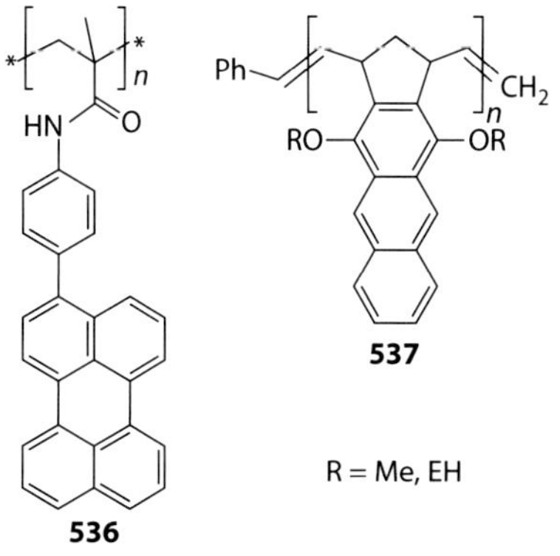

CHART 2.122
Chemical structures of polymers containing polycyclic aromatic hydrocarbons as pendant groups.

Generally, the fluorescence wavelength and quantum yield of PPP polymers are very sensitive to the dihedral angle between the phenylene units in the chain. As the energy barrier of this rotation is low, the PL (and EL) maxima depend on the film fabrication conditions. This ambiguity can be eliminated by designing the polymer, where the phenylene rings are fixed at a certain angle. Prominent examples of such systems, where the phenylene units are bound in fully planar pairs, are the PF materials discussed above. In 1995, Kreyenschmidt and coworkers [619] reported poly(tetrahydropyrene) **538** (Chart 2.123). The six-member unsaturated cyclic bridges fix the biphenylene pairs at a dihedral angle of 15°–20°, which is higher than that in PFs (~0°), but somewhat lower than in other PPP (23° for unsubstituted **500**). In contrast to PPP, the ethylene bridge offers an opportunity to introduce solubilizing substituents into polymer **538** without affecting the dihedral angle and, thus, the emission properties could remain constant (λ_{PL} = 425 nm in solution, 457 nm in films). In this line, the same group later reported a poly(indenofluorene), **539**, with three phenylene units under planarization (λ_{PL} = 432 nm) [620]. This indenofluorene unit was also exploited in the Yamamoto synthesis of a random indenofluorene–anthracene copolymer, **540**, which emitted a deep-blue color (λ_{PL} = 435 nm; CIE: x = 0.21, y = 0.23) and demonstrated significantly higher color stability (in solid-state PL and EL experiments) than homopolymer **539** [621].

As the next step in this way, Jacob and coworkers [622] have reported PPP-type polymers **541**, containing planar pentaphenylene blocks. As expected, the emission maximum (λ_{PL} = 445 nm) of **541** was found between those of indenofluorene **539** (432–434 nm) and completely planar ladder-type PPP (450 nm) (see Section 2.5.2) [622]. Single-layer PLEDs ITO/PEDOT/**541b**/Ca/Al showed stable pure-blue emission with brightness in excess of 200 cd/m² (at 7 V) (Chart 2.124).

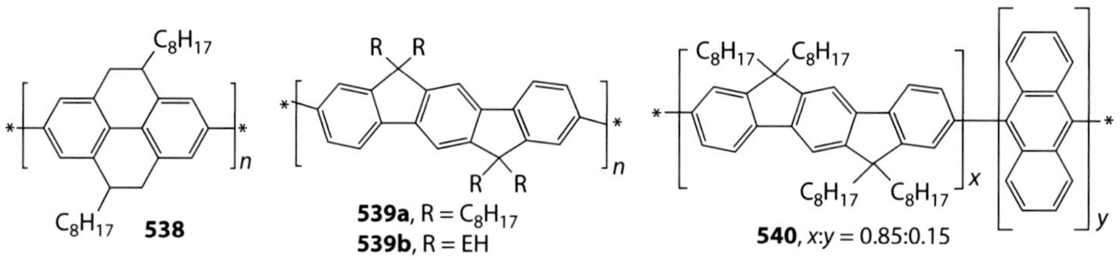

CHART 2.123
Chemical structures of poly(tetrahydropyrene), poly(indenofluorene), and random indenofluorene–anthracene copolymers.

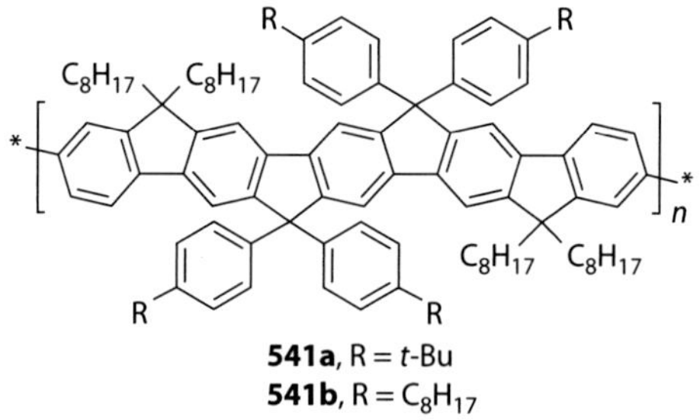

CHART 2.124
Chemical structures of PPPs containing planar pentaphenylene blocks.

An ultimate correction of the intrachain dihedral angle was achieved in LPPP, first synthesized by Scherf and Müllen (**542, 543**) [623] (Chart 2.125). The synthesis of these and other types of conjugated ladder-type polymers was reviewed by Scherf and Müllen [624] in 1992, and more recently, by Scherf [625]. Owing to planarization of the backbone, the band gap of LPPP is decreased to 2.6 eV, and the solution emission maximum is shifted to ~450 nm. The first LPPP-based EL device was reported by Grem and Leising [597] for polymer **542b** using Al, Ca, and In anodes and ITO as a cathode. The EL spectra confirmed the tendency of the LPPP to aggregate in the solid state, which resulted in yellow emission color. Considering that blue-light EL was one of the objectives for exploration of PPP-type materials, this tendency of a red shift of the EL maximum stands as a major obstacle. The early studies claimed an excimer formation (rather than ground state aggregates) as an origin of the low-energy emission [626]. On the other hand, in the light of recent findings that ketone (fluorenone) defect quenching is responsible for the red shift of the emission in PF polymers (see also Section 2.3) [392,393], the same mechanism of blue-emission quenching can be anticipated for LPPP materials [393,627]. This is particularly true for polymer **542a** because the hydrogen-terminated methylene bridge in this compound should be more sensitive to oxidation. In fact, it was shown that replacing the hydrogen atom in the methylene bridge with CH_3 reduces the luminescence quenching in the solid state (cf. **542b**: Φ_{PL} = 72% [in solution] and 10% [in films]; **543b**: Φ_{PL} = 84% [in solution] and 24% [in films] [626]; Φ_{PL} as high as 40% for **543b** in films was claimed later [625]). The long-wavelength emission in the PL and EL spectra of pure **543b** in films, although not completely eliminated, was dramatically suppressed compared with **542b** [628,629]. The problem of low-energy emission in ladder-type PPP was the subject of recent studies by UV–vis and IR spectroscopic annealing experiments, complemented with quantum chemical calculations. Three

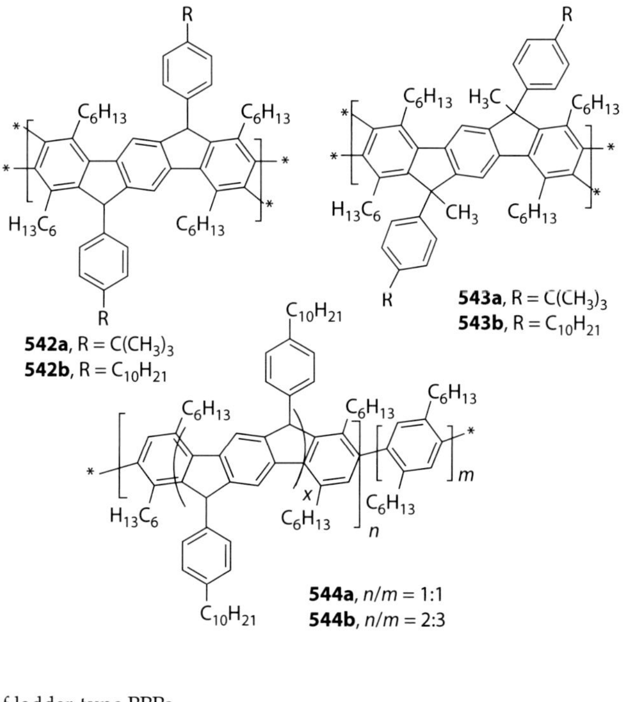

542a, R = $C(CH_3)_3$
542b, R = $C_{10}H_{21}$

543a, R = $C(CH_3)_3$
543b, R = $C_{10}H_{21}$

544a, *n*/*m* = 1:1
544b, *n*/*m* = 2:3

CHART 2.125
Chemical structures of ladder-type PPPs.

types of oxidative defects (ketonic, phenolate, and phenol defects) have been discussed as responsible for the low-energy emission [627].

Whereas the initial measurements of the freshly fabricated PLED ITO/**543b**/Al showed $\Phi_{EL}^{ex} = 0.1\%$, maturing the device by passing 1 mA current increases Φ_{EL}^{ex} up to 4% (at 1 µA current). The PLED emitted blue–green light (CIE: $x = 0.23$, $y = 0.33$ [533]) with a turn-on voltage of ~10 V and a maximum brightness of 2000 cd/m². This record value QE of 4% (as well as typical values of 1.5%–2% [533]) for a simple single-layer PLED using Al cathode is really outstanding and has attracted much attention to this class of LEPs. Other related applications such as polymer lasers [630,631], photovoltaics, and semiconductors have been also demonstrated for polymers **543** [625]. Recently, the first nanosphere-based PLED was fabricated by spin coating a water emulsion of **543b** (stabilized with poly(styrene sulfonate) [PSS]) on an ITO/PEDOT anode [632]. The device emitted blue–green light with an EL efficiency of 0.5 cd/A (with Al cathode), which compares favorably to the efficiency obtained using traditional organic solution deposition technique (0.3 cd/A in the same device structure).

An efficient white-emitting PLED was fabricated by blending LPPP **543b** with a phenylene ethynylene–pyrene copolymer (**554**, Chart 2.126) [633]. Blends of blue-emitting **543b**, with a very small amount of red-emitting **554**, demonstrated efficient energy transfer, emitting rather pure white light (CIE: $x = 0.31$, $y = 0.33$) at all driving potentials. The Φ_{EL}^{ex} as high as 1.2% (the highest for a white PLED at that time) was reported for the device ITO/**543b**:**554**:PMMA/Al. An efficient yellow-emitting PLED was fabricated with **543b** by blending in small amounts (1%) of PT **425** [532,533]. The precise emission color of this device can be tuned by adjusting the amount of the PT component and Φ_{EL}^{ex} as high as 4.2% was achieved. The PLED fabricated under the same conditions with neat **543b** showed a QE of 2%.

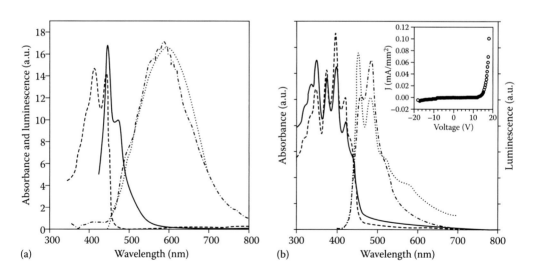

FIGURE 2.34
(a) Absorption (dashed line), solution PL (solid line), PL in films (dotted–dashed line), and EL (dotted line) spectra for **542b**. (b) Absorption spectra of **544a** (dashed line), **544b** (solid line), PL (dotted line), and EL (dotted–dashed line) spectra of **544b**. Inset: I–V characteristic of ITO/**544b**/Al device. (From Grem, G., Paar, C., Stampfl, J., Leising, G., Huber, J., and Scherf, U., *Chem. Mater.*, 7, 2, 1995. With permission.)

It is noteworthy that a lower QE was reported for **543b** using a Ca cathode [70]. A Φ_{EL}^{ex} of ITO/**543b**/Ca was only 0.4%, although it can be improved to 1.3% by modifying the ITO anode with PANI or SiO_2 nanoparticle layer. For the latter, a very high brightness of >40,000 cd/m² was demonstrated.

As mentioned above, in spite of the origin of the long-wavelength emission in PF and related materials due to ketone defects, aggregation is responsible for the emission shift in the solid state. This is because strong inter- and intrachain interactions allow an efficient quenching of the excitons in the entire material by a relatively few defect sites. To minimize the formation of aggregates as observed in fully planar LPPP, segmented so-called steplad-der copolymers **544**, where the completely flat LPPP blocks are separated by distorted PPP segments, have been designed [634,635]. Indeed, whereas the homopolymer **542b** reveals a dramatic red shift of the emission when going from a solution to a film, the emission of seg-mented copolymer **544** stays almost unaltered (Figure 2.34). Sandwiching **544b** between ITO and Al electrodes affords a pure blue-light emitting PLED with an average QE of 0.2% [634].

A significant suppression of the yellow emission band of LPPP was also achieved by dilution in PVK matrix, resulting in 2–3 times increase in EL efficiency, compared with the device prepared with neat LPPP (ITO/polymer/Al) [636].

2.5.2 Poly(Phenylene Ethynylene)s

After the first successful synthesis of a soluble poly(phenylene ethynylene) (PPE) deriva-tive by Giesa and coworkers, and elaboration of elegant methods for preparation of dif-ferent derivatives of this polymer, PPE materials have played important roles in many electronic applications, from sensors and molecular wires to polarized displays [637–639]. The first EL from PPE was reported in 1993 by Swanson et al. [640] for dialkoxy-PPE **545** sandwiched between ITO and Ca electrodes [640] (Chart 2.126). The polymer **545** has a low band gap of ~2.25 eV and emits orange–red light with λ_{PL} ~ 585 nm and λ_{EL} ~ 605 nm (with a broad tail at ~800 nm). A very detailed study of a series of dialkoxy-PPE polymers and copolymers was conducted by Wrighton and coworkers [641–643]. A PLQY of up to 86% and 36% was demonstrated for polymers **546** in solution and in films, respectively [643]. It was demonstrated that the polymer band gap and the emission maxima are shifted from the solution (E_g ~ 2.6 eV, λ_{PL} = 473 nm) to the solid state (E_g ~2.25–2.45 eV, λ_{PL} ~530–590 nm). These shifts, as well as the observed decrease in the PLQY in films, are related to the degree of crystalline order and can be controlled by the substitution pattern. Some PLED devices ITO/polymer/Al were reported for copolymers **546** (R = $C_{18}H_{37}$, R′ = EH) [644]. The EL performance of the device with neat polymer was rather low (Φ_{EL}^{ex} = 0.004%, maximum brightness 4 cd/m²), but it can be essentially improved by blending with poly-TPD hole-transporting material (Φ_{EL}^{ex} = 0.02%, maximum brightness 146 cd/m²).

Hanssen and coworkers [645] reported epicholestanoxy PPE **545c**, where the steroid substituents were supposed to suppress crystallization in the solid state and, therefore, to maximize the PL efficiency. PPEs **547**, synthesized by Swager's group [646,647], con-tained triptycene units, which acted as an "insulating cover," providing a good separa-tion between the PPE chains. Consequently, the PL spectra in solution and in films were almost identical (λ_{PL} = 455 nm). Owing to the nanoporous structure formed by the tripty-cene molecular shape, polymers **547** act as efficient sensors for trinitrotoluene (TNT), an explosive. The latter is based on fluorescence quenching by the nitroaromatic electron-accepting TNT molecules [646]. Targeting possible biosensor applications, Tan and cowork-ers [648] reported water-soluble PPE polyelectrolyte **548**, similar to the PPV-type material **166**. Aggregation of **548** was observed on changing the solvent from MeOH to water.

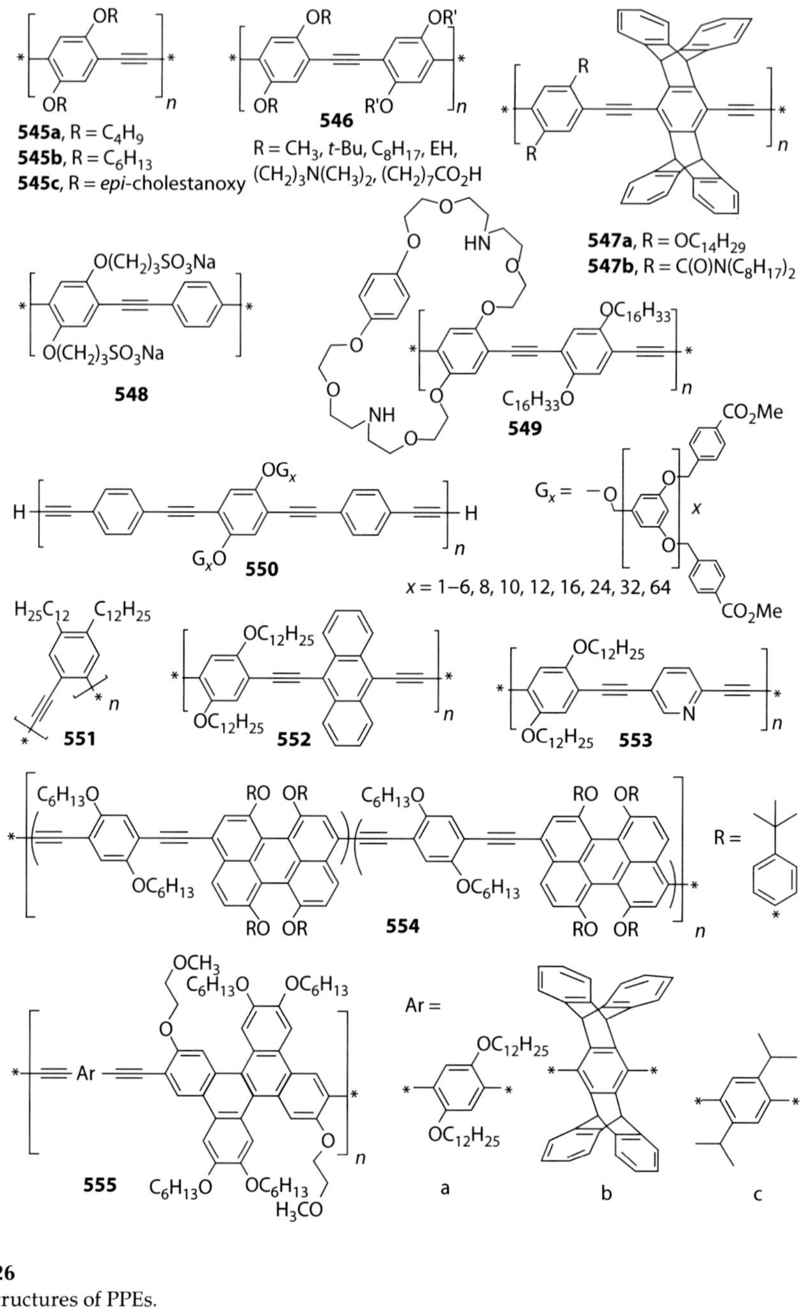

CHART 2.126
Chemical structures of PPEs.

Significant red shift and broadening of the emission spectrum occurred. Swager's group [649] reported a cyclophane-type PPE **549**, for which an interesting aggregation-related phenomenon was observed. While the PL of **549** in solution and in Langmuir–Blodgett films has a narrow emission band (λ_{PL} = 470–480 nm, band half-width ~30 nm) and very low quantum yield (5%–6%), aggregation of the polymer in spin-coated films increases the quantum yield to 21%, and the aggregate emission is seen as a very wide band (λ_{PL} = 520 nm, band half-width ~140 nm), contrary to expectation.

An interesting example of extremely long monodisperse phenylene ethynylene "wires" **550** insulated (wrapped) by dendritic substituents was recently reported by Li and coworkers [650]. For the oligomers with the first-generation dendrimer substituents (G_1), the PLQY decreased with increasing "wire" length (n), from 80%–90% to <60%, whereas steadily high (>80%) Φ_{PL} was observed for all the oligomers containing bulkier G_3 substituents. Furthermore, in contrast to the G_1-substituted homologs, increasing the concentration of **550-G_3** (in THF solution) does not decrease the fluorescence yield.

The EL of poly(o-phenylene ethynylene) **551** was reported by Yoshino and coworkers [651]. As expected, the optical band gap of **551** (3.1 eV) is larger than that of *para*-connected polymer **545**, making the former a blue–purple-emitting material ($\lambda_{PL} \sim 400$). Interestingly, **551** possess a low LUMO energy and can be n-doped electrochemically (partially reversible wave at $E_{pc} = -1.2$ V vs. Ag/Ag$^+$ [~ -0.9 V vs. SCE]). The PLED ITO/**551**/Al emits blue light with two peaks at $\lambda_{EL} \sim 410$ and 430 nm.

Just as with other conjugated polymers, the emission color of PPE can be tuned by introducing different conjugated fragments in the polymer chain (copolymers **552** and **553**) [641,652]. Thus, red–orange EL ($\lambda_{EL} \sim 590$ nm and a shoulder at 530 nm) was reported for the anthracene-based polymer **552** (in PLED ITO/**552**/Al), whereas introducing a pyridine moiety into the polymer chain results in a significant blue shift of the emission, and a blue–green-emitting ($\lambda_{EL} \sim 480$ nm) PLED was fabricated with this material [652].

Higher polycyclic aromatic units such as perylene (**554** [653]) and dibenzocrysene moieties (**555a–c** [654]) were introduced into the PPE backbone. The former are red-emitting materials with $\lambda_{PL} = 580$–590 nm, and were used to create a white-emitting PLED with LPPP [633]. Significantly higher-energy, blue-color emission was observed from the crysene derivatives ($\lambda_{PL} = 474$–480, $\Phi_{PL} = 25\%$–34%) [654]. The latter were employed as fluorescent sensor materials and demonstrated better sensitivity than other similar iptycene-containing LEPs (e.g., **547**) [654].

2.5.3 Polyacetylenes

Polyacetylene (PA), the first and the most conducting polymer, was traditionally regarded as a nonemissive material. Although it is certainly true for the unsubstituted PA in the doped and neutral states, strong fluorescence has recently been demonstrated from disubstituted (diphenyl and phenylalkyl) PAs [655–657]. Sun and coworkers [656] fabricated red, green, and blue EL devices by sandwiching monophenyl-PA (**556**), diphenyl-PA (**557**), and phenylalkyl-PA (**558**), respectively, between ITO and Mg:Al electrodes (Chart 2.127). As expected from a low PLQY (<0.1%) of monophenyl-PA, the efficiency of the red-emitting PLED based on polymer **556** was negligibly low ($\sim 10^{-4}\%$). Better results were demonstrated by green- (**557**) and blue-emitting (**558**) disubstituted PAs $\left(\Phi_{EL}^{ex} \sim 0.01\%\right)$, in accordance with their high PL efficiency (>60%). The structural variations included biphenyl- and carbazole-PAs **559** [658,659] and **560** [660]. The former is UV- and violet-emitting material ($\lambda_{PL} = 350$–370 nm), where the emitting sites (biphenyl chromophores) are separated from the PA backbone by a long alkyl chain. In contrast, strong green fluorescence ($\lambda_{PL} = 525$ nm) of the carbazole **560** cannot be attributed to the individual carbazolyl substituents. A remarkably efficient PLED ($\Phi_{EL}^{ex} = 2\%$, maximum brightness 2700 cd/m^2) was fabricated from **560** as a bilayer device ITO/**560**/Alq$_3$/Mg/Al, where Alq$_3$ acts as an ETL and the observed EL is exclusively due to PA [660].

Inspired by very high EL efficiency of OLEDs with molecular siloles, Chen and coworkers [661] reported an interesting series of silole-substituted PAs **561–563** (Chart 2.128). Whereas the fluorescence of all three polymers in solution was very low ($\Phi_{PL} < 0.5\%$), a significant increase of the PLQY (20–50 times) was observed in the solid state of **562** and

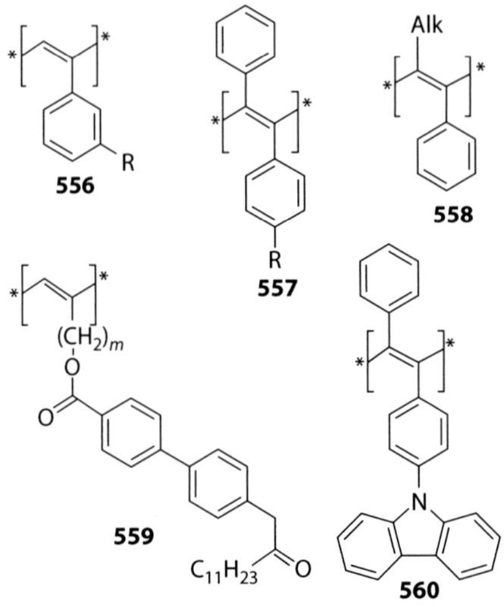

CHART 2.127
Chemical structures of PAs.

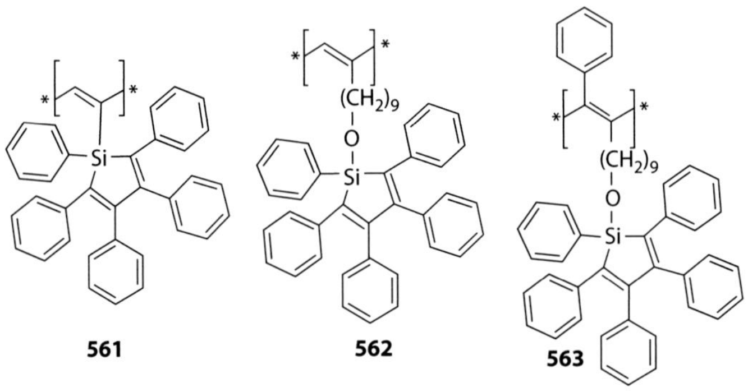

CHART 2.128
Chemical structures of silol-substituted PAs.

563. The phenomenon was observed earlier in molecular siloles, as well as in cyclophane–PPE **549**. While an efficient energy transfer from silole to the PA backbone renders **561**, a red-emitting material (λ_{EL} = 664 nm), the longer backbone-to-silole bridged **562** and **563** exhibited blue–green fluorescence (λ_{EL} = 512 nm). The EL efficiency of single-layer devices with all three polymers was similar (~0.013%), in spite of different PL properties, suggesting that the device performance was limited by charge-transport processes. Indeed, the multilayer device ITO/**563**:(PVK **564**)/BCP/Alq$_3$/Al emitted blue light (λ_{EL} = 496 nm) with a relatively high current and quantum efficiencies (1.45 cd/A and 0.55%, respectively) and a maximum brightness of 1118 cd/m^2 [661].

2.5.4 Carbazole-Containing Polymers

The high hole mobility and excellent photoconductive properties of carbazole-containing polymers, such as PVK (**564**) and poly(*N*-epoxyprolylcarbazole) (PEPK, **565**), place them among the most studied polymers for optoelectronic application (Chart 2.129). They have

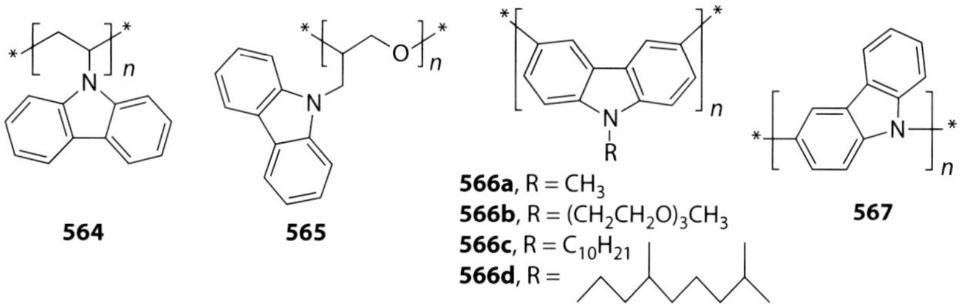

CHART 2.129
Chemical structures of carbazole-containing polymers.

been commercialized in a number of devices and processes (photocopiers, laser print-ers, holographic security stamps, etc.) [585]. The first application of carbazole-containing polymer (PVK, **564**) in an EL device was described in 1983 by Partridge [3], although the reported blue EL was due to molecular fluorescent dyes, incorporated into a PVK matrix. In 1994, Hu and coworkers [215] reported blue EL from pure PVK material (devices ITO/PVK/Ca and ITO/PVK/Al) as well as from a mixture of PVK with PPV block copoly-mer **189b**. Using PBD/PMMA blend as an additional ETL, a brightness of 200 cd/m^2 was achieved. Generally, the EL properties of nonconjugated carbazoles are rather poor. On the other hand, as an excellent hole-transporting material, PVK has been extensively used as an HTL [106,230,280,296,298,299] or as a hole-transporting material in blends with other conjugated EL materials in PLEDs [312,528,552]. Chapter 4 of this book gives many exam-ples of PVK as a host material for high-efficiency phosphorescent PLEDs, using organome-tallic phosphorescent dopants.

Several carbazole homopolymers (**566a** [662], **566b** [663], **566c,d** [664], and **567** [665]) have been synthesized as blue-fluorescent materials. The PL maximum of **566c** in solution is ~420 nm, and it shifts to ~490 nm in films [664]. A blue-emitting single-layer LEC (λ_{EL} ~ 440 and 490 nm) with $\Phi_{EL}^{ex} = 0.02\%$ was reported for polymer **566b** [663]. Interestingly, mixing **566b** with polyquinoxaline (**615b**) results in emission of a new color corresponding to an energy difference between the LUMO of polyquinoxaline and the HOMO of polycarba-zole. A significantly improved Φ_{EL}^{ex} of an LEC (1%) was demonstrated with this polymer blend material [663].

3,6-Carbazole–diacetylene copolymer **568** was used as an HTL in bilayer devices with macrocyclic carbazole oligomer **569**, which acted as an electron-transporting and light-emitting material [666] (Chart 2.130). Copolymer **568** itself is a blue-emitting material ($\lambda_{PL} = 400$ nm), whereas the oligomers **569** exhibit green fluorescence with $\lambda_{PL} = 520$ nm. Rather remarkably, the cyclooctamer **569** could be prepared in 45% yield by simple Knoevenagel coupling of carbazole-3,6-dicarbaldehyde with 3,6-*bis*(cyanoacetoxymethyl)carbazole. The PLED fabricated as ITO/**568**/**569**/Al demonstrated $\Phi_{EL}^{ex} = 0.44\%$, although the device maxi-mum brightness was rather low (60 cd/m^2).

A series of carbazole-3,6-diyl polymers, including homopolymer **566c**, alternating carbazole–oxadiazole and carbazole–fluorene copolymers **570–574**, and analogous ran-dom copolymers containing all three (carbazole, fluorene, and oxadiazole) units were pre-pared and studied as host materials for electrophosphorescent LEDs [667] (Chart 2.131). In line with previous observations, the performance of carbazole-3,6-diyl polymers in PLED was rather low (0.2 cd/A for ITO/PEDOT/**566c**/Ba diode); however, using a triplet-emitter Ir-SC4 (Covion Organic Semiconductors GmbH), very high-efficiency green-phosphorescent PLEDs (up to 23 cd/A) were fabricated. It was shown that small structural

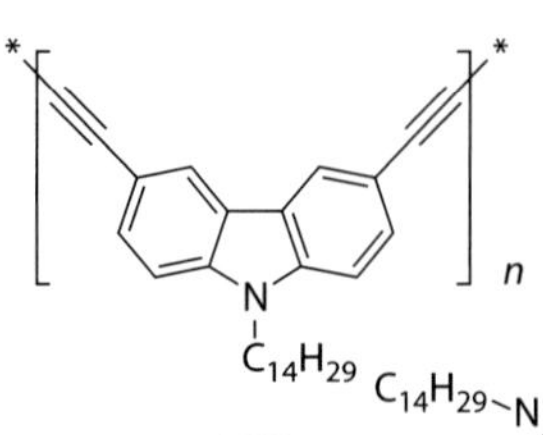

568

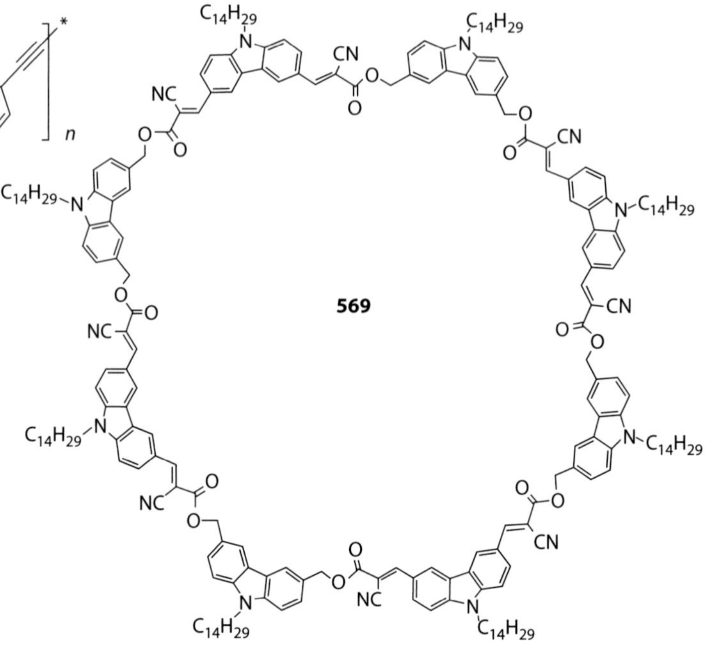

569

CHART 2.130
Chemical structures of 3,6-carbazole-diacetylene copolymer and macrocyclic carbazole oligomer.

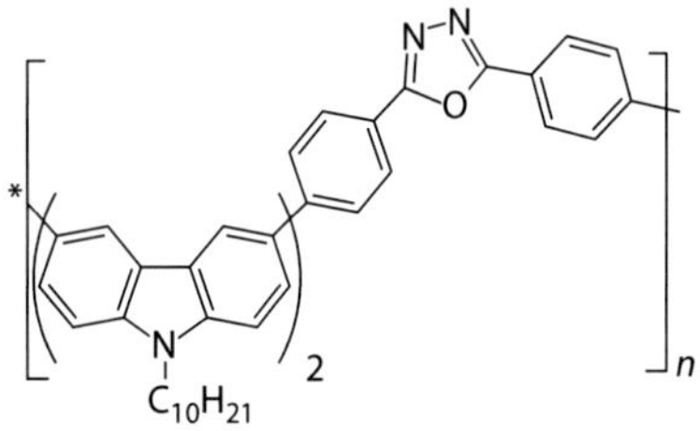

570

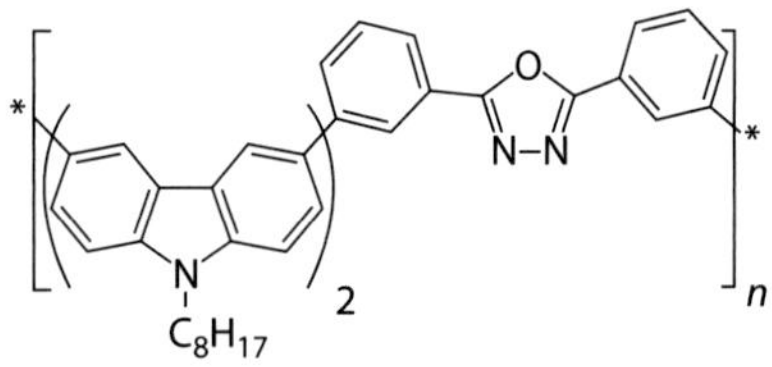

571

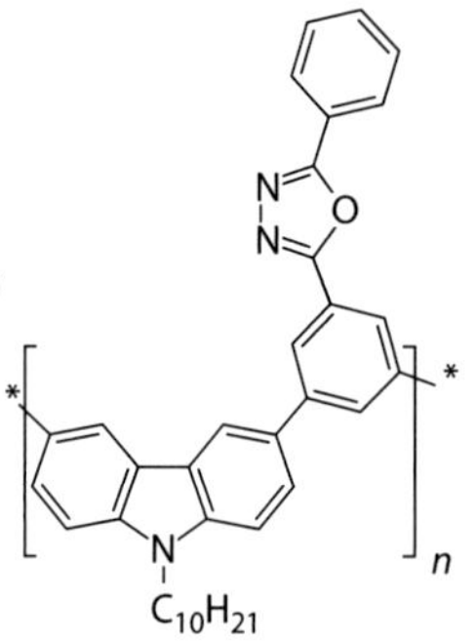

572

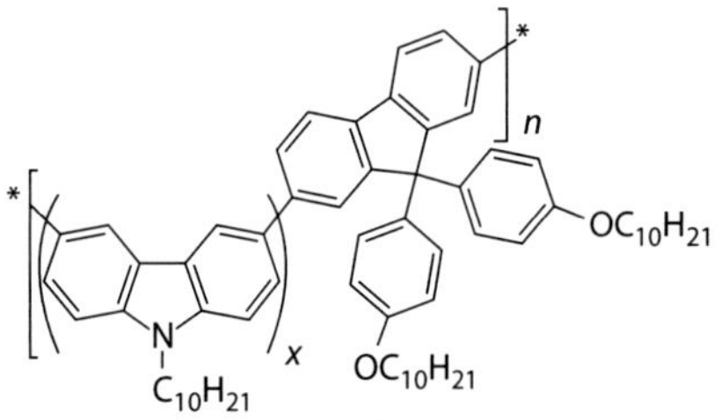

573, *x* = 1 or 2

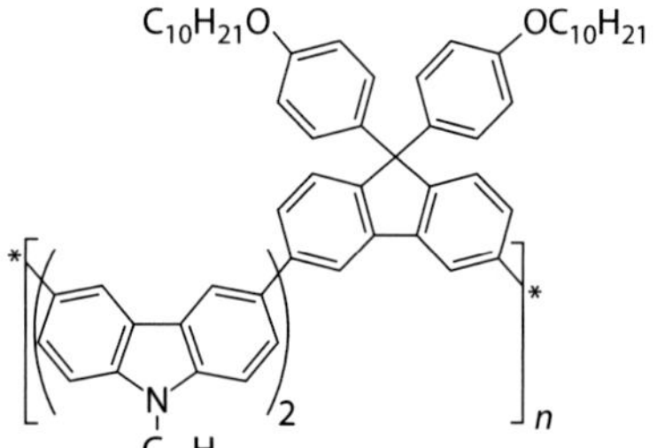

574

CHART 2.131
Chemical structures of carbazole-3,6-diyl polymers.

variations such as substitution position in the fluorene unit allows control of the triplet energy without disturbing the HOMO–LUMO levels of the polymers.

Cao and coworkers [668,669] prepared random 3,6-carbazole–BT copolymers **575** and **576** by Suzuki coupling (Chart 2.132). PL and EL missions of the carbazole segment were completely quenched for copolymers with BT concentrations as low as 1%, showing an efficient

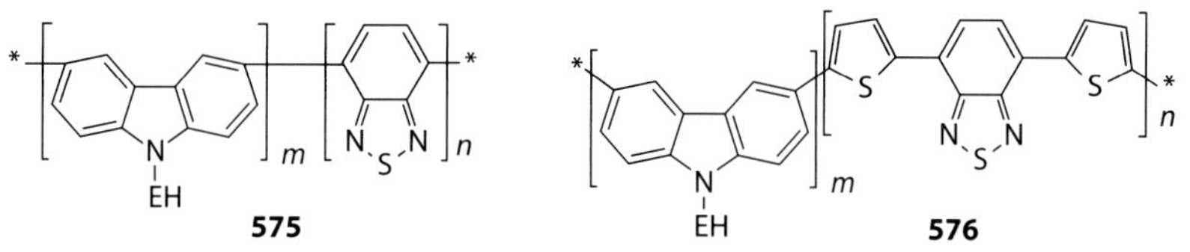

CHART 2.132
Chemical structures of random 3,6-carbazole-BT copolymers.

energy transfer on the narrow band-gap BT sites. Copolymers **576** emitted a saturated red light (from 660 to 730 nm, depending on stoichiometry) with a luminance of 70–631 cd/m^2 and Φ_{EL}^{ex} = 0.55%–1.48%. Also, a very-high-efficiency red-emitting PLED (λ_{EL} = 680 nm; CIE: x = 0.67, y = 0.33) was fabricated by blending small amounts of **575** (*m:n* = 4:1) into MEH–PPV. The device ITO/PEDOT/MEH–PPV:**576** (240:1)/CsF/Al showed Φ_{EL}^{ex} = 3.8% [75]. The emission of copolymers **575** was hypsochromically shifted to 570–620 nm and showed a lower Φ_{EL}^{ex} ~ 0.01%–0.48%.

It is noteworthy that the polymer involving linking via positions 3 and 6 of the carbazole ring hinders the conjugation due to a *"meta*-type" connection. The poor conjugation might be responsible for low PLQY of polymers **566** (15% for **566b** in THF solution). On the other hand, functionalization of carbazole in positions 2 and 7 is a synthetically challenging task. In contrast to fluorene, electrophilic halogenation of carbazole occurs exclusively at positions 3 and 6 (and then at 1 and 8) due to strong *para-* and *ortho*-directing effect of the nitrogen atom. To prepare fully conjugated polycarbazoles, Leclerc and coworkers synthesized 2,7-dihalocarbazoles via reductive cyclization of 4,4′-dichloro-2-nitrobiphenyl (in refluxing triethylphosphite) or thermal decomposition of 4,4′-dinitro-2-azidobiphenyl (followed by multistep conversion of the nitro groups into iodine substituents) (Scheme 2.72). The polymerization was performed via Yamamoto or Suzuki protocols to afford conjugated polymers **577a–c** [421,562,670]. Later, polycarbazoles **577b,d–f** were prepared by Pd-catalyzed cross-coupling of Grignard derivatives of 2,7-dibromocarbazoles [671]. In contrast to 3,6-linked polycarbazoles **566**, polymers **577** showed pure blue emission (λ_{PL} [CHCl$_3$] = 417–420 nm, λ_{PL} [film] = 439–442 nm) with high PLQY of 76%–80% (in solution). For **577**, a blue-emitting PLED was fabricated but an initial test device showed a rather low performance (maximum brightness below 100 cd/m^2) [268].

As demonstrated in previous sections, the carbazole unit was introduced as a pendant group or as a chain member in major classes of EL polymers such as PPVs (**103**–**114, 152, 195, 208**) and PFs (**69, 70, 255–258**). A variety of 2,7-carbazole-derived polymers with different conjugated units, such as 2-alkoxy- and 2,5-dialkoxy-1,4-phenylene (**578**) and 1,1′-binaphthalene-6,6′-diyl (**579** [672]), 2,5-pyridine (**580**), 2,7-fluorene (**258** [268,421]), 2,5′-bithiophene (**583** [421]), 5,8-quinoxaline (**581**), quinquethiophene-*S,S*-dioxide (**476** [562]), 2,5-thiophene (**582**), 2,5-furan (**584**), and acetylene (**585** [673]) were reported by Leclerc and coworkers (Chart 2.133). Taking advantage of a wide range of electronic properties of the comonomers, an efficient tuning of the HOMO (approximately −5.0 to −5.5 eV) and LUMO (approximately −1.85 to −3.5 eV) energies, and emission color (spanning from blue to red, λ_{PL} ~ 400 to 700 nm) was achieved in this series.

An interesting hybrid carbazole–fluorene ladder polymer, **586**, was synthesized recently by Patil and coworkers [674] via Suzuki coupling of carbazole diboronic ester with dibromodibenzoylbenzene followed by conversion of the ketone into a tertiary alcohol and

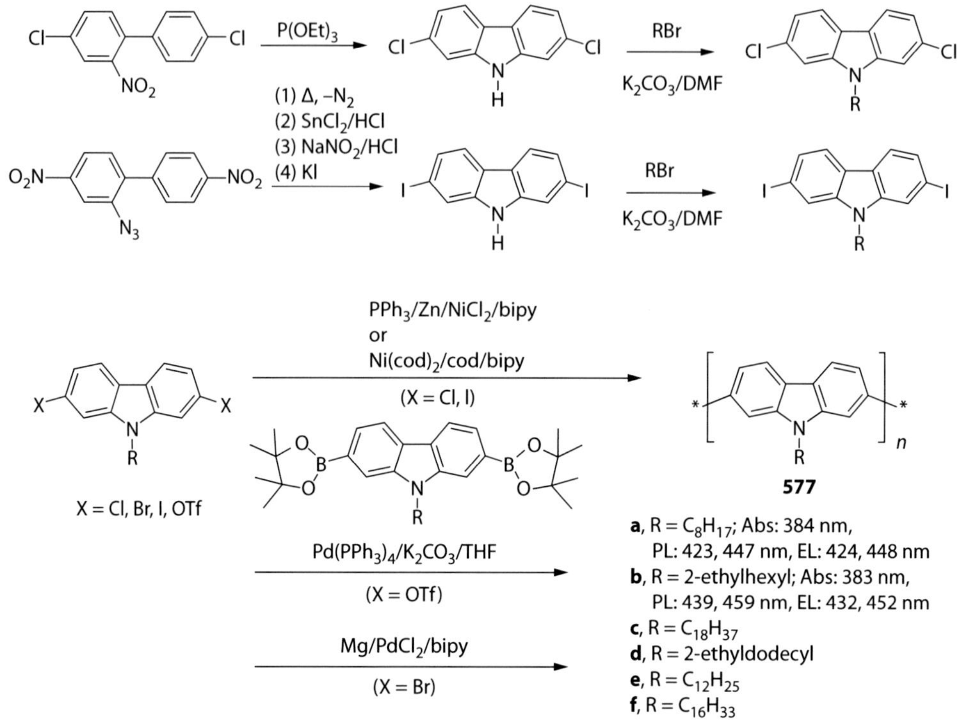

SCHEME 2.72
Synthesis of 2,7-dihalocarbazole monomers and their polymerization by Yamamoto and Suzuki protocols. (From Morin, J.-F. and Leclerc, M., *Macromolecules*, 34, 4680, 2001; Morin, J.-F. and Leclerc, M., *Macromolecules*, 35, 8413, 2002; Zotti, G., Schiavon, G., Zecchin, S., Morin, J.-F., and Leclerc, M., *Macromolecules*, 35, 2122, 2002; Iraqi, A. and Wataru, I., *Chem. Mater.*, 16, 442, 2004.)

Lewis acid–catalyzed cyclization (Scheme 2.73). Upon photoexcitation, **586** emits blue light (λ_{PL} = 470 nm) with very small Stokes shift (10 nm, 0.057 eV), expected from its rigid geometry. Photophysical studies suggest a nearly defect-free intrachain structure of this material, although practical applications of **586** in LEDs are still to be explored.

Morin and coworkers [675] synthesized a series of 2,7-carbazolylene–vinylene copolymers **587–590** via Horner–Emmons reaction. Although all the copolymers are fluorescent in solution (λ_{PL} ~ 460–505 nm, Φ_{PL} ~ 16%–67%), only **588** retains its fluorescence in the solid state (λ_{PL} = 656, 701 nm). A device ITO/PEDOT/**588**/LiF/Al/Ag emitted orange–red light (λ_{EL} = 640 nm, CIE 1976 coordinates: u' = 0.40, v' = 0.54) with a low turn-on voltage of 3 V, an efficiency of 0.17 cd/A (at 5 V), and a maximum brightness of 245 cd/m^2 (at 10 V) (Chart 2.134).

Polymers containing carbazol fragments were also explored as hole-transporting materials for PLEDs. Xie and coworkers reported carbazole-based polymers with aromatic main-chain segments connected via alkylene spacers synthesized by Ni(0)-catalyzed Yamamoto-type aryl–aryl coupling reactions [676]. The polymers (Chart 2.135) emit blue light in solution with maxima emission peak of about 410 nm. The hole-transporting properties of synthesized polymers were tested in PLED devices as the HTL with Alq$_3$ as EL material. The device with polymer **591a** exhibits the best overall performance (turn-on voltage: 4 V; maximum photometric efficiency: 2.6 cd/A; maximum brightness: 2200 cd/m^2).

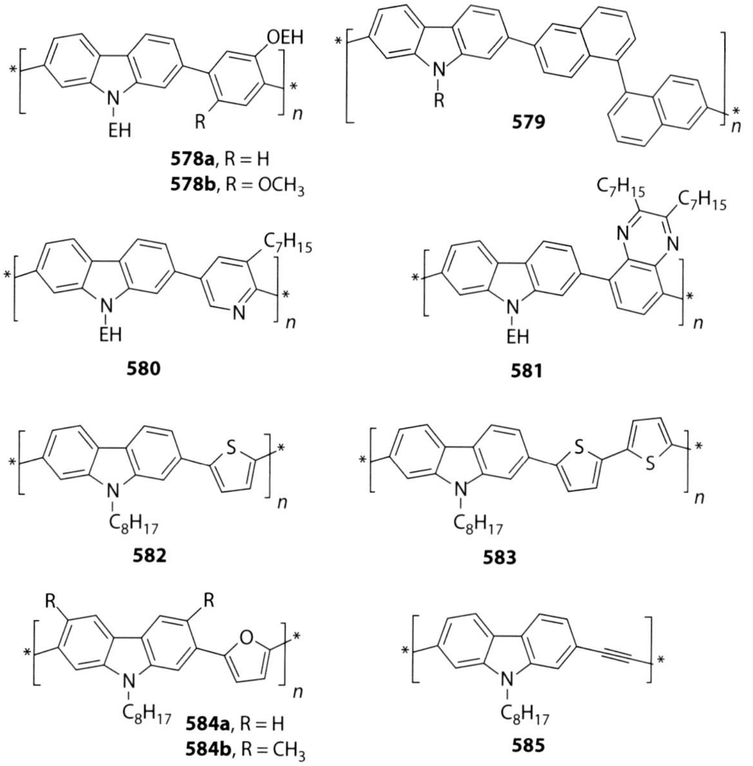

CHART 2.133
Chemical structures of 2,7-carbazole–derived polymers.

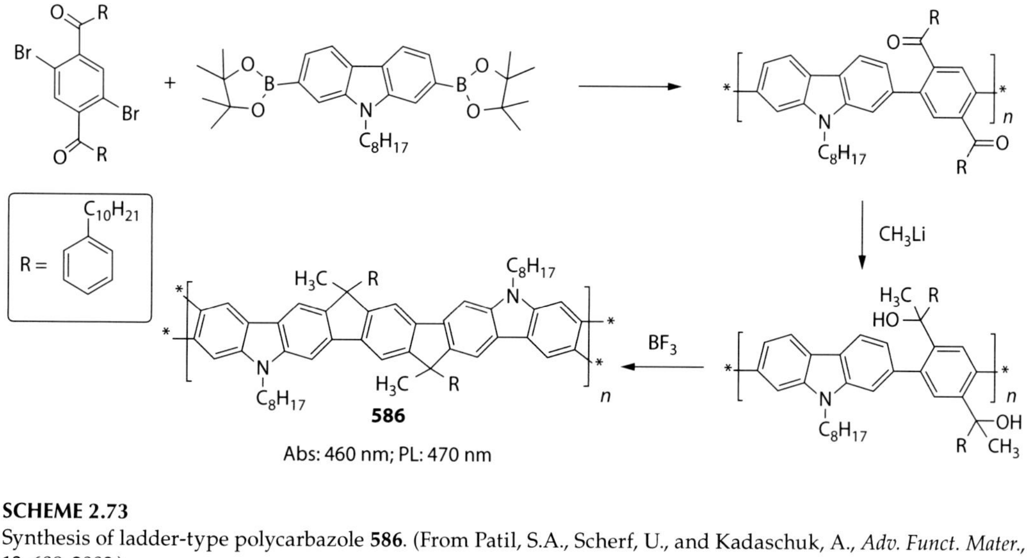

SCHEME 2.73
Synthesis of ladder-type polycarbazole **586**. (From Patil, S.A., Scherf, U., and Kadaschuk, A., *Adv. Funct. Mater.*, 13, 609, 2003.)

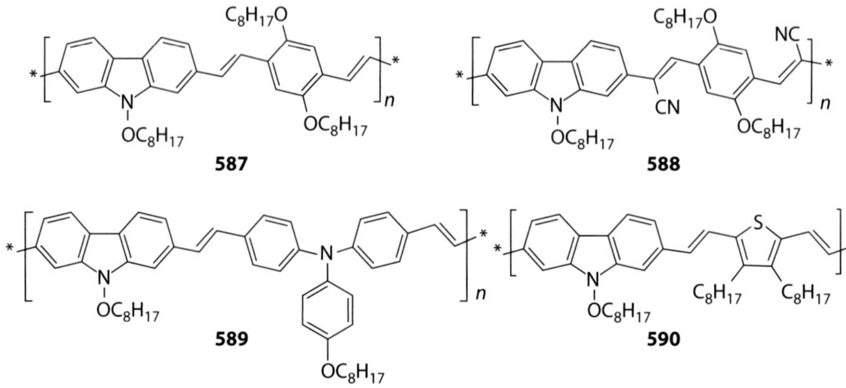

CHART 2.134
Chemical structures of 2,7-carbazolylene–vinylene copolymers.

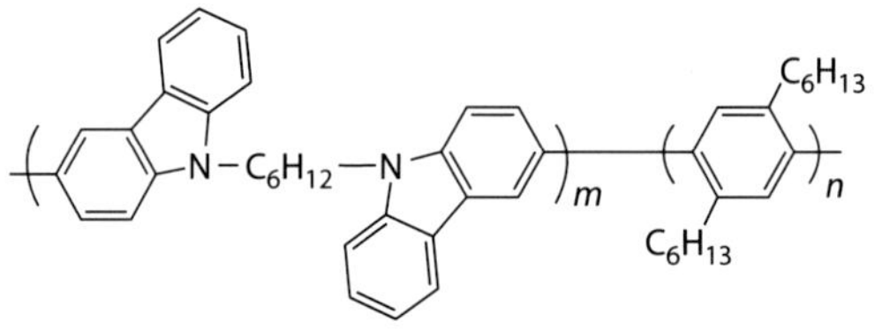

	Mole fraction of phenyl (%)
591a	0
591b	49
591c	64
591d	78

CHART 2.135
Chemical structures of polymers containing hole-transporting carbazol units.

2.5.5 Poly(Pyridine)s and Related Poly(N-Heterocycle)s

The first EL from polypyridine was reported by Gebler and coworkers [677] in 1995. The polymer **592**, prepared by Yamamoto Ni-mediated reductive coupling of 2,5-dibromopyridine [678] was soluble in formic acid (due to formation of a protonated form **593**) and could be spin coated on an ITO glass substrate for preparation of PLED (Chart 2.136). It had a rather high PL in the solid state (37 ± 3% [679], 30 ± 3% [680]). The device ITO/**592**/Al emitted green light (λ_{EL} ~ 500 nm) at voltages above 8 V. Introduction of an additional HTL (PANI) reduced the turn-on voltage to 4 V [677]. Unfortunately, the efficiency of the device ITO/**592**/Al is very low (Φ_{EL}^{ex} ~ 0.001% [679] to 0.002% [680]). Also, the issue of deprotonation of the pyridine units (which might occur during vacuum deposition of a second electrode) and the possible influence of residual protonated species has not been addressed.

EL pyridine–phenylene copolymers **594** [681] and **595** [682] have been synthesized and studied by Bryce and coworkers. Although a rather low Φ_{EL}^{ex} (<0.1%) was reported for the devices, an interesting phenomenon was found for polymers **595**. When the PLED (ITO/PEDOT/**595**/Ca/Al) was fabricated using acidic solutions, a strong red shift in the EL band compared with that obtained with the neutral solution (from 510 to 575 nm) was observed. The authors explained this concept by planarization of the protonated polymer chain as

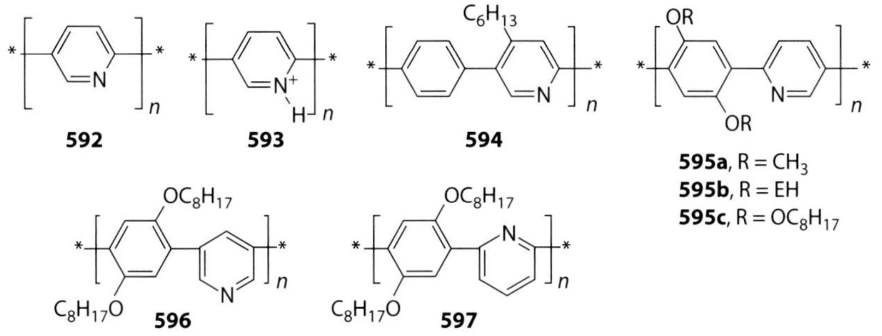

CHART 2.136
Chemical structures of polypyridines.

a result of intramolecular hydrogen bonding N···H···O. Variation of pyridine linkage in copolymers **595**, **596**, and **597** affects the PL and EL emissions (λ_{EL} = 444, 432, and 428 nm, respectively), although the details of the efficiency of LEDs based on these polymers have not been reported [683].

Owing to the electronegative atom, polypyridines are good electron acceptors: from UPS and UV–vis absorption spectra, Miyamae and coworkers [684] estimated E_A = 3.5 eV, I_P = 6.3 eV. Hwang and coworkers [685] reported E_A = 2.9 eV, I_P = 5.7 eV based on electrochemical measurements. For the double-layer ITO/PPV/**592**/Al device, in which **592** acts as electron-transport and hole-blocking layer, Hwang and coworkers [685] reported a EL efficiency of 0.12 cd/A that is 17 times higher than for an single-layer PPV-based PLED. The improvement in Φ_{EL}^{ex} by a factor of 60 (from 0.004% to 0.25%) for this device configuration was demonstrated by Dailey and coworkers [680].

Several poly(*p*-pyridine vinylene)s (PPyV, **598**) have been reported as isoelectronic analogs of PPVs (see also PPyV–fluorenevinylene copolymer **325** [310]) (Chart 2.137) [686–688]. The neutral and the protonated (**599**) forms showed strong luminescence from excimeric states that contributed to the observed low PLQYs and red shift in the emission spectra from solution (from 470 to >600 nm). A device fabricated as ITO/**598**/Al showed an Φ_{EL}^{ex} = 0.02%, which can be somewhat improved to 0.05% by introduction of PPV **1** as an HTL [688]. An efficiency as high as 0.3% was reported for copolymer **600** on a bilayer PLED with a PANI-networked electrode [689]. Several studies of pyridine homopolymers and copolymers (**592**, **598**, **600**) as light-emitting materials for symmetrically configured alternating-current light-emitting (SCALE) devices [690] and inverted LEDs [689] using high-work-function electrodes (Au) have been performed.

Several light-emitting copolymers with pyrazines in the backbone are known. In 1996, a Japanese group reported fluorescent polyimide **601** [691] (Chart 2.138). An LED device based

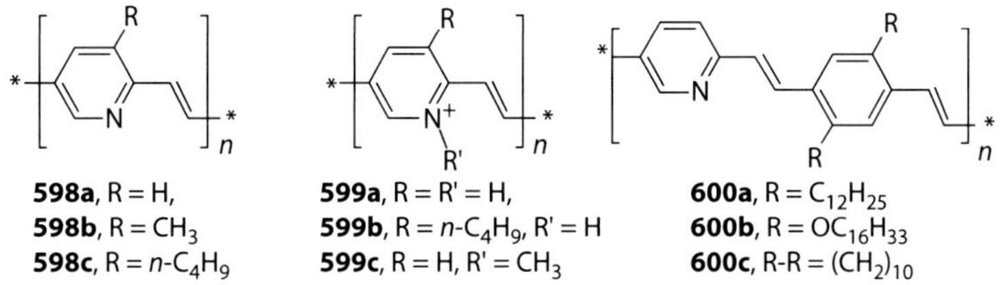

CHART 2.137
Chemical structures of poly(*p*-pyridine vinylene)s.

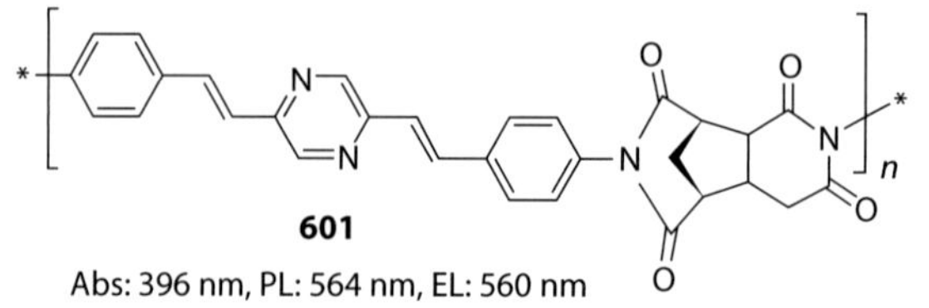

601

Abs: 396 nm, PL: 564 nm, EL: 560 nm

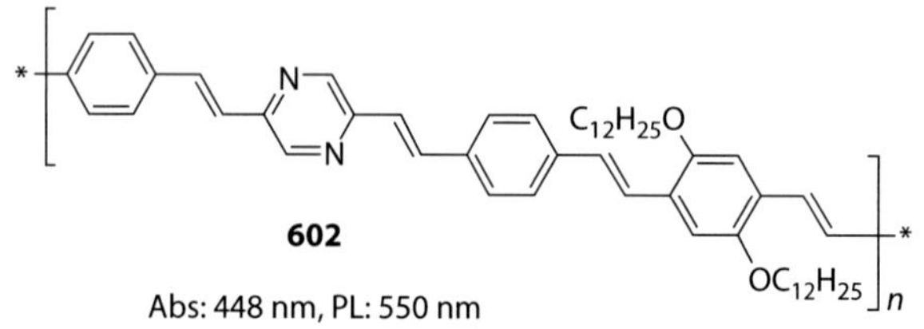

602

Abs: 448 nm, PL: 550 nm

CHART 2.138
Chemical structures of pyridine–pyrazine copolymers.

on a Langmuir–Blodgett film of **601** was fabricated as (ITO/**601**/Mg:Al) and showed orange–red emission (λ_{EL} = 560 nm) with a turn-on voltage of 7 V. Peng and Galvin [168] demonstrated orange emission from the PLED based on pyrazine-containing PPV **602** (Φ_{EL}^{ex} = 0.012% and a turn-on voltage of 10 V for ITO/**602**/Al).

A solid study of PL and EL properties of a series of poly(quinoline) homopolymers **603**–**605** and quinoline and anthrazoline copolymers **606**–**608** and **495a–e** was reported by Jenekhe and coworkers [579,580,692–694] (Chart 2.139). Changing the substituent R' and a colink R, the emission color was tuned from blue to red (λ_{EL} = 410–622 nm) (Figure 2.35). The solid-state PLQY of these polymers was usually on the level of 10%, although a higher efficiency (20%–30%) was observed for the *tert*-butyl-substituted polymers **604** and **608a** [692]. The PLED devices fabricated with **603**–**608** in ITO/HTL/polymer/Al configuration showed moderately high Φ_{EL}^{ex} ~ 0.1%–1%, with values repeating the trend of PL efficiencies (the highest Φ_{EL}^{ex} = 0.92% and 1.08% were due to polymers **604** and **608a**, respectively). As mentioned before, emissive polymers **495a–e** (Scheme 2.69) showed good electron-transport properties and a weak green EL (Φ_{EL}^{ex} = 0.004 – 0.06% in ITO/**495a–e**/Al). A large improvement in the performance of MEH–PPV-based PLEDs was achieved using polymers **495a–e** as the electron-transport materials (Φ_{EL}^{ex} was up to 1.4% and brightness up to 2311 cd/m²) [579,580]. Considering a relatively low EL turn-on voltage of 5–10 V, these polymers present a promising class of polyheterocyclic materials for LED displays.

Jen and coworkers [695,696] reported PLEDs based on variations of the above-mentioned 6,6'-bisquinoline copolymers (**609**–**612**), containing conjugated and nonconjugated linkers (Chart 2.140). For copolymers with nonconjugated units, the EL efficiency was very low (0.002% for ITO/**612**/Al [696]), although a similar value was demonstrated for MEH–PPV in the same device configuration. Introducing conjugated arylamine or phenylene vinylene linkers improves Φ_{EL}^{ex} to 0.018% (for triarylamine copolymer ITO/**609**/Al) and 0.06% (for phenylene vinylene copolymer ITO/**610**/Al) [695]. Also, nonconjugated, relatively more electronegative copolymers **611** and **612** showed a lower turn-on voltage (~6 V).

Several other 6,6'-bisquinoline-based copolymers (**613**, **614**) have been reported [697] (Chart 2.141). The absorption and PL spectra of polymers **614a,b** are red shifted, compared with the analog with *p*-phenylene fragment **613**. The effect is much more pronounced in films than in solution (films: λ_{PL} ≈ 495 nm for **613** and a broad structureless band at ≈520–650 nm for **614a,b**). Single-layer (ITO/PEDOT/polymer/Ca/Ag) devices fabricated with

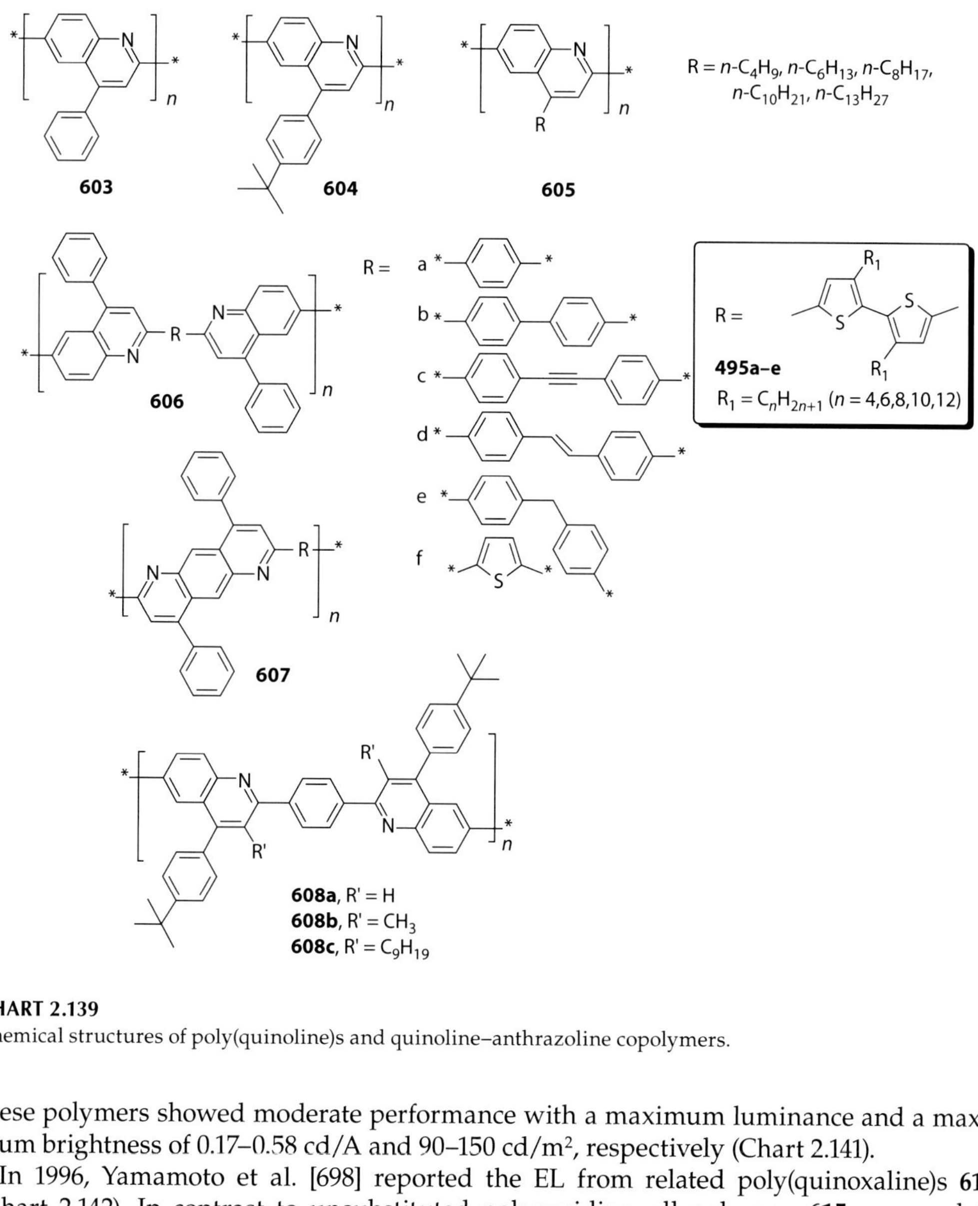

CHART 2.139
Chemical structures of poly(quinoline)s and quinoline–anthrazoline copolymers.

these polymers showed moderate performance with a maximum luminance and a maximum brightness of 0.17–0.58 cd/A and 90–150 cd/m², respectively (Chart 2.141).

In 1996, Yamamoto et al. [698] reported the EL from related poly(quinoxaline)s **615** (Chart 2.142). In contrast to unsubstituted polypyridine, all polymers **615** appeared to be soluble in organic solvents. For all polymers of the series, single-layer (ITO/polymer/Mg(Ag)) and double-layer devices with different HTL emitted green–blue light at $\lambda_{max} = 490$ nm; however, the device brightness was rather low (on the order of 1–10 cd/m²).

A better PLED performance was observed by Grice and coworkers [182] for ITO/PEDOT/polymer/Al devices with quinoxaline–phenylene vinylene copolymers **616** and **617** as emitting layers. The Φ_{EL}^{ex} and maximum brightness were measured as 0.012% and 0.01%, and 120 and 35 cd/m², respectively. The turn-on voltages of these devices were reasonably low, 6.0 and 4.0 V, respectively. The performance of PLEDs with polymer **616** was further improved by blending with 5 wt.% of a hole-transport material, 1,1-*bis*(di-4-tolylaminophenyl)cyclohexane, that enhanced the Φ_{EL}^{ex} to 0.06% and the maximum brightness to 450 cd/m².

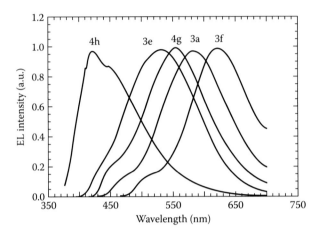

FIGURE 2.35
Tuning the electroluminescence color in polyquinoline derivatives (device structure ITO/HTL/EL polymer/Al, where HTL is triarylamine-based molecular material (for polymers **606**) or PVK **564** (for polymer **608**); EL polymer: **606a** (3a), **606e** (3e), **606f** (3f), **608a** (4g), and **608b** (4h). (From Zhang, X., Shetty, A.S., and Jenekhe, S.A., *Macromolecules*, 32, 7422, 1999. With permission.)

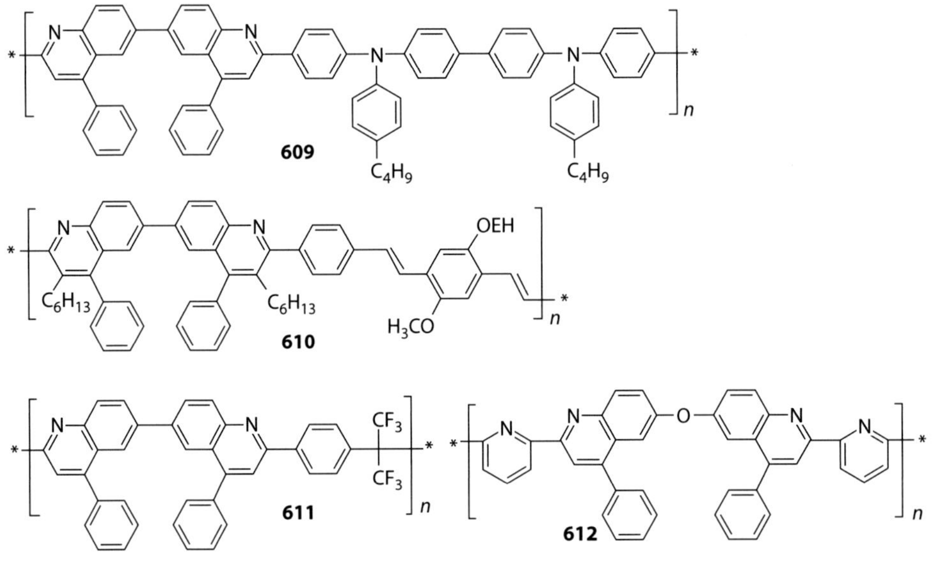

CHART 2.140
Chemical structures of pyridine-6,6′-bisquinoline copolymers.

Excellent electron-transporting properties of quinoxaline (also demonstrated for non-conjugated quinoxaline-containing polymer **618** [699] and quinoxaline-based polyether **619** [700]) resulted in a substantially decreased turn-on voltage of PPV/**620** PLED (3.6 V), which is much lower than that of pure PPV in the same conditions (7 V). These diodes showed a maximum luminance of 710 cd/m² (ca. 40 times brighter than the PPV diode at the same current density and voltage) [701].

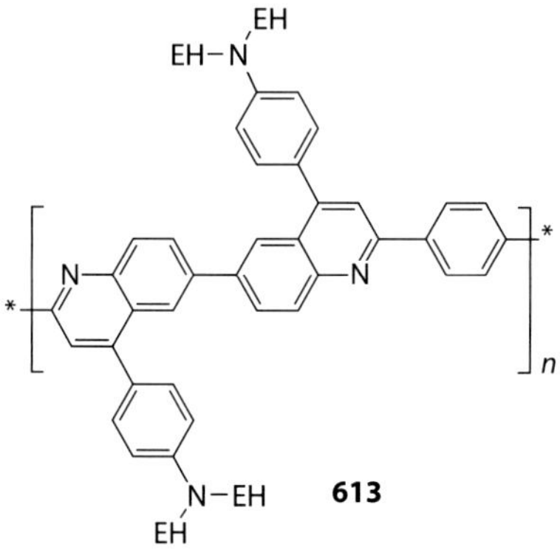
613

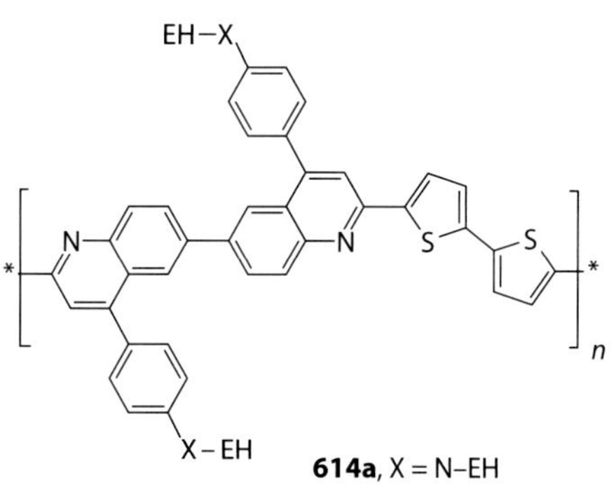
614a, X = N–EH
614b, X = O

CHART 2.141
Chemical structures of 6,6′-bisquinoline–based copolymers.

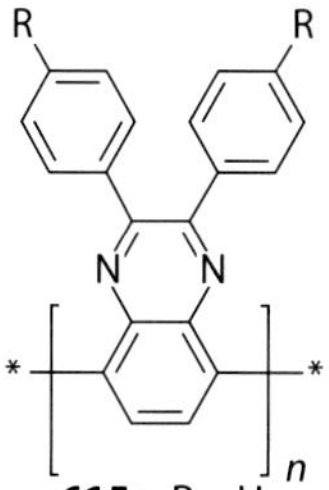
615a, R = H
615b, R = CH_3
615c, R = OCH_3
615d, R = Ph

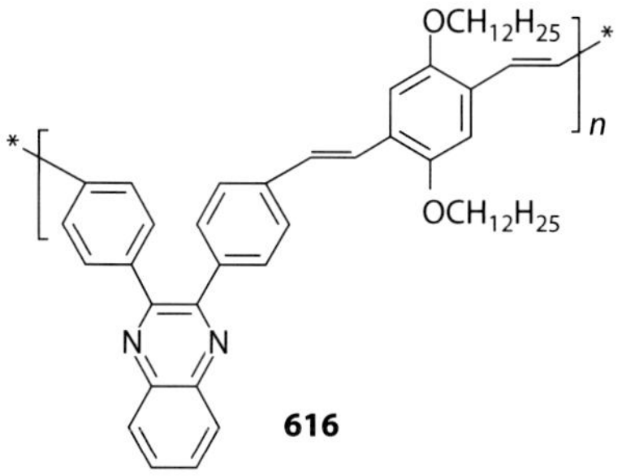
616

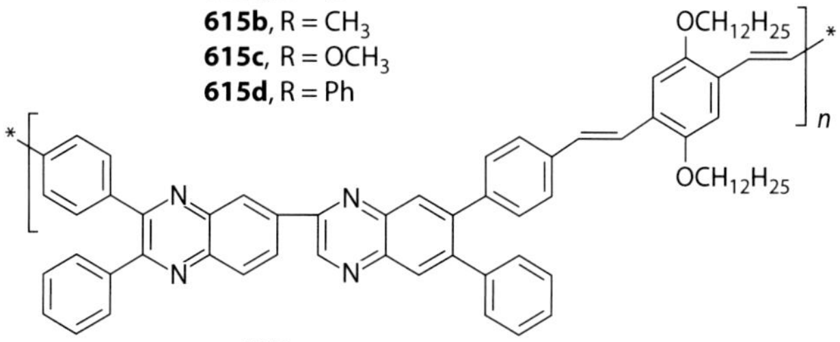
617

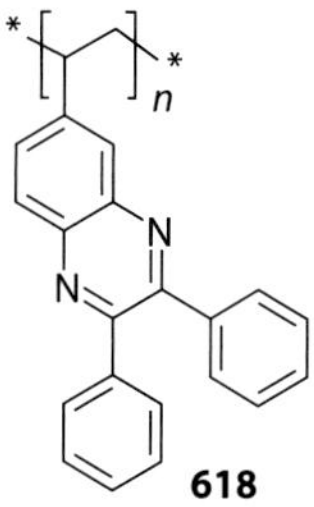
618

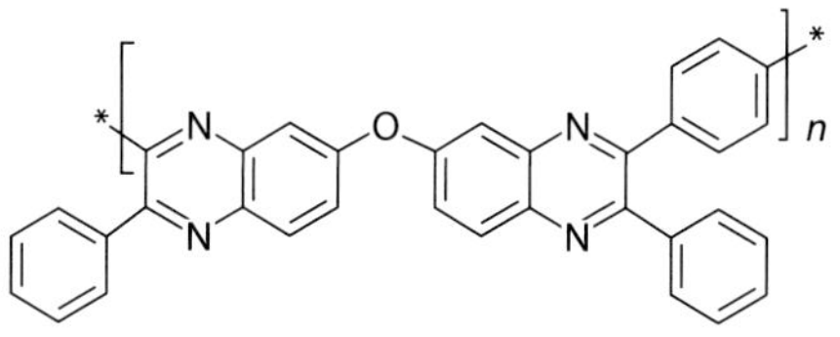
619

Abs: 381 nm, PL: 431 nm, LUMO: –2.76 eV

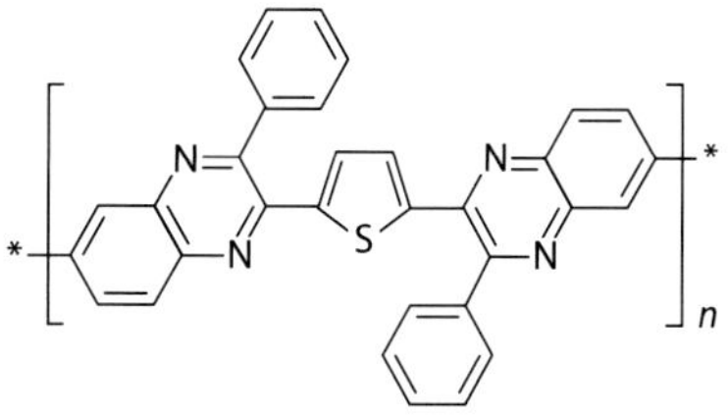
620

CHART 2.142
Chemical structures of polyquinoxalines.

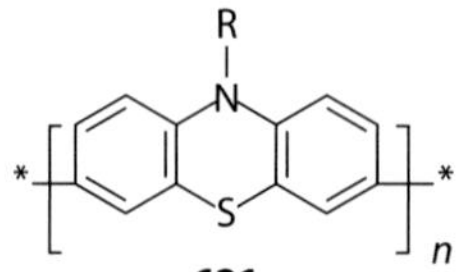

621

a, R = C$_6$H$_{13}$; Abs; 287 nm, PL: 490 nm, EL: 480 nm,
HOMO: −5.0 eV, LUMO: −2.24 eV
b, R = 2-ethylhexyl; Abs: 285 nm, PL: 478 nm, CIE (0.16, 0.32),
HOMO: −5.0 eV, LUMO: −2.23 eV

CHART 2.143
Chemical structures of fluorene–phenothiazine copolymers.

Incorporation of phenothiazine was successful in tuning the emission characteristics and LUMO energy level of fluorene–phenothiazine copolymers **368–370**, as discussed earlier [326,327] (Chart 2.89 and Chart 2.90). Phenothiazine homopolymers **621a,b** have also been prepared and characterized [252,326]. Although they emitted blue–greenish light with reasonably high Φ_{PL} in solution (43% for **621a**), the performance of the devices was rather low. Thus, ITO/PEDOT/**621b**/Ca diode showed a maximum brightness of only 9.0 cd/m^2, and a maximum current efficiency of 0.002 cd/A [252] (Chart 2.143).

2.5.6 Oxadiazole, Oxazole, and Thiadiazole Polymers

Owing to relatively high electron affinity and very good PL efficiency, molecular materials based on oxadiazole, particularly, PBD (**21**), are among the most popular electron-transport materials for OLEDs. The oxadiazole moieties, including PBD, were introduced as pendant groups in many EL polymers to reduce the electron-injection barrier and improve the EL efficiency of the device. We have already described oxadiazole-containing PPVs (**68, 72–75, 112–113, 146–153**), PFs (**270–273, 275–282**), and PTs (**488–494**) in previous parts.

Blends of PBD with EL polymers were employed to balance the charge-injection and -transport characteristics of PLEDs [71,112,528,594]. To avoid phase separation and molecular PBD crystallization, several groups introduced the PBD moieties as pendant groups in nonconjugated polymethacrylate [702–705] and polyethylene [699] chains. Polymers **622–624** have been studied as materials for PLEDs [702,703], in single-layer devices and in combination with PPV as HTL (Chart 2.144). The authors mentioned that the device instability is a great problem for these systems, and that engineering a more robust polymer backbone would be necessary. Later, Jiang and coworkers [706] synthesized and studied related polymers **625** and **626** based on polystyrene backbones. Having observed the immiscibility of oxadiazole polymer **625** with PVK, the authors designed the copolymer **626** containing electron- and hole-transporting units. The device ITO/**626**/Mg:Al showed improved Φ_{EL}^{ex} of 0.3%, although the turn-on voltage was still high (16 V). Through doping the polymer with different molecular dyes, the emission color was tuned from blue to orange [706]. The use of polymers **627a–c** as electron-transport layers in a two-layer LED (ITO/PPV/**627a–c**/Ag) increased the intensity of light emission by a factor of 100, compared with the single-layer ITO/PPV/Ag device; reaching Φ_{EL}^{ex} of up to 0.1% and a maximum brightness of 300 cd/m^2 at 30 V (turn-on voltage 6.5–8.5 V) [705].

In 1995, Pei and Yang reported polymers **628** [707,708] and **629** [709] with a short conjugation length and wide π–π^* energy gap (Chart 2.145). They were not fluorescent but could be used as an electron-injection layer to significantly improve the QE of PLEDs based on

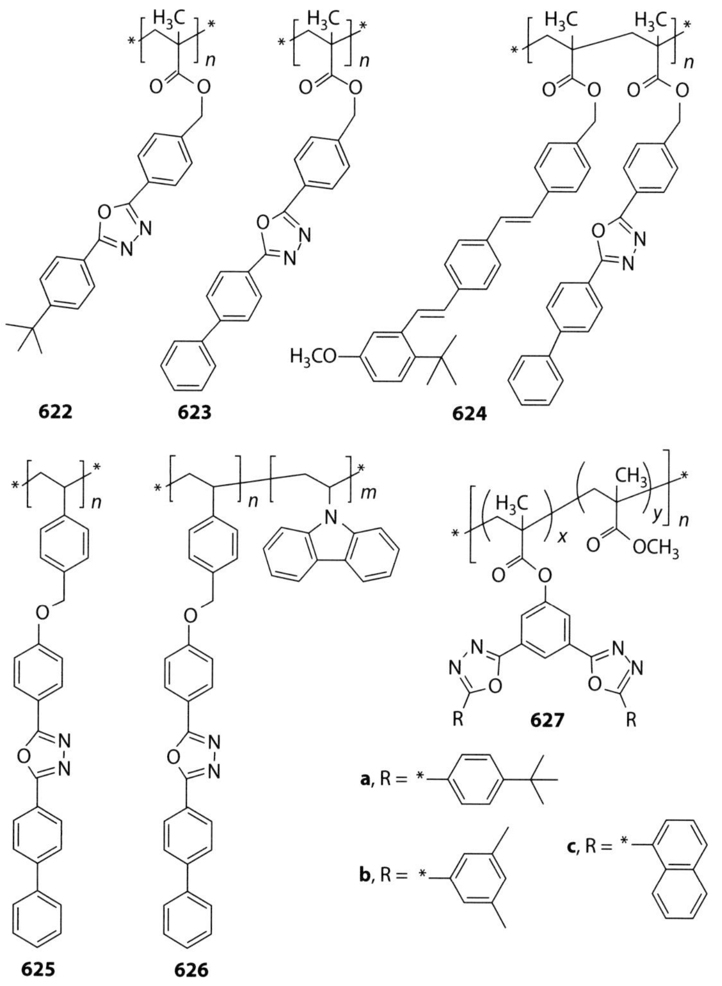

CHART 2.144
Chemical structures of polymethacrylates with PBD as pendant groups.

PPVs (from 0.002% to 0.08% for MEH–PPV **13** and from 0.03% to 0.30% for BCHA–PPV **25**) [709]. The additional oxadiazole moiety in **629** further enhances the electron-transport properties of the polymer and decreases the operation voltage of LEDs. Polymer **630** with a longer conjugation length has an efficient blue fluorescence (λ_{PL} = 554, 465 nm, Φ_{PL} [film] = 35%) and the PLED ITO/PANI/**630**/**628**/Ca emits a bright-blue light with Φ_{EL}^{ex} = 0.1% (turn-on voltage 4.5 V) [709]. Some other examples of electron-transport and hole-blocking oxadiazole–phenylene copolymers with short conjugation length interrupted by C(CF$_3$)$_2$ groups (**631** [710]) or by ether links (**632** [700]) have been reported in the literature.

Holmes and coworkers [710,711] introduced oxadiazole moieties into formally conjugated (although having one *m*-phenylene linkage) polymer **633** (Chart 2.146). The polymerization was performed by condensation of terephthaldihydrazide with 5-*tert*-butyl-1,3-benzenedicarboxylic acid, followed by dehydration and cyclization of the resulting polyhydrazide. On the basis of CV studies, the authors concluded that the electron affinity of **633**, although

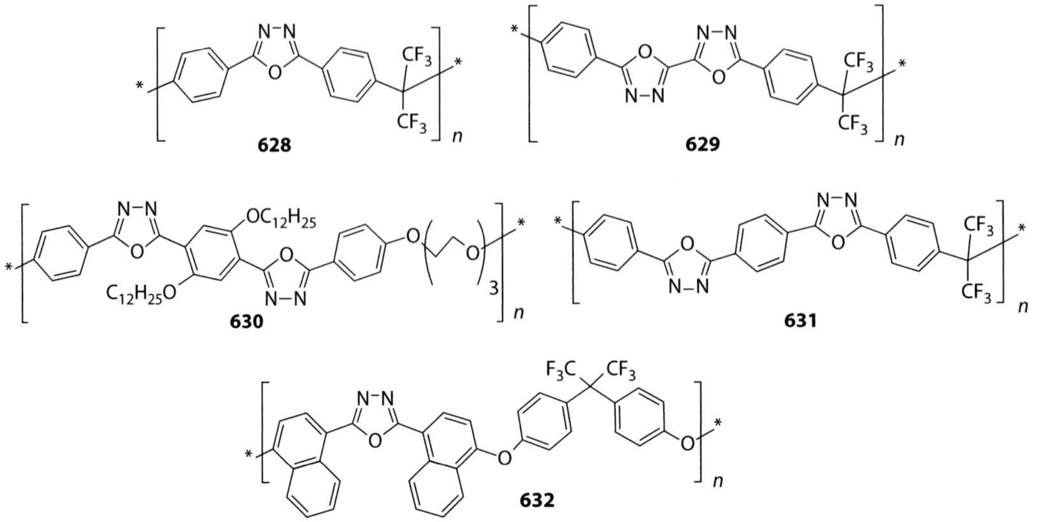

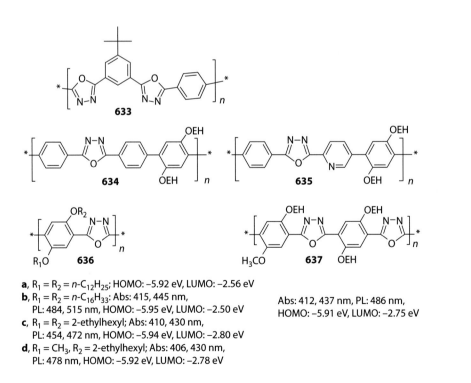

CHART 2.145
Chemical structures of polymers containing oxadiazole with a short conjugation length.

a, $R_1 = R_2 = n\text{-}C_{12}H_{25}$; HOMO: −5.92 eV, LUMO: −2.56 eV
b, $R_1 = R_2 = n\text{-}C_{16}H_{33}$: Abs: 415, 445 nm,
    PL: 484, 515 nm, HOMO: −5.95 eV, LUMO: −2.50 eV
c, $R_1 = R_2 = $ 2-ethylhexyl; Abs: 410, 430 nm,
    PL: 454, 472 nm, HOMO: −5.94 eV, LUMO: −2.80 eV
d, $R_1 = CH_3$, $R_2 = $ 2-ethylhexyl; Abs: 406, 430 nm,
    PL: 478 nm, HOMO: −5.92 eV, LUMO: −2.78 eV

Abs: 412, 437 nm, PL: 486 nm,
HOMO: −5.91 eV, LUMO: −2.75 eV

CHART 2.146
Chemical structures of conjugated polymers containing oxadiazole units.

higher than that of MEH–PPV, is below that of, e.g., CN–PPVs, and only a moderate PLQY (11%) was found.

Inspired by good electron-transport properties and high PL of PBD and, particularly, a claim by Cao and coworkers [71] of exceptional performance of PBD/MEH–PPV mixtures (EL of up to 50% of the PL yield), Wang and coworkers [712] reported the first poly(PBD) homopolymer (**634**) and its aza-derivative (**635**). The device ITO/PEDOT/MEH–PPV:**634**/Al showed Φ_{EL}^{ex} of 0.26%, compared with 0.01% obtained with MEH–PPV alone in an identically prepared device.

Janietz and coworkers [713,714] reported blue emission from fully conjugated alternating oxadiazole–phenylene polymers **636a–d** and **637**. Electrochemical estimation of their LUMO energy levels (–2.50 to –2.80 eV) characterizes them as potentially good electron-transport materials. A rectification ratio of 10^4 to 10^6 was reported for ITO/**636c**/Al diode in the negative bias region; however, no light emission was demonstrated.

Polybenzobisthiazoles **638** and polybenzobisoxazole **639** have been used as efficient electron-transport materials in PLEDs [74] (Chart 2.147). Although these polymers show poor fluorescence quantum yields in thin films likely due to excimer formation [715], double-layer devices ITO/PEDOT/polymer/ETL/Al with PPV or MEH–PPV as emissive polymers and **638a** or **639** as ETL had a turn-on voltage as low as 2.8 V, a luminance of up to 1400 cd/m², and Φ_{EL}^{ex} of up to 2.5%. Very recently, benzoxazole-based polymers have been studied as LEP. The fully conjugated polymer **639** was used in oriented fibers to create polarized PLEDs [716]. The resulting device ITO/**639**/Al can be turned on at ~5 V emitting red light (λ_{EL} = 620 nm) with polarization ratio of 1.6. The same group reported EL from nonconjugated benzoxazole copolymers **640** [717]. While the ratio *x*:*y* decreases from 1:0 to 0:1, the emission band of **640** broadens significantly (while keeping the maximum at ~520 nm), so that at *x*:*y* = 0:1 the emission band covers almost the entire visible region. The emission color of PLED fabricated as ITO/**640**/Al can be tuned from green (CIE: *x* = 0.25, *y* = 0.53) to the color reported as white, although the reported CIE coordinates (*x* = 0.24, *y* = 0.30) are relatively far from the definition of white (CIE: *x* = 0.33, *y* = 0.33).

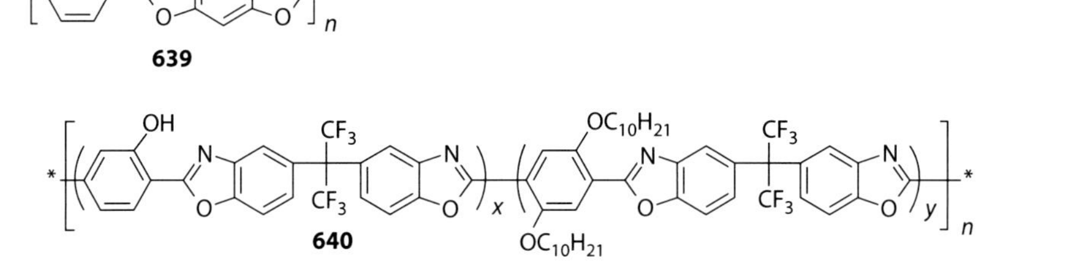

CHART 2.147
Chemical structures of polybenzobisthiazoles and polybenzobisoxazole.

2.5.7 Boron-, Phosphorus-, and Silicon-Containing Polymers

Introducing heteroatoms such as O, N, S in the backbone of conjugated polymers is routinely used to modify their electronic properties, and particularly, the HOMO–LUMO energies. Other heteroatoms only recently have been studied in this aspect [718].

Several relatively stable boron-containing conjugated polymers have been synthesized by Chujo and coworkers by high-yield hydroboration of C–C bonds (for polymers 641 [719,720]) or C–N triple bonds (for polymers 643 [721]), or by Grignard-type reaction of phenylenedibromide with substituted dimethoxyborane in the presence of Mg (for polymers 642 [722]) (Chart 2.148). The conjugation in these polymers is provided by interaction of the carbon π-system with the vacant p-orbital of boron, which should greatly enhance the electron-transporting properties of the polymers. In solution, 642 and 643 reveal strong blue or blue–green PL with λ_{PL} = 440–496 nm, while 641 has three PL maxima of 416 nm (blue), 495 nm (green), and 493 nm (orange–red), thus emitting white light [720]. No EL devices have been reported thus far, and the stability issue is likely to be a problem for practical applications of these materials [718].

Makioka and coworkers [723] prepared phosphorus-containing polymers (2,7-(9-oxo-9-phosphafluorenylene)-*alt-co*-(1,4-arylene))s 644, which emit blue light in solution (413–433 nm) with rather high Φ_{PL} (68%–81%), similar to PFs. In films, however, their emission is red shifted by 32–44 nm becoming green–blue, which is at a much higher wavelength than that for fluorene-*alt-co*-phenylene analogs [280,281,294,295] (Chart 2.149). The introduction of phosphorus-containing moieties into conjugated thiophene polymers has been investigated since the first polymeric system containing a phosphole unit (645) was reported [724]. The maximum emission peak of 645 is located at 470 nm with a quantum yield of 0.092 [724]. Baumgartner's group has investigated the dithieno[3,2-*b*:2′,3′-*d*] phosphole systems and explored their optoelectronic properties [725–727]. Copolymer 646 containing dithieno[3,2-*b*:2′,3′-*d*]phosphole was synthesized and exhibits a very strong blue photoluminescence (λ_{em} = 424) due to the great distance between the emitting centers, reducing the potential for quenching processes that could occur at higher densities [725].

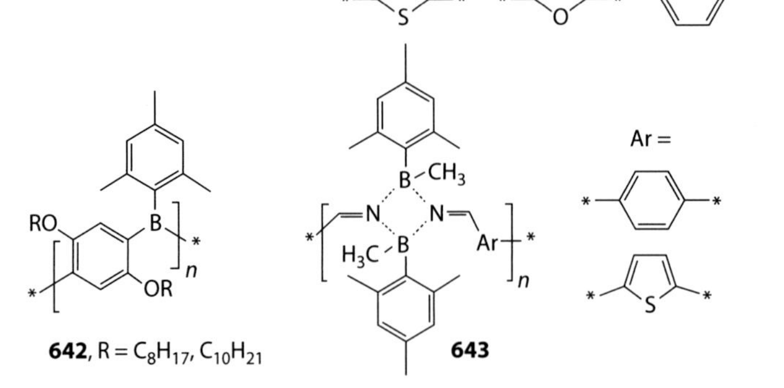

CHART 2.148
Chemical structures of boron-containing polymers.

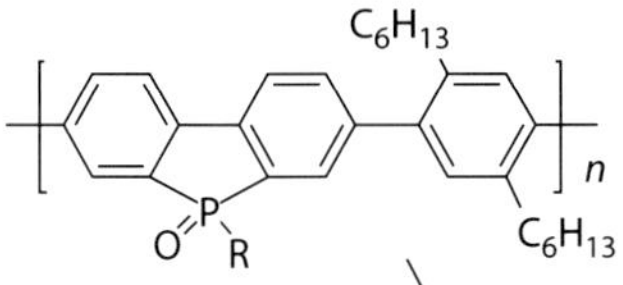

644a, R = n-C_9H_{19}
644b, R = n-C_3H_7
644c, R = $CH(CH_3)C_6H_{13}$

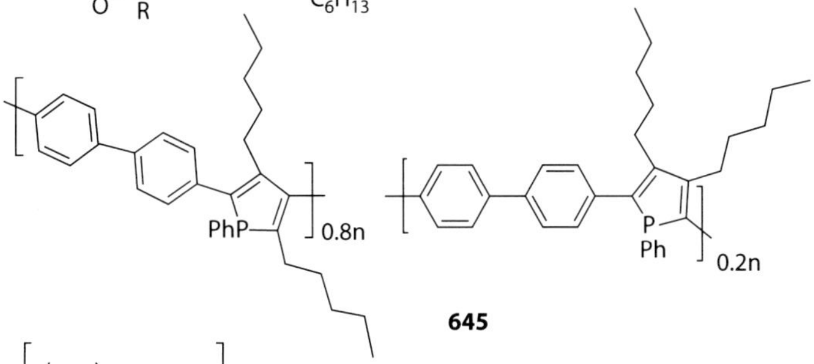

645

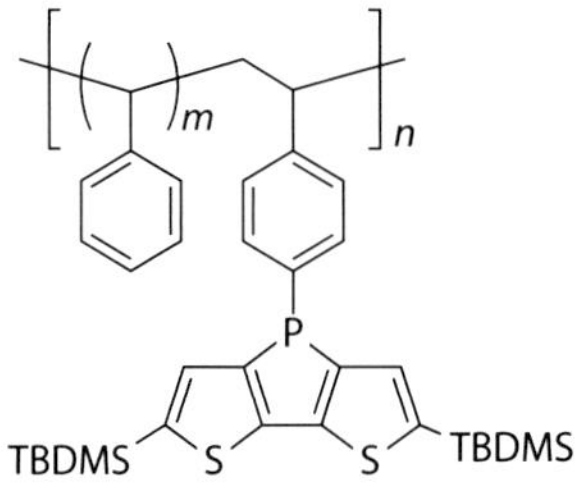

646

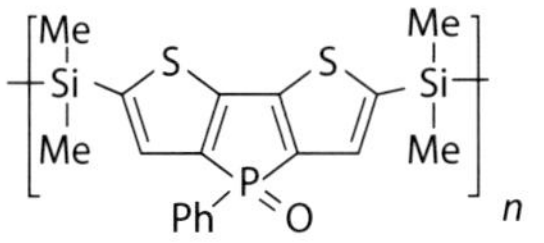

647

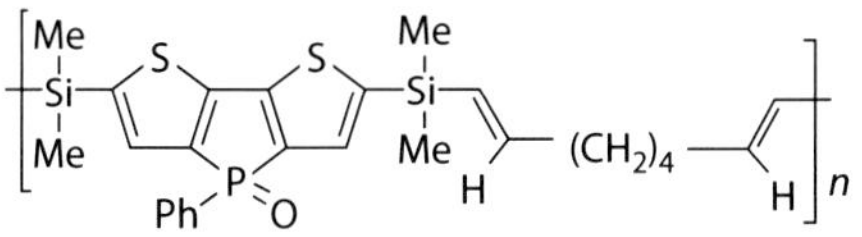

648

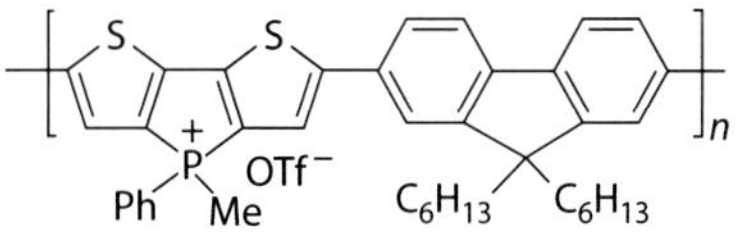

649

CHART 2.149
Chemical structures of phosphorus-containing polymers.

Later, Baumgartner's group reported the blue light–emitting Si–H functionalized dithieno[3,2-*b*:2′,3′-*d*]phosphole polymers (**647** and **648**) [726]. The emission properties of polymers **647** and **648** are close to those of the monomer (λ_{em} = 459). A copolymer (**649**) containing cationic dithieno[3,2-*b*:2′,3′-*d*]phosphole synthesized via the Suzuki–Miyaura cross-coupling reaction was reported for luminescent, conjugated polyelectrolytes application by the same group [727]. The emission and absorption (λ_{ex} [π–π*] = 485 nm) maxima of CPE**649** (CPE = conjugate polyelectrolytes) are strongly red shifted from those of the corresponding 9,9-dehexyl-fluorene homopolymer (λ_{ex} = 388 nm; λ_{em} = 445 nm) [342] and related fluorine copolymers containing 2,2-bithienyl (λ_{ex} = 398 nm; λ_{em} = 483 nm) [296], indicating a significantly lowered LUMO level by the dithienophosphole moiety. Yamada and coworkers [728] have prepared the luminescent silole polymers **650a–e**. The blue emission of the homopolymers **650a,b** can be shifted into the red region by copolymerization with other conjugated units (but for the price of lowered PLQY) (Chart 2.150). Although no device studies have been reported yet, excellent electron-transport properties are expected from such materials [729]. Other polymers containing silole units (**651–657**) were reported (Chart 2.151). Liu et al. [730] synthesized two random silole-containing copolymers via a Suzuki coupling reaction. A double-layer device (ITO/BTPD–PFCB/**651a**/Ca)-based polymer **651a** with a small amount of 4,4-diphenyldithienosilole unit (10 mol%) shows bright-blue emission with maximum brightness of 25,900 cd/m^2 and a maximum

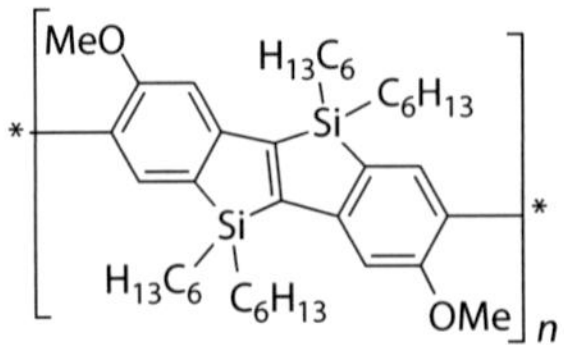

650a, Abs: 420 nm, PL: 476 nm, Φ_{PL}: 36%

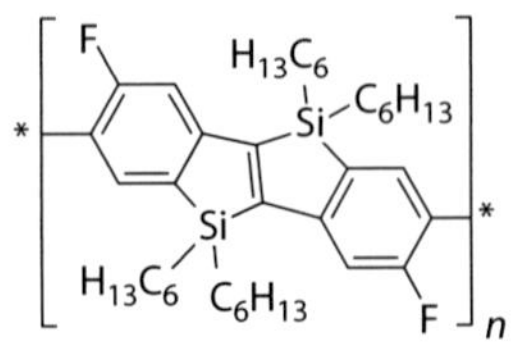

650b, Abs: 418 nm, PL: 467 nm, Φ_{PL}: 51%

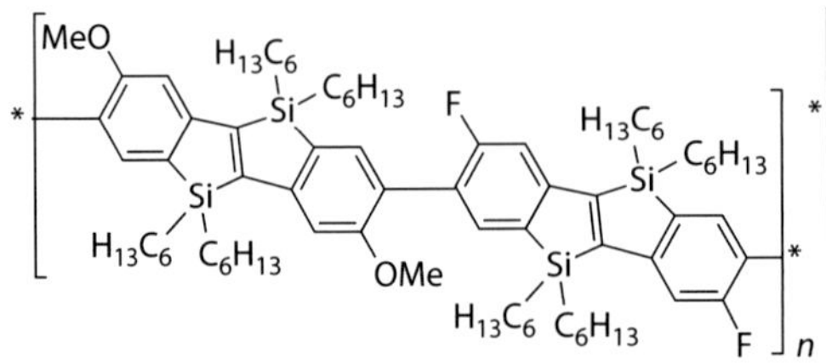

650c, Abs: 422 nm, PL: 454 nm, Φ_{PL}: 63%

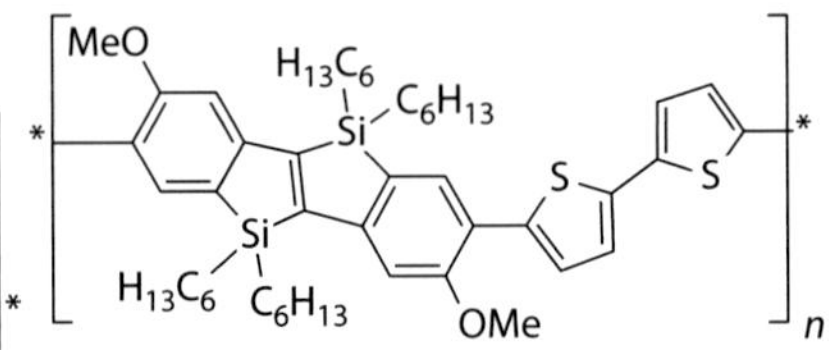

650d, Abs: 494 nm, PL: 547 nm, Φ_{PL}: 2.5%

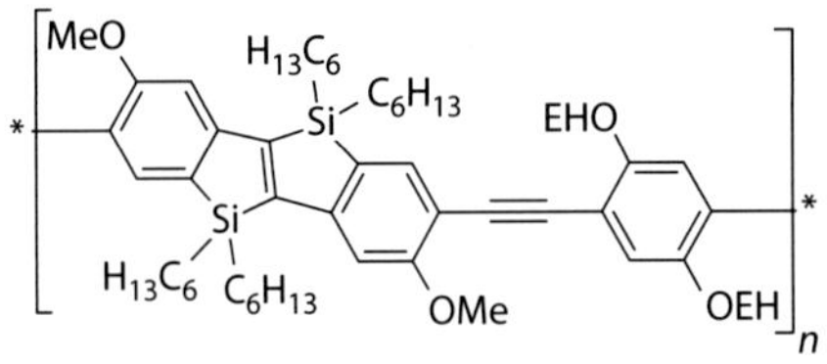

650e, Abs: 482 nm, PL: 506 nm, Φ_{PL}: 15%

CHART 2.150
Chemical structures of silole polymers.

external QE of 1.64% due to both the improved charge injection and recombination in this polymer [730]. Copolymers **652** and **653** were reported by Ohshita's group [731]; however, the single-layer devices (ITO/polymer/Mg–Ag) with **652** and **653** emit weak red color. A series of 3,6-silafluorene-based copolymers **654** were reported as blue EL materials [732]. Poly(3,6-silafluorene) is a typical wide-band-gap conjugated polymer with ultraviolet light emission. By introducing different functional moieties into the polymers, the resulting polymers (**654**) can emit efficient deep-blue light due to the intrachain energy transfer [732]. Pang's group reported optical properties of light-emitting π-conjugated polymers containing biphenyl and dithienosilole [733]. The presence of 5% dithienosilole in the copolymers of **656** and **657** had little influence on the absorption and emission. Double-layer light-emitting devices using polymers as emissive layers had low turn-on voltage (3.5–4 V) and moderate external quantum efficiencies (0.14%–0.30%) (Chart 2.151) [733].

2.5.8 Summary

Although the major research activities in PLEDs are now concentrated in three major classes of LEPs—PFs, PPVs, and, to a lesser extent, PTs, the study of other classes of conjugated and nonconjugated polymers may give rise to high-performance EL materials with tunable emission color. Importantly, the charge-transport and charge-injection properties can be tuned in a wider region than is accessible by structural modification of PF, PPV, and PT materials. The very high EL performance achieved in PPP-type polymers (including ladder-type PPPs) and very high chemical and thermal stability of the PPP materials render them as very promising candidates for commercial PLEDs. On the basis

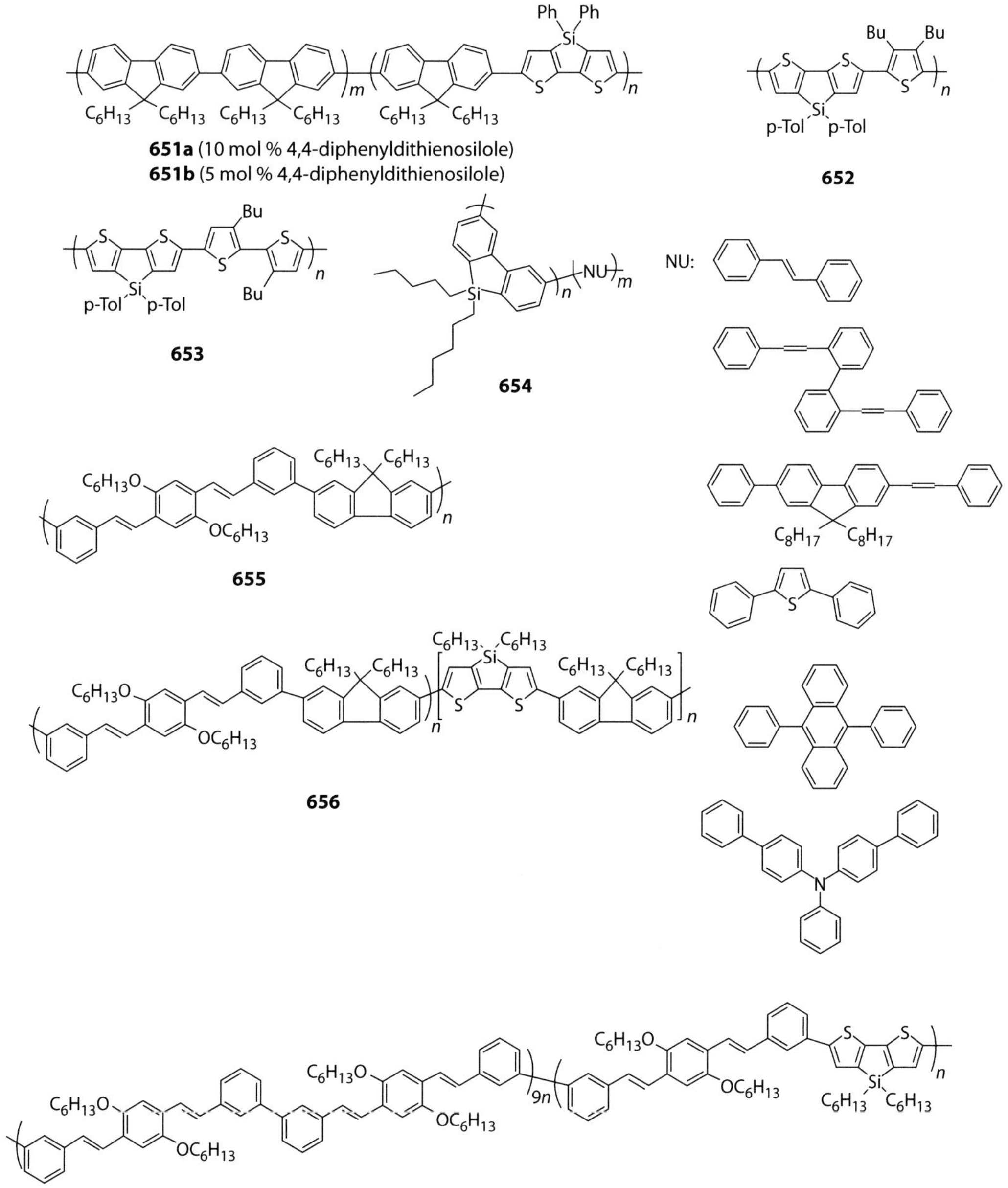

CHART 2.151
Chemical structures of polymers containing silole units.

of LPPP **543b**, blue–green-emitting PLEDs with Φ_{EL}^{ex} up to 4% [628] and brightness as high as 40,000 cd/m^2 [70] was achieved by different engineering techniques. An efficient yellow-emitting PLED (Φ_{EL}^{ex} = 4.2% [533]) and a white-emitting PLED (Φ_{EL}^{ex} = 1.2% [633]) were fabricated by blending LPPP **543b** with PT- and PPE-based polymers, respectively.

Introducing heteroaromatic moieties (mainly with N and, to a lesser extent, with O and S) in the backbone of the polymer or as a pendant group can substantially modify the

LUMO level of the materials, improving their electron-transport properties and facilitating electron injection in PLEDs; however, the efficiencies still lag behind the other systems.

2.6 Conclusions and Outlook

In the previous four sections, we described the progress in the development of LEPs, from the beginning of the 1990s up to 2014. Practical applications in PLEDs became a field of major research activity worldwide, in academia and in industrial laboratories. More than 600 different LEPs described in this book (and, most probably, >1000 reported LEPs) have been developed through the efforts of many researchers with different scientific background. At one extreme, this fact is reflected in unsystematic and often-illogical approaches the science took in the development of new materials (which was the biggest frustration for the authors, who were trying to integrate all materials in one story). At the other extreme, it allowed an element of serendipity that continues to keep the scientific part of the field alive. Although at the moment, the commercial success of PLEDs probably depends more on engineering issues, it would certainly not be possible without several important scientific breakthroughs. Besides the pioneering work of Friend and coworkers [1] on PPV-based PLEDs, the following major discoveries, while certainly not an exhaustive list, should be considered:

- Design of soluble LEPs by introduction of solubilizing substituents
- Prevention of interchain interactions (and resulting luminescence quenching) by sterically hindering substituents
- Fine tuning the emission of the polymers via copolymerization
- Control of the charge injection and transport by introduction of electron donor–acceptor substituents (or comonomers and end-capping groups)
- Control of the interface at the anode and cathode
- Control of the film morphology as a function of spin-casting solvent

Following these principles, a number of EL polymers with an external QE of >3%–5% have been demonstrated. However, there are still a number of issues to be solved for successful competition with small-molecule OLEDs. Since one of the main targets for LEPs is display technology, pure red, green, and blue emitters are of particular importance. Among these, green color is the easiest to achieve with a traditional class of green-emitting polymers, such as phenyl-PPV and PF-based polymers. The achieved photometric current efficiency of green-emission polymer materials is much higher than other color–emission polymers, which is also due to the higher sensitivity of the human eye to the green color. High-efficiency red-emitting polymers are considerably less common, and, in terms of technology, this niche probably belongs to electrophosphorescent PLEDs (which are based on blends of conventional LEPs with phosphorescent metal complexes, or their covalent conjugates; see Chapter 7). Design of a highly efficient polymer emitter giving a pure blue color is probably one of the major challenges to the LEP field. High-energy excited states (~3 eV) generally impose higher reactivity (and lower stability) to blue emitters. Undoubtedly, PFs, as easily functionalizable polymers, are the most promising materials for blue PLEDs. Color purity and color stability are the major concerns with blue-emitting

TABLE 2.10

Current PLED Performance by Sumitomo Chemical and CDT

Spin/BE Data at 1000 cd/m²	Red			Green			Blue	
Efficiency (cd/A)	31	22	17	75	75	12.4	8.6	7.3
Color (CIE x, y)	0.62, 0.38	0.65, 0.35	0.65, 0.35	0.32, 0.63	0.32, 0.63	0.14, 0.12	0.14, 0.12	0.14, 0.13
T_{50} lifetime (h)	350,000	350,000	>150,000	>300,000	1,900,000	16,000	–	–
V_d (V)	4.2	4.1	3.6	5.1	4.5	3.9	3.6	4.1

polymers. A recent demonstration of the origin of the "parasitic" green band in PFs arising from ketone defects opens a number of possible (and already partially realized) solutions for this problem. Table 2.10 shows state-of-the-art performance of PLEDs developed by Sumitomo Chemical and CDT. Green PLED exhibits the highest efficiency of 75 cd/A, and red PLED has the best half-lifetime of 350,000 h. However, the performance of blue PLED is still far behind that of red and green PLEDs in terms of efficiency and lifetime. More efforts are needed to develop novel, efficient, and stable blue light–emitting polymer materials.

The next most promising application for EL polymers is general illumination, where high efficiency, stability, and low cost are the major criteria. Although the sufficiently high-efficiency yellow-emitting polymers (e.g., oxadiazole-containing PPV **75**, photometric efficiency of 21.1 cd/A [129]) are already available, the true need is in much less developed white-emitting LEPs. To date, true white color in LEP can be achieved through a single-polymer system containing different chromophores or by blending several types of materials. A pure white emission (CIE: $x = 0.35$, $y = 0.32$) was demonstrated by a molecular design of polyfluorene bearing orange dopant on the main chain (**395**) with high luminance efficiency of 7.30 cd/A [450].

Overall, the stability in terms of color, and even more in terms of operation lifetime, is the major issue for PLEDs. A very significant role in device degradation is played by material impurities that are very difficult to control in polymers, as in any polydisperse system. Rigorous monomer purification and a careful choice of the polymerization method are just as important for the EL performance as the specific polymer structure. At the same time, the device architecture is also directly responsible for the device efficiency and stability. A careful adjustment of charge-injection barriers can minimize the driving voltage and improve the device efficiency and stability. To conclude, the future progress in PLEDs will depend on a close collaboration between organic and polymer chemists on one side and surface scientists and engineers on the other.

2.7 Appendix

2.7.1 Syntheses of Poly(*p*-Phenylene Vinylene)s

2.7.1.1 Wessling–Zimmerman (Thermoconversion) Precursor Route to PPV

2.7.1.1.1 *Poly(2-Fluoro-5-(n-Hexyloxy)-1,4-Phenylene Vinylene) (84) (Chart 2.152)*

From Gurge, R.M., Sarker, A.M., Lahti, P.M., Hu, B., and Karasz, F.E., *Macromolecules*, 30: 8286–8292, 1997. Copyright 1997, American Chemical Society, Washington, DC. With permission.

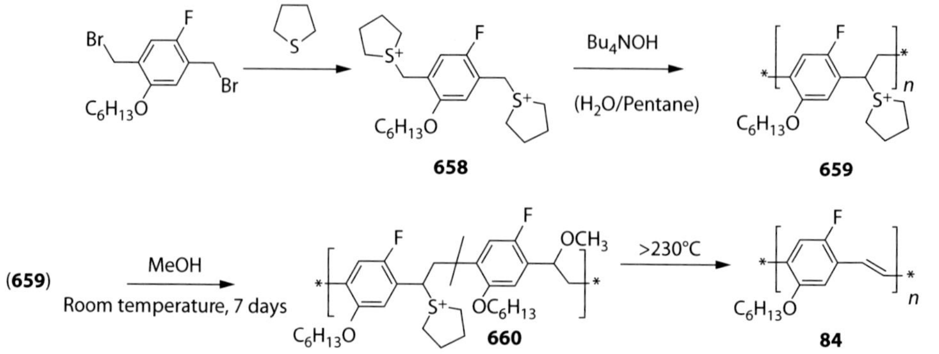

CHART 2.152
Synthesis of poly(2-fluoro-5-(*n*-hexyloxy)-1,4-phenylene vinylene).

Bis(tetrahydrothiophenium) salt **658**: 1-Fluoro-4-hexyloxy-2,5-*bis*(chloromethyl)benzene (1.22 g, 3.19 mmol) was suspended in a solution of tetrahydrothiophene (1.4 g, 15.9 mmol) in 35 ml of dry methanol. The mixture was stirred at 55°C for 24 h. The solvent and excess tetrahydrothiophene were removed by distillation to give an off-white residue, which was dissolved in a minimum amount of methanol and precipitated into 200 ml of dry acetone. After vacuum drying, white powder of **658** (1.25 g, 70%) was obtained with melting point at 163°C–166°C (dec). This compound is stable in a freezer under inert atmosphere. An analytical sample was obtained by 3-fold precipitation from methanol into acetone. Anal. Calcd for $C_{22}H_{35}Br_2FOS_2$: C, 47.32; H, 6.31; Br, 28.62; F, 3.40; S, 11.48. Found: C, 47.38; H, 6.27; Br, 29.1; F, 3.50; S, 11.52. ^{1}H NMR (80 MHz, CD_3OD): 7.64 (d, 1H, J = 8.8 Hz), 7.52 (d, 1H, J = 4.5 Hz), 4.77 (s, 2H), 4.67 (s, 2H), 4.29 (t, 2H, J = 6.2 Hz), 3.66 (m, 8H), 2.48 (m, 8H), 1.54 (m, 8H), 1.02 (asym t, 3H, J = 7.5 Hz).

Polyelectrolyte **659**: The *bis*(sulfonium) salt **658** (600 mg, 1.08 mmol) was dissolved in 25 ml distilled water. The solution was filtered through a glass frit and placed in a 100 ml round-bottom flask. An equal volume of pentane was added to the flask, and the two-phase system was cooled to 0°C under argon. A solution of tetramethylammonium hydroxide was also cooled to 0°C under argon and pentane. Both solutions were thoroughly purged with argon gas for 1 h. Then, the base (0.47 ml, 1.30 mmol) was added swiftly by syringe. The polymerization was allowed to proceed for 1 h at 0°C. The excess base was neutralized with 6 M HCl solution to a phenolphthalein end point. The resulting yellow–green solution was dialyzed against distilled water (Spectropore 1 filters, M_w cutoff 6000–8000) for 3 days to give a uniform green solution. This solution can be used to cast films, which are soluble in methanol but insoluble in THF and $CHCl_3$. This material is appropriate for conversion to polyether **660**.

Polyether **660**: The sulfonium polyelectrolyte **659** was dissolved in a minimal amount of methanol. The solution was stirred for 1 week under argon, during which pendant polyether **660** precipitated from the solution. The tacky yellow material was completely soluble in $CHCl_3$ and THF. Polymer **660** can be dissolved in THF and passed through a 0.2-mm filter for GPC, or for use in film casting and device fabrication. Cast films can be completely redissolved into $CHCl_3$ or THF. GPC in THF (polystryrene standards) gave M_w = 91,500 and M_n = 33,800.

PPV **84**: A film of polyether **660** was clamped between two 0.125-in-thick Teflon plates, and then heated at 230°C for 6 h under vacuum (<0.01 mm Hg) to give a red film of PPV **84**. The final polymer films are insoluble in $CHCl_3$, THF, and MeOH. Anal. Calcd for $C_{14}H_{17}OF$:

C, 76.33; H, 7.78; F, 8.62; S, 0.0. Found: C, 72.02; H, 7.44; F, 8.3; S, 0.30. IR (neat film, cm^{-1}): 962 (*trans* HC=CH). UV–vis (neat film on quartz): λ_{max} = 455 nm. PL (neat film on quartz, excitation at 390 nm): λ_{max} = 630 nm.

2.7.1.2 Gilch Polymerization Procedure

2.7.1.2.1 Preparation of Poly[2-(3,7-Dimethyloctyloxy)-5-Methoxy-p-Phenylene Vinylene] (14) (Chart 2.153)

From Becker, H., Spreitzer, H., Ibrom, K., and Kreuder, W., *Macromolecules*, 32: 4925–4932, 1999. Copyright 1999, American Chemical Society, Washington, DC. With permission.

A 4-l, four-neck flask fitted with mechanical (Teflon) stirrer, reflux condenser, thermometer, and dropping funnel was dried (stream of hot air) and flushed with N$_2$. The reactor was then charged with 2.3 l of dry 1,4-dioxane, and the solvent was degassed by passing N$_2$ through it for about 15 min. The solvent was heated to 98°C with an oil bath, and 14.0 g (38.7 mmol) of 2,5-*bis*(chloromethyl)-1-(3,7-dimethyloctyloxy)-4-methoxybenzene was added as a solid. (The solid was rinsed with about 10 ml of dry 1,4-dioxane.) A 11.3 g (100 mmol, 2.6 equivalent) sample of potassium *tert*-butoxide, dissolved in 100 ml of 1,4-dioxane, was added dropwise to the reaction solution from the dropping funnel during a period of 5 min. During this addition, the reaction mixture changed color from colorless to greenish to yellow–orange, and the viscosity increased significantly. After the addition was complete, the mixture was stirred further for about 5 min at 98°C; 8.70 g of potassium *tert*-butoxide (77 mmol, 2 equivalent) in 77 ml of dry 1,4-dioxane was then added during a period of 1 min, and stirring was continued for 2 h at 96°C–98°C. The solution was then cooled to 50°C for a period of about 2 h. The reaction mixture was finally mixed with 15 ml (260 mmol, 1.5 equivalent based on the base) of acetic acid (diluted with the same amount of 1,4-dioxane) and stirred further for 20 min. The solution was then deep orange, and the viscosity increased. For the workup, the reaction solution was slowly poured into 2.5 l of intensively stirred water. The resulting mixture was stirred further for 10 min, 200 ml of methanol was added, and the precipitated polymer was filtered off. This was washed with 200 ml of methanol and dried under reduced pressure at room temperature. A 10.04 g (34.8 mmol, 90%) sample of crude polymer was obtained as red fibers. The polymer was purified by dissolving it in 1.1 l of THF (60°C), cooling the solution to 40°C, and precipitating the polymer by dropwise addition of 1.2 l of methanol. After washing with 200 ml of methanol, it was dried at room temperature under reduced pressure. This

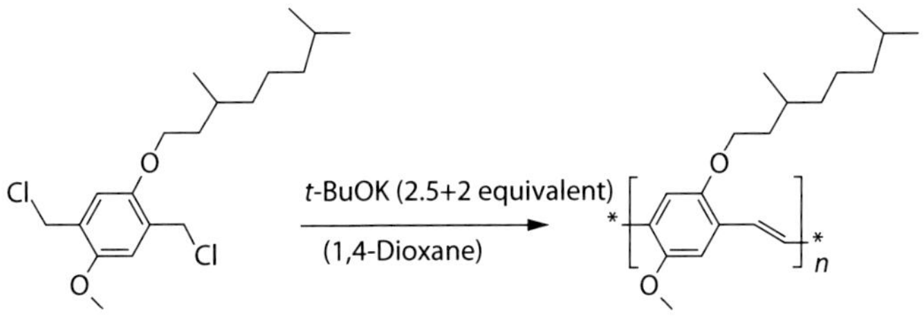

CHART 2.153
Synthesis of poly[2-(3,7-dimethyloctyloxy)-5-methoxy-*p*-phenylene vinylene].

procedure was repeated once more using 1.0 l of THF and 1.0 l of methanol. A 6.03 g (20.9 mmol, 54%) sample of poly[2-(3,7-dimethyloctyloxy)-5-methoxy-*p*-phenylene vinylene] (**14**) was obtained as a dark-orange, fibrous polymer. ^{1}H NMR (400 MHz, CDCl$_3$): 7.7–6.5 (br m, 4H; H$_{arom}$, olefin-H); 4.5–3.6 (br m, 5H; OCH$_3$, OCH$_2$); 2.1–0.6 (br m, 19H; aliph-H). GPC (THF + 0.25% of oxalic acid; column set SDV500, SDV1000, SDV10000 (from PSS), 35°C, UV detection at 254 nm, polystyrene standard): M_w = 1.5 × 10^6 g/mol, M_n = 3.1 × 10^5 g/mol. Elemental analysis: Calcd: C, 79.12%; H, 9.78%; O, 11.09%. Found: C, 78.88%; H, 9.82%; O, 11.00%; Cl, 25 ppm; K, 41 ppm; Na, 23 ppm.

2.7.1.3 Chlorine (Bromine) Precursor Route

2.7.1.3.1 Preparation of Poly{2-[2′-Ethylhexyloxy]-5-[4″-Methylsulfonyl Phenyl]-1,4-Phenylenvinylene} (77) (Chart 2.154)

From Boardman, F.H., Grice, A.W., Rüther, M.G., Sheldon, T.J., Bradley, D.D.C., and Burn, P.L., *Macromolecules*, 32: 111–117, 1999. Copyright 1999, American Chemical Society, Washington, DC. With permission.

A solution of potassium *tert*-butoxide dissolved in dry THF (0.3 M, 2.6 ml, 0.8 mmol) was added to a solution of *bis*(bromomethyl)methylsulfonylbiphenyl (0.5 g, 0.9 mmol) in dry THF (3 ml) cooled in an acetone or ice bath under nitrogen. A bright-yellow viscous solution was formed, and dry THF (4 ml) was added. The reaction mixture was allowed to warm to room temperature after 10 min, and stirred for 2 h. A small amount of a precipitate was formed. The reaction mixture was filtered through a plug of cotton wool, and polymer **661** was precipitated by adding the reaction mixture dropwise to ice-cold 2-propanol (50 ml). The mixture was centrifuged for 10 min at 4000 rpm, and the supernatant was removed. The crude polymer **661** was briefly dried under vacuum, dissolved in dry THF, and then precipitated by addition to an excess of 2-propanol. The solid was collected after centrifugation, and the process was repeated a further 2 times. The residue was finally collected by dissolution in a minimum of THF, and the solvent was removed to leave **661** as an orange–yellow solid (200 mg, 40%). λ_{max} (thin film)/cm^{-1} 955 (HC=C), 1154 (SO$_2$), 1315 (SO$_2$); λ_{max} (CH$_2$Cl$_2$)/nm 300sh; ^{1}H NMR (500 MHz; CDCl$_3$) 0.74–1.00 (br m, CH$_3$), 1.14–1.90 (br m, CH and CH$_2$), 2.94–3.25, 3.30–4.10 (br m, CH$_3$SO$_2$, OCH$_2$, and ArCH$_2$), 5.05–5.23 (br s, CHBr), 6.27–7.63 (br m, Ar H or vinyl H), 7.67–8.05 (br m, ArH); GPC of equilibrated sample, M_w = 2.7 × 10^5 and M_n = 1.3 × 10^5, PDI = 2.1.

Thin films of **661** were heated at 160°C under a dynamic vacuum for 14 h to yield **77**. λ_{max} (film on KBr disk) (cm^{-1}) 955 (C=CH), 1154 (SO$_2$), 1315 (SO$_2$); λ_{max} (thin film) (nm) 285sh, 437.

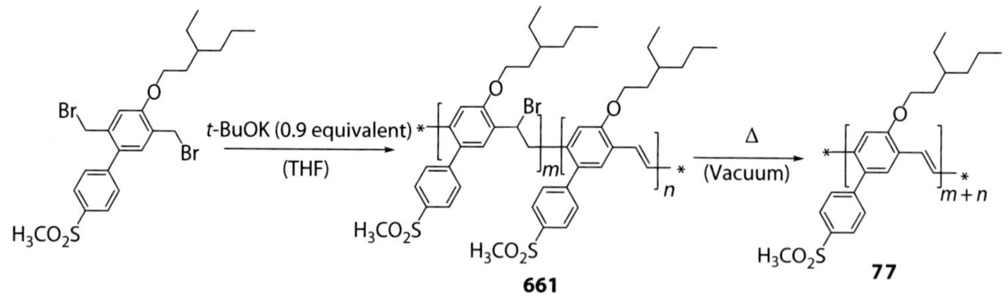

CHART 2.154
Synthesis of poly{2-[2′-ethylhexyloxy]-5-[4″-methylsulfonyl phenyl]-1,4-phenylenvinylene}.

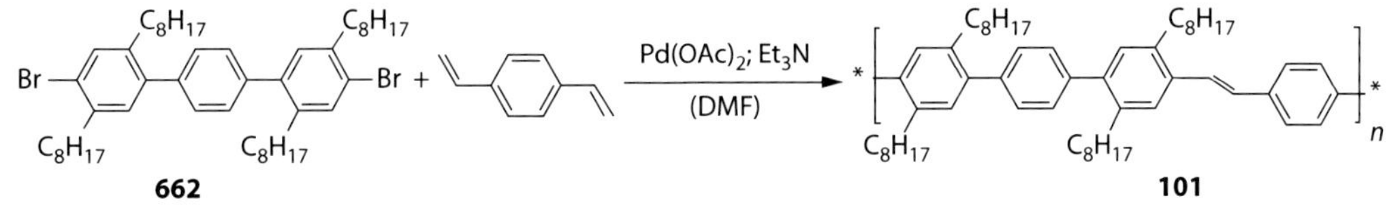

CHART 2.155
Synthesis of poly[2,5,2,5″-tetraoctyl-*p*-terphenyl-4,4-ylenevinylene-*p*-phenylene vinylene].

2.7.1.4 Heck-Coupling Route

2.7.1.4.1 Preparation of Poly[2,5,2,5″-Tetraoctyl-p-Terphenyl-4,4-Ylenevinylene-p-Phenylene Vinylene] (101) (Chart 2.155)

From Hilberer, A., Brouwer, H.-J., van der Scheer, B.-J., Wildeman, J., and Hadziioannou, G., *Macromolecules*, 28: 4525–4529, 1995. Copyright 1995, American Chemical Society, Washington, DC. With permission.

A mixture of monomer **662** (1.0 g, 1.2 mmol), *p*-divinylbenzene (0.155 g, 1.2 mmol), Pd(OAc)$_2$ (0.011 g, 0.05 mmol), tri-*o*-tolylphosphine (0.071 g, 0.23 mmol), triethylamine (2 ml), and DMF (5 ml) was placed in a heavy wall pressure tube. The tube was degassed, closed (Teflon bushing), and heated to 100°C. After 40 h of reaction, thin-layer chromatography showed that the monomers were consumed. The reaction mixture was then poured into methanol (75 ml). The precipitated material was filtered off and dried under vacuum, giving a brown–yellow polymer (0.88 g, conversion = 91%). The crude polymer was dissolved in chloroform and then filtered through a small column of Kiesel gel to remove traces of catalyst. The resulting solution was concentrated and precipitated in methanol (75 ml). The yellow polymer **101** was collected by filtration and thoroughly dried under vacuum. ^{1}H NMR (broad signals): 0.89 (t, CH$_3$, 12H), 1.25/1.29/1.5 (m, CH$_2$, 48H), 2.64 (m, CH$_2$, 8H), 7.1–7.6 (3 main peaks, arom. CH and vinyl CH, 16H); ^{13}C NMR: 126.8, 128.8, 129.2, 130.1, 131.5 (arom CH), 134.7, 137.2, 137.6, 138.1, 138.2, 140.1 (arom C). Anal. Calcd for C$_{60}$H$_{84}$ (repeating unit): C, 89.49; H, 10.51. Found: C, 88.02; H, 10.51; Br, 0.89.

2.7.1.5 Knoevenagel-Coupling Route

2.7.1.5.1 Preparation of CN–PPV Copolymer (133) (Chart 2.156)

From Kim, D.-J., Kim, S.-H., Zyung, T., Kim, J.-J., Cho, I., and Choi, S.K., *Macromolecules*, 29: 3657–3660, 1996. Copyright 1996, American Chemical Society, Washington, DC. With permission.

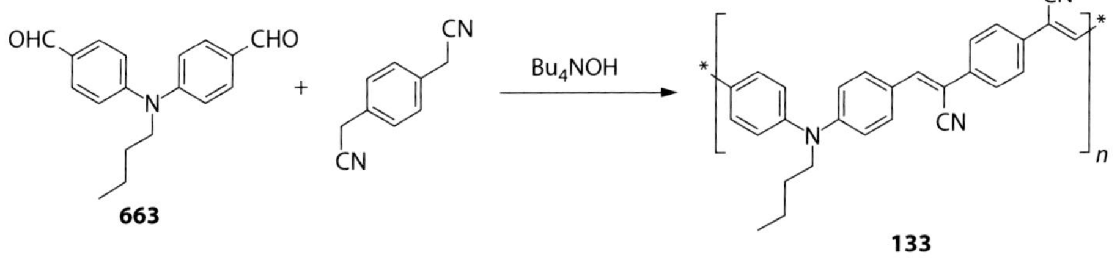

CHART 2.156
Synthesis of CN–PPV copolymer.

Under an argon atmosphere, to a stirred solution of equimolar quantities of dialdehyde **663** and 1,4-phenylenediacetonitrile in THF and *tert*-butyl alcohol (1:1) was added dropwise 5 mol% of Bu$_4$NOH (1 M in methanol) at 45°C for 20 min. The resulting paste-like polymeric product that precipitated from the solution during polymerization was collected and thoroughly washed with methanol to remove ionic species and unreacted compounds. The scarlet polymeric product was dried in a vacuum oven at 40°C for 2 days. The polymer yield was 93%. GPC (THF, polystyrene standard): $M_w = 1.5 \times 10^6$ g/mol, $M_n = 3.1 \times 10^5$ g/mol.

2.7.2 Syntheses of Polyfluorenes

2.7.2.1 Synthesis of PF Monomers

2.7.2.1.1 2,7-Dibromofluorene (664) (Chart 2.157)

From Perepichka, I.I., Perepichka, I.F., Bryce, M.R., and Pålsson, L.O., *Chem. Commun.*, 3397–3399, 2005.

Fluorene (292 g, 1.76 mol) was dissolved in acetic acid (2.6 l) at ~70°C, and H$_2$SO$_4$ (96%, 25 ml) was added slowly to this solution. The reaction mixture was allowed to cool to ~50°C with stirring, and a solution of bromine (150 ml, 2.92 mol) in acetic acid (200 ml) was added dropwise for 2–3 h, keeping the temperature at 40°C–55°C to avoid crystallization of the fluorene. When about 1/3–1/2 of bromine was added, dibromofluorene started to crystallize. Simultaneously, with addition of a second half of bromine, KBrO$_3$ (100 g, 0.60 mol) was added in small portions (caution: add slowly, exothermic reaction) at 40°C–55°C with intense stirring, which promotes intensive precipitation of dibromofluorene. The mixture was stirred for 3–4 h, allowing it to cool gradually to room temperature. After cooling the mixture to 10°C, the solid was filtered off, washed with 70% AcOH (0.5 l) and water until pH 7, and dried, affording crude product as creme-colored solid (481 g, 85%) of >95% purity (by ^{1}H NMR). To further purify the product, it was stirred in AcOH (~1 l) at reflux (no full dissolution) for 4 h, cooled, filtered off, washed with AcOH, and dried, affording 455 g (80%) of 2,7-dibromofluorene **664**. ^{1}H NMR (400 MHz, CDCl$_3$): δ 7.66 (2H, d, J_{1-3} = 1.8 Hz, H-1,8), 7.59 (2H, d, J_{3-4} = 8.0 Hz, H-4,5), 7.50 (2H, dd, J_{3-4} = 8.0 Hz, J_{1-3} = 1.8 Hz, H-3,6), 3.89 (2H, s, CH$_2$). ^{13}C NMR (100 MHz, CDCl$_3$): δ 144.79, 139.69, 130.15, 128.31, 121.19, 120.94, 36.56.

2.7.2.1.2 2,7-Dibromo-9,9-Dihexylfluorene (665) (Chart 2.158)

From Perepichka, I.I., Perepichka, I.F., Bryce, M.R., and Pålsson, L.O., *Chem. Commun.*, 3397–3399, 2005.

Under argon, 3-l, three-neck flask was charged with 2,7-dibromofluorene **664** (130.0 g, 0.40 mol), 1-bromohexane (220 ml, 1.57 mol), and dry THF (1.0 l). After the full dissolution, the mixture was cooled to 0°C and a solution of potassium *tert*-butoxide (100.6 g, 0.90 mol) in dry THF (1.0 l) was added dropwise at 0°C to +5°C with vigorous stirring during 1.5 h. Upon adding the *tert*-butoxide solution, the reaction mixture become orange (generation of fluorene anion) and then the color vanished to light pink (at the end of *tert*-butoxide addition, no orange color is produced). The mixture was stirred at room temperature for 4 h,

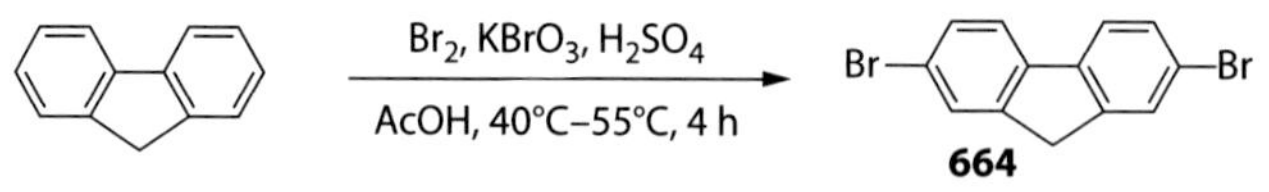

CHART 2.157
Synthesis of 2,7-dibromofluorene.

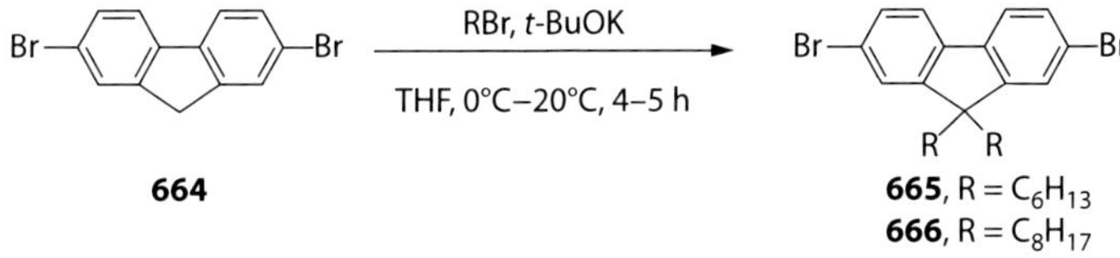

CHART 2.158
Synthesis of 2,7-dibromo-9,9-dihexylfluorene.

filtered off from the KBr precipitate, and the solid was washed on a filter with DCM. The filtrate was evaporated on a rotavapor, the residue was dissolved in DCM (1.5 l), washed with water, dried over MgSO$_4$, and the solvent was evaporated. Excess of 1-bromohexane was removed *in vacuo* (80°C, 1 mbar) yielding crude product (196.7 g, 99.6%) as yellow crystals. This was purified by column chromatography (7 × 17 cm column, silica gel, eluent—petroleum ether with a boiling point of 40°C–60°C) to afford 2,7-dibromo-9,9-dihexylfluorene **665** (179.5 g, 90.9%) as colorless plates. The material can also be additionally recrystallized from hexane or ethanol. ^{1}H NMR (500 MHz, CDCl$_3$): δ 7.51 (2H, d, *J* = 7.8 Hz, H-1,8), 7.45 (2H, dd, *J* = 1.8 Hz and 7.8 Hz, H-3,6), 7.44 (2H, d, *J* = 1.8 Hz, H-4,5), 1.96–1.87 (4H, m, *CH$_2$C$_5$H$_{11}$*), 1.16–1.08 (4H, m, CH$_2$CH$_2$CH$_2$C$_3$H$_7$), 1.08–0.98 (8H, m, [CH$_2$]$_3$*CH$_2$CH$_2$CH$_3$*), 0.78 (6H, t, *J* = 7.4 Hz, CH$_3$), 0.62–0.53 (4H, m, CH$_2$CH$_2$C$_4$H$_9$).

2,7-Dibromo-9,9-dioctylfluorene **666** was obtained as described in the above procedure, and after the column chromatography was additionally recrystallized from ethanol, yielding pure product in 78% yield. ^{1}H NMR (400 MHz, CDCl$_3$): δ 7.52 (2H, dd, *J$_{3-4}$* = 8.0 Hz, *J$_{1-4}$* = 0.8 Hz, H-4,5), 7.45 (2H, dd, *J$_{3-4}$* = 8.0 Hz, *J$_{1-3}$* = 1.6 Hz, H-3,6), 7.44 (2H, br s, H-1,8), 1.93–1.80 (4H, m, *CH$_2$C$_7$H$_{15}$*), 1.26–1.00 [20H, m, (CH$_2$)$_2$(CH$_2$)$_5$CH$_3$], 0.83 (6H, t, *J* = 7.2 Hz, CH$_3$), 0.63–0.53 (4H, m, CH$_2$CH$_2$C$_6$H$_{13}$).

^{13}C NMR (100 MHz, CDCl$_3$): δ 152.57, 139.07, 130.15, 126.18, 121.47, 121.13, 55.69, 40.15, 31.76, 29.86, 29.18, 29.15, 23.62, 22.60, 14.08.

2.7.2.1.3 9,9-Dihexylfluorene-2,7-Diboronic Acid (**667**) (Chart 2.159)

From Perepichka, I.I., Perepichka, I.F., Bryce, M.R., and Pålsson, L.O., *Chem. Commun.*, 3397–3399, 2005.

To a stirred solution of 2,7-dibromo-9,9-dihexylfluorene **665** (30.0 g, 60.9 mmol) in dry THF (1.0 l) under argon, a solution of BuLi in hexane (2.5 M; 54 ml, 135 mmol) was added dropwise at −78°C. The mixture was stirred at this temperature for 6 h to give a white suspension. Triisopropylborate (60 ml, 258 mmol) was added quickly and the mixture was stirred overnight, allowing the temperature to rise gradually to room temperature. Water (300 ml) was added, and the mixture was stirred at room temperature for 4 h. Organic solvents were removed on a rotavapor (35°C, 40 mbar); water (1.1 l) was added, and the mixture was acidified with concentrated HCl. The product was extracted into diethyl ether (7 × 300 ml), the organic layer was dried over MgSO$_4$, and the solvent was removed

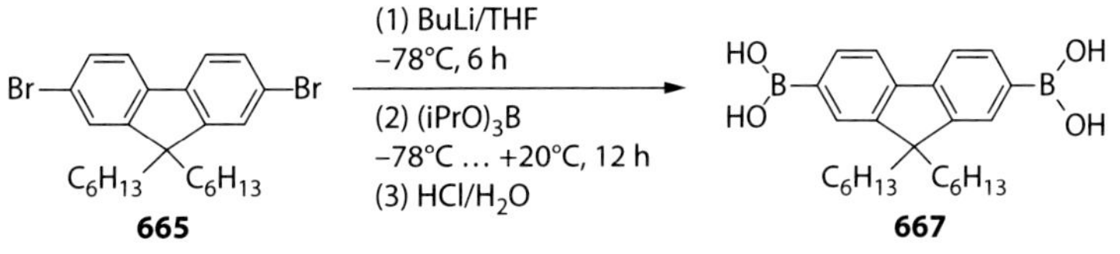

CHART 2.159
Synthesis of 9,9-dihexylfluorene-2,7-diboronic acid.

on the rotavapor. The residue was dissolved in acetone (110 ml) and reprecipitated into a mixture of water (130 ml) and concentrated HCl (70 ml), affording desired product **667** (24.3 g, 94.5%) as white powder. The product can be additionally purified by dissolution in acetone (100 ml) and addition of hexane (200 ml) to this solution. ^{1}H NMR (400 MHz, acetone-d_6): δ 7.99 (2H, dd, H-1,8), 7.90 (2H, dd, J_{3-4} = 7.6 Hz, J_{1-3} = 1.3 Hz, H-3,6), 7.80 (2H, dd, J_{3-4} = 7.6 Hz, J_{1-4} = 0.6 Hz, H-4,5), 7.19 (4H, s, OH), 2.12–2.00 (4H, m, $CH_2C_5H_{11}$), 1.2–0.9 [12H, m, $(CH_2)_2(CH_2)_3CH_3$], 0.74 (6H, t, J = 7.2 Hz, CH_3), 0.64–0.54 (4H, m, $CH_2CH_2C_4H_9$). ^{13}C NMR (100 MHz, acetone-d_6): δ 150.87, 144.09, 133.87, 129.39, 119.92, 55.50, 41.09, 32.27, 30.39, 24.57, 23.16, 14.21.

2.7.2.1.4 2,7-Bis(4,4,5,5-Tetramethyl-1,3,2-Dioxaborolan-2-yl)- 9,9-Dioctylfluorene (**668**) (Chart 2.160)

From Sonar, P., Zhang, J., Grimsdale, A.C., Müllen, K., Surin, M., Lazzaroni, R., Leclère, P., Tierney, S., Heeney, M., and McCulloch, I., *Macromolecules*, 37: 709–715, 2004. Copyright 2004, American Chemical Society, Washington, DC. With permission.

To a stirred solution of 2,7 dibromo-9,9-dioctylfluorene **666** (5.0 g, 9.1 mmol) in THF (70 ml) at –78°C was added dropwise *n*-butyllithium in hexanes (7.6 ml, 2.5 M, 19 mmol) at –78°C. The mixture was warmed to 0°C for 15 min and then cooled back to –78°C. 2-Isopropoxy-4,4,5,5-tetramethyl-1,3,2-dioxaborolane (4.0 g, 21.5 mmol) was added rapidly to the solution, and the resulting mixture was warmed to room temperature and stirred for 24 h. The mixture was then poured into water and extracted with diethyl ether. The organic extract was washed with brine and dried over magnesium sulfate. The solvent was removed under reduced pressure, and the crude product was purified by column chromatography eluting with 2% ethyl acetate and hexane to give 2,7-*bis*(4,4,5,5-tetramethyl-1,3,2-dioxaborolan-2-yl)-9,9-dioctylfluorene **668** as a pale yellow solid (3.80 g, 65%). HRMS: Calcd for $C_{41}H_{64}B_2O_4$: 642.59. Found: 642.37. ^{1}H NMR (250 MHz, CD$_2$Cl$_2$): δ 7.83 (d, 2H), 7.76 (s, 2H), 7.73 (d, 2H), 2.05 (m, 4H), 1.44 (s, 24H), 1.25–1.09 (m, 20H), 0.82 (t, 6H), 0.59 (m, 4H). ^{13}C NMR (62.5 MHz, CD$_2$Cl$_2$): δ 150.35, 143.82, 133.84, 128.74, 119.23, 83.51, 54.96, 39.36, 31.68, 29.89, 29.08, 29.00, 24.87, 23.50, 22.49, 13.96.

2.7.2.2 Suzuki-Coupling Polymerization (Chart 2.161)

From Ranger, M., Rondeau, D., and Leclerc, M., *Macromolecules*, 30: 7686, 1997. Copyright 1997, American Chemical Society, Washington, DC. With permission.

Carefully purified 2,7-dibromofluorene derivative (1 equivalent), 2,7-*bis*(4,4,5,5-tetramethyl-1,3,2-dioxaborolan-2-yl)-9,9-dioctylfluorene **668** (1 equivalent), and Pd(PPh$_3$)$_4$ (1.5–0.5 mol%) were dissolved in a mixture of toluene ([monomer] = 0.5 M) and aqueous 2 M Na$_2$CO$_3$ (or K$_2$CO$_3$) (1:1.5 toluene). The solution was first put under a nitrogen atmosphere and was refluxed with vigorous stirring for 48 h. The whole mixture was then poured into methanol (150 ml). The precipitated material was recovered by filtration through a

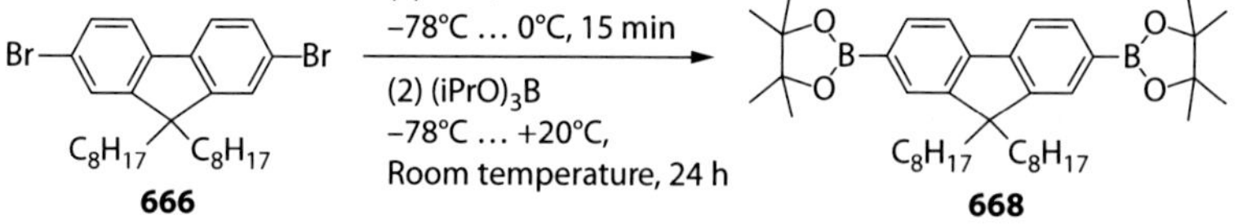

CHART 2.160
Synthesis of 2,7-bis(4,4,5,5-tetramethyl-1,3,2-dioxaborolan-2-yl)-9,9-dioctylfluorene.

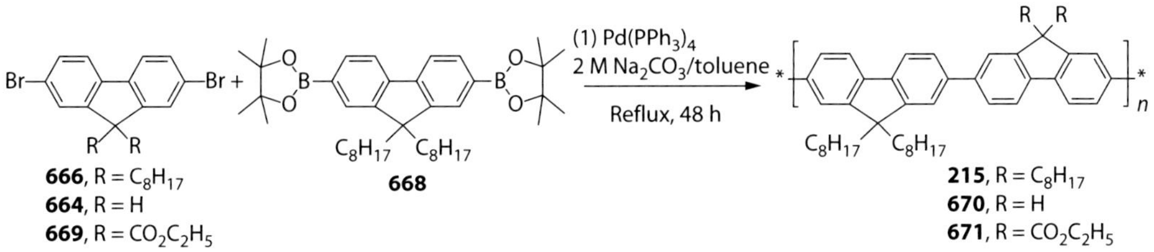

CHART 2.161
Suzuki-coupling polymerization.

Büchner funnel and washed with dilute HCl. The solid material was washed for 24 h in a Soxhlet apparatus using acetone to remove oligomers and catalyst residues. The resulting polymers were soluble in THF and CHCl$_3$. Yields: ~65%–90%.

Poly[2,7-(9,9-dioctylfluorene)] **215**. ^{1}H NMR (400 MHz, CDCl$_3$): δ (ppm) 7.85 (2H, d), 7.68 (4H, m), 2.15 (4H, m), 1.2 (24H, m), 0.8 (6H, t).

Poly[2,7'-(9,9-dioctyl-2',7'-bifluorene)] **670**. ^{1}H NMR (400 MHz, CDCl$_3$): δ (ppm) 7.1–8.0 (12H, m), 4.0 (2H, s), 2.1 (4H, m), 1.1 (24H, m), 0.8 (6H, t).

Poly-2,7'-(diethyl 9,9-dioctyl-2,7'-bifluorene-9',9'-dicarboxylate)] **671**. ^{1}HNMR (400 MHz, CDCl$_3$): δ (ppm) 8.2 (2H, s), 7.75 (5H, m), 7.65 (5H, m), 4.45–4.2 (4H, m), 2.1 (4H, m), 1.7 (1.5H, m), 1.4–1.0 (24H, m), 0.9–0.6 (6H, 2t).

*2.7.2.2.1 Fluorene Copolymer (**323**) (Chart 2.162)*

From Sonar, P., Zhang, J., Grimsdale, A.C., Müllen, K., Surin, M., Lazzaroni, R., Leclère, P., Tierney, S., Heeney, M., and McCulloch, I., *Macromolecules*, 37: 709, 2004. Copyright 2004, American Chemical Society, Washington, DC. With permission.

A solution of 2,6-dibromo-4-hexylbithieno [3,2-*b*:2'3'-*e*]pyridine **672** (0.300 g, 0.466 mmol), 9,9-dioctylfluorene-2,7-di(ethyleneboronate) **668** (0.202 g, 0.466 mmol), and Pd(PPh$_3$)$_4$ (0.020 g, 0.5 mol%) in a mixture of toluene (15 ml) and aqueous 2 M K$_2$CO$_3$ (10 ml) was refluxed with vigorous stirring for 72 h under N$_2$. The cooled mixture was poured into methanol (150 ml), and the precipitate was recovered by filtration and washed with dilute HCl. The precipitate was then extracted with acetone in a Soxhlet apparatus for 24 h. The dried residue was dissolved in dichloromethane and stirred with EDTA solution overnight. The residue of crude polymer was twice dissolved in chloroform and reprecipitated from methanol to give the copolymer **323** as a yellow solid (0.2 g, 67%). Elemental analysis: Found: C 79.09, H 3.15, N 2.11, S 5.23%. Calculated: C 79.58, H 8.65, N 2.11, S 9.66%. GPC: M_n = 10,577 g/mol, M_w = 31,297 g/mol, PDI = 2.95. ^{1}H NMR (250 MHz, CD$_2$Cl$_2$): δ 7.88 (s, 2H), 7.82 (m, 2H), 7.59 (m, 2H), 7.24 (d, 2H), 3.10 (t, 2H), 2.1 (m, 6H), 0.9 (m, 40H). ^{13}C NMR (62.5 MHz, CD$_2$Cl$_2$): δ 155.54, 152.60, 142.00, 141.86, 138.89, 128.40, 121.17, 120.99, 120.67, 40.58, 34.64, 32.13, 31.93, 30.29, 29.54, 25.11, 22.95, 14.18.

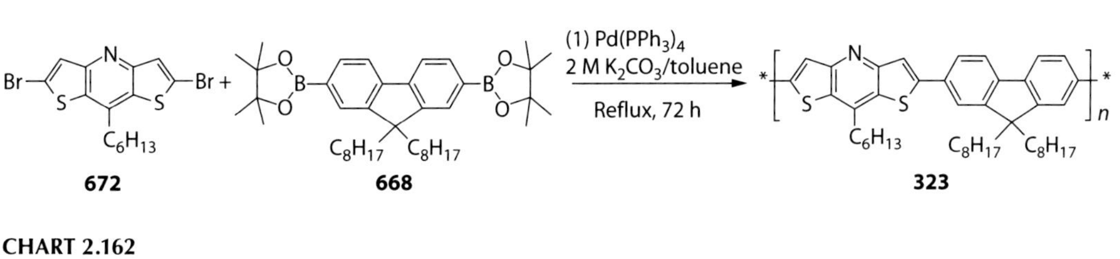

CHART 2.162
Synthesis of fluorene copolymer.

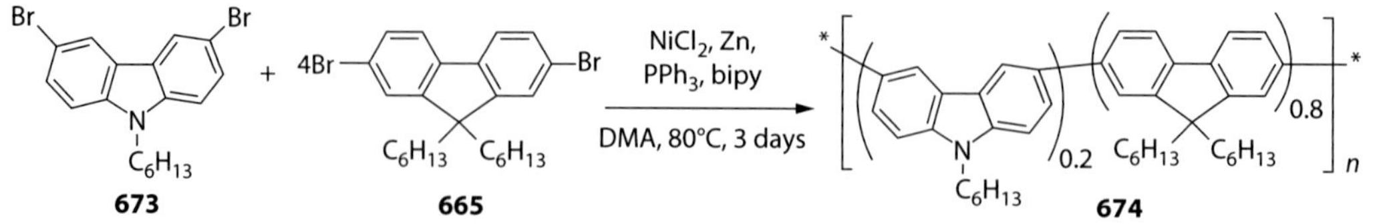

CHART 2.163
Synthesis of poly[(9,9-dihexylfluorene-2,7-diyl)-*co*-(*N*-hexylcarbazole-3,6-diyl)].

2.7.2.3 Yamamoto Polymerization

2.7.2.3.1 Poly[(9,9-Dihexylfluorene-2,7-Diyl)-co-(N-Hexylcarbazole-3,6-Diyl)] (674) (Chart 2.163)

From Stéphan, O. and Vial, J.-C., *Synth. Met.*, 106: 115, 1999. Copyright 1999, Elsevier, Amsterdam. With permission.

A mixture of 2,7-dibromo-9,9-dihexylfluorene **665** and 3,6-dibromo-*N*-hexylcarbazole **673** (molar ratio 4:1, total 2.5 mmol), triphenylphosphine (65.6 mg, 0.25 mmol), zinc powder 100 mesh, 99.998% (506 mg, 7.75 mmol), 2,2'-dipyridyl (19.5 mg, 0.125 mmol), and nickel chloride (16.2 mg, 0.125 mmol) were charged in a 20 ml flask. *N*,*N*-dimethylacetamide (3 ml) was added via syringe, and the mixture was stirred at 80°C for 3 days. After cooling, the polymer is precipitated by pouring the solution in a methanol-concentrated HCl mixture. The solid is collected by filtration and purified by subsequent precipitation in methanol–acetone mixture. The number-average molecular weight of the copolymer poly(DHF-*co*-NHK) **674**, evaluated by GPC calibrated with polystyrene standards, has been estimated at 4200 g/mol (polydispersity of about 2.5).

2.7.2.3.2 Poly[9,9-Bis(4-Diphenylaminophenyl)-2,7-Fluorene] (241) (Chart 2.164)

From Ego, C., Grimsdale, A.C., Uckert, F., Yu, G., Srdanov, G., and Müllen, K., *Adv. Mater.*, 14: 809, 2002. Copyright 2002, Wiley-VCH, Weinheim. With permission.

A solution of Ni(cod)$_2$ (132 mg, 0.48 mmol), 2,2'-bipyridine (75 mg, 0.48 mmol), and 1,5-cyclooctadienyl (52 mg, 0.48 mmol) in dry DMF (3 ml) was heated at 75°C for 30 min under an argon atmosphere. A solution of the monomer **675** (355 mg, 0.44 mmol) in dry toluene (5 ml) was added, and the mixture was heated at 75°C for a further 24 h and then poured into a methanol/HCl (2:1) mixture. The crude product was collected, dissolved in CHCl$_3$, and then precipitated from methanol/acetone (4:1). Residual impurities were removed by extraction with acetone in a Soxhlet apparatus to give polymer **241** (277 mg, 97%). GPC (THF) M_n: 12,060 g/mol, M_w: 25,240 g/mol, PDI 2.1 (polystyrene

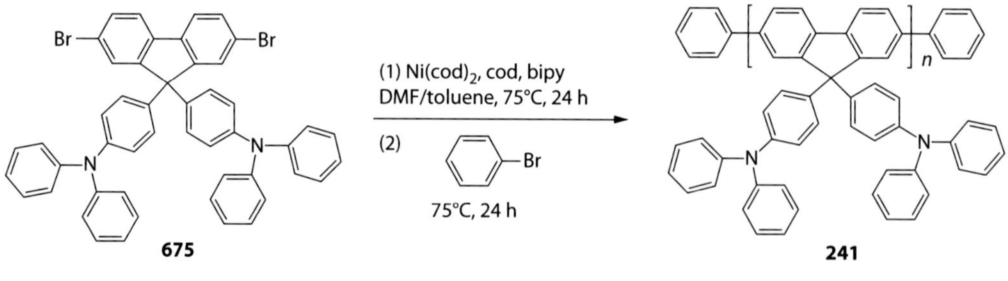

CHART 2.164
Synthesis of poly[9,9-bis(4-diphenylaminophenyl)-2,7-fluorene].

standards)/^{1}H-NMR (300 MHz, C$_2$D$_2$Cl$_2$): 6.60–6.72 (m, 28H, aryl H), 7.40–7.69 (b, 4H, fluorenyl H), 7.70–7.96 (b, 2H, fluorenyl H). ^{13}C-NMR (126 MHz, C$_2$D$_2$Cl$_2$): 64.6, 120.6, 123.1, 123.3, 124.7, 125.3, 129.1, 129.4, 139.7, 146.6, 146.9, 148.0, 152.8. λ_{max} (CHCl$_3$): 308, 384 nm. PL (CHCl$_3$) 441, 419 nm; (film) 428, 452 nm.

2.7.2.3.3 Poly(Dialkylfluorene-co-Anthracene)s (253a–c) (Chart 2.165)

From Klärner, G., Davey, M.H., Chen, E.-D., Scott, J.C., and Miller, R.D., *Adv. Mater.*, 13: 993–997, 1998. Copyright 1998, Wiley-VCH, Weinheim. With permission.

Into a Schlenk tube was placed *bis*(1,5-cyclooctadiene)-nickel(0) (2.6 mmol), 2,2′-bipyridyl (2.6 mmol), 1,5-cyclooctadiene (0.2 ml), DMF (4 ml), and toluene (8 ml). The reaction mixture was heated to 80°C for 0.5 h under argon. The dibromide comonomers **665** and **676** dissolved in degassed toluene (8 ml; molar ratio of dibromides to nickel complex: 0.65) were added under argon to the DMF–toluene solution, and the polymerization was maintained at 80°C for 3 days in the dark. 2-Bromofluorene (molar ratio of dibromides to monobromide: 0.1) dissolved in degassed toluene (1 ml) was added, and the reaction continued for 12 h. The polymers were precipitated by addition of the hot solution dropwise to an equivolume mixture of concentrated HCl, methanol, and acetone. The isolated polymers were then dissolved in toluene or dichloromethane and reprecipitated with methanol/acetone (1:1). The copolymers were dried at 80°C *in vacuo*. The isolated yields of copolymers **253a–c** were 79%–85%.

2.7.3 Syntheses of Polythiophenes

2.7.3.1 Polymerization of Thiophene Monomers with FeCl$_3$ (Chart 2.166)

From Pomerantz, M., Tseng, J.J., Zhu, H., Sproull, S.J., Reynolds, J.R., Uitz, R., Amott, H.G., and Haider, M.I., *Synth. Met.*, 41–43: 825, 1991. Copyright 1991, Elsevier, Amsterdam. With permission.

Into a 100-ml, three-neck flask equipped with a magnetic stirring bar, condenser, dropping funnel, and an inlet for dry air was added 3-alkylthiophene (7 mmol). A solution of anhydrous FeCl$_3$ (1.0 g) in chloroform (50 ml) was added to the alkylthiophene for about

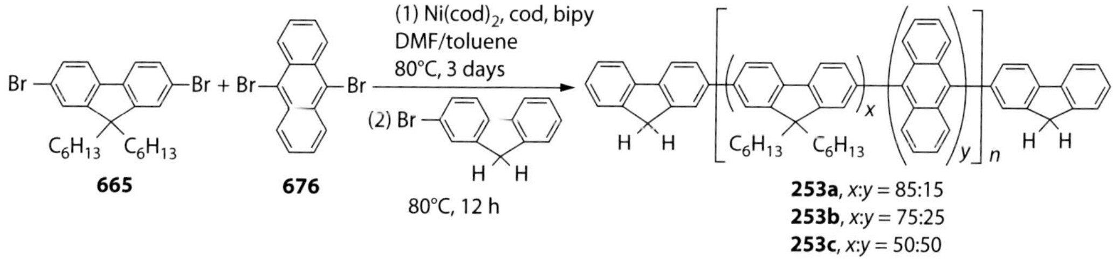

CHART 2.165
Synthesis of poly(dialkylfluorene-*co*-anthracene)s.

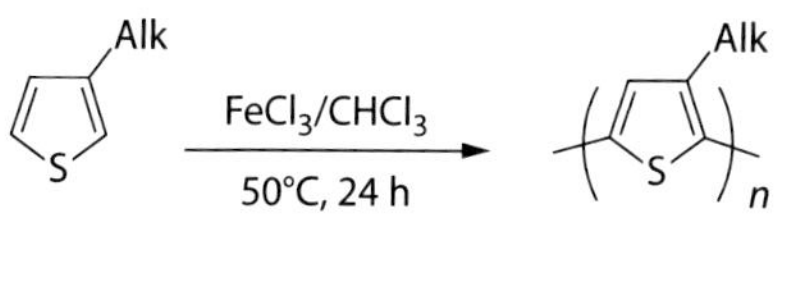

CHART 2.166
Polymerization of thiophene monomers with FeCl$_3$.

20 min, and the solution became dark. The mixture was then warmed to 50°C and stirred for 24 h at this temperature, while dry air was bubbled through to remove HCl from the reaction mixture. The black mixture was washed with water to remove excess $FeCl_3$ to give a dark-green–black mixture. This was stirred with concentrated aqueous ammonia (20 ml) and chloroform for 30 min at room temperature to produce a yellowish–red chloroform solution of dedoped polymer, which was washed several times with water and dried over $MgSO_4$. Removal of the solvent gave 60%–97% yield of the dark-purple poly(3-alkylthiophene). Soxhlet extraction with methanol was used to remove the low-molecular-weight fractions from the bulk polymer.

2.7.3.2 Yamamoto Polymerization (Chart 2.167)

From Yamamoto, T., Morita, A., Miyazaki, Y., Maruyama, T., Wakayama, H., Zhou, Z.H., Nakamura, Y., Kanbara, T., Sasaki, S., and Kubota, K., *Macromolecules*, 25: 1214, 1992. Copyright 1992, American Chemical Society, Washington, DC. With permission.

$Ni(cod)_2$ (1.59 g, 6.00 mmol), 1,5-cyclooctadiene (531 mg, 5.00 mmol), and 2,2'-bipyridyl (937 mg, 6.00 mmol) were dissolved in DMF (20 ml), in a Schlenk tube under argon. To the solution was added 2,5-dibromothiophene (1.21 g, 5.00 mmol) at room temperature. The reaction mixture was stirred at 60°C for 16 h to yield a reddish–brown precipitate. The reaction mixture was then poured into HCl–acidic methanol, and the precipitate of PT was separated by filtration. The precipitate was washed with HCl–acidic methanol, ethanol, hot toluene, a hot aqueous solution of EDTA (pH 3.80), a hot aqueous solution of EDTA (pH 9), and distilled water, in this order, and dried under vacuum to yield a reddish–brown powder of PT.

When 3(4)-substituted thiophenes (e.g., 3-hexylthiophene) are used as monomers, the polymers are partly or completely soluble in low polar organic solvents like toluene, chloroform, dichloromethane, and THF. Therefore, after washing with ethanol, the polymer is dried, then dissolved in chloroform and reprecipitated into methanol or acetone to remove low-molecular-weight fractions. The solid is collected by filtration, washed with methanol, and dried *in vacuo*.

2.7.3.3 McCullough Method of Preparation of Regioregular HT Poly(3-Alkylthiophenes) (Chart 2.168)

From McCullough, R.D., Lowe, R.D., Jayaraman M., and Anderson, D.L., *J. Org. Chem.*, 58: 904, 1993. Copyright 1993, American Chemical Society, Washington, DC. With permission.

2.7.3.3.1 Preparation of Monomers: 2-Bromo-3-n-Butylthiophene (**415**, R = n-C_4H_9)

Into a dry round-bottom flask was placed 68.5 ml (0.8 M) of acetic acid, which was then sparged with argon (5 min). Then 7.7 g (0.055 mol) of freshly distilled 3-*n*-butylthiophene was added. The mixture was cooled to 10°C, whereupon a 2.5 M solution of bromine (2.8 ml, 0.055 mol) in acetic acid was added dropwise from an addition funnel for a period of

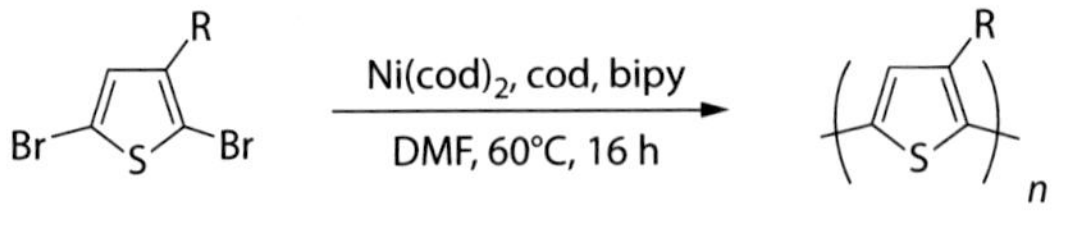

CHART 2.167
Yamamoto polymerization.

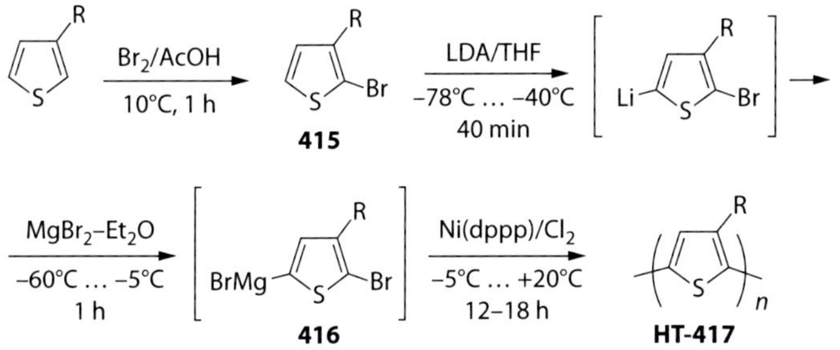

CHART 2.168
McCullough method of preparation of regioregular HT poly(3-alkylthiophenes).

30 min, while the temperature was maintained at 10°C–15°C. The material was then stirred in an ice bath for 30 min and was then poured onto ice. The mixture was then extracted into CHCl₃, the CHC1₃ layer washed with NaOH until pH 6 and dried over MgSO₄, and the solvent removed by rotary evaporation. The product was twice distilled (80°C/1.8 mm Hg) to yield 5.8 g (48%) of **386** (R = *n*-C₄H₉). In a similar manner, the monomers with R = *n*-C₆H₁₃ (49%), *n*-C₈H₁₇ (40%), and *n*-C₁₂H₂₅ (41%) have been obtained.

2.7.3.3.2 Polymerization of **415** (R = n-C₆H₁₃) Using 0.5 mol% Catalyst

Into a dry round-bottom flask was placed dry diisopropylamine (2.11 ml, 15 mmol) and freshly distilled, dry THF (75 ml, 0.2 M). To the mixture was added 6.0 ml of 2.5 M *n*-BuLi (15 mmol) at room temperature. The mixture was cooled to −40°C and stirred for 40 min. The reaction mixture, containing LDA, was then cooled to −78°C, and 2-bromo-3-hexylthiophene (**415**, R = *n*-C₆H₁₃) (3.7 g, 15 mmol) was added. The mixture was stirred for 40 min at −40°C. The mixture was then cooled to −60°C, MgBr₂·Et₂0 (3.87 g, 15 mmol) was added, and the reaction was stirred at −60°C for 20 min. The reaction was then warmed to −40°C and stirred for 15 min. The reaction was then allowed to slowly warm to −5°C, whereupon all of the MgBr₂·Et₂O had reacted. At −5°C, Ni(dppp)Cl₂ (39 mg, 0.072 mmol, 0.48 mol%) was added. The mixture was allowed to warm to room temperature overnight (≈12–18 h). The polymer was then precipitated with MeOH (300 ml), and the resulting red precipitate was then filtered and washed with MeOH, H₂O, and MeOH again. The solid was then dried under vacuum. Removal of oligomers and impurities was achieved by subjecting the solid to Soxhlet extractions with MeOH first, followed by hexanes. The polymer was then dissolved in CHCl₃ using a Soxhlet extractor, the CHCl₃ was removed, and the residue was dried under vacuum to yield 760 mg (36% yield) of 95% HT-coupled P3HT **HT-417** (R = *n*-C₆H₁₃). The most recent preparation (precipitation with hexane) resulted in 98% HT–HT-coupled polymer. GPC analysis (THF-soluble fraction): $M_w ≈ 10,000$ and PDI = 1.6.

2.7.3.3.3 Polymerization of **415** (R = n-C₈H₁₇) Using 2 × 0.5 mol% Catalyst

The exact procedure was performed as listed above except on a 18 mmol scale; however, after stirring for 15 h, 0.41 mol% of Ni(dppp)Cl₂ was added (40 mg, 0.074 mmol) at 25°C. The solution was then stirred an additional 18 h. The polymer was then precipitated with MeOH (400 ml) and allowed to remain for 2 days in MeOH, and the red precipitate was allowed to settle. The solution was decanted and the solid filtered and washed with MeOH, H₂O, and MeOH again. The solid was dried under vacuum and Soxhlet extracted with MeOH and hexanes. The polymer was then dissolved in CHCl₃ using a Soxhlet extractor, the CHCl$_3$ was removed, and the residue was dried and yielded 2.28 g (65% yield) of 96% HT poly(3-octylthiophene). The most recent preparation gave 97% HT–HT coupling. GPC analysis (THF-soluble fraction): M_w = 24,424 and PDI = 1.98.

2.7.3.4 Rieke Zinc (Zn*)-Mediated Polymerization

2.7.3.4.1 Typical Preparation of Rieke Zinc (Zn*)

From Chen, T.-A., Wu, X., and Rieke, R.D., *J. Am. Chem. Soc.*, 117: 233, 1995. Copyright 1995, American Chemical Society. Washington, DC. With permission.

Procedure A. Finely cut (ca. 5 × 5 × 0.75 mm) lithium (0.15 g, 22.0 mmol) and a catalytic amount (10 mol%) of naphthalene (0.28 g, 2.20 mmol) were weighed into a 100-ml, two-neck, round-bottom flask equipped with an elliptical Teflon stir bar in an argon dry box; the flask was sealed with a septum and a condenser topped with stopcock outlet. Similarly, ZnCl$_2$ (1.50 g, 11.0 mmol) was weighed into a 50 ml, two-neck, round-bottom flask, equipped with a stir bar; the flask was sealed with a septum and stopcock. The flasks were then transferred to the manifold system and the argon inlet fitted. THF (15 ml) was added to the flask with lithium and naphthalene, while ZnCl$_2$ was dissolved in THF (25 ml). To the flask with lithium and naphthalene, the THF solution of ZnCl$_2$ was transferred via cannula dropwise so as addition was completed in ca. 1.5 h under moderate stirring. The reaction mixture was further stirred until the lithium was consumed (ca. 30 min), and the resulting black suspension of active zinc thus prepared was ready for use.

Procedure B. Finely cut (0.15 g, 22.0 mmol) and a stoichiometric amount of naphthalene (2.80 g, 22.0 mmol) were weighed into a 100-ml flask, and ZnCl$_2$ (1.5 g, 11.0 mmol) was weighed into a 50-ml flask. The lithium and naphthalene were dissolved in THF (20 ml) in ca. 2 h. ZnCl$_2$ was dissolved in THF (20 ml), and the solution was transferred into the flask with lithium naphthalide via cannula for 10 min. The reaction mixture was further stirred for 1 h, and the resulting black suspension of active zinc thus prepared was ready for use.

2.7.3.4.2 Pd-Catalyzed Preparation of Regiorandom Poly(3-Alkylthiophenes) (HT-417)
    from 2,5-Dibromoalkylthiophenes (418) and Rieke Zinc (Zn*) (Chart 2.169)

From Chen, T.-A., Wu, X., and Rieke, R.D., *J. Am. Chem. Soc.*, 117: 233, 1995. Copyright 1995, American Chemical Society, Washington, DC. With permission.

2,5-Dibromo-3-alkylthiophene (alkyl = n-C$_6$H$_{13}$ or n-C$_8$H$_{17}$) (10.0 mmol) in THF (20 ml) was added to the flask with newly prepared Zn* (11.0 mmol in 40 ml of THF) via cannula at 0°C, and the mixture was stirred for 1 h at room temperature. A 0.2 mol% amount of Pd(PPh$_3$)$_4$ (23.1 mg, 0.02 mmol, in 20 ml of THF) was added via cannula. The mixture was then stirred for 24 h at room temperature (or reflux for 6 h). The polymer was precipitated with a solution of MeOH (100 ml) and 2 N HCl (50 ml), and purified by reprecipitation from

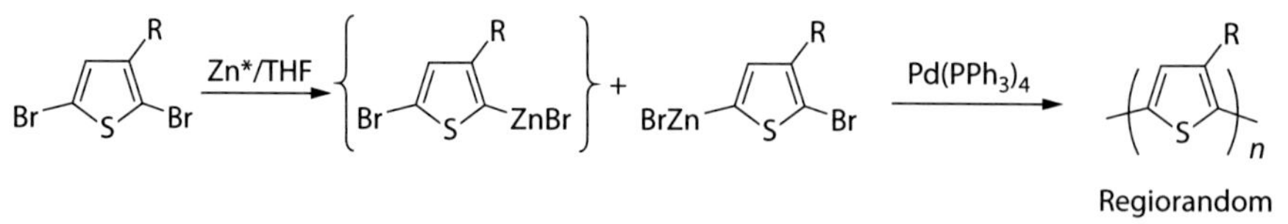

Regiorandom
417

CHART 2.169
Pd-catalyzed preparation of regiorandom poly(3-alkylthiophenes) (**HT-417**) from 2,5-dibromoalkylthiophenes (**418**) and Rieke zinc (Zn*).

the polymer solution of chloroform upon addition of MeOH. After drying under vacuum, red–brown, rubber-like polymers of regiorandom PATs were obtained in 97%–99% yields.

2.7.3.4.3 Ni-Catalyzed Preparation of Regioregular HT Poly(3-Alkylthiophenes) (HT-417) from 2,5-Dibromoalkylthiophenes (418) and Rieke Zinc (Zn*) (Chart 2.170)

From Chen, T.-A., Wu, X., and Rieke, R.D., J. Am. Chem. Soc., 117: 233, 1995. Copyright 1995, American Chemical Society, Washington, DC. With permission.

2,5-Dibromo-3-alkylthiophene (**418**, alkyl = n-C$_4$H$_9$, n-C$_6$H$_{13}$, n-C$_8$H$_{17}$, n-C$_{10}$H$_{21}$, n-C$_{12}$H$_{25}$, n-C$_{14}$H$_{29}$) (10.0 mmol, in 20 ml of THF) was added via a cannula to the newly prepared Zn* (11.0 mmol, in 40 ml of THF) at –78°C. The mixture was stirred for 1 h at this temperature and allowed to warm to 0°C naturally in ca. 3 h; 0.2 mol% of Ni(dppe)C1$_2$ (11.0 mg, 0.02 mmol, in 20 ml of THF) was added via cannula at 0°C. The mixture was stirred for 24 h at room temperature. A dark-purple precipitate was formed gradually in this period. The mixture was poured into a solution of MeOH (100 ml) and 2 N HCl (50 ml); the resulting dark precipitate was filtered and washed with MeOH and 2 N HCl solution, and then dried. Reprecipitation of polymer from chloroform solution upon addition of MeOH and drying under vacuum gave dark polymer of regioregular HT poly(3-alkylthiophenes). Purification of polymer by Soxhlet extractions with MeOH for 24 h and then with hexane for 24 h afforded regioregular **HT-417** in 67%–82% yields. The regioregularity of the polymers, according to NMR analysis, was from 97% to >98.5% of HT linkage.

2.7.3.5 Rieke Nickel-Catalyzed Polymerization

2.7.3.5.1 Poly(Alkyl Thiophene-3-Carboxylates) (429 and 430) (Chart 2.171)

From Pomerantz, M., Cheng, Y., Kasim, R.K., and Elsenbaumer, R.L., *J. Mater. Chem.*, 9: 2155–2163, 1999. Copyright 1999, The Royal Society of Chemistry, Cambridge. With permission.

A 25-ml one-necked flask was charged with NiI$_2$ (1.563 g, 4.994 mmol), freshly cut lithium (0.080 g, 11 mmol), naphthalene (0.064 g, 0.50 mmol), and THF (10 ml), and the mixture was stirred vigorously at room temperature for 12 h under argon. To the precipitated black-nickel powder, alkyl 2,5-dibromo-3-carboxylate (R = n-C$_6$H$_{13}$ or n-C$_8$H$_{17}$) (2.0 mmol) in THF (5 ml) was added directly via a syringe. The mixture was refluxed for 60 h under argon.

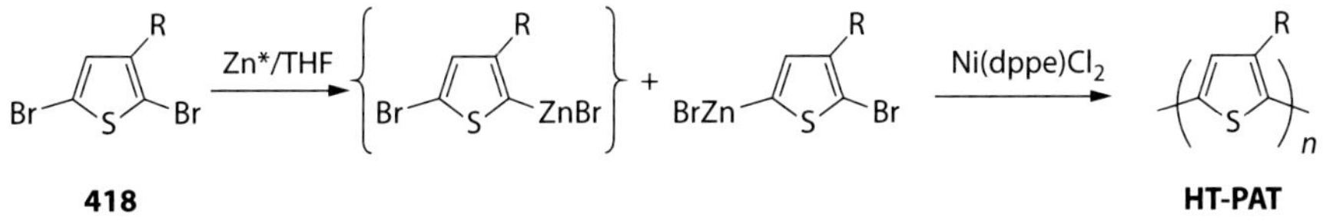

418　　　　　　　　　　　　　　　　　　　　　　　　　　　**HT-PAT**

CHART 2.170
Ni-Catalyzed preparation of regioregular HT poly(3-alkylthiophenes) (**HT-417**) from 2,5-dibromoalkylthiophenes (**418**) and Rieke zinc (Zn*).

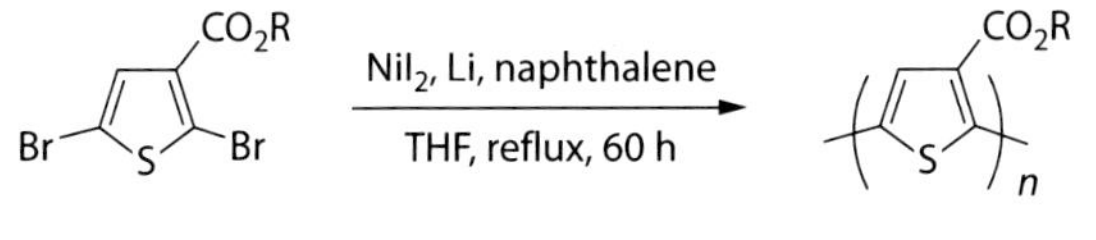

CHART 2.171
Synthesis of poly(alkyl thiophene-3-carboxylates).

The reaction mixture was diluted with diethyl ether to 100 ml, and then filtered to remove the metal powder. The dark-red organic phase was washed with water (3 × 50 ml) and dried over $MgSO_4$. The ether was removed with rotary evaporator, and a dark-red solid polymer was obtained. The polymer was extracted with methanol in a Soxhlet extractor for 48 h to remove the low-molecular-weight material, and, after drying under vacuum at room temperature, a red solid was obtained. The yields are 32%–33%.

2.7.4 Commercial Availability of Light-Emitting Polymers

To address new researchers entering the field of PLED, especially those not trained in organic synthesis, we should mention that many LEP are now available commercially from several companies, such as Aldrich (www.sigma-aldrich.com), American Dyes Source (www.adsdyes.com), H.W. Sands (http://www.hwsands.com/), and Dow Chemicals (http://www.dow.com/pled/).

In Aldrich, the widest collection belongs to PPV derivatives, which includes dialkoxy-PPVs (e.g., MEH–PPV **13, 14**), phenyl-substituted PPV **45**, *meta–para*-linked PPV copolymer **177**, and others. Many monomer precursors for PPV synthesis by Gilch and Wittig procedure as well as the Wessling–Zimmerman precursor **2** for unsubstituted PPV are also available.

Acknowledgments

Dr. Shidi Xun has written the updates for this chapter. We thank Dr. Weishi Wu at DuPont Display and Dr. Gary A. Johansson at DuPont CRD for proofreading of the manuscript. We thank Dr. Jeremy Burroughes at Sumitomo Chemical Co., Ltd., for providing the valuable information of current PLED performance (Table 2.10). DFP thanks the National Science and Engineering Research Council for support through the Discovery and AGENO grants.

References

1. J.H. Burroughes, D.D.C. Bradley, A.R. Brown, R.N. Marks, K. Mackay, R.H. Friend, P.L. Burn, and A.B. Holmes, Light-emitting diodes based on conjugated polymers, *Nature*, 347: 539–541, 1990.
2. R.H. Partridge, Radiation sources, U.S. Patent 3,995,299, November 30, 1976.
3. R.H. Partridge, Electroluminescence from polyvinylcarbazole films (parts 1–4), *Polymer*, 24: 733–762, 1983.
4. R.J. Visser, Application of polymer light-emitting materials in light-emitting diodes, backlights and displays, *Philips J. Res.*, 51: 467–477, 1998.
5. R.H. Friend, D.D.C. Bradley, and P.D. Townsend, Photo-excitation in conjugated polymers, *J. Phys. D: Appl. Phys.*, 20: 1367–1384, 1987.
6. D.R. Baigent, N.C. Greenham, J. Gruener, R.N. Marks, R.H. Friend, S.C. Moratti, and A.B. Holmes, Light-emitting diodes fabricated with conjugated polymers—Recent progress, *Synth. Met.*, 67: 3–10, 1994.
7. A. Greiner, B. Bolle, P. Hesemann, J.M. Oberski, and R. Sander, Preparation and structure–property relationships of polymeric materials containing arylenevinylene segments—Perspectives for new light-emitting materials, *Macromol. Chem. Phys.*, 197: 113–134, 1996.

8. J.L. Segura, The chemistry of electroluminescent organic materials, *Acta Polym.*, 49: 319–344, 1998.
9. R.H. Friend, R.W. Gymer, A.B. Holmes, J.H. Burroughes, R.N. Marks, C. Taliani, D.D.C. Bradley, D.A. Dos Santos, J.L. Brédas, M. Lögdlung, and W.R. Salaneck, Electroluminescence in conjugated polymers, *Nature*, 397: 121–127, 1999.
10. U. Mitschke and P. Bäuerle, The electroluminescence of organic materials, *J. Mater. Chem.*, 10: 1471–1507, 2000.
11. R.E. Martin, F. Geneste, and A.B. Holmes, Synthesis of conjugated polymers for application in light-emitting diodes (PLEDs), *C.R. Acad. Sci. Paris, t. 1*, Série IV: 447–470, 2000.
12. A. Greiner, Design and synthesis of polymers for light-emitting diodes, *Polym. Adv. Technol.*, 9: 371–389, 1998.
13. L. Dai, B. Winkler, L. Dong, L. Tong, and A.W.H. Mau, Conjugated polymers for light-emitting applications, *Adv. Mater.*, 13: 915–925, 2001.
14. I.D. Rees, K.L. Robinson, A.B. Holmes, C.R. Towns, and R. O'Deil, Recent developments in light-emitting polymers, *MRS Bull.*, 27: 451–455, 2002.
15. F. Babudri, G.M. Farinola, and F. Naso, Synthesis of conjugated oligomers and polymers: The organometallic way, *J. Mater. Chem.*, 14: 11–34, 2004.
16. L. Akcelrud, Electroluminescent polymers, *Prog. Polym. Sci.*, 28: 875–962, 2003.
17. N. Tessler, G.J. Denton, and R.H. Friend, Lasing from conjugated-polymer microcavities, *Nature*, 382: 695–697, 1996.
18. F. Hide, M. Díaz-García, M.R. Andersson, Q. Pei, and A.J. Heeger, Semiconducting polymers: A new class of solid-state laser materials, *Science*, 273: 1833–1836, 1996.
19. G. Yu, J. Gao, J.C. Hummelen, F. Wudl, and A.J. Heeger, Polymer photovoltaic cells: Enhanced efficiencies via a network of internal donor–acceptor heterojunctions, *Science*, 270: 1789–1791, 1995.
20. A. Kraft, A.C. Grimsdale, and A.W. Holmes, Electroluminescent conjugated polymers—Seeing polymers in a new light, *Angew. Chem. Int. Ed.*, 37: 402–428, 1998.
21. R.A. Wessling and R.G. Zimmerman, Polyelectrolytes from bis sulfonium salts, U.S. Patent 3,401,152, 1968.
22. A. Beerden, D. Vanderzande, and J. Gelan, The effect of anions on the solution behaviour of poly(xylylene tetrahydrothiophenium chloride) and on the elimination to poly(*p*-phenylene vinylene), *Synth. Met.*, 52: 387–394, 1992.
23. R.O. Garay, U. Baier, C. Bubeck, and K. Müllen, Low-temperature synthesis of poly(*p*-phenylene vinylene) by the sulfonium salt route, *Adv. Mater.*, 5: 561–564, 1993.
24. P.L. Burn, A.B. Holmes, A. Kraft, D.D.C. Bradley, A.R. Brown, R.H. Friend, and R.W. Cymer, Chemical tuning of electroluminescent copolymers to improve emission efficiency and allow patterning, *Nature*, 356: 47–49, 1992.
25. D.A. Halliday, P.L. Burn, D.D.C. Bradley, R.H. Friend, O.M. Gelsen, A.B. Holmes, A. Kraft, J.H.F. Martens, and K. Pichler, Large changes in optical response through chemical pre-ordering of poly(*p*-phenylene vinylene), *Adv. Mater.*, 5: 40–43, 1993.
26. S. Son, A. Dodabalapur, A.J. Lovinger, and M.E. Galvin, Luminescence enhancement by the introduction of disorder into poly(*p*-phenylene vinylene), *Science*, 269: 376–378, 1995.
27. W.J. Mitchell, C. Pena, and P.L. Burn, Thermal routes to low HOMO–LUMO energy gap poly(arylene vinylene)s, *J. Mater. Chem.*, 12: 200–205, 2002.
28. A. Marletta, F.A. Castro, C.A.M. Borges, O.N. Oliveira, Jr., R.M. Faria, and F.E.G. Guimarães, Enhanced optical properties of layer-by-layer films of poly(*p*-phenylene vinylene) alternated with a long chain counterion and converted at low temperatures, *Macromolecules*, 35: 9105–9109, 2002.
29. S. Iwatsuki, M. Kubo, and T. Kumeuchi, New method for preparation of poly(phenylene vinylene) film, *Chem. Lett.*, 2: 1971–1974, 1991.
30. E.G.J. Staring, D. Braun, G.L.J.A. Rikken, R.J.C.E. Demandt, Y.A.R.R. Kessener, M. Bauwmans, and D. Broer, Chemical vapour deposition of poly(1,4-phenylene vinylene) films, *Synth. Met.*, 67: 71–75, 1994.

31. K.M. Vaeth and K.F. Jenson, Selective growth of poly(*p*-phenylene vinylene) prepared by chemical vapor deposition, *Adv. Mater.*, 11: 814–820, 1999.
32. K.M. Vaeth and K.F. Jenson, Chemical vapor deposition of poly(*p*-phenylene vinylene) based light emitting diodes with low turn on voltage, *Appl. Phys. Lett.*, 71: 2091–2093, 1997.
33. Y.-J. Miao and G.C. Bazan, Paracyclophene route to poly(*p*-phenylene vinylene), *J. Am. Chem. Soc.*, 116: 9379–9380, 1994.
34. V.P. Conticello, D.L. Gin, and R.H. Grubbs, Ring-opening metathesis polymerization of substituted bicyclo[2.2.2]octadienes: A new precursor route to poly(*p*-phenylene vinylene), *J. Am. Chem. Soc.*, 114: 9708–9710, 1992.
35. H.G. Gilch and W.L. Wheelwright, Polymerization of α-halogenated *p*-xylenes with base, *J. Polym. Sci. Part A*, 4: 1337–1349, 1966.
36. J. Wiesecke and M. Rehahn, [2,2]Paracyclophanes with defined substitution pattern—Key compounds for the mechanistic understanding of the Gilch reaction to poly(*p*-phenylene vinylene), *Angew. Chem. Int. Ed.*, 42: 567–570, 2003.
37. B.R. Hsieh, Y. Yu, A.C. van Laeken, and H. Lee, General methodology toward soluble poly(*p*-phenylene vinylene) derivatives, *Macromolecules*, 30: 8094–8095, 1997.
38. C.J. Neef and J.P. Ferraris, MEH-PPV: Improved synthetic procedure and molecular weight control, *Macromolecules*, 33: 2311–2314, 2000.
39. J.P. Ferraris and C.J. Neef, Methods for the synthesis and polymerization of α,α′-dihalo-*p*-xylenes, U.S. Patent 6,426,399 B1, July 30, 2002.
40. A. Issaris, D. Vanderzande, and J. Gelan, Polymerization of a *p*-quinodimethane derivative to a precursor of poly(*p*-phenylene vinylene)—Indications for a free radical mechanism, *Polymer*, 38: 2571–2574, 1997.
41. H. Becker, H. Spreitzer, K. Ibrom, and W. Kreuder, New insights into the microstructure of Gilch-polymerized PPVs, *Macromolecules*, 32: 4925–4932, 1999.
42. W.J. Swatos and B. Gordon, III, Polymerization of 2,2-di-*n*-hexyloxy-α,α′-dichloro-*p*-xylene with potassium *tert*-butoxide: A novel route to poly(2,5-di-*n*-hexyloxy-*p*-phenylene vinylene), *Polym. Prepr.*, 31(1): 505–506, 1990.
43. D.M. Johansson, M. Theander, G. Srdanov, G. Yu, O. Inganäs, and M.R. Andersson, Influence of polymerization temperature on molecular weight, photoluminescence, and electroluminescence for a phenyl-substituted poly(*p*-phenylene vinylenes), *Macromolecules*, 34: 3716–3719, 2001.
44. B.R. Hsieh and W.R. Feld, A dehydrochlorination (DHCL) route to poly(2,3-diphenyl-1,4-phenylene vinylene) (DP-PPV), *Polym. Prepr.*, 34(2): 410–411, 1993.
45. L. Lutsen, P. Adriaensens, H. Becker, A.J. Van Breemen, D. Vanderzande, and J. Gelan, New synthesis of a soluble high molecular weight poly(arylene vinylene): Poly[2-methoxy-5-(3,7-dimethyloctyloxy)-*p*-phenylene vinylene]. Polymerization and device properties, *Macromolecules*, 32: 6517–6525, 1999.
46. R.W. Lenz and C.E. Handlovitis, Thermally stable hydrocarbon polymers: Polyterephthalylidenes, *J. Org. Chem.*, 25: 813–817, 1960.
47. S.C. Moratti, R. Cervini, A.B. Holmes, D.R. Baigent, R.H. Friend, N.C. Greenham, J. Grüner, and P.J. Hamer, High electron affinity polymers for LEDs, *Synth. Met.*, 71: 2117–2120, 1995.
48. J. Liao and Q. Wang, Ruthenium-catalyzed Knoevenagel condensation: A new route toward cyano-substituted poly(*p*-phenylene vinylene)s, *Macromolecules*, 37: 7061–7063, 2004.
49. A. Hilberer, H.-J. Brouwer, B.-J. van der Scheer, J. Wildeman, and G. Hadziioannou, Synthesis and characterization of a new efficient blue-light-emitting copolymer, *Macromolecules*, 28: 4525–4529, 1995.
50. A. Greiner and W. Heitz, New synthetic approach to poly(1,4-phenylene vinylene) and its derivatives by palladium catalyzed arylation of ethylene, *Macromol. Chem. Rapid Commun.*, 9: 581–588, 1988; and references therein.
51. R.N. McDonald and T.W. Campbell, The Wittig reaction as a polymerization method, *J. Am. Chem. Soc.*, 82: 4669–4671, 1960.
52. B. Francois, S. Izzillo, and P. Iratçabal, Substituted PPV block copolymer from anionically prepared precursor, *Synth. Met.*, 102: 1211–1212, 1999.

53. M. Aguiar, M.C. Fugihara, I.A. Hümmelgen, L.O. Péres, J.R. Garcia, J. Gruber, and L. Akcelrud, Interchain luminescence in poly(acetoxy-*p*-phenylene vinylene), *J. Lumin.*, 96: 219–225, 2002.

54. D.D.C. Bradley, Conjugated polymer electroluminescence, *Synth. Met.*, 54: 401–415, 1993.

55. M. Strukelj, F. Papadimitrakopoulos, T.M. Miller, and L.J. Rothberg, Design and application of electron-transporting organic materials, *Science*, 267: 1969–1972, 1995.

56. N. Tessler, N.T. Harrison, and R.H. Friend, High peak brightness polymer light-emitting diodes, *Adv. Mater.*, 10: 64–68, 1998.

57. M. Alvaro, A. Corma, B. Ferrer, M.S. Galletero, H. García, and E. Peris, Increasing the stability of electroluminescent phenylene vinylene polymers by encapsulation in nanoporous inorganic materials, *Chem. Mater.*, 16: 2142–2147, 2004.

58. P. Kumar, A. Mehta, S.M. Mahurin, S. Dai, M.D. Dadmun, B.G. Sumpter, and M.D. Barnes, Formation of oriented nanostructures from single molecules of conjugated polymers in microdroplets of solution: The role of solvent, *Macromolecules*, 37: 6132–6140, 2004.

59. K.-Y. Jen, L.W. Shackletter, and R. Elsenbaumer, Synthesis and conductivity of poly(2,5-dimethoxy-1,4-phenylene vinylene), *Synth. Met.*, 22: 179–183, 1987; and references therein.

60. I. Murase, T. Ohnishi, T. Noguchi, and M. Hirooka, Highly conducting poly(phenylene vinylene) derivatives via soluble precursor process, *Synth. Met.*, 17: 639–644, 1987.

61. S.H. Askari, S.D. Rughooputh, and F. Wudl, Soluble substituted-polyphenylene vinylene (PPV) conducting polymers: Spectroscopic studies, *Synth. Met.*, 29: E129–E134, 1989.

62. F. Wudl and G. Srdanov, Conducting Polymer Formed of Poly(2-methoxy-5-(2′-ethylhexyloxy)-*p*-phenylene vinylene), U.S. Patent 5,189,136, February 23, 1993.

63. D.W.J. McCallien, A.C. Thomas, and P.L. Burn, A study of the molecular weight of the chloro-precursor polymer to MEHPPV, *J. Mater. Chem.*, 9: 847–849, 1999.

64. P.L. Burn, A.W. Grice, A. Tajbakhsh, D.D.C. Bradley, and A.C. Thomas, Insoluble poly [2-(2′-ethylhexyloxy)-5-methoxy-1,4-phenylene vinylene] for use in multiplayer light-emitting diodes, *Adv. Mater.*, 9: 1171–1176, 1997.

65. D. Braun and A.J. Heeger, Visible light emission from semiconducting polymer diodes, *Appl. Phys. Lett.*, 58: 1982–1984, 1991.

66. Q. Pei, G. Yu, C. Zhang, Y. Yang, and A.J. Heeger, Polymer light-emitting electrochemical cells, *Science*, 269: 1086–1088, 1995.

67. Q. Pei, Y. Yang, C. Zhang, and A.J. Heeger, Polymer light-emitting electrochemical cells: *In situ* formation of a light-emitting p–n junction, *J. Am. Chem. Soc.*, 118: 3922–3929, 1996.

68. S.A. Carter, M. Angelopoulos, S. Karg, P.J. Brock, and J.C. Scott, Polymeric anodes for improved polymer light-emitting diode performance, *Appl. Phys. Lett.*, 70: 2067–2069, 1997.

69. Y. Cao, G. Yu, and A.J. Heeger, Efficient, low operating voltage polymer light-emitting diodes with aluminum as the cathode material, *Adv. Mater.*, 10: 917–920, 1998.

70. V. Blyznyuk, B. Ruhstaller, P.J. Brock, U. Scherf, and S.A. Carter, Self-assembled nanocomposite polymer light-emitting diodes with improved efficiency and luminance, *Adv. Mater.*, 11: 1257–1261, 1999.

71. Y. Cao, I.D. Parker, G. Yu, C. Zhang, and A.J. Heeger, Improved quantum efficiency for electroluminescence in semiconducting polymers, *Nature*, 397: 414–417, 1999.

72. T.-W. Lee, O.O. Park, J.-J. Kim, J.-M. Hong, and Y.C. Kim, Efficient photoluminescence and electroluminescence from environmentally stable polymer/clay nanocomposites, *Chem. Mater.*, 13: 2217–2222, 2001.

73. B.J. Matterson, J.M. Lupton, A.F. Safonov, M.G. Salt, W.L. Barnes, and I.D.W. Samuel, Increased efficiency and controlled light output from a microstructured light-emitting diode, *Adv. Mater.*, 13: 123–127, 2001.

74. M.M. Alam and S.A. Jenekhe, Polybenzobisazoles as efficient electron-transport materials for improving the performance and stability of polymer light-emitting diodes, *Chem. Mater.*, 14: 4775–4780, 2002.

75. Y-H. Niu, J. Huang, and Y. Cao, High-efficiency polymer light-emitting diodes with stable saturated red emission: Use of carbazole-based copolymer blends in a poly(*p*-phenylene vinylene) derivative, *Adv. Mater.*, 15: 807–811, 2003.

76. X.Y. Deng, W.M. Lau, K.Y. Wong, K.H. Low, H.F. Chow, and Y. Cao, High efficiency low operating voltage polymer light-emitting diodes with aluminum cathode, *Appl. Phys. Lett.*, 84: 3522–3524, 2004.

77. H. Spreitzer, H. Becker, E. Kluge, W. Kreuder, H. Schenk, R. Demandt, and H. Schoo, Soluble phenyl-substituted PPVs—New materials for highly efficient polymer LEDs, *Adv. Mater.*, 10: 1340–1343, 1998.

78. H. Frohne, D.C. Müller, and K. Meerholz, Continuously variable hole injection in organic light emitting diodes, *Chemphyschem*, 707–711, 2002.

79. I.D. Parker, Y. Cao, and C.Y. Yang, Lifetime and degradation effects in polymer light-emitting diodes, *J. Appl. Phys.*, 85: 2441–2447, 1999.

80. E. Peeters, M.P.T. Christiaans, R.A.J. Janssen, H.F.M. Schoo, H.P.J.M. Dekkers, and E.W. Meijer, Circularly polarized electroluminescence from a polymer light-emitting diode, *J. Am. Chem. Soc.*, 119: 9909–9910, 1997.

81. B.S. Chuah, F. Cacialli, D.A. dos Santos, N. Feeder, J.E. Davies, S.C. Moratti, A.B. Holmes, R.H. Friend, and J.L. Brédas, A highly luminescent polymer for LEDs, *Synth. Met.*, 102: 935–936, 1999.

82. R.E. Martin, F. Geneste, B.S. Chuah, C. Fischmeister, Y. Ma, A.B. Holmes, R. Riehn, F. Cacialli, and R.H. Friend, Versatile synthesis of various conjugated aromatic homo- and copolymers, *Synth. Met.*, 122: 1–5, 2001.

83. S.-C. Lo, A.K. Sheridan, I.D.W. Samuel, and P.L. Burn, Comparison of the electronic properties of poly[2-(2'-ethylhexyloxy)-1,4-phenylene vinylene] prepared by different precursor routes, *J. Mater. Chem.*, 9: 2165–2170, 1999.

84. F. Wudl and H.M. Peters, Highly soluble, conductive, luminescent polyphenylene vinylenes, and products and uses thereof, PCT Patent Appl. WO 94/20589, September 15, 1994.

85. P.L. Burn, D.D.C. Bradley, R.H. Friend, D.A. Halliday, A.B. Holmes, R.W. Jackson, and A. Kraft, Precursor route chemistry and electronic properties of poly(p-phenylene vinylene), poly[(2,5-dimethyl-p-phenylene)vinylene] and poly[(2,5-dimethoxy-p-phenylene)vinylene], *J. Chem. Soc. Perkin Trans.*, 1: 3225–3231, 1992.

86. G.J. Sarnecki, P.L. Burn, A. Kraft, R.H. Friend, and A.B. Holmes, The synthesis and characterization of some poly(2,5-dialkoxy-1,4-phenylene vinylenes), *Synth. Met.*, 55: 914–917, 1993.

87. S. Wang, J. Yang, Y. Li, H. Lin, Z. Guo, S. Xiao, Z. Shi, D. Zhu, H.-S. Woo, D.L. Carroll, I.-S. Kee, and J.-H. Lee, Composites of C60 based poly(phenylene vinylene) dyad and conjugated polymer for polymer light-emitting devices, *Appl. Phys. Lett.*, 80: 3847–3849, 2002.

88. J. Vandenbergh, I.V. Severen, L. Lutsen, P. Adriaensens, H.J. Bolink, T.J. Cleij, and D. Vanderzande, Tetra-alkoxy substituted PPV derivatives: A new class of highly soluble liquid crystalline conjugated polymers, *Polym. Chem.*, 2: 1279–1286, 2011.

89. F. Wudl, S. Höger, C. Zhang, P. Pakbaz, and A.J. Heeger, Conjugated polymers for organic LEDs: Poly[2,5-*bis*(3a,5b-cholestanoxy)phenylene vinylene] (BCHA-PPV); a processible, yellow light emitter, *Polym. Prepr.*, 34(1): 197–198, 1993.

90. C.L. Gettinger, A.J. Heeger, J.H. Drake, and D.J. Pine, A photoluminescence study of poly (phenylene vinylene) derivatives: The effect of intrinsic persistence length, *J. Chem. Phys.*, 101: 1673–1678, 1994.

91. F. Wudl and S. Hoger, Highly organic solvent soluble, water insoluble electroluminescent polyphenylene vinylenes having pendant steroid group and products and uses thereof, U.S. Patent 5,679,757, October 21, 1997.

92. M.R. Andersson, G. Yu, and A.J. Heeger, Photoluminescence and electroluminescence of films from soluble PPV-polymers, *Synth. Met.*, 85: 1275–1276, 1997.

93. R.O. Garay, B. Mayer, F.E. Karasz, and R.W. Lenz, Synthesis and characterization of poly[2,5-*bis*(triethoxy)-1,4-phenylene vinylene], *J. Polym. Sci. A*, 33: 525–531, 1995.

94. B.S. Chuah, D.-H. Hwang, S.T. Kim, S.C. Moratti, A.B. Holmes, J.C. De Mello, and R.H. Friend, New luminescent polymers for LEDs, *Synth. Met.*, 91: 279–282, 1997.

95. M. Fahlman, M. Lögdlund, S. Stafström, W.R. Salaneck, R.H. Friend, P.L. Burn, A.B. Holmes, K. Kaeriyama, Y. Sonnoda, O. Lhost, F. Meyers, and J.L. Brédas, Experimental and theoretical

studies of the electronic structure of poly(*p*-phenylene vinylene) and some ring-substituted derivatives, *Macromolecules*, 28: 1959–1965, 1995.

96. L.M. Leung and G.L. Chik, Phase-transfer catalysed synthesis of disubstituted poly(phenylene vinylene), *Polymer*, 34: 5174–5179, 1993.

97. Y. Sonoda, Y. Nakao, and K. Kaeriyama, Preparation and properties of poly(1,4-phenylene vinylene) derivatives, *Synth. Met.*, 55–57: 918–923, 1993.

98. A.G.J. Staring and R.J.C.E. Demandt, Light-emitting diode comprising an active layer of 2,5-substituted poly(*p*-phenylene vinylene), PCT Patent Appl. WO 32526, November 30, 1995.

99. B.J. Schwartz, F. Hide, M.R. Andersson, and A.J. Heeger, Ultrafast studies of stimulated emission and gain in solid films of conjugated polymers, *Chem. Phys. Lett.*, 265: 327–333, 1997.

100. S. Höger, J. McNamara, S. Schricker, and F. Wudl, Novel silicon-substituted, soluble poly(phenylene vinylene)s: Enlargement of the semiconductor bandgap, *Chem. Mater.*, 6: 171–173, 1994.

101. C. Zhang, S. Hoeger, K. Pakbaz, F. Wudl, and A.J. Heeger, Improved efficiency in green polymer light-emitting diodes with air-stable electrodes, *J. Electron. Mater.*, 23: 453–458, 1994.

102. S.T. Kim, D.-H. Hwang, X.C. Li, J. Grüner, R.H. Friend, A.B. Holmes, and H.K. Shim, Efficient green electroluminescent diodes based on poly(2-dimethyloctylsilyl-1,4-phenylene vinylene), *Adv. Mater.*, 8: 979–982, 1996.

103. D.-H. Hwang, S.T. Kim, H.-K. Shim, A.B. Holmes, S.C. Moratti, and R.H. Friend, Green light-emitting diodes from poly(2-dimethyloctylsilyl-1,4-phenylene vinylene), *Chem. Commun.* 19: 2241–2242, 1996.

104. H.Y. Chu, D. Hwang, L. Do, J. Chang, H. Shim, A.B. Holmes, and T. Zyung, Electroluminescence from silyl-disubstituted PPV derivative, *Synth. Met.*, 101: 216–217, 1999.

105. Z. Chen, W. Huang, L. Wang, E. Kang, B.J. Chen, C.S. Lee, and S.T. Lee, A family of electroluminescent silyl-substituted poly(*p*-phenylene vinylene)s: Synthesis, characterization, and structure–property relationship, *Macromolecules*, 33: 9015–9025, 2000.

106. T. Ahn, S. Ko, J. Lee, and H. Shim, Novel cyclohexylsiyl—Or phenylsilyl-substituted poly(*p*-phenylene vinylene)s via the halogen precursor route and Gilch polymerization, *Macromolecules*, 35: 3495–3505, 2002.

107. D.-H. Hwang, J.-I. Lee, N.-S. Cho, and H.-K. Shim, Light-emitting properties of germyl-substituted PPV derivative synthesized via soluble precursor, *J. Mater. Chem.*, 14: 1026–1030, 2004.

108. B.R. Hsieh, Y. Yu, E.W. Forsythe, G.M. Schaaf, and W.A. Feld, A new family of highly emissive soluble poly(*p*-phenylene vinylene) derivatives. A step toward fully conjugated blue-emitting poly(*p*-phenylene vinylenes), *J. Am. Chem. Soc.*, 120: 231–232, 1998.

109. H. Spreitzer, W. Kreuder, H. Becker, H. Schoo, and R. Demandt, Aryl-substituted poly(*p*-arylene vinylenes), method for the production and use thereof in electroluminescent components, PCT Patent Appl. WO 98/27136, June 25, 1998 (in German).

110. H. Becker, H. Spreitzer, W. Kreuder, E. Kluge, H. Schenk, I. Parker, and Y. Cao, Soluble PPVs with enhanced performance—A mechanistic approach, *Adv. Mater.*, 12: 42–48, 2000.

111. H. Becker and P. Stössel, Substituted poly(arylene vinylenes), method for the production thereof and their use in electroluminescent devices, PCT Patent Appl. WO 01/34722 A1, May 17, 2001 (in German).

112. M.R. Andersson, O. Thomas, W. Mammo, M. Svensson, M. Theander, and O. Inganäs, Substituted polythiophenes designed for optoelectronic devices and conductors, *J. Mater. Chem.*, 9: 1933–1940, 1999.

113. D.M. Johansson, G. Srdanov, G. Yu, M. Theander, O. Inganäs, and M.R. Andersson, Synthesis and characterization of highly soluble phenyl-substituted poly(*p*-phenylene vinylenes), *Macromolecules*, 33: 2525–2529, 2000.

114. D.M. Johansson, X. Wang, T. Johansson, O. Inganäs, G. Yu, G. Srdanov, and M.R. Andersson, Synthesis of soluble phenyl-substituted poly(*p*-phenylene vinylenes) with a low content of structural defects, *Macromolecules*, 35: 4997–5003, 2002.

115. Z. Chen, N.H.S. Lee, W. Huang, Y. Xu, and Y. Cao, New phenyl-substituted PPV derivatives for polymer light-emitting diodes—Synthesis, characterization and structure–property relationship study, *Macromolecules*, 36: 1009–1020, 2003.

116. N.H.S. Lee, Z.-K. Chen, W. Huang, Y.-S. Xu, and Y. Cao, Synthesis and characterization of naphthyl-substituted poly(*p*-phenylene vinylene)s with few structural defects for polymer light-emitting diodes, *J. Polym. Sci., Part A: Polym. Chem.*, 42: 1647–1657, 2004.
117. M. Kimura, M. Sato, N. Adachi, T. Fukawa, E. Kanbe, and H. Shirai, Poly(*p*-phenylene vinylene)s wrapped with 1,3,5-phenylene-based rigid dendrons, *Chem. Mater.* 19: 2809–2815, 2007.
118. S. Jin, M. Jang, and H. Suh, Synthesis and characterization of highly luminescent asymmetric poly(*p*-phenylene vinylene) derivatives for light-emitting diodes, *Chem. Mater.*, 14: 643–650, 2002.
119. Y. Jin, S. Song, S. H. Park, J.-A Park, J. Kim, H. Y. Woo, K. Lee, and H. Suh, Synthesis and properties of various PPV derivatives with phenyl substituents, *Polymer* 49: 4559–4568, 2008.
120. S. Chung, J. Jin, C.H. Lee, and C.E. Lee, Improved-efficiency light-emitting diodes prepared from organic-soluble PPV derivatives with phenylanthracene and branched alkoxy pendants, *Adv. Mater.*, 10: 684–688, 1998.
121. S.H. Lee, B. Jang, and T. Tsutsui, Sterically hindered fluorenyl-substituted poly(*p*-phenylene vinylenes) for light-emitting diodes, *Macromolecules*, 35: 1356–1364, 2002.
122. S.H. Lee, T. Yasuda, and T. Tsutsui, Charge carrier mobility in blue–green emitting fluorenyl-substituted poly(*p*-phenylene vinylene)s, *J. Appl. Phys.*, 95: 3825–3827, 2004.
123. D. Shin, Y. Kim, H. You, and S. Kwon, Sterically hindered and highly thermal stable spriobifluorenyl-substituted poly(*p*-phenylene vinylene) for light-emitting diodes, *Macromolecules*, 36: 3222–3227, 2003.
124. S. Chung, K. Kwon, S. Lee, J. Jin, C.H. Lee, C.E. Lee, and Y. Park, Highly efficient light-emitting diodes based on an organic-soluble poly(*p*-phenylene vinylene) derivatives carrying the electron-transporting PBD moiety, *Adv. Mater.*, 10: 1112–1116, 1998.
125. K. Kim, Y. Hong, S. Lee, J. Jin, Y. Park, B. Sohn, W. Kim, and J. Park, Synthesis and luminescence properties of poly(*p*-phenylene vinylene) derivatives carrying directly attached carbazole pendants, *J. Mater. Chem.*, 11: 3023–3030, 2001.
126. J.D. Stenger-Smith, P. Zarras, L.H. Merwin, S.E. Shaheen, B. Kippelen, and N. Peyghambarian, Synthesis and characterization of poly(2,5-*bis*(*N*-methyl-*N*-hexylamino)phenylene vinylene), a conjugated polymer for light-emitting diodes, *Macromolecules*, 31: 7566–7569, 1998.
127. H. Meng, W. Yu, and W. Huang, Facile synthetic route to a novel electroluminescent polymer— Poly(*p*-phenylene vinylene) containing a fully conjugated aromatic oxadiazole side chain, *Macromolecules*, 32: 8841–8847, 1999.
128. Z. Chen, H. Meng, Y. Lai, and W. Huang, Photoluminescent poly(*p*-phenylene vinylene)s with an aromatic oxadiazole moiety as the side chain: Synthesis, electrochemistry, and spectroscopy study, *Macromolecules*, 32: 4351–4358, 1999.
129. D.W. Lee, K.-Y. Kwon, J.-I. Jin, Y. Park, Y.-R. Kim, and I.-W. Hwang, Luminescence properties of structurally modified PPVs: PPV derivatives bearing 2-(4-*tert*-butylphenyl)-5-phenyl-1,3,4-oxadizole pendants, *Chem. Mater.*, 13: 565–574, 2001.
130. S. Jin, M. Kim, J.Y. Kim, K. Lee, and Y. Gal, High-efficiency poly(*p*-phenylene vinylene)-based copolymers containing an oxadiazole pendant group for light-emitting diodes, *J. Am. Chem. Soc.*, 126: 2474–2480, 2004.
131. J. Gordon, T.J. Sheldon, D.D.C. Bradley, and P.L. Burn, The synthesis of an electronically asymmetric substituted poly(arylene vinylene); poly{2-(2'-ethylhexyloxy)-5-[(*E*)-4''-nitrostyryl]-1,4-phenylene vinylene}, *J. Mater. Chem.*, 6: 1253–1258, 1996.
132. F.H. Boardman, A.W. Grice, M.G. Rüther, T.J. Sheldon, D.D.C. Bradley, and P.L. Burn, A new electron-withdrawing group containing poly(1,4-phenylene vinylene), *Macromolecules*, 32: 111–117, 1999.
133. S.W. Ko, B. Jung, T. Ahn, and H. Shim, Novel poly(*p*-phenylene vinylene)s with an electron-withdrawing cyanophenyl group, *Macromolecules*, 35: 6217–6223, 2002.
134. A.C. Grimsdale, F. Cacialli, J. Grüner, X. Li, A.B. Holmes, S.C. Moratti, and R.H. Friend, Novel poly(arylene vinylene)s carrying donor and acceptor substituents, *Synth. Met.*, 76: 165–167, 1996.
135. I. Kang, H. Shim, and T. Zyung, Yellow-light-emitting fluorine-substituted PPV derivative, *Chem. Mater.*, 9: 746–749, 1997.

136. R.M. Gurge, A.M. Sarker, P.M. Lahti, B. Hu, and F.E. Karasz, Light emitting properties of fluorine-substituted poly(1,4-phenylene vinylenes), *Macromolecules*, 30: 8286–8292, 1997.
137. M. Jang, S. Song, and H. Shim, Efficient green light-emitting polymer by balanced injection of electron and holes: New electron accepting perfluorinated substituent, *Polymer*, 41: 5675–5679, 2000.
138. Y. Jin, J. Kim, S. Lee, J.Y. Kim, S.H. Park, K. Lee, and H. Suh, Novel electroluminescent polymers with fluoro groups in vinylene units, *Macromolecules*, 37: 6711–6715, 2004.
139. H. Antoniadis, M.A. Abkowitz, and B.R. Hsieh, Carrier deep-trapping mobility-lifetime products in poly(*p*-phenylene vinylene), *Appl. Phys. Lett.*, 65: 2030–2032, 1994.
140. P.L. Burn, A. Kraft, D.R. Baigent, D.D.C. Bradley, A.R. Brown, R.H. Friend, R.W. Gymer, A.B. Holmes, and R.W. Jackson, Chemical tuning of the electronic properties of poly(*p*-phenylene vinylenes)-based copolymers, *J. Am. Chem. Soc.*, 115: 10117–10124, 1993.
141. Z. Bao, Z. Peng, M.E. Galvin, and E.A. Chandross, Novel oxadiazole side chain conjugated polymers as single-layer light-emitting diodes with improved quantum efficiencies, *Chem. Mater.*, 10: 1201–1204, 1998.
142. Z. Peng and J. Zhang, New oxadiazole-containing conjugated polymer for single-layer light-emitting diodes, *Chem. Mater.*, 11: 1138–1143, 1999.
143. C. Huang, W. Huang, J. Guo, C.-Z. Yang, and E.-T. Kang, A novel rigid-rod alternating poly (*p*-phenylene vinylene) derivative with oligo(ethylene oxide) side chains, *Polymer*, 42: 3929–3938, 2004.
144. B.S. Chuah, F. Geneste, A.B. Holmes, R.E. Martin, H. Rost, F. Cacialli, R.H. Friend, H.-H. Hörhold, S. Pfeiffer, and D.-H. Hwang, The copolymer route to new luminescent materials for LEDs, *Macromol. Symp.*, 154: 177–186, 2000.
145. H. Beck, H. Spreitzer, W. Kreuder, E. Kluge, H. Vestweber, H. Schenk, and K. Treacher, Advances in polymers for PLEDs: From a polymerization mechanism to industrial manufacturing, *Synth. Met.*, 122: 105–110, 2001.
146. P.K.H. Ho, J.-S. Kim, J.H. Burroughes, H. Becker, S.F.Y. Li, T.M. Brown, F. Cacialli, and R.H. Friend, Molecular-scale interface engineering for polymer light-emitting diodes, *Nature*, 404: 481–484, 2000.
147. D.-W. Kim, H.-K. Kim, J.-M. Son, I.-S. Kee, D.-H. Hwang M.-C. Chung, and J.-H. Lee, Highly alkoxylated PPV copolymers for PLEDs: Syntheses optical properties and enhancement of device lifetimes, *J. Lumin.*, 131: 1288–1293, 2011.
148. B. Sohn, K. Kim, D.S. Choi, Y.K. Kim, S.C. Jeoung, and J. Jin, Synthesis and luminescence properties of poly[2-(9,9-dihexylfluoren-2-yl)-1,4-phenylene vinylene] and its copolymers containing 2-(2-ethylhexyloxy)-5-methoxy-1,4-phenylene vinylene units, *Macromolecules*, 35: 2876–2881, 2002.
149. Z. Peng, J. Zhang, and B. Xu, New poly(*p*-phenylene vinylene) derivatives exhibiting high photoluminescence quantum efficiencies, *Macromolecules*, 32: 5162–5164, 1999.
150. J.A. Mikroyannidis, Synthesis by Heck coupling of soluble, blue-light-emitting fully conjugated poly(*p*-phenylene vinylene)s with highly phenylated side groups, *Macromolecules*, 35: 9289–9295, 2002.
151. W.J. Feast, I.S. Millichamp, R.H. Friend, M.E. Horton, D. Phillips, S.D.D.V. Rughooputh, and G. Rumbles, Optical absorption and luminescence in poly(4,4'-diphenylenediphenylvinylene), *Synth. Met.*, 10: 181–191, 1985.
152. F. Cacialli, R.H. Friend, N. Haylett, R. Daik, W.J. Feast, D.A. dos Santos, and J.L. Brédas, Efficient green light-emitting diodes from a phenylated derivative of poly(*p*-phenylene vinylene), *Appl. Phys. Lett.*, 69: 3794–3796, 1996.
153. T. Ahn, S. Song, and H. Shim, Highly photoluminescent and blue–green electroluminescent polymers: New silyl- and alkoxy-substituted poly(*p*-phenylene vinylene) related copolymers containing carbazole or fluorene groups, *Macromolecules*, 33: 6764–6771, 2000.
154. H. Li, Y. Zhang, Y. Hu, D. Ma, L. Wang, X. Jing, and F. Wang, Novel soluble *N*-phenyl-carbazole-containing PPV for light-emitting devices: Synthesis, electrochemical, optical and electroluminescent properties, *Macromol. Chem. Phys.*, 205: 247–255, 2004.

155. F. Liang, T. Kurata, H. Nishide, and J. Kido, Synthesis and electroluminescent properties of N-phenylcarbazole-substituted poly(*p*-phenylenevinylene), *J. Polym. Sci., Part A: Polym. Chem.*, 43: 5765–5773, 2005.

156. J.J. Kim, K.-S. Kim, H.C. Kim, and M. Ree, Synthesis and properties of photoluminescent polymers bearing electron-facilitating oxadiazole derivative side groups, *J. Polym. Sci., Part A: Polym. Chem.*, 40: 1173–1183, 2002.

157. S.-H. Jin, J.-E. Jung, D.-K. Park, B.-C. Jeon, S.-K. Kwon, Y.-H. Kim, D.-K. Moon, S.-H. Kim, and Y.-S. Gal, Synthesis and characterization of color tunable electroluminescent polymer by blending oxadiazole containing polymer, *Eur. Polym. J.*, 37: 921–925, 2001.

158. J. Shi and S. Zheng, Conjugated polymer containing arylamine pendants for light-emitting diodes, *Macromolecules*, 34: 6571–6576, 2001.

159. Y.-J. Pu, M. Soma, J. Kido, and H. Nishide, A novel triphenylamine-substituted poly(*p*-phenylene vinylene): Improved photo- and electroluminescent properties, *Chem. Mater.*, 13: 3817–3819, 2001.

160. F. Bai, M. Zheng, G. Yu, and D. Zhu, The photo- and electroluminescence of some novel light emitting copolymers, *Thin Solid Films*, 363: 118–121, 2000.

161. A. Kimoto, K. Masachika, J.-S. Cho, M. Higuchi, and K. Yamamoto, Novel poly(*p*-phenylene vinylene)s with a phenylazomethine dendron as a metal-collecting site, *Org. Lett.*, 6: 1179–1182, 2004.

162. Y. Zhang, Y. Hu, H. Li, L. Wang, X. Jing, F. Wang, and D. Ma, Polymer light-emitting diodes based on a bipolar transporting luminescent polymer, *J. Mater. Chem.*, 13: 773–777, 2003.

163. Z. Tan, R. Tang, F. Xi, and Y. Li, Synthesis, characterization, and electroluminescence of new conjugated PPV derivatives bearing triphenylamine side-chain through a vinylene bridge, *Polym. Adv. Technol.*, 18: 963–970, 2007.

164. N.C. Greenham, S.C. Moratti, D.D.C. Bradley, R.H. Friend, and A.B. Holmes, Efficient light-emitting diodes based on polymers with high electron affinities, *Nature*, 365: 628–630, 1993.

165. Y. Liu, G. Yu, Q. Li, and D. Zhu, Synthesis and characterization of new poly(cyanoterephthalylidene)s for light-emitting diodes, *Synth. Met.*, 122: 401–408, 2001.

166. D.-J. Kim, S.-H. Kim, T. Zyung, J.-J. Kim, I. Cho, and S.K. Choi, Synthesis of a new class of processible electroluminescent poly(cyanoterephthalidene) derivative with a tertiary amine linkage, *Macromolecules*, 29: 3657–3660, 1996.

167. J.-H. Lee, J.-W. Park, and S.-K. Choi, Synthesis and electroluminescent property of a new conjugated polymer based on carbazole derivative: Poly(3,6-*N*-2-ethylhexyl carbazolyl cyanoterephthalidene), *Synth. Met.*, 88: 31–35, 1997.

168. Z. Peng and M.E. Galvin, Polymers with high electron affinities for light-emitting diodes, *Chem. Mater.*, 10: 1785–1788, 1998.

169. X. Wu, Y. Liu, and D. Zhu, Synthesis and characterization of a new conjugated polymer containing cyano substituents for light-emitting diodes, *J. Mater. Chem.*, 11: 1327–1331, 2001.

170. M. Hohloch, J.L. Segura, S.E. Döttinger, D. Honhholz, E. Steinhuber, H. Spreitzer, and M. Hanack, Design, synthesis and study of photoluminescence and electroluminescence of new poly(2,6-naphthylene vinylene) derivatives, *Synth. Met.*, 84: 319–322, 1997.

171. Y. Xiao, W. Yu, S. Chua, and W. Huang, A novel series of copolymers containing 2,5-dicyano-1,4-phenylene vinylene—Synthetic tuning of the HOMO and LUMO energy levels of conjugated polymers, *Chem. Eur. J.*, 6: 1318–1321, 2000.

172. Y. Xiao, W.-L. Yu, Z.-K. Chen, N.H.S. Lee, Y.-H. Lai, and W. Huang, Synthesis and characterization of a novel light-emitting copolymer with improved charge-balancing property, *Thin Solid Films*, 363: 102–105, 2000.

173. M.S. Liu, X. Jiang, S. Liu, P. Herguth, and A.K.Y. Jen, Effect of cyano substituents on electron affinity and electron-transporting properties of conjugated polymers, *Macromolecules*, 35: 3532–3538, 2002.

174. I. Benjamin, E.Z. Faraggi, Y. Avny, D. Davydov, and R. Neumann, Fluorinated poly(*p*-phenylene vinylene) copolymers: Preparation and use in light-emitting materials, *Chem. Mater.*, 8: 352–355, 1996.

175. R. Riehn, J. Morgado, R. Iqbal, S.C. Moratti, A.B. Holmes, S. Volta, and F. Cacialli, Fluorine substituted poly(*p*-phenylene vinylenes) copolymers, *Synth. Met.*, 124: 67–69, 2001.

176. Z. Peng and J. Zhang, Novel oxadiazole-containing conjugated polymers as efficient single-layer light-emitting diodes, *Synth. Met.*, 105: 73–78, 1999.

177. J.H. Kim, J.H. Park, and H. Lee, Highly efficient novel poly(*p*-phenylene vinylene) derivative with 1,3,4-oxadiazole pendant on a vinylene unit, *Chem. Mater.*, 15: 3414–3416, 2003.

178. Z. Peng, Z. Bao, and M.E. Galvin, Polymers with bipolar carrier transport abilities for light emitting diodes, *Chem. Mater.*, 10: 2086–2090, 1998.

179. J.A. Mikroyannidis, I.K. Spiliopoulos, T.S. Kasimis, A.P. Kulkarni, and S.A. Jenekhe, Synthesis, photophysics, and electroluminescence of conjugated poly(*p*-phenylene vinylene) derivatives with 1,3,4-oxadiazoles in the backbone, *Macromolecules*, 36: 9295–9302, 2003.

180. Z. Peng, Z. Bao, and M.E. Galvin, Oxadiazole-containing conjugated polymers for light-emitting diodes, *Adv. Mater.*, 10: 680–682, 1998.

181. S.-Y. Zhang, F. Kong, R. Sun, R.-K. Yuan, X.-Q. Jiang, and C.-Z. Yang, Synthesis, characterization, and electro-optical properties of a soluble conjugated polymer containing an oxadiazole unit in the main chain, *J. Appl. Polym. Sci.*, 89: 2618–2623, 2003.

182. A.W. Grice, A. Tajbakhsh, P.L. Burn, and D.D.C. Bradley, A blue-emitting triazole-based conjugated polymer, *Adv. Mater.*, 9: 1174–1178, 1997.

183. P. Karastatiris, J.A. Mikroyannidis, I.K. Spiliopoulos, A.P. Kulkarni, and S.A. Jenekhe, Synthesis, photophysics and electroluminescence of new quinoxaline-containing poly(*p*-phenylene vinylene)s, *Macromolecules*, 37: 7867–7878, 2004.

184. L.-S. Yu and S.A. Chen, Full-range tunability of electron and hole carrier mobilities and density ratios via incorporation of highly electron-deficient moieties in poly(phenylene vinylene) side chains, *Adv. Mater.*, 16: 744–748, 2004.

185. J.H. Kim and H. Lee, Synthesis, electrochemistry, and electroluminescence of novel red-emitting poly(*p*-phenylene vinylene) derivative with 2-pyran-4-ylidene-malononitrile obtained by the Heck reaction, *Chem. Mater.*, 14: 2270–2275, 2002.

186. B. Jiang, S.-W. Yang, and W.E. Jones, Jr., Conjugated porphyrin polymers: Control of chromophore separation by oligophenylene vinylene bridges, *Chem. Mater.*, 9: 2031–2034, 1997.

187. B. Jiang, S.-W. Yang, R. Niver, and W.E. Jones, Jr., Metalloporphyrin polymers bridged with conjugated cyano-substituted stilbene units, *Synth. Met.*, 94: 205–210, 1998.

188. P. Karastatiris, J. A. Mikroyannidis, I. K. Spiliopoulos, Bipolar poly(*p*-phenylene vinylene)s bearing electron-donating triphenylamine or carbazole and electron-accepting quinoxaline moieties, *J. Polym. Sci., Part A: Polym. Chem.*, 46: 2367–2378, 2008.

189. S. Li, P. Zhao, Y. Huang, T. Li, C. Tang, R. Zhu, L. Zhao, Q. Fan, S. Huang, Z. Xu, and W. Huang, Poly-(*p*-phenylene vinylenes) with pendent 2,4-difluorophenyl and fluorenyl moieties: Synthesis, characterization, and device performance, *J. Polym. Sci., Part A: Polym. Chem.*, 47: 2500–2508, 2009.

190. H. Li, Y. Hu, Y. Zhang, D. Ma, L. Wang, X. Jing, F. Wang, Soluble, saturated-red-light-emitting poly-(*p*-phenylene vinylene) containing triphenylamine units and cyano groups, *J. Polym. Sci., Part A: Polym. Chem.*, 42: 3947–3953, 2004.

191. H. Li, Y. Geng, S. Tong, H. Tong, R. Hua, G. Su, L. Wang, X. Jing, and F. Wang, Synthesis and characterization of alternating copolymers containing triphenylamine as hole-transporting units, *J. Polym. Sci., Part A: Polym. Chem.*, 39: 3278–3286, 2001.

192. S. Shi and F. Wudl, Synthesis and characterization of a water-soluble poly(*p*-phenylene vinylene) derivative, *Macromolecules*, 23: 2119–2124, 1990.

193. H.-L. Wang, D.W. McBranch, V.I. Klimov, R. Helgeson, and F. Wudl, Controlled unidirectional energy transfer in luminescent self-assembled conjugated polymer superlattices, *Chem. Phys. Lett.*, 315: 173–180, 1999.

194. S.-C. Chang, J. Bharathan, Y. Yang, R. Helgeson, F. Wudl, M.B. Ramey, and J.R. Reynolds, Dual-color polymer light-emitting pixels processed by hybrid inkjet printing, *Appl. Phys. Lett.*, 73: 2561–2563, 1998.

195. L. Chen, D.W. McBranch, H.-L. Wang, R. Helgeson, F. Wudl, and D.G. Whitten, Highly sensitive biological and chemical sensors based on reversible fluorescence quenching in a conjugated polymer, *Proc. Natl. Acad. Sci. U. S. A.*, 96: 12287–12292, 1999.

196. F. Cacialli, J.S. Wilson, J.J. Michels, C. Daniel, C. Silva, R.H. Friend, N. Severin, P. Samorì, J.P. Rabe, M.J. O'Connell, P.N. Taylor, and H.L. Anderson, Cyclodextrin-threaded conjugated polyrotaxanes, as insulated molecular wires with reduced interstrand interaction, *Nat. Mater.*, 1: 160–164, 2002.

197. J.J. Michels, M.J. O'Connell, P.N. Taylor, J.S. Wilson, F. Cacialli, and H.L. Anderson, Synthesis of conjugated polyrotaxanes, *Chem. Eur. J.*, 9: 6167–6176, 2003.

198. J. Terao, A. Tang, J.J. Michels, A. Krivokapic, and H.L. Anderson, Synthesis of poly(*p*-phenylene vinylene) rotaxanes by aqueous Suzuki coupling, *Chem. Commun.*, 1: 56–57, 2004.

199. L. Zheng, R.C. Urian, Y. Liu, A.K. Jen, and L. Pu, A binaphthyl-based conjugated polymer for light-emitting diodes, *Chem. Mater.*, 12: 13–15, 2000.

200. A.M. Sarker, L. Ding, P.M. Lahti, and F.E. Karasz, Synthesis and luminescent studies of poly(*p*-phenylene vinylene)s containing a biphenyl moiety, *Macromolecules*, 35: 223–230, 2002.

201. T. Ahn, M.S. Jang, H.-K. Shim, D.-H. Hwang, and T. Zyung, Blue electroluminescent polymers: Control of conjugation length by kink linkages and substituents in the poly(*p*-phenylene vinylene)-related copolymers, *Macromolecules*, 32: 3279–3285, 1999.

202. Y. Pang, J. Li, B. Hu, and F.E. Karasz, A highly luminescent poly[(*m*-phenylene vinylene)-*alt*-(*p*-phenylene vinylene)] with defined conjugation length and improved solubility, *Macromolecules*, 32: 3946–3950, 1999.

203. F. Cacialli, B.S. Chuah, R.H. Friend, S.C. Moratti, and A.B. Holmes, Blue-emitting diodes from a *m*-linked 2,3-substituted alkoxy poly(*p*-phenylene vinylene), *Synth. Met.*, 111: 155–158, 2000.

204. L. Ding, F.E. Karasz, Z. Lin, M. Zheng, L. Liao, and Y. Pang, Effect of Förster energy transfer and hole transport layer on performance of polymer light-emitting diodes, *Macromolecules*, 34: 9183–9188, 2001.

205. H. Schlick, F. Stelzer, F. Meghdadi, and G. Leising, Synthesis and optical characterization of new highly luminescent poly(*m,p*-phenylene vinylene) derivatives, *Synth. Met.*, 119: 529–530, 2001.

206. G. Brizius, N.G. Pschirer, W. Steffen, K. Stitzer, H.-C. zur Loye, and U.H.F. Bunz, Alkyne metathesis with simple catalyst systems: Efficient synthesis of conjugated polymers containing vinyl groups in main side chain, *J. Am. Chem. Soc.*, 122: 12435–12440, 2000.

207. D.A.M. Egbe, S. Sell, C. Ulbricht, E. Birckner, and U.-W. Grummt, Mixed alkyl- and alkoxy-substituted poly[(phenylene ethynylene)-*alt*-(phenylene vinylene)] hybrid polymers: Synthesis and photophysical properties, *Macromol. Chem. Phys.*, 205: 2105–2115, 2004.

208. L. Ding, D.A.M. Egbe, and F.E. Karasz, Photophysical and optoelectronic properties of green-emitting alkoxy-substituted PE/PV hybrid conjugated polymers, *Macromolecules*, 37: 6124–6131, 2004.

209. D.A.M. Egbe, B. Carbonnier, L. Ding, D. Mühlbacher, E. Birckner, T. Pakula, F.E. Karasz, and U.-W. Grummt, Supramolecular ordering, thermal behaviour, and photophysical, electrochemical, and electroluminescent properties of alkoxy-substituted yne-containing poly(phenylene-vinylene)s, *Macromolecules*, 37: 7451–7463, 2004.

210. Q. Chu, Y. Pang, L. Ding, and F.E. Karasz, Green-emitting PPE–PPV hybrid polymers: Efficient energy transfer across the *m*-phenylene bridge, *Macromolecules*, 36: 3848–3853, 2003.

211. H. Liang, J. Yan, and J. Lu, Synthesis and optical characterization of a novel blue luminescent polymer: Regioregular poly(1-alkoxy-2,4-*m*-phenylene vinylene), *Synth. Met.*, 142: 143–145, 2004.

212. P.L. Burn, A.B. Holmes, A. Kraft, D.D.C. Bradley, A.R. Brown, and R.H. Friend, Synthesis of a segmented conjugated polymer-chain giving a blue-shifted electroluminescence and improved efficiency, *J. Chem. Soc., Chem. Commun.*, 1: 32–34, 1992.

213. J.C. Carter, I. Grizzi, S.K. Heeks, D.J. Lacey, S.G. Latham, P.G. May, O. Ruiz de los Paños, K. Pichler, C.R. Towns, and H.F. Wittmann, Operating stability of light-emitting polymer diodes based on poly(*p*-phenylene vinylene), *Appl. Phys. Lett.*, 71: 34–36, 1997.

214. Z. Yang, I. Sokolik, and F.E. Karasz, A soluble blue-light-emitting polymer, *Macromolecules*, 26: 1188–1190, 1993.

215. B. Hu, Z. Yang, and F.E. Karasz, Electroluminescence of pure poly(*N*-vinylcarbazole) and its blends with a multiblock copolymer, *J. Appl. Phys.*, 76: 2419–2422, 1994.

216. R.G. Sun, Y.Z. Wang, D.K. Wang, Q.B. Zheng, E.M. Kyllo, T.L. Gustafson, F. Wang, and A.J. Epstein, High PL quantum efficiency of poly(phenylene vinylene) systems through exciton confinement, *Synth. Met.*, 111–112: 595–602, 1999.

217. T. Zyung, D.-H. Hwang, I.-N. Kang, H.-K. Shim, W.-Y. Hwang, and J.-J. Kim, Novel blue electroluminescent polymers with well-defined conjugation length, *Chem. Mater.*, 7: 1499–1503, 1995.

218. N. Benfaremo, D.J. Sandman, S. Tripathy, J. Kumar, K. Yang, M.F. Rubner, and C. Lyons, Synthesis and characterization of luminescent polymers of distyrylbenzenes with oligo(ethylene glycol) spacers, *Macromolecules*, 31: 3595–3599, 1998.

219. F. Cacialli, R.H. Friend, W.J. Feast, and P.W. Lovenich, Poly(distyrylbenzene-*block*-sexi(ethylene oxide)), a highly luminescent processable derivative of PPV, *Chem. Commun.*, 18: 1778–1779, 2001.

220. L. Duan, Y. Qiu, and H. Wang, Blue electroluminescence from a processible derivative of PPV based copolymer with tri(ethylene oxide) segments in the backbone, *Synth. Met.*, 137: 1133–1135, 2003.

221. H. Wang, L. Duan, Y. Qui, X. Wang, and D. Liu, Synthesis and electroluminescence properties of a novel poly(*p*-phenylene vinylene)-based copolymer with tri(ethylene oxide) segments on the backbone, *J. Appl. Polym. Sci.*, 83: 2195–2200, 2002.

222. E.-N. Chang and S.-A. Chen, Cyano-containing phenylene vinylene-based copolymer as blue luminescent and electron transport material in polymer light-emitting diodes, *J. Appl. Phys.*, 85: 2057–2061, 1999.

223. M. Zheng, L. Ding, E.E. Gürel, P.M. Lahti, and F.E. Karasz, Oxadiazole containing conjugated–nonconjugated blue and blue–green light-emitting copolymers, *Macromolecules*, 34: 4124–4129, 2001.

224. J.A. Mikroyannidis and J.K. Spiliopoulos, Synthesis and optical properties of novel blue-light emitting poly(*p*-phenylene vinylene) derivatives with pendant oxadiazole and cyano groups, *J. Polym. Sci., Part A: Polym. Chem.*, 42: 1768–1778, 2004.

225. A.M. Sarker, E.E. Gürel, M. Zheng, P.M. Lahti, and F.E. Karasz, Synthesis, characterization, and photophysical studies of new blue-emitting segmented copolymers, *Macromolecules*, 34: 5897–5901, 2001.

226. S. Zheng, J. Shi, and R. Mateu, Novel blue light emitting polymer containing an adamantine moiety, *Chem. Mater.*, 12: 1814–1817, 2000.

227. M. Zheng, A.M. Sarker, E.E. Gürel, P.M. Lahti, and F.E. Karasz, Structure–property relationships in light-emitting polymers: Optical, electrochemical and thermal studies, *Macromolecules*, 33: 7426–7430, 2000.

228. H.-K. Kim, M.-K. Ryu, K.-D. Kim, S.-M. Lee, S.-W. Cho, and J.-W. Park, Tunable electroluminescence from silicon-containing poly(*p*-phenylene vinylene)-related copolymers with well-defined structures, *Macromolecules*, 31: 1114–1123, 1998.

229. K.-D. Kim, J.-S. Park, H.-K. Kim, T.B. Lee, and K.T. No, Blue electroluminescence from novel silicon-containing poly(cyanoterephthalylidene) copolymers, *Macromolecules*, 31: 7267–7272, 1998.

230. M.R. Robinson, S. Wang, A.J. Heeger, and G.C. Bazan, A tetrahedral oligo(phenylene vinylene) molecule of intermediate dimensions: Effect of molecular shape on the morphology and electroluminescence of organic glasses, *Adv. Funct. Mater.*, 11: 413–419, 2001.

231. R. Zhang, G. Zhang, and J. Shen, A new approach for the synthesis of conjugated–non-conjugated poly(phenylene vinylene)—Polyacrylamide copolymers, *Chem. Commun.*, 10: 823–824, 2000.

232. J.-K. Lee, R.R. Schrock, D.R. Baigent, and R.H. Friend, A new type of blue-emitting electroluminescent polymer, *Macromolecules*, 28: 1966–1971, 1995.

233. J. Rault-Berthelot, Polyfluorenes, a family of versatile conjugated polymers. Anodic synthesis, physicochemical properties, electrochemical behaviour and application fields, *Recent Res. Dev. Macromol. Res.*, 3: 425–437, 1998.

234. M. Bernius, M. Inbasekaran, E. Woo, W. Wu, and L. Wujkowski, Fluorene-based polymers—Preparation and applications, *J. Mater. Sci.: Mater. Electron.*, 11: 111–116, 2000.
235. M.T. Bernius, M. Inbasekaran, J. O'Brien, and W. Wu, Progress in light-emitting polymers, *Adv. Mater.*, 12: 1737–1750, 2000.
236. M. Leclerc, Polyfluorenes: Twenty years of progress, *J. Polym. Sci., Part A: Polym. Chem.*, 39: 2867–2873, 2001.
237. D. Neher, Polyfluorene homopolymers: Conjugated liquid-crystalline polymers for bright emission and polarized elecroluminescence, *Macromol. Rapid Commun.*, 22: 1365–1385, 2001.
238. U. Scherf and E.J.W. List, Semiconducting polyfluorenes—Towards reliable structure–property relationships, *Adv. Mater.*, 14: 477–487, 2002.
239. W. Wu, M. Inbasekaran, M. Hudack, D. Welsh, W. Yu, Y. Cheng, C. Wang, S. Kram, M. Tacey, M. Bernius, R. Fletcher, K. Kiszka, S. Munger, and J. O'Brien, Recent development of polyfluorene-based RGB materials for light emitting diodes, *Microelectron. J.*, 35: 343–348, 2004.
240. D.Y. Kim, H.N. Cho, and C.Y. Kim, Blue light emitting polymers, *Prog. Polym. Sci.*, 25: 1089–1139, 2000.
241. J. Rault-Berthelot and J. Simonet, The anodic oxidation of fluorene and some of its derivatives, *J. Electrochem. Soc.*, 182: 187–192, 1985.
242. J. Rault-Berthelot and J. Simonet, The polyfluorenes: A family of versatile electroactive polymers (I). Electropolymerization of fluorenes, *Nouv. J. Chim.*, 10: 169–177, 1986.
243. H.-J. Cho, B.-J. Jung, N.S. Cho, J. Lee, and H.-K. Shim, Synthesis and characterization of thermally stable blue light-emitting polyfluorenes containing siloxane bridges, *Macromolecules*, 36: 6704–6710, 2003.
244. B. Liu, W.-L. Yu, Y.-H. Lai, and W. Huang, Blue-light-emitting fluorene-based polymers with tunable electronic properties, *Chem. Mater.*, 13: 1984–1991, 2001.
245. P. Lu, H. Zhang, F. Shen, B. Yang, D. Li, Y. Ma, X. Chen, and J. Li, A wide-bandgap semiconducting polymer for ultraviolet and blue light emitting diodes, *Macromol. Chem. Phys.*, 204: 2274–2280, 2003.
246. W. Yang, J. Huang, C. Liu, Y. Niu, Q. Hou, R. Yang, and Y. Cao, Enhancement of color purity in blue-emitting fluorene–pyridine-based copolymers by controlling the chain rigidity and effective conjugation length, *Polymer*, 45: 865–872, 2004.
247. W. Yang, Q. Hou, C. Liu, Y. Niu, J. Huang, R. Yang, and Y. Cao, Improvement of color purity in blue-emitting polyfluorene by copolymerization with dibenzothiophene, *J. Mater. Chem.*, 13: 1351–1355, 2003.
248. E. Lim, B. Jung, and H. Shim, Synthesis and characterization of a new light-emitting fluorene–thieno[3,2-*b*]thiophene-based conjugated copolymer, *Macromolecules*, 36: 4288–4293, 2003.
249. C. Xia and R.C. Advincula, Decreased aggregation phenomena in polyfluorenes by introducing carbazole copolymer units, *Macromolecules*, 34: 5854–5859, 2001.
250. N.S. Cho, D.-H. Hwang, B.-J. Jung, E. Lim, J. Lee, and H.-K. Shim, Synthesis, characterization, and electroluminescence of new conjugated polyfluorene derivatives containing various dyes as comonomers, *Macromolecules*, 37: 5265–5273, 2004.
251. N.S. Cho, D.-H. Hwang, J.-I. Lee, B.-J. Jung, and H.-K. Shim, Synthesis and color tuning of new fluorene-based copolymers, *Macromolecules*, 35: 1224–1228, 2002.
252. D.-H. Hwang, S.-K. Kim, M.-J. Park, J.-H. Lee, B.-W. Koo, I.-N. Kang, S.-H. Kim, and T. Zyung, Conjugated polymers based on phenothiazine and fluorene in light-emitting diodes and field effect transistors, *Chem. Mater.*, 16: 1298–1303, 2004.
253. S. Beaupré and M. Leclerc, Optical and electrical properties of π-conjugated polymers based on electron-rich 3,6-dimethoxyfluorene unit, *Macromolecules*, 36: 8986–8991, 2003.
254. E.P. Woo, W.R. Shiang, M. Inbasekaran, and G.R. Roof, 2,7-Diaryl-9-substituted fluorenes and 9-substituted fluorene oligomers and polymers, U.S. Patent 5,708,130, January 13, 1998.
255. J.-H. Lee and D.-H. Hwang, Alkoxyphenyl-substituted polyfluorene: A stable blue-light-emitting polymer with good solution processability, *Chem. Commun.*, 22: 2836–2837, 2003.

256. S. Setayesh, A.C. Grimsdale, T. Weil, V. Enkelmann, K. Müllen, F. Meghdadi, E.J.W. List, and G. Leising, Polyfluorenes with polyphenylene dendron side chains: Toward non-aggregating, light-emitting polymers, *J. Am. Chem. Soc.*, 123: 946–953, 2001.

257. A. Pogantsch, F.P. Wenzl, E.J.W. List, G. Leising, A.C. Grimsdale, and K. Müllen, Polyfluorenes with dendron side chains as the active materials for polymer light-emitting devices, *Adv. Mater.*, 14: 1061–1064, 2002.

258. D. Marsitzky, R. Vestberg, P. Blainey, B.T. Tang, C.J. Hawker, and K.R. Carter, Self-encapsulation of poly-2,7-fluorenes in a dendrimer matrix, *J. Am. Chem. Soc.*, 123: 6965–6972, 2001.

259. C.-H. Chou and C.-F. Shu, Synthesis and characterization of dendronized polyfluorenes, *Macromolecules*, 35: 9673–9677, 2002.

260. Y. Fu, J. Li, S. Yan, and Z. Bo, High molecular weight dendronized poly(fluorene)s with peripheral carbazole groups: Synthesis, characterization, and properties, *Macromolecules*, 37: 6395–6400, 2004.

261. D. Vak, C. Chun, C.L. Lee, J.-J. Kim, and D.-Y. Kim, A novel spiro-functionalized polyfluorene derivative with solubilizing side chains, *J. Mater. Chem.*, 14: 1342–1346, 2004.

262. Y. Wu, J. Li, Y. Fu, and Z. Bo, Synthesis and extremely stable blue light emitting poly(spirobifluorene)s with Suzuki polycondensation, *Org. Lett.*, 6: 3485–3487, 2004.

263. C. Ego, A.C. Grimsdale, F. Uckert, G. Yu, G. Srdanov, and K. Müllen, Triphenylamine-substituted polyfluorene—A stable blue-emitter with improved charge injection for light-emitting diodes, *Adv. Mater.*, 14: 809–811, 2002.

264. J. Pei, X.-L. Liu, Z.-K. Chen, X.-H. Zhang, Y.-H. Lai, and W. Huang, First hydrogen-bonding-induced self-assembled aggregates of a polyfluorene derivatives, *Macromolecules*, 36: 323–327, 2003.

265. G. Klärner, M.H. Davey, E.-D. Chen, J.C. Scott, and R.D. Miller, Colorfast blue-light-emitting random copolymers derived from di-*n*-hexylfluorene and anthracene, *Adv. Mater.*, 13: 993–997, 1998.

266. O. Stéphan and J.-C. Vial, Blue light electroluminescent devices based on a copolymer derived from fluorene and carbazole, *Synth. Met.*, 106: 115–119, 1999.

267. Y. Li, J. Ding, M. Day, Y. Tao, J. Lu, and M. D'Iorio, Synthesis and properties of random and alternating fluorene/carbazole copolymers for use in blue light-emitting devices, *Chem. Mater.*, 16: 2165–2173, 2004.

268. J.-F. Morin, S. Beaupré, M. Leclerc, I. Lévesque, and M. D'Iorio, Blue light-emitting devices from new conjugated poly(*N*-substituted-2,7-carbazole) derivatives, *Appl. Phys. Lett.*, 80: 341–343, 2002.

269. Q. Fang and T. Yamamoto, New alternative copolymer constituted of fluorene and triphenylamine units with a tunable -CHO group in the side chain. Quantitative transformation of the -CHO group to -CH=CHAr groups and optical and electrochemical properties of the polymers, *Macromolecules*, 37: 5894–5899, 2004.

270. B.-J. Jung, J.-I. Lee, H.Y. Chu, L.-M. Do, and H.-K. Shim, Synthesis of novel fluorene-based poly(iminoarylene)s and their application to buffer layer in organic light-emitting diodes, *Macromolecules*, 35: 2282–2287, 2002.

271. F.-I. Wu, S. Reddy, and C.-F. Shu, Novel oxadiazole-containing polyfluorene with efficient blue electroluminescence, *Chem. Mater.*, 15: 269–274, 2003.

272. C. Shu, R. Dodda, F. Wu, M.S. Liu, and A.K. Jen, Highly efficient blue-light-emitting diodes from polyfluorene containing bipolar pendant groups, *Macromolecules*, 36: 6698–6703, 2003.

273. H.-H. Sung and H.-C. Lin, Novel alternating fluorene-based conjugated polymers containing oxadiazole pendants with various terminal groups, *Macromolecules*, 37: 7945–7954, 2004.

274. J. Ding, M. Day, G. Robertson, and J. Roovers, Synthesis and characterization of alternating copolymers of fluorene and oxadiazole, *Macromolecules*, 35: 3474–3483, 2002.

275. S. Janietz, J. Barche, A. Wedel, and D. Sainova, *n*-Type copolymers with fluorene and 1,3,4-heterodiazole moieties, *Macromol. Chem. Phys.*, 205: 187–198, 2004.

276. X. Zhan, Y. Liu, X. Wu, S. Wang, and D. Zhu, New series of blue-emitting and electron-transporting copolymers based on fluorene, *Macromolecules*, 35: 2529–2537, 2002.

277. N.C. Yang, Y.H. Park, and D.H. Suh, Synthesis and properties of new ultraviolet-blue-emissive fluorene-based aromatic polyoxadiazoles with confinement moieties, *J. Polym. Sci., Part A: Polym. Chem.*, 41: 674–683, 2003.

278. X.-H. Zhou, J.-C. Yan, and J. Pei, Exploiting an imidazole-functionalized polyfluorene derivative as a chemosensory material, *Macromolecules*, 37: 7078–7080, 2004.

279. M.S. Liu, Z. Jiang, P. Herguth, and A.K.-Y. Jen, Efficient cyano-containing electron-transporting polymers for light-emitting diodes, *Chem. Mater.*, 13: 3820–3822, 2001.

280. W.-L. Yu, J. Pei, Y. Cao, W. Huang, and A.J. Heeger, New efficient blue light emitting polymer for light emitting diodes, *Chem. Commun.* 18: 1837–1838, 1999.

281. W.-L. Yu, Y. Cao, J. Pei, W. Huang, and A.J. Heeger, Blue polymer light-emitting diodes from poly(9,9-dihexylfluorene-*alt-co*-2,5-didecyloxy-*p*-phenylene), *Appl. Phys. Lett.*, 75: 3270– 3272, 1999.

282. W.-L. Yu, J. Pei, W. Huang, and A.J. Heeger, Spiro-functionalized polyfluorene derivatives as blue light-emitting materials, *Adv. Mater.*, 12: 828–831, 2000.

283. D. Marsitzky, J. Murray, J.C. Scott, and K.R. Carter, Amorphous poly-2,7-fluorene networks, *Chem. Mater.*, 13: 4285–4289, 2001.

284. X.-M. Liu, C. He, X.-T. Hao, L.-W. Tan, Y. Li, and K.S. Ong, Hyperbranched blue-light-emitting alternating copolymers of tetrabromoarylmethane/silane and 9,9-dihexylfluorene-2,7-diboronic acid, *Macromolecules*, 37: 5965–5970, 2004.

285. Q. Peng, M. Xie, Y. Huang, Z. Lu, and D. Xiao, New series of highly phenyl-substituted poly-fluorene derivatives for polymer light-emitting diodes, *J. Polym. Sci., Part A: Polym. Chem.*, 42: 2985–2993, 2004.

286. N.G. Pschirer and U.H.F. Bunz, Poly(fluorenylene ethynylene)s by alkyne metathesis: Optical properties and aggregation behavior, *Macromolecules*, 33: 3961–3963, 2000.

287. X. Jiang, S. Liu, H. Ma, and A.K. Jen, High-performance blue light-emitting diode based on a binaphthyl-containing polyfluorene, *Appl. Phys. Lett.*, 76: 1813–1815, 2000.

288. W. Wang, J. Xu, and Y.-H. Lai, Alternating conjugated and transannular chromophores: Tunable property of fluorene–paracyclophane copolymers via transannular π–π interaction, *Org. Lett.*, 5: 2765–2768, 2003.

289. W. Wang, J. Xu, Y.-H. Lai, and F. Wang, Alternating aromatic and transannular chromophores with and without linker: Effect of transannular π–π interaction on the optical property of dithi-aparacyclophane-based copolymers, *Macromolecules*, 37: 3546–3553, 2004.

290. B. Liu, W.-L. Yu, J. Pei, S.-Y. Liu, Y.-H. Lai, and W. Huang, Design and synthesis of bipyri-dyl-_containing conjugated polymers: Effects of polymer rigidity on metal ion sensing, *Macromolecules*, 34: 7932–7940, 2001.

291. T. Zhang, R. Wang, H. Ren, Z. Chen, and J. Li, Deep blue light-emitting polymers with fluori-nated backbone for enhanced color purity and efficiency, *Polymer*, 53: 1529–1534, 2012.

292. U. Giovanella, C. Botta, F. Galeotti, B. Vercelli, S. Battato, and M. Pasini, Perfluorinated polymer with unexpectedly efficient deep blue electroluminescence for full-colour OLED displays and light therapy applications, *J. Mater. Chem. C*, 1: 5322–5329, 2013.

293. F.-I. Wu, R. Dodda, D.S. Reddy, and C.-F. Shu, Synthesis and characterization of spiro-linked poly(terfluorene): A blue-emitting polymer with controlled conjugated length, *J. Mater. Chem.*, 12: 2893–2897, 2002.

294. A. Donat-Bouillud, I. Lévesque, Y. Tao, M. D'Iorio, S. Beaupré, P. Blondin, M. Ranger, J. Bouchard, and M. Leclerc, Light-emitting diodes from fluorene-based π-conjugated poly-mers, *Chem. Mater.*, 12: 1931–1936, 2000.

295. M. Ranger and M. Leclerc, Optical and electrical properties of fluorene-based π-conjugated polymers, *Can. J. Chem.*, 76: 1571–1577, 1998.

296. B. Liu, W.-L. Yu, Y.-H. Lai, and W. Huang, Synthesis, characterization, and structure–property relationship of novel fluorene–thiophene-based conjugated copolymers, *Macromolecules*, 33: 8945–8952, 2000.

297. B. Liu, Y.-H. Niu, W.-L. Yu, Y. Cao, and W. Huang, Application of alternating fluorene and thio-phene copolymers in polymer light-emitting diodes, *Synth. Met.*, 129: 129–134, 2002.

298. J. Pei, W.-L. Yu, W. Huang, and A.J. Heeger, The synthesis and characterization of an efficient green electroluminescent conjugated polymer: Poly[2,7-*bis*(4-hexylthienyl)-9,9-dihexylfluorene], *Chem. Commun.* 17: 1631–1632, 2000.

299. J. Pei, W.-L. Yu, J. Ni, Y.-H. Lai, W. Huang, and A.J. Heeger, Thiophene-based conjugated polymers for light-emitting diodes: Effect of aryl groups on photoluminescence efficiency and redox behavior, *Macromolecules*, 34: 7241–7248, 2001.

300. Y. He, S. Gong, R. Hottori, and J. Kanicki, High performance organic polymer light-emitting heterostructure devices, *Appl. Phys. Lett.*, 74: 2265–2267, 1999.

301. C.I. Wilkinson, D.G. Lidzey, L.C. Palilis, R.B. Fletcher, S.J. Martin, X.H. Wang, and D.D.C. Bradley, Enhanced performance of pulse driven small area polyfluorene light emitting diodes, *Appl. Phys. Lett.*, 79: 171–173, 2001.

302. A.J. Campbell, D.D.C. Bradley, and H. Antoniadis, Dispersive electron transport in an electroluminescent polyfluorene copolymer measured by the current integration time-of-flight method, *Appl. Phys. Lett.*, 79: 2133–2135, 2001.

303. I.S. Millard, High-efficiency polyfluorene polymers suitable for RGB applications, *Synth. Met.*, 111–112: 119–123, 2000.

304. Q. Peng, Z.-Y. Lu, Y. Huang, M.-G. Xie, D. Xiao, S.-H. Han, J.-B. Peng, and Y. Cao, Novel efficient green electroluminescent conjugated polymers based on fluorene and triarylpyrazoline for light-emitting diodes, *J. Mater. Chem.*, 14: 396–401, 2004.

305. P. Sonar, J. Zhang, A.C. Grimsdale, K. Müllen, M. Surin, R. Lazzaroni, P. Leclère, S. Tierney, M. Heeney, and I. McCulloch, 4-Hexylbithieno[3,2-*b*:2',3'-*e*]pyridine: An efficient electron-accepting unit in fluorene and indenofluorene copolymers for light-emitting devices, *Macromolecules*, 37: 709–715, 2004.

306. K. Nomura, H. Morimoto, Y. Imanishi, Z. Ramhani, and Y. Geerts, Synthesis of high molecular weight *trans*-poly(9,9-di-*n*-octylfluorene-2,7-vinylene) by the acyclic diene metathesis polymerization using molybdenum catalysts, *J. Polym. Sci., Part A: Polym. Chem.*, 39: 2463–2470, 2001.

307. S. Jin, S.Y. Kang, M. Kim, Y.U. Chan, J. Young, K. Lee, and Y. Gal, Synthesis and electroluminescence properties of poly(9,9-di-*n*-octylfluorenyl-2,7-vinylene) derivatives for light-emitting display, *Macromolecules*, 36: 3841–3847, 2003.

308. D.-H. Hwang, J.-D. Lee, J.-M. Kang, S. Lee, C.-H. Lee, and S.-H. Jin, Synthesis and light-emitting properties of poly(9,9-di-*n*-octylfluorenyl-2,7-vinylene) and PPV copolymers, *J. Mater. Chem.*, 13: 1540–1545, 2003.

309. Y. Jin, J. Ju, J. Kim, S. Lee, J.Y. Kim, S.H. Park, S.-M. Son, S.-H. Jin, K. Lee, and H. Suh, Design, synthesis, and electroluminescent property of CN-poly(dihexylfluorenevinylene) for LEDs, *Macromolecules*, 36: 6970–6975, 2003.

310. J.K. Kim, J.W. Yu, J.M. Hong, H.N. Cho, D.Y. Kim, and C.Y. Kim, An alternating copolymer consisting of light emitting and electron transporting units, *J. Mater. Chem.*, 9: 2171–2176, 1999.

311. Q. Peng, Z. Lu, Y. Huang, M. Xie, D. Xiao, and D. Zou, Novel light-emitting polymers derived from fluorene and maleimide, *J. Mater. Chem.*, 13: 1570–1574, 2003.

312. X. Zhan, S. Wang, Y. Liu, X. Wu, and D. Zhu, New series of blue-emitting and electron-transporting copolymers based on cyanostilbene, *Chem. Mater.*, 15: 1963–1969, 2003.

313. S. Beaupré and M. Leclerc, Fluorene-based copolymers for red-emitting diodes, *Adv. Funct. Mater.*, 12: 192–196, 2002.

314. S.-H. Jin, M.-Y. Kim, D.-S. Koo, and Y.-I. Kim, Synthesis and properties of poly(fluorene-*alt*-cyanophenylene vinylene)-based alternating copolymers for light-emitting diodes, *Chem. Mater.*, 16: 3299–3307, 2004.

315. Q. Peng, Z.-Y. Lu, Y. Huang, M.-G. Xie, S.-H. Han, J.-B. Peng, and Y. Cao, Synthesis and characterization of new red-emitting polyfluorene derivatives containing electron-deficient 2-pyran-4-ylidene-malononitrile moieties, *Macromolecules*, 37: 260–266, 2004.

316. M. Chen, E. Perzon, M.R. Andersson, S.K.M. Jönsson, M. Fahlman, and M. Berggren, 1 Micron wavelength photo- and electroluminescence from π-conjugated polymer, *Appl. Phys. Lett.*, 84: 3570–3572, 2004.

317. H. Ye, B. Zhao, M. Liu, X. Zhou, Y. Li, D. Li, S.-J. Su, W. Yang, and Y. Cao, Dual-functional conjugated polymers based on trifluoren-2-yl-amine for RGB organic light-emitting diodes, *J. Mater. Chem.*, 21:17454–17461, 2011.

318. G. Klärner, J.-I. Lee, M.H. Davey, and R.D. Miller, Exciton migration and trapping in copolymers based on dialkylfluorenes, *Adv. Mater.*, 11: 115–119, 1999.

319. J.-I. Lee, T. Zyung, R.D. Miller, Y.H. Kim, S.C. Jeoung, and D. Kim, Photoluminescence study on exciton migration and trapping in a copolymer based on poly(fluorene), *J. Mater. Chem.*, 10: 1547–1550, 2000.

320. D.-H. Hwang, N.S. Cho, B.-J. Jung, H.-K. Shim, J.-I. Lee, L.-M. Do, and T. Zyung, Band gap tuning of new light emitting conjugated polymers, *Opt. Mater.*, 21: 199–203, 2002.

321. N. Nemoto, H. Kameshima, Y. Okano, and T. Endo, Synthesis of novel π-conjugating polymers based on dibenzothiophene, *J. Polym. Sci., Part A: Polym. Chem.*, 41: 1521–1526, 2003.

322. R. Yang, R. Tian, Q. Hou, W. Yang, and Y. Cao, Synthesis and optical and electroluminescent properties of novel conjugated copolymers derived from fluorene and benzoselenadiazole, *Macromolecules*, 36: 7453–7460, 2003.

323. J. Yang, C. Jiang, Y. Zhang, R. Yang, W. Yang, Q. Hou, and Y. Cao, High-efficiency saturated red emitting polymers derived from fluorene and naphtoselenadiazole, *Macromolecules*, 37: 1211–1218, 2004.

324. Q. Hou, Y. Xu, W. Yang, M. Yuan, J. Peng, and Y. Cao, Novel red-emitting fluorene-based copolymers, *J. Mater. Chem.*, 12: 2887–2892, 2002.

325. Q. Hou, Q. Zhou, Y. Zhang, W. Yang, R. Yang, and Y. Cao, Synthesis and electroluminescent properties of high-efficiency saturated red emitter based on copolymers from fluorene and 4,7-di(4-hexylthien-2-yl)-2,1,3-benzothiadiazole, *Macromolecules*, 37: 6299–6305, 2004.

326. X. Kong, A.P. Kulkarni, and S.A. Jenekhe, Phenothiazine-based conjugated polymers: Synthesis, electrochemistry, and light-emitting properties, *Macromolecules*, 36: 8992–8999, 2003.

327. D.-H. Hwang, M.-J. Park, S.-K. Kim, N.-H. Lee, C. Lee, Y.-B. Kim, and H.-K. Shim, Characterization of white electroluminescent devices fabricated using conjugated polymer blends, *J. Mater. Res.*, 19: 2081–2086, 2004.

328. B. Qu, L. Feng, H. Yang, Z. Gao, C. Gao, Z. Chen, L. Xiao, and Q. Gong, Color-stable deep red-emitting OLEDs based on a soluble terpolymer containing fluorine, thiophene and benzothiadiazole units, *Synth. Met.*, 162: 1587–1593, 2012.

329. I. Osken, A.S. Gundogan, E. Tekin, M.S. Eroglu, and T. Ozturk, Fluorene-dithienothiophene-*S,S*-dioxide copolymers. Fine-tuning for OLED applications, *Macromolecules*, 46: 9202–9210, 2013.

330. M. Fukuda, K. Sawada, and K. Yoshino, Fusible conducting poly(9-alkylfluorene) and poly(9,9-dialkylfluorene) and their characteristics, *Jpn. J. Appl. Phys., Pt. 2 Lett.*, 28: L1433–L1435, 1989.

331. M. Fukuda, K. Sawada, S. Morita, and K. Yoshino, Novel characteristics of conducting poly (9-alkylfluorene), poly(9,9-dialkylfluorene) and poly(1,10-*bis*(9′-alkylfluorenyl)alkane), *Synth. Met.*, 41–43: 855–858, 1991.

332. M. Fukuda, K. Sawada, and K. Yoshino, Synthesis of fusible and soluble conducting polyfluorene derivatives and their characteristics, *J. Polym. Sci., Part A: Polym. Chem.*, 31: 2465–2471, 1993.

333. Y. Ohmori, A. Uchida, K. Muro, and K. Yoshino, Blue electroluminescent diodes utilizing poly(alkylfluorene), *Jpn. J. Appl. Phys.*, 30: L1941–L1943, 1991.

334. Q. Pei and Y. Yang, Efficient photoluminescence and electroluminescence from a soluble polyfluorene, *J. Am. Chem. Soc.*, 118: 7416–7417, 1996.

335. Q. Pei, G. Yu, and Y. Yang, Polyfluorenes as materials for photoluminescence and electroluminescence, PCT Patent WO 97/33323, September 12, 1997.

336. Q. Pei, G. Yu, and Y. Yang, Polyfluorenes as materials for photoluminescence and electroluminescence, U.S. Patent 5,900,327, May 4, 1999.

337. E.P. Woo, W.R. Shiang, M. Inbasekaran, and G.R. Roof, 2,7-Diaryl-9-substituted fluorenes and 9-substituted fluorene oligomers and polymers, U.S. Patent 5,962,631, October 5, 1999.

338. E.P. Woo, M. Inbasekaran, W.R. Shiang, G.R. Roof, M.T. Bernius, and W. Wu, Fluorene-containing polymers and compounds useful in the preparation thereof, U.S. Patent 6,169,163 B1, January 2, 2001.

339. E.P. Woo, W.R. Shiang, M. Inbasekaran, G.R. Roof, M.T. Bernius, and W. Wu, Fluorene-containing polymers and compounds useful in the preparation thereof, U.S. Patent 6,512,083, January 28, 2003.

340. E.P. Woo, W.R. Shiang, M. Inbasekaran, G.R. Roof, M.T. Bernius, and W. Wu, Fluorene-containing polymers and compounds useful in the preparation thereof, U.S. Patent 6,514,632, February 4, 2003.

341. T. Yamamoto, Electrically conducting and thermally stable π-conjugated poly(arylene)s prepared by organometallic processes, *Prog. Polym. Sci.*, 17: 1153–1205, 1992.

342. G. Klaerner and R.D. Miller, Polyfluorene derivatives: Effective conjugation lengths from well-defined oligomers, *Macromolecules*, 31: 2007–2009, 1998.

343. M. Kreyenschmidt, G. Klaerner, T. Fuhrer, J. Ashenhurst, S. Karg, W.D. Chen, V.Y. Lee, J.C. Scott, and R.D. Miller, Thermally stable blue-light-emitting copolymers of poly(alkylfluorene), *Macromolecules*, 31: 1099–1103, 1998.

344. H.-G. Nothofer, A. Meisel, T. Miteva, D. Neher, M. Forster, M. Oda, G. Liester, D. Sainova, A. Yasude, D. Lupo, W. Knoll, and U. Scherf, Liquid crystalline polyfluorenes for blue polarized electroluminescence, *Macromol. Symp.*, 154: 139–148, 2000.

345. M. Ranger and M. Leclerc, Novel base-dopable poly(2,7-fluorenylene) derivatives, *Chem. Commun.*, 17: 1597–1598, 1997.

346. M. Ranger, D. Rondeau, and M. Leclerc, New well-defined poly(2,7-fluorene) derivatives: Photoluminescence and base doping, *Macromolecules*, 30: 7686–7691, 1997.

347. M. Inbasekaran, W. Wu, and E.P. Woo, Process for preparing conjugated polymers, U.S. Patent 5,777,070, July 7, 1998.

348. C.R. Towns and R. O'Dell, Polymer preparation, PCT Patent WO 00/53656, September 14, 2000.

349. M. Grell, D.D.C. Bradley, X. Long, T. Chamberlain, M. Inbasekaran, E.P. Woo, and M. Soliman, Chain geometry, solution aggregation and enhanced dichroism in the liquid-crystalline conjugated polymer poly(9,9-dioctylfluorene), *Acta Polym.*, 49: 439–444, 1998.

350. G. Fytas, H.G. Nothofer, U. Scherf, D. Vlassopoulos, and G. Meier, Structure and dynamics of nondilute polyfluorene solutions, *Macromolecules*, 35: 481–488, 2002.

351. M. Grell, D.D.C. Bradley, M. Inbasekaran, and E.P. Woo, A glass-forming conjugated main-chain liquid crystal polymer for polarized electroluminescence applications, *Adv. Mater.*, 9: 798–802, 1997.

352. B. Schartel, V. Wachtendorf, M. Grell, D.D.C. Bradley, and M. Hennecke, Polarized fluorescence and orientational order parameters of a liquid-crystalline conjugated polymer, *Phys. Rev. B*, 60: 277–283, 1999.

353. M. Grell, W. Knoll, D. Lupo, A. Miesel, T. Miteva, D. Neher, H.-G. Nothofer, U. Scherf, and A. Yasuda, Blue polarized electroluminescence from a liquid crystalline polyfluorene, *Adv. Mater.*, 11: 671–675, 1999.

354. M. Grell, M. Redecker, K.S. Whitehead, D.D.C. Bradley, M. Inbasekaran, E.P. Woo, and W. Wu, Monodomain alignment of thermotropic fluorene copolymers, *Liq. Cryst.*, 26: 1403–1407, 1999.

355. M. Grell and D.D.C. Bradley, Polarized luminescence from oriented molecular materials, *Adv. Mater.*, 11: 895–905, 1999.

356. K.S. Whitehead, M. Grell, D.D.C. Bradley, M. Jandke, and P. Strohriegl, Highly polarized blue electroluminescence from homogeneously aligned films of poly(9,9-dioctylfluorene), *Appl. Phys. Lett.*, 76: 2946–2948, 2000.

357. S. Kawana, M. Durrel, J. Lu, J.E. Macdonald, M. Grell, D.D.C. Bradley, P.C. Jules, R.A.L. Jones, and S.L. Bennett, X-ray diffraction study of the structure of thin polyfluorene films, *Polymer*, 43: 1907–1913, 2002.

358. S.H. Chen, H.L. Chou, A.C. Su, and S.A. Chen, Molecular packing in crystalline poly(9,9-di-*n*-octyl-2,7-fluorene), *Macromolecules*, 37: 6833–6838, 2004.

359. M. Grell, D.D.C. Bradley, G. Ungar, J. Hill, and K.S. Whitehead, Interplay of physical structure and photophysics for a liquid crystalline polyfluorene, *Macromolecules*, 32: 5810–5817, 1999.

360. J. Teetsov and M.A. Fox, Photophysical characterization of dilute solution and ordered thin films of alkyl-substituted polyfluorenes, *J. Mater. Chem.*, 9: 2117–2122, 1999.

361. M. Misaki, Y. Ueda, S. Nagamatsu, Y. Yoshida, N. Tanigaki, and K. Yase, Formation of single-crystal-like poly(9,9-dioctylfluorene) thin films by the friction-transfer technique with subsequent thermal treatments, *Macromolecules*, 37: 6926–6931, 2004.

362. M. Oda, H.-G. Nothofer, G. Lieser, U. Scherf, S.C.J. Meskers, and D. Neher, Circularly polarized electroluminescence from liquid-crystalline chiral polyfluorenes, *Adv. Mater.*, 12: 362–365, 2000.

363. H.-Z. Tang, M. Fujiki, and T. Sato, Thermodriven conformational transition of optically active poly[2,7-{9,9-*bis*[(S)-2-methyloctyl]}fluorene] in solution, *Macromolecules*, 35: 6439–6445, 2002.

364. H.-Z. Tang, M. Fujiki, and M. Motonaga, Alkyl side chain effects of optically active polyfluorenes on their chiroptical absorption and emission properties, *Polymer*, 43: 6213–6220, 2002.

365. L. Wu, T. Sato, H.-Z. Tang, and M. Fujiki, Conformation of a polyfluorene derivative in solution, *Macromolecules*, 37: 6183–6188, 2004.

366. M.J. Banach, R.H. Friend, and H. Sirringhaus, Influence of the casting solvent on the thermotropic alignment of thin liquid crystalline polyfluorene copolymer films, *Macromolecules*, 37: 6079–6085, 2004.

367. X. Gong, P.K. Iyer, D. Moses, G.C. Bazan, A.J. Heeger, and S.S. Xiao, Stabilized blue emission from polyfluorene-based light-emitting diodes: Elimination of fluorenone defects, *Adv. Funct. Mater.*, 13: 325–329, 2003.

368. M. Ariu, D.G. Lidzey, M. Sims, A.J. Cadby, P.A. Lane, and D.D.C. Bradley, The effect of morphology on the temperature-dependent photoluminescence quantum efficiency of the conjugated polymer poly(9,9-dioctylfluorene), *J. Phys.: Condens. Matter*, 14: 9975–9986, 2002.

369. Y. Koizumi, S. Seki, A. Acharya, A. Saeki, and S. Tagawa, Delocalization of positive and negative charge carriers on oligo- and poly-fluorenes studied by low-temperature matrix isolation technique, *Chem. Lett.*, 33: 1290–1291, 2004.

370. J. Jo, C. Chi, S. Höger, G. Wegner, and D.Y. Yoon, Synthesis and characterization of monodisperse oligofluorenes, *Chem. Eur. J.*, 10: 2681–2688, 2004.

371. S. Janietz, D.D.C. Bradley, M. Grell, C. Giebeler, M. Inbasekaran, and E.P. Woo, Electrochemical determination of the ionization potential and electron affinity of poly(9,9-dioctylfluorene), *Appl. Phys. Lett.*, 73: 2453–2455, 1998.

372. T. Miteva, A. Meisel, W. Knoll, H.G. Nothofer, U. Scherf, D.C. Müller, K. Meerholz, A. Yasuda, and D. Neher, Improving the performance of polyfluorene-based organic light-emitting diodes via end-capping, *Adv. Mater.*, 13: 565–570, 2001.

373. L.S. Liao, M.K. Fung, C.S. Lee, S.T. Lee, M. Inbasekaran, E.P. Woo, and W.W. Wu, Electronic structure and energy band gap of poly(9,9-dioctylfluorene) investigated by photoelectron spectroscopy, *Appl. Phys. Lett.*, 76: 3582–3584, 2000.

374. L.S. Liao, L.F. Cheng, M.K. Fung, C.S. Lee, S.T. Lee, M. Inbasekaran, E.P. Woo, and W.W. Wu, Oxygen effect on the interface formation between calcium and a polyfluorene film, *Phys. Rev. B*, 62: 10004–10007, 2000.

375. D. Poplavsky, J. Nelson, and D.D.C. Bradley, Ohmic hole injection in poly(9,9-dioctylfluorene) polymer light-emitting diodes, *Appl. Phys. Lett.*, 83: 707–709, 2003.

376. T. van Woudenbergh, J. Wildeman, P.W.M. Blom, J.J.A.M. Bastiaansen, and B.M.W. Langeveld-Voss, Electron-enhanced hole injection in blue polyfluorene-based polymer light-emitting diodes, *Adv. Funct. Mater.*, 14: 677–683, 2004.

377. A.J. Campbell, D.D.C. Bradley, H. Antoniadis, M. Inbasekaran, W.W. Wu, and A.P. Woo, Transient and steady-state space-charge-limited currents in polyfluorene copolymer diode structures with ohmic hole injection contacts, *Appl. Phys. Lett.*, 76: 1734–1736, 2000.

378. M. Redecker, D.D.C. Bradley, M. Inbasekaran, and E.P. Woo, Nondispersive hole transport in an electroluminescent polyfluorene, *Appl. Phys. Lett.*, 73: 1565–1567, 1998.

379. G. Greczynski, M. Fahlman, W.R. Salaneck, N. Johansson, D.A. dos Santos, A. Dkhissi, and J.L. Brédas, Electronic structure of poly(9,9-dioctylfluorene) in the pristine and reduced state, *J. Chem. Phys.*, 116: 1700–1706, 2002.

380. J. Cornil, I. Gueli, A. Dkhissi, J.C. Sancho-Garcia, E. Hennebicq, J.P. Calbert, V. Lemaur, D. Beljonne, and J.L. Brédas, Electronic and optical properties of polyfluorene and fluorene-based copolymers: A quantum-chemical characterization, *J. Chem. Phys.*, 118: 6615–6623, 2003.

381. E. Zojer, A. Pogantsch, E. Hennebicq, D. Beljonne, J. Brédas, and E.J.W. List, Green emission from poly(fluorene)s: The role of oxidation, *J. Chem. Phys.*, 117: 6794–6802, 2002.

382. X.H. Yang, D. Neher, C. Spitz, E. Zojer, J.L. Brédas, R. Günther, and U. Scherf, On the polarization of the green emission of polyfluorenes, *J. Chem. Phys.*, 119: 6832–6839, 2003.

383. I. Franco and S. Tretiak, Electron-vibrational relaxation of photoexcited polyfluorenes in the presence of chemical defects: A theoretical study, *Chem. Phys. Lett.*, 372: 403–408, 2003.

384. I. Franco and S. Tretiak, Electron-vibrational dynamics of photoexcited polyfluorenes, *J. Am. Chem. Soc.*, 126: 12130–12140, 2004.

385. U. Lemmer, S. Heun, R.F. Mahrt, U. Scherf, M. Hopmeier, U. Siegner, E.O. Göbel, K. Müllen, and H. Bässler, Aggregate fluorescence in conjugated polymers, *Chem. Phys. Lett.*, 240: 373–378, 1995.

386. V.N. Bliznyuk, S.A. Carter, J.C. Scott, G. Klärner, R.D. Miller, and D.C. Miller, Electrical and photoinduced degradation of polyfluorene based films and light-emitting devices, *Macromolecules*, 32: 361–369, 1999.

387. J.-I. Lee, G. Klaerner, and R.D. Miller, Oxidative stability and its effect on the photoluminescence of poly(fluorene) derivatives: End group effects, *Chem. Mater.*, 11: 1083–1088, 1999.

388. L.M. Herz and R.T. Phillips, Effects of interchain interactions, polarization anisotropy, and photo-oxidation on the ultrafast photoluminescence decay from a polyfluorene, *Phys. Rev. B*, 61: 13691–13697, 2000.

389. H.-Z. Tang, M. Fujiki, Z.-B. Zhang, K. Torimitsu, and M. Motonaga, Nearly pure blue photoluminescent poly(2,7-[9-{3,5-*bis*[3,5-*bis*(benzyloxy)benzyloxy]benzyl]-9-(3,6-dioxaoctyl)]fluorene} in film, *Chem. Commun.*, 23: 2426–2427, 2001.

390. G. Zeng, W.-L. Yu, S.-J. Chua, and W. Huang, Spectral and thermal stability study for fluorene-based conjugated polymers, *Macromolecules*, 35: 6907–6914, 2002.

391. S. Panozzo, J.-C. Vial, Y. Kervela, and O. Stéphan, Fluorene–fluorenone copolymer: Stable end efficient yellow-emitting material for electroluminescent devices, *J. Appl. Phys.*, 92: 3495–3502, 2002.

392. E.J.W. List, R. Guentner, P.S. de Freitas, and U. Scherf, The effect of keto defect sites on the emission properties of polyfluorene-type materials, *Adv. Mater.*, 14: 374–378, 2002.

393. M. Gaal, E.J.W. List, and U. Scherf, Excimers or emissive on-chain defects? *Macromolecules*, 36: 4236–4237, 2003.

394. L. Romaner, A. Pogantsch, P.S. de Freitas, U. Scherf, M. Gaal, E. Zojer, and E.J.W. List, The origin of green emission in polyfluorene-based conjugated polymers: On-chain defect fluorescence, *Adv. Funct. Mater.*, 13: 597–601, 2003.

395. S. Gamerith, C. Gadermaier, U. Scherf, and E.J.W. List, Emission properties of pristine and oxidatively degraded polyfluorene type polymers, *Phys. Status Solidi A*, 201: 1132–1151, 2004.

396. P. Scanducci de Freitas, U. Scherf, M. Collon, and E.J.W. List, (9,9-Dialkylfluorene-*co*-fluorenone) Copolymers containing low fluorenone fractions as model systems for degradation-induced changes in polyfluorene-type semiconducting materials, *e-Polymers*, Abstr. No. 009, 2002.

397. W. Zhao, T. Cao, and J.M. White, On the origin of green emission in polyfluorene polymers: The roles of thermal oxidation degradation and crosslinking, *Adv. Funct. Mater.*, 14: 783–790, 2004.

398. M. Sims, D.D.C. Bradley, M. Ariu, M. Koeberg, A. Asimakis, M. Grell, and D.G. Lidzey, Understanding of the origin of the 535 nm emission band in oxidized poly(9,9-dioctylfluorene): The essential role of inter-chain/inter-segment interactions, *Adv. Funct. Mater.*, 14: 765–781, 2004.

399. M. Stoessel, G. Wittmann, J. Staudigel, F. Steuber, J. Blässing, W. Roth, H. Klausmann, W. Rogler, J. Simmerer, A. Winnacker, M. Inbasekaran, and E.P. Woo, Cathode-induced luminescence quenching in polyfluorenes, *J. Appl. Phys.*, 87: 4467–4475, 2000.

400. M.R. Craig, M.M. de Kok, J.W. Hofstraat, A.P.H.J. Schenning, and E.W. Meijer, Improving color purity and stability in a blue emitting polyfluorene by monomer purification, *J. Mater. Chem.*, 13: 2861–2862, 2003.

401. G. Hughes, C. Wang, A.S. Batsanov, M. Fern, S. Frank, M.R. Bryce, I.F. Perepichka, A.P. Monkman, and B.P. Lyons, New pyrimidine- and fluorine-containing oligo(arylene)s: Synthesis, crystal structures, optoelectronic properties and a theoretical study, *Org. Biomol. Chem.*, 1: 3069–3077, 2003.

402. J.-I. Lee, G. Klaerner, and R.D. Miller, Structure–property relationship for excimer formation in poly(alkylfluorene) derivatives, *Synth. Met.*, 101: 126, 1999.
403. Y.-S. Suh, S.W. Ko, B.-J. Jung, and H.-K. Shim, Synthesis and electroluminescent properties of cyclohexyl-substituted polyfluorenes, *Opt. Mater.*, 21: 109–118, 2002.
404. M. Oda, H.-G. Nothofer, U. Scherf, V. Šunjić, D. Richter, W. Regenstein, and D. Neher, Chiroptical properties of chiral substituted polyfluorenes, *Macromolecules*, 35: 6792–6798, 2002.
405. A. Rieman and W. Ude, Method for making 9,9-*Bis*(4-hydroxyphenyl)-fluorene, U.S. Patent 4,675,458, June 23, 1987.
406. W. Orth, E. Pastorek, W. Weiss, and H.W. Kleffner, Preparation of 9,9-*bis*-(4-hydroxyphenyl)-fluorene, U.S. Patent 5,149,886, September 22, 1992.
407. W. Orth, E. Pastorek, W. Weiss, and H.W. Kleffner, Preparation of 9,9-*bis*-(4-hydroxyphenyl)-fluorene, U.S. Patent 5,169,990, December 8, 1992.
408. M. Yamada, J. Sun, Y. Suda, and T. Nakaya, Synthesis of fluorenebisphenoxy derivatives by acid-sulfur compound catalyzed condensation reaction, *Chem. Lett.*, 28: 1055–1056, 1998.
409. D.-H. Hwang, M.-J. Park, and J.-H. Lee, EL properties of stable blue light-emitting polyfluorene copolymers, *Mater. Sci. Eng. C*, 24: 201–204, 2004.
410. D. Sainova, T. Miteva, H.G. Nothofer, U. Scherf, I. Glowacki, J. Ilanski, H. Fujikawa, and D. Neher, Control of color and efficiency of light-emitting diodes based on polyfluorenes blended with hole-transport molecules, *Appl. Phys. Lett.*, 76: 1810–1812, 2000.
411. A. Grice, D.D.C. Bradley, M. Grell, M.D. Bradley, M.T. Bernius, M. Inbasekaran, W.W. Wu, and E.P. Woo, High brightness and efficiency blue light-emitting polymer diodes, *Appl. Phys. Lett.*, 73: 629–631, 1998.
412. J.I. Lee, D.-H. Hwang, H. Park, L.-M. Do, H.Y. Chu, T. Zyung, and R.D. Miller, Light-emitting electrochemical cells based on poly(9,9-*bis*(3,6-dioxaheptyl)-fluorene-2,7-diyl), Vol. 1 *Synth. Met.*, 111–112: 195–197, 2000.
413. D.C. Müller, T. Braig, H.-G. Nothofer, M. Arnoldi, M. Gross, U. Scherf, O. Nuyken, and K. Meerholz, Efficient blue organic light-emitting diodes with graded hole-transport layers, *Chemphyschem*, 1: 207–211, 2000.
414. S. Xiao, M. Nguyen, X. Gong, Y. Cao, H. Wu, D. Moses, and A.J. Heeger, Stabilization of semi-conducting polymers with silsesquioxane, *Adv. Funct. Mater.*, 13: 25–29, 2003.
415. G. Klärner, J.-I. Lee, V.Y. Lee, E. Chan, J.-P. Chen, A. Nelson, D. Markiewicz, R. Siemens, J.C. Scott, and R.D. Miller, Cross-linkable polymers based on dialkyfluorenes, *Chem. Mater.*, 11: 1800–1805, 1999.
416. J.P. Chen, G. Klaerner, J.-I. Lee, D. Markiewicz, V.Y. Lee, R.D. Miller, and J.C. Scott, Efficient, blue light-emitting diodes using cross-linked layers of polymeric arylamine and fluorene, *Synth. Met.*, 107: 129–135, 1999.
417. X. Gong, W. Ma, J.C. Ostrowski, K. Bechgaard, G.C. Bazan, A.J. Heeger, S. Xiao, and D. Moses, End-capping as a method for improving carrier injection in electrophosphorescent light-emitting diodes, *Adv. Funct. Mater.*, 14: 393–397, 2004.
418. J.P. Chen, D. Markiewicz, V.Y. Lee, G. Klaerner, R.D. Miller, and J.C. Scott, Improved efficiencies of light-emitting diodes through incorporation of charge transporting components in tri-block polymers, *Synth. Met.*, 107: 203–207, 1999.
419. J.K. Kim, S.I. Hong, H.N. Cho, D.Y. Kim, and C.Y. Kim, An alternating copolymer for a blue light-emitting diode, *Polym. Bull.*, 38: 169–176, 1997.
420. J. Lu, Y. Tao, M. D'iorio, Y. Li, J. Ding, and M. Day, Pure deep blue light-emitting diodes from alternating fluorene/carbazole copolymers by using suitable hole-blocking materials, *Macromolecules*, 37: 2442–2449, 2004.
421. J.-F. Morin and M. Leclerc, Synthesis of conjugated polymers derived from *N*-alkyl-2,7-carbazoles, *Macromolecules*, 34: 4680–4682, 2001.
422. M. Redecker, D.D.C. Bradley, M. Inbasekaran, W.W. Wu, and E.P. Woo, High mobility hole transport fluorene–triarylamine copolymers, *Adv. Mater.*, 11: 241–246, 1999.

423. M. Redecker, D.D.C. Bradley, K.J. Baldwin, D.A. Smith, M. Inbasekaran, W.W. Wu, and E.P. Woo, An investigation of the emission solvatochromism of a fluorene–triarylamine copolymer studied by time resolved spectroscopy, *J. Mater. Chem.*, 9: 2151–2154, 1999.

424. E.P. Woo, M.T. Bernius, M. Inbasekaran, and W. Wu, Fluorene-containing polymers and electroluminescent devices therefrom, U.S. Patent 6,309,763, October 30, 2001.

425. M. Inbasekaran, E. Woo, W. Wu, M. Bernius, and L. Wujkowski, Fluorene homopolymers and copolymers, *Synth. Met.*, 111–112: 397–401, 2000.

426. S. Hamai and F. Hirayama, Actinometric determination of absolute fluorescence quantum yields, *J. Phys. Chem.*, 87: 83–89, 1983.

427. S. Liu, X. Jiang, H. Ma, M.S. Liu, and A.K.-Y. Jen, Triarylamine-containing poly(perfluorocyclobutane) as hole-transporting materials for polymer light-emitting diodes, *Macromolecules*, 33: 3514–3517, 2000.

428. X. Kong and S.A. Jenekhe, Block copolymers containing conjugated polymer and polypeptide sequences: Synthesis and self-assembly of electroactive and photoactive nanostructures, *Macromolecules*, 37: 8180–8183, 2004.

429. T. Virgili, D. Lidzey, and D.D.C. Bradley, Efficient energy transfer from blue to red in tetraporphyrin-doped poly(9,9-dioctylfluorene) light-emitting diodes, *Adv. Mater.*, 12: 58–62, 2000.

430. I. Lévesque, A. Donat-Bouillud, Y. Tao, M. D'Iorio, S. Beaupré, P. Blondin, M. Ranger, J. Bouchard, and M. Leclerc, Organic tunable diodes from polyfluorene derivatives, *Synth. Met.*, 122: 79–81, 2001.

431. P. Blondin, J. Bouchard, S. Beupré, M. Belletête, G. Duroche, and M. Leclerc, Molecular design and characterization of chromic polyfluorene derivatives, *Macromolecules*, 33: 5874–5879, 2000.

432. B. Liu, W.-L. Yu, J. Pei, Y.-H. Lai, W. Huang, Y.-H. Niu, and Y. Cao, Synthesis and characterization of novel fluorene–thiophene-based conjugated copolymers, *Mater. Sci. Eng. B*, 85: 232–235, 2001.

433. G. Vamvounis and S. Holdcroft, Enhancing solid-state emission from conjugated polymers via self-forming host–guest systems, *Adv. Mater.*, 16: 716–719, 2004.

434. A.R. Buckley, M.D. Rahn, J. Hill, J. Cabanillas-Gonzalez, A.M. Fox, and D.D.C. Bradley, Energy transfer dynamics in polyfluorene-based polymer blends, *Chem. Phys. Lett.*, 339: 331–336, 2001.

435. S.-H. Jin, H.-J. Park, J.Y. Kim, K. Lee, S.-P. Lee, D.-K. Moon, H.-J. Lee, and Y.-S. Gal, Polyfluorenevinylene derivative by Gilch polymerization for light emitting diode applications, *Macromolecules*, 35: 7532–7534, 2002.

436. Commercially available from American Dye Source, Inc. See at: www.adsdyes.com.

437. J. Pei, X.-L. Liu, W.-L. Yu, Y.-H. Lai, Y.-H. Niu, and Y. Cao, Efficient energy transfer to achieve narrow bandwidth red emission from Eu^{3+}-grafting conjugated polymers, *Macromolecules*, 35: 7274–7280, 2002.

438. Q.D. Ling, E.T. Kang, K.G. Neoh, and W. Huang, Synthesis and nearly monochromatic photoluminescence properties of conjugated copolymers containing fluorene and rare earth complexes, *Macromolecules*, 36: 6995–7003, 2003.

439. J.-I. Lee, G. Klaerner, M.H. Davey, and R.D. Miller, Color tuning in polyfluorenes by copolymerization with low band gap comonomers, *Synth. Met.*, 107: 1087–1088, 1999.

440. H.N. Cho, J.K. Kim, D.Y. Kim, C.Y. Kim, N.W. Song, and D. Kim, Statistical copolymers for blue-light-emitting diodes, *Macromolecules*, 32: 1476–1481, 1999.

441. Y. Niu, Q. Hou, and Y. Cao, Thermal annealing below the glass transition temperature: A general way to increase performance of light-emitting diodes based on copolyfluorenes, *Appl. Phys. Lett.*, 81: 634–636, 2002.

442. O. Stéphan, F. Tran-Van, and C. Chevrot, New organic materials for light emitting devices based on dihexylfluorene-*co*-ethylenedioxythiophene copolymers exhibiting improved hole-injecting properties, *Synth. Met.*, 131: 31–40, 2002.

443. B.W. D'Andrade and S.R. Forrest, White organic light-emitting devices for solid-state lighting, *Adv. Mater.*, 16: 1585–1595, 2004.

444. C. Ego, D. Marsitzky, S. Becker, J. Zhang, A.C. Grimsdale, K. Müllen, J.D. MacKenzie, C. Silva, and R.H. Friend, Attaching perylene dyes to polyfluorene: Three simple, efficient methods for facile color tuning of light-emitting polymers, *J. Am. Chem. Soc.*, 125: 437–443, 2003.

445. P. Herguth, X. Jiang, M.S. Liu, and A.K.-Y. Jen, Highly efficient fluorene and benzothiadiazole-based conjugated copolymers for polymer light-emitting diodes, *Macromolecules*, 35: 6094–6100, 2002.

446. C.D. Müller, A. Falcou, N. Reckefuss, M. Rojahn, V. Widerhirn, P. Rudati, H. Frohne, O. Nuyken, H. Becker, and K. Meerholz, Multicolour organic light-emitting displays by solution processing, *Nature*, 421: 829–833, 2003.

447. G. Tu, Q. Zhou, Y. Cheng, L. Wang, D. Ma, X. Jing, and F. Song, White electroluminescence from polyfluorene chemically doped with 1,8-napthalimide moieties, *Appl. Phys. Lett.*, 12: 2172–2174, 2004.

448. J. Liu, Q. Zhou, Y. Cheng, Y. Geng, L. Xiang, D. Ma, X. Jing, and F. Wang, The first single polymer with simultaneous blue, green, and red emission for white electroluminescence, *Adv. Mater.*, 17: 2974–2978, 2005.

449. G. Tu, C. Mei, Q. Zhou, Y. Cheng, Y. Geng, L. Wang, D. Ma, X. Jing, and F. Wang, Highly efficient pure-white-light-emitting diodes from a single polymer: Polyfluorene with naphthalimide moieties, *Adv. Funct. Mater.*, 16: 101–106, 2006.

450. J. Liu, Q. Zhou, Y. Cheng, Y. Geng, L. Wang, D. Ma, X. Jing, and F. Wang, White electroluminescence from a single-polymer system with simultaneous two-color emission: Polyfluorene as the blue host and a 2,1,3-benzothiadiazole derivative as the orange dopant on the main chain, *Adv. Funct. Mater.*, 16: 957–965, 2006.

451. B. Liu, W.-L. Yu, Y.-H. Lai, and W. Huang, Synthesis of a novel cationic water-soluble efficient blue photoluminescent conjugated polymer, *Chem. Commun.*, 7: 551–552, 2000.

452. B. Liu, W.-L. Yu, Y.-H. Lai, and W. Huang, Blue-light-emitting cationic water-soluble polyfluorene derivatives with tunable quaternization degree, *Macromolecules*, 35: 4975–4982, 2002.

453. F. Huang, H. Wu, D. Wang, W. Yang, and Y. Cao, Novel electroluminescent polyelectrolytes based on polyfluorene, *Chem. Mater.*, 16: 708–716, 2004.

454. F. Huang, L. Hou, H. Wu, X. Wang, H. Shen, W. Cao, W. Yang, and Y. Cao, High-efficiency, environment-friendly electroluminescent polymers with stable high work function metal as a cathode: Green- and yellow-emitting conjugated polyfluorene polyelectrolytes and their neutral precursors, *J. Am. Chem. Soc.*, 126: 9845–9853, 2004.

455. B. Liu, S. Wang, G.C. Bazan, and A. Mikhailovsky, Shape-adaptable water-soluble conjugated polymers, *J. Am. Chem. Soc.*, 125: 13306–13307, 2003.

456. B. Liu and G.C. Bazan, Interpolyelectrolyte complexes of conjugated copolymers and DNA: Platforms for multicolor biosensors, *J. Am. Chem. Soc.*, 126: 1942–1943, 2004.

457. H.D. Burrows, V.M.M. Lobo, J. Pina, M.L. Ramos, J.S. de Melo, A.J.M. Velente, M.J. Tapia, S. Pradhan, and U. Scherf, Fluorescence enhancement of the water-soluble poly{1,4-phenylene-[9,9-*bis*(4-phenoxybutylsulfonate)]fluorene-2,7-diyl}copolymer in *n*-dodecylpentaoxyethylene glycol ether micelles, *Macromolecules*, 37: 7425–7427, 2004.

458. L. Ying, P. Zalar, S.D. Collins, Z. Chen, A.A. Mikhailovsky, T.-Q. Nguyen, and G.C. Bazan, All-conjugated triblock polyelectrolytes, *Adv. Mater.*, 24: 6496–6501, 2012.

459. J. Roncali, Conjugated poly(thiophenes): Synthesis, functionalization, and applications, *Chem. Rev.*, 92: 711–738, 1992.

460. D. Fichou, Ed., *Handbook of Oligo- and Polythiophenes*, Wiley, Weinheim, 1999, p. 534.

461. R.D. McCullough, The chemistry of conducting polythiophenes, *Adv. Mater.*, 10: 93–116, 1998.

462. J. Roncali, Electrogenerated functional conjugated polymers as advanced electrode materials, *J. Mater. Chem.*, 9: 1875–1893, 1999.

463. M. Theander, O. Inganäs, W. Mammo, T. Olinga, M. Svensson, and M. Andersson, Photophysics of substituted polythiophenes, *J. Phys. Chem. B*, 103: 7771–7780, 1999.

464. N.C. Greenham, I.D.W. Samuel, G.R. Hayes, R.T. Phillips, Y.A.R.R. Kessener, S.C. Moratti, A.B. Holmes, and R.H. Friend, Measurements of absolute photoluminescence quantum efficiencies in conjugated polymers, *Chem. Phys. Lett.*, 241: 89–96, 1995.

465. F. Chen, P.G. Mehta, L. Takiff, and R.D. McCullough, Improved electroluminescence perfor- mance of poly(3-alkylthiophenes) having a high head-to-tail (HT) ratio, *J. Mater. Chem.*, 6: 1763– 1766, 1996.

466. F. Garnier, Organic-based electronics *à la Carte*, *Acc. Chem. Res.*, 32: 209–215, 1999.

467. H. Saadeh, T. Goodson, III, and L. Yu, Synthesis of a polyphenylene-*co*-furan and polyphenyl- ene-*co*-thiophene and comparison of their electroluminescent properties, *Macromolecules*, 30: 4608–4612, 1997.

468. B. Kraabel, D. Moses, and A.J. Heeger, Direct observation of the intersystem crossing in poly(3- octylthiophene), *J. Chem. Phys.*, 103: 5102–5108, 1995.

469. M. Leclerc and K. Faïd, Conformation-induced chromism in conjugated polymers, in: *Handbook of Conducting Polymers*, T.A. Skotheim, R.L. Elsenbaumer, and J.R. Reynolds, Eds., Marcel Dekker, New York, 1998, pp. 695–706.

470. G. Zerbi, B. Chierichetti, and O. Inganäs, Vibrational spectra of oligothiophenes as model of polythiophenes, *J. Chem. Phys.*, 94: 4637–4645, 1991.

471. G. Zerbi, B. Chierichetti, and O. Inganäs, Thermochromism in polyalkylthiophenes: Molecular aspects from vibrational spectroscopy, *J. Chem. Phys.*, 94: 4646–4658, 1991.

472. Y. Yamamoto, K. Sanechika, and A. Yamamoto, Preparation of thermostable and electric- conducting poly(2,5-thienylene), *J. Polym. Sci., Polym. Lett. Ed.*, 18: 9–12, 1980.

473. J.W.-P. Lin and L.P. Dudek, Synthesis and properties of poly(2,5-thienylene), *J. Polym. Sci., Polym. Chem. Ed.*, 18: 2869–2873, 1980.

474. M. Kobayashi, J. Chen, T.-C. Chung, F. Moraes, A.J. Heeger, and F. Wudl, Synthesis and proper- ties of chemically coupled poly(thiophene), *Synth. Met.*, 9: 77–86, 1984.

475. T. Yamamoto, A. Morita, Y. Miyazaki, T. Maruyama, H. Wakayama, Z.H. Zhou, Y. Nakamura, T. Kanbara, S. Sasaki, and K. Kubota, Preparation of π-conjugated poly(thiophene-2,5-diyl), poly(*p*-phenylene), and related polymers using zerovalent nickel complexes. Linear structure and properties of the π-conjugated polymers, *Macromolecules*, 25: 1214–1223, 1992.

476. K. Yoshino, S. Hayashi, and R. Sugimoto, Preparation and properties of conducting heterocyclic polymer films by chemical method, *Jpn. J. Appl. Phys., Part 2*, 23: L899–L900, 1984.

477. M. Leclerc, F.M. Diaz, and G. Wegner, Structural analysis of poly(3-alkylthiophenes), *Makromol. Chem.*, 190: 3105–3116, 1989.

478. M. Pomerantz, J.J. Tseng, H. Zhu, S.J. Sproull, J.R. Reynolds, R. Uitz, H.G. Amott, and M.I. Haider, Processable polymers and copolymers of 3-alkylthiophenes and their blends, *Synth. Met.*, 41–43: 825–833, 1991.

479. K. Tamao, S. Kodama, I. Nakajima, M. Kumada, A. Minato, and K. Suzuki, Nickel–phosphine complex-catalyzed Grignard coupling—II: Grignard coupling of heterocyclic compounds, *Tetrahedron*, 38: 3347–3354, 1982.

480. S. Amou, O. Haba, K. Shirato, T. Hayakawa, M. Ueda, K. Takeuchi, and M. Asai, Head-to-tail regioregularity of poly(3-hexylthiophene) in oxidative coupling polymerization with $FeCl_3$, *J. Polym. Sci., Part A: Polym. Chem.*, 37: 1943–1948, 1999.

481. M.R. Andersson, D. Selse, M. Berggren, H. Järvinen, T. Hjertberg, O. Inganäs, O. Wennerström, and J.-E. Österholm, Regioselective polymerization of 3-(4-octylphenyl)thiophene with $FeCl_3$, *Macromolecules*, 27: 6503–6506, 1994.

482. R.D. McCullough and R.D. Lowe, Enhanced electrical conductivity in regioselectively synthe- sized poly(3-alkylthiophenes), *J. Chem. Soc., Chem. Commun.*, 1: 70–72, 1992.

483. R.D. McCullough, R.D. Lowe, M. Jayaraman, and D.L. Anderson, Design, synthesis, and con- trol of conducting polymer architectures: Structurally homogeneous poly(3-alkylthiophenes), *J. Org. Chem.*, 58: 904–912, 1993.

484. R.S. Loewe, S.M. Khersonsky, and R.D. McCullough, A simple method to prepare head-to- tail coupled, regioregular poly(3-alkylthiophenes) using Grignard metathesis, *Adv. Mater.*, 11: 250–253, 1999.

485. R.S. Loewe, P.C. Ewbank, J. Liu, L. Zhai, and R.D. McCullough, Regioregular, head-to-tail cou- pled poly(3-alkylthiophenes) made easy but the GRIM method: Investigation of the reaction and the origin of regioselectivity, *Macromolecules*, 34: 4324–4333, 2001.

486. R.D. McCullough and R.S. Loewe, Method of forming poly-(3-substituted) thiophenes, U.S. Patent 6,166,172, December 26, 2000.
487. R.D. McCullough, P.C. Ewbank, and R.S. Loewe, Self-assembly and disassembly of regioregular, water soluble polythiophenes: Chemoselective ionchromatic sensing in water, *J. Am. Chem. Soc.*, 119: 633–634, 1997.
488. S. Guillerez and G. Bidan, New convenient synthesis of highly regioregular poly(3-octylthiophene) based on the Suzuki coupling reaction, *Synth. Met.*, 93: 123–126, 1998.
489. T.A. Chen and R.D. Rieke, The first regioregular head-to-tail poly(3-hexylthiophene-2,5-diyl) and a regiorandom isopolymer: Nickel versus palladium catalysis of 2(5)-bromo-5(2)-(bromozincio)-3-hexylthiophene polymerization, *J. Am. Chem. Soc.*, 114: 10087–10088, 1992.
490. T.-A. Chen, X. Wu, and R.D. Rieke, Regiocontrolled synthesis of poly(3-alkylthiophenes) mediated by Rieke zinc: Their characterization and solid-state properties, *J. Am. Chem. Soc.*, 117: 233–244, 1995.
491. Y. Ohmori, M. Uchida, K. Muro, and K. Yoshino, Visible-light electroluminescent diodes utilizing poly(3-alkylthiophenes), *Jpn. J. Appl. Phys., Part 2*, 30: L1938–L1940, 1991.
492. Y. Ohmori, M. Uchida, K. Muro, and K. Yoshino, Effects of alkyl chain lengths and carrier confinement layer on characteristics of poly(3-alkylthiophene) electroluminescent diodes, *Solid State Commun.*, 80: 605–608, 1991.
493. M. Uchida, Y. Ohmori, C. Morishima, and K. Yoshino, Visible and blue electroluminescent diodes utilizing poly(3-alkylthiophene)s and poly(alkylfluorene)s, *Synth. Met.*, 55–57: 4168–4173, 1993.
494. D. Braun, G. Gustafsson, D. McBranch, and A.J. Heeger, Electroluminescence and electrical transport in poly(3-octylthiophene) diodes, *J. Appl. Phys.*, 72: 564–568, 1992.
495. N.C. Greenham, A.R. Brown, D.D.C. Bradley, and R.H. Friend, Electroluminescence in poly(3-alkylthienylene)s, *Synth. Met.*, 57: 4134–4138, 1993.
496. A. Bolognesi, C. Botta, Z. Geng, C. Flores, and L. Denti, Modified poly(3-alkylthiophene) for LED preparation, *Synth. Met.*, 71: 2191–2192, 1995.
497. A. Bolognesi, W. Porzio, G. Bajo, G. Zannoni, and L. Fanning, Highly regioregular poly (3-alkylthiophenes): A new synthetic route and characterization of the resulting polymers, *Acta Polym.*, 50: 151–155, 1999.
498. A. Bolognesi, C. Botta, and L. Cecchinato, Optical properties and electroluminescence of poly(3-alkylmethoxy-thiophene) single- and double-layer structures, *Synth. Met.*, 111–112: 187–189, 2000.
499. A. Bolognesi, W. Porzio, F. Provasoli, and T. Ezquerra, The thermal behavior of low-molecular-weight poly(3-decylthiophene), *Makromol. Chem.*, 194: 817–827, 1993.
500. M. Pomerantz, H. Yang, and Y. Cheng, Poly(alkyl thiophene-3-carboxylates). Synthesis and characterization of polythiophenes with a carbonyl group directly attached to the ring, *Macromolecules*, 28: 5706–5708, 1995.
501. T. Yamamoto and H. Hayashi, π-Conjugated soluble and fluorescent poly(thiophene-2,5-diyl)s with phenolic, hindered phenolic and p-$C_6H_4OCH_3$ substituents. Preparation, optical properties, and redox reaction, *J. Polym. Sci., Part A: Polym. Chem.*, 35: 463–474, 1997.
502. S.H. Jin, B.U. Yoo, S.Y. Kang, Y.S. Gal, and D.K. Moon, Synthesis and electro-optical properties of polythiophene derivatives for electroluminescence display, *Opt. Mater.*, 21: 153–157, 2002.
503. S.-H. Ahn, M.-Z. Czae, E.-R. Kim, H. Lee, S.-H. Han, J. Noh, and M. Hara, Synthesis and characterization of soluble polythiophene derivatives containing electron-transporting moiety, *Macromolecules*, 34: 2522–2527, 2001.
504. B. Xu and S. Holdcroft, Molecular control of luminescence from poly(3-hexylthiophenes), *Macromolecules*, 26: 4457–4460, 1993.
505. M. Pomerantz, Y. Cheng, R.K. Kasim, and R.L. Elsenbaumer, Poly(alkylthiophene-3-carboxylates). Synthesis, properties and electroluminescence studies of polythiophenes containing carbonyl group directly attached to the ring, *J. Mater. Chem.*, 9: 2155–2163, 1999.
506. P. Barta, F. Cacialli, R.H. Friend, and M. Zagórska, Efficient photo- and electroluminescence of regioregular poly(alkylthiophene)s, *J. Appl. Phys.*, 84: 6279–6284, 1998.

507. M. Zagórska and B. Krische, Chemical synthesis and characterization of soluble poly(4,4'-dialkyl-2,2'-bithiophenes), *Polymer*, 31: 1379–1383, 1990.

508. P. Barta, P. Dannetun, S. Stafström, M. Zagórska, and A. Proñ, Temperature evolution of the electronic band structure of the undoped and doped regioregular analog of poly(3-alkylthiophenes): A spectroscopic and theoretical study, *J. Chem. Phys.*, 100: 1731–1741, 1994.

509. R.E. Gill, G.G. Malliaras, J. Wildeman, and G. Hadziioannou, Tuning of photo- and electroluminescence in alkylated polythiophenes with well-defined regioregularity, *Adv. Mater.*, 6: 132–135, 1994.

510. J.L. Brédas, R. Silbey, D.S. Boudreaux, and R.R. Chance, Chain-length dependence of electronic and electrochemical properties of conjugated systems: Polyacetylene, polyphenylene, polythiophene, and polypyrrole, *J. Am. Chem. Soc.*, 105: 6555–6559, 1983.

511. M. Boman and S. Stafström, Interpretation of anomalous absorption spectra. A theoretical study of the geometric, electronic and optical properties of poly[3-(4-octylphenyl)thiophene], *Mol. Cryst. Liq. Cryst. Sci. Technol.*, 256: 705–710, 1994.

512. O. Inganäs, Making polymer light emitting diodes with polythiophenes, in: *Organic Electroluminescent Materials and Devices*, S. Miyata and H.S. Nalwa, Eds., Gordon & Breach, Langhorne, 1997, pp. 147–175.

513. M.R. Andersson, M. Berggren, O. Inganäs, G. Gustafsson, J.C. Gustafsson-Carlberg, D. Selse, T. Hjertberg, and O. Wennerström, Electroluminescence from substituted poly(thiophenes): From blue to near-infrared, *Macromolecules*, 28: 7525–7529, 1995.

514. M.R. Andersson, W. Mammo, T. Olinga, M. Svensson, M. Theander, and O. Inganäs, Synthesis of regioregular phenyl substituted polythiophenes with FeCl$_3$, *Synth. Met.*, 101: 11–12, 1999.

515. Q. Pei, H. Järvinen, J.E. Österholm, O. Inganäs, and J. Laakso, Poly[3-(4-octylphenyl)thiophene], a new processible conducting polymer, *Macromolecules*, 25: 4297–4301, 1992.

516. M. Berggren, G. Gustafsson, O. Inganäs, M.R. Andersson, O. Wennerström, and T. Hjertberg, Green electroluminescence in poly-(3-cyclohexylthiophene) light-emitting diodes, *Adv. Mater.*, 6: 488–490, 1994.

517. M.R. Andersson, M. Berggren, T. Olinga, T. Hjertberg, O. Inganäs, and O. Wennerström, Improved photoluminescence efficiency of films from conjugated polymers, *Synth. Met.*, 85: 1383–1384, 1997.

518. A. Ruseckas, E.B. Namadas, T. Ganguly, M. Theander, M. Svensson, M.R. Andersson, O. Inganäs, and V. Sundström, Intra- and interchain luminescence in amorphous and semicrystalline films of phenyl-substituted polythiophene, *J. Phys. Chem. B*, 105: 7624–7631, 2001.

519. M. Berggren, G. Gustafsson, O. Inganäs, M.R. Andersson, O. Wennerström, and T. Hjertberg, Thermal control of near-infrared and visible electroluminescence in alkyl–phenyl substituted polythiophenes, *Appl. Phys. Lett.*, 65: 1489–1491, 1994.

520. M. Berggren, P. Bergman, J. Fagerström, O. Inganäs, M.R. Andersson, H. Weman, M. Granström, S. Stafström, O. Wennerström, and T. Hjertberg, Controlling inter-chain and intra-chain excitations of a poly(thiophene) derivative in thin films, *Chem. Phys. Lett.*, 304: 84–90, 1999.

521. Y. Li, G. Vamvounis, and S. Holdcroft, Tuning optical properties and enhancing solid-state emission of poly(thiophene)s by molecular control: A postfunctionalization approach, *Macromolecules*, 35: 6900–6906, 2002.

522. Y. Li, G. Vamvounis, J. Yu, and S. Holdcroft, A novel and versatile methodology for functionalization of conjugated polymers. Transformation of poly(3-bromo-4-hexylthiophene) via palladium-catalyzed coupling chemistry, *Macromolecules*, 34: 3130–3132, 2001.

523. V. Saxena and V.S. Shirodkar, A study of light-emitting diodes constructed with copolymers having cyclohexyl thiophene and hexyl thiophene units, *J. Appl. Polym. Sci.*, 77: 1051–1055, 2000.

524. M. Berggren, O. Inganäs, G. Gustafsson, J. Rasmusson, M.R. Andersson, T. Hjertberg, and O. Wennerström, Light-emitting diodes with variable colours from polymer blends, *Nature*, 372: 444–446, 1994.

525. O. Inganäs, M. Berggren, M.R. Andersson, G. Gustafsson, T. Hjertberg, O. Wennerström, P. Dyreklev, and M. Granström, Thiophene polymers in light emitting diodes: Making multicolor devices, *Synth. Met.*, 71: 2121–2124, 1995.

526. M. Granström and O. Inganäs, White light emission from a polymer blend light emitting diode, *Appl. Phys. Lett.*, 68: 147–149, 1996.

527. M. Granström, M. Berggren, D. Pede, O. Inganäs, M.R. Andersson, T. Hjertberg, and O. Wennerström, Self-organizing polymer films—A route to novel electronic devices based on conjugated polymers, *Supramol. Sci.*, 4: 27–34, 1997.

528. S. Destri, U. Giovanella, A. Fazio, W. Porzio, B. Gabriele, and G. Zotti, A new soluble poly(bithiophene)-co-3,4-di(methoxycarbonyl)methylthiophene for LED, *Org. Electron.*, 3: 149–156, 2002.

529. M. Berggren, G. Gustafsson, O. Inganäs, M.R. Andersson, T. Hjertberg, and O. Wennerström, White light from an electroluminescent diode made from poly[3(4-octylphenyl)-2,2'-bithiophene] and an oxadiazole derivative, *Appl. Phys. Lett.*, 76: 7530–7534, 1994.

530. T. Granlund, L.A.A. Petterson, M.R. Andersson, and O. Inganäs, Interference phenomenon determines the color in an organic light emitting diode, *J. Appl. Phys.*, 81: 8097–8103, 1997.

531. G. Yu, H. Nishino, A.J. Heeger, T.-A. Chen, and R.D. Rieke, Enhanced electroluminescence from semiconducting polymer blends, *Synth. Met.*, 72: 249–252, 1995.

532. E.J.W. List, L. Holzer, S. Tasch, G. Leising, M. Catellani, and S. Luzzati, Efficient, single layer yellow light emitting diodes made of a blend of a ladder-type poly(*p*-phenylene) and poly-alkylthiophene, *Opt. Mater.*, 12: 311–314, 1999.

533. E.J.W. List, L. Holzer, S. Tasch, G. Leising, U. Scherf, K. Müllen, M. Catellani, and S. Luzzati, Efficient single-layer yellow-light emitting-diodes with ladder-type poly(*p*-phenylene/poly (decyl-thiophene) blends, *Solid State Commun.*, 109: 455–459, 1999.

534. A. Bolognesi, G. Bajo, J. Paloheimo, T. Östergård, and H. Stubb, Polarized electroluminescence from an oriented poly(3-alkylthiophene) Langmuir–Blodgett structure, *Adv. Mater.*, 9: 121–124, 1997.

535. P. Dyreklev, M. Beggren, O. Inganäs, M.R. Andersson, O. Wennerström, and T. Hjertberg, Polarized electroluminescence from an oriented substituted polythiophene in a light emitting diode, *Adv. Mater.*, 7: 43–45, 1995.

536. M. Berggren, O. Inganäs, T. Granlund, S. Guo, G. Gustafsson, and M.R. Andersson, Polymer light-emitting diodes placed in microcavities, *Synth. Met.*, 76: 121–123, 1996.

537. M. Granström, Novel polymer light-emitting diode designs using poly(thiophenes), *Polym. Adv. Technol.*, 8: 424–430, 1997.

538. M. Granström, M. Berggren, and O. Inganäs, Micrometer- and nanometer-sized polymeric light emitting diodes, *Science*, 267: 1479–1481, 1995.

539. M. Granström and O. Inganäs, Flexible arrays of sub-micron sized polymeric light-emitting diodes, *Adv. Mater.*, 7: 1012–1015, 1995.

540. F. Geiger, M. Stoldt, H. Schweizer, P. Bäuerle, and E. Umbach, Electroluminescence from oligothiophene-based light-emitting devices, *Adv. Mater.*, 5: 922–925, 1993.

541. S.A. Lee, Y. Yoshida, M. Fukuyama, and S. Hotta, Phenyl-capped oligothiophenes: Novel light-emitting materials with different molecular alignments in thin films, *Synth. Met.*, 106: 39–43, 1999.

542. S. Hotta, Y. Ichino, Y. Yoshida, and M. Yoshida, Spectroscopic features of thin films of thiophene/phenylene co-oligomers with vertical molecular alignment, *J. Phys. Chem. B*, 104: 10316–10320, 2000.

543. S.A. Lee, S. Hotta, and F. Nakanishi, Spectroscopic characteristics and intermolecular interactions of thiophene/phenylene co-oligomers in solutions, *J. Phys. Chem. A*, 104: 1827–1833, 2000.

544. K. Uchiyama, H. Akimichi, S. Hotta, H. Noge, and H. Sakaki, Electroluminescence from thin film of a semiconducting oligothiophene deposited in ultrahigh vacuum, *Synth. Met.*, 63: 57–59, 1994.

545. T. Noda, H. Ogawa, N. Noma, and Y. Shirota, A novel yellow-emitting material, 5,5"-bis{4-[bis(4-methylphenyl)amino]phenyl}-2,2':5',2"-terthiophene, for organic electroluminescent devices, *Appl. Phys. Lett.*, 70: 699–701, 1997.

546. T. Noda, H. Ogawa, N. Noma, and Y. Shirota, A novel family of amorphous molecular materials containing an oligothiophene moiety as color-tunable emitting materials for organic electroluminescent devices, *Adv. Mater.*, 9: 720–722, 1997.

547. K. Chondroudis and D.B. Mitzi, Electroluminescence from an organic–inorganic perovskite incorporating a quaterthiophene dye within lead halide perovskite layer, *Chem. Mater.*, 11: 3028–3030, 1999.

548. J. Khunchalee, R. Tarsang, N. Prachumrak, S. Jungsuttiwong, T. Keawin, T. Sudyoadsuk, and V. Promarak, Synthesis and properties of oligofluorene-thiophenes as emissive materials for organic electroluminescent devices: Color-tuning from deep blue to orange, *Tetrahedron*, 68: 8416–8423, 2012.

549. A.W. Freeman, S.C. Koene, P.R.L. Malenfant, M.E. Thompson, and J.M.J. Fréchet, Dendrimer-containing light-emitting diodes: Toward site-isolation of chromophores, *J. Am. Chem. Soc.*, 122: 12385–12386, 2000.

550. G. Barbarella, L. Favaretto, M. Zambianchi, O. Pudova, C. Arbizzani, A. Bongini, and M. Mastragostino, From easily oxidized to easily reduced thiophene-based materials, *Adv. Mater.*, 10: 551–554, 1998.

551. G. Barbarella, L. Favaretto, G. Sotgiu, M. Zambianchi, L. Antolini, O. Pudova, and A. Bongini, Oligothiophene *S,S*-dioxides. Synthesis and electronic properties in relation to the parent oligothiophenes, *J. Org. Chem.*, 63: 5497–5506, 1998.

552. G. Gigli, M. Ani, G. Barbarella, L. Favaretto, F. Cacialli, and R. Cingolani, High photo and electroluminescence efficiency oligothiophenes, *Physica E*, 7: 612–615, 2000.

553. L. Antolini, E. Tadesco, G. Barbarella, L. Favaretto, G. Sotgiu, M. Zambianchi, D. Casarini, G. Gigli, and R. Cingolani, Molecular packing and photoluminescence efficiency in odd-membered oligothiophene *S,S*-dioxides, *J. Am. Chem. Soc.*, 122: 9006–9013, 2000.

554. G. Barbarella, L. Favaretto, G. Sotgiu, M. Zambianchi, A. Bongini, C. Arbizzani, M. Mastragostino, M. Anni, G. Gigli, and R. Cingolani, Tuning solid-state photoluminescence frequencies and efficiencies of oligomers containing one central thiophene-*S,S*-dioxide unit, *J. Am. Chem. Soc.*, 122: 11971–11978, 2000.

555. G. Gigli, G. Barbarella, L. Favaretto, F. Cacialli, and R. Cingolani, High-efficiency oligothiophene-based light-emitting diodes, *Appl. Phys. Lett.*, 75: 439–441, 1999.

556. A. Berlin, G. Zotti, S. Zecchin, G. Schiavon, M. Cocchi, D. Virgili, and C. Sabatini, 3,4-Ethylenedioxy-substituted bithiophene-*alt*-thiophene-*S,S*-dioxide regular copolymers. Synthesis and conductive, magnetic and luminescence properties, *J. Mater. Chem.*, 13: 27–33, 2003.

557. A. Charas, J. Morgado, J.M.G. Martinho, L. Alcácer, S.F. Lim, R.H. Friend, and F. Cacialli, Synthesis and luminescence properties of three novel polyfluorene copolymers, *Polymer*, 44: 1843–1850, 2003.

558. A. Charas, J. Morgado, J.M.G. Martinho, L. Alcácer, and F. Cacialli, Electrochemical and luminescent properties of poly(fluorene) derivatives for optoelectronic applications, *Chem. Commun.* 13: 1216–1217, 2001.

559. A. Charas, J. Morgado, J.M.G. Martinho, A. Fedorov, L. Alcácer, and F. Cacialli, Excitation energy transfer and spatial exciton confinement in polyfluorene blends for application in light-emitting diodes, *J. Mater. Chem.*, 12: 3523–3527, 2002.

560. M. Passini, S. Destri, W. Porzio, C. Botta, and U. Giovanella, Electroluminescent poly(fluorene-*co*-thiophene-*S,S*-dioxide): Synthesis, characterisation and structure–property relationships, *J. Mater. Chem.*, 13: 807–813, 2003.

561. S. Destri, M. Pasini, W. Porzio, G. Gigi, D. Pisignano, and C. Capolupo, Emission properties and solid-state aggregation in poly(fluorene–thiophene-*S,S*-dioxide) and in its model oligomer, *Synth. Met.*, 138: 289–293, 2003.

562. J.-F. Morin and M. Leclerc, 2,7-Carbazole-based conjugated polymers for blue, green, and red light emission, *Macromolecules*, 35: 8413–8417, 2002.

563. G.G. Malliaras, J.K. Herrema, J. Wildeman, R.H. Wieringa, R.E. Gill, S.S. Lampoura, and G. Hadziioannou, Tuning of the photo- and electroluminescence in multi-block copolymers of poly[(silanylene)-thiophene]s via exciton confinement, *Adv. Mater.*, 5: 721–723, 1993.

564. J.K. Herrema, P.F. van Hutten, R.E. Gill, J. Wildeman, R.H. Wieringa, and G. Hadziioannou, Tuning of the luminescence in multiblock alternating copolymers. 1. Synthesis and spectroscopy of poly[(silanylene)thiophene]s, *Macromolecules*, 28: 8102–8116, 1995.

565. K. Yoshino, M. Hirohata, T. Sonoda, R. Hidayat, A. Fujii, A. Naka, and M. Ishikawa, Electroluminescence and photoluminescence characteristics of poly(disilanylene oligophenylene)s and poly(disilanylene oligothienylene)s, *Synth. Met.*, 102: 1158, 1999.

566. M. Fahlman, J. Birgersson, K. Kaeriyama, and W.R. Salaneck, Poly(2,5-diheptyl-1,4-phenylene-*alt*-2, 5-thienylene): A new material for blue-light-emitting diodes, *Synth. Met.*, 75: 223–228, 1995.

567. J.-L. Brédas and A.J. Heeger, Influence of donor and acceptor substituents on the electronic characteristics of poly(*p*-phenylene vinylene) and poly(*p*-phenylene), *Chem. Phys. Lett.*, 217: 507–512, 1994.

568. J.-L. Brédas, R.L. Elsenbaumer, R.R. Chance, and R. Silbey, Electronic properties of sulfur containing conjugated polymers, *J. Chem. Phys.*, 78: 5656–5662, 1983.

569. J. Birgersson, K. Kaeriyama, P. Barta, P. Bröms, M. Fahlman, T. Granlund, and W.R. Salaneck, Efficient blue-light emitting devices from conjugated polymer blends, *Adv. Mater.*, 8: 982–985, 1996.

570. J. Pei, W.-L. Yu, W. Huang, and A.J. Heeger, A novel series of efficient thiophene-based light-emitting conjugated polymers and application in polymer light-emitting diodes, *Macromolecules*, 33: 2462–2471, 2000.

571. A.-L. Ding, J. Pei, Y.-H. Lai, and W. Huang, Phenylene-functionalized polythiophene derivatives for light-emitting diodes: Their synthesis, characterization and properties, *J. Mater. Chem.*, 11: 3082–3086, 2001.

572. K.Y. Musick, Q.-S. Hu, and L. Pu, Synthesis of binaphthyl–oligothiophene copolymers with emissions of different colors: Systematically tuning the photoluminescence of conjugated polymers, *Macromolecules*, 31: 2933–2942, 1998.

573. Y. Liu, G. Yu, A.K.-Y. Jen, Q.-S. Hu, and L. Pu, A binaphthyl–bithiophene copolymer for light-emitting devices, *Macromol. Chem. Phys.*, 203: 37–40, 2002.

574. S.-Y. Song and H.-K. Shim, Synthesis of poly(phenylene-*alt*-thiophene) polymer and characterization of its emitting properties, *Synth. Met.*, 111–112: 437–439, 2000.

575. W. Huang, H. Meng, W.-L. Yu, J. Gao, and A.J. Heeger, A new blue light-emitting polymer containing substituted thiophene and an arylene-1,3,4-oxadiazole moiety, *Adv. Mater.*, 10: 593–596, 1998.

576. W. Huang, W.-L. Yu, H. Meng, J. Pei, and S.F.Y. Li, New series of blue-light-emitting polymers constituted of 3-alkylthiophenes and 1,4-di(1,3,4-oxadiazolyl)phenylene, *Chem. Mater.*, 10: 3340–3345, 1998.

577. W. Huang, H. Meng, W.-L. Yu, J. Pei, Z.-K. Chen, and Y.-H. Lai, A novel series of *p–n* diblock light-emitting copolymers based on oligothiophenes and 1,4-*bis*(oxadiazolyl)-2,5-dialkyloxy-benzenes, *Macromolecules*, 32: 118–126, 1999.

578. H. Meng and W. Huang, Novel photoluminescent polymers containing oligothiophene and *m*-phenylene-1,3,4-oxadiazole moieties: Synthesis and spectroscopic and electrochemical studies, *J. Org. Chem.*, 65: 3894–3901, 2000.

579. C.J. Tonzola, M.M. Alam, and S.A. Jenekhe, New soluble *n*-type conjugated copolymer for light-emitting diodes, *Adv. Mater.*, 14: 1086–1090, 2002.

580. C.J. Tonzola, M.M. Alam, B.A. Bean, and S.A. Jenekhe, New soluble *n*-type conjugated polymers for use as electron transport materials in light-emitting diodes, *Macromolecules*, 37: 3554–3563, 2004.

581. A. Dhanabalan, J.K.J. van Duren, P.A. van Hal, J.L.J. van Dongen, and R.A.J. Janssen, Synthesis and characterization of a low bandgap conjugated polymer for bulk heterojunction photovoltaic cells, *Adv. Funct. Mater.*, 11: 255–262, 2001.

582. C.J. Brabec, C. Winder, N.S. Sariciftci, J.C. Hummelen, A. Dhanabalan, P.A. van Hal, and R.A.J. Janssen, A low-bandgap semiconducting polymer for photovoltaic devices and infrared emitting diodes, *Adv. Funct. Mater.*, 12: 709–721, 2002.

583. J. Roncali, Oligothienylenevinylenes as a new class of multinanometer linear π-conjugated systems for micro- and nanoelectronics, *Acc. Chem. Res.*, 33: 147–156, 2000.

584. I. T. Kim, S. W. Lee, S. Y. Kim, J. S. Lee, G. B. Park, S. H. Lee, S. K. Kang, J.-G. Kang, C. Park, S.-H. Jin, Synthesis, characterization, and properties of a new thiophene-benzobisthiazole copolymer, *Syn. Met.*, 156: 38–41, 2006.

585. J.V. Grazulevicius, P. Strohriegl, J. Pielichowski, and K. Pielichowski, Carbazole-containing polymers: Synthesis, properties and applications, *Prog. Polym. Sci.*, 29: 1297–1353, 2003.

586. G. Grem, G. Leditzky, B. Ullrich, and G. Leising, Realization of a blue-light emitting device using poly(*p*-phenylene), *Adv. Mater.*, 4: 36–37, 1992.

587. G. Grem, G. Leditzky, and G. Leising, Blue electroluminescent device based on conjugated polymer, *Synth. Met.*, 51: 383–389, 1992.

588. A.-D. Schlüter and G. Wegner, Palladium and nickel catalyzed polycondensation—The key to structurally defined polyarene and other aromatic polymers, *Acta Polym.*, 44: 59–69, 1993.

589. M. Rahahn, A.-D. Schlüter, G. Wegner, and W.J. Feast, Soluble poly(*p*-phenylene)s. 1. Extension of the Yamamoto synthesis to dibromobenzenes substituted with flexible side chains, *Polymer*, 30: 1054–1059, 1989.

590. N. Tanigaki, H. Masude, and K. Kaeriyama, Substituted poly(*p*-phenylene)s prepared from 2,5-diheptylbenzene-1,4-*bis*(trimethylene boronate), *Polymer*, 38: 1221–1226, 1997.

591. Y. Yang, Q. Pei, and A.J. Heeger, Efficient blue polymer light-emitting diodes from a series of soluble poly(*p*-phenylene)s, *J. Appl. Phys.*, 79: 934–936, 1996.

592. S.-A. Chen and C.-I. Chao, Poly(2-alkoxy-*p*-phenylene)s as deep-blue light-emitting polymers, *Synth. Met.*, 79: 93–96, 1996.

593. A. Edwards, S. Blumstengel, I. Sokolik, R. Dorsinville, H. Yun, T.K. Kwei, and Y. Okamoto, Blue photo- and electroluminescence from poly(benzoyl-1,4-phenylene), *Appl. Phys. Lett.*, 70: 298–300, 1997.

594. J. Huang, H. Zhang, W. Tian, J. Hou, Y. Ma, J. Shen, and S. Liu, Violet-blue electroluminescent diodes utilizing conjugated polymer blends, *Synth. Met.*, 87: 105–108, 1997.

595. W.-X. Jing, A. Kraft, S.C. Moratti, J. Grüner, F. Cacialli, P.J. Hamer, A.B. Holmes, and R.H. Friend, Synthesis of a polyphenylene light-emitting copolymer, *Synth. Met.*, 67: 161–163, 1994.

596. M. Hamaguchi and K. Yoshino, Blue electroluminescence from poly(2,5-diheptyloxy-1,4-phenylene), *Jpn. J. Appl. Phys.*, 34: L587–L589, 1995.

597. G. Grem and G. Leising, Electroluminescence of "wide-bandgap" chemically tunable cyclic conjugated polymers, *Synth. Met.*, 57: 4105–4110, 1993.

598. Y. Yang, Q. Pei, and A.J. Heeger, Efficient blue light-emitting diodes from a soluble poly(*p*-phenylene): Internal field emission measurement of the energy gap in semiconducting polymers, *Synth. Met.*, 78: 263–267, 1996.

599. J. Birgerson, M. Fahlman, P. Bröm, and W.R. Salneck, Conjugated polymer surfaces and interfaces: A mini-review and some new results, *Synth. Met.*, 80: 125–130, 1996.

600. M. Hamaguchi, H. Sawada, J. Kyokane, and K. Yoshino, Blue electroluminescence from poly(*p*-phenylene) solubilized by perfluoropropylation, *Chem. Lett.*, 25: 527–528, 1996.

601. V. Cimrová, W. Schmidt, R. Rulkens, M. Schulze, W. Meyer, and D. Neher, Efficient blue light-emitting devices based on rigid-rod polyelectrolytes, *Adv. Mater.*, 8: 585–588, 1996.

602. S. Kim, J. Jackiw, E. Robinson, K.S. Schanze, J.R. Reynolds, J. Baur, M.F. Rubner, and D. Boils, Water soluble photo- and electroluminescent alkoxy-sulfonated poly(*p*-phenylenes) synthesized via palladium catalysis, *Macromolecules*, 31: 964–974, 1998.

603. J.W. Baur, S. Kim, P.B. Balanda, J.R. Reynolds, and M.F. Rubner, Thin-film light-emitting devices based on sequentially adsorbed multilayers of water-soluble poly(*p*-phenylene)s, *Adv. Mater.*, 10: 1452–1455, 1998.

604. M. Remmers, D. Neher, J. Grüner, R.H. Friend, G.H. Gelinck, J.M. Warman, C. Quattrocchi, D.A. dos Santos, and J.-L. Brédas, The optical, electronic, and electroluminescent properties of novel poly(*p*-phenylene)-related polymers, *Macromolecules*, 29: 7432–7445, 1996.

605. V. Cimrová, D. Neher, M. Remmers, and I. Kmínek, Blue light-emitting devices based on novel polymer blends, *Adv. Mater.*, 10: 676–680, 1998.

606. Y.-H. Kim, J.-H. Ahn, D.-C. Shin, and S.-K. Kwon, Synthesis and characterization of poly (terphenylenevinylene) derivatives containing alkoxy substituents and (or) phenyl pendant group, *Polymer*, 45: 2525–2532, 2004.

607. S.J. Park, S.-H. Jung, J.M. Kim, and H.-N. Cho, Highly phenyl-substituted fluorene copolymers for light-emitting diode, *Mater. Sci. Eng. C*, 24: 99–102, 2004.

608. M.A. Keegstra, V. Cimrová, D. Neher, and U. Scherf, Synthesis and electroluminescent properties of quaterphenyl and sexiphenyl containing copolymers, *Macromol. Chem. Phys.*, 197: 2511–2519, 1996.

609. F.D. Konstandakopoulou, K.G. Gravalos, and J.K. Kallitsis, Synthesis and characterization of processible aromatic–aliphatic polyethers with quinquephenyl segments in the main chain for light-emitting applications, *Macromolecules*, 31: 5264–5271, 1998.

610. J.K. Kallitsis, K.G. Gravalos, A. Hilberer, and G. Hadziioannou, Soluble polymers with laterally attached oligophenyl units for potential use as blue luminescent materials, *Macromolecules*, 30: 2989–2996, 1997.

611. V. Deimede, J.K. Kallitsis, and T. Pakula, Synthesis and properties of amorphous blue-light-emitting polymers with high glass-transition temperatures, *J. Polym. Sci., Part A: Polym. Chem.*, 39: 3168–3179, 2001.

612. K. Kaeriyama, Y. Tsukahara, S. Negoro, N. Tanigaki, and H. Masuda, Preparation and properties of soluble polyphenylenes, *Synth. Met.*, 84: 263–264, 1997.

613. P. Hodge, G.A. Power, and M.A. Rabjohns, Synthesis of poly(2,6-anthracene-2,6-diyl) and a copolymer containing alternately anthracene-2,6-diyl and *p*-phenylene units, *Chem. Commun.* 1: 73–74, 1996.

614. A.K.-Y. Jen, Y. Liu, Q.-S. Hu, and L. Pu, Efficient light-emitting diodes based on a binaphthalene-containing polymer, *Appl. Phys. Lett.*, 75: 3745–3747, 1999.

615. F.D. Konstandakopoyloy and J.K. Kallitsis, Soluble rigid–flexible polyethers containing *bis*(biphenyl)anthracene or *bis*(styryl)anthracene units in the main chain for light-emitting applications, *J. Polym. Sci., Part A: Polym. Chem.*, 37: 3826–3837, 1999.

616. S. Zheng and J. Shi, Novel blue-emitting polymers containing dinaphthylanthracene moiety, *Chem. Mater.*, 13: 4405–4407, 2001.

617. C.H. Lee, S.H. Ryu, and S.Y. Oh, Characteristics of a single-layered organic electroluminescent device using a carrier-transporting copolymer and a non-conjugated light-emitting polymer, *J. Polym. Sci., Part B: Polym. Phys.*, 41: 2733–2743, 2003.

618. M.-L. Tsai, C.-Y. Liu, Y.-Y. Wang, J.-Y. Chen, T.-C. Chou, H.-M. Lin, S.-H. Tsai, and T.J. Chow, Preparation and luminescent properties of polymer containing dialkoxyacenes, *Chem. Mater.*, 16: 3373–3380, 2004.

619. M. Kreyenschmidt, F. Uckert, and K. Müllen, A new soluble poly(*p*-phenylene) with tetrahydropyrene repeating units, *Macromolecules*, 28: 4577–4582, 1995.

620. S. Setayesh, D. Marsitzky, and K. Müllen, Bridging the gap between polyfluorene and ladder-poly-*p*-phenylene: Synthesis and characterization of poly-2,8-indenofluorene, *Macromolecules*, 33: 2016–2020, 2000.

621. D. Marsitzky, J.C. Scott, J.-P. Chen, V.Y. Lee, R.D. Miller, S. Setayesh, and K. Müllen, Poly-2,8-(indenofluorene-*co*-anthracene)—A colorfast blue-light-emitting random copolymer, *Adv. Mater.*, 13: 1096–1099, 2001.

622. J. Jacob, S. Sax, T. Piok, E.J.W. List, A.C. Grimsdale, and K. Müllen, Ladder-type pentaphenylenes and their polymers: Efficient blue-light emitters and electron-accepting materials via a common intermediate, *J. Am. Chem. Soc.*, 126: 6987–6995, 2004.

623. U. Scherf and K. Müllen, Polyarylenes and poly(arylene vinylenes), 7. A soluble ladder polymer via bridging of functionalized poly(*p*-phenylene)-precursors, *Makromol. Chem., Rapid Commun.*, 12: 489–497, 1991.

624. U. Scherf and K. Müllen, Design and synthesis of extended π-systems: Monomer, oligomers, polymers, *Synthesis*, 23–28, 1992.

625. U. Scherf, Ladder-type materials, *J. Mater. Chem.*, 9: 1853–1864, 1999.

626. J. Stampfil, S. Tasch, G. Leising, and U. Scherf, Quantum efficiencies of electroluminescent poly(*p*-phenylenes), *Synth. Met.*, 71: 2125–2128, 1995.

627. L. Romaner, G. Heimel, H. Weisenhofer, P.S. de Freitas, U. Scherf, J.-L. Brédas, E. Zojer, and E.J.W. List, Ketonic defects in ladder-type poly(*p*-phenylene)s, *Chem. Mater.*, 16: 4667–4674, 2004.

628. S. Tasch, A. Niko, G. Leising, and U. Scherf, Highly efficient electroluminescence of new wide band gap ladder-type poly(*p*-phenylenes), *Appl. Phys. Lett.*, 68: 1090–1092, 1996.

629. G. Leising, S. Tasch, F. Meghdadi, L. Athouel, G. Froyer, and U. Scherf, Blue electroluminescence with ladder-type poly(*para*-phenylene) and *para*-hexaphenyl, *Synth. Met.*, 81: 185–189, 1996.

630. C. Kallinger, M. Hilmer, A. Haugeneder, M. Perner, W. Spirkl, U. Lemmer, J. Feldmann, U. Scherf, K. Müllen, A. Gombert, and V. Wittwer, A flexible conjugated polymer laser, *Adv. Mater.*, 10: 920–923, 1998.

631. C. Bauer, H. Giessen, B. Schnabel, E.-B. Kley, C. Schmitt, U. Scherf, and R.F. Mahrt, A surface-emitting circular grating polymer laser, *Adv. Mater.*, 13: 1161–1164, 2001.

632. T. Piok, S. Gamerith, C. Gadermaier, H. Plank, F.P. Wenzl, S. Patil, R. Montenegro, T. Tietzke, D. Neher, U. Scherf, K. Landfester, and E.J.W. List, Organic light-emitting devices fabricated from semiconducting nanospheres, *Adv. Mater.*, 15: 800–804, 2003.

633. S. Tasch, E.J.W. List, O. Ekström, W. Graupner, G. Leising, P. Schlichting, U. Rohr, Y. Geerts, U. Scherf, and K. Müllen, Efficient white light-emitting diodes realizes with new processible blends of conjugated polymers, *Appl. Phys. Lett.*, 71: 2883–2885, 1997.

634. G. Grem, C. Paar, J. Stampfl, G. Leising, J. Huber, and U. Scherf, Soluble segmented stepladder poly(*p*-phenylenes) for blue-light-emitting diodes, *Chem. Mater.*, 7: 2–4, 1995.

635. J. Grüner, P.J. Hamer, R.H. Friend, H.-J. Huber, U. Scherf, and A.B. Holmes, A high efficiency blue-light-emitting diode based on novel ladder poly(*p*-phenylene)s, *Adv. Mater.*, 6: 748–752, 1994.

636. Y. Xiaohui, H. Yanbing, W. Zhenjia, C. Xiaohong, X. Zheng, and X. Xurong, Blue polymer light emitting diode based on ladder poly(*p*-phenylene), *Thin Solid Films*, 363: 211–213, 2000.

637. R. Giesa, Synthesis and properties of conjugated poly(arylene ethynylene)s, *J. Macromol. Sci., C, Rev. Macromol. Chem. Phys.*, 36: 631–670, 1996.

638. U.H.F. Bunz, Poly(arylene ethynylene)s: Syntheses, properties, structures, and applications, *Chem. Rev.*, 100: 1605–1644, 2000.

639. C. Weder, S. Sarwa, A. Montali, C. Bastiaansen, and P. Smith, Incorporation of photoluminescent polarizers into liquid crystal displays, *Science*, 279: 835–837, 1998.

640. L.S. Swanson, J. Shinar, Y.W. Ding, and T.J. Barton, Photoluminescence, electroluminescence, and optically detected magnetic resonance study of 2,5-dialkoxy derivatives of poly(*p*-phenylene acetylene) (PPA) and PPA-based light-emitting diodes, *Synth. Met.*, 55: 1–6, 1993.

641. T.M. Swager, C.J. Gil, and M.S. Wrighton, Fluorescence studies of poly(*p*-phenylene)s: The effect of anthracene substitution, *J. Phys. Chem.*, 99: 4886–4893, 1995.

642. D. Ofer, T.M. Swager, and M.S. Wrighton, Solid-state ordering and potential dependence of conductivity in poly(2,5-dialkoxy-*p*-phenylene ethylene), *Chem. Mater.*, 7: 418–425, 1995.

643. C. Weder and M.S. Wrighton, Efficient solid-state photoluminescence in new poly(2,5-dialkoxy-*p*-phenylene ethynylene)s, *Macromolecules*, 29: 5157–5165, 1996.

644. C. Schmitz, P. Pösch, M. Thelekkat, H.-W. Schmidt, A. Montali, K. Feldman, P. Smith, and C. Weder, Polymeric light-emitting diodes based on poly(*p*-phenylene ethynylene), poly(triphenyldiamine), and spiroquinoxaline, *Adv. Funct. Mater.*, 11: 41–46, 2001.

645. R.A.J. Hanssen, N.S. Sariciftci, K. Pakbaz, J.J. McNamara, S. Schricker, A.J. Heeger, and F. Wudl, Photoinduced absorption of π-conjugated polymers in solution, *Synth. Met.*, 69: 441–442, 1995.

646. J.-S. Yang and T.M. Swager, Fluorescent porous polymer films as TNT chemosensors: Electronic and structural effects, *J. Am. Chem. Soc.*, 120: 11864–11873, 1998.

647. V.E. Williams and T.M. Swager, Iptycene-containing poly(arylene ethynylene)s, *Macromolecules*, 33: 4069–4073, 2000.

648. C. Tan, M.R. Pinto, and K.S. Schanze, Photophysics, aggregation and amplified quenching of a water-soluble poly(phenylene ethynylene), *Chem. Commun.* 5: 446–447, 2002.

649. R. Deans, J. Kim, M.R. Machacek, and T.M. Swager, A poly(*p*-phenylene ethynylene) with a highly emissive aggregated phase, *J. Am. Chem. Soc.*, 122: 8565–8566, 2000.

650. W.-S. Li, D.-L. Jiang, and T. Aida, Photoluminescence properties of discrete conjugated wires wrapped within dendrimeric envelopes: "Dendrimer effects" on π-electronic conjugation, *Angew. Chem. Int. Ed.*, 43: 2943–2947, 2004.

651. K. Yoshino, K. Tada, and M. Onoda, Optical properties of poly(3,4-dialkyl-1,6-phenylene ethynylene), *Jpn. J. Appl. Phys.*, 33: L1785–L1788, 1994.

652. K. Tada, M. Onoda, M. Hirohata, T. Kawai, and K. Yoshino, Blue–green electroluminescence in copolymer based on poly(1,4-phenylene ethynylene), *Jpn. J. Appl. Chem.*, 35: L251–L253, 1996.

653. H. Quante, P. Schlichting, U. Rohl, Y. Geerts, and K. Müllen, Novel perylene-containing polymers, *Macromol. Chem. Phys.*, 197: 4029–4044, 1996.

654. S. Yamaguchi and T.M. Swager, Oxidative cyclization of *bis*(biaryl)acetylenes: Synthesis and photophysics of dibenzo[*g,p*]chrysene-based fluorescent polymers, *J. Am. Chem. Soc.*, 123: 12087–12088, 2001.

655. S.V. Frolov, A. Fujii, D. Chinn, M. Hirohata, R. Hidayat, M. Taraguchi, T. Masuda, K. Yoshino, and Z.V. Vardeny, Microlasers and micro-LEDs from disubstituted polyacetylene, *Adv. Mater.*, 10: 869–872, 1998; and references therein.

656. R. Sun, Q. Zheng, X. Zhang, T. Masuda, and T. Kobayashi, Light-emitting substituted polyacetylenes, *Jpn. J. Appl. Phys.*, 38: 2017–2023, 1999; and references therein.

657. R. Hidayat, S. Tatsuhara, D.W. Kim, M. Ozaki, K. Yoshino, M. Teraguchi, and T. Masuda, Time-resolved study of luminescence in highly luminescent disubstituted polyacetylene and its blend with poorly luminescent monosubstituted polyacetylene, *Phys. Rev. B*, 61: 10167–10173, 2000; and references therein.

658. Y.M. Huang, J.W.Y. Lam, K.K.L. Cheuk, W. Ge, and B.Z. Tang, Strong luminescence from poly(1-alkynes), *Macromolecules*, 32: 5976–5978, 1999.

659. J.W.Y. Lam, Y. Dong, K.K.L. Cheuk, J. Luo, Z. Xie, H.S. Kwok, Z. Mo, and B.Z. Tang, Liquid crystalline and light-emitting polyacetylenes: Synthesis and properties of biphenyl-containing poly(1-alkynes) with different functional bridges and spacer length, *Macromolecules*, 35: 1229–1240, 2002.

660. Q. Zheng, R. Sun, X. Zhang, T. Masuda, and T. Kobayashi, Electroluminescent devices based on poly(diphenylacetylene) with carbazolyl side groups, *Jpn. J. Appl. Phys.*, 36: L1508–L1510, 1997.

661. J. Chen, Z. Xie, J.W.Y. Chen, C.C.W. Law, and B.Z. Tang, Silole-containing polyacetylenes. Synthesis, thermal stability, light-emission, nanodimensional aggregation and restricted intramolecular rotation, *Macromolecules*, 36: 1108–1117, 2003.

662. S.T. Wellinghoff, D. Zhi, T.J. Kedrowski, S.A. Dick, S.A. Jenekhe, and H. Ishida, Electronic conduction mechanism in polycarbazole iodine complexes, *Mol. Cryst. Liq. Cryst.*, 106: 289–304, 1984.

663. Y. Yang and Q. Pei, Light-emitting electrochemical cells from a blend of *p*- and *n*-type luminescent conjugated polymers, *Appl. Phys. Lett.*, 70: 1926–1928, 1997.

664. Z.-B. Zhang, M. Fujiki, H.-Z. Tang, M. Motonaga, and K. Torimitsu, The first high molecular weight poly(*N*-alkyl-3,6-carbazole)s, *Macromolecules*, 35: 1988–1990, 2002.

665. S. Grigalevicius, J.V. Grazulevicius, V. Gaidelis, and V. Jankauskas, Synthesis and properties of poly(3,9-carbazole) and low-molecular-mass glass-forming carbazole compounds, *Polymer*, 43: 2603–2608, 2002.

666. S. Maruyama, X.-T. Tao, H. Hokari, T. Noh, Y. Zhang, T. Wada, H. Sasabe, H. Suzuki, T. Watanabe, and S. Miyata, Electroluminescent applications of a cyclic carbazole oligomers, *J. Mater. Chem.*, 8: 893–898, 1999.

667. A. van Dijken, J.J.A.M. Bastiaansen, N.M.M. Kiggen, B.M.W. Langeveld, C. Rothe, A. Monkman, I. Bach, P. Stössel, and K. Brunner, Carbazole compounds as host materials in light-emitting diodes: Polymer hosts for high-efficiency light-emitting diodes, *J. Am. Chem. Soc.*, 126: 7718–7727, 2004.

668. J. Huang, Y. Niu, W. Yang, Y. Mo, M. Yuan, and Y. Cao, Novel electroluminescent polymers derived from carbazole and benzothiadiazole, *Macromolecules*, 35: 6080–6082, 2002.

669. J. Huang, Y. Xu, Q. Hou, W. Yang, M. Yuan, and Y. Cao, Novel red electroluminescent polymers derived from carbazole and 4,7-*bis*(2-thienyl)-2,1,3-benzothiadiazole, *Macromol. Rapid Commun.*, 23: 709–712, 2002.

670. G. Zotti, G. Schiavon, S. Zecchin, J.-F. Morin, and M. Leclerc, Electrochemical, conductive, and magnetic properties of 2,7-carbazole-based conjugated polymers, *Macromolecules*, 35: 2122–2128, 2002.

671. A. Iraqi and I. Wataru, Preparation and properties of 2,7-linked *N*-alkyl-9*H*-carbazole main-chain polymers, *Chem. Mater.*, 16: 442–448, 2004.

672. J.-F. Morin, P.-L. Boudreault, and M. Leclerc, Blue-light-emitting conjugated polymers derived from 2,7-carbazoles, *Macromol. Rapid Commun.*, 23: 1032–1036, 2002.

673. J.-F. Morin and M. Leclerc, Solvatochromic properties of 2,7-carbazole-based conjugated polymers, *Macromolecules*, 36: 4624–4630, 2003.

674. S.A. Patil, U. Scherf, and A. Kadaschuk, New conjugated ladder polymer containing carbazole moieties, *Adv. Funct. Mater.*, 13: 609–614, 2003.

675. J.-F. Morin, N. Drolet, Y. Tao, and M. Leclerc, Syntheses and characterization of electroactive and photoactive 2,7-carbazolenevinylene-based conjugated oligomers and polymers, *Chem. Mater.*, 16: 4619–4626, 2004.

676. S. Grigalevicius, J. Simokaitiene, J.V. Grazulevicius, L. Ma, and Z. Xie, Hole-transporting polymers containing carbazol-3,9-diyl and 1,4-phenylene fragments in the main chain, *Synth. Met.*, 158: 739–743, 2008.

677. D.D. Gebler, Y.Z. Wang, J.W. Blatchford, S.W. Jessen, L.-B. Lin, T.L. Gustafson, H.L. Wang, T.M. Swager, A.G. MacDiarmid, and A.J. Epstein, Blue electroluminescent devices based on soluble poly(p-pyridine), *J. Appl. Phys.*, 78: 4264–4266, 1995.

678. T. Yamamoto, T. Murauyama, Z.-H. Zhou, T. Ito, T. Fukuda, Y. Yoneda, F. Begum, T. Ikeda, S. Sasaki, H. Takezoe, A. Fukuda, and K. Kubota, π-Conjugated poly(pyridine-2,5-diyl), poly(2,2′-bipyridine-5,5′-diyl), and their alkyl derivatives. Preparation, linear structure, function as a ligand to form their transition metal complexes, catalytic reactions, n-type electrically conducting properties, optical properties, and alignment on substrates, *J. Am. Chem. Soc.*, 116: 4832–4845, 1994.

679. M. Halim, I.D.W. Samuel, J.N.G. Pillow, A.P. Monkman, and P.L. Burn, Control of colour and charge injection in conjugated dendrimer/polypyridine bilayer LEDs, *Synth. Met.*, 102: 1571–1574, 1999.

680. S. Dailey, M. Halim, E. Rebourt, L.E. Horsburgh, I.D.W. Samuel, and A.P. Monkman, An efficient electron-transporting polymer for light-emitting diodes, *J. Phys.: Condens. Matter*, 10: 5171–5178, 1998.

681. C. Wang, M. Kilitziraki, J.A.H. MacBride, M.R. Bryce, L.E. Horsburgh, A.K. Sheridan, A.P. Monkman, and I.D.W. Samuel, Tuning the optoelectronic properties of pyridine-containing polymers for light-emitting devices, *Adv. Mater.*, 12: 217–222, 2000.

682. A.P. Monkman, L.-O. Pålsson, R.W.T. Higgins, C. Wang, M.R. Bryce, A.S. Batsanov, and J.A.K. Howard, Protonation and subsequent intramolecular hydrogen bonding as a method to control chain structure and tune luminescence in heteroatomic conjugated polymers, *J. Am. Chem. Soc.*, 124: 6049–6055, 2002.

683. S.-C. Ng, H.-F. Lu, H.S.O. Chan, A. Fujii, T. Laga, and K. Yoshino, Blue electroluminescence from a novel donor/acceptor polymer structure, *Adv. Mater.*, 12: 1122–1125, 2000.

684. T. Miyamae, D. Yoshimura, H. Ishii, Y. Ouchi, K. Seki, T. Miyazaki, T. Koike, and T. Yamamoto, Ultraviolet photoelectron spectroscopy of poly(pyridine-2,5-diyl), poly(2,2′-bipyridine-5,5′-diyl), and their K-doped states, *J. Chem. Phys.*, 103: 2738–2744, 1995.

685. M.-Y. Hwang, M.-Y. Hua, and S.-A. Chen, Poly(pyridine-2,5-diyl) as electron-transport/hole blocking layer in poly(phenylene vinylene) light-emitting diode, *Polymer*, 40: 3233–3235, 1999.

686. M.J. Marsella, D.-K. Fu, and T.M. Swager, Synthesis of regioregular poly(methyl pyridinium vinylene): An isoelectric analogue to poly(phenylene vinylene), *Adv. Mater.*, 7: 145–147, 1995.

687. J. Tiuan, C.-C. Wu, W.E. Thompson, J.C. Sturm, R.A. Register, M.J. Marsella, and T.M. Swager, Electroluminescent properties of self-assembled polymer thin films, *Adv. Mater.*, 7: 395–398, 1995.

688. J. Tian, C.-C. Wu, M.E. Thompson, J.C. Strum, and R.A. Register, Photophysical properties, self-assembled films, and light-emitting diodes of poly(p-pyridylvinylene)s and poly(p-pyridinium vinylene)s, *Chem. Mater.*, 7: 2190–2198, 1995.

689. Y.Z. Wang, D.D. Gebler, D.K. Fu, T.M. Swager, A.G. MacDiarmid, and A.J. Epstein, Light-emitting devices based on pyridine-containing conjugated polymers, *Synth. Met.*, 85: 1179–1182, 1995.

690. A.J. Epstein, J.W. Blatchford, Y.Z. Wang, S.W. Jessen, D.D. Gebler, L.B. Lin, T.L. Gustafson, H.-L. Wang, Y.W. Park, T.M. Swager, and A.G. MacDiarmid, Poly(*p*-pyridine)- and poly (*p*-pyridyl vinylene)-based polymers: Their photophysics and application to SCALE devices, *Synth. Met.*, 78: 253–261, 1996; and references therein.

691. A. Wu, T. Akagi, M. Jikei, M. Kakimoto, Y. Imai, S. Ukishima, and Y. Takahashi, New fluorescent polyimides for electroluminescent devices based on 2,5-distyrylpyrazine, *Thin Solid Films*, 273: 214–217, 1996.

692. X. Zhang, A.S. Shetty, and S.A. Jenekhe, Electroluminescence and photophysical properties of polyquinolines, *Macromolecules*, 32: 7422–7429, 1999.

693. Y. Zhu, M.M. Alam, and S.A. Jenekhe, Regioregular head-to-tail poly(4-alkylquinoline)s: Synthesis, characterization, self-organization, photophysics, and electroluminescence of new *n*-type conjugated polymers, *Macromolecules*, 36: 8958–8968, 2003.

694. X. Zhang and S.A. Jenekhe, Electroluminescence of multicomponent conjugated polymers. 1. Roles of polymer/polymer interfaces in emission enhancement and voltage-tunable multicolor emission in semiconducting polymer/polymer heterojunction, *Macromolecules*, 33: 2069–2082, 2000.

695. M.S. Liu, Y. Liu, R.C. Urian, H. Ma, and A.K.-Y. Jen, Synthesis and characterization of poly-quinolines for light-emitting diodes, *J. Mater. Chem.*, 9: 2201–2204, 1999.

696. Y. Liu, H. Ma, and A.K.-Y. Jen, Synthesis and characterization of quinoline-based copolymers for light-emitting diodes, *J. Mater. Chem.*, 11: 1800–1804, 2001.

697. H. Krüger, S. Janietz, D. Sainova, and A. Wedel, New organo-soluble conjugated polyquino-lines, *Macromol. Chem. Phys.*, 204: 1607–1615, 2003.

698. T. Yamamoto, K. Sugiyama, T. Kushida, T. Inoue, and T. Kanbara, Preparation of new electron-accepting π-conjugated polyquinoxalines. Chemical and electrochemical reduction, electrically conducting properties, and use in light-emitting diodes, *J. Am. Chem. Soc.*, 118: 3930–3937, 1996.

699. S. Dailey, W.J. Feast, R.J. Peace, I.C. Sage, S. Till, and E.L. Wood, Synthesis and device character-ization of side-chain polymer electron transport materials for organic semiconducting applica-tions, *J. Mater. Chem.*, 11: 2238–2243, 2001.

700. P. Pösch, R. Fink, M. Thelakkat, and H.-W. Schmidt, A comparison of hole blocking/electron transport polymers in organic LEDS, *Acta Polym.*, 47: 487–494, 1998.

701. Y. Cui, X. Zhang, and S.A. Jenekhe, Thiophene-linked polyphenylquinoxaline: A new electron transport conjugated polymer for electroluminescent devices, *Macromolecules*, 32: 3824–3826, 1999.

702. F. Cacialli, X.-C. Li, R.H. Friend, S.C. Moratti, and A.B. Holmes, Light-emitting diodes based on poly(methacrylates) with distyrylbenzene and oxadiazole side chains, *Synth. Met.*, 75: 161–168, 1995.

703. X.-C. Li, F. Cacialli, M. Giles, J. Grüner, R.H. Friend, A.B. Holmes, S.C. Moratti, and T.M. Yong, Charge transport polymers for light emitting diodes, *Adv. Mater.*, 7: 898–900, 1995.

704. M. Strukelj, T.M. Miller, F. Papadimitrakopoulos, and S. Son, Effects of polymeric electron trans-porters and the structure of poly(*p*-phenylene vinylene) on the performance of light-emitting diodes, *J. Am. Chem. Soc.*, 117: 11976–11983, 1995.

705. M. Greczmiel, P. Strohriegl, M. Meier, and W. Brütting, Polymethacrylates with pendant oxa-diazole units synthesis and application in organic LEDs, *Macromolecules*, 30: 6042–6046, 1997.

706. X. Jiang, R.A. Register, K.A. Killeen, M.E. Thompson, F. Pschenitzka, and J.C. Sturm, Statistical copolymers with side-chain hole and electron transport groups for single-layer electrolumines-cent device applications, *Chem. Mater.*, 12: 2542–2549, 2000.

707. Q. Pei and Y. Yang, Bright blue electroluminescence from an oxadiazole-containing copolymer, *Adv. Mater.*, 7: 559–561, 1995.

708. Y. Yang and Q. Pei, Electron injection polymer for polymer light-emitting diodes, *J. Appl. Phys.*, 77: 4807–4809, 1995.

709. Q. Pei and Y. Yang, 1,3,4-Oxadiazole-containing polymers as electron-injection and blue elec-troluminescent materials in polymer light-emitting diodes, *Chem. Mater.*, 7: 1568–1575, 1995.

710. R. Cervini, X.-C. Li, G.W.C. Spencer, A.B. Holmes, S.C. Morath, and R.H. Friend, Electrochemical and optical studies of PPV derivatives and poly(aromatic oxadiazoles), *Synth. Met.*, 84: 359–360, 1997.

711. X.-C. Li, G.C.W. Spencer, A.B. Holmes, S.C. Moratti, F. Cacialli, and R.H. Friend, The synthesis, optical and charge-transport properties of poly(aromatic oxadiazole)s, *Synth. Met.*, 76: 153–156, 1996.

712. C. Wang, M. Kilitziraki, L.-O. Pålsson, M.R. Bryce, A.P. Monkman, and I.D.W. Samuel, Polymeric alkoxy PBD [2-(4-biphenyl)-5-phenyl-1,3,4-oxadiazole] for light-emitting diodes, *Adv. Funct. Mater.*, 11: 47–50, 2001.

713. S. Janietz and S. Anlauf, A new class of organosoluble rigid-rod fully aromatic poly(1,3,4-oxadiazole)s and their solid-state properties, 1., *Macromol. Chem. Phys.*, 203: 427–432, 2002.

714. S. Janietz, S. Anlauf, and A. Wedel, A new class of organosoluble rigid-rod, fully aromatic poly(1,3,4-oxadiazole)s and their solid-state properties, 2., *Macromol. Chem. Phys.*, 203: 433–538, 2002.

715. S.A. Jenekhe and J.A. Osaheni, Excimers and exciplexes of conjugated polymers, *Science*, 265: 765–768, 1994.

716. C.C. Wu, P.Y. Tsay, H.Y. Cheng, and S.J. Bai, Polarized luminescence and absorption of highly oriented, fully conjugated, heterocyclic aromatic rigid-rod polymer poly-*p*-phenylenebenzoxazole, *J. Appl. Phys.*, 95: 417–423, 2004.

717. S.J. Bai, C.C. Wu, T.D. Dang, F.E. Arnold, and B. Sakaran, Tunable and white light-emitting diodes of monolayer fluorinated benzoxazole graft copolymer, *Appl. Phys. Lett.*, 84: 1656–1658, 2004.

718. C.D. Entwistle and T.B. Marder, Application of three-coordinate organoboron compounds and polymers in optoelectronics, *Chem. Mater.*, 16: 4574–4585, 2004.

719. M. Matsumi, K. Naka, and Y. Chujo, Extension of π-conjugation length via the vacant p-orbital of the boron atom. Synthesis of novel electron deficient π-conjugated systems by hydroboration polymerization and their blue light emitting, *J. Am. Chem. Soc.*, 120: 5112–5113, 1998.

720. N. Matsumi, M. Miyata, and Y. Chujo, Synthesis of organoboron π-conjugated polymers by hydroboration polymerization between heteroaromatic diynes and mesitylborane and their light emitting properties, *Macromolecules*, 32: 4467–4469, 1999.

721. M. Matsumi, K. Naka, and Y. Chujo, π-Conjugated poly(cyclodiborazane)s with intramolecular charge transferred structure, *Macromolecules*, 33: 3956–3957, 2000.

722. M. Matsumi, K. Naka, and Y. Chujo, Poly(*p*-phenylene-borane)s, novel organoboron π-conjugated polymers via Grignard reagent, *J. Am. Chem. Soc.*, 120: 10776–10777, 1998.

723. Y. Makioka, T. Hayashi, and M. Tanaka, Poly[2,7-(9-oxo-9-phosphafluorenylene)-*alt-co*-(1,4-arylene)]s: Phosphorus-containing π-conjugated polymers, *Chem. Lett.*, 33: 44–45, 2004.

724. S. S. H. Mao, and T. D. Tilley, A versatile, transition-metal mediated route to blue-light-emitting polymers with chemically tunable luminescent properties, *Macromolecules*, 30: 5566–5569, 1997.

725. T. Baumgartner, T. Neumann, and B. Wirges, The dithieno[3,2-*b*:2′,3′-*d*]phosphole system: A novel building block for highly luminescent π-conjugated materials, *Angew. Chem. Int. Ed.* 43: 6197–6201, 2004.

726. T. Baumgartner and W. Wilk, Synthesis and unexpected reactivity of Si–H functionalized dithieno[3,2-*b*:2′,3′-*d*]phospholes, *Org. Lett.*, 8: 503–606, 2006.

727. S. Durben, Y. Dienes, and T. Baumgartner, Cationic dithieno[3,2-*b*:2′,3′-*d*]phospholes: A new building black for luminescent, conjugated polyelectrolytes, *Org. Lett.*, 8: 5893–5896, 2006.

728. C. Xu, H. Yamada, A. Wakamiya, S. Yamaguchi, and K. Tamao, Ladder *bis*-silicon-bridged stilbenes as a new building unit for fluorescent π-conjugated polymers, *Macromolecules*, 37: 8978–8983, 2004.

729. S. Yamaguchi and K. Tamao, A key role of orbital interaction in the main group element-containing π-electron systems, *Chem. Lett.*, 34: 2–7, 2005.

730. M. S. Liu, J. Luo, and A. K.-Y. Jen, Efficient green-light-emitting diodes from silole-containing copolymers, *Chem. Mater.*, 15: 3496–3500, 2003.

731. J. Ohshita, K. Kimura, K.-H. Lee, A. Kunai, Y.-W. Kwak, E.-C. Son, and Y. Kunugi, Synthesis of silicon-bridged polythiophene derivatives and their applications to EL device materials, *J. Polym. Sci., Part A: Polym. Chem.*, 45: 4588–4596, 2007.

732. Y.-Q. Mo, X.-Y. Deng, X. Jiang, and Q.-H. Cui, Blue electroluminescence from 3,6-silafluorene-based copolymers, *J. Polym. Sci., Part A: Polym. Chem.*, 47: 3286–3295, 2009.

733. L. Liao, A. Cirpan, Q. Chu, F. E. Karasz, and Y. Pang, Synthesis and optical properties of light-emitting π-conjugated polymers containing biphenyl and dithienosilole, *J. Polym. Sci., Part A: Polym. Chem.*, 45: 2048–2058, 2007.

3

Organic Small-Molecule Materials for Organic Light-Emitting Diodes

Shijian Su, Norman Herron, and Hong Meng

CONTENTS

3.1 Introduction

Small-molecule organic light-emitting diodes (SMOLEDs), inspired by the search for blue light–emitting devices based on organic crystals such as anthracene, can be traced back to the early work of Pope et al. in the 1960s [1]. The development of thin-film organic electroluminescent devices with relatively low driving voltages (<30 V direct current) by Vincett et al. at Xerox Canada in 1982 was a major step forward in this field [2]. A significant breakthrough in achieving high-electrical-efficiency organic light-emitting diodes (OLEDs) using small-molecule-based organic materials was the discovery of OLEDs reported by

Kodak scientists in 1987 [3]. In that publication, a double layer consisting of thin films of a hole-transporting triarylamine and a light-emitting and electron-transporting aluminum tris-8-hydroxyquinoline (Alq_3) layer, sandwiched between a transparent indium–tin oxide (ITO) electrode and an Al/Mg electrode, emitted green light under applied direct-current voltage. Although the quantum efficiency (QE) of such fluorescent material–based SMOLEDs is limited by spin statistics to only ~25%, phosphorescent SMOLEDs developed by the Princeton and University of Southern California (USC) groups achieved QEs approaching 100% [4,5]. These efficiencies far exceed those of any previously described devices. Such a fundamental breakthrough in device engineering and materials selection has ignited progress in the OLED field. Undoubtedly, device construction, device engineering, and, particularly, new materials design continue to drive advances in this field.

Numerous articles have covered research in the SMOLED field, and many have focused on various aspects of the design, synthesis, and applications of many classes of materials in the past three decades. Several major companies such as Kodak, Sony, Samsung, DuPont, LG, as well as newly born R&D companies such as Universal Display Corporation (UDC), Novaled GmbH, e-Ray Optoelectronics Technology Co. Ltd., America Organic Semiconductor, LLC, etc., have targeted the development of high-efficiency, long-lifetime SMOLED materials and devices.

Like Chapter 2 that reviews polymeric light-emitting materials, this chapter will attempt to review most of the important small-molecular materials used in OLEDs. In many cases, where such data is available, we will also describe the related device structures and electroluminescence (EL) performance associated with those materials. One of the difficulties in evaluating the relative merits of any given set of materials in OLEDs is the strong interplay of the materials stack and the device architecture chosen for the test device. This is further complicated by the diverse measures of device performance, which are often used to describe the test devices—combinations of luminance and power efficiencies, brightness and voltage stability expressed as lifetimes under diverse luminance and current conditions, defect and black-spot appearance and growth, etc. In such an undertaking, it is often tempting to make value judgments and critical assessments of the various data under review; however, we have attempted to refrain from such analysis, fraught with controversy as it is, and have tried to simply present an unbiased selection of representative data from the original authors' works. It is left to the reader to evaluate the merits and conclusions of this work by a careful reading of the primary publications.

3.1.1 Organic Light-Emitting Diode Device Structure

OLED devices are fabricated on a glass, plastic, metal, or ceramic substrate as a multilayer-stacked structure. The simplest manifestation of an OLED is a sandwich structure consisting of an emission layer (EML) between an anode and a cathode. More typical is an increased-complexity OLED structure consisting of an anode, an anode buffer or hole-injection layer (HIL), a hole-transport layer (HTL), a light-emitting layer, an electron-transport layer (ETL), a cathode interfacial or electron-injection layer (EIL), and a cathode. In some devices, a hole-blocking layer (HBL), an electron-blocking layer (EBL), and a stabilizer layer are also applied to achieve the desired performance. Although the structure of a typical OLED can contain many layers, not all of these layers are necessarily present in all OLED architectures. Indeed, much of the current research on OLEDs focuses on the development of the simplest possible and most easily processed architecture that can deliver the optimal combination of device properties.

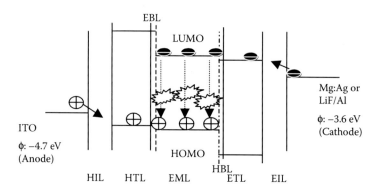

FIGURE 3.1
Function of each layer in a typical OLED.

3.1.2 Organic Light-Emitting Diode Operating Mechanism

The light-generating mechanism of OLEDs can be summarized by the following processes and is illustrated in Figure 3.1.

The function of each layer can be summarized as follows.

1. *Electron and hole injection from the cathode and the anode*

   When an electric field is applied between the anode and the cathode, electrons and holes are injected from the cathode and the anode, respectively, into the organic layers. With a matched energy barrier between the electron and the HILs and the cathode and the anode, electrons and holes are efficiently injected into the ETL and HTL.

2. *Electron and hole migration through the ETL and HTL*

   Once the electrons and holes have been injected, they migrate into the ETL and HTL to form excited states referred to as polarons by physicists or radical ions by chemists. These polarons or radical ions move by means of a so-called charge-hopping mechanism, through the electron- and hole-transport materials (ETMs and HTMs), which typically possess good charge-mobility properties, and eventually into the EML.

3. *Charge recombination*

   The charges then meet at the organic EML, and the device is optimized by fine tuning so as to match the number of electrons and holes coming through the EBL or HBL. Once the opposite charges recombine, an exciton is formed, and depending on the nature of the emission materials (EMs) and according to appropriate selection rules, singlet fluorescence or triplet phosphorescence is emitted.

3.2 Anode and Cathode Materials

In the following sections, materials appropriate for inclusion into all of the layers shown will be described; however, creative device physicists and material chemists continue to manipulate these materials into ever more intricate and elegant architectures. This includes the addition of new layers of new materials with new functions such that, as is usually the case, any review of this type that attempts to cover such a rapidly developing area will be outdated even before it is published.

3.2.1 Anode Materials

The anode material is, most typically, transparent ITO coated onto a glass or plastic substrate. Chapter 5 describes the details of such transparent anode materials. The general requirements for an anode material are as follows:

1. Highly conductive so as to reduce contact resistance
2. High work function (WF) ($\phi > 4.1$ eV) to promote efficient hole injection
3. Good film-forming and wetting properties of applied organic materials so as to ensure good contact with these adjacent organic layers
4. Good stability, both thermal and chemical
5. Transparent, or else highly reflective if used in top-emitting OLEDs

Clearly, in any light-emitting device, the light must escape from the device, and in OLEDs the window through which this occurs is typically provided by the anode ITO. ITO is a highly degenerate n-type semiconductor with high conductivity. It is transparent in the visible range owing to its large band gap of >4.0 eV [6]. Although other transparent and conductive electrode materials certainly exist (e.g., fluorine-doped tin oxide [7] and aluminum-doped zinc oxide [8]), the ready availability of glass substrates, precoated and (as needed) prepatterned with ITO as an item of commerce (owing to its extensive use in liquid crystal display screens) make ITO the most common material of choice. Although ITO has many desirable properties, it does suffer from several shortcomings. Most notably, ITO has a relatively high resistivity ($\sim 2 \times 10^{-4}$ Ω cm), moderate surface roughness (typically $\sim$2 nm), a chemically reactive surface (which may result in ion migration into the device), and a low and variable WF (4.5–4.8 eV), leading to hole-injection difficulties with some materials. The WF of ITO is rather sensitive to cleaning procedures (ozone or plasma treatments) used during device fabrication. Deposition techniques used for generating thin ($\sim$100 nm) coatings of ITO typically require sputtering (e-beam, pulsed laser [9], etc.) from an ITO target (typically 0–14% SnO_2 in In_2O_3) or from an In/Sn alloy in a reactive oxygen ($Ar–O_2$) atmosphere. In many instances, such deposition must be followed by annealing of the film at rather high temperature (>200°C) to reduce the resistivity to acceptable levels. Such high-temperature annealing processes are precluded for plastic substrates leading to even poorer resistivity in such systems. Attempts have been made to develop solution-coating approaches to generate ITO anodes but still require high-temperature annealing to reduce the resistance [10]. It is most often the poor conductivity of ITO that limits the size of passive-matrix OLED displays. Nevertheless, recently, new techniques have been developed to deposit ITO at low temperature to produce ITO-coated plastic substrate such as polyethylene terephthalate (PET) or polyethylene naphthalate (PEN), and commercial ITO-coated plastic substrates are now available [11].

Other materials such as gold ($\phi = 4.9$ eV), aluminum ($\phi = 4.2$ eV), indium-doped zinc oxide, magnesium indium oxide, nickel tungsten oxide, or other transparent conductive oxide materials have been studied as anodes in OLEDs. Furthermore, the WF of ITO can be varied by surface treatments such as application of a very thin layer of Au, Pt, Pd, or C, acid or base treatments, self-assembly of active surface molecules, or plasma treatment. As an example, Helander et al. recently demonstrated a chlorinated ITO transparent electrode with a WF of >6.1 eV that provides a direct match to the energy levels of the active light-emitting materials in state-of-the art OLEDs [12].

Anode materials are most typically deposited by evaporation, sputtering, or chemical vapor deposition methods. Other methods such as screen printing, laser ablation, and electrochemical deposition have also been used.

3.2.2 Cathode Materials

Unlike the constraints on anode material, the constraints on cathode materials are usually lower because typically they do not need to constitute the transparent electrode material. In certain instances, where a completely transparent OLED is needed (windshield and heads-up displays), ITO may also be used as the cathode with suitable modification [13]. In general, cathode materials are pure metals or metal alloys. The requirements for cathode materials are as follows:

1. High conductivity
2. Low WF to promote electron injection
3. Good film-forming and wetting properties to ensure good contact with adjacent organic layers
4. Good stability
5. Highly reflective or transparent if used in top-emitting OLEDs

Typically, the cathode is a low-WF metal or alloy system, such as Mg, Ca, Ba, and Al. Clearly, a low WF facilitates electron injection into the lowest unoccupied molecular orbital (LUMO) level of the ETM. However, low WF also implies high chemical reactivity, and problems can occur with direct chemical reduction of organic materials in contact with such low-WF metals. Although production of such species may be detrimental to device performance, it may, in favorable instances, actually assist in charge injection. Ease of oxidation of low-WF metals and alloys can also lead to difficulties in processing devices that become very sensitive to moisture and oxygen contaminants. A very popular solution to the problem of low WF, yet readily processable cathodes, is the two-layer cathode comprising a vapor-deposited thin (<5 nm) layer of LiF and subsequent Al. This cathode owes its discovery to work from Kodak, which showed that this combination, especially in contact with reducible species in the ETL, such as Alq_3, leads to the production of anionic species (e.g., Alq_3^- at the electrode surface, possibly with an Li^+ countercation and with cogeneration of AlF_3) [14]. This selective doping at the electrode and ETL interface then leads to improved charge injection similar to when lower WF cathode metals are used. This same principle has been used to generate other hybrid cathode systems, although all typically use an alkali-metal, alkali-earth, or rare-earth salt or oxide in conjunction with Al [15]. It is often the reactivity of the cathode metals that demands the high-quality hermetic seals used in OLED devices, and, in many cases of device failure due to inadequate sealing, black-spot defects appear in the device due to the effect of air and moisture on the cathode metal. A common approach to controlling this problem is to include a sacrificial getter material inside the encapsulated device to scavenge water and oxygen before they can corrode the cathode. The most popular cathode materials are Al (ϕ = 4.2 eV), LiF/Al (ϕ = 3.6–3.8 eV), Ca/Al, Mg/Ag (ϕ = 2.90 eV), and Ba/Al (2.60 eV). Although even lower WF can be achieved with, e.g., Yb (ϕ = 2.4 eV), the low reflectivity index of the latter makes it less suitable for OLED applications. The active metal Ca (ϕ = 2.60 eV) often has to be accompanied with other metals such as Al to increase the device lifetime. It is worth noting that the WF of the metals can be affected by their purity, their deposition method, and the surface structure, and the crystal orientation of the deposited films.

Research on cathode materials focuses on reduction of the high chemical activity of the lower-WF metals (e.g., Ca/Al), the increase of the chemical stability, and improvement of the sticking coefficient of the interlayer materials (e.g., LiF/Al).

3.3 Hole-Injection and -Transport Materials

3.3.1 Hole-Injection Materials

Closely related to the anode modifications described above, the use of a hole-injection material (HIM) to improve charge injection into the OLED device has spawned a number of materials that have been shown to provide benefits, particularly in terms of lower operating voltages and extended lifetimes of devices.

Since an OLED is a multilayer device structure, the interfacial electronic structure at the organic–metal and organic–organic interfaces plays an important role in the devices. The interface structure and energy-level alignment of organic–inorganic and organic–organic has been investigated by Kahn and Seki [16,17]. These studies are very helpful in elucidating the role of the HIM as well as the electron-injection material (EIM), although much work is still needed to fully understand this field. A book edited by Salaneck et al. emphasizes the interface issues, ubiquitous in organic electronics, and interested readers may find additional detailed information there [18,19]. Hung and Chen combined organic–metal interface energy diagrams and ultraviolet (UV) photoelectron spectroscopy (UPS) results, as shown in Figure 3.2 [20].

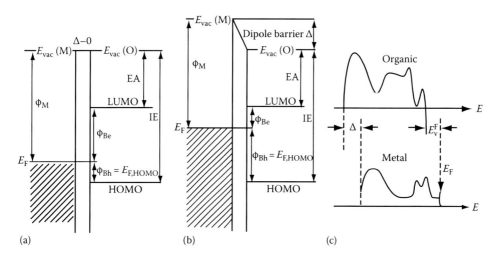

FIGURE 3.2
Schematic of an organic–metal interface energy diagram (a) without and (b) with an interface dipole and (c) UPS spectra of metal and organic. (From Hung, L.S. and Chen, C.H., *Mater. Sci. Eng.*, R39, 143, 2002. With permission.)

By careful examination of the various work dealing with organic–metal interfaces, they concluded the following:

1. In general, the dipolar difference between the metal and organic is negative (vacuum level lowered), with the exception that when electron-withdrawing groups such as fluorine atoms are attached to the metal surface, this value is positive.
2. The higher the WF of the metal, the larger the shift of the dipolar energy.
3. The interface dipole in organic–organic junctions is negligible, with the exception of strong donor–acceptor interfaces where a barrier of 0.2–0.3 eV may exist due to the charge-transfer process.
4. The dipolar difference is very complex in organic–metal interfaces (metal may deposit into the organic layer).

The HIL acts as an interface connection layer between the anode and the HTL so as to improve the film-forming property of the subsequent organic layer and to facilitate efficient hole injection. HIMs should have good adhesion to the anode and should serve to smooth the anode surface. The most common HIMs are

1. Porphyrinic metal complexes
2. Conducting polymers
3. Self-assembly compounds
4. Fluorocarbon polymers
5. Electron donors doped with electron acceptors
6. Inorganic HIMs
7. Organic HIMs

3.3.1.1 Porphyrinic Metal Complexes

In 1996, the Kodak group achieved a highly stable OLED by introducing a thin layer (~15 nm) of copper phthalocyanine (CuPc, 1) (Scheme 3.1) between the anode (ITO) layer and the hole-transport 4,4′-bis[N-(1-naphthyl-1)-N-phenyl-amino]-biphenyl (α-NPD) layer [21]. The function of the CuPc as an HIM has two possible controversial mechanisms. One study showed that an optimized thin layer of CuPc actually lowers the driving voltage by reducing the hole-injection barrier compared with an α-NPD–ITO interface as was

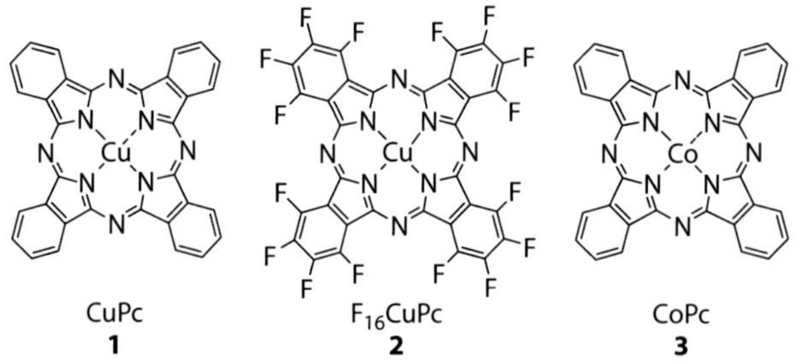

CuPc
1

F_{16}CuPc
2

CoPc
3

SCHEME 3.1
Chemical structures of porphyrinic metal complex HIMs.

explained by the Kodak group. Another study, however, showed that inserting the CuPc helps balance the hole and electron currents (by way of sacrificing the hole-injection efficiency), but results in an increased operating voltage [22]. Tadayyon et al., using UPS, studied the interfacial band energies of CuPc inserted between ITO and α-NPD, and concluded that the good wetting of CuPc on ITO may also partially contribute to the enhancement of the device stability, in addition to balancing the hole and electron injection [23]. Hill and Kahn also studied the interface of ITO–CuPc–NPD by UPS as well as the characteristics of related hole-only devices, and concluded that the effects of the CuPc layer depend on the WF of the underlying ITO [24]. These researchers measured the offset between their ITO and CuPc at about 0.70 eV. α-NPD and CuPc are offset by about 0.5 eV. In this case, the CuPc layer improves the hole injection.

CuPc is a readily available pigment material that has very high thermal stability and is a modest semiconductor. When applied to ITO as a thin overlayer (usually by high-vacuum evaporation), the highest occupied molecular orbital (HOMO) level of the CuPc effectively pins the ITO WF at ~5.1 eV, which can lead to improved charge injection at lower voltages and much less sensitivity of performance upon ITO cleaning protocols [25]. Although CuPc has found widespread use, other phthalocyanines, including the metal-free version [26], fluorinated version of F_{16}CuPc (**2**) [27], and cobalt complex of CoPc (**3**) [28], have also been used for HIL applications. The structurally similar porphyrin materials have also found applicability for this purpose.

Further improvement of the device efficiency was achieved by adapting so-called quantum well layers. In the report by Qiu's group, an OLED device with an optimum quadriquantum-well structure of four alternating layers of CuPc and NPD gives 3 times the efficiency compared with the conventional structure [29]. The authors explained that this observation is due to the improved balance between holes and electrons.

3.3.1.2 Conducting Polymers

The performance of OLED devices employing CuPc as an HIL is unstable due to thermally induced HTM crystallization on the CuPc surface [30]. One approach to improve the hole injection and enhance the device stability is to overcoat the CuPc or else to directly deposit onto the ITO itself with a buffer layer of a conductive polymer such as acid-doped (typically using polystyrene sulfonic acid, PSS) PEDOT (poly-3,4-ethylenedioxythiophene) [31], PANI (polyaniline) [32], or polypyrrole [33]. Such coatings can typically be deposited from water solutions or suspensions of the polymer by spin coating, ink-jet printing, etc. Such layers can result in improved surface smoothness and compatibility with subsequent organic layers, a somewhat higher WF (>5.0 eV) for better hole injection, and improved barrier properties by protecting the EMLs of the device from the reactive ITO surface. The smoothing effect of a conductive polymer layer should not be underestimated, as conductive spikes on a native ITO surface are often large enough to contribute to serious electrical shorting to the cathode (given that the typical organic layers used in OLEDs are very thin). On the negative side, however, most of the available conductive polymer materials used for this application are strongly acidic due to the doping necessary to induce conductivity. The acidity can, in many instances, lead to corrosion problems (the ITO surface may actually partially dissolve during coating) and also problems with EMLs, which may be acid sensitive (e.g., many are based on amine-containing host or guest materials that can react with acid functionality). This effect of the material acidity can manifest itself as a short device lifetime, including poor shelf life.

3.3.1.3 Self-Assembled Monolayers (SAMs)

Surface modification by chemically attaching a monolayer (or thicker layers) of organic species is an efficient way to control the chemical and electronic properties of ITO electrodes at the molecular level. This approach has the potential for exercising exquisite control over the dipoles on the ITO surface and, thereby, the charge-injection properties. Such an approach, utilizing attachment of organotin, phosphorus, and phenolic species, has been shown to be effective in controlling the anode WF [34]. Other work has claimed silane monolayer species for a similar purpose [35]. Overall, this approach may offer a chemical solution to many of the issues that arose due to the device performance of ITO and has become a very active area of investigation.

Marks' group introduced two new HIMs, 4,4'-bis[(*p*-trichlorosilylpropyl-phenyl) phenyl-amino]biphenyl (TPD-Si$_2$Cl, **4**) and 4,4'-bis[(*p*-methyloxylsilylpropylphenyl) phenylamino] biphenyl (TPD-Si$_2$OMe, **5**), which combine the hole-transporting efficiency of *N,N*-diphenyl-*N,N*-bis(3-methylphenyl)-1,1-biphenyl)-4,4-diamine) (TPD) and the strong cross-linking and densification tendencies of organosilanol groups (Scheme 3.2) [36]. Covalent chemical bonding of TPD-Si$_2$ to the anode (e.g., ITO) and its self-cross-linking as a hole-injection and adhesion interlayer can enhance OLED hole-injection and device stability. High-efficiency and high-luminance SMOLEDs are fabricated by introduction of the TPD-Si$_2$ interlayer, which significantly increases OLED current and light output by a factor of ~50–80 as well as improves the external quantum efficiencies (EQEs) [37]. The HTLs enhance polymer light-emitting diode (PLED) anode hole injection and exhibit significantly greater electron-blocking capacity than PEDOT:PSS [38]. The organosiloxane HIM approach offers the convenience of fabrication, flexibility in choosing HTL components, and reduced HTL-induced luminescence quenching, and can be applied as a general strategy to enhance PLED performance. The same group also reported a class of triphenylamine (TPA)-based molecules (**4, 6–8**) with various contents of TPA and trichlorosilane groups attached, and studied the hole-injection characteristics of their SAMs on ITO [39]. The large observed interfacial molecular structure effects offer an approach to tuning OLED hole-injection flux over 1–2 orders of magnitude, resulting in up to 3-fold variations in OLED brightness at identical bias and up to a 2 V driving voltage reduction at identical brightness.

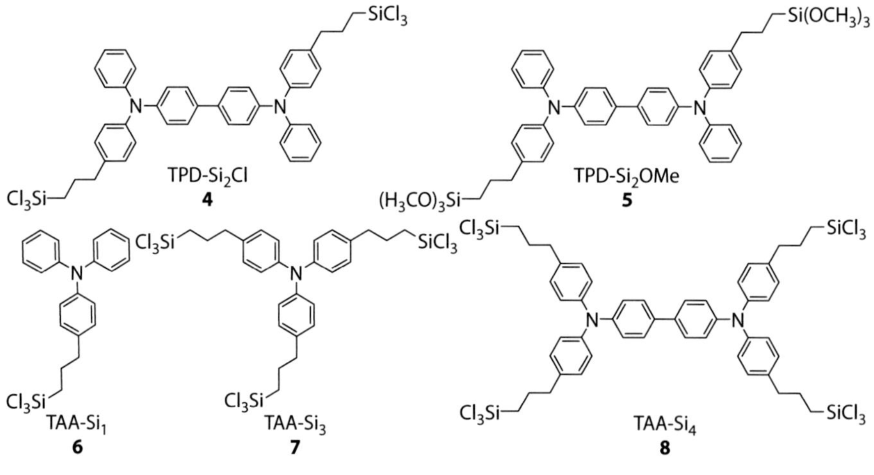

SCHEME 3.2
Chemical structures of triphenylamine-containing silane HIMs.

Kaji's group systematically investigated the relationships between the preparation conditions of SAM and the resultant device properties by utilizing pentyltriethoxysilane (PTES) [40]. The performances of OLEDs and hole-only devices reveal that the catalytic condition, the value of H_2O/PTES ratio (r), and the concentration of PTES (C_{PTES}) have a great effect on the hole-injection property. When SAM formation is performed under $r = 15$ and $C_{PTES} = 250$ mM, with 0.01 M HNO_3 as a catalyst, the OLEDs exhibit excellent device performance with the highest hole injection.

Although the effect of using silane SAMs on ITO to improve hole injection in OLEDs is rather outstanding, the drawback of a tendency to self-condense in solution, resulting in a difficult formation of uniform monolayers, restricts their wide application. Moreover, the amount of silane SAMs on ITO is determined by the concentration of surface hydroxyl groups. By comparison, phosphonic acid SAMs with better surface loading and greater hydrolytic stability can easily form a metal–O–P bond on the ITO surface through their own hydroxyl (–OH) groups, which is not limited by the surface hydroxyl content [41]. As a result of the excellent self-assembly property of phosphonic acid, some studies utilizing their SAMs on ITO have been conducted to enhance hole injection of OLEDs.

Hanson et al. reported that the hole-injection barrier at the ITO/HTL junction could be greatly reduced by binding a 4TPA (9) (Scheme 3.3) SAM onto the ITO surface [42]. The device of ITO/F_4-TCNQ-4TPA/NPD/Alq$_3$/Al using F_4-TCNQ-doped 4TPA SAM as an HIL shows a 10,000-fold increase for both current density and luminance versus the device without this doped SAM. Rampulla et al. investigated the charge-injection properties of three monolayers of 4TPA (9), 2TPA (10), and DPA (11) bonded to ITO electrodes using transition voltage spectroscopy [43]. At the ITO electrode, there is no large difference among these three molecules for electron injection; however, the hole-injection barrier of the aromatic compounds (2TPA and 4TPA) is higher than that of the saturated compound (DPA). Jen's group developed a series of phosphonic acid SAMs based on TPA and used them to modify the surface of an ITO anode [44]. The current density for hole-only devices reveals that the hole injection of these TPA-based compounds shows the following trend: TPA-PA

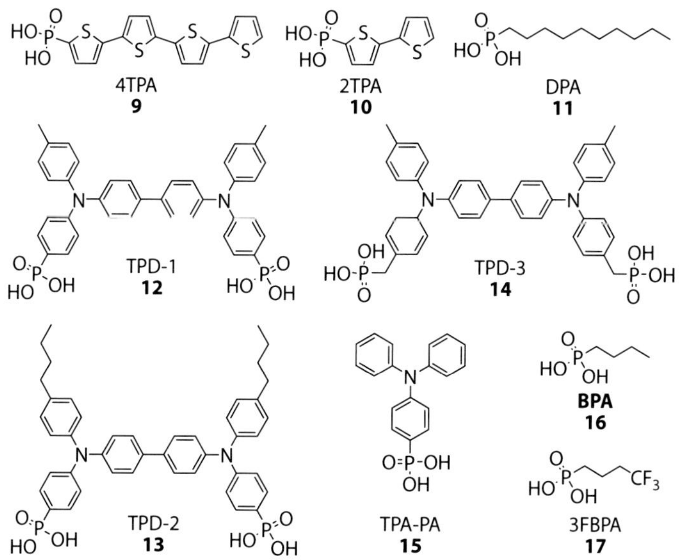

SCHEME 3.3
Chemical structures of phosphonic acid HIMs.

(15) << TPD-1 (**12**) ≤ TPD-2 (**13**) << TPD-3 (**14**), which is correlated with the trend in their HOMO energy levels. All of the SAMs with good hole-transporting ability have dipole moments orienting away from the ITO surface, leading to an effective change in ITO WF. Compared with the devices with bare ITO, devices using SAM-modified anodes exhibit significant higher current density and brightness at 10 V. Recently, Yu et al. reported SAMs of binary mixtures of 1-butylphosphonic acid (BPA, **16**) and the trifluoromethyl-terminated analogue 4,4,4-trifluoro-1-butylphosphonic acid (3FBPA, **17**) formed on ITO surfaces could tune the WF of ITO over a range of 5.0–5.75 eV by varying the mixing ratio of the two adsorbents [45]. The performances of OLED devices with structure of ITO/mixed-SAMs/HTL/Alq$_3$/EIL/Al, which utilize proper HTL (NPD or BPAPF) and EIL (LiF or Cs$_2$CO$_3$) to approach charge-carrier balance, can be greatly improved.

3.3.1.4 Fluorocarbon Polymers

Hung et al. reported using a plasma polymerization of CHF$_3$-coated ITO as an anode in OLEDs, and demonstrated dramatically improved device performance and stability [46]. The high ionization potential of polymerized fluorocarbon (9.5 eV), coupled with its ability to impede indium diffusion into the organic layers, stabilizes the device lifetime. In addition, the unusually low resistivity of this thin layer of plasma-polymerized fluorocarbon lowers the device operating voltage as well. The same group later explained that the enhanced performance and hole injection is attributed to the dipolar layer formed by the negatively charged fluorine [47]. The WF of ITO is strongly influenced by the electrostatic conditions at the surface. The introduction of the plasma-treated polyfluorocarbon creates a dipolar interface between ITO and the polymer layer, which leads to a lower barrier between the ITO and the HTL as depicted in Figure 3.3.

Instead of using plasma-polymerized polyfluorocarbon as HIL, Qiu et al. utilized a thermally deposited Teflon (polytetrafluoroethylene) thin layer as an HIL, which results in the same effect [48].

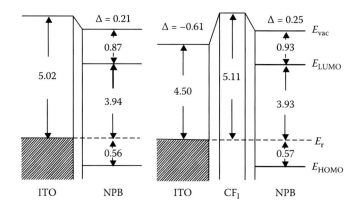

FIGURE 3.3
UPS spectra of NPB on a UV–ozone-treated ITO substrate (left) and NPB on an untreated ITO substrate overlaid by a 0.3-nm-thick CF$_x$ film (right), and schematic energy-level diagrams of the corresponding interfaces. The positions of E_{HOMO} and vacuum level (E_{vac}) are derived from UPS measurements, and the position of the lowest unoccupied molecular orbital of NPB (E_{LUMO}). (From Tang, J.X., Li, Y.Q., Zheng, L.R., and Hung, L.S., *J. Appl. Phys.*, 95, 4397, 2004. With permission.)

3.3.1.5 Inorganic Hole-Injection Materials

It has been reported that a thin interfacial inorganic layer such as SiO_2, SiO_xN_y, or TiO_2 can improve device performance [49–51]. Effective hole injection was achieved by Gao et al. using thermally evaporated ultrathin layer of $InCl_3$ on ITO anode [52]. The improved hole injection is evidenced by the formation of an In–Cl dipole, which improves the ITO WF (by 0.48 eV) and reduces the hole-injection barrier (by 0.15 eV). Phosphorescent OLEDs with the device configuration of $ITO/InCl_3/CBP/CBP:Ir(ppy)_2(acac)/TPBI/LiF/Al$ give the current efficiency of 76 cd/A and the maximum power efficiency of 42.4 lm/W at 20 mA/cm^2, >30% improved compared with the devices based on UV–ozone-treated ITO anode. More than a 3-fold improvement in half-life time (estimated at initial luminance of 1000 cd/cm^2) is also observed.

The thin layers of metal oxides, like MoO_3, WO_3, and Ni_2O_3, have also been used as HIMs in OLEDs. By using UPS and inverse photoelectron spectroscopy (IPES), Kröger et al. discovered that the electron affinity (EA), WF, and ionization energy (IE) of MoO_3 thin film were 6.7, 6.9, and 9.7 eV, respectively [53]. In green devices of $ITO/HIL/NPD/2,2'$-bi-(9,10-diphenyl-anthracene) (TPBA):9,10-bis[phenyl(m-tolyl)-amino]anthracene (TPA)/Alq$_3$/LiF/Al, MoO_x was investigated as an HIM to improve the device performance [54]. Owing to the enhanced hole injection, the stable interface between MoO_x/NPD, and an improved balance of holes and electrons, the MoO_x-based device displays much higher maximum power efficiency and stability than those of devices using CuPc and m-MTDATA as HILs.

Ni_2O_3 was utilized as an HIL in single-layer OLEDs by Liu and coworkers [55,56]. By inserting a 1-nm ultrathin HIL of Ni_2O_3 and MoO_3, the 2,3,6,7-tetrahydro-1,1,7,7-tetramethyl-1H,5H,11H-10-(2-benzothiazolyl)quinolizino-[9,9a,1gh] coumarin (C545T)-doped Alq$_3$ emitting system–based single-layer OLED gives a current efficiency of 6.4 cd/A for Ni_2O_3 HIL and 12.8 cd/A for MoO_3 HIL at 100 cd/m^2 at a doping concentration of 1%. The improved performance is attributed to the balanced charge due to the modified hole-injection and electron-blocking (for MoO_3 devices) characteristics at the anode/EML interface. Their later study on phosphorescent OLEDs (PHOLEDs) also showed largely improved efficiency of 22 and 25.5 cd/A for Ni_2O_3 and MoO_3 HIL, respectively, in a device configuration of $ITO/HIL/FIrpic/FIrpic:TPBI/LiF/Al$, >6-fold better than the reference devices without the Ni_2O_3 or MoO_3 HIL. Raised anode WF (5.3 and 5.4 eV for Ni_2O_3 and MoO_3, respectively) and lowered hole-injection barrier contribute to the efficiency improvement and lowered turn-on voltage, as evidenced by both x-ray photoemission spectroscopy (XPS) and UPS measurements. Further higher performance is achieved in dual-EML devices $ITO/MoO_3/FIrpic:CBP/FIrpic:TPBI/LiF/Al$, in which hole injection is further eased by the blend layer with CBP.

WO_3 was utilized as an HIL in inverted OLEDs by Kim et al. [57]. By insertion of an ultrathin layer between the low WF Al anode and NPD HTL, efficient hole transfer is achieved owing to the strong band bending at the WO_3/NPD interface, which makes the NPD HOMO level approach the WO_3 CB, giving largely reduced hole-injection barrier (0.15 eV). The authors also demonstrated that the major role of the WO_3 film is to help charge transport, forming a charge-generation layer between NPD and WO_3. Devices with Alq$_3$ EML exhibit improvement in current efficiency by a 3-order magnitude and a decrease in turn-on voltage compared with those without a WO_3 HIL as a result of favorable energy-level alignment.

Yook and Lee developed a device structure of $ITO/DNTPD/MoO_3/NPD/Alq_3/LiF/Al$ having MoO_3 as an interlayer between HIL and HTL [58]. Owing to the enhanced hole injection by interfacial doping effect stemming from interfacial charge complex between MoO_3 and NPD, the driving voltage of the MoO_3-based devices with high current densities is reduced by 1.3 V. However, in the study of Murata's group, they discovered that the interfacial charge-generation mechanism (composed of charge transfer, separation, and transit) is more dominant than the hole-injection mechanism for improvement of OLED performances using MoO_3 as a buffer layer [59]. Furthermore, Meyer et al. demonstrated that MoO_3 films formed by spin coating possess electronic properties nearly identical to those MoO_3 films thermally evaporated under vacuum [60]. From the current–voltage characteristics of hole-only devices with a structure of ITO/HIL/poly(9,9′-dioctylfluorene-co-bis-*N,N*′-(4-butylphenyl)diphenyl-amine) (TFB)/Hg using spin-coated MoO_3, thermally evaporated MoO_3 and PEDOT:PSS as HILs, it is found that spin-coated MoO_3 films exhibit similar hole-injection efficiency as thermally evaporated MoO_3, and better performances than PEDOT:PSS.

Recently, Vasilopoulou et al. studied the influence of hydrogenation on the WF of molybdenum oxides [61]. The study discovered that incorporation of hydrogen within the molybdenum oxide lattice resulted in a high density of occupied gap states near the Fermi level and improved n-type conductivity. Unlike the oxygen vacancy formation counterparts, which exhibit a large decline in WF, hydrogenated molybdenum oxides (H_yMoO_x) only showed a slight WF decrease while possessing a tuned favorable energy-level alignment with organic layers. OLEDs using H_yMoO_x as an HIL achieve a maximum current efficiency of 10.5 cd/A and power efficiency of 7.3 lm/W in an ITO/Mo oxide/F8BT/POM (12-phosphotungstic acid)/Al device structure, >3-fold of those using stoichiometric molybdenum oxides.

Aside from the mechanisms thus far studied, the enhanced performance may also arise from the improved smoothness of the surface of ITO, which leads to more homogeneous adhesion of the HTL. In addition, the optimized thickness of the buffer layer also helps balance the device charges due to reduced hole injection.

3.3.1.6 Doping Hole-Transport Materials

To obtain low operating voltage in OLEDs, the method of using p-type doping or n-type doping structure in charge injection/-transport layers is mostly studied. A doping strategy was first explored using PLEDs where a hole-injecting conducting polymer such as polythiophene was doped with an oxidizing agent, such as $FeCl_3$ [62], or MEH–PPV was doped with iodine [63]. To improve the hole injection from the anode, hole-transporting polymers of TPDPES (**18**) [64] and PC–TPB–DEG (**19**) [65] were doped with tris(4-bromophenyl) aminium hexachloroantimonate (TBPAH, **20**) to be an HIL (Scheme 3.4). In SMOLEDs, an HTL doped with a strong electron acceptor compound has been used as an HIL; for example, vanadyl phthalocyanine (VOPc) HTL has been doped with a very strong acceptor tetrafluorotetracyanoquinodimethane (F_4-TCNQ) (**21**) (Scheme 3.5) [66]. These HILs dramatically lower the driving voltage and improve the device efficiency.

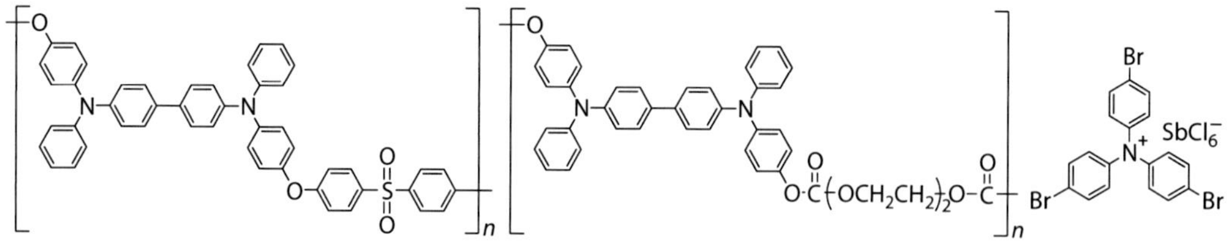

TPDPES
18

PC-TPB-DEG
19

TBPAH
20

SCHEME 3.4
Chemical structures of p-doping polymer HIMs.

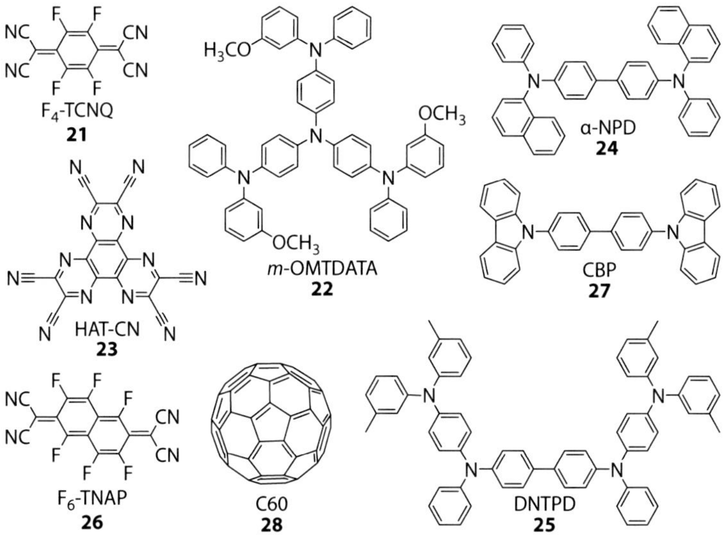

SCHEME 3.5
Chemical structures of p-type materials and p-doped HIMs.

A series of p-doped aromatic diamines have been reported by Pfeiffer et al. as excellent HIMs, e.g., 4,4′,4″-tris(N, N-diphenylamino)triphenylamine (TDATA) or 1, 4-benzenediamine,N-(3-methoxyphenyl)-$N′,N′$-bis[4-[(3-methoxyphenyl)phenylamino] phenyl]-N-phenyl (m-OMTDATA, **22**) doped with F$_4$-TCNQ (**21**) by controlled coevaporation [67,68]. Multilayered OLEDs consisting of ITO/F$_4$-TCNQ (2%):TDATA (100 nm)/TPD (10 nm)/Alq$_3$ (65 nm)/LiF (1 nm)/Al achieved a very low operating voltage of 3.4 V, giving 100 cd/m^2 at 9.1 mA/cm^2.

Besides F$_4$-TCNQ, a strong p-type material, HAT-CN (**23**), was also used as a dopant in α-NPD (**24**) HTL to enhance the hole-injection property [69]. The LUMO level of HAT-CN (−5.7 eV) was similar to the HOMO level of α-NPD (−5.5 eV), so that extra hole carriers can be generated by interlayer electron transfer from α-NPD to HAT-CN. Lowered turn-on voltage, improved EQE (from 10.8% to 16.8%) and power efficiency (from 17 to 47 lm/W) are obtained at 1000 cd/m^2 in a green PHOLED of ITO/α-NPD:HAT-CN/TCTA/ TCTA:TPBI:Ir(ppy)$_3$/BCP/Alq$_3$/LiF/Al due to a better charge balance brought by the enhanced hole injection. Cho et al. developed two different types of green OLEDs with HAT-CN as a p-dopant and interlayer by itself [70]. Although two types of devices display a similar low operating voltage (~4.0 V), the devices of ITO/DNTPD/HAT-CN/DNTPD/ GGH01:GGD01/LGC201:LiQ/LiQ/Mg:Ag using HAT-CN as an interconnecting layer exhibit higher maximum current efficiency (51.7 cd/A) and power efficiency (39.5 lm/W) than the devices of ITO/DNTPD:HAT-CN (3%)/DNTPD/GGH01:GGD01/LGC201:LiQ/ LiQ/Mg:Ag with HAT-CN-doped DNTPD (**25**) as HIL (49.1 cd/A, 38.7 lm/W). Liao et al. replaced the p-type doped HTL by a nondoped HAT-CN layer as an intermediate connector in tandem OLED devices [71]. Comparing with traditional devices using F$_4$-TCNQ-doped α-NPD intermediate layer and MoO$_3$/WO$_3$-doped Alq$_3$:Li layer, the HAT-CN layer–based tandem device shows almost ideal charge-injection and -transport properties, as the turn-on voltage was slightly smaller than that of the single-layer reference device. Lifetime measurement also indicates improved voltage stability as a result of better interface condition.

Koech and coworkers reported a strong electron-withdrawing p-type dopant, 1,3,4,5,7,8-hexafluorotetracyanonaphthoquinodimethane (F$_6$-TNAP, **26**) [72]. Obvious decrease in turn-on voltage is found in a bulk-doped device, while the interface-doped device exhibited no enhancement in operating voltage, indicating that F$_6$-TNAP offers an improved film conductivity other than a better hole injection. Blue PHOLED devices in an ITO/F$_6$-TNAP:α-NPD/TCTA/FIrpic:mCP/PO15/LiF/Al structure show a >20% reduction of operating voltage at 10 mA/cm^2 at the doping concentration of 2 wt.%, as a result of charge-transfer-complex formation between F$_6$-TNAP and the α-NPD bulk.

The method of p-type doping with MoO$_3$ is also effective to improve performances of OLEDs. In the report of Yoon's group, MoO$_3$ was studied as a p-dopant in α-NPD [73]. The devices of ITO/MoO$_3$-doped α-NPD/α-NPD/C545T-doped Alq$_3$/Alq$_3$/LiF/Al using the MoO$_3$-doped α-NPD as an HIL have an improved device efficiency with low driving voltage because of the interface effects and decrease of hole injection barrier.

Besides HTMs, a general host material of 4,4'-bis(carbazol-9-yl)biphenyl (CBP, **27**) was also doped with MoO$_x$ as an HIL [74]. Impedance spectroscopy shows that increase of MoO$_x$ content from 5% to 20% results in a large decrease in hole-injection barrier from 0.63 to 0.18 eV. Devices based on an ITO/CBP:MoO$_x$/CBP:Ir(ppy)$_2$(acac)/TPBI/LiF/Al structure exhibit a gradual increase in power efficiency with the increment of doping concentration from 5% to 20%, twice as high as the devices with MoO$_x$/CBP bilayer (63 vs. 28 lm/W).

An MoO$_3$-doped C$_{60}$ (**28**) layer was used as an HIL in OLEDs by Zou et al. [75]. By way of co-evaporation, a 5-nm-thick, 1:1 in volume codeposited layer is inserted in a device structure of ITO/HIL/NPD/Alq$_3$/LiF/Al. The driving voltage at luminance of 100 cd/m^2 is reduced by 35% to be 5.7 eV compared with reference devices without an HIL, which leads to a >40% increased power efficiency at 10 mA/cm^2. The generation of charge-transfer complexes, i.e., $\left[C_{60}^+:MoO_3^- \right]$, as evidenced by UV/visible (vis)/near-infrared absorption spectra, leads to the lowered turn-on voltage as a result of the increase in charge-carrier density. A similar study was carried out by Li et al. using an MoO$_3$ doped CuPc thin film [76]. Best performance is obtained in devices with 50% doping concentration, giving a driving voltage as low as 4.4 V and power efficiency of 4.3 lm/W at luminance of 100 cd/m^2. In addition to the formation of charge transfer complex $\left[CuPc^+:MoO_3^- \right]$, the performance enhancement is also attributed to the better film morphology.

3.3.1.7 Organic Hole-Injection Materials

Besides doping an HTM, a nondoped HTM with high-lying HOMO level can be also used as an HIL. Lee et al. reported a new HIM containing a rigid diphenylbinaphthyl core in the structure, namely PPBN (**29**) (Scheme 3.6) [77]. This molecule has excellent thermal stability (T_d > 480°C) and high glass transition temperature (T_g) (148°C) with a HOMO level of −5.10 eV. Owing to the increase of hole injection and transport, the device of ITO/HIL/NPD/Alq$_3$/Al:Li using PPBN as the HIL can reach an efficiency of 3.38 cd/A, which is higher than the control device (2.14 cd/A) with a commercially available material, 4,4',4"-tris-(N-(naphthalen-2-yl)-N-phenyl-amino) triphenylamine (2-TNATA, **30**) as the HIL.

Three DDP derivatives of B-DDP (**31**), T-DDP (**32**), and BT-DDP (**33**) were synthesized by Kim et al. [78]. OLEDs containing these materials as an HIL demonstrate a significant reduction in driving voltage compared with those using α-NPD, which has been believed to be a result of the decrease in hole-injection barrier at the anode/HIL interface induced by an enhanced horizontal orientation of the planar molecules. It is interesting to notice that these materials reduce the hole-injection barrier regardless of the HIL thickness, but show little influence on the overall hole mobility of the devices.

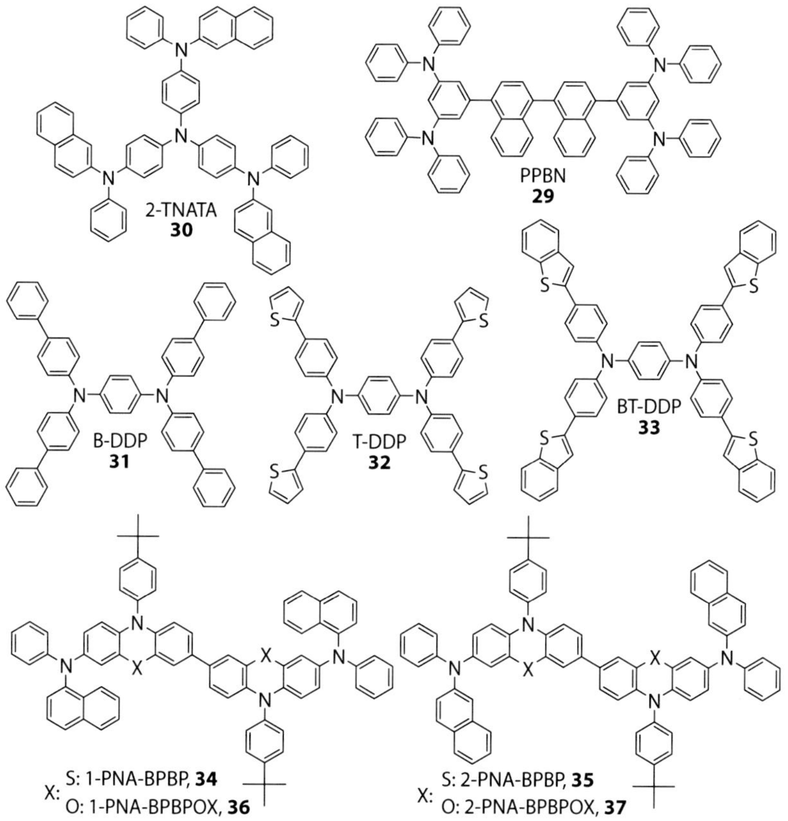

SCHEME 3.6
Chemical structures of nondoped HIMs.

Park et al. developed a new class of HIMs based on dimeric phenothiazine and phenoxazine [1-PNA-BPBP (**34**), 2-PNA-BPBP (**35**), 1-PNA-BPBPOX (**36**), and 2-PNA-BPBPOX (**37**)], which exhibit excellent thermal stability and T_gs as high as 160°C [79]. The thin film of 1-PNA-BPBPOX remains almost unchanged even after exposure to high temperatures for a 3-day period. Using these materials as an HIL in green OLEDs featuring the structure of ITO/HIL/NPD/Alq$_3$/LiF/Al, the highest power efficiency of 2.8 lm/W is achieved with 1-PNA-BPBPOX, >30% improved compared with the reference devices using 2-TNATA as the HIL.

3.3.2 Hole-Transport Materials

The HTL materials are very common in small-molecule-based OLED devices but are less common in polymer-based devices because conjugated polymers are usually good hole conductors themselves. They serve to provide a hole-conductive (via charge hopping) pathway for positive charge carriers to migrate from the anode into the EML. On the basis of this requirement, HTMs are usually easily oxidized and are fairly stable in the one-electron oxidized (radical–cation) form. This further translates into the materials having a fairly shallow HOMO energy level—preferably isoenergetic with the anode/HIL WF and somewhat lower in energy than the HOMO energy level of the EML. This latter property improves the chances of charge flow into the EML with minimal charge trapping. As the main function of the HTL is to conduct the positive charge-carrier holes, hole traps (higher-energy HOMO materials) should be avoided either in the bulk of the material (i.e., hole-trapping impurity levels of ~0.1% are typically required) or at interfaces. Another, perhaps less commonly appreciated, function of the HTL is that it should act as an EBL to prevent the flow of electrons from the EML and ultimately to the anode. For this purpose, a very shallow LUMO level is desirable. With these properties in mind, the commonly used HTL materials fall into several simple chemical classes:

1. Triarylamines
2. Triphenylmethanes
3. Phenylazomethines
4. Polyacene and carbazole derivatives
5. Spiro fluorene and TPA derivatives
6. Indole derivatives
7. Furan derivatives

3.3.2.1 Triarylamines

The most commonly used HTL materials are triarylamine compounds. These compounds were developed as HTMs for photoconductive applications such as xerography [80]. They naturally have been selected as HTMs for OLED applications largely because of their ready availability and their good electrochemical and thermal stabilities. The hole mobilities of these materials are also adequate for OLED applications. In addition, high purity, so as to ensure low hole-trap contamination, is believed necessary for long-lived OLED performance, and such materials may often be train sublimed to very high purity.

Two of the most widely used HTMs of the triarylamine family in OLEDs are N,N'-(3-methylphenyl)-1,1'-biphenyl-4,4'-diamine (TPD, **38**) (Scheme 3.7) and α-NPD (**24**). TPD and α-NPD have modestly high hole drift mobility, and were initially developed as charge-transport layers in xerography. However, TPD and α-NPD, which have low T_g of 65°C and 95°C, respectively, tend to crystallize or expand during device operation. For comparison, the T_g of Alq$_3$ is >170°C. It is commonly believed that a good HTM should have both a low-energy barrier from the anode and a relatively high T_g. These properties will improve the hole-injection efficiency, reduce the crystallization, and thus increase the lifetime of the device. Adachi et al. showed that the ionization potential of HTLs was found to be the dominant factor for obtaining high durability in organic EL devices [81]. The formation of the small energy barrier at the interface of an HTL and anode was required for high durability. However, their results showed that there are no straightforward relationships between melting point, T_g of the HTMs, and durability of the EL devices.

Fujikawa et al. studied a series of TPA oligomers from the dimer TPD up to the related pentamer and used them as HTMs [82]. Their results indicated that the thermal stability of the OLEDs was dramatically improved using an HTM TPTE (**39**), a tetramer of TPA. The resulting OLED devices show uniform light emission in continuous operation up to 140°C without breakdown [83].

Using the same biphenyl backbone as TPD and α-NPD, Thompson's group synthesized a series of triarylamines with high T_gs ranging from 85°C to 155°C as HTMs [84]. The OLED devices fabricated by using PPD (**40**) or ISB (**41**) showed comparable device performance

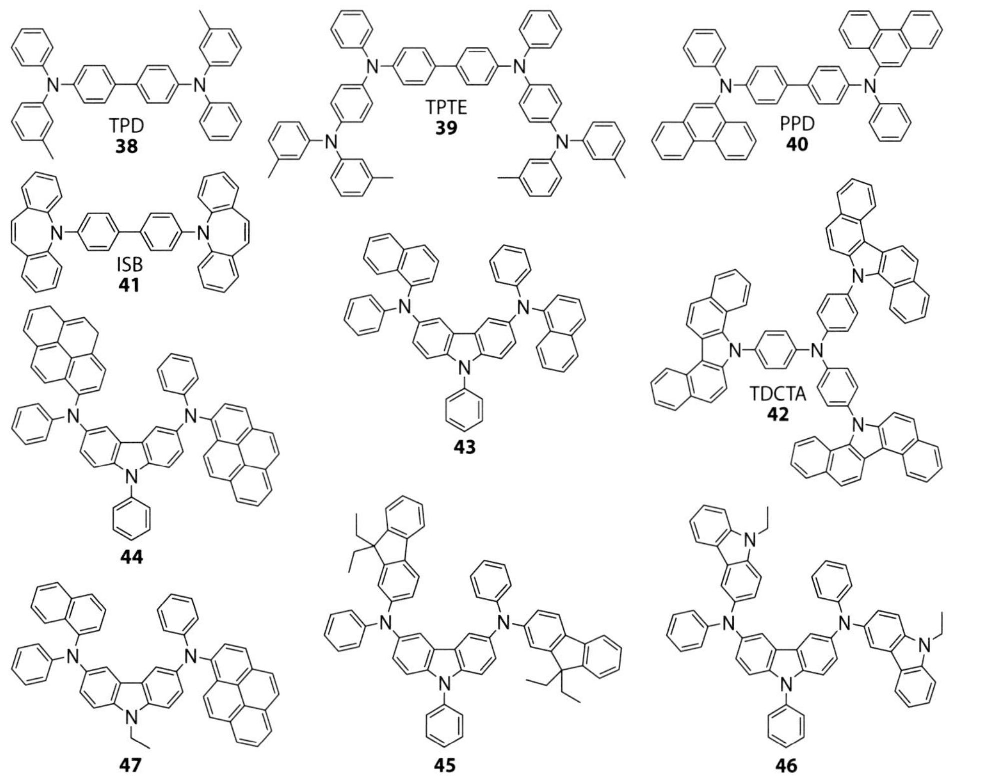

SCHEME 3.7
Chemical structures of triarylamine-based HTMs.

to the TPD or NPD devices, while maintaining the high T_g. The HTL can be composed of one compound or a mixture of several HTMs. Interestingly, in contrast to the reports of Giebeler and coworkers, who reported that the energetics between ITO and HTL interface was critical to the operating voltage and the QE of an OLED [85], Thompson's group found that there is no relationship between the HOMO energy and the device QE or turn-on voltage. Instead, they have shown that an asymmetrical substitution of the amine group hinders charge transport, thereby raising the turn-on and operating voltages.

Replacement of the biphenyl bridge with a fluorene unit can moderately change the T_g (15–20°C higher in fluorene containing species) and electronic absorption spectra (red shift of about 30 nm for fluorene species). The fluorene unit–based and biphenyl compounds–based triarylamine compounds have the same range of hole mobilities. Hreha et al. studied the details of these changes related to their optical and electronic properties, and related these to the device performance [86]. Their results showed that the fluorene series has similar OLED performance as the biphenyl species. Introducing electron-withdrawing fluorine atoms or electron-donating methoxy groups can tune the ionization potential, with little effect on the T_g. Later studies by Wong et al. showed that bulky 9,9-diarylfluorene-substituted triarylamine increased T_g up to 134°C [87].

Chen et al. synthesized a star-shaped triarylamine with a very high T_g as an HTM, 4,4′,4″-tri(N-dibenzo[a,g]carbazoyl)triphenylamine (TDCTA, **42**) [88]. The device fabricated using TDCTA as an HTL showed good hole-transporting properties, although the hole barrier is higher owing to the deeper HOMO energy level of TDCTA (–5.72 eV).

Thomas et al. reported a series of light-emitting peripheral diarylamine derivatives containing carbazole units, which possess dual functions, as both active emitting materials and HTMs (**43–47**) [89]. These luminescent materials are amorphous, with high T_g (120–194°C) and high thermal stability ($T_d > 450$°C), and with HOMO energy levels of 4.86–5.04 eV are also suitable for HTMs. OLED devices fabricated using these materials appear to be very stable. Using high T_g (>150°C) tri- and tetra-substituted carbazole derivatives as HTMs, Zhang et al. achieved high-performance OLEDs, comparable to the same device structure using NPD as the HTL [90].

Earlier work by Shirota et al. focused on the investigation of starburst triarylamine amorphous glasses with high T_g as HTMs. Scheme 3.8 shows the chemical structures of two representative starburst HTMs 4,4′,4″-tris(3-methylphenylphenylamino)-triphenylamine (*m*-MTDATA, **48**) ($T_g = 75$°C) and 4,4′,4″-tri(N-carbazolyl)triphenylamine (TCTA, **49**) ($T_g = 150$°C) [91,92]. OLEDs fabricated using such materials showed dramatically improved thermal stability and high efficiency. Tao's group also reported a series of star-like hole-transport and-emitting materials based on carbazole or diphenylthienylamine units [93,94].

A new branched carbazole derivative with phenyl ethylene moieties attached, 1,3,5-tris(2-(9-ethylcarbazyl-3)ethylene)benzene (TECEB, **50**), was prepared as an HTM for OLEDs [95]. TECEB has a HOMO energy level of −5.2 eV and hole-drift mobility of 10^{-4} cm^2/(V s), comparable to NPD. The device performance (maximum luminance of about 10,000 cd/m^2 and current efficiency of 3.27 cd/A) in a standard HTL/Alq$_3$ double-layer device is also comparable to NPD, but TECEB has a higher T_g (130°C) and its ease of synthesis is superior to NPD. Distyryl units linked to a TPD derivative, N,N′-bis(4-(2,2-diphenylethenyl)-phenyl)-N,N′-di(p-tolyl)-bendidine (DPS, **51**), reported by Yamashita and coworkers, showed good hole-transport properties and improved thermal stability compared with the parent TPD [96].

He et al. synthesized an HTM with two TPA groups at the 3,3′-positions of binaphthyl, namely TPA–BN–TPA (**52**) [97] (Scheme 3.9). The high T_g (130°C) and T_d (439°C) with a HOMO level of −5.26 eV make this material a promising substitute for NPD. When using TPA–BN–TPA as an HTL in ITO/TPA–BN–TPA/Alq$_3$/Mg:Ag device, a maximum current efficiency of 3.85 cd/A was achieved, which is better than the device with NPD as an HTL. 3,3′,5,5′-tetrakis(p-tolyldiamino)biphenyl (TTAB, **53**) was synthesized by Kim and

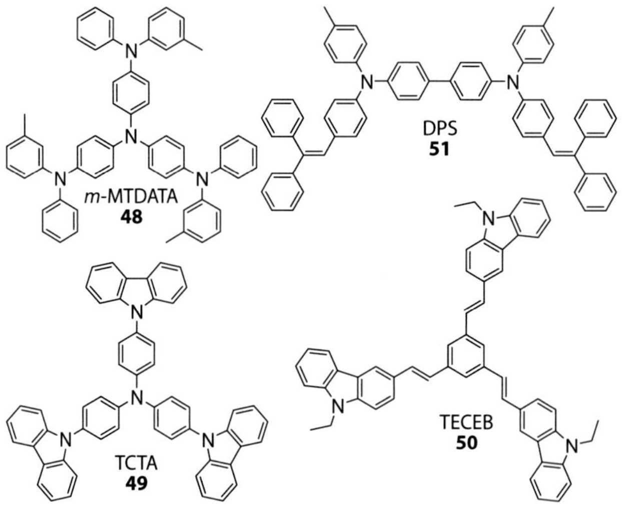

SCHEME 3.8
Chemical structures of star-shaped HTMs.

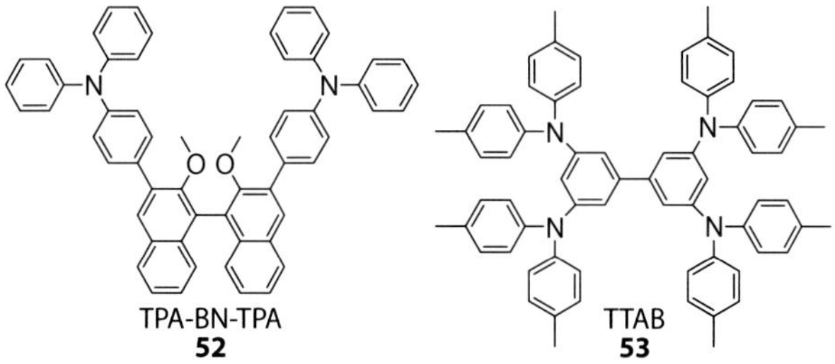

SCHEME 3.9
Chemical structures of HTMs.

Hwang [98]. Because of its rigid and near-spherical structure, this molecule has high thermal stability with T_g of 108°C and T_d of 414°C and good solid film-forming capability. Using TTAB as an HTL instead of NPD, the device with a structure of ITO/2-TNATA/TTAB/Alq$_3$:C545T/Alq$_3$/LiF/Al shows a remarkable improvement of the current density and luminance. Owing to the good hole-transporting ability and suitable HOMO energy level (–5.3 eV) of TTAB, a highly efficient OLED in a device configuration of ITO/TTAB/Alq$_3$:C545T/Alq$_3$/LiF/Al having TTAB as an HIM and HTM exhibits a current efficiency of 14.55 cd/A, superior to the standard device (11.66 cd/A) with 2-TNATA/NPD two layers.

As TPA and its derivatives with excellent hole-injection-/-transporting ability have been widely used in OLEDs and other optoelectronic devices, it is important to understand the relationship between molecular structure and properties of TPA derivatives for designing efficient HTMs and devices. Cias et al. performed theoretical calculation on TPA derivatives with various *p*-substitution on phenyl groups [99]. Density functional theory (DFT) computations reveal that electron-donating substituents such as amino and methoxy groups increase both HOMO and LUMO energy of the TPA-cored molecule, and make the molecule more flexible, while electron-withdrawing ones lead to a lowered energy level and rigid molecular framework. As the modification of HOMO and LUMO level results in favorable contact in the anode/organic layer interface and efficient electron blocking, it is demonstrated from this study that effective TPA-based HTM may be achieved by alternating the appropriate substitution. Moreover, tuning of the flexibility of molecular framework can offer controlled hole mobility, and subsequently the balanced charge transport in OLED devices. By introducing various fluorine groups onto TPA, a series of TPA fluorinated derivatives (TPAFs, **54**) (Scheme 3.10) were synthesized by Li and coworkers [100]. They tune both the energy level and hole-transport ability by changing the position and species of fluorine groups and devices of ITO/TPAFs/Alq$_3$/LiF/Al fabricated with the

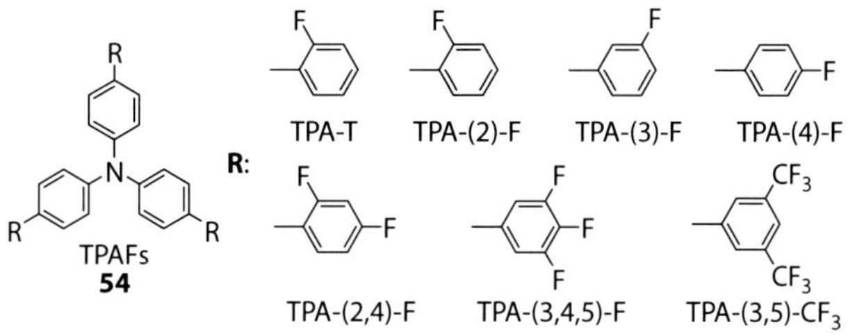

SCHEME 3.10
Chemical structures of fluorinated TPA HTMs.

o-fluorine molecule (TPA-(2)-F) as HTL show the highest current efficiency of 4.7 cd/A, nearly 20% higher than that of NPD-based devices.

Attracted by the good chemical and thermal stabilities and great potential applications for OLEDs of oligofluorenes, Li and Wong developed a series of oligofluorenes OFn-EG (**55**) (Scheme 3.11) with TPA and carbazole moieties. The HOMO energy level of these oligofluorenes with high T_gs ranging from 192°C to 254°C is about −5.2 eV, which can greatly reduce the hole-injection barrier from ITO to the active layer [101] to be used as hole-injection/ -transport materials.

Two HTMs based on the triphenylene core, namely NPAPT (**56**) and NPAPPT (**57**), were also developed by the same group [102]. Compared with an HTM of α-NPD, using NPAPT and NPAPPT as an HTL in ITO/2-TNATA/HTL/Alq₃/LiF/Al devices shows better power efficiencies as a result of their good hole-injection and -transport properties. Further, high T_gs and suitable HOMO levels of NPAPT (153°C, −5.36 eV) and NPAPPT (157°C, −5.39 eV) suggest that they can be potentially good candidates as HTMs after fine optimizing the device structure. Similarly, Hwang et al. developed a new group of triarylamine derivatives (**58–60**) based on the phenylnaphthyldiamine [103]. When serving as an HTL in the device of structure ITO/2-TNATA/HTL/IDE215:3% IDE118/Alq₃/LiF/Al, the devices have better current efficiency than the device based on the analogue **61** due to the efficient hole transport resulting from higher radical cation stability as well as the steric effect of the naphthalene moiety. Kwak et al. synthesized two HTMs containing phenanthrene (NphenD, **62**) or anthracene (NAD, **63**) functionalities with high T_gs (about 200°C) [104]. Although the efficiencies of ITO/*m*-MTDATA/HTL/Alq₃/LiF/Al devices with NPhenD or NAD as the HTL show no obvious improvement, the NPhenD- and NAD-based devices

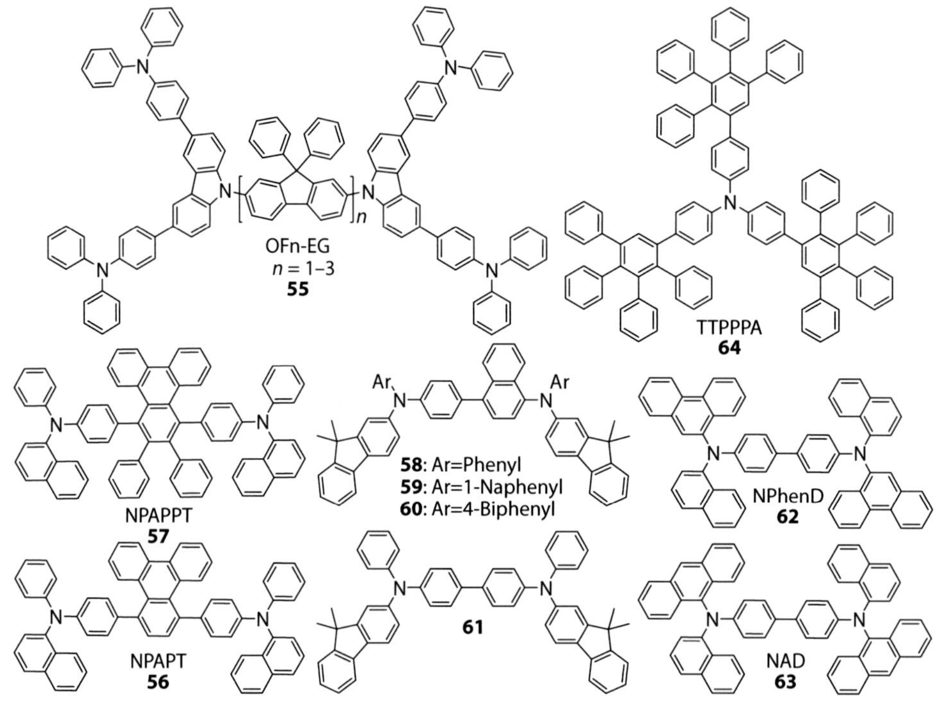

SCHEME 3.11
Chemical structures of HTMs.

exhibit improved stability at high temperature (up to 420 K) as well as better operational device lifetime in comparison with the device having α-NPD as HTL.

On the basis of the TPA core, a starburst material, 4,4′,4″-tris[(2,3,4,5-tetraphenyl)phenyl] phenylamine (TTPPPA, **64**), was synthesized and used as an HTM in OLEDs by Tong et al. [105]. Compared with the standard ITO/NPD/Alq₃/LiF/Al device (3.0 cd/A and 2.9 lm/W), the device of ITO/TTPPPA/Alq₃/LiF/Al has better efficiency of 5.3 cd/A and 4.3 lm/W. This significant enhancement in efficiency is attributed to the moderate hole mobility [5 × 10^{-5} cm²/(V s)] of TTPPPA resulting in better balanced hole and electron recombination in the emitting layer. Moreover, TTPPPA with high T_g (202°C) and good film-forming property could be an alternative material to NPD.

Although small molecule–based triarylamines are commonly used as HTMs, few of them can function both as emitting as well as hole-transporting materials with high T_gs. Tao et al. reported a bipolar star-shaped molecule, 4,4′,4″-tris(8-quinoline)-triphenylamine (TQTPA, **65**) and investigated it as an HTM (Scheme 3.12) [106]. The single-layer device of TQTPA gave sky-blue emission and a very low turn-on voltage of 2.8 V, indicating a balanced hole/electron transport within the EML. Typical bilayer devices with the configuration of ITO/TQTPA/Alq₃/LiF/MgAg yield a maximum current efficiency of 5.6 cd/A, nearly 50% better than those with an NPD HTL. Interestingly, the TQTPA-based devices gave better performance although their hole drift mobility is lower than that of NPD. It is believed that the slightly lower hole mobility leads to a further balanced charge recombination in the EML, and thus an improved efficiency. Two bimesityl-based triarylamines, TPAMeTy (**66**) and NPAMeTy (**67**), have been developed by Moorthy et al. [107]. Owing to the rigid noncoplanar bimesityls and triarylamine group, both of them have high T_gs (>120°C) and exhibit excellent blue emission and hole-transport ability. The

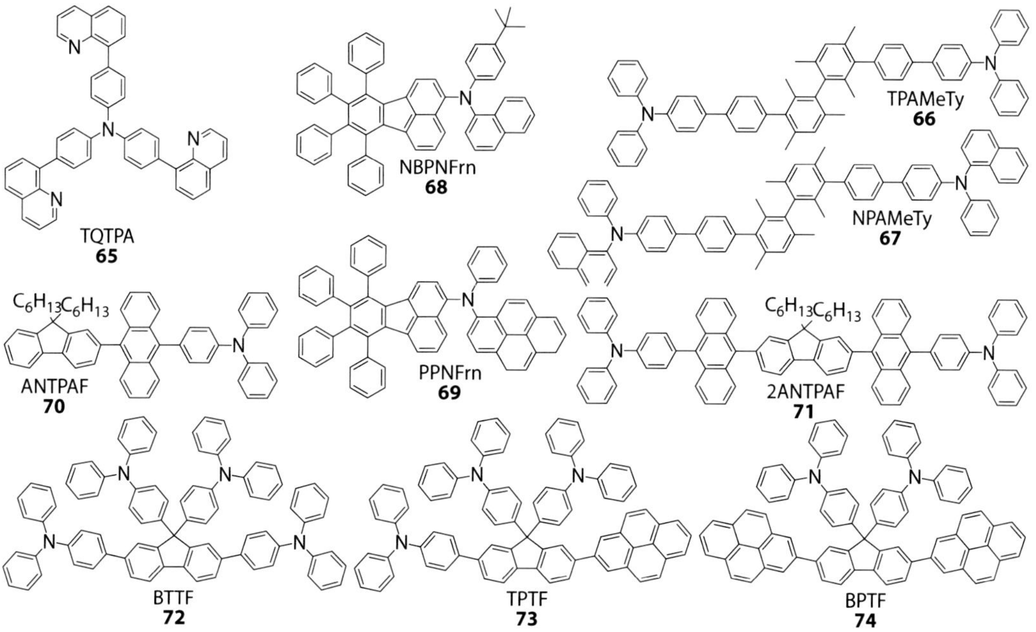

SCHEME 3.12
Chemical structures of dual-function HTMs.

NPAMeTy-based device of ITO/NPAMeTy/TPBI/LiF/Al with NPAMeTy serving as both hole-transporting and emissive material yields an efficiency of 1.94 cd/A with blue emission of a very high color purity (Commission Internationale de l'Eclairage [CIE] coordinates [0.15, 0.10]). Thomas's group synthesized NBPNFrn (**68**) and PPNFrn (**69**) containing fluoranthene and triarylamine segments [108]. Owing to the emitting character of fluoranthene moiety and hole-transporting ability of triarylamine functionality, these two materials can be used as emitting hole transporters in OLEDs. The devices with NBPNFrn and PPNFrn as hole-transporting and emitting layer exhibit good performances with bright greenish–yellow emission. Particularly, excellent current efficiency of 6.73 cd/A and EQE of 1.86% were achieved for the PPNFrn-based device by using TPBI as an ETL. Anthracene (**70,71**) [109] and fluorene (**72–74**) [110] derivatives with TPA moiety were also studied as emitters and hole transporters for OLEDs.

Besides a neat HTL, a general HTM can also be doped with another material to be an HTL. Liao et al. reported a highly efficient blue LED using a composite hole-transport material (c-HTM) having a CuPc-doped α-NPD [111]. This c-HTL reduces the hole current and thus balances the hole–electron charge recombination, although a slight increase of the driving voltage has been observed for such c-HTM-based device. Ha et al. demonstrated that dispersed TPD in a fluorine-containing polyimide matrix as an HTM has significantly improved device performance [112]. Flexible and fluorinated polyimide as a matrix exhibits the lowest turn-on voltage and a high EL efficiency [113].

Triarylamine-containing cross-linkable HTMs were also investigated as an HTL. Grubbs' group reported a series of cross-linkable triarylamine-containing poly(norbornenes) and investigated them as the HTMs in a bilayer OLED [114]. However, cross-linking was found to decrease the device performance due to the low T_g of the polymers and the poor film quality after UV irradiation.

Improved cross-linkable HTMs have been investigated. These include photo-cross-linkable and thermo-cross-linkable polymers. A new type of low-molecular-weight HTM based on TPAs bearing cross-linkable oxetane functionalities has been developed [115,116]. This material can be easily photo-patterned by exposure to a very short-wavelength UV light source to form a patterned and insoluble HTL. Similar photo-cross-linkable HTMs based on side-chain acrylate or styryl functional groups have been synthesized [117]. A series of spiro-linked and photo-cross-linkable HTMs have been patented by the Canon group [118].

Jen's group synthesized a series of high-T_g fluorinated polymers with a hole-transporting TPA group covalently attached as side chains [119]. This type of polymer can be processed into thin films by simple thermal cyclopolymerization without introducing any by-products [120]. The polymers are insoluble in most organic solvents and can be conveniently streamlined into a multilayer device fabrication process [121]. The HOMO energy level of the polymer is −5.32 eV. It was demonstrated that the EL performance of super yellow PPV and blue polyfluorene derivatives using this hole-transport polymer is comparable to or better than that using PEDOT:PSS as an HTL [122].

3.3.2.2 Triphenylmethanes

As with the triarylamines, the triphenylmethanes were first developed for xerographic and photoconductor applications [123]. The prototypical example of a material of this type is MPMP. This compound has one of the highest hole mobilities known for amorphous organic materials but is prone to crystallization and has a low T_g, making it less useful for long-lived OLED architectures. The HOMO and LUMO energy levels of MPMP are −5.53 and −1.88 eV, respectively, and the hole mobility of MPMP is in the range of 10^{-3} to

10^{-4} cm^2/(V s), as measured by time-of-flight (TOF) method [124]. This value is comparable to the best TPA compounds. Experimental results confirmed that there is little difference in the energetics and the mobility between the MPMP and NPD HTMs. However, when using MPMP as an HTM with a green iridium complex phosphorescent-emitter-based OLED, the device showed improved QE compared with the equivalent NPD-based device, especially when using neat films of the iridium complex as the emitter [125]. The authors explained that this difference in device performance is due to the high triplet energy level of MPMP (>3 eV) compared with that of NPD, which is 2.55 eV. The lower triplet energy level of NPD contributes to the energy-transfer-quenching processes from the excited state of the iridium emitter to the triplet state of the HTM [125].

3.3.2.3 Phenylazomethines

Yamamoto et al. have designed and synthesized a series of diphenylamine-substituted phenylazomethine dendrimers (DP-G$_n$, **75**–**78**) (Scheme 3.13) as HTMs [126–129]. These dendrimers showed a relatively high thermal stability, a multiredox system due to the terminal amine moiety, and a stepwise metal complexation with metal ions. The EL performance of the double-layer devices utilizing these dendrimers as HTMs (HOMO: −5.2 to −5.4 eV) and Alq$_3$ as the emitting and ETM increased with higher dendrimer generations. Using the metal ion–complexed (0.5 equiv. SnCl$_2$) DP-G$_2$ dendrimers, the luminance and EL efficiency of the devices were increased by more than double and >30%, respectively. These phenylazomethine dendrimers, capable of forming metal complexes, are also promising materials for highly efficient OLEDs [130,131].

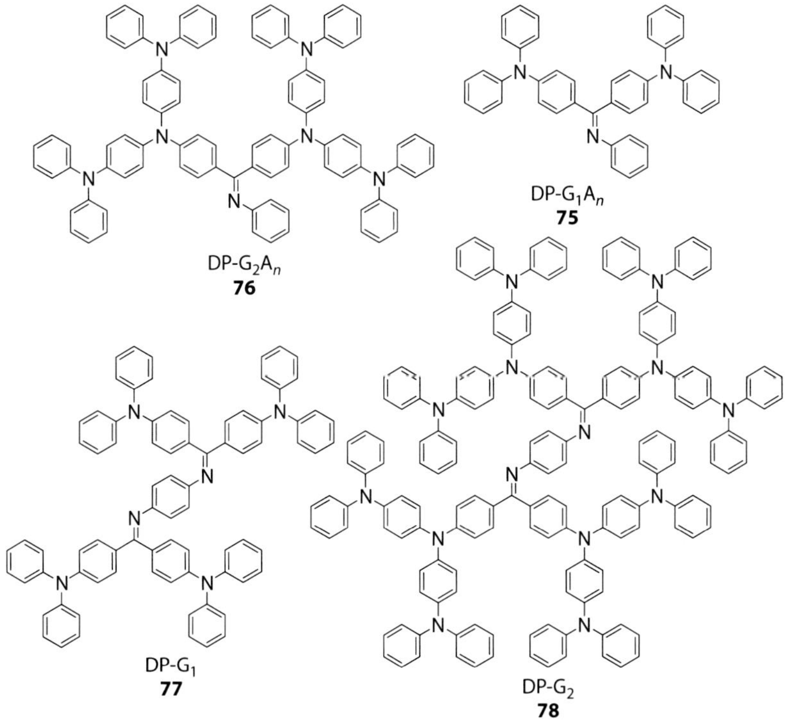

SCHEME 3.13
Diphenylamine-substituted phenylazomethine dendrimers (DP-G$_n$, n = 1, 2).

3.3.2.4 Polyacene and Carbazole Derivatives

To develop HTMs with moderate hole-transporting mobility for a balanced hole and electron injection into the EML, Xia et al. designed and synthesized a novel class of HTMs without triarylamine, namely TMOADS (**79**), TMOADN (**80**), TMOADP-1-N (**81**), and TMOADP-2-N (**82**), having tetra(methoxy)anthracene as a molecular platform (Scheme 3.14) [132]. These non-amine-based HTMs have similar HOMO energy levels (–5.25 eV) and high-lying LUMO energy levels (–2.11 to –2.28 eV) as well as good thermal stability ($T_d > 323°C$), which suggest that they are suitable for use as hole transporters with good electron-blocking abilities. In the standard devices of ITO/CF$_x$/HTMs/Alq$_3$/LiF/Al, the devices based on TMOADN, TMOADN, TMOADP-1-N, and TMOADP-2-N display comparable to or better performance than those of devices using NPD as HTL. These results indicate that these non-amine-based HTMs can be ideal substitutes for NPD.

Okumoto et al. demonstrated that three polyacene derivatives, rubrene (**83**), TBADN (**84**), and TPBA (**85**), can also be used as HTMs in OLEDs [133]. Owing to their suitable HOMO energy levels and good hole-injection properties, the performance of the devices with the structure of ITO/CF$_x$/HTL/rubrene + 1% DBP/Alq$_3$/LiF/Al using rubrene, TBADN, and TPBA as HTLs is greatly improved. The device with TBADN as HTL displays significantly higher current efficiency than that of the device using NPD as HTL. When rubrene is investigated as HTM in the above standard device, the luminance lifetime (time for 10% reduction of the luminance) is 280 h with driving voltage of 3.4 V, better than that of the NPD-based device (lifetime of 29 h, driving voltage of 3.9 V).

A class of binaphthyl (BN) derivatives, namely BN1 (**86**), BN2 (**87**), and BN3 (**88**) (Scheme 3.15), were studied as HTMs in OLEDs by Gao's group [134]. These molecules decompose at temperatures as high as 485–545°C and can form thermally stable amorphous

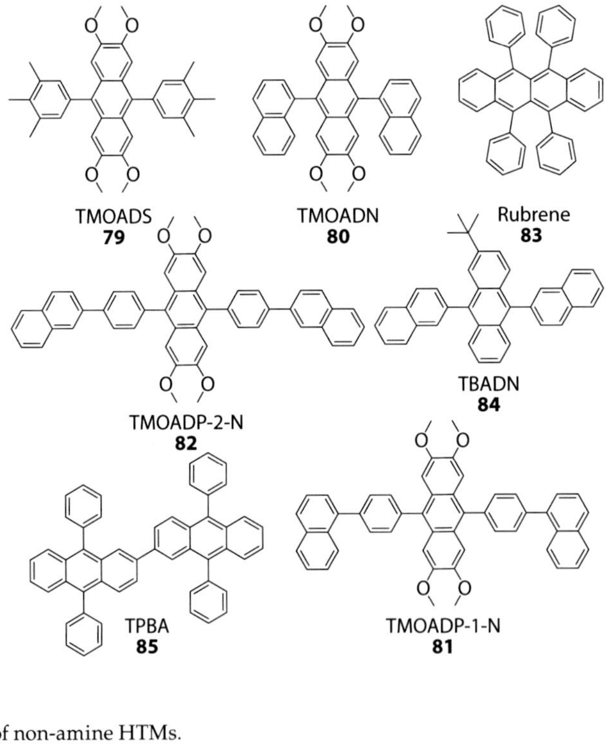

TMOADS **79** TMOADN **80** Rubrene **83**

TMOADP-2-N **82** TBADN **84**

TPBA **85** TMOADP-1-N **81**

SCHEME 3.14
Chemical structures of non-amine HTMs.

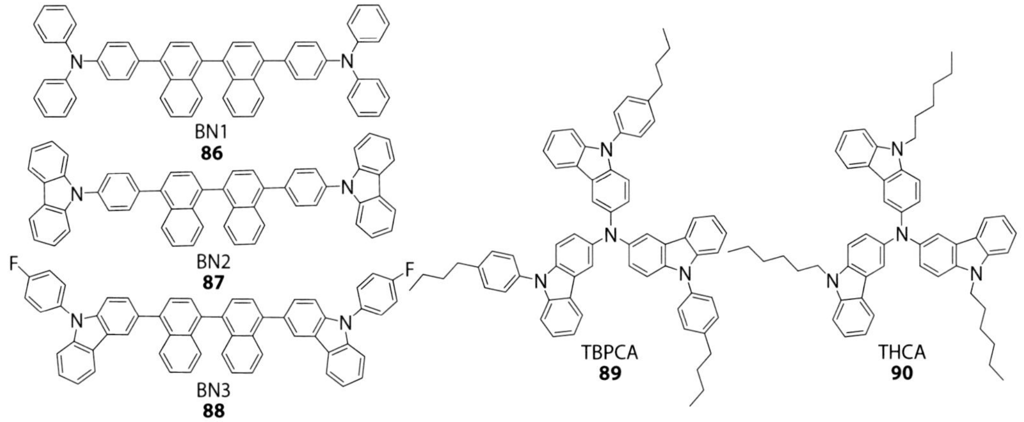

SCHEME 3.15
Chemical structures of carbazole-based HTMs.

films. At a constant current density of 25 mA/cm^2, a single-layer device of ITO/BN3/Al has a much longer lifetime (>80 h) than that of a single-layer device of ITO/NPB/Al (8 h). In addition, in the devices of ITO/BN1, BN2, and BN3/Alq$_3$/LiF/Al, the best performance is achieved in the BN1-based device with a current efficiency of 8.03 cd/A, 3 times higher than that of the device with NPD as the HTL (2.39 cd/A).

Two star-shaped carbazole-based molecules were synthesized by Michaleviciute et al. as HTL and blue-EML in bilayer white OLEDs [135]. Both compounds exhibit excellent thermal stability; TBPCA (**89**) showed an extremely high decomposition temperature of 458°C. The hole mobility of THCA (**90**) is measured by the TOF method to be as high as 10^{-3} cm^2/(V s) at an electric field of 3.6 × 10^5 V/cm.

3.3.2.5 Spiro Fluorene Derivatives

Although triarylamine derivatives with hole mobilities exceeding 10^{-4} cm^2/(V s) are commonly used as HTMs for OLEDs, the low T_gs and poor thermal stability of most molecules have great effect on the device efficiency and long-term stability. To overcome this problem, the development of materials having spiro-annulated structures [136,137], which can efficiently reduce crystallization tendencies and suppress intermolecular aggregations without significantly changing charge-transport properties, is one of the most promising approaches. Moreover, the T_gs of triarylamine or fluorene-based spiro-annulated materials will also be improved. Compared with using NPD and TPD HTMs, using **91** in ITO/HTM/Alq$_3$/LiF/Al devices showed very high luminescent efficiency [138] (Scheme 3.16).

Liao et al. reported a class of spirobifluorene-based triarylamine derivatives, namely 27DPSF (**92**), 22DPASF (**93**), 22DPSF (**94**), 22DTSF (**95**), and 22DBPSF (**96**) [139]. Owing to their three-dimensional structural feature and suitable HOMO levels (−5.51 to −5.26 eV), the devices of ITO/*m*-MTDATA/spiro fluorene derivatives/Alq$_3$/LiF/Al yield a maximum current efficiency of 5.6 cd/A and EQE of 1.8%. 27DPSF and 22DPASF exhibit the highest hole mobilities [up to 2 × 10^{-3} cm^2/(V s)], and the hole mobilities decrease from 22DPASF, 22DPSF, 22DTSF to 22DBPSF stage by stage as a result of the crucial influence of spatial hindrance.

Novel bifunctional materials, namely DFSTPA (**97**), TFSTPA (**98**), and TFSDTC (**99**), having spiro-annulated triarylamine structures were synthesized by Liu's group [140]. These materials have similar HOMO energy levels of −5.32 eV and high T_gs (197–253°C) because of

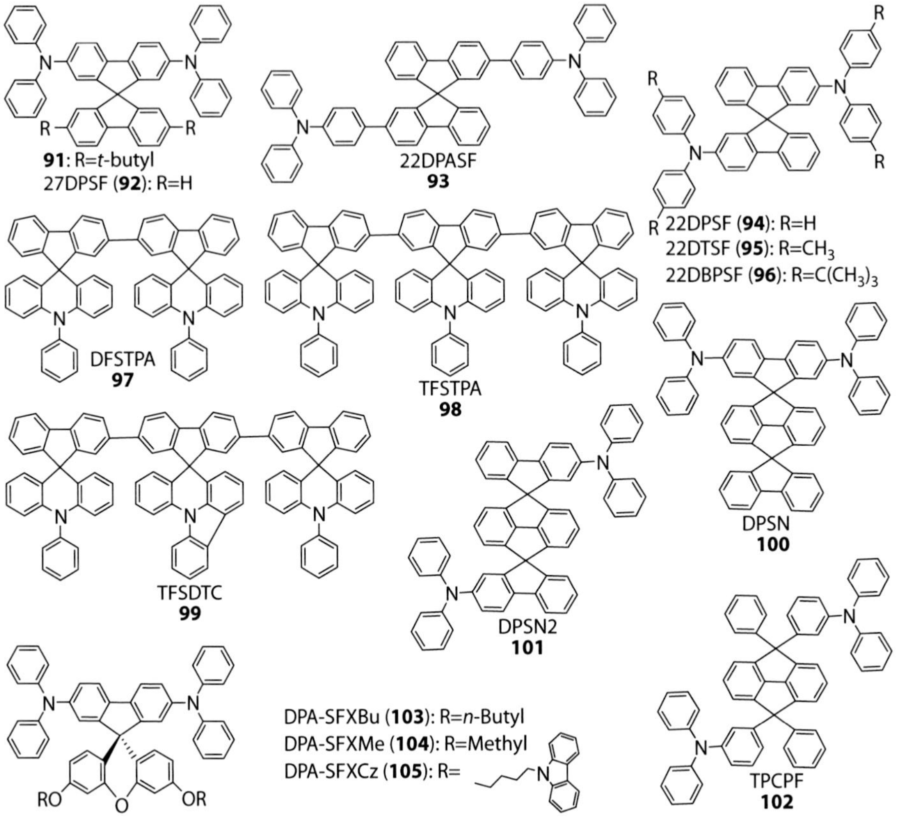

SCHEME 3.16
Chemical structures of spiro-fluorene–based HTMs.

the spiro-configuration structures, which is beneficial to improve the efficiency and lifetime of OLEDs. The devices using TFSTPA and TFSDTC as the EML as well as the hole-injection and -transport layer display deep-blue emission with maximum current efficiency of 1.91 cd/A and CIE coordinates of (0.16, 0.07). Owing to the relatively low hole mobility of DFSTP, the device with structure of ITO/DFSTPA/Alq$_3$/LiF/Al having DFSTPA as the HIL and HTL gives higher current efficiency (5.56 cd/A) than the device with NPD as the HTL (3.97 cd/A).

Cho et al. synthesized a double spirobifluorene derivative with high T_g (152°C), DSPN (**100**), having two aromatic amine units attached to the structure [141]. This HTM has a suitable HOMO energy level (−5.56 eV) for hole injection into host material and a relatively high triplet energy (2.44 eV) to effectively block the triplet excitons of Ir(ppy)$_3$ (2.40 eV). The maximum EQE of the device with structure of ITO/DNTPD/DSPN/BSFM:Ir(ppy)$_3$/ TSPO1/LiF/Al is 16.5%, which is much higher than that of the device using NPD as HTL (10.4%). DSPN2 (**101**), which has structural similarity to DSPN, was also developed and investigated as an HTM by this group [142]. Although T_g of DSPN2 (142°C) is lower than DSPN, the more suitable HOMO energy level (−5.76 eV) and higher triplet energy (2.53 eV) makes the device of ITO/DNTPD/DSPN2/BSFM:Ir(ppy)$_3$/TSPO1/LiF/Al exhibit a maximum EQE of 17.4%. By changing the structure of DSPN2, they further synthesized a high-triplet-energy HTM, namely TPCPF (**102**), based on cyclopenta[*def*]fluorene [143]. As TPCPF possesses a high triplet energy of 2.81 eV and low-lying LUMO of −2.10 eV, this compound can effectively suppress triplet exciton quenching and block electron leakage. Compared with the device of ITO/PEDOT:PSS/TCTA/CPBDC:FIrpic/TSPO1/LiF/Al with TCTA as HTL (~11.5%), the device using TPCPF as HTL shows better maximum EQE (~14%).

On the basis of modification to the spirobifluorene, Zou's group developed a series of HTMs with a spiro(fluorene-9,9′-xanthene) core (**103–105**) [144]. Because of the electron-rich xanthene ring with electron-donor alkyloxy pendant groups and diphenylamine moiety, these materials exhibit high HOMO energy levels close to the WF of ITO and good hole-transporting property.

3.3.2.6 Indole Derivatives

An interesting series of high-T_g HTMs based on indolo[3,2-*b*]carbazoles (**106**) (Scheme 3.17) has been discovered by the Xerox group [145]. These compounds not only showed the desired hole-transport properties and high T_g of 164°C but also display an unusual atropisomerism with two discrete *trans-* and *cis-*rotational isomers, which greatly improves their tendency to form stable amorphous glasses. Recently, new indolo[3,2-*b*]carbazole derivatives were reported by Simokaitiene et al. [146] with differently substituted phenyl groups at nitrogen atoms. These derivatives all have excellent thermal stability with decomposition temperatures exceeding 420°C. The best hole mobility of 2.8 × 10⁻³ cm²/(V s) is obtained at an electric field of 6.4 × 10⁵ V/cm in the amorphous layer of compound. It is interesting to notice that the position of the methoxy group on phenyl groups shows few influences on the hole-transporting characteristics of these series of compounds, which is contrary to previous studies on 2-carbazole diphenylamine derivatives.

A highly phenylated isoindole HTM, 1,3,4,5,6,7-hexaphenyl-2-[3′-(9-ethylcarbazolyl)]-isoindole (HPCzI, **107**), was designed and synthesized by Lee's group [147]. The hole

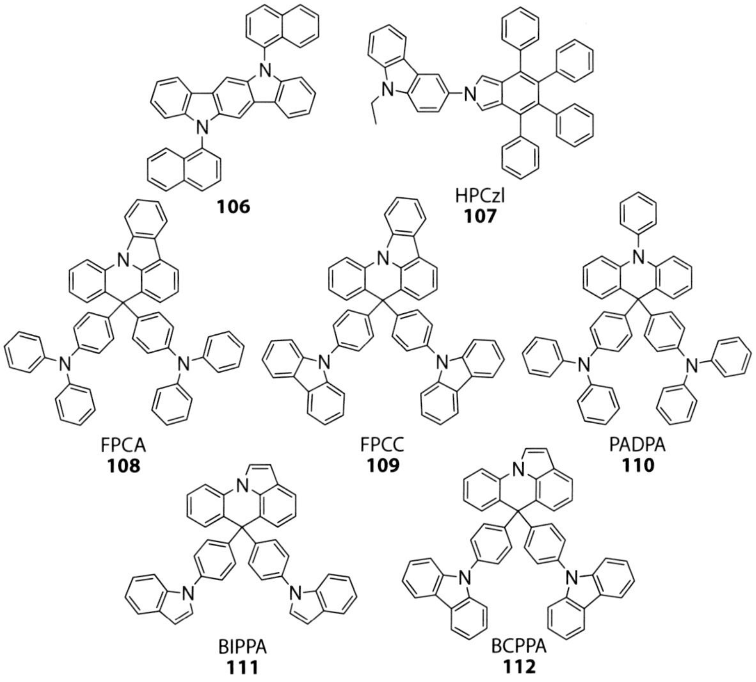

106

HPCzI
107

FPCA
108

FPCC
109

PADPA
110

BIPPA
111

BCPPA
112

SCHEME 3.17
Chemical structures of indole-based HTMs.

mobility of an HPCzI thin film was 4.3–6.0 × 10^{-5} cm²/(V s) (studied by using transient EL with an applied electric field in the range of 7.6 × 10^5 to 1.4 × 10^6 V/cm), which is of the same order of magnitude as that of NPD [7.8–9.9 × 10^{-5} cm²/(V s)]. The bilayer device ITO/HPCzI/Alq₃/MgAg showed comparable luminescent efficiency to the same device structure where NPD was used as the HTL. The benefit of HPCzI as an HTM is that the current efficiency is rather stable when the device is operated under higher current density. Two indolo acridine-based HTMs, namely FPCA (**108**) and FPCC (**109**), were reported by Park and Lee [148] for deep-blue PHOLED. Both compounds possess excellent thermal stability with T_g >140°C. Deep HOMO and high triplet energy level is simultaneously observed for FPCA and FPCC, indicating these materials can provide exciton blocking and hole-injection properties in a deep-blue OLED device. Best performance is achieved for the FPCA-based device in a configuration of ITO/DNTPD/NPD/(FPCC or FPCA)/mCPPO1:FCNIrpic/TSPO1/LiF/Al with a high EQE of 22% and maximum power efficiency of 25.2 lm/W. Relatively high hole mobility of 1.4 × 10^{-3} cm²/(V s) is also found in a hole-only device. The high device performances are attributed to improved recombination efficiency and effective triplet exciton blocking. Kim and Lee synthesized a high-triplet-energy HTM based on the acridine core [149]. Compound PADPA (**110**), with similar structure to FPCA, possesses high triplet energy of 2.89 eV and HOMO level of −5.96 eV. Owing to the efficient triplet exciton blocking and hole injection into the EML, PADPA- and FPCA-based devices of ITO/DNTPD/NPD/PADPA or FPCA/mCPPO1:FCNIrpic/TSPO1/LiF/Al show almost the same performance with maximum power efficiency of 21.1 lm/W, which is much better than that of the device of ITO/DNTPD/NPD/TAPC/mCPPO1:FCNIrpic/TSPO1/LiF/Al (14.4 lm/W).

Recently, Park et al. synthesized two novel HTMs with high T_gs (>120°C) containing a fused indole core structure, namely BIPPA (**111**) and BCPPA (**112**) [150]. The HOMO energy levels of BIPPA and BCPPA are −5.90 and −6.04 eV, respectively. Owing to the reduced energy barrier for hole injection from BIPPA or BCPPA into the host material (HOMO: −6.13 eV) and the high triplet energy (>2.90 eV), the devices of ITO/PEDOT:PSS/NPD/BIPPA or BCPPA/mCPPO1:FCNIrpic/TSPO1/LiF/Al exhibit a good performance, with a maximum EQE of 24.7% and a maximum current efficiency of 30.7 cd/A.

3.3.2.7 Furan Derivatives

Tsuji et al. developed a series of HTMs, namely TPBDF (**113**), MPBDF (**114**), DPABDF (**115**), and DMPABDF (**116**), based on a benzodifuran (BDF) core (Scheme 3.18) [151]. Except for the compound MPBDF, these materials with T_gs ranging from 90°C to 135°C display higher mobilities [5.6 × 10^{-4}–2.8 × 10^{-3} cm²/(V s)] than α-NPD [3.6 × 10^{-4} cm²/(V s)]. However, the devices of ITO/PEDOT:PSS/HTL/Alq₃/Liq/Al having TPBDF and MPBDF with no arylamino groups as HTLs exhibit significantly better power efficiency (2.6 lm/W both for TPBDF and MPBDF) and current efficiency (4.1 cd/A for TPBDF, 4.9 cd/A for MPBDF) than the device with α-NPD as HTL (2.0 lm/W, 3.6 cd/A). Owing to the synergetic effect of the BDF core and the arylamine substituents, the device with the structure ITO/PEDOT:PSS/DMPABDF/Alq₃/Liq/Al displays the highest power efficiency of 2.7 lm/W. It was not long before Tsuji et al. again synthesized an ambipolar BDF derivative, bis(carbazolyl)benzodifuran (CZBDF, **117**), and investigated it as host material in p–i–n homojunction OLEDs [152]. As the excellent ambipolar CZBDF with high T_g of 162°C possesses a wide band gap (3.3 eV) and well-balanced hole [3.7 × 10^{-3} cm²/(V s)] and electron [4.4 × 10^{-3} cm²/(V s)] drift mobility, the performance of the CZBDF-based homojunction devices using appropriate materials as emissive dopants are comparable to, or better than that of complicated heterojunction devices.

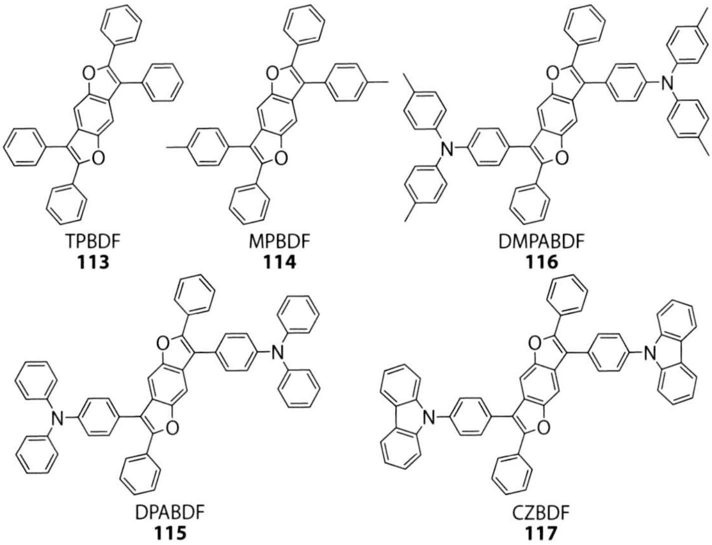

SCHEME 3.18
Chemical structures of benzodifuran-based HTMs.

3.4 Electron-Injection and -Transport Materials

3.4.1 Electron-Injection Materials

Owing to the relatively high mobility of holes compared with the mobility of electrons in organic materials, holes are often the major charge carriers in OLED devices. It is widely realized that improvement of electron injection is of critical importance for obtaining highly efficient low-voltage OLEDs. To better balance holes and electrons, one approach is to use low-WF metals, such as Ca (2.9 eV), Mg (3.7 eV), Li (2.9 eV), or Cs (2.1 eV), protected by a stable metal, such as Al or Ag, to reduce the electron-injection barrier and to improve the device efficiency. However, the problem with such an approach is that the long-term stability of the device is poor because the reactive metals with a low WF are always susceptible to atmospheric moisture and oxygen. Another approach is to lower the electron-injection barrier by introducing an EIM layer between the cathode material and the organic layer. The optimized thickness of the EIM layer is usually about 0.3–1.0 nm. The function of the EIM is to lower the WF of the cathode and/or to enhance the electron injection by reducing the high chemical reactivity of the low-WF metals, thereby increasing the chemical stability and improving the sticking coefficient of the interlayer materials. The typical EIMs are

1. LiF, CsF
2. Metal oxides: (ZnO, MnO, MnO_2, TiO_2, HfO_2, MgO, ZrO_2)
3. Cs_2CO_3, $(CH_3)_3CCOOCs$, $(CH_3COO)_2Zn$, Na_3PO_4, Li_3PO_4, NaSt, $LiCoO_2$
4. n-Doped EIMs
5. Water-/alcohol-soluble materials
6. Other EIMs

3.4.1.1 LiF, CsF

LiF is usually inserted between the cathode and the organic layer to enhance the electron injection because it is stable in air and the OLED performance is easily reproducible, and the interfacial chemical reaction is considered as the mechanism of the buffer-enhanced carrier injection in Al-cathode device [153]. It has been reported that metallic lithium is liberated from the interfacial chemical reaction of Al and LiF in the presence of Alq_3, which would, in turn, n-dope the underlying Alq_3 layer [153]. The chemical reaction such as $3LiF + Al + 3Alq_3 \rightarrow AlF_3 + 3Li^+Alq_3^-$ exists in the conventional $Al/LiF/Alq_3$ triple-layer system in previous work based on XPS and UPS (Figure 3.4) [14]. Sun and coworkers [154] found that the performance of Au-cathode OLEDs can be also efficiently improved by introducing an LiF buffer layer. The structures of their devices were ITO/NPD (40 nm)/Alq_3 (80 nm)/LiF (x nm)/Au (80 nm), and there is no chemical shift occurring in the $Au/LiF/Alq_3$ system, indicating a rather different mechanism of the LiF thin layer from that of Al-cathode device. The device with Au/LiF (3 nm) combined cathode has a prominent improvement in both current and luminance. The luminescence and the current efficiency of the device can be even comparable with conventional Al-cathode device. This nonchemical reaction effect could be attributed to tunneling and partially to interface dipole and the forbiddance of metal penetration into organic layer. Thus, LiF could be viewed as a kind of "interfacial reactive precursor" to low-WF alkali metal because its effectiveness is sensitive to the adjacent materials used [155].

The use of CsF and LiF as an EIL has the same effect. However, unlike LiF, CsF reacts directly with Al and releases Cs metal, whereas the dissociation of LiF in the presence of Al is thermodynamically disallowed and proceeds only in the presence of suitable reducible organic materials such as Alq_3. Thus, CsF is more generally applicable to many organic materials.

3.4.1.2 Metal Oxides: ZnO, MnO, MnO₂, TiO₂, HfO₂, MgO, ZrO₂

Metal oxides usually can be used as efficient EIMs because of their good charge-injection property, high transparency, and low resistance compared with organic materials. Titanium

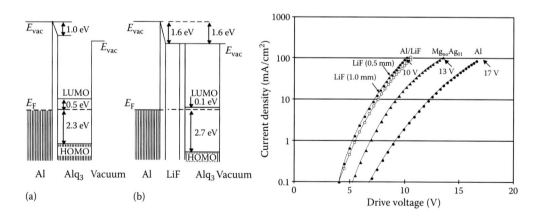

FIGURE 3.4
Energy diagrams of the (a) Alq_3–Al interface and (b) Alq_3–LiF–Al interface (left). (From Mori, T., *Appl. Phys. Lett.*, 73, 2763, 1998. With permission.) Current–voltage characteristics of three EL devices using an Al, $Mg_{0.9}Ag_{0.1}$, and Al/LiF electrode, respectively (right). (From Hung, L.S., Tang, C.W., and Mason, M.G., *Appl. Phys. Lett.*, 70, 152, 1997. With permission.)

oxide (TiO$_2$) [conduction band (CB): −3.8 eV, valence band (VB): −7.0 eV] was applied as an EIL in a class of hybrid organic–inorganic-based light-emitting diodes (HyLEDs) by Bolink et al. [156]. Lee et al. fabricated highly efficient inverted blue phosphorescent OLEDs based on FIrpic as an emitter and zinc oxide (ZnO) nanoparticles (NPs) (CB: −4.0 eV, VB: −7.5 eV) as an EIM, which showed peak efficiencies of 16.5 cd/A and 8.2%, about 3 times higher than those of the device without ZnO NPs [157]. Tokmoldin et al. reported using zirconium dioxide (ZrO$_2$) (CB: −3.0 eV, VB: −8.7 eV) as EIL in HyLEDs; the luminance of the poly(9,9′-dioctylfluorene-co-benzothiadiazole) (F8BT) devices is 5 times greater than that previously reported using TiO$_2$ or ZnO EIL [158]. Two wide-band-gap metal oxides, hafnium dioxide (HfO$_2$) and magnesium oxide (MgO), were also used as EIL in HyLEDs of ITO/EIL/F8BT/MoO$_3$/Au compared with ZnO. The current efficacy of 3.3 cd/A accompanied with high luminance exceeding 10,000 cd/m^2 have been obtained for the HyLEDs based on MgO [159].

The air-stable Mott insulator manganese mono-oxide (MnO) (CB: −2.6 eV, VB: −6.5 eV) was proved to be an efficient electron injecting and transporting material instead of the reactive alkali metals [160], which eliminated the problem of the oxidation of reactive dopants. In Alq$_3$-based OLEDs, a power efficiency of 1.1 lm/W was obtained by inserting a 3-nm-thick MnO as EIL, >0.8 lm/W for the reference device with 0.5-nm-thick LiF [161].

Recently, an n-type semiconducting manganese dioxide (MnO$_2$) was used as an EIL in the Alq$_3$-based OLEDs. The authors carried out *in situ* photoelectron spectroscopy experiments, which showed remarkable reduction of the electron-injection barrier without significant chemical reactions between Alq$_3$ and MnO$_2$. The reduction of the electron-injection barrier is due to the n-type doping effect, and the lack of strong interfacial reaction is advantageous with regard to more efficient electron injection than a conventional LiF EIL [162].

3.4.1.3 Cs$_2$CO$_3$, (CH$_3$)$_3$CCOOCs, (CH$_3$COO)$_2$Zn, Na$_3$PO$_4$, Li$_3$PO$_4$, NaSt, LiCoO$_2$

Cesium carbonate (Cs$_2$CO$_3$) has been recently reported to be more effective than LiF in terms of facilitating electron injection [163]. Cs$_2$CO$_3$ would decompose during vacuum vapor evaporation to liberate metallic cesium, which would be responsible for the injection improvement [164]. Therefore, Cs$_2$CO$_3$ can be viewed as a "thermal-decomposable precursor" to alkali metal. As cesium is liberated during the deposition process and no interfacial chemical reaction is required afterward, devices with a Cs$_2$CO$_3$ EIL have a wider choice of adjacent materials [155]. However, Chen and Wu found that Cs$_2$CO$_3$ molecules were deposited on the substrates without decomposition, regardless of the evaporation rates, based on the signature features of carbonate groups and ionization energies measured in UPS spectra and the binding energy shifts of core level electrons. The reaction between Cs$_2$CO$_3$ and Alq$_3$ forms complex of Cs$^+$ ions with Alq$_3^-$ radical anions, which increases the electron concentration in the organic films, with oxygen and carbon dioxide released from the surfaces. Besides, they also found that 0.5-Å-thick Cs$_2$CO$_3$ is sufficient to reduce the electron-injection barrier of Alq$_3$; the thickness needed to convert the ITO surface to low-WF cathode is about 10 Å [165].

Shangguan et al. evaluated the electron-injection efficacy of cesium pivalate [(CH$_3$)$_3$CCOOCs] in the Alq$_3$-based OLEDs, which showed better electron injection than LiF/Al cathode and the power efficiency was improved by about 19% at current density of 50 mA/cm^2, indicating that the nonaromatic alkali metal complex can also have good matching with the chemically stable compound and exhibit good electron-injection properties [166]. Furthermore, zinc acetate [(CH$_3$COO)$_2$Zn] was also reported [167].

The metal salt of lithium phosphate (Li$_3$PO$_4$) [168] or sodium triphosphate (Na$_3$PO$_4$) [169] was deposited between the organic semiconductor and an Al cathode; the bilayer cathode enabled a device performance of OLEDs competitive to the benchmark cathode LiF/Al and increased the device lifetime. In addition, Zhang et al. reported the device with Alq$_3$ doped with 10-(2-benzothiazolyl)-1,1,7,7-tetramethyl-2,3,6,7-tetrahydro-1H,5H,11H[1]benzopyrano[6,7,8-ij]quinolizin-11-one as the EML and lithium cobalt oxide (LiCoO$_2$) as the EIL showed promising efficiency (10.74 cd/A at 11V) and longer lifetime (2.8 times as much as LiF/Al control device). Lithium cobalt oxide was proved to be thermally decomposed in vacuum to form lithium oxide, which was responsible for the enhanced electron injection [170]. Siemund et al. first introduced the molecule sodium stearate (NaSt) as EIL in combination with the fluorescent polymer phenylene substituted poly(*para*-phenylenevinylene) (Ph-PPV) in OLEDs, the fabricated devices show current efficiencies up to 8.4 cd/A, indicating that the employed NaSt/Al bilayer cathode has adequate electron-injection capabilities in conjunction with Ph-PPV and, therefore, NaSt has the potential to become a nontoxic alternative to the widely-used alkali halide LiF [171].

3.4.1.4 n-Doped EIMs

Just as p-doping HIMs has been exploited, n-doping EIMs have also been explored. Kido et al. reported using an Li-doped Alq$_3$ layer as an EIM, which generates the radical anions of Alq$_3$ that, in turn, serve as intrinsic electron carriers and lead to improved device performance [172]. Lithium salts such as acetate or benzoate can also enhance the electron injection by a similar mechanism [173].

With a p-doped HIL and 4,7-diphenyl-1,10-phenanthroline (BPhen) doped via coevaporation of Cs metal as an EIM in a PHOLED device, He et al. have achieved a power efficiency of ca. 77 lm/W and an EQE of 19.3% at 100 cd/m^2 with an operating voltage of only 2.65 V. More important, the efficiency decays weakly with increasing brightness, and a power efficiency of 50 lm/W is obtained even at 4000 cd/m^2 [174]. Such a p–i–n device features efficient carrier injection from both contacts into the doped transport layers and low ohmic losses in these highly conductive layers and low operating voltages are obtained compared with conventional nondoped OLEDs.

Many other inorganic compounds, like MnO [161], lithium carbonate (Li$_2$CO$_3$) [175], cesium hydroxide (CsOH) [176], and LiF [177] were also reported as dopants in Alq$_3$ for an efficient electron-injection and electron-transport layer. The bilayer structure [178], a mixed layer of Alq$_3$ and 8-hydroxyquinolinato lithium (Liq) and a Liq layer, showed high efficiency of 11.6 cd/A compared with 9.8 cd/A for LiF EIL, and the lifetime of OLED was improved by 40% at 1000 cd/m^2 with the bilayer structure. Besides, BPhen doped with rubidium carbonate (Rb$_2$CO$_3$) [179] or cesium oxide (C$_{s2}$O) [180] or cesium azide (CsN$_3$) [181] were also investigated as n-doped EILs in OLEDs.

3.4.1.5 Water-/Alcohol-Soluble Materials

Water-/alcohol-soluble conjugated polymers and molecular materials have received great attention because they could be processed from water or other polar solvents, which offer good opportunities to avoid interfacial mixing upon fabrication of multilayer polymer optoelectronic devices by solution processing. This can dramatically improve charge injection from a high-WF metal cathode, resulting in great enhancement of the device performance [182]. It was proposed that their distinguished electron-injection properties were attributable to the dipole formation between the EIL and the metal cathode, while

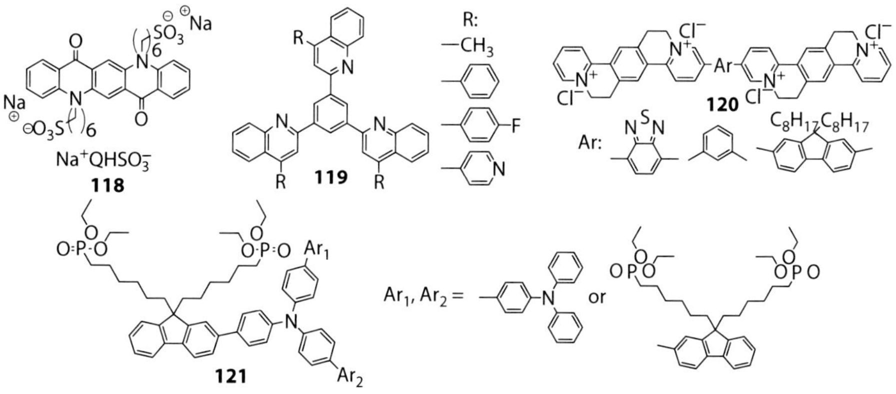

SCHEME 3.19
Chemical structures of water-/alcohol-soluble small-molecular EIMs.

the ionic charge redistribution by the moving of the counterions was also found to play an important role on the excellent electron-injection properties [182]. Thus far, most of the water-/alcohol-soluble materials are conjugated polyelectrolytes or their neutral precursors with polar pendant groups on their side chains, and the skeleton of the main chains are limited to polyfluorenes [183–188], polycarbazoles [189,190], and polythiophenes [191]. Jo and coworkers [192] developed three nonconjugated polyviologen derivatives with different alkyl chain length and counteranion as an interfacial layer. Pho and coworkers [193] reported a water-/alcohol-soluble small molecule based on the commercially available pigment quinacridone (**118**) (Scheme 3.19). In addition, dendritic oligoquinolines (**119**) [194], fully conjugated electron-deficient pyridinium salt–based molecules (**120**) [195], and neutral alcohol-soluble small-molecule materials composed of electron-rich TPA and fluorene featuring phosphonate side chains (**121**) [196] were also reported as EILs.

3.4.1.6 Other EIMs

Heavily doping the metal oxide as cathode enhances the electron-injection properties, such as with Cs-doped ZnO in HyLEDs [197] and Cs-doped TiO$_2$ in OLEDs [198]. Recently, Brine et al. introduced ionic liquids (IL, **122–124**) (Scheme 3.20) directly to the precursor solution for the ZnO, and a heavily doped ZnO:IL cathode was obtained to give improved electron-injection properties [199].

Duan et al. reported thermally decomposable lithium nitride (Li$_3$N) as the EIL for highly efficient low-voltage OLEDs. They used a quartz crystal microbalance method to investigate the decomposition process of Li$_3$N during vacuum thermal evaporation *in situ*. A thermodynamics study also proves that Li$_3$N decomposes in vacuum to form metallic lithium

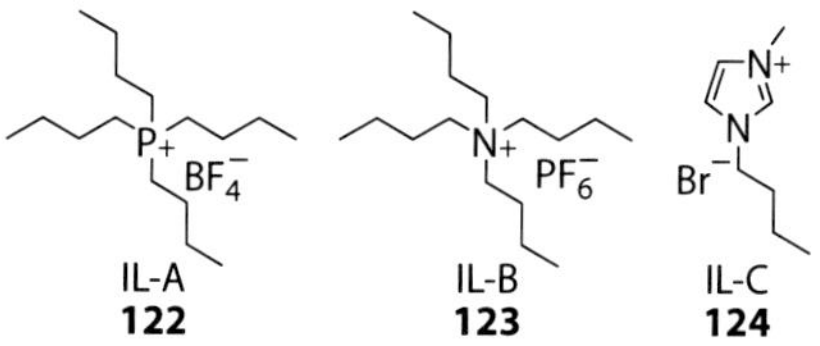

SCHEME 3.20
Chemical structures of ionic liquid EIMs.

and nitrogen. OLEDs with Li_3N as the EIL outperform those with conventional LiF in every respect. An optimized green OLED with an Li_3N EIL shows 35% higher efficiency and more than doubled lifetime compared with the control device with an LiF EIL [155].

The use of inorganic alkali metal salts as EIMs has been demonstrated to improve electron injection and to enhance the EL efficiency of OLEDs. The fact that using Li or Cs metal as the interface between aluminum and the organic layer (such as Alq_3) lowers the driving voltage indicates that modification of the Alq_3–Al interface with either elemental lithium or lithium compounds causes doping, which results in the creation of quinolate radical anions that then favor electron injection. This leads to the question of whether the properties of improved electron injection and good EL efficiency can be realized together in a simple lithium–quinolate complex. Lithium–quinolate complexes, 8-hydroxyquinolinolatolithium (Liq, **125**), 8-hydroxyquinolinolato Cs (Csq) [200], 2-methyl-8-hydroxyquinolinolatolithium (LiMeq, **126**), 4-phenanthridinolatolithium (Liph, **127**), and 2-(5-phenyl-1,3,4-oxadiazolyl) phenolatolithium (LiOXD, **128**) (Scheme 3.21), have been synthesized and investigated as EIMs used between the cathode Al and Alq_3 [201–205]. The results confirm that the function of these lithium–quinolate complexes is the same as that of LiF. A very thin layer (0.5–5.0 nm) of the complex deposited on the Alq_3 layer enhances the QE of the devices and reduces the driving voltage. In addition, the advantage of using Liq over LiF as an injection layer is that the efficiency is less sensitive to the Liq thickness.

Lithium phenolate complexes, lithium 2-(2-pyridyl)phenolate (LiPP, **129**), lithium 2-(2′,2″-bipyridine-6′-yl)phenolate (LiBPP, **130**), and lithium 2-(isoquinoline-1′-yl)phenolate (LiIQP, **131**), showed lower sublimation temperatures of 305–332°C compared with 717°C of LiF and exhibited high efficiencies compared with that of the device using LiF in OLEDs [206].

Pentacene films were also used as an EIL between the Al and Alq_3 layers in inverted OLEDs; the devices were turned on at 4.7 V and exhibited a current efficiency of 9.5 cd/A without any dopants or reactive metals [207]. Lee et al. reported a diimide-type bis-[1,10] phenanthrolin-5-yl-pyromellitic diimide (BPhen-PMDI, **132**) (Scheme 3.22) as EIL that vastly improved the device current density and the luminance [208].

LiF mixed with subphthalocyanine (SubPc, **133**) as EIL was demonstrated by Chen et al. The photoemission spectra indicated that the lithium released from LiF would remove chlorine atoms from SubPc molecules and created a large density of gap states at the interfaces [209].

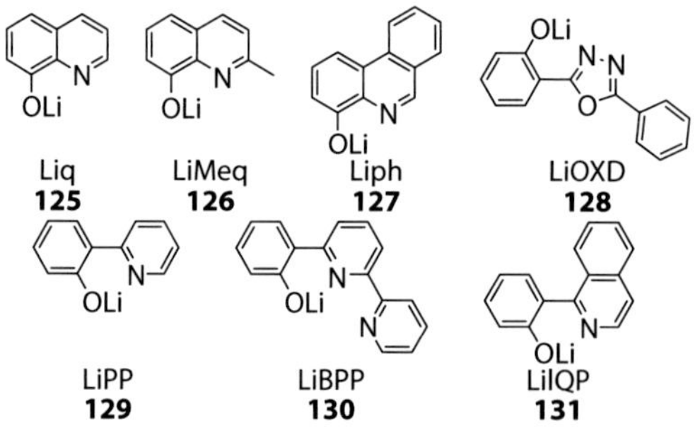

SCHEME 3.21
Chemical structures of lithium complex EIMs.

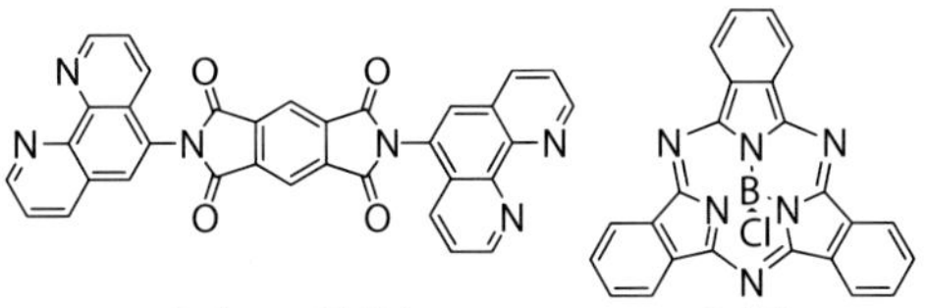

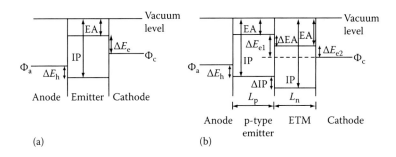

Bphene-PMDI
132

SubPc
133

SCHEME 3.22
Chemical structures of EIMs.

3.4.2 Electron-Transport Materials

Perhaps the most widely investigated layer in an OLED device is the ETL. This layer functions as a conducting material to help transport electrons from the cathode and into the organic layers of the device—ideally transporting the electrons via a hopping mechanism involving transitory production of anion radicals of the molecules involved. Therefore, the material needs to have a LUMO level close in energy to the WF of the cathode material used as to aid charge injection. It also needs to be composed of a material that is relatively stable in its one-electron reduced form. As with all organic layers, it should form good amorphous films and have a high T_g to favor stable operation over extended periods.

Since most of the high-efficiency organic emitters have p-type character and mainly a hole-transporting behavior, to achieve high-efficiency device performance an ETM is necessary to balance the charge injection and transport. In fact, it is documented that introducing an ETM into OLEDs results in orders-of-magnitude improvement in the device performance. The functions of the ETMs are to reduce the energy barrier between the cathode and the emitter, and to help the electrons be easily transported to the emitter. Two reviews of ETMs have been published within the past few years [210,211]. Figure 3.5 shows the energy-level diagrams of a single-layer OLED and a double-layer OLED after the introduction of the ETM layer. The electron-injection energy barrier (ΔE_e) is determined by the EA or the LUMO level and the WF of the cathode (Φ_c), while the hole-injection energy barrier (ΔE_h) is determined by the ionization potential or the HOMO level and the WF of the anode (Φ_a). Introducing an ETM lowers the energy barrier for electron injection ($\Delta E_{e2} < \Delta E_{e1}$). Meanwhile, most ETMs also serve as an HBL to efficiently confine the exciton formation in the EML and thus balance charge injection.

(a)

(b)

FIGURE 3.5
Energy-level diagrams of a (a) single-layer OLED and (b) two-layer OLED based on a p-type emitter and an ETM. (From Kulkarni, A.P., Tonzola, C.J., Babel, A., and Jenekhe, S.A., *Chem. Mater.*, 16, 4556, 2004. With permission.)

A good ETM should have the following properties:

1. High EA (<3.2 eV): This will match the WF of the cathode and reduce the energy barrier difference between the cathode and the emitter.
2. Reasonably good electron-transport mobility [$\mu_e > 10^{-5}$ cm^2/(V s)]: This aids transporting electrons to the emitter layer and efficiently confines the exciton in the EML.
3. High thermal stability ($T_g > 120°C$): The materials should not easily crystallize and should withstand joule heating during device operation.
4. Stable electrochemistry and electric-field stability (reversible one-electron reduction): Withstanding a high electric field is a key issue for the lifetime of the device.
5. Match the optical band gap of the emitters: The materials should avoid light absorption and scattering to maximize light output and increase the efficiency.
6. Phase compatibility and processability: The materials should be processable and compatible with neighboring materials to get pin hole–free and uniform films.

On the basis of these criteria, the chemical structures of useable ETMs include

1. Metal chelates
2. N=C (imine)-containing compounds
3. Cyano- and F-substituted compounds
4. Silole compounds
5. Boron compounds
6. Phosphine oxides
7. Doped ETMs
8. Other ETMs

Methods to determine or justify the utility of the electron-transport properties of ETMs are TOF electron mobility and electron-only diode device measurement, as well as the overall OLED performance.

3.4.2.1 Metal Chelate Electron-Transport Materials

The most commonly used material of this type is certainly Alq$_3$ (**134**) (Scheme 3.23), and almost all long-lived OLED devices include Alq$_3$ as the ETL. Interestingly, Alq$_3$ was the first emission and ETM explored by the Kodak group in their pioneering papers, and, thus far, it is still one of the best ETMs, EMs, and host materials. It is worthwhile mentioning here that, in addition to the use of metal chelates as ETMs, most of them can also be used as EMs. In other words, many are electron-transporting and emission materials.

Alq$_3$ has risen to a prominent position among OLED materials and remains the most widely studied metal chelate material. In the Alq$_3$ crystal structure, the distorted octahedral geometry of the 8-hydroxyquinoline ligands surrounding the Al^{3+} ion center makes it less prone to photoluminescence (PL) quenching in the solid state. It is thermally stable, has a T_g of 172°C [212], and can easily be thermally deposited to form pin hole–free amorphous thin films due to its intrinsic polymorphic phase behavior [213]. The electron mobility of Alq$_3$ is 1.4 × 10^{-6} cm^2/(V s), far higher than its hole mobility of 2.0 × 10^{-8} cm^2/(V s) as estimated by TOF

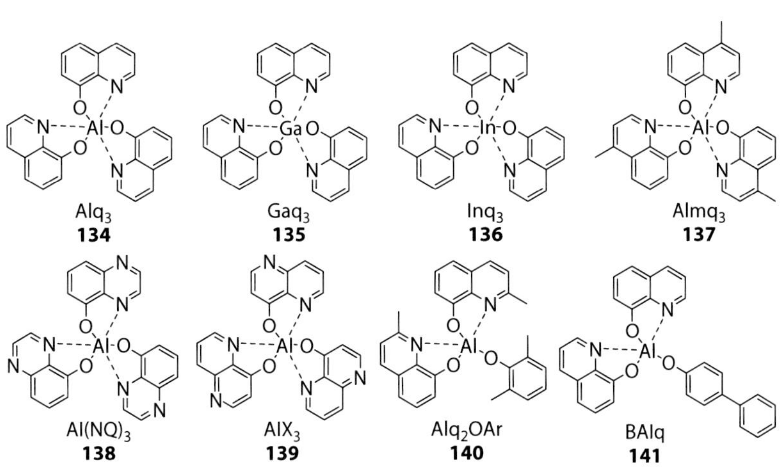

SCHEME 3.23
Chemical structures of aluminum, gallium, and indium chelate ETMs.

measurements [214]. The HOMO energy level is −5.95 eV, and its LUMO energy level is −3.00 eV [215]. These optical and electronic data indicate that Alq_3 is an electron acceptor, consistent with its use as an efficient ETM. For example, using Alq_3 as an ETM has enabled an >100 times improvement in EQE of MEH–PPV bilayer OLEDs compared with the MEH–PPV single-layer OLEDs [216].

In addition to Alq_3, the same group metal chelates Gaq_3 (**135**) and Inq_3 (**136**) have also been investigated as ETMs and EMs. Chen et al. studied the OLED performance using these metal chelates and found that as the size of the metal ion increased, the luminescent efficiency decreased and the electron mobility increased [217]. The observed performance has been confirmed by PL QE measurements, UPS, and UV–visible (UV–vis) spectral data. Alq_3 has a solid-state fluorescence QE of 25–32% with the thickness of the films from 0.01 to 1 μm [218], which is 4 times higher than the fluorescence efficiency of Gaq_3 and Inq_3 [219]. The emission spectrum shifts from 528 nm (Alq_3) to 548 nm (Gaq_3) and 556 nm (Inq_3) when the size of the central metal ion increases. The electron affinities of Alq_3, Gaq_3, and Inq_3 are −3.1, −2.9, and −3.4 eV, respectively.

By varying the substituents at the 5 position of Alq_3 chelates using electron-withdrawing or electron-donating groups attached to aryl moieties, a new class of electroluminescent compounds with tunable emission colors ranging from bluish–green to orange–red have been synthesized. Their OLED performances have been investigated; however, the luminescence and QE of the devices were lower than that of the parent Alq_3-based device [220,221]. The results indicate that the emission color shifts from blue to red when the substituents change from strong electron-withdrawing to strong electron-donating groups. At the same time, the QE decreases accordingly. The electron mobility of this class of materials has not been reported. Kido and Iizumi studied the OLED performance using tris(4-methyl-8-quinolinolate) Al(III) chelates ($Almq_3$, **137**) as EML, host material, and ETM, and found that the OLED EQE of $Almq_3$ is twice that of the Alq_3-based device [222].

Theoretical modeling work predicting the emission color of Alq_3 derivatives has shown that the emission properties of the ligand dominate the fluorescence of the complexes [223,224]. The electronic π–π* transitions in Alq_3 are localized on the quinolate ligands with the filled π orbitals (HOMOs) located on the phenoxide side of the quinolate ligand, and the unfilled π* orbitals (LUMOs) are on the pyridyl side. Substitution of an electron-withdrawing substituent at the C-5 or C-7 position of the phenoxide side of the quinolate ligand will cause a blue shift of the absorption spectrum relative to the parent unsubstituted Alq$_3$. An electron-withdrawing substituent at the C-4 or C-2 position of the pyridyl side will, conversely, cause a red shift. This trend will be reversed when using electron-donating substituents. This general rule helps the rational design of any desired emission color in the Alq$_3$ series.

Another example is where the Kodak group claim in a patent that replacing CH at the 5 position of the quinolate ligand with an N atom results in a compound, Al(NQ)$_3$ (**138**), with a hypsochromically shifted Alq$_3$ emission at 440 nm, a 90 nm blue shift from the parent Alq$_3$ emission [225]. A red-shifted emission compound, AlX$_3$ (**139**), is attributed to the reverse effect, when a nitrogen atom replaces the CH at the 4 position of the quinolate ligand.

Interestingly, methyl substitution at the C-2 position of the quinoline ligand hinders the formation of stable tris-chelates of aluminum. However, a phenolic ligand bearing bulky substituents, such as 2,5-dimethylphenol or 4-phenylphenol, used as an ancillary ligand can effectively shield the Al^{3+} from nucleophilic attack and has been found to improve stability [226]. Alq$_2$OAr (**140**) compounds show greenish–blue emission with a peak of ~490 nm. BAlq (**141**) has an emission wavelength peak of 476 nm. The blue shifts of C-2 methyl-substituted Al complexes are due to the steric hindrance enforced by the 2-methyl group with the bulky ligands, which reduces the conjugation overlap. OLEDs made using such materials show poor efficiency however. Alternately, these materials are suitable as HBLs owing to their large band gap coupled with their high EA.

Despite the ubiquity of aluminum hydroxyquinolinate chelates as ETMs, other metal chelates, such as group II metal ions of Zn^{2+} and Be^{2+} have also been used as the ETM in OLEDs (Scheme 3.24) [227–229].

Donze et al. studied a series of electron-withdrawing-group substituted bis-(8-hydroxy-quinoline)zinc(II) (Znq$_2$, **142**) derivatives as ETMs [230]. They found that these Znq$_2$ derivatives showed excellent electron-transport properties compared with Alq$_3$. This is due to the better π–π overlap of molecular orbitals of the Znq$_2$ derivatives and the extended electronic states in its tetrameric form with respect to Alq$_3$ [231]. Higher electron mobilities compared with Alq$_3$ may also contribute to better electron-transport properties of such Zn complexes, when used as ETMs in OLEDs.

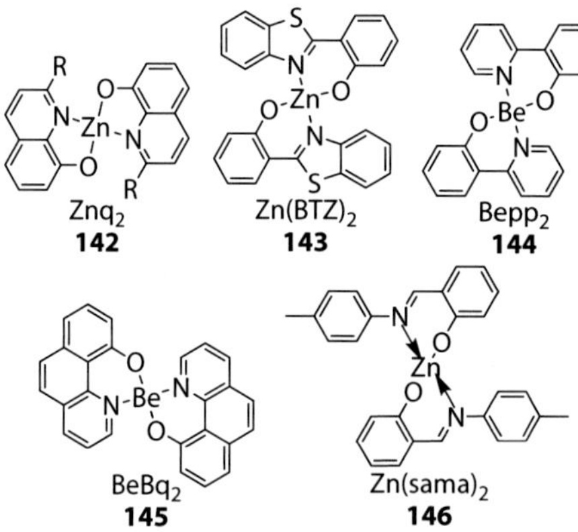

SCHEME 3.24
Chemical structures of zinc and beryllium chelate ETMs.

Bis(2-(2-hydroxphenyl)benzothiazolate)zinc(II), [Zn(BTZ)$_2$, **143**], is an excellent white emitter. The HOMO and LUMO energy levels of Zn(BTZ)$_2$ are −5.41 and −2.65 eV, respectively. Just as was found by Donze et al. for Znq$_2$ derivatives, Yu et al. [232] found that the electron transport of Zn(BTZ)$_2$ is better than that of Alq$_3$, although the electron-injection barrier is higher for Zn(BTZ)$_2$. This has been explained by the strong intermolecular interaction of Zn(BTZ)$_2$ molecules. This same group has examined the use of Zn(BTZ)$_2$ as an ETM in PLEDs, and the results are consistent with those with SMOLEDs [233].

Beryllium chelates bis(2-(2-hydroxyphenyl)-pyridine)beryllium (BePP$_2$, **144**) and bis(10-hydroxybenzo-quinolinato) beryllium (Bebq$_2$, **145**) have also been investigated as electron-transport emitters having blue or white emission. In some cases, when using Bebq$_2$ as an ETM, it shows higher electron-transport properties than Alq$_3$. The improved performance may be explained by the matched energy level of the ETM and the EM [234–236].

Chen and coworkers reported a series of zinc complexes with salicylidene–aniline and its derivatives as ligands, and found that the substitution of CH$_3$, OCH$_3$, CN, and N(CH$_3$)$_2$ on aniline ring of ligands can finely tune the properties of the corresponding zinc complexes. The devices with Zn(sama)$_2$ (**146**) as the ETL show a better performance than the Alq$_3$-based devices [237]. Tanaka et al. reported a series of oxadiazole metal chelate materials. However, these complexes suffer stability issues due to the intrinsic instability of the excited state of the molecules. Therefore, the lifetimes of OLEDs fabricated using these compounds are fairly short [238,239].

3.4.2.2 N=C (Imine)-Containing Electron-Transport Materials

A dendritic molecular species, 1,3,5-tris(*N*-phenylbenzimidizol-2-yl)benzene (TPBI, **147**) (Scheme 3.25), and its derivatives were patented by Kodak as blue emitters and ETMs [240]. The reason to add TPBI as a key ETM is because this material has received much attention recently, and there are hundreds of papers that use TPBI as the ETM or host material in the literature. It can be easily synthesized by heating a mixture of *N*-phenyl-1,2-phenylenediamine (3 equiv.) and 1,3,5-benzenetricarbonyl chloride (1 equiv.). The electron mobility of TPBI is of the order of 10^{-6}–10^{-5} cm^2/(V s), which is slightly higher than that of Alq$_3$, and its mobility is also electric field dependent as measured by transient EL spectroscopy [241,242]. Compared with Alq$_3$, TPBI has a large band gap and its LUMO level is −2.7 eV while its HOMO level is in the range of −6.2 to −6.7 eV, making it also function as a hole-blocking and possible host material [243,244]. Using TPBI as an ETL and HBL, high-efficiency (55 cd/A) devices based on a solution-processable phosphorescent green emitter have been demonstrated [245].

Jenekhe's group has recently explored a series of anthrazoline compounds as n-type semiconductors [246]. These polycyclic anthrazolines have a relatively high EA (2.9–3.1 eV) and high electron mobility owing to their rigid and planar structure. In addition, these compounds are thermally stable with T_gs >300°C. They can also form amorphous thin films by either spin coating or vapor deposition. MEH–PPV-based OLEDs, fabricated using ATZL (**148**) as the ETM, show up to a 50-fold improvement in brightness with EQE as high as 3.1%, when compared with single-layer PLEDs.

Tris(phenylquinoxaline) compounds containing two imine nitrogens on a phenyl ring possess high EA (2.6–2.8 eV), high mobility [10^{-4} cm^2/(V s) at 10^6 V/cm], and good thermal stability. Star-shaped TPQ (**149**) was investigated for application as an ETM in a PPV-based PLED, which resulted in a 5-fold enhancement in brightness compared with single-layer devices, with EQEs of 0.01% to 0.11% [247].

Oxadiazole-based ETMs are perhaps the most widely investigated organic ETMs. The initial work studying 5(4-biphenyl)-2-(4-*tert*-butyl-phenyl)-1,3,4-oxadiazole (PBD, **150**) as an

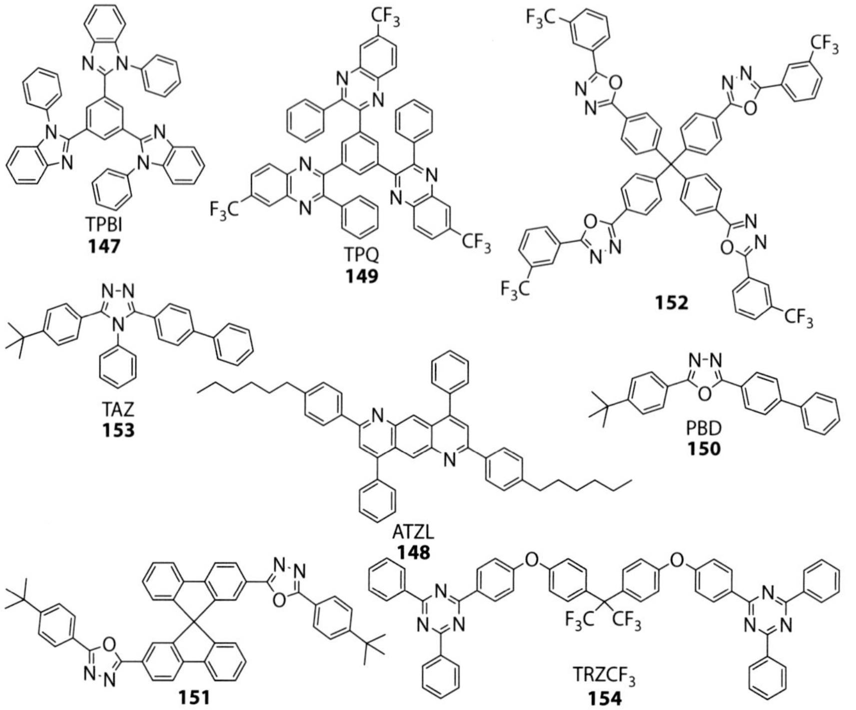

SCHEME 3.25
Chemical structures of ETMs.

ETM in OLEDs is a result of its high EA and its excellent thermal stability [248]. The EA value of PBD is 2.16 eV and ionization potential is 6.06 eV [249]. While the device efficiency is improved by introducing the PBD layer, the thin film of vacuum evaporated PBD, owing to its low T_g (60°C), tends to crystallize during device operation due to joule heating. Later work focused on the design of amorphous oxadiazole compounds for OLED applications. Spiro-linked oxadiazole compounds (e.g., **151**) and star-shaped tetraphenylmethane-based oxadiazole (e.g., **152**) were developed to reduce crystallinity and yet maintain solubility and stability. By introducing CF_3 groups, it can be tailored to have an even higher EA (2.26 eV) than PBD. However, no long-lived stable OLEDs based on this class of ETMs have yet been demonstrated.

1,3,5-Triazines are well-known compounds with high thermal stability and higher EA than 1,3,4-oxadiazoles and 1,2,4-triazoles (e.g., TAZ, **153**). Fink et al. studied a series of dimeric 1,3,5-triazine ethers (e.g., TRZCF$_3$, **154**) for application as ETMs for OLEDs [250]. However, despite their high EA, the efficiency of the OLEDs improved only modestly. One possible explanation is due to their rather poor electron mobilities.

Yokoyama et al. investigated two oxadiazole derivatives (Scheme 3.26). One of them has bulky *tert*-butyl terminals and forms an isotropic film, whereas the other has planar bipyridyl terminals and forms a highly anisotropic film. The very large optical anisotropy of the latter means that the planar molecules stack horizontally, leading to large overlaps of π orbitals and >30 times higher electron mobility. For vacuum-deposited organic amorphous films, molecular orientation is a concept of vital importance for understanding and improving the fundamental properties of the films [251].

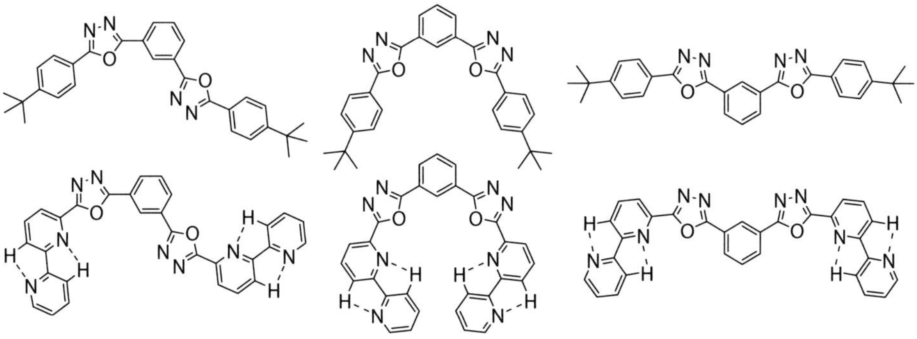

SCHEME 3.26
Intramolecular hydrogen bonding formation of ETMs containing oxadiazole and pyridine.

Recently, pyridine-containing ETMs have received much attention owing to their excellent electron-transport ability, low LUMO and HOMO energy levels, and high triplet energy levels. Su et al. designed and synthesized a series of four-pyridylbenzene-armed biphenyl derivatives (**155** and **156**) (Scheme 3.27) [252] as an electron-transport and exciton- and hole-blocking layer for the *fac*-tris(2-phenylpyridine)iridium (Ir(ppy)$_3$)–based green phosphorescent OLEDs, giving significantly improved efficiency in comparison to that with both the ETL of Alq$_3$ and the exciton- and hole-blocking layer of 2,9-dimethyl-4,7-diphenylphenanthroline (BCP). They also demonstrated two pyridine-containing triphenylbenzene derivatives of 1,3,5-tri(*m*-pyrid-3-yl-phenyl)benzene (Tm3PyPB, **157**) (HOMO: 6.68 eV, LUMO: 2.73 eV) and 1,3,5-tri(*p*-pyrid-3-yl-phenyl)benzene (TpPyPB, **158**) (HOMO: 6.65 eV, LUMO: 3.04 eV) with high electron mobility of 1.0×10^{-3} and 7.9×10^{-3} cm^2/(V s) and high triplet energy level of 2.78 and 2.57 eV, respectively. Highly efficient blue and green phosphorescent OLEDs were achieved by using Tm3PyPB and TpPyPB as an ETL,

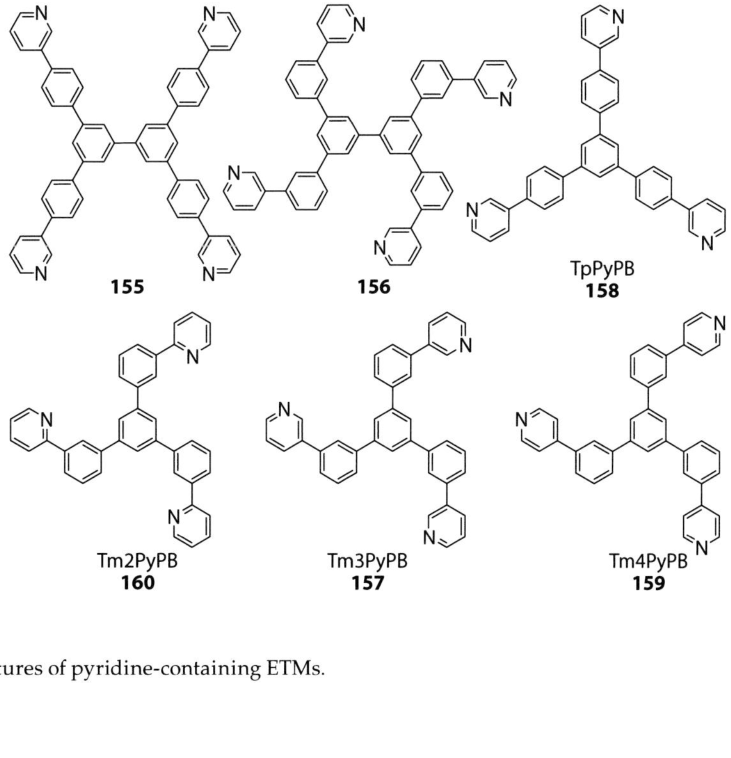

SCHEME 3.27
Chemical structures of pyridine-containing ETMs.

respectively, owing to improved electron injection and transport and good confinement of carriers and excitons within the emissive layer [253]. Besides, Su et al. further thoroughly studied the structure–property relationships of the pyridine-containing triphenylbenzene ETMs based on the different nitrogen atom positions of the pyridine ring for highly efficient blue phosphorescent OLEDs. It was found that their HOMO and LUMO energy levels and confinement of FIrpic triplet excitons are strongly dependent on the nitrogen atom orientation, decreasing as the nitrogen atom of the pyridine ring moves from position 2 to 3 and 4. This series of ETMs exhibit a high electron mobility of 10^{-4}–10^{-3} cm²/(V s) and a high triplet energy level of 2.75 eV. Reduced driving voltages are obtained when the nitrogen atom position changes because of improved electron injection as a result of the reduced LUMO energy level, but a better carrier balance is achieved for the Tm3PyPB (**157**)-based device. An EQE >93% of maximum EQE was achieved for the 1,3,5-tri(*m*-pyrid-4-yl-phenyl)benzene (Tm4PyPB, **159**)-based device at an illumination-relevant luminance of 1000 cd/m², indicating reduced efficiency roll-off due to better confinement of FIrpic triplet excitons by Tm4PyPB in contrast to 1,3,5-tri(*m*-pyrid-2-yl-phenyl)benzene (Tm2PyPB, **160**) and Tm3PyPB [254].

Togashi and coworkers [255] reported a series of pyridine-containing triphenylene-based ETMs (Bpy-TP1–Bpy-TP4, **161–164**) (Scheme 3.28) with a coplanar molecular structure. These ETMs exhibit optical anisotropy, and thus they adopt an orientation that is almost parallel to the substrate. The OLED containing Bpy-TP2 (**162**) showed a significantly lower driving voltage than the OLED with Alq₃, and the lifetime of the blue OLED containing Bpy-TP2 is equivalent to that of the OLED with TPBI. The increase in the driving voltage of the device containing Bpy-TP2 is significantly suppressed compared with that of the device with TPBI. Jeon et al. [256] synthesized a pyridine-substituted spirofluorene derivative (SPBP, **165**) as an ETM, giving >20% enhanced EQE and power efficiency for blue OLEDs compared with Alq₃ due to efficient electron-transport and hole-blocking

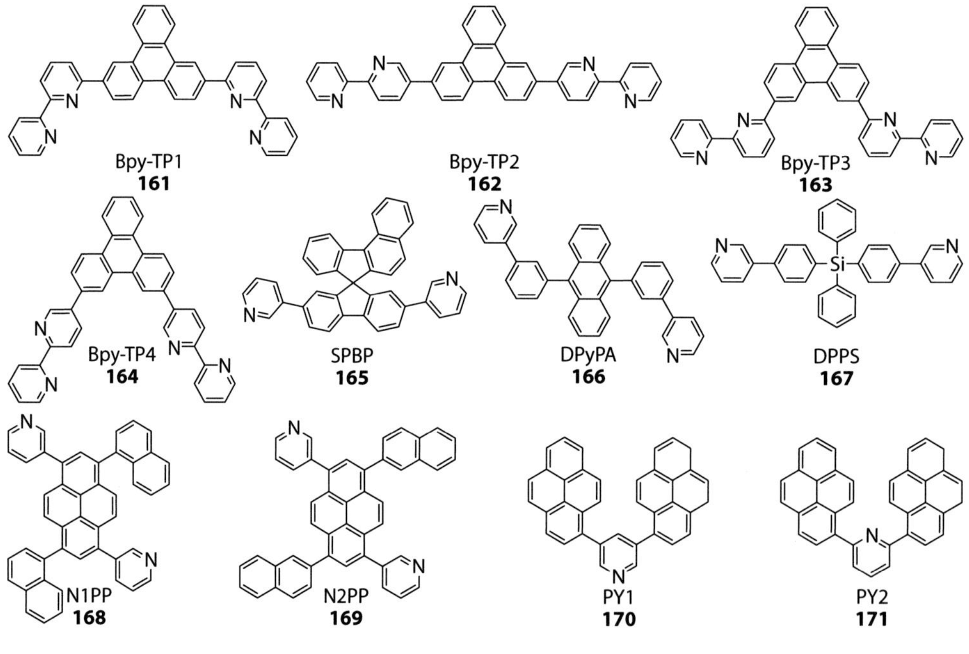

SCHEME 3.28
Chemical structures of pyridine-containing ETMs.

properties. A maximum EQE of 7.72% and a maximum power efficiency of 9.17 lm/W were achieved. Sun and coworkers [257] developed a pyridine-containing anthracene derivative (DPyPA, **166**) that exhibits ambipolar transport properties, with both electron and hole mobilities of around 10^{-3} cm²/(V s), about 2 orders of magnitude higher than that of Alq₃. The nitrogen atom in the pyridine ring of DPyPA coordinates to lithium cations, which leads to efficient electron injection when LiF/Al is used as the cathode. The power efficiencies of the devices based on DPyPA are greater by 80–140% relative to those of the Alq₃-based devices. Xiao et al. [258] reported nearly 100% internal QE in an organic blue-light electrophosphorescent device by using DPPS (**167**) as an ETL with a wide energy gap.

Oh et al. [259] reported two kinds of pyrene-based ETM, N1PP (**168**) and N2PP (**169**). The EQEs of the devices with these ETMs increase by >50% at 1 mA/cm² compared with those of the devices with Alq₃ as an ETM. Electron mobilities in N1PP and N2PP films are 3 times higher than that in Alq₃. Pu et al. [260] developed two dipyrenylpyridines, PY1 (**170**) and PY2 (**171**), as ETMs for OLEDs; both of the compounds show much higher electron mobilities than that of Alq₃, and have a similar ionization potential and EA. Probably due to the steric hindrance to the central pyridine group, the barrier height of electron injection from cathode to PY1 in OLED is much smaller than that to PY2, indicating that the chemical affinity of ETM with cathode is more important than their own EA to improve the electron injection.

Li et al. [261] reported two series of n-type triphenylpyridine derivatives (**172–179**) (Scheme 3.29) with good thermal properties, and their applications to efficient deep-blue OLEDs as electron-transport and hole-blocking materials. The devices show higher efficiency (2.54 cd/A) and better color purity (0.15, 0.10) compared with those of similarly structured blue OLEDs using state-of-the-art ETMs.

Pyrimidine-based molecules also have excellent electron-transport capability. Kido's group first developed 2-phenylpyrimidine skeleton-based ETMs [262] of B3PyPPM (**180**) and B4PyPPM (**181**) (Scheme 3.30) with HOMO levels of 7.15 eV and LUMO levels of 3.41 eV (B3PyPPM) and 3.44 eV (B4PyPPM), respectively. Their electron mobilities measured by TOF technology were in the order of 10^{-7} cm²/(V s) for B3PyPPM and 10^{-6} cm²/(V s) for B4PyPPM, respectively. These values were almost the same as that of conventional ETM, Alq₃ and 3-(4-biphenylyl)-4-phenyl-5-(4-*tert*-butylphenyl)-1,2,4-triazole (*t*-BuTAZ).

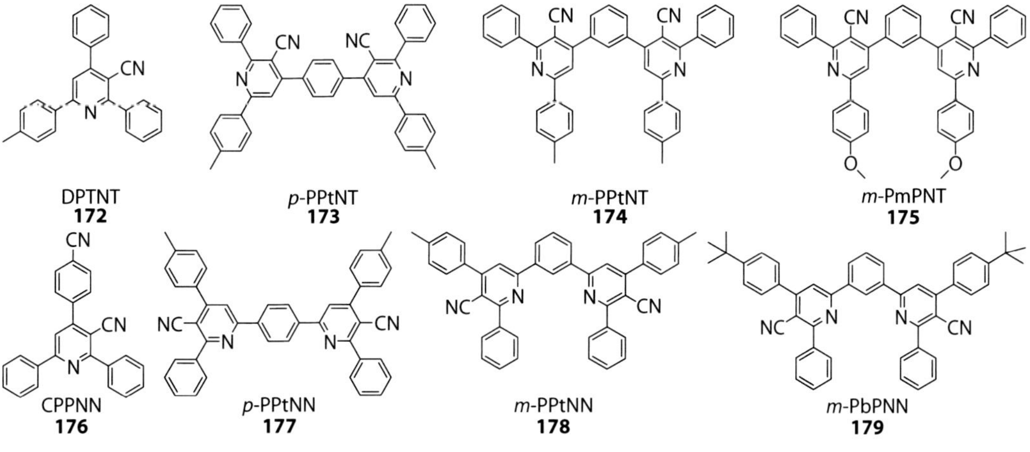

SCHEME 3.29
Chemical structures of pyridine-containing ETMs.

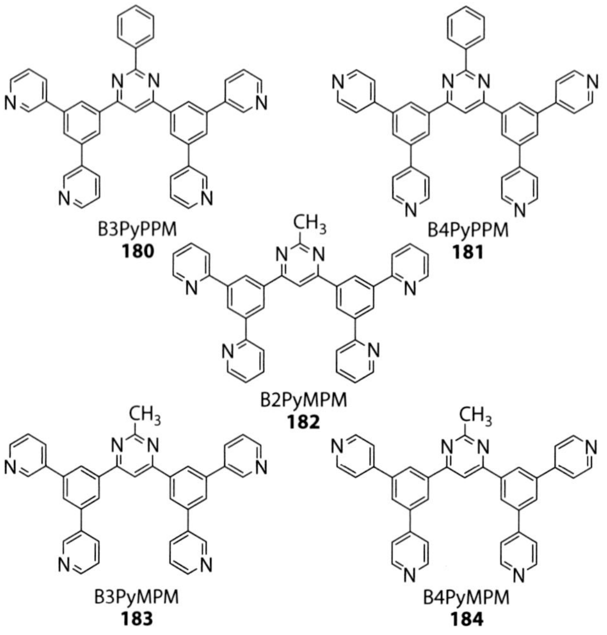

SCHEME 3.30
Chemical structures of pyrimidine- and pyridine-containing ETMs.

They further designed and synthesized a series of ETMs based on the 2-methylpyrimidine skeleton, namely B2PyMPM (**182**), B3PyMPM (**183**), and B4PyMPM (**184**), to investigate the influence of substituted pyridine rings on the physical properties and electron mobilities. The ionization potential is observed to increase in the order B2PyMPM (6.62 eV) < B3PyMPM (6.97 eV) < B4PyMPM (7.30 eV). The electron mobility of B4PyMPM was measured to be 10 times higher than that of B3PyMPM and 100 times higher than that of B2PyMPM. From both thin layer chromatography and differential scanning calorimetry analyses, the degree of the H-bonding interactions is considered to increase in the order B2PyMPM << B3PyMPM < B4PyMPM. Using Bässler's disorder formalism, the degree of energetic disorder is estimated to decrease in the order B2PyMPM (91 meV) > B3PyMPM (88 meV) > B4PyMPM (76 meV), and the positional disorder is 2.7 for B2PyMPM and <1.5 for B3PyMPM and B4PyMPM. These results clearly indicate that the position of substituted pyridine rings is critically important to adjust the fundamental physical properties and the electron mobilities [263].

In addition, Liu and coworkers synthesized a series of hybrid heterocycle-containing ETMs (**185–190**) (Scheme 3.31) with different pyridine substitution positions in a single molecule by regioselective sequential palladium-catalyzed Suzuki cross-coupling reactions. The developed ETMs have lower-lying HOMO (6.42–6.48 eV) and LUMO energy levels (2.42–3.21 eV). Extremely low turn-on voltage for EL, which is 0.2–0.3 V lower than the minimum value of $h\nu$/e, was experimentally achieved by utilizing the developed ETMs as an electron-transport and hole-/exciton-blocking layer for the classical Ir(ppy)$_3$-based green phosphorescent OLEDs [264].

As is well known, 1,3,5-triazine has a higher EA than pyrimidine and pyridine. Recently, Su et al. developed a series of 1,3,5-triazine-core-containing ETMs (**191–193**) (Scheme 3.32) with low-lying HOMO (6.48–6.73 eV) and LUMO (2.90–3.13 eV) energy levels [265].

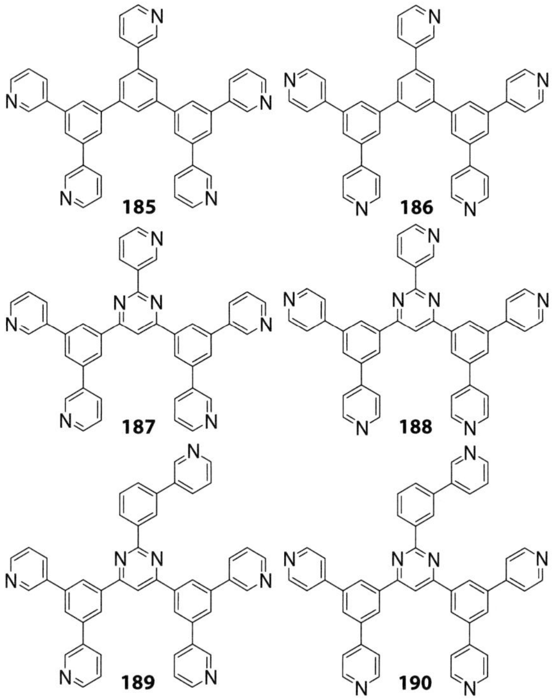

SCHEME 3.31
Chemical structures of hybrid heterocyle-containing ETMs.

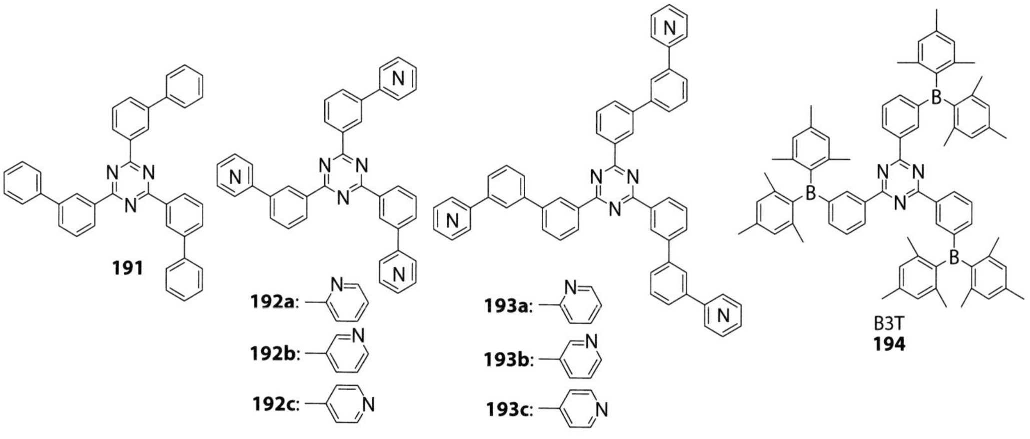

SCHEME 3.32
Chemical structures of 1,3,5-triazine-containing ETMs.

Extremely low threshold voltage for EL, which is 0.1–0.2 V lower than the minimum value of *hv*/e, was experimentally achieved without using any n-doped ETL for the Ir(ppy)$_3$-based green phosphorescent OLEDs [265]. In addition, triarylboron compounds can undergo reversible reduction while maintaining a high chemical and thermal stability. Therefore, a triarylboron-based ETM of B3T **(194)** with a triazine core was developed by Sun and coworkers. B3T has a low LUMO (3.25 eV) and a deep HOMO (6.73 eV) energy level and a high triplet energy level (3.07 eV) [266].

3.4.2.3 Fluorine-Substituted Electron-Transport Materials

A perfluorinated, *para*-conjugated oligophenylene with high EA exhibited improved electron-transport properties and was investigated as an ETM (**195–198**) (Scheme 3.33) [267]. The electron mobility of NPF-6 (**195**), determined by the TOF technique, is much higher than that of Alq_3 under the same conditions. In fact, an OLED fabricated using NPF-6 as the ETL showed higher EQE than an equivalent Alq_3-based device.

3.4.2.4 Silole Electron-Transport Materials

In contrast to lowering the LUMO energy levels via introduction of electron-withdrawing groups (such as C=N double bond–containing aromatic heterocycles, metal–quinolinol complexes, and CN- or F-substituted conjugated systems) to achieve high-EA compounds for ETMs, the high electron-accepting ability of the silole ring ascribed to the σ*–σ* conjugation between the σ* orbital of the exocyclic σ bonds on silicon and the σ* orbital of the butadiene moiety in the ring makes silole compounds with a low LUMO energy level particularly suitable for use as ETMs. Scheme 3.34 shows some examples of silole ETMs (**199–201**) investigated by several groups [268–272]. 2,5-Bis-(2′,2″-bipyridin-6-yl)-1,1-dimethyl-3,4-diphenylsilacyclopentadiene (PyPySPyPy, **200**) showed an electron mobility of 2×10^{-4} cm²/(V s) at 0.64 MV/cm measured by the TOF technique, higher than that of

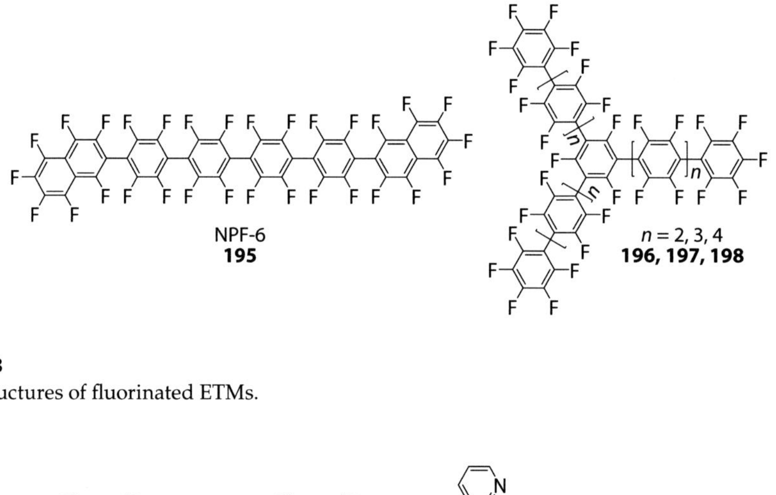

SCHEME 3.33
Chemical structures of fluorinated ETMs.

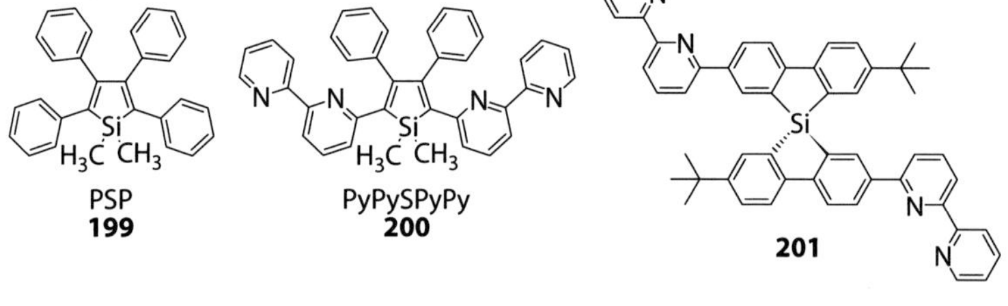

SCHEME 3.34
Chemical structures of silole compounds as ETMs.

Alq$_3$ [273]. Compared with Alq$_3$, the EAs of these compounds are slightly higher (3.3 eV), which in part explains their superior electron-transport properties.

Improved EL efficiencies were obtained when using these compounds as the ETL [274]. Besides their use as ETMs, some siloles are also being explored as emissive materials or host materials for OLEDs [275]. However, it was also reported that the stability of devices using silole compounds is lower compared with equivalent Alq$_3$-based devices [276]. Further improvement in such device performance by designing new silole compounds is necessary.

3.4.2.5 Boron-Based Electron-Transport Materials

The electron-withdrawing dimesitylboryl substituted compound 5,5″-bis-(dimesitylboryl)-2, 2′:5′,2″-terthiophene (BMB-3T, **202**) (Scheme 3.35) was reported as an ETM [277]. The molecule showed reversible two-peak reductions with high EA (3.05 eV) and can form amorphous films by vacuum evaporation. Using BMB-3T as the ETM for the Alq$_3$-based OLEDs, a brightness of 21,400 cd/m^2 and an EQE of 1.1% were obtained, compared with 13,000 cd/m^2 and 0.9% for the OLEDs without BMB-3T.

Kido's group developed an ETM of tris-[3-(3-pyridyl)mesityl]borane (3TPYMB, **203**) with high triplet excited energy level. Larger current density and higher luminance were achieved for the OLED using 3TPYMB as an ETL, compared with the OLED with a conventional ETM of Alq$_3$ [278].

3.4.2.6 Phosphine Oxide Electron-Transport Materials

Organic phosphine oxide (PO) materials exhibit excellent electron-injection and hole- and exciton-blocking capabilities, as well as good morphological stability. Von Ruden et al. [279] reported two PO electron-transport/hole-blocking materials, BM-A10 (**204**) and BM-A11 (**205**) (Scheme 3.36). The attachment of the phenyl rings to the central pyridyl ring is different in the two molecules, and this difference causes the two molecules to pack differently, which could very well influence the π-orbital overlap and lead to differences in electron-transporting ability and overall device properties.

Lee and coworkers developed a universal ETM (SPPO2, **206**) with LUMO level of 2.4 eV and HOMO level of 6.0 eV. SPPO2 shows a high T_g of 118°C and good film-forming property. SPPO2 reduced the driving voltage of OLEDs irrespective of the energy level of the host materials due to efficient electron injection from SPPO2 to host material. In particular, SPPO2 reduced the driving voltage of the blue device as a wide-band-gap host material by >3 V. Therefore, SPPO2 can be used as a universal ETL for OLEDs [280].

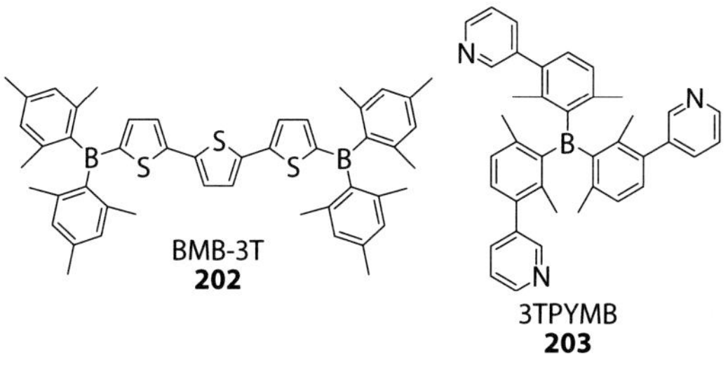

SCHEME 3.35
Chemical structures of boron-based ETMs.

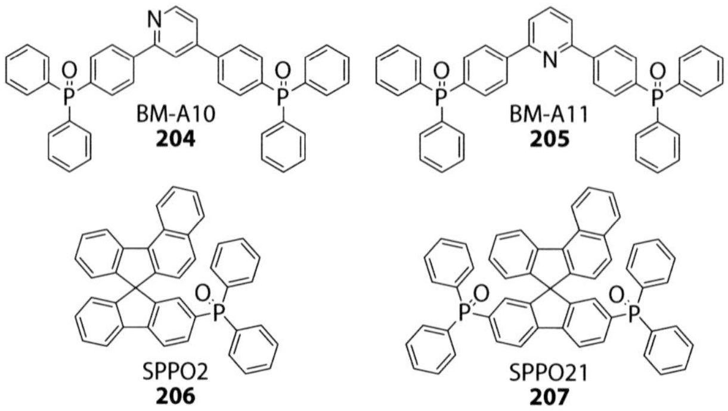

SCHEME 3.36
Chemical structures of phosphine oxide ETMs.

Jeon et al. developed high-efficiency blue fluorescent OLEDs without an LiF layer using a spirobenzofluorene-based phosphine oxide (SPPO21, **207**) as the ETM and direct deposition of Al cathode on it. The power efficiency of the device was improved from 0.9% to 6.1% by using SPPO21 instead of Alq_3. Using SPPO21/Al instead of Alq_3/LiF/Al would simplify the device structure and the fabrication process [281].

3.4.2.7 Doped Electron-Transport Materials

A universal approach to improve electron injection/transport is by doping the ETM with organic [282–284], inorganic [285], or low-WF metal [286] n-type dopants to modify the interface electronic structure and/or to enhance bulk conductivity of the ETMs. Ma et al. reported Liq-doped BPhen as an ETL combined with a p-doped HTL; high current efficiency of 5.90 cd/A and power efficiency of 4.51 lm/W were achieved for the Alq_3-based p–i–n OLEDs [287]. By inserting a bis(ethylenedithio)-tetrathiafulvalene (BEDT-TTF)-doped BPhen layer as an ETL, enhanced electron injection and luminance as well as decreased operating voltage of the OLEDs were achieved [288]. Lee et al. investigated the effect of the n-ETL/ETL interface on the performance of inverted OLEDs and found that B3PYMPM resulted in the lowest turn-on voltage of 2.4 V and the highest EQE of 19.8% at low voltage [289]. Eom and coworkers [290] reported iridium(III)bis(40,60difluorophe nylpyridinato)tetrakis (1-pyrazolyl)borate (FIr6)-based blue phosphorescent OLEDs with 3TPYMB as the ETL and 3TPYMB:Cs as the EIL, achieving an EQE of 20 ± 1% and peak power efficiency of 36 ± 2 lm/W, which could be attributed to the high triplet energy of 3TPYMB and the increased conductivity of 3TPYMB:Cs. Besides, LiF [291], Li-doped aluminum triquinolate [292], and Cs-doped phenyldipyrenylphosphine oxide (POPy2) [293] as the ETL were also reported.

Recently, Earmme and Jenekhe reported that the FIrpic-based light-blue devices with solution-processed TmPyPB doped with Cs_2CO_3 as an ETL have a current efficiency of 37.7 cd/A with an EQE of 19.0%, which was the best performance observed to date in all solution-processed blue phosphorescent OLEDs. The results show that a small-molecule ETL doped with alkali metal salt can be realized by solution-processing device to enhance overall device performance and orthogonal solution processing of metal salt-doped ETM is a promising strategy for applications in various solution-processed multilayered organic electronic devices [294].

3.4.2.8 Other Electron-Transport Materials

Huang et al. [295] reported a series of 2,8-disubstituted dibenzothiophene and 2,8-disubstituted dibenzothiophene-S,S-dioxide derivatives containing quinoxaline and pyrazine moieties (**208–211**) (Scheme 3.37). These materials are amorphous, with T_g ranging from 132°C to 194°C. The compounds exhibit high electron mobilities and serve as effective ETM for OLED. Double-layer devices are fabricated with the structure of ITO/Qn/ETL/LiF/Al, where yellow-emitting 2,3-bis[4-(N-phenyl-9-ethyl-3-carbazolylamino)phenyl]quinoxaline (Qn) serves as the emitting layer. An EQE of 1.41%, a power efficiency of 4.94 lm/W, and a current efficiency of 1.62 cd/A were achieved at a current density of 100 mA/cm².

Jenekhe's group [296] reported bisindenoanthrazolines with a novel π-conjugated heptacyclic framework (**212–215**) (Scheme 3.38) that show LUMO energy levels of 3.65–3.72 eV. Single-crystal x-ray diffraction showed that the novel heptacyclic framework is planar and leads to a slipped face-to-face π-stacking. The high EA (3.7 eV) and high electron mobility [1×10^{-7} to 1×10^{-4} cm²/(V s)] of the bisindenoanthrazolines make them attractive electron-transport candidates in organic electronics. DADF and DADK were shown to assemble in solution into single-crystal nanowires with widths of 100–250 nm and lengths of 10–200 μm. DADF are excellent ETMs in realizing high-performance phosphorescent OLEDs; a brightness of 62,000 cd/m² and current efficiency of 39.2 cd/A at a brightness of 4270 cd/m² was achieved in green PHOLEDs.

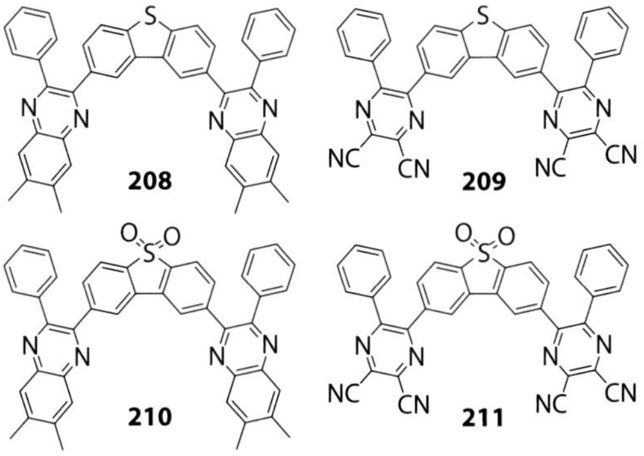

SCHEME 3.37
Chemical structures of dibenzothiophene and dibenzothiophene-S,S-dioxide ETMs.

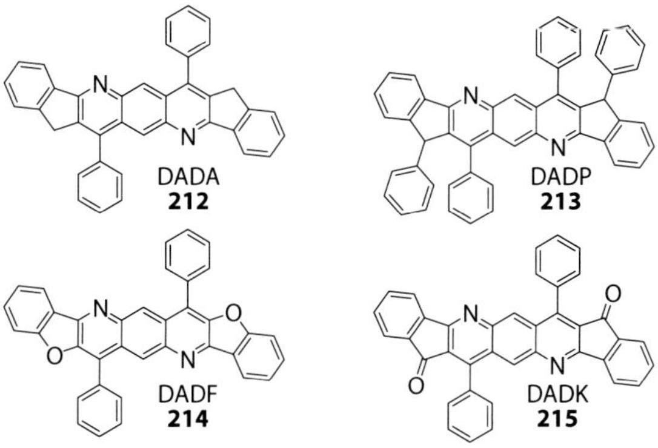

SCHEME 3.38
Chemical structures of bisindenoanthrazoline ETMs.

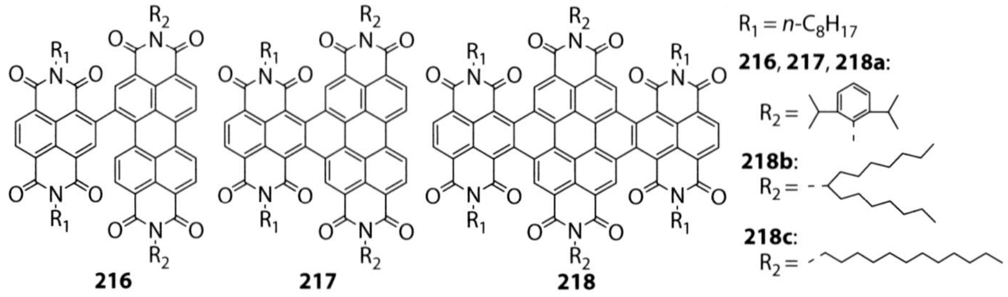

SCHEME 3.39
Chemical structures of hybrid perylene ETMs.

Yue et al. [297] reported hybrid perylene arrays (**216–218**) (Scheme 3.39) via a combination of Stille coupling and C–H transformation. The ability to extend the π system along the equatorial axis of arylenes not only leads to broadened light absorption but also increases the EA, which can facilitate electron injection and transport with ambient stability.

3.5 Light-Emitting Materials

The material that gets most of the glory in OLED devices is naturally that which generates the light output. In many cases, however, the so-called EML is actually a mixture of two or more materials wherein there is at least one electroluminescent emissive material in conjunction with a charge-transporting host material. Such guest–host systems are extremely common in SMOLED devices, whereas in polymeric OLED devices the EML is usually composed of a single polymer comprising multiple distinct monomer units, which combine to produce the necessary emissive and charge-transporting properties into a single-phase material. This is a broad generalization, of course, and there are certainly examples of SMOLEDs that use a single material as the EML, and of polymer OLEDs that use multiple phases (e.g., polymer blends or doped polymers) as the EML.

Considering the roles that must be fulfilled by the EML, many properties must be effectively combined. The layer must be able to transport charge—both holes and electrons—in order that the charge carriers are able to move through the layer and find each other. The recombined charges must then create an excited state in the material, which can collapse from this exciton state back to its ground state and in doing so emit a photon of light. All of this must occur efficiently with little to none of the input energy being dissipated as heat or electrochemical transformations of the materials themselves. Furthermore, mixtures of materials, if used for long-lived devices, must remain uniformly dispersed (a high T_g and have good film-forming properties as a solid solution) and not be subject to materials migration under an applied electric field (no electrophoresis). This latter requirement tends to exclude ionic materials from this application.

The chemical and photo-physical characteristics of the emissive material itself lead to categorization of OLEDs containing them into two main types:

1. SMOLEDs contain small-molecule emissive materials that can be processed by either vacuum deposition (evaporative) techniques or solution coating. The emissive small molecule may be a fluorescent (singlet excited state) or a phosphorescent (triplet excited state) emitter.

2. PLEDs contain polymeric emissive materials that are almost exclusively processed by solution coating (spin coating or ink-jetting). This has been discussed in Chapter 2. While most polymer work uses fluorescent emissive materials, there are a few examples of phosphorescent materials being incorporated into a polymer chain and being used as phosphorescent emitters. This part of the materials discussion will be covered in Chapter 4.

The phenomenon of organic EL was first demonstrated using a small-molecule fluorescent emitter in a vapor-deposited OLED device. The Kodak group first used metal oxinoid materials such as the octahedral complex Alq_3 (discussed above as an ETM) as the fluorescent green emitter in their pioneering work on OLED architectures [298].

To this day, Alq_3 is a stable emitter material in SMOLED devices, and metallic complexes of this type have many of the desired stability and film-forming properties necessary to provide useful performance. The complex can exist as both *fac* and *mer* isomers, and little attention is usually given to which isomer is present in the thin films of devices; however, the starting material for the evaporation is usually predominantly *mer*. A large body of work using other oxinoid complexes, particularly with zinc and magnesium, has shown similar performances. These materials have good electron-transporting abilities in addition to their luminescent properties and consequently also find great utility in the ETL (see Section 3.4.2.1). At this stage, the oxinoid materials, especially Alq_3 and BAlq, are also, more typically, used as host materials for other more efficient dopant materials of lower emission energy.

A bewildering array of materials has been used as emitters in SMOLEDs since this early work on Alq_3. In the following sections, we will present a brief review of light-emitting materials, especially host–guest emitter systems, and give a perspective description of all the current state-of-the-art small-molecule materials for emission at the three primary colors needed for full-color display applications.

The principle of the electronic processes in molecules can be schematically illustrated with the classical Jablonski diagram, which was first proposed by Prof. A. Jablonski in 1935 to describe absorption and emission of light. Figure 3.6 illustrates the electronic processes of the host–guest molecules.

When a host molecule is excited from the ground state by either absorbing light energy or being driven by electric energy to a higher vibrational energy level, it is subjected to collisions with the surrounding molecules. It can directly release its energy through radiative decay or nonradiative decay processes to the ground state, or in the presence of a suitable guest molecule, energy transfer processes will occur. The latter event, depicted in the left side of the diagram as an energy transfer transition from the host molecule to the guest molecule, occurs through Förster, Dexter, or radiative energy transfer processes. At this point, the radiative decay processes will occur from the luminescent guest molecules. It may be noted that the emission spectrum observed is sometimes the emission from only the guest molecules due to complete energy transfer processes, but sometimes it combines the guest and host molecule emission due to incomplete energy transfer.

Because molecular excited states may also transfer from molecule to molecule while conserving their spin and energy, one can treat them as quasiparticles named excitons. The highly localized excited states are known as Frenkel excitons, having radii of a few angstroms. One can treat the Frenkel exciton as the hop of charge carriers (electron, hole) to a neighboring molecule. Because the rate of exciton hopping is given by the multiplication between the rate of electron transfer and the rate of hole transfer, the theory of electron transfer can shed light on the understanding of exciton hopping.

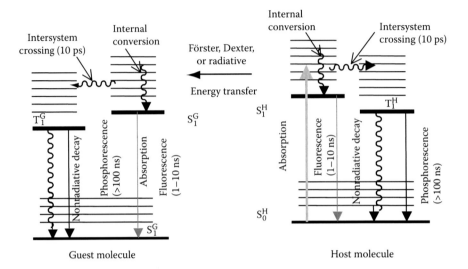

FIGURE 3.6
Electronic processes of host–guest molecules, in which the guest molecules can emit light through both singlet and triplet states. S_0^H, S_0^G: a singlet ground state of the host and guest molecules, respectively. S_1^H, S_1^G: a first excited singlet state of the host and guest molecules, respectively. T_1^H, T_1^G: a first excited triplet state of the host and guest molecules, respectively.

During OLED operation, singlet and, in some cases, triplet excitations may first be created in the host material. Then through charge or energy transfer from the host to the guest, singlet or triplet excited states are formed in the guest. For an effective guest–host system, several factors have to be considered, such as the phase compatibility of the host and guest, the aggregation of the molecules, and the host–guest energy level and orbital alignment. Thoms and coworkers have studied efficient EL in a host–guest system and, using computation methods, predicted suitable host carbazole molecules for phosphorescent Ir guest complexes [299]. In this charge-transfer process, the band gap of the guest should fall within the band gap of the host to favor transport of electrons and holes from the host to the guest, where they should then recombine. If such energy transfer processes dominate, efficient energy transfer requires that the energy of the excited state of the host should be higher than that of the emissive excited state of the guest.

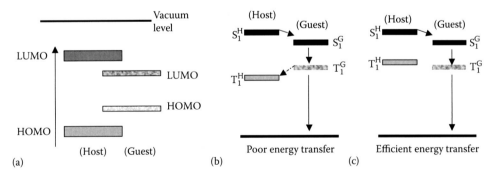

FIGURE 3.7
Energy-level relationship in a phosphorescent guest–host system. (a) The band gap of the guest falls within the band gap of the host. (b) The energy of the excited state of the host is lower than that of the emissive excited state of the guest. (c) The energy of the excited state of the host is higher than that of the emissive excited state of the guest.

This applies to both singlet excited states and triplet excited states of the host and the guest as shown in Figure 3.7.

The efficiency of charge or energy transfer for the singlet excited (fluorescent) state is easy to verify if there is an overlap between the emission spectrum of the host and the absorption spectrum of the guest. Beyond this requirement, for an efficient energy transfer from the host to the guest of the triplet state (phosphorescent), the excited triplet state of the host should be higher than that of the guest. In the following section, we will discuss some widely used fluorescent and phosphorescent light-emitting systems, including host materials and appropriate guest or dopant materials.

3.5.1 Fluorescent Light-Emitting Materials

3.5.1.1 Fluorescent Host Materials

The basic requirements for the host materials dictate that they should have the following properties:

1. Good electron and hole conduction with thermal, chemical, and electrochemical stabilities.
2. Matching HOMO and LUMO energy levels with the guest materials. For efficient energy transfer processes, the LUMO energy level of the host normally should be shallower than the corresponding LUMO energy level of the guest. Likewise, the HOMO energy level of the host should be deeper than that of the guest.
3. For a triplet emissive guest (phosphorescent) dopant, the triplet energy level of the host normally should be higher than that of the guest.
4. Fast energy transfer processes.
5. Phase compatibility with the guest materials.

A well-known ETM and green light–emitting material, Alq$_3$ (**134**), was also the first OLED host material reported by Tang et al. [300]. It is one of the most widely studied electron-transport host materials and has a HOMO energy level of −5.7 eV and a LUMO energy level of −3.0 eV with a band gap of 2.7 eV. The triplet energy level of Alq$_3$ was determined to be 2.0 eV. Singlet emission peaks centered at 560 nm make it suitable as a host for green and red emission.

TPBI (**147**) is another electron-transport host material for both fluorescent and phosphorescent dopants. It has a large band gap of 3.5–4.0 eV with HOMO energy level of −6.2 to −6.7 eV and LUMO energy level of −2.7 eV, making it suitable for singlet green and red emission and in some case for blue dopants [243]. TPBI also possesses good hole-blocking properties owing to its deep HOMO energy level.

DPF (**219**) (Scheme 3.40) was used as a host material for a red dye of DCJTB. A large overlap between its PL spectrum and the absorption spectrum of DCJTB ensures efficient energy transfer to DCJTB. Red OLED with a configuration of ITO/NPB (50 nm)/DPF:2 wt.% DCJBT (10 nm)/TPBI (30 nm/)/LiF (0.5 nm)/Mg:Ag (200 nm) exhibits maximum efficiencies of 4.2 cd/A and 3.9 lm/W, and these values are much higher than those of the Alq$_3$-based device [301]. In addition to the good performance, the DPF-based device also shows better color stability and lower efficiency roll-off than the Alq$_3$-based device.

The most widely used host materials for fluorescent blue light–emitting materials are anthracene and distyryl derivatives, also shown in Scheme 3.40. These materials have large band gaps, are thermally stable with high T_g, and have good phase-compatibility

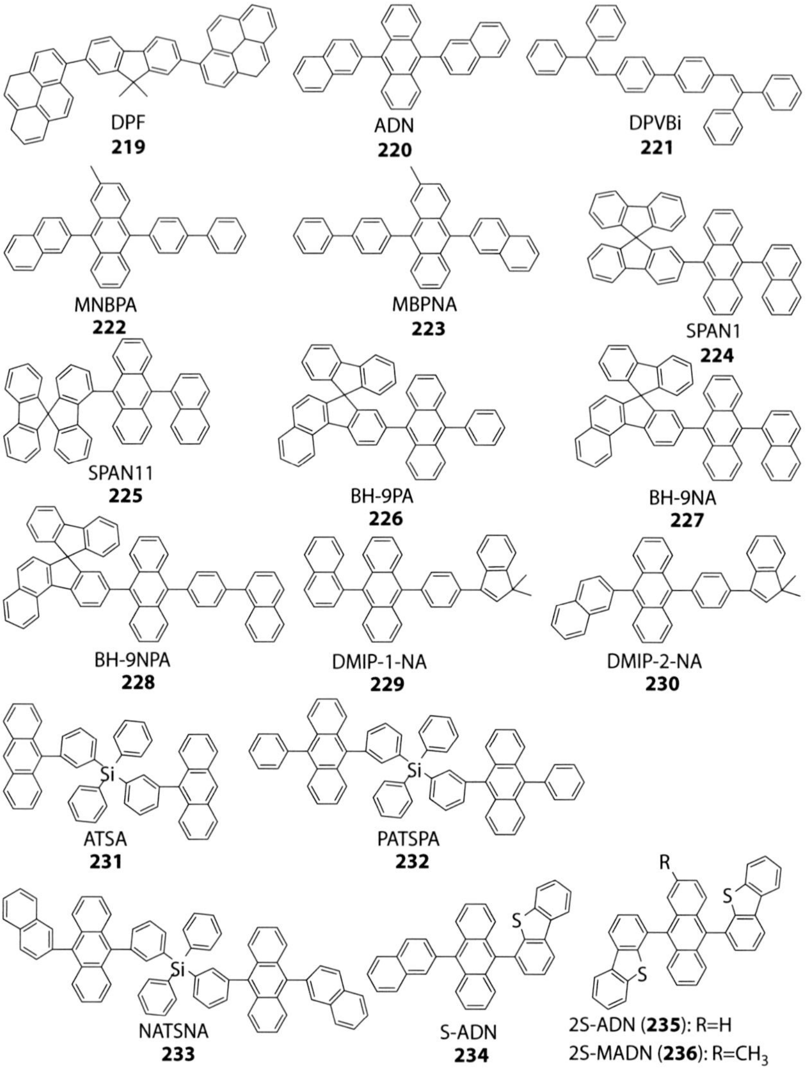

SCHEME 3.40
Chemical structures of host materials for fluorescent emitters.

with their guest blue emitters, which in turn also belong to the anthracene, distyryl amine compounds, perylene, and fluorene derivatives.

Besides symmetrical anthracene derivatives like ADN (**220**), some unsymmetrical 9,10-disubstituted anthracene derivatives were also recently developed as host materials for fluorescent blue emitters. MNBPA (**222**) and MBPNA (**223**) were developed by Huang et al. [302], and they have similar T_gs of 132.5°C and 133.8°C, respectively, both of which are higher than that of MADN (120°C). This can be ascribed to the unsymmetrical nature of MNBPA and MBPNA with substitution of biphenyl instead of one naphthyl group in MADN to hamper the facile packing of the molecules. Light-emitting device with a configuration of ITO/CF$_x$/2-TNATA (60 nm)/NPB (10 nm)/MBPNA:3% BUBD-1 (40 nm)/TPBI (20 nm)/LiF (1 nm)/Al exhibits the best performance with a maximum EQE of 6.34% and CIE coordinates of (0.152, 0.222).

Spirobifluorene was also introduced onto the C-9 position of naphthylanthracene as a terminal group, resulting in two host materials, SPAN1 (**224**) and SPAN11 (**225**), in which naphthylanthracene was linked with the 2 or 4 position of spirobifluorene, respectively. It has been found that the 4-spirobifluorene substitution was better than 2-spirobifluorene in terms of thermal stability and wide band gap. However, the 4-spirobifluorene substituted derivative with wide band gap has a negative effect on current density and efficiency of the devices using these materials as host and BD-1 as dopant [303].

A series of spiro[benzo[*c*]fluorene-7,9′-fluorene] substituted anthracene derivatives with different conjugation length, namely BH-9PA (**226**), BH-9NA (**227**), and BH-9NPA (**228**), were developed by Gong et al. All these derivatives were used as host materials in a device structure of ITO/DNTPD/NPB/host:5% dopant/Alq$_3$/LiF/Al. BH-9PA-based devices composed of DSA-Ph and BD-6MDPA as dopants show blue EL spectra at 468 and 464 nm at 7 V and current efficiencies of 7.03 and 6.60 cd/A, respectively [304].

Indene was also adopted as a substituent of 10-naphthylanthracene derivatives to develop two novel unsymmetrical host materials, DMIP-1-NA (**229**) and DMIP-2-NA (**230**). Both DMIP-1-NA and DMIP-2-NA exhibit intense fluorescence and good thermal stability. A maximum power efficiency of 8.39 lm/W was achieved by using DMIP-2-NA as host and BD-1 as dopant [305].

Lyu et al. developed a group of silylanthracene derivatives with a nonplanar structure, namely ATSA (**231**), PATSPA (**232**), and NATSNA (**233**), which have high thermal stability and good amorphous film-forming capability. A moderate current efficiency of 1 cd/A was achieved by using these compounds as the blue emitter. Moreover, they can be also used as host materials for DPAVBi, giving high color purity with CIE coordinates of (0.142, 0.149) and high efficiency up to 7.5 cd/A (6.3%) [306].

Wang et al. developed a series of anthracene derivatives, S-ADN (**234**), 2S-ADN (**235**) and 2S-MADN (**236**), comprising π-electron-rich dibenzothiophene moiety as a side group to adjust the HOMO/LUMO levels and to enhance the hole-transport and electron-injection ability. Efficient blue-emitting devices could be obtained by using these compounds as host materials [307].

3.5.1.2 Fluorescent Dopant Materials

To match up with host materials, the basic requirements for the guest or dopant materials are as follows:

1. Highly fluorescent.
2. For green pixel color, the standard of CIE 1931 color chromaticity coordinates are (0.30, 0.60). For red color, the standard of CIE 1931 color chromaticity coordinates are (≥0.62, ≤0.37), the standard red CRT Phosphors of the Society of Motion Picture and Television Engineers (SMPTE-C) is (0.64, 0.34), and the European Broadcasting Union (EBU) is (0.64, 0.33). For blue pixel color, the standard of CIE 1931 color chromaticity coordinates are (0.14–0.16, 0.11–0.15). For white pixel color, the standard of CIE 1931 color chromaticity coordinates are (0.313, 0.329).
3. Matched HOMO and LUMO energy levels with host materials.
4. For a triplet emissive guest (phosphorescent) dopant, the triplet energy level of the dopant normally should be lower than that of the host.
5. Fast energy transfer processes.
6. The dopant materials should have phase compatibility with the host materials.

3.5.1.2.1 Red Fluorescent Dopants

A series of red fluorescent dyes has been reviewed by Chen [308]. The red fluorescent dopants, according to their structural characteristics, are summarized as

1. Push–pull red emitters
2. Polyacene red emitters
3. Metal chelate red emitters

3.5.1.2.1.1 Push–Pull Red Emitters　The most widely studied push–pull fluorescent red emitters are pyran-containing compounds. Among them, 4-(dicyanomethylene)-2-methyl-6-[p-(dimethylamino)styryl]-4H-pyran (DCM, **237**) (Scheme 3.41) is a bright and efficient red arylidene laser dye invented in 1974 [309], and was the first dopant in host–guest system OLEDs introduced by Kodak researchers in 1989. The device ITO/HTL/Alq$_3$:DCM/Mg:Ag showed an EQE of 2.3%. The emission is orange–red with peak emission in the range of 570–620 nm for DCM and 610–650 nm for DCJ (**238**) depending on the concentration of the dopants. High-concentration doping results in more saturated red emission; however, low efficiency due to concentration quenching becomes a factor at these high dopant levels.

The DCM molecule has an electron-donor π-acceptor (so-called push–pull) structure. By introducing a more rigid julolidine ring into the donor moiety, the Kodak group prepared a DCJ molecule, which has about a 10-nm red shift of the emission spectrum. The molecule shows high QE and a more saturated red emission compared with DCM [310,311]. However, both the DCM and DCJ series have an active methyl group, which can suffer further condensation reactions to form undesired bis-condensation by-products. This by-product shows a broad, very weak fluorescence. Consequently, during the synthesis of materials in the DCM or DCJ series, careful control of the reaction condition and repeated purification of the product are needed to obtain high-performance devices. The drawbacks of such a complex, synthetic protocol and tedious purification processes make for

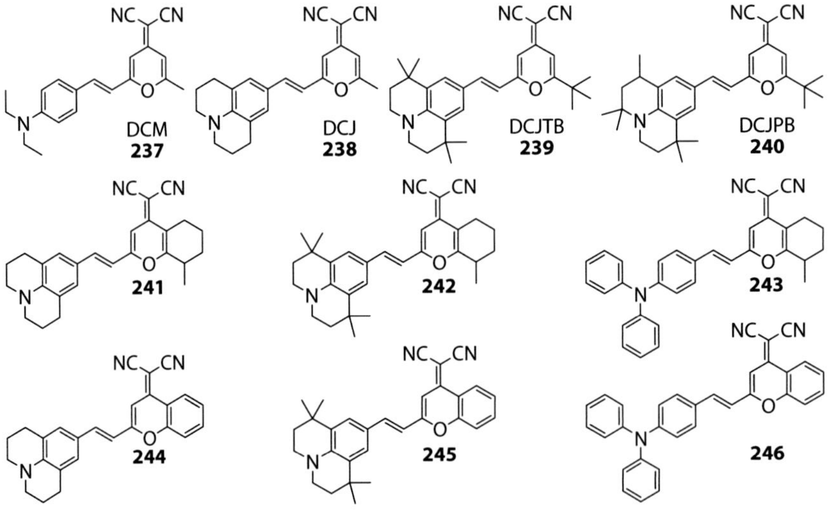

SCHEME 3.41
Chemical structures of push–pull red emitters.

difficulties in large-scale manufacturing. This problem can be solved by substituting bulky and sterically demanding substituents such as *iso*-propyl and *tert*-butyl group for the parent methyl group. This leads to the DCJTB series (**239**), which exhibit improved high QE while suppressing the concentration quenching. More importantly, the bulky *tert*-butyl or *iso*-propyl substituent at the pyran ring avoids further condensation with aldehyde groups and improves the purity of the materials [310,312]. However, the red OLEDs using the DCJTB series have the drawback of requiring higher doping concentrations, of the order of 2–4%, to tune the CIE to a standard red. Unlike the symmetric DCJTB series, the unsymmetrically substituted DCJPB series (**240**) has five methyl groups instead of four and offers an excellent red emission with comparable performance, yet with relatively lower required doping level of 0.5–2.5% [313].

While the DCJTB series replaced the active methyl group with *tert*-butyl or *iso*-propyl substituents to avoid bis-Knoevenagel condensation reactions during the synthesis of DCM or DCJ series, Ma et al. came up with the idea of using substituted cyclohexane rings to block the reactive site of the pyran ring. They then synthesized a series of 4*H*-benzopyran-based red emitters (**241–243**) [314].

Yet another strategy was developed by Zhang et al. using a bis-condensation inactive chromene ring instead of the pyran ring so as to generate a series of chromene-based red dopants (**244–246**) [315]. The merit of these chromene dopants is their relatively long emission wavelength peaks compared with DCM or DCJTB materials due to the more conjugated chromene moiety, and this contributes to the more saturated red emission. In fact, the EL spectra of OLED devices of ITO/TPD/Alq$_3$:chromene-dopants/Alq$_3$/Mg:Ag exhibited satisfactory red emission color. However, these chromene-based red emitters showed lower fluorescent quantum yield compared with the DCM (>70%) or DCJTB series (78%). As a result, the efficiency of the OLEDs of these compounds is, in general, not competitive with that of devices made from DCM or DCJTB. Nevertheless, the convenient synthetic procedures and the possibility of further chemical modification of the chromene series provide the possibility to yet prepare high-efficiency red dopants.

The impurity issues of DCM materials have been solved by the above-mentioned methods through the elimination of the active methyl group by introducing unsymmetrical pyran moieties. Another issue for the DCM series is color purity. The emission spectra of the DCM red emitters demonstrate poor red color saturation having a broad emission band centered at 590–620 nm and a large full width at half maximum (FWHM) (100 nm). Tao et al. reported a DCDDC dopant that shows a bright red emission with a peak wavelength of 650 nm and an FWHM of only 70 nm (Scheme 3.42) [316]. The DCDDC molecules (**247**) can be easily synthesized in high yield and purity [317].

The main backbone of DCDDC is the same as that of DCM, except that the former lacks a double bond at the pyrene ring but rather has an additional methyl substituent on the

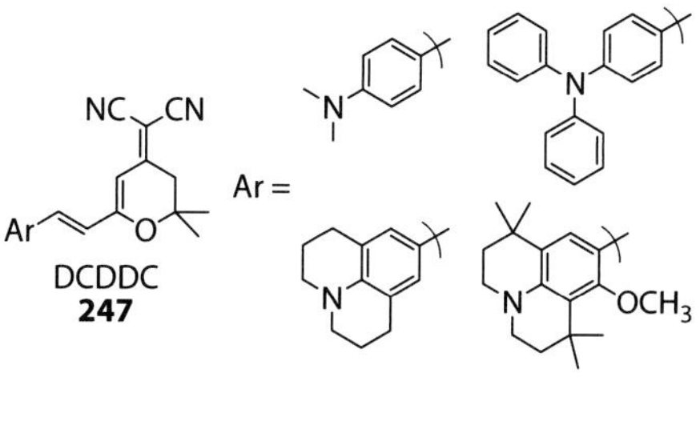

SCHEME 3.42
Chemical structures of isophorone-based red emitters.

6 position of the ring. This minor difference leads to DCDDC having a narrow emission band due to the simple one donor–acceptor system, while in the DCM series the mixture of two donor–acceptor effects produces broad emission bands. OLEDs fabricated with the architecture ITO/PVK:TPD/Alq$_3$:1% DCDDC/Alq$_3$/Mg:Ag have a turn-on voltage of 5 V, peak luminance of 5600 cd/m^2 at 15 V, and the highest EL power efficiency of 1.6 lm/W. The device exhibits red emission with a peak maximum at 630 nm and an additional shoulder emission of 530 nm from Alq$_3$, which disappears at 2% doping but at the price of a lower power efficiency. This is common to the red emission OLEDs due to inefficient energy transfer processes, and this can be improved when using additional assistant dopants. The authors also pointed out that by further modification of the chemical structure of DCDCC, it is possible to achieve even better red dopants. Following DCDCC, Li et al. reported a series of isophorone-based red emitters [318]. However, the QE of these red emitters is in the range of 10–20%, significantly lower than the DCM or DCJTB series. The OLED devices using these materials as dopants showed saturated red emission along with stable current efficiency at high driving voltages.

Instead of eliminating the bis-condensed by-products by the above strategy, Jung et al. have found that some bis-condensation products themselves show high fluorescence, if properly designed. A series of bis-condensed DCM derivatives has been synthesized (**248**) (Scheme 3.43) [319]. Knoevenagel condensation of 1:2.1 stoichiometric ratio of the pyran and aldehyde generates a pure bis-condensation product. Because of the extended π-conjugated system, these bis-condensed red emitters are about 40–50 nm red shifted compared with their respective monosubstituted DCM analogues.

Using a 5.2% doping level of compound **248b**, EL with color close to the NTSC red standard with CIE (0.658, 0.337) was achieved. Owing to the inefficiency of the low-energy transfer processes, emission from the host or the electron transporter is often observed. To overcome this, Hamada et al. [320] reported improved red emission by using a codopant such as rubrene as a sensitizer to assist the energy transfer processes between the host and the red dopant; however, this process involves complex deposition control. To simplify the device fabrication processes while keeping the high efficiency of the device, several strategies have been applied. Qiu's group used a new host and an ETM composed of the binuclear complex bis(salicylidene-*o*-aminophenolato)-bis(8-quinolinoato)-bisgallium(III) (Ga$_2$(saph)$_2$q$_2$), which replaced Alq$_3$ and achieved a pure red emission based on a DCJTB red emitter with CIE (0.67, 0.33), and an efficiency of 2.04 cd/A, which is double that of the Alq$_3$-based device [321].

By optimization of device structures and by using different hole and electron-injection or transport materials, Liu et al. achieved an excellent red OLED with a very high efficiency based on the DCJTB molecule. The OLED structure is glass (0.7 mm)/SiO$_2$ (20 nm) ITO/CF$_x$/NPD (110 nm)/Alq$_3$:5% rubrene:2% DCJTB (30 nm)/Alq$_3$ (55 nm)/LiF (0.1 nm)/ Al (200 nm). A luminescent efficiency of 3.24 cd/A with CIE (0.643, 0.354) and a power efficiency of 1.19 lm/W at a drive current density of 20 mA/cm^2 at 8.53 V is reported [322].

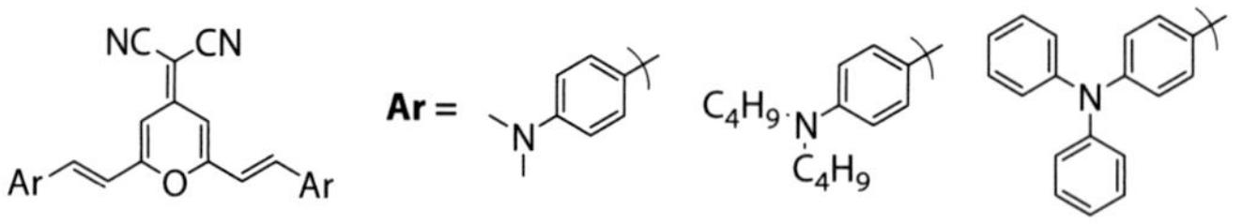

SCHEME 3.43
Chemical structures of bis-condensed DCM red emitters.

In 2007, a new DCM-type dye (DCQTB, **249**) (Scheme 3.44) with a molecular structure similar to DCJTB was reported [323]. Compared with DCJTB, DCQTB shows red-shifted emission but blue-shifted absorption, and the absorption of DCQTB has a better overlap with the emission of Alq$_3$. As a result, DCQTB exhibits higher luminous efficiency in a device using Alq$_3$ as a host. A device with a configuration of ITO/NPB (60 nm)/Alq$_3$:2.3 wt.% DCQBT (7 nm)/BCP (12 nm)/Alq$_3$ (45 nm)/LiF (0.3 nm)/Al (300 nm) was fabricated, giving a maximum brightness of 6201 cd/m^2 and a maximum current efficiency of 4.41 cd/A with CIE (0.65, 0.35), in contrast to a maximum luminance of 3453 cd/m^2 and a maximum current efficiency of 3.01 cd/A with CIE (0.62, 0.38) obtained for the DCJTB-based device in the same structure.

OJ1 (**250**) and OJ2 (**251**) are highly efficient dopants that possess a dipolar skeleton analogous to that of DCJTB and can be prepared in high yield by using commercially available raw material [324]. The T_gs of OJ1 and OJ2 are 107°C and 103°C , respectively, and are higher than that of DCJTB (98°C), indicating a better morphological stability in the solid state. Devices with a typical configuration of ITO/NPB (40 nm)/Alq$_3$:x% dye (40 nm)/TPBI (40 nm)/LiF (1 nm)/Al (150 nm) were fabricated, and an optimal performance with an EQE of 2.4% and a maximum current efficiency of 5.6 cd/A was achieved for the 1% doped OJ2-based device. When replacing the TPBI layer with 30 nm Alq3 as an ETL and a 10 nm BCP as an HBL, the device reached the highest brightness of 43,400 cd/m^2 with a current efficiency of 6.46 cd/A, displaying a pure red color with CIE (0.62, 0.38) and a narrow bandwidth (FWHM = 72 nm). These values are significantly better than those reported for DCJTB.

Efficient red fluorescent compounds, Red 1 (**252**) and Red 2 (**253**), based on bulky groups–substituted DCJTB derivatives were synthesized and characterized [325]. Introducing bulky bicycle(2,2,2)octane groups in the pyran moiety and *tert*-butyl or isopropyl group in the julolidine moiety could suppress the aggregation between the emitters. By coupling with the utilization of a rubrene-Alq$_3$ cohost system, complete energy transfer between the dopants (Red 1 and Red 2) and Alq$_3$ could be realized and high device efficiencies could be achieved. The device configuration is as follows: ITO/NPB (40 nm)/Alq$_3$:rubrene (50%):red dopant (1%) (20 nm)/Alq$_3$ (40 nm)/Liq (1 nm)/Al. The Red 1–based device showed a current efficiency of 6.89 cd/A and a power efficiency of 3.09 lm/W at 20 mA/cm^2 with CIE

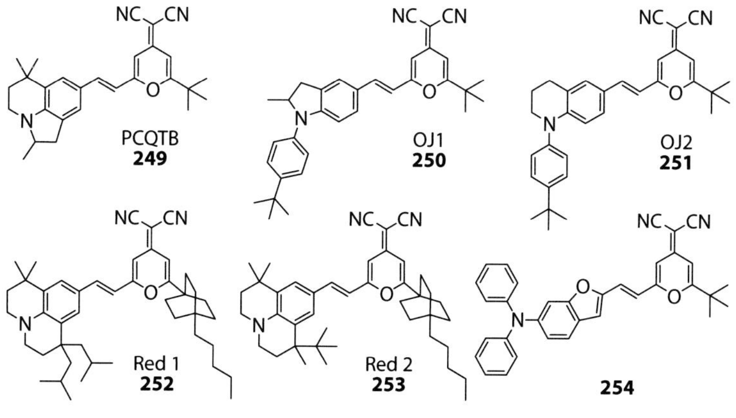

SCHEME 3.44
Chemical structures of push–pull red emitters.

coordinates (0.64, 0.36) at 7.0 V. The Red 2–based exhibited a current efficiency of 6.86 cd/A and a power efficiency of 3.08 lm/W at 20 mA/cm² with CIE coordinates (0.64, 0.36) at 7.0 V.

Leung et al. designed and synthesized a new red emitter **254** in which 6-*N,N*-(diphenylamino)benzofuran replaces the dialkylaminoaryl group of DCJTB to stabilize the cationic intermediates [326]. Red OLED of ITO/NPB (70 nm)/Bebq₂:1% red dopant (7 nm)/BCP (5 nm)/Bebq₂ (33 nm)/LiF (1 nm)/Al (150 nm) exhibited red EL with a current efficiency of 2.9 cd/A at 100 cd/m² and a maximum brightness of 62,000 cd/m².

Besides pyran-containing compounds, squaraine dyes, widely applied in the fields of organic solar cells, electrophotography, optical storage systems, imaging materials, and laser dyes, also emit sharp red emission with a rather small Stokes shift. Using Alq₃ as the host, red emissive OLEDs with an architecture of ITO/NPD/Alq₃:Sq/Mg:Ag can be prepared. However, green emission from the Alq₃ host cannot be avoided because of the inefficient energy transfer process [327]. Using NPD as the host material can, however, lead to pure red emission [328].

3.5.1.2.1.2 Polyacene Red Emitters   Scheme 3.45 lists some examples of polyacene-based red emitters. 6,13-Diphenylpentacene (DPP, **255**), doped in an Alq₃ host, emits pure red light centered at 625 nm with CIE (0.63, 0.34) and an EQE of 1.3% at 100 A/m² [329]. Benzo[*a*]aceanthrylene- and perylene-substituted arylamine compounds [Acen, (PPA) (PSA)Pe-1, **256**] have also been investigated as red dopants [330]. In an OLED device of ITO/NPD/Acen/BCP/Alq₃/Mg:Ag, red emission with CIE (0.64, 0.34) has been observed.

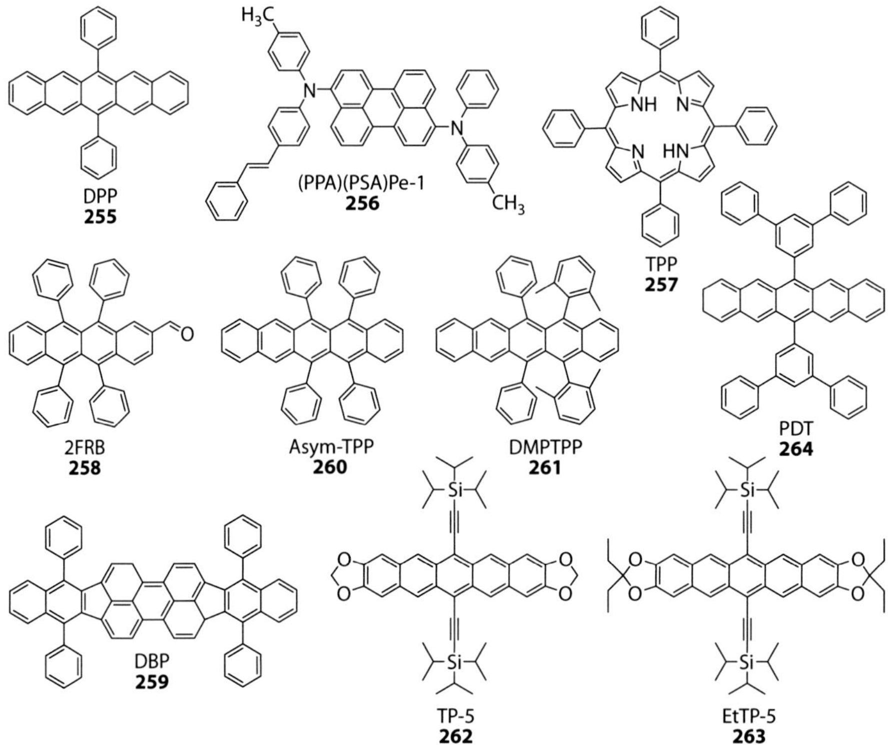

SCHEME 3.45
Chemical structures of polyacene red emitters.

Porphyrin-macrocyclic compound 5,10,15,20 tetraphenyl-21*H*,23*H*-porphine (TPP, **257**) shows a very narrow red emission along with a longer wavelength emission shoulder; this red color is too red for display application. None of these red emitters demonstrated high efficiency in red OLEDs; therefore, they are not good candidates for display applications.

Doping a red-emitting rubrene derivative, 2FRB (**258**), into different types of hosts to fabricate OLEDs has been examined by Li et al. [331]. Two devices were prepared with configurations of ITO/2-TNATA (10 nm)/NPB (40 nm)/NPB:*x*% 2FRB (10 nm)/TPBI (60 nm)/ LiF (0.5 nm)/Al (200 nm) and ITO/2-TNATA (10 nm)/NPB (40 nm)/Alq$_3$:*x*% 2FRB (10 nm)/ Alq$_3$ (50 nm)/LiF (0.5 nm)/Al (200 nm). The emission can be tuned from 580 to 607 nm in NPB host and 560 to 622 nm in Alq$_3$ host. The Alq$_3$-hosted devices exhibited higher efficiencies with a maximum current efficiency of 2.42 cd/A and a maximum luminance of 3100 cd/m², and showed more saturated red emission peaking at 622 nm with CIE (0.65, 0.35) by doping 3.2% 2FRB. The results indicate that the rubrene derivative is a promising bipolar dopant for red light-emitting device.

Red OLEDs in configuration of ITO/fluorocarbon (CF$_x$)/NPB (60 nm)/rubrene:1 wt.% DBP (40 nm)/DBzA (20 nm)/Li (1 nm)/Al (200 nm) were reported by using tetraphenyl-dibenzoperiflanthene (DBP, **259**) as a red dopant and rubrene as host [332]. The red OLED using an ETL consisting of DBzA instead of Alq$_3$ exhibited a high power efficiency of 5.3 lm/W at a current density of 20 mA/cm² and gave a saturated red emission with CIE (0.66, 0.34).

Two asymmetric pentacene derivatives, asym-TPP (**260**) and DMPDPP (**261**), have been prepared and spectroscopically characterized by Kafafi's group [333]. Although the maximum absolute PL quantum yields of asym-TPP and DMPDPP were measured to be as low as 21% and 20%, respectively, at low concentrations in Alq$_3$, red OLEDs based on the fluorescent DMPDPP still yield an EQE approximately equal to 1%, which is close to the theoretical limit.

TP-5 (**262**) and EtTP-5 (**263**) were dioxolane-substituted pentacene derivatives reported by Wolak et al. [334]. They exhibit sharp red emission in solution with a peak at 625–630 nm and an FWHM of 21 nm. Remarkably, EtTP-5 shows a PL quantum yield of 72%, which is a 4-fold increase relative to 6,13-diaryl pentacene derivatives. Devices based on either TP-5 or EtTP-5 doped into Alq$_3$ with a structure of ITO/NPD (50 nm)/Alq$_3$:TP-5 or EtTP-5 (35 nm)/Alq$_3$ (35 nm)/Mg:Ag (15:1/120 nm) were fabricated and characterized. The optimized EtTP-5–based device with a doping concentration of 0.25 mol% exhibited a maximum EQE of 3.3%, which is close to the theoretical limit considering its PL quantum yield (3.8%).

Luo et al. demonstrated a new type of light-emitting layer based on codoping a pentacene derivative, PDT (**264**), and rubrene into Alq$_3$ [335]. Red OLEDs with structure of ITO/TOD (50 nm)/Alq$_3$:3 mol% PDT/*x* mol% rubrene (*y* nm)/BPhen (25 nm)/LiF (1 nm)/Al were fabricated and characterized. The optimized device with 60 nm EML and 1 mol% rubrene has a maximum current efficiency of 2.4 cd/A at 120 mA/cm² and high brightness of 2894 cd/m².

3.5.1.2.1.3 Metal Chelates It is well known that rare-earth complexes such as europium complexes emit sharp spectral bands due to electronic transitions between inner d and f orbitals of the central rare-earth metal ions and are expected to show high luminescent efficiency since both singlet and triplet excitons are involved in the luminescence process. A pure red light with a peak of 612 nm and a half bandwidth of 3 nm, was observed in the OLED employing Eu(Tmphen)(TTA)$_3$ (**265**) complex as a red dopant [336] (Scheme 3.46). The devices show a maximum luminance up to 800 cd/m², an EQE of 4.3%, a current

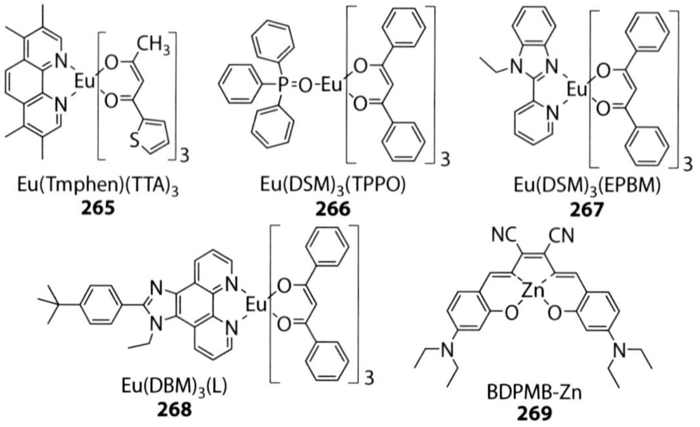

SCHEME 3.46
Chemical structures of red emitters of metal chelates.

efficiency of 4.7 cd/A, and a power efficiency of 1.6 lm/W. However, the EQE decreases markedly with increasing current due to the triplet–triplet annihilation in the device. Other europium complexes such as Eu(DBM)$_3$(TPPO) (**266**), Eu(DBM)$_3$(EPBM) (**267**), and Eu(DBM)$_3$(L) (**268**) were also investigated as red dopants in OLEDs [337]. Although rare-earth complexes have good color purity, their efficiency and chemical stability, thus far, fall short of the requirements for commercial applications.

Schiff bases with intramolecular charge-transfer complexes, such as 2,3-bis[(4-diethyl-amino-2-hydroxybenzylidene)amino]but-2-enedinitrile zinc (II) (BDPMB-Zn, **269**), emit red fluorescence with fluorescent quantum yields up to 67%. OLEDs with a structure of ITO/TPD/TPD:BDPMB-Zn/Alq$_3$:BDPMB-Zn/Alq$_3$/Mg-Ag showed very bright saturated red emission with CIE (0.67, 0.32) with a luminance of 2260 cd/m^2 at 20 V and a current efficiency of 0.46 cd/A at 20 mA/cm^2. In addition, the EL spectra do not change with the doping concentration in the range of 0.5–3% [338].

3.5.1.2.2 Green Fluorescent Dopants

3.5.1.2.2.1 Coumarin Dyes The first class of green dopants is the coumarin dyes (Scheme 3.47). Coumarin laser dyes such as (10-(2-benzothiazolyl)-1,1,7,7-tetramethyl-2,3,6,7-tetra-hydro-1*H*,5*H*,11*H*-[*l*]benzo-pyrano[6,7,8-*ij*]quinolizin-11-one) (C-545T, **270**), its bulky *t*-butyl group–substituted derivative C-545TB (**271**), and the methyl group–substituted C-545MT (**272**) have been widely investigated as green fluorescent dopants. These green couma-rin dopants generally have a high fluorescent QE in their dilute solutions (up to 90%). When these green dyes are doped in electroluminescent devices, the QEs of the devices are similarly high. To match the requirements of high color purity, stability, and high QE, substituted coumarins have been designed and synthesized. Chen et al. of Kodak patented a series of green dopants based on versions of the coumarin molecule [339]. For example, C-545 (**273**) shows a high fluorescent QE; however, it tends to aggregate in the solid state because of its nearly planar chemical structure. As a result, even 1% doping in an OLED device can cause an undesirable shift in the hue attributed to an emission shoulder at long wavelength [340]. The design of green dopant C-545T judiciously introduces the sterically hindered tetra-methyl groups at the julolidyl ring and, in doing so, dramatically reduces the dye quenching and keeps the desirable green color and high luminescent properties [341]. Further improvement of the coumarin dyes has led to the discovery of the thermally

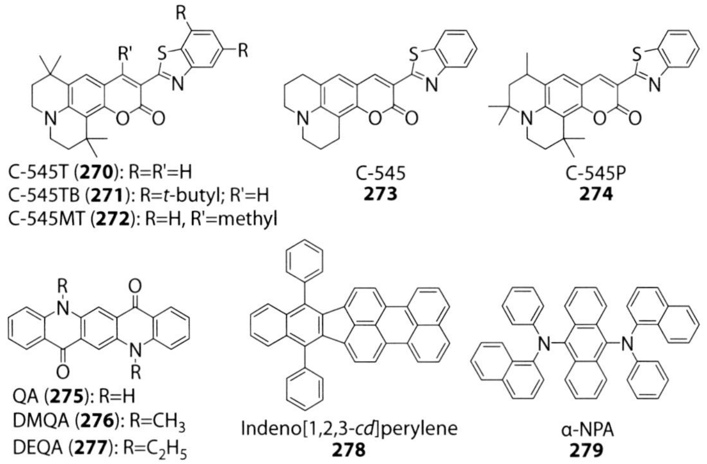

SCHEME 3.47
Chemical structures of green fluorescent emitters.

stable green dopants C-545MT and C-545TB [342]. Devices fabricated using C-545TB in the structure ITO/CHF$_3$ plasma/NPB/Alq$_3$:1% C-545TB/Alq$_3$/Mg:Ag gave a saturated green emission CIE (0.30, 0.64) with an output of 2585 cd/m^2, and a current efficiency of 12.9 cd/A with a power efficiency of 3.5 lm/W at driving current density of 20 mA/cm^2.

Chen's group has synthesized a new green dopant C-545P (**274**) by introducing asymmetric tetra-methyl steric spacers on the julolidyl ring [343]. C-545P has good thermal stability and photostability; when used as a dopant in an Alq$_3$-hosted OLED, it shows better device performance than that of C-545T. This is attributed to the asymmetric substituents, which minimize aggregation. The authors, however, did not compare the performance with C-545TB.

3.5.1.2.2.2 Quinacridones The second class of green dopants is the family of quinacridone (**275–277**) (Scheme 3.47) fluorescent dyes. Patents on derivatives of these compounds have been filed by Kodak for use in OLEDs [344].

Using DMQA as a green dopant, Shi and Tang of the Kodak group fabricated a green OLED with a device structure of ITO/CuPc/NPD/Alq$_3$:0.8% DMQA/Alq$_3$/Mg:Ag. A current efficiency of 7.3 cd/A and a maximum light output of 1462 cd/m^2 have been achieved with a half-life time of 7500 h [345]. They found that using quinacridone as a dopant, the device is very unstable. This is presumably due to the intermolecular hydrogen bonding in the unsubstituted quinacridone. The N–H moiety contributes to hydrogen bonding of neighboring quinacridone molecules, which favors quenching of the fluorescence. Wakimoto et al. also studied the stability of the OLED devices based on quinacridone compounds as dopants [346]. Their study supports that the steric hindrance invoked by the bulky substituents of quinacridone prevents excimer formation and prolongs the lifetime of the devices. For example, using DEQA as a green dopant in Alq$_3$, Murata et al. demonstrated efficient and thermally stable OLED performance [347]. In a paper published in the proceedings of a Society for Information Display conference in 2004, Wang et al. demonstrated a very high efficiency OLED using DMQA as a green dopant. Surprisingly, in their simple OLED structure, ITO/Teflon/Alq$_3$:0.7% DMQA/BAlq/Alq$_3$/Mg:Ag, they achieved a luminance of >88,000 cd/m^2 with an EQE of 5.4% (21.1 cd/A) at a voltage of 19.8 V and current density of 418 mA/cm^2 [348]. In a published patent

[349], Kodak disclosed a highly stable and longer-lifetime OLED using DMQA as the green dopant in a double host material (aminoanthracene and Alq$_3$) in the device structure.

3.5.1.2.2.3 Indeno[1,2,3-cd]Perylenes   The third class of green dopants is indeno[1,2,3-*cd*] perylene compounds. Kodak applied for a patent on a series of green dopants based on the [1,2,3-*cd*]perylene skeleton (**278**) (Scheme 3.47) [350–352]. They claim using the [1,2,3-*cd*] perylene dopant improves the current efficiency (7.4 cd/A) and stability (lifetime >2200 h at 70°C) compared with pure Alq$_3$ or Alq$_3$ doped with quinacridone green dopant devices. However, the color purity (λ_{max} 500–550 nm) is still an issue. Their device structure is ITO/CHF$_3$ plasma (1 nm)/NPD (75 nm)/TBADN (38 nm):0–2.5%-[1,2,3-*cd*]perylene/BAlq (10 nm)/Alq$_3$ (28 nm)/Mg:Ag (220 nm).

3.5.1.2.2.4 Diaminoanthracene Compounds   The fourth class of green emitters is the diami-noanthracene compounds (**279**) (Scheme 3.47). Very high-efficiency and bright green OLEDs have been fabricated with diaminoanthracene derivatives as either HTMs or both hole-transport and emitting materials [353]. The optimized devices emit narrow FWHM of 52-nm green light with a remarkable maximum EQE of 3.68%, a current efficiency of 14.79 cd/A, a power efficiency of 7.76 lm/W, and a maximum brightness of 64,991 cd/m^2. Tokito et al. also reported blue–green hole-transport emitting materials using related dibenzochrysene derivatives [354].

3.5.1.2.2.5 Other Heterocyclic Green Dopants   Other green dopants, **280–282** (Scheme 3.48), have been investigated; however, the efficiency is not high. Using a high doping level of ca. 16% of heterocyclic nitrogen-containing compound PAQ-NEt$_2$ as a dopant in the device ITO/NPD/NPD:16% PAQ-NEt$_2$/TPBI/Mg:Ag, Tao et al. fabricated a device that gave a sharp, bright, and efficient green EL peaked at 530 nm with an FWHM of 60 nm [355]. The maximum luminance is 37,000 cd/m^2 at 10.0 V with a maximum power efficiency of 4.2 lm/W, a current efficiency of 6.0 cd/A, and an EQE of 1.6%. Interestingly, whereas the PL of the doped film showed emission of the host NPD material even at the high doping concen-tration of 20%, the host emission was not observed in the EL spectrum. This is presumed to be due to charge-trapping processes, which occur in competition with the energy transfer process. The biologically active oxazolone compound GFP has also been investigated as a green dopant in OLEDs; however, the efficiency is very low [356].

   Adachi's group disclosed the photophysical and electroluminescent properties of a series of hetero-annulated π-conjugated dithieno[3,2-*b*:2′3′*d*]metallole derivatives **283–286** incor-porating Ge, Si, P, and S atoms as a bridging center (Scheme 3.49) [357]. Multilayer OLEDs have been fabricated with the following device configuration: ITO/α-NPD (40 nm)/3 wt.% dithienometallole:TBADN (30 nm)/BPhen (50 nm)/LiF (0.8 nm)/Al (80 nm). Among the devices, the SO$_2$–BT-based one shows the highest EQE of 6.1% with CIE coordinates of

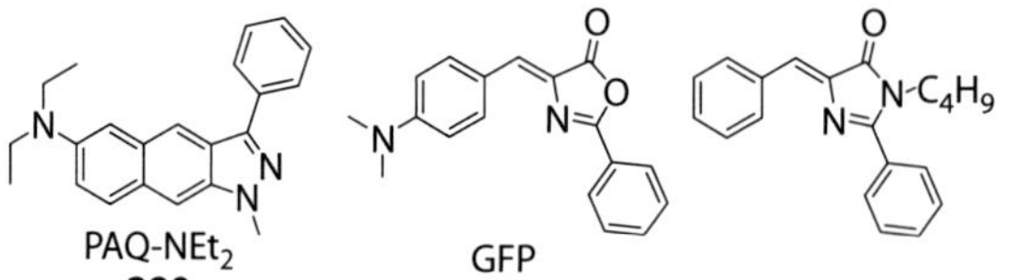

PAQ-NEt$_2$
280

GFP
281

282

SCHEME 3.48
Chemical structures of green fluorescent emitters.

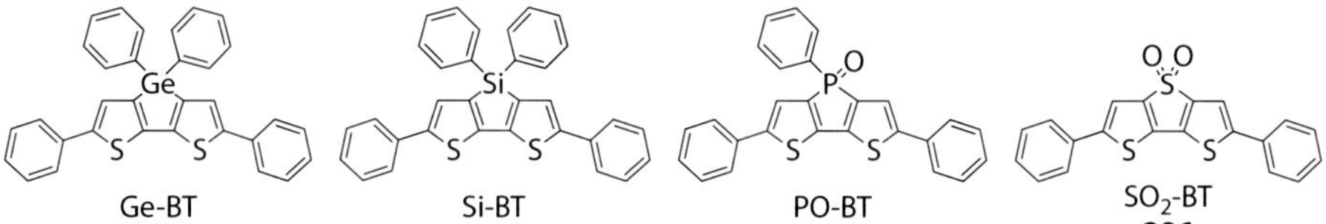

SCHEME 3.49
Chemical structures of green fluorescent emitters.

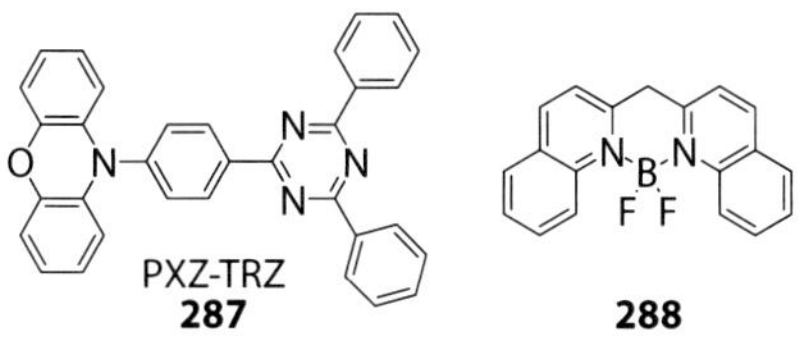

SCHEME 3.50
Chemical structures of green fluorescent emitters.

(0.28, 0.55). Notably, even when the luminance reaches 10,000 cd/m², the EQE of the SO₂–BT-based device still remains nearly 6%, indicating a balanced injection and transport of holes and electrons in the device at high current densities. They also proposed an efficient thermally activated delayed fluorescence emitter PXZ–TRZ (**287**) (Scheme 3.50) based on a phenoxazine (PXZ) electron donor unit and a 2,4,6-triphenyl-1,3,5-triazine (TRZ) electron acceptor unit [358]. An OLED containing PXZ–TRZ as an emitter layer was fabricated, exhibiting a maximum EQE of 12.5% with green emission.

3.5.1.2.2.6 Metal Chelates   The sixth class of green emitters is metal organic complexes. Alq₃ was the first green emitter. Alq₃ emission exhibits relatively saturated green color [CIE 1931 coordinates (0.32, 0.55)] and, thus far, it is still one of the best green emitters available.

Other metal complexes such as those of boron also shown in Scheme 3.50 (**288**) have also been reported to be viable green emitters.

3.5.1.2.3 Blue Fluorescent Dopants

For the blue pixel, the standard CIE 1931 color chromaticity coordinates are (0.14–0.16, 0.11–0.15). Since a relatively large band gap is required for blue emitters, the appropriate blue host materials with even larger band gap are needed to optimize the energy transfer requirements. The main challenge in designing the blue emitter or its host is the device stability.

Many large band-gap organic materials have been explored for blue emission. To summarize, they are the distyrylarylene (DSA) series, anthracenes, chrysenes, pyrenes, perylenes, fluorenes, heterocyclic compounds, and metal complexes.

3.5.1.2.3.1 Distyrylarylene Series   The most efficient fluorescent blue emitters yet reported belong to the DSA series (Scheme 3.51).

Hosokawa et al. first reported the use of a DSA host DPVBI (**221**) and amino-substituted DSAs such as BCzVB (**289**) and BCzVBI (**290**) as dopants in a blue OLED with an OLED structure of ITO/CuPc/TPD/DPVBI:BCzVB or BCzVBI/Alq₃/Mg:Ag [359]. Both amino-substituted DSAs

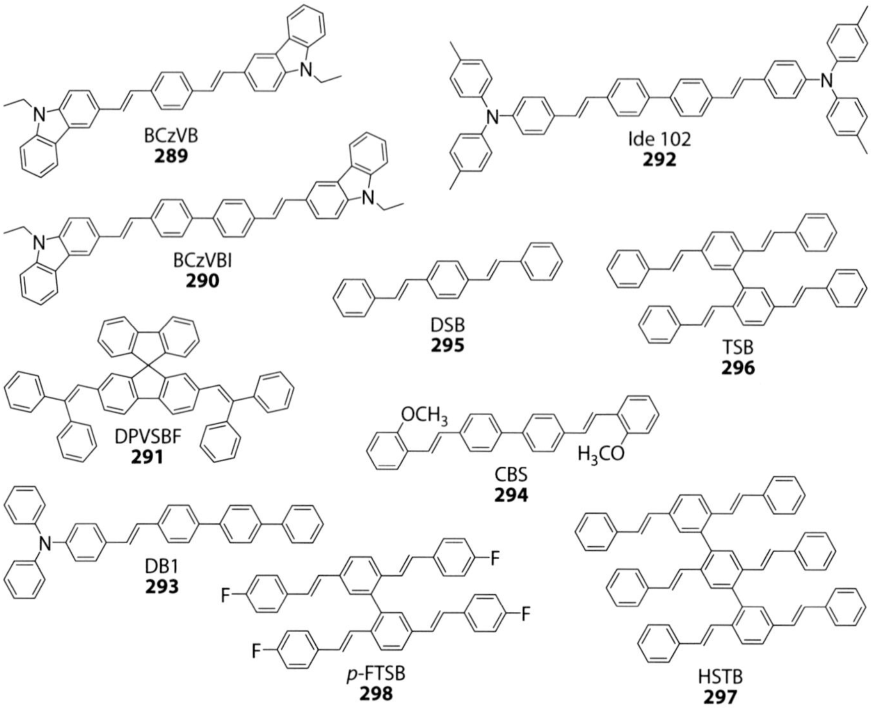

SCHEME 3.51
Chemical structures of DSA-based blue fluorescent emitters.

gave almost identical EL emission with a peak maximum centered at 468 nm with two shoulders at 445 and 510 nm. The highest power efficiency was observed in a BCzVBI-doped device that gave a power efficiency of 1.5 lm/W with an EQE of 2.4%. The highest luminance, >10,000 cd/m², in the blue region was obtained at 14 V. The efficiency is twice that of the nondoped device, which gives a power efficiency of 0.7–0.8 lm/W. The nonplanar host DPVBI has a blue emission with a band gap of 3.08 eV. The authors also pointed out that the EL characteristics are attributable to the dopant owing to the efficient energy-transfer mechanism and not to a charge-trapping mechanism. This material also has nice film-forming properties. The initial half-life time of the above device was measured to be 500 h at an initial luminance of 100 cd/m². Later, the same group used an improved HTL with a DSA host and a DSA–amine dopant that gave a half-life time of >5000 h.

Using the familiar dicarbazole material CBP as a host material and introducing BAlq as a hole-blocking material and LiQ as an EIL in the BCzVB-doped OLED device with a structure of ITO/TPD/CBP:BCzVB/BAlq/Alq₃/LiQ/Al, Wu et al. achieved a maximum luminance of 11,000 cd/m² with an EQE of 3.3% [360]. It is interesting to note that the emission spectra of CBP doped with BCzVB either in PL or in EL are rather different from the neat BCzVB- or DPVBI-doped BCzVB in PL or EL, with the former blue shifted and having better color purity. The EL emission of the CBP doped with BCzVB exhibits a dominant peak at 448 nm with an additional peak at 476 nm of FWHM of 60 nm, while the emission color corresponds to CIE (0.15, 0.16), where the neat BCzVB or DPVBI doped with BCzVB has CIE (0.16, 0.21). The authors explain such color differences between the CBP-doped and neat BCzVB or DPVBi-doped BCzVB devices to be due to the different excitation mechanisms of the two different guest–host doped systems. The excitation mechanism of the

CBP-doped-with-BCzVB device is attributed to both Förster energy transfer and carrier trapping, whereas the DPVBI-doped-with-BCzVB device is only a Förster energy transfer–dominated process, as mentioned by Hosokawa et al. This conclusion has been deduced from the HOMO and LUMO energy levels of the hosts and the dopant, and confirmed by an experimental study. The emission spectra of the neat BCzVB and DPVBi doped with BCzVB are essentially the same and can be attributed to an energy transfer process. The emission spectrum of CBP doped with BCzVB is rather different; it comes from emission contributed from both CBP and BCzVB molecules owing to both charge trapping as well as a partial energy transfer process.

The emitter DPVBI has a low T_g of ~64°C. This can be increased to a higher T_g by introducing more robust substituents. Tao's group designed a DPVBI analogue using a spirobifluorene unit instead of a biphenyl [361]. This rigid and orthogonally shaped spirobifluorene DPVSBF (**291**) exhibits a T_g of ~115°C. Improved performance has been observed both in blue OLEDs and in doped white OLEDs. In addition, the lifetime of DPVSBF-doped devices has been improved, which is also attributed to the higher morphological stability of the spirobifluorene DPVSBF host.

Suzuki et al. studied a series of different triaryl- or tetraaryl-benzene host materials and found that using triarylbenzene (TPB3) as a host material with a DSA–amine (Ide 102, **292**)-doped OLED showed a peak luminance of 142,000 cd/m² at 12 V. This device also showed a luminescent efficiency of 6.0 lm/W at 5 V and 820 cd/m², and an EQE of 2.4%. The lifetime of the device is also better than that of a device with DPVBI as the host [362].

Chen's group reported a deep-blue OLED based on an asymmetric mono(styryl) amine derivative DB1 (**293**). The PL spectrum of this deep-blue dopant in toluene solution showed a peak emission of 438 nm, which is about 20 nm hypsochromic shift compared with DSA–amine symmetric dopant, due to the shorter chromophoric conjugated length of the mono(styryl) amine. The OLED device based on this blue dopant achieved a very high efficiency of 5.4 cd/A, with CIE coordinates of (0.14, 0.13) [363].

By introducing methoxy groups, Huang et al. designed the more soluble DSA derivative CBS (**294**) [364], and this compound is easy to purify. In the triple-layer devices ITO/TPD/CBS/Alq₃ or Tb complex/Al, the multipeak emission with band peaks centered at 505 nm was observed when Alq₃ was used as the ETM, due to emission from the Alq₃ itself in the ETL. However, a blue emission was observed when the ETM was changed to a Tb complex.

Same as arylamine DSA derivatives used as guest materials in blue OLEDs, Geise et al. synthesized a series of alkyloxy-substituted biphenyl vinyl compounds as dopant materials [365]. These authors studied OLEDs fabricated using PVK as a host polymer and hole-transporting PBD as a codopant with a PLED structure of ITO/PEDOT:PSS/PVK:23OVBP:PBD/LiF/Al, which gave an optimized EQE of 0.7% and brightness of 1600 cd/m² at 100 mA/cm² with emission peak at 450 nm.

A notable design of oligo-phenyl vinyl blue emitters by Ma et al. is also shown in Scheme 3.51 (**295–298**) [366,367]. By linking two or three distyrylbenzene molecule units through a phenyl–phenyl bond, the dimers **296** and trimer **297** exhibit very interesting optical properties. In dilute toluene solution, distyrylbenzene molecule **295** shows deep-blue emission with maximum emission peaks at 403 and 420 nm, and its dimer and trimer show a slight red-shift emission (10–20 nm) compared with **295**. However, in the solid state, the emission spectrum properties are rather different, with both dimer and trimer blue shifted compared with **295**. The trimer emits a pure blue color with a maximum emission peak at 450 nm; the dimer emits at a maximum peak emission of 454 nm; and **295** red shifts itself about 54 nm from its dilute solution with a peak emission at 474 nm. Besides, the trimer also exhibits the highest solid-state photoluminescent efficiency. The authors explained that these are due to the weak intermolecular interactions of **297** in the solid states. The fluorine-substituted dimer **298** shows a slight blue-shift emission in the solid state compared with **295**. OLED fabricated using trimer as emitter emitted a blue light with CIE (0.19, 0.22) with a highest luminescent efficiency of 3.5 cd/A (1.6 lm/W) and a brightness of about 3837 cd/m^2.

3.5.1.2.3.2 Polyacene Series Besides DSA as a blue host or dopant in blue OLED application, anthracene derivatives with high QE and emission color in the blue range make them attractive materials.

In realizing the poor film-forming property of 9,10-(diphenyl)anthracene, the Kodak group improved this property by designing a series of blue emitters based on further substituted anthracene derivatives. The chemical structures of these materials were patented in a U.S. Patent in 1999 [368]. In their patent, Kodak also reported the EL data using one of these compounds as a host material and using 2,5,8,11-tetra-*tert*-butylperylene (TBP, **299**) as a blue dopant (Scheme 3.52). The device structure is ITO/CuPc/NPD/anthracene compound:0.5% TBP/Alq$_3$/Mg:Ag. The EL of the device showed blue emission with CIE color coordinates of (0.144, 0.196). Without the dopant, the EL emission with CIE color coordinates are (0.162, 0.107). No efficiency data or device lifetime data are reported.

Perylene (**300**) and its derivatives have been widely used as blue dopant materials owing to their excellent color purity and stability. Efficient blue emitters with excellent CIE coordinates are found in biaryl compound 2,2'-bistriphenylenyl (BTP, **301**) [243]. A device of structure ITO/TPD/BTP/TPBI/Mg:Ag emits bright blue emission with CIE (0.14, 0.11). A maximum brightness of 21,200 cd/m^2 at 13.5 V with a maximum EQE of 4.2% (4.0 cd/A) and a power efficiency of 2.5 lm/W have been achieved.

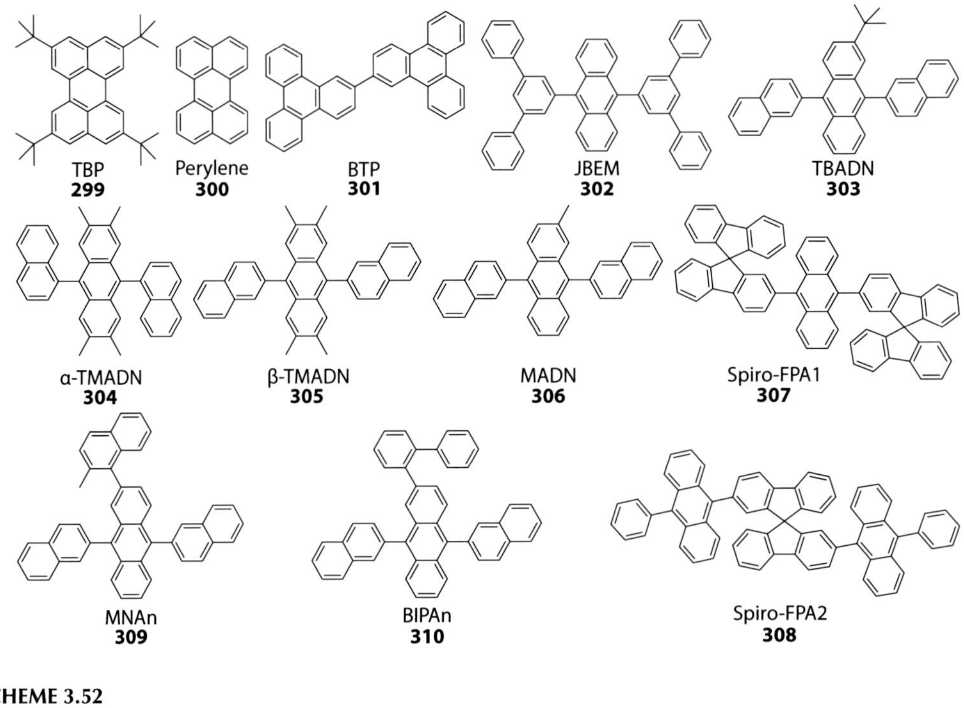

SCHEME 3.52
Chemical structures of polyacene-based blue fluorescent emitters.

Jiang et al. were the first to report a relatively stable blue OLED based on the anthracene derivative JBEM (**302**) [369]. With the similar OLED structure as that used above by Kodak of ITO/CuPc/NPD/JBEM:perylene/Alq$_3$/Mg:Ag and using JBEM as a blue host material, the device shows a maximum luminance of 7526 cd/m^2 and a luminance of 408 cd/m^2 at a current density of 20 mA/cm^2. The maximum power efficiency is 1.45 lm/W with CIE (0.14, 0.21). A half-life of >1000 h at an initial luminance of 100 cd/m^2 has been achieved. The authors also compared the device performance using DPVBI as a host, which gave them a less stable device.

Several groups have studied naphthalene-substituted anthracene derivatives as hosts or emitter materials in blue OLEDs. The Kodak group used ADN (**220**) as a host and TBP as a dopant in ITO/CuPc/NPD/ADN:TBP/Alq$_3$/Mg:Ag [370]. They achieved a narrow vibronic emission centered at 465 nm with CIE (0.154, 0.232) and a luminescent efficiency as high as 3.5 cd/A. In comparison, the nondoped device shows a broad and featureless bluish–green emission centered at 460 nm with CIE (0.197, 0.257) and an EL efficiency <2.0 cd/A. The operational lifetimes of the doped device and the nondoped device were 4000 and 2000 h at an initial luminance of 636 and 384 cd/m^2, respectively.

Normally, using ADN as an EML inserted between the HTL and ETL in the multilayer device results in broad emission as was observed in the above Kodak device. This is due to emission from the ETL material Alq$_3$. Increasing the thickness of the ADN layer cannot completely avoid this undesired emission, and, in addition, the operating voltage is high and the device is not stable. Hung et al., by introducing an additional HBL and confining the exciton emission in the ADN layer, achieved a pure blue emission color with CIE (0.1566, 0.1076). The device structure is ITO/NPD/ADN/BCP/Alq$_3$/Mg:Ag. However, the drawback with this device is the operating voltage that is high owing to the high energy barrier between the BCP and the Alq$_3$ layers. By replacing both BCP and Alq$_3$ with a combined high-electron-mobility ETM and hole-blocking material TPBI, and by using the doped perylene device structure ITO/NPD/ADN:0.5% perylene/TPBI/LiF/Mg:Ag, a low driving voltage and a pure blue emission device is achieved. The device exhibits a luminescent efficiency of 3.6 cd/A with CIE (0.1451, 0.1446). It is envisioned that, if an HIL is incorporated into such a device, the efficiency will be improved further. Using this device structure but employing an unsymmetrical anthracene derivative, TBADN (**303**), as the blue emitter in a nondoped device, Tao et al. achieved a blue device with excellent color purity with CIE (0.14, 0.10) and a maximum current efficiency of 2.6 cd/A [371]. The EL emission of the device is also stable compared with ADN devices when operated at high current. The authors attributed the color purity of the TBADN device to the bulky *tert*-butyl group on ADN, which prevents charged complex formation at the interface.

Qiu's group modified the ADN structure and synthesized α-TMADN (**304**) and β-TMADN (**305**) that contain substituted *tetra*-methyl anthracene [372]. This sterically congested nonplanar structure hinders the close spacing of the molecules and improves the film-forming properties. The OLED device structure is ITO/NPD/TMADN/BPhen/Mg:Ag. The device with β-TMADN as the EM exhibits blue emission with CIE (0.163, 0.221). A high EQE of 2.8% and a brightness up to 10,000 cd/m^2 at 12 V with the maximum efficiency of 4.5 cd/A (2.51 lm/W) have been achieved. The performance of the device using a blend of the two molecules α-TMADN and β-TMADN (9:1) demonstrates even better results with a brightness >12,000 cd/m^2 at 15 V and a maximum efficiency of 5.2 cd/A (2.72 lm/W) and color coordinates of (0.152, 0.229) with emission color centered at 466 nm.

Chen's group designed MADN (**306**) through introduction of one methyl substituent at the C-2 position of the anthracene ring. This material can form nice films without the problem of crystallization while keeping the HOMO and LUMO energy of ADN. The efficiency

(1.4 cd/A) of MADN-based blue OLEDs with CIE (0.15, 0.10) is slightly lower than the equivalent ADN device. Later, using a styrylamine as a dopant, and an optimized device structure (ITO/CF$_x$/NPD/MADN:DSA-Ph/Alq$_3$/LiF/Al), Lee et al. achieved a very high-efficiency OLED with an EL efficiency of 9.7 cd/A at 20 mA/cm^2 and 5.7 V; while the lifetime of such a device is projected to have a half-decay lifetime of 46,000 h at an initial brightness of 100 cd/m^2, the color purity is sacrificed with an emission peak shift from 452 nm with CIE (0.15, 0.10) to a major peak at 464 nm with a shoulder centered at 490 nm with CIE (0.16, 0.32) [373].

Other derivatives have been reported, such as the spiro-linked fluorene-anthracenes, which preserve the optical and electrochemical properties of anthracene while reducing the tendency for crystallization and enhancing the solubility and T_g. Highly efficient deep blue OLEDs have been demonstrated by using spiro-FPA1 (**307**) as an EM in a p–i–n type OLED structure: ITO/MeO-TPD:2% F4-TCNQ/spiro-TAD/spiro-FPA1/TPBI/BPhen:Cs/Al. A very low operating voltage of 3.4 V at luminance of 1000 cd/m^2 was obtained, which is the lowest value reported for either small-molecule or polymer blue electroluminescent devices. Pure blue color with CIE coordinates (0.14, 0.14) has been measured with very high current (4.5 cd/A) and QEs (3.0% at 100 cd/m^2 at 3.15 V) [374]. In another paper, spiro-FPA2 (**308**) was used as a host material with an OLED device structure of ITO/CuPc/NPD/spiro-FPA2:1% TBP/Alq$_3$/LiF that produces a high luminescent efficiency of 4.9 cd/A [375].

By introducing the hole-transport arylamine as an end cap for an anthracene backbone, Danel et al. designed a series of novel materials [376]. The aim of these dual-function materials is to combine the emitting property of the blue anthracene luminophore with the hole-transporting property of the triarylamine to simplify the device fabrication steps. Although the introduction of the arylamino moieties produces moderate QE ($\Phi_f \sim 20\%$) for these materials, the OLEDs using them as emitters as well as HTMs demonstrate only moderate EL performance with a maximum luminance of 12,922 cd/m^2 and 1.8 lm/W with CIE (0.15, 0.15).

Shin et al. developed two ADN derivatives, MNAn (**309**) and BIPAn (**310**), by introducing highly twisted *ortho*-methylated naphthyl or *ortho*-phenylated phenyl groups at the 2 position of ADN. It has been found that these derivatives have good thermal stability and pure blue EL. A multilayered device using MNAn as a doping emitting material exhibits a maximum EQE of 3.61% (3.55 cd/A) and CIE coordinates of (0.15, 0.13). Similar device performance could be obtained by using BIPAn as a doping emitting material [377].

Karatsu et al. developed a family of anthracene derivatives by substituting silyl or silylethynyl and arylsilyl groups into the 9,10 position of anthracene. They found that substitution of the bulky substituents, such as the triphenylsilyl groups, at the 9,10 positions of anthracene could remarkably increase the fluorescent quantum yield. The substituents may also control the intermolecular stacking by diminishing the electroplex formation and inter- and intramolecular photochemical reactions. OLEDs composed of these molecules as dopants in CBP showed that pure blue EL was obtained [378].

3.5.1.2.3.3 Heterocyclic Compounds Pyrazole-containing compounds are useful blue dyes. Tao's group studied a series of pyrazolo [3,4-*b*]quinoline derivatives as blue emitters (Scheme 3.53) [379]. An EQE about 3.5% using one of the dopants 4-(*p*-methoxyphenyl)-3-methyl-1-phenyl-1*H*-pyrazolo[3,4-*b*]quinoline (MeOPAQ-H) (**311**) in an OLED with a configuration of ITO/NPD/CBP/TPBI:MeOPAQ-H/TPBI/Mg:Ag was achieved. A brightness of 13,000 cd/m^2 with a blue emission CIE (0.19, 0.16) was demonstrated. Other pyrazole derivatives such as Pyzo-1 (**312**) and Pyzo-2 (**313**) also emitted blue light; however, the EL performance is rather poor [380]. Phenanthrolines, e.g., BCP, widely used as hole-blocking materials, were used as blue-emitting materials in the late 1990s [381,382]. The pure blue emission in this case may come from the HTM NPD emission.

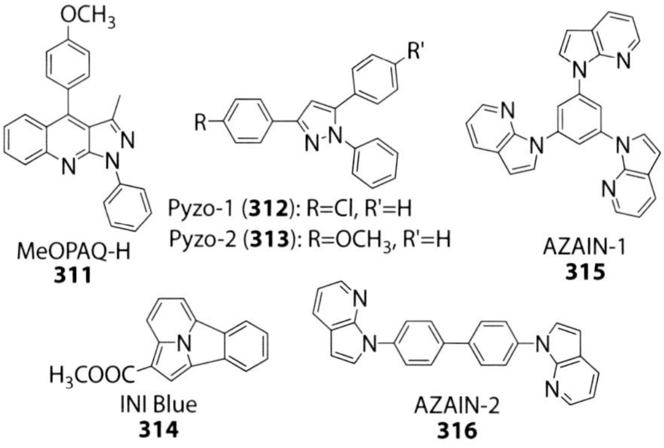

SCHEME 3.53
Chemical structures of heterocyclic blue fluorescent emitters.

Wudl's group synthesized a new indolizino[3,4,5-*ab*]isoindole derivative INI Blue (**314**) as a blue guest material for OLEDs [383]. An OLED with the structure of ITO/CuPc/CBP/ INI Blue/Salq/Alq$_3$/MgF2 generates a very bright blue emission with CIE (0.165, 0.188). The maximum luminance is 5674 cd/m² with a power efficiency of 0.94 lm/W and current efficiency of 2.7 cd/A. The device was claimed to be rather stable; however, the lifetime has not been reported.

Wang's group systematically studied a series of blue emitters based on 7-azaindole compounds [384]. Interestingly, blue emitters such as 1,3,5-tri(*N*-7-azaindolyl)-benzene (AZAIN-1, **315**) and 4,4′-di(*N*-7-azaindolyl)-biphenyl (AZAIN-2, **316**), when used as the emitting layers in OLEDs, can function as both hole-transporter and emitter materials. The presence of an HTL such as NPD or TPD in these OLED devices decreases the luminescent efficiency. Devices fabricated with the structure ITO/CuPc/AZAIN-1 or 2/PBD/ LiF/Al emitted deep-blue emission.

3.5.1.2.3.4 Spiro-Linked Blue Emitters Spiro-linked organic glasses, which improve the processability and morphology of the films while keeping the electronic properties intact, are very interesting materials for OLEDs. Salbeck et al. designed a series of spiro-linkage fluorene compounds with electron-donating and electron-withdrawing groups (Scheme 3.54) [136,385].

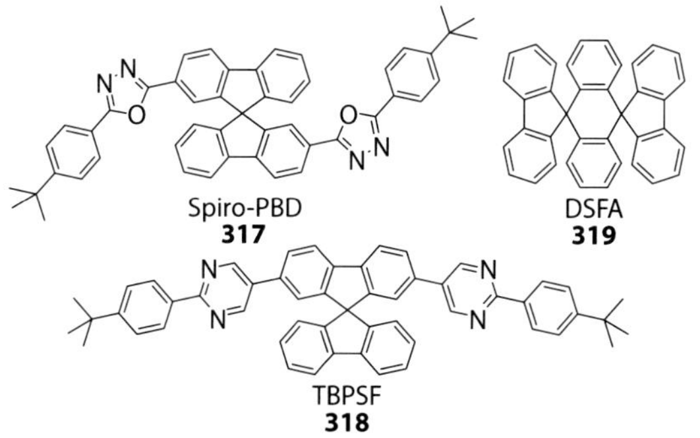

SCHEME 3.54
Chemical structures of spiro-linked blue fluorescent emitters.

A blue OLED with a very low turn-on voltage of 2.7 V and a luminance of 500 cd/m² at 5 V with the structure ITO/spiro-TAD/spiro-PBD (**317**)/Al:Mg has been reported. The robust and morphologically stable spirobifluorene-cored pyrimidine oligoaryl blue emitter 2,7-bis[2-(4-*tert*-butylphenyl)pyrimidine-5yl]-9,9′-spirobifluorene (TBPSF, **318**) with very high PL QE (80%) has been reported by Wu et al. [386]. When using this material as an emitter or a host layer in ITO/PEDOT:PSS/NCB/TBPSF:0–1% perylene/Alq₃/LiF/Al, the nondoped and doped devices exhibit unusual longevity even to currents up to 5000 mA/cm², with maximal brightness >30,000 and ~80,000 cd/m² for nondoped and doped devices, respectively. The EQEs of nondoped and doped devices are 2.3% (1.6 cd/A) and 4% (5.2 cd/A), respectively.

Qiu's group investigated the spirofluorene linked dihydroanthracene compound di-spiro-9,9′-di-fluorene-9″,9‴-(9,10-dihydro-anthracene) (DSFA, **319**), originally developed in the 1930s, as a blue emitter in ITO/*m*-MTDATA/NPD/DSFA/Mg:Ag [387]. The device exhibited a pure blue emission CIE (0.17, 0.15) at a low turn-on voltage of 6 V with a maximum power efficiency of 4.73 lm/W (15.1 cd/A).

3.5.1.2.3.5 Organosilicon Blue Emitters Aromatic chromophores consisting of linked organosilicon compounds possess excellent charge-transport properties due to the interaction between the Si 3d orbitals and the ligands. These organosilicon compounds have been proven to be good EMs owing to their excellent film-forming properties and their high T_g. Chan et al. reported an efficient blue emitter based on the tetraphenylsilane compound shown in Scheme 3.55 [388].

OLED ITO/NPD/Ph₃Si(PhTPAOXD) (**320**)/Alq₃/Mg:Ag emits pure blue light with an EL emission band centered at 460 nm (FWHM: 75 nm) and CIE (0.17, 0.17). The maximum luminance exceeds 20,000 cd/m² at 15 V, with an EQE of 1.7% and a power efficiency of 0.9 lm/W. Later, the same group optimized the device structures by introducing an HTL of an organosilicon compound and achieved a much higher performance [389]. An optimized OLED with a structure of ITO/Ph₂Si(Ph(NPA)₂)₂ (**321**)/NPD/Ph₃Si(PhTPAOXD)/ Alq₃/Mg:Ag possesses good stability and color purity with EL emission maximum at 460 nm, corresponding to CIE (0.16, 0.18). The device shows a maximum luminance of 19,000 cd/m² and an EQE of 2.4% (3.1 cd/A) with a power efficiency of 1.1 lm/W.

Compound 2PSP (**322**) exhibits a bright blue–greenish emission with a solid-state PL QE of 97%. OLEDs using 2PSP as the EML showed a very low operating voltage, an EQE of 4.8%, and luminous power efficiency of 9 lm/W at a brightness of 100 cd/m². This value is

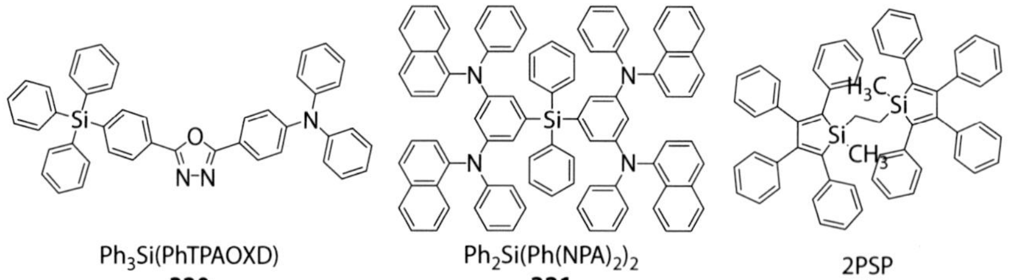

Ph₃Si(PhTPAOXD)
320

Ph₂Si(Ph(NPA)₂)₂
321

2PSP
322

SCHEME 3.55
Chemical structures of organosilicon blue fluorescent emitters.

the best efficiency achieved for SMOLED using nondoped emissive and carrier-transport layers, and is close to the theoretical limit for a device using a fluorescent emitter [273].

3.5.1.2.3.6 Metal Chelates Owing to its outstanding stability and luminescent efficiency, Alq_3 is the most widely used and studied emitting material. By structural modifications of Alq_3, several blue emitters based on Alq_3 structures have been investigated, and Kodak has patented the basic structures of several possible blue emitters based on Alq_3 [225, 390]. The first blue aluminum chelate structure contains a phenolato ligand and two R_2-8-quinolinolato ligands. The second blue aluminum chelate structure comprises a bis(R_2-8-quinolinolato) aluminum(III)-μ-oxo-bis(R_2-8-quinolinolato)-aluminum(III).

Yu et al. synthesized two methyl–substituted Alq_3, named tris(2,3-dimethyl-8-hydroxy-quinoline) aluminum complex ($Alm_{23}q_3$, **323**) (Scheme 3.56) [391]. This compound emits blue color with an emission peak centered at 470 nm and FWHM of 90 nm. OLEDs with a structure of ITO/TPD/$Alm_{23}q_3$/Mg:Ag emit blue light, and the luminous efficiency is 0.62 lm/W with a maximum luminance of 5400 cd/m^2 at 19 V.

Bis(2-methyl-8-quinolinolato)aluminum hydroxide ($AlMq_2OH$, **324**) with only two quinolate ligands emits blue color with the maximum peak emission at 485 nm and FWHM of 80 nm [392]. Devices fabricated with a structure of ITO/CuPc/NPD/$AlMq_2OH$/LiF/Al give a maximum brightness of 14,000 cd/m^2 at 480 mA/cm^2.

Beryllium chelates such as bis[2-(2-hydroxyphenyl)-pyridine]beryllium ($Bepp_2$, **144**) emit pure blue light with an emission peak centered at 465 nm [393]. ITO/NPD/$Bepp_2$/LiF/Al exhibited a maximum luminance of 15,000 cd/m^2 and a maximum luminescent efficiency of 3.43 lm/W (3.8 cd/A). The emission color may have contributions from both NPD and $Bepp_2$, as stated by the authors.

Other metal chelate materials of ligands such as benzimidazole have been disclosed in a Kodak patent as their beryllium or aluminum complexes [394]. These materials generate very pure blue color within CIE range of (0.15–0.16, 0.12–0.17) [239,395].

3.5.1.2.3.7 Other Blue Emitters Some aromatic amine HTMs [396], like α-NPD (**24**), PPD (**40**), and tri(p-terphenyl-4-yl)amine (p-TTA, **325**) (Scheme 3.57) [397], can also emit efficient blue light if used with an appropriate HBL; however, their performance is generally very poor.

5,5′-Bis(dimesitylboryl)-2,2′-bithiophene (BMB-2T, **326**) forms a stable amorphous glass and emits pure blue color with a high fluorescence QE of 86% in THF solution [398]. However, an OLED with ITO/m-MTDATA/TPD/BMB-2T/Mg:Ag emits with a broad emission due to an exciplex with TPD. The exciplex can be prevented by insertion of a thin layer of 1,3,5-tris(biphenyl-4-yl)benzene (TBB) between TPD and BMB-2T, leading to a pure blue

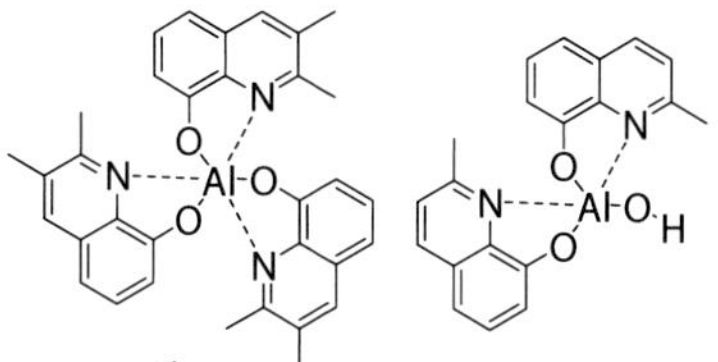

$Alm_{23}q_3$
323

$AlMq_2OH$
324

SCHEME 3.56
Chemical structures of metal chelate blue fluorescent emitters.

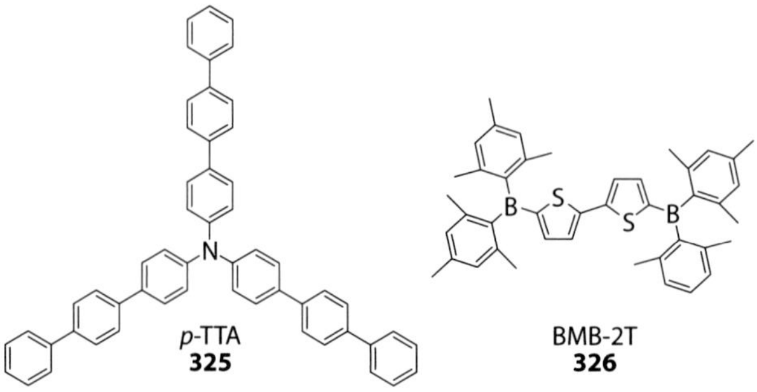

SCHEME 3.57
Chemical structures of blue fluorescent emitters.

emission. It seems that the boron complex or boron-containing compounds easily form an exciplex with common HTMs. Other similar blue emitter materials also demonstrate such behavior.

3.5.2 Phosphorescent Light-Emitting Materials

3.5.2.1 Host Materials for Phosphorescent Emitters

3.5.2.1.1 Red Phosphorescent Hosts

BAlq (**141**), one of the very good hole-blocking materials, has recently been used as a host material for phosphorescent OLEDs [399]. BAlq has HOMO and LUMO energy levels of −5.90 and −3.0 eV, respectively [400]. Its triplet energy level has not been reported. A theory modeling study of this molecule predicted the triplet energy is about 2.2 eV, which is 0.2 eV higher than that of Alq_3 [401]. This property of BAlq makes it suitable as a host material for red phosphorescent dopants, such as bis(2-(2′-benzo[4,5-*a*]thienyl)pyridinato-*N*,C3′) iridium (acetylacetonate) [(Btp)$_2$Iracac], which has a triplet energy level of 2.02 eV [402]. In fact, long lifetime and high efficiency of red phosphorescent OLEDs in such an architecture have been reported.

Zn(BTP)$_2$ [the same structure as Zn(BTZ)$_2$, **143**] is another electron-transport host material in red phosphorescent OLEDs [403]. Zn(BTP)$_2$ has HOMO and LUMO energy levels of −5.41 and −2.65 eV, respectively, and its triplet energy is 2.50 eV, making it suitable as a host material for red phosphorescent dopants. A low driving voltage of 3.2 V, a power efficiency of 8.8 lm/W, and a current efficiency of 8.6 cd/A were obtained for the device with the structure of ITO/NPB (70 nm)/Zn(BTP)$_2$:5% Ir(piq)$_3$ (50 nm)/DBzA (20 nm)/LiF (1 nm)/Al (100 nm).

Tsuzuki and Tokito reported red OLEDs using a green light–emitting iridium complex (ppy)$_2$Ir(acac) (**327**) (Scheme 3.58) as the host and Ir(piq)$_3$ as a red phosphorescent guest in a structure of ITO/PEDOT:PSS (35 nm)/NPD (40 nm)/(ppy)$_2$Ir(acac):x% Ir(piq)$_3$ (35 nm)/BCP (10 nm)/Alq$_3$ (35 nm)/LiF (0.5 nm)/Al (100 nm) [404]. The triplet energy of (ppy)$_2$Ir(acac) was estimated to be 2.3 eV, and it is higher than that of Ir(piq)$_3$ (2.0 eV). A maximum EQE of 9.2% and a maximum power efficiency of 11 lm/W were obtained when the concentration of Ir(piq)$_3$ was only 0.3 wt.%, and driving voltage was much lower than the CBP:Ir(piq)$_3$ devices. The reduced driving voltage was attributed to the 0.7 eV narrower E_g of (ppy)$_2$Ir(acac) than CBP.

Besides metal chelates, organic compounds containing carbazole, phosphine oxide, and heterocycles are also mostly investigated as host materials for red phosphorescent emitters.

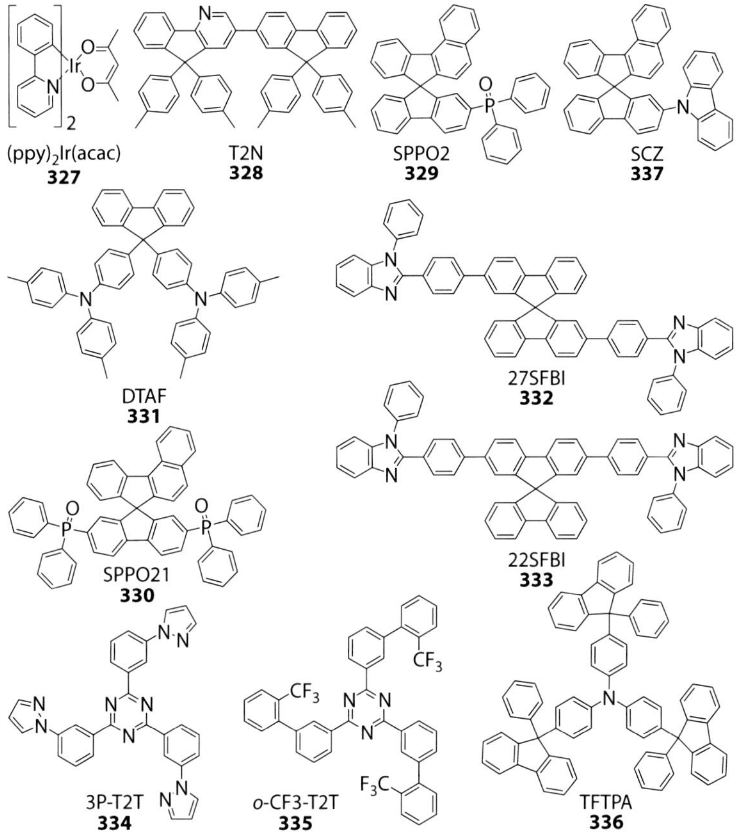

SCHEME 3.58
Chemical structures of host materials for red phosphorescent emitters.

These host materials can be categorized as electron-transport host, hole-transport host, and bipolar host, according to their building blocks and carrier-transport properties.

T2N (**328**), which contains a pyridyl moiety, can serve as both a host and an efficient electron-transporting material blended with various heavy metal–containing red (Ir, Ru, Os, and Pt) phosphors for highly efficient phosphorescent OLEDs [405]. The HOMO and LUMO energy levels of T2N are –5.75 and –2.5 eV, respectively, and its triplet energy is 2.45 eV. Devices with architectures of ITO/PEDOT:PSS (30 nm)/TCTA (40 nm)/T2N:dopant (25 nm)/T2N (35 nm)/LiF (0.5 nm)/Al (100 nm) were prepared. When using 9 wt.% $Os(fptz)_2(PPhMe_2)_2$ as the dopant, the highest EQE (15.5%) and highest brightness of saturated red emission (29,700 cd/m^2) with CIE coordinates (0.68, 0.32) were obtained.

Phosphine oxide derivative is another kind of electron-transport host, like SPPO2 (**329**) [406], which possesses low-lying HOMO (–6.0 eV) and LUMO (–2.4 eV) energy levels. Its triplet energy (2.4 eV) is suitable for triplet energy transfer to red-emitting phosphors such as $Ir(piq)_2acac$ with a triplet energy of 2.2 eV. Red OLEDs fabricated using SPPO2 as the host material in the device structure of ITO/PEDOT:PSS (60 nm)/NPB (20 nm)/TCTA (10 nm)/SPPO2:10 wt.% $Ir(pq)_2acac$ (30 nm)/BCP (5 nm)/Alq$_3$ (20 nm)/LiF (1 nm)/Al (200 nm) gave a high EQE of 14.3% and a current efficiency of 20.4 cd/A. The EML where SPPO2 serves as the host material also showed a smooth surface morphology with an average surface roughness <1 nm and a high T_g of 119°C, indicating SPPO2 is a promising host for red and deep-red phosphorescent devices. Its analogue, SPPO21 (**330**), developed by Lee's group has a triplet energy level of 2.32 eV, a LUMO energy level of 2.57 eV, and a HOMO energy level of 6.12 eV [407]. LiF-free devices with a configuration of ITO/PEDOT:PSS (60 nm)/NPB (30 nm)/SPPO21:10% Ir(pq)$_2$acac (30 nm)/SPPO21 or Alq$_3$ (25 nm)/Al (200 nm) were fabricated. Although there is an obvious energy barrier (0.43 eV) between Al and SPPO21, the coordination between phosphine oxide group and the metal surface and the resultant charge transfer complex may lead to a much more efficient electron injection than that from the Al cathode to Alq$_3$. It is not difficult to understand that the device using SPPO21 as an ETL showed much higher efficiencies than the Alq3 counterpart. It is worth noting that the device did not have any separate electron-transport, and LiF layer with a configuration of ITO/PEDOT:PSS (60 nm)/NPB (30 nm)/SPPO21:10% Ir(pq)$_2$acac (55 nm)/Al (200 nm) still gave a satisfactory performance with a maximum EQE of 10.5% and a maximum power efficiency of 10.2 lm/W. In a further study, single layer OLED using Ir(pq)$_2$acac as the red phosphorescent emitter and a cohost system of SPPO21 (n-type) and NPB (p-type) were developed [408]. Electrons could be directly injected from the cathode through the SPPO21 layer and the device structure of the PHOLEDs could be simplified without sacrificing the device performances by using the phosphine oxide–type SPPO21 host material.

A new hole-transporting host, DTAF (**331**), and electron-transporting hosts, 27SFBI (**332**) and 22SFBI (**333**), were reported by Kao et al. [409]. Because the orthogonal configuration leads to intermolecular spatial hindrance, and 22SFBI shows significantly less electron mobility than 27SFBI. Thus, 27SFBI was chosen as an electron-transporting host together with DTAF (hole-transporting host) to fabricate a doubly doped red phosphorescent OLED. Owing to the different charge-transport behavior and the mismatched HOMO/LUMO energy levels of DTAF and 22SFBI, the charge carriers could accumulate at the hole-transport and electron-transport interfacial regions. A device with configuration of ITO/PEDOT:PSS/DTAF (15 nm)/DTAF:6 wt.% Os1 (5 nm)/27SFBI:6 wt.% Os1 (20 nm)/27SFBI (55 nm)/LiF/Al exhibited a maximum brightness of 32,700 cd/m^2, a maximum power efficiency of 16.5 lm/W, and an EQE of 10.9% at 1000 cd/m^2.

A series of electron-transport host materials of T2T (the same structure as **191**), 3P-T2T (**334**), 3N-T2T (the same structure as **192b**), and *o*-CF3-T2T (**335**), were developed by Chen et al. [410]. The electron-transport ability of these materials was investigated in electron-only devices, and the results indicate that the introduction of N-heterocyclic polar peripheries, pyridyl or pyrazolyl groups, is advantageous to the electron-transport properties of the bulk material over electron-withdrawing groups such as CF$_3$. The best efficiencies (EQE of 16.4%, power efficiency of 20.0 lm/W, and current efficiency of 15.8 cd/A) were obtained for the 3P-T2T–based device with a configuration of ITO/PEDOT:PSS (30 nm)/NPB (20 nm)/TCTA (5 nm)/3P-T2T:10 wt.% (Mpq)$_2$Ir(acac) (25 nm)/3P-T2T (50 nm)/LiF (0.5 nm)/Al (100 nm).

Shu's group developed a new hole-transport host of tris(4-(9-phenylfluoren-9-yl)pheny-l) amine (TFTPA, **336**) [411]. Introduction of the sterically bulky 9-phenyl-9-fluorenyl groups gives a high T_g of 186°C and electrochemical stability. OLEDs with typical multilayer architecture of ITO/NPB (30 nm)/TFTPA:x wt.% of dopant (40 nm)/TPBI (40 nm)/Mg:Ag (100 nm)/Ag (100 nm) were fabricated. Devices using Ir(piq)$_2$acac as a red phosphorescent dopant with doping concentration of 21 wt.% showed the best performance with a low turn-on voltage of 2.6 V. The maximum values of EQE and power efficiency reached 9.6% and 9.0 lm/W, respectively; the devices exhibited a much reduced efficiency roll-off at high brightness.

SCZ (**337**) is a carbazole-type hole-transport host material. To investigate the electroluminescent properties of SCZ as a host [412], red phosphorescent OLEDs were prepared with the structure of ITO/DNTPD (60 nm)/NPB (20 nm)/SCZ:*x* wt.% Ir(pq)$_2$acac (30 nm)/ BCP (5 nm)/Alq$_3$ (20 nm)/LiF (1 nm)/Al (200 nm). The doping concentration was optimized to be 1 wt.% and the device showed <10% decrease of EQE at 10,000 cd/m^2 from the maximum EQE.

POAPF (**338**) (Scheme 3.59) containing both electron-transport phosphine oxide and hole-transport TPA is an efficient bipolar host material for osmium phosphor Os(fptz)$_2$(PPh$_2$Me)$_2$ [413]; it has HOMO and LUMO energy levels of −5.24 and −2.48 eV, respectively, and triplet energy (*E*$_T$) of 2.72 eV. A maximum EQE of 18.6% and a power efficiency of 34.5 lm/W were achieved by using POAPF as a host, and both of these data were obviously higher than the CBP-based device with the same configuration of ITO/TPD (30 nm)/host:7 wt.% Os(fptz)$_2$(PPh$_2$Me)$_2$ (30 nm)/BPhen (30 nm)/LiF (1.5 nm)/Al (100 nm).

Replacement of the two diphenylphosphoryl groups with phenylsulfone groups yields a new host material, SAF (**339**) [414]. Red OLED was fabricated in a configuration of ITO/ TPD (30 nm)/SAF:7 wt.% Ir(piq)$_3$ (30 nm)/BPhen (30 nm)/LiF (1.5 nm)/Al (100 nm), giving a maximum EQE of 15.8% and current efficiency of 19.6 cd/A. These values are much higher than that of the device using CBP as a host material. The superior performance of the SAF-based device can be attributed to the extremely balanced bipolar transport ability of SAF [3.4 × 10^{-5} cm^2/(V s) for electron and 3.5 × 10^{-5} cm^2/(V s) for hole] measured by TOF transient photocurrent techniques. The HOMO and LUMO energy levels of SAF are −5.29 and −2.88 eV, respectively, which are suitable for the injection of electrons and holes. In addition, its triplet energy (2.71 eV) is very high to avoid reverse energy transfer from the triplet dopant to the host.

Su et al. have designed and synthesized a series of bicarbazole-substituted bipolar molecules (**340**) containing various heterocyclic cores, such as pyridine, pyrimidine, and pyrazine (Scheme 3.60) [415]. The energy levels of these materials can be tuned by the

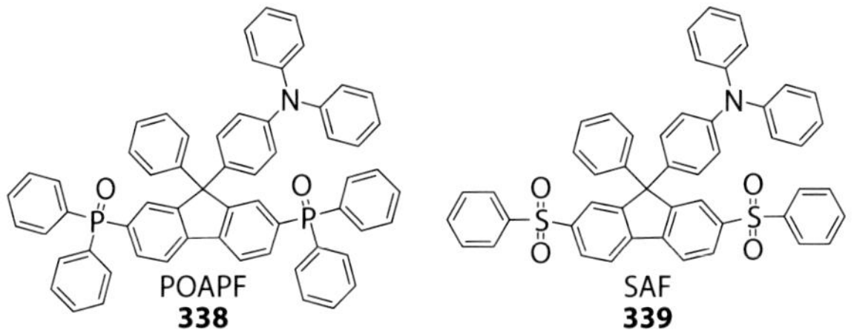

POAPF
338

SAF
339

SCHEME 3.59
Chemical structures of bipolar host materials for red phosphorescent emitters.

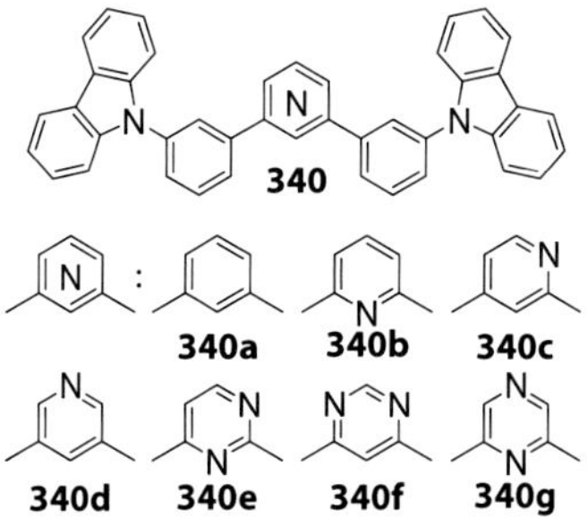

340

340a   **340b**   **340c**

340d   **340e**   **340f**   **340g**

SCHEME 3.60
Chemical structures of bicarbazole-substituted bipolar host materials.

change of heterocyclic cores and their nitrogen atom orientations, and decrease of singlet-triplet exchange energy (ΔE_{ST}) was achieved by introducing one or two nitrogen atoms into the central arylene. The device in a configuration of ITO/TPDPES:10% TBPAH (20 nm)/TAPC (35 nm)/host:4 wt.% Ir(piq)$_3$ (10 nm)/TPyBPZ (65 nm)/LiF (0.5 nm)/Al (100 nm) using 46DCzPPm (**340f**) as the host showed the highest efficiencies, which can be attributed to its low-lying LUMO energy level and the smallest ΔE_{ST}, giving excellent electron injection from ETL and thus improved carrier balance. Some of these host materials were also investigated as the host for RGB p–i–n homojunction OLEDs comprising an MoO$_3$-doped host as an HTL and a Cs$_2$CO$_3$-doped host as an ETL [416]. The device with an EML consisting of 46DCzPPm (**340f**) doped with Ir(piq)$_3$ showed EQE of 8.5% at 100 cd/m^2.

Tao et al. reported a series of 1,2,4-triazole-cored TPA derivatives (**341** and **342**) with various linkages between the triazole and TPA moieties (Scheme 3.61) [417]. Red phosphorescent OLEDs in a configuration of ITO/MoO$_3$ (10 nm)/NPB (80 nm)/host:6 wt.% (piq)$_2$Ir(acac) (20 nm)/BCP (10 nm)/Alq$_3$ (30 nm)/LiF (1 nm)/Al (100 nm) were fabricated by using these compounds as the host material and Ir(piq)$_2$acac as the red phosphor. Among these devices, the best performance was achieved for the device hosted by *o*-TPA-*m*-TPTAZ (**342**), which showed a maximum current efficiency of 12.4 cd/A and a maximum EQE of 16.4%; these values are close to the best results for deep-red electrophosphorescence.

Su et al. reported a series of host materials, TCPB (**343**), TCPY (**344**), TCPM (**345**), and TCPZ (**346**), containing a carbazole periphery and a triphenyl arylene core. Gradually reduced singlet-triplet exchange energy from 0.60 to 0.14 eV was obtained with increasing of the N atom amount in the aromatic core [418,419]. A narrow energy band-gap of 2.95 eV, a low-lying LUMO energy level of 3.23 eV, and a HOMO energy level of 6.18 eV were achieved for TCPZ (**346**) by introducing 1,3,5-triazine as an electron-withdrawing unit of the host. Thanks to the relatively lower electron-injection barrier from the ETL, the Ir(piq)$_3$-based red phosphorescent device gave a very low turn-on voltage (V_{on}) of 2.3 V and driving voltages of 3.02 and 4.25 V achieved at 100 and 1000 cd/m^2. In addition, a very

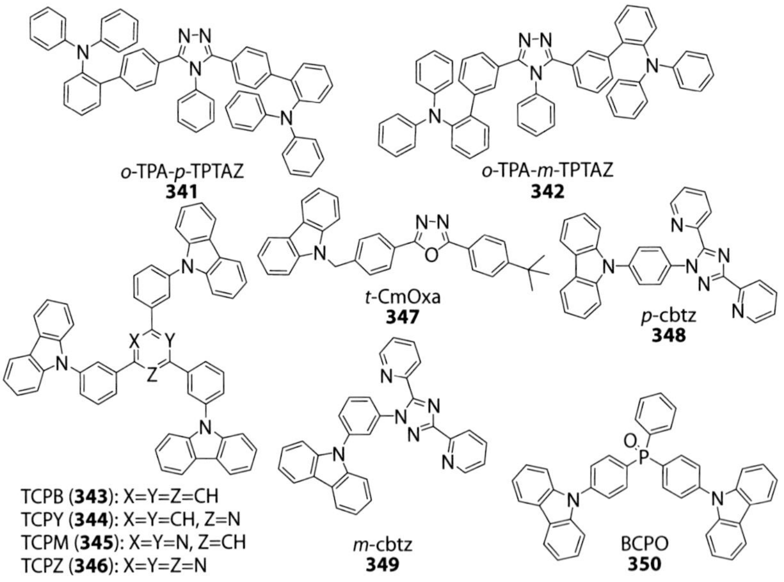

o-TPA-*p*-TPTAZ
341

o-TPA-*m*-TPTAZ
342

t-CmOxa
347

p-cbtz
348

TCPB (**343**): X=Y=Z=CH
TCPY (**344**): X=Y=CH, Z=N
TCPM (**345**): X=Y=N, Z=CH
TCPZ (**346**): X=Y=Z=N

m-cbtz
349

BCPO
350

SCHEME 3.61
Chemical structures of bipolar host materials for red phosphorescent emitters.

high maximum EQE of 19.1% and power efficiencies of 15.8 and 8.85 lm/W were achieved at 100 and 1000 cd/m^2.

t-CmOxa (**347**) is an oxadiazole-substituted carbazole derivative [420] that exhibits a relatively lower HOMO energy level (6.14 eV) than that of CBP, and the introduction of the oxadiazole will not only lower the LUMO energy level (2.62 eV) but will also make this host material possess good electron-transport capability. For the devices with a structure of ITO/NPB (30 nm)/*t*-CmOxa:4 wt.% Ir(DBQ)$_2$acac (30 nm)/Alq$_3$ (30 nm)/Mg$_{0.9}$:Ag$_{0.1}$ without using any HBL, a turn-on voltage of 4 V and a maximum brightness of 13,790 cd/m^2 were achieved, and the power efficiency and EQE are 2.6 lm/W and 4.52%, respectively, at 100 cd/m^2.

Two bipolar host materials, *p*-cbtz (**348**) and *m*-cbtz (**349**), containing phenylcarbazole and 3,5-bis(2-pyridyl)-1,2,4-triazole moieties were synthesized and tested for the fabrication of various phosphorescent OLEDs by Hung et al. [421] in a configuration of ITO/PEDOT:PSS (30 nm)/NPB (20 nm)/TCTA (5 nm)/*p*-cbtz or *m*-cbtz:10 wt.% Os1 (20 nm)/TPBI (50 nm)/LiF(0.5 nm)/Al (100 nm). The *p*-cbtz–based device exhibited a maximum brightness of 29,100 cd/m^2 at 10.5 V and outstanding EQE, current efficiency, and power efficiency of 16.5%, 20.3 cd/A, and 18.8 lm/W, respectively; with CIE coordinates of (0.63, 0.34), while the *m*-cbtz–based device gave a maximum brightness of 36,700 cd/m^2 at 13 V and excellent EL efficiencies of 16.7%, 18.2 cd/A, and 12.5 lm/W, with CIE coordinates of (0.64, 0.35). Compared with *p*-cbtz, the *m*-cbtz counterpart exhibited better performance since the LUMO energy level of *p*-cbtz (2.12 eV) is slightly higher than that of *m*-cbtz (2.15 eV), which is caused by the distinctive bonding mode of the phenylene linker. *p*-cbtz and *m*-cbtz possess almost the same HOMO energy levels of 5.68 and 5.69 eV, respectively. Triplet energies of *p*-cbtz and *m*-cbtz were determined to be 2.82 and 2.75 eV. Because of these characteristics, *p*-cbtz and *m*-cbtz can be considered as two potential host materials for phosphorescent OLEDs.

Cheng's group developed a bipolar host material, BCPO (**350**), containing phosphine oxide and carbazole, which can be seen as a modified molecule by inserting a phosphine oxide (PhPO) moiety into CBP [422]. It can be expected that the introduction of PhPO moiety is helpful for increasing T_g, triplet energy, and electron-transporting ability. BCPO has a HOMO energy level of –5.26 eV, a LUMO energy level of –2.16 eV, and a triplet energy of 3.01 eV. Devices with a common structure of ITO/NPB (10 nm)/TCTA (20 nm)/BCPO:7 wt.% Ir(piq)$_3$ (30 nm)/BCP (10 nm)/Alq$_3$ (50 nm)/LiF (1 nm)/Al (100 nm) were prepared and exhibited a low turn-on voltage of 2.7 V, maximum EQE of 17.0%, current efficiency of 19.4 cd/A, and power efficiency of 20.4 lm/W.

Tsuji et al. reported a well-balanced ambipolar material, CZBDF (**117**), as a universal host to fabricate efficient p–i–n homojunction devices that emit light across the full visible color range [152]. CZBDF possesses a high T_g (162°C), which is high enough for practical application. Its HOMO and LUMO energy levels were estimated to be –5.52 and –2.20 eV. Red phosphorescent OLEDs with a configuration of ITO/CZBDF:V$_2$O$_5$ (20 nm)/CZBDF (10 nm)/CZBDF:3 wt.% Ir(piq)$_3$ (80 nm)/CZBDF (10 nm)/CZBDF:Cs (10 nm)/Al (80 nm) gave a maximum EQE of 5.6% at 10 V.

3.5.2.1.2 Green Phosphorescent Hosts

Compared with the hosts for red phosphorescent emitters, larger energy band-gaps are necessary for the hosts for green phosphorescent emitters. A general large band-gap electron-transport host is 3-phenyl-4-(1'-naphthyl)-5-phenyl-1,2,4-triazole (TAZ, **351**) (Scheme 3.62), which has a HOMO energy level of –6.6 eV and a LUMO energy level of –2.6 eV. Using TAZ as the host, a maximum EQE (Φ_{ext}) of 15.5% and a luminous power efficiency of

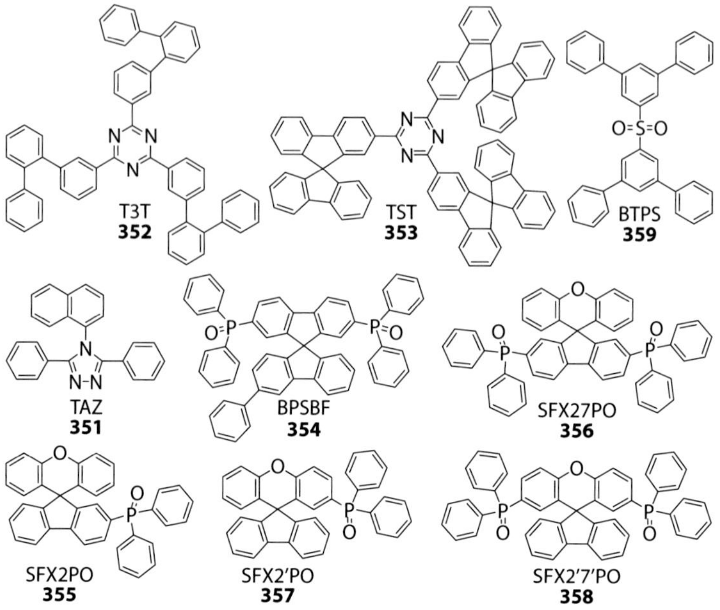

SCHEME 3.62
Chemical structures of electron-transport host materials for green phosphorescent emitters.

40 lm/W can be achieved in a green phosphorescent OLED; the value of Φ_{ext} is almost double compared with that using a CBP host device [423]. The authors explain that this is because the phosphorescent decay lifetime of 7% Ir(ppy)$_3$ in the TAZ (τ = 650 ns) is longer than that in CBP (τ = 380 ns), and the phosphorescence efficiency is approximately proportional to the excited state lifetime [424].

Three star-shaped 1,3,5-triazine derivatives of 2,4,6-tris(biphenyl-3-yl)-1,3,5-triazine (T2T, the same structure as **191**), 2,4,6-tris(triphenyl-3-yl)-1,3,5-triazine (T3T, **352**), and 2,4,6-tris(9,9'-spirobifluoren-2-yl)-1,3,5-triazine (TST, **353**) were developed as electron transport–type host materials for green phosphorescent OLEDs. The HOMO/LUMO energy levels for T2T, T3T, and TST are −5.64/−2.08 eV, −5.71/−2.15 eV, and −5.69/−2.47 eV, respectively. Triplet energies of T2T, T3T, and TST were estimated to be 2.80, 2.69, and 2.54 eV, respectively. The device incorporating T2T as the host, (PPy)$_2$Ir(acac) as the phosphorescent dopant, and TBPI as the ETL, achieved a high EQE (η_{ext}) of 17.5% and a power efficiency (η_p) of 59.0 lm/W. For the same device configuration, the T3T-based device provided η_{ext} and η_p values of 14.4% and 50.6 lm/W, respectively; the TST-based device provided values of 5.1% and 12.3 lm/W, respectively [425]. Chen et al. also reported a systematic comparison of the physical properties and the dual-role (host and electron transport) applications of star-shaped 1,3,5-triazine–based electron transport–type hosts of T2T, 3N-T2T (the same structure as **192b**), 3P-T2T (**334**), and oCF3-T2T (**335**) with different peripheral groups [426]. Among these hosts, 3P-T2T can serve both as a promising host and ETM for various Ir-based electrophosphorescent devices. These phosphorescent OLEDs with the same device architecture exhibited low operation voltages with a maximum η_{ext} of 8%, 15.7%, 16.9%, 16.4%, and 10.8% for sky blue (FIrpic), green [(PPy)$_2$Ir(acac)], yellow [(Bt)$_2$Ir(acac)], red [(Mpq)$_2$Ir(acac)], and white [FIrpic + 0.5 wt.%(Mpq)$_2$Ir(acac)] electrophosphorescence, respectively.

Phosphine oxide derivatives can also be used as an electron-transport host. Lee's group developed a phosphine oxide–based host material, BPSBF (**354**), for highly efficient green

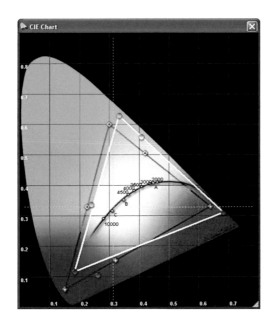

FIGURE 1.15
CIE coordinates of red, green, and blue polymer emitters.

FIGURE 1.24
PM displays in 96 × 64 format made at DuPont Display.

FIGURE 1.25
Photo of an FC AMOLED in 4″ diagonal QVGA format (a) and 14.1″ diagonal WXGA format (b).

FIGURE 1.26
LG's OLED TV displayed at Society of Information Display 2014 meeting.

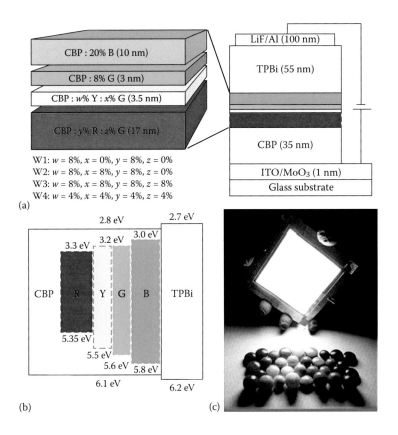

FIGURE 3.13
Device configurations (a) and energy-level diagrams (b) for WOLEDs W1–W4. The dopants employed are FIrpic for blue (B), Ir(ppy)₂(acac) for green (G), Ir(BT)₂(acac) for yellow (Y), and Ir(MDQ)₂(acac) for red (R). (From Chang, Y.-L., Song, Y., Wang, Z., Helander, M.G., Qiu, J., Chai, L., Liu, Z., Scholes, G.D. and Lu, Z., *Adv. Funct. Mater.*, 23, 705, 2012. With permission.)

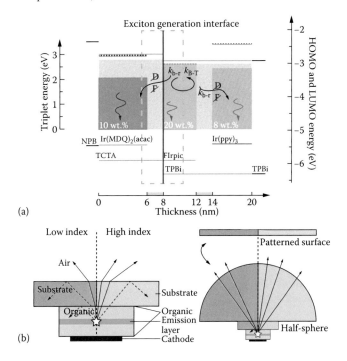

FIGURE 3.14
Energy-level diagram (a) and light modes (b) in an OLED. (From Reineke, S., Lindner, F., Schwartz, G., Seidler, N., Walzer, K., Luessem, B. and Leo, K., *Nature*, 459, 234, 2009. With permission.)

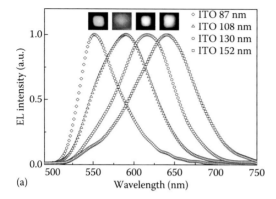
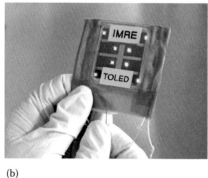

(a)   (b)

FIGURE 6.15
(a) EL spectra of a set of structurally identical OLEDs having a bilayer anode of metal/ITO. Inset pictures are the actual photographs taken for the microcavity OLEDs demonstrating color tuning with graded ITO thickness. (b) Top-emitting OLEDs with microcavity architectures to emit variable color and to enhance light output using a single emissive material.

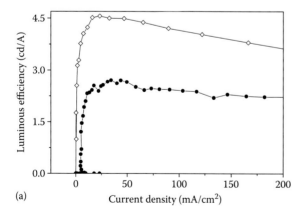
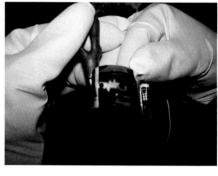

(a)   (b)

FIGURE 6.24
(a) Luminous efficiency of two top-emitting OLEDs with a configuration of glass/Ag (200 nm)/ITO (130 nm)/ PEDOT:PSS (80 nm)/Ph-PPV (80 nm)/semitransparent cathode (closed circles), and Al-PET/acrylic layer/Ag (200 nm)/CF$_x$ (0.3 nm)/Ph-PPV (110 nm)/semitransparent cathode (open diamonds). (b) Photo showing a flexible top-emitting electroluminescent device on Al-PET substrate.

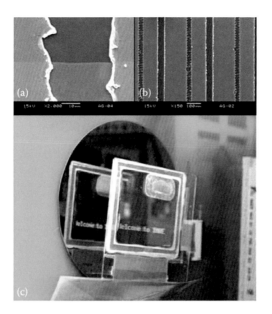

FIGURE 6.30
Microscopic images of cathode lines: (a) image taken for a laser-ablated cathode edges before the optimization, showing burr and jagged shapes; (b) typical cathode edge profile formed using an optimized ablation process; and (c) photograph of a prototype semitransparent OLED passive matrix display.

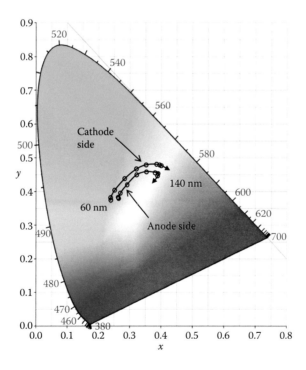

FIGURE 6.33
Effect of organic layer thickness variation on CIE color coordinates of anode and cathode emissions in semi-transparent WOLED. In this design, a pair of 40-nm-thick NPB HTL and B3PyPB ETL is chosen by considering an optimal combination of high power efficiency and preferable CIE color coordinates.

FIGURE 8.12
Flexible OLED display made by South China University of Technology. (Courtesy of Dr. Lei Wang.)

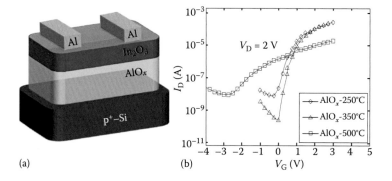

(a) (b)

FIGURE 9.2
(a) Schematic of the TFT structure, (b) transfer characteristic curves of the In$_2$O$_3$ TFTs fabricated using AlO$_x$ films prepared at different temperatures as the gate dielectrics. (From Nayak, P.K. et al., *Appl. Phys. Lett.*, 103, 033518, 2013.)

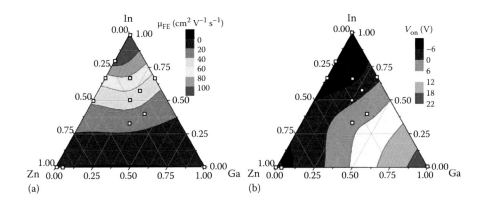

FIGURE 9.4
(a) μ_{FE} and (b) V_{on} obtained for TFTs with different oxide semiconductor compositions, in the indium–gallium–zinc oxide system. (From Jang, J. et al., *Adv. Mater.*, 25, 1042, 2013.)

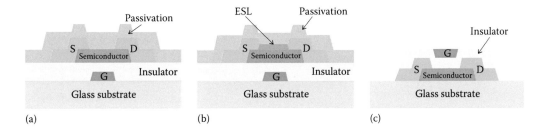

FIGURE 9.5
Typical device structures used to fabricate amorphous oxide semiconductor TFTs with (a) back-channel etch, (b) etch-stopper layer, and (c) top-gate structures.

FIGURE 9.15
Photograph of 4.8″ full-color AMOLED display.

phosphorescent OLEDs [427]. Its triplet energy was estimated to be 2.71 eV, which is high enough for energy transfer to green phosphorescent dopants with a triplet energy around 2.40 eV. The HOMO and LUMO energy levels of BPSBF are −6.53 and −2.87 eV, respectively. The maximum EQE of the BPSBF-based OLEDs using Ir(ppy)$_3$ as a green phosphorescent dopant was 25.3%, achieved at 10% doping concentration, and the EQE slightly decreased to 22.0% at 1000 cd/m^2.

Huang's group designed and synthesized a series of phosphine oxide (PO) hosts based on diphenylphosphine oxide and spiro[fluorene-9,9′-xanthene] (SFX) moieties, SFX2PO (**355**), SFX27PO (**356**), SFX2′PO (**357**), and SFX2′7′PO (**358**) [428]. It was shown that the device based on SFX27PO with the strongest carrier injecting ability had the impressively high maximum efficiencies of 70.0 cd/A, 77.0 lm/W, and 19.2% EQE, which were the highest among the green light–emitting devices with the configuration of ITO/MoO$_x$ (2 nm)/*m*-MTDATA:MoO$_x$ (15 wt.%, 30 nm)/*m*-MTDATA (10 nm)/Ir(ppz)$_3$ (10 nm)/spiro-PO host:Ir(ppy)$_2$acac (10 wt.%, 10 nm)/BPhen (40 nm)/KBH4 (1 nm)/Al.

Kido's group introduced a *m*-terphenyl–modified sulfone derivative, 5,5″″-sulfonyl-di-1,1′:3′,1″-terphenyl (BTPS, **359**), as a host material for blue and green phosphorescent OLEDs [429]. Its triplet energy was estimated as 2.79 eV. A green phosphorescent OLED with a structure of ITO (130 nm)/TAPC (30 nm)/5–15 wt.% Ir(ppy)$_3$-doped BTPS (10 nm)/B3PyPB (50 nm)/Liq (1 nm)/Al (100 nm) was fabricated. The 10 wt.%-doped device showed a very high $\eta_{p,100}$ of 105 lm/W ($\eta_{c,100}$: 100 cd/A, $\eta_{ext,100}$: 28%) at 100 cd/m^2 and a $\eta_{p,1000}$ of 82 lm/W ($\eta_{c,1000}$: 75 cd/A, $\eta_{ext,1000}$: 26%) at 1000 cd/m^2, and the average performance was 100 lm/W at 100 cd/m^2. These performances are some of the highest levels in the scientific literature. It indicates that a high-efficiency phosphorescent OLED can be realized by using guest charge trapping, although an EML exhibits a low η_{PL} in an exciton-confining high E_T host material.

A widely used hole-transport host material for triplet emitters is 4,4′-bis(9-carbazolyl)-biphenyl (CBP, **27**). The HOMO and LUMO energy levels of CBP are −6.30 and −3.0 eV, respectively. The triplet energy level of CBP is 2.67 eV. Interestingly, CBP was reported to have a bipolar transport character [430]. These properties enable CBP to act as a good host material for green, yellow, and red triplet emitters. The triplet energy level is too low to yield highly efficient devices with blue triplet emitters, although a blue phosphorescent device has been fabricated using CBP as a host and a turquoise blue triplet emitter FIrpic as a guest. FIrpic has a higher triplet energy level (2.75 eV) compared with CBP, and therefore this endothermic host–guest energy transfer produces a low-efficiency device [431].

Shu's group reported a host material that contains a TPA core and three 9-phenyl-9-fluorenyl peripheries, tris[4-(9-phenylfluoren-9-yl)phenyl]amine (TFTPA, **360**) (Scheme 3.63) [432]. Doped with a green light–emitting dopant of Ir(ppy)$_3$, CIE coordinates of (0.26, 0.66) were achieved for the device at a concentration of 21 wt.%, and the maximum values of η_{ext} and η_p were 12.0% (44.1 cd/A at 11.5 mA/cm^2) and 21.0 lm/W, respectively.

Kido's group developed a simple *m*-terphenyl modified carbazole derivative, 9-phenyl-3,6-bis-[1,1′;3′1″]terphenyl-50-yl-9 H-carbazole (CzTP, **361**), as a host material [433]. The triplet energy of CzTP is 2.70 eV. The green PHOLEDs using CzTP as a host material showed a maximum power efficiency ($\eta_{p,max}$) of 113 lm/W with a structure of ITO (110 nm)/TPDPES:10 wt.% TBPAH (20 nm)/TAPC (30 nm)/CzTP: 8 wt.% Ir(ppy)$_3$ (10 nm)/B3PyPB (50 nm)/LiF (0.5 nm)/Al (100 nm).

Fan et al. reported two diarylmethylene-bridged TPA/fluorene hybrids, BTPAF1 (**362**) and BTPAF2 (**363**) [434], which have HOMO/LUMO energy levels of 5.19/1.89 and 5.15/1.85 eV and triplet energies of 2.86 and 2.86 eV, respectively. Devices with BTPAF1

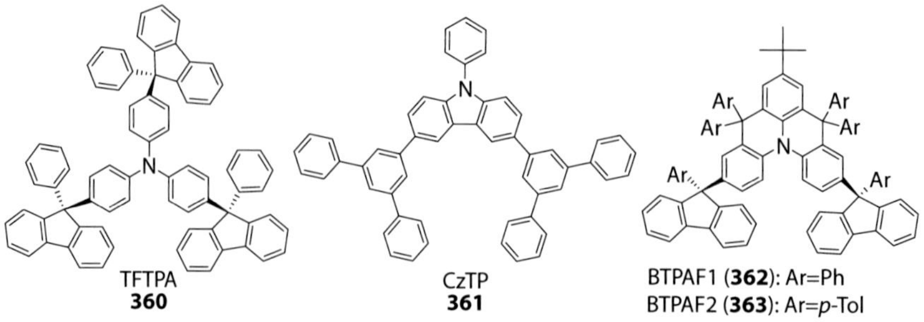

SCHEME 3.63
Chemical structures of hole-transport host materials for green phosphorescent emitters.

as host materials showed maximum EQE as high as 9.4% for blue and 19.5% for green electrophosphorescence.

Chi et al. synthesized an amorphous host material, SInF3 (**364**) (Scheme 3.64), featuring a coplanar pure-hydrocarbon *p*-terphenylene (indenofluorene) backbone and exhibiting a high triplet energy, excellent thermal stability, and high hole mobility, making it suitable for use in green phosphorescent OLEDs [435]. The value of E_T of SInF3, estimated from the phosphorescence peak (492 nm) at 77 K, was 2.52 eV. The HOMO and LUMO energy levels are −5.83 and −2.34 eV, respectively. The EL spectrum reveals that the emission arose purely from ppy$_2$Ir(acac) with CIE coordinates of (0.32, 0.64); that is, no emission occurred from the neighboring materials. The maximum values of η_{ext} [15.8% (60 cd/A)] and η_p (63 lm/W) were achieved at a practical brightness of 50 cd/m² at 3 V. A device incorporating this tailor-made host material exhibited a brightness of 10,000 cd/m² at an extremely low operating voltage (5 V), with efficiencies of 14% (53.5 cd/A) and 33.6 lm/W.

Yang's group constructed two novel oligo-9,9′-spirobifluorenes through *ortho*-linkage 44BSF (**365**) and 24TSF (**366**) [436]. The triplet energy of ortho-linked 24TSF was determined to be 2.55 eV by the highest-energy vibronic subband of the phosphorescence spectrum at 77 K. Devices with 24TSF as the full-hydrocarbon host material and Ir(ppy)$_3$ as the triplet emitter in a configuration of ITO/MoO$_3$ (10 nm)/NPB (40 nm)/24TSF:8% Ir(ppy)$_3$ (30 nm)/BCP (10 nm)/Alq$_3$ (30 nm)/LiF (1 nm)/Al (100 nm) showed a maximum EQE of 12.6% (48.2 cd/A) and power efficiency of 26.8 lm/W.

Bipolar host materials have aroused considerable interest recently because they can significantly provide balanced electron and hole flows compared with conventional host materials. With more balanced carrier injection and transport, the bipolar host materials could not only lessen driving voltages, improve efficiencies, and simplify device structures in phosphorescent OLEDs but also make the exciton recombination zone in the EML less confined near the interfaces between the EML and the neighboring layers, which will

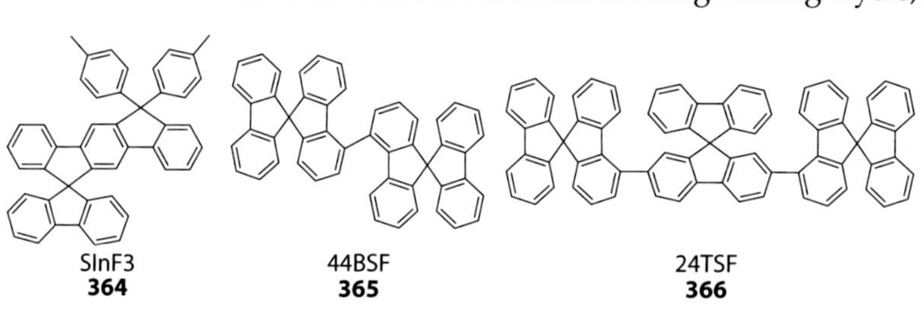

SCHEME 3.64
Chemical structures of pure hydrocarbon host materials for green phosphorescent emitters.

alleviate the triplet–triplet annihilation and lower the efficiency roll-off. Compared with the conventional phosphorescent hosts, both hole-transporting (electron-donating structure) and electron-transporting (electron-accepting structure) moieties are usually incorporated in the molecular structures of the bipolar host materials.

Ma and Yang's group have synthesized a series of carbazole/oxadiazole hybrid molecules, CzOXD (**367–370**) (Scheme 3.65) [437,438]. These four compounds exhibit similar LUMO energy levels (2.55–2.59 eV), whereas their HOMO energy levels vary in a range from 5.55 to 5.69 eV, depending on the linkage modes of the electronic donor and acceptor moieties. The HOMO and LUMO energy levels of *o*-CzOXD (**367**) are 5.55 and 2.56 eV, respectively, and its triplet energy (2.68 eV) is higher than that of CBP (2.56 eV). Devices hosted by *o*-CzOXD achieved maximum current efficiencies as high as 77.9 cd/A for Ir(ppy)$_3$ and 64.2 cd/A for (ppy)$_2$Ir(acac). Devices containing the *o*-CzOXD host showed maximum EQEs as high as 20.2% for green and 18.5% for deep-red electrophosphorescence, which are much higher than those of identical devices with CBP as the host. They also developed a TPA/oxadiazole hybrid, namely *m*-TPA-*o*-OXD (**371**), as both host and exciton-blocking material in the phosphorescent OLEDs [439]. Its HOMO/LUMO energy levels are −5.21 and −2.23 eV relative to vacuum level, respectively. The HOMO energy level of this compound is even higher than that of the most widely used NPB (5.4 eV). The triplet energy of this compound in film state is 2.51 eV. A maximum EQE of 23.7% and a maximum power efficiency of 105 lm/W were achieved for green electrophosphorescence. A series of tetraarylsilane compounds, namely *p*-BISiTPA (**372**), *m*-BISiTPA (**373**), *p*-OXDSiTPA (**374**), and *m*-OXDSiTPA (**375**), were also developed by incorporating electron-donating arylamine and electron-accepting benzimidazole or oxadiazole into one molecule via a silicon-bridge linkage mode [440]. These compounds have high triplet energies of 2.69, 2.73, 2.70, and 2.72 eV, respectively. Phosphorescent OLEDs hosted by *p*-BISiTPA achieved a maximum EQE of 22.7%. In addition, the EQE is still as high as 22.4% at a high luminance of 1000 cd/m^2.

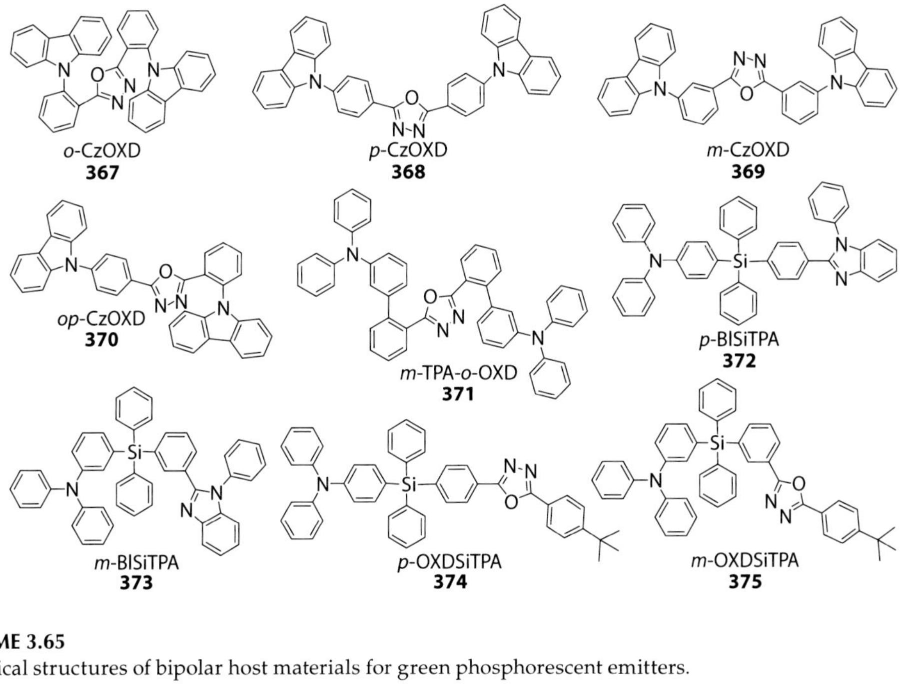

SCHEME 3.65
Chemical structures of bipolar host materials for green phosphorescent emitters.

Thompson's group also developed a series of silicon-containing compounds like (9,90-dimethylfluoren-2-yl)$_4$Si (SiFl4, **376**) (Scheme 3.66) as a host material for red, green, and blue phosphorescent devices, giving a peak efficiency of 3% [441].

Tao et al. synthesized a series of 1,2,4-triazole-cored TPA derivatives [417]. The best EL performance was achieved for the green electrophosphorescent device hosted by *o*-TPA-*m*-PTAZ (**377**) (E_T = 2.57 eV, HOMO = 5.31 eV) in a configuration of ITO/MoO$_3$ (10 nm)/NPB (80 nm)/host:9 wt.% Ir(ppy)$_3$ (20 nm)/BCP (10 nm)/Alq$_3$ (30 nm)/LiF (1 nm)/Al (100 nm), giving maximum current efficiency of 50.7 cd/A and maximum EQE of 14.2%. Chen et al. synthesized a series of bipolar hosts of CbzCBI (**378**), mCPCBI (**379**), CbzNBI (**380**), and mCPNBI (**381**) containing hole-transporting carbazole and electron-transporting benz-imidazole moieties [442]. Notably, the triplet energies (E_T) of CbzCBI and mCPCBI (2.49 and 2.50 eV) are significantly lower than those of CbzNBI and mCPNBI (2.72 and 2.73 eV). A green phosphorescent device incorporating CbzCBI as the host doped with (PBi)$_2$Ir(acac) in a configuration of ITO/(PEDOT:PSS) (30 nm)/DTAF (40 nm)/host:dopant 8 wt.% (30 nm)/TPBI (30 nm)/LiF/Al achieved a maximum EQE, current efficiency, and power efficiency of 20.1%, 70.4 cd/A, and 63.2 lm/W, respectively.

Hung's group reported two bipolar host materials, CBzIm (**382**) and COxaPh (**383**), comprising a hole-transport carbazole core functionalized with electron-transport moieties of benzimidazole/oxadiazole at C3 and C6 positions, respectively [443]. The triplet energies (E_T) of CBzIm and COxaPh were determined to be 2.76 and 2.65 eV, respectively. CBzIm with a high E_T (2.76 eV) is suitable to serve as a blue phosphor host, and a sky-blue phosphor (DFPPM)$_2$Irpic that exhibits superior properties than those of the popular blue emitter FIrpic was used as the dopant to give a maximum EQE of 15.7%. The better π-delocalization of COxaPh led to a lower triplet energy (E_T = 2.65 eV), which can be used to accommodate green and red phosphors, providing excellent device performance with

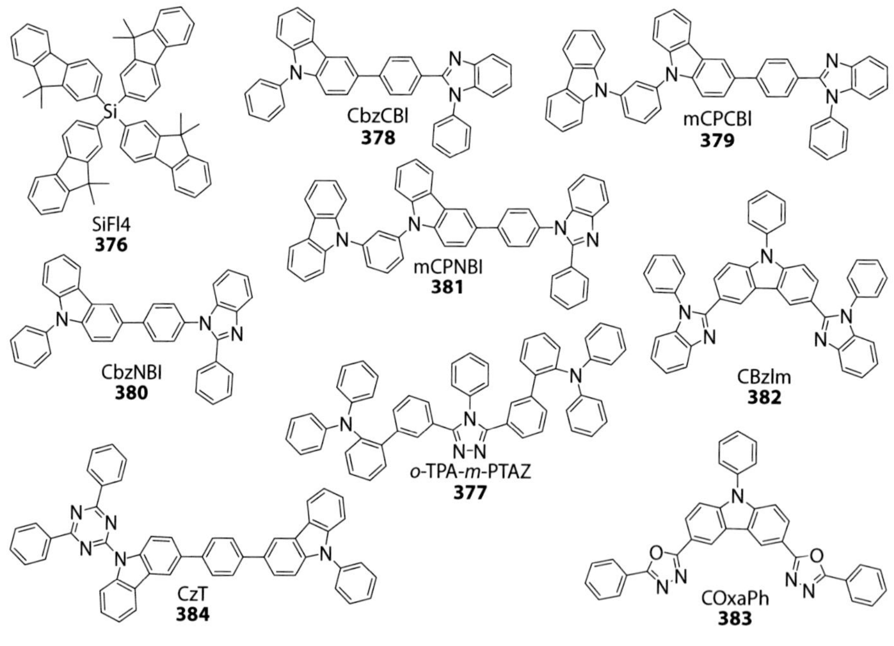

SCHEME 3.66
Chemical structures of host materials for green phosphorescent emitters.

an EQE as high as 17.7% for green [(ppy)$_2$Ir(acac)] and 20.6% for red [Os(bpftz)$_2$(PPh$_2$Me)$_2$] phosphors, respectively.

Chang et al. synthesized a novel bipolar host material 9-(4,6-diphenyl-1,3,5-triazin-2-yl)-9'-phenyl-3,3'-bicarbazole (CzT, **384**) comprising a dicarbazole donor linking to an electron deficient 1,3,5-triazine acceptor [444]. CzT has sufficient triplet energy of 2.67 eV, and its HOMO and LUMO energy levels were estimated to be −5.49 and −2.77 eV, respectively. It was used as a host for an efficient yellowish–green iridium complex, (TPm)$_2$Ir(acac); the as-fabricated trilayer and bilayer phosphorescent OLEDs demonstrated maximum efficiencies of 20.0% (75.7 cd/A and 71.3 lm/W) and 20.1% (76.3 cd/A and 72.7 lm/W), respectively.

Yang's group synthesized a series of hybrid bipolar phosphorescent host by conjugating a carbazole moiety to the different positions of the rigid skeleton of 1,2-diphenyl-1*H*-phenanthro[9,10-*d*]imidazole (**385**) (Scheme 3.67) [445]. Highly efficient green and orange phosphorescent OLEDs were achieved by employing these compounds as host. For the device of structure ITO/MoO$_3$ (10 nm)/NPB (80 nm)/TCTA (5 nm)/mPhBINCP (**385d**):9 wt.% Ir(ppy)$_3$ (20 nm)/TmPyPB (45 nm)/LiF (1 nm)/Al (100 nm), a maximum current efficiency ($\eta_{c,max}$) of 77.6 cd/A, maximum power efficiency ($\eta_{p,max}$) of 80.3 lm/W, and maximum EQE ($\eta_{EQE,max}$) of 21% were obtained. Furthermore, these hosts are also applicable for the orange phosphorescent emitter (fbi)$_2$Ir(acac). $\eta_{c,max}$ of 57.2 cd/A, $\eta_{EQE,max}$ of 19.3%, and $\eta_{p,max}$ of 59.8 lm/W were achieved for a yellow phosphorescent OLED with pPhBICP (**385a**) as host. These results demonstrated that the phenanthroimidazole unit is an excellent electron-transporting group for constructing bipolar phosphorescent host.

Wang's group designed two host materials, 4,5'-*N*,*N*'-dicarbazolyl-(2-phenylpyridine) (CPPY, **386**) and 4,5'-*N*,*N*'-dicarbazolyl-(2-phenylpyrimidine) (CPHP, **387**), as host for single-layer OLEDs (Scheme 3.68) [446]. The triplet energies of these materials are 2.62 and 2.61 eV, respectively. Single-layer OLEDs with performance comparable to those of conventional multilayer devices can be achieved by careful control of charge transport within the host and the energy-level alignment of the host material at metal/organic interfaces. Devices based on these hosts using Ir(ppy)$_2$(acac) as emitter exhibit maximum EQEs of 21.5% and 26.8%, respectively, which are the highest values reported to date for a simplified single-layer device.

Su and Kido and coworkers reported a series of three-carbazole-armed host materials (**343–346**) composed of carbazole as the periphery and triphenyl arylene as the core [418, 419]. Their triplet energies are 2.66, 2.63, 2.64, and 2.52 eV, respectively. For the green phosphorescent OLEDs in a structure of ITO/TPDPES:TBPAH (20 nm)/TAPC (30 nm)/host:8 wt.% Ir(ppy)$_3$ (10 nm)/TmPyBPZ (50 nm)/LiF (0.5 nm)/Al (100 nm), reduced driving voltages were achieved for the devices based on TCPY (**344**) and TCPM (**345**) in comparison with the device based on TCPB (**343**), especially at a brightness <1000 cd/m^2. The highest efficiency was achieved for TCPM, exhibiting η_{ext} of 25.6% and η_P of 109 lm/W at 100 cd/m^2. In

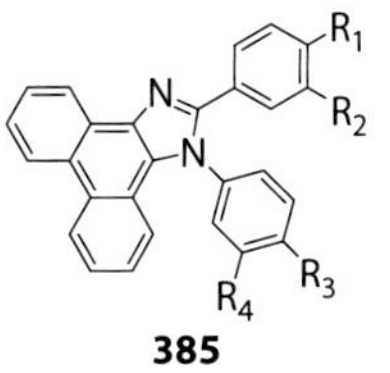

pPhBICP (**385a**): R$_1$=Carbazole,R$_2$=R$_3$=R$_4$=H
mPhBICP (**385b**): R$_2$=Carbazole,R$_1$=R$_3$=R$_4$=H
pPhBINCP (**385c**): R$_3$=Carbazole,R$_1$=R$_2$=R$_4$=H
mPhBINCP (**385d**): R$_4$=Carbazole,R$_1$=R$_2$=R$_3$=H

385

SCHEME 3.67
Chemical structures of phenanthroimidazole-containing bipolar host materials for green phosphorescent emitters.

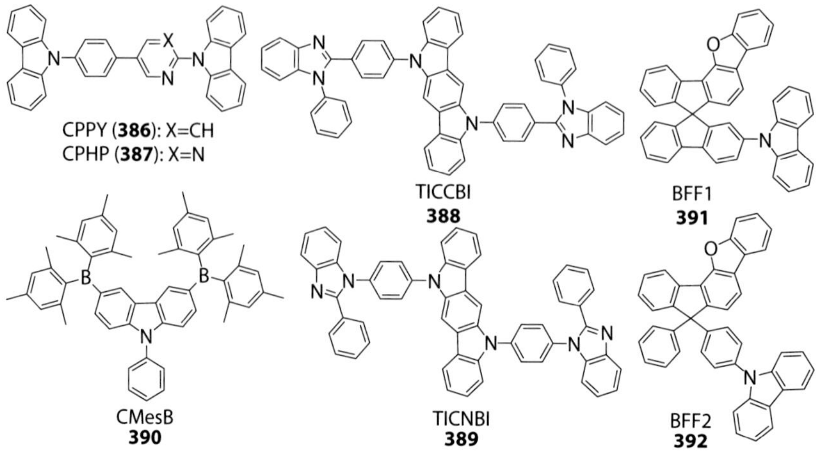

SCHEME 3.68
Chemical structures of host materials for green phosphorescent emitters.

addition, besides the high efficiency achieved by the host of TCPM, its η_{ext} rolls very slightly from a display-relevant luminance of 100 cd/m^2 to an illumination-relevant luminance of 1000 cd/m^2. Even at a much brighter luminance of 10,000 cd/m^2, η_{ext} remains 21.9% for the device based on TCPM, compared with 12.1% and 19.6% for the devices based on TCPB and TCPY, respectively. A bicarbazole-substituted bipolar host material (**340b**) was also used for green electrophosphorescence. η_{ext} remains 23.4% (corresponding to 83.9 cd/A) with a power efficiency of 62.3 lm/W at a much brighter luminance of 10,000 cd/m^2, and both are some of the highest ever values for the Ir(ppy)$_3$-based green phosphorescent OLEDs [415].

Wong's group synthesized two donor–acceptor bipolar host materials, TICCBI (**388**) and TICNBI (**389**), by incorporating electron-accepting benzimidazole and electron-donating indolo[3,2-*b*]carbazole into one molecule [447]. The triplet energies (E_T) of TICCBI and TICNBI are identical as 2.61 eV. The devices, in a typical multilayer architecture of ITO/ PEDOT:PSS (30 nm)/NPB (20 nm)/TCTA (5 nm)/emitter (25 nm)/TPBI (50 nm)/LiF (0.5 nm)/ Al (100 nm) with (PPy)$_2$Ir(acac) as green dopant and TICNBI as host, exhibited higher EQE of 15%, current efficiency of 55.8 cd/A, and power efficiency of 56 lm/W compared with that (14%, 52.4 cd/A, and 31.2 lm/W, respectively) of the device using TICCBI as the host. It is worth noting that the η_{ext} of the device using TICNBI still remains as high as 14.9% at a brightness of 1000 cd/m^2 without significant decay.

Wu's group reported a bipolar host material containing dimesityl borane/carbazole hybrid, CMesB (**390**) [448]. The HOMO/LUMO energy levels of CMesB were elucidated to be −5.69/−2.36 eV. Considering its high triplet energy (2.94 eV), CMesB is a promising universal host material and has been applied to phosphorescent OLEDs of various colors. Red/green/blue/white (RGBW) OLEDs based on CMesB all show high EQEs (20.7% for red, 20.0% for green, 16.5% for blue, and 15.7% for white) at practical brightnesses.

Lee's group developed a series of fluorenobenzofuran (BFF) derivatives as the host materials for green electrophosphorescence [449]. The triplet energies are 2.70 and 2.71 eV for BFF1 (**391**) and BFF2 (**392**), respectively. Maximum EQE of 19.7% was achieved for the BFF1-based device in a structure of ITO (50 nm)/DNTPD (60 nm)/TAPC (30 nm)/BFF1:Ir(ppy)$_3$ (30 nm, 3%)/TSPO1 (25 nm)/LiF (1 nm)/Al (200 nm), and the EQE at 1000 cd/m^2 was 19.5%. In particular, the EQE at 5000 cd/m^2 was as high as 18.3%.

3.5.2.1.3 Blue Phosphorescent Hosts

The challenge for achieving exothermic energy transfer in blue electrophosphorescence is in selecting a higher triplet energy host. The carbazole-based host materials typically have triplet energies <3 eV, and this value is close to the band gap of deep-blue dopants. Obviously, to achieve a deep-blue phosphorescent OLED, host materials need to have much larger triplet energies, so as to maintain efficient exothermic energy transfer from host to guest. This is a difficult task, not only because of the limited availability of such large band-gap materials but also because of the difficulty of charge injection from charge-transport materials to such large band-gap host materials, which normally have deeper HOMO and shallower LUMO energy levels. To circumvent these limitations, Forrest's group proposed an idea to employ a host molecule that has a large band gap and higher triplet energy levels as an inert matrix, leaving the guest molecules to both conduct and trap charges, allowing for direct exciton formation on the guest phosphor [450]. This process allows exciton formation, and recombination at the guest molecular sites while eliminating the need for an electrically active host. Bear in mind that an exothermic path from host to guest must be maintained for an efficient charge transfer. This also requires large triplet energy level host materials so as to avoid back flow of the transferred excitons formed on guest molecules to the lower-energy host molecules.

To improve this, Tokito et al. reported a CBP derivative, 4,4'-bis(9-carbazolyl)-2,2'-dimethyl-biphenyl (CDBP, **393**) (Scheme 3.69), in which two methyl groups are grafted onto the biphenyl unit, which enforces a nonplanar structure and generates a significant blue shift of the triplet state (from 2.67 to 3.0 eV) [451]. A dramatically improved external efficiency of blue phosphorescent and white phosphorescent devices has been achieved employing CDBP as a host, indicating an efficient triplet energy confinement on the phosphorescent guests [451–453].

One example of a large band-gap host material reported by Forrest's group is *N,N'*-dicarbazolyl-3,5-benzene (mCP, **394**) [454]. mCP and CBP have similar charge injection and transport properties, but mCP has a triplet energy of 3.0 eV. This large triplet energy allows efficient energy transfer to the triplet emitter guest. In fact, using mCP as a host and FIrpic as the blue phosphorescent dopant, the EQE of the resulting PHOLED is 7.5%. This value is 50% higher than the equivalent CBP-containing device (η_{ext} = 5%), clearly demonstrating the beneficial effect of a higher triplet energy level for mCP. Another carbazole compound is *N,N'*-dicarbazolyl-1,4-dimethene-benzene (DCB, **395**), which has two carbazoles linked with a phenyl moiety decoupled by two methylene groups [455]. DCB is a large band-gap molecule with a band gap of 3.5 eV, while its triplet energy is 2.95 eV. The expected exothermic charge-transfer process is observed in blue and white devices, where FIrpic is used as the blue phosphorescent dopant [456].

Arysilane compounds have been found suitable for such applications. Scheme 3.70 shows the chemical structures of several arysilane compounds: diphenyldi(*o*-tolyl)silane (UGH1,

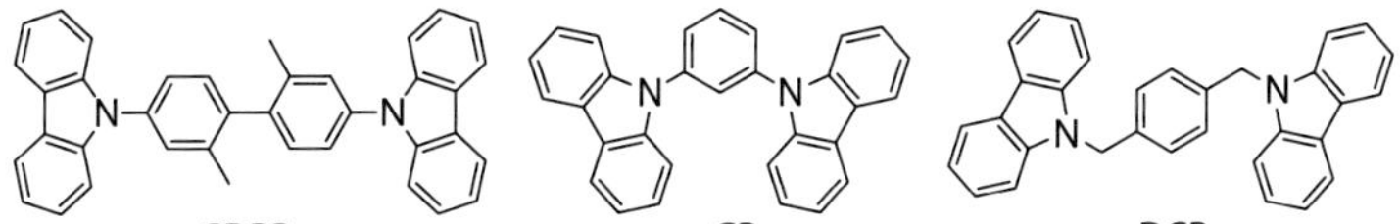

CDBP
393

mCP
394

DCB
395

SCHEME 3.69
Chemical structures of carbazole-based host materials for blue phosphorescent emitters.

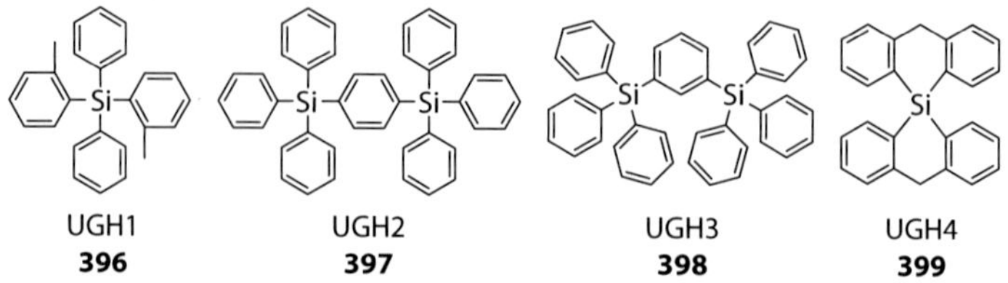

SCHEME 3.70
Chemical structures of arylsilane host materials for blue phosphorescent emitters.

396), *p*-bis(triphenylsilyl)benzene (UGH2, **397**), *m*-bis(triphenylsilyl)benzene (UGH3, **398**), and 9,9′-spirobisilaanthracene (UGH4, **399**) [457]. These host materials have large band gaps in the range of 4.5–5.0 eV, and their triplet energies are all >3 eV. The large energy gaps of these materials may be attributed to the isolation of phenyl groups by Si atoms, which preclude direct conjugation between aromatic phenyl–phenyl rings.

These UGH series materials have electronic structures that are similar in their ground and excited states, indicating that there is little or no conjugation between the arylsilicon groups. The singlet energy band gap and triplet energies of all these molecules are ~4.4 and ~3.5 eV, respectively. They do, however, have different glass transition temperatures and melting points associated with their differing structural configurations. The morphology of the thin films prepared using UGH1 is crystalline and very rough, which limits its application as a host material. UGH2, UGH3, and UGH4 give smooth, pin hole–free thin films. However, UGH4 solutions and thin films are not air-stable and thus have not been considered practical. High-EQE PHOLED devices using UGH2 and UGH3 as host materials and the turquoise phosphorescent dopant FIrpic (E_T = 2.72 eV) as a guest have been achieved.

A hybrid material of mCP and UGH1, 3,5-bis(9-carbazoyl)tetraphenylsilane (SimCP, **400**) (Scheme 3.71), has been developed as a host material. When using SimCP in place of mCP, an improved device performance has been obtained. This is probably due to the high T_g of the SimCP host with a more branched structure, preventing molecular aggregation and thus diminishing triplet–triplet annihilation. Two host materials, BTPASi1 (**401**) and BTPASi2 (**402**), with bridged TPA and triphenylsilane moieties were reported as hosts for blue and green phosphors by Jiang et al., giving a maximum EQE of as high as 15.4% for blue and 19.7% for green electrophosphorescence [458].

Considering that the carbazole monomer has a high E_T around 3.04 eV, but the materials with a molecular size as small as a monomer often suffer from the tendency of

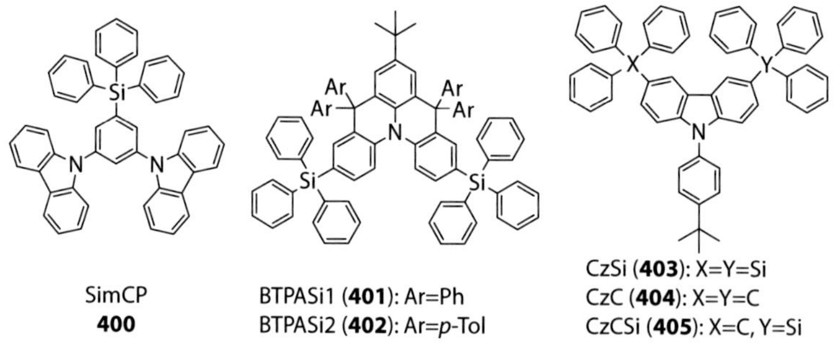

SCHEME 3.71
Chemical structures of arylsilane-containing host materials for blue phosphorescent emitters.

crystallization and thermal instability, and furthermore carbazole derivatives without groups to protect C3, C6 position could be electrochemically unstable, Tsai et al. developed efficient phosphorescent host materials, namely CzSi (**403**), based on 3,6-bis(triphenylsilyl) carbazole. It was found that CzSi materials have a high T_g of 131°C and good electrochemical stability. The increase of T_g originated from its nonplanar structure, which could prevent molecular aggregation. Good electrochemical stability, with respect to mCP, was due to block effect on active C3, C6 position of carbazole moieties. CzSi possesses a high E_T of 3.02 eV, which is almost identical with that of unsubstituted carbazole. The tetrahedral structures of Si atoms serve as a spacer, blocking π-conjugation extending from carbazole cores to the peripheral substituents. Therefore, the high triplet energy level of carbazole monomer could be retained [459].

To understand why CzSi (**403**) as host is successful in giving highly efficient devices, Tsai et al. further synthesized two novel CzSi analogues, CzC (**404**) and CzCSi (**405**). All these three compounds have high E_T in the range of 2.97–3.02 eV, together with a high T_g of 131–163°C and superior electrochemical stability. However, the carrier-transport properties show rather significant dependence on different substitutions. The CzSi host, which has more suitable carrier-transport properties, renders broadened distributions of the triplet excitons in phosphorescent devices, reducing the quenching associated with triplet–triplet annihilation and giving larger resistance against efficiency roll-off at higher brightnesses [460].

Tetraphenylsilane-based phosphine oxide derivatives were also synthesized as the host materials to obtain high triplet energy; these compounds have high triplet energy due to the Si disconnecting the conjugation of the phenyl unit, but have poor charge-transport properties. Diphenyl[4-(triphenylsilyl)phenyl] phosphine oxide (EMPA1, **406**) (Scheme 3.72) has a wide HOMO–LUMO gap (4.1 eV), high singlet (4.3 eV) and triplet (3.4 eV) energies, and an electron-dominated charge transport that follows a trap-free space charge limited model with an average electron mobility of 5.7×10^{-6} cm²/(V s) and a hole mobility of 1.1×10^{-6} cm²/(V s). The EMPA1 device doped with FIrpic showed an EQE of 21% and a current efficiency of 45 cd/A; however, the driving voltage is high (>6 V) [461]. Tetraphenylsilane modified with two phosphine oxide units was also synthesized [bis(4-(diphenylphosphoryl) phenyl)diphenylsilane (BDDS, **407**)] [462]. BDDS shows stronger electron-transport property than EMPA1, and a maximum EQE of 18.1% and an EQE of 12.2% at 1000 cd/m² with CIE coordinates of (0.137, 0.191) were achieved for the BDDS-based device doped with bis((3,5-difluoro-4-cyanophenyl)pyridine) iridium picolinate (FCNIrpic) [462].

Another series of wide-band-gap phosphine oxide/triphenylsilane hybrid materials, like 4-diphenylphosphine oxide-4′-9H-carbazol-9-yl-tetraphenylsilane (CSPO, **408**), 4-diphenylphosphine oxide-4′-[3-(9H-carbazol-9-yl)-carbazol-9-yl]-tetraphenylsilane (DCSPO, **409**), 4-diphenylphosphine oxide-4′,4″-di(9H-carbazol-9-yl)-tetraphenylsilane (pDCSPO, **410**), 4-diphenylphosphine oxide-4′-[3,6-di(9H-carbazol-9-yl)-9H-carbazol-9-yl]-tetraphenylsilane (TCSPO, **411**), and 4-diphenylphosphine oxide-4′,4″,4‴-tri (9H-carbazol-9-yl)-tetraphenylsilane (pTCSPO, **412**), were also developed as host materials [463]. This series of compounds containing different contents of p-type carbazole and n-type phosphine oxide units has balanced charge-injection and transport properties. An EQE of 27.5% and a maximum current efficiency of 49.4 cd/A were achieved for the DCSPO/FIrpic-based device, and the efficiencies are as high as 41.2 cd/A and 23.0% even at 10,000 cd/m². TFTPA (**360**) with a TPA core was also used as a host material for blue electrophosphorescence by Shih et al., and the devices showed better performance than those using mCP or CBP as host. At a practical brightness of 100 cd/A, the efficiency of the FIrpic-based device reached 12% (26 cd/A) [411].

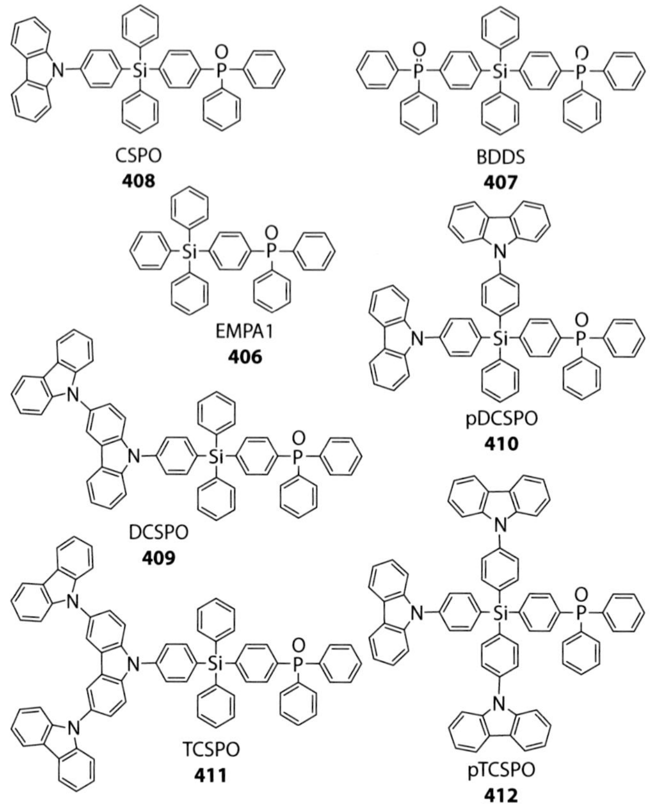

SCHEME 3.72
Chemical structures of tetraphenylsilane/phosphine oxide hybrid host materials for blue phosphorescent emitters.

Zheng et al. developed two ambipolar host materials, MCAF (**413**) and PCAF (**414**) (Scheme 3.73), comprising electron-donating carbazole and electron-accepting 4,5-azafluorene units. MCAF and PCAF have high E_T values of 2.82 and 2.83 eV, respectively, probably due to the combinatorial effects of incorporation of high intrinsic E_T fragments and isolation of sp³-hybridized C9 atom of 4,5-azafluorene to invalidate the E_T loss originating from the extended conjugation length. Owing to their high nonplanar twisted structures, high T_g values of 187°C and 188°C were obtained for MCAF and PCAF, respectively. OLEDs, in which MCAF served as host and FIrpic was used as dopant, exhibited a rather low turn-on voltage of 2.6 eV and a high power efficiency of 31.3 lm/W (17.9%). PCAF showed a rather inferior performance when compared with MCAF; this fact could be ascribed to the inferior balanced carrier-transporting capacity of PCAF [464].

With carbazole moieties as the electron donors and phosphine oxide moieties as electron acceptors, Wang et al. reported two star-shaped host materials, TCTP (**415**) and TPCz (**416**), which have different topology structures. With respect to TCTA (**49**), these two compounds have ambipolar transporting properties; however, TPCz shows a more predominately electron-transporting feature, while TCTP has been demonstrated to possess balanced bipolar hole and electron-transporting features. The star-shaped topology structures impart both TCTP and TPCz with high T_g and result in the formation of an amorphous glass state. Phosphine oxide leakages disrupt the conjugation and ensure TCTP and TPCz with high

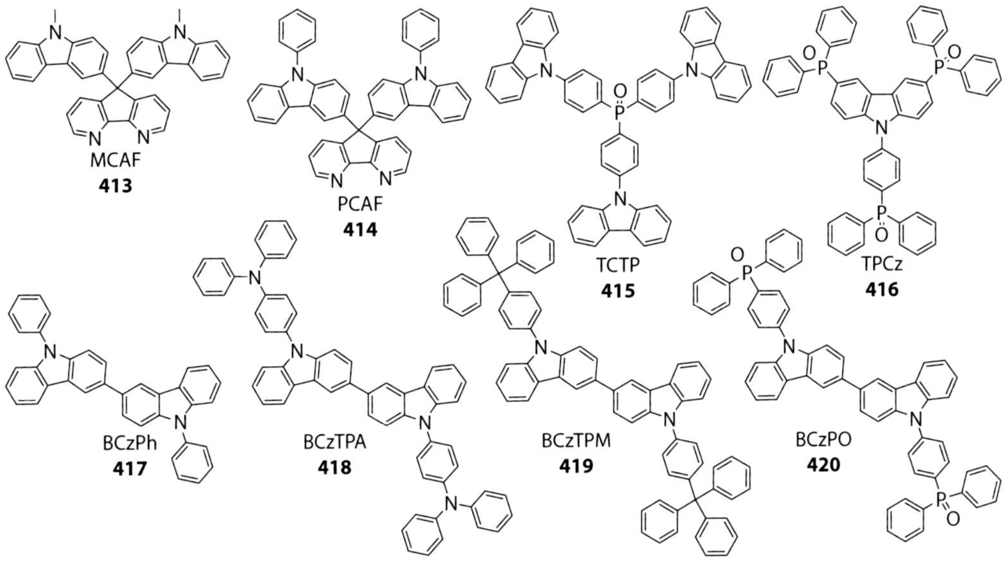

SCHEME 3.73
Chemical structures of bipolar host materials for blue phosphorescent emitters.

E_T >3.0 eV. With reasonable device configurations, high-performance blue electrophosphorescent devices with comparable efficiencies of 15.9% and 16.7% (35.0–36.4 cd/A) have been realized using TCTP and TPCz as the host for the blue phosphor FIrpic, respectively [465].

Su et al. reported a molecular design strategy of combining a carbazole electron donor with high triplet energy and a pyridine electron acceptor with high EA to give bipolar host materials of 2,6-bis(3-(carbazol-9-yl)phenyl)pyridine (26DCzPPy, **340b**) and 3,5-bis(3-(carbazol-9-yl)phenyl)pyridine (35DCzPPy, **340d**). An EQE of 24% and a power efficiency of 46 lm/W were achieved at the practical brightness of 100 cd/m² by using 26DCzPPy as the host for the FIrpic-based blue phosphorescent OLEDs. Even at a brighter emission of 1000 cd/m², the efficiencies remain >22% and >34 lm/W [466].

Sasabe et al. reported a series of 3,3′-bicarbazole–based host materials, which are modified by phenyl (BCzPh, **417**), TPA (BCzTPA, **418**), tetraphenylmethane (BCzTPM, **419**), and triphenylphosphine oxide (BCzPO, **420**), respectively. It was reported that 3,3′-bicarbazole, a dimer of carbazole, has not only a high E_T but also a small ΔE_{ST} of 0.46 eV [467]. Owing to the sterically twisted structures, BCz can greatly reduce host–host aggregation, which may cause excimer formation. The developed host materials have high E_T around 2.87–2.91 eV, together with high fluorescent quantum yield of an 11 wt.% FIrpic-doped film in the range of 62–64%. When these compounds were used as host materials for FIrpic, OLEDs exhibited superior performance; for instance, BCzTPM exhibited high power efficiency of 45.5 lm/W (44.5 cd/A, EQE 19.6%) with an extremely low driving voltage at 3.1 V at 100 cd/m² [467]. They also reported a *m*-terphenyl modified carbazole host material CzTP (**361**). CzTP has an E_{T1} of 2.7 eV, which is 0.07 eV smaller than that of FIrpic (2.77 eV), the decrease of E_T could be attributed to the extended π-conjugation of 3, 6-position on carbazole. However, it has been found that CzTP is still a good host for FIrpic, and the FIrpic-based OLED devices containing CzTP as a host showed maximum power efficiency of 55 lm/W [433].

Since the first report of 4,4′-bis (diphenylphosphine oxide) biphenyl (PO1, **421**) (Scheme 3.74) as a host material for a blue PHOLED using FIrpic as phosphor [468], the phosphine

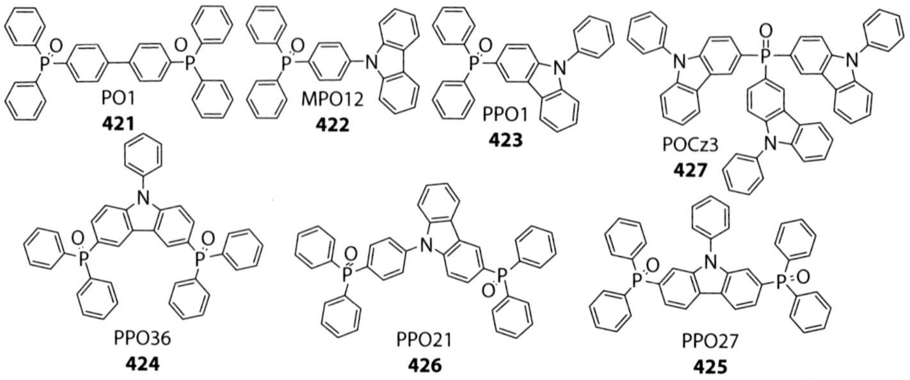

SCHEME 3.74
Chemical structures of phosphine oxide–containing host materials for blue phosphorescent emitters.

oxide (PO) groups have attracted great attention as a functional group for the host because the PO groups have unique properties to meet the basic requirements in host applications. Many colorful PO host materials have been reported, and some of them are gathered as follows.

The most widely studied PO-based host materials are carbazole-based PO compounds. Hole-transporting carbazole can be combined with PO to obtain bipolar charge-transport properties. The first reported phenylcarbazole-based PO host material may be *N*-(4-diphenylphosphoryl phenyl) carbazole (MPO12, **422**) [469], which has a substituted PO unit at the 4 position of the phenyl unit of 9-phenylcarbazole. Such a synthetic strategy offers the potential for materials with high triplet exciton energy and ambipolar charge-transporting capability. MPO12 showed a high E_T of 3.10 eV and HOMO/LUMO levels of 5.45 and 1.03 eV, respectively. The EQE of the FIrpic-based blue PHOLED fabricated using MPO12 as host was only 9.1% due to poor charge balance in the emitting layer. Carbazole was further modified at various positions. PO was introduced at the 3 position [3-(diphenylphosphoryl)-9-phenyl-9*H*-carbazole (PPO1, **423**)] and the 3,6 positions [3,6-bis(diphenylphosphoryl)-9-phenyl-9*H*-carbazole (PPO36, **424**)] of the carbazole unit to improve electron-transport properties [470].

To study the position of the PO substitution on carbazole units, PO introduced at the 2,7 position [2,7-bis(diphenylphosphoryl)-9-phenyl-9*H*-carbazole (PPO27, **425**)] of the carbazole units was synthesized, and it was compared with PPO36 whose PO was introduced at the 3,6 position of the carbazole units [471]. It was demonstrated that PPO27 reduced the driving voltage and improved the QE of blue PHOLEDs due to the reduced HOMO–LUMO gap and charge balance. This was because the LUMO of carbazole is localized at the 2,7 position, with little distribution at the 3,6 position, leading to substitution of an electron-withdrawing PO group at the 2,7 position greatly affecting the LUMO. The reduced band gap in PPO27 lowered the required driving voltage of FIrpic-based blue PHOLEDs, leading to improved QE. A high EQE of 23.9% and power efficiency of 37.5 lm/W were obtained using PPO27.

3-(Diphenylphosphoryl)-9-(4-(diphenylphosphoryl)phenyl)-9-carbazole (PPO21, **426**) has also been synthesized as a phenylcarbazole-based PO derivative [472]. PO was introduced at the 4 position of the phenyl unit and one PO at the 3 position of the carbazole unit. Because of the presence of two PO moieties, PPO21 has good electron-transporting capability. A high E_T of 3.01 eV was obtained due to the disrupted conjugation via the PO group.

The device using PPO21 as host with tris((3,5-difluoro-4-cyanophenyl)pyridine) iridium (FCNIr) as emitter revealed a maximum EQE of 19.2%.

3,3′,3″-Phosphoryl tris(9-phenyl-9*H*-carbazole) (POCz3, **427**) has also been tested as a phenylcarbazole-based PO derivative [473]. With the carbazoles acting as hole transporters/donors and the triphenylphosphine oxide groups as electron transporters/acceptors, it serves as a bipolar host material for FIrpic to realize highly efficient single-layer blue PHOLED, achieving a maximum EQE up to 9% (10.4 lm/W, 21.3 cd/A) with low efficiency roll-off.

Fluorene-based structures are also popular as the core structures of PO-type host materials. The fluorene units have weak electron-transporting properties, and modification by PO units can strengthen the electron-transport properties; thus, the fluorene-based PO host materials have typically been used as electron transport–type host materials in PHOLEDs. The first fluorene-based PO derivative was 2,7-bis(diphenyl-phosphine oxide)-9,9-dimethylfluorene (PO6, **428**) (Scheme 3.75) [474], which shows a high E_T of 2.72 eV. Its application as host and ETMs in FIrpic blue PHOLEDs resulted in a maximum EQE of 8.1%.

The spirobifluorene-based PO derivatives were developed to improve the poor thermal stability of fluorene-base derivatives. 2-Diphenylphosphoryl-9,9′-spirobi[fluorene] (SPPO1, **429**) was synthesized as a spirobifluorene-based host material for blue PHOLEDs [475]. Unexpectedly, its T_g is as low as 96°C. The SPPO1-based blue PHOLEDs with FIrpic as a blue phosphorescent dopant showed a high EQE of 16.3% and a current efficiency of 31.4 cd/A when compared with 10 cd/A of the mCP device.

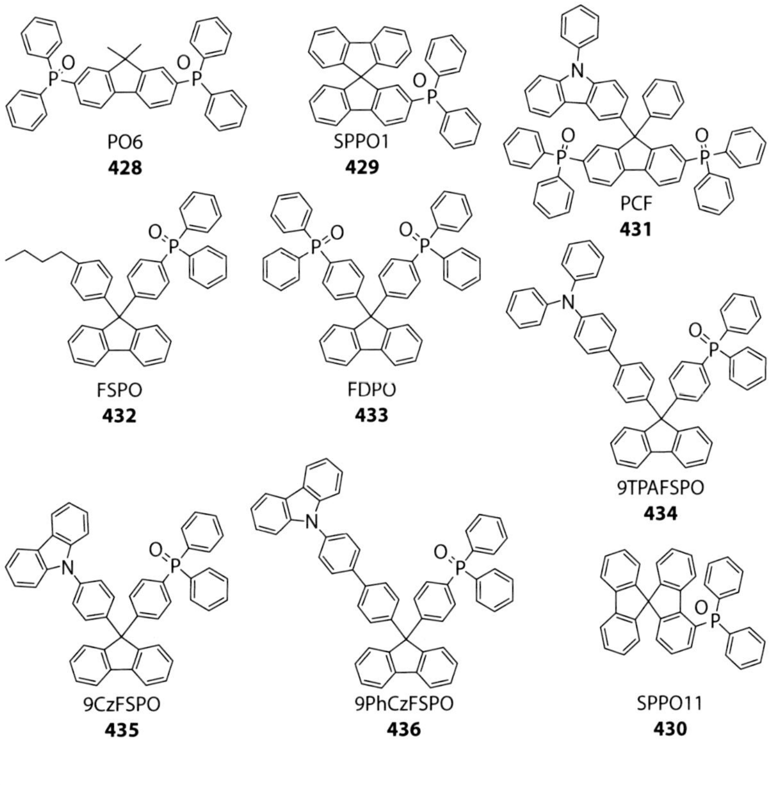

SCHEME 3.75
Chemical structures of phosphine oxide–containing host materials for blue phosphorescent emitters.

4-(Diphenyl-phosphoryl)-9,9'-spirobifluorene (SPPO11, **430**) was synthesized, which was different from SPPO1 in that the PO was attached at the 4 position of the spirobifluorene core [476]. By changing the substitution position of the PO unit, the T_g of SPPO11 was improved from 96°C to 127°C, and the high T_g stabilized the film morphology at high temperature.

Spiro[fluorene-9,9'-xanthene] (SFX) PO derivatives **355–358** were also reported as host materials for blue phosphorescence [428]. The power efficiency of the SFX2PO (**355**)- and SFX27PO (**356**)-based devices was 17.6 and 18.2 lm/W, respectively, both exceeding the performance of the SFX2'PO (**357**)- and SFX2'7'PO (**358**)-based devices (17.2 and 6.9 lm/W).

Although fluorene-based PO host materials have shown high QEs in red, green, and blue PHOLEDs, significant efficiency roll-off was observed in all devices because of the strong electron transport properties of the host materials. The problem of poor hole-transporting properties of the fluorene-based PO derivatives was solved by introducing electron-rich moieties, such as aromatic amine or carbazole functional groups. 2,7-Bis(diphenylphosphine oxide)-9-(9-phenylcarbazol-3-yl)-9-phenylfluorene (PCF, **431**) was synthesized as a carbazole-modified fluorene compound with two PO units [477]. By appending diphenylphosphine oxide groups onto a carbazole/fluorene hybrid, the PCF-based device doped with FIrpic displays both electron- and hole-transporting characteristics, resulting in a low turn-on voltage (2.6 V) and greatly improved power efficiencies (maximum power efficiency, 26.2 lm/W).

PO moieties bonded to the 9 position of fluorene would achieve a high E_T. 9-(4'-Butylphen-yl)-9-(diphenylphosphorylphenyl)fluorene (FSPO, **432**) and 9,9-bis(diph enylphosphorylphenyl) fluorene (FDPO, **433**) have an E_T of 2.99 and 3.01 eV, respectively [478], which is 0.27 eV higher than that of the 2,7-PO–substituted fluorene. However, the device performances of those host materials were not high owing to poor hole injection into the emitting layer.

Introducing hole-transporting TPA/carbazole and electron-transporting PO moieties at the 9 position of fluorene would preserve the high E_T and improve the carrier injecting/transporting ability. (4'-{9-[4-(Diphenyl-phosphinoyl)-phenyl]-9*H*-fluoren-9-yl} c-biphenyl-4-yl)-diphenyl-amine (9TPAFSPO, **434**), 9-(4-(9-(4-(diphenylphosphoryl) phenyl)-9*H*-fluoren-9-yl)phenyl)-9*H*-carbazole (9CzFSPO, **435**), and 9-(4'-(9-(4-(diphe-nylphosphoryl)phenyl)-9*H*-fluoren-9-l)biphenyl-4-yl)-9*H*-carbazole (9PhCzFSPO, **436**) were synthesized, and the E_T of these three PO compounds is as high as 3.01 eV. The ΔE_{ST} of 9TPAFSPO is extremely small at 0.3 eV, which is 0.15 eV smaller than that of mCP. 9CzFSPO and 9PhCzFSPO also obtained a very small ΔE_{ST} of only 0.45 eV, similar to mCP, through a much bigger conjugated area. The maximum power efficiency of the 9CzFSPO-based device doped with FIrpic is 35.6 lm/W, with corresponding current efficiency of 35.5 cd/A and EQE of 14.4% [479].

Other than fluorene- and carbazole-based host materials, dibenzothiophene-, TPA-, pyridine-, and arylsilane-based PO host materials were also reported.

By incorporating a TPA moiety into the otherwise electron-transporting aromatic phosphine oxide, a simple bipolar host material, 4-(diphenylphosphoryl)-*N,N*-diphenylaniline (HM-A1, **437**) (Scheme 3.76), was reported. HM-A1 had a higher triplet energy of 2.84 eV than FIrpic (2.65 eV). The power efficiency and EQE of the HM-A1–based devices doped with FIrpic is 26 lm/W and 15.4% at the brightness of 800 cd/m², respectively [480]. Other TPA-modified PO derivatives, 4-(diphenylphosphoryl)-*N*-(4-(diphenylphosphoryl)phenyl)-*N*-phenylaniline (DDPP, **438**) and tris(4-(diphenylphosphoryl) phenyl)amine (TDPA, **439**), were also reported. By incorporating two or three PO units to the TPA monomer core, the hole-transporting ability became stronger than that of HM-A1. The solution-processed

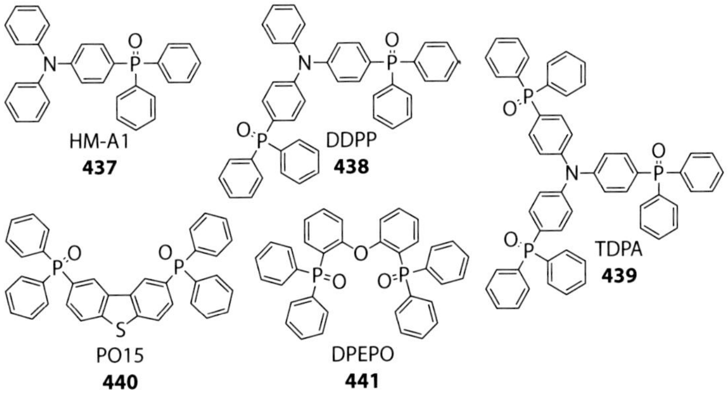

SCHEME 3.76
Chemical structures of phosphine oxide–containing host materials for blue phosphorescent emitters.

devices using DDPP and TDPA as the hosts for the phosphorescence emitter FCNIrpic showed the maximum current efficiency of 9.7 and 6.6 cd/A, respectively [481].

Dibenzothiophene is an electron-deficient core, and its modification with PO units results in electron transport–type host materials. 2,8-Bis(diphenylphosphoryl)dibenzothiophene (PO15, **440**) that possesses two PO units at the 2,8 positions of the dibenzothiophene core was reported, and showed a high E_T suitable for application in blue PHOLEDs. Although the EQE was not high in the resulting FIrpic blue PHOLEDs, it was enhanced in FCNIrpic blue PHOLEDs by optimizing the device structure to confine triplet excitons in the emitting layer [482].

PO derivatives containing an insulating core was also synthesized as host materials for blue PHOLEDs, namely bis{2-[di(phenyl)phosphino]-phenyl}ether oxide (DPEPO, **441**). In DPEPO, there are two different insulating linkages—an electron-drawing P=O bond and an electron-donating –O– bond. As a result, DPEPO achieved a high E_T close to 3.0 eV, but the resulting EQE of the corresponding device was only 6.9% [483].

3.5.2.2 Phosphorescent Dopants

Some of the highest-efficiency phosphorescent dopant materials are the iridium organometallic complexes—particularly when compared with lanthanide [484] and platinum complexes. The triplet lifetime of this class of materials is short, normally around 1–100 μs. The phosphorescent QE (ϕ_p) is high at room temperature, and the color of the dopants can be easily tuned simply by modification of the ligand chemical structures. In addition, in contrast to platinum complex–based devices, the iridium complex–based PHOLEDs can exhibit only a slow drop-off in QE when operated at high current—in large part because of the reduced triplet–triplet quenching that results from the short excited state lifetimes.

The first iridium complex used in PHOLED devices was *fac*-tris(2-phenylpyridine) iridium (Ir(ppy)$_3$) complex [485]. It has a short triplet lifetime (~1 μs) and high phosphorescent efficiency (ϕ_p = 40% at room temperature in solution) [486]. However, in the solid state, most iridium complexes showed very low phosphorescent QE due to aggregate quenching. In most cases, the complexes have to be diluted in host materials to avoid reducing the phosphorescence efficiency. For example, Ir(ppy)$_3$ suffers aggregate quenching when the doping level is >8 wt.% in a CBP host [485]. In neat thin films, ϕ_p decreases to <1%, while in the doped film of 7.9% CBP host, ϕ_p reaches >60% and can become as high as 90% when it is doped at 1% in polystyrene films [487]. Recent studies have demonstrated that the ϕ_p of 2% Ir(ppy)$_3$ doped into CBP gave 100% phosphorescence QE [5]. PHOLEDs fabricated with the structure ITO/NPD/CBP:Ir(ppy)$_3$/BCP/Alq$_3$/Mg:Ag showed a QE and a power efficiency of 8% (28 cd/A) and 31 lm/W, respectively [485]. Later, using the same device configuration but with a slightly higher dopant concentration of 6.5% Ir(ppy)$_3$ in CBP and Li$_2$O/Al as the cathode instead of Mg:Ag, a Japanese group reproduced the device performance with 1.8 times higher QE compared with the original Princeton group data [488].

3.5.2.2.1 Red Phosphorescent Dopants

3.5.2.2.1.1 Iridium Complexes 2-Benzo[*b*]thiophen-2-yl-pyridine is a general ligand for red light–emitting iridium complexes, such as Ir(btp)$_2$(acac) (**442**) (Scheme 3.77). Introducing dendrimer substituents, a solution-processable iridium red complex, (btp$_2$-D1)Ir(acac) (**443**), gave pure red emission.

To create pure red phosphorescent emission, a systematic study of the ligand structure and the emission properties was carried out by Tsuboyama et al. [489]. It was found that the red shift of the phosphorescence is attributable to introduction of more conjugated ligands.

Su et al. reported a series of highly efficient red-emissive phenylisoquinoline iridium complexes [490]. These iridium complexes showed triplet lifetimes (1.2–2.5 μs in CH$_2$Cl$_2$ and 0.15–0.56 μs in solid-film states) considerably shorter than that of Ir(btp)$_2$(acac) (6 μs in THF). The phosphorescent QEs of these complexes are also higher. Devices fabricated using the configuration ITO/NPD/6% Ir complex:CBP/BCP/Alq$_3$/Mg:Ag exhibited bright orange–red or orange emission. It is interesting to note that the emission color is independent of the applied voltage. The EQE decreased very slowly with increasing current, in contrast with the Ir(btp)$_2$(acac)-based device, ranking this class of materials as attractive candidates for red emission display applications.

Following on from the isoquinoline iridium complexes, various modifications of the ligands on iridium have been designed in order to achieve high efficiency as well as highly stable devices. Yang et al. reported a series of phenyl and naphthalene ligand red iridium complexes [491]. Red phosphorescent iridium dendrimers have been prepared by Burn's group [492]. Fang et al. developed a series derivatives of (piq)$_2$Ir(acac) (**445**) (Scheme 3.78) [493]. It is interesting that the 6′-substituted complexes show a red-shift tendency in their PL spectra compared with (piq)$_2$Ir(acac) but the 4′-substituted complexes exhibit a slightly blue-shift tendency. The quantum yields of the 6′-substituted complexes are rather lower than those of the 4′-substituted complexes. When the isoquinoline was replaced with

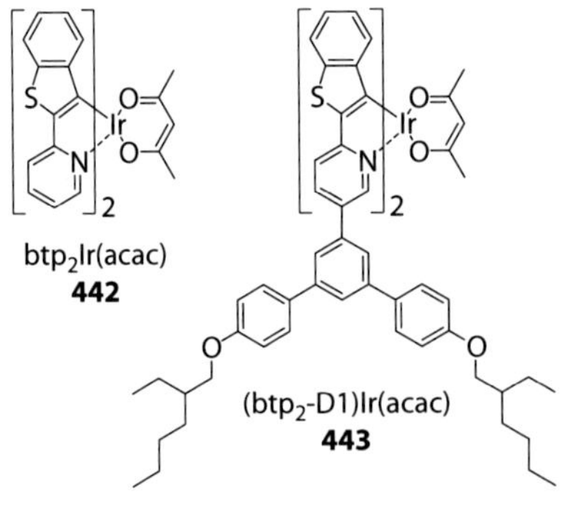

btp$_2$Ir(acac)
442

(btp$_2$-D1)Ir(acac)
443

SCHEME 3.77
Chemical structures of red iridium complexes.

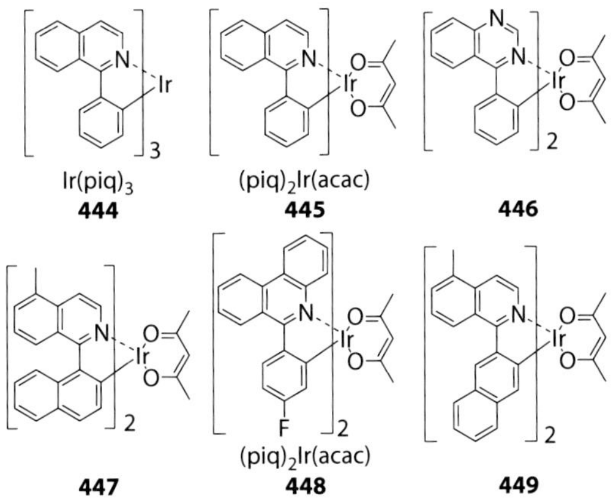

SCHEME 3.78
Chemical structures of red iridium complexes.

quinazolinato (**446**) or the phenyl was simultaneously expanded to naphthyl (**447,449**), a red shift of the PL spectra was found [494].

Schneidenbach et al. reported three orange–red phosphorescent iridium(III) complexes bearing 2-(4-fluorophenyl)-3-methyl-quinoxaline (fpmqx) cyclometalated ligands combined with three different ancillary ligands (**450–452**) (Scheme 3.79) [495]. They found that using iridium(III) complexes with different ancillary ligands as phosphorescent dopants will not only affect the emission colors and efficiencies but also the device lifetimes. With the simple test architecture ITO/NPD (20 nm)/TCTA:8% iridium(III) complex dopants (20 nm)/TPBi (45 nm)/LiF/Al, *mer*-(fpmqx)$_2$Ir(pic) (**451**)-based devices emit saturated red light of CIE of (0.64, 0.35) and show higher efficiencies with a power efficiency of 10.4 lm/W and a current efficiency of 13.9 cd/A at a current density of $J = 1$ mA/cm^2. However, the *mer*-(fpmqx)$_2$Ir(trz) (**452**)-based devices exhibited an apparent longer device lifetime, which

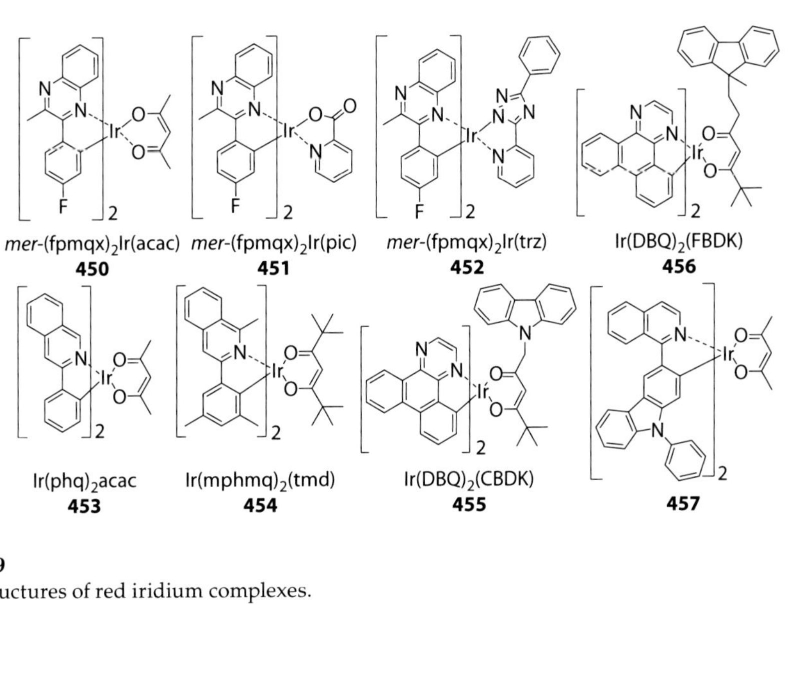

SCHEME 3.79
Chemical structures of red iridium complexes.

can be attributed to the stabilizing effect of the strongly electron-accepting character of the triazolylpyridine ligand.

Kim et al. designed and synthesized a series of iridium(III) complexes with methylated phenyl ring and quinoline ring and have ancillary ligands, such as acac and tmd groups [496]. They found that addition of an electron-donating methyl group to the metallated phenyl ring (mphq) exhibited a bathochromic shift compared with Ir(phq)$_2$acac (453), while the additional introduction of methyl group to the quinoline ring (mphmq ligand) led to a hypsochromic shift compared with the spectral range of the mphq ligand. In addition, the change of ancillary ligand to a tmd moiety from an acac moiety results in a significant improvement of device efficiency. However, it is interesting that all metallated products exhibited higher QE than Ir(phq)$_2$acac. The device structure in this research is ITO/DNTPD (40 nm)/Bebq$_2$:3 wt.% dopants (30 nm)/Bebq$_2$ (20 nm)/LiF (0.5 nm)/Al (100 nm). Ir(mphmq)$_2$acac (454)-based devices gave high efficiencies with a current efficiency of 29.9 cd/A, a power efficiency of 25.4 lm/W, and CIE of (0.65, 0.35) at a brightness of 1000 cd/m^2.

Huang's group developed two red iridium complexes bearing a carbazole or fluorene functionalized β-diketonate, Ir(DBQ)$_2$(CBDK) (455) and Ir(DBQ)$_2$(FBDK) (466) [497]. Time-resolved PL spectra revealed that Ir(DBQ)$_2$(CBDK) and Ir(DBQ)$_2$(FBDK) have a reduction of the triplet–triplet annihilation compared with Ir(DBQ)$_2$(acac). Nondoped OLEDs were also fabricated by using these iridium complexes as a nondoped emitting layer in a configuration ITO/NPB (20 nm)/Ir complex (20 nm)/BCP (5 nm)/Alq$_3$ (30 nm)/Mg$_{0.9}$Ag$_{0.1}$ (200 nm)/Ag (80 nm). The Ir(DBQ)$_2$(CBDK)-based device gave an emission peak at 634–644 nm under different driving voltages with CIE of (0.64, 0.36), and exhibited the highest brightness of 9106 cd/m^2 at 14 V, a maximum EQE of 6.2%, and a power efficiency of 3.49 lm/W.

Ma et al. reported a series of iridium(III) complexes with ligands bearing carbazole chromophores [498,499]. In this research, they found that integration of a carbazole unit into cyclometalated ligands leads to improvement of hole-transporting ability in the resulting complexes, and the difference in the ligation position (3 or 2 position C atom of the carbazole unit coordinates to the metal center) results in significant shifts in the emission spectra with the changes in wavelength. Highly efficient devices were fabricated by using some of these iridium(III) complexes as phosphorescent dopants in a configuration ITO/NPB (40 nm)/CBP:x wt.% dopant (20 nm)/BCP (10 nm)/Alq$_3$ (30 nm)/LiF(1 nm)/Al (100 nm). The device based on 4 wt.% (457) gave a maximum EQE of 11.8% and a maximum current efficiency of 10.2 cd/A with CIE of (0.68, 0.33).

Ha's group developed two iridium complexes with 2,3-diphenyl-4-methyl-quinoline rings (Scheme 3.80) [500]. They suggested that the methoxy group at the 4 position of the phenyl ring contributes to increase of the LUMO as well as the HOMO energy levels, leading to increased energy band gap. Devices were fabricated using the two Ir complexes as phosphorescent dopants in a configuration of ITO/2-TNATA (60 nm)/NPB (20 nm)/CBP:10 wt.% Ir complex (30 nm)/BCP (10 nm)/Alq$_3$ (20 nm)/Liq (2 nm)/Al (10 nm). The maximum current efficiencies of the devices based on (4-Me-2,3-dpq)$_2$Ir(acac) (458) and (4-Me-2,3-dpq-OMe$_2$)$_2$Ir(acac) (459) are 9.1 and 12.2 cd/A with CIE coordinates of (0.64, 0.35) and (0.61, 0.37), respectively. A red phosphorescent analogue, (2,3-dpq)$_2$Ir(acac) (460), without methyl and methoxy groups was developed by Chuang et al., and a current efficiency of 11.4 cd/A was obtained at 15 V with CIE of (0.66, 0.34) for the device with a configuration of ITO/2-TNATA (5 nm)/NPB (40 nm)/BAlq:5 wt.% (2,3-dpq)$_2$Ir(acac) (60 nm)/AlQ$_3$ (30 nm)/LiF (1 nm)/Al (100 nm) [501].

Seo et al. reported a series of red phosphorescent emitters based on a ligand of 2,4-diphenylquinoline (2,4-dpq) and its derivatives [502]. Compared with Ir(pq)$_2$(acac),

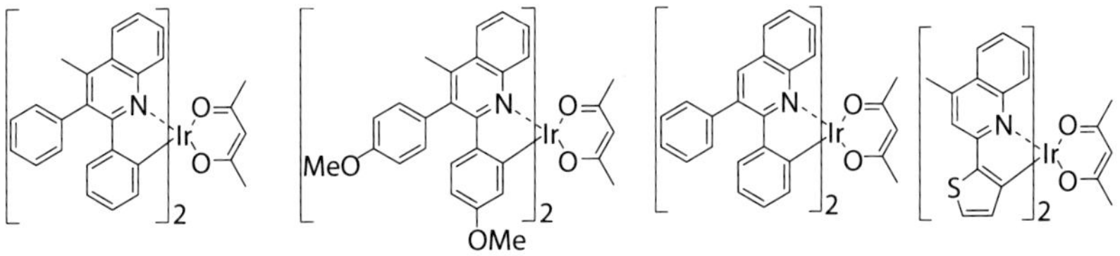

$(4\text{-Me-}2,3\text{-dpq})_2Ir(acac)$ **458** — $(4\text{-Me-}2,3\text{-dpq(OMe)}_2)_2Ir(acac)$ **459** — $(2,3\text{-dpq})_2Ir(acac)$ **460** — $(tmq)_2Ir(acac)$ **465**

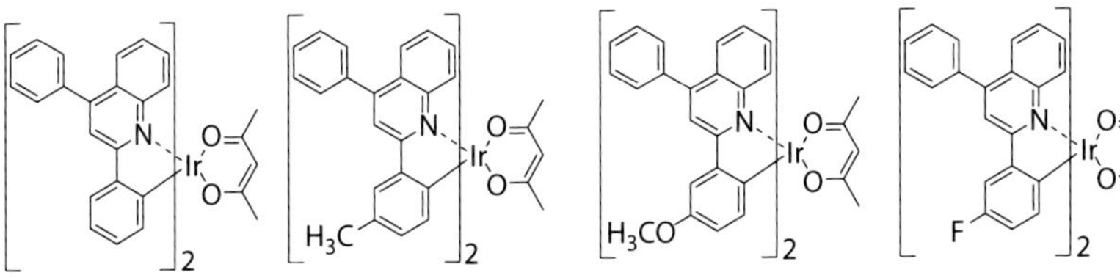

$(2,4\text{-dpq})_2Ir(acac)$ **461** — $(2,4\text{-dpq-5-}CH_3)_2Ir(acac)$ **462** — $(2,4\text{-dpq-5-}OCH_3)_2Ir(acac)$ **463** — $(2,4\text{-dpq-5-F})_2Ir(acac)$ **464**

SCHEME 3.80
Chemical structures of red iridium complexes.

$Ir(2,4\text{-dpq})_2(acac)$ (**461**) shows 15-nm bathochromic shift in the emission peak because of its longer conjugation length. However, the OLEDs with the emitter containing the dpq ligands show low efficiency due to the weak metal-to-ligand charge transfer (MLCT) characteristics. Methyl, methoxy, and fluoride were introduced onto the 2,4-dpq ligand to give Ir complexes of $Ir(2,4\text{-dpq-5-}CH_3)_2(acac)$ (**462**), $Ir(2,4\text{-dpq-5-}OCH_3)_2(acac)$ (**463**), and $Ir(2,4\text{-dpq-5-F})_2(acac)$ (**464**) to improve the efficiency of OLEDs. Maximum current efficiencies of 8.20, 10.20, and 15.80 cd/A were achieved for the devices with $Ir(dpq\text{-5}OCH_3)2(acac)$, $Ir(dpq\text{-5}CH_3)2(acac)$, and $Ir(2,4\text{-dpq-5-F})_2(acac)$ as a red phosphorescent emitter, respectively.

Cheng's group reported a deep-red phosphorescent emitter, $(tmq)_2Ir(acac)$ (**465**) [503]. Red OLEDs were fabricated by using an EML containing $(tmq)_2Ir(acac)$ as a dopant and BIQS as a host. Owing to the existence of two 6*H*-indolo(2,3-*b*)quinoxaline moieties, BIQS possesses a low-lying LUMO energy level of −2.88 eV and a HOMO energy level of −5.93 eV. Its triplet energy was estimated to be 2.34 eV, which is suitable for efficient energy transfer to the deep-red emitters. The device with a configuration of ITO/NPB (20 nm)/TCTA (10 nm)/BIQS:*x* wt.% $(tmq)_2Ir(acac)$ (30 nm)/BCP (15 nm)/Alq_3 (50 nm)/LiF (1 nm)/Al (100 nm) gave a brightness of 58,688 cd/m^2, a maximum EQE of 25.9%, a current efficiency of 37.3 cd/A, and a power efficiency of 32.9 lm/W. It is worth mentioning that the BIQS/$(tmq)_2Ir(acac)$ is the most efficient system for deep-red PHOLEDs reported to date.

Just as nondoped red fluorescent dyes, nondoped phosphorescent iridium complexes consisting of two chelating phenyl-substituted quinazoline and one (2-pyridyl) pyrazolate or triazolate have been reported by Song et al. (**466**) (Scheme 3.81) [504]. All of

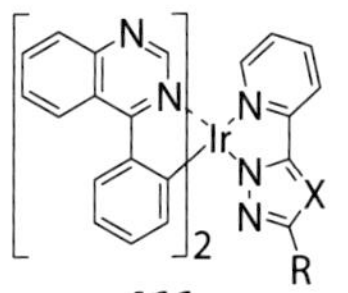

466

466a: X=CH, R=CF_3
466b: X=CH, R=*t*-Bu
466c: X=N, R=CF_3

SCHEME 3.81
Chemical structures of red iridium complexes.

these complexes exhibited bright red phosphorescence with relatively short excited state lifetimes of 0.4–1.05 μs. PHOLEDs fabricated using the compounds **466a** and **466b** with relatively higher doping level up to 21% gave saturated red emission 626 and 652 nm corresponding to CIE (0.66, 0.34) and (0.69, 0.31), respectively. The EQE of a doped device is ~6.3%. The nondoped device fabricated by using complex **466a** showed an EQE of 5.5% at 8 V, and a current density of 20 mA/cm² with luminance as high as 5780 cd/m². The high efficiency of phosphorescence under a high doping level is attributed to the very short emission lifetime.

3.5.2.2.1.2 Europium Complexes　Wang et al. reported a simple triple-layered pure red OLED with low turn-on voltage of 4.0 V, high efficiency (maximal current efficiency of 5.21 cd/A and power efficiency of 1.6 lm/W), and a slow efficiency roll-off [505], where Eu(DBM)₃(sbf) (**467**) (Scheme 3.82) was used as a dopant in a configuration of ITO/TPD (40 nm)/BCP:8 wt.% Eu(DBM)₃(sbf) (30 nm)/Alq₃ (30 nm)/LiF (1 nm)/Al (100 nm). The device gave pure red emission 612 nm corresponding to CIE (0.66, 0.33) and (0.66, 0.34) at 100 and 1365 cd/m², respectively, and it is one of the best europium complex–based devices.

Red phosphorescent OLEDs wherein a series of europium β-diketonates served as phosphorescent emitters were fabricated by De Silva et al. [506] in a structure of ITO/NPB (60 nm)/CBP:x% Eu complex (60 nm)/BCP (20 nm)/Alq₃ (40 nm)/LiF (0.5 nm)/Al (100 nm). The device containing 5% Eu(TTA)₃(TPTZ) (**468**) in the EML exhibited the highest current efficiency of 8.98 cd/A, but showed dramatically efficient roll-off at high brightness. Strangely, the highly photoluminescent complex Eu(BTFA)₃(TPTZ) (**469**) (quantum yield of 69.7%) only displayed a low EL of 7.25 cd/m².

Li's group fabricated high-efficiency, pure red OLEDs by doping a europium complex Eu(DBM)₃(pyzphen) (**470**) into an electron-transporting/hole-blocking host BPhen [507]. Red OLED with a structure of ITO/TPD (50 nm)/BPhen:25 wt.% Eu(DBM)₃(pyzphen) (25 nm)/BPhen (35 nm)/LiF (1 nm)/Al (100 nm) showed a maximum current efficiency of

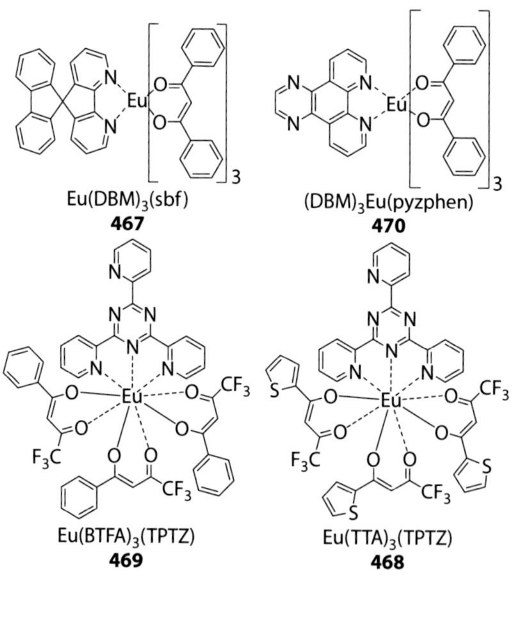

Eu(DBM)₃(sbf)
467

(DBM)₃Eu(pyzphen)
470

Eu(BTFA)₃(TPTZ)
469

Eu(TTA)₃(TPTZ)
468

SCHEME 3.82
Chemical structures of red europium complexes.

5.1 cd/A, corresponding to an EQE of 2.4%. It is exciting that the efficiency roll-off in this device is much smaller than that of Eu complex OLEDs reported previously, which can be attributed to the hole-injection barrier at the TPD/BPhen interface that is high enough to prevent the excess holes injecting into EML and then quenching the luminescence. For example, the current efficiency has a slight decrease from 4.7 to 3.8 cd/A, while the brightness increases from 10 to 100 cd/m^2. The result means that the efficiency roll-off, a major obstacle to the development of Eu-complex organic light-emitting devices, has been greatly alleviated.

3.5.2.2.1.3 Platinum Complexes Pt(II) octaethylporphine (PtOEP, **471**) (Scheme 3.83) was one of the first nonlanthanide phosphorescent organometallic complexes used in OLED devices [4]. It showed a very nice narrow red emission band centered at 650 nm. The drawback of such platinum complexes is their long triplet lifetime, which leads to low QE especially in high-current conditions, rendering them poor candidates for phosphorescent OLED applications. Tsuzuki et al. developed efficient OLEDs using phosphorescent (ppy)$_2$Ir(acac) as a host and Pt-SA-1 (**472**) as a guest [508]. Owing to good overlap between the PL spectra of (ppy)$_2$Ir(acac) and the absorption band of Pt-SA-1 in the region from 500 to 600 nm, efficient energy transfer from host to guest might be achieved. The devices in a configuration of ITO/PEDOT:PSS (35 nm)/NPD (40 nm)/(ppy)$_2$Ir(acac):1 wt.% Pt-SA-1 (35 nm)/BCP (10 nm)/Alq$_3$ (35 nm)/LiF (0.5 nm)/Al (100 nm) exhibited a maximum EQE of 8.3% and a maximum current efficiency of 4.9 cd/A with red emission λ_{max} of 645 nm and CIE of (0.68, 0.32).

Ikai et al. demonstrated a series of platinum complexes **473** and **474** and found that there is an interesting enhancement of EQE and increase of triplet excited-state lifetimes with introducing facially encumbered and bulky meso-aryl substituents into the Pt(II) porphyrin complexes [509]. In other words facial encumbrance and steric bulkiness could suppress nonradiative decay. A maximum EQE of 8.2% was achieved for the device based on **474** at a current density of 0.1 mA/cm^2.

Fukagawa's group reported highly efficient red phosphorescent OLEDs based on a Pt(II) complex, TLEC-025 (**475**) [510]. TLEC-025 exhibits a quantum yield of 0.58, which is comparable to that for some efficient red iridium complexes. A red phosphorescent OLED with a

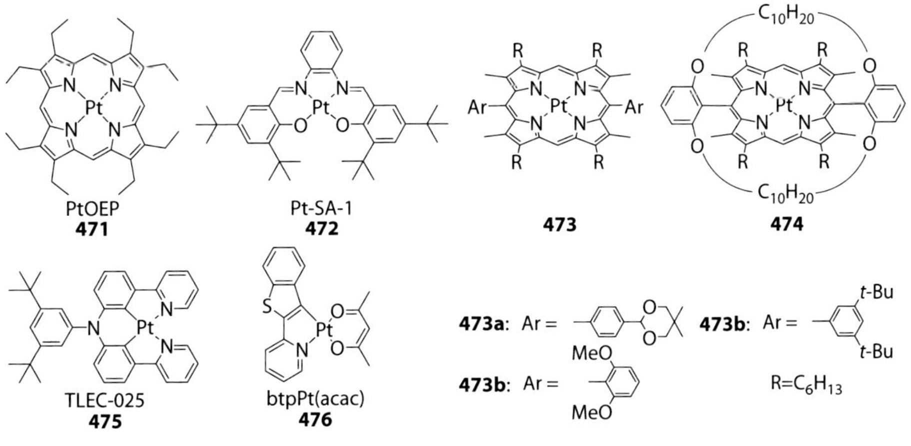

SCHEME 3.83
Chemical structures of red platinum complexes.

structure of ITO/PEDOT:PSS (30 nm)/NPD (40 nm)/Bebq$_2$:6 wt.% TLEC-025 (35 nm)/TPBI (40 nm)/LiF (1 nm)/Al (100 nm) gave an EQE of 17.1%, a power efficiency of 13.6 lm/W, a driving voltage of 4.4 V, and a pure-red-light emission with CIE of (0.66, 0.34) at a current density of 100 cd/m^2, and the device using TLEC-025 as a dopant shows higher efficiency and longer lifetime than the PHOLED using Ir(piq)$_3$.

3.5.2.2.1.4 Osmium Complexes   A high-efficiency red organic light-emitting device using a highly sterically hindered red phosphorescent dye, Os(fptz)$_2$(PPh$_2$Me)$_2$ (**477**) (Scheme 3.84), which possesses a relatively short phosphorescent lifetime (τ = 96 ns), was reported [511]. Researchers found that the Os(fptz)$_2$(PPh$_2$Me)$_2$ doped in n-type host materials (such as BAlq) produced higher current efficiency at lower doping concentration than in p-type hosts (such as CBP), which can be attributed to a narrow recombination zone near the HTL/EML interface resulting in more efficient recombination. The EQE and current efficiency reached 11.6% and 12.1 cd/A in the devices with a structure of ITO/CuPc (15 nm)/ NPB (60 nm)/BAlq:15 wt.% Os(fptz)$_2$(PPh$_2$Me)$_2$ (40 nm)/BAlq (15 nm)/Alq$_3$ (20 nm)/LiF (1 nm)/Al (200 nm). In a further research, highly efficient red light–emitting devices based on a TFTPA (**360**) host doped with Os(fptz)$_2$(PPh$_2$Me)$_2$ was reported [512]. Since the driving voltage dramatically decreased with the increasing of the doping concentration of Os(fptz)$_2$(PPh$_2$Me)$_2$ from 7 to 21 wt.%, improvement of power efficiency in these devices has been achieved. The device with a multilayer configuration consisting of ITO/NPB (30 nm)/ TFTPA:21 wt.% Os(fptz)$_2$(PPh$_2$Me)$_2$ (40 nm)/TPBI (40 nm)/LiF (1 nm)/Al (100 nm) gave maximum current efficiency and power efficiency of 29.9 cd/A and 25.2 lm/W, respectively, and the efficiencies remained 29.2 cd/A and 22.2 lm/W at 1000 cd/m^2.

3.5.2.2.1.5 Copper and Ruthenium Complexes   Wang et al. reported a series of orange–red to red phosphorescent heteroleptic Cu(I) complexes [513]. A red light–emitting device using Cu(mdpbq)(DPEphos)(BF$_4$) (**481c**) (Scheme 3.85) as phosphorescent dopant was prepared with a structure of ITO/PEDOT:PSS/TCCz:15% Cu(mdpbq)(DPEphos)(BF$_4$)/TPBI/LiF/Al, emitting efficient red phosphorescence with a current efficiency of 6.4 cd/A and an EQE of 4.5%.

A breakthrough in the field of OLEDs using less expensive Ru(II) metal complexes has been [514] as potential phosphorescent dopants. Researchers have found that the emission QEs and radiative lifetimes of neutral Ru(II) emitters are comparable to those of Os(II) or Ir(III) complexes. The devices with a structure consisting of ITO/PEDOT:PSS/NPB (40 nm)/CBP:3 wt.% Ru A (**482**) (30 nm)/BCP (10 nm)/Alq$_3$ (30 nm)/LiF (1 nm)/Mg:Ag (10:1) (50 nm) gave an EQE of 7.03%, a current efficiency of 8.02 cd/A, and a power efficiency of

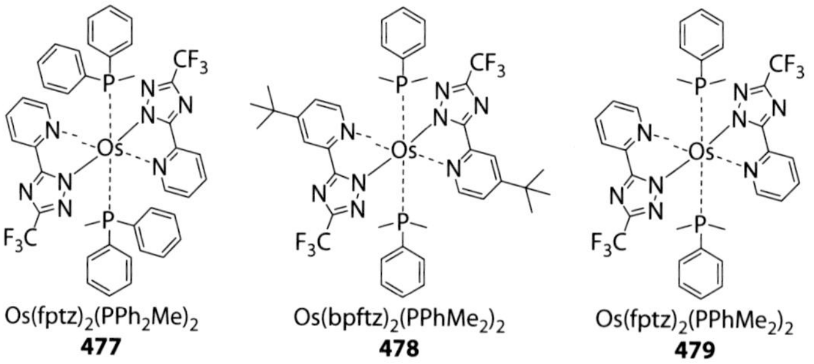

Os(fptz)$_2$(PPh$_2$Me)$_2$
477

Os(bpftz)$_2$(PPhMe$_2$)$_2$
478

Os(fptz)$_2$(PPhMe$_2$)$_2$
479

SCHEME 3.84
Chemical structures of red osmium complexes.

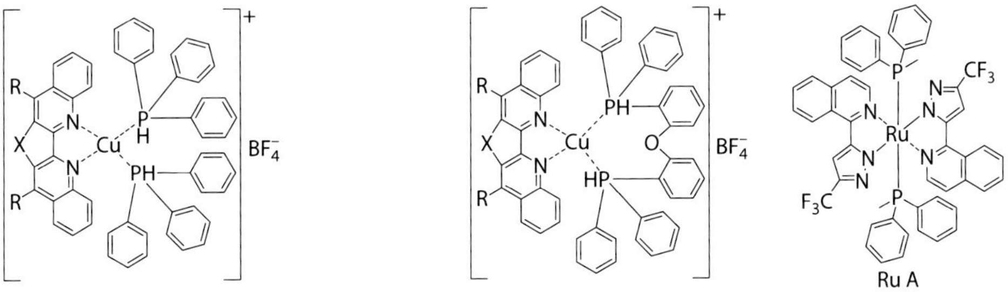

480
Cu(bq)(PPh$_3$)$_2$](BF$_4$) (**480a**): R=H, X=H,H
Cu(dpbq)(PPh$_3$)$_2$](BF$_4$) (**480b**): R=C$_6$H$_5$, X=H,H
Cu(mdpbq)(PPh$_3$)$_2$](BF$_4$) (**480c**): R=C$_6$H$_5$, X=CH$_2$

481
Cu(bq)(DPEphos)](BF$_4$) (**481a**): R=H, X=H,H
Cu(dpbq)(DPEphos)](BF$_4$) (**481b**): R=C$_6$H$_5$, X=H,H
Cu(mdpbq)(DPEphos)](BF$_4$) (**481c**): R=C$_6$H$_5$, X=CH$_2$

482
Ru A

SCHEME 3.85
Chemical structures of red copper and ruthenium complexes.

2.74 lm/W at 20 mA/cm^2. These values demonstrate that ruthenium complexes can be potential candidates for red phosphorescent dopants of OLEDs.

3.5.2.2.2 Green Phosphorescent Dopants

3.5.2.2.2.1 Iridium Complexes Scheme 3.86 lists some iridium complexes used in green PHOLEDs. Ir(ppy)$_3$ (**483**) is the original green dopant as mentioned earlier. The peak wavelength of Ir(ppy)$_3$ thin films is 510 nm with an FWHM of 70 nm. The color of the Ir(ppy)$_3$-based device corresponds to CIE coordinates of (0.27, 0.63). The doped device of 6% Ir(ppy)$_3$ in a CBP host exhibited high QE performance (~8%); however, the neat thin film of Ir(ppy)$_3$ only gave ~0.8% QE, almost 10 times lower than the doped device. By introducing electron-withdrawing fluorine groups, the DuPont team was able to demonstrate that the thin neat film of Ir–2h (**484**) could give high-efficiency device performance [515]. The result is in accordance with the photoluminescent QE measurement. In the thin film, the QE of Ir–2h is 10 times higher than that of Ir(ppy)$_3$. This, in part, is due to less pronounced self-quenching processes in the bulky substituted Ir–2h complex. The bulky substituent effect to suppress the self-quenching of the iridium dopant was clearly demonstrated by introducing even more bulky substituents such as in the iridium complex Ir(mmppy)$_3$ (**485**). This iridium complex also showed higher solid state QE.

Another way to increase the QE of iridium complexes is by elaboration into dendrimeric structures, which usually consist of surface groups, dendrons, and cores (**486**, **487**). Dendrimers have some advantages, such as a convergent synthesis that allows for a modular approach to different generations of the dendrimers, the fine tuning of

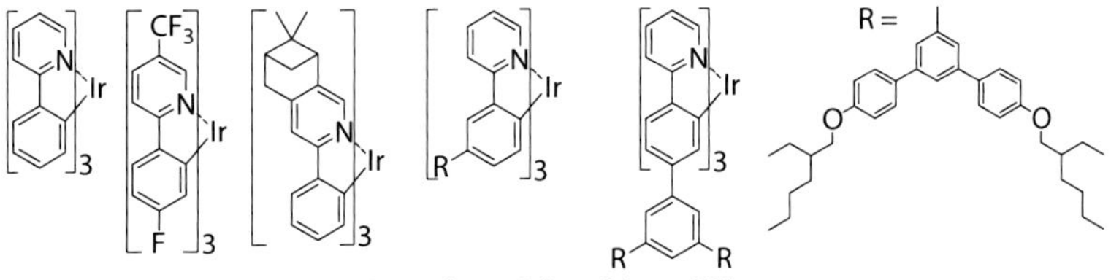

Ir(ppy)$_3$
483

Ir-2h
484

Ir(mmppy)$_3$
485

Ir(ppy-G1)$_3$
486

Ir(ppy-G2)$_3$
487

SCHEME 3.86
Chemical structures of green iridium complexes.

the electronic properties and processability, the flexibility to control the intermolecular interactions, and the ease of the purification process. Markham et al. reported a series of iridium dendrimer complexes and have successfully fabricated solution-processable PHOLEDs by spin coating of the host and guest mixture solution to the ITO substrate followed by cathode deposition [516]. The dendrimer complexes also exhibit high photo-luminescent QE compared with nondendritic iridium complexes. It is interesting to note that the higher the generation number of the dendrimers, the higher the QE it has [517]. Spin coating 20% Ir(ppy-G2)$_3$ (487) in CBP as an active layer, the maximum EQE and power efficiency were 8.1% (28 cd/A) and 6.9 lm/W, respectively. Similar devices made using Ir(ppy)$_3$ demonstrated 10 times lower QE. High-efficiency PHOLEDs have been achieved in double-layer devices with the device configuration of ITO/Ir(ppy-G1)$_3$:TCTA/TPBI/LiF/Al by selecting TCTA as the host material and using TPBI as the ETL or HBL. The best devices have a maximum efficiency of 40 lm/W (55 cd/A) at 4.5 V and 400 cd/m^2. The low turn-on voltage of 3.0 V and a maximum brightness of 12,000 cd/m^2 at 7 V have been observed. The devices are modestly stable but not yet comparable with the more complex vacuum-evaporated Ir(ppy)$_3$-based devices containing both HTL and ETL [245]. The high efficiency is attributed to the excellent uniform film-forming properties of the dendrimer Ir complex in the host layer, and the balanced charge injection.

Color tuning the green emission of Ir(ppy)$_3$ has been achieved by substitution of donor or acceptor groups either on the phenyl or on the pyridine ring. Another method to tune the color of emission is through ligand changes. While the phosphorescent QE of Ir(ppy)$_3$ in solution at room temperature is about 40–50%, a Swiss group could increase the QE close to 100% by changing the ligand configuration [518]. By introducing ligands such as CN-, NCS-, and NCO-, which have strong ligand field stabilization energy, a large gap between the E_g and LUMO of the ligand is created. This approach may allow for tuning the emission color to the blue range. The QE is also increased. Unfortunately, such materials are unstable for vacuum deposition due to the nonvolatile nature of these iridium complex salts. Although the materials may be solution-processable, their ionic charge makes it unlikely that long-lived devices can be achieved.

Ma et al. at PPG applied for patents on a series of iridium star-like bidentate complexes [519]. OLEDs fabricated using the dopants showed green emission with higher EQE and enhanced stability compared with a similar Ir(ppy)$_3$-based device.

Another high-efficiency series of iridium complexes is based on benzimidazole ligand complexes such as in Ir(bim)$_2$acac (488) (Scheme 3.87) [520]. Ir(bim)$_2$acac exhibits green phosphorescence in dilute CH$_2$Cl$_2$ solution with a maximum peak emission at 509 nm and a QE of about 40%. It can be doped as high as 12% without losing its QE. Disclosed by PPG and the UDC companies, this green dopant also showed improved EL performance as compared with an equivalent Ir(ppy)$_3$ device [521].

Kang et al. reported iridium complexes with cyclometalated 2-cycloalkenyl-pyridine ligands Ir(chpy)$_3$ (489) and Ir(mchpy)$_3$ (490) as highly efficient emitters for OLEDs [522]. The CIE coordinates of the devices in a configuration of ITO/NPB (40 nm)/CBP:iridium complex (30 nm)/BCP or BAlq (10 nm)/Alq$_3$ (40 nm)/LiF (1 nm)/Al (100 nm) were (0.40, 0.59). The device with Ir(chpy)$_3$ (6%) as an emitter showed a maximum EQE of 18.7%, a current efficiency of 69 cd/A, and a power efficiency of 62 lm/W, which are much higher than the Ir(ppy)$_3$-based device. They also developed a silyl-containing homoleptic Ir(III) complex, Ir(disppy)$_3$ (491) [523]. A power efficiency of 17.3 lm/W was achieved at 10 mA/cm^2 with Ir(disppy)$_3$ as a dopant compared with 11.7 lm/W for the equivalent Ir(ppy)$_3$ device.

Ha's group reported two heteroleptic phosphorescent blue–green emitters, Ir(ppy)$_2$(F$_2$-ppy) (492) and Ir(ppy)(F$_2$-ppy)$_2$ (493), which exhibit emission peaks at 502 and 495 nm with

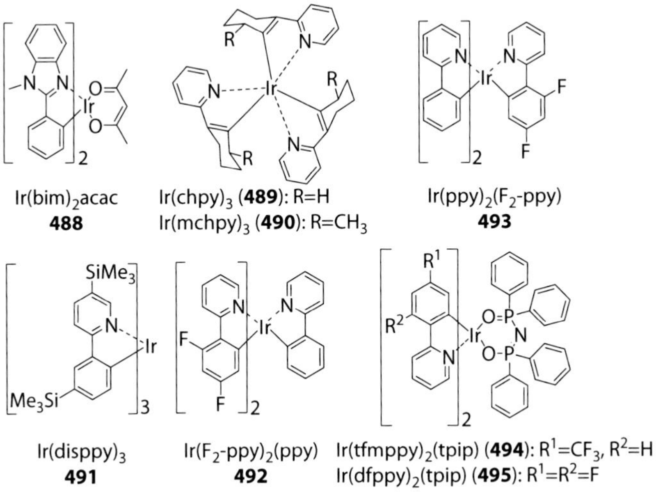

SCHEME 3.87
Chemical structures of green iridium complexes.

CIE coordinates of (0.26, 0.55) and (0.35, 0.54), giving maximum current efficiencies of 8.93 and 13.80 cd/A, respectively [524]. The EQE of the device with Ir(ppy)(F$_2$-ppy)$_2$ as a dopant is 3.63% at 10 mA/cm^2. It was expected that blue–green phosphors can be applied as prospective components for white OLEDs with suitable color complements.

Zhu et al. successfully prepared two bis-cyclometalated iridium complexes of Ir(tfmppy)$_2$(tpip) (**494**, tfmppy = 4-trifluoromethylphenylpyridine, tpip = tetraphe-nyl-imido-diphosphinate) and Ir(dfppy)$_2$(tpip) (**495**, dfppy = 4,6-difluorophenylpyridine) [525]. For the **494**-based device, a maximum luminance of 64,351 cd/m^2 and maximum efficiencies of 67.95 cd/A and 69.90 lm/W were achieved, and the efficiency roll-off from 100 to 1000 cd/m^2 was only 11.88%. For the **495**-based device, only a 6.7% efficiency roll-off from 100 to 1000 cd/m^2 was observed. The results suggest that these iridium complexes are good green and blue–green phosphorescent materials for potential applications in OLEDs, particularly at high luminance and high current density.

3.5.2.2.2.2 Gold Complexes Chan's group reported the discovery of an entirely new class of luminescent neutral cyclometalated alkynylgold(III) complexes, **496** (Scheme 3.88), that exhibits strong PL in various media at both ambient and low temperatures [526]. The EL spectra of the devices based on the developed gold complexes as triplet emitters peaked at ~528 nm with FWHM of 90 nm (3106 cm^{-1}) and CIE coordinates of (0.34, 0.57), exhibiting a maximum current efficiency of 37.4 cd/A and a maximum power efficiency of 26.2 lm/W, which is comparable to that of the optimized Ir(ppy)$_3$:CBP devices reported.

3.5.2.2.2.3 Copper Complexes Osawa's group developed a series of highly emissive three-coordinate copper(I) complexes, (dtpb)CuX (**497**) [X = Cl, Br, or I; dtpb = 1,2-bis(o-ditolylphosphino)benzene] [527]. Conventional OLEDs containing (dtpb)CuBr as the emitter exhibited bright green luminescence with a current efficiency of 65.3 cd/A and a maximum EQE of 21.3% in a three-layer structure of ITO (110 nm)/TAPC (30 nm)/mCP + 10% (dtpb)CuBr (25 nm)/3TPYMB (50 nm)/LiF (0.5 nm)/Al (100 nm). This efficiency is comparable to that of cyclometalated iridium(III)-based devices, and the present work suggests

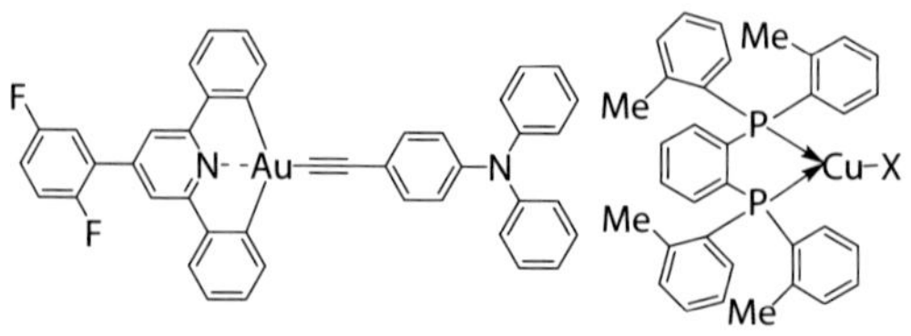

AuIII
496

(dtpb)CuX (X =
Cl, Br, or I)
497

SCHEME 3.88
Chemical structures of green gold and copper complexes.

that three-coordinate copper(I) complexes are also promising EL materials in terms of efficiency and thermal stability.

3.5.2.2.3 Blue Phosphorescent Dopants

There are still some obstacles to the development of blue phosphorescent OLEDs suitable for commercial applications. One particular problem is finding a material with a high-enough triplet state to correspond to a blue emission wavelength. Furthermore, finding a host into which this material may be doped without quenching the emissive state is a major challenge since the triplet state of the host must be even higher in energy than the emitter triplet state. The high-energy triplet state of a blue phosphorescent emitter appears to be difficult to maintain without degradation; therefore, lifetimes of phosphorescent blues are a problem. Most work reported in this area has focused on a material dubbed FIrpic and developed by the USC and Princeton groups (Scheme 3.89) [431,454].

Heteroleptic iridium(III) bis[(4,6-di-fluorophenyl)-pyridinato-*N*,*C*2′-picolinate] (FIrpic, **498**) and bis(4′,6′-difluorophenylpyridinato)tetrakis-1-pyrazolyl)borate (FIr6, **499**) use fluoro-substituted phenylpyridine ligands and an anionic 2-picolinic acid or tetra(pyrazolyl) borate as an auxiliary ligand, respectively [528]. Devices fabricated by using ITO/CuPc/ NPD/CBP or mCP:6% FIrpic/BAlq/LiF/Al exhibited a maximum QE of 6.1% and a luminous power efficiency of 7.7 lm/W with a peak luminance of 6400 cd/m². The EL performance has been improved by using a graded doped EML in a host DCB, which shows a peak power efficiency of 15.4 cd/A and a maximum brightness of 35,000 cd/m², a sharp increase compared with the conventional PHOLED, which has a peak power efficiency of 8.7 cd/A and a maximum brightness of 17,000 cd/m² [529]. This material, in a suitably high-energy carbazole host, has given up to 20 cd/A but with poor blue chromaticity, CIE (0.17, 0.34), which is a common problem with such materials resulting from the long wavelength emission tail that seems to arise from vibronic structures. Improvement of color purity and QE using FIr6 as blue dopant and Si-based large band-gap material as a host

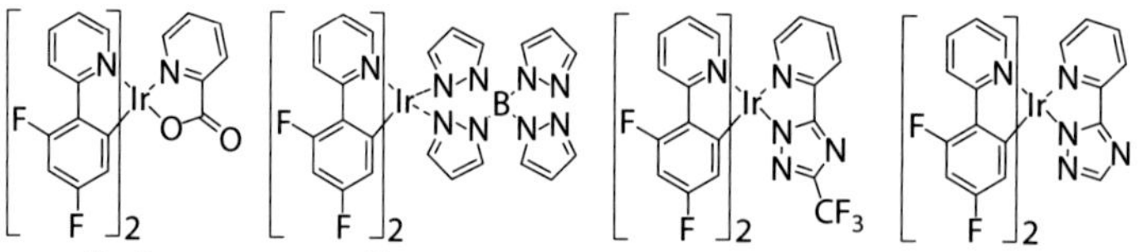

FIrpic
498

FIr6
499

FIrtaz
500

FIrN4
501

SCHEME 3.89
Chemical structures of blue iridium complexes.

gave peak quantum and power efficiencies of 12% and 14 lm/W in UGH2, with CIE of (0.16, 0.26) [450]. Again, the lifetimes of such systems are still problematic with only hundreds of hours of operation being claimed.

Yeh and coworkers have developed two new phosphorescent blue emitters that have two identical 2-(2,4-difluorophenyl)pyridine ligands and are derivatives of the FIrpic compound, iridium(III) bis(4,6-difluorophenylpyridinato)-3-(trifluoromethyl)-5-(pyridin-2-yl)-1, 2,4-triazolate (FIrtaz, **500**) and iridium(III) bis(4,6-difluorophenylpyridinato)-5-(pyridin-2-yl)-1*H*-tetrazolate (FIrN4, **501**) [530]. These two blue emitters both show a 10-nm blue shift of the emission compared with FIrpic. Unfortunately, the efficiency of such blue emitters is inferior to those of FIrpic and FIr6. There is no lifetime data reported for such devices.

In addition to using pyrazole as the auxiliary ligand to blue shift the emission of phenylpyridine homoleptic iridium complexes, replacing the pyridine ring of phenylpyridine ligands with a pyrazole ring can also widen the HOMO and LUMO band gap of the complexes as a consequence of lowering the HOMO and raising the LUMO energy levels of the complexes, respectively. Thompson's group synthesized a series of blue homoleptic phenylpyrazolyl iridium complexes (**502–506**) (Scheme 3.90) [531]. However, these homoleptic complexes showed strong ultrapure blue-to-blue emission (390–440 nm) only at very low temperature (77 K); however, unfortunately, they all showed very weak emission at room temperature, rendering them unsuitable for PHOLED applications. Although heteroleptic phenylpyrazolyl–phenylpyridine complexes showed moderate emission at room temperature, the emission color is bluish–green, and there is no device data reported for these complexes [532].

UDC claimed that they have successfully achieved a luminescent efficiency of 22 cd/A for a sky-blue PHOLED with CIE (0.16, 0.37) with >15,000 h operating lifetime at 200 cd/m^2 [533]. The possible chemical structure of this sky-blue emitter probably involves replacing the phenyl ring of phenylpyrazolyl with an extended fluorenyl unit, which has effective emission at room temperature, as presented at an American Chemical Society meeting by Dr. Forrest [534].

Orselli et al. investigated neutral heteroleptic mononuclear iridium complexes (**507**) with (2,4-difluoro)phenylpyridine and different pyridine-1,2,4-triazole ligands (Scheme 3.91). Various electron-poor/-rich substituents have been introduced into the 5 position of triazole ring to test their substituent effects. It was found that increasing the electron-withdrawing capabilities generally leads to a lowering of the HOMO level with a consequent slight widening of the HOMO–LUMO gap and a blue shift in emission. **507a** and **507e** were characterized by a device configuration of ITO/*m*-MTDATA:F4-TCNQ (40 nm)/*m*-MTDATA (10 nm)/TCTA:dopant (25 nm)/HBL (10 nm)/TPBI (40 nm)/LiF (1 nm)/Al(100 nm), and a high EQE of 7% together with a good CIE of (0.17, 0.27) were obtained by these devices [535].

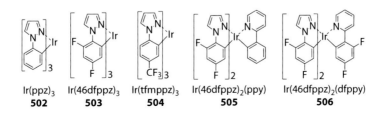

Ir(ppz)$_3$ **502**   Ir(46dfppz)$_3$ **503**   Ir(tfmppz)$_3$ **504**   Ir(46dfppz)$_2$(ppy) **505**   Ir(46dfppz)$_2$(dfppy) **506**

SCHEME 3.90
Chemical structures of blue iridium complexes.

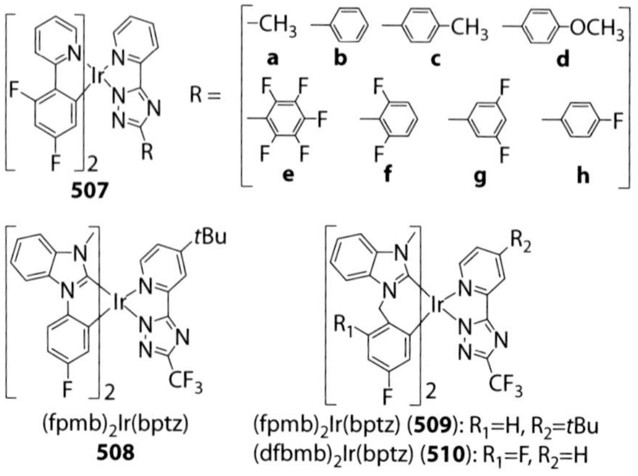

SCHEME 3.91
Chemical structures of blue iridium complexes.

Chang et al. reported the use of a different functionalized benzyl carbene chelate ligands and their coordination to a different iridium source reagent could achieve blue-emitting Ir(III) metal complexes. A prototype phosphorescent OLED with CIE coordinates (0.16, 0.13) has been fabricated with **510**; the CIE coordinates of the obtained device are much better than for other devices fabricated from some well-known blue phosphorescent dopants, like FIrpic, FIr6, FIrtaz, and FIrN4 [536].

Wang et al. synthesized blue light–emitting bisorthometalated Ir(III) complex called Ir(ppy)$_2$(PPh$_3$)Cl with substitutional pyridine and phosphine ligands. The electrophosphorescent polymer light-emitting diodes based on Ir(ppy)$_2$(PPh$_3$)Cl have low turn-on voltages of <4 V. The brightness and EQE for the single-layer device reaches 1190 cd/m^2 at 12 V and 2.1% at a current density of 9.85 mA/cm^2, respectively [537].

The progress with phosphorescent blue emitters suggests that it may be rather possible to achieve high-efficiency blue phosphorescent candidates by carefully designing the proper ligands coupled with appropriate selection of auxiliary ligands.

3.5.3 Nondoped Organic Light-Emitting Materials

3.5.3.1 Nondoped Red Light–Emitting Materials

Red emission chomophores having a long wavelength emission band are usually polar, such as the above DCM series. The push–pull red emitters are normally prone to aggregation in the solid state owing to dipole–dipole interactions or through intermolecular π–π stacking, especially when the molecules are flat, as is the case for DCM. As a consequence, the push–pull red emitter is used mainly as the dopant in guest–host systems. However, there are some push–pull molecules that have a twisted molecular structure and can be used as nondoped red emitters. This class of red emitters demonstrates better performance as compared with the arylamine-substituted PAH series (**511–517**) (Scheme 3.92).

All of these molecules have an electron-withdrawing group in the center and two electron-donating arylamine groups attached at each end. They all demonstrate strong red fluorescence in the solid state, and they can also be used as nondoped OLED materials. Nondoped OLEDs avoid reproducibility problems of reliably achieving the optimum doping concentration during processing and are easily adapted to a mass production line.

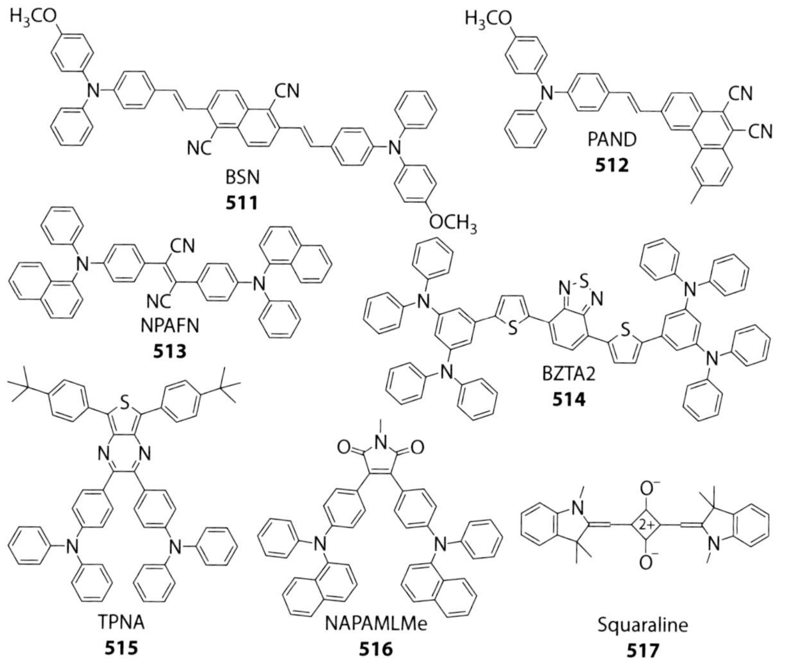

SCHEME 3.92
Chemical structures of nondoped red emitters.

The 1,10-dicyano-substituted bis-styrylnaphthalene derivative (BSN, **511**) reported by Sony has a very high photoluminescent QE up to 80% with emission wavelength of 630 nm. In a device of ITO/2-TNATA/NPD/BSN/Alq$_3$/Li$_2$O/Al, BSN as a red emitter without dopants displays an impressive high-color stability with luminous efficiency of 2.8 cd/A at 500 cd/m^2 with CIE (0.63, 0.37) [538]. In a recent patent, Sony disclosed another high-efficiency, stable, and bright red emitter 3-(2-{4-[(4-methoxy-phenyl)-phenyl-amino]-phenyl}-vinyl)-6-methyl-phenanthrene-9,10-dicarbonitrile (PAND, **512**) [539]. This compound emits bright red emission with maximum wavelength emission at 645 nm.

The nondoped red emitter bis(4-(N-(1-naphthyl)phenylamino)phenyl)fumaronitrile (NPAFN, **513**), reported by Chen's group, showed very interesting photoluminescent properties: in solution, the compound barely emits red emission; however, in the solid state, it exhibits strong orange–red emission at the maximum wavelength of 616 nm, and in the OLED of ITO/NPAFN/BCP/TPBI/Mg:Ag, an EL had a pure red emission with a maxima wavelength at 636 nm, corresponding to CIE (0.64, 0.35).

BTZA2 (**514**) containing a benzo[1,2,5]thiadiazole core and peripheral diarylamines groups exhibits strong red fluorescence in the solid state. Single-layer OLEDs fabricated using this compound demonstrated a maximum brightness of >5000 cd/m^2 [540].

Compound NAPAMLMe (**516**) is an amorphous glass material reported by Yeh et al. in a nondoped OLED [541]. An optimized OLED device, with a structure of ITO/NPD/NAPAMLMe/BCP/Alq$_3$/Mg:Ag, emits pure red light with CIE (0.66, 0.33), a brightness of 4600 cd/m^2, and an EQE as high as 1.6%. In addition, the color of the red emission is rather stable and is also voltage independent, which has been attributed to the very well-confined excitonic zone. A saturated red emission with CIE (0.64, 0.33) has been observed using a star-shaped thieno-[3,4-*b*]-pyrazine (TPNA, **515**) red emitter in a simple double-layer OLED,

ITO/TPNA/TPBI/Mg:Ag, in which the TPNA compound functions as a hole-transporting red emitter [542].

Lee et al. developed two red fluorophores, PhSPFN (**518**) and FPhSPFN (**519**) (Scheme 3.93), and nondoped red OLEDs with the structure of ITO/NPB (30 nm)/PhSPFN or FPhSPFN (30 nm)/BCP (20 nm)/TPBI (30 nm)/LiF (1 nm)/Al (150 nm) were fabricated [543]. The PhSPFN-based device showed unsatisfactory performance because the red EL of PhSPFN is partially beyond the visible range to humans. On the other hand, the FPhSPFN-based device exhibited EQE of 2.5% and luminance as high as 500 cd/m^2 at 20 mA/cm^2. Such performance of FPhSPFN is obviously higher than that of the previously reported NPAFN.

A dicyanovinyl-functionalized chromophore TPD-FCN (**520**) was reported as a red emitter by Yang's group [544]. The nondoped device with a configuration of ITO/CuPc (8 nm)/DBTPA (20 nm)/TPD-FCN (35 nm)/BCP (15 nm)/Alq$_3$ (5 nm)/LiF (1 nm)/Al (100 nm) reached a high brightness of 1399 cd/m^2 and a current efficiency of 2.85 cd/A at a low current density of 20 mA/cm^2.

Huang et al. reported a dithienylbenzothiadiazole derivative (**521**) that exhibits a PL quantum yield of 0.58 and shows a maximum fluorescence emission at 613 nm [545]. Since there is preferential electron transport in the red-emitting compound, the device without an HBL demonstrated a satisfactory current efficiency of 2.2 cd/A at a brightness of 402 cd/m^2.

Qiu's group developed a naphtha(2,3-*c*)(1,2,5)thiadiazole derivative BDPNTD (**522**) for nondoped red OLEDs [546]. BDPNTD has a low-lying LUMO energy level of −3.58 eV, indicating a low electron-injection barrier. A double-layer device, with a structure of ITO/NPB (40 nm)/BDPNTD (80 nm)/Mg:Ag (150 nm)/Ag (50 nm) without using any ETL, displays a maximum EQE 2.52% (corresponding to 2.05 cd/A) at the current density of 22.7 A/m^2, an emission wavelength at 644 nm, and CIE coordinates of (0.65, 0.35).

3.5.3.2 Nondoped Green Light–Emitting Materials

An efficient ambipolar pure green emitter, 2,2-bis(triphenylamine)-3,3-diphenyl-[6,6]biquinoxaline (TPA-2PQx, **523**), was developed by Jenekhe's group (Scheme 3.94) [547]. The

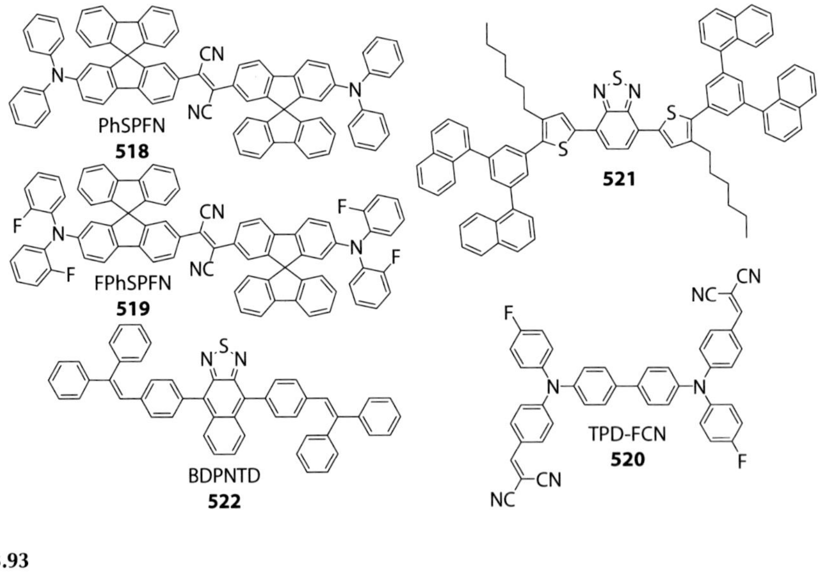

SCHEME 3.93
Chemical structures of nondoped red emitters.

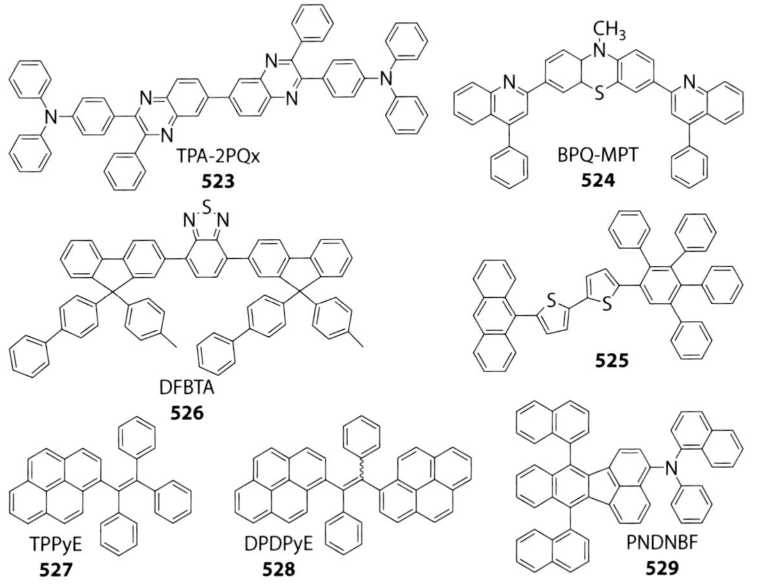

SCHEME 3.94
Chemical structures of nondoped green emitters.

single-layer OLEDs with TPA-2PQx as a nondoped emitter emitted pure green light with an EL maximum at 505 nm and CIE coordinates of (0.19, 0.55), giving bright green EL up to 15,330 cd/m^2 and high efficiency up to 7.9 cd/A. A donor–acceptor molecule, 3,7-[bis(4-phenyl-2-quinolyl)]-10-methylphenothiazine (BPQ-MPT, **524**), was also reported by the same group [548]. Green-light emission with a maximum at ca. 540 nm and CIE coordinates of (0.35, 0.60) were achieved by using BPQ-MPT as a nondoped EML, giving high brightness of >40,000 cd/m^2 and high efficiency of 21.9 cd/A at 1140 cd/m^2.

A nondoped emitter of 5-(2,3,4,5-tetraphenyl)phenyl-5-(9-anthracenyl)-2,2-bithiophene (**525**) was developed by Liu's group [549], achieving a maximum current efficiency of 4.9 cd/A and an EQE of 1.85% with an EL peak centered at 516 nm and CIE coordinates of (0.27, 0.51).

Ku et al. reported a highly efficient nondoped OLED by using 9,9-diarylfluorene-terminated 2,1,3-benzothiadiazole (DFBTA, **526**) as a green emitter, giving a green light emission with CIE coordinates of (0.38, 0.58), a maximum EQE (η_{ext}) of 3.7% (12.9 cd/A), and a maximum brightness at 168,000 cd/m^2 [550].

Two ethene derivatives, 1,2,2-tripheny-1-pyrenylethene (TPPyE, **527**) and 1,2-diphenyl-1,2-di-pyrenylethene (DPDPyE, **528**), were developed by Tang's group [551]. In contrast to the general observation that excimer formation quenches the light emission of fluorophores, TPPyE and DPDPyE exhibit efficient excimer emissions in the solid state with high fluorescence quantum yield up to 100%. Green-light emission with maximum current efficiency of 10.2 cd/A and maximum EQE of 3.3% was achieved by using DPDPyE as a nondoped EML, making the present luminogens promising materials for the construction of efficient nondoped green OLEDs.

A series of benzo[*k*]fluoranthene derivatives incorporating a diarylamine group was developed by Tian's group as green electroluminescent emitters [552]. Two-layer electroluminescent devices employing them as HTL and EML achieved efficient green-light emission with CIE coordinates of (0.28, 0.61) under a low driving voltage. A further optimized three-layer device based on one of these compounds, 3-(1-naphthylphenylamino)-7,12-diphenylbenzo[*k*]fluo-ranthene (PNDPBF, **529**), exhibited a maximum current efficiency of

10.2 cd/A (6.7 lm/W at 4.70 V) in a device structure of ITO/CF$_x$/2-TNAT (40 nm)/PNDNBF (40 nm)/TPBI (20 nm)/LiF (1 nm)/Al.

3.5.3.3 Nondoped Blue Light–Emitting Materials

Nondoped blue light–emitting materials are the most widely investigated, and most of them are anthracene derivatives. Wu et al. reported a blue light–emitting material, BFAn (**530**), containing an anthracene core and end capped with 9-*p*-tolyl-9-fluorenyl groups (Scheme 3.95). Introduction of the sterically congested fluorene groups provides BFAn with a decomposition temperature (T_d) of 510°C and a superior T_g value of 227°C, making BFAn form a high-quality amorphous film with good morphological stability. OLEDs using BFAn as a nondoped emitter exhibited an excellent EQE of 5.1% (5.6 cd/A) with CIE coordinates of (0.15, 0.12) [553].

Zhang et al. developed two blue light–emitting materials, namely TBMFA (**531**) and TBMFPA (**532**), containing an anthracene core and end capped with 9,9-dimethylfluorene. TBMFPA possesses a high T_g of 165°C, and OLED with TBMFPA as a nondoped EML exhibited an EQE of 4.76% (2.67 cd/A) with CIE coordinates of (0.147, 0.096) [554].

Shih et al. reported a blue light–emitting material containing an anthracene core and end capped with tetraphenylenethylene, TPVAn (**533**), which has a high T_g of 155°C, indicating excellent thermal stability. While used as a nondoped emitter, the device showed a high efficiency of 5.3 cd/A with CIE coordinates of (0.14, 0.12) [555].

Through introducing phenyl or naphthyl group on the *ortho*-position of the phenyl substituent in 9,10-diphenyleneanthracene, Wang et al. synthesized two anthracene derivatives, TBBPA (**534**) and TBNPA (**535**). OLEDs based on TBBPA as the nondoped blue emitter exhibited deep-blue emission with CIE coordinates of (0.15, 0.07) and a high efficiency of 3.0% [556].

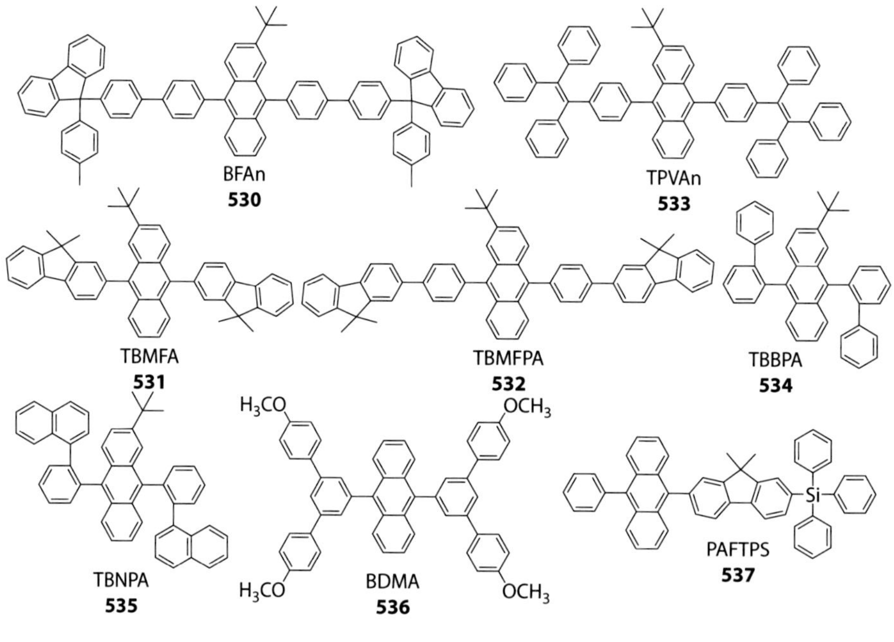

SCHEME 3.95
Chemical structures of nondoped blue emitters.

Park et al. reported a di-(4-methoxyphenyl)-phenyl–substituted anthracene derivative, BDMA (**536**). BDMA has a high fluorescence yield and good thermal stability with a high T_g of 206°C. OLED with BDMA as a nondoped EML shows a maximum current efficiency of 2.2 cd/A with a turn-on voltage of 4.4 V. The device emits blue light stably even at high voltages of 5–18 V with CIE coordinates of (0.16, 0.12) [557].

Kim et al. reported a series of anthracene derivatives with (9,9-dimethyl-9H-fluoren-7-yl) triphenylsilyl or triphenylsilyl group end capped on various positions of the anthracene skeleton. Introduction of a triphenylsilane group renders these derivatives with a nonplanar structure, which could suppress the close packing of molecules in the solid state and improve the morphological stability of thin films. An OLED incorporating PAFTPS (**537**) as a nondoped EML exhibited a high EQE of 2.02% with CIE coordinates of (0.152, 0.072) [558].

Tao et al. developed a series of anthracene derivatives end capped with TPA, phenyl, naphthyl, and pyrenyl. Owing to good hole-transport property of anthracene–TPA derivatives, highly efficient OLEDs were achieved by using these derivatives as both emitter and hole transporter [559].

Cho et al. developed an anthracene-based deep blue–emitting material, PCAN (**538**), which is asymmetrically functionalized with *N*-phenylcarbazole and naphthalene. The vacuum-deposited, nondoped EL device exhibited excellent current efficiency as high as 3.64 cd/A (4.61%) with a saturated deep-blue CIE coordinates of (0.151, 0.086) [560].

Chien et al. reported a highly efficient blue emitter, POAn (**539**) (Scheme 3.96), composed of anthracene core and terminal triphenylphosphine oxide groups. Owing to its noncoplanar

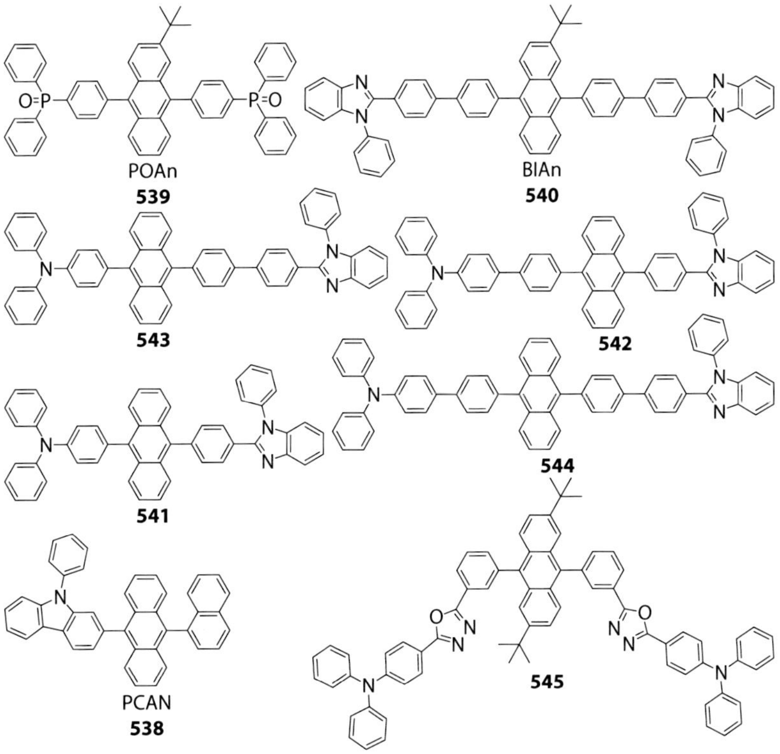

SCHEME 3.96
Chemical structures of nondoped blue emitters.

configuration that reduces the tendency to crystalize and the weaker intermolecular inter-action, POAn possesses pronounced morphological stability. Simple double-layer devices incorporating POAn as the emitting, electron-transporting, and electron-injecting material produce bright deep-blue light with CIE coordinates of (0.15, 0.07) and maximum efficiency of 4.3% (2.9 cd/A) [561]. Wu et al. also developed an efficient anthracene-based ETM BIAn (540) containing 1-phenylbenzoimidazyl end groups. When BIAn was used as an ETL instead of TPBI, the operating voltage was further decreased to give improved power efficiency [553].

Huang et al. reported a series of bipolar anthracene derivatives containing hole-transporting TPA and electron-transporting benzimidazole. OLEDs comprising these anthracene derivatives as nondoped blue emitter were fabricated. The devices based on 541, 543, and 544 emit blue light, while the 542-based device emits white light at a low current density due to the electromer formation. A single-layer OLED device based on 544 exhibited a current efficiency of 3.33 cd/A with CIE coordinates of (0.16, 0.16) at 20 mA/cm^2 and a maximum brightness of 8472 cd/m^2 at 8.7 V [562].

Reddy et al. reported a series of anthracene-based blue emitters in which 1,3,4-oxadiazole groups were introduced to improve their electron-transport properties. The electron affinity of these derivatives are around 3.7 eV, and the ionization potential are around 6.7–6.8 eV, which is higher than the most commonly used electron-transport emitting material, Alq$_3$. These materials can be used as a nondoped EML as well as an efficient ETM [563].

Zhu et al. reported four blue fluorescent emitters with bifunctional charge-transport groups, TPA and oxadiazole, appended to the 9 and 10 positions of the anthracene core. These four compounds have high T_g >154°C and possess noncoplanar structures to sup-press intermolecular interaction within the films. OLED with 545 as an emitter achieves a maximum power efficiency of 2.0 lm/W and CIE coordinates of (0.16, 0.10) [564].

Besides anthracene, fluorene is also a widely investigated unit for nondoped blue emit-ters. Kreger et al. reported an efficient OLED by using 546 with a TPA core and three fluorene arms as a blue emitter (Scheme 3.97). Owing to the hole-transporting property

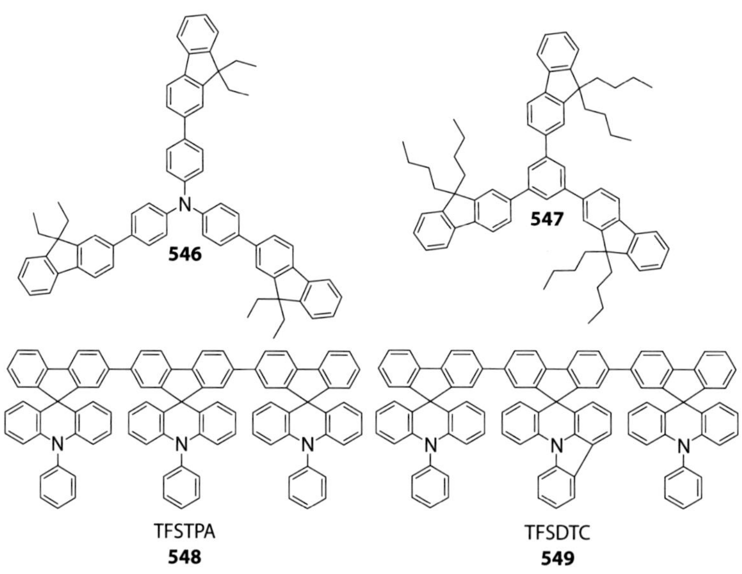

SCHEME 3.97
Chemical structures of nondoped blue emitters.

contributed by electron-rich TPA core, it could also be treated as an HTL. A hole-blocking material with a low-lying HOMO of −6.0 eV, referred to as **547**, with a triphenylbenzene core and three fluorene side arms was also developed in their work. The blue light–emitting device showed pure blue emission with a current efficiency of 2 cd/A and CIE coordinates of (0.15, 0.15) [565].

Jiang et al. developed a series of fluorene-based oligomers with novel spiro-annulated triarylamine structures. Simple double-layer devices with a configuration of ITO/TFSTPA (**548**) or TFSDTC (**549**)/TPBI/LiF/Al, where TFSTPA and TFSDTC serve as hole-transporting blue light–emitting materials, show a deep-blue emission with CIE coordinates of (0.17, 0.12) and (0.16, 0.07) and current efficiencies of 1.63 and 1.91 cd/A for the TFSTPA- and TFSDTC-based devices, respectively [140].

Phosphorescent emitters can be also used as a nondoped EML. As an example, Liu et al. reported two iridium complexes containing carbazole-functionalized β-diketonate, namely $Ir(ppy)_2(CBDK)$ and $Ir(dfppy)_2(CBDK)$, to be used as nondoped phosphorescent emitters in blue–green and green OLEDs. The results show that the nondoped four-layer device for $Ir(ppy)_2(CBDK)$ achieves maximum power efficiency of 4.54 lm/W, which is higher than that of $Ir(ppy)_2(acac)$ (0.53 lm/W) [566].

Liao et al. reported a series of blue light–emitting analogues of the well-known green fluorophore Alq_3. These metal chelates of 4-hydroxy-1,5-naphthyridine exhibit very deep-blue fluorescence, wide band gap energy, high charge-carrier mobility, and superior thermal stability. When these complexes are used as emitters, maximum EQE of 4.2% and CIE of (0.15, 0.07) could be obtained. They are also a qualified wide-band-gap host material for efficient blue perylene [CIE (0.14, 0.17) and maximum EQE 3.8%] or deep-blue 9,10-diphenylanthracene [CIE (0.15, 0.06) and maximum EQE 2.8%]. Besides, these chelates can also be utilized as electron-transporting materials [567].

3.6 White Organic Light-Emitting Diodes

3.6.1 Fluorescent White Organic Light-Emitting Diodes

White organic light-emitting diodes (WOLEDs) have many applications, including as backlights in liquid crystal displays, serving as light sources in fabricating full-color OLED displays using color-filtering techniques and as general lighting sources. According to the emitters used in WOLEDs, WOLEDs can be classified as fluorescent WOLEDs, phosphorescent WOLEDs, and fluorescent/phosphorescent hybrid WOLEDs.

3.6.1.1 Fluorescent WOLEDs with Primary Colors

The combination of three primary emission colors of blue, green, and red is the simplest method to achieve WOLEDs. Kido et al. reported using three emitter layers with different carrier-transport properties to produce a white emission [568]. The multilayer structure of such an OLED is ITO/TPD/*p*-EtTAZ/Alq_3/Alq_3:Nile red/Alq_3/Mg:Ag, in which a blue emission from the TPD layer, a green emission from the Alq_3 layer, and a red emission from the Nile red dye are produced (Figure 3.8). These three primary-color emissions produce the white light by carefully controlling the recombination zone and modulating the doping level. A maximum luminance >2200 cd/m² is achieved at 16 V. This strategy,

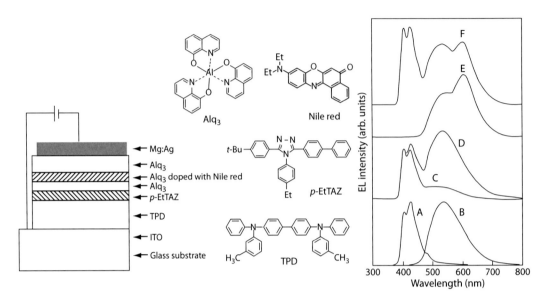

FIGURE 3.8
Configuration of the white OLED structure and the chemical structures of the materials (left). EL spectra of
(A) ITO/TPD/*p*-EtTAZ/Mg:Ag, (B) ITO/TPD/Alq/Mg:Ag, (C) ITO/TPD/*p*-EtTAZ (5 nm)/Alq/Mg:Ag, (D) ITO/
TPD/*p*-EtTAZ (3 nm)/Alq/Mg:Ag, (E) ITO/TPD/Alq/Alq:1% Nile red/Alq/Mg:Ag, and (F) ITO/TPD/*p*-EtTAZ
(3 nm)/Alq/Alq:1% Nile red/Alq/Mg:Ag (right). (From Kido, J., Kimura, M., and Nagai, K., *Science*, 267, 1332,
1995. With permission.)

although it can be used to achieve white emission, leads to a tedious device fabrication
process, and the emission color is sensitive to the operating voltage and device structure
parameters such as active layer thickness and doping concentration.

WOLEDs with four wavelengths were fabricated by Jeong et al. [569]. They used three
doped layers, which were obtained by separating recombination zones into three emissive layers. The white light could be emitted by simultaneously controlling the emitter
thickness and concentration of fluorescent dyes in each emissive layer, resulting in partial
excitations among those three emissive layers. The EL spectra of the device obtained in
this study were not sensitive to the driving voltage of the device. Also, the maximum
luminance for the white OLED with CIE coordinates of (0.34, 0.34) was 56,300 cd/m² at an
applied bias voltage of 11.6 V. Also, its EQE and power efficiency at 100 cd/m² were 1.68%
and 2.41 lm/W, respectively.

Seo et al. had demonstrated highly efficient WOLEDs by using two emissive materials as
dopant [570]. It was found that the WOLEDs fabricated in this study emitted a white color
consisting of three primary colors: red, green, and blue. The luminance–voltage (*L–V*)
characteristics of the WOLEDs showed a maximum luminance of 30,500 cd/m² at 14 V and
a maximum current efficiency of 38.0 cd/A. The CIE coordinates of the WOLEDs are (0.33,
0.40) at 10 V.

3.6.1.2 Fluorescent WOLEDs with Complementary Colors

White emission can also be achieved by the combination of complementary colors like blue
and orange–red emission, and this is the most widely investigated route to achieve white
emission.

Cheon and Shinar demonstrated that by deposition of a thin layer of the blue emitter DPVBI on the DCM-2–doped NPD device, an efficient WOLED with a brightness of >50,000 cd/m² and a power efficiency of 4.1 lm/W (EQE of 3.0%) could be achieved [571].

Chien et al. reported a very bright WOLED with a much extended lifetime using 2,7-bis(2,2-diphenylvinyl)-9,9′-spirobifluorene (DPVSBF) doped with DCJTB as the EML. The device ITO/NPD/DPVSBF:0.2% DCJTB/Alq₃/LiF/Al showed a brightness of 1575 cd/m² with an EQE of 3.31%, a current efficiency of 8 cd/A, and a power efficiency of 5.35 lm/W at a current density of 20 mA/cm² at 4.7 V. The CIE stayed constant at (0.32–0.35, 0.34–0.36) when driving voltages changed from 6 to 12 V [361].

A highly efficient and chromatically stable WOLED based on an anthracene derivative blue emitter doped with yellow–orange 5,6,11,12-tetraphenylnaphthacene (rubrene) was reported by Qiu's group [572]. By simple deposition of the mixture of the predoped rubrene and two anthracene derivatives in the host material NPD, a maximum brightness of 20,100 cd/m² with a peak EQE of 2.4% (5.6 cd/A) at 9 V and luminance-independent CIE coordinates of (0.32, 0.34) has been achieved in an OLED with the device structure of ITO/NPD/ADN:2.5% TBADN:0.025% rubrene/Alq₃/Mg:Ag. The results also indicate that using two anthracene derivatives improves the morphology of the doped films and depresses the crystallization of the dopant, which in turn contributes to the high performance and stability of the device. This strategy has also been applied to achieve a high-efficiency blue OLED by the same group.

A series of highly fluorescent blue-emitting materials based on fluorene and anthracene hybrids (550–552, Scheme 3.98) were designed and synthesized for OLEDs by Ye et al. [573]. These materials feature a phenyl-substituted fluorene dimer as a bulky and rigid core, and anthracene as a functional active group. They fabricated highly efficient fluorescent WOLEDs employing an interesting emission in the longer wavelength of DNAPFB combined with DPAVBi emission due to incomplete energy transfer between DNAPFB and DPAVBi to achieve stable white-light emission in a binary host–guest single emissive layer with high efficiency of 14.8 cd/A (5.3 lm/W).

van Gemmern et al. investigated the influence of carrier conductivity and injection on efficiency and chromaticity in small-molecule WOLEDs based on a spirobiphenyl

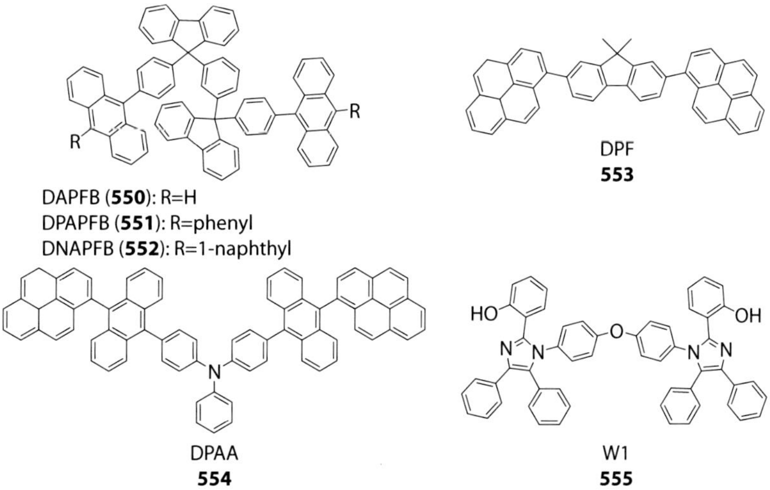

DAPFB (**550**): R=H
DPAPFB (**551**): R=phenyl
DNAPFB (**552**): R=1-naphthyl

DPF
553

DPAA
554

W1
555

SCHEME 3.98
Chemical structures of materials used for fluorescent WOLEDs.

derivative [4,4′-bis(2,2′-diphenylvinyl)-1,1′-spirobiphenyl] and rubrene [574]. Devices with warm-white emission with CIE coordinates of (0.43, 0.42) were produced with power and current efficiencies of 5 lm/W and 10.9 cd/A, respectively, at a luminance of 1000 cd/m². The maximum EQE at a current density of 20 mA/cm² was 4.6%.

Choukri et al. regulated the thickness and the position of the rubrene layer, allowing fine chromaticity tuning from deep blue to pure yellow via bright white with CIE coordinates of (0.33, 0.32) [575]. Jou et al. fabricated efficient, chromaticity-stable fluorescent white, including pure white, OLEDs by having an effective exciton-confining device architecture with a single EML via vapor deposition of the solution-mixed targets of a blue light–emitting host doped with a trace amount of a red and/or a green dye [576]. The best power efficiency of the resulting two-spectrum fluorescent devices was 7.5 lm/W (10.8 cd/A) at 11 cd/m² with CIE of (0.423, 0.426), or 6.5 lm/W (9.6 cd/A) at 12 cd/m² with pure white light with coordinates of (0.346, 0.343). The best power efficiency of the three-spectrum counterparts was 6.7 lm/W (9.9 cd/A) at 15 cd/m² with a nearly pure white light with CIE coordinates of (0.325, 0.374). All color variations were less than (0.007, 0.006) between 100 and 10,000 cd/m². On the basis of mixing orange and blue emission, Shen et al. [577] realized high-brightness WOLEDs in which the maximum luminance was 15,000 cd/m² at 20 V and 8010 cd/m² at 12 V. Su et al. [578] used copper complexes as an HIL and Cu(bis(2-(diphenylphosphino) phenyl)ether(6,7-dicyanodipyrido(2,2-*d*:2′,3′-*f*)quinoxalin))-doped CBP as a yellow EML to give WOLEDs. Wu's group [579] used a thin layer of ADN [9,10-bis(2-naphthyl)anthracene] doped with rubrene as the source of white emission. The device with the structure of ITO/NPB/ADN:0.5% rubrene/Alq$_3$/Mg:Ag showed a maximum current efficiency of 3.7 cd/A with CIE coordinates of (0.33, 0.43).

Extremely low-voltage WOLEDs in p–i–n structure with fluorescent emitters were realized by Duan et al. [580]. White light was obtained by a two-complementary-colors system, in which the yellow- and the blue-emitting components are based on rubrene and spiro-DPVBi, respectively. The effects on the device performances of various EBLs and hosts for rubrene were discussed. The best device showed a luminance of 1000 cd/m² at bias as low as 2.9 V, and 10,000 cd/m² at 4.7 V with a maximum power efficiency of 8.7 lm/W. The CIE chromaticity coordinates changed from (0.36, 0.45) at 1000 cd/m² to (0.33, 0.42) at 10,000 cd/m², showing high color stability.

Highly efficient p–i–n two-component WOLEDs had been fabricated by Ho et al. [581], with a thin dual EML composed of one codeposited emitting layer with dyes and one emitting layer, which gave rise to a balanced white emission. The p–i–n white device achieved a current efficiency of 10 cd/A and a power efficiency of 9.3 lm/W at 1000 cd/m² and a low voltage of 3.4 V with CIE coordinates of (0.32, 0.43). Otherwise, the electroluminescent color of this p–i–n WOLED had been shown to be immune to drive current density variations.

Hsiao et al. [582] reduced the driving voltage in WOLEDs by selectively doping into an ambipolar blue-emitting layer. A thin codoped layer served as a probe for detecting the position of maximum recombination rate in the dopant layer. Owing to the energy barrier and bipolar carrier transport, the maximum recombination rate was found to be close to, but not exactly at, the interface of the hole-transporting layer and the EML. The device driving voltage decreased by 21.7%, nearly 2 V in reduction, due to the increased recombination current from the faster exciton relaxation. They are able to successfully generate 112 cd/m² at 4 V from their WOLEDs simply by engineering the structure of the EML.

Jou et al. [583] fabricated efficient fluorescent WOLEDs with the use of an efficient blue–green host material doped with red dye. One resulting two-wavelength white-emission device showed a maximum EQE of 4.8% and a high power efficiency of 14.8 lm/W with a luminance of 100 cd/m² at 3.8 V. The high efficiency might be attributed to the high EL character of the host, relatively high host-to-guest energy transfer efficiency, and effective device architecture.

High-efficiency and low-operating-voltage fluorescent WOLEDs had been realized by doping either BPhen or NPB into the blue light–emissive layer [584]. Devices doped with BPhen (or NPB) exhibited a maximum power efficiency of 8.7 lm/W (7.6 lm/W). Such performance improvement was ascribed to the incorporation of a better electron-transporting layer and an improved carrier transport through the emissive layer by mixing with the higher-drift-mobility materials. This provided a simple and general means to improve the power efficiency of WOLEDs.

A relatively high-efficiency, fluorescent pure white OLED was fabricated by Jou et al. [585]; they used a polysilicic acid (PSA) nanodot-embedded polymeric HTL. The diode employed a mixed host in the single emissive layer, which comprised 0.5 wt.% yellow 5,6,11,12-tetra-phenylnaphthacene doped in the mixed host of 50% 2-(*N,N*-diphenyl-amino)-6-(4-(*N,N*-diphenylamino)styryl)naphthalene and 50% *N,N*'-bis-(1-naphthyl)-*N,N*'-diphenyl-1,10-biphenyl-4-4'-diamine. By incorporating 7 wt.% 3-nm PSA nanodots into the HTL of poly(3,4-ethylene-dioxythiophene)-poly(styrenesulfonate), the efficiency at 100 cd/m^2 was increased from 13.5 lm/W (14.7 cd/A; EQE: 7.2%) to 17.1 lm/W (17.6 cd/A; EQE: 8.3%). The marked efficiency improvement may be attributed to the introduction of the PSA nanodot, leading to a better carrier-injection balance.

Wang et al. [586] demonstrated a fluorescent WOLED with double emissive layers. The yellow and blue dyes were doped into the same conductive host materials. The maximum luminance and power efficiency of the WOLED were 14.6 cd/A and 9.5 lm/W at 0.01 mA/cm^2, with a maximum brightness of 20,100 cd/m^2 at 17.8 V. The CIE coordinates changed lightly from (0.27, 0.37) to (0.28, 0.36), as the applied voltage increased from 6 to 16 V. The high efficiencies can be attributed to the balance between holes and electrons.

Lin et al. had developed a WOLED with stable color chromaticity with respect to drive current density [587]. The device achieved an EL efficiency of 17.1 cd/A and 7.9 lm/W at 20 mA/cm^2. From 112 to 32,010 nits, the CIE$_{x,y}$ coordinates variation was $\Delta(x,y) < \pm(0.01, 0.01)$, which was superior to common WOLEDs that tended to change color with drive conditions. Moreover, the device reached a half-decay $t_{1/2}$ lifetime of >40,000 h at an initial luminance of 300 cd/m^2.

Tsai and Jou [588] fabricated a long-lifetime, high-efficiency WOLED with a mixed host in one or double EMLs. This device exhibited the longest lifetime, 5 times longer than that of its pure NPB counterpart. The resulting efficiency was 6.0 lm/W (10.9 cd/A) at 10 mA/cm^2, 33% better than that of the NPB counterpart. They attributed the improvements to the mixed-host structure, which effectively dispersed carriers and gave a good charge balance.

WOLEDs with mixed interfaces between emitting layers (MI-EML WOLEDs) were developed by Lee et al. [589], giving luminance and efficiency as high as 26,213 cd/m^2 and 9.85 cd/A. Efficiencies of the MI-EML WOLEDs are about 1.5 times better than those of the conventional three-EML WOLEDs for luminance of 1000–5000 cd/m^2, and their half-decay lifetime shows 3.1 times improvement. Note that if one operates typical active-matrix mobile-phone displays based on combination of WOLED and color filters to produce standard white emission for high-definition televisions and illumination sources, MI-EML WOLEDs will have advantages of 25% less power consumption and 2.8 times longer lifetime over conventional three-EML WOLEDs.

Jou et al. fabricated efficient fluorescent WOLEDs with low carrier-injection barriers [590]. By blending in the blue host in the emissive layer an efficient electroluminescent cohost, with the doping of a trace amount of red dye, bright and color-stable white emission with high power efficiency of 14.6 lm/W at 100 cd/m^2 or current efficiency of 19.2 cd/A at 300 cd/m^2 or 18.7 cd/A at 10,000 cd/m^2 was obtained. The resulting synergistic increase in brightness and efficiency may be attributed to the presence of cascading new routes with comparatively lower electron-injection barrier.

A series of group III metal chelates have been synthesized and characterized for the versatile application of OLEDs by Liao et al. [567]. These metal chelates were based on 4-hydroxy-1,5-naphthyridine derivatives as chelating ligands, and they are the blue-version analogues of well-known green fluorophore Alq$_3$. They had successfully demonstrated that the application of these unusual hydroxynaphthyridine metal chelates can be very versatile and effective. First, they had solved or alleviated the problem of exciplex formation that took place between the hole-transporting layer and hydroxynaphthyridine metal chelates, of which OLED application has been prohibited to date. Second, these deep-blue materials could play various roles in OLED application. They could be a highly efficient nondopant deep-blue emitter: maximum EQE η_{ext} of 4.2%, CIE$_{x,y}$ = (0.15, 0.07). For solid-state lighting application, these compounds are desirable as a host material for a yellow dopant (rubrene) in achieving high-efficiency (η_{ext} 4.3% and η_P 8.7 lm/W at a luminance of 100 cd/m^2 or η_{ext} 3.9% and η_P 5.1 lm/W at a luminance of 1000 cd/m^2) white EL (CIE$_{x,y}$ = 0.30, 0.35).

Joly et al. reported on the synthesis and characterization of mixed thiophene–phosphole–fluorene derivatives that can be used as dopants for DPVBi [591]. The increase of the doping rate had a small impact on the CIE coordinates and on the EQE. These results were very appealing toward the development of "easy-to-make" WOLEDs. A series of orange light–emitting dopants incorporating a phosphole ring had been developed by Fadhel et al. [592], for use in WOLEDs with simple structures. The coevaporation of DPVBi and phosphole-based dopants was explored, to construct devices with the structure of ITO/CuPc/α-NPB/doped DPVBi/Alq$_3$/LiF/Al. The possibility of easily chemically manipulating the structure of the dopants allowed for the straightforward optimization of their structures, which gave access to WOLEDs exhibiting excellent performance for devices based on fluorescent materials.

The lifetime of the organic devices remains a major challenge that must be overcome before the wide application of WOLEDs technology. Duan et al. presented a strategy to achieve WOLEDs (Figure 3.9) with an extremely long lifetime by wisely controlling the recombination zone [593]. The key feature of their OLED layer structure was the utilization of double blue-emitting layers to stabilize the blue emission. A mixed-host blue-emitting layer consisting of 78% α,β-ADN:20% NPB:2% ENPN was utilized to broaden the recombination zone and dilute the concentration of any degradation-related quenching species. A second blue-emitting layer of 98% α,β-ADN:2% ENPN was then deposited onto the mixed-host blue-emitting layer to achieve better charge confinement. Combined with a mixed-host yellow-emitting layer, they achieved a record high lifetime of >150,000 h at an initial brightness of 1000 cd/m^2 in a WOLED together with a stable color over the whole lifespan.

Jou et al. [594] had demonstrated a sunlight-style color-temperature tunable as well as color-temperature span tunable OLED (Figure 3.10), the color temperature of which ranges between 2300 and 8200 K, fully covering those of the entire daylight at different times and regions, achieved by employing daylight complementary emissive constituents with the use of a hole-modulating layer to regulate the distribution of holes.

The fabrication of nondoped multilayer WOLEDs did not involve the difficult-to-control doping process. To obtain OLEDs with high-efficiency and full-spectrum white-light emission, Yang et al. [595] proposed a nondoped-type structure. A white emission with CIE of (0.3556, 0.3117) at 5 V and (0.282, 0.2658) at 15 V was obtained. Its luminance was 11,497 cd/m^2 at 15 V, and the maximum efficiency was 4.8 lm/W at 5 V. Yeh et al. [596]

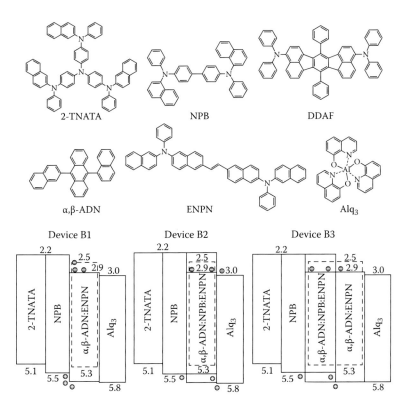

FIGURE 3.9
Schematic energy-level diagram of the blue OLEDs and the molecules used. (From Duan, L., Zhang, D., Wu, K., Huang, X., Wang, L. and Qiu, Y., *Adv. Funct. Mater.*, 21, 3540, 2011. With permission.)

composed all nondopant WOLEDs. The CIE coordinates of fairly white EL of devices varied only little from (0.34, 0.39) to (0.34, 0.38) at driving voltages between 6 and 14 V. The devices exhibited relatively high CRIs in the range of 74–81, which are essentially voltage independent. The other WOLEDs showed even better color purity of white EL with CIE of (0.34, 0.31) along with higher color rendering index (CRI) of 83 at 8 V, although higher voltage deteriorated the color quality of these WOLEDs.

Excellent bipolar carrier-transport properties of DPF (**553**, Scheme 3.98) had been elucidated by using different device structures by Tao et al. [597]. A nondoped device using DPF as emitter showed highly efficient blue emission with a maximum efficiency of 6.0 cd/A and CIE coordinates of (0.15, 0.19). Another device based on rubrene-doped DPF as the EML gave high-efficiency pure white emission with good color stability, a maximum efficiency of 10.5 cd/A, and CIE coordinates of (0.28, 0.35). The excellent bipolar transport capability and high performance as both emitter and host suggest that DPF was an efficient and versatile material for various applications in OLEDs.

High-efficiency nondoped WOLEDs were demonstrated by Tong et al. [598]; they used both the intrinsic and exciplex emissions from a single electroluminescent material, TPyPA. The simple device structure exhibited a luminance of 10,000 cd/m² at a low driving voltage of 4.5 V, and high current and power efficiencies of 9.4 cd/A and 9.0 lm/W, respectively. Such WOLED showed excellent color stability and purity with the CIE coordinates of (0.31, 0.35), which remained unchanged over a wide range of luminance from 100 to 20,000 cd/m².

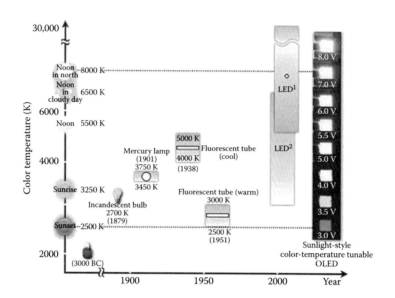

FIGURE 3.10
Sunlight-style color-temperature tunable OLED with color-temperature span from 2300 to 9700 K. (From Jou, J.-H., Wu, M.-H., Shen, S.-M., Wang, H.-C., Chen, S.-Z., Chen, S.-H., Lin, C.-R. and Hsieh, Y.-L., *Appl. Phys. Lett.*, 95, 013307, 2009. With permission.)

A nondoped-type pure white OLED with high CRI had been reported by Yang et al. [599]. The structure of the device was ITO/m-MTDATA/NPB/DPVBi/DCJTB (x nm)/Alq$_3$/ QAD (y nm)/Alq$_3$/BPhen/Alq$_3$/LiF/Al. Through the optimization of x and y, white emission with CIE coordinates of (0.33, 0.33) at 8 V was obtained, and the color temperature and CRI were 6712 K and 91 at 8 V, respectively. The CIE coordinates of the device changed from (0.37, 0.36) at 4 V to (0.32, 0.33) at 15 V, which are well in the white region. Its maximum luminance was 27,360 cd/m^2 at 15 V and maximum power efficiency was 3.16 lm/W at 7 V. Moreover, the current efficiency was largely insensitive to the current density.

Zhao et al. [600] reported on high-efficiency orange and white OLEDs by introducing an ultrathin nondoped orange-emitting layer (<1 nm) adjacent to the electron-transporting layer. The orange OLEDs showed a high EQE of 16.4% at 1000 cd/m^2, and the resulting four-color white OLEDs emitted an EQE of 15.5% and a maximum CRI of 87 at a luminance of 1000 cd/m^2. The used white device structure was ITO/PEDOT:PSS/NPB/TCTA/TCTA:8% (piq)$_2$Ir(acac)/TCTA:8% Ir(ppy)$_3$/TCTA:15% FIrpic/(fbi)$_2$Ir(acac)(0.5 Å)/TmPyPB/LiF/Al.

Nondoped white OLEDs based on bluish–green light-emitting material [1,3,6,8-tetrakis (4-(1,2,2-triphenylvinyl)phenyl)pyrene (TTPEPy)], red light–emitting material [4-(4-(1,2,2-tri-phenylvinyl)phenyl)-7-(5-(4-(1,2,2-triphenylvinyl)phenyl)thiophen-2-yl)benzo(c)(1,2,5) thiadiazole (BTPETTD)], and blue light–emitting material [4,4'-bis(1-phenyl-1H-phenanthro(9,10-d)imidazol-2-yl)biphenyl (DDPi)] had been demonstrated by Chen et al. [601,602].

3.6.1.3 Fluorescent WOLEDs with Single Emissive Material

A further strategy to achieve white emission uses rare-earth complexes. For example, a dysprosium complex emits two band emissions: a yellow band (580 nm) corresponding to the $^4F_{9/2} \rightarrow {}^6H_{13/2}$ transition and a blue band (480 nm) corresponding to $^4F_{9/2} \rightarrow {}^6H_{15/2}$

transition of the Dy^{3+} ion in the complex. Hong et al. reported Dy complex white-emission OLEDs of a structure of ITO/PVK:Dy complex/Mg:Ag device [603].

Stable white emission with CIE coordinates of (0.3519, 0.3785) was obtained in such a rare earth–based OLED device. The authors mentioned that the EQE of the device was not good, possibly due to the inefficient energy transfer process between the ligand and the rare-earth metal. A suitable choice of the ligand may improve this type of device performance.

The broad PL emission spectra of some metal chelates match the requirements for white emission. Hamada et al. investigated a series of Zn complexes and found that bis(2-(2-hydroxyphenyl)benzothiazolate)zinc [Zb(BTZ)$_2$, **143**] is the best white-emission candidate. An OLED with a structure of ITO/TPD/Zn(BTZ)$_2$/OXD-7/Mg:In showed greenish–white emission with CIE (0.246, 0.363) with a broad emission spectrum (FWHM: 157 nm) consisting of two emission peaks centered at 486 and 524 nm [604]. A maximum luminance of 10,190 cd/m^2 at 8 V was achieved.

A WOLED fabricated from a single-emitting component has been reported by Li et al. [605]. An OLED with a structure of three-layer ITO/TECEB/BCP/Alq$_3$/Mg:Ag emits a broad emission band from blue to red–orange with CIE coordinates of (0.298–0.304, 0.31–0.327). Interestingly, the long-wavelength emissions do not come from the interface exciplex nor from the Alq$_3$. They are produced by electronic excitation of TECEB (**50**) as evidenced by a single-layer device with an OLED structure of ITO/TECEB/Mg:Ag, which emits a similar spectrum as the three-layer device. This identical EL spectrum implies that TECEB is the exclusive component responsible for the white emission. The authors propose that the longer wavelength emissions come from the singlet electromer (TECEB$^+$/TECEB$^-$)* rather than the excimer as supported by its PL spectra in solution and in the solid state. In addition, and more likely, phosphorescence of TECEB may contribute to the longer wavelength emission as well, as confirmed by the transient luminescence lifetime decay test.

A TPA-based compound, DPAA (**554**, Scheme 3.98), had been designed and synthesized for application in WOLEDs by Tao et al. [606]. The PL of the compound in thin film showed a white emission due to the combination of exciton and excimer emissions. Taking advantage of this property, a WOLED was fabricated by using the compound as the sole emitter. The WOLED exhibited highly efficient white emission with a low turn-on voltage of 3 V, a maximum brightness of 12,320 cd/m^2, a maximum current efficiency of 7.0 cd/A, and power efficiency of 7.1 lm/W at 8 V with CIE coordinates of (0.29, 0.34).

The unique and unprecedented EL behavior of the white-emitting molecule 3-(1-(4-(4-(2-(2-hydroxyphenyl)-4,5-diphenyl-1H-imidazol-1-yl)phenoxy)phenyl)-4,5-diphenyl-1H-imidazol-2-yl)naphthalen-2-ol (W1, **555**) (Scheme 3.98), fluorescence emission from which was controlled by the excited-state intramolecular proton transfer (ESIPT) was investigated by Kim et al. [607]. Finally, the device structure to create a color-stable, color-reproducible, and simple-structured WOLED using W1 was investigated. The maximum luminance efficiency, power efficiency, and luminance of the WOLED are 3.10 cd/A, 2.20 lm/W, and 1092 cd/m^2, respectively.

3.6.1.4 Fluorescent WOLEDs with Excimer or Exciplex Emissions

Another interesting WOLED based on a boron hydroxyphenylpyridine complex was reported by Liu et al. [608]. The PL emission of such a material in fluid solution as well as in the solid state is blue (450 nm). However, the EL spectrum of ITO/NPD/(mdppy)BF/LiF/Al gives a broad emission band with a stable CIE coordinate of (0.30, 0.36). The authors explained that this white emission is due to the formation of the exciplex between NPD and (mdppy)BF. A maximum efficiency of 3.6 lm/W (6.5 cd/A) with a luminance of 110 cd/m² at 5.5 V was achieved.

Kalinowski et al. reported an efficient white device with particularly high CRI of 90 by using electron donor–electron acceptor emitter layers, in which the broad emission band of an exciplex mixes with excimer emission (Figure 3.11) [609]. The CRI can be tuned by varying the proportions of the monomer triplet, phosphor excimer, and exciplex mixture emission components of the system, and its further improvement by varying the electron donor–electron acceptor ratio and/or introducing electronically active spacers between hole-transporting and emissive layers of the device. They concluded that the use of electron donor and electron acceptor phosphor single-dopant systems may allow obtaining highly efficient WOLEDs with a color rendition approaching CRI = 100 for ideal white light.

Kim et al. broadened the EL spectra of WOLEDs based on complementary colors [610]. They reported the optical and EL properties of WOLEDs that had two emitters of 1,1,4,4-tetraphenyl-1,3-butadiene and (4-dicyanomethylene)-2-methyl-6-(*p*-dimethylaminostyryl)-4*H*-pyran with emission peaks of 400 and 580 nm, respectively. The WOLEDs showed broadening EL spectra through the visual range from 400 to 780 nm. This spectral broadening was related to an exciplex emission at the organic solid interface.

3.6.1.5 Fluorescent WOLEDs with Color-Change Material

Li et al. fabricated high chromatic-stability and efficient WOLEDs by using red-dye doped 2-TNATA as an inside color-change material (CCM) as well as an HIL in the devices [611], which absorbed the blue emission from emissive active layer and then gave red emission. One part of transmitted blue emission in the EML in combination with red emission in HIL produced a white emission. They found that the CCM film had good electric and optical characteristics, and was independent of temperature and electric field. They fabricated three WOLEDs with different thicknesses, and observed unchanged CIE coordinates under varied current density ranging from 0.001 to 1.0 A/cm² or after long-time continuous operation at 0.37 A/cm².

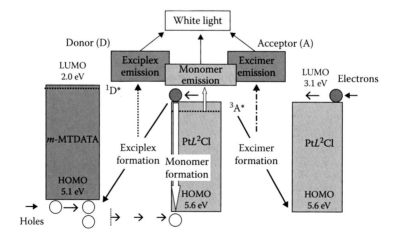

FIGURE 3.11
Material and device structures using excimer/exciplex to obtain WOLEDs with high CRIs. (From Kalinowski, J., Cocchi, M., Virgili, D., Fattori, V., and Williams, J.A.G., *Adv. Mater.*, 19, 4000, 2007. With permission.)

3.6.2 Phosphorescent White Organic Light-Emitting Diodes

As an alternative backlight source, with a potential for high efficiency, low cost, and large area fabrication, white phosphorescent OLEDs have received great attention for their lighting applications. White emission can be achieved either through multilayer OLED structures combining different EMLs or a single EML containing different emitting materials to cover the entire visible spectrum, or through blue or bluish–green emitters in combination with complementary emitters or excimer emissions to achieve the desired white-light emission.

3.6.2.1 Phosphorescent WOLEDs with Primary Colors

The three primary-color phosphors can be doped into either one or several layers to form WOLEDs with single or multiple EMLs. It is simple to make WOLEDs by doping RGB phosphors into a single EML. Doping three emitters into a single layer would probably induce energy transfer processes from the short-wavelength emitters to the longer-wavelength ones, which would result in a color shift with respect to driving voltage or brightness, and thus impair the quality of the white light. Thus, the doping level of the emitters in this kind of WOLED must be precisely controlled to achieve decent white-light quality at the acceptable brightness level. Generally, the doping level for blue phosphor is much higher than that of the green and red counterparts in order to maintain a more balanced RGB light component in the white EL spectrum.

By using a large band-gap host material UGH2, selecting different phosphorescent dopants, and controlling the doping concentration and thickness, a very high-efficiency white PHOLED was presented by Forrest's group. The devices have various white-emission colors with CIE coordinates varying from (0.43, 0.45) at 0.1 mA/cm^2 to (0.38, 0.45) at 10 mA/cm^2, with a CRI of 80 and a maximum EQE up to 46 lm/W [612].

Different from doping three phosphors into single-EML, R–G–B primary colors from phosphorescent emitters can be set into different layers individually to make WOLEDs with multi-EML configuration. Despite the complexity in device fabrication, WOLEDs with multi-EMLs would serve as a good platform with wider scope for device optimization. As with fluorescent white OLEDs, the first attempt to achieve white PHOLEDs was described by the Forrest group using multilayer device structures integrating blue (FIrpic), yellow (bt$_2$Ir(acac)), and orange–red (btp$_2$Ir(acac)) phosphorescent dopants (Figure 3.12) [613]. Through control of the dopant concentration and thickness, white emission with the desired color can be achieved.

Sun and Forrest [614] demonstrated a white PHOLED with a three-section EML where excitons were formed in the multiple-emission regions. The EML consisted of a stepped progression of HOMO and LUMO energies of the ambipolar hosts. Analysis showed that $36 \pm 6\%$ of the excitons form in the blue-emitting region, while $64 \pm 6\%$ form in the green-emitting region at 100 mA/cm^2. The doping of the red, green, and blue phosphors allowed for efficient utilization of excitons formed in these multiple regions. On the basis of this architecture, the WOLED had an internal QE close to unity. The WOLED had total external quantum and power efficiencies of $\eta_{\text{ext},t} = 26 \pm 1\%$ and $\eta_{p,t} = 63 \pm 3$ lm/W at 12 cd/m^2, decreasing to $\eta_{\text{ext},t} = 23 \pm 1\%$ and $\eta_{p,t} = 37 \pm 2$ lm/W at 500 cd/m^2. When an nondoped ETL was used, the peak efficiency was $\eta_{\text{ext},t} = 28 \pm 1\%$.

An all-phosphor, four-color WOLED (Figure 3.13) is presented by Chang et al. [615], employing a novel device design principle utilizing molecular energy transfer or, specifically, triplet exciton conversion within common organic layers in a cascaded emissive zone

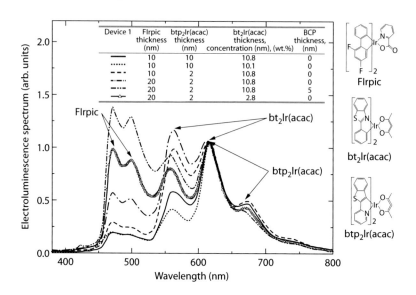

FIGURE 3.12

Variation, at 10 mA/cm², in the electroluminescence spectra with layer thickness, dopant concentration, and the insertion of an exciton or hole-blocking layer between the FIrpic and btp₂Ir(acac) doped layers for Device 1. Molecular structural formulas of btp₂Ir(acac), FIrpic, and bt₂Ir(acac) (right). (From D'Andrade, B.W., Thompson, M.E., and Forrest, S.R., *Adv. Mater.*, 14, 147, 2002. With permission.)

configuration to achieve exceptional performance: an EQE of 24.5% at 1000 cd/m² with a CRI of 81, and an EQE of 20.4% at 5000 cd/m² with a CRI of 85, using FIrpic for blue phosphor, $Ir(ppy)_2(acac)$ for green phosphor, $Ir(BT)_2(acac)$ for yellow, and $Ir(MDQ)_2(acac)$ for red phosphor.

Efficient phosphorescent WOLEDs were realized by Song et al. [616]. The device emission color was controlled by varying dopant concentrations and the thicknesses of blue and green–red layers, as well as tuning the thickness of an exciton-blocking layer. The maximum luminance and power efficiency of the WOLED were 42,700 cd/m² at 17 V and 8.48 lm/W at 5 V, respectively. The CIE coordinates changed from (0.41, 0.42) to (0.37, 0.39) when the luminance ranged from 1000 to 30,000 cd/m².

Cheng et al. fabricated a phosphorescent WOLED with multiple emissive layers, in which a blue emissive layer is sandwiched between red and green ones [617]. This WOLED had a maximum luminance of 48,000 cd/m² at 17 V, a maximum power efficiency of 9.9 lm/W at 4 V, and a CRI of 82. In addition, the emission color of this device was fairly stable at high luminance: its CIE coordinates slightly changed from (0.431, 0.436) to (0.400, 0.430) when the luminance ranged from 2000 to 40,000 cd/m².

Sun and Forrest [618] demonstrated high-efficiency WOLEDs employing three adjacent phosphorescent EMLs. This structure distributed the exciton generation region across the three hosts to form a stepped progression of HOMO and LUMO. The three-EML WOLED had a CRI of 81 and peak forward-viewing external quantum (EQE) and power efficiencies of 16.6 ± 0.8% and 32 ± 1 lm/W, respectively, corresponding to a total EQE = 28 ± 1% and a total power efficiency = 54 ± 3 lm/W. When an n-doped electron-transporting layer was used, the total power efficiency peaked at 64 ± 3 lm/W, and rolled off to 34 ± 2 lm/W at 1000 cd/m².

Reineke et al. [619] reported a white PHOLED (Figure 3.14) containing FIrpic (blue), $Ir(ppy)_3$ (green), and iridium(III) bis(2-methyldibenzo(*f,h*)quinoxaline) (acetylacetonate)

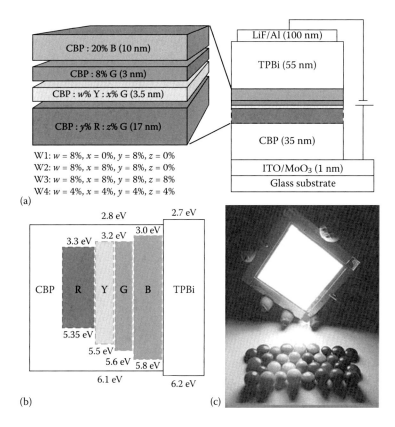

FIGURE 3.13
(See color insert.) Device configurations (a) and energy-level diagrams (b) for WOLEDs W1–W4. The dopants employed are FIrpic for blue (B), Ir(ppy)₂(acac) for green (G), Ir(BT)₂(acac) for yellow (Y), and Ir(MDQ)₂(acac) for red (R). (From Chang, Y.-L., Song, Y., Wang, Z., Helander, M.G., Qiu, J., Chai, L., Liu, Z., Scholes, G.D., and Lu, Z., *Adv. Funct. Mater.*, 23, 705, 2012. With permission.)

(Ir(MDQ)₂ (acac)) (red) as the emitting phosphors, where the blue EML was surrounded by red and green sublayers (2 nm of sublayer to suppress Förster-type transfer) of the EML to harvest unused excitons from the blue EML. They achieved 90 lm/W at 1000 cd/m² through combining a high-refractive-index (1.78) glass substrate with a periodic light-outcoupling structure, with CIE coordinates of (0.41, 0.49) owing to less contribution from blue emission, which could be enhanced by changing to a wide-gap host and transporting material as mentioned in Section 3.5.2.1.3, and even 124 lm/W when using a 3D light extraction system, surpassing the efficiency of fluorescent tube. This indicates a way for white organic light with efficiency beyond 100 lm/W.

Sasabe et al. reported on the development of high-efficiency blue and white OLEDs incorporating a blue iridium carbene complex [620]. They used *mer*-tris (*N*-dibenzofuranyl-*N*′-methylimidazole)iridium(III)(Ir(dbfmi)) as blue phosphor, and Ir(ppy)₃ and iridium bis(2-phenyl-quinoly-*N*,C²′)dipivaloylmethane (PQ₂Ir(dpm) as green and orange–red phosphors, respectively. White OLED with a structure of ITO/TAPC/TCTA/PQ₂Ir(dpm) 2 wt.% doped CBP/Ir(ppy)₃ 6 wt.% doped CBP (1 nm)/Ir(dbfmi) 10 wt.% doped PO9/B3PyPB/LiF/Al) was fabricated, showing a maximum power efficiency of 59.9 lm/W and a power efficiency of 43.3 lm/W at 1000 cd/m² and an illumination-acceptable CRI >80 without any light-outcoupling enhancement.

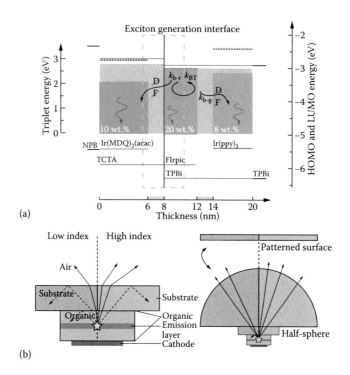

(a)

(b)

FIGURE 3.14
(See color insert.) Energy-level diagram (a) and light modes (b) in an OLED. (From Reineke, S., Lindner, F., Schwartz, G., Seidler, N., Walzer, K., Luessem, B., and Leo, K., *Nature*, 459, 234, 2009. With permission.)

A yellowish–green triplet emitter, bis(5-(trifluoromethyl)-2-*p*-tolylpyridine) (acetylacetonate)iridium(III), was conveniently synthesized and used in the fabrication of both monochromatic and WOLEDs [621]. By combining this compound with a phosphorescent sky-blue emitter, FIrpic, and a red emitter, bis(2-benzo(*b*)thiophen-2-yl-pyridine)(acetylaceto-nate)iridium(III) (Ir(btp)$_2$(acac)), the resulting electrophosphorescent WOLEDs show three evenly separated main peaks and give a high efficiency of 34.2 cd/A (13.2% and 18.5 lm/W) at a luminance of 100 cd/m^2.

Weichsel et al. investigated a host–guest–system containing four phosphorescent emitters and two matrix materials with different transport properties [622]. They found that FIr6 promotes an energy back-transfer to the TCTA and SPPO1 host. While the back-transfer on TCTA leads to an enhanced nonradiative recombination, the excitation of SPPO1 could be used by transferring it to Ir(ppy)$_3$. Furthermore, thin interlayers have the possibility to suppress Förster and Dexter energy transfer. Nevertheless, Dexter energy transfer over the matrix is possible if the energy levels of the interlayer material are not much higher than those of the emitter molecules and if the triplet lifetimes of the interlayers are high.

Hsiao et al. investigated the strong influence of the thickness of FIrpic doped *N,N'*-dicarbazolyl-3,5-benzene (mCP) and diphenylbis(4-(9-carbazoyl)phenyl)silane (SiCa) blue-emitting layer on color stability [623,624]. They fabricated a white OLED with high color stability, which exhibited nearly invariant CIE coordinates throughout the practical luminance range from 1050 (0.310, 0.441) to 9120 cd/m^2 (0.318, 0.446) and maximum efficiencies of 26.4 cd/A and 19.8 lm/W. Jou et al. [625,626] reported the fabrication of a highly efficient, very high-CRI white OLEDs using five organic dyes doped into two different phospho-rescent and fluorescent emissive layers separated by a high-triplet-energy interlayer. The resulting white OLED achieves a 93 CRI with a power efficiency of 23.3 lm/W at 1000 cd/m^2, or 14.3 lm/W at 10,000 cd/m^2. This high CRI was attributed to the five dyes employed in this design, which together emit a relatively wide spectrum that nearly covers the entire range of visible light. At the proper thickness, the interlayer enabled the device to balance the distribution of carriers in the two emissive zones and achieve a maximum power effi-ciency while maintaining high CRI.

3.6.2.2 Phosphorescent WOLEDs with Complementary Colors

Besides R–G–B phosphorescent emitters, phosphors showing complementary colors, such as blue (B) and orange (O), can also be utilized to produce white-light emission in the devices. By reducing the number of phosphors used, the device fabrication process can be generally simplified. Owing to the absence of green-emitting phosphor, the CRI is usu-ally moderate for the white light emitted from the WOLEDs with B–O complementary colors. By reducing the number of emitters employed, B–O complementary-color phos-phorescent WOLEDs still attract much research attention for their relatively simple device configuration.

Like RGB single-EML WOLEDs, the phosphors in a B–O complementary-color configu-ration can be doped in a small-molecular host to form EML by a vacuum deposition pro-cess. Similarly, the doping level for the phosphors also needs to be carefully controlled to achieve a desirable white EL spectrum.

Reineke et al. [627] investigated phosphorescent OLEDs comprising mixed films of TCTA as the EML. On the basis of the results of PL experiments, they intermitted the EML with thin neat layers of TCTA acting as an exciton blocking layer inside the EML, which suppresses triplet–triplet annihilation. They showed that this EML structure leads to an improved roll-off behavior. Starting at the initial EQEs of 15.8% and 14.4% at low bright-ness for the reference and interlayer device, respectively, those structures yield critical cur-rent densities j_c of 140 and 270 mA/cm^2 defining the current density of half-value internal EQE. Their present study reported on a novel device concept rendering it possible to also harvest the triplet excitons from a fluorescent blue emitter for light generation in white OLEDs, thus giving rise to a high QE [628]. By incorporating electrically doped charge-transport layers, a maximum power efficiency of 28.0 lm/W at a brightness of 1000 cd/m^2 was obtained in the forward direction. Finally, the device had a good CRI of 86, making it well suited for lighting applications.

Su et al. [629] reported an iridium(III) bis(4,6-(difluorophenyl) pyridinato-*N,C*20) picolinate (FIrpic)-based blue OLED with a carrier- and exciton-confining structure (Figure 3.15). The device has a power efficiency of 55 lm/W at luminance of 100 cd/m^2. A very efficient white OLED was achieved with the insertion of ultrathin layers of orange light–emitting iridium(III) bis-(2-phenylquinoly-*N,C*2′)dipivaloylmethane (PQ2Ir) in the host material used for FIrpic. A power efficiency of 53 lm/W at a display-relevant lumi-nance of 100 cd/m^2 was obtained in the forward direction, and rolls off slightly to 44 lm/W at an illumination-relevant luminance of 1000 cd/m^2 without the use of any outcoupling techniques. Several strategies were employed to achieve the efficient light emission. First, since low operating voltage is necessary to achieve a high power efficiency, thin emissive layers were employed, and a stepped progression of the HOMO and LUMO energy levels of the HTM, host materials, and ETM was incorporated to facilitate hole and electron injec-tion into the emissive layers. Second, a double-EML and a bipolar host were employed to

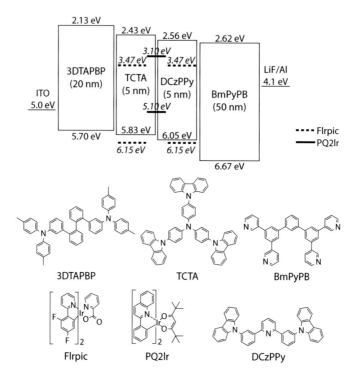

FIGURE 3.15
Top: Proposed energy-level diagrams of the blue and white OLEDs, showing the highest occupied and lowest unoccupied molecular orbital energies relative to vacuum. The thickness of the constituents used is also shown. Bottom: Molecular structures and acronyms of the materials used. (From Su, S.-J., Gonmori, E., Sasabe, H., and Kido, J., *Adv. Mater.*, 20, 4189, 2008. With permission.)

broaden the exciton-formation zone, and thus reduce efficiency roll off compared with conventional single-EML OLEDs. Third, HTMs and ETMs with wide energy gaps and high-lying LUMO and low-lying HOMO energy levels, respectively, were employed adjacent to the emissive layers to confine carriers and triplet excitons.

Hung et al. reported the synthesis, characterization, and various OLED applications of a bipolar material, CPhBzIm [630]. The device was investigated by a simple configuration: ITO/PEDOT:PSS/DTAF/emitter layer/TPBI/LiF/Al. A two-color WOLED under the same device structure by utilizing 12 wt.% FIrpic and 0.3 wt.% Os(bpftz)$_2$(PPh$_2$Me)$_2$ codoped into CNBzIm as a single emitting layer was fabricated. The maximum values of efficiencies (η_{ext}, η_c, and η_p) were 15.7%, 35 cd/A, and 36.6 lm/W, respectively. Even at a practical brightness of 1000 cd/m^2, the η_{ext} remained at 12.4% [631,632].

Lin et al. reported the synthesis and physical properties of a bipolar host material, CMesB, containing the dimesitylborane/carbazole hybrid [448]. The white-emitting device had the structure of ITO/TAPC/CMesB doped with 20 wt.% of FIrpic and 1 wt.% of Os(bpftz)$_2$(PPh$_2$Me)$_2$/TmPyPB/LiF/Al. The device achieved peak efficiencies of 15.7%, 22.6 lm/W, and 34.8 cd/A at practical brightnesses.

Seidel et al. [633] reported on highly efficient white OLEDs comprising the two phosphorescent emitters FPt1 and FIrpic. Investigating the influence of the dopants in the EMLs on the device performance, a series of devices with different FIrpic and FPt1 concentrations had been processed. After optimizing the device with respect to efficiency and white color coordinates, the influence of a thickness variation of the EMLs was investigated. The experiments revealed that the device performance is more sensitive to the parameters of the EML containing FPt1. Additional optical simulations showed that interference effects play a crucial role in this device setup. The optimized device resulted in an EQE of 10%, a power efficiency of 16 lm/W, near-warm-white color coordinates of (0.39, 0.41), and a CRI of 81 at 1000 cd/m^2.

Lee et al. [634] demonstrated high-efficiency white PHOLEDs based on yellow and composite blue emitters. The composite blue emitter was constructed from a wide-band-gap host, organometallic iridium dopant, and a carrier-transporting material. Under the same driving current density and emissive color, the current efficiency of the white PHOLEDs can be enhanced by a factor of 1.4 comparing to that of using typical blue emitter composed of host and dopant only. Attaching an outcoupling enhancement film onto glass substrate, the white PHOLEDs with a current efficiency of 47 cd/A, a power efficiency of 32 lm/W, and a CIE of (0.40, 0.44) at a practical brightness of 1000 cd/m^2 can be achieved.

Lee et al. reported a significant enhancement and a reduced efficiency roll-off in white phosphorescent OLEDs based on a mixed-host structure [635]. The combination of two hole transport–type host materials, mCP and TCTA, with an electron transport–type host, UGH3 [636], as a mixed-host emissive layer resulted in an effective charge-carrier injection and distribution of the recombination zone, and therefore highly efficient white PHOLEDs with a mixed-host structure showed a power efficiency of 37 lm/W, which is >4 times higher than 8.7 lm/W in a single-host structure device at a luminance of 1000 cd/cm^2.

On the basis of the unipolar TCTA, Yang et al. synthesized a bipolar host material, TCTAPO, (9-(4-(bis(4-(9H-carbazol-9-yl)phenyl)amino)phenyl)-9H-carbazol-3-yl) diphenylphosphine oxide, by directly imparting an electron-transporting diphenylphosphine oxide moiety to TCTA [637]. The single-EML WOLED based on the emitter of TCTAPO:FIrpic:(fbi)$_2$Ir(acac) exhibited a maximum current efficiency of 43.7 cd/A, a maximum power efficiency of 47.0 lm/W, and a maximum EQE (η_{ext}) of 17.2%. Compared with the PHOLEDs with TCTA as the host, the maximum power efficiencies increased about 81% for white devices.

Recently, Han et al. reported several hosts with carbazole–phosphine oxide (PO) and dibenzothiophene (DBT) structures for high-efficiency phosphorescent WOLEDs [638,639]. Those hosts revealed strong electron-injecting/-transporting ability without interference in hole injection/transportation and T1. The corresponding white PHOLEDs based on DBTDPO showed excellent performance, including driving voltages of <2.8 V for 100 cd/m^2 and 3.2 V for 1000 cd/m^2, and maximum efficiencies of 43.5 lm/W for white-emitting devices.

Highly efficient phosphorescent white OLEDs were developed by Seo et al. using a deep-blue phosphorescent emitter doped into a mixed host of high triplet energy host materials, 2,8-di(9H-carbazol-9-yl)dibenzo(b,d)furan (DFCZ) and 2,8-bis(diphenylphosphoryl) dibenzo(b,d)furan (DFPO), respectively [640,641]. A high QE of 19.5% with color coordinates of (0.29, 0.38) and 19.8% with color coordinates of (0.39, 0.46) were achieved in the phosphorescent WOLEDs using the mixed-host emitting layer doped with a deep-blue phosphorescent dopant.

Gong et al. reported a series of tetraarylsilane compounds that were designed and synthesized by incorporating electron-donating arylamine or carbazole and electron-accepting benzimidazole or oxadiazole into one molecule via a silicon-bridge linkage mode [440,642–644]. The thermal, photophysical, and electrochemical properties of those compounds can be finely tuned through the different groups and linking topologies. The silicon-interrupted conjugation of the electron-donating and electron-accepting segments gives these materials the following advantages: (i) relative high triplet energies; (ii) the HOMO and LUMO levels of the compounds mainly depend on the electron-donating and electron-accepting groups, respectively; and (iii) bipolar transporting feature as indicated by hole-only and electron-only devices. In terms of the applicability of the host materials for multicolor triplet emitters, they demonstrated two-color and all-phosphor WOLED with the configuration of ITO/MoO$_3$/NPB/TCTA/p-OXDSiTPA:8 wt.% FIrpic:0.67 wt.% (fbi)$_2$Ir(acac)/TPBI/LiF/Al. FIrpic and (fbi)$_2$Ir(acac) were codoped into the bipolar host p-OXDSiTPA as the single emitting layer. The device with p-OXDSiTPA as host achieved $\eta_{c.max}$ of 51.4 cd/A, $\eta_{p.max}$ of 51.9 lm/W, and $\eta_{ext.max}$ of 18.3%. The high efficiencies and low-efficiency roll-off at high luminance for devices can be attributed to the use of the bipolar hosts, which may result in balanced charge fluxes and a broad distribution of recombination regions within the emitting layer.

High-efficiency single-layer WOLEDs were fabricated through thermal vapor deposition technique by Yin et al. [645]. A bipolar transport material with high triplet energy (2.75 eV), namely 2,7-bis(diphenylphosphoryl)-9-(4-(N,N-dipheny-lamino)phenyl)-9-phenylfluorene (POAPF) [646], was used as the host material in their study. The single-layer WOLED based on FIrpic and PO-01 exhibited high efficiency of 30.9 cd/A and 20.9 lm/W at 1000 cd/cm^2. The efficiency of the single-layer OLED was comparable to that of the multilayer OLED, while the efficiency roll-off was much slower than that of the multilayer OLED. Yook et al. also reported phosphorescent white OLEDs were developed by using a spirofluorene based phosphine oxide (SPPO1) as a host material in a deep blue–emitting layer [647–650].

By codoping FIrpic for blue emission and (fbi)$_2$Ir(acac) for orange emission into a single host material, Wang et al. reported a high-efficiency WOLED with excellent color stability [651,652]. They had successfully realized a high-efficiency phosphorescent WOLED with a peak power efficiency of 42.5 lm/W and an EQE of 19.3% by codoping two kinds of iridium complexes into mCP, respectively; the maximum power efficiencies reached 33.0 lm/W and an EQE of 13.3% with TCTA host material. Comprehensive experimental and theoretical evidence had demonstrated that the realization of such high efficiency results from the coexistence of two parallel channels, i.e., direct exciton formation following hole trapping and direct electron injection onto (fbi)$_2$Ir(acac) molecules, and effective energy transfer from mCP to FIrpic molecules to harvest excitons for white emission.

Besides single-layer phosphorescent WOLEDs, WOLEDs with multiple EMLs can be fabricated by doping each of the two phosphors exhibiting B–O complementary colors into different EMLs. Despite the relatively complicated device manufacture compared with the single-EML WOLEDs, such double- or multiple-EML devices can give more flexibility to improve EL performance through fine optimization of every single EML and the careful energy matching of the EMLs.

Wang et al. reported a series of iridium complexes containing a strong electron-withdrawing CF$_3$ or F as an orange phosphorescent dopant [653,654]. These orange iridium complexes were used as an emitter in combination with the traditional blue-emitting FIrpic to fabricate two-element WOLEDs, and highly efficient white-light emission was realized with a peak forward-viewing η_c of 68.6 cd/A and a peak η_p of 34 lm/W.

Chang et al. [655–658] had shown that the unique multifunctional capabilities of the orange–red phosphorescent Os complex are highly useful for achieving excellent internal and external efficiencies of WOLEDs. They reported an efficient phosphorescent WOLED structure that would provide improved color stability versus biases. The two-component phosphorescent white OLEDs yielded a CRI of 73 and CIE coordinates close to the warm-white emission (0.39, 0.40) at 1000 cd/m^2, a performance adequate for applications in domestic lighting. Furthermore, these devices exhibited stable colors over a wide brightness range of 100 to 1000 cd/m^2 and yielded EL efficiencies of up to 15.3%, 33.3 cd/A, and 22.7 lm/W in the forward directions.

Lai et al. reported the synthesis of a new biscyclometalated yellow phosphorescent dopant, iridium(III) bis(2-(2-naphthyl)pyridine)(acetylacetonate) (Ir(npy)2acac), and studies of its application in WOLEDs by combining blue emission from FIrpic or F3Irpic [659,660]. The optimized devices demonstrated high peak forward-viewing η_c, η_p, and EQE of 34 cd/A, 22 lm/W, and 11%, respectively, along with CIE coordinates of (0.45, 0.54).

Besides generally doped emissive layers, nondoped emissive layers can also be used for WOLED fabrication. Lee et al. [661] demonstrated a WOLED by dispersing a host-free, yellow phosphorescent material in between double blue phosphorescent emitters. The device performance achieved a comparable value to that of using a complicated host–guest doping system to form the yellow emitter in WOLEDs. Zhao et al. [662,663] investigated WOLEDs by inserting a thin layer of nondoped yellow phosphorescent (tbt)$_2$Ir(acac) between doped blue-emitting layer and electron-transporting layer.

3.6.2.3 Phosphorescent WOLEDs with Excimer or Exciplex Emissions

The above strategy, using three primary colors as phosphorescent dopants, has a large problem in controlling the efficiency of the energy transfer, which eventually will result in unbalanced brightness of the colors. In addition, the difficult manufacturing processes of such an approach will be difficult to render the approach cost-effective. One alternative approach to achieve white emission is by using a phosphorescent excimer (similarly to what was described in white fluorescent OLEDs). The excimer dopant can emit a broad spectrum, and high-quality white emission may simultaneously arise from its monomer and aggregate states.

By a careful study of the optical and electronic properties of phosphorescent dopants with host material CBP, Forrest's group selected two blue phosphorescent dopants, FIrpic and platinum(II) [2-(4,6-difluorophenyl)pyridinato-*N*,C2'] (2,4-pentanedionato) (FPt1), as excimers in their white PHOLED structure (Figure 3.16) [664]. Although the planar

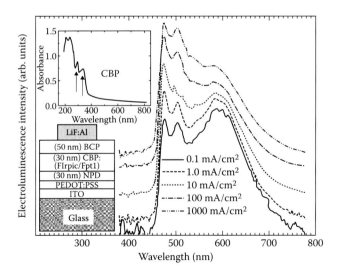

FIGURE 3.16
Normalized EL spectra of the white PHOLED at several current densities. Top inset: absorption of neat CBP films (100 nm); bottom inset: white PHOLED structure. (From D'Andrade, B.W., Brooks, J., Adamovich, V., Thompson, M.E., and Forrest, S.R., *Adv. Mater.*, 14, 1032, 2002. With permission.)

structure of the dopant FPt1 forms a broad excimer emission, the codopant of a blue dopant of FIrpic (FIrpic does not form excimer emission) with FPt1 balances the emission color and produces the desired white emission with CIE (0.33, 0.44). Both of these blue dopants have similar triplet energy and overlapped emission spectra, which are good for energy transfer when using CBP as a host. Since the triplet energy level of CBP is above the triplet levels of the two dopants, an endothermal triplet energy transfer process is expected between the host and the dopants. Nevertheless, a white PHOLED fabricated using two dopants where each doping concentrations is 6% in CBP produces a bright white emission with a maximum EQE of 4.4% (10.1 cd/A) and a luminance of 34,000 cd/m² at 16.6 V with a power efficiency of 4.8 lm/W and a CRI of 78.

As described by the same group, efficient white electrophosphorescence has been achieved with a single emissive dopant by screening a series of dopant molecules based on platinum(II) [2-(4,6-difluorophenyl)pyridinato-*N*,C2'] β-diketonates, which have a blue monomer emission (λ_{max}: 468, 500, 540 nm) and a broad orange aggregate emission (λ_{max}: 580 nm) in the doped host system. Since the intensity of the orange band increases relative to the blue monomer emission as the doping level increases, by judicious control of the ratio of monomer to aggregate emission, the doping concentration, the degree of steric bulk on the dopant, and the choice of the host material, a pure white emission spectrum close to standard white illumination has been achieved in a white PHOLED using a single phosphorescent excimer dopant platinum(II) [2-(4,6-difluorophenyl)pyridinato-*N*,C2'] (6-methyl-2,4-heptandionato-*O*,O) (FPt2). A high-efficiency white PHOLED with a structure of ITO/NPD/Irppz/mCBP:10–20% FPt2/BCP/Alq₃/LiF/Al emits white emission with CIE (0.36, 0.44) and CRI value >67 at both low (1 cd/m²) and high luminescence (500 cd/m²) levels. The devices gave a peak EQE of 6.4% (12.2 lm/W, 17.0 cd/A) [665].

The triplet phosphorescent excimer approach to a white PHOLED requires only energy transfer from the host to the monomer dopant or charge trapping on the dopant, but lacks energy transfer to the excimer owing to the zero ground state of the excimer. This strategy avoids the cascade energy transfer processes in different color dopants and allows simple optimization of doping levels to achieve the desired white color.

3.6.2.4 Phosphorescent WOLEDs with Single Phosphorescent Emitter

A new phosphorescent platinum(II)–pyridyltriazolate complex, bis(3,5-bis(2-pyridyl)-1,2,4-triazolato)platinum(II) [Pt(ptp)₂, 556] (Scheme 3.99), was synthesized and incorporated into OLEDs for evaluation as an electrophosphorescent dopant by Li et al. [666]. OLEDs with the structure glass/ITO/NPB/*x*% Pt(ptp)2:CBP/TPBI/Mg:Ag, 1:10, (200 nm) were fabricated. The emissions for the 7.5–15% Pt(ptp)₂:CBP devices are characterized by CIE coordinates of (0.3, 0.5), rendering near-white EL with green hue. Bhansali et al. [667] reported high-efficiency single-emitter white OLEDs from the same phosphorescent platinum complex,

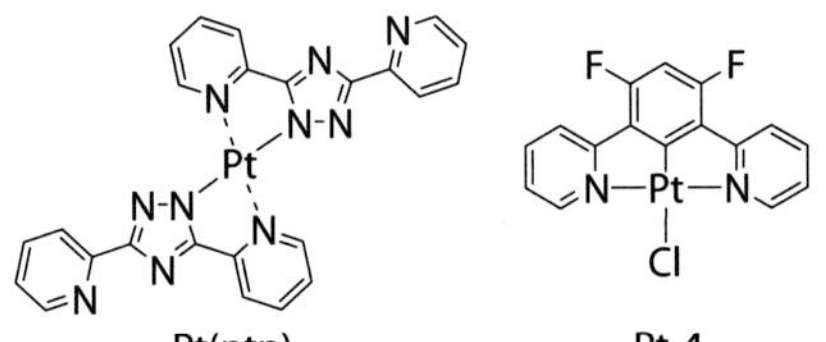

Pt(ptp)₂
556

Pt-4
557

SCHEME 3.99
Chemical structures of platinum complexes used for phosphorescent WOLEDs.

Pt(ptp)$_2$. They had demonstrated good control of the recombination zone and its impact on device performance by adjusting the number of emissive layers, their thicknesses, and relative positions in the device stack. Device optimization has resulted in single-emitter WOLEDs with a peak power efficiency of 30.4 ± 1.3 lm/W, EQE of 17.0 ± 0.1%, and correlated color temperature of 3450 K, within acceptable warm-white range despite the CRI being only 62.

Yang et al. [668] reported a novel blue-emitting Pt complex, Pt(II) (1,3-difluoro-4,6-di(2-pyridinyl)benzene)chloride (Pt-4, **557**) (Scheme 3.99). The photophysical studies of Pt-4 suggested that the high quantum yield and narrow emission spectra of Pt-4 can be mainly attributed to strong mixing of MLCT character to the lowest excited state. They had demonstrated, for the first time, highly efficient blue-emitting electrophosphorescent devices employing Pt complexes with a peak EQE of 16%, power efficiency of 20 lm/W, and CIE coordinates of (0.15, 0.26), which suggested that the utilization of Pt complexes is a viable approach for the development of blue OLEDs. In addition, a single-doped white-emitting device with a peak EQE of ca. 9.3% ph/d with CIE coordinates of (0.33, 0.36) was also reported. The performance of blue and white OLEDs using Pt-4 can be further enhanced by employing new matrix and carrier-transporting materials, and further characterization of the lifetime of the devices based on the Pt complexes is currently under way.

3.6.3 Hybrid White Light–Emitting Diodes

Fully phosphorescent WOLEDs have the highest reported efficiencies owing to potentially up to 100% internal QE of triplet emitters. However, these devices introduce an intrinsic exchange energy loss mainly arising from energy transfer from the host's singlets to the phosphorescent guest's triplets. Moreover, their development is limited by the availability of high-performance blue hosts/emitters, which are not stable enough to ensure a long device lifetime. Therefore, a combination of blue fluorescence and phosphorescence of other colors is considered to be an ideal solution to approach high-efficiency and long lifetime WOLEDs.

Forrest et al. introduced this device concept that exploits a blue fluorescent molecule in exchange for a phosphorescent dopant, in combination with green and red phosphor dopants, to yield high power efficiency and stable color balance, while maintaining the potential for unity internal QE. Two distinct modes of energy transfer within this device serve to channel nearly all of the triplet energy to the phosphorescent dopants, retaining the singlet energy exclusively on the blue fluorescent dopant (Figure 3.17). Additionally, eliminating the exchange energy loss to the blue fluorophore allows for roughly 20% increased power efficiency compared with a fully phosphorescent device. The device challenges incandescent sources by exhibiting total external quantum and power efficiencies that peak at 18.7% and 37.6 lm/W, respectively, decreasing to 18.4% and 23.8 lm/W at a high luminance of 500 cd/m^2 [669].

A WOLED was produced upon systematic introduction of deep-blue fluorescence from 4,4′-bis(9-ethyl-3-carbazovinylene)-1,1′-biphenyl (BCzVBi) to broad-band yellow phosphorescence from bis(3,5-bis(2-pyridyl)-1,2,4-triazolato)platinum(II) Pt(ptp)$_2$ in the common host CBP [670]. The WOLED exhibits striking stability of color and efficiency, as manifested by parameters at high brightness of 1000 cd/m^2, sustaining 94–122% of their values at 50 cd/m^2.

Baek and Lee [671] demonstrated that simple-structure WOLEDs without an interlayer could be achieved using the combination of phosphor-sensitized-fluorescent red and phosphorescent blue EMLs. In addition, the main cause of the color shift with increasing

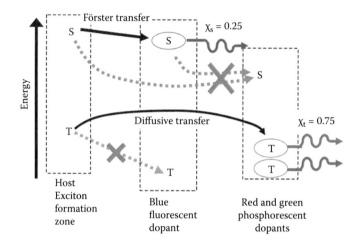

FIGURE 3.17
Proposed energy transfer mechanisms in the fluorescent/phosphorescent WOLED. (From Sun, Y.R., Giebink, N.C., Kanno, H., Ma, B.W., Thompson, M.E., and Forrest, S.R., *Nature*, 440, 908, 2006. With permission.)

current density was identified, and the color shift of the WOLEDs was successfully suppressed by properly balancing emission from the red and blue EMLs. Consequently, a maximum EQE of 6.2% (a current efficiency of 14.3 cd/A) and very stable CIE coordinates of (0.32 ± 0.01, 0.42 ± 0.002) were achieved. However, the elimination of an interlayer for the combination with a fluorescent blue EML causes about 50% decrease in the efficiency and a large change in the CIE coordinates with the driving current density.

Lei et al. [672] fabricated a WOLED by hybrid emissive layers that combined phosphorescence with fluorescence. The WOLED showed stable CIE and a high efficiency of 9.6 cd/A at 1.8 A/m². Maximum luminance of 17,400 cd/m² was achieved at 3000 A/m².

Lee et al. [673] used a blue iridium complex to sensitize a red fluorescent dye, giving a white emission device with a CRI of 70. The maximum luminance and current efficiency of the device are 16,220 cd/m² and 9.28 cd/A, respectively. Chang et al. [674] investigated WOLEDs, making use of both blue-phosphor-sensitized orange–red fluorescence and the residual blue phosphorescence. By doping a green phosphor into the poorly emitting ETL to recycle excitons formed there, the EL efficiencies can be up to 12.1%, 35.3 cd/A, and 23.9 lm/W.

By introducing a thin interlayer between the phosphorescent and the fluorescent region, Schwartz et al. [675] demonstrated devices reaching a current efficiency of 16.3 cd/A at 100 cd/m² and a CRI of 85 at warm-white CIE coordinates of (0.47, 0.42). They [676] also investigated the electron and hole mobility in mixed layers with different mix ratios, using both space-charge-limited currents of single-carrier devices with electrically doped charge-transport layers and TOF measurements. Both experimental methods yielded consistent results. The 1:1 blend showed balanced ambipolar charge-carrier transport, which was advantageous for the application as exciton-blocking interlayer in hybrid WOLEDs. The EL spectrum was rather stable against changes in interlayer thickness and driving current. Moreover, the EQE was enhanced by a factor of 2.5 as compared with a device without interlayer.

Chen et al. [677] fabricated WOLEDs with a bipolar transport layer between blue fluorescent and orange phosphorescent emitting layers. Pure white emission with CIE coordinates of (0.33, 0.34) and a maximum luminance of 40,960 cd/m^2 were obtained. The maximum current efficiency of the device is 13.4 cd/A with CRI of 71. Wang et al. [678] modulated EL spectra from interfacial exciplex emissions by varying the ratios of two acceptors (TPA and phenanthroline derivative) of exciplex-type devices in which the emissive wavelengths were tuned from 530 to 656 nm. The white device has CIE coordinates of (0.32, 0.35) and CRI of 90.4 with high color stability at various biases.

WOLEDs were fabricated by Wang et al. [679] by using phosphorescent material (*t*-bt)$_2$Ir(acac) doped in a CBP matrix as a yellow light–emitting layer and a thin layer DPVBi as the blue light–emitting layer. The light color of the OLEDs can be adjusted by changing doping concentration and the thickness of the DPVBi thin layer. The maximum luminance and power efficiency of 5% doped device reached 15,460 cd/m^2 and 8.1 lm/W, respectively. The 3% doped device showed CIE coordinates of (0.344, 0.322) at 8 V and a maximum power efficiency of 5.7 lm/W at 4.5 V.

Zhao et al. [680] demonstrated highly efficient triplet multiple-quantum-well-structure WOLEDs by introducing phosphorescent Ir complex material of bis(2-phenylbenzo-thiozo-lato-*N*,C2′)iridium(III)(acetylacetonate) (Ir(bt)$_2$(acac)) as orange dopant and fluorescent dye BCzVBi as blue dopant, and narrow band-gap semiconductor CBP as host material. The maximum luminance, current efficiency, and power efficiency are 19,000 cd/m^2, 14.5 cd/A, and 5.4 lm/W, respectively.

High-efficiency hybrid WOLEDs with a double light-emitting layer structure for high CRI were fabricated [681,682]. The hybrid WOLEDs with a double light-emitting layer structure was composed of a red phosphorescent layer and a blue–green layer of a blue light–emitting host doped with a green phosphor. A bipolar spacer by blending an HTM TCTA and an ETM bis(2-(2-hydroxy-phenyl)-pyridine)beryllium (Bepp2) was inserted between the red-emitting layer and blue–green-emitting layer. The resulting hybrid WOLEDs achieved a CRI of 90 and kept rather stable spectral emission with CIE coordinates of (0.42, 0.44) independent of driving voltages. Furthermore, the hybrid WOLEDs also exhibited a high efficiency where the maximum current efficiency, EQE, and power efficiency reached 29.4 cd/A, 13.8%, and 34.2 lm/W, and still remained at 25.4 cd/A, 11.9%, and 23.0 lm/W at 1000 cd/m^2, respectively.

Wan et al. [683] synthesized and characterized a series of indolizine derivatives and demonstrated their applications as electron-transporting hosts and deep-blue emitters. Of the indolizine derivatives, a blue fluorescent material, BPPI (3-(4,4′-biphenyl)-2-diphenylin-dolizine), was found to have (i) blue emission with high quantum yields, (ii) good morpho-logical and thermal stabilities, (iii) electron-transporting properties, and (iv) a sufficiently high triplet energy level to act as a host for red or yellow–orange phosphorescent dopants. They demonstrated a fluorescent/phosphorescent hybrid WOLED with a simplified struc-ture of ITO/NPB/TCTA/BPPI/BPPI:2% Ir(2-phq)$_3$/BPPI/LiF/Al, giving stable emissions, and current efficiency and EQE of 17.8 cd/A and 10.7%.

High-efficiency white OLED with a maximum power efficiency of 50.6 lm/W was obtained by combined hybrid blue device and ultrathin phosphorescent yellow emitter by Piao et al. [684]. At the practical brightness of 1000 cd/m^2, the power efficiency of the white device was 28.3 lm/W with a low voltage of 3.37 V and CIE coordinates of (0.40, 0.44). A highly efficient deep blue–emitting material with a diphenylaminofluorene emit-ting core unit (DAF) was synthesized and used as a deep blue–emitting material in pure white OLEDs [648]. The WOLEDs were fabricated by combining the deep blue–emitting material with phosphorescent red- and green-emitting materials. A high QE of 12.3% was demonstrated by optimizing the device structure of the WOLEDs using the DAF dopant.

A series of compounds containing arylamine and 1,2-diphenyl-1*H*-benz(*d*)imidazole moieties were developed as ambipolar blue-emitting materials by Chen et al. [685], with tunable blue-emitting wavelengths, tunable ambipolar carrier-transport properties and tunable triplet energy gaps. When these ambipolar blue-emitting materials are lightly doped with a yellow–orange-emitting iridium complex, white OLEDs can be achieved, as well as by the use of the incomplete energy transfer between the host and the dopant.

A very high-efficiency WOLED fabricated by the combination of a fluorescent blue emitter and a phosphorescent red emitter was reported by Qin and Tao [686]. The device structure is ITO/NPD/TCTA:2% TPP/BCP:0.4% Ir(piq)$_3$/Alq$_3$/Mg:Ag. The white emission contributed from both fluorescent blue light and phosphorescent red light was observed between 10 and 15 V, and a maximum white light luminance of 1076 cd/m^2 with CIE of (0.27, 0.24) and current efficiency of 1.35 cd/A was achieved. This practical approach to white-light emission is a clever strategy considering the currently available, relatively stable, and high-efficiency blue fluorescent, and very high-efficiency red phosphorescent materials; however, further work still needs to be done to optimize the device performance.

Recently, a unique multiple-EML white OLED was fabricated by Su et al. by utilizing FIrpic and tris(1-phenylisoquinolinolato-C2,*N*)iridium(III) (Ir(piq)$_3$) as the complementary triplet emitters and a bipolar material TCPZ (**346**) as the host. Different from most of the thus far reported white OLEDs fabricated with blue/red complementary triplet emitters that exhibit CRI <70, a high CRI of 82 is achieved owing to the combination of blue and red phosphorescence emissions from FIrpic and Ir(piq)$_3$, and the emerging green fluorescence emission from TCPZ. It also shows an alternative route to achieve high-quality white-light emission by the utilization of the guest, host, and exciplex emission [419].

3.6.4 Stacked and Tandem White Light–Emitting Diodes

While the above methods have been widely used in white OLED research, other strategies can be envisioned, such as (i) using stacked OLED structures, where blue, green, and red OLEDs operate independently to achieve the combined white color emission [687]; (ii) using one blue OLED coupled with green and red emitters resulting in white emission [688]; and (iii) employing multimode optical cavities to capture white emission, which has been investigated in Alq$_3$-based OLED [689]. However, there are pros and cons to using all of these methods.

To improve the lifetime and reduce thermal degradation of OLEDs due to the excessive current in practical use, it is necessary to obtain high brightness and high efficiency at lower current density. A promising approach is via stacked OLEDs consisting of vertically stacked multiple emissive units separated by charge-generating layers. Stacked OLEDs attract particular interest because the current efficiency and luminance can scale linearly with the number of emitting units.

Khan et al. [690] demonstrated WOLED with a RBG stacked multilayer structure. They obtained a white OLED with current efficiency of 5.60 cd/A and CIE coordinates of (0.34, 0.34) at 200 mA/cm^2 by adjusting the order and thickness of emitting layer in RBG structure. Its maximum luminance is 20,700 cd/m^2 at current density of 400 mA/cm^2. The results were explained on the basis of the theory of excitons generation and diffusion. According to the theory of excitons generation and diffusion, an equation had been set up that relates EL spectra to the thickness of every layer and to the exciton diffusion length.

Ho et al. had fabricated highly efficient tandem p–i–n white OLEDs [691]. A current efficiency of 23.9 cd/A and a power efficiency of 7.8 lm/W were achieved at 20 mA/cm^2

with CIE coordinates of (0.30, 0.43) by utilizing an optical transparent bilayer with doped organic p–n junction as the connecting layer.

Chen et al. [692] demonstrated two kinds of stacked WOLEDs employing Alq_3:20 wt.% Mg/MoO_3 as a charge-generation layer (Figure 3.18). White-light emission can be obtained by mixing blue fluorescence and orange phosphorescence. Stacked WOLEDs with individual blue fluorescent and orange phosphorescent emissive units had better color stability and higher efficiency than that with double white emissive units, which was attributed to the avoidance of the movement of the charge recombination zone and elimination of the Dexter energy transfer between blue and orange EMLs occurring in the latter. The efficiency of the stacked WOLED was 35.9 cd/A at 1000 cd/m².

Kim et al. found that the color stability of WOLEDs could be improved by using a charge-confining device structure with an emitting layer sandwiched between emitting layers [693]. A high EQE of 8.2% was achieved at 500 cd/m², and there was little change of color coordinate from 100 to 10,000 cd/m².

Lee et al. [694] reported efficient tandem WOLEDs by using bathocuproine:Cs_2CO_3/MoO_3 as an effective interconnecting layer (Figure 3.19). They utilized two primary colors of sky-blue and orange fluorescent emitters to obtain efficient white EL. Although single WOLEDs using two adjacent emitting layers showed a maximum current efficiency of 7.96 cd/A with CIE coordinates of (0.28, 0.34), the tandem WOLED device made by stacking two single-color OLEDs in series demonstrated doubled maximum current efficiency of 17.14 cd/A with CIE coordinates of (0.28, 0.41).

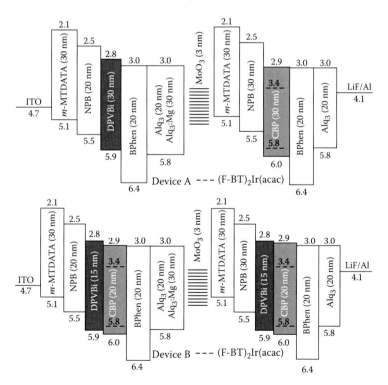

FIGURE 3.18
Device structures of stacked OLEDs. (From Chen, P., Xue, Q., Xie, W., Duan, Y., Xie, G., Zhao, Y., Hou, J., Liu, S., Zhang, L., and Li, B., *Appl. Phys. Lett.*, 93, 153508, 2008. With permission.)

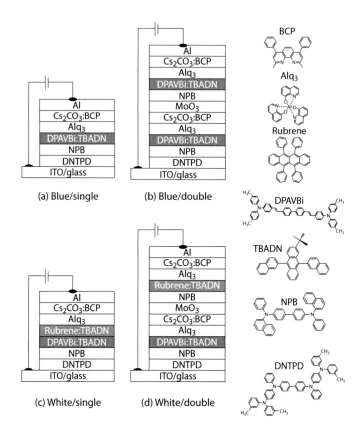

FIGURE 3.19
Stacked OLEDs using Cs₂CO₃/MoO₃ as an effective interconnecting layer. (From Lee, T.-W., Noh, T., Choi, B.-K., Kim, M.-S., Shin, D.W., and Kido, J., *Appl. Phys. Lett.*, 92, 043301, 2008. With permission.)

Stacked WOLEDs comprising blue fluorescent and orange phosphorescent emissive units employing Alq₃:Mg/MoO₃ as charge-generation layer were fabricated by Chen et al. [695]. They demonstrated that charge-carrier separation takes place only in the MoO₃ layer. Stacked WOLEDs with better performance was obtained by adjusting the thickness of MoO₃. The stacked WOLED with efficiency of 39.2 cd/A had excellent color stability with the CIE coordinates only changing from (0.407, 0.405) to (0.398, 0.397) when luminance increased from 22 to 10,000 cd/m².

The significant improvement of power efficiency and efficiency roll-off in tandem WOLEDs was achieved using an interface-modified C60/pentacene organic heterojunction (Figure 3.20) as the charge-generation layer by Chen et al. [696]. The resulting tandem WOLED exhibited a maximum power efficiency of 53.8 lm/W and a maximum current efficiency of 101.5 cd/A without any outcoupling techniques. Furthermore, they presented a high-efficiency tandem white OLED based on a C60/5,5‴-bis(naphth-2-yl)-2,2′:5′,2″:5″,2‴-quaterthiophene (NaT4) organic heterojunction as a charge-generation layer, giving a maximum current efficiency of 111.3 cd/A, maximum power efficiency of 50.5 lm/W, and maximum external QE of 38.7% [697].

By utilizing 2,9-dimethyl-4,7-diphenyl-1,10-phenanthroline (BCP):Li/MoO₃ as an effective charge-generation layer, Wang et al. [698] demonstrated that extremely high-efficiency tandem WOLEDs could be realized by using single-EML device configurations. This stacked device achieved maximum forward-viewing current efficiency of 110.9 cd/A and EQE of 43.3% at 1 μA/cm². Meanwhile, they demonstrated a high-efficiency white top-emitting

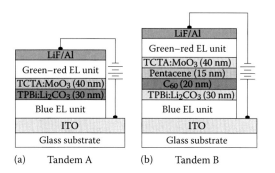

FIGURE 3.20
Schematic diagrams of the tandem WOLEDs using TPBI:Li$_2$CO$_3$/TCTA:MoO$_3$ (tandem A; a) and C$_{60}$/pentacene (tandem B; b) as charge-generation layers. (From Chen, Y., Chen, J., Ma, D., Yan, D., and Wang, L., *Appl. Phys. Lett.*, 99, 103304, 2011. With permission.)

OLED by introducing a tandem structure [699]. The microcavity effects were used to finalize the configuration of the tandem white top-emitting OLED. The maximum forward-viewing EQE and current efficiency of the device were 16.9% and 41.1 cd/A, respectively.

Among the previously reported WOLEDs, white PHOLEDs may become the most effective light source in the future owing to their 100% internal emission efficiency, which raises the possibility to approach the upper limit of the maximum external power efficiency of 80 lm/W with CRI >80 [700]. Currently, the major challenges facing such white PHOLEDs are their lifetime and color stability under operating conditions.

3.7 Conclusion and Remarks

While the individual materials described in the preceding sections are all valuable and in many cases essential for the realization of good, long-lived performance in OLED devices, it is equally true that the materials themselves are only a small part of the OLED story. As frequently alluded to above, the precise architecture into which the materials are placed is crucially important to device performance. Furthermore, while the bulk materials properties themselves, and especially their purities, are important, the interfaces formed between different materials are equally critical. Both chemical and morphological effects at any of the materials' interfaces in a device can have profound effects on the overall device performance. It is the precise interplay of materials design, processing, and ultimate device architecture that must be controlled if the remarkable phenomenon of efficient EL in organic materials is ever to be transformed into the display technology of the future.

Acknowledgments

SJS would like to thank all the students of the group at South China University of Technology for their contribution regarding the contents of this review. He also greatly appreciates the financial support from the National Natural Science Foundation of China (51073057 and 91233116), the Ministry of Science and Technology (2011AA03A110), and the Ministry of Education (NCET-11-0159).

References

1. M. Pope, H. P. Kallmann, P. Magnante, "Electroluminescence in organic crystals," *J. Chem. Phys.*, 38:2042–2043 (1963).
2. R. S. Vincett, W. A. Barlow, R. A. Hann, G. G. Roberts, "Electron conduction and low voltage blue electroluminescence in vacuum-deposited organ films," *Thin Solid Films*, 94:171–183 (1982).
3. C. W. Tang, S. A. VanSlyke, "Organic electroluminescent diodes," *Appl. Phys. Lett.*, 51:913–915 (1987).
4. M. A. Baldo, D. F. O'Brien, Y. You, A. Shoustikov, S. Sibley, M. E. Thompson, S. R. Forrest, "Highly efficient phosphorescent emission from organic electroluminescent devices," *Nature*, 395:151–154 (1998).
5. Y. Kawamura, K. Goushi, J. Brooks, J. J. Brown, H. Sasabe, C. Adachi, "100% Phosphorescence quantum efficiency of Ir(III) complexes in organic semiconductor films," *Appl. Phys. Lett.*, 86:071104 (2005).
6. T. Nagamoto, Y. Maruta, O. Omoto, "Electrical and optical properties of vacuum-evaporated indium–tin oxide films with high electron mobility," *Thin Solid Films*, 192:17–25 (1990).
7. T. Kawashima, H. Matsui, N. Tanabe, "New transparent conductive films: FTO coated ITO," *Thin Solid Films*, 445:241–244 (2003).
8. N. W. Schmidt, T. S. Totushek, W. A. Kimes, D. R. Callender, J. R. Doyle, "Effects of substrate temperature and near-substrate plasma density on the properties of dc magnetron sputtered aluminum doped zinc oxide," *J. Appl. Phys.*, 94:5514–5521 (2003).
9. Y. Wu, C. H. M. Maree, R. F. Haglund, J. D. Hamilton, M. A. M. Paliza, M. B. Huang, L. C. Feldman, R. A. Weller, "Resistivity and oxygen content of indium tin oxide films deposited at room temperature by pulsed-laser ablation," *J. Appl. Phys.*, 86:991–994 (1999).
10. H. Tomonaga, T. Morimoto, "Indium–tin oxide coatings via chemical solution deposition," *Thin Solid Films*, 392:243–248 (2001).
11. O. Marcovitch, Z. Klein, I. Lubezky, "Transparent conductive indium oxide films deposited on low temperature substrates by activated reactive evaporation," *Appl. Opt.*, 28:2792–2795 (1989).
12. M. G. Helander, Z. B. Wang, J. Qiu, M. T. Greiner, D. P. Puzzo, Z. W. Liu, Z. H. Lu, "Chlorinated indium tin oxide electrodes with high work function for organic device compatibility," *Science*, 332:944–947 (2011).
13. J. K. Mahon, T. X. Zhou, P. E. Burrows, S. R. Forrest, M. E. Thompson, "Small-molecule organic light emitting devices in flat panel display applications," in: *Light-Emitting Diodes: Research, Manufacturing, and Applications, Second SPIE Proceedings*, 3279:87–92 (1998).
14. M. G. Mason, C. W. Tang, L. S. Hung, P. Raychaudhuri, J. Madathil, D. J. Giesen, L. Yan, Q. T. Le, Y. Gao, S. T. Lee, L. S. Liao, L. F. Cheng, W. R. Salaneck, D. A. dos Santos, J. L. Bredas, "Interfacial chemistry of Alq$_3$ and LiF with reactive metals," *J. Appl. Phys.*, 89:2756–2765 (2001).
15. Y. Cao, "Thin metal–oxide layer as stable electron-injecting electrode for light emitting diodes," PCT Int. Appl., WO 2000022683, p. 37 (2000).
16. I. G. Hill, A. Rajagopal, A. Kahn, Y. Hu, "Molecular level alignment at organic semiconductor–metal interfaces," *Appl. Phys. Lett.*, 73:662–664 (1999).
17. H. Ishii, K. Sugiyama, E. Ito, K. Seki, "Energy level alignment and interfacial electronic structures at organic/metal and organic/organic interfaces," *Adv. Mater.*, 11:605–625 (1999).
18. W. R. Salaneck, S. Stafstrom, J. L. Bredas, *Conjugated Polymer Surfaces and Interfaces: Electronic and Chemical Structure of Interfaces for Polymer Light Emitting Devices*, Cambridge University Press, London (2003).

19. W. R. Salaneck, K. Seki, A. Kahn, J. Pireaux, *Conjugated Polymer and Molecular Interfaces*, 1st ed., Marcel Dekker, New York (2001).
20. L. S. Hung, C. H. Chen, "Recent progress of molecular organic electroluminescent materials and devices," *Mater. Sci. Eng. R.*, 39:143–222 (2002).
21. S. A. Van Slyke, C. H. Chen, C. W. Tang, "Organic electroluminescent devices with improved stability," *Appl. Phys. Lett.*, 69:2160–2162 (1996).
22. E. W. Forsythe, M. A. Abkowitz, Y. L. Gao, "Tuning the carrier injection efficiency for organic light-emitting diodes," *J. Phys. Chem. B*, 104:3948–3952 (2000).
23. S. M. Tadayyon, H. M. Grandin, K. Griffiths, P. R. Norton, H. Aziz, Z. D. Popovic, "CuPc buffer layer role in OLED performance: A study of the interfacial band energies," *Org. Electron.*, 5:157–166 (2004).
24. I. G. Hill, A. Kahn, "Combined photoemission/in vacuo transport study of the indium tin oxide/copper phthalocyanine/N,N'-diphenyl-N,N'-bis(l-naphthyl)-1,1'biphenyl-4,4"diamine molecular organic semiconductor system," *J. Appl. Phys.*, 86:2116–2122 (1999).
25. F. Nuesch, M. Carrara, M. Schaer, D. B. Romero, L. Zuppiroli, "The role of copper phthalocyanine for charge injection into organic light emitting devices," *Chem. Phys. Lett.*, 347:311–317 (2001).
26. S. H. Jung, J. H. Choi, S. M. Yang, W. J. Cho, C. S. Ha, "Syntheses and characterization of soluble phthalocyanine derivatives for organic electroluminescent devices," *Mater. Sci. Eng. B-Solid State Mater. Adv. Technol.*, 85:160–164 (2001).
27. M. Ichikawa, K. Kobayashi, T. Koyama, Y. Taniguchi, "Intense and efficient ultraviolet electroluminescence from organic light-emitting devices with fluorinated copper phthalocyanine as hole injection layer," *Thin Solid Films*, 515:3932–3935 (2007).
28. H. Lee, J. Lee, K. Jeong, Y. Yi, J. H. Lee, J. W. Kim, S. W. Cho, "Hole injection enhancements of a CoPc and CoPc:NPB mixed layer in organic light-emitting devices," *J. Phys. Chem. C*, 116:13210–13216 (2012).
29. Y. Qiu, Y. D. Gao, P. Wei, L. D. Wang, "Organic light-emitting diodes with improved hole–electron balance by using copper phthalocyanine/aromatic diamine multiple quantum wells," *Appl. Phys. Lett.*, 80:2628–2630 (2002).
30. J. Cui, Q. L. Huang, J. G. C. Veinot, H. Yan, T. J. Marks, "Interfacial microstructure function in organic light-emitting diodes: Assembled tetraaryldiamine and copper phthalocyanine interlayers," *Adv. Mater.*, 14:565–569 (2002).
31. W. H. Kim, A. J. Makinen, N. Nikolov, R. Shashidhar, H. Kim, Z. H. Kafafi, "Molecular organic light-emitting diodes using highly conducting polymers as anodes," *Appl. Phys. Lett.*, 80:3844–3846 (2002).
32. Y. Yang, A. J. Heeger, "Polyaniline as a transparent electrode for polymer light-emitting diodes: Lower operating voltage and higher efficiency," *Appl. Phys. Lett.*, 64:1245–1247 (1994).
33. J. Gao, A. J. Heeger, J. Y. Lee, C. Y. Kim, "Soluble polypyrrole as the transparent anode in polymer light-emitting diodes," *Synth. Met.*, 82:221–223 (1996).
34. J. Schwartz, E. L. Bruner, N. Koch, A. R. Span, S. L. Bernasek, A. Kahn, "Controlling the work function of indium tin oxide: Differentiating dipolar from local surface effects," *Synth. Met.*, 138:223–227 (2003).
35. J. Lee, B. J. Jung, J. I. Lee, H. Y. Chu, L. M. Do, H. K. Shim, "Modification of an ITO anode with a hole-transporting SAM for improved OLED device characteristics," *J. Mater. Chem.*, 12:3494–3498 (2002).
36. Q. L. Huang, J. Cui, H. Yan, J. G. C. Veinot, T. J. Marks, "Small molecule organic light-emitting diodes can exhibit high performance without conventional hole transport layers," *Appl. Phys. Lett.*, 81:3528–3530 (2002).
37. Q. L. Huang, J. Cui, J. G. C. Veinot, H. Yan, T. J. Marks, "Realization of high-efficiency/high-luminance small-molecule organic light-emitting diodes: Synergistic effects of siloxane anode functionalization/hole-injection layers, and hole/exciton-blocking/electron-transport layers," *Appl. Phys. Lett.*, 82:331–333 (2003).

38. H. Yan, P. Lee, N. R. Armstrong, A. Graham, G. A. Evmenenko, P. Dutta, T. J. Marks, "High-performance hole-transport layers for polymer light-emitting diodes. Implementation of organosiloxane cross-linking chemistry in polymeric electroluminescent devices," *J. Am. Chem. Soc.*, 127:3172–3183 (2005).

39. Q. L. Huang, J. F. Li, G. A. Evmenenko, P. Dutta, T. J. Marks, "Systematic investigation of nanoscale adsorbate effects at organic light-emitting diode interfaces. Interfacial structure–charge injection–luminance relationships," *Chem. Mater.*, 18:2431–2442 (2006).

40. Y. Tokudome, T. Fukushima, A. Goto, H. Kaji, "Enhanced hole injection in organic light-emitting diodes by optimized synthesis of self-assembled monolayer," *Org. Electron.*, 12:1600–1605 (2011).

41. H. Ma, H. L. Yip, F. Huang, A. K. Y. Jen, "Interface engineering for organic electronics," *Adv. Funct. Mater.*, 20:1371–1388 (2010).

42. E. L. Hanson, J. Guo, N. Koch, J. Schwartz, S. L. Bernasek, "Advanced surface modification of indium tin oxide for improved charge injection in organic devices," *J. Am. Chem. Soc.*, 127:10058–10062 (2005).

43. D. M. Rampulla, C. M. Wroge, E. L. Hanson, J. G. Kushmerick, "Charge transport across phosphonate monolayers on indium tin oxide," *J. Phys. Chem. C*, 114:20852–20855 (2010).

44. J. A. Bardecker, H. Ma, T. Kim, F. Huang, M. S. Liu, Y. J. Cheng, G. Ting, A. K. Y. Jen, "Self-assembled electroactive phosphonic acids on ITO: Maximizing hole-injection in polymer light-emitting diodes," *Adv. Funct. Mater.*, 18:3964–3971 (2008).

45. S. Y. Yu, D. C. Huang, Y. L. Chen, K. Y. Wu, Y. T. Tao, "Approaching charge balance in organic light-emitting diodes by tuning charge injection barriers with mixed monolayers," *Langmuir*, 28:424–430 (2012).

46. L. S. Hung, L. R. Zheng, M. G. Mason, "Anode modification in organic light-emitting diodes by low-frequency plasma polymerization of CHF_3," *Appl. Phys. Lett.*, 78:673–675 (2001).

47. J. X. Tang, Y. Q. Li, L. R. Zheng, L. S. Hung, "Anode/organic interface modification by plasma polymerized fluorocarbon films," *J. Appl. Phys.*, 95:4397–4403 (2004).

48. Y. Qiu, Y. D. Gao, L. D. Wang, D. Q. Zhang, "Efficient light emitting diodes with Teflon buffer layer," *Synth. Met.*, 130:235–237 (2002).

49. Z. B. Deng, X. M. Ding, S. T. Lee, W. A. Gambling, "Enhanced brightness and efficiency in organic electroluminescent devices using SiO_2 buffer layers," *Appl. Phys. Lett.*, 74:2227–2229 (1999).

50. C. O. Poon, F. L. Wong, S. W. Tong, R. Q. Zhang, C. S. Lee, S. T. Lee, "Improved performance and stability of organic light-emitting devices with silicon oxy-nitride buffer layer," *Appl. Phys. Lett.*, 83:1038–1040 (2003).

51. Z. F. Zhang, Z. B. Deng, C. J. Liang, M. X. Zhang, D. H. Xu, "Organic light-emitting diodes with a nanostructured TiO_2 layer at the interface between ITO and NPB layers," *Displays*, 24:231–234 (2003).

52. C. H. Gao, S. D. Cai, W. Gu, D. Y. Zhou, Z. K. Wang, L. S. Liao, "Enhanced hole injection in phosphorescent organic light-emitting diodes by thermally evaporating a thin indium trichloride layer," *ACS Appl. Mater. Interfaces*, 4:5211–5216 (2012).

53. M. Kroger, S. Hamwi, J. Meyer, T. Riedl, W. Kowalsky, A. Kahn, "P-type doping of organic wide band gap materials by transition metal oxides: A case-study on molybdenum trioxide," *Org. Electron.*, 10:932–938 (2009).

54. X. Y. Jiang, Z. L. Zhang, J. Cao, W. Q. Zhu, "High stability and low driving voltage green organic light emitting diode with molybdenum oxide as buffer layer," *Solid State Electron.*, 52:952–956 (2008).

55. Z. W. Liu, M. G. Helander, Z. B. Wang, Z. H. Lu, "Band alignment at anode/organic interfaces for highly efficient simplified blue-emitting organic light-emitting diodes," *J. Phys. Chem. C*, 114:16746–16749 (2010).

56. Z. W. Liu, M. G. Helander, Z. B. Wang, Z. H. Lu, "Efficient single-layer organic light-emitting diodes based on C545T–Alq_3 system," *J. Phys. Chem. C*, 114:11931–11935 (2010).

57. Y. H. Kim, S. Kwon, J. H. Lee, S. M. Park, Y. M. Lee, J. W. Kim, "Hole injection enhancement by a WO$_3$ interlayer in inverted organic light-emitting diodes and their interfacial electronic structures," *J. Phys. Chem. C*, 115:6599–6604 (2011).

58. K. S. Yook, J. Y. Lee, "Low driving voltage in organic light-emitting diodes using MoO$_3$ as an interlayer in hole transport layer," *Synth. Met.*, 159:69–71 (2009).

59. T. Matsushima, G. H. Jin, Y. Kanai, T. Yokota, S. Kitada, T. Kishi, H. Murata, "Interfacial charge transfer and charge generation in organic electronic devices," *Org. Electron.*, 12:520–528 (2011).

60. J. Meyer, R. Khalandovsky, P. Gorrn, A. Kahn, "MoO$_3$ films spin-coated from a nanoparticle suspension for efficient hole-injection in organic electronics," *Adv. Mater.*, 23:70–73 (2011).

61. M. Vasilopoulou, A. M. Douvas, D. G. Georgiadou, L. C. Palilis, S. Kennou, L. Sygellou, A. Soultati, I. Kostis, G. Papadimitropoulos, D. Davazoglou, P. Argitis, "The influence of hydrogenation and oxygen vacancies on molybdenum oxides work function and gap states for application in organic optoelectronics," *J. Am. Chem. Soc.*, 134:16178–16187 (2012).

62. D. B. Romero, M. Schaer, L. Zuppiroli, B. Cesar, B. François, "Effects of doping in polymer light-emitting diodes," *Appl. Phys. Lett.*, 67:1659–1661 (1995).

63. F. Huang, A. G. MacDiarmid, B. R. Hsieh, "An iodine-doped polymer light-emitting diode," *Appl. Phys. Lett.*, 71:2415–2417 (1997).

64. Y. Sato, T. Ogata, J. Kido, "Organic electroluminescent devices with polymer buffer layer," in: *Proc. SPIE, Organic Light-Emitting Materials and Devices IV*, 4105:134–142 (2000).

65. A. Yamamori, C. Adachi, T. Koyama, Y. Taniguchi, "Doped organic light emitting diodes having a 650-nm-thick hole transport layer," *Appl. Phys. Lett.*, 72:2147–2149 (1998).

66. J. Blochwitz, M. Pfeiffer, T. Fritz, K. Leo, "Low voltage organic light emitting diodes featuring doped phthalocyanine as hole transport material," *Appl. Phys. Lett.*, 73:729–731 (1998).

67. X. Zhou, M. Pfeiffer, J. Blochwitz, A. Werner, A. Nollau, T. Fritz, K. Leo, "Very-low-operating-voltage organic light-emitting diodes using a p-doped amorphous hole injection layer," *Appl. Phys. Lett.*, 78:410–412 (2001).

68. G. F. He, O. Schneider, D. S. Qin, X. Zhou, M. Pfeiffer, K. Leo, "Very high-efficiency and low voltage phosphorescent organic light-emitting diodes based on a p–i–n junction," *J. Appl. Phys.*, 95:5773–5777 (2004).

69. K. S. Yook, S. O. Jeon, J. Y. Lee, "Efficient hole injection by doping of hexaazatriphenylene hexacarbonitrile in hole transport layer," *Thin Solid Films*, 517:6109–6111 (2009).

70. S. H. Cho, S. W. Pyo, M. C. Suh, "Low voltage top-emitting organic light emitting devices by using 1,4,5,8,9,11-hexaazatriphenylene-hexacarbonitrile," *Synth. Met.*, 162:402–405 (2012).

71. L. S. Liao, W. K. Slusarek, T. K. Hatwar, M. L. Ricks, D. L. Comfort, "Tandem organic light-emitting mode using hexaazatriphenylene hexacarbonitrile in the intermediate connector," *Adv. Mater.*, 20:324–329 (2008).

72. P. K. Koech, A. B. Padmaperuma, L. A. Wang, J. S. Swensen, E. Polikarpov, J. T. Darsell, J. E. Rainbolt, D. J. Gaspar, "Synthesis and application of 1,3,4,5,7,8-hexafluorotetracyanonaphthoquinodimethane (F6-TNAP): A conductivity dopant for organic light-emitting devices," *Chem. Mater.*, 22:3926–3932 (2010).

73. W. J. Shin, J. Y. Lee, J. C. Kim, T. H. Yoon, T. S. Kim, O. K. Song, "Bulk and interface properties of molybdenum trioxide-doped hole transporting layer in organic light-emitting diodes," *Org. Electron.*, 9:333–338 (2008).

74. S. D. Cai, C. H. Gao, D. Y. Zhou, W. Gu, L. S. Liao, "Study of hole-injecting properties in efficient, stable, and simplified phosphorescent organic light-emitting diodes by impedance spectroscopy," *ACS Appl. Mater. Interfaces*, 4:312–316 (2012).

75. Y. Zou, Z. B. Deng, D. H. Xu, J. Xiao, M. Y. Zhou, H. L. Du, Y. S. Wang, "Enhanced performance in organic light-emitting diode by utilizing MoO$_3$-doped C$_{60}$ as effective hole injection layer," *Synth. Met.*, 161:2628–2631 (2012).

76. L. S. Li, M. Guan, G. H. Cao, Y. Y. Li, Y. P. Zeng, "Low operating-voltage and high power-efficiency OLED employing MoO$_3$-doped CuPc as hole injection layer," *Displays*, 33:17–20 (2012).

77. J. Y. Lee, J. Y. Park, S. H. Min, K. W. Lee, Y. G. Baek, "A thermally stable hole injection material for use in organic light-emitting diodes," *Thin Solid Films*, 515:7726–7731 (2007).

78. J. Y. Kim, D. Yokoyama, C. Adachi, "Horizontal orientation of disk-like hole transport molecules and their application for organic light-emitting diodes requiring a lower driving voltage," *J. Phys. Chem. C*, 116:8699–8706 (2012).

79. Y. Park, B. Kim, C. Lee, A. Hyun, S. Jang, J. H. Lee, Y. S. Gal, T. H. Kim, K. S. Kim, J. Park, "Highly efficient new hole injection materials for OLEDs based on dimeric phenothiazine and phenoxazine derivatives," *J. Phys. Chem. C*, 115:4843–4850 (2011).

80. A. M. Horgan, "Composite layered imaging member for electrophotography," U.S. Patent 4,047,948, p. 14 (1977).

81. C. Adachi, K. Nagai, N. Tamoto, "Molecular design of hole transport materials for obtaining high durability in organic electroluminescent diodes," *Appl. Phys. Lett.*, 66:2679–2681 (1995).

82. H. Fujikawa, S. Tokito, Y. Taga, "Energy structures of triphenylamine oligomers," *Synth. Met.*, 91:161–162 (1997).

83. S. Tokito, H. Tanaka, A. Okada, Y. Taga, "High-temperature operation of an electroluminescent device fabricated using a novel triphenylamine derivative," *Appl. Phys. Lett.*, 69:878–880 (1996).

84. D. F. O'Brien, P. E. Burrows, S. R. Forrest, B. E. Koene, D. E. Loy, M. E. Thompson, "Hole transporting materials with high glass transition temperatures for use in organic light-emitting devices," *Adv. Mater.*, 10:1108–1112 (1998).

85. C. Giebeler, H. Antoniadis, D. D. C. Bradley, Y. Shirota, "Influence of the hole transport layer on the performance of organic light-emitting diodes," *J. Appl. Phys.*, 85:608–615 (1999).

86. R. D. Hreha, C. P. George, A. Haldi, B. Domercq, M. Malagoli, S. Barlow, J. L. Bredas, B. Kippelen, S. R. Marder, "2,7-Bis(diarylamino)-9,9-dimethylfluorenes as hole-transport materials for organic light-emitting diodes," *Adv. Funct. Mater.*, 13:967–973 (2003).

87. K. T. Wong, Z. J. Wang, Y. Y. Chien, C. L. Wang, "Synthesis and properties of 9,9-diarylfluorene-based triaryldiamines," *Org. Lett.*, 3:2285–2288 (2001).

88. J. P. Chen, H. Tanabe, X. C. Li, T. Thoms, Y. Okamura, K. Ueno, "Novel organic hole transport material with very high T-g for light-emitting diodes," *Synth. Met.*, 132:173–176 (2003).

89. K. R. J. Thomas, J. T. Lin, Y. T. Tao, C. W. Ko, "Light-emitting carbazole derivatives: Potential electroluminescent materials," *J. Am. Chem. Soc.*, 123:9404–9411 (2001).

90. Q. Zhang, J. S. Chen, Y. X. Cheng, L. X. Wang, D. G. Ma, X. B. Jing, F. S. Wang, "Novel hole-transporting materials based on 1,4-bis(carbazolyl)benzene for organic light-emitting devices," *J. Mater. Chem.*, 14:895–900 (2004).

91. Y. Kuwabara, H. Ogawa, H. Inada, N. Noma, Y. Shirota, "Thermally stable multilayered organic electroluminescent devices using novel starburst molecules, 4,4′,4″-tri(N-carbazolyl) triphenyl-amine (TCTA) and 4,4′,4″-tris(3-methylphenylphenylamino)triphenylamine (*m*-MTDATA), as hole-transport materials," *Adv. Mater.*, 6:677–679 (1994).

92. Y. Shirota, "Organic materials for electronic and optoelectronic devices," *J. Mater. Chem.*, 10:1–25 (2000).

93. K. R. J. Thomas, J. T. Lin, Y. T. Tao, C. W. Ko, "New star-shaped luminescent triarylamines: Synthesis, thermal, photophysical, and electroluminescent characteristics," *Chem. Mater.*, 14:1354–1361 (2002).

94. I. Y. Wu, J. T. Lin, Y. T. Tao, E. Balasubramaniam, Y. Z. Su, C. W. Ko, "Diphenylthienylamine-based star-shaped molecules for electroluminescence applications," *Chem. Mater.*, 13:2626–2631 (2001).

95. J. Y. Li, D. Liu, Y. Q. Li, C. S. Lee, H. L. Kwong, S. T. Lee, "A high T_g carbazole-based hole-transporting material for organic light-emitting devices," *Chem. Mater.*, 17:1208–1212 (2005).

96. K. Yamashita, T. Mori, T. Mizutani, H. Miyazaki, T. Takeda, "EL properties of organic light-emitting-diode using TPD derivatives with diphenylstylyl groups as hole transport layer," *Thin Solid Films*, 363:33–36 (2000).

97. Q. G. He, H. Z. Lin, Y. F. Weng, B. Zhang, Z. M. Wang, G. T. Lei, L. D. Wang, Y. Qiu, F. L. Bai, "A hole-transporting material with controllable morphology containing binaphthyl and triphenylamine chromophores," *Adv. Funct. Mater.*, 16:1343–1348 (2006).

98. Y. K. Kim, S. H. Hwang, "Highly efficient organic light-emitting diodes using novel hole-transporting materials," *Synth. Met.*, 156:1028–1035 (2006).

99. P. Cias, C. Slugovc, G. Gescheidt, "Hole transport in triphenylamine based OLED devices: From theoretical modeling to properties prediction," *J. Phys. Chem. A*, 115:14519–14525 (2011).

100. Z. F. Li, Z. X. Wu, W. Fu, P. Liu, B. Jiao, D. D. Wang, G. J. Zhou, X. Hou, "Versatile fluorinated derivatives of triphenylamine as hole-transporters and blue–violet emitters in organic light-emitting devices," *J. Phys. Chem. C*, 116:20504–20512 (2012).

101. Z. H. Li, M. S. Wong, "Synthesis and functional properties of end-dendronized oligo(9,9-diphenyl)fluorenes," *Org. Lett.*, 8:1499–1502 (2006).

102. J. Y. Park, J. M. Kim, H. Lee, K. Y. Ko, K. S. Yook, J. Y. Lee, Y. G. Baek, "Thermally stable triphenylene-based hole-transporting materials for organic light-emitting devices," *Thin Solid Films*, 519:5917–5923 (2011).

103. S. H. Hwang, Y. K. Kim, Y. Kwak, C. H. Lee, J. Lee, S. Kim, "Improved performance of organic light-emitting diodes using advanced hole-transporting materials," *Synth. Met.*, 159:2578–2583 (2009).

104. J. Kwak, Y. Y. Lyu, S. Noh, H. Lee, M. Park, B. Choi, K. Char, C. Lee, "Hole transport materials with high glass transition temperatures for highly stable organic light-emitting diodes," *Thin Solid Films*, 520:7157–7163 (2012).

105. Q. X. Tong, S. L. Lai, M. Y. Chan, K. H. Lai, J. X. Tang, H. L. Kwong, C. S. Lee, S. T. Lee, "High T_g triphenylamine-based starburst hole-transporting material for organic light-emitting devices," *Chem. Mater.*, 19:5851–5855 (2007).

106. S. L. Tao, L. Li, J. S. Yu, Y. D. Jiang, Y. C. Zhou, C. S. Lee, S. T. Lee, X. H. Zhang, O. Y. Kwon, "Bipolar molecule as an excellent hole-transporter for organic-light emitting devices," *Chem. Mater.*, 21:1284–1287 (2009).

107. J. N. Moorthy, P. Venkatakrishnan, D. F. Huang, T. J. Chow, "Blue light-emitting and hole-transporting amorphous molecular materials based on diarylaminobiphenyl-functionalized bimesitylenes," *Chem. Commun.*, 2146–2148 (2008).

108. N. Kapoor, K. R. J. Thomas, "Fluoranthene-based triarylamines as hole-transporting and emitting materials for efficient electroluminescent devices," *New J. Chem.*, 34:2739–2748 (2010).

109. A. M. Thangthong, D. Meunmart, N. Prachumrak, S. Jungsuttiwong, T. Keawin, T. Sudyoadsuk, V. Promarak, "Bifunctional anthracene derivatives as non-doped blue emitters and hole-transporters for electroluminescent devices," *Chem. Commun.*, 47:7122–7124 (2011).

110. A. M. Thangthong, N. Prachumrak, S. Namuangruk, S. Jungsuttiwong, T. Keawin, T. Sudyoadsuk, V. Promarak, "Synthesis, properties and applications of biphenyl functionalized 9,9-bis(4-diphenylaminophenyl)fluorenes as bifunctional materials for organic electroluminescent devices," *Eur. J. Org. Chem.*, 27:5263–5274 (2012).

111. C. H. Liao, M. T. Lee, C. H. Tsai, C. H. Chen, "Highly efficient blue organic light-emitting devices incorporating a composite hole transport layer," *Appl. Phys. Lett.*, 86:203507 (2005).

112. C. S. Ha, J. H. Shin, H. Lim, W. J. Cho, "Characterization of organic electroluminescent devices introducing fluorine-containing polyimide to hole-transporting layer," *Mater. Sci. Eng. B-Solid State Mater. Adv. Technol.*, 85:195–198 (2001).

113. J. H. Shin, J. W. Park, W. K. Lee, N. J. Jo, W. J. Cho, C. S. Ha, "Organic electroluminescent devices using fluorine-containing polyimides as a hole transporting layer," *Synth. Met.*, 137:1017–1018 (2003).

114. E. Bellmann, S. E. Shaheen, S. Thayumanavan, S. Barlow, R. H. Grubbs, S. R. Marder, B. Kippelen, N. Peyghambarian, "New triarylamine-containing polymers as hole transport materials in organic light-emitting diodes: Effect of polymer structure and cross-linking on device characteristics," *Chem. Mater.*, 10:1668–1676 (1998).

115. M. S. Bayerl, T. Braig, O. Nuyken, D. C. Muller, M. Gross, K. Meerholz, "Crosslinkable hole-transport materials for preparation of multilayer organic light emitting devices by spin-coating," *Macromol. Rapid Commun.*, 20:224–228 (1999).

116. E. Bacher, S. Jungermann, M. Rojahn, V. Wiederhirn, O. Nuyken, "Photopatterning of crosslinkable hole-conducting materials for application in organic light-emitting devices," *Macromol. Rapid Commun.*, 25:1191–1196 (2004).

117. B. Domercq, R. D. Hreha, N. Larribeau, J. N. Haddock, S. R. Marder, B. Kippelen, "Photo-crosslinkable polymers as hole-transport materials for organic light-emitting diodes," in: *Proc SPIE, Organic Photonic Materials and Devices IV*, 4642:88–96 (2002).

118. B. R. Hsieh, X. C. Li, A. Sellinger, G. E. Jabbour, "Semiconducting hole injection materials for organic light emitting devices," U.S. Patent 6,830,830 (2004).

119. S. Liu, X. Z. Jiang, H. Ma, M. S. Liu, A. K. Y. Jen, "Triarylamine-containing poly(perfluorocy-clobutane) as hole-transporting material for polymer light-emitting diodes," *Macromolecules*, 33:3514–3517 (2000).

120. J. Ji, S. Narayan-Sarathy, R. H. Neilson, J. D. Oxley, D. A. Babb, N. G. Rondan, D. W. Smith, "[p-((Trifluorovinyl)oxy)phenyl]lithium: Formation, synthetic utility, and theoretical support for a versatile new reagent in fluoropolymer chemistry," *Organometallics*, 17:783–785 (1998).

121. X. Z. Jiang, M. S. Liu, A. K. Y. Jen, "Bright and efficient exciplex emission from light-emitting diodes based on hole-transporting amine derivatives and electron-transporting polyfluorenes," *J. Appl. Phys.*, 91:10147–10152 (2002).

122. X. Gong, D. Moses, A. J. Heeger, S. Liu, A. K. Y. Jen, "High-performance polymer light-emitting diodes fabricated with a polymer hole injection," *Appl. Phys. Lett.*, 83:183–185 (2003).

123. P. M. Borsenberger, "Hole transport in bis(4-*N,N*-diethylamino-2-methylphenyl)-4-methyl-phenylmethane doped polymers," *Phys. Status Solidi Basic Solid State Phys.*, 173:671–680 (1992).

124. P. M. Borsenberger, W. T. Gruenbaum, E. H. Magin, "Hole transport in vapor-deposited triphe-nylmethane glasses," *Jpn. J. Appl. Phys.*, 35:2698–2703 (1996).

125. Y. Wang, "Dramatic effects of hole transport layer on the efficiency of iridium-based organic light-emitting diodes," *Appl. Phys. Lett.*, 85:4848–4850 (2004).

126. M. Higuchi, S. Shiki, K. Ariga, K. Yamamoto, "First synthesis of phenylazomethine dendrimer ligands and structural studies," *J. Am. Chem. Soc.*, 123:4414–4420 (2001).

127. M. Higuchi, M. Tsuruta, H. Chiba, S. Shiki, K. Yamamoto, "Control of stepwise radial complex-ation in dendritic polyphenylazomethines," *J. Am. Chem. Soc.*, 125:9988–9997 (2003).

128. A. Kimoto, J. S. Cho, M. Higuchi, K. Yamamoto, "Novel carbazole dendrimers having a metal coordination site as a unique hole-transport material," *Macromol. Symp.*, 209:51–65 (2004).

129. K. Yamamoto, M. Higuchi, S. Shiki, M. Tsuruta, H. Chiba, "Stepwise radial complexation of imine groups in phenylazomethine dendrimers," *Nature*, 415:509–511 (2002).

130. A. Kimoto, J. S. Cho, M. Higuchi, K. Yamamoto, "Synthesis of asymmetrically arranged den-drimers with a carbazole dendron and a phenylazomethine dendron," *Macromolecules*, 37:5531–5537 (2004).

131. J. S. Cho, A. Kimoto, M. Higuchi, K. Yamamoto, "Synthesis of diphenylamine-substituted phenyl-azomethine dendrimers and the performance of organic light-emitting diodes," *Macromol. Chem. Phys.*, 206:635–641 (2005).

132. Z. Y. Xia, J. H. Su, W. Y. Wong, L. Wang, K. W. Cheah, H. Tian, C. H. Chen, "High performance organic light-emitting diodes based on tetra(methoxy)-containing anthracene derivatives as a hole transport and electron-blocking layer," *J. Mater. Chem.*, 20:8382–8388 (2010).

133. K. Okumoto, H. Kanno, Y. Hamada, H. Takahashi, K. Shibata, "Organic light-emitting devices using polyacene derivatives as a hole-transporting layer," *J. Appl. Phys.*, 100:044507 (2006).

134. B. Wei, J. Z. Liu, Y. Zhang, J. H. Zhang, H. N. Peng, H. L. Fan, Y. B. He, X. C. Gao, "Stable, glassy, and versatile binaphthalene derivatives capable of efficient hole transport, hosting, and deep-blue light emission," *Adv. Funct. Mater.*, 20:2448–2458 (2010).

135. A. Michaleviciute, E. Gurskyte, D. Y. Volyniuk, V. V. Cherpak, G. Sini, P. Y. Stakhira, J. V. Grazulevicius, "Star-shaped carbazole derivatives for bilayer white organic light-emitting diodes combining emission from both excitons and exciplexes," *J. Phys. Chem. C*, 116:20769–20778 (2012).

136. T. Spehr, R. Pudzich, T. Fuhrmann, J. Salbeck, "Highly efficient light emitters based on the spiro concept," *Org. Electron.*, 4:61–69 (2003).

137. U. Bach, K. De Cloedt, H. Spreitzer, M. Gratzel, "Characterization of hole transport in a new class of spiro-linked oligotriphenylamine compounds," *Adv. Mater.*, 12:1060–1063 (2000).

138. S. Tokito, K. Noda, K. Shimada, S. Inoue, M. Kimura, Y. Sawaki, Y. Taga, "Influence of hole transporting material on device performance in organic light-emitting diode," *Thin Solid Films*, 363:290–293 (2000).

139. Y. L. Liao, W. Y. Hung, T. H. Hou, C. Y. Lin, K. T. Wong, "Hole mobilities of 2,7-and 2,2'-disubstituted 9,9'-spirobifluorene-based triaryldiamines and their application as hole transport materials in OLEDs," *Chem. Mater.*, 19:6350–6357 (2007).

140. Z. Q. Jiang, Z. Y. Liu, C. L. Yang, C. Zhong, J. G. Qin, G. Yu, Y. Q. Liu, "Multifunctional fluorene-based oligomers with novel spiro-annulated triarylamine: Efficient, stable deep-blue electroluminescence, good hole injection, and transporting materials with very high T_g," *Adv. Funct. Mater.*, 19:3987–3995 (2009).

141. Y. J. Cho, O. Y. Kim, J. Y. Lee, "Synthesis of an aromatic amine derivative with novel double spirobifluorene core and its application as a hole transport material," *Org. Electron.*, 13:351–355 (2012).

142. Y. J. Cho, J. Y. Lee, "Thermally stable aromatic amine derivative with symmetrically substituted double spirobifluorene core as a hole transport material for green phosphorescent organic light-emitting diodes," *Thin Solid Films*, 522:415–419 (2012).

143. Y. J. Cho, J. Y. Lee, "Cyclopenta[*def*]fluorene based high triplet energy hole transport material for blue phosphorescent organic light-emitting diodes," *Org. Electron.*, 13:1044–1048 (2012).

144. Z. Z. Chu, D. Wang, C. Zhang, F. Z. Wang, H. W. Wu, Z. B. Lv, S. C. Hou, X. Fan, D. C. Zou, "Synthesis of spiro[fluorene-9,9'-xanthene] derivatives and their application as hole-transporting materials for organic light-emitting devices," *Synth. Met.*, 162:614–620 (2012).

145. N. X. Hu, S. Xie, Z. Popovic, B. Ong, A. M. Hor, S. N. Wang, "5,11-Dihydro-5,11-di-1-naphthylindolo[3,2-*b*]carbazole: Atropisomerism in a novel hole-transport molecule for organic light-emitting diodes," *J. Am. Chem. Soc.*, 121:5097–5098 (1999).

146. J. Simokaitiene, E. Stanislovaityte, J. V. Grazulevicius, V. Jankauskas, R. Gu, W. Dehaen, Y. C. Hung, C. P. Hsu, "Synthesis and properties of methoxyphenyl-substituted derivatives of indolo[3,2-*b*]carbazole," *J. Org. Chem.*, 77:4924–4931 (2012).

147. B. X. Mi, P. F. Wang, M. W. Liu, H. L. Kwong, N. B. Wong, C. S. Lee, S. T. Lee, "Thermally stable hole-transporting material for organic light-emitting diode: An isoindole derivative," *Chem. Mater.*, 15:3148–3151 (2003).

148. M. S. Park, J. Y. Lee, "Indolo acridine-based hole-transport materials for phosphorescent OLEDs with over 20% external quantum efficiency in deep blue and green," *Chem. Mater.*, 23:4338–4343 (2011).

149. M. Kim, J. Y. Lee, "Improved power efficiency in deep blue phosphorescent organic light-emitting diodes using an acridine core based hole transport material," *Org. Electron.*, 13:1245–1249 (2012).

150. M. S. Park, D. H. Choi, B. S. Lee, J. Y. Lee, "Fused indole derivatives as high triplet energy hole transport materials for deep blue phosphorescent organic light-emitting diodes," *J. Mater. Chem.*, 22:3099–3104 (2012).

151. H. Tsuji, C. Mitsui, L. Ilies, Y. Sato, B. Nakamura, "Synthesis and properties of 2,3,6,7-tetraarylbenzo[1,2-b:4,5-*b'*]difurans as hole-transporting material," *J. Am. Chem. Soc.*, 129:11902–11903 (2007).

152. H. Tsuji, C. Mitsui, Y. Sato, E. Nakamura, "Bis(carbazolyl)benzodifuran: A high-mobility ambipolar material for homojunction organic light-emitting diode devices," *Adv. Mater.*, 21:3776–3779 (2009).

153. L. S. Hung, R. Q. Zhang, P. He, G. Mason, "Contact formation of LiF/Al cathodes in Alq-based organic light-emitting diodes," *J. Phys. D Appl. Phys.*, 35:103–107 (2002).

154. Z. Y. Sun, B. F. Ding, B. Wu, Y. T. You, X. M. Ding, X. Y. Hou, "LiF layer at the interface of Au cathode in organic light-emitting devices: A nonchemical induced carrier injection enhancement," *J. Phys. Chem. C*, 116:2543–2547 (2012).

155. L. Duan, Q. Liu, Y. Li, Y. D. Gao, G. H. Zhang, D. Q. Zhang, L. D. Wang, Y. Qiu, "Thermally decomposable lithium nitride as an electron injection material for highly efficient and stable OLEDs," *J. Phys. Chem. C*, 113:13386–13390 (2009).

156. H. J. Bolink, E. Coronado, D. Repetto, M. Sessolo, E. M. Barea, J. Bisquert, G. Garcia-Belmonte, J. Prochazka, L. Kavan, "Inverted solution processable OLEDs using a metal oxide as an electron injection contact," *Adv. Funct. Mater.*, 18:145–150 (2008).

157. H. Lee, I. Park, J. Kwak, D. Y. Yoon, C. Lee, "Improvement of electron injection in inverted bottom-emission blue phosphorescent organic light emitting diodes using zinc oxide nanoparticles," *Appl. Phys. Lett.*, 96:153306 (2010).

158. N. Tokmoldin, N. Griffiths, D. D. C. Bradley, S. A. Haque, "A hybrid inorganic–organic semiconductor light-emitting diode using ZrO_2 as an electron-injection layer," *Adv. Mater.*, 21:3475–3478 (2009).

159. H. J. Bolink, H. Brine, E. Coronado, M. Sessolo, "Hybrid organic–inorganic light emitting diodes: Effect of the metal oxide," *J. Mater. Chem.*, 20:4047–4049 (2010).

160. J. X. Luo, L. X. Xiao, Z. J. Chen, B. Qu, Q. H. Gong, "Insulator MnO: Highly efficient and air-stable n-type doping layer for organic photovoltaic cells," *Org. Electron.*, 11:664–669 (2010).

161. J. X. Luo, L. X. Xiao, Z. J. Chen, Q. H. Gong, "Highly efficient organic light emitting devices with insulator MnO as an electron injecting and transporting material," *Appl. Phys. Lett.*, 93:133301 (2008).

162. H. Lee, J. Lee, P. Jeon, K. Jeong, Y. Yi, T. G. Kim, J. W. Kim, J. W. Lee, "Highly enhanced electron injection in organic light-emitting diodes with an n-type semiconducting MnO2 layer," *Org. Electron.*, 13:820–825 (2012).

163. J. S. Huang, Z. Xu, Y. Yang, "Low-work-function surface formed by solution-processed and thermally deposited nanoscale layers of cesium carbonate," *Adv. Funct. Mater.*, 17:1966–1973 (2007).

164. Y. Li, D. Q. Zhang, L. Duan, R. Zhang, L. D. Wang, Y. Qiu, "Elucidation of the electron injection mechanism of evaporated cesium carbonate cathode interlayer for organic light-emitting diodes," *Appl. Phys. Lett.*, 90:012119 (2007).

165. C. Mei-Hsin, I. W. Chih, "The roles of thermally evaporated cesium carbonate to enhance the electron injection in organic light emitting devices," *J. Appl. Phys.*, 104:113713 (2008).

166. R. G. Shangguan, G. Y. Mu, X. F. Qiao, L. Wang, K. W. Cheah, X. J. Zhu, C. H. Chen, "Low sublimation temperature cesium pivalate complex as an efficient electron injection material for organic light-emitting diode devices," *Org. Electron.*, 12:1957–1962 (2011).

167. Z. Y. Lu, Z. B. Deng, J. J. Zheng, Y. Zou, H. L. Du, Z. Chen, Y. S. Wang, "Efficient organic light-emitting diodes with zinc acetate as an effective electron injection layer," *Phys. E—Low-Dimens. Syst. Nanostruct.*, 41:1733–1737 (2009).

168. A. Gassmann, C. Melzer, H. von Seggern, "The Li_3PO_4/Al bilayer: An efficient cathode for organic light emitting devices," *J. Appl. Phys.*, 105:084513 (2009).

169. Y. H. Yin, Z. B. Deng, R. S. Zeng, Z. Y. Lu, J. C. Lun, Z. Chen, Y. Zou, H. L. Du, Y. S. Wang, "Sodium triphosphate as an efficient electron injection layer in organic light-emitting diodes," *Synth. Met.*, 162:1804–1808 (2012).

170. D. Q. Zhang, Y. Li, G. H. Zhang, Y. D. Gao, L. Duan, L. D. Wang, Y. Qiu, "Lithium cobalt oxide as electron injection material for high performance organic light-emitting diodes," *Appl. Phys. Lett.*, 92:073301 (2008).

171. H. Siemund, F. Brocker, H. Gobel, "Enhancing the electron injection in polymer light-emitting diodes using a sodium stearate/aluminum bilayer cathode," *Org. Electron.*, 14:335–343 (2013).

172. J. Kido, T. Matsumoto, "Bright organic electroluminescent devices having a metal-doped electron-injecting layer," *Appl. Phys. Lett.*, 73:2866–2868 (1998).

173. C. Ganzorig, M. Fujihira, "A lithium carboxylate ultrathin film on an aluminum cathode for enhanced electron injection in organic electroluminescent devices," *Jpn. J. Appl. Phys. Pt. 2 Lett.*, 38:L1348–L1350 (1999).

174. G. F. He, M. Pfeiffer, K. Leo, M. Hofmann, J. Birnstock, R. Pudzich, J. Salbeck, "High-efficiency and low-voltage p–i–n electrophosphorescent organic light-emitting diodes with double-emission layers," *Appl. Phys. Lett.*, 85:3911–3913 (2004).

175. P. C. Kao, J. H. Lin, J. Y. Wang, C. H. Yang, S. H. Chen, "Li_2CO_3 as an n-type dopant on Alq$_3$-based organic light emitting devices," *J. Appl. Phys.*, 109:094505 (2011).

176. T. Xiong, F. X. Wang, X. F. Qiao, D. G. Ma, "Cesium hydroxide doped tris-(8-hydroxyquinoline) aluminum as an effective electron injection layer in inverted bottom-emission organic light emitting diodes," *Appl. Phys. Lett.*, 92:263305 (2008).

177. K. R. Choudhury, J. H. Yoon, F. So, "LiF as an n-dopant in tris(8-hydroxyquinoline) aluminum thin films," *Adv. Mater.*, 20:1456–1461 (2008).

178. J. H. Kwon, J. Y. Lee, "High efficiency and long lifetime in organic light-emitting diodes using bilayer electron injection structure," *Synth. Met.*, 159:1292–1294 (2009).

179. C. Mei-Hsin, C. Yu-Hung, L. Chang-Tin, L. Guan-Ru, I. W. Chih, L. Dong-Seok, K. Jang-Joo, P. Tun-Wen, "Electronic and chemical properties of cathode structures using 4,7-diphenyl-1,10-phenanthroline doped with rubidium carbonate as electron injection layers," *J. Appl. Phys.*, 105:113714 (2009).

180. S. Y. Chen, T. Y. Chu, J. F. Chen, C. Y. Su, C. H. Chen, "Stable inverted bottom-emitting organic electroluminescent devices with molecular doping and morphology improvement," *Appl. Phys. Lett.*, 89:053518 (2006).

181. Y. Kyoung Soo, J. Soon Ok, M. Sung-Yong, L. Jun Yeob, Y. Ha-Jin, N. Taeyong, K. Sung-Kee, L. Tae-Woo, "Highly efficient p–i–n and tandem organic light-emitting devices using an air-stable and low-temperature-evaporable metal azide as an n-dopant," *Adv. Funct. Mater.*, 20:1797–1802 (2010).

182. F. Huang, H. Wu, Y. Cao, "Water/alcohol soluble conjugated polymers as highly efficient electron transporting/injection layer in optoelectronic devices," *Chem. Soc. Rev.*, 39:2500–2521 (2010).

183. F. Huang, L. T. Hou, H. B. Wu, X. H. Wang, H. L. Shen, W. Cao, W. Yang, Y. Cao, "High-efficiency, environment-friendly electroluminescent polymers with stable high work function metal as a cathode: Green- and yellow-emitting conjugated polyfluorene polyelectrolytes and their neutral precursors," *J. Am. Chem. Soc.*, 126:9845–9853 (2004).

184. H. B. Wu, F. Huang, Y. Q. Mo, W. Yang, D. L. Wang, J. B. Peng, Y. Cao, "Efficient electron injection from a bilayer cathode consisting of aluminum and alcohol-/water-soluble conjugated polymers," *Adv. Mater.*, 16:1826–1830 (2004).

185. R. Yang, H. Wu, Y. Cao, G. C. Bazan, "Control of cationic conjugated polymer performance in light emitting diodes by choice of counterion," *J. Am. Chem. Soc.*, 128:14422–14423 (2006).

186. S.-H. Oh, D. Vak, S.-I. Na, T.-W. Lee, D.-Y. Kim, "Water-soluble polyfluorenes as an electron injecting layer in PLEDs for extremely high quantum efficiency," *Adv. Mater.*, 20:1624–1629 (2008).

187. C. Hoven, R. Yang, A. Garcia, A. J. Heeger, T.-Q. Nguyen, G. C. Bazan, "Ion motion in conjugated polyelectrolyte electron transporting layers," *J. Am. Chem. Soc.*, 129:10976–10977 (2007).

188. F. Huang, Y. Zhang, M. S. Liu, A. K. Y. Jen, "Electron-rich alcohol-soluble neutral conjugated polymers as highly efficient electron-injecting materials for polymer light-emitting diodes," *Adv. Funct. Mater.*, 19:2457–2466 (2009).

189. X. Xu, W. Cai, J. Chen, Y. Cao, "Conjugated polyelectrolytes and neutral polymers with poly(2,7-carbazole) backbone: Synthesis, characterization, and photovoltaic application," *J. Polym. Sci. Polym. Chem.*, 49:1263–1272 (2011).

190. X. Xu, B. Han, J. Chen, J. Peng, H. Wu, Y. Cao, "2,7-Carbazole-1,4-phenylene copolymers with polar side chains for cathode modifications in polymer light-emitting diodes," *Macromolecules*, 44:4204–4212 (2011).

191. J. H. Seo, A. Gutacker, Y. Sun, H. Wu, F. Huang, Y. Cao, U. Scherf, A. J. Heeger, G. C. Bazan, "Improved high-efficiency organic solar cells via incorporation of a conjugated polyelectrolyte interlayer," *J. Am. Chem. Soc.*, 133:8416–8419 (2011).

192. M. Y. Jo, Y. E. Ha, J. H. Kim, "Polyviologen derivatives as an interfacial layer in polymer solar cells," *Sol. Energy Mater. Sol. Cells*, 107:1–8 (2012).

193. T. V. Pho, H. Kim, J. H. Seo, A. J. Heeger, F. Wudl, "Quinacridone-based electron transport layers for enhanced performance in bulk-heterojunction solar cells," *Adv. Funct. Mater.*, 21:4338–4341 (2011).

194. E. Ahmed, T. Earmme, S. A. Jenekhe, "New solution-processable electron transport materials for highly efficient blue phosphorescent OLEDs," *Adv. Funct. Mater.*, 21:3889–3899 (2011).
195. H. Ye, X. W. Hu, Z. X. Jiang, D. C. Chen, X. Liu, H. Nie, S. J. Su, X. Gong, Y. Cao, "Pyridinium salt-based molecules as cathode interlayers for enhanced performance in polymer solar cells," *J. Mater. Chem. A*, 1:3387–3394 (2013).
196. D. C. Chen, H. Zhou, M. Liu, W. M. Zhao, S. J. Su, Y. Cao, "Novel cathode interlayers based on neutral alcohol-soluble small molecules with a triphenylamine core featuring polar phosphonate side chains for high-performance polymer light-emitting and photovoltaic devices," *Macromol. Rapid Commun.*, 34:595–603 (2013).
197. H. J. Bolink, H. Brine, E. Coronado, M. Sessolo, "Phosphorescent hybrid organic–inorganic light-emitting diodes," *Adv. Mater.*, 22:2198–2201 (2010).
198. M.-H. Park, J.-H. Li, A. Kumar, G. Li, Y. Yang, "Doping of the metal oxide nanostructure and its influence in organic electronics," *Adv. Funct. Mater.*, 19:1241–1246 (2009).
199. H. Brine, J. F. Sánchez-Royo, H. J. Bolink, "Ionic liquid modified zinc oxide injection layer for inverted organic light-emitting diodes," *Org. Electron.*, 14:164–168 (2013).
200. K. Xie, J. Qiao, L. Duan, Y. Li, D. Zhang, G. Dong, L. Wang, Y. Qiu, "Organic cesium salt as an efficient electron injection material for organic light-emitting diodes," *Appl. Phys. Lett.*, 93:183302 (2008).
201. C. Schmitz, H. W. Schmidt, M. Thelakkat, "Lithium-quinolate complexes as emitter and interface materials in organic light-emitting diodes," *Chem. Mater.*, 12:3012–3019 (2000).
202. Z. G. Liu, O. V. Salata, N. Male, "Improved electron injection in organic LED with lithium quinolate/aluminium cathode," *Synth. Met.*, 128:211–214 (2002).
203. A. Fukase, J. Kido, "Organic electroluminescent devices having self-doped cathode interface layer," *Jpn. J. Appl. Phys. Pt. 2 Lett.*, 41:L334–L336 (2002).
204. F. S. Liang, J. S. Chen, L. X. Wang, D. G. Ma, X. B. Jing, F. S. Wang, "A hydroxyphenyloxadiazole lithium complex as a highly efficient blue emitter and interface material in organic light-emitting diodes," *J. Mater. Chem.*, 13:2922–2926 (2003).
205. X. Y. Zheng, Y. Z. Wu, R. G. Sun, W. Q. Zhu, X. Y. Jiang, Z. L. Zhang, S. H. Xu, "Efficiency improvement of organic light-emitting diodes using 8-hydroxy-quinolinato lithium as an electron injection layer," *Thin Solid Films*, 478:252–255 (2005).
206. Y. J. Pu, M. Miyamoto, K. Nakayama, T. Oyama, Y. Masaaki, J. Kido, "Lithium phenolate complexes for an electron injection layer in organic light-emitting diodes," *Org. Electron.*, 10:228–232 (2009).
207. Y. Changhun, C. Hyunsu, K. Hyeseung, L. Young Mi, P. Yongsup, Y. Seunghyup, "Electron injection via pentacene thin films for efficient inverted organic light-emitting diodes," *Appl. Phys. Lett.*, 95:053301 (2009).
208. H. Lee, G. Cho, S. Woo, S. Nam, J. Jeong, H. Kim, Y. Kim, "Phenanthroline diimide as an organic electron-injecting material for organic light-emitting devices," *RSC Adv.*, 2:8762–8767 (2012).
209. C. Yu-Hung, C. Yu-Jen, L. Guan-Ru, I. W. Chih, P. Tun-Wen, "Improvements of electron injection efficiency using subphthalocyanine mixed with lithium fluoride in cathode structures of organic light emitting diodes," *Org. Electron.*, 12:562–565 (2011).
210. A. P. Kulkarni, C. J. Tonzola, A. Babel, S. A. Jenekhe, "Electron transport materials for organic light-emitting diodes," *Chem. Mater.*, 16:4556–4573 (2004).
211. G. Hughes, M. R. Bryce, "Electron-transporting materials for organic electroluminescent and electrophosphorescent devices," *J. Mater. Chem.*, 15:94–107 (2005).
212. K. A. Higginson, X. M. Zhang, F. Papadimitrakopoulos, "Thermal and morphological effects on the hydrolytic stability of aluminum tris(8-hydroxyquinoline) (Alq$_3$)," *Chem. Mater.*, 10:1017–1020 (1998).
213. M. Brinkmann, G. Gadret, M. Muccini, C. Taliani, N. Masciocchi, A. Sironi, "Correlation between molecular packing and optical properties in different crystalline polymorphs and amorphous thin films of mer-tris(8-hydroxyquinoline)aluminum(III)," *J. Am. Chem. Soc.*, 122:5147–5157 (2000).
214. R. G. Kepler, P. M. Beeson, S. J. Jacobs, R. A. Anderson, M. B. Sinclair, V. S. Valencia, P. A. Cahill, "Electron and hole mobility in tris(8-hydroxyquinolinolato-N1,O8) Aluminum," *Appl. Phys. Lett.*, 66:3618–3620 (1995).

215. J. D. Anderson, E. M. McDonald, P. A. Lee, M. L. Anderson, E. L. Ritchie, H. K. Hall, T. Hopkins, E. A. Mash, J. Wang, A. Padias, S. Thayumanavan, S. Barlow, S. R. Marder, G. E. Jabbour, S. Shaheen, B. Kippelen, N. Peyghambarian, R. M. Wightman, N. R. Armstrong, "Electrochemistry and electrogenerated chemiluminescence processes of the components of aluminum quinolate/triarylamine, and related organic light-emitting diodes," *J. Am. Chem. Soc.*, 120:9646–9655 (1998).

216. K. Kim, D. W. Lee, J. I. Jin, "Electroluminescence properties of poly[2-(2'-ethylhexyloxy)-5-methoxy-1,4-phenylenevinylene]/tris(8-hydroxyquinoline) aluminum two-layer devices," *Synth. Met.*, 114:49–56 (2000).

217. B. J. Chen, X. W. Sun, Y. K. Li, "Influences of central metal ions on the electroluminescence and transport properties of tris-(8-hydroxyquinoline) metal chelates," *Appl. Phys. Lett.*, 82:3017–3019 (2003).

218. D. Z. Garbuzov, V. Bulović, P. E. Burrows, S. R. Forrest, "Photoluminescence efficiency and absorption of aluminum–tris-quinolate (Alq_3) thin films," *Chem. Phys. Lett.*, 249:433–437 (1996).

219. L. S. Burrows, L. S. Sapochak, D. M. McCarty, S. R. Forrest, M. E. Thompson, "Metal ion dependent luminescence effects in metal trisquinolate organic heterojunction light emitting devices," *Appl. Phys. Lett.*, 64:2718–2720 (1994).

220. R. Pohl, V. A. Montes, J. Shinar, P. Anzenbacher, "Red–green–blue emission from tris(5-aryl-8-quinolinolate)Al(III) complexes," *J. Org. Chem.*, 69:1723–1725 (2004).

221. V. A. Montes, G. Li, R. Pohl, J. Shinar, P. Anzenbacher, "Effective color tuning in organic light-emitting diodes based on aluminum tris(5-aryl-8-hydroxyquinoline) complexes," *Adv. Mater.*, 16:2001–2003 (2004).

222. J. Kido, Y. Iizumi, "Fabrication of highly efficient organic electroluminescent devices," *Appl. Phys. Lett.*, 73:2721–2723 (1998).

223. M. Sugimoto, M. Anzai, K. Sakanoue, S. Sakaki, "Modulating fluorescence of 8-quinolinolato compounds by functional groups: A theoretical study," *Appl. Phys. Lett.*, 79:2348–2350 (2001).

224. M. Sugimoto, S. Sakaki, K. Sakanoue, M. D. Newton, "Theory of emission state of tris(8-quinolinolato)aluminum and its related compounds," *J. Appl. Phys.*, 90:6092–6097 (2001).

225. S. A. Vanslyke, P. S. Bryan, F. V. Lovecchio, "Blue emitting internal junction organic electroluminescent device (II)," U.S. Patent 5,150,006 (1992).

226. P. S. Bryan, F. V. Lovecchio, S. A. Vanslyke, "Mixed ligand 8-quinolinolato aluminum chelate luminophors," U.S. Patent 5,141,671 (1992).

227. L. S. Sapochak, F. E. Benincasa, R. S. Schofield, J. L. Baker, K. K. C. Riccio, D. Fogarty, H. Kohlmann, K. F. Ferris, P. E. Burrows, "Electroluminescent zinc(II) bis(8-hydroxyquinoline): Structural effects on electronic states and device performance," *J. Am. Chem. Soc.*, 124:6119–6125 (2002).

228. C. H. Chen, J. M. Shi, "Metal chelates as emitting materials for organic electroluminescence," *Coord. Chem. Rev.*, 171:161–174 (1998).

229. Y. Hamada, T. Sano, H. Fujii, Y. Nishio, H. Takahashi, K. Shibata, "Organic light-emitting diodes using 3- or 5-hydroxyflavone–metal complexes," *Appl. Phys. Lett.*, 71:3338–3340 (1997).

230. N. Donze, P. Pechy, M. Gratzel, M. Schaer, L. Zuppiroli, "Quinolinate zinc complexes as electron transporting layers in organic light-emitting diodes," *Chem. Phys. Lett.*, 315:405–410 (1999).

231. Y. Kai, M. Morita, N. Yasuoka, N. Kasai, "The crystal and molecular structure of anhydrous zinc 8-quinolinolate complex," *Bull. Chem. Soc. Jpn.*, 58:1631–1635 (1985).

232. G. Yu, S. W. Yin, Y. Q. Liu, Z. G. Shuai, D. B. Zhu, "Structures, electronic states, and electroluminescent properties of a zinc(II) 2-(2-hydroxyphenyl)benzothiazolate complex," *J. Am. Chem. Soc.*, 125:14816–14824 (2003).

233. X. J. Xu, G. Yu, Y. Q. Liu, R. P. Tang, F. Xi, D. B. Zhu, "Effect of metal chelate layer on electroluminescent and current–voltage characteristics," *Appl. Phys. Lett.*, 86:202109 (2005).

234. S. W. Pyo, S. P. Lee, H. S. Lee, O. K. Kwon, H. S. Hoe, S. H. Lee, Y. K. Ha, Y. K. Kim, J. S. Kim, "White-light-emitting organic electroluminescent devices using new chelate metal complexes," *Thin Solid Films*, 363:232–235 (2000).

235. T. Sano, Y. Nishio, Y. Hamada, H. Takahashi, T. Usuki, K. Shibata, "Design of conjugated molecular materials for optoelectronics," *J. Mater. Chem.*, 10:157–161 (2000).

236. Y. Q. Li, Y. Liu, J. H. Guo, F. Wu, W. J. Tian, B. F. Li, Y. Wang, "Photoluminescent and electroluminescent properties of phenol–pyridine beryllium and carbonyl polypyridyl Re(I) complexes codeposited films," *Synth. Met.*, 118:175–179 (2001).

237. L. Chen, J. Qiao, J. F. Xie, L. Duan, D. Q. Zhang, L. D. Wang, Y. Qiu, "Substituted azomethine–zinc complexes: Thermal stability, photophysical, electrochemical and electron transport properties," *Inorg. Chim. Acta*, 362:2327–2333 (2009).

238. H. Tanaka, S. Tokito, Y. Taga, A. Okada, "Novel metal-chelate emitting materials based on polycyclic aromatic ligands for electroluminescent devices," *J. Mater. Chem.*, 8:1999–2003 (1998).

239. S. Tokito, K. Noda, H. Tanaka, Y. Taga, T. Tsutsui, "Organic light-emitting diodes using novel metal–chelate complexes," *Synth. Met.*, 111:393–396 (2000).

240. J. Shi, C. W. Tang, C. H. Chen, "Blue organic electroluminescent devices," U.S. Patent 5,645,948, p. 22 (1997).

241. T. C. Wong, J. Kovac, C. S. Lee, L. S. Hung, S. T. Lee, "Transient electroluminescence measurements on electron-mobility of N-arylbenzimidazoles," *Chem. Phys. Lett.*, 334:61–64 (2001).

242. Y. Q. Li, M. K. Fung, Z. Y. Xie, S. T. Lee, L. S. Hung, J. M. Shi, "An efficient pure blue organic light-emitting device with low driving voltages," *Adv. Mater.*, 14:1317–1321 (2002).

243. H. T. Shih, C. H. Lin, H. H. Shih, C. H. Cheng, "High-performance blue electroluminescent devices based on a biaryl," *Adv. Mater.*, 14:1409–1412 (2002).

244. C. Adachi, M. A. Baldo, S. R. Forrest, S. Lamansky, M. E. Thompson, R. C. Kwong, "High-efficiency red electrophosphorescence devices," *Appl. Phys. Lett.*, 78:1622–1624 (2001).

245. S. C. Lo, N. A. H. Male, J. P. J. Markham, S. W. Magennis, P. L. Burn, O. V. Salata, I. D. W. Samuel, "Green phosphorescent dendrimer for light-emitting diodes," *Adv. Mater.*, 14:975–979 (2002).

246. C. J. Tonzola, M. M. Alam, W. Kaminsky, S. A. Jenekhe, "New n-type organic semiconductors: Synthesis, single crystal structures, cyclic voltammetry, photophysics, electron transport, and electroluminescence of a series of diphenylanthrazolines," *J. Am. Chem. Soc.*, 125:13548–13558 (2003).

247. M. Redecker, D. D. C. Bradley, M. Jandke, P. Strohriegl, "Electron transport in starburst phenylquinoxalines," *Appl. Phys. Lett.*, 75:109–111 (1999).

248. C. Adachi, T. Tsutsui, S. Saito, "Organic electroluminescent device having a hole conductor as an emitting layer," *Appl. Phys. Lett.*, 55:1489–1491 (1989).

249. J. Pommerehne, H. Vestweber, W. Guss, R. F. Mahrt, H. Bassler, M. Porsch, J. Daub, "Efficient two layer LEDs on polymer blend basis," *Adv. Mater.*, 7:551–554 (1995).

250. R. Fink, Y. Heischkel, M. Thelakkat, H. W. Schmidt, C. Jonda, M. Huppauff, "Synthesis and application of dimeric 1,3,5-triazine ethers as hole-blocking materials in electroluminescent devices," *Chem. Mater.*, 10:3620–3625 (1998).

251. D. Yokoyama, A. Sakaguchi, M. Suzuki, C. Adachi, "Enhancement of electron transport by horizontal molecular orientation of oxadiazole planar molecules in organic amorphous films," *Appl. Phys. Lett.*, 95:243303 (2009).

252. S.-J. Su, D. Tanaka, Y.-J. Li, H. Sasabe, T. Takeda, J. Kido, "Novel four-pyridylbenzene-armed biphenyls as electron-transport materials for phosphorescent OLEDs," *Org. Lett.*, 10:941–944 (2008).

253. S. J. Su, T. Chiba, T. Takeda, J. Kido, "Pyridine-containing triphenylbenzene derivatives with high electron mobility for highly efficient phosphorescent OLEDs," *Adv. Mater.*, 20:2125–2130 (2008).

254. S. Shi-Jian, Y. Takahashi, T. Chiba, T. Takeda, J. Kido, "Structure–property relationship of pyridine-containing triphenyl benzene electron-transport materials for highly efficient blue phosphorescent OLEDs," *Adv. Funct. Mater.*, 19:1260–1267 (2009).

255. K. Togashi, S. Nomura, N. Yokoyama, T. Yasuda, C. Adachi, "Low driving voltage characteristics of triphenylene derivatives as electron transport materials in organic light-emitting diodes," *J. Mater. Chem.*, 22:20689–20695 (2012).

256. S. O. Jeon, K. S. Yook, J. Y. Lee, "Pyridine substituted spirofluorene derivative as an electron transport material for high efficiency in blue organic light-emitting diodes," *Thin Solid Films*, 519:890–893 (2010).

257. Y. Sun, L. Duan, D. Zhang, J. Qiao, G. Dong, L. Wang, Y. Qiu, "A pyridine-containing anthracene derivative with high electron and hole mobilities for highly efficient and stable fluorescent organic light-emitting diodes," *Adv. Funct. Mater.*, 21:1881–1886 (2011).

258. L. X. Xiao, S. J. Su, Y. Agata, H. L. Lan, J. Kido, "Nearly 100% internal quantum efficiency in an organic blue-light electrophosphorescent device using a weak electron transporting material with a wide energy gap," *Adv. Mater.*, 21:1271–1274 (2009).

259. H. Y. Oh, C. Lee, S. Lee, "Efficient blue organic light-emitting diodes using newly-developed pyrene-based electron transport materials," *Org. Electron.*, 10:163–169 (2009).

260. Y. J. Pu, M. Yoshizaki, T. Akiniwa, K. I. Nakayama, J. Kido, "Dipyrenylpyridines for electron-transporting materials in organic light emitting devices and their structural effect on electron injection from LiF/Al cathode," *Org. Electron.*, 10:877–882 (2009).

261. N. Li, S. L. Lai, W. M. Liu, P. F. Wang, J. J. You, C. S. Lee, Z. T. Liu, "Synthesis and properties of n-type triphenylpyridine derivatives and applications in deep-blue organic light-emitting devices as electron-transporting layer," *J. Mater. Chem.*, 21:12977–12985 (2011).

262. H. Sasabe, T. Chiba, S. J. Su, Y. J. Pu, K. I. Nakayama, J. Kido, "2-Phenylpyrimidine skeleton-based electron-transport materials for extremely efficient green organic light-emitting devices," *Chem. Commun.*, 5821–5823 (2008).

263. H. Sasabe, D. Tanaka, D. Yokoyama, T. Chiba, Y. J. Pu, K. Nakayama, M. Yokoyama, J. Kido, "Influence of substituted pyridine rings on physical properties and electron mobilities of 2-methylpyrimidine skeleton-based electron transporters," *Adv. Funct. Mater.*, 21:336–342 (2011).

264. M. Liu, S. J. Su, M. C. Jung, Y. B. Qi, W. M. Zhao, J. Kido, "Hybrid heterocycle-containing electron-transport materials synthesized by regioselective Suzuki cross-coupling reactions for highly efficient phosphorescent OLEDs with unprecedented low operating voltage," *Chem. Mater.*, 24:3817–3827 (2012).

265. S. J. Su, H. Sasabe, Y. J. Pu, K. Nakayama, J. Kido, "Tuning energy levels of electron-transport materials by nitrogen orientation for electrophosphorescent devices with an 'ideal' operating voltage," *Adv. Mater.*, 22:3311–3316 (2010).

266. C. Sun, Z. M. Hudson, M. G. Helander, Z. H. Lu, S. N. Wang, "A polyboryl-functionalized triazine as an electron transport material for OLEDs," *Organometallics*, 30:5552–5555 (2011).

267. Y. Sakamoto, T. Suzuki, A. Miura, H. Fujikawa, S. Tokito, Y. Taga, "Synthesis, characterization, and electron-transport property of perfluorinated phenylene dendrimers," *J. Am. Chem. Soc.*, 122:1832–1833 (2000).

268. K. Tamao, M. Uchida, T. Izumizawa, K. Furukawa, S. Yamaguchi, "Silole derivatives as efficient electron transporting materials," *J. Am. Chem. Soc.*, 118:11974–11975 (1996).

269. G. Yu, S. W. Yin, Y. Q. Liu, J. S. Chen, X. J. Xu, X. B. Sun, D. G. Ma, X. W. Zhan, Q. Peng, Z. G. Shuai, B. Z. Tang, D. B. Zhu, W. H. Fang, Y. Luo, "Structures, electronic states, photoluminescence, and carrier transport properties of 1,1-disubstituted 2,3,4,5-tetraphenylsiloles," *J. Am. Chem. Soc.*, 127:6335–6346 (2005).

270. X. W. Zhan, C. Risko, F. Amy, C. Chan, W. Zhao, S. Barlow, A. Kahn, J. L. Bredas, S. R. Marder, "Electron affinities of 1,1-diaryl-2,3,4,5-tetraphenyisiloles: Direct measurements and comparison with experimental and theoretical estimates," *J. Am. Chem. Soc.*, 127:9021–9029 (2005).

271. S. H. Lee, B. B. Jang, Z. H. Kafafi, "Highly fluorescent solid-state asymmetric spirosilabifluorene derivatives," *J. Am. Chem. Soc.*, 127:9071–9078 (2005).

272. H. Murata, G. G. Malliaras, M. Uchida, Y. Shen, Z. H. Kafafi, "Non-dispersive and air-stable electron transport in an amorphous organic semiconductor," *Chem. Phys. Lett.*, 339:161–166 (2001).

273. H. Murata, Z. H. Kafafi, M. Uchida, "Efficient organic light-emitting diodes with undoped active layers based on silole derivatives," *Appl. Phys. Lett.*, 80:189–191 (2002).

274. M. Uchida, T. Izumizawa, T. Nakano, S. Yamaguchi, K. Tamao, K. Furukawa, "Structural optimization of 2,5-diarylsiloles as excellent electron-transporting materials for organic electroluminescent devices," *Chem. Mater.*, 13:2680–2683 (2001).

275. B. Z. Tang, X. W. Zhan, G. Yu, P. P. S. Lee, Y. Q. Liu, D. B. Zhu, "Efficient blue emission from siloles," *J. Mater. Chem.*, 11:2974–2978 (2001).

276. J. Ohshita, M. Nodono, H. Kai, T. Watanabe, A. Kunai, K. Komaguchi, M. Shiotani, A. Adachi, K. Okita, Y. Harima, K. Yamashita, M. Ishikawa, "Synthesis and optical, electrochemical, and electron-transporting properties of silicon-bridged bithiophenes," *Organometallics*, 18:1453–1459 (1999).

277. T. Noda, Y. Shirota, "5,5'-bis(dimesitylboryl)-2,2'-bithiophene and 5,5"-bis(dimesitylboryl)-2,2':5',2"-terthiophene as a novel family of electron-transporting amorphous molecular materials," *J. Am. Chem. Soc.*, 120:9714–9715 (1998).

278. D. Tanaka, T. Takeda, T. Chiba, S. Watanabe, J. Kido, "Novel electron-transport material containing boron atom with a high triplet excited energy level," *Chem. Lett.*, 36:262–263 (2007).

279. A. L. Von Ruden, L. Cosimbescu, E. Polikarpov, P. K. Koech, J. S. Swensen, L. Wang, J. T. Darsell, A. B. Padmaperuma, "Phosphine oxide based electron transporting and hole blocking materials for blue electrophosphorescent organic light emitting devices," *Chem. Mater.*, 22:5678–5686 (2010).

280. S. O. Jeon, K. S. Yook, C. W. Joo, J. Y. Lee, "A phosphine oxide derivative as a universal electron transport material for organic light-emitting diodes," *J. Mater. Chem.*, 19:5940–5944 (2009).

281. S. O. Jeon, K. S. Yook, J. Y. Lee, S. M. Park, J. W. Kim, J. H. Kim, J. A. Hong, Y. Park, "Mechanism for the direct electron injection from Al cathode to the phosphine oxide type electron transport layer," *Appl. Phys. Lett.*, 98:073306 (2011).

282. A. G. Werner, F. Li, K. Harada, M. Pfeiffer, T. Fritz, K. Leo, "Pyronin B as a donor for n-type doping of organic thin films," *Appl. Phys. Lett.*, 82:4495–4497 (2003).

283. C. K. Chan, W. Zhao, S. Barlow, S. Marder, A. Kahn, "Decamethylcobaltocene as an efficient n-dopant in organic electronic materials and devices," *Org. Electron.*, 9:575–581 (2008).

284. N. Cho, H.-L. Yip, S. K. Hau, K.-S. Chen, T.-W. Kim, J. A. Davies, D. F. Zeigler, A. K. Y. Jen, "n-Doping of thermally polymerizable fullerenes as an electron transporting layer for inverted polymer solar cells," *J. Mater. Chem.*, 21:6956–6961 (2011).

285. J. Meyer, M. Kroeger, S. Hamwi, F. Gnam, T. Riedl, W. Kowalsky, A. Kahn, "Charge generation layers comprising transition metal-oxide/organic interfaces: Electronic structure and charge generation mechanism," *Appl. Phys. Lett.*, 96:193302 (2010).

286. K. S. Yook, S. O. Jeon, S. Y. Min, J. Y. Lee, H. J. Yang, T. Noh, S. K. Kang, T. W. Lee, "Highly efficient p–i–n and tandem organic light-emitting devices using an air-stable and low-temperature-evaporable metal azide as an n-dopant," *Adv. Funct. Mater.*, 20:1797–1802 (2010).

287. J. W. Ma, W. Xu, X. Y. Jiang, Z. L. Zhang, "Organic light-emitting diodes based on new n-doped electron transport layer," *Synth. Met.*, 158:810–814 (2008).

288. K. S. Lee, D. H. Kim, D. U. Lee, T. W. Kim, "Organic light-emitting devices with an n-type bis(ethylenedithio)-tetrathiafulvalene-doped 4,7-diphenyl-1,10-phenanthroline electron transport layer operating at low voltage," *Thin Solid Films*, 521:193–196 (2012).

289. J. H. Lee, P. S. Wang, H. D. Park, C. I. Wu, J. J. Kim, "A high performance inverted organic light emitting diode using an electron transporting material with low energy barrier for electron injection," *Org. Electron.*, 12:1763–1767 (2011).

290. S. H. Eom, Y. Zheng, E. Wrzesniewski, J. Lee, N. Chopra, F. So, J. Su, "Effect of electron injection and transport materials on efficiency of deep-blue phosphorescent organic light-emitting devices," *Org. Electron.*, 10:686–691 (2009).

291. Q. Yang, Y. Y. Hao, Z. G. Wang, Y. F. Li, H. Wang, B. S. Xu, "Double-emission-layer green phosphorescent OLED based on LiF-doped TPBi as electron transport layer for improving efficiency and operational lifetime," *Synth. Met.*, 162:398–401 (2012).

292. V. V. Jarikov, K. P. Klubek, L. S. Liao, C. T. Brown, "Operating lifetime recovery in organic light-emitting diodes having an azaaromatic hole-blocking/electron-transporting layer," *J. Appl. Phys.*, 104:074914 (2008).

293. T. Matsushima, C. Adachi, "Extremely low voltage organic light-emitting diodes with p-doped alpha-sexithiophene hole transport and n-doped phenyldipyrenylphosphine oxide electron transport layers," *Appl. Phys. Lett.*, 89:253506 (2006).

294. T. Earmme, S. A. Jenekhe, "Solution-processed, alkali metal-salt-doped, electron-transport layers for high-performance phosphorescent organic light-emitting diodes," *Adv. Funct. Mater.*, 22:5126–5136 (2012).

295. T. H. Huang, W. T. Whang, J. Y. Shen, Y. S. Wen, J. T. Lin, T. H. Ke, L. Y. Chen, C. C. Wu, "Dibenzothiophene/oxide and quinoxaline/pyrazine derivatives serving as electron-transport materials," *Adv. Funct. Mater.*, 16:1449–1456 (2006).

296. E. Ahmed, T. Earmme, G. Q. Ren, S. A. Jenekhe, "Novel n-type conjugated ladder heteroarenes: Synthesis, self-assembly of nanowires, electron transport, and electroluminescence of bisindenoanthrazolines," *Chem. Mater.*, 22:5786–5796 (2010).

297. W. Yue, A. Lv, J. Gao, W. Jiang, L. Hao, C. Li, Y. Li, L. E. Polander, S. Barlow, W. Hu, S. Di Motta, F. Negri, S. R. Marder, Z. Wang, "Hybrid rylene arrays via combination of Stille coupling and C–H transformation as high-performance electron transport materials," *J. Am. Chem. Soc.*, 134:5770–5773 (2012).

298. S. A. Vanslyke, C. W. Tang, L. C. Roberts, "Electroluminescent device with organic luminescent medium," U.S. Patent 4,720,432A, p. 16 (1988).

299. T. Thoms, S. Okada, J. P. Chen, M. Furugori, "Improved host material design for phosphorescent guest–host systems," *Thin Solid Films*, 436:264–268 (2003).

300. C. W. Tang, S. A. VanSlyke, C. H. Chen, "Electroluminescence of doped organic thin films," *J. Appl. Phys.*, 65:3610–3616 (1989).

301. S. L. Tao, L. B. Niu, J. S. Yu, Y. D. Jiang, X. H. Zhang, "High-performance organic red-light-emitting device based on DCJTB and a new host material," *J. Lumin.*, 130:70–73 (2010).

302. J. H. Huang, B. Xu, M. K. Lam, K. W. Cheah, C. H. Chen, J. H. Su, "Unsymmetrically amorphous 9,10-disubstituted anthracene derivatives for high-efficiency blue organic electroluminescence devices," *Dyes Pigments*, 89:155–161 (2011).

303. J. Y. Lee, S. E. Jang, C. W. Joo, K. S. Yook, J. W. Kim, C. W. Lee, "Thermally stable fluorescent blue organic light-emitting diodes using spirobifluorene based anthracene host materials with different substitution position," *Synth. Met.*, 160:1184–1188 (2010).

304. M. S. Gong, H. S. Lee, Y. M. Jeon, "Highly efficient blue OLED based on 9-anthracene-spirobenzofluorene derivatives as host materials," *J. Mater. Chem.*, 20:10735–10746 (2010).

305. Z. Y. Xia, Z. Y. Zhang, J. H. Su, Q. Zhang, K. M. Fung, M. K. Lam, K. F. Li, K. W. Cheah, W. Y. Wong, H. Tian, C. H. Chen, "Robust and highly efficient blue light-emitting hosts based on indene-substituted anthracene," *J. Mater. Chem.*, 20:3768–3774 (2010).

306. Y. Y. Lyu, J. Kwak, O. Kwon, S. H. Lee, D. Kim, C. Lee, K. Char, "Silicon-cored anthracene derivatives as host materials for highly efficient blue organic light-emitting devices," *Adv. Mater.*, 20:2720–2729 (2008).

307. L. Wang, Z. Y. Wu, W. Y. Wong, K. W. Cheah, H. Huang, C. H. Chen, "New blue host materials based on anthracene-containing dibenzothiophene," *Org. Electron.*, 12:595–601 (2011).

308. C. T. Chen, "Evolution of red organic light-emitting diodes: Materials and devices," *Chem. Mater.*, 16:4389–4400 (2004).

309. F. G. Webster, W. C. McColgin, U.S. Patent 3,852,683 (1974).

310. C. H. Chen, K. P. Klubek, J. Shi, "Red organic electroluminescent materials," U.S. Patent 5,908,581 (1999).

311. C. H. Chen, C. W. Tang, J. Shi, "Red organic electroluminescent devices," U.S. Patent 5,935,720 (1999).

312. C. H. Chen, C. W. Tang, J. Shi, K. P. Klubek, "Recent developments in the synthesis of red dopants for Alq(3) hosted electroluminescence," *Thin Solid Films*, 363:327–331 (2000).

313. M. Chang, W. Huang, W. Huang, "Red organic electroluminescent materials," U.S. Patent 6,649,089 (2003).

314. C. Q. Ma, Z. Liang, X. S. Wang, B. W. Zhang, Y. Cao, L. D. Wang, Y. Qiu, "A novel family of red fluorescent materials for organic light-emitting diodes," *Synth. Met.*, 138:537–542 (2003).

315. X. H. Zhang, B. J. Chen, X. Q. Lin, O. Y. Wong, C. S. Lee, H. L. Kwong, S. T. Lee, S. K. Wu, "A new family of red dopants based on chromene-containing compounds for organic electroluminescent devices," *Chem. Mater.*, 13:1565–1569 (2001).

316. X. T. Tao, S. Miyata, H. Sasabe, G. J. Zhang, T. Wada, M. H. Jiang, "Efficient organic red electroluminescent device with narrow emission peak," *Appl. Phys. Lett.*, 78:279–281 (2001).

317. R. Lemke, "Knoevenagel condensation in dimethylformamide," *Synthesis*, 5:359–361 (1974).

318. J. Y. Li, D. Liu, Z. R. Hong, S. W. Tong, P. F. Wang, C. W. Ma, O. Lengyel, C. S. Lee, H. L. Kwong, S. Lee, "A new family of isophorone-based dopants for red organic electroluminescent devices," *Chem. Mater.*, 15:1486–1490 (2003).
319. B. J. Jung, C. B. Yoon, H. K. Shim, L. M. Do, T. Zyung, "Pure-red dye for organic electroluminescent devices: Bis-condensed DCM derivatives," *Adv. Funct. Mater.*, 11:430–434 (2001).
320. Y. Hamada, H. Kanno, T. Tsujioka, H. Takahashi, T. Usuki, "Red organic light-emitting diodes using an emitting assist dopant," *Appl. Phys. Lett.*, 75:1682–1684 (1999).
321. J. Qiao, Y. Qiu, L. D. Wang, L. Duan, Y. Li, D. Q. Zhang, "Pure red electroluminescence from a host material of binuclear gallium complex," *Appl. Phys. Lett.*, 81:4913–4915 (2002).
322. T. H. Liu, C. Y. Iou, S. W. Wen, C. H. Chen, "4-(Dicyanomethylene)-2-*t*-butyl-6-(1,1,7,7-tetramethyljulolidyl-9-enyl)-4*H*-pyran doped red emitters in organic light-emitting devices," *Thin Solid Films*, 441:223–227 (2003).
323. Y. S. Yao, Q. X. Zhou, X. S. Wang, Y. Wang, B. W. Zhang, "A DCM-type red-fluorescent dopant for high-performance organic electroluminescent devices," *Adv. Funct. Mater.*, 17:93–100 (2007).
324. Y. J. Chang, T. J. Chow, "Highly efficient red fluorescent dyes for organic light-emitting diodes," *J. Mater. Chem.*, 21:3091–3099 (2011).
325. K. H. Lee, M. H. Park, C. S. Kim, Y. K. Kim, S. S. Yoon, "Modified 4-(dicyanomethylene)-2-*tert*-butyl-6-(1,1,7,7-tetra-methyljulolidy-19-enyl)-4*H*-pyran-containing red fluorescent emitters for efficient organic light-emitting diodes," *Thin Solid Films*, 520:510–514 (2011).
326. M. K. Leung, C. C. Chang, M. H. Wu, K. H. Chuang, J. H. Lee, S. J. Shieh, S. C. Lin, C. F. Chiu, "6-*N,N*-diphenylaminobenzofuran-derived pyran containing fluorescent dyes: A new class of high-brightness red-light-emitting dopants for OLED," *Org. Lett.*, 8:2623–2626 (2006).
327. T. Mori, K. Miyachi, T. Kichimi, T. Mizutani, "Electrical and luminescent properties of color-changeable organic electroluminescent diode using squarylium dyes," *Jpn. J. Appl. Phys.*, 33:6594–6598 (1994).
328. B. W. Zhang, W. Zhao, Y. Cao, X. S. Wang, Z. L. Zhang, X. Y. Jiang, S. H. Xu, "Photoluminescence and electroluminescence of squarylium cyanine dyes," *Synth. Met.*, 91:237–241 (1997).
329. L. C. Picciolo, H. Murata, Z. H. Kafafi, "Organic light-emitting devices with saturated red emission using 6,13-diphenylpentacene," *Appl. Phys. Lett.*, 78:2378–2380 (2001).
330. T. H. Huang, J. T. Lin, Y. T. Tao, C. H. Chuen, "Benzo[*a*]aceanthrylene derivatives for red-emitting electroluminescent materials," *Chem. Mater.*, 15:4854–4862 (2003).
331. T. L. Li, X. Li, W. L. Li, B. Chu, Z. S. Su, L. L. Han, Z. Z. Hu, Z. Q. Zhang, "Tunable red emission by incorporation of a rubrene derivative in p-type and n-type hosts in organic light emitting devices," *Thin Solid Films*, 517:4629–4632 (2009).
332. K. Okumoto, H. Kanno, Y. Hamada, H. Takahashi, K. Shibata, "High efficiency red organic light-emitting devices using tetraphenyldibenzoperiflanthene-doped rubrene as an emitting layer," *Appl. Phys. Lett.*, 89:013502 (2006).
333. B. B. Jang, S. H. Lee, Z. H. Kafafi, "Asymmetric pentacene derivatives for organic light-emitting diodes," *Chem. Mater.*, 18:449–457 (2006).
334. M. A. Wolak, J. Delcamp, C. A. Landis, P. A. Lane, J. Anthony, Z. Kafafi, "High-performance organic light-emitting diodes based on dioxolane-substituted pentacene derivatives," *Adv. Funct. Mater.*, 16:1943–1949 (2006).
335. X. D. Luo, H. Gu, B. D. Ding, L. Wang, X. W. Zhang, W. Q. Zhu, X. Y. Jiang, Z. L. Zhang, "Red organic light-emitting devices based on a pentacene derivative," *Curr. Appl. Phys.*, 11:844–848 (2011).
336. J. F. Fang, D. G. Ma, "Efficient red organic light-emitting devices based on a europium complex," *Appl. Phys. Lett.*, 83:4041–4043 (2003).
337. L. Huang, K. Z. Wang, C. H. Huang, F. Y. Li, Y. Y. Huang, "Bright red electroluminescent devices using novel second-ligand-contained europium complexes as emitting layers," *J. Mater. Chem.*, 11:790–793 (2001).
338. P. F. Wang, Z. R. Hong, Z. Y. Xie, S. W. Tong, O. Y. Wong, C. S. Lee, N. B. Wong, L. S. Hung, S. T. Lee, "A bis-salicylaldiminato Schiff base and its zinc complex as new highly fluorescent red dopants for high performance organic electroluminescence devices," *Chem. Commun.*, 1664–1665 (2003).

339. C. H. Chen, C. W. Tang, J. Shi, K. P. Klubek, "Green organic electroluminescent devices," U.S. Patent 6,020,078 (2000).
340. T. Wakimoto, Y. Yonemoto, Japan Patent 6,240,243 (1994).
341. J. L. Fox, C. H. Chen, U.S. Patent 4,736,032 (1988).
342. C. H. Chen, C. W. Tang, "Efficient green organic light-emitting diodes with sterically hindered coumarin dopants," *Appl. Phys. Lett.*, 79:3711–3713 (2001).
343. M. T. Lee, C. K. Yen, W. P. Yang, H. H. Chen, C. H. Liao, C. H. Tsai, C. H. Chen, "Efficient green coumarin dopants for organic light-emitting devices," *Org. Lett.*, 6:1241–1244 (2004).
344. J. Shi, C. W. Tang, "Organic electroluminescent devices with high operational stability," U.S. Patent 5,593,788, p. 13 (1997).
345. J. M. Shi, C. W. Tang, "Doped organic electroluminescent devices with improved stability," *Appl. Phys. Lett.*, 70:1665–1667 (1997).
346. T. Wakimoto, Y. Yonemoto, J. Funaki, M. Tsuchida, R. Murayama, H. Nakada, H. Matsumoto, S. Yamamura, M. Nomura, "Stability characteristics of quinacridone and coumarin molecules as guest dopants in the organic LEDs," *Synth. Met.*, 91:15–19 (1997).
347. H. Murata, C. D. Merritt, H. Inada, Y. Shirota, Z. H. Kafafi, "Molecular organic light-emitting diodes with temperature-independent quantum efficiency and improved thermal durability," *Appl. Phys. Lett.*, 75:3252–3254 (1999).
348. L. Wang, Y. Gao, P. Wei, Y. Qiu, "Novel structure organic light-emitting diodes with high performance," *SID Dig. Tech. Pap.*, 35:703–705 (2004).
349. K. P. Klubek, C. W. Tang, "Stable organic light emitting devices using aminoanthracenes," U.S. Patent 0,153,163 (2005).
350. C. T. Brown, D. Y. Kondakov, "Efficient electroluminescent device," Eur. Pat. Appl., EP1357613A2 (2003).
351. J. D. Debad, J. C. Morris, V. Lynch, P. Magnus, A. J. Bard, "Dibenzotetraphenylperiflanthene: Synthesis, photophysical properties, and electrogenerated chemiluminescence," *J. Am. Chem. Soc.*, 118:2374–2379 (1996).
352. R. Adams, M. H. Gold, "Synthesis of 1,3-diphenyldihydroisobenzofurans, 1,3-diphenylisobenzofurans and *o*-dibenzoylbenzenes from the diene addition products to dibenzoylethylene," *J. Am. Chem. Soc.*, 62:56–61 (1940).
353. M. X. Yu, J. P. Duan, C. H. Lin, C. H. Cheng, Y. T. Tao, "Diaminoanthracene derivatives as high-performance green host electroluminescent materials," *Chem. Mater.*, 14:3958–3963 (2002).
354. S. Tokito, K. Noda, H. Fujikawa, Y. Taga, M. Kimura, K. Shimada, Y. Sawaki, "Highly efficient blue–green emission from organic light-emitting diodes using dibenzochrysene derivatives," *Appl. Phys. Lett.*, 77:160–162 (2000).
355. Y. T. Tao, E. Balasubramaniam, A. Danel, B. Jarosz, P. Tomasik, "Sharp green electroluminescence from 1*H*-pyrazolo[3,4-*b*]quinoline-based light-emitting diodes," *Appl. Phys. Lett.*, 77:1575–1577 (2000).
356. Y. J. You, Y. K. He, P. E. Borrows, S. R. Forrest, N. A. Petasis, M. E. Thompson, "Fluorophores related to the green fluorescent protein and their use in optoelectronic devices," *Adv. Mater.*, 12:1678–1681 (2000).
357. R. Kondo, T. Yasuda, Y. S. Yang, J. Y. Kim, C. Adachi, "Highly luminescent pi-conjugated dithienometalloles: Photophysical properties and their application in organic light-emitting diodes," *J. Mater. Chem.*, 22:16810–16816 (2012).
358. H. Tanaka, K. Shizu, H. Miyazaki, C. Adachi, "Efficient green thermally activated delayed fluorescence (TADF) from a phenoxazine–triphenyltriazine (PXZ–TRZ) derivative," *Chem. Commun.*, 48:11392–11394 (2012).
359. C. Hosokawa, H. Higashi, H. Nakamura, T. Kusumoto, "Highly efficient blue electroluminescence from a distyrylarylene emitting layer with a new dopant," *Appl. Phys. Lett.*, 67:3853–3855 (1995).
360. Y. Z. Wu, X. Y. Zheng, W. Q. Zhu, R. G. Sun, X. Y. Jiang, Z. L. Zhang, S. H. Xu, "Highly efficient pure blue electroluminescence from 1,4-bis[2-(3-*N*-ethylcarbazoryl)vinyl]benzene," *Appl. Phys. Lett.*, 83:5077–5079 (2003).

361. C. H. Chuen, Y. T. Tao, F. I. Wu, C. F. Shu, "White organic light-emitting diodes based on 2,7-bis(2,2-diphenylvinyl)-9,9'-spirobifluorene: Improvement in operational lifetime," *Appl. Phys. Lett.*, 85:4609–4611 (2004).

362. K. Suzuki, A. Seno, H. Tanabe, K. Ueno, "New host materials for blue emitters," *Synth. Met.*, 143:89–96 (2004).

363. M. T. Lee, C. H. Liao, C. H. Tsai, C. H. Chen, "Highly efficient, deep-blue doped organic light-emitting devices," *Adv. Mater.*, 17:2493–2497 (2005).

364. L. Huang, H. Tian, F. Y. Li, D. Q. Gao, Y. Y. Huang, C. H. Huang, "Blue organic electroluminescent devices based on a distyrylarylene derivative as emitting layer and a terbium complex as electron-transporting layer," *J. Lumin.*, 97:55–59 (2002).

365. X. B. Ding, J. G. Zheng, Y. D. Jin, G. Aerts, B. X. Peng, P. L. Heremans, G. Borghs, H. J. Geise, "Development and characterization of alkyloxy-substituted biphenyl compounds as guest in blue OLEDs," *Synth. Met.*, 142:267–273 (2004).

366. F. He, H. Xu, B. Yang, Y. Duan, L. L. Tian, K. K. Huang, Y. G. Ma, S. Y. Liu, S. H. Feng, J. C. Shen, "Oligomeric phenylenevinylene with cross dipole arrangement and amorphous morphology: Enhanced solid-state luminescence efficiency and electroluminescence performance," *Adv. Mater.*, 17:2710–2718 (2005).

367. Y. Duan, F. He, G. Cheng, J. Li, Y. G. Ma, S. Y. Liu, "Blue electroluminescent devices based on a trimeric phenylenvinylene derivative as emitting layer," *Thin Solid Films*, 492:275–278 (2005).

368. J. Shi, C. H. Chen, K. P. Klubek, "Organic electroluminescent elements for stable blue electroluminescent devices," U.S. Patent 5,972,247 (1999).

369. X. Y. Jiang, Z. L. Zhang, X. Y. Zheng, Y. Z. Wu, S. H. Xu, "A blue organic emitting diode from anthracene derivative," *Thin Solid Films*, 401:251–254 (2001).

370. J. M. Shi, C. W. Tang, "Anthracene derivatives for stable blue-emitting organic electroluminescence devices," *Appl. Phys. Lett.*, 80:3201–3203 (2002).

371. S. L. Tao, Z. R. Hong, Z. K. Peng, W. G. Ju, X. H. Zhang, P. F. Wang, S. K. Wu, S. T. Lee, "Anthracene derivative for a non-doped blue-emitting organic electroluminescence device with both excellent color purity and high efficiency," *Chem. Phys. Lett.*, 397:1–4 (2004).

372. Y. Kan, L. D. Wang, L. Duan, Y. C. Hu, G. S. Wu, Y. Qiu, "Highly-efficient blue electroluminescence based on two emitter isomers," *Appl. Phys. Lett.*, 84:1513–1515 (2004).

373. M. T. Lee, H. H. Chen, C. H. Liao, C. H. Tsai, C. H. Chen, "Stable styrylamine-doped blue organic electroluminescent device based on 2-methyl-9,10-di(2-naphthyl)anthracene," *Appl. Phys. Lett.*, 85:3301–3303 (2004).

374. D. Gebeyehu, K. Walzer, G. He, M. Pfeiffer, K. Leo, J. Brandt, A. Gerhard, P. Stossel, H. Vestweber, "Highly efficient deep-blue organic light-emitting diodes with doped transport layers," *Synth. Met.*, 148:205–211 (2005).

375. W. J. Shen, R. Dodda, C. C. Wu, F. I. Wu, T. H. Liu, H. H. Chen, C. H. Chen, C. F. Shu, "Spirobifluorene-linked bisanthracene: An efficient blue emitter with pronounced thermal stability," *Chem. Mater.*, 16:930–934 (2004).

376. K. Danel, T. H. Huang, J. T. Lin, Y. T. Tao, C. H. Chuen, "Blue-emitting anthracenes with end-capping diarylamines," *Chem. Mater.*, 14:3860–3865 (2002).

377. M. G. Shin, S. O. Kim, H. T. Park, S. J. Park, H. S. Yu, Y. H. Kim, S. K. Kwon, "Synthesis and characterization of *ortho*-twisted asymmetric anthracene derivatives for blue organic light emitting diodes (OLEDs)," *Dyes Pigments*, 92:1075–1082 (2012).

378. T. Karatsu, R. Hazuku, M. Asuke, A. Nishigaki, S. Yagai, Y. Suzuri, H. Kita, A. Kitamura, "Blue electroluminescence of silyl substituted anthracene derivatives," *Org. Electron.*, 8:357–366 (2007).

379. Y. T. Tao, E. Balasubramaniam, A. Danel, A. Wisla, P. Tomasik, "Pyrazoloquinoline derivatives as efficient blue electroluminescent materials," *J. Mater. Chem.*, 11:768–772 (2001).

380. Z. Y. Lu, Q. Jiang, W. G. Zhu, M. G. Xie, Y. B. Hou, X. H. Chen, Z. J. Wang, D. C. Zou, T. Tsutsui, "Efficient blue emission from pyrazoline organic light emitting diodes," *Synth. Met.*, 111:425–427 (2000).

381. Y. Kijima, N. Asai, S. Tamura, "A blue organic light emitting diode," *Jpn. J. Appl. Phys. Pt. 1 Reg. Pap. Short Notes Rev. Pap.*, 38:5274–5277 (1999).

382. T. Tachikawa, S. Terazono, S. Tokita, "Blue light-emitting electroluminescent devices with 2,9-disubstituted 1,10-phenanthrolines," *Synth. Met.*, 91:247–248 (1997).

383. T. Mitsumori, M. Bendikov, O. Dautel, F. Wudl, T. Shioya, H. Sato, Y. Sato, "Synthesis and properties of highly fluorescent indolizino[3,4,5-*ab*]isoindoles," *J. Am. Chem. Soc.*, 126:16793–16803 (2004).

384. Q. G. Wu, J. A. Lavigne, Y. Tao, M. D'Iorio, S. N. Wang, "Novel blue luminescernt/electroluminescent 7-azaindole derivatives: 1,3-di(*N*-7-azaindolyl)benzene, 1-bromo-3,5-di(*N*-7-azaindolyl)benzene, 1,3,5-tri(*N*-7-azaindolyl)benzene, and 4,4'-di(*N*-7-azaindolyl)biphenyl," *Chem. Mater.*, 13:71–77 (2001).

385. J. Salbeck, N. Yu, J. Bauer, F. Weissortel, H. Bestgen, "Low molecular organic glasses for blue electroluminescence," *Synth. Met.*, 91:209–215 (1997).

386. C. C. Wu, Y. T. Lin, H. H. Chiang, T. Y. Cho, C. W. Chen, K. T. Wong, Y. L. Liao, G. H. Lee, S. M. Peng, "Highly bright blue organic light-emitting devices using spirobifluorene-cored conjugated compounds," *Appl. Phys. Lett.*, 81:577–579 (2002).

387. B. Li, Y. Qiu, L. D. Wang, Y. D. Gao, "Spiro-annulated compound as stable and high-efficiency blue host light-emitter," *Jpn. J. Appl. Phys. Pt. 1 Reg. Pap. Short Notes Rev. Pap.*, 41:5599–5601 (2002).

388. L. H. Chan, H. C. Yeh, C. T. Chen, "Blue light-emitting devices based on molecular glass materials of tetraphenylsilane compounds," *Adv. Mater.*, 13:1637–1641 (2001).

389. L. H. Chan, R. H. Lee, C. F. Hsieh, H. C. Yeh, C. T. Chen, "Optimization of high-performance blue organic light-emitting diodes containing tetraphenylsilane molecular glass materials," *J. Am. Chem. Soc.*, 124:6469–6479 (2002).

390. S. A. Vanslyke, "Blue emitting internal junction organic electroluminescent device(I)," U.S. Patent 5,151,629 (1992).

391. J. S. Yu, Z. J. Chen, Y. Sakuratani, H. Suzuki, M. Tokita, S. Miyata, "A novel blue light emitting diode using tris(2,3-methyl-8-hydroxyquinoline) aluminum(III) as emitter," *Jpn. J. Appl. Phys. Pt 1 Reg. Pap. Short Notes Rev. Pap.*, 38:6762–6763 (1999).

392. L. M. Leung, W. Y. Lo, S. K. So, K. M. Lee, W. K. Choi, "A high-efficiency blue emitter for small molecule-based organic light-emitting diode," *J. Am. Chem. Soc.*, 122:5640–5641 (2000).

393. Y. Liu, J. H. Guo, J. Feng, H. D. Zhang, Y. Q. Li, Y. Wang, "High-performance blue electroluminescent devices based on hydroxyphenyl–pyridine beryllium complex," *Appl. Phys. Lett.*, 78:2300–2302 (2001).

394. J. Shi, C. H. Chen, K. P. Klubek, "Blue luminescent materials for organic electroluminescent devices," U.S. Patent 5,755,999 (1998).

395. J. Shi, C. H. Chen, "Efficient blue organic electroluminescent devices," U.S. Patent 5,928,802 (1999).

396. Y. Sato, S. Ichinosawa, T. Ogata, M. Fugono, Y. Murata, "Blue-emitting organic EL devices with a hole blocking layer," *Synth. Met.*, 111:25–29 (2000).

397. H. Ogawa, K. Ohnishi, Y. Shirota, "Tri(*p*-terphenyl-4-yl)amine as a novel blue-emitting material for organic electroluminescent devices," *Synth. Met.*, 91:243–245 (1997).

398. T. Noda, H. Ogawa, Y. Shirota, "A blue-emitting organic electroluminescent device using a novel emitting amorphous molecular material, 5,5'-bis(dimesitylboryl)-2,2'-bithiophene," *Adv. Mater.*, 11:283–285 (1999).

399. T. Tsuji, S. Kawami, S. Miyaguchi, T. Naijo, T. Yuki, S. Matsuo, H. Miyazaki, "Red-phosphorescent OLEDs employing bis(8-quinolinolato)-phenolato-aluminum(III) complexes as emission-layer hosts," *SID Symp. Dig. Tech. Pap.*, 35(1):900 (2004).

400. C. F. Qiu, Z. L. Xie, H. Y. Chen, B. Z. Tang, M. Wong, H. S. Kwok, "Polymer electrophosphorescent light-emitting diode using aluminum bis(2-methyl-8-quinolinato) 4-phenylphenolate as an electron-transport layer," *IEEE J. Sel. Top. Quantum Electron.*, 10:101 (2004).

401. T. Y. Chu, Y. S. Wu, J. F. Chen, C. H. Chen, "Characterization of electronic structure of aluminum (III) bis(2-methyl-8-quninolinato)-4-phenylphenolate (BAlq) for phosphorescent organic light emitting devices," *Chem. Phys. Lett.*, 404:121–125 (2005).

402. F. C. Chen, G. F. He, Y. Yang, "Triplet exciton confinement in phosphorescent polymer light-emitting diodes," *Appl. Phys. Lett.*, 82:1006–1008 (2003).

403. H. Kanno, K. Ishikawa, Y. Nishio, A. Endo, C. Adachi, K. Shibata, "Highly efficient and stable red phosphorescent organic light-emitting device using bis 2-(2-benzothiazoyl)phenolato zinc(II) as host material," *Appl. Phys. Lett.*, 90:123509 (2007).
404. T. Tsuzuki, S. Tokito, "Highly efficient, low-voltage phosphorescent organic light-emitting diodes using an iridium complex as the host material," *Adv. Mater.*, 19:276–280 (2007).
405. T. Y. Hwu, T. C. Tsai, W. Y. Hung, S. Y. Chang, Y. Chi, M. H. Chen, C. I. Wu, K. T. Wong, L. C. Chi, "An electron-transporting host material compatible with diverse triplet emitters used for highly efficient red- and green-electrophosphorescent devices," *Chem. Commun.*, 4956–4958 (2008).
406. S. O. Jeon, K. S. Yook, C. W. Joo, H. S. Son, S. E. Jang, J. Y. Lee, "High efficiency red phosphorescent organic light-emitting diodes using a spirobenzofluorene type phosphine oxide as a host material," *Org. Electron.*, 10:998–1000 (2009).
407. S. O. Jeon, K. S. Yook, C. W. Joo, J. Y. Lee, "Simple high efficiency red phosphorescent organic light-emitting diodes without LiF electron injection layer," *J. Phys. D Appl. Phys.*, 42:225103 (2009).
408. C. W. Joo, S. O. Jeon, K. S. Yook, J. Y. Lee, "Red phosphorescent organic light-emitting diodes with indium tin oxide/single organic layer/Al simple device structure," *Org. Electron.*, 11:36–40 (2010).
409. M. T. Kao, W. Y. Hung, Z. H. Tsai, H. W. You, H. F. Chen, Y. Chi, K. T. Wong, "Using a double-doping strategy to prepare a bilayer device architecture for high-efficiency red PhOLEDs," *J. Mater. Chem.*, 21:1846–1851 (2011).
410. H. F. Chen, T. C. Wang, S. W. Lin, W. Y. Hung, H. C. Dai, H. C. Chiu, K. T. Wong, M. H. Ho, T. Y. Cho, C. W. Chen, C. C. Lee, "Peripheral modification of 1,3,5-triazine based electron-transporting host materials for sky blue, green, yellow, red, and white electrophosphorescent devices," *J. Mater. Chem.*, 22:15620–15627 (2012).
411. P. I. Shih, C. H. Chien, T. I. Wu, C. F. Shu, "A novel fluorene-triphenylamine hybrid that is a highly efficient host material for blue-, green-, and red-light-emitting electrophosphorescent devices," *Adv. Funct. Mater.*, 17:3514–3520 (2007).
412. S. E. Jang, S. O. Jeon, Y. J. Cho, K. S. Yook, J. Y. Lee, "Stable efficiency roll-off in red phosphorescent organic light-emitting diodes using a spirofluorene-benzofluorene based carbazole type host material," *J. Lumin.*, 130:2184–2187 (2010).
413. C. H. Chien, F. M. Hsu, C. F. Shu, Y. Chi, "Efficient red electrophosphorescence from a fluorene-based bipolar host material," *Org. Electron.*, 10:871–876 (2009).
414. F. M. Hsu, C. H. Chien, Y. J. Hsieh, C. H. Wu, C. F. Shu, S. W. Liu, C. T. Chen, "Highly efficient red electrophosphorescent device incorporating a bipolar triphenylamine/bisphenylsulfonyl-substituted fluorene hybrid as the host," *J. Mater. Chem.*, 19:8002–8008 (2009).
415. S. J. Su, C. Cai, J. Kido, "RGB phosphorescent organic light-emitting diodes by using host materials with heterocyclic cores: Effect of nitrogen atom orientations," *Chem. Mater.*, 23:274–284 (2011).
416. C. Cai, S. J. Su, T. Chiba, H. Sasabe, Y. J. Pu, K. Nakayama, J. Kido, "High-efficiency red, green and blue phosphorescent homojunction organic light-emitting diodes based on bipolar host materials," *Org. Electron.*, 12:843–850 (2011).
417. Y. T. Tao, Q. Wang, L. Ao, C. Zhong, C. L. Yang, J. G. Qin, D. G. Ma, "Highly efficient phosphorescent organic light-emitting diodes hosted by 1,2,4-triazole-cored triphenylamine derivatives: Relationship between structure and optoelectronic properties," *J. Phys. Chem. C*, 114:601–609 (2010).
418. S. J. Su, C. Cai, J. Kido, "Three-carbazole-armed host materials with various cores for RGB phosphorescent organic light-emitting diodes," *J. Mater. Chem.*, 22:3447–3456 (2012).
419. S. J. Su, C. Cai, J. Takamatsu, J. Kido, "A host material with a small singlet–triplet exchange energy for phosphorescent organic light-emitting diodes: Guest, host, and exciplex emission," *Org. Electron.*, 13:1937–1947 (2012).
420. M. Guan, Z. Q. Chen, Z. Q. Bian, Z. W. Liu, Z. L. Gong, W. Baik, H. J. Lee, C. H. Huang, "The host materials containing carbazole and oxadiazole fragment for red triplet emitter in organic light-emitting diodes," *Org. Electron.*, 7:330–336 (2006).

421. W. Y. Hung, G. M. Tu, S. W. Chen, Y. Chi, "Phenylcarbazole–dipyridyl triazole hybrid as bipolar host material for phosphorescent OLEDs," *J. Mater. Chem.*, 22:5410–5418 (2012).

422. H.-H. Chou, C.-H. Cheng, "A highly efficient universal bipolar host for blue, green, and red phosphorescent OLEDs," *Adv. Mater.*, 22:2468–2471 (2010).

423. C. Adachi, M. A. Baldo, S. R. Forrest, M. E. Thompson, "High-efficiency organic electrophosphorescent devices with tris(2-phenylpyridine)iridium doped into electron-transporting materials," *Appl. Phys. Lett.*, 77:904–906 (2000).

424. M. Klessinger, J. Michl, *Excited States and Photochemistry of Organic Molecules*, VCH Publishers, New York (1995).

425. H.-F. Chen, S.-J. Yang, Z.-H. Tsai, W.-Y. Hung, T.-C. Wang, K.-T. Wong, "1,3,5-Triazine derivatives as new electron transport-type host materials for highly efficient green phosphorescent OLEDs," *J. Mater. Chem.*, 19:8112–8118 (2009).

426. H.-F. Chen, T.-C. Wang, S.-W. Lin, W.-Y. Hung, H.-C. Dai, H.-C. Chiu, K.-T. Wong, M.-H. Ho, T.-Y. Cho, C.-W. Chen, C.-C. Lee, "Peripheral modification of 1,3,5-triazine based electron-transporting host materials for sky blue, green, yellow, red, and white electrophosphorescent devices," *J. Mater. Chem.*, 22:15620–15627 (2012).

427. S. O. Jeon, J. Y. Lee, "Above 20% external quantum efficiency in green and white phosphorescent organic light-emitting diodes using an electron transport type green host material," *Org. Electron.*, 12:1893–1898 (2011).

428. J. Zhao, G.-H. Xie, C.-R. Yin, L.-H. Xie, C.-M. Han, R.-F. Chen, H. Xu, M.-D. Yi, Z.-P. Deng, S.-F. Chen, Y. Zhao, S.-Y. Liu, W. Huang, "Harmonizing triplet level and ambipolar characteristics of wide-gap phosphine oxide hosts toward highly efficient and low driving voltage blue and green PHOLEDs: An effective strategy based on spiro-systems," *Chem. Mater.*, 23:5331–5339 (2011).

429. H. Sasabe, Y. Seino, M. Kimura, J. Kido, "A *m*-Terphenyl-modified sulfone derivative as a host material for high-efficiency blue and green phosphorescent OLEDs," *Chem. Mater.*, 24:1404–1406 (2012).

430. H. Kanai, S. Ichinosawa, Y. Sato, "Effect of aromatic diamines as a cathode interface layer," *Synth. Met.*, 91:195–196 (1997).

431. C. Adachi, R. C. Kwong, P. Djurovich, V. Adamovich, M. A. Baldo, M. E. Thompson, S. R. Forrest, "Endothermic energy transfer: A mechanism for generating very efficient high-energy phosphorescent emission in organic materials," *Appl. Phys. Lett.*, 79:2082–2084 (2001).

432. I. S. Ping, C. Chen-Han, W. Fang-Iy, S. Ching-Fong, "A novel fluorene-triphenylamine hybrid that is a highly efficient host material for blue-, green-, and red-light-emitting electrophosphorescent devices," *Adv. Funct. Mater.*, 17:3514–3520 (2007).

433. H. Sasabe, Y.-J. Pu, K.-I. Nakayama, J. Kido, "*m*-Terphenyl-modified carbazole host material for highly efficient blue and green PHOLEDS," *Chem. Commun.*, 6655–6657 (2009).

434. C. Fan, Y. H. Chen, Z. Q. Jiang, C. L. Yang, C. Zhong, J. G. Qin, D. G. Ma, "Diarylmethylene-bridged triphenylamine derivatives encapsulated with fluorene: Very high Tg host materials for efficient blue and green phosphorescent OLEDs," *J. Mater. Chem.*, 20:3232–3237 (2010).

435. L.-C. Chi, W.-Y. Hung, H.-C. Chiu, K.-T. Wong, "A high-efficiency and low-operating-voltage green electrophosphorescent device employing a pure-hydrocarbon host material," *Chem. Commun.*, 3892–3894 (2009).

436. Z. Jiang, H. Yao, Z. Zhang, C. Yang, Z. Liu, Y. Tao, J. Qin, D. Ma, "Novel oligo-9,9'-spirobifluorenes through *ortho*-linkage as full hydrocarbon host for highly efficient phosphorescent OLEDs," *Org. Lett.*, 11:2607–2610 (2009).

437. Y. T. Tao, Q. Wang, C. L. Yang, Z. Q. Zhang, T. T. Zou, J. G. Qin, D. G. Ma, "A simple carbazole/oxadiazole hybrid molecule: An excellent bipolar host for green and red phosphorescent OLEDs," *Angew. Chem. Int. Ed.*, 47:8104–8107 (2008).

438. Y. Tao, Q. Wang, C. Yang, C. Zhong, K. Zhang, J. Qin, D. Ma, "Tuning the optoelectronic properties of carbazole/oxadiazole hybrids through linkage modes: Hosts for highly efficient green electrophosphorescence," *Adv. Funct. Mater.*, 20:304–311 (2010).

439. Y. Tao, Q. Wang, C. Yang, C. Zhong, J. Qin, D. Ma, "Multifunctional triphenylamine/oxadiazole hybrid as host and exciton-blocking material: High efficiency green phosphorescent OLEDs using easily available and common materials," *Adv. Funct. Mater.*, 20:2923–2929 (2010).

440. S. L. Gong, Y. H. Chen, J. J. Luo, C. L. Yang, C. Zhong, J. G. Qin, D. G. Ma, "Bipolar tetraarylsilanes as universal hosts for blue, green, orange, and white electrophosphorescence with high efficiency and low efficiency roll-off," *Adv. Funct. Mater.*, 21:1168–1178 (2011).

441. W. Wei, P. I. Djurovich, M. E. Thompson, "Properties of fluorenyl silanes in organic light emitting diodes," *Chem. Mater.*, 22:1724–1731 (2010).

442. Y. M. Chen, W. Y. Hung, H. W. You, A. Chaskar, H. C. Ting, H. F. Chen, K. T. Wong, Y. H. Liu, "Carbazole–benzimidazole hybrid bipolar host materials for highly efficient green and blue phosphorescent OLEDs," *J. Mater. Chem.*, 21:14971–14978 (2011).

443. H.-F. Chen, L.-C. Chi, W.-Y. Hung, W.-J. Chen, T.-Y. Hwu, Y.-H. Chen, S.-H. Chou, E. Mondal, Y.-H. Liu, K.-T. Wong, "Carbazole and benzimidazole/oxadiazole hybrids as bipolar host materials for sky blue, green, and red PhOLEDs," *Org. Electron.*, 13:2671–2681 (2012).

444. C.-H. Chang, M.-C. Kuo, W.-C. Lin, Y.-T. Chen, K.-T. Wong, S.-H. Chou, E. Mondal, R. C. Kwong, S. Xia, T. Nakagawa, C. Adachi, "A dicarbazole–triazine hybrid bipolar host material for highly efficient green phosphorescent OLEDs," *J. Mater. Chem.*, 22:3832–3838 (2012).

445. H. Huang, Y. Wang, S. Zhuang, X. Yang, L. Wang, C. Yang, "Simple phenanthroimidazole/carbazole hybrid bipolar host materials for highly efficient green and yellow phosphorescent organic light-emitting diodes," *J. Phys. Chem. C*, 116:19458–19466 (2012).

446. Z. M. Hudson, Z. Wang, M. G. Helander, Z.-H. Lu, S. Wang, "N-heterocyclic carbazole-based hosts for simplified single-layer phosphorescent OLEDs with high efficiencies," *Adv. Mater.*, 24:2922–2928 (2012).

447. H.-C. Ting, Y.-M. Chen, H.-W. You, W.-Y. Hung, S.-H. Lin, A. Chaskar, S.-H. Chou, Y. Chi, R.-H. Liu, K.-T. Wong, "Indolo 3,2-b carbazole/benzimidazole hybrid bipolar host materials for highly efficient red, yellow, and green phosphorescent organic light emitting diodes," *J. Mater. Chem.*, 22:8399–8407 (2012).

448. M. S. Lin, L. C. Chi, H. W. Chang, Y. H. Huang, K. C. Tien, C. C. Chen, C. H. Chang, C. C. Wu, A. Chaskar, S. H. Chou, H. C. Ting, K. T. Wong, Y. H. Liu, Y. Chi, "A diarylborane-substituted carbazole as a universal bipolar host material for highly efficient electrophosphorescence devices," *J. Mater. Chem.*, 22:870–876 (2012).

449. S. O. Jeon, J. Y. Lee, "Fluorenobenzofuran as the core structure of high triplet energy host materials for green phosphorescent organic light-emitting diodes," *J. Mater. Chem.*, 22:10537–10541 (2012).

450. R. J. Holmes, B. W. D'Andrade, S. R. Forrest, X. Ren, J. Li, M. E. Thompson, "Efficient, deep-blue organic electrophosphorescence by guest charge trapping," *Appl. Phys. Lett.*, 83:3818–3820 (2003).

451. S. Tokito, T. Iijima, Y. Suzuri, H. Kita, T. Tsuzuki, F. Sato, "Confinement of triplet energy on phosphorescent molecules for highly-efficient organic blue-light-emitting devices," *Appl. Phys. Lett.*, 83:569–571 (2003).

452. S. Tokito, T. Iijima, T. Tsuzuki, F. Sato, "High-efficiency white phosphorescent organic light-emitting devices with greenish–blue and red-emitting layers," *Appl. Phys. Lett.*, 83:2459–2461 (2003).

453. I. Tanaka, Y. Tabata, S. Tokito, "Energy-transfer and light-emission mechanism of blue phosphorescent molecules in guest–host systems," *Chem. Phys. Lett.*, 400:86–89 (2004).

454. R. J. Holmes, S. R. Forrest, Y. J. Tung, R. C. Kwong, J. J. Brown, S. Garon, M. E. Thompson, "Blue organic electrophosphorescence using exothermic host–guest energy transfer," *Appl. Phys. Lett.*, 82:2422–2424 (2003).

455. G. T. Lei, L. D. Wang, L. Duan, J. H. Wang, Y. Qiu, "Highly efficient blue electrophosphorescent devices with a novel host material," *Synth. Met.*, 144:249–252 (2004).

456. G. T. Lei, L. D. Wang, Y. Qiu, "Blue phosphorescent dye as sensitizer and emitter for white organic light-emitting diodes," *Appl. Phys. Lett.*, 85:5403–5405 (2004).

457. X. F. Ren, J. Li, R. J. Holmes, P. I. Djurovich, S. R. Forrest, M. E. Thompson, "Ultrahigh energy gap hosts in deep blue organic electrophosphorescent devices," *Chem. Mater.*, 16:4743–4747 (2004).

458. Z. Q. Jiang, Y. H. Chen, C. Fan, C. L. Yang, Q. Wang, Y. T. Tao, Z. Q. Zhang, J. G. Qin, D. G. Ma, "Bridged triphenylamines as novel host materials for highly efficient blue and green phosphorescent OLEDs," *Chem. Commun.*, 3398–3400 (2009).

459. M. H. Tsai, H. W. Lin, H. C. Su, T. H. Ke, C. C. Wu, F. C. Fang, Y. L. Liao, K. T. Wong, C. I. Wu, "Highly efficient organic blue electrophosphorescent devices based on 3,6-bis(triphenylsilyl) carbazole as the host material," *Adv. Mater.*, 18:1216–1220 (2006).

460. M. H. Tsai, T. H. Ke, H. W. Lin, C. C. Wu, S. F. Chiu, F. C. Fang, Y. L. Liao, K. T. Wong, Y. H. Chen, C. I. Wu, "Triphenylsilyl- and trityl-substituted carbazole-based host materials for blue electrophosphorescence," *ACS Appl. Mater. Interfaces*, 1:567–574 (2009).

461. M. Mamada, S. Ergun, C. Perez-Bolivar, P. Anzenbacher, "Charge transport, carrier balance, and blue electrophosphorescence in diphenyl[4-(triphenylsilyl)phenyl]phosphine oxide devices," *Appl. Phys. Lett.*, 98:073305 (2011).

462. O. Y. Kim, J. Y. Lee, "High efficiency deep blue phosphorescent organic light-emitting diodes using a tetraphenylsilane based phosphine oxide host material," *J. Ind. Eng. Chem.*, 18:1029–1032 (2012).

463. H. Liu, G. Cheng, D. H. Hu, F. Z. Shen, Y. Lv, G. N. Sun, B. Yang, P. Lu, Y. G. Ma, "A highly efficient, blue-phosphorescent device based on a wide-bandgap host/FIrpic: Rational design of the carbazole and phosphine oxide moieties on tetraphenylsilane," *Adv. Funct. Mater.*, 22:2830–2836 (2012).

464. C. J. Zheng, J. Ye, M. F. Lo, M. K. Fung, X. M. Ou, X. H. Zhang, C. S. Lee, "New ambipolar hosts based on carbazole and 4,5-diazafluorene units for highly efficient blue phosphorescent OLEDs with low efficiency roll-off," *Chem. Mater.*, 24:643–650 (2012).

465. L. X. Wang, J. Q. Ding, Q. Wang, L. Zhao, D. G. Ma, X. B. Jing, F. S. Wang, "Design of star-shaped molecular architectures based on carbazole and phosphine oxide moieties: Towards amorphous bipolar hosts with high triplet energy for efficient blue electrophosphorescent devices," *J. Mater. Chem.*, 20:8126–8133 (2010).

466. S. J. Su, H. Sasabe, T. Takeda, J. Kido, "Pyridine-containing bipolar host materials for highly efficient blue phosphorescent OLEDs," *Chem. Mater.*, 20:1691–1693 (2008).

467. H. Sasabe, N. Toyota, H. Nakanishi, T. Ishizaka, Y. J. Pu, J. Kido, "3,3′-Bicarbazole-based host materials for high-efficiency blue phosphorescent OLEDs with extremely low driving voltage," *Adv. Mater.*, 24:3212–3217 (2012).

468. P. E. Burrows, A. B. Padmaperuma, L. S. Sapochak, P. Djurovich, M. E. Thompson, "Ultraviolet electroluminescence and blue–green phosphorescence using an organic diphosphine oxide charge transporting layer," *Appl. Phys. Lett.*, 88:183503 (2006).

469. X. Y. Cai, A. B. Padmaperuma, L. S. Sapochak, P. A. Vecchi, P. E. Burrows, "Electron and hole transport in a wide bandgap organic phosphine oxide for blue electrophosphorescence," *Appl. Phys. Lett.*, 92:083308 (2008).

470. S. K. Kim, B. Yang, Y. G. Ma, J. H. Lee, J. W. Park, "Exceedingly efficient deep-blue electroluminescence from new anthracenes obtained using rational molecular design," *J. Mater. Chem.*, 18:3376–3384 (2008).

471. H. S. Son, C. W. Seo, J. Y. Lee, "Correlation of the substitution position of diphenylphosphine oxide on phenylcarbazole and device performances of blue phosphorescent organic light-emitting diodes," *J. Mater. Chem.*, 21:5638–5644 (2011).

472. J. Y. Lee, S. O. Jeon, K. S. Yook, C. W. Joo, "High-efficiency deep-blue-phosphorescent organic light-emitting diodes using a phosphine oxide and a phosphine sulfide high-triplet-energy host material with bipolar charge-transport properties," *Adv. Mater.*, 22:1872–1876 (2010).

473. H. H. Chang, W. S. Tsai, C. P. Chang, N. P. Chen, K. T. Wong, W. Y. Hung, S. W. Chen, "A new tricarbazole phosphine oxide bipolar host for efficient single-layer blue PhOLED," *Org. Electron.*, 12:2025–2032 (2011).

474. A. B. Padmaperuma, L. S. Sapochak, P. E. Burrows, "New charge transporting host material for short wavelength organic electrophosphorescence: 2,7-bis(diphenylphosphine oxide)-9,9-dimethylfluorene," *Chem. Mater.*, 18:2389–2396 (2006).

475. J. Y. Lee, S. O. Jeon, K. S. Yook, C. W. Joo, H. S. Son, "A high triplet energy phosphine oxide derivative as a host and exciton blocking material for blue phosphorescent organic light-emitting diodes," *Thin Solid Films*, 518:3716–3720 (2010).

476. J. Y. Lee, S. E. Jang, C. W. Joo, S. O. Jeon, K. S. Yook, "The relationship between the substitution position of the diphenylphosphine oxide on the spirobifluorene and device performances of blue phosphorescent organic light-emitting diodes," *Org. Electron.*, 11:1059–1065 (2010).

477. F. M. Hsu, C. H. Chien, P. I. Shih, C. F. Shu, "Phosphine-oxide-containing bipolar host material for blue electrophosphorescent devices," *Chem. Mater.*, 21:1017–1022 (2009).

478. H. Xu, D. H. Yu, Y. B. A. Zhao, C. M. Han, D. G. Ma, Z. P. Deng, S. Gao, P. F. Yan, "Fluorene-based phosphine oxide host materials for blue electrophosphorescence: An effective strategy for a high triplet energy level," *Chem.-Eur. J.*, 17:2592–2596 (2011).

479. D. H. Yu, F. C. Zhao, C. M. Han, H. Xu, J. Li, Z. Zhang, Z. P. Deng, D. G. Ma, P. F. Yan, "Ternary ambipolar phosphine oxide hosts based on indirect linkage for highly efficient blue electrophosphorescence: Towards high triplet energy, low driving voltage and stable efficiencies," *Adv. Mater.*, 24:509–514 (2012).

480. E. Polikarpov, J. S. Swensen, N. Chopra, F. So, A. B. Padmaperuma, "An ambipolar phosphine oxide-based host for high power efficiency blue phosphorescent organic light emitting devices," *Appl. Phys. Lett.*, 94:223304 (2009).

481. W. Jiang, W. Yang, X. X. Ban, B. Huang, Y. Q. Dai, Y. M. Sun, L. Duan, Y. Qiu, "Synthesis of new bipolar materials based on diphenylphosphine oxide and triphenylamine units: Efficient host for deep-blue phosphorescent organic light-emitting diodes," *Tetrahedron*, 68:9672–9678 (2012).

482. U. S. Bhansali, E. Polikarpov, J. S. Swensen, W. H. Chen, H. P. Jia, D. J. Gaspar, B. E. Gnade, A. B. Padmaperuma, M. A. Omary, "High-efficiency turquoise-blue electrophosphorescence from a Pt(II)-pyridyltriazolate complex in a phosphine oxide host," *Appl. Phys. Lett.*, 95:233304 (2009).

483. H. Xu, C. M. Han, Y. B. Zhao, J. S. Chen, Z. P. Deng, D. G. Ma, Q. Li, P. F. Yan, "A simple phosphine-oxide host with a multi-insulating structure: High triplet energy level for efficient blue electrophosphorescence," *Chem.-Eur. J.*, 17:5800–5803 (2011).

484. F. S. Liang, L. X. Wang, D. G. Ma, X. B. Jing, F. S. Wang, "Oxadiazole-containing material with intense blue phosphorescence emission for organic light-emitting diodes," *Appl. Phys. Lett.*, 81:4–6 (2002).

485. M. A. Baldo, S. Lamansky, P. E. Burrows, M. E. Thompson, S. R. Forrest, "Very high-efficiency green organic light-emitting devices based on electrophosphorescence," *Appl. Phys. Lett.*, 75:4–6 (1999).

486. K. A. King, P. J. Spellane, R. J. Watts, "Excited-state properties of a triply *ortho*-metalated iridium(III) complex," *J. Am. Chem. Soc.*, 107:1431–1432 (1985).

487. W. Holzer, A. Penzkofer, T. Tsuboi, "Absorption and emission spectroscopic characterization of Ir(ppy)(3)," *Chem. Phys.*, 308:93–102 (2005).

488. T. Tsutsui, M. J. Yang, M. Yahiro, K. Nakamura, T. Watanabe, T. Tsuji, Y. Fukuda, T. Wakimoto, S. Miyaguchi, "High quantum efficiency in organic light-emitting devices with iridium-complex as a triplet emissive center," *Jpn. J. Appl. Phys. Pt. 2 Lett. Express Lett.*, 38:L1502–L1504 (1999).

489. A. Tsuboyama, H. Iwawaki, M. Furugori, T. Mukaide, J. Kamatani, S. Igawa, T. Moriyama, S. Miura, T. Takiguchi, S. Okada, M. Hoshino, K. Ueno, "Homoleptic cyclometalated iridium complexes with highly efficient red phosphorescence and application to organic light-emitting diode," *J. Am. Chem. Soc.*, 125:12971–12979 (2003).

490. Y. J. Su, H. L. Huang, C. L. Li, C. H. Chien, Y. T. Tao, P. T. Chou, S. Datta, R. S. Liu, "Highly efficient red electrophosphorescent devices based on iridium isoquinoline complexes: Remarkable external quantum efficiency over a wide range of current," *Adv. Mater.*, 15:884–888 (2003).

491. C. H. Yang, C. C. Tai, I. W. Sun, "Synthesis of a high-efficiency red phosphorescent emitter for organic light-emitting diodes," *J. Mater. Chem.*, 14:947–950 (2004).

492. M. J. Frampton, E. B. Namdas, S. C. Lo, P. L. Burn, I. D. W. Samuel, "The synthesis and properties of solution processable red-emitting phosphorescent dendrimers," *J. Mater. Chem.*, 14:2881–2888 (2004).

493. K. H. Fang, L. L. Wu, Y. T. Huang, C. H. Yang, I. W. Sun, "Color tuning of iridium complexes—Part I: Substituted phenylisoquinoline-based iridium complexes as the triplet emitter," *Inorg. Chim. Acta*, 359:441–450 (2006).

494. C. H. Yang, K. H. Fang, W. L. Su, S. P. Wang, S. K. Su, I. W. Sun, "Color tuning of iridium complexes for organic light-emitting diodes: The electronegative effect and pi-conjugation effect," *J. Organomet. Chem.*, 691:2767–2773 (2006).

495. D. Schneidenbach, S. Ammermann, M. Debeaux, A. Freund, M. Zollner, C. Daniliuc, P. G. Jones, W. Kowalsky, H. H. Johannes, "Efficient and long-time stable red iridium(III) complexes for organic light-emitting diodes based on quinoxaline ligands," *Inorg. Chem.*, 49:397–406 (2010).

496. D. H. Kim, N. S. Cho, H. Y. Oh, J. H. Yang, W. S. Jeon, J. S. Park, M. C. Suh, F. H. Kwon, "Highly efficient red phosphorescent dopants in organic light-emitting devices," *Adv. Mater.*, 23:2721–2726 (2011).

497. Z. W. Liu, M. Guan, Z. Q. Bian, D. B. Nie, Z. L. Gong, Z. B. Li, C. H. Huang, "Red phosphorescent iridium complex containing carbazole-functionalized beta-diketonate for highly efficient nondoped organic light-emitting diodes," *Adv. Funct. Mater.*, 16:1441–1448 (2006).

498. C. L. Ho, W. Y. Wong, Z. Q. Gao, C. H. Chen, K. W. Cheah, B. Yao, Z. Y. Xie, Q. Wang, D. G. Ma, L. X. Wang, X. M. Yu, H. S. Kwok, Z. Y. Lin, "Red-light-emitting iridium complexes with hole-transporting 9-arylcarbazole moieties for electrophosphorescence efficiency/color purity trade-off optimization," *Adv. Funct. Mater.*, 18:319–331 (2008).

499. C. L. Yang, X. W. Zhang, H. You, L. Y. Zhu, L. Q. Chen, L. N. Zhu, Y. T. Tao, D. G. Ma, Z. G. Shuai, J. G. Qin, "Tuning the energy level and photophysical and electroluminescent properties of heavy metal complexes by controlling the ligation of the metal with the carbon of the carbazole unit," *Adv. Funct. Mater.*, 17:651–661 (2007).

500. G. Y. Park, Y. Ha, "Red phosphorescent iridium(III) complexes containing 2,3-diphenylquinoline derivatives for OLEDs," *Synth. Met.*, 158:120–124 (2008).

501. T. H. Chuang, C. H. Yang, P. C. Kao, "Efficient red-emitting cyclometalated iridium(III) complex and applications of organic light-emitting diode," *Inorg. Chim. Acta*, 362:5017–5022 (2009).

502. J. H. Seo, S. C. Lee, Y. K. Kim, Y. S. Kim, "Substituent effects of iridium (III) complexes with 2,4-diphenylquinoline for efficient red organic light-emitting diodes," *Thin Solid Films*, 517:1346–1348 (2008).

503. C.-H. Fan, P. Sun, T.-H. Su, C.-H. Cheng, "Host and dopant materials for idealized deep-red organic electrophosphorescence devices," *Adv. Mater.*, 23:2981–2985 (2011).

504. Y. H. Song, S. J. Yeh, C. T. Chen, Y. Chi, C. S. Liu, J. K. Yu, Y. H. Hu, P. T. Chou, S. M. Peng, G. H. Lee, "Bright and efficient, non-doped, phosphorescent organic red-light-emitting diodes," *Adv. Funct. Mater.*, 14:1221–1226 (2004).

505. S. M. Wang, J. Y. Zhang, Y. H. Hou, C. X. Du, Y. J. Wu, "4,5-Diaza-9,9'-spirobifluorene functionalized europium complex with efficient photo- and electro-luminescent properties," *J. Mater. Chem.*, 21:7559–7561 (2011).

506. C. R. De Silva, F. Y. Li, C. H. Huang, Z. P. Zheng, "Europium beta-diketonates for red-emitting electroluminescent devices," *Thin Solid Films*, 517:957–962 (2008).

507. Q. Xin, W. L. Li, G. B. Che, W. M. Su, X. Y. Sun, B. Chu, B. Li, "Improved electroluminescent performances of europium–complex based devices by doping into electron-transporting/hole-blocking host," *Appl. Phys. Lett.*, 89:223524 (2006).

508. T. Tsuzuki, Y. Nakayama, J. Nakamura, T. Iwata, S. Tokito, "Efficient organic light-emitting devices using an iridium complex as a phosphorescent host and a platinum complex as a red phosphorescent guest," *Appl. Phys. Lett.*, 88:243511 (2006).

509. M. Ikai, F. Ishikawa, N. Aratani, A. Osuka, S. Kawabata, T. Kajioka, H. Takeuchi, H. Fujikawa, Y. Taga, "Enhancement of external quantum efficiency of red phosphorescent organic light-emitting devices ices with facially encumbered and bulky Pt-II porphyrin complexes," *Adv. Funct. Mater.*, 16:515–519 (2006).

510. H. Fukagawa, T. Shimizu, H. Hanashima, Y. Osada, M. Suzuki, H. Fujikake, "Highly efficient and stable red phosphorescent organic light-emitting diodes using platinum complexes," *Adv. Mater.*, 24:5099–5103 (2012).

511. T. H. Liu, S. F. Hsu, M. H. Ho, C. H. Liao, Y. S. Wu, C. H. Chen, Y. L. Tung, P. C. Wu, Y. Chi, "Phosphorescence of red Os(fptz)(2)(PPh2Me)(2) doped organic light-emitting devices with n and p hosts," *Appl. Phys. Lett.*, 88:063508 (2006).

512. C. H. Wu, P. I. Shih, C. F. Shu, Y. Chi, "Highly efficient red organic light-emitting devices based on a fluorene-triphenylamine host doped with an Os(II) phosphor," *Appl. Phys. Lett.*, 92:233303 (2008).

513. Q. S. Zhang, J. Q. Ding, Y. X. Cheng, L. X. Wang, Z. Y. Xie, X. B. Jing, F. S. Wang, "Novel hetero-leptic Cu–I complexes with tunable emission color for efficient phosphorescent light-emitting diodes," *Adv. Funct. Mater.*, 17:2983–2990 (2007).

514. Y. L. Tung, L. S. Chen, Y. Chi, P. T. Chou, Y. M. Cheng, E. Y. Li, G. H. Lee, C. F. Shu, F. I. Wu, A. J. Carty, "Orange and red organic light-emitting devices employing neutral Ru(II) emitters: Rational design and prospects for color tuning," *Adv. Funct. Mater.*, 16:1615–1626 (2006).

515. Y. Wang, N. Herron, V. V. Grushin, D. LeCloux, V. Petrov, "Highly efficient electrolumines-cent materials based on fluorinated organometallic iridium compounds," *Appl. Phys. Lett.*, 79:449–451 (2001).

516. J. P. J. Markham, S. C. Lo, S. W. Magennis, P. L. Burn, I. D. W. Samuel, "High-efficiency green phosphorescence from spin-coated single-layer dendrimer light-emitting diodes," *Appl. Phys. Lett.*, 80:2645–2647 (2002).

517. S. C. Lo, E. B. Namdas, P. L. Burn, I. D. W. Samuel, "Synthesis and properties of highly efficient electroluminescent green phosphorescent iridium cored dendrimers," *Macromolecules*, 36:9721–9730 (2003).

518. M. K. Nazeeruddin, R. Humphry-Baker, D. Berner, S. Rivier, L. Zuppiroli, M. Graetzel, "Highly phosphorescence iridium complexes and their application in organic light-emitting devices," *J. Am. Chem. Soc.*, 125:8790–8797 (2003).

519. B. Ma, R. Walters, R. Kwong, "OLEDs utilizing multidentate ligand systems," U.S. Patent 200,5170,206, p. 35 (2005).

520. B. Ma, D. B. Knowles, C. S. Brown, D. Murphy, M. E. Thompson, "Organic light emitting mate-rials and devices," PCT Int. Appl., WO 045002 A1 (2004).

521. B. Ma, D. B. Knowles, C. S. Brown, D. Murphy, M. E. Thompson, "Organic light emitting mate-rials and devices," U.S. Patent 6,687,266 B1, p. 17 (2004).

522. D. M. Kang, J.-W. Kang, J. W. Park, S. O. Jung, S.-H. Lee, H.-D. Park, Y.-H. Kim, S. C. Shin, J.-J. Kim, S.-K. Kwon, "Iridium complexes with cyclometalated 2-cycloalkenylpyridine ligands as highly efficient emitters for organic light-emitting diodes," *Adv. Mater.*, 20:2003–2007 (2008).

523. S. O. Jung, Q. Zhao, J. W. Park, S. O. Kim, Y. H. Kim, H. Y. Oh, J. Kim, S. K. Kwon, Y. Kang, "A green emitting iridium(III) complex with narrow emission band and its application to phos-phorescence organic light-emitting diodes (OLEDs)," *Org. Electron.*, 10:1066–1073 (2009).

524. J. H. Seo, Y. K. Kim, Y. Ha, "Efficient blue–green organic light-emitting diodes based on hetero-leptic tris-cyclometalated iridium(III) complexes," *Thin Solid Films*, 517:1807–1810 (2009).

525. Y.-C. Zhu, L. Zhou, H.-Y. Li, Q.-L. Xu, M.-Y. Teng, Y.-X. Zheng, J.-L. Zuo, H.-J. Zhang, X.-Z. You, "Highly efficient green and blue–green phosphorescent OLEDs based on iridium complexes with the tetraphenylimidodiphosphinate ligand," *Adv. Mater.*, 23:4041–4046 (2011).

526. V. K.-M. Au, K. M.-C. Wong, D. P.-K. Tsang, M.-Y. Chan, N. Zhu, V. W.-W. Yam, "High-efficiency green organic light-emitting devices utilizing phosphorescent bis-cyclometalated alkynylgold(III) complexes," *J. Am. Chem. Soc.*, 132:14273–14278 (2010).

527. M. Hashimoto, S. Igawa, M. Yashima, I. Kawata, M. Hoshino, M. Osawa, "Highly efficient green organic light-emitting diodes containing luminescent three-coordinate copper(i) com-plexes," *J. Am. Chem. Soc.*, 133:10348–10351 (2011).

528. J. Li, P. I. Djurovich, B. D. Alleyne, I. Tsyba, N. N. Ho, R. Bau, M. E. Thompson, "Synthesis and characterization of cyclometalated Ir(III) complexes with pyrazolyl ancillary ligands," *Polyhedron*, 23:419–428 (2004).

529. G. T. Lei, L. D. Wang, Q. Yong, "Improved performance of electrophosphorescent organic light-emitting diode by graded doped emissive layer," *Jpn. J. Appl. Phys. Pt. 2 Lett. Express Lett.*, 43:L1226–L1228 (2004).

530. S. J. Yeh, M. F. Wu, C. T. Chen, Y. H. Song, Y. Chi, M. H. Ho, S. F. Hsu, C. H. Chen, "New dopant and host materials for blue-light-emitting phosphorescent organic electroluminescent devices," *Adv. Mater.*, 17:285–289 (2005).

531. A. B. Tamayo, B. D. Alleyne, P. I. Djurovich, S. Lamansky, I. Tsyba, N. N. Ho, R. Bau, M. E. Thompson, "Synthesis and characterization of facial and meridional tris-cyclometalated iridium(III) complexes," *J. Am. Chem. Soc.*, 125:7377–7387 (2003).

532. K. Dedeian, J. M. Shi, N. Shepherd, E. Forsythe, D. C. Morton, "Photophysical and electrochemical properties of heteroleptic tris-cyclometalated iridium(III) complexes," *Inorg. Chem.*, 44:4445–4447 (2005).

533. Available at http://www.universaldisplay.com.

534. T. Sajoto, A. B. Tamayo, P. I. Djurovich, M. Yousufuddin, R. Bau, M. E. Thompson, R. J. Holmes, S. R. Forrest, "Recent progress in blue phosphorescent iridium(III) complexes and their application to organic light emitting devices (OLEDS)," *Abstr. Pap. Am. Chem. Soc.*, 229:U1031 (2005).

535. E. Orselli, G. S. Kottas, A. E. Konradsson, P. Coppo, R. Frohlich, R. Frtshlich, L. De Cola, A. van Dijken, M. Buchel, H. Borner, "Blue-emitting iridium complexes with substituted 1,2,4-triazole ligands: Synthesis, photophysics, and devices," *Inorg. Chem.*, 46:11082–11093 (2007).

536. C. F. Chang, Y. M. Cheng, Y. Chi, Y. C. Chiu, C. C. Lin, G. H. Lee, P. T. Chou, C. C. Chen, C. H. Chang, C. C. Wu, "Highly efficient blue-emitting iridium(III) carbene complexes and phosphorescent OLEDs," *Angew. Chem. Int. Ed.*, 47:4542–4545 (2008).

537. Y. M. Wang, F. Teng, L. H. Gan, H. M. Liu, X. H. Zhang, W. F. Fu, Y. S. Wang, X. R. Xu, "Blue light-emitting bisorthometalated Ir(III) complex: Origin of blue emission and application in electrophosphorescent devices," *J. Phys. Chem. C*, 112:4743–4747 (2008).

538. K. Takada, M. Ichimura, T. Ishibashi, S. Tamura, "Bis(aminostyryl)stilbene-type compound, synthetic intermediate for the compound, manufacture of the intermediate and the compound, and organic electroluminescent device," Jpn. Kokai Tokkyo Koho, JP2001131128, p. 44 (2001).

539. T. Ishibashi, M. Ichimura, S. Tamura, N. Ueda, "Organic electroluminescent device or display using styryl compound," PCT Int. Appl., WO 2004003104, p. 142 (2004).

540. K. R. J. Thomas, J. T. Lin, M. Velusamy, Y. T. Tao, C. H. Chuen, "Color tuning in benzo[1,2,5]thiadiazole-based small molecules by amino conjugation/deconjugation: Bright red-light-emitting diodes," *Adv. Funct. Mater.*, 14:83–90 (2004).

541. H. C. Yeh, L. H. Chan, W. C. Wu, C. T. Chen, "Non-doped red organic light-emitting diodes," *J. Mater. Chem.*, 14:1293–1298 (2004).

542. K. R. J. Thomas, J. T. Lin, Y. T. Tao, C. H. Chuen, "Star-shaped thieno-[3,4-*b*]-pyrazines: A new class of red-emitting electroluminescent materials," *Adv. Mater.*, 14:822–826 (2002).

543. Y. T. Lee, C. L. Chiang, C. T. Chen, "Solid-state highly fluorescent diphenylaminospirobifluorenylfumaronitrile red emitters for non-doped organic light-emitting diodes," *Chem. Commun.*, 217–219 (2008).

544. X. B. Cao, Y. G. Wen, Y. L. Guo, G. Yu, Y. Q. Liu, L. M. Yang, "Undoped, red organic light-emitting diodes based on a *N,N,N',N'*-tetraphenylbenzidine (TPD) derivative as red emitter with a triphenylamine derivative as hole-transporting layer," *Dyes Pigments*, 84:203–207 (2010).

545. J. Huang, X. F. Qiao, Y. J. Xia, X. H. Zhu, D. G. Ma, Y. Cao, J. Roncali, "A dithienylbenzothiadiazole pure red molecular emitter with electron transport and exciton self-confinement for nondoped organic red-light-emitting diodes," *Adv. Mater.*, 20:4172–4175 (2008).

546. Y. Qiu, P. Wei, D. Q. Zhang, J. Qiao, L. Duan, Y. K. Li, Y. D. Gao, L. D. Wang, "Novel naphtho[2,3-*c*][1,2,5]thiadiazole derivative for non-doped small molecular organic red-light-emitting diodes," *Adv. Mater.*, 18:1607–1611 (2006).

547. J. M. Hanock, A. P. Gifford, Z. Yan, L. Yun, S. A. Jenekhe, "n-Type conjugated oligoquinoline and oligoquinoxaline with triphenylamine endgroups: Efficient ambipolar light emitters for device applications," *Chem. Mater.*, 18:4924–4932 (2006).

548. A. P. Kulkarni, X. X. Kong, S. A. Jenekhe, "High-performance organic light-emitting diodes based on intramolecular charge-transfer emission from donor–acceptor molecules: Significance of electron–donor strength and molecular geometry," *Adv. Funct. Mater.*, 16:1057–1066 (2006).

549. H. Zhang, X. Xu, W. Qiu, T. Qi, X. Gao, Y. Liu, K. Lu, C. Du, G. Yu, Y. Liu, "Unsymmetrical dendrimers as highly efficient light-emitting materials: Synthesis, photophysics, and electroluminescence," *J. Phys. Chem. C*, 112:13258–13263 (2008).

550. S.-Y. Ku, L.-C. Chi, W.-Y. Hung, S.-W. Yang, T.-C. Tsai, K.-T. Wong, Y.-H. Chen, C.-I. Wu, "High-luminescence non-doped green OLEDs based on a 9,9-diarylfluorene-terminated 2,1,3-benzo-thiadiazole derivative," *J. Mater. Chem.*, 19:773–780 (2009).

551. Z. Zhao, S. Chen, J. W. Y. Lam, Z. Wang, P. Lu, F. Mahtab, H. H. Y. Sung, I. D. Williams, Y. Ma, H. S. Kwok, B. Z. Tang, "Pyrene-substituted ethenes: Aggregation-enhanced excimer emission and highly efficient electroluminescence," *J. Mater. Chem.*, 21:7210–7216 (2011).

552. Z.-Y. Xia, J.-H. Su, H.-H. Fan, K.-W. Cheah, H. Tian, C. H. Che, "Multifunctional diarylamine-substituted benzo[*k*]fluoranthene derivatives as green electroluminescent emitters and nonlinear optical materials," *J. Phys. Chem. C*, 114:11602–11606 (2010).

553. C. H. Wu, C. H. Chien, F. M. Hsu, P. I. Shih, C. F. Shu, "Efficient non-doped blue-light-emitting diodes incorporating an anthracene derivative end-capped with fluorene groups," *J. Mater. Chem.*, 19:1464–1470 (2009).

554. X. H. Zhang, C. J. Zheng, W. M. Zhao, Z. Q. Wang, D. Huang, J. Ye, X. M. Ou, C. S. Lee, S. T. Lee, "Highly efficient non-doped deep-blue organic light-emitting diodes based on anthracene derivatives," *J. Mater. Chem.*, 20:1560–1566 (2010).

555. P. I. Shih, C. Y. Chuang, C. H. Chien, E. W. G. Diau, C. F. Shu, "Highly efficient non-doped blue-light-emitting diodes based on an anthrancene derivative end-capped with tetraphenylethylene groups," *Adv. Funct. Mater.*, 17:3141–3146 (2007).

556. Z. Q. Wang, C. Xu, W. Z. Wang, L. M. Duan, Z. Li, B. T. Zhao, B. M. Ji, "High-color-purity and high-efficiency non-doped deep-blue electroluminescent devices based on novel anthracene derivatives," *New J. Chem.*, 36:662–667 (2012).

557. J. W. Park, Y. H. Kim, S. Y. Jung, K. N. Byeon, S. H. Jang, S. K. Lee, S. C. Shin, S. K. Kwon, "Efficient and stable blue organic light-emitting diode based on an anthracene derivative," *Thin Solid Films*, 516:8381–8385 (2008).

558. Y. S. Kim, K. H. Lee, J. K. Park, J. H. Seo, S. W. Park, Y. K. Kim, S. S. Yoon, "Efficient deep-blue and white organic light-emitting diodes based on triphenylsilane-substituted anthracene derivatives," *J. Mater. Chem.*, 21:13640–13648 (2011).

559. S. Tao, Y. C. Zhou, C. S. Lee, S. T. Lee, D. Huang, X. H. Zhang, "Highly efficient nondoped blue organic light-emitting diodes based on anthracene-triphenylamine derivatives," *J. Phys. Chem. C*, 112:14603–14606 (2008).

560. I. Cho, S. H. Kim, J. H. Kim, S. Park, S. Y. Park, "Highly efficient and stable deep-blue emitting anthracene-derived molecular glass for versatile types of non-doped OLED applications," *J. Mater. Chem.*, 22:123–129 (2012).

561. C. H. Chien, C. K. Chen, F. M. Hsu, C. F. Shu, P. T. Chou, C. H. Lai, "Multifunctional deep-blue emitter comprising an anthracene core and terminal triphenylphosphine oxide groups," *Adv. Funct. Mater.*, 19:560–566 (2009).

562. J. H. Huang, J. H. Su, X. Li, M. K. Lam, K. M. Fung, H. H. Fan, K. W. Cheah, C. H. Chen, H. Tian, "Bipolar anthracene derivatives containing hole- and electron-transporting moieties for highly efficient blue electroluminescence devices," *J. Mater. Chem.*, 21:2957–2964 (2011).

563. M. A. Reddy, G. Mallesham, A. Thomas, K. Srinivas, V. J. Rao, K. Bhanuprakash, L. Giribabu, R. Grover, A. Kumar, M. N. Kamalasanan, R. Srivastava, "Synthesis and characterization of novel 2,5-diphenyl-1,3,4-oxadiazole derivatives of anthracene and its application as electron transporting blue emitters in OLEDs," *Synth. Met.*, 161:869–880 (2011).

564. M. R. Zhu, Q. A. Wang, Y. Gu, X. S. Cao, C. Zhong, D. G. Ma, J. G. Qin, C. L. Yang, "Efficient deep-blue emitters comprised of an anthracene core and terminal bifunctional groups for non-doped electroluminescence," *J. Mater. Chem.*, 21:6409–6415 (2011).

565. K. Kreger, M. Bate, C. Neuber, H. W. Schmidt, P. Strohriegl, "Combination development of blue OLEDs based on star shaped molecules," *Adv. Funct. Mater.*, 17:3456–3461 (2007).

566. Z. W. Liu, Z. Q. Bian, L. Ming, F. Ding, H. Y. Shen, D. B. Nie, C. H. Huang, "Green and blue–green phosphorescent heteroleptic iridium complexes containing carbazole-functionalized beta-diketonate for non-doped organic light-emitting diodes," *Org. Electron.*, 9:171–182 (2008).

567. S. H. Liao, J. R. Shiu, S. W. Liu, S. J. Yeh, Y. H. Chen, C. T. Chen, T. J. Chow, C. I. Wu, "Hydroxynaphthyridine-derived group III metal chelates: Wide band gap and deep blue analogues of green Alq$_3$ (tris(8-hydroxyquinolate)aluminum) and their versatile applications for organic light-emitting diodes," *J. Am. Chem. Soc.*, 131:763–777 (2009).

568. J. Kido, M. Kimura, K. Nagai, "Multilayer white light emitting organic electroluminescent device," *Science*, 267:1332–1334 (1995).

569. C. H. Jeong, J. T. Lim, M. S. Kim, J. H. Lee, J. W. Bae, G. Y. Yeom, "Four-wavelength white organic light-emitting diodes using 4,4′-bis- carbazoyl-(9)-stilbene as a deep blue emissive layer," *Org. Electron.*, 8:683–689 (2007).

570. J. H. Seo, J. H. Seo, J. H. Park, Y. K. Kim, J. H. Kim, G. W. Hyung, K. H. Lee, S. S. Yoon, "Highly efficient white organic light-emitting diodes using two emitting materials for three primary colors (red, green, and blue)," *Appl. Phys. Lett.*, 90:203507 (2007).

571. K. O. Cheon, J. Shinar, "Bright white small molecular organic light-emitting devices based on a red-emitting guest-host layer and blue-emitting 4,4(′)-bis(2,2(′)-diphenylvinyl)-1,1′-biphenyl," *Appl. Phys. Lett.*, 81:1738–1740 (2002).

572. L. D. Wang, L. Duan, G. T. Lei, Y. Qiu, "Bright white organic light-emitting diodes based on anthracene derivatives and rubrene," *Jpn. J. Appl. Phys. Pt. 2 Lett. Express Lett.*, 43:L560–L562 (2004).

573. S. Ye, J. Chen, C.-A. Di, Y. Liu, K. Lu, W. Wu, C. Du, Y. Liu, Z. Shuai, G. Yu, "Phenyl-substituted fluorene-dimer cored anthracene derivatives: Highly fluorescent and stable materials for high performance organic blue- and white-light-emitting diodes," *J. Mater. Chem.*, 20:3186–3194 (2010).

574. P. van Gemmern, V. van Elsbergen, S. P. Grabowski, H. Boerner, H.-P. Loebl, H. Becker, H. Kalisch, M. Heuken, R. H. Jansen, "Influence of carrier conductivity and injection on efficiency and chromaticity in small-molecule white organic light-emitting diodes based on 4,4′-bis(2,2′-diphenylvinyl)-1,1′-spirobiphenyl and rubrene," *J. Appl. Phys.*, 100:123707 (2006).

575. H. Choukri, A. Fischer, S. Forget, S. Chenais, M.-C. Castex, D. Ades, A. Siove, B. Geffroy, "White organic light-emitting diodes with fine chromaticity tuning via ultrathin layer position shifting," *Appl. Phys. Lett.*, 89:183513 (2006).

576. J. H. Jou, Y. S. Chiu, R. Y. Wang, H. C. Hu, C. P. Wang, H. W. Lin, "Efficient, color-stable fluorescent white organic light-emitting diodes with an effective exciton-confining device architecture," *Org. Electron.*, 7:8–15 (2006).

577. W.-C. Shen, Y.-K. Su, L.-W. Ji, "High bright white organic light-emitting diode based on mixing orange and blue emission," *J. Cryst. Growth*, 293:48–51 (2006).

578. Z. S. Su, G. B. Che, W. L. Li, W. M. Su, M. T. Li, B. Chu, B. Li, Z. Q. Zhang, Z. Z. Hu, "White-electrophosphorescent devices based on copper complexes using 2-(4-biphenylyl)-5-(4-*tert*-butyl-phenyl)-1,3,4-oxadiazole as chromaticity-tuning layer," *Appl. Phys. Lett.*, 88:213508 (2006).

579. S. Tao, Z. Peng, X. Zhang, S. Wu, "Efficient and stable single-dopant white OLEDs based on 9,10-bis (2-naphthyl) anthracene," *J. Lumin.*, 121:568–572 (2006).

580. Y. Duan, M. Mazzeo, V. Maiorano, F. Mariano, D. Qin, R. Cingolani, G. Gigli, "Extremely low voltage and high bright p–i–n fluorescent white organic light-emitting diodes," *Appl. Phys. Lett.*, 92:113304 (2008).

581. M.-H. Ho, S.-F. Hsu, J.-W. Ma, S.-W. Hwang, P.-C. Yeh, C. H. Chen, "White p–i–n organic light-emitting devices with high power efficiency and stable color," *Appl. Phys. Lett.*, 91:113518 (2007).

582. C.-H. Hsiao, C.-F. Lin, J.-H. Lee, "Driving voltage reduction in white organic light-emitting devices from selectively doping in ambipolar blue-emitting layer," *J. Appl. Phys.*, 102:094508 (2007).

583. J.-H. Jou, C.-P. Wang, M.-H. Wu, P.-H. Chiang, H.-W. Lin, H. C. Li, R.-S. Liu, "Efficient fluorescent white organic light-emitting diodes with blue–green host of di(4-fluorophenyl)amino-di(styryl)biphenyl," *Org. Electron.*, 8:29–36 (2007).

584. S. L. Lai, M. Y. Chan, M. K. Fung, C. S. Lee, S. T. Lee, "Efficiency enhancement and voltage reduction in white organic light-emitting devices," *Appl. Phys. Lett.*, 90:203510 (2007).

585. J.-H. Jou, C.-C. Chen, Y.-C. Chung, M.-T. Hsu, C.-H. Wu, S.-M. Shen, M.-H. Wu, W.-B. Wang, Y.-C. Tsai, C.-P. Wang, J.-J. Shyue, "Nanodot-enhanced high-efficiency pure-white organic light-emitting diodes with mixed-host structures," *Adv. Funct. Mater.*, 18:121–126 (2008).

586. Y. Wang, Y.-L. Hua, X.-M. Wu, L. Zhang, Q.-C. Hou, Q. Liu, "High-efficiency fluorescent white organic light-emitting device with double emissive layers," *Org. Electron.*, 9:273–278 (2008).

587. M.-F. Lin, L. Wang, W.-K. Wong, K.-W. Cheah, H.-L. Tam, M.-T. Lee, M.-H. Ho, C. H. Chen, "Highly efficient and stable white light organic light-emitting devices," *Appl. Phys. Lett.*, 91:073517 (2007).

588. Y.-C. Tsai, J.-H. Jou, "Long-lifetime, high-efficiency white organic light-emitting diodes with mixed host composing double emission layers," *Appl. Phys. Lett.*, 89:243521 (2006).

589. Y. G. Lee, I. S. Kee, H. S. Shim, I. H. Ko, S. Lee, K. H. Koh, "White organic light-emitting devices with mixed interfaces between light emitting layers," *Appl. Phys. Lett.*, 90:243508 (2007).

590. J.-H. Jou, M.-H. Wu, C.-P. Wang, Y.-S. Chiu, P.-H. Chiang, H.-C. Hu, R.-Y. Wang, "Efficient fluorescent white organic light-emitting diodes using cohost/emitter dual-role possessed di(triphenyl-amine)-1,4-divinyl-naphthalene," *Org. Electron.*, 8:735–742 (2007).

591. D. Joly, D. Tondelier, V. Deborde, W. Delaunay, A. Thomas, K. Bhanuprakash, B. Geffroy, M. Hissler, R. Reau, "White organic light-emitting diodes based on quench-resistant fluorescent organophosphorus dopants," *Adv. Funct. Mater.*, 22:567–576 (2012).

592. O. Fadhel, M. Gras, N. Lemaitre, V. Deborde, M. Hissler, B. Geffroy, R. Reau, "Tunable organophosphorus dopants for bright white organic light-emitting diodes with simple structures," *Adv. Mater.*, 21:1261–1265 (2009).

593. L. Duan, D. Zhang, K. Wu, X. Huang, L. Wang, Y. Qiu, "Controlling the recombination zone of white organic light-emitting diodes with extremely long lifetimes," *Adv. Funct. Mater.*, 21:3540–3545 (2011).

594. J.-H. Jou, M.-H. Wu, S.-M. Shen, H.-C. Wang, S.-Z. Chen, S.-H. Chen, C.-R. Lin, Y.-L. Hsieh, "Sunlight-style color-temperature tunable organic light-emitting diode," *Appl. Phys. Lett.*, 95:013307 (2009).

595. H. Yang, Y. Zhao, J. Hou, S. Liu, "White organic light-emitting devices with non-doped-type structure," *Displays*, 27:183–186 (2006).

596. S. J. Yeh, H. Y. Chen, M. F. Wu, L. H. Chan, C. L. Chiang, H. C. Yeh, C. T. Chen, J. H. Lee, "All non-dopant red–green–blue composing white organic light-emitting diodes," *Org. Electron.*, 7:137–143 (2006).

597. S. Tao, C. S. Lee, S.-T. Lee, X. Zhang, "Efficient blue and white organic light-emitting devices based on a single bipolar emitter," *Appl. Phys. Lett.*, 9:013507 (2007).

598. Q.-X. Tong, S.-L. Lai, M.-Y. Chan, J.-X. Tang, H.-L. Kwong, C.-S. Lee, S.-T. Lee, "High-efficiency nondoped white organic light-emitting devices," *Appl. Phys. Lett.*, 91:023503 (2007).

599. H. Yang, Y. Shi, Y. Zhao, Y. Meng, W. Hu, J. Hou, S. Liu, "High colour rendering index white organic light-emitting devices with three emitting layers," *Displays*, 29:327–332 (2008).

600. Y. Zhao, J. Chen, D. Ma, "Realization of high efficiency orange and white organic light emitting diodes by introducing an ultra-thin undoped orange emitting layer," *Appl. Phys. Lett.*, 99:163303 (2011).

601. S. Chen, Z. Zhao, B. Z. Tang, H. S. Kwok, "Non-doped white organic light-emitting diodes based on aggregation-induced emission," *J. Phys. D Appl. Phys.*, 43:095101 (2010).

602. S. Chen, Z. Zhao, Z. Wang, P. Lu, Z. Gao, Y. Ma, B. Z. Tang, H.-S. Kwok, "Bi-layer non-doped small-molecular white organic light-emitting diodes with high colour stability," *J. Phys. D Appl. Phys.*, 44:145101 (2011).

603. Z. R. Hong, W. L. Li, D. X. Zhao, C. J. Liang, X. Y. Liu, J. B. Peng, D. Zhao, "White light emission from OEL devices based on organic dysprosium-complex," *Synth. Met.*, 111:43–45 (2000).

604. Y. Hamada, T. Sano, H. Fujii, Y. Nishio, H. Takahashi, K. Shibata, "White light emitting materials for organic electroluminescent devices," *Jpn. J. Appl. Phys.*, 35:L1339–L1341 (1996).

605. J. Y. Li, D. Liu, C. W. Ma, O. Lengyel, C. S. Lee, C. H. Tung, S. Lee, "White-light emission from a single-emitting-component organic electroluminescent device," *Adv. Mater.*, 16:1538–1541 (2004).

606. S. Tao, Y. Zhou, C.-S. Lee, S.-T. Lee, D. Huang, X. Zhang, "A triphenylamine derivative as a single-emitting component for highly-efficient white electroluminescent devices," *J. Mater. Chem.*, 18:3981–3984 (2008).

607. S. H. Kim, S. Park, J. E. Kwon, S. Y. Park, "Organic light-emitting diodes with a white-emitting molecule: Emission mechanism and device characteristics," *Adv. Funct. Mater.*, 21:644–651 (2011).

608. Y. Liu, J. H. Guo, H. D. Zhang, Y. Wang, "Highly efficient white organic electroluminescence from a double-layer device based on a boron hydroxyphenylpyridine complex," *Angew. Chem. Int. Ed.*, 41:182–184 (2002).

609. J. Kalinowski, M. Cocchi, D. Virgili, V. Fattori, J. A. G. Williams, "Mixing of excimer and exciplex emission: A new way to improve white light emitting organic electrophosphorescent diodes," *Adv. Mater.*, 19:4000–4005 (2007).

610. Y. M. Kim, Y. W. Park, J. H. Choi, B. K. Ju, J. H. Jung, J. K. Kim, "Spectral broadening in electroluminescence of white organic light-emitting diodes based on complementary colors," *Appl. Phys. Lett.*, 90:033506 (2007).

611. C. Li, H. Kimura, T. Saito, K. Sakurai, B. Wei, M. Ichikawa, Y. Taniguchi, "Study of color-conversion-materials in chromatic-stability white organic light-emitting diodes," *Opt. Express*, 15:14422–14430 (2007).

612. B. W. D'Andrade, R. J. Holmes, S. R. Forrest, "Efficient organic electrophosphorescent white-light-emitting device with a triple doped emissive layer," *Adv. Mater.*, 16:624–628 (2004).

613. B. W. D'Andrade, M. E. Thompson, S. R. Forrest, "Controlling exciton diffusion in multilayer white phosphorescent organic light emitting devices," *Adv. Mater.*, 14:147–151 (2002).

614. Y. Sun, S. R. Forrest, "Multiple exciton generation regions in phosphorescent white organic light emitting devices," *Org. Electron.*, 9:994–1001 (2008).

615. Y.-L. Chang, S. Yin, Z. Wang, M. G. Helander, J. Qiu, L. Chai, Z. Liu, G. D. Scholes, Z. Lu, "Highly efficient warm white organic light-emitting diodes by triplet exciton conversion," *Adv. Funct. Mater.*, 23:705–712 (2012).

616. R. Song, Y. Duan, S. Chen, Y. Zhao, J. Hou, S. Liu, "White organic light-emitting devices employing phosphorescent iridium complex as RGB dopants," *Semicond. Sci. Technol.*, 22:728–731 (2007).

617. G. Cheng, Y. Zhang, Y. Zhao, Y. Lin, C. Ruan, S. Liu, T. Fei, Y. Ma, Y. Cheng, "White organic light-emitting devices with a phosphorescent multiple emissive layer," *Appl. Phys. Lett.*, 89:043504 (2006).

618. Y. Sun, S. R. Forrest, "High-efficiency white organic light emitting devices with three separate phosphorescent emission layers," *Appl. Phys. Lett.*, 91:263503 (2007).

619. S. Reineke, F. Lindner, G. Schwartz, N. Seidler, K. Walzer, B. Luessem, K. Leo, "White organic light-emitting diodes with fluorescent tube efficiency," *Nature*, 459:234–239 (2009).

620. H. Sasabe, J.-I. Takamatsu, T. Motoyama, S. Watanabe, G. Wagenblast, N. Langer, O. Molt, E. Fuchs, C. Lennartz, J. Kido, "High-efficiency blue and white organic light-emitting devices incorporating a blue iridium carbene complex," *Adv. Mater.*, 22:5003–5007 (2010).

621. S. Chen, G. Tan, W.-Y. Wong, H.-S. Kwok, "White organic light-emitting diodes with evenly separated red, green, and blue colors for efficiency/color-rendition trade-off optimization," *Adv. Funct. Mater.*, 21:3785–3793 (2011).

622. C. Weichsel, S. Reineke, M. Furno, B. Luesem, K. Leo, "Organic light-emitting diodes for lighting: High color quality by controlling energy transfer processes in host–guest–systems," *J. Appl. Phys.*, 111:033102 (2012).

623. C.-H. Hsiao, Y.-H. Lan, P.-Y. Lee, T.-L. Chiu, J.-H. Lee, "White organic light-emitting devices with ultra-high color stability over wide luminance range," *Org. Electron.*, 12:547–555 (2011).

624. C.-H. Hsiao, S.-W. Liu, C.-T. Chen, J.-H. Lee, "Emitting layer thickness dependence of color stability in phosphorescent organic light-emitting devices," *Org. Electron.*, 11:1500–1506 (2010).

625. J.-H. Jou, Y.-C. Chou, S.-M. Shen, M.-H. Wu, P.-S. Wu, C.-R. Lin, R.-Z. Wu, S.-H. Chen, M.-K. Wei, C.-W. Wang, "High-efficiency, very-high color rendering white organic light-emitting diode with a high triplet interlayer," *J. Mater. Chem.*, 21:18523–18526 (2011).

626. J.-H. Jou, S.-M. Shen, C.-R. Lin, Y.-S. Wang, Y.-C. Chou, S.-Z. Chen, Y.-C. Jou, "Efficient very-high color rendering index organic light-emitting diode," *Org. Electron.*, 12:865–868 (2011).

627. S. Reineke, G. Schwartz, K. Walzer, K. Leo, "Reduced efficiency roll-off in phosphorescent organic light emitting diodes by suppression of triplet–triplet annihilation," *Appl. Phys. Lett.*, 91:123508 (2007).

628. G. Schwartz, M. Pfeiffer, S. Reineke, K. Walzer, K. Leo, "Harvesting triplet excitons from fluorescent blue emitters in white organic light-emitting diodes," *Adv. Mater.*, 19:3672–3676 (2007).

629. S. J. Su, E. Gonmori, H. Sasabe, J. Kido, "Highly efficient organic blue- and white-light-emitting devices having a carrier- and exciton-confining structure for reduced efficiency roll-off," *Adv. Mater.*, 20:4189–4194 (2008).

630. W.-Y. Hung, L.-C. Chi, W.-J. Chen, Y.-M. Chen, S.-H. Chou, K.-T. Wong, "A new benzimidazole/carbazole hybrid bipolar material for highly efficient deep-blue electrofluorescence, yellow–green electrophosphorescence, and two-color-based white OLEDs," *J. Mater. Chem.*, 20:10113–10119 (2010).

631. W.-Y. Hung, L.-C. Chi, W.-J. Chen, E. Mondal, S.-H. Chou, K.-T. Wong, Y. Chi, "A carbazole–phenylbenzimidazole hybrid bipolar universal host for high efficiency RGB and white PhOLEDs with high chromatic stability," *J. Mater. Chem.*, 21:19249–19256 (2011).

632. W.-Y. Hung, Z.-W. Chen, H.-W. You, F.-C. Fan, H.-F. Chen, K.-T. Wong, "Efficient carrier- and exciton-confining device structure that enhances blue PhOLED efficiency and reduces efficiency roll-off," *Org. Electron.*, 12:575–581 (2011).

633. S. Seidel, R. Krause, A. Hunze, G. Schmid, F. Kozlowski, T. Dobbertin, A. Winnacker, "High efficient two color white phosphorescent organic light emitting diode," *J. Appl. Phys.*, 104:064505 (2008).

634. M.-T. Lee, J.-S. Lin, M.-T. Chu, M.-R. Tseng, "Improvement in carrier transport and recombination of white phosphorescent organic light-emitting devices using a composite blue emitter," *Appl. Phys. Lett.*, 93:133306 (2008).

635. J. Lee, J.-I. Lee, J. Y. Lee, H. Y. Chu, "Enhanced efficiency and reduced roll-off in blue and white phosphorescent organic light-emitting diodes with a mixed host structure," *Appl. Phys. Lett.*, 94:193305 (2009).

636. J. Lee, J.-I. Lee, H. Y. Chu, "Efficient and color stable phosphorescent white organic light-emitting devices based on an ultra wide band-gap host," *Synth. Met.*, 159:991–994 (2009).

637. X. Yang, H. Huang, B. Pan, M. P. Aldred, S. Zhuang, L. Wang, J. Chen, D. Ma, "Modified 4,4',4"-tri(*N*-carbazolyl)triphenylamine as a versatile bipolar host for highly efficient blue, orange, and white organic light-emitting diodes," *J. Phys. Chem. C*, 116:15041–15047 (2012).

638. C. Han, Z. Zhang, H. Xu, S. Yue, J. Li, P. Yan, Z. Deng, Y. Zhao, P. Yan, S. Liu, "Short-axis substitution approach selectively optimizes electrical properties of dibenzothiophene-based phosphine oxide hosts," *J. Am. Chem. Soc.*, 134:19179–19188 (2012).

639. C. Han, G. Xie, H. Xu, Z. Zhang, L. Xie, Y. Zhao, S. Liu, W. Huang, "A single phosphine oxide host for high-efficiency white organic light-emitting diodes with extremely low operating voltages and reduced efficiency roll-off," *Adv. Mater.*, 23:2491–2496 (2011).

640. C. W. Seo, J. Y. Lee, "High efficiency in two color and three color phosphorescent white organic light-emitting diodes using a 2,7-substituted 9-phenylcarbazole derivative as the host material," *Org. Electron.*, 12:1459–1464 (2011).

641. C. W. Seo, J. Y. Lee, "Highly efficient white phosphorescent organic light emitting diodes using a mixed host structure in deep blue emitting layer," *Thin Solid Films*, 520:5075–5079 (2012).

642. S. Gong, Q. Fu, Q. Wang, C. Yang, C. Zhong, J. Qin, D. Ma, "Highly efficient deep-blue electrophosphorescence enabled by solution-processed bipolar tetraarylsilane host with both a high triplet energy and a high-lying HOMO level," *Adv. Mater.*, 23:4956–4959 (2011).

643. S. Gong, Y. Chen, X. Zhang, P. Cai, C. Zhong, D. Ma, J. Qin, C. Yang, "High-performance blue and green electrophosphorescence achieved by using carbazole-containing bipolar tetraarylsilanes as host materials," *J. Mater. Chem.*, 21:11197 (2011).

644. S. L. Gong, C. L. Yang, J. G. Qin, "Efficient phosphorescent polymer light-emitting diodes by suppressing triplet energy back transfer," *Chem. Soc. Rev.*, 41:4797–4807 (2012).

645. Y. Yin, X. Piao, Y. Li, Y. Wang, J. Liu, K. Xu, W. Xie, "High-efficiency and low-efficiency-roll-off single-layer white organic light-emitting devices with a bipolar transport host," *Appl. Phys. Lett.*, 101:063306 (2012).

646. F. M. Hsu, C. H. Chien, C. F. Shu, C. H. Lai, C. C. Hsieh, K. W. Wang, P. T. Chou, "A bipolar host material containing triphenylamine and diphenylphosphoryl-substituted fluorene units for highly efficient blue electrophosphorescence," *Adv. Funct. Mater.*, 19:2834–2843 (2009).

647. K. S. Yook, S. O. Jeon, C. W. Joo, J. Y. Lee, M. S. Kim, H. S. Choi, S. J. Lee, C.-W. Han, Y. H. Tak, "Highly efficient pure white phosphorescent organic light-emitting diodes using a deep blue phosphorescent emitting material," *Org. Electron.*, 10:681–685 (2009).

648. K. S. Yook, S. O. Jeon, J. Y. Lee, K. H. Lee, Y. S. Kwon, S. S. Yoon, J. H. Yoon, "High efficiency pure white organic light-emitting diodes using a diphenylaminofluorene-based blue fluorescent material," *Org. Electron.*, 10:1378–1381 (2009).

649. K. S. Yook, S. O. Jeon, J. Y. Lee, "Pure white phosphorescent organic light-emitting diodes using a phosphine oxide derivative as a high triplet energy host material," *Thin Solid Films*, 518:5827–5831 (2010).

650. K. S. Yook, S. O. Jeon, C. W. Joo, J. Y. Lee, "High efficiency phosphorescent white organic light-emitting diodes using a spirofluorene based phosphine oxide host material," *Thin Solid Films*, 518:4462–4466 (2010).

651. Q. Wang, J. Ding, D. Ma, Y. Cheng, L. Wang, X. Jing, F. Wang, "Harvesting excitons via two parallel channels for efficient white organic LEDs with nearly 100% internal quantum efficiency: Fabrication and emission–mechanism analysis," *Adv. Funct. Mater.*, 19:84–95 (2009).

652. Q. Wang, J. Ding, D. Ma, Y. Cheng, L. Wang, "Highly efficient single-emitting-layer white organic light-emitting diodes with reduced efficiency roll-off," *Appl. Phys. Lett.*, 94:103503 (2009).

653. R. Wang, D. Liu, H. Ren, T. Zhang, H. Yin, G. Liu, J. Li, "Highly efficient orange and white organic light-emitting diodes based on new orange iridium complexes," *Adv. Mater.*, 23:2823–2827 (2011).

654. R. Wang, D. Liu, R. Zhang, L. Deng, J. Li, "Solution-processable iridium complexes for efficient orange–red and white organic light-emitting diodes," *J. Mater. Chem.*, 22:1411 (2012).

655. C.-H. Chang, Y.-H. Lin, C.-C. Chen, C.-K. Chang, C.-C. Wu, L.-S. Chen, W.-W. Wu, Y. Chi, "Efficient phosphorescent white organic light-emitting devices incorporating blue iridium complex and multifunctional orange–red osmium complex," *Org. Electron.*, 10:1235–1240 (2009).

656. C.-H. Chang, C.-C. Chen, C.-C. Wu, S.-Y. Chang, J.-Y. Hung, Y. Chi, "High-color-rendering pure-white phosphorescent organic light-emitting devices employing only two complementary colors," *Org. Electron.*, 11:266–272 (2010).

657. C.-H. Chang, K.-C. Tien, C.-C. Chen, M.-S. Lin, H.-C. Cheng, S.-H. Liu, C.-C. Wu, J.-Y. Hung, Y.-C. Chiu, Y. Chi, "Efficient phosphorescent white OLEDs with high color rendering capability," *Org. Electron.*, 11:412–418 (2010).

658. C.-H. Chang, Y.-S. Ding, P.-W. Hsieh, C.-P. Chang, W.-C. Lin, H.-H. Chang, "Highly efficient phosphorescent blue and white organic light-emitting devices with simplified architectures," *Thin Solid Films*, 519:7992–7997 (2011).

659. S.-L. Lai, S.-L. Tao, M.-Y. Chan, M.-F. Lo, T.-W. Ng, S.-T. Lee, W.-M. Zhao, C.-S. Lee, "Iridium(III) bis 2-(2-naphthyl)pyridine (acetylacetonate)-based yellow and white phosphorescent organic light-emitting devices," *J. Mater. Chem.*, 21:4983–4988 (2011).

660. S. L. Lai, S. L. Tao, M. Y. Chan, T. W. Ng, M. F. Lo, C. S. Lee, X. H. Zhang, S. T. Lee, "Efficient white organic light-emitting devices based on phosphorescent iridium complexes," *Org. Electron.*, 11:1511–1515 (2010).

661. M.-T. Lee, M.-T. Chu, J.-S. Lin, M.-R. Tseng, "Host-free, yellow phosphorescent material in white organic light-emitting diodes," *J. Phys. D Appl. Phys.*, 43:442003 (2010).
662. J. Zhao, J. Yu, X. Hu, M. Hou, Y. Jiang, "High-efficiency white organic light-emitting devices with a non-doped yellow phosphorescent emissive layer," *Thin Solid Films*, 520:4003–4007 (2012).
663. J. Zhao, J. Yu, L. Zhang, J. Wang, "Non-doped phosphorescent white organic light-emitting devices with a quadruple-quantum-well structure," *Phys. B Condens. Matter*, 407:2753–2757 (2012).
664. B. W. D'Andrade, J. Brooks, V. Adamovich, M. E. Thompson, S. R. Forrest, "White light emission using triplet excimers in electrophosphorescent organic light-emitting devices," *Adv. Mater.*, 14:1032–1036 (2002).
665. V. Adamovich, J. Brooks, A. Tamayo, A. M. Alexander, P. I. Djurovich, B. W. D'Andrade, C. Adachi, S. R. Forrest, M. E. Thompson, "High efficiency single dopant white electrophosphorescent light emitting diodes," *New J. Chem.*, 26:1171–1178 (2002).
666. M. Li, W.-H. Chen, M.-T. Lin, M. A. Omary, N. D. Shepherd, "Near-white and tunable electrophosphorescence from bis 3,5-bis(2-pyridyl)-1,2,4-triazolato platinum(II)-based organic light emitting diodes," *Org. Electron.*, 10:863–870 (2009).
667. U. S. Bhansali, H. Jia, I. W. H. Oswald, M. A. Omary, B. E. Gnade, "High efficiency warm-white organic light emitting diodes from a single emitter in graded-doping device architecture," *Appl. Phys. Lett.*, 100:183305 (2012).
668. X. Yang, Z. Wang, S. Madakuni, J. Li, G. E. Jabbour, "Efficient blue- and white-emitting electrophosphorescent devices based on platinum(II) 1,3-difluoro-4,6-di(2-pyridinyl)benzene chloride," *Adv. Mater.*, 20:2405–2409 (2008).
669. Y. R. Sun, N. C. Giebink, H. Kanno, B. W. Ma, M. E. Thompson, S. R. Forrest, "Management of singlet and triplet excitons for efficient white organic light-emitting devices," *Nature*, 440:908–912 (2006).
670. U. S. Bhansali, H. Jia, M. A. Quevedo Lopez, B. E. Gnade, W.-H. Chen, M. A. Omary, "Controlling the carrier recombination zone for improved color stability in a two-dopant fluorophore/phosphor white organic light-emitting diode," *Appl. Phys. Lett.*, 94:203501 (2009).
671. H.-I. Baek, C. Lee, "Simple white organic light emitting diodes with improved color stability and efficiency using phosphorescent and fluorescent emitters," *J. Appl. Phys.*, 103:124504 (2008).
672. G. Lei, X. Chen, L. Wang, M. Zhu, W. Zhu, L. Wang, Y. Qiu, "White organic light emitting devices with hybrid emissive layers combining phosphorescence and fluorescence," *J. Phys. D Appl. Phys.*, 41:105114 (2008).
673. W. F. Xie, S. L. Chew, C. S. Lee, S. T. Lee, P. F. Wang, H. L. Kwong, "High-efficiency white organic light-emitting devices using a blue iridium complex to sensitize a red fluorescent dye," *J. Appl. Phys.*, 100:096114 (2006).
674. C.-H. Chang, Y.-J. Lu, C.-C. Liu, Y.-H. Yeh, C.-C. Wu, "Efficient white OLEDs employing phosphorescent sensitization," *J. Display Technol.*, 3:193–199 (2007).
675. G. Schwartz, K. Fehse, M. Pfeiffer, K. Walzer, K. Leo, "Highly efficient white organic light emitting diodes comprising an interlayer to separate fluorescent and phosphorescent regions," *Appl. Phys. Lett.*, 89:083509 (2006).
676. G. Schwartz, T.-H. Ke, C.-C. Wu, K. Walzer, K. Leoc, "Balanced ambipolar charge carrier mobility in mixed layers for application in hybrid white organic light-emitting diodes," *Appl. Phys. Lett.*, 93:073304 (2008).
677. P. Chen, W. Xie, J. Li, T. Guan, Y. Duan, Y. Zhao, S. Liu, C. Ma, L. Zhang, B. Li, "White organic light-emitting devices with a bipolar transport layer between blue fluorescent and orange phosphorescent emitting layers," *Appl. Phys. Lett.*, 91:023505 (2007).
678. D. Wang, W. L. Li, Z. S. Su, T. L. Li, B. Chu, D. F. Bi, L. L. Chen, W. M. Su, H. He, "Broad wavelength modulating and design of organic white diode based on lighting by using exciplex emission from mixed acceptors," *Appl. Phys. Lett.*, 89:033511 (2006).
679. J. Wang, J. Yu, L. Li, X. Tang, Y. Jiang, "Efficient white organic light-emitting devices using a thin 4,4'-bis(2,2'-diphenylvinyl)-1,1'-diphenyl layer," *J. Phys. D Appl. Phys.*, 41:045104 (2008).

680. B. Zhao, Z. Su, W. Li, B. Chu, F. Jin, X. Yan, F. Zhang, D. Fan, T. Zhang, Y. Gao, J. Wang, "High efficient white organic light-emitting diodes based on triplet multiple quantum well structure," *Appl. Phys. Lett.*, 101:053310 (2012).

681. F. Zhao, N. Sun, H. Zhang, J. Chen, D. Ma, "Hybrid white organic light-emitting diodes with a double light-emitting layer structure for high color-rendering index," *J. Appl. Phys.*, 112:084504 (2012).

682. F. Zhao, Z. Zhang, Y. Liu, Y. Dai, J. Chen, D. Ma, "A hybrid white organic light-emitting diode with stable color and reduced efficiency roll-off by using a bipolar charge carrier switch," *Org. Electron.*, 13:1049–1055 (2012).

683. J. Wan, C.-J. Zheng, M.-K. Fung, X.-K. Liu, C.-S. Lee, X.-H. Zhang, "Multifunctional electron-transporting indolizine derivatives for highly efficient blue fluorescence, orange phosphorescence host and two-color based white OLEDs," *J. Mater. Chem.*, 22:4502–4510 (2012).

684. X. Piao, Y. Yin, J. Liu, Y. Li, K. Xu, Y. Wang, W. Xie, "High-efficiency blue and white organic light-emitting devices by combining fluorescent and phosphorescent blue emitters," *Org. Electron.*, 13:2412–2416 (2012).

685. C.-H. Chen, W.-S. Huang, M.-Y. Lai, W.-C. Tsao, J. T. Lin, Y.-H. Wu, T.-H. Ke, L.-Y. Chen, C.-C. Wu, "Versatile, benzimidazole/amine-based ambipolar compounds for electroluminescent applications: Single-layer, blue, fluorescent OLEDs, hosts for single-layer, phosphorescent OLEDs," *Adv. Funct. Mater.*, 19:2661–2670 (2009).

686. D. S. Qin, Y. Tao, "White organic light-emitting diode comprising of blue fluorescence and red phosphorescence," *Appl. Phys. Lett.*, 86:113507 (2005).

687. L. S. Liao, K. P. Klubek, C. W. Tang, "High-efficiency tandem organic light-emitting diodes," *Appl. Phys. Lett.*, 84:167–169 (2004).

688. A. R. Duggal, J. J. Shiang, C. M. Heller, D. F. Foust, "Organic light-emitting devices for illumination quality white light," *Appl. Phys. Lett.*, 80:3470–3472 (2002).

689. A. Dodabalapur, L. J. Rothberg, T. M. Miller, "Color variation with electroluminescent organic semiconductors in multimode resonant cavities," *Appl. Phys. Lett.*, 65:2308–2310 (1994).

690. M. A. Khan, W. Xu, J. Cao, Y. Bai, W. Q. Zhu, X. Y. Jiang, Z. L. Zhang, "Spectral studies of white organic light-emitting devices based on multi-emitting layers," *Displays*, 28:26–30 (2007).

691. M.-H. Ho, T.-M. Chen, P.-C. Yeh, S.-W. Hwang, C. H. Chen, "Highly efficient p–i–n white organic light emitting devices with tandem structure," *Appl. Phys. Lett.*, 91:233507 (2007).

692. P. Chen, Q. Xue, W. Xie, Y. Duan, G. Xie, Y. Zhao, J. Hou, S. Liu, L. Zhang, B. Li, "Color-stable and efficient stacked white organic light-emitting devices comprising blue fluorescent and orange phosphorescent emissive units," *Appl. Phys. Lett.*, 93:153508 (2008).

693. S. H. Kim, J. Jang, J. Y. Lee, "Improved color stability in white phosphorescent organic light-emitting diodes using charge confining structure without interlayer," *Appl. Phys. Lett.*, 91:123509 (2007).

694. T.-W. Lee, T. Noh, B.-K. Choi, M.-S. Kim, D. W. Shin, J. Kido, "High-efficiency stacked white organic light-emitting diodes," *Appl. Phys. Lett.*, 92:043301 (2008).

695. P. Chen, Q. Xue, W. Xie, G. Xie, Y. Duan, Y. Zhao, S. Liu, L. Zhang, B. Li, "Influence of inter-layer on the performance of stacked white organic light-emitting devices," *Appl. Phys. Lett.*, 95:123307 (2009).

696. Y. Chen, J. Chen, D. Ma, D. Yan, L. Wang, "Tandem white phosphorescent organic light-emitting diodes based on interface-modified C-60/pentacene organic heterojunction as charge generation layer," *Appl. Phys. Lett.*, 99:103304 (2011).

697. Y. Chen, H. Tian, J. Chen, Y. Geng, D. Yan, L. Wang, D. Ma, "Highly efficient tandem white organic light-emitting diodes based upon C-60/NaT$_4$ organic heterojunction as charge generation layer," *J. Mater. Chem.*, 22:8492–8498 (2012).

698. Q. Wang, J. Ding, Z. Zhang, D. Ma, Y. Cheng, L. Wang, F. Wang, "A high-performance tandem white organic light-emitting diode combining highly effective white-units and their interconnection layer," *J. Appl. Phys.*, 105:076101 (2009).

699. Q. Wang, Y. Chen, J. Chen, D. Ma, "White top-emitting organic light-emitting diodes employing tandem structure," *Appl. Phys. Lett.*, 101:133302 (2012).

700. B. W. D'Andrade, S. R. Forrest, "White organic light-emitting devices for solid-state lighting," *Adv. Mater.*, 16:1585–1595 (2004).

4

Phosphorescent Polymer Light-Emitting Diodes

Xu Han, Dmitrii F. Perepichka, Hong Meng, and Mang-Mang Ling

CONTENTS

4.1 Introduction

Light-emitting diodes (LEDs) based on luminescent small organic molecules (OLEDs) and polymers (PLEDs) have become one of the major areas in the science of organic materials. A number of commercial OLED-based displays have been in the market for a couple of decades, and commercialization has already been initiated for PLED displays. Despite the technological problems of efficiency and stability of PLEDs as compared with OLEDs (which is, first of all, a material purity issue), the former promises to revolutionize the display-manufacturing technology as it provides the possibility of inexpensive solution fabrication (see also Chapter 2). Indeed, ambient temperature and pressure fabrication conditions (spin coating, bar coating, ink-jet printing, etc., of PLED-based large-area screens), enabled by good film-forming properties of polymers, are particularly attractive for industrial applications. However, most of the PLEDs show an external quantum efficiency (EQE) of not more than 15%, which limits the achieved power efficiency below ~40 lm/W (compared with ~90 lm/W for the best inorganic light-emitting devices) [1]. Besides the energy consumption issue, low efficiency also poses a problem of heat dissipation, which affects device stability. Tailoring the efficiency of OLEDs and PLEDs is a very complex material and device-engineering task; however, one of the major improvement potentials follows from a simple consideration of the basic photophysics.

The principle of operation of all LEDs is based on radiative decay of excited states (excitons) created in the molecule (polymer, inorganic material) by injecting electrons and holes in the lowest unoccupied molecular orbital (LUMO) and highest occupied molecular orbital (HOMO), respectively. As these electrons and holes are generally injected with random spins, their recombination would give both symmetric (triplet) and antisymmetric (singlet) states in the ratio of 3:1 (based on statistical consideration). Only one of these, a singlet, has an allowed transition to the ground state, which means that only 25% of the created excitons can produce light in typical fluorescent materials. Therefore, to increase the efficiency of a LED, one would need either to control the spin of the injected electrons and holes or to find a way to harvest the formed triplet excitons. There is a limited understanding on how the spin control can be achieved [2]. Recently, Uoyama and coworkers [3] reported a novel pathway to attain the greatest possible electroluminescence (EL) efficiency from simple aromatic compounds that exhibit efficient thermal-activated delayed fluorescence with high photoluminescence (PL) efficiency. The energy gap between the singlet and triplet excited states of these metal-free organic molecules is minimized, therefore promoting highly efficient spin up-conversion from nonradiative triplet states to radiative singlet states while maintaining high radiative decay rates. Alternatively, radiative decay of triplet states, known as phosphorescence, is known to be an efficient process in many heavy-metal complexes with efficient spin-orbital coupling.

At the same time, the early reports on OLED devices fabricated with Eu^{3+} complexes did not discuss the triplet-harvesting issue and also showed rather low quantum efficiency, owing to the low solid-state PL quantum yield of these emitters [4,5]. Only in 1998, Baldo and coworkers [6] reported phosphorescent OLEDs with very high efficiency using energy transfer from the fluorescent organic host material (Alq_3) to the triplet-emitting dopant (Pt–porphine complex), which breaks the 25% theoretical efficiency limit of the electrofluorescent emission. Since then, a number of high-efficiency electrophosphorescent OLEDs have been fabricated (and already been commercialized), where a nearly 100% internal quantum efficiency has been achieved [7].

There is no reason why the same principle cannot be applied for light-emitting polymers as host materials to pave the way to high-efficiency solution-processable LEDs. In fact, polymer-based electrophosphorescent LEDs (PPLEDs) based on polymer fluorescent hosts and transition metal complexes were reported only a year after the phosphorescent OLED was reported [8]. In spite of a relatively limited research activity in PPLEDs, as compared with phosphorescent OLEDs, it is hoped that 100% internal quantum efficiency can also be achieved for polymer LEDs. In this chapter, we will give a brief description of the photophysics beyond the operation of electrophosphorescent devices, followed by examples of the materials, devices, and processes experimentally studied in the field till the beginning of 2013.

4.2 Photophysical Aspects of Electrophosphorescence

4.2.1 Singlet–Triplet Excitons Ratio

Before describing the phosphorescent light emission in PLEDs, we will briefly review the relevant basic photophysical principles. The Pauli exclusion principle states that no two electrons in an atom can have identical quantum numbers. A simple quantum mechanical calculation can show that the total spin of two electrons has a value of either 0, where the spin state is antisymmetric (called a singlet) under electron exchange, or 1, where there are three possible symmetric spin states under electron exchange (called triplets) [9]. The optical transitions are allowed only from singlet to singlet and from triplet to triplet, but not between singlet and triplet states.

During photoexcitation, all the formed excitons are singlets. The radiative decay back to a singlet ground state is allowed and the PL quantum yield of 100% is observed in many molecules. In an EL process, however, the injection of electrons from the cathode and holes from the anode in the device is generally spin independent and random. Their recombination results in the formation of four possible spin states (usually with equal cross section), one singlet and three triplets. As the radiative transition from the triplet excited state to the singlet ground state is prohibited, the theoretical maximum quantum yield of electrofluorescence is limited to 25%. Having said this, we should mention that in the last few years, an increasing number of theoretical [10,11] and experimental [12–16] studies suggest that the singlet exciton yield (h_S) in some materials such as conjugated polymers can be significantly higher than 25% owing to the spin-dependent exciton formation process [17]. Vardeny and coworkers [13] demonstrated a nonmonotonic dependence between the ratio of singlet–triplet exciton formation cross sections (s_S/s_T) and the band gap of conjugated polymers. The minimum value of s_S/s_T (~2) was reported for the polymer with the band gap around 2.5 eV, and this can be increased to as high as 5 for polymers with either higher or lower band gap.

There is a certain amount of controversy in some of these reports; for example, a h_S from 20 ± 4% [18] to ~45% [13] was reported for thin films of poly(2-methoxy-5-(2'-ethylhexyloxy)–1,4-phenylene vinylene) (MEH–PPV). For small molecules, however, a h_S close to 25% is invariably observed [13,19]. Whatever is the case, a significant amount of triplets is formed in PLEDs, and harvesting these can result in a tremendous increase of device efficiency. The triplet emission in organic polymers (as well as small molecules) is prohibited and usually cannot be observed at room temperature. However, many transition metals show high-yield triplet emission (phosphorescence) owing to the effective intersystem crossing enabled by strong spin–orbital interaction. Therefore, transferring the formed triplets to triplet-emitting dopants can potentially increase the internal quantum efficiency of the EL devices to 100%. A practical realization of this goal for PLEDs is given in the following sections; however, before that, we would like to describe briefly the basic principles of the exciton transfer processes relevant to the operation of phosphorescent LEDs.

4.2.2 Exciton Transfer Processes

The electronic processes in the host–guest molecular system are best illustrated by a classical Jablonski diagram [20], which was first proposed in 1933 to describe the absorption and emission of light (Figure 4.1).

When a host molecule in its ground state $\left(S_0^H\right)$ absorbs light, it is excited to a higher-energy singlet state $\left(S_1^H\right)$. At this point, it can directly release the absorbed energy via radiative decay, generating a photon with energy equal to $S_1^H - S_0^H$, or the molecule could be subject to collisions with the surrounding molecules and undergo nonradiative decay processes to the ground state (or even decompose). In the presence of the guest molecules, energy transfer processes may occur, which brings the guest molecule to its first excited state and returns the host to the ground state. This energy transfer may occur via Förster,

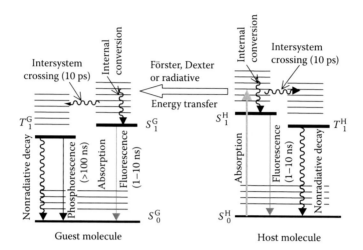

FIGURE 4.1
Electronic processes in fluorescent host–phosphorescent guest molecular systems. S_0^H, S_0^G: singlet ground state of host and guest molecules, respectively. S_1^H, S_1^G: first excited singlet state of host and guest molecules, respectively. T_1^H, T_1^G: first excited triplet state of host and guest molecules, respectively.

Dexter, or radiative energy transfer mechanisms, as will be described later. Then, radiative decay processes will occur in the guest molecules. In the real host–guest systems, one can observe the emission of the host molecule, the guest molecule, or a combination of both, depending on the efficiency of the energy transfer processes.

As these excited states may transfer from molecule to molecule while conserving their spin, one can treat them as quasiparticles called excitons. The highly localized excited states with radius of a few angstroms are known as Frenkel excitons. The rate of exciton transfer (hopping) relates to the electron transfer rate. According to the Marcus equation, the rate of electron transfer is in exponential relation with the change in free energy (driving force) [21,22]:

$$k = k_{max} \exp\left(-\frac{(\Delta G_0 + \lambda)^2}{4\lambda k_B T}\right) \tag{4.1}$$

where k_{max} is an electronic coupling coefficient, ΔG_0 is the change in free energy, λ is the reorganization energy depending on the distance between molecules, k_B is the Boltzmann constant, and T is the temperature.

In most cases, when the driving force is lower than the reorganization energy, the electron transfer rate increases when the driving force $|\Delta G_0|$ is increased. There is, however, the so-called "Marcus inversion region" $(-\Delta G_0 > \lambda)$, such that the larger the driving force, the lower the electron transfer rate.

The principles of the Marcus theory are most directly related to the Dexter energy transfer process, which is based on direct electron exchange between the host and the guest (Figure 4.2) [23]. Both the HOMO and the LUMO of the guest must be within the HOMO–LUMO gap of the host molecule to allow such electron transfer processes, and its efficiency depends on the driving force (determined by the difference in HOMO and LUMO energies of the host and the guest). The rate constant of the Dexter energy transfer decreases exponentially with the distance between the host and the guest molecules:

$$k = K \cdot J \cdot \exp\left(-2r/L\right) \tag{4.2}$$

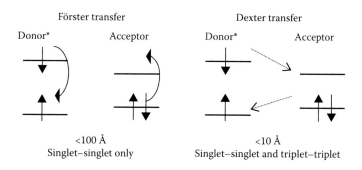

FIGURE 4.2
Comparison of Förster and Dexter energy transfer processes.

where r is the distance, L is the characteristic length on the order of 1–5 Å at which the transfer is effective, K is a prefactor related to electron exchange, and J is the spectral overlap integral.

Therefore, Dexter energy transfer is a short-range process that is significant only at distances <10 Å. As this can be logically concluded from Figure 4.2, the Dexter mechanism allows both singlet–singlet and triplet–triplet energy transfers.

Another major energy transfer process, the so-called Förster transfer mechanism, is based on a dipole–dipole interaction between the host excited state and the guest ground state (Figure 4.2) [24]. It does not include the transfer of electrons and may occur over significantly larger distances. The rate constant of the Förster energy transfer is inversely proportional to the sixth power of the distance R between the molecules:

$$ k = \left(\frac{R_0}{R} \right)^6 \frac{1}{\tau_D} \qquad (4.3) $$

where τ_D is the lifetime of the donor excited state and the constant R_0 is the critical quenching radius (usually between 50 and 100 Å).

Under this mechanism, a long-range (up to 100 Å) energy transfer can occur without the emission of a photon; however, it requires the emission spectrum of the host to overlap with the absorption spectrum of the guest molecule. Naturally, the excited state of the guest $\left(S_1^G \right)$ should be lower in energy than the excited state of the host $\left(S_1^H \right)$, but otherwise the relative position of HOMO and LUMO orbitals of the host and the guest is not important. Considering that the ground state of the guest is (always) a singlet, the Förster mechanism allows for energy transfer only from singlet to singlet. As shown later, the Förster and Dexter transfer processes often coexist in the real host–guest system, and both of them play a role in phosphorescent LEDs.

In real phosphorescent LED devices, a high yield of excitons localized on the guest molecule (usually present in low concentration) is determined not only by energy transfer from the host matrix but also by a charge trapping on these molecules, which then leads to preferential exciton formation on these species. If the HOMO (LUMO) of the guest lies above (below) that of the host material, a hole (electron) generated in the host during LED operation may be trapped on the guest energy well, until meeting its counter particle and forming an exciton. The higher the difference between the HOMO (LUMO) orbitals of the host and the guest, the higher is the efficiency of hole (electron) trapping, and sometimes direct charge trapping could be the prime mechanism of the exciton generation on the guest molecules. However, the charge trapping also creates a barrier for charge transport across the device, resulting in significant increase of the operating voltage.

Once the excitons have been generated and transferred onto the guest molecule, the emissive properties of the latter determine the luminescence quantum yield. For phosphorescent guests, a major nonradiative process responsible for decreased quantum efficiency is the so-called triplet–triplet annihilation, which results in the formation of one singlet ground state and one excited state (i.e., one exciton is lost) (Equation 4.4). This annihilation is particularly pronounced for the molecules with long phosphorescence lifetime, and it is held responsible for a strong decrease of the efficiency in PPLED at high current densities [25]:

$$T_1 + T_1 \rightarrow S_1 + S_0 \tag{4.4}$$

4.3 Nonconjugated Polymers as Host Materials

Following the encouraging results demonstrated by metal complex-based phosphorescent OLEDs [5,6,26], several groups started investigating the possibility of attaining electrophosphorescence in solution-processable polymer-based LEDs. The first report on using triplet-emitting metal complexes blended in semiconducting polymers, acting as a host, appeared in 1999 [8,27,28]. Thus, Cleave and coworkers [8] used *N*-arylnaphthalimide-containing polystyrene (PNP **2**) as a host material and platinum octaethylporphyrin (PtOEP **1**) as a triplet-emissive dopant (phosphor) (Chart 4.1) [8]. An overlap between the host emission and the guest absorption allows an efficient Förster energy transfer from the host to the guest (Figure 4.3). On the basis of time-resolved PL and EL experiments, the authors suggest that the transfer of both singlet and triplet states of the PNP polymer onto platinum complex is taking place, although the direct charge trapping and recombination on PtOEP may also be responsible for the increased red emission.

4.3.1 Poly(9-Vinylcarbazole)s

To date, the most widely investigated host polymer material for PPLEDs is poly(9-vinylcarbazole) (PVK **4**, Chart 4.2), which is because of its high-energy blue-emissive excited state, excellent film-forming properties, high thermal stability, and reasonable hole mobility (~10^{-5} cm^2/(V s)) [29,30]. The first report mentioning the energy transfer from the PVK onto a triplet-emitting dye (Eu–phenanthroline complex) appeared in 1999, although the authors have chosen another polymer, poly(*p*-phenylene), to fabricate a PPLED (see Section 4.4.2) [31]. Lee et al. [32] reported a PLED device using the PVK host material and a green-emitting

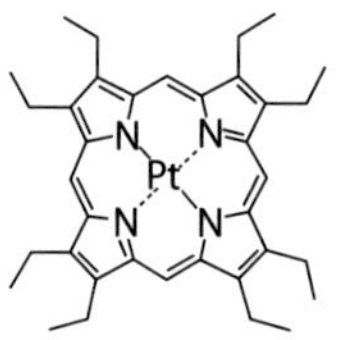

PtOEP (**1**) 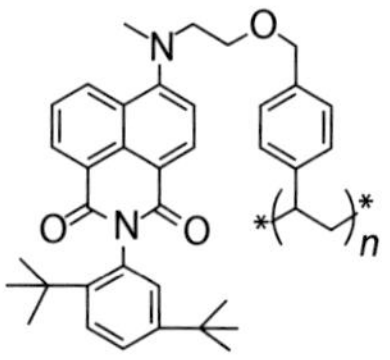   PNP (**2**)

CHART 4.1
Chemical structure of PtOEP and PNP.

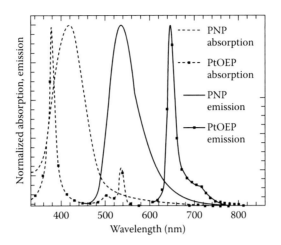

FIGURE 4.3
Absorption and emission spectra of the host PNP (**2**) and the guest PtOEP (**1**) (5% in poly(methylmethacrylate) excited at 514.5 nm). (From Cleave, V., Yahioglu, G., Barny, P.L., Friend, R.H., and Tessler, N., *Adv. Mater.*, 11, 285, 1999. With permission.)

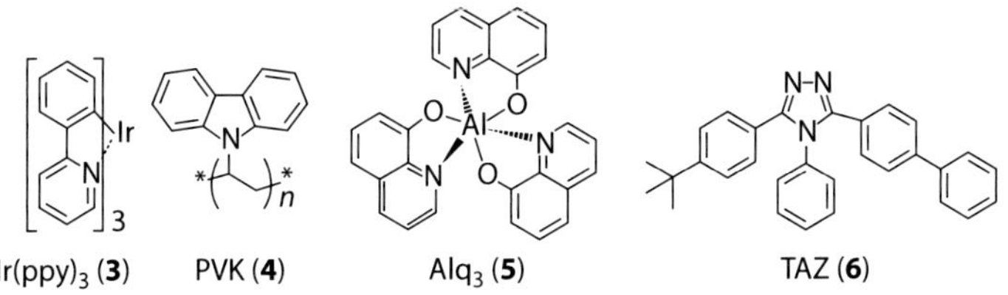

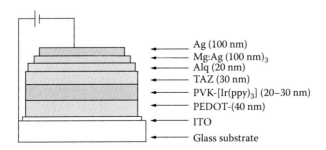

Ir(ppy)$_3$ (**3**)   PVK (**4**)   Alq$_3$ (**5**)   TAZ (**6**)

CHART 4.2
Chemical structure of Ir(ppy)$_3$, PVK, Alq$_3$, and TAZ.

tris(2-phenylpyridine)iridium complex Ir(ppy)$_3$ (**3**) (for the synthesis and photophysics of a large series of Ir complexes, see Refs. [33,34]) as a phosphorescent dopant. The multilayer device built with 8% Ir(ppy)$_3$ and PVK emissive layer showed the maximum EQE of 1.9% (with only little decrease upon 100 times increase of the operating current) and light output in excess of 2500 cd/m^2 (Figure 4.4). Although the efficiency was lower than that of the best small molecule–based electrophosphorescent LED, it was significantly higher than that of a similar structure device without a phosphorescent dopant [30].

Ag (100 nm)
Mg:Ag (100 nm)$_3$
Alq (20 nm)
TAZ (30 nm)
PVK-[Ir(ppy)$_3$] (20–30 nm)
PEDOT-(40 nm)
ITO
Glass substrate

FIGURE 4.4
Multilayer PPLED configuration. (From Lee, C., Lee, K.B., and Kim, J., *Appl. Phys. Lett.*, 77, 2280, 2000. With permission.)

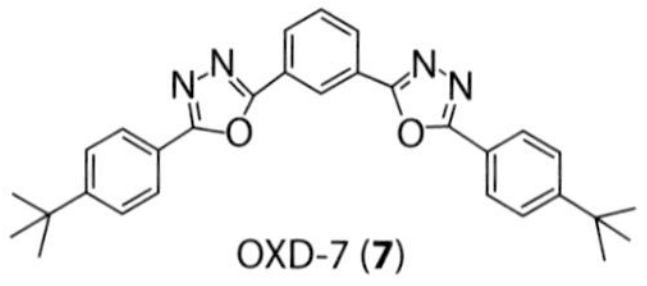

OXD-7 (**7**)

CHART 4.3
Chemical structure of OXD-7.

A higher-efficiency, yet simpler-structure PPLED device fabricated with the same dopant and host materials was almost simultaneously reported by Yang and Tsutsui [35]. The highest EQE of their device, ITO/PVK:5% Ir(ppy)$_3$/OXD-7/Mg:Ag (where ITO is indium tin oxide) (using OXD-7 (**7**) as an electron-transporting layer (ETL), Chart 4.3), reached the value of 7.5%, which was the first reported PLED with an external efficiency of >5%, an upper limit of the fluorescent PLEDs. The power efficiency was 5.8 lm/W at the luminance of 106 cd/m^2.

A detailed study of the influence of film morphology, guest concentration, and device structure on the performance of (Ir(ppy)$_3$/PVK)-based PPLEDs has been undertaken by Vaeth and Tang [36] at Kodak. They evaluated different solvents for spin-coating deposition of polymer films and found that only relatively low-volatile solvents such as toluene or chlorobenzene can reproducibly form the high-quality pinhole-free films. The efficiency of a single-layer device ((C$_n$F$_{2n}$-modified ITO)/PVK:Ir(ppy)$_3$/Mg/Ag) measured at 20 mA/cm^2 was 5 cd/A at 1.5% concentration of Ir(ppy)$_3$ and reached the value of 8.7 cd/A when the dopant concentration was increased to 3.5% (limited by the solubility in chlorobenzene) (Figure 4.5). Balancing the charge transport and injection via blending with electron-transporting material, 2-(4-biphenyl)-5-(4-*tert*-butylphenyl)-1,3,4-oxadiazole (PBD) (**8**), and introducing additional hole-blocking and electron-injection layers (TPBI **9**, ADN **10**, tBP **11**, and Alq$_3$ **5**) significantly improved the device efficiency; the very high current efficiency of 30 cd/A and the EQE of 8.5% were reported for the multilayer device (C$_n$F$_{2n}$-modified ITO)/PVK:PBD:3.5% Ir(ppy)$_3$/TPBI/Alq$_3$/Mg/Ag (Chart 4.4, Figure 4.5).

FIGURE 4.5
PPLED structure for a single-layer (a) and multilayered (b) devices. (From Vaeth, K.M. and Tang, C.W., *J. Appl. Phys.*, 92, 3447, 2002. With permission.)

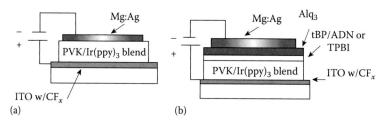

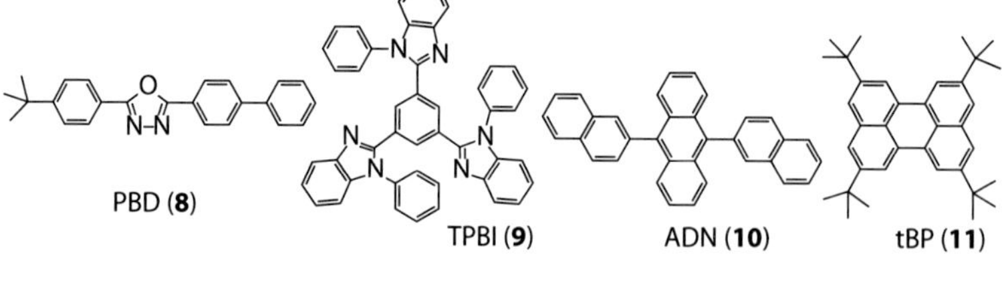

PBD (**8**)   TPBI (**9**)   ADN (**10**)   tBP (**11**)

CHART 4.4
Chemical structure of PBD, TPBI, ADN, and tBP.

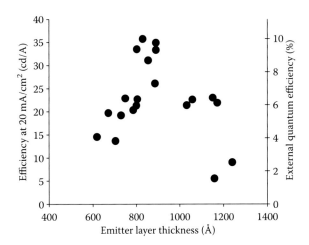

FIGURE 4.6
Dependence of the current efficiency on the thickness of the electroluminescent layer for PPLED ITO/PEDOT/
PVK (**4**):3% Ir(ppy)$_3$ (**3**)/CsF/Mg:Ag. (From Vaeth, K.M. and Dicillo, J., *J. Polym. Sci.: Part B: Polym. Phys.*, 41, 2715, 2003. With permission.)

A further optimization of the performance of the PVK-/Ir(ppy)$_3$-based LEDs was reported in 2003 [37]. The authors undertook a systematic variation of a number of factors affecting the charge injection and recombination, cf. concentration of the dopant and thickness of the EL layer, electron- and hole-transporting layers, hole-blocking layer, and cathode material. Small changes (ca. 50%) in the thickness of EL layer can increase the EQE of the device by more than a factor of 3 (Figure 4.6), although, very likely, in lieu of a reduced lifetime. The best performance was achieved for a simple device with poly(3,4-ethylenedioxythiophene) (PEDOT)-modified anode (ITO/PEDOT/PVK:3% Ir(ppy)$_3$/CsF/Mg:Ag). An unprecedented maximum current efficiency of 37.3 cd/A (10.5%) for PLED devices was achieved at the high brightness of 2240 cd/A (at 10 V). The efficiency changes very little even at higher brightness, showing 36 cd/A (10.1%) at 7200 cd/A (at 12 V).

Using a methylated analog of Ir(ppy)$_3$ (Ir(mppy)$_3$ **12**, Chart 4.5), Yang and coworkers [38] have recently prepared another highly efficient PPLED operating at very low voltage (5.5 V); a single-layer device, ITO/PEDOT/Ir(mppy)$_3$–PBD–PVK/CsF/Al, showed a peak power efficiency of 14.1 lm/W (27 cd/A, 7.6%).

Chen and coworkers [39] suggested that the long lifetime of the triplet excitons may cause the concentration quenching (due to triplet–triplet annihilation) and thus account for the observed decrease of the quantum efficiency at high current densities. On the other hand, the long lifetime of the triplet state of the host materials would facilitate the energy transfer [40]. To compare the mutual influence of these factors, the authors studied the

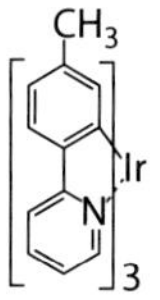

Ir(mppy)$_3$ (**12**)

CHART 4.5
Chemical structure of Ir(mppy)$_3$.

PPLEDs fabricated with PVK (**4**) and poly(dioctylfluorene) PDOF (**13**) host materials (triplet exciton lifetime of 100 and 2.5 ms, respectively), and the platinum (PtOX **14**) or iridium (Btp$_2$Ir **15**) complexes as guest phosphorescent materials (triplet lifetime of 80 and 5 μs, respectively) (Chart 4.6, Table 4.1) [39]. To balance the charge transport and injection for the polymers with different HOMO and LUMO levels, an electron transport material, PBD (**8**), was blended in PVK-based PPLEDs, whereas a hole-transporting PVK layer was introduced into polyfluorene-based devices (Table 4.2). As expected from the triplet-state lifetimes, PVK-based devices showed better performance than polyfluorene-based devices. The Btp$_2$Ir dopant appeared to be more efficient than PtOX, and also showed a higher stability. The PVK–Btp$_2$Ir device emitted very pure saturated red light (λ_{max} = 614 nm, Commission Internationale de l'Eclairage (CIE) coordinates: 0.66, 0.33), revealing a maximum EQE of 3.3% and a brightness of 1400 cd/m^2.

The same group investigated the degradation mechanism of PVK–PtOX-based PPLED [41]. The results suggest that electron trapping with the formation of anionic PtOX (**14**) species is responsible for the device's instability, and thus blending the electron transport material PBD (**8**), which competes with the triplet emitter in electron-trapping processes and improves the device half-lifetime by approximately 40 times (to 45 h); however, it is still very far from a technologically acceptable level.

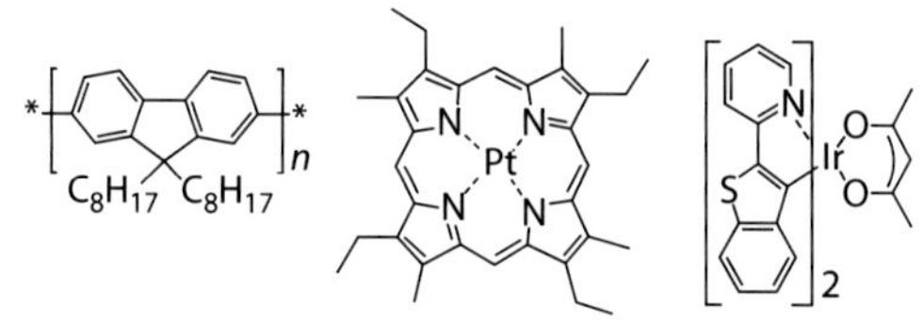

CHART 4.6
Chemical structure of PDOF, PtOX, and Btp$_2$Ir.

TABLE 4.1

Photophysical Parameters of Some Polymer Hosts and Phosphorescent Dopants

Compound	HOMO (eV)	LUMO (eV)	E_g (eV)	E_T (eV)[a]	l_{PL} (nm)	F_{PL} (%)	T_1 (Lifetime)	Ref.
PVK **4**	−5.54	−2.04	3.5	2.50	410	5	(100 ms)	66 (65)
PDOF **13**	−5.77	−2.08	3.69	2.15	420	50	2.5 ms	66
PF3CNP1 **51**	−5.74	−2.89	3.15		475			69
PtOX **14**	−5.3	−2.9	2.4	1.91	650		80 ms	41, 65
Ir(ppy)$_3$ **3**	−5.12	−2.11	3.01	2.41	516	(40)	(2 ms)	66 (31)
PPIr **19**	−5.20	−2.19	3.01	2.41	516	(34)	(1.5 ms)	66 (32)
Ir(Ocppy)$_3$ **52**	−5.04	−2.20	2.84	2.41	518			66
Bt$_2$Ir **17**	−5.36	−2.65	2.69	2.23	560	(26)	(1.8 ms)	66 (32)
Btp$_2$Ir **15**	−5.16	−2.38	2.78	2.02	614	(21)	(5.8 ms)	66 (32)
Bzq$_2$Ir **20a**				2.26	548	27	4.5 ms	32
Bsn$_2$Ir **20b**				2.05	606	22	1.8 ms	32
Ir(DPF)$_3$ **22**	−4.69	−2.10	2.59	2.27	546			50
Ir(HFP)$_3$ **24**	−4.73	−2.25	2.48	2.07	600			50
Ir(DPPF)$_3$ **23**	−4.80	−2.21	2.59	2.25	550			50

[a] Above HOMO.

TABLE 4.2

Performance of ITO/PEDOT/PVK (**4**):PBD (**8**):Dopant/Ca/Al (I) and ITO/PEDOT/PVK/PDOF (**13**):Dopant/Ca/Al (II) PPLEDs, Doped with Btp$_2$Ir (**15**) and with PtOX (**14**) Complexes

Host	Dopant (wt.%)	Maximum QE at mA/cm^2 Brightness (cd/m^2)	QE at High-Current Density at mA/cm^2 QE/Maximum QE	Maximum Brightness (cd/m^2)
PVK	BtpIr (4%)	3.3%	2.2%	
		5.7	80	1400
		147	65%	
	PtOX (4%)	2.5%	0.79%	
		3.1	60	270
		28	32%	
	BtpIr (5%)	2.4%	1.0%	
		1.3	80	800
		25	43%	
PF	PtOX (5%)	1.2%	0.26%	
		1.7	80	125
		12	21%	

Source: Chen, F., Yang, Y., Thompson, M.E., and Kido, J., *Appl. Phys. Lett.*, 80, 2308, 2002. With permission.

Kawamura et al. [42] used a PVK host polymer doped with a series of phosphorescent iridium complexes, FIrpic (**16**), Ir(ppy)$_3$ (**3**), Bt$_2$Ir (**17**), and Btp$_2$Ir (**15**), to achieve a full-color spectrum, from blue to red, and their mixture for white emission (Figure 4.7). The multi-layer PPLED device ITO/PEDOT/PVK:complex/BCP/Alq$_3$/Mg:Ag was built with batho-cuproine (BCP **18**) as a hole- or exciton-blocking layer to confine long-living triplet states in the luminescent region [43]. The EL characteristics of the prepared PPLEDs containing 5 wt.% of iridium complexes are listed in Table 4.3. All single-component PPLEDs exhibited pure color emission originating from the guest molecules, suggesting very efficient energy transfer from the host. The highest EQE (5.1%) was demonstrated by green PPLED using

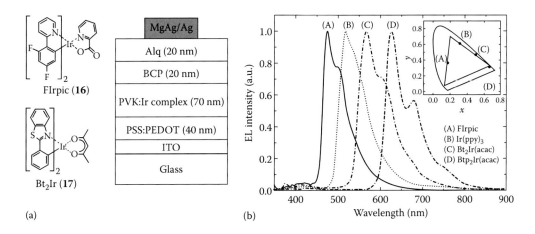

(a)   (b)

FIGURE 4.7

(a) Chemical structures of PPLED materials. (b) Device structure and the electroluminescence (EL) spectra of the PPLEDs prepared with PVK (**4**) host polymer and the above dopants (5 wt.%). (From Kawamura, Y., Yanagida, S., and Forrest, S.R., *J. Appl. Phys.*, 92, 87, 2002. With permission.)

TABLE 4.3

Characteristics of PPLEDs with PVK (**4**)–Ir Complex (5 wt.%) Blends

	FIrpic[a]	Ir(ppy)$_3$	Bt$_2$Ir	Btp$_2$Ir
λ_{max} (nm)	474	517	565	623
CIE (*x*,*y*)	(0.18, 0.36)	(0.32, 0.61)	(0.50, 0.47)	(0.67, 0.31)
h_{ext} (%) [h_p (lm/W)] at 0.1 mA/cm^2	0.96 (0.69)	4.9 (6.3)	1.5 (1.1)	2.2 (0.60)
1	1.2 (0.70)	5.1 (5.2)	2.0 (1.3)	2.1 (0.44)
10	1.2 (0.60)	4.8 (3.8)	1.8 (1.0)	1.8 (0.32)
100	0.96 (0.39)	3.5 (2.2)	1.2 (0.59)	1.3 (0.18)
J (mA/cm^2) for 100 cd/m^2	3.4	0.55	1.7	6.1
J_0 (mA/cm^2)	410	260	190	190

Source: Kawamura, Y., Yanagida, S., and Forrest, S.R., *J. Appl. Phys.*, 92, 87, 2002. With permission.

Note: External QE (h_{ext}) and power efficiency (h_p) are given as functions of current density; λ_{max} is the EL emission peak; J_0 is the current at which the h_{ext} falls to 50% of its peak value and is characteristic of a triplet–triplet annihilation rate.

[a] Device doped at 10 wt.%.

Ir(ppy)$_3$ as a dopant, whereas essentially lower efficiencies were observed for blue FIrpic-, yellow Bt$_2$Ir-, and red Btp$_2$Ir-based PPLEDs. This can be due to the poor energy overlap between the PVK host and those dopants. Interestingly, for the red PPLED, the efficiency can be increased from 2.2% to 3.3% by employing a sensitizer. Adding an additional green dopant Ir(ppy)$_3$ as a sensitizer allows for the energy transfer from PVK, through Ir(ppy)$_3$, to Btp$_2$Ir. Blending several dopants with different emission color into a PVK host matrix leads to white-emitting devices, although the ratio of dopants must be carefully controlled to avoid a complete energy transfer to the lowest-energy (most red) emitter. For example, PPLED in the configuration ITO/PEDOT/PVK:10% FIrpic (**16**):0.25% Bt$_2$Ir (**17**):0.25% Btp$_2$Ir (**15**)/BCP (**18**)/Alq$_3$/Mg:Ag emitted white color (CIE: 0.33, 0.41) with the maximum h_{ext} = 2.1% and h_p = 1.4 lm/W at 10 V.

These (Bt$_2$Ir **17**, Btp$_2$Ir **15**) and other (PPIr **19**, Bzq$_2$Ir **20a**, Bsn$_2$Ir **20b**) (Chart 4.7) Ir-based dopants have been simultaneously studied by Lamansky and coworkers [44], revealing similar good efficiencies in PVK-based PPLEDs.

Even higher-efficiency red, blue, and green PPLEDs have been fabricated with the above complexes and PVK using Cs as a cathode (device ITO/PEDOT/PVK:OXD-7 (**7**):Ir complex/Cs/Al) [45]. Using Btp$_2$Ir (**15**), Ir(ppy)$_3$ (**3**), and FIrpic (**16**) phosphorescent dopants, red (CIE: 0.66, 0.33), green (CIE: 0.31, 0.60), and blue (CIE: 0.17, 0.30) emission with efficiency as high as 4.3, 31, and 14 cd/A, respectively, was achieved.

He and coworkers [46] tried to improve the efficiency of PLEDs using the phosphorescent sensitizer to transfer all singlet and triplet excitons from the host polymer (PVK) to

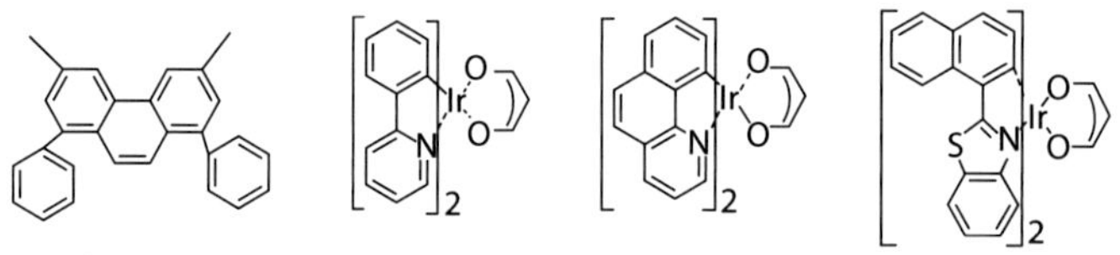

Bcp (**18**)   PPIr (**19**)   Bzq$_2$Ir (**20a**)   Bsn$_2$Ir (**20b**)
Green (0.28, 0.65)   Yellow (0.41, 0.57)   Red (0.62, 0.34)

CHART 4.7
Chemical structure of BCP, PPIr, Bzq$_2$Ir, and Bsn$_2$Ir.

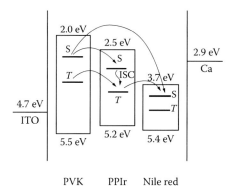

FIGURE 4.8
Energy diagram and proposed energy transfer mechanism in the blend system. (From He, G., Chang, S., Chen, F., Li, Y., and Yang, Y., *Appl. Phys. Lett.*, 81, 1509, 2002. With permission.)

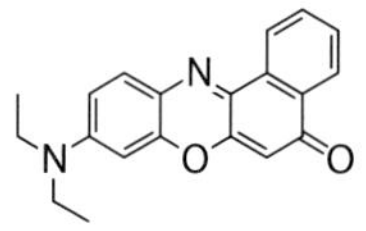

Nile red (**21**)

CHART 4.8
Chemical structure of Nile red.

the singlet fluorescent dye, as illustrated in Figure 4.8. They used PPIr (**19**) (Chart 4.7) as the phosphorescent dopant and Nile red (**21**) (Chart 4.8) as a fluorescent dye. Because of the very low concentration of PPIr and Nile red, their absorption could hardly be observed. However, when the blend film was excited at 300 nm, the main emission came from PPIr and Nile red, indicating a complete energy transfer from the host to the dye materials. Furthermore, the EL spectrum of PPLED built as ITO/PEDOT/PVK:PBD:PPIr:Nile red (100:100:1:1)/Ca/Al showed a single band at 600 nm corresponding to emission of Nile red. The authors suggest that the energy transfer proceeds through a combination of singlet–singlet (PVK → Nile red) and triplet–singlet (PPIr → Nile red) transitions (Figure 4.8). The Förster mechanism was claimed for both processes on the basis of a very low concentration (and thus a large separation) of the dye molecules, although it should be prohibited for the latter process (triplet–singlet transfer). Nevertheless, comparing the prepared PPLED with those of control devices fabricated without PPIr (**19**) sensitizer or with a fluorescent green sensitizer Alq$_3$ (**5**) showed approximately threefold higher EL efficiency for the former (6.4 cd/A). This fact strongly supports the authors' claim that the triplets formed or transferred on PPIr phosphor undergo further transfer onto Nile red dopant.

Gong and coworkers [47] at University of California, Santa Barbara (UCSB), reported high-efficiency PPLEDs based on PVK polymer doped with a new p-extended iridium complex, Ir(DPF)$_3$ (**22**) (blended with PBD (**8**) as electron transporter) (Chart 4.9). The good overlap of the emission of the host polymer with the absorption of the dopant Ir(DPF)$_3$ allows an efficient Förster energy transfer between the host and the guest materials (Figure 4.9). Indeed, studying the concentration dependence of the PL spectra reveals an almost complete transfer of the PVK short-wavelength emission into the red emission of the Ir complex at concentrations above ~1% (although total suppression of the PVK–PBD PL emission was achieved only at 8% dopant concentration). In EL devices, there was no

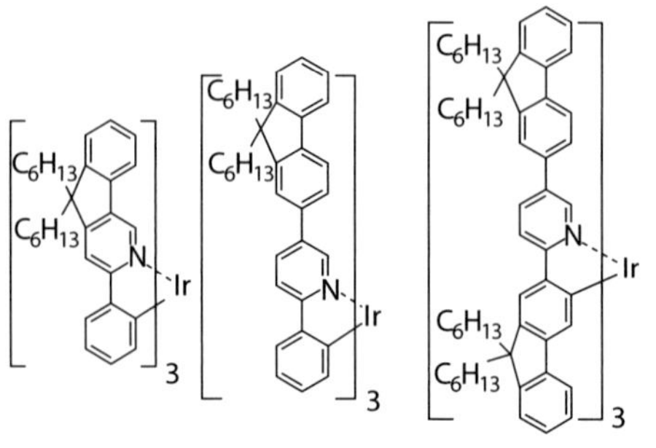

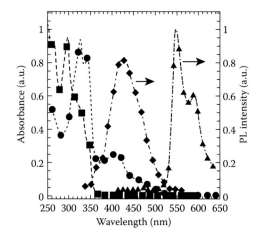

CHART 4.9
Chemical structure of Ir(DPF)$_3$, Ir(DPPF)$_3$, and Ir(HFP)$_3$.

FIGURE 4.9
Normalized absorption or emission spectra of the neat films of Ir(DPF)$_3$ (**22**) (absorption: ●, emission: ▲) and the blend films of PVK (**4**)–PBD (**8**) (40 wt.%) (absorption: ■, emission: ◆). (From Gong, X., Robinson, M.R., Ostrowski, J.C., Moses, D., Bazan, G.C., and Heeger, A.J., *Adv. Mater.*, 14, 581, 2002. With permission.)

short-wavelength emission even at extremely low concentration of Ir(DPF)$_3$ (0.001 mol% per repeat unit of PVK), which suggests that the charge trapping on the metal complex, rather than just Förster energy transfer, is responsible for the EL of this PPLED. The device built as ITO/PEDOT/PVK–PBD (40%):Ir(DPF)$_3$/Ca/Ag emits yellowish-green light at λ_{max} = 550 nm (shoulder at 590 nm) with an EQE of 10%, luminous efficiencies as high as 36 cd/A, and brightness in excess of 8000 cd/m^2 at 75 mA/cm^2 (55 V). Although the operating voltage was very high, the device efficiency for the first time approached that of small molecule–based phosphorescent LEDs, yet at a very low doping level.

A somewhat lower operating voltage was reported by the same group for a related guest dopant, Ir(DPPF)$_3$ (**23**) [48]. A green–yellow-emitting (λ_{max} = 550 nm, shoulder at 590 nm) single-layer device, ITO/PEDOT/PVK:PBD:Ir(DPPF)$_3$/Ca/Ag, achieved the maximum brightness of 3500 cd/m^2 at 30 V (cf. 45 V for an Ir(DPF)$_3$) [47]. A high EQE of 8% (29 cd/A) was achieved in this device; however, the high operating voltage resulted in relatively low power efficiency of ~3.3 lm/W.

As already demonstrated, the efficient color tuning in the PVK–Ir complex systems can be achieved by the modification of the organic ligands. An extended conjugation in another member of the fluorene–pyridine complexes, Ir(HFP)$_3$ (**24**), shifts the emission maximum into the red by ca. 30 nm. A bright-red electrophosphorescent device was fabricated with Ir(HFP)$_3$ as a guest material in the configuration ITO/PEDOT/PVK:PBD:Ir(HFP)$_3$/Ca/Ag (λ_{max} ~580 nm) [49]. The maximum EQE of 5% (7.2 cd/A) was achieved at a brightness of 170 cd/m^2.

The same group later published an account comparing the performance of the three fluorene-containing dopants described above with PVK host polymer (as well as with several polyfluorene hosts; see Section 4.4.1) [50]. Analyzing the PL and EL properties of different host–guest combinations, they have reached the conclusion that the Förster energy transfer plays only a minor role in the operation of such PPLEDs and that the direct charge trapping by dopants is the major mechanism of the electrophosphorescent process.

Although platinum and, particularly, iridium complexes are the two most frequently used classes of dopants for PPLEDs, other metals such as gold, copper, rhenium, ruthenium, europium, and osmium have also been studied as dopants. Ma et al. [27] have studied two phosphorescent complexes, Au2MDP (**25**) and Cu4MDP (**26**) (Chart 4.10), as dopants in a PVK host polymer. The solution PL quantum yields for these two phosphors (23% and 42%, respectively) are rather high compared with the other gold and copper complexes. The PPLEDs fabricated in the configuration ITO/PVK:Au2MDP/Al and ITO/PVK:Cu4MDP/Al can be turned on at 10 and 12 V, respectively. However, the EQE of these devices appeared to be very low (0.1% for Cu4MDP), which can be partially attributed to the unoptimized device structure, with very high charge injection barriers. Comparison of the PL and EL spectra of the device suggests a significant increase in the ratio of triplet–singlet emission in the EL process, suggesting that the charge-trapping mechanism occurs also for these dopants.

Li et al. [28] and Kan et al. [51] also studied PVK-based PPLEDs using a rhenium complex. They have prepared the PPLED in the configuration ITO/PVK:bpyRe (**27**):DCM (**28**)/Al (DCM is 4-(dicyanomethylene)-2-methyl-6-[*p*-(dimethylamino)styryl]-4*H*-pyran), where the bpyRe complex (Chart 4.10) acts as a coupler of energy transfer between the PVK host and the DCM-fluorescent emitter, which allows harvesting both singlet and triplet excitons [51]. The device emitted red light with λ_{max} ~570 nm (CIE: 0.45, 0.53). The highest EL efficiency (0.42 cd/A) was shown by the devices containing 10 wt.% of bpyRe complex and 1 wt.% of DCM dye, which is a reasonably good value for the unoptimized single-layer device. This efficiency is 10 times higher than that of the similar device without the phosphorescent bpyRe dopant, and is twice higher than that of the PVK-based device containing the bpyRe (**27**) but not the DCM dye (0.2 cd/A) [28]. Later, a detailed study of PL and EL of PVK films doubly doped with green-emitting Ir(ppy)$_3$ (**3**) and a red-emitting ruthenium dye Ru(Ph$_2$phn)$_3$ (**29**) was reported [52].

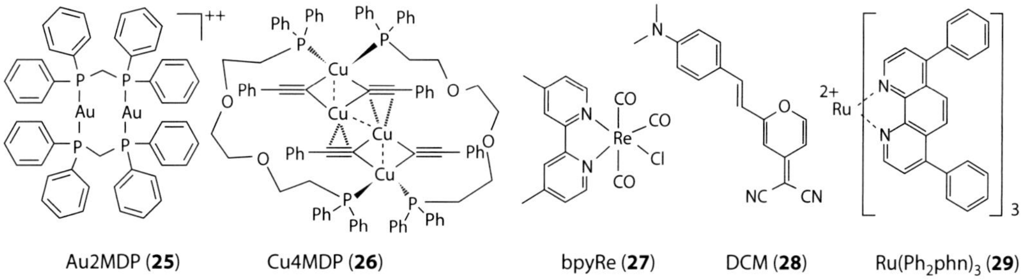

Au2MDP (**25**)   Cu4MDP (**26**)   bpyRe (**27**)   DCM (**28**)   Ru(Ph$_2$phn)$_3$ (**29**)

CHART 4.10
Chemical structure of Au2MDP, Cu4MDP, bpyRe, DCM, and Ru(Ph$_2$phn)$_3$.

Shih et al. [53] demonstrated the fabrication of white PLEDs in a single emitting layer containing a PVK (4) host blended with 30 wt.% of electron-transporting PBD (8) and two dopants: a blue fluorescent dye, 4,4′-bis[2-{4-(*N*,*N*-diphenylamino)phenyl}vinyl]biphenyl (DPAVBi **30**), and a phosphorescent orange light–emitting osmium complex, Os(bpftz) (**31**, Chart 4.11). The device configuration consists of an ITO/polystyrenesulfonate-doped PEDOT, (PEDOT:PSS)/PVK:PBD:DPAVBi (**30**):Os(bpftz) (**31**)/TPBI/Mg:Ag/Ag. This doubly doped device exhibited an intense white emission having CIE coordinates of (0.33, 0.34) with a high EQE of 6.12% (13.2 cd/A) and a maximum brightness of 11,306 cd/m². At a high luminance of 10,000 cd/m², only a minor color shift was observed from (0.33, 0.34) to (0.33, 0.32), representing a very stable white emission.

Wu and coworkers [54] reported efficient white PLEDs in a single emitting layer containing a host matrix of PVK/OXD-7 (7) and two or three phosphorescent iridium complexes as dopants. They employed iridium bis(1-phenylisoquinoline) (acetylacetonate) ((piq)₂Ir(acac) **32**, Chart 4.12, Ir(mppy)₃ (**12**), and FIrpic (**16**) as the red–green–blue (RGB) emitter, respectively. The RGB monochromatic and white light–emitting devices have identical single emitting layer configuration of ITO/PEDOT:PSS/polymer blend/Ba/Al. A series of devices with various dopant concentrations and ratios were fabricated and characterized, as shown in Table 4.4. The double-doped devices exhibited an intense white emission having CIE coordinates of (0.33, 0.36) at 12 mA/cm², peak power efficiency of 6.3 lm/W, and a peak luminous efficiency of 16.1 cd/A for forward viewing. The whole triple-doped device exhibited white emission having CIE coordinates of (0.34, 0.47), a peak power efficiency of 9.5 lm/W, and a peak luminous efficiency of 24.3 cd/A.

The same group later published a study [55] of a series of white PLEDs in a single emitting layer containing a host matrix of PVK/OXD-7 (7) and two phosphorescent iridium complexes as dopants. They employed blue-emitting FIrpic (**16**) and newly laboratory-made efficient yellow iridium complexes **33** and **34** (Chart 4.13). Both monochromatic and white light–emitting devices have the same identical single emitting layer configuration of ITO/PEDOT:PSS/polymer blend/Ba/Al as reported before [54]. The double-doped devices exhibited a peak EQE of 19.1%, a peak power efficiency of 20.3 lm/W (can be up to 40 lm/W

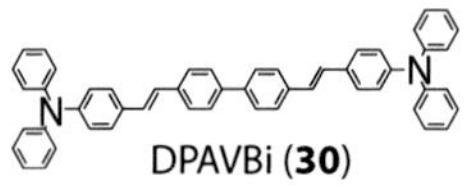

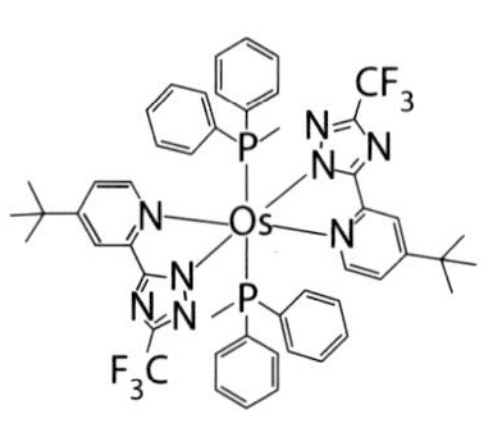

CHART 4.11
Chemical structure of DPAVBi and Os(bpftz).

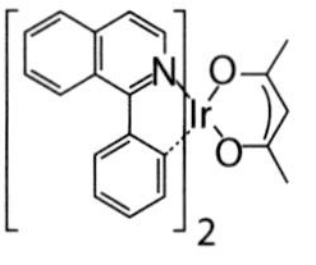

CHART 4.12
Chemical structure of (piq)₂Ir(acac).

TABLE 4.4

Device Performance of RGB Iridium Complexes–Doped White-Emitting PLEDs

Device	Composition	CIE[e]		Max LE[f] (cd/A)	Max EQE[g] (%)	Max PE[h] (lm/W)	CCT (K)	CRI
A[a]	40:1[c]	0.255	0.357	16.0	12.9	7.6	9083	65
B[a]	30:1[c]	0.319	0.352	14.6	11.4	5.6	6131	46
C[a]	20:1[c]	0.329	0.362	16.1	10.0	6.3	5896	52
D[a]	10:1[c]	0.465	0.339	4.9	5.0	2.2	2061	28
E[b]	50:1:1[d]	0.301	0.467	21.9	13.8	10.0	6297	62
F[b]	20:1:1[d]	0.343	0.468	24.3	14.4	9.5	5010	77
G[b]	30:0.1:1[d]	0.382	0.469	18.9	12.9	8.2	3891	75
H[b]	50:0.1:1[d]	0.257	0.395	16.3	12.1	5.4	8267	62

Source: Wu, H., Zou, J., Liu, F., Wang, L., Mikhailovsky, A., Bazan, G.C., Yang, W., and Cao, Y., *Adv. Mater.*, 20, 696, 2008. With permission.

Note: CCT, correlated color temperature; CRI, color rendering index.

[a] Doubly doped devices.
[b] Triply doped devices.
[c] FIrpic:Ir(piq) ratio; FIrpic is fixed at 10 wt.% of the active layer blend.
[d] FIrpic:Ir(mppy)$_3$:Ir(piq) ratio; FIrpic is fixed at 10 wt.% of the active layer blend.
[e] Observer: 2 degrees; obtained at 12 mA/cm^2.
[f] Maximal front viewing luminous efficiency in cd/A.
[g] Maximal external quantum efficiency in %.
[h] Maximal front viewing power efficiency in lm/W.

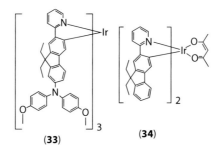

(**33**)   (**34**)

CHART 4.13
Chemical structure of yellow phosphors **33** and **34**.

if used as a lighting source), and a peak luminous efficiency of 42.9 cd/A with CIE coordinates of (0.395, 0.452). At a practical luminance of 1000 cd/m^2, the power efficiency is still retained at 16.8 lm/W (corresponding to a total power efficiency of 30 lm/W), while the luminous efficiency is as high as 41.7 cd/A. The high efficiency of the white PLEDs, together with the simplicity of the device architecture and fabrication process, renders the devices promising candidates for large-area, low-cost, solid-state lighting sources.

Recently, the same group [56] demonstrated highly power-efficient white PLEDs with a single emitting layer containing two newly synthesized yellow-emitting iridium complexes **35** and **36** functionalized with the sterically hindered diarylfluorene chromophores and a saturated red iridium(III)-based dendrimer Ir-G2 (**37**) (Chart 4.14). The white PLEDs were fabricated by a simple solution-processing method. The device configuration is ITO/ PEDOT:PSS/PVK:OXD-7:FIrpic:Ir(mppy)$_3$:**35** (or **36**):Ir-G2/Ba/Al. PPLEDs were achieved with peak forward-viewing power efficiency close to 40 lm/W, corresponding to an EQE of 28.8% and a luminous efficiency of 60 cd/A. At a luminance of 100 or 1000 cd/m^2, the

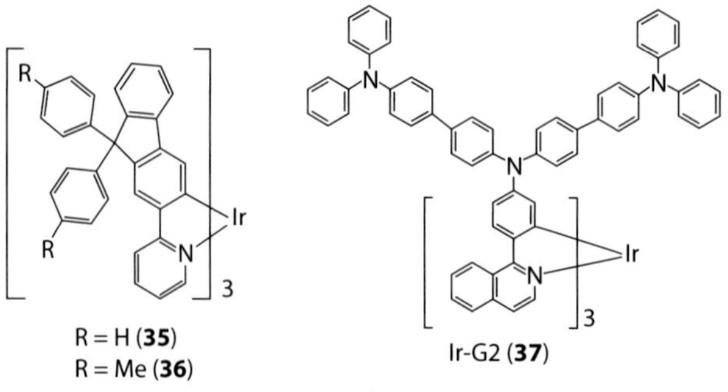

CHART 4.14
Chemical structure of the new yellow phosphors **35** and **36**, and a saturated red-emitting dendrimer Ir-G2.

power efficiencies were found to be 30.7 and 20.7 lm/W, respectively. The remarkable improvement in efficiency significantly outperforms other white light sources such as incandescent light bulbs.

Yin and coworkers [57] proposed a strategy of steric hindrance functionalization of π-stacked polymers to control their electronic structures and phase behaviors. A solution-processable π-stacked homopolymer host, poly(N-vinyl-3-(9-phenylfluoren-9-yl)carbazole) (PVPFK **38**) (Chart 4.15), was synthesized. PVPFK (**38**) bearing bulky 9-phenylfluorenyl moieties with steric hindrance effect exhibits high triplet energy levels (E_T levels) of 2.80 eV and better electron-transporting ability than PVK (**4**). The device architecture was ITO/PEDOT:PSS/PVPFK:30% FIrpic/TPBI/Ca:Ag. The PVPFK-based devices exhibit lower operation voltage (the turn-on voltage is 6.1 V), higher luminous efficiency (14.2 cd/A), and better power efficiency (2.8 lm/W) than the PVK (**4**) device (turn-on voltage, 7.4 V; luminous efficiency, 13.3 cd/A; and power efficiency, 2.4 lm/W) under the same conditions.

Very recently, Cheng et al. [58] reported [Pt(O^N^C^N)]-type complexes **39a–g** (Chart 4.16) used as single emitters in solution-processed PLEDs. Both monochromic and white PLEDs studied in this work were fabricated with a simple architecture of ITO/PEDOT:PSS/PVK:OXD-7:Pt(II) complex/1,3,5-tri(m-pyrid-3-yl-phenyl)benzene (TmPyPb)/TPBI (**9**)/LiF/Al. PPLEDs exhibit a maximum EQE of 15.55%, current efficiency of 53.63 cd/A, and power efficiency of 25.99 lm/W for green. The white PLEDs show a maximum EQE of 12.73%, current efficiency of 27.44 cd/A, and power efficiency of 17.01 lm/W. These EL values are the highest values ever achieved for PLEDs based on Pt(II) complexes.

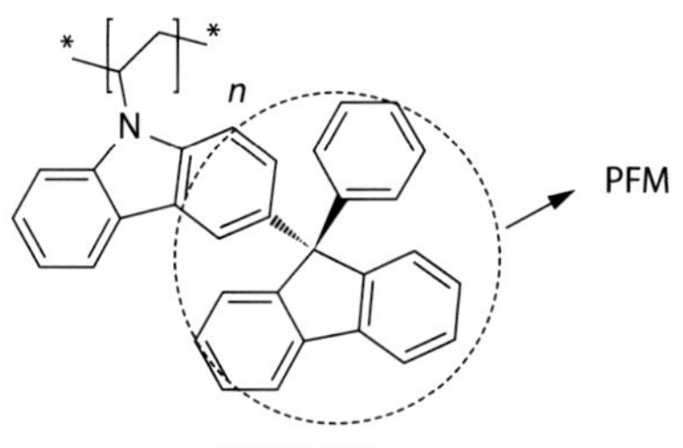

CHART 4.15
Chemical structure of PVPFK.

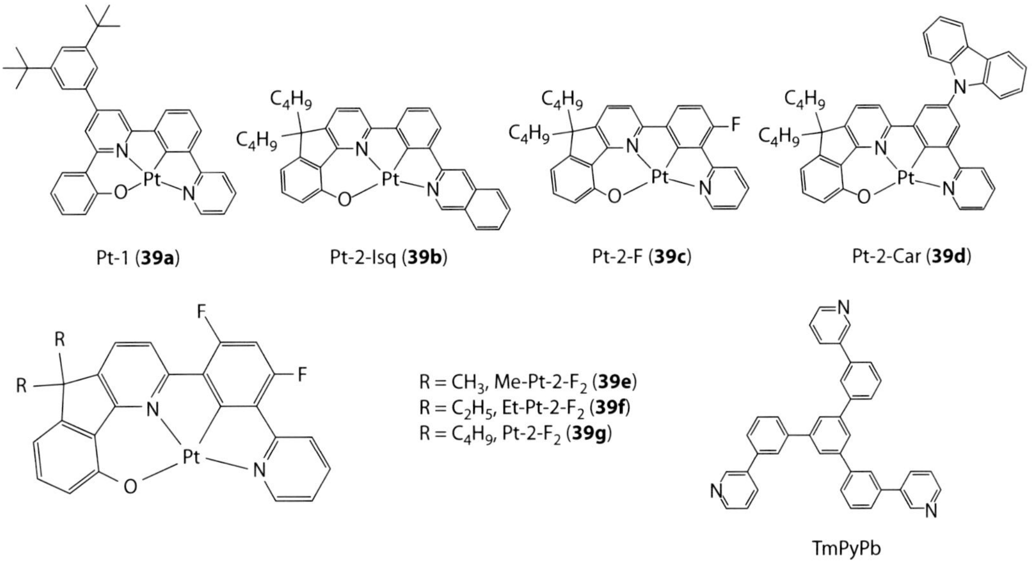

CHART 4.16
Chemical structure of Pt(II) complexes Pt-1, Pt-2-Isq, Pt-2-F, Pt-2-Car, Me-Pt-2-F$_2$, Et-Pt-2-F$_2$, Pt-2-F$_2$, and TmPyPb.

4.3.2 Other Types of Nonconjugated Polymers

Wang et al. [59] synthesized two new types of nonconjugated polymers, poly[*N*-(2-ethyl-hexyl)carbazole-*alt*-*N*-(4-aminophenyl)carbazole] (PECAC **40**) and poly[*N*-(2-ethylhexy-loxy-phenyl) carbazole-*alt*-*N*-(4-aminophenyl)carbazole] (PEPCAC **41**) (Chart 4.17). The carbazole groups are designed to generate wide band gap, and a triarylamine moiety is incorporated to provide good hole-transporting property. Green and red PPLEDs were fabricated with the configuration ITO/PEDOT/polymers:iridium complex (10 wt.%)/BAlq (**42**)/Alq$_3$ (**5**)/LiF/Al. The emitting layer consists of either PECAC (**40**) or PEPCAC (**41**) as

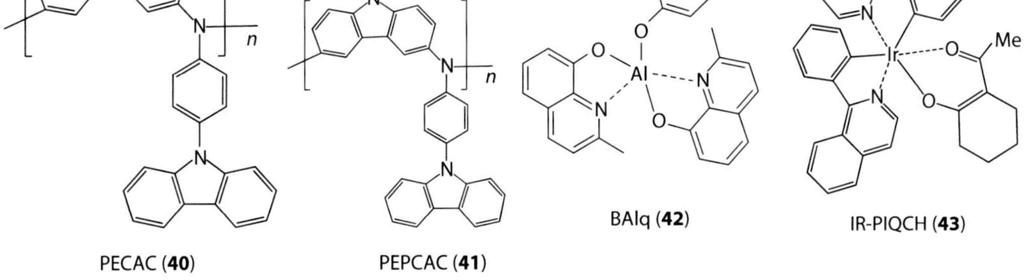

CHART 4.17
Chemical structure of PECAC, PEPCAC, BAlq, and IR-PIQCH.

a host matrix with suitable iridium complexes, Ir(ppy)$_3$ (**3**) as a green dopant, or iridium (III) bis(2-phenyl-isoquinoline)(2-acetyl-cyclohexone) (IR-PIQCH, CAS 1309777-03-4, **43**) as a red dopant. For devices doped with Ir(ppy)$_3$ (**3**), the maximum luminescence and driving voltage are 453 cd/cm^2 and 8 V for PECAC (**40**), and 77 cd/cm^2 and 14 V for PEPCAC (**41**). In the case of IR-PIQCH (**43**)–doped PPLEDs, the maximum luminescence and driving voltage are 576 cd/cm^2 and 6 V for PECAC (**40**), and 59 cd/cm^2 and 12 V for PEPCAC (**41**), respectively.

Recently, Thesen et al. [60] designed a series of novel host polymers containing styrene-derived monomers with triphenylamine-based units, functionalized at the *para* positions with the following side groups: diphenylamine, 3-methylphenyl-aniline, 1- and 2-naphthylamine, carbazole, and phenothiazine (**44a–f**, Chart 4.18). Green and red PPLEDs were fabricated with the configuration ITO/PEDOT:PSS/polymer blend/CsF/ Ca/Ag. Ir(mppy)$_3$ (**12**) was used as a phosphorescent dopant, and PBD (**8**) was used as electron-transporting material. For the carbazole-substituted polymer **44e** (T$_g$ = 246°C), a luminous efficiency of 35 cd/A and a brightness of 6700 cd/m^2 at 10 V are accessible. The phenothiazine-functionalized polymer **44f** (T$_g$ = 220°C) shows nearly the same outstanding PLED behavior. The authors believe that it can be attributed to a good balance of electron and hole mobility. In addition, these two polymers exhibit relatively high glass transition temperatures of >200°C, which is a favorable property for devices with stable lifetimes.

Shao and coworkers [61] prepared a novel, bipolar polymeric host PCzPO (**45**, Chart 4.19) based on a nonconjugated poly(aryl ether) containing triarylphosphine oxide units in the main chain and carbazole units in the side chains. Double-layer blue-emitting PPLEDs fabricated with the configuration ITO/PEDOT:PSS/PCzPO:FIrpic/TPCz (**46**)/LiF/Al achieved an efficiency significantly higher than those using PVK (**4**) as a host. Here, TPCz stands for 3,6-bis(diphenylphosphoryl)-9-(4-(diphenylphosphoryl)phenyl)-9*H*-carbazole, which acts as a hole/exciton-blocking material. This polymeric host possesses a bipolar character and a high E$_T$ level of 2.96 eV. With a FIrpic (**16**) concentration of 5 wt.%, the maximum

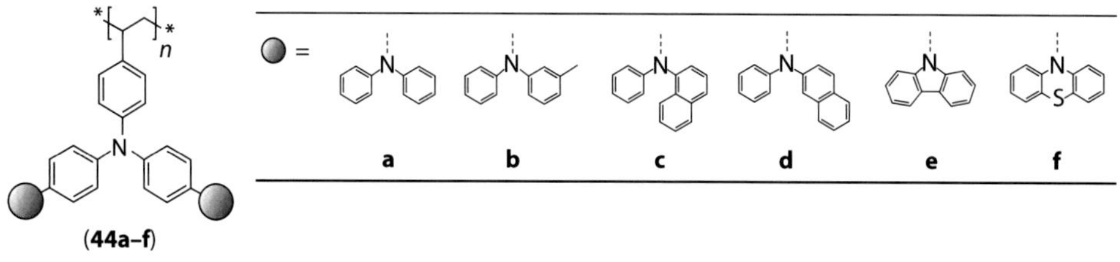

CHART 4.18
Chemical structure of polymers **44a–f**.

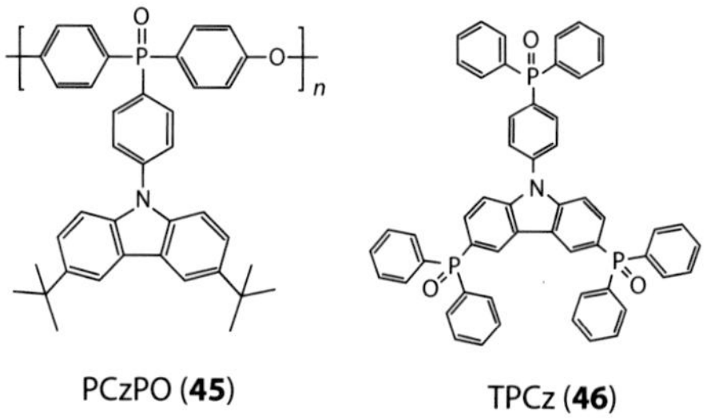

PCzPO (**45**)   TPCz (**46**)

CHART 4.19
Chemical structure of PCzPO and TPCz.

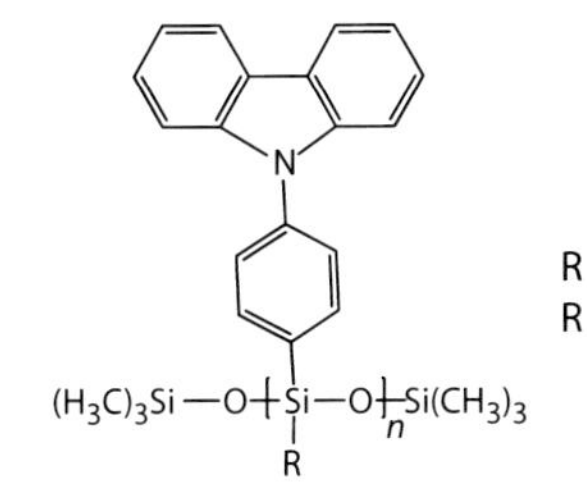

R = methyl, PCzMSi (**47**)
R = phenyl, PCzPhSi (**48**)

CHART 4.20
Chemical structure of PCzMSi and PCzPhSi.

luminous efficiency, power efficiency, and EQE of blue-emitting PPLEDs reach 23.3 cd/A, 10.6 lm/W, and 10.8%, respectively, at a practical brightness of 116 cd/m². At a brightness level of 1000 cd/m², the EQE still remained as high as 9.1%.

Very recently, Sun et al. [62] reported a novel series of efficient nonconjugated host polymers PCzMSi (**47**) and PCzPhSi (**48**) containing carbazole moieties linked to the backbone of polysiloxane through a phenyl bridge (Chart 4.20). The silicon–oxygen linkage disrupts their conjugation and results in a sufficiently high E_T level of 3.0 eV. Blue light–emitting PPLEDs fabricated with the configuration ITO/PEDOT:PSS/host polymer:10 wt.% FIrpic/TmPyPb/TPBI/LiF/Al show good overall performance with low efficiency roll-off. The device using PCzMSi (**47**) as the host demonstrates the best performance with a maximum current efficiency of 22.8 cd/A, a maximum power efficiency of 9.4 lm/W, and a maximum EQE of 11.9% at a practical luminance of 1165 cd/m². Even at a brightness of 5000 cd/m², the EQE still remains as high as 10%, suggesting a gentle roll-off of device efficiency at high current density.

4.4 Conjugated Polymers as Host Materials

4.4.1 Polyfluorenes

Guo and coworkers [63] were the first to use polyfluorene as a host material in PPLED application. They demonstrated an electrophosphorescent device using conjugated polyfluorene PDOF (**13**) as the host material and PtOX (**14**) as a dopant. The emission spectrum of PDOF overlaps well with the absorption of the PtOX dopant, suggesting an efficient energy transfer from PDOF to PtOX. The device in the configuration ITO/PEDOT/PVK/PDOF:PtOX/Ca:Al, having only 1% of the dopant, showed deep-red emission at λ_{max} = 656 nm with an EQE of 2.3% at 11 cd/m², which is twice higher than that of the fluorescent PDOF-based PLED.

O'Brien et al. [64] employed a similar Pt complex (PtOEP **1**) in PDOF-based PPLEDs. The light-emitting device was built with *N,N*-diphenyl-*N',N'*-di(3-carboxyphenyl)-4,4'-diaminobiphenyl (BFA) as a hole-transporting or injecting layer (ITO/BFA/PDOF:4% PtOEP/Ca), which allowed the maximal EQE of 3.5% and a peak brightness >200 cd/m² to be reached. The operating voltage (~25 V), however, is much higher than that of polyfluorene-based fluorescent PLEDs and is certainly above the technologically acceptable level. The absorption and PL spectra of PtOEP/PDOF blends suggest an efficient Förster energy transfer (Figure 4.10); however, measurements of triplet-state dynamics in PDOF

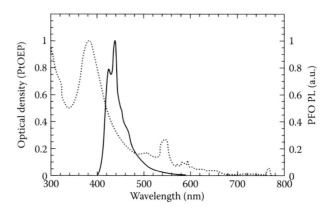

FIGURE 4.10
Emission spectra of the host PDOF (**13**) (solid line) and the absorption of the guest PtOEP (**1**) (broken line). (From O'Brien, D.F., Giebelerb, C., Fletcherb, R.B., Cadbyb, A.J., Palilisb, L.C., Lidzeyb, D.G., Laneb, P.A., Bradley, D.D.C., and Blau, W., *Synth. Met.*, 116, 379, 2001. With permission.)

(by photoinduced absorption spectroscopy) show that virtually no Dexter energy transfer takes place in this system [65]. Accordingly, the PDOF triplets cannot be transferred onto the dopant and are not expected to contribute to the emission, and there must be some other mechanism responsible for high observed EL efficiency. Indeed, increase of the driving voltage with increasing concentration of PtOEP (**1**) and the fact that the ionization potential of PtOEP is 0.5 eV lower than that of PDOF suggest that a direct trapping of holes (followed by electron trapping and recombination) occurs on PtOEP. As a result, a higher electric field (i.e., a higher operating voltage) is needed to transport the required amount of charge through the device.

As in conventional, electrofluorescent devices, the turn-on voltage of electrophosphorescent devices can be essentially decreased in light-emitting electrochemical cell (LEC). Chen and coworkers [66] have prepared red-emitting LEC using a hydrophilic polyfluorene BDOH-PF (**49**) host (Chart 4.21) together with the triplet emitter Btp$_2$Ir (**15**) [66]. The turn-on voltage was as low as ~3 V, and a power efficiency of 1.0 lm/W was achieved, whereas the corresponding conventional PPLED device (without electrolyte) showed six times lower power efficiency and higher turn-on voltage.

A very detailed study of the energy transfer processes in PPLEDs has been recently reported by Kim and coworkers [67]. They compared the behavior of the nonconjugated PVK and conjugated fluorene–phenylene host copolymer poly[9,9'-di-*n*-hexyl-2,7-fluorene-*alt*-1,4-(2,5-di-*n*-hexyloxy)phenylene (PFHP **50**) (Chart 4.21) while using Ir(ppy)$_3$ (**3**) and PtOEP (**1**) as the guest dopants. The emission of both polymers is in good overlap with the absorption spectra of the dopants, suggesting efficient Förster energy transfer. The experimental results indicated efficient singlet and triplet energy transfers in PVK/Ir(ppy)$_3$ and PVK/PtOEP

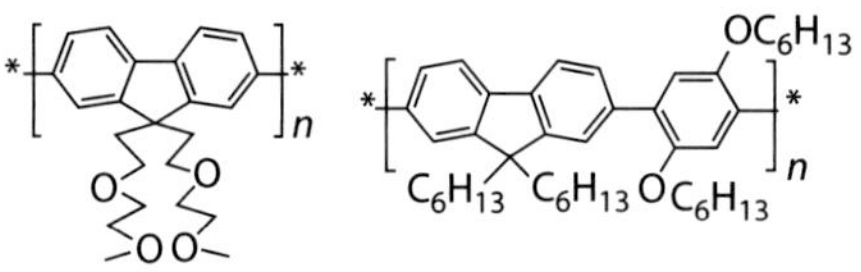

CHART 4.21
Chemical structure of BDOH-PF and PFHP.

systems due to large Förster radius, high energy (2.46 eV), and long lifetime of the host trip-let state. In the case of the conjugated PFHP, similar criteria do not result in efficient energy transfer, which the authors explain by the formation of the aggregates in the conjugated polymer. However, the actual triplet energy of PFHP (**50**) (2.3 eV) is lower than that of Ir(ppy)$_3$ (2.4 eV), which might well explain the observed behavior (Table 4.1). The PPLED fabricated as ITO/PEDOT/PVK–Ir(ppy)$_3$/TAZ (**6**)/Alq$_3$ (**5**)/LiF/Al showed the maximum EQE up to 6%, which is 15 times higher than that of a corresponding device using PFHP as a host polymer. Lower efficiency of conjugated PDOF-based PPLED as compared with nonconjugated PVK in PPLEDs was also mentioned by Yang and coworkers [39,68], who ascribed this fact to quenching the phosphorescence of Ir complexes by low-energy triplet state of the polymer.

Bazan and coworkers [50,69] have studied the fluorene copolymer PF3CNP1 (**51**), contain-ing electron-deficient dicyanophenylene fragment as a host material for PPLEDs (Chart 4.22). They used p-extended fluorene-containing dopants Ir(DPF)$_3$ (**22**), Ir(DPPF)$_3$ (**23**), and Ir(HFP)$_3$ (**24**), which have previously shown excellent results in PVK-based devices. The spectral overlap of the guest absorption and host emission suggests an efficient Förster energy transfer (Figure 4.11). The PPLED ITO/PEDOT/PF3CNP1:Ir(HFP)$_3$/Ca/Ag with only 1 wt.% of the dopant exhibited a maximum luminescence of 2200 cd/m^2 and a maxi-mum EQE of 1.5% and EL efficiency of 3 cd/A (achieved at 15 V) [69]. These values are essentially lower than those for the similar PVK-based device, although the turn-on volt-age was brought down to 4.5 V, as compared with 10 V for a similar PVK-based device [49]. In addition, the stability of polyfluorene-based device was also higher. Recently, even higher-efficiency red-emitting device made from Ir(HFP)$_3$ (**24**) and PDOF (**13**) was reported

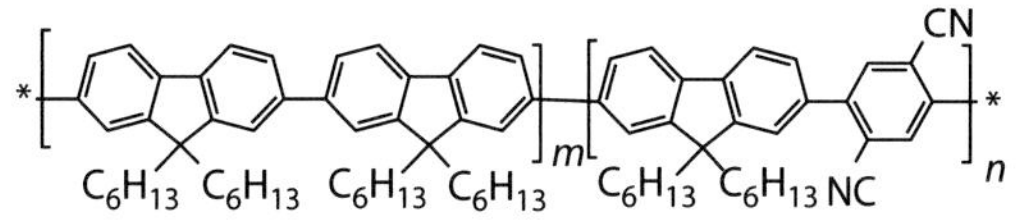

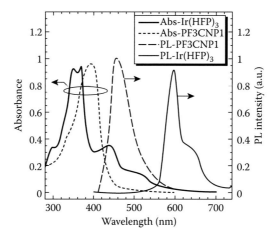

PF3CNP1 (**51**) (*m* = 0.75, *n* = 0.25)

CHART 4.22
Chemical structure of PF3CNP1.

FIGURE 4.11
Absorption or emission spectra of the guest Ir(HFP)$_3$ (**24**) and host PF3CNP1 (**51**) materials. (From Gong, X., Ostrowski, J.C., Bazan, G.C., Moses, D., Heeger, A.J., Liu, M.S., and Jen, A.K.Y., *Adv. Mater.*, 15, 45, 2003. With permission.)

by the same group. PPLED fabricated with the configuration ITO/PEDOT/PDOF:Ir(HFP)$_3$/ Ca/Ag showed an EQE of 4.5%, a current efficiency of 6.2 cd/A, a maximum brightness of >1000 cd/A, and a turn-on voltage of ~5 V [50].

Chen and coworkers [70] studied the energy transfer from the PDOF host polymer onto a series of iridium complexes [Ir(ppy)$_3$ **3**, PPIr **19**, Bt$_2$Ir **17**, Btp$_2$Ir **15**] with triplet energy levels positioned below and above that of the host polymer (Table 4.1). They found a relationship between the triplet energy of the dopants and the PL emission. When the triplet exciton (T_1) of the dopant (PPIr **19**) is higher in energy than that of the host polymer, it can be quenched via back energy transfer from dopant onto the polymer, and only the emission of the latter is observed (Figure 4.12). This is not the case if the triplet energy of the dopant (Btp$_2$Ir **15**) is lower than that of the host polymer, and the triplet excitons, thus confined on the dopant, can decay radiatively with high quantum yield. When the triplet energy levels of the dopant (Bt$_2$Ir **17**) and the host polymer are close, the competition between the energy transfer processes from the host to the dopant and from the dopant back to the host polymer will take place. However, in the PPLED devices, the charge-trapping process often results in domination of the dopant EL, regardless of the energy-level difference. Nevertheless, the device efficiency correlates with the above-mentioned charge transfer processes: the dopant Btp$_2$Ir with the lowest triplet energy showed the highest EQE (2%).

Chen and coworkers [68] also compared the EL performance of these phosphorescent dopants as well as the alkyl-substituted Ir(Ocppy)$_3$ (**52**) (Chart 4.23) blended in PDOF (**13**) and PVK (**4**). The study once again confirmed a relatively poor performance (in terms of brightness and efficiency) of the PDOF host as compared with nonconjugated PVK (Table 4.5). Interestingly, the "plasticizing" alkyl substituents in the complex Ir(Ocppy)$_3$ (**52**) improve the film morphology (as demonstrated by atomic force microscopy) and give the best current efficiency, although the turn-on voltage of the PPLED with this dopant also tends to be the highest (which is explained by more efficient hole trapping on this complex) [68].

Thus, to achieve high EL efficiency, the triplet energy of a host material should be above that of the dopant. For this reason, conjugated polyfluorenes are not expected to perform

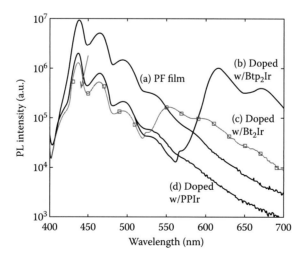

FIGURE 4.12
PL spectra of the (a) undoped PDOF (**13**) film and PDOF doped with (b) 11 wt.% Btp$_2$Ir (**15**), (c) 10 wt.% Bt$_2$Ir (**17**), and (d) 10 wt.% PPIr (**19**), pumping at 382 nm. (From Chen, F., He, G., and Yang, Y., *Appl. Phys. Lett.*, 82, 1006, 2003. With permission.)

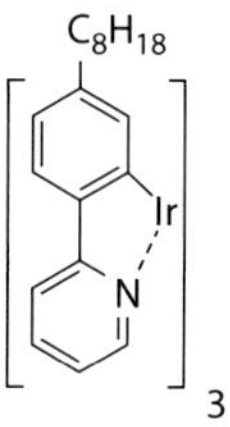

C_8H_{18}

Ir(Ocppy)$_3$ (**52**)

CHART 4.23
Chemical structure of Ir(Ocppy)$_3$.

TABLE 4.5

Comparison of the Performance of PVK (**4**) (ITO/PEDOT/PVK:PBD:Dopant/Ca/Al) and PDOF (**13**) (ITO/PEDOT/PVK/PDOF:Dopant/Ca/Al)-Based PPLED with Different Dopants

Host	Dopant (wt.% in Host)	Turn-On Voltage (V)	Efficiency (cd/A)	λ_{max} (nm)
	Ir(ppy)$_3$ (3%)	5.5	12.7	516
	PPIr (3%)	5.2	13.0	516
PVK	Ir(Ocppy)$_3$ (3%)	9.6	16.0	518
	BtIr (3%)	5.5	8.0	560
	BtpIt (4%)	6.5	2.6	614
	Ir(ppy)$_3$ (3%)	4.5	3.9	520
	PPIr (3%)	4.5	4.1	526
PF	Ir(Ocppy)$_3$ (2%)	11.0	6.2	518
	BtIr (5%)	5.2	3.0	560
	BtpIr (5%)	5.0	1.9	614

Source: Chen, F.C., Yang, Y., and Pei, Q., *Appl. Phys. Lett.*, 81, 4278, 2002. With permission.

well with green dopants owing to their low triplet energy, although it can be used as an efficient host material for red-emitting PPLEDs. A red-emitting PLED having highest efficiency based on polyfluorene was reported by Jiang and coworkers [71]. Blending silsesquioxane end-capped PDOF and triplet emitter (piq)$_2$Ir(acac) (**32**), they have produced saturated red light with CIE-coordinated values of 0.67, 0.33. The device ITO/PEDOT/PVK/PDOF:PBD(30%):(piq)$_2$Ir(acac)(2%)/Ba/Al showed an EQE as high as 12% and power efficiency of 5.2 cd/A (at 15 V).

An interesting approach to white-emitting polyfluorene-based PPLEDs was reported by Moses and coworkers [72,73]. As described in detail in Chapter 2, fluorenone defects in the polyfluorene ring result in efficient quenching of the blue fluorene emission producing green color. The group at UCSB deliberately introduced a low concentration (1%) of fluorenone units into the PDOF (**13**) chain and blended this green-emitting material with blue-emitting pristine PDOF and a small amount (~1%) of the red-emitting phosphorescent dopant Ir(HFP)$_3$ (**24**) to produce pure white light with current efficiency of up to 3 cd/A (turn-on voltage 5 V, maximal brightness 6100 cd/m^2) (Figure 4.13, device II). The color of such a device (CIE: 0.32, 0.33) is almost independent of the current density.

Su and coworkers [74] used the Os(fppz) (**53**) triplet emitter to improve the EL of the fluorene copolymer PF-Q (**54**) (Chart 4.24) [74]. A red-emitting PPLED, ITO/PEDOT/2.4%Os(fppz) (**53**):PF-Q (**54**)/TPBI (**9**)/Mg:Ag, showed an EQE of 6.6% and a maximum brightness of 10,400 cd/m^2, which is almost 10 times higher than that in a similar device without a triplet emitter.

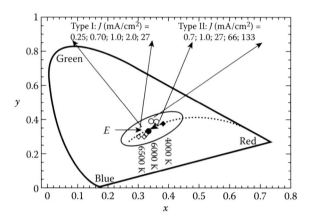

FIGURE 4.13
CIE chromaticity diagram of white PPLEDs based on PDOF (**13**) and Ir(HFP)$_3$ (**24**). (From Gong, X., Ma, W., Ostrowski, J.C., Bazan, G.C., Moses, D., and Heeger, A.J., *Adv. Mater.*, 16, 615, 2004.)

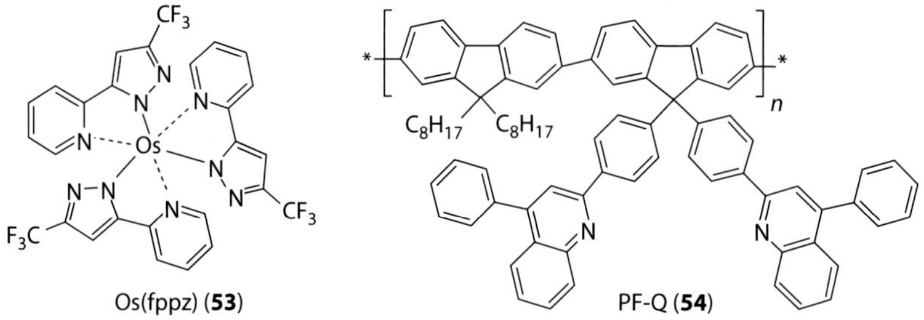

Os(fppz) (**53**)   PF-Q (**54**)

CHART 4.24
Chemical structure of PF-Q and Os(fppz).

King et al. [75] investigated the reduction of the rate of triplet energy transfer from the dopant to the host by making inert *t*-butyl substitutions to the ligands of the widely used phosphorescent dopant Ir(ppy)$_3$ (**3**) to get its derivatives, **55a**, **55b**, and *fac-tris*-[2-(4′-*tert*-butyl)phenylpyridine]iridium(III) Ir(Bu-ppy)$_3$ (**56**) (Chart 4.25). PPLEDs were constructed with the configuration ITO/PEDOT:PSS/poly(9,9′-spirobifluorene) (PSBF **57**):Ir complex/Ca/Al. The rate of triplet back-transfer for Ir(ppy)$_3$ (**3**), **55a**, **55b**, and Ir(Bu-ppy)$_3$ (**56**) (triplet energy of ca. 2.5 eV) in a PSBF host with a significantly lower triplet energy of 2.1 eV (Figure 4.14) was measured. Transient photoinduced absorption measurements of the doped films showed that the time constant for the build-up of triplets on the polymer host follow the sequence of 7.1 ns (**55a**) > 6.3 ns (**55b**) > 5.3 ns (**56**) > 2.8 ns (**3**), indicating that the bulky substitutions introduce steric bulk to the dopant that inhibit triplet energy back-transfer to the polymer host. This is attributed to the strong distance dependence of the Dexter-type energy transfer involved. Introducing bulky groups into the ligand of the phosphor could suppress triplet energy back-transfer by reducing the orbital overlap of the phosphor and polymer host necessary for Dexter transfer.

Kim et al. [76] demonstrated highly efficient white PPLEDs using blue-emitting PFs as a host and two phosphorescent iridium complexes as dopants. They employed the blue emitter BlueJ (**58**) from Dow Chemical Company; a green phosphorescent emitter, *fac-tris*[2-(2-pyridinyl-kN)-5-(3,4-di-*sec*-butoxyphenyl)phenyl-kC]iridium(III) (Ir(PBPP)$_3$ **59**), and

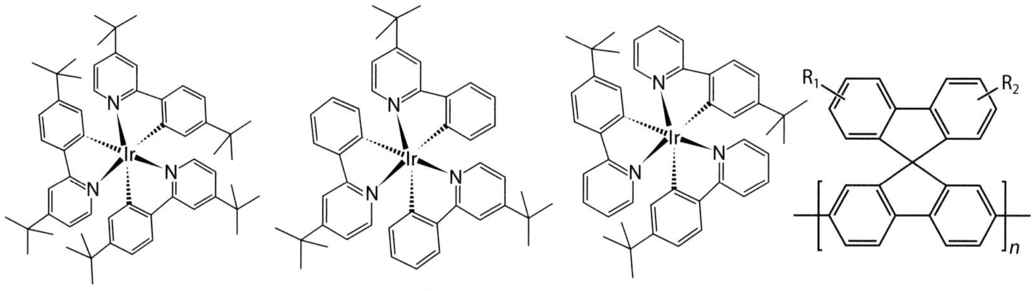

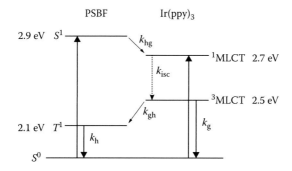

CHART 4.25
Chemical structure of the derivatives of phosphorescent dopant Ir(ppy)$_3$ (**3**) **55a**, **55b**, and Ir(Bu-ppy)$_3$ (**56**), and host polymer PSBF (**57**).

FIGURE 4.14
Energy-level diagram for the polymer–dopant system used in the study. k_{hg}, rate of host–guest singlet energy (Föster) transfer; k_{isc}, intersystem crossing rate of the polymer; k_{gh}, rate of triplet energy transfer to the polymer triplet; k_h, radiative decay rate of the host polymer; k_g, rate of phosphorescence of the complex; PSBF (**57**), poly(9,9′-spirobifluorene); *S*, singlet energy level; *T*, triplet energy level. (From King, S.M., Al-Attar, H.A., Evans, R.J., Congreve, A., Beeby, A., and Monkman, A.P., *Adv. Funct. Mater.*, 16, 1043, 2006. With permission.)

a red phosphorescent emitter, *tris*(1-phenylisoquinoline)iridium(III) (Ir(PIQ)$_3$ **60**) (Chart 4.26). PPLEDs were fabricated with a single active layer architecture of ITO/PEDOT:PSS/BlueJ:PVK:Ir(PBPP)$_3$:Ir(PIQ)$_3$/BCP/Li:Al. The BCP (**18**) layer acted as a hole-blocking layer. PVK (**4**) was used as co-host to improve the overall miscibility, facilitate hole transport, and reduce phosphorescence quenching by raising the triplet energy levels of the polymer blend at the PVK/PF interface. The energy-level diagram of PVK, BlueJ (**58**), Ir(PBPP)$_3$ (**59**), Ir(PIQ)$_3$ (**60**), and BCP (**18**) is shown in Figure 4.15. Nearly pure white light with CIE coordinates in the region of (0.34, 0.34) to (0.32, 0.32) was obtained. The peak luminance efficiency of the BlueJ 65:PVK 25:Ir(PBPP)$_3$ 9.7:Ir(PIQ)$_3$ 0.3 (wt.%) device is 12.52 cd/A at a current density of 7.23 mA/cm^2, a voltage of 6.2 V, and a brightness of 905 cd/m^2, corresponding to an EQE of 3.2%.

Xu and coworkers [77] reported efficient white PLEDs containing a highly efficient blue emitter, polyhedral oligomeric silsesquioxane terminated poly(9,9′-dioctylfluorene) (PFO-poss **61**) as a host, and two phosphorescent iridium complexes as dopants. The authors employed green-emitting Ir(Bu-ppy)$_3$ (**56**) and red-emitting *bis*-(1-phenylisoquinolyl) iridium(III) (1-trifluoro) acetylacetonate (Piq)$_2$Ir(acaF) (**62**) (Chart 4.27). The PPLED is fabricated with an architecture of ITO/PEDOT:PSS/PVK/PFO-poss:Ir(Bu-ppy)$_3$:(Piq)$_2$Ir(acaF)/Ba/Al. The PVK layer was used to block the electron penetrating from PFO-poss to PEDOT.

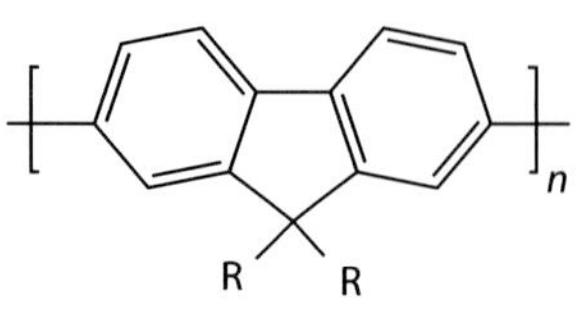

BlueJ (**58**)

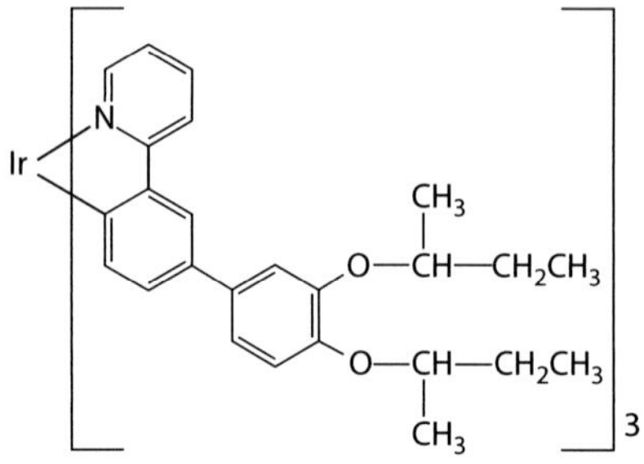

Ir(PBPP)$_3$ (**59**)

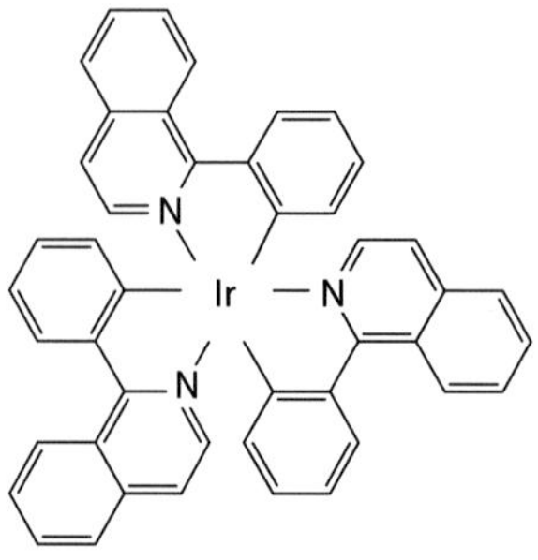

Ir(PIQ)$_3$ (**60**)

CHART 4.26
Chemical structure of BlueJ, Ir(PBPP)$_3$, and Ir(PIQ)$_3$.

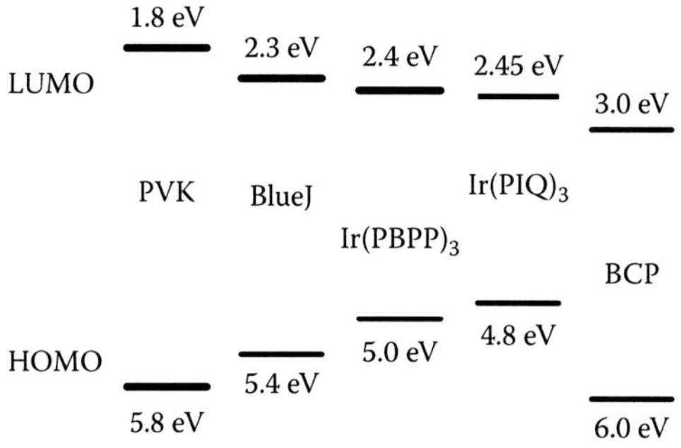

FIGURE 4.15
Energy-level diagram of PVK (**4**), BlueJ (**58**), Ir(PBPP)$_3$ (**59**), Ir(PIQ)$_3$ (**60**), and BCP (**18**). (From Kim, T.H., Lee, H.K., Park, O.O., Chin, B.D., Lee, S.H., and Kim, J.K., *Adv. Funct. Mater.*, 16, 611, 2006. With permission.)

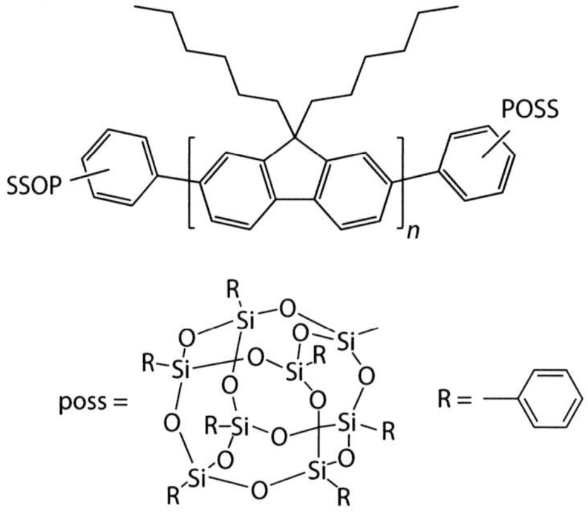

PFO-poss (**61**)

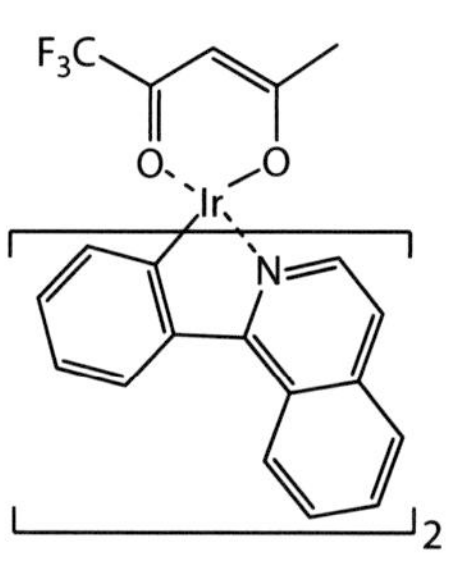

(Piq)$_2$Ir(acaF) (**62**)

CHART 4.27
Chemical structure of PFO-poss and (Piq)$_2$Ir(acaF).

For the green emission, the trap mechanism is the main exciton recombination process, while for the red emission, both the energy transfer and the trap effect are primary mechanisms. A pure white light emission with CIE coordinates of (0.33, 0.33) was obtained with the doping concentrations of both red and green triple emitters at 0.14 wt.%. The PPLEDs exhibit peak luminous efficiency of 9.0 cd/A and power efficiency of 5.5 lm/W at 5.6 V. The maximum luminance of 10,200 cd/m^2 was gained at 11 V.

The same group [78] developed a series of morphology-stable carbazole-based iridium(III) complexes **63a–f** (Chart 4.28) with green to red emission. The nonconjugated PVK (**4**) and conjugated PFO-poss (**61**) were applied as the host matrixes. The device configurations were ITO/PEDOT:PSS/PVK:40% PBD:Ir complex/Ba/Al and ITO/PEDOT:PSS/PVK/PFO-poss:30% PBD:Ir complex/Ba/Al. With the device structure of ITO/PEDOT:PSS/PVK/PFO-poss:30% PBD:2% **63b**/Ba/Al, a maximum EQE of 6.4% and a maximum luminous efficiency of 6.0 cd/A with red emission at 608 nm were achieved. With the device configuration of ITO/PEDOT:PSS/PVK/PFO-poss:30% PBD:4% **63f**/Ba/Al, a maximum EQE of 9.9% and a maximum luminous efficiency of 22.4 cd/A with yellow–green emission at 544 nm were obtained. By introducing the long and bulky alkyl chain in the ligand frame, the morphological stability of the complexes and the compatibility between the complexes and polymer host can be improved. It was observed that **63b** and **63d** imparted by the *N*-decyl long chains at the *N* atom of carbazole showed a significantly improvement of device performance than their short-chain analogues **63a** and **63c** under identical device configurations.

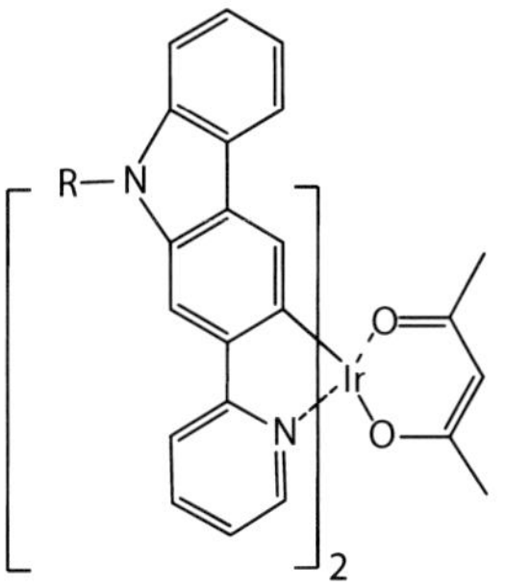

(**63a**) R = C$_2$H$_5$ Ir(2-PyEtCz)$_2$(acac)
(**63b**) R = *n*-C$_{10}$H$_{21}$ Ir(2-PyDeCz)$_2$(acac)

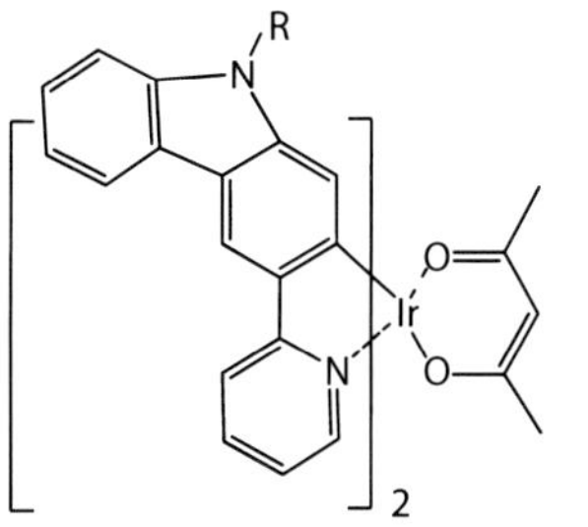

(**63c**) R = C$_2$H$_5$ Ir(3-PyEtCz)$_2$(acac)
(**63d**) R = *n*-C$_{10}$H$_{21}$ Ir(3-PyDeCz)$_2$(acac)

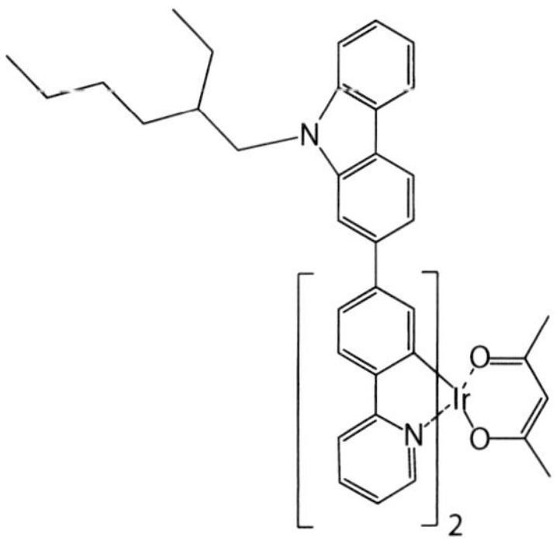

(**63e**) Ir(2-PhPyCz)$_2$(acac)

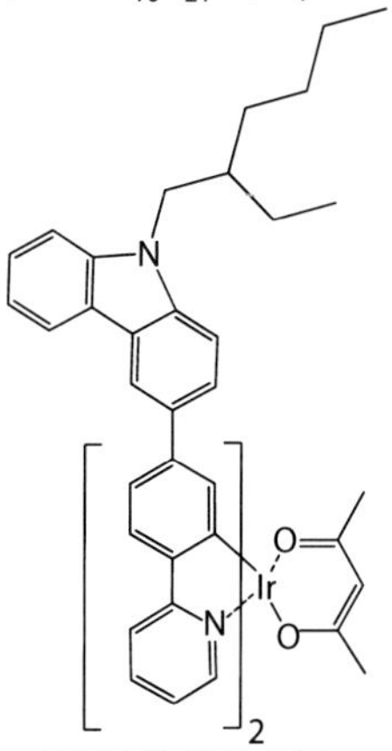

(**63f**) Ir(3-PhPyCz)$_2$(acac)

CHART 4.28
Chemical structure of carbazole-based Ir(III) complexes **63a–f**.

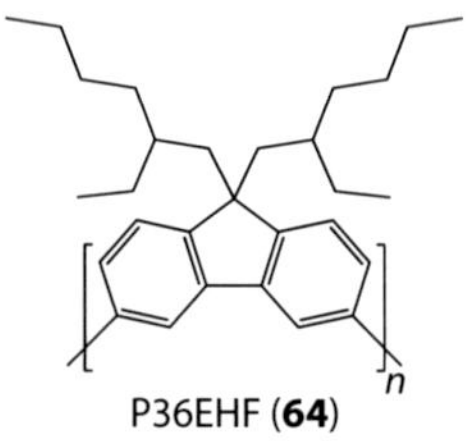

CHART 4.29
Chemical structure of P36EHF.

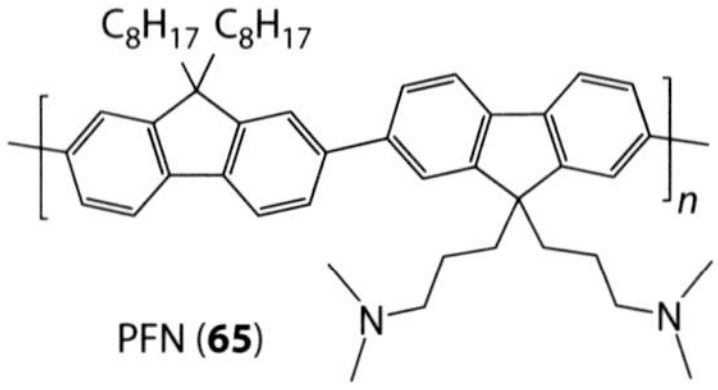

CHART 4.30
Chemical structure of PFN.

Later, the same group [79] reported the synthesis of a novel fluorene homopolymer poly(9,9'-bis(2-ethylhexyl)-3,6-fluorene) (P36EHF **64**) with the fluorene units linked at the 3,6-positions (Chart 4.29). The triplet energy of P36EHF (**64**) is 2.58 eV, which is much higher than that of poly(2,7-fluorene)s (2.15 eV); therefore, P36EHF (**64**) is a suitable polymer host for green- and even blue-emitting phosphors. PPLEDs were fabricated with an architecture of ITO/PEDOT:PSS/PVK/P36EHF:FIrpic/Ba/Al. The triplet energy back-transfer from FIrpic (**16**) to P36EHF (**64**) was significantly reduced because of the relatively high triplet energy level of P36EHF (**64**). P36EHF-based green and blue PPLEDs achieved modest device performance (0.4 cd/A, at 10 wt.% FIrpic (**16**) doping concentration). The design of this type of conjugated polymer showed an example to achieve homopolymers with high triplet energies.

Recently, Chen et al. [80] studied efficient green light-emitting PPLEDs based on neutral aminoalkyl substituted polyfluorene, poly(9,9-bis(3'-(*N,N*-dimethylamino)propyl)-2,7-fluorene)-*alt*-2,7-(9,9-dioctylfluorene) (PFN **65**, Chart 4.30) as the host, an Ir complex Ir(Bu-PPy)₃ (**56**) as the dopant, and Al as the cathode. PPLEDs were fabricated with an architecture of ITO/PEDOT:PSS/PVK/PFN: Ir(Bu-PPy)₃/Al, wherein PFN plays a dual role of both the host and electron-injection material. High luminous efficiency of 35.7 cd/A, EQE of 14.2%, and maximum luminance of 17,161 cd/m² were achieved, with the optimized thickness of the PVK and EL layers. The device performance was comparable to those of the devices using PVK as the host, Ir(Bu-PPy)₃ (**56**) as dopant, and a low-work-function metal as cathode (cf. barium or calcium). The authors believe that the effective electron injection from the PFN/Al bilayer cathode improved charge balance, and the significant reduction of back-transfer of excitons from the Ir complex to the low-lying PFN (**65**) triplet level in the PVK/PFN interface is responsible for the high device performance.

4.4.2 Poly(*p*-Phenylenes)

The first report on increasing efficiency of a conjugated PLED via doping with a phosphorescent dye was described in 1999 by McGehee and coworkers [31], who added triplet-emitting europium complexes to substituted poly(*p*-phenylene) CN-PPP (**66**). Comparing

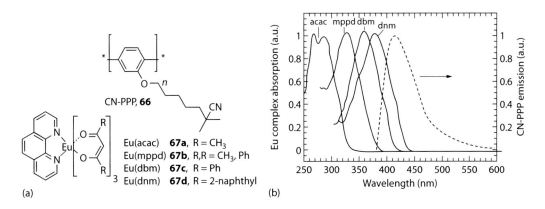

FIGURE 4.16
(a) Molecular structure of Eu triplet emitters (**67a–d**) and CN-PPP (**66**) host polymer. (b) Emission spectrum of CN-PPP (broken line) and absorption of the Eu complexes with different ligands (solid lines). (From McGehee, M.D., Bergstedt, T., Zhang, C., Saab, A.P., O'Regan, M.B., Bazan, G.C., Srdanov, V.I., and Heeger, A.J., *Adv. Mater.*, 11, 1349, 1999. With permission.)

the absorption spectra of a series of Eu–phenanthroline complexes (**67**) with the emission of the CN-PPP, they have shown that only phenyl- or naphthyl-substituted dopants can efficiently accept the energy transfer from the host polymer (Figure 4.16). As expected from the overlap of the polymer emission and the dye absorption spectra, the most efficient energy transfer was observed for naphthyl-substituted Eu (dnm) (**67d**), as manifested in complete replacement of the CN-PPP (**66**) PL by the dopant emission (l_{PL} ~610 nm) at 5 wt.% concentration. The PLED device fabricated with PVK (**4**) as a hole-transporting layer (ITO/PVK/CN-PPP:5% Eu (dnm) (**67d**)/Ca) showed an EQE of 1.1%. Although this is not a great value even for fluorescent PLEDs, the paper presented an important demonstration that adding a triplet-emitting dye increases the efficiency of the device (cf. 0.3% for PLED without the Eu complex). The authors also mention achieving a very narrow emission spectral width (<4 nm) as a route to improve color purity. It is noteworthy that the increase of the device efficiency occurs in spite of a very low F_{PL} of the Eu complex (2% cf. 80% for CN-PPP polymer), which is due to a very long phosphorescent lifetime of ca. 0.5 ms. The role of harvesting the triplets was mentioned in the paper, although, at that time, only as one of the hypotheses.

Clearly, using a triplet emitter with a shorter excited state lifetime should improve the device efficiency. In 2002, Zhu and coworkers [81] used the iridium complex Ir(ppy)$_3$ (**3**) and its derivatives with shorter decay time to dope poly(p-phenylene)s EHO-PPP (**68**) (Chart 4.31) and CN-PPP (**66**), and demonstrated significantly improved device efficiency. The good overlap between the absorption band of the dopant Ir(ppy)$_3$ (**3**) and the emission

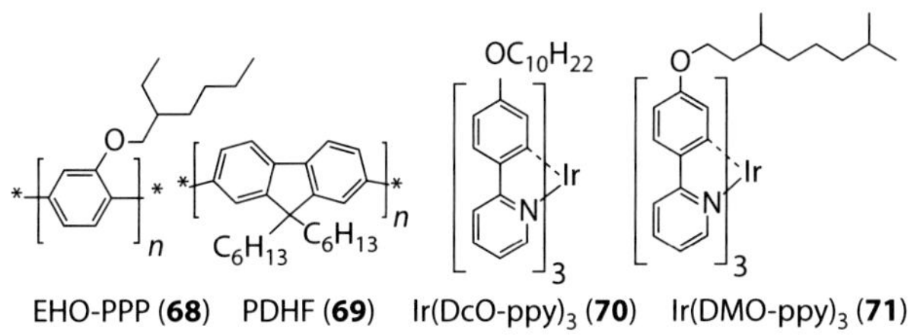

EHO-PPP (**68**)   PDHF (**69**)   Ir(DcO-ppy)$_3$ (**70**)   Ir(DMO-ppy)$_3$ (**71**)

CHART 4.31
Chemical structure of EHO-PPP, PDHF, Ir(DcO-ppy)$_3$, and Ir(DMO-ppy)$_3$.

of the host EHO-PPP and CN-PPP polymers meets the requirement for the Förster energy transfer. PPLEDs were fabricated using PVK (**4**) as a hole-transporting layer as ITO/PVK/ host polymer:iridium complex/Ba/Al. The best device was made with 4 wt.% Ir(Bu-ppy)₃ (**56**)-doped CN-PPP. The maximum EQE and EL efficiencies were 5.1% and 12 cd/A, respectively, observed at a brightness of 800 cd/m² and current density of 6.8 mA/cm². It changes only scarcely in the brightness range of 120–2500 cd/m², in contrast to many small-molecule phosphorescent LEDs [26]. The authors compared the quantum efficiency of EL devices made with different host materials (PVK **4**, PDHF **69**, CN-PPP **66**, and EHO-PPP **68**) using 2% of the iridium dopants, at a current density of 13.3 mA/cm² (Table 4.6) [81].

All the devices emitted green light with a peak at 515 nm (CIE: 0.33, 0.58). The EQE of CN-PPP (**66**) and EHO-PPP (**68**) were close, with the former a little bit higher. However, very low quantum efficiency was found in devices made with the host polymers PDHF (**69**) (Chart 4.31) and PVK (**4**) in spite of the good overlap of the emission spectra of these two polymers with the guest materials. Interestingly, the iridium complex with *t*-butyl group Ir(Bu-ppy)₃ (**56**) as the guest material showed higher quantum efficiency than Ir(ppy)₃ (**3**). The authors attribute this improvement to the more homogeneous distribution of the guest molecules in the host polymer matrix, owing to the alkyl substituent. However, the same group later demonstrated that using longer alkyl chains in the ligand (complexes Ir(DcO-ppy)₃ (**70**) and Ir(DMO-ppy)₃ (**71**)), on the contrary, decreases the efficiency of the device (with CN-PPP host) [82]. The *t*-butyl-substituted complex still remains the most efficient in the series, providing a high EQE even at a very high brightness (4.2%, 10 cd/A at 2500 cd/m²). At the same time, extending the conjugation in the ligand can be used to tune the emission color, and a red-emitting PPLED (l_{EL} ~600 nm, CIE: 0.59, 0.38) was fabricated with CN-PPP host and isoquinoline–Ir complex ((piq)₂Ir(acac) **32**) [83]. The two-layer device ITO/PVK/CN-PPP:4%(piq)₂Ir(acac)/Ba/Al showed an external efficiency of 1.3% (0.47 cd/A).

Yang and coworkers [84] suggested that replacing PDOF (**13**) with a similar host polymer, ladder poly(*p*-phenylene) (LPPP) (**72**) (Chart 4.32), may reduce the hole-trapping process in PtOEP (**1**)-doped PPLEDs owing to lower hole trap depth (i.e., the difference of host–guest HOMO levels), which is only 0.2 eV, compared with 0.5 eV for PDOF/PtOEP. Indeed, a single-layer PPLED fabricated as ITO/PEDOT/LPPP+PtOEP/Ca/Al had a switch-on voltage of only 4 V, which, in contrast to polyfluorene PPLEDs, was independent of the dopant concentration. The maximum efficiency (achieved at ~7 V) was quite low (0.06 cd/A), although it was

TABLE 4.6

Device Performance of PPLEDs with Ir(ppy)₃ (**3**) and Ir(Bu-ppy)₃ (**56**) Dopants and Different Host Polymers

Host Polymer			Device Parameter of 2% Ir Complex-Doping Conc.					
Polymer	λ_{max}	QE (%)[a]	IrR₃	V	I (mA/cm²)	cd/m²	cd/A	QE (%)
PDHF	420	0.50	Ir(ppy)₃	14.5	13.3	9	0.07	0.04
PVK	410	0.005	Ir(ppy)₃	16.0	13.3	5	0.02	0.01
EHO-PPP	415	0.20	Ir(ppy)₃	22.0	13.3	334	2.51	1.06
EHO-PPP	415	0.20	Ir(Bu-ppy)₃	22.7	13.3	521	3.91	1.65
PPP	430	0.22	Ir(ppy)₃	33.0	13.3	597	4.48	2.15
PPP	430	0.22	Ir(Bu-ppy)₃	28.3	13.3	942	7.07	2.99

Source: Zhu, W., Mo, Y., Yuan, M., Yang, W., and Cao, Y., *Appl. Phys. Lett.*, 80, 2045, 2002. With permission.

[a] External QE of device with pure host blue polymer in the same device configuration as doped polymer ITO/PVK/host polymer/Ba/Al.

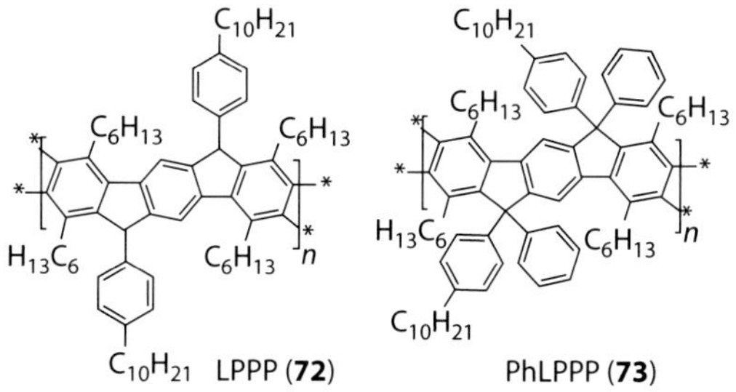

CHART 4.32
Chemical structure of LPPP and PhLPPP.

significantly improved in double-layer devices using PBD (**8**) or Alq$_3$ (**5**) electron-injecting layers (0.5 cd/A at 8 V) and triple-layer devices using both electron- and hole (PVK)-injecting layers (1.2 cd/A, 2.5%). For the latter, the brightness of 100 cd/m^2 is achieved at a voltage of ~13 V, which is significantly lower than that for other PtOEP (**1**)-based LEDs. The relatively low efficiency of the devices was explained by inefficient Förster energy transfer (due to a poor host–guest spectral overlap), while the offset between the HOMOs of the polymer and the Pt complex is still too high for the Dexter energy transfer. The authors conclude that an alternative concept of reducing the operating voltage must be developed.

Lupton et al. [85] studied delayed PL and EL of another ladder polyphenylene, PhLPPP (**73**). A pronounced phosphorescence at ca. 600 nm was observed at room temperature. Elemental analysis revealed the presence of ~80 ppm of Pd (i.e., one Pd atom per 1700 polymer units), as an unintentional impurity originating from the polymerization catalyst, presumably covalently bound to the polymer backbone. This finding adds an additional scrutiny for the purification of EL-conjugated polymers, most of which are prepared with the help of transition metal catalysis.

Harrison et al. [86] demonstrated a near-infrared PPLED based on polyphenylene PPP-OR11 (**74**) and lanthanide complex Yb(TPP)Tp (**75**) (Chart 4.33). Complete quenching of the polymer emission was observed at the dye concentration of 5%, producing near-infrared emission of the dopant at l_{EL} = 977 nm. The PPLED device can be turned on at 4 V (cf. 8 V for corresponding PPLEDs with nonconjugated polymers), although the EQE is very low (~0.01%).

Chen and coworkers [87] examined the phosphorescence quenching of high-triplet-energy phosphor guests bis(2-phenylpyridinato-*N,C*$^{2'}$)iridium(acetylacetonate) (ppy)$_2$Ir(acac) (**76**, green emitter), FIrpic (**16**, blue emitter), and FIr6 (**77**, blue emitter) by low-triplet-energy poly(*p*-phenylene)s C$_8$OPPP (**78**) and CzPPP (**79**) as host (Chart 4.34). The shielding effect provided by the side chain in the polymer host affects the triplet energy back-transfer.

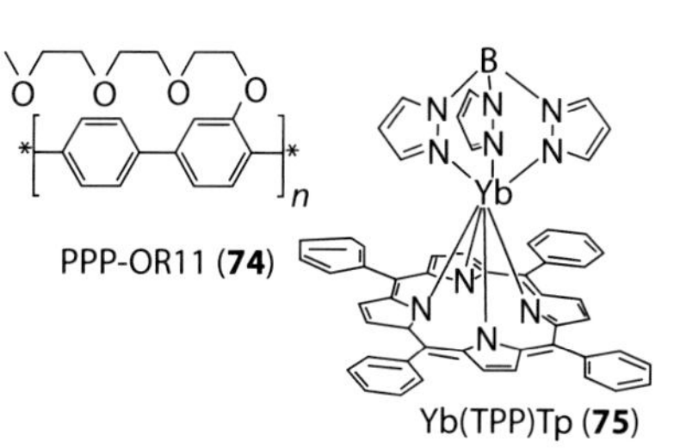

PPP-OR11 (**74**)

Yb(TPP)Tp (**75**)

CHART 4.33
Chemical structure of PPP-OR11 and Yb(TPP)Tp.

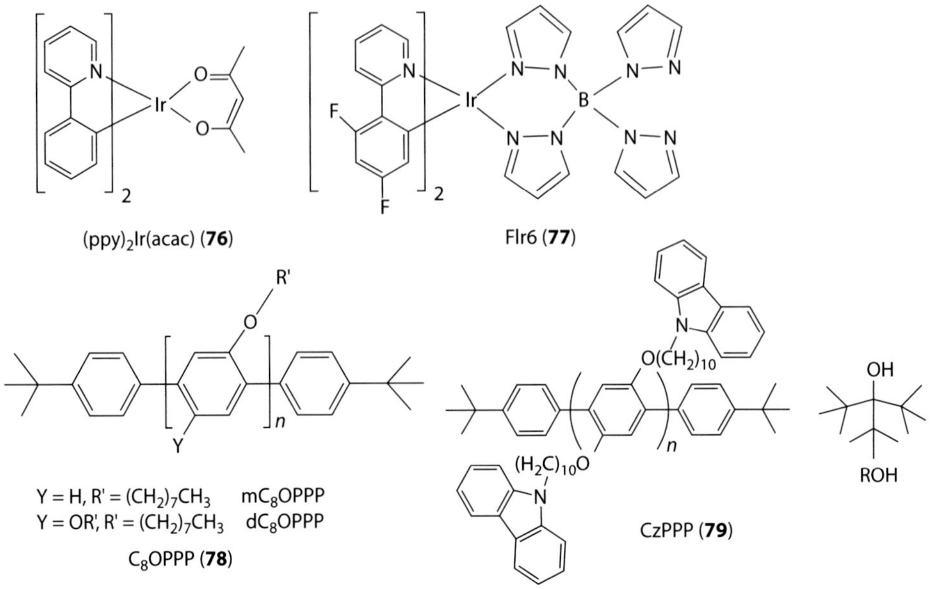

CHART 4.34
Chemical structure of (ppy)$_2$Ir(acac), FIr6, C$_8$OPPP, CzPPP, and ROH.

PPLEDs were fabricated with the architecture ITO/CF$_x$/polymer:dopant/TPBI/CsF/Ca/Al. CF$_x$ thin film (work function 5.3–5.7 eV) was used as a hole injection layer. The energy-level diagram of the materials used in the devices is shown in Figure 4.17. Capping the dialkoxyl-susbstituents with a carbazole (Cz) moiety, CzPPP (**79**) provides enhanced extent of shielding. An excellent device efficiency of 30 cd/A (8.25%) for a green electrophosphorescent device can be achieved with CzPPP (**79**) as a host, which is higher than that of dC$_8$OPPP (**78**) as a host (15 cd/A).

The same group [88] later published a similar system using a branched alcohol 3-*tert*-butyl-2,2,4,4-tetramethylpentan-3-ol (ROH) (Chart 4.34) into green emission phosphor-doped dialkoxyl-substituted poly(*p*-phenylene)s as host and high-triplet-energy (ppy)$_2$Ir(acac)

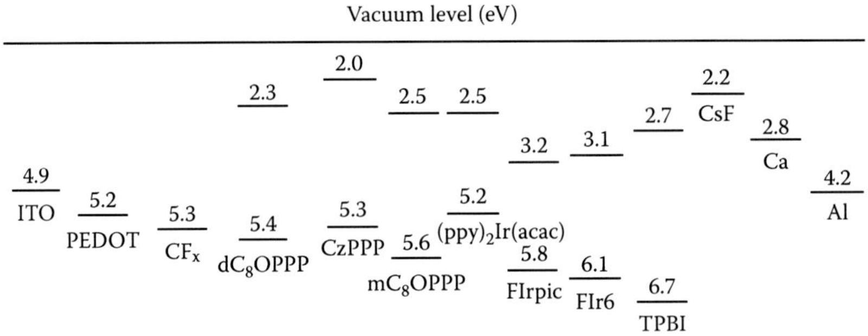

FIGURE 4.17
Energy-level diagram of the materials used in the devices. (From Huang, S.P., Jen, T.H., Chen, Y.C., Hsiao, A.E., Yin, S.H., Chen, H.Y., and Chen, S.A., *J. Am. Chem. Soc.*, 130, 4699, 2008. With permission.)

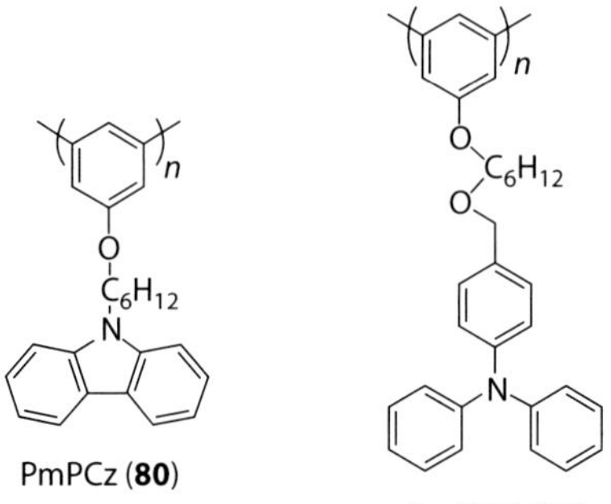

CHART 4.35
Chemical structure of PmPCz and PmPTPA.

(**76**) (2.41 eV) as the guest. PPLEDs were fabricated with the architecture ITO/PEDOT:PSS/ polymer:dopant/TPBI/CsF/Ca/Al. The performance of the green PPLEDs can be improved from 7.1 to 25.1 cd/A for C_8OPPP (**78**), and from 32.2 to 42 cd/A for CzPPP (**79**), respectively. This finding suggests that a promotion of specific electro-optical properties for phosphor-doped polymer systems can be obtained through self-assembling interactions in addition to chemical structure modification.

Recently, Liu and coworkers [89,90] reported two poly(*m*-phenylene) derivatives tethering carbazole unit (PmPCz **80**) or triphenylamine unit (PmPTPA **81**) (Chart 4.35). The E_T levels of PmPCz (**80**) and PmPTPA (**81**) reached 2.64 and 2.65 eV, respectively, which resulted from the twisted linkage of the phenyl rings in the polymer backbone. The E_T is sufficiently high to prevent triplet energy back-transfer and endows the polymers to host blue or white electrophosphorescence. In addition, the functional side groups can facilitate hole injection. PmPCz-based single-layer blue PPLED with a device architecture of ITO/ PEDOT:PSS/PmPCz:FIrpic (1 wt.%)/CsF/Al exhibited a luminous efficiency of 4.69 cd/A. The high-lying HOMO energy level of PmPTPA (**81**, 5.35 eV) facilitates hole injection. Blue PPLEDs were fabricated with the configuration ITO/PEDOT:PSS/PmPTPA:FIrpic (15 wt.%)/OXD-7/CsF/Al. OXD-7 (**7**) was used as an electron-transporting layer to boost electron injection. The PmPTPA-based devices with FIrpic (**16**) exhibit a luminous efficiency of 17.9 cd/A and EQE of 9.3% at a brightness of 1000 cd/m². White PPLEDs with blue, green, and red phosphors dispersed in PmPTPA (**81**) show a luminous efficiency of 22.1 cd/A and EQE of 10.6% at a brightness of 1000 cd/m². For both the blue and white PPLEDs based on conjugated polymer host PmPTPA (**81**), the EL performances are fairly comparable to those of the state-of-the-art nonconjugated polymer host PVK (**4**), indicating that conjugated polymers can be suitable host materials for PPLEDs with all emission colors.

4.4.3 Polycarbazoles

As we already mentioned, the efficiency of PPLEDs based on conjugated polymers is usually lower than of those based on nonconjugated polymers (as PVK). Although high-efficiency red PPLED based on polyfluorene as the host material has been recently demonstrated [71], the use of the conjugated polymers to achieve green electrophosphorescence is very difficult and it is practically impossible for blue color emission. This is due to quenching of the triplet emission of the transition metal complex by the lower-energy triplet state of the conjugated polymers [91]. At the same time, high-triplet-energy polymers usually also possess a high band gap, which increases the charge injection barriers and the operating voltage of the device and reduces the power efficiency. The polymers with the singlet–triplet gap ($E_{S \to T}$) less than ~0.5 eV are rare [92]; $E_{S \to T}$ of 0.62 eV (between zero phonon fluorescence and phosphorescence peaks) was recently reported, and in most EL polymers it is higher [93,94]. Therefore, high triplet energy (at least ≥2.5 eV) results from a high band gap (>3 eV).

To address this problem, a series of carbazole-3,6-diyl conjugated homo- and copolymers **82–91** with fluorene and oxadiazole units have been studied as host materials for the triplet emitter Ir-SC4 (**92**) (Figure 4.18) [95]. Controlling the conjugation in the polymer by

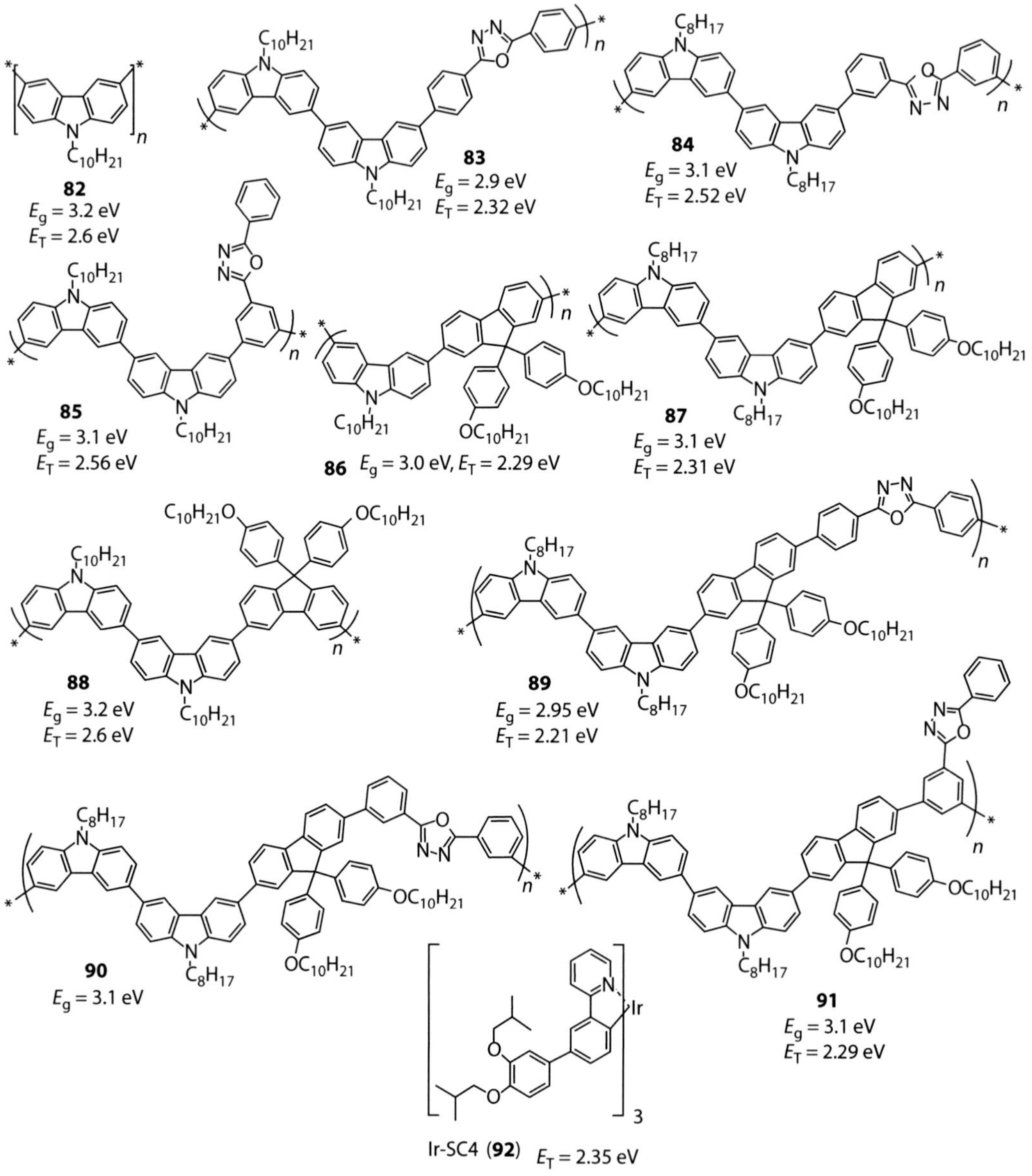

FIGURE 4.18
Structure, band gap (E_g), and triplet energy (E_T) of carbazole-3,6-diyl polymers (**82–91**) used as guest materials for triplet emitter Ir-SC4 (**92**). (From van Dijken, A., Bastiaansen, J.J.A.M., Kiggen, N.M.M., Langeveld, B.M.W., Rothe, C., Monkman, A., Bach, I., Stössel, P., and Brunner, K., *J. Am. Chem. Soc.*, 126, 7718, 2004. With permission.)

changing the linkage position of the fluorene and oxadiazole units, the authors could tune the triplet energy from 2.3 to 2.6 eV without affecting the polymer band gap. Conjugation in the homopolymer **82** is reduced by a "meta" (3,6-) connection, and introducing highly conjugated fluorene or oxadiazole units with "para" connection decreases the energy of the triplet by ca. 0.3 eV. Only polymers **82**, **84**, **85**, and **88** with all "meta" connections have the sufficiently high triplet energy to offset the green triplet emitter Ir-SC4 (E_T = 2.35 eV). Accordingly, the current efficiency of PPLEDs fabricated with these materials was significantly higher compared with other polymers in the series, cf. 23 cd/A for **85** versus 0.45 cd/A for **83** (in device ITO/PEDOT/polymer: Ir-SC4/Ba/Al).

Later, Chen and coworkers [96] developed a modified 4,4′-bis(9-carbazolyl)biphenyl (CBP) as the basic building block for a conjugated homopolymer P(*t*Bu-CBP) (**93**), and two copolymers P(*t*Bu-CBPF) (**94**) and P(*t*Bu-CBPP) (**95**) (Chart 4.36). P(*t*Bu-CBP) (**93**) possesses a relatively high E_T level of 2.53 eV, suitable HOMO level of 5.3 eV, and LUMO level of 2.04 eV. The E_T levels of copolymers P(*t*Bu-CBPF) (**94**) and P(*t*Bu-CBPP) (**95**) are 2.28 and 2.30 eV, respectively, owing to the incorporation of dialkyl-substituted fluorene and dialkoxy-substituted phenylene with CBP. PPLEDs were fabricated using P(*t*Bu-CBP) (**93**), P(*t*Bu-CBPF) (**94**), and P(*t*Bu-CBPP) (**95**) as the hosts, and $(ppy)_2Ir(acac)$ (**76**) and $(piq)_2Ir(acac)$ (**32**)

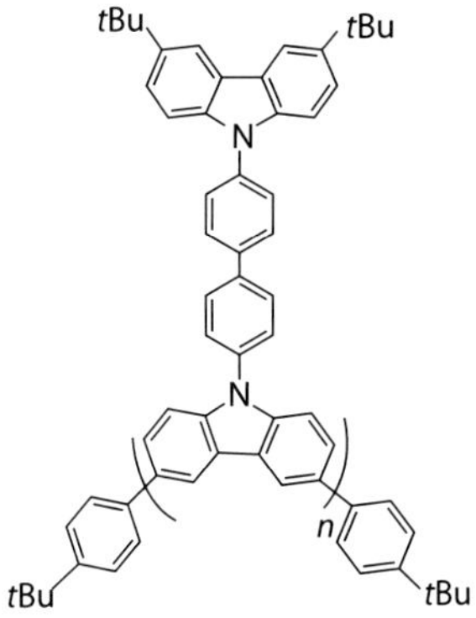

P(*t*Bu-CBP) (**93**)

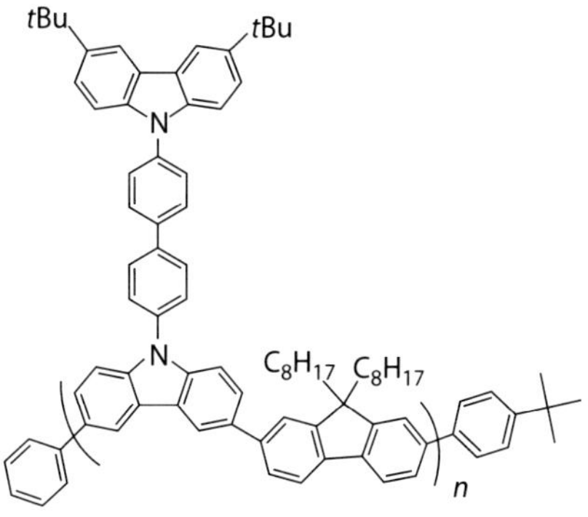

P(*t*Bu-CBPF) (**94**)

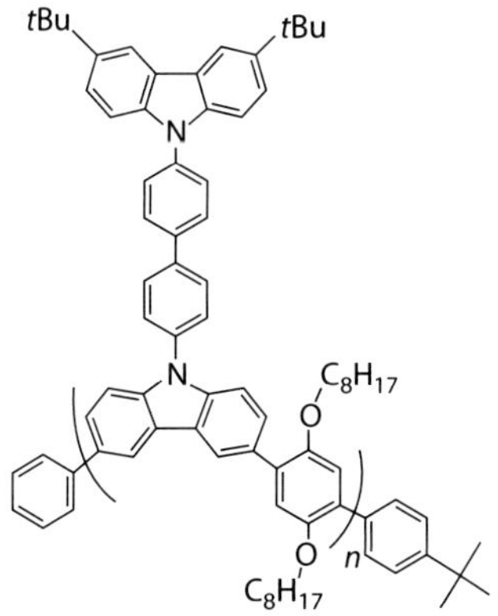

P(*t*Bu-CBPP) (**95**)

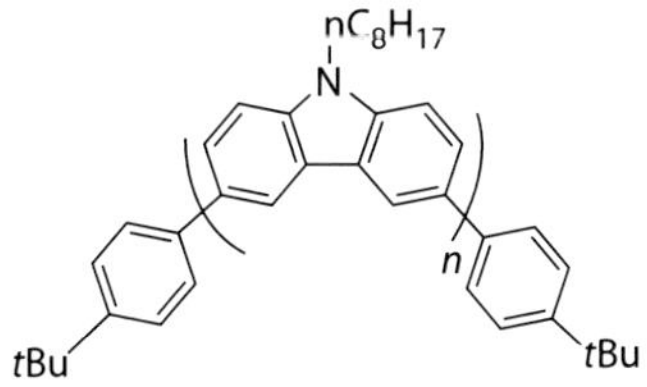

P(3,6-Cz) (**96**)

CHART 4.36
Chemical structure of P(*t*Bu-CBP), P(*t*Bu-CBPF), P(*t*Bu-CBPP), and P(3,6-Cz).

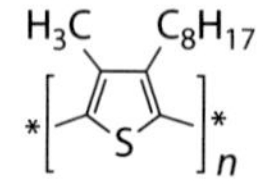

$$H_3C \quad C_8H_{17}$$

PMOT (**97**)

CHART 4.37
Chemical structure of PMOT.

as the green and red emitters, respectively. The device architecture was fabricated as ITO/ PEDOT:PSS/polymer:Ir complex/TPBI (**9**)/CsF/Al. The best device performances were achieved by the homopolymer P(*t*Bu-CBP)-based devices, with high luminous efficiency and EQE of 23.7 cd/A, 6.57% for the green device and 5.1 cd/A, 4.23% for the red device, respectively. For both devices, the efficiencies were higher than those of the control devices with P(3,6-Cz) (**96**) as the host by a factor of 4. The authors conclude that the difference in E_T levels of host and guest is a key factor in determining the device efficiency. In addition, the HOMO and LUMO levels of the host polymer, which provide more balance in electron and hole fluxes and prevent the formation of an excimer, is another important factor in determining device performance.

4.4.4 Polythiophenes

Polythiophene, because of its low band gap and low triplet energy, is not expected to be a good host material for PPLEDs. Nevertheless, recently, Wang et al. [97] demonstrated an efficient energy transfer from a poly(dialkylthiophene) (PMOT **97**) (Chart 4.37) to a PtOX (**14**) (Chart 4.6) dopant. Two alkyl substituents on thiophene ring result in significant twist of the polymer, increasing the band gap to ~3.2 eV (S_1 at 3.77 eV), which becomes essentially high for energy transfer to red-emitting phosphors. Indeed, while PMOT itself shows blue emission at l_{PL} ~480 nm, its blend with 5% PtOX emits a single band at 650 nm. The PPLED ITO/PEDOT/8%PtOX:PMOT/Al showed an EL efficiency of 0.7 cd/A, which is a significant improvement from a similar device without phosphorescent dye (0.05 cd/A). It is worth mentioning that the triplet energy of polymer **97** is extremely low (2.2 eV) for such a wide band-gap polymer. Therefore, only few phosphors possessing lower triplet energy (cf. PtOX:T_1 = 1.9 eV) may perform well with this host material.

4.5 Metallorganic Electrophosphorescent Polymers

Most reported PPLEDs were fabricated by doping a polymer with a phosphorescent dye. However, aggregation and phase separation effects may cause serious problems for device performance and aging. In this section, we describe the very recent progress in intrinsically electrophosphorescent polymers containing triplet-emitting complexes either as pendant substituents or as a part of a backbone.

The EL from a metallorganic polymer PPEP (**98**) (Chart 4.38) containing transition metal (Pt) as a part of the polymer backbone has been studied by Wilson and coworkers [14]. Both triplet (at ~750 nm) and singlet (at ~560 nm) emissions have been observed in PL and EL spectra. Although no performance data for the PPLEDs have been reported, the key conclusion of the work was that there is a spin-dependent exciton formation in the conjugated

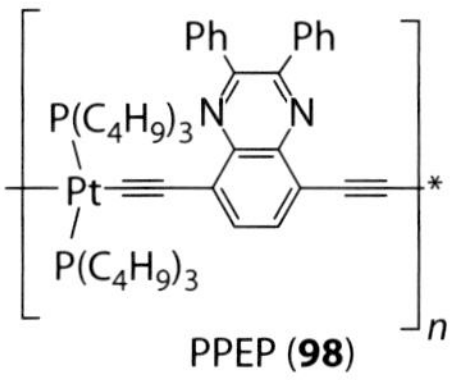

CHART 4.38
Chemical structure of PPEP.

polymers. The singlet–triplet ratio for PPEP determined from the ratio of the singlet and triplet emission in PL and EL spectra was 57%, whereas the usual 1:4 ratio (22%) was found for an analog monomer. The authors suggest that not always the phosphorescent materials have to be used in order to achieve a quantum efficiency >25%.

Nevertheless, in practice, the presence of a triplet-emitting dopant does significantly improve the PLED performance, often giving an EQE well above 5%, which could not be achieved in fluorescent LEDs. The Korean group was the first to publish the nonconjugated PVK copolymer **99** containing iridium complex as a comonomer (Chart 4.39) [98]. The polymer was prepared by copolymerization of vinylcarbazole with vinylphenylpyridine followed by metallization to afford ~8 wt.% concentration of the Ir complex. Green PL and EL was demonstrated for the films of **99** ($l_{PL/EL}$ ~520 nm). The multilayer PPLED ITO/PEDOT/**99**/TAZ (**6**)/Alq$_3$ (**5**)/LiF/Al showed a relatively low (as for phosphorescent LEDs) turn-on voltage of 8 V, a high efficiency of 4.4% (5.0 lm/W), and a maximum brightness of 12,900 cd/m^2 at 24 V. Interestingly, PVK triplet harvesting occurs by intramolecular rather than intermolecular energy transfer: the emission of diluted (10^{-4} M) solutions of **99** occurs solely from the carbazole units (l_{PL} ~375 nm), whereas at higher concentrations (10^{-2} M) strong Ir complex emission (l_{PL} ~520 nm) is observed.

A Japanese group reported a series of similar iridium-containing polymers **100**, **101**, and **102** [99,100], and **103** and **104** [101] (Chart 4.40). By employing complexation ligands with different electron-donating abilities, pure green (**100**, $l_{PL/EL}$ ~523 nm), blue (**101**, $l_{PL/EL}$ ~475 nm), and red (**102**, $l_{PL/EL}$ ~620 nm) phosphorescent polymers were obtained. The role of different electron transport materials (blended with the polymer at 30 wt.%) as well as the iridium complex ratio was investigated, and the best performance was achieved for the device ITO/PEDOT/polymer:OXD-7/Ca/Al (OXD-7 (**7**) used as an electron-transporting material) and 0.2 mol% (EQE 5.5%, current efficiency 7.1 cd/A), 0.6 mol% (9%, 30 cd/A), and 1.0 mol% (3.5%, 4.1 cd/A) of Ir for the red-, green-, and blue-emitting polymers, respectively [99]. The performance was further increased in a multilayer device containing additional ETLs (BAlq **42**, Chart 4.17), even with an aluminum cathode. The device ITO/PEDOT/polymer blend:OXD-7/BAlq/LiF/Al showed an extraordinarily high EQE of 6.6% (**102**: red,

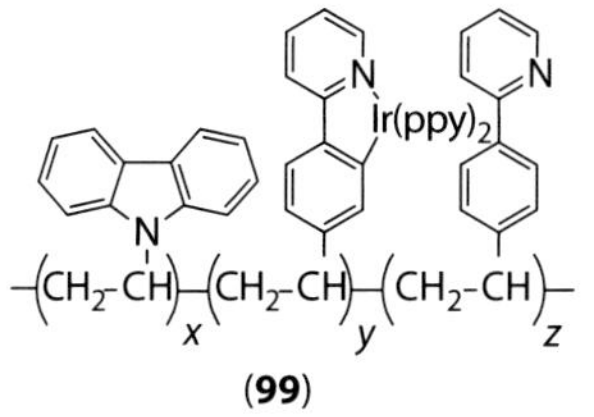

CHART 4.39
Chemical structure of a copolymer **99**.

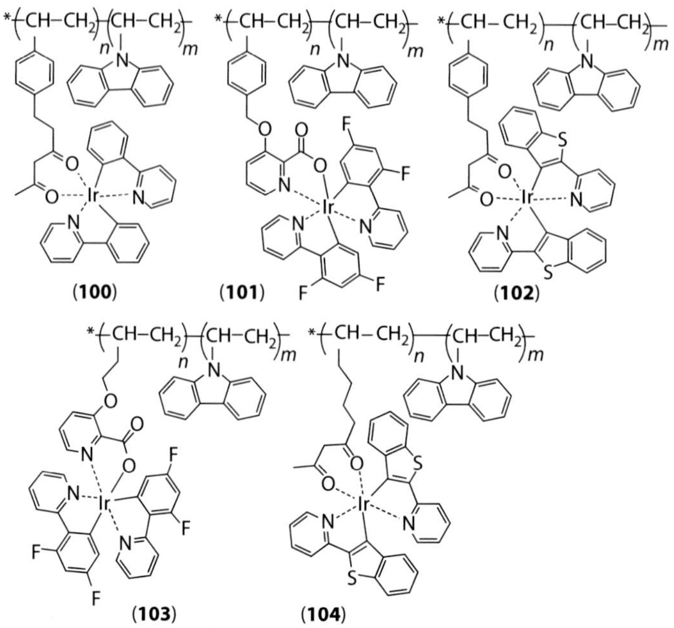

CHART 4.40
Chemical structure of iridium-containing polymers **100–104**.

current efficiency 14.5 cd/A, maximum brightness 1600 cd/m²), 11% (**100**: green, 40.3 cd/A, 15,000 cd/m²), and 6.9% (**101**: blue, 5.5 cd/A, 3000 cd/m²) [100].

Furthermore, efficient (4.5%) pure white-emitting PPLED (CIE: 0.34, 0.36) was fabricated from a 10:1 blend of a blue-emitting **101** with red-emitting **102** polymers [99]. Recently, even a higher-efficiency white-emitting PPLED was fabricated in a similar manner by blending a blue-emitting **103** with red-emitting **104** polymers [101]. The device ITO/PEDOT/polymer blend:OXD-7/BAlq/LiF/Al can be turned on at ~5 V to emit pure white light (CIE: 0.34, 0.36) with an EQE as high as 6.0% (5.2 lm/W at 100 cd/m²) and a maximum brightness of 2000 cd/m².

An interesting new approach to phosphorescent polymers has been reported by Furuta and coworkers [102]. Using a living polymerization reaction (with alkoxyamine catalyst), they have prepared the polymer **105** (Chart 4.41), which contains the electron transport oxadiazole units, hole-transporting triphenylamine units, and triplet-emitting Pt complex,

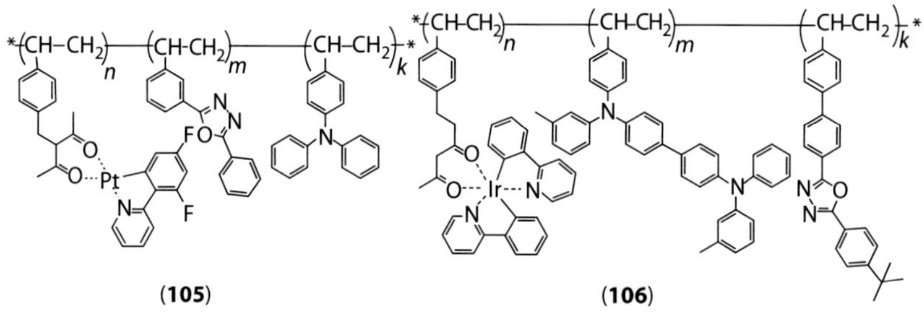

CHART 4.41
Chemical structure of polymers **105** and **106** containing electron transport oxadiazole units, hole-transporting triphenylamine units, and triplet-emitting Pt complex.

all grafted on a polystyrene chain. On the basis of the known wide emission spectrum of the Pt complex (which consists of blue emission of isolated complex and red emission of the aggregate), near-white-emitting PPLED ($x = 0.30$–0.38, $y = 0.43$–0.50, depending on the ratio between comonomers) was fabricated and an EQE of 4.6% was reported.

A high-efficiency PPLED device has been fabricated with polymer **106** [103]. The authors mentioned that the relatively low HOMO of the carbazole (lower than the anode work function, -4.7 eV for ITO) in previously studied polymers **100–104** results in a high operating voltage. Furthermore, the device efficiency is compromised by the low electron mobility of PVK (**4**). Therefore, they prepared polymer **106** with triarylamine hole-transporting and oxadiazole electron-transporting units, together with an Ir emitter. Tuning the first two components, a good balance between the electron and the hole injection in the device was achieved, which resulted in an EQE of 11.8% and a power efficiency of 38.6 lm/W in the device ITO/PEDOT/**106**/Cs (green emitter).

The first example of triplet-emitting complexes grafted onto polyfluorene backbone was reported in 2002 by Pei et al. [104]. Copolymers **107a–c** containing a chelating 2,2′-bipyridyl moiety in a side chain can be prepared with different Eu content (Chart 4.42). Their emission is governed by a moderately efficient energy transfer from polyfluorene onto the Eu center, and can be tuned by changing the complex ligands and the content of Eu [104]. The most effective energy transfer manifested in a single red emission band at 612 nm was observed for the complex **107a** and required an Eu content as high as ~25 mol%. However, the PLED fabricated with PVK (**4**) hole-transporting layer (ITO/PVK/**107a**/Ba/Al) showed a very low EQE of 0.07%, which is a reflection of a very long phosphorescence lifetime and a low quantum yield generally observed in Eu complexes [31]. A related polyfluorene–Eu complex **108c** with carboxylic-chelating moiety of the polymer ligand was reported by Ling and coworkers [105]. Again, the efficiency of the energy transfer was relatively low, confirming poor performance of Eu complexes as triplet-emitting dopants.

The first efficient phosphorescent fluorene polymer was reported by Chen et al. [106], who synthesized a series of polyfluorenes containing both the triplet-emitting iridium complex and the charge-transporting carbazole moieties as pendant groups on the polymer chain. By simply changing the comonomer feeding ratio in Suzuki or Yamamoto polymerization reactions, the copolymers **109** with different content of phosphorescent

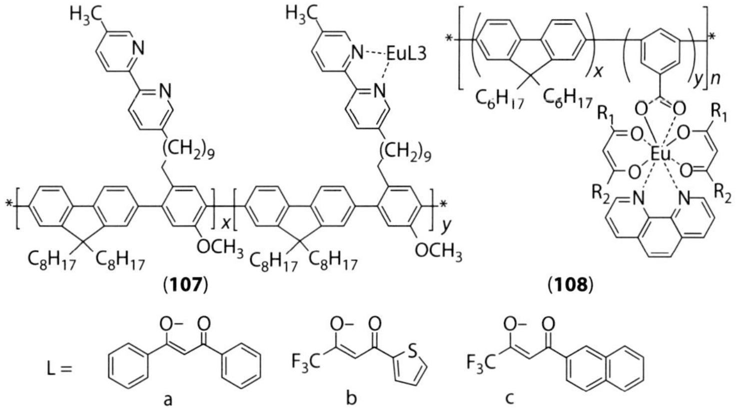

CHART 4.42
Chemical structure of copolymers **107** and **108** prepared with different Eu content to contain a chelating 2,2′-bipyridyl moiety in a side chain.

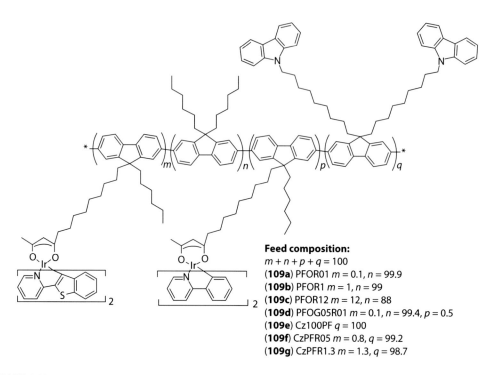

CHART 4.43
Chemical structure of copolymers **109a–g** with different contents of phosphorescent and hole-transporting units.

and hole-transporting units have been prepared (Chart 4.43). The performance of the PPLEDs fabricated with these materials in the configuration ITO/PEDOT/polymer/Ca/Al is given in Table 4.7. The charge mobility of the copolymers containing iridium complex moieties was up to two orders of magnitude lower than that of PDOF (**13**) homopolymers, which indicates a strong charge trapping on the phosphorescent moieties. However, the efficiency of the Ir-containing copolymers was substantially higher than those without the Ir moieties, and in the case of copolymer CzPFR13 (**109g**), the maximum EQE of 1.59%

TABLE 4.7

Characteristics of PLEDs Fabricated with Polymers Containing
Phosphorescent Moieties as Pendant Groups (PFO Stands for PDOF **13**)

Polymer	Turn-On Voltage[a] (V/100 nm)	Maximum Efficiency (cd/A) (h_{max}, %) (V)	Maximum Brightness (cd/m²) (V)
PFO	5.7	0.049 (0.05) (9)	258 (11)
PFOR01	6.9	0.037 (0.017) (8)	57 (11)
PFOR1	6.9	0.88 (0.57) (10)	1479 (13)
PFOR12	8.0	1.0 (0.43) (17.5)	508 (23)
PFOG05R01	6.7	0.23 (0.16) (15)	335 (18)
Cz100PF	3.3	1.28 (0.74) (5)	5029 (7)
CzPFR08	4.3	2.16 (1.32) (9)	3735 (10)
CzPER1.3	4.9	2.8 (1.59) (7)	4321 (15)

Source: Chen, X., Liao, J., Liang, Y., Ahmed, M.O., Tseng, H., and Chen, S., *J. Am. Chem. Soc.*, 125, 636, 2003. With permission.

[a] Brightness over 0.2 cd/m².

(EL efficiency 2.8 cd/A) was achieved at a low voltage of 7 V (and brightness 65 cd/m^2). Moreover, the efficiency stays relatively high (1.6 cd/A) at a high brightness of 4320 cd/m^2 (achieved at 15 V).

Later, Jiang and coworkers [107] reported similar electrophosphorescent fluorene copolymers **110** and **111**, where the hole-transporting carbazole units have been introduced in the polymer backbone (Chart 4.44). Optimizing the polymer structure (comonomer ratio) and the device structure (blending with electron-transporting material PBD **8**), an EQE of 4.9% was achieved.

A different type of phosphorescent polyfluorenes (green-emitting **112** and red-emitting **113**) (Chart 4.45) was recently prepared by Sandee and coworkers [108]. The triplet-emitting iridium complex was made a part of a conjugated polymer backbone, which should provide more efficient energy transfer. The dopant concentration was adjusted by controlling the length of the polymer ($n \sim 5, 10, 20, 30,$ and 40). The triplet-state emission was observed in all polymers, and the energy transfer was more efficient than that in corresponding blends. The PPLED devices, fabricated as ITO/PEDOT/polymer/Ca/Al, showed a moderately high EL efficiency of 1.5% (for red-emitting **113**, $n = 40$).

Schulz et al. [109] developed two series of copolymers, PFPyIr (**114**) and PFTIr (**115**), containing the fluorene-*alt*-pyridine or fluorene-*alt*-thiophene building blocks and iridium complex unit in the main chain (Chart 4.46). The E_T levels of 2,5-linked pyridine-based and 3,4-linked thiophene-based copolymers are 2.13 eV (PFPyIr **114**) and 2.88 eV (PFTIr **115**), respectively. The E_T level of PFTIr (**115**) is about 0.7 eV higher than the E_T of the embedded phosphor (2.22 eV). PPLEDs were fabricated with the architecture ITO/PEDOT:PSS/copolymer/CsF/Al or ITO/PEDOT:PSS/copolymer/TPBI/CsF/Al. The phosphorescent quantum yields of PFPyIr (**114**) and PFTIr (**115**) PPLEDs were 0.05 and 0.20, and the EQEs were 0.32% and 0.84%, respectively.

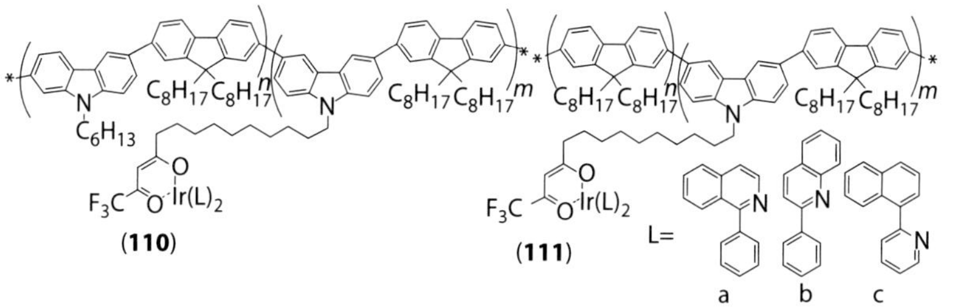

CHART 4.44
Chemical structure of electrophosphorescent fluorene copolymers **110** and **111** where the hole-transporting carbazole units have been introduced into the polymer backbone.

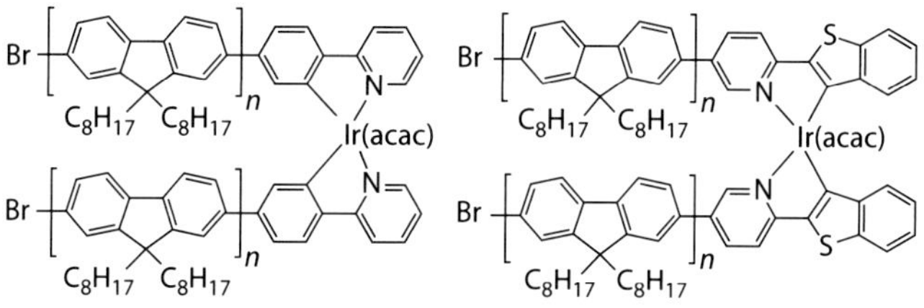

CHART 4.45
Chemical structure of green-emitting and red-emitting phosphorescent polyfluorenes **112** and **113**.

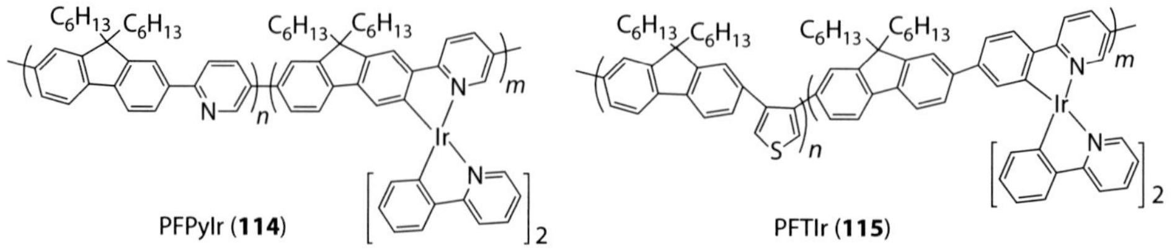

CHART 4.46
Chemical structure of PFPyIr and PFTIr.

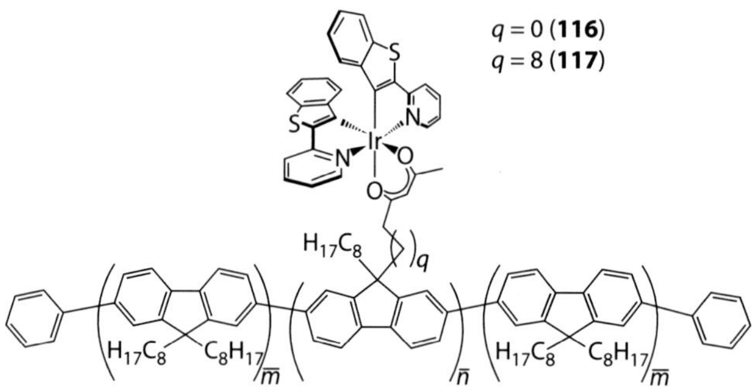

CHART 4.47
Chemical structure of the copolymers **116** and **117** with and without alkyl spacers (tethers) between the polyfluorene chains and pendant red phosphorescent iridium complexes.

Evans and coworkers [110] developed a series of phosphorescent copolymers with a red-emitting phosphor Btp2Ir (**15**) unit attached either directly or through an octamethylene-tethered –(CH$_2$)$_8$– chain at the 9-position of 9-octylfluorene hosts **116** and **117** (Chart 4.47). The E_T of the attached phosphor (2.0 eV) is close to that of the polyfluorene backbone (2.1 eV); thus, the triplet energy back-transfer cannot be avoided. PPLEDs were fabricated with an architecture of ITO/PEDOT:PSS/copolymer/Ca/Al. The phosphorescent quantum yield of **116** and **117** were measured as 22% and 12%, respectively. EQEs of 2.0% and 1.1% were obtained from **116**- and **117**-based devices, respectively.

Liu and coworkers [111] reported a series of π-conjugated chelating polymers (Chart 4.48) with charged iridium complex units based on 1,10-phenanthroline in the backbones. 3,8-Dibromo-1,10-phenanthroline served as an N^N ligand to form a charged Ir complex monomer with 1-(9,9′-dioctylfluoren-2-yl)isoquinoline (Fiq) as the cyclometalated ligand. Chelating polymers showed almost complete energy transfer when the feed ratio of the Ir complex monomer was 4 mol%. Red light–emitting PPLEDs using chelating polymer without carbazole units phen-PFOIr2 (**118c**) and chelating polymer with carbazole units phen-PFOCzIr2 (**118e**) as the light-emitting layers were fabricated in the architecture ITO/PEDOT:PSS/chelating polymer/BCP/LiF/Al. The current density and luminance are 153.2 mA/cm^2 and 83 cd/m^2 at 20 V for the device based on polymer phen-PFOIr2 (**118c**) and 265.8 mA/cm^2 and 100 cd/m^2 at 21 V for the device based on polymer phen-PFOCzIr2 (**118e**), respectively.

Zhen and coworkers [112] developed a series of chelating polymers, PFCzMppyIr (**119**) and PFCzMppyIrhm (**120**) (Chart 4.49), containing the fluorene-*alt*-carbazole (PFCz) building block and alkyl-substituted ligands of iridium complex unit. The chelating polymers showed highly efficient energy transfer of excitons from the PFCz host segment to the

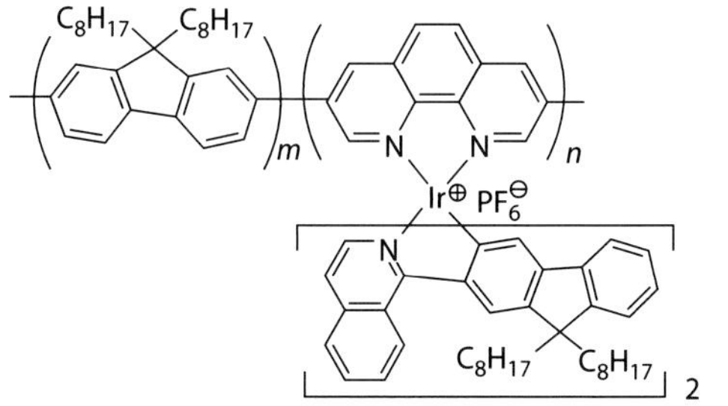

(**118a**) phen-PFOIr01 $m = 99.5$, $n = 0.5$
(**118b**) phen-PFOIr05 $m = 99.9$, $n = 0.1$
(**118c**) phen-PFOIr2 $m = 98$, $n = 2$
(**118d**) phen-PFOIr4 $m = 96$, $n = 4$

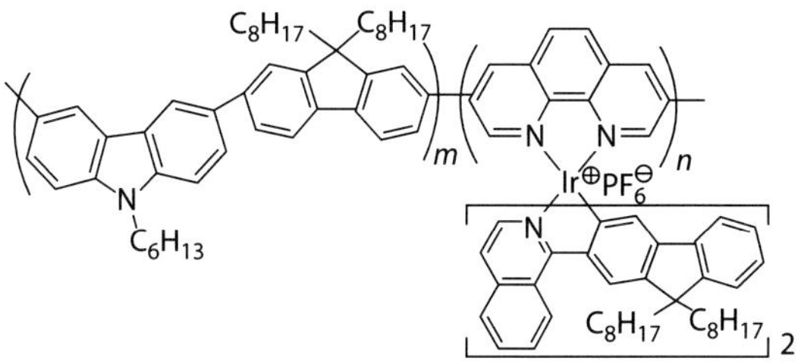

$m = 98$, $n = 2$, phen-PFOCzIr2 (**118e**) 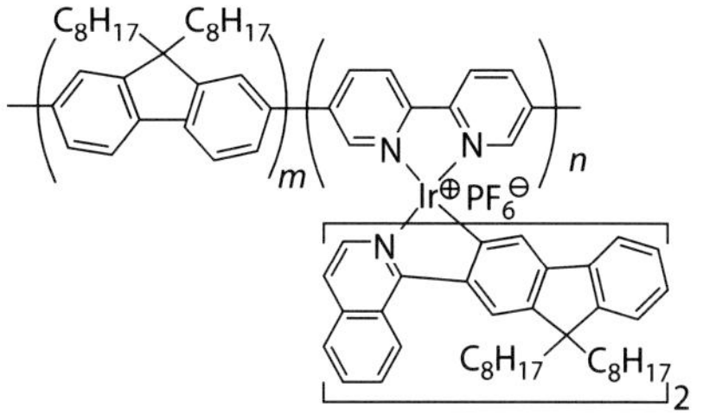 $m = 98$, $n = 2$, bpy-PFOIr2 (**118f**)

CHART 4.48
Chemical structure of chelating polymers **118a–f**.

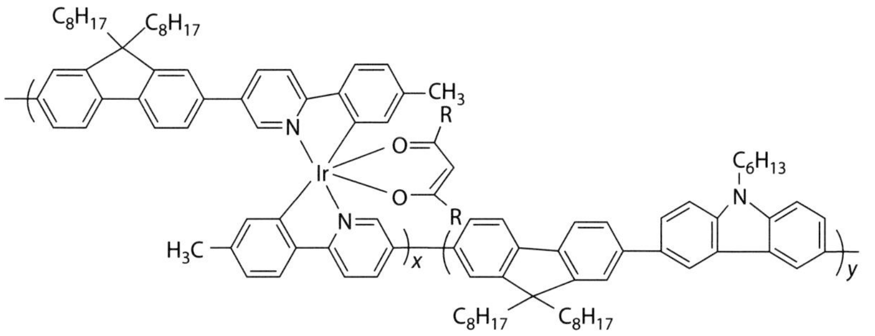

R = CH$_3$, PFCzMppyIr (**119**)   R = C(CH$_3$)$_3$, PFCzMppyIrhm (**120**)

CHART 4.49
Chemical structure of the chelating copolymers PFCzMppyIr and PFCzMppyIrhm.

Ir complex by an intramolecular trapping mechanism. PPLEDs were fabricated with an architecture of ITO/PEDOT:PSS/copolymer:PBD (30 wt.%)/Ba/Al. The EQE and luminous efficiency of a device made with copolymer PFCzMppyIrhm (**120**) reached 4.1% and 5.4 cd/A, respectively, with a luminance of 1730 cd/m^2. The emission peak was 577 nm, at a current density of 32.2 mA/cm^2. Moreover, the PPLEDs exhibited long-term stability. No notable efficiency decay was observed while increasing current density, which may be attributed to the reduced concentration quenching and triplet–triplet annihilation.

The same group [113] studied another series of phosphorescent chelating copolymers based on the linkage isomers of 2-(1-naphthalene)pyridine- and 2-(2-naphthalene)pyridine–bicycloiridium complexes PF1–NpyIrm (**121a–d**) and PF2–NpyIrm (**122a–d**) (Chart 4.50). The double-layer saturated red light–emitting PPLEDs were fabricated in the configuration ITO/PEDOT:PSS/PVK/polymer/Ba/Al. The EL characteristics of the devices are

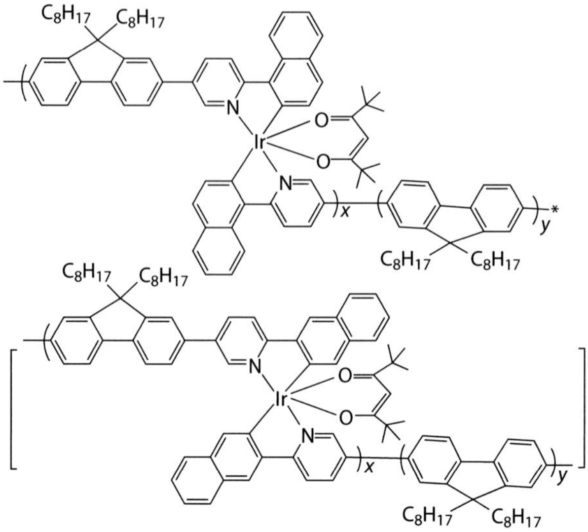

PF1–NpyIrm05 (**121a**)
PF1–NpyIrm1 (**121b**)
PF1–NpyIrm2 (**121c**)
PF1–NpyIrm5 (**121d**)

PF2–NpyIrm1 (**122a**)
PF2–NpyIrm2 (**122b**)
PF2–NpyIrm5 (**122c**)
PF2–NpyIrm7 (**122d**)

CHART 4.50
Chemical structure of chelating polymers PF1–NpyIrm and PF2–NpyIrm.

TABLE 4.8
Device Performances of the Chelating Copolymers

	J = 30 mA/cm²				*J* = 100 mA/cm²				CIE Coordinates	
Polymer	**Bias (V)**	***L* (cd/m²)**	**LE (cd/A)**	**QE_{ext} (%)**	**Bias (V)**	***L* (cd/m²)**	**LE (cd/A)**	**QE_{ext} (%)**	*x*	*y*
1NpyIrm/PFO	17.8	396	1.3	2.1	20.0	900	0.9	1.4	0.62	0.37
PF1-NpyIrm05	14.7	451	1.5	3.9	16.2	1242	1.3	3.3	0.65	0.31
PF1-NpyIrm1	15.1	732	2.5	6.5	16.5	2010	2.0	5.3	0.65	0.31
PF1-NpyIrm2	18.0	570	1.9	4.9	20.0	1560	1.6	4.1	0.66	0.30
PF1-NpyIrm5	15.6	602	2.0	5.2	18.0	1507	1.5	4.0	0.67	0.31
PF2-NpyIrm1	17.3	334	1.1	1.5	19.4	800	0.8	1.0	0.56	0.40
PF2-NpyIrm2	17.8	396	1.3	1.6	20.0	600	0.6	1.0	0.59	0.40
PF2-NpyIrm5	18.4	430	1.4	2.5	15.6	1136	1.1	1.9	0.56	0.43

Source: Zhen, H., Luo, C., Yang, W., Song, W., Du, B., Jiang, J., Jiang, C., Zhang, Y., and Cao, Y., *Macromolecules*, 39, 1693, 2006. With permission.
Note: Bias, bias voltage; *J*, current density; *L*, luminance; LE, luminous efficiency; QE_{ext}, external quantum efficiency. Device structure: ITO/PEDOT:PSS/PVK/polymer/Ba/Al.

listed in Table 4.8. The extraordinary device performances were achieved from copolymer PF1–NpyIrm1 (**121b**). At the current density of 5.2 mA/cm², the maximal EQE of 7.0% was obtained with the luminance of 138 cd/m². The EQE value reached 6.5% at the current density of 38 mA/cm², with the emission peak at 630 nm and the luminance of 926 cd/m². The EQE remained as high as 5.3% at a high current density of 100 mA/cm².

The same group later reported another series of novel copolymers (**123a–d**) (Chart 4.51) with fluorene-*alt*-carbazole segments and β-diketonate moieties coordinating to iridium $(piq)_2$Irdbm (where piq is 1-phenylisoquinolin and dbm is dibenzoylmethane) [114]. PPLEDs were fabricated with an architecture of ITO/PEDOT:PSS/copolymer:40% PBD/Ba/Al. A saturated red-emitting PPLED based on PFCzIrpiq3 (**123d**) was achieved with

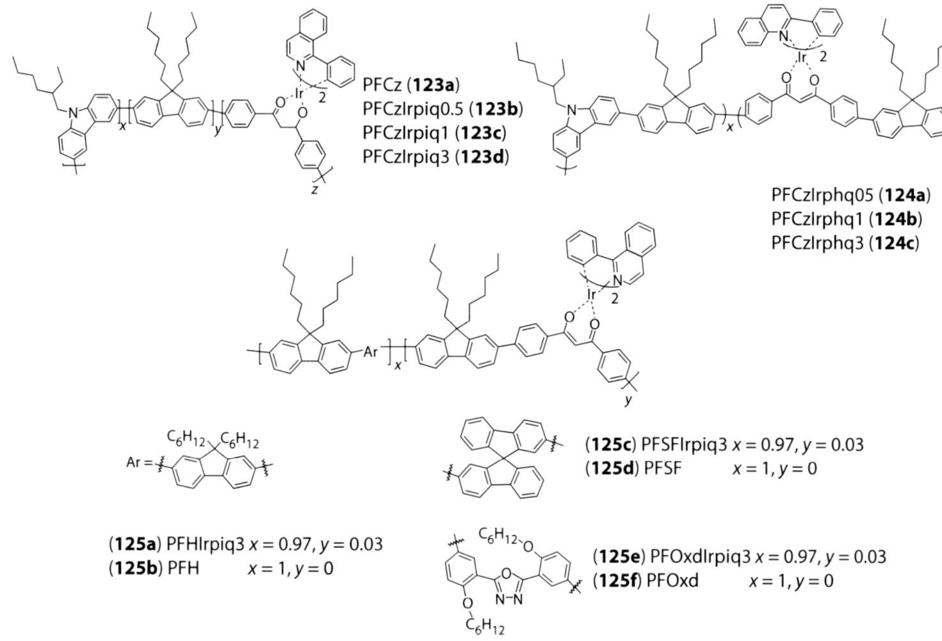

CHART 4.51
Chemical structure of copolymers **123–125**.

emission peak at 628 nm, a maximum EQE of 0.6% at the current density of 38.5 mA/cm^2, and a maximum luminance of 541 cd/m^2 at 15.8 V. Then, the authors [115] compared the EL properties of fluorene-carbazole-based copolymers incorporated with 2-phenylquinoline- or 1-phenylisoquinoline-based Ir complexes through a β-diketonate ancillary ligand. The copolymers PFCzIrphq (**124a–c**) (Chart 4.51) emit orange–red light with an emission peak at ca. 590 nm, while PFCzIrpiq (**123a–d**) emit saturated red light with an emission maximum at ca. 620 nm, with a slight red shift compared with the emissions for the corresponding monomeric iridium complexes, respectively. They further investigated the relationship between the structure of copolymers **125a–f** and the EL properties by incorporation of an oxadiazole or spirobifluorene unit to the backbone [116]. To optimize the device performance, three types of PPLEDs were fabricated with the following configurations: device I, ITO/PEDOT:PSS/copolymer/Ba/Al; device II, ITO/PEDOT:PSS/copolymer:PBD/Ba/Al; and device III, ITO/PEDOT:PSS/PVK/copolymer/Ba/Al. PPLEDs based on PFOxdIrpiq3 (**125e**) (E_T = 2.32 eV) and PFSFIrpiq3 (**125c**) (E_T = 2.18 eV) exhibited maximum EQE of 2.93% and 3.21%, respectively. The EL characteristics of the devices are listed in Table 4.9. The effective inhibition of interchain triplet energy transfer because of sterically hindered spirobifluorene units on the backbone contributed to the improved efficiency.

The authors [117] also compared two types of phosphorescent copolymers with phosphors incorporated into the polymer backbone by either embedding (**126a** and **126b**) or end-capping via an ancillary ligand β-diketone (**127**) (Chart 4.52). PPLEDs were fabricated with the configuration ITO/PEDOT:PSS/PVK/copolymer/Ba/Al. The EL device using end-capping copolymer **127** as the emitting layer exhibited higher efficiency than the copolymers **126a** and **126b** with the same phosphor embedded in the polymer main chain, since **127** reduced triplet exciton back-transfer from the iridium complex to the polyfluorene

TABLE 4.9

Device Performance Data

Polymer	Device	L_{max} (cd/m²)/ Bias (V)[a]	J (mA/cm²)[b]	η_{max} (%)	LE_{max} (cd/A)	η at 20 mA/cm² (%)	η at 100 mA/cm² (%)	η at 100 cd/m² (%)
PFHIrpiq3	I	160/16.0	68.6	0.21	0.10	0.20	0.20	0.20
	II	389/20.0	56.8	0.49	0.24	0.44	0.48	0.46
	III	556/18.7	34.7	0.74	0.35	0.71	0.67	0.66
PFCzIrpiq3	I	222/24.0	33.0	0.24	0.14	0.23	0.22	0.23
	II	541/14.8	38.5	0.60	0.34	0.54	0.59	0.59
	III	224/30.0	33.2	0.28	0.16	0.28	0.26	0.27
PFSFIrpiq3	I	962/18.4	11.9	1.97	1.04	1.89	1.63	1.89
	II	967/16.3	26.9	1.96	1.04	1.95	1.60	1.96
	III	2140/24.1	8.24	3.21	1.70	3.03	2.55	2.93
PFOxdIrpiq3	I	420/26.9	24.5	0.85	0.43	0.80	0.61	0.80
	II	399/25.4	14.5	0.80	0.40	0.80	0.64	0.79
	III	1179/30	4.30	2.93	1.46	2.64	1.86	2.92

Source: Zhang, K., Chen, Z., Zou, Y., Gong, S., Yang, C., Qin, J., and Cao, Y., *Chem. Mater.*, 21, 3306, 2009. With permission.

[a] Maximum luminance and corresponding bias.

[b] Current intensity for maximum external quantum efficiency and luminance efficiency.

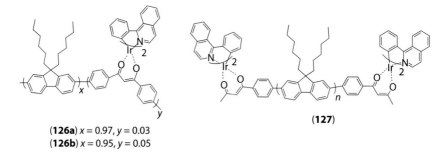

(**126a**) x = 0.97, y = 0.03
(**126b**) x = 0.95, y = 0.05

(**127**)

CHART 4.52
Chemical structure of copolymers **126a**, **126b**, and **127**.

backbone than iridium complex embedded in the polymer main chain of **126a** and **126b**. A red-emitting PPLED shows an emission peak at 633 nm, a maximum EQE of 1.70% at a current density of 3.58 mA/cm², and a maximum luminance of 706 cd/m² at 18 V.

Similarly, another two series of fluorene-*alt*-carbazole or fluorene-based copolymers were prepared with a new iridium complex with the cyclometalated ligand 2-*p*-tolyl-benzothiazole incorporated into the polymer backbone by either embedding (**128a–c**) or end-capping via an ancillary ligand β-diketone (**129**) (Chart 4.53) [118]. PPLEDs were fabricated with two configurations of ITO/PEDOT:PSS/PVK/copolymer/Ba/Al and ITO/PEDOT:PSS/PVK/copolymer:40% PBD/Ba/Al. It is also observed that the EL device using end-capping copolymer **129** as the emitting layer exhibited higher efficiency than the copolymers **128a–c** with the same phosphor embedded in the polymer main chain, which is also attributed to the fact that **129** suffers less from triplet exciton back-transfer from the iridium complex to the polymer backbone than **128a–c** do. An orange–red-emitting PPLED shows an emission peak at 599 nm, a maximum EQE of 2.19% at a current density

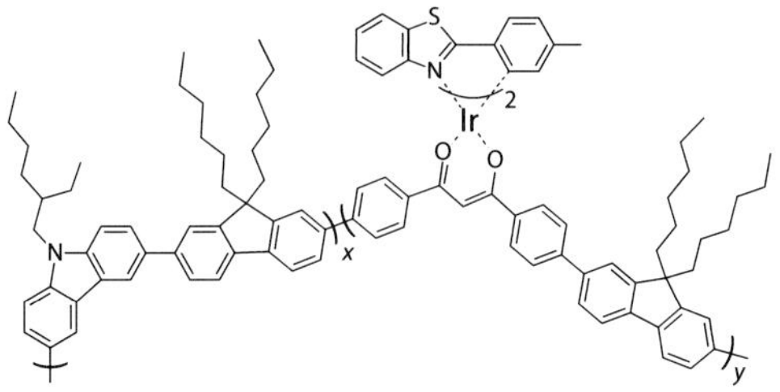

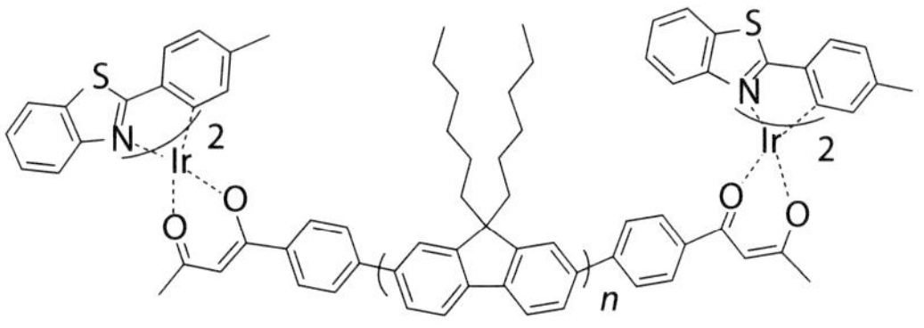

CHART 4.53
Chemical structure of copolymers **128a–c** and **129**.

of 0.3 mA/cm², and a maximum luminance of 2347 cd/m² at 17 V. This approach opens a door toward the molecular design of EL polymers for efficient PPLEDs.

The same group later published a novel poly(fluorene-*alt*-carbazole) (PFCz)-based copolymer with 3,6-carbazole-*N*-alkyl grafted iridium complex using 2,3-diphenylpyrazine as ligand (IrBpz). Ying et al. [119] reported the EL study of the copolymer–iridium complex PFCzIrBpq (**130**) (Chart 4.54). The orange–red light–emitting PPLEDs was fabricated with a configuration of ITO/PEDOT:PSS/copolymer:30% PBD/Ba/Al. The PPLEDs showed a maximum luminous efficiency of 5.58 cd/A and a maximal luminance of 8625 cd/m². White light with CIE coordinates of (0.33, 0.27) was observed from white PPLEDs made by the copolymer containing 0.4 mol% iridium complex. A luminous efficiency of 2.30 cd/A with a luminance of 2068 cd/m² was observed from white PPLEDs.

The authors [120] also designed and synthesized another series of efficient and almost pure white light–emitting copolymers PFBT-Phq (**131a–h**) containing a benzothiadiazole group attached to a fluorene backbone and an iridium complex attached to the side chain (Chart 4.55). In this system, both fluorescent and phosphorescent moieties were used in a single copolymer. By tuning the ratio of the green light–emitting benzothiadiazole group to the red emitter 2-phenylquinoline iridium complex (Phq), the EL spectra can be adjusted to achieve white light emission. The PPLEDs were fabricated with the configuration ITO/PEDOT:PSS/PVK/copolymer/CsF/Al. The devices from PFBT1-Phq2 (**131b**) and PFBT3-Phq2 (**131c**) exhibited white light with CIE coordinates of (0.34, 0.33) and (0.32, 0.33), respectively. A maximum luminous efficiency of 6.1 cd/A at a current density of 2.2 mA/cm² and

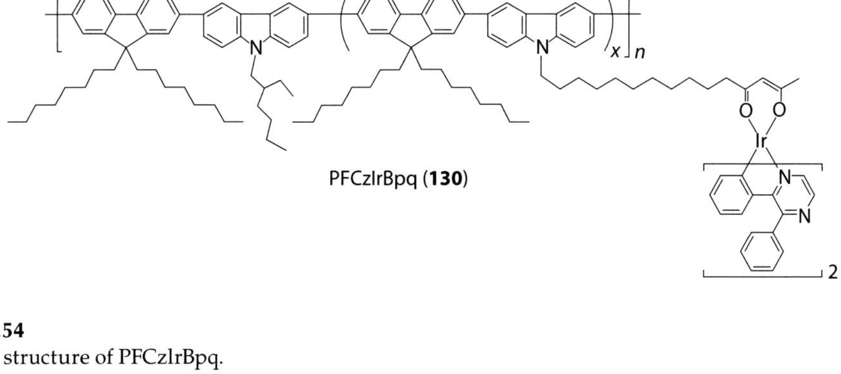

PFCzIrBpq (**130**)

CHART 4.54
Chemical structure of PFCzIrBpq.

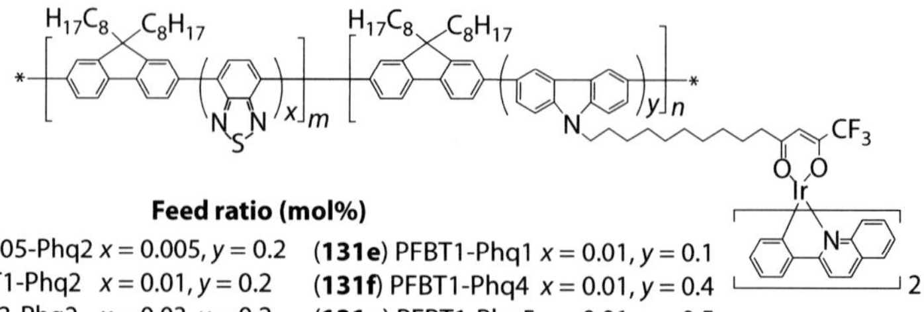

Feed ratio (mol%)

(**131a**) PFBT05-Phq2 $x = 0.005, y = 0.2$
(**131b**) PFBT1-Phq2 $x = 0.01, y = 0.2$
(**131c**) PFBT3-Phq2 $x = 0.03, y = 0.2$
(**131d**) PFBT5-Phq2 $x = 0.05, y = 0.2$
(**131e**) PFBT1-Phq1 $x = 0.01, y = 0.1$
(**131f**) PFBT1-Phq4 $x = 0.01, y = 0.4$
(**131g**) PFBT1-Phq5 $x = 0.01, y = 0.5$
(**131h**) PFBT3-Phq3 $x = 0.03, y = 0.3$

CHART 4.55
Chemical structure of copolymers PFBT-Phq.

a maximum luminance of 10,110 cd/m^2 at a current density of 345 mA/cm^2 were achieved from PPLEDs of PFBT5-Phq2 (**131d**).

In another work, the authors [121] studied iridium complex iridium(III) bis(2-(2′-benzo[4,5-α]-thienyl) pyridinato-*N*,C$^{3'}$)2,2,6,6-tetramethyl-3,5-heptanedioe (btp)$_2$Ir(tmd) incorporated into a fluorene-benzothiadiazole copolymer backbone (**132a–e**, Chart 4.56). By tuning the ratio of the green light–emitting benzothiadiazole group to the red emitter (btp)$_2$Ir(tmd), the EL spectra can be adjusted to achieve white light emission. The PPLEDs were fabricated with the configuration ITO/PEDOT:PSS/PVK/copolymer/CsF/Al. The devices from PFIrR1G03 (**132b**) exhibited white light with a maximum EQE of 3.7%, a maximum luminous efficiency of 3.9 cd/A at the current density of 1.6 mA/cm^2 with the CIE coordinates of (0.33, 0.34), and a maximum luminance of 4180 cd/m^2 at the current density of 268 mA/cm^2 with the CIE coordinates of (0.31, 0.32). The white light emission of devices from the copolymers is stable over the whole white light region at different applied voltages.

Recently, the same group [122] reported another iridium complex, iridium(III) bis(2-(1-naphthalene)pyridine-C$^{2'}$,*N*)-2,2,6,6-tetramethyl-3,5-heptanedione (1-npy)$_2$Ir(tmd), incorporated into a fluorene-benzothiadiazole copolymer backbone (**133a–d**, Chart 4.57). By tuning the ratio of green light–emitting benzothiadiazole group to the red emitter (1-npy)$_2$Ir(tmd), the EL spectra can be adjusted to achieve white light emission. The PPLEDs were fabricated with the same configuration of ITO/PEDOT:PSS/PVK/copolymer/CsF/Al. The devices from PFG03–IrR07 (**133d**) exhibited a maximum luminous efficiency of 5.3 cd/A, a maximum luminance of 9900 cd/m^2 at a current density of 453 mA/cm^2, and CIE coordinates of (0.32, 0.34). The white emission of devices based on these materials showed very good color quality, with a high color rendering index between 84 and 89.

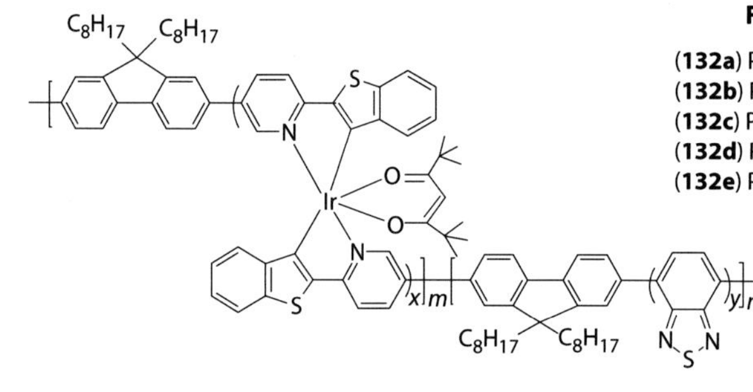

Feed ratio (mol%)

(**132a**) PFIrR05G02 $x = 0.05, y = 0.02$
(**132b**) PFIrR1G03 $x = 0.1, y = 0.03$
(**132c**) PFIrR1G04 $x = 0.1, y = 0.04$
(**132d**) PFIrR1G05 $x = 0.1, y = 0.05$
(**132e**) PFIrR3G03 $x = 0.3, y = 0.03$

CHART 4.56
Chemical structure of copolymers **132a–e**.

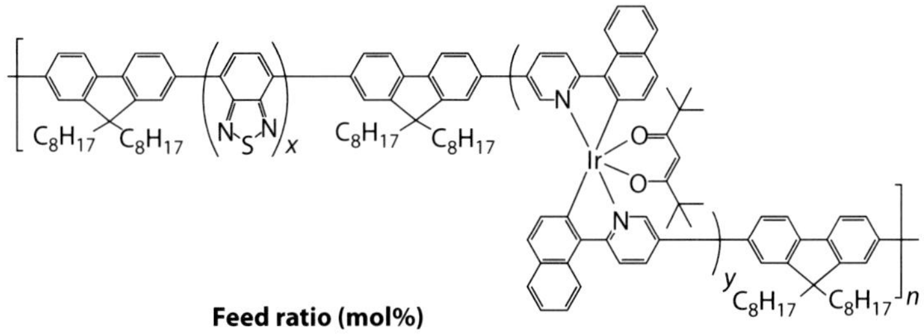

Feed ratio (mol%)

(**133a**) PFG02-IrR1 $x = 0.02, y = 0.1$ (**133c**) PFG04-IrR1 $x = 0.04, y = 0.1$
(**133b**) PFG05-IrR1 $x = 0.05, y = 0.1$ (**133d**) PFG03-IrR07 $x = 0.03, y = 0.075$

CHART 4.57
Chemical structure of copolymers **133a–d**.

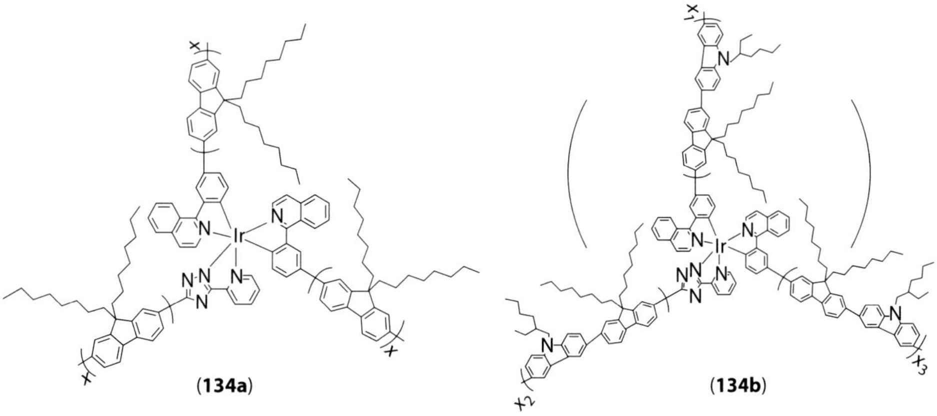

CHART 4.58
Chemical structure of copolymers **134a** and **134b**.

Very recently, the authors [123] investigated a series of red-emitting hyperbranched polymers (**134a** and **134b**, Chart 4.58) utilizing polyfluorene (PF) and poly(fluorene-*alt*-carbazole) (PFCz) as the branches and the phenyl-isoquinoline iridium complex (Piq)₂Ir(Pytz) (where Piq is 1-phenylisoquinoline and Pytz is 3-(pyridin-2-yl)-1*H*-1,2,4-triazole) as the core. These copolymers were synthesized through a Suzuki polycondensation, using the "$A_2 + A_2' + B_3$" approach. The HOMO and LUMO levels of the PFCz branch are enhanced compared with those of the PF branches, indicating that the incorporation of the Cz moiety into the fluorene branches can improve hole injection in the polymers, while the potential barrier of electron injection is enlarged. The hyperbranched polymer **134a** using fluorene branches exhibited a better device performance in the configuration ITO/PEDOT:PSS/PVK/polymer/CsF/Al. A maximum luminous efficiency of 6.54 cd/A (peak EQE = 4.88%) at 5.4 V with CIE coordinates of (0.65, 0.34) was obtained from copolymer **134a**. The device efficiencies from the copolymers showed a reduced roll-off upon an increase of the current density.

Besides the employment of iridium complexes as emitters for phosphorescent emission, Chien et al. [124] investigated another kind of heavy metal osmium complex with a shorter radiative lifetime than Ir complexes. A series of polyfluorene copolymers TOF-Os(bpftz)$_x$ (**135a–d**, Chart 4.59), containing red-emitting osmium complex Os(bpftz) (**31**), bearing two 3-trifluoromethyl-5-(4-*tert*-butyl-2-pyridyl)triazolate (bpftz) cyclometalated ligands, were introduced into the backbone of the copolymer. PPLEDs were fabricated with the configuration

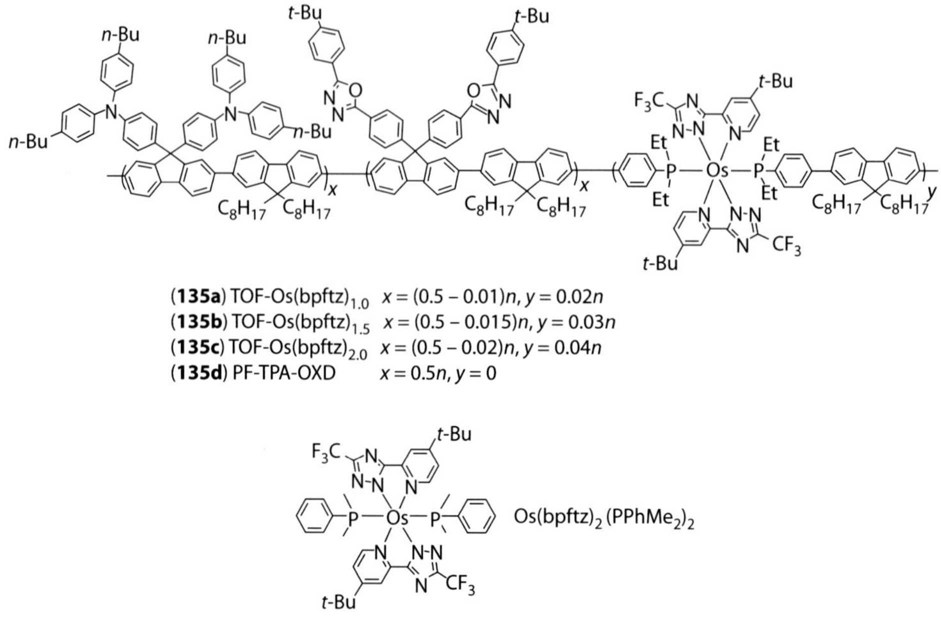

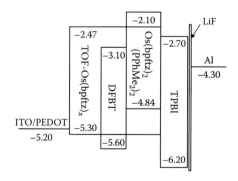

(135a) TOF-Os(bpftz)$_{1.0}$ $x = (0.5 - 0.01)n$, $y = 0.02n$
(135b) TOF-Os(bpftz)$_{1.5}$ $x = (0.5 - 0.015)n$, $y = 0.03n$
(135c) TOF-Os(bpftz)$_{2.0}$ $x = (0.5 - 0.02)n$, $y = 0.04n$
(135d) PF-TPA-OXD $x = 0.5n$, $y = 0$

Os(bpftz)$_2$ (PPhMe$_2$)$_2$

CHART 4.59
Chemical structure of copolymers **135a–d** and model compound Os(bpftz)$_2$(PPhMe$_2$)$_2$.

ITO/PEDOT:PSS/TOF-Os(bpftz)$_x$/TPBI/LiF/Al. The energy-level diagram of the materials involved in the EL devices is shown in Figure 4.19. The optimized TOF-Os(bpftz)$_x$ PPLEDs ($x = 1.5$) exhibited a maximum EQE of 18.0% (luminous efficiency of 26.3 cd/A), a maximum brightness of 38,000 cd/m^2, and CIE chromaticity coordinates of (0.64, 0.35). The maximum luminous efficiency was 10.7 cd/A (EQE of 5.4%) with a brightness of 442 cd/m^2, obtained at a current density of 4.1 mA/cm^2 for white light–emitting PPLEDs with CIE coordinates (0.37, 0.30).

Wu et al. [125] developed a series of copolymers (**136a–c**, Chart 4.60) containing a green fluorophore and a red phosphor in the backbone and the side chains, respectively. PPLEDs based on **136c** were fabricated with the configuration ITO/PEDOT:PSS/**136c**/CsF/Al. The optimized device exhibited a low turn-on voltage of 2.8 V and a luminance of 1000 cd/m^2 at below 6 V. The peak luminous efficiency and power efficiency were 8.2 cd/A and 7.2 lm/W, respectively.

FIGURE 4.19
Energy-level diagram of materials involved in EL devices. (From Chien, C.H., Liao, S.F., Wu, C.H., Shu, C.F., Chang, S.Y., Chi, Y., Chou, P.T., and Lai, C.H., *Adv. Funct. Mater.*, 18, 1430, 2008. With permission.)

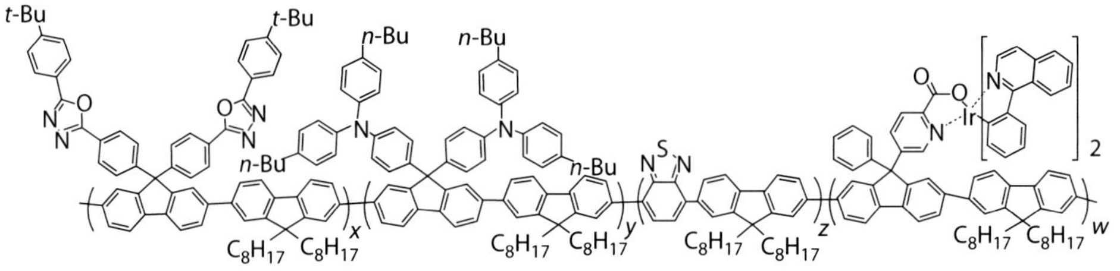

CHART 4.60
Chemical structure of copolymers **136a–c**.

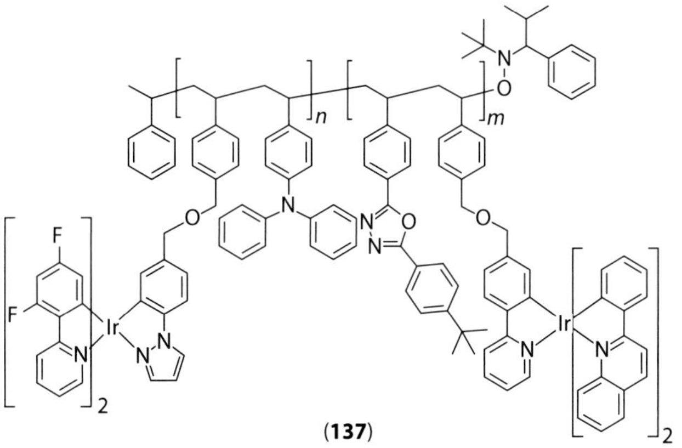

$$(\textbf{136a})\ x = y = 0.247n,\ z = 0.002n,\ w = 0.004n$$
$$(\textbf{136b})\ x = y = 0.2485n,\ z = 0.0015n,\ w = 0.0015n$$
$$(\textbf{136c})\ x = y = 0.2496n,\ z = 0.0004n,\ w = 0.0004n$$

(**137**)

CHART 4.61
Chemical structure of copolymer **137**.

In addition, the **136c** device showed relatively stable white emission. The CIE chromaticity coordinates of the devices changed from (0.35, 0.38) at 10 mA/cm^2 to (0.33, 0.36) at 100 mA/cm^2, with an almost constant color render index value of 82 at all measured current densities.

Poulsen and coworkers [126] employed triarylamine (TPA) and oxadiazole (OXA) building blocks to be incorporated into diblock TPA-b-OXA copolymer **137** (Chart 4.61). This approach minimized energy transfer between the two-colored species of green–blue Ir(dfppy)$_2$(tpzs) and orange–red Ir(pq)$_2$(tpys) iridium complexes by site isolation through morphology control. A higher concentration of red emitters was achieved. PPLEDs were fabricated with the configuration ITO/PEDOT:PSS/copolymer/LiF/Al. However, the device efficiency was rather low (EQE >1.5%).

4.6 Electrophosphorescent Dendrimers

Triplet-emitting metals can be also used as multidentate sites to synthesize solution-processable phosphorescent dendrimers. Anthopoulos and coworkers [127] reported a

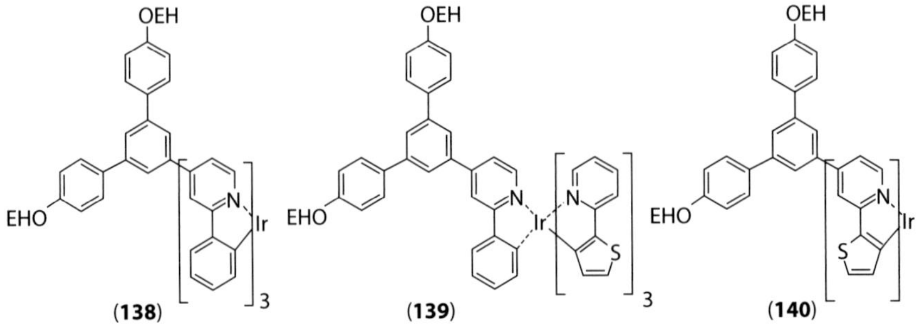

CHART 4.62
Chemical structure of a first-generation dendrimers **138–140**.

high-efficiency single-layer LED fabricated from a first-generation dendrimer **138** (Chart 4.62) blended with electron-transporting (TPBI, **9**) and hole-transporting (4,4′-bis(9-carbazolyl)diphenyl (CBP)) materials [127]. The green-emitting (l_{EL} = 518 nm) device with ITO and LiF/Al electrodes showed an EQE of 10.4% and high power efficiency of 12.8 lm/W (at 8 V), whereas the operating voltage (at 100 cd/m²) was as low as 6 V. Later, the same group reported red-emitting phosphorescent dendrimers **139** and **140** [128]. The resulting multilayer OLEDs, fabricated by spin coating, showed an EQE of 4.6% (4.5 lm/W) and saturated red emission (CIE: 0.67, 0.33). Furthermore, the precise color of the device can be conveniently adjusted by blending green-emitting dendrimer **138** with red-emitting **139** to afford high-efficiency devices (current efficiency of up to 31 cd/A) [129].

The same group [130] also demonstrated how to control intermolecular interactions of light-emitting dendrimers **141** (Chart 4.63) to achieve efficient EL devices through solution processing without using any host materials. Bilayer devices were fabricated using **141** as the green-emitting layer and a TPBI (**9**) as the electron-transporting layer. The turn-on voltage was 2.5 V with an EQE value of 13.6% (30 lm/W, 47 cd/A, 110 cd/m²) reached at 4.8 V. Host-free dendrimer-based OLEDs are a significant step toward the easy fabrication of low-cost, high-performance organic light-emitting displays.

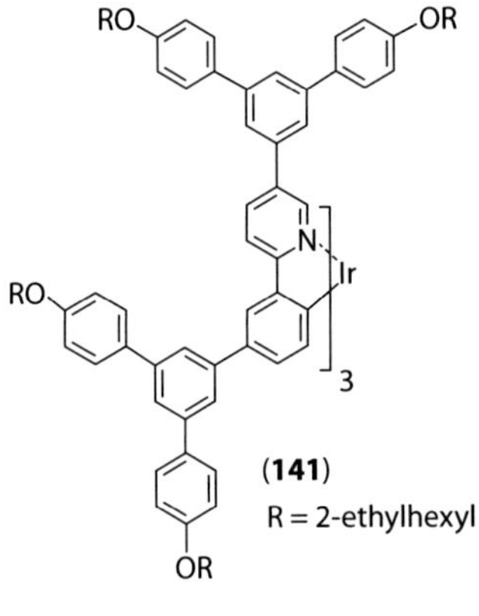

CHART 4.63
Chemical structure of dendrimer **141**.

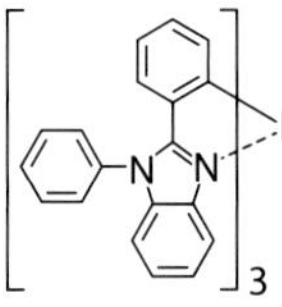

CHART 4.64
Chemical structure of dendrimers **142–144**.

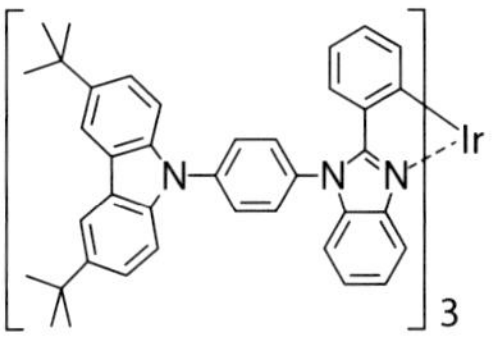

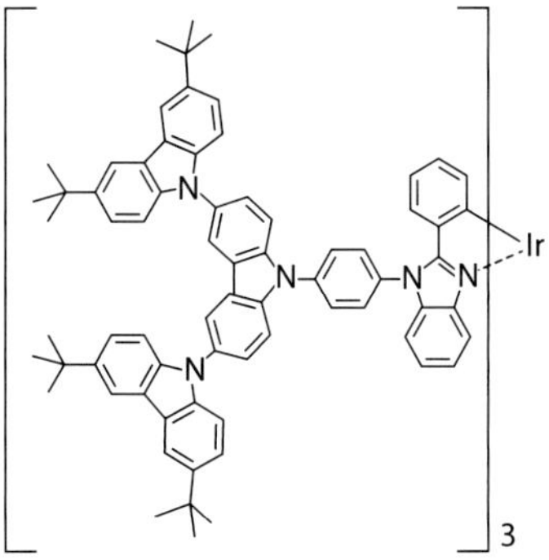

State-of-the-art light-emitting dendrimers can reach a maximum luminous efficiency of 34.7 cd/A (~20 cd/A at 1000 cd/m^2) for green emitters [131], 9.2 cd/A (5.7 cd/A at 1000 cd/m^2) for red emitters [132], and 5.3 cd/A (4.9 cd/A at 1000 cd/m^2) for blue emitters [133].

Ding et al. [131] reported green-emitting iridium dendrimers (**142–144**, Chart 4.64) with rigid hole-transporting carbazole dendrons. Light-emitting devices were fabricated with the structure ITO/PEDOT:PSS/neat dendrimer/TPBI/LiF/Al. A device based on **144** showed a maximum EQE of 10.3% and a maximum luminous efficiency of 34.7 cd/A. By doping the dendrimers into a carbazole-based host, the maximum EQE can be further increased to 16.6% and a maximum luminous efficiency of 57.9 cd/A.

Liang and coworkers [132] reported a series of red-emitting iridium dendrimers (**145–148**, Chart 4.65) with 1-phenylisoquinoline derivatives encapsulated with peripheral aryl-amines as dendrons. Devices were fabricated from dendritic iridium complexes with a small-molecule host solution using solution processing in various device configurations. Dendrimers **147** and **148** showed similar device performances with a maximum EQE of 12.8% and 11.8% and luminous efficiency of 9.2 and 8.5 cd/A at 0.1 mA/cm^2, respectively.

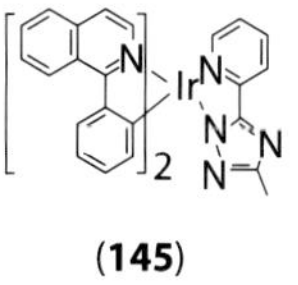

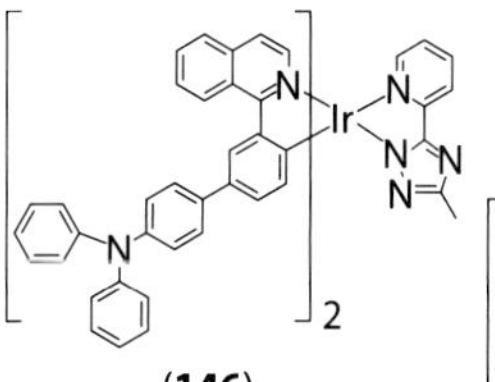

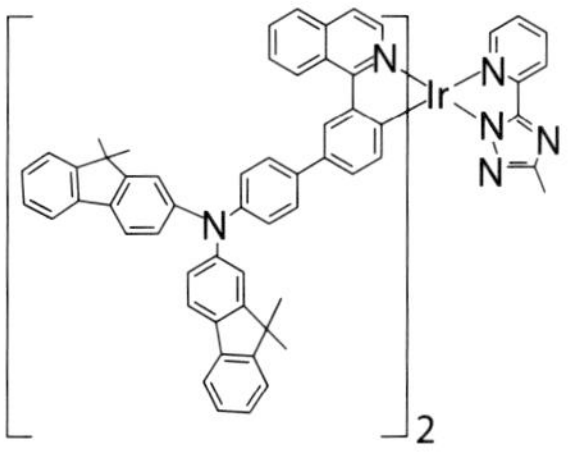

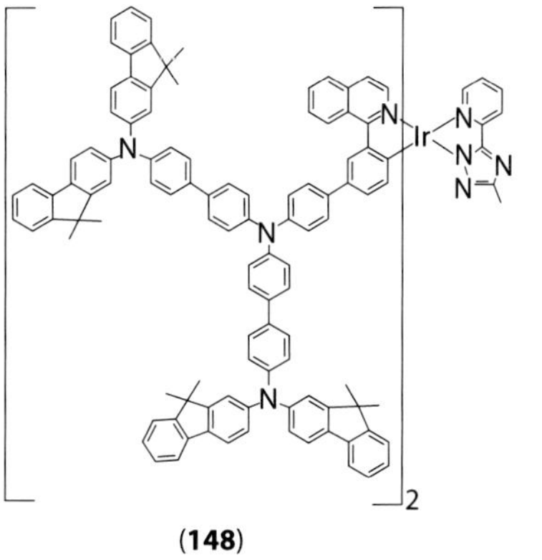

CHART 4.65
Chemical structure of dendrimers **145–148**.

Devices based on host material PDOF (30% PBD (**8**)) showed a slightly higher efficiency for **147**, with a maximum EQE of 13.9% at a higher current density of 6.4 mA/cm² and luminance of 601 cd/m².

The same group [133] later published a study of a series of color-stable blue OLEDs utilizing a solution-processable fluorescent π-conjugated dendrimer (**149**, Chart 4.66), with a maximum luminous efficiency of 5.3 cd/A. Efficient green, red, and white OLEDs were also demonstrated by doping **149** with phosphorescent dyes. Combining high device

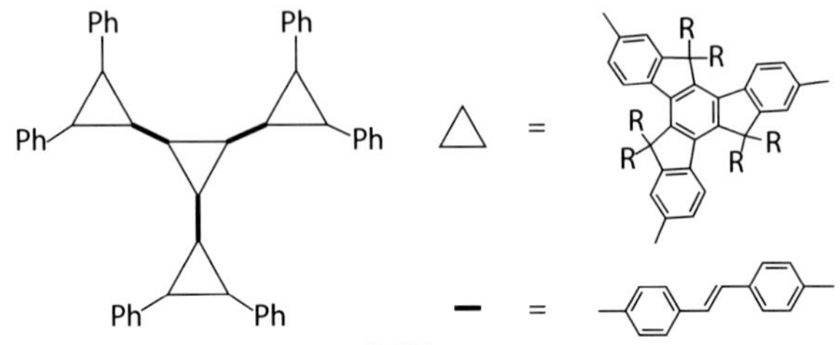

(**149**)

CHART 4.66
Chemical structure of dendrimer **149**.

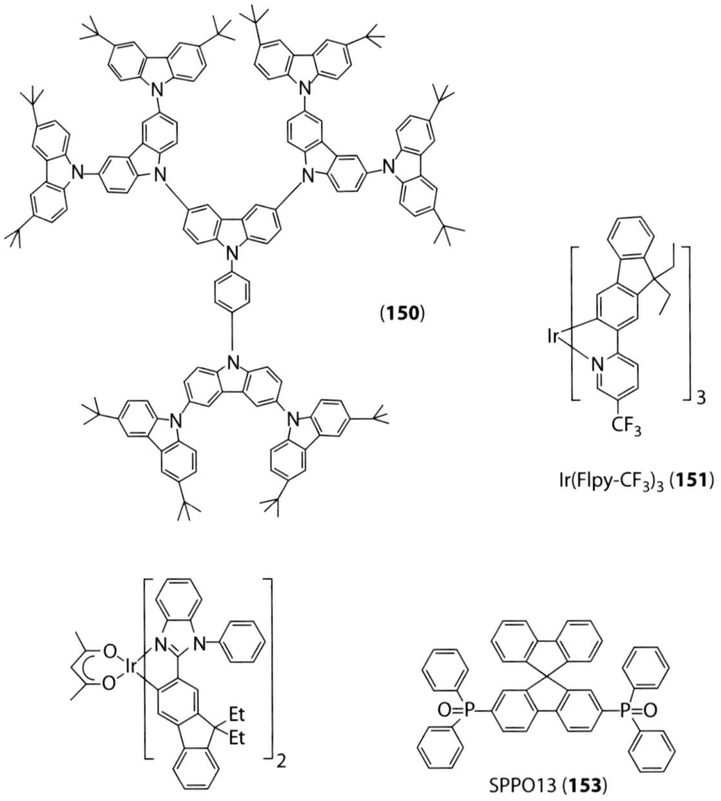

(**150**)

(fbi)$_2$Ir(acac) (**152**)

SPPO13 (**153**)

CHART 4.67
Chemical structure of dendrimer **150**, Ir(Flpy-CF$_3$)$_3$, (fbi)$_2$Ir(acac), and SPPO13.

efficiency with solution processability, **149** becomes one of the best blue-emitting materials for fabricating flat-panel displays and white light lighting panels via solution processing.

Very recently, Zhang et al. [134] reported an extremely high-efficiency solution-processed white OLED using dendritic host material **150** (Chart 4.67) and a novel efficient orange phosphorescent iridium complex 5-trifluoromethyl-2-(9,9-diethylfluoren-2-yl)pyridine (Ir(Flpy-CF$_3$)$_3$ **151**) and bis(2-(9,9-diethyl-9*H*-fluoren-2-yl)-1-phenyl-1*H*-benzoimidazol-*N*,*C*^{3})-iridium(acetylacetonate) ((fbi)$_2$Ir(acac) **152**), which show good miscibility with the host matrix. The white OLEDs were fabricated with a structure of ITO/PEDOT:PSS/dendritic host material **150**:FIrpic:Ir(Flpy-CF$_3$)$_3$/2,7-bis(diphenylphosphoryl)-9,9′-spirobi(fluorene) (SPPO13 **153**)/LiF/Al. The solution-processable carbazole-based dendritic host not only provides efficient hole injection from the PEDOT:PSS layer because of its high-lying HOMO level but also possesses a very promising high triplet energy level (2.89 eV), which effectively reduces the quenching of triplet excitons on the blue phosphor. The optimized device exhibited a forward-viewing luminous efficiency of 70.6 cd/A, EQE of 26.0%, and power efficiency of 47.6 lm/W at a luminance of 100 cd/m^2. The light-emitting efficiency of the solution-processed white OLEDs approached that of the competitive fluorescent lamp (40–70 lm/W).

4.7 Conclusions and Remarks

Polymer-based electrophosphorescent LEDs are a new research area explored only since 1999. Nevertheless, the practical achievements reported for the last 13 years suggest a strong potential of this technology to pioneer the manufacturing of large-area low-cost displays in the future by using solution techniques. The EL efficiency demonstrated by PPLEDs (EQE up to 29%, current efficiency up to 60 cd/A), although still a little lower than that of the best phosphorescent small molecules OLEDs, overrides conventional, fluorescent polymer LEDs. Most of the phosphorescent polymeric materials studied consist of a fluorescent host polymer doped by a low-molecular phosphorescent dye. However, many prominent examples of metallorganic phosphorescent polymers have been recently demonstrated.

Although the empirical search was a major contributor toward new host materials and guest dopants, the rational design based on consideration of energy levels and energy transfer criteria is playing a very important role in the creation of specific host–guest systems. The well-developed theory of electron and exciton transfer allows scientists to work out the basic material requirements. Thus, the spectral overlap between the host emission and the guest absorption is needed for effective Förster transfer of singlet exciton, and the HOMO and LUMO levels of the dopant should preferably be within the band gap of the host materials (to facilitate the Dexter transfer of triplets). Thus, one needs a wide band-gap polymer (yet highly fluorescent and possessing substantial charge mobility) and a smaller gap efficient phosphor dopant. At the same time, the triplet energy of the polymer should be above the triplet of the dopant to preclude back-transfer of the triplet excitons on the polymer. This latter condition is difficult to fulfill [92] and the structure–property relationships are not well established in this area. Probably, a high-level computational study should be implemented to help in the design of such novel host materials.

Thus far, poly(*N*-vinylcarbazole) PVK has been shown as the most efficient host polymer for PPLEDs. In combination with green-emitting [Pt(O^N^C^N)]-type complexes (**39a–g**), the champion EQE of 15.55% for green and 12.73% for white devices have been achieved, respectively [58]. The record white PPLEDs were achieved using PVK:OXD-7:FIrpic:Ir(mppy)$_3$:yellow-emitting iridium complexes **35** (or **36**):Ir-G2 as the emitting layer with peak forward-viewing power efficiency close to 40 lm/W, corresponding to an EQE of 28.8% and a luminous efficiency of 60 cd/A [56]. The conjugated polymers are usually poor hosts for high-energy blue and green emitters, but can perform rather well with red-emitting phosphors. Thus, an EQE of 18.0% (luminous efficiency of 26.3 cd/A) was reported for a red-emitting PPLED built with polyfluorene copolymers TOF-Os(bpftz)$_x$ (**135a–d**), containing red-emitting osmium complex Os(bpftz) (**31**) [124]. Recently, a breakthrough of an EQE of 14.2% (maximum luminance of 17,161 cd/m^2) was reported for a green-emitting PPLED built with poly(9,9-bis(3'-(*N,N*-dimethylamino)propyl)-2,7-fluorene)-*alt*-2,7-(9,9-dioctylfluorene) (PFN **65**) as the host, Ir(Bu-PPy)$_3$ (**56**) as the dopant, and Al as the cathode [80]. A blue-emitting PPLED based on PmPTPA (**81**) and FIrpic (**16**) exhibited an EQE of 9.3% (luminous efficiency of 17.9 cd/A) [90]. These achievements indicate that conjugated polymers are suitable host materials for PPLEDs with all emission colors.

At the same time, a rather high operating voltage, as compared with fluorescent PLEDs, results in a relatively low power conversion efficiency (particularly for nonconjugated hosts). There were, thus far, only a couple of reports on PPLED device with high power efficiency close to 40 lm/W (cf. 38.6 lm/W [103] and 40 lm/W [56]), while for the majority of PPLEDs it does not exceed 30 lm/W. This is caused by direct charge trapping on the phosphor dopants, which also might be responsible for the device degradation.

The most important problem for commercialization of PPLEDs is their short operation lifetime. The lower operation stability of phosphorescent LEDs as compared with fluorescent devices might be expected from the longer lifetime of emitting states (triplet excitons), which facilitates side chemical processes. Surprisingly, there are only scarce studies of this issue in the literature [41], and, unfortunately, device stability has rarely been the main objective in the synthesis of new host and guest materials. Clearly, systematic studies on device degradation and accompanying chemical processes are needed for a rational design of new phosphors and polymer hosts with practically acceptable stability.

References

1. SR Forrest, The road to high efficiency organic light emitting devices, *Org. Electron.*, 4: 45–48, 2003.
2. CH Li, G Kioseoglou, OMJ van't Erve, AT Hanbicki, BT Jonker, R Mallory, M Yasar, and A Petrou, Spin injection across (110) interfaces: Fe/GaAs(110) spin-light-emitting diodes, *Appl. Phys. Lett.*, 85: 1544–1546, 2004.
3. H Uoyama, K Goushi, K Shizu, H Nomura, and C Adachi, Highly efficient organic light-emitting diodes from delayed fluorescence, *Nature*, 492: 234–238, 2012.
4. J Kido, H Haromichi, K Hongawa, K Nagai, and K Okuyama, Bright red light-emitting organic electroluminescent devices having a europium complex as an emitter, *Appl. Phys. Lett.*, 65: 2124–2126, 1994.
5. T Sano, M Fujita, T Fujii, Y Hamada, K Shibata, and K Kuroki, Novel europium complex for electroluminescent devices with sharp red emission, *Jpn. J. Appl. Phys.*, 34 (Part 1): 1883–1887, 1995.
6. MA Baldo, DF O'Brien, Y You, A Shoustikov, S Sibley, ME Thompson, and SR Forrest, Highly efficient phosphorescent emission from organic electroluminescent devices, *Nature*, 395: 151–154, 1998.

7. C Adachi, MA Baldo, ME Thompson, and SR Forrest, Nearly 100% internal phosphorescence efficiency in an organic light emitting device, *J. Appl. Phys.*, 90: 5048–5051, 2001.
8. V Cleave, G Yahioglu, PL Barny, RH Friend, and N Tessler, Harvesting singlet and triplet energy in polymer LEDs, *Adv. Mater.*, 11: 285–288, 1999.
9. PW Atkins, and RS Friedman, *Molecular Quantum Mechanics*, 3rd ed., Oxford University Press, Oxford, 1997, pp. 109–119.
10. Z Shuai, D Beljonne, RJ Silbey, and JL Bredas, Singlet and triplet exciton formation rates in conjugated polymer light-emitting diodes, *Phys. Rev. Lett.*, 24: 131–134, 2000.
11. S Ramasesha, S Mazumdar, K Tandon, and M Das, Electron correlation effects in electron–hole recombination and triplet–triplet scattering in organic light emitting diodes, *Synth. Met.*, 139: 917–920, 2003.
12. Y Cao, ID Parker, G Yu, C Zhang, and AJ Heeger, Improved quantum efficiency for electroluminescence in semiconducting polymers, *Nature*, 397: 414–417, 1999.
13. M Wohlgenannt, K Tandon, S Mazumdar, S Ramasesha, and ZV Vardeny, Formation cross-sections of singlet and triplet excitons in p-conjugated polymers, *Nature*, 409: 494–497, 2001.
14. JS Wilson, AS Dhoot, AJAB Seeley, MS Khan, A Köhler, and RH Friend, Spin-dependent exciton, formation in p-conjugated compounds, *Nature*, 413: 828–831, 2001.
15. AS Dhoot, and NC Greenham, Triplet formation in polyfluorene derivatives, *Adv. Mater.*, 24: 1834–1837, 2002.
16. W Wohlgenannt, XM Jiang, C Yang, OJ Korovyanko, and ZV Vardeny, Spin-dependent polaron pair recombination in p-conjugated polymers: Enhanced singlet exciton densities, *Synth. Met.*, 139: 921–924, 2003.
17. A Köhler, and J Wilson, Phosphorescence and spin-dependent exciton formation in conjugated polymers, *Org. Electron.*, 4: 179–189, 2003.
18. M Segal, MA Baldo, RJ Holmes, SR Forrest, and ZG Soos, Excitonic singlet–triplet ratios in molecular and polymeric organic materials, *Phys. Rev. B*, 68: 075211, 2003.
19. MA Baldo, DF O'Brien, ME Thompson, and SR Forrest, Excitonic singlet–triplet ratio in a semiconducting organic thin film, *Phys. Rev. B*, 60: 14422, 1999.
20. A Jablonski, Efficiency of anti-stokes fluorescence in dyes, *Nature*, 131: 839–841, 1933.
21. RA Marcus, On the theory of oxidation–reduction reactions involving electron transfer. I, *J. Chem. Phys.*, 24: 966–979, 1956.
22. GL Closs, and JR Miller, Intramolecular long-distance electron transfer in organic materials, *Science*, 240: 440–447, 1988.
23. DL Dexter, A theory of sensitized luminescence in solids, *J. Chem. Phys.*, 21: 836–850, 1953.
24. T Förster, Transfer mechanisms of electronic excitation, *Discuss. Faraday Soc.*, 27: 7–17, 1959.
25. MA Baldo, C Adachi, and SR Forrest, Transient analysis of electrophosphorescence II. Transient analysis of triplet–triplet annihilation, *Phys. Rev. B*, 62: 10967–10977, 2000.
26. C Adachi, MA Baldo, SR Forrest, and ME Thompson, High-efficiency organic electrophosphorescent devices with tris(2-phenylpyridine)iridium doped into electron-transporting materials, *Appl. Phys. Lett.*, 77: 904–906, 2000.
27. Y Ma, CM Che, HY Chao, X Zhou, WH Chan, and J Shen, High luminescence gold(I) and copper(I) complexes with a triplet excited state for use in light-emitting diodes, *Adv. Mater.*, 11: 852–857, 1999.
28. Y Li, Y Wang, Y Wu, and J Shen, Carbonyl polypyridyl Re(I) complexes as organic electroluminescent materials, *Synth. Met.*, 99: 257–260, 1999.
29. WD Gill, Drift mobilities in amorphous charge-transfer complexes of trinitrofluorenone and poly-*n*-vinylcarbazole, *J. Appl. Phys.*, 43: 5033–5040, 1972.
30. J Kido, K Hongawa, K Okuyama, and K Nagai, Bright blue electroluminescence from poly (*N*-vinylcarbazole), *Appl. Phys. Lett.*, 63: 2627–2629, 1993.
31. MD McGehee, T Bergstedt, C Zhang, AP Saab, MB O'Regan, GC Bazan, VI Srdanov, and AJ Heeger, Narrow bandwidth luminescence from blends with energy transfer from semiconducting conjugated polymers to europium complexes, *Adv. Mater.*, 11: 1349–1354, 1999.

32. C Lee, KB Lee, and J Kim, Polymer phosphorescent light-emitting devices doped with tris (2-phenylpyridine) iridium as a triplet emitter, *Appl. Phys. Lett.*, 77: 2280–2282, 2000.

33. A Tsuboyama, H Iwawaki, M Furugori, T Mukaide, J Kamatani, S Igawa, T Moriyama, S Muira, T Takiguchi, S Oakada, M Hoshino, and K Ueno, Homoleptic cyclometalated iridium complexes with highly efficient red phosphorescence and application to organic light-emitting diode, *J. Am. Chem. Soc.*, 125: 12971–12979, 2003.

34. S Lamansky, P Djurovich, D Murphy, F Abdel-Razzaq, HE Lee, C Adachi, PE Burrows, SR Forrest, and ME Thompson, Highly phosphorescent bis-cyclometalated iridium complexes: Synthesis, photophysical characterization, and use in organic light-emitting diodes, *J. Am. Chem. Soc.*, 123: 4304–4312, 2001.

35. M Yang, and T Tsutsui, Use of poly(9-vinylcarbazole) as host material for iridium complexes in high-efficiency organic light-emitting devices, *Jpn. J. Appl. Phys.*, 39: L828–L829, 2000.

36. KM Vaeth, and CW Tang, Light-emitting diodes based on phosphorescent guest/polymeric host systems, *J. Appl. Phys.*, 92: 3447–3453, 2002.

37. KM Vaeth, and J Dicillo, High-efficiency doped polymeric organic light-emitting diodes, *J. Polym. Sci.: Part B: Polym. Phys.*, 41: 2715–2725, 2003.

38. X Yang, D Neher, D Hertel, and TK Däubler, Highly efficient single-layer polymer electrophosphorescent devices, *Adv. Mater.*, 16: 161–166, 2004.

39. F Chen, Y Yang, ME Thompson, and J Kido, High-performance polymer light-emitting diodes doped with a red phosphorescent iridium complex, *Appl. Phys. Lett.*, 80: 2308–2310, 2002.

40. MA Baldo, S Lamansky, PE Burrows, ME Thompson, and SR Forrest, Very high-efficiency green organic light-emitting devices based on electrophosphorescence, *Appl. Phys. Lett.*, 75: 4–6, 1999.

41. SC Chang, G He, FC Chen, TF Guo, and Y Yang, Degradation mechanism of phosphorescent-dye-doped polymer light-emitting diodes, *Appl. Phys. Lett.*, 79: 2088–2090, 2001.

42. Y Kawamura, S Yanagida, and SR Forrest, Energy transfer in polymer electrophosphorescent light emitting devices with single and multiple doped luminescent layers, *J. Appl. Phys.*, 92: 87–93, 2002.

43. DF O'Brien, MA Baldo, ME Thompson, and SR Forrest, Improved energy transfer in electrophosphorescent devices, *Appl. Phys. Lett.*, 74: 442–444, 1999.

44. S Lamansky, PI Djurovich, F Abdel-Razzaq, S Garon, DL Murphy, and ME Thompson, Cyclometalated Ir complexes in polymer organic light-emitting diodes, *J. Appl. Phys.*, 92: 1570–1572, 2002.

45. A Nakamura, T Tada, M Mizukami, and S Yagyu, Efficient electrophosphorescent polymer light-emitting devices using a Cs/Al cathode, *Appl. Phys. Lett.*, 84: 130–132, 2004.

46. G He, S Chang, F Chen, Y Li, and Y Yang, Highly efficient polymer light-emitting devices using a phosphorescent sensitizer, *Appl. Phys. Lett.*, 81: 1509–1511, 2002.

47. X Gong, MR Robinson, JC Ostrowski, D Moses, GC Bazan, and AJ Heeger, High-efficiency polymer-based electrophosphorescent devices, *Adv. Mater.*, 14: 581–585, 2002.

48. X Gong, JC Ostrowski, D Moses, GC Bazan, and AJ Heeger, Electrophosphorescence from a polymer guest–host system with an iridium complex as guest: Förster energy transfer and charge trapping, *Adv. Funct. Mater.*, 13: 439–444, 2003.

49. X Gong, JC Ostrowski, GC Bazan, D Moses, and AJ Heeger, Red electrophosphorescent from polymer doped with iridium complex, *Appl. Phys. Lett.*, 81: 3711–3713, 2002.

50. X Gong, JC Ostrowski, D Moses, GC Bazan, and AJ Heeger, High-performance polymer-based electrophosphorescent light-emitting diodes, *J. Polym. Sci. B: Polym. Phys.*, 41: 2691–2705, 2003.

51. S Kan, X Liu, F Shen, J Zhang, Y Ma, Y Wang, and J Shen, Improved efficiency of single-layer polymer light-emitting devices with poly(vinylcarbazole) doubly doped with phosphorescent and fluorescent dyes as the emitting layer, *Adv. Funct. Mater.*, 13: 603–608, 2003.

52. F Shen, H Xia, C Zhang, D Lin, X Liu, and Y Ma, Spectral investigation for phosphorescent polymer light-emitting devices with doubly doped phosphorescent dyes, *Appl. Phys. Lett.*, 84: 55–57, 2004.

53. PI Shih, CF Shu, YL Tung, and Y Chi, Efficient white-light-emitting diodes based on poly(*N*-vinylcarbazole) doped with blue fluorescent and orange phosphorescent materials, *Appl. Phys. Lett.*, 88: 251110, 2006.

54. H Wu, J Zou, F Liu, L Wang, A Mikhailovsky, GC Bazan, W Yang, and Y Cao, Efficient single active layer electrophosphorescent white polymer light-emitting diodes, *Adv. Mater.*, 20: 696–702, 2008.

55. H Wu, G Zhou, J Zou, C Ho, WY Wong, W Yang, J Peng, and Y Cao, Efficient polymer white-light-emitting devices for solid-state lighting, *Adv. Mater.*, 21: 4181–4184, 2009.

56. J Zou, H Wu, CS Lam, C Wang, J Zhu, C Zhong, S Hu, CL Ho, GJ Zhou, H Wu, WCH Choy, J Peng, Y Cao, and WY Wong, Simultaneous optimization of charge-carrier balance and luminous efficacy in highly efficient white polymer light-emitting devices, *Adv. Mater.*, 23: 2976–2980, 2011.

57. CR Yin, SH Ye, J Zhao, MD Yi, LH Xie, ZQ Lin, YZ Chang, F Liu, H Xu, NE Shi, Q Yan, and W Huang, Hindrance-functionalized π-stacked polymer host materials of the cardo-type carbazole-fluorene hybrid for solution-processable blue electrophosphorescent devices, *Macromolecules*, 44: 4589–4595, 2011.

58. G Cheng, PK Chow, SCF Kui, CC Kwok, and CM Che, High-efficiency polymer light-emitting devices with robust phosphorescent platinum(II) emitters containing tetradentate dianionic O^N^C^N ligands, *Adv. Mater.*, 25: 6765–6770, 2013.

59. H Wang, JT Ryu, DU Kim, YS Han, LS Park, HY Cho, SJ Lee, and Y Kwon, Green and red electrophosphorescent devices consisting of cabazole/triarylamine-based polymers doped with iridium complexes, *Mol. Cryst. Liq. Cryst.*, 471: 279–291, 2007.

60. MW Thesen, B Höfer, M Debeaux, S Janietz, A Wedel, A Köhler, HH Johannes, and H Krueger, Hole-transporting host–polymer series consisting of triphenylamine basic structures for phosphorescent polymer light-emitting diodes, *J. Polym. Sci. A: Polym. Chem.*, 48: 3417–3430, 2010.

61. S Shao, J Ding, T Ye, Z Xie, L Wang, X Jing, and F Wang, A novel, bipolar polymeric host for highly efficient blue electrophosphorescence: A non-conjugated poly(aryl ether) containing triphenylphosphine oxide units in the electron-transporting main chain and carbazole units in hole-transporting side chains, *Adv. Mater.*, 23: 3570–3574, 2011.

62. D Sun, Q Fu, Z Ren, W Li, H Li, D Ma, and S Yan, Carbazole-based polysiloxane hosts for highly efficient solution-processed blue electrophosphorescent devices, *J. Mater. Chem. C: Mater. Opt. Electron. Devices*, 1: 5344–5350, 2013.

63. TF Guo, SC Chang, Y Yang, RC Kwong, and ME Thompson, Highly efficient electrophosphorescent polymer light-emitting devices, *Org. Electron.*, 1: 15–20, 2000.

64. DF O'Brien, C Giebelerb, RB Fletcherb, AJ Cadbyb, LC Palilisb, DG Lidzeyb, PA Laneb, DDC Bradley, and W Blau, Electrophosphorescence from a doped polymer light emitting diode, *Synth. Met.*, 116: 379–383, 2001.

65. PA Lane, LC Palilis, DF O'Brien, C Giebeler, AJ Cadby, DG Lidzey, AJ Campbell, W Blau, and DDC Bradley, Origin of electrophosphorescence from a doped polymer light emitting diode, *Phys. Rev. B*, 63: 235206, 2001.

66. FC Chen, Y Yang, and Q Pei, Phosphorescent light-emitting electrochemical cell, *Appl. Phys. Lett.*, 81: 4278–4280, 2002.

67. Y Noh, C Lee, J Kim, and K Yase, Energy transfer and device performance in phosphorescent dye doped polymer light emitting diodes, *J. Chem. Phys.*, 118: 2853–2864, 2003.

68. FC Chen, SC Chang, G He, S Pyo, Y Yang, M Kurotaki, and J Kido, Energy transfer and triplet exciton confinement in polymeric electrophosphorescence devices, *J. Polym. Sci. B: Polym. Phys.*, 41: 2681–2690, 2003.

69. X Gong, JC Ostrowski, GC Bazan, D Moses, AJ Heeger, MS Liu, and AKY Jen, Electrophoresence from a conjugated copolymer doped with an iridium complex: High brightness and improved operational stability, *Adv. Mater.*, 15: 45–49, 2003.

70. F Chen, G He, and Y Yang, Triplet exciton confinement in phosphorescent polymer light-emitting diodes, *Appl. Phys. Lett.*, 82: 1006–1008, 2003.

71. C Jiang, W Yang, J Peng, S Xiao, and Y Cao, High-efficiency, saturated red-phosphorescent polymer light-emitting diodes based on conjugated polymers doped with an Ir complex, *Adv. Mater.*, 16: 537–541, 2004.

72. X Gong, W Ma, JC Ostrowski, GC Bazan, D Moses, and AJ Heeger, White electrophosphorescence from semiconducting polymer blends, *Adv. Mater.*, 16: 615–619, 2004.
73. X Gong, D Moses, AJ Heeger, and S Xiao, White light electrophosphorescence from polyfluorene-based light-emitting diodes: Utilization of fluorenone defects, *J. Phys. Chem. B*, 108: 8601–8605, 2004.
74. HJ Su, FI Wu, CF Shu, YL Tung, Y Chi, and GH Lee, Polyfluorene containing diphenylquinoline pendants and their applications in organic light emitting diodes, *J. Polym. Sci. A: Polym. Chem.*, 43: 859–869, 2005.
75. SM King, HA Al-Attar, RJ Evans, A Congreve, A Beeby, and AP Monkman, The use of substituted iridium complexes in doped polymer electrophosphorescent devices: The influence of triplet transfer and other factors on enhancing device performance, *Adv. Funct. Mater.*, 16: 1043–1050, 2006.
76. TH Kim, HK Lee, OO Park, BD Chin, SH Lee, and JK Kim, White-light-emitting diodes based on iridium complexes via efficient energy transfer from a conjugated polymer, *Adv. Funct. Mater.*, 16: 611–617, 2006.
77. YH Xu, JB Peng, JX Jiang, W Xu, W Yang, and Y Cao, Efficient white-light-emitting diodes based on polymer codoped with two phosphorescent dyes, *Appl. Phys. Lett.*, 87: 193502, 2005.
78. K Zhang, Z Chen, C Yang, X Zhang, Y Tao, L Duan, L Chen, L Zhu, J Qin, and Y Cao, Improving the performance of phosphorescent polymer light-emitting diodes using morphology-stable carbazole-based iridium complexes, *J. Mater. Chem.*, 17: 3451–3460, 2007.
79. Z Wu, Y Xiong, J Zou, L Wang, J Liu, Q Chen, W Yang, J Peng, and Y Cao, High-triplet-energy poly(9,9′-bis(2-ethylhexyl)-3,6-fluorene) as host for blue and green phosphorescent complexes, *Adv. Mater.*, 20: 2359–2364, 2008.
80. Z Chen, QL Niu, Y Zhang, L Ying, JB Peng, and Y Cao, Efficient green electrophosphorescence with Al cathode using an effective electron-injecting polymer as the host, *ACS Appl. Mater. Interfaces*, 1: 2785–2788, 2009.
81. W Zhu, Y Mo, M Yuan, W Yang, and Y Cao, Highly efficient electrophosphorescent devices based on conjugated polymers doped with iridium complexes, *Appl. Phys. Lett.*, 80: 2045–2047, 2002.
82. W Zhu, C Liu, L Su, W Yang, M Yuan, and Y Cao, Synthesis of new iridium complexes and their electrophosphorescent properties in polymer light-emitting diodes, *J. Mater. Chem.*, 13: 50–55, 2003.
83. W Zhu, M Zhu, Y Ke, L Su, M Yuan, and Y Cao, Synthesis and red electrophosphorescence of a novel cyclometalated iridium complex in polymer light-emitting diodes, *Thin Solid Films*, 446: 128–131, 2004.
84. XH Yang, D Neher, U Scherf, SA Bagnich, and H Bässler, Polymer electrophosphorescent devices utilizing a ladder-type poly(para-phenylene) host, *J. Appl. Phys.*, 93: 4413–4419, 2003.
85. JM Lupton, A Pogantsch, T Piok, EJW List, S Patil, and U Scherf, Intrinsic room-temperature electrophosphorescence from a π-conjugated polymer, *Phys. Rev. Lett.*, 89: 167401, 2002.
86. BS Harrison, TJ Foley, AS Knefely, JK Mwaura, GB Cunningham, TS Kang, M Bougettaya, JM Boncella, JR Reynolds, and KS Schanze, Near-infrared photo- and electroluminescence of alkoxy-substituted poly(p-phenylene) and nonconjugated polymer/lanthanide tetraphenylporphirin blends, *Chem. Mater.*, 16: 2938–2947, 2004.
87. SP Huang, TH Jen, YC Chen, AE Hsiao, SH Yin, HY Chen, and SA Chen, Effective shielding of triplet energy transfer to conjugated polymer by its dense side chains from phosphor dopant for highly efficient electrophosphorescence, *J. Am. Chem. Soc.*, 130: 4699–4707, 2008.
88. YK Huang, TH Jen, YT Chang, NJ Yang, HH Lu, and SA Chen, Enhancing shielding of triplet energy transfer to poly(p-phenylene)s from phosphor dopant by addition of branched alcohol for highly efficient electrophosphorescence, *ACS Appl. Mater. Interfaces*, 2: 1094–1099, 2010.
89. J Liu, and Q Pei, Poly(m-phenylene): Conjugated polymer host with high triplet energy for efficient blue electrophosphorescence, *Macromolecules*, 43: 9608–9612, 2010.
90. J Liu, L Li, and Q Pei, Conjugated polymer as host for high efficiency blue and white electrophosphorescence, *Macromolecules*, 44: 2451–2456, 2011.
91. M Sudhakar, PI Djurovich, TE Hogen-Esch, and ME Thompson, Phosphorescence quenching by conjugated polymers, *J. Am. Chem. Soc.*, 125: 7796–7797, 2003.

92. A Köhler, and D Beljonne, The singlet–triplet exchange energy in conjugated polymers, *Adv. Funct. Mater.*, 14: 11–18, 2004.

93. YV Romanovskii, A Gerhard, B Schweitzer, U Scherf, RI Personov, and H Bässler, Phosphorescence of p-conjugated oligomers and polymers, *Phys. Rev. Lett.*, 84: 1027–1030, 2000.

94. D Hertel, S Setayesh, HG Nothofer, U Scherf, K Müllen, and H Bässler, Phosphorescence in conjugated poly(*para*-phenylene-derivatives), *Adv. Mater.*, 13: 65–70, 2001.

95. A van Dijken, JJAM Bastiaansen, NMM Kiggen, BMW Langeveld, C Rothe, A Monkman, I Bach, P Stössel, and K Brunner, Carbazole compounds as host materials in light-emitting diodes: Polymer hosts for high-efficiency light-emitting diodes, *J. Am. Chem. Soc.*, 126: 7718–7727, 2004.

96. YC Chen, GS Huang, CC Hsiao, and SA Chen, High triplet energy polymer as host for electrophosphorescence with high efficiency, *J. Am. Chem. Soc.*, 128: 8549–8558, 2006.

97. X Wang, MR Andersson, ME Thompson, and O Inganäs, Electrophosphorescence from substituted poly(thiophene) doped with iridium or platinum complex, *Thin Solid Films*, 468: 226–233, 2004.

98. CL Lee, NG Kang, YS Cho, JS Lee, and JJ Kim, Polymer electrophosphorescent device: Comparison of phosphorescent dye doped and coordinated systems, *Opt. Mater.*, 21: 119–123, 2002.

99. S Tokito, M Suzuki, F Sato, M Kamachi, and K Shirane, High-efficiency phosphorescent polymer light-emitting devices, *Org. Electron.*, 4: 105–111, 2003.

100. S Tokito, M Suzuki, and F Sato, Improvement of emission efficiency in polymer light-emitting devices based on phosphorescent polymers, *Thin Solid Films*, 445: 353–357, 2003.

101. M Suzuki, T Hatekayama, S Tokito, and F Sato, High-efficiency white phosphorescent polymer light-emitting devices, *IEEE J. Sel. Top. Quantum Electron.*, 10: 115–120, 2004.

102. PT Furuta, L Deng, S Garon, ME Thompson, and JMJ Fréchet, Platinum-functionalized random copolymers for use in solution-processible, efficient, near-white organic light-emitting diodes, *J. Am. Chem. Soc.*, 126: 15388–15389, 2004.

103. M Suzuki, S Tokito, F Sato, T Igarashi, K Kondo, T Koyama, and T Yamaguchi, Highly efficient polymer light-emitting devices using ambipolar phosphorescent polymers, *Appl. Phys. Lett.*, 86: 103507, 2005.

104. J Pei, XL Liu, WL Yu, YH Lai, YH Niu, and Y Cao, Efficient energy transfer to achieve narrow bandwidth red emission from Eu^{3+}-grafting conjugated polymers, *Macromolecules*, 35: 7274–7280, 2002.

105. QD Ling, ET Kang, KG Neoh, and W Huang, Synthesis and nearly monochromatic photoluminescence properties of conjugated copolymers containing fluorene and rare earth complexes, *Macromolecules*, 36: 6995–7003, 2003.

106. X Chen, J Liao, Y Liang, MO Ahmed, H Tseng, and S Chen, High-efficiency red-light emission from polyfluorenes grafted with cyclometalated iridium complexes and charge transport moiety, *J. Am. Chem. Soc.*, 125: 636–637, 2003.

107. J Jiang, C Jiang, W Yang, H Zhen, F Huang, and Y Cao, High-efficiency electrophosphorescent fluorene-*alt*-carbazole copolymers *N*-grafted with cylometalated Ir complexes, *Macromolecules*, 38: 4072–4080, 2005.

108. AJ Sandee, CK Williams, NR Evans, JE Davies, CE Boothby, A Köhler, RH Friend, and AB Holmes, Solution-processible conjugated electrophosphorescent polymers, *J. Am. Chem. Soc.*, 126: 7041–7048, 2004.

109. GL Schulz, X Chen, SA Chen, and S Holdcroft, Enhancement of phosphorescence of Ir complexes bound to conjugated polymers: Increasing the triplet level of the main chain, *Macromolecules*, 39: 9157–9165, 2006.

110. NR Evans, LS Devi, CSK Mak, SE Watkins, SI Pascu, A Köhler, RH Friend, CK Williams, and AB Holmes, Triplet energy back transfer in conjugated polymers with pendant phosphorescent iridium complexes, *J. Am. Chem. Soc.*, 128: 6647–6656, 2006.

111. SJ Liu, Q Zhao, Y Deng, YJ Xia, J Lin, QL Fan, LH Wang, and W Huang, π-Conjugated chelating polymers with a charged iridium complex in the backbones: Toward saturated-red phosphorescent polymer light-emitting diodes, *J. Phys. Chem. C*, 111: 1166–1175, 2007.

112. H Zhen, C Jiang, W Yang, J Jiang, F Huang, and Y Cao, Synthesis and properties of electrophosphorescent chelating polymers with iridium complexes in the conjugated backbone, *Chem. Eur. J.*, 11: 5007–5016, 2005.

113. H Zhen, C Luo, W Yang, W Song, B Du, J Jiang, C Jiang, Y Zhang, and Y Cao, Electrophosphorescent chelating copolymers based on linkage isomers of naphthylpyridine-iridium complexes with fluorene, *Macromolecules*, 39: 1693–1700, 2006.

114. K Zhang, Z Chen, C Yang, S Gong, J Qin, and Y Cao, Saturated red-emitting electrophosphorescent polymers with iridium coordinating to β-diketonate units in the main chain, *Macromol. Rapid Commun.*, 27: 1926–1931, 2006.

115. K Zhang, Z Chen, Y Zou, C Yang, J Qin, and Y Cao, Synthesis, characterization, and photophysics of electroluminescent copolymers with a quinoline-based iridium complex in the main chain: A versatile method for constructing metal-containing copolymers, *Organometallics*, 26: 3699–3707, 2007.

116. K Zhang, Z Chen, Y Zou, S Gong, C Yang, J Qin, and Y Cao, Effective suppression of intra- and interchain triplet energy transfer to polymer backbone from the attached phosphor for efficient polymeric electrophosphorescence, *Chem. Mater.*, 21: 3306–3314, 2009.

117. K Zhang, Z Chen, C Yang, Y Zou, S Gong, J Qin, and Y Cao, First iridium complex end-capped polyfluorene: Improving device performance for phosphorescent polymer light-emitting diodes, *J. Phys. Chem. C*, 112: 3907–3913, 2008.

118. K Zhang, Z Chen, C Yang, Y Zou, S Gong, Y Tao, J Qin, and Y Cao, Iridium complexes embedded into and end-capped onto phosphorescent polymers: Optimizing PLED performance and structure–property relationships, *J. Mater. Chem.*, 18: 3366–3375, 2008.

119. L Ying, J Zou, A Zhang, B Chen, W Yang, and Y Cao, Novel orange-red light-emitting polymers with cyclometaled iridium complex grafted in alkyl chain, *J. Organomet. Chem.*, 694: 2727–2734, 2009.

120. J Jiang, Y Xu, W Yang, R Guan, Z Liu, H Zhen, and Y Cao, High-efficiency white-light-emitting devices from a single polymer by mixing singlet and triplet emission, *Adv. Mater.*, 18: 1769–1773, 2006.

121. H Zhen, W Xu, W Yang, Q Chen, Y Xu, J Jiang, J Peng, and Y Cao, White-light emission from a single polymer with singlet and triplet chromophores on the backbone, *Macromol. Rapid Commun.*, 27: 2095–2100, 2006.

122. Q Chen, N Liu, L Ying, W Yang, H Wu, W Xu, and Y Cao, Novel white-light-emitting polyfluorenes with benzothiadiazole and Ir complex on the backbone, *Polymer*, 50: 1430–1437, 2009.

123. T Guo, L Yu, B Zhao, Y Li, Y Tao, W Yang, Q Hou, H Wu, and Y Cao, Highly efficient, red-emitting hyperbranched polymers utilizing a phenyl-isoquinoline iridium complex as the core, *Macromol. Chem. Phys.*, 213: 820–828, 2012.

124. CH Chien, SF Liao, CH Wu, CF Shu, SY Chang, Y Chi, PT Chou, and CH Lai, Electrophosphorescent polyfluorenes containing osmium complexes in the conjugated backbone, *Adv. Funct. Mater.*, 18: 1430–1439, 2008.

125. FI Wu, XH Yang, D Neher, R Dodda, YH Tseng, and CF Shu, Efficient white-electrophosphorescent devices based on a single polyfluorene copolymer, *Adv. Funct. Mater.*, 17: 1085–1092, 2007.

126. DA Poulsen, BJ Kim, B Ma, CS Zonte, and JMJ Frèchet, Site isolation in phosphorescent bichromophoric block copolymers designed for white electroluminescence, *Adv. Mater.*, 22: 77–82, 2010.

127. TD Anthopoulos, JPJ Markham, EB Namdas, IDW Samuel, SC Lo, and PL Burn, Highly efficient single-layer dendrimer light-emitting diodes with balanced charge transport, *Appl. Phys. Lett.*, 82: 4824–4826, 2003.

128. TD Anthopoulos, MJ Frampton, EB Namdas, PL Burn, and IDW Samuel, Solution-processable red phosphorescent dendrimers for light-emitting device applications, *Adv. Mater.*, 16: 557–560, 2004.

129. EB Namdas, TD Anthopoulos, IDW Samuel, MJ Frampton, SC Lo, and PL Burn, Simple color tuning of phosphorescent light-emitting diodes, *Appl. Phys. Lett.*, 86: 161104–161106, 2005.

130. SC Lo, TD Anthopoulos, EB Namdas, PL Burn, and IDW Samuel, Encapsulated cores: Host-free organic light-emitting diodes based on solution-processible electrophosphorescent dendrimers, *Adv. Mater.*, 17: 1945–1948, 2005.
131. J Ding, J Gao, Y Cheng, Z Xie, L Wang, D Ma, X Jing, and F Wang, Highly efficient green-emitting phosphorescent iridium dendrimers based on carbazole dendrons, *Adv. Funct. Mater.*, 16: 575–581, 2006.
132. B Liang, L Wang, Y Xu, H Shi, and Y Cao, High-efficiency red phosphorescent iridium dendrimers with charge-transporting dendrons: Synthesis and electroluminescent properties, *Adv. Funct. Mater.*, 17: 3580–3589, 2007.
133. L Wang, Y Jiang, J Luo, Y Zhou, J Zhou, J Wang, J Pei, and Y Cao, Highly efficient and color-stable deep-blue organic light-emitting diodes based on a solution-processible dendrimer, *Adv. Mater.*, 21: 4854–4858, 2009.
134. B Zhang, G Tan, CS Lam, B Yao, CL Ho, L Liu, Z Xie, WY Wong, J Ding, and L Wang, High-efficiency single emissive layer white organic light-emitting diodes based on solution-processed dendritic host and new orange-emitting iridium complex, *Adv. Mater.*, 24: 1873–1877, 2012.

5

Polarized Light Emission from Organic Light-Emitting Diodes

Daniel Steiger and Christoph Weder

CONTENTS

5.1 Introduction

Uniaxially oriented, form-anisotropic photoluminescent (PL) molecules usually exhibit anisotropic optical characteristics, i.e., linearly polarized absorption and emission. The phenomenon of linearly polarized PL from inorganic crystals has been known for more than a century [1] and was reported for oriented blends of ductile polymers and low-molecular-weight organic PL molecules as early as the 1930s [2]. The emission of linearly polarized light has since been observed for a plethora of materials that were oriented by a broad variety of orientation methods [3]. A similar effect is the emission of circularly polarized (CP) light, which can be observed in the case of chiral PL molecules [4] and also in the case of nonchiral PL molecules that are embedded in a chiral matrix [5]. The emission of linearly polarized or CP light from PL matter has recently attracted a great deal of interest, and PL polarizers [6–10] may be useful for a variety of applications that range from security features [11] to liquid crystal displays (LCDs) [12].

Polarized chromatic light, which is essential for a variety of devices and applications, is usually generated by the use of an isotropic light source in combination with a polarizer [13]. Probably the technologically most important application of this combination is in LCDs, which currently represent the dominating flat-panel display technology [14]. In the case of backlit LCDs, a white light source is employed, and the three primary colors—red, green, and blue—are generated by means of absorbing color filters. However, dichroic sheet polarizers and conventional color filters rely on the principle of light absorption and are extremely energy-inefficient elements [15]. As a matter of fact, in conventional LCDs, about 90% of the light generated by the backlight is absorbed by the back polarizers and the color filters, and consequently these devices exhibit severe limitations in brightness and energy efficiency [15]. One approach of improving the yield of polarized light is the replacement of dichroic polarizers with elements that reflect or scatter light [16–19]. By using appropriate supplementary elements to recycle reflected or scattered energy, the ultimate efficiency of these systems can, in principle, approach unity (twice that of dichroic polarizers). In practice, efficiencies up to 80% are achieved [15]. Rather than recycling wrongly polarized light, an alternative way to improve the overall efficiency of the system is to directly generate polarized light. In this context, the combination of PL polarizers with a monochromatic, an ultraviolet (UV), or a blue light source [7,12] has been proposed. This approach is attractive as it not only eliminates an absorbing polarizer but also makes the terribly inefficient color filters obsolete. The PL materials act as active color filters, and in principle, isotropic-to-polarized conversion efficiencies that approach the PL quantum efficiency of the emitting species could be attainable by this concept. Taking this approach one step further, one arrives at light sources that directly emit polarized chromatic light, a feature that had hitherto been unique to lasers. Although the general idea had already been proposed in the patent literature [20], Dyreklev et al. [21] were the first to demonstrate that the concept of polarized light emission can also be exploited in the case of polymer light-emitting diodes (PLEDs). Most conventional PLEDs and also light-emitting diodes (LEDs) based on low-molecular-weight organic emitters (OLEDs) comprise an isotropic (i.e., unoriented, nonchiral) emitting layer, and concomitantly their emission is essentially unpolarized [22–24]. However, as the singlet excited states generated through the recombination of electrons and holes are identical with those generated through photoexcitation [25], light-emitting devices comprising a uniaxially oriented emitting layer and an otherwise conventional architecture emit linearly polarized light (Figure 5.1a). In their pioneering study, Dyreklev et al. employed a modestly oriented conjugated polymer, and as a result, the emission of the device was only moderately polarized, as evidenced by the reported ratio of 2.4 of the electroluminescence (EL) intensities observed through a linear polarizer oriented parallel and perpendicular to the direction of uniaxial alignment (Figure 5.1b) [21].

Beyond their potential application as a backlight in LCDs, light sources that generate polarized chromatic light are technologically important for a variety of applications and devices that range from optical data storage, to communications systems, to LCD projectors, to medical applications. CP LEDs allow display configurations in which the reflection of ambient light is reduced and which therefore exhibit significantly improved contrast. Thus, the concept of polarized LEDs generated much excitement, and considerable research and development activities focused on this subject have evolved worldwide [26–28]. The general architecture of an LED emitting linearly polarized light (and similarly CP light; see Section 5.2) is rather straightforward and only differs from conventional LEDs in that a uniaxially oriented emitting layer is used (cf. Figure 5.1a). However, as will become apparent later in this chapter, the implementation of the concept is a nontrivial exercise. The fabrication of highly oriented emitting layers requires special materials and processing

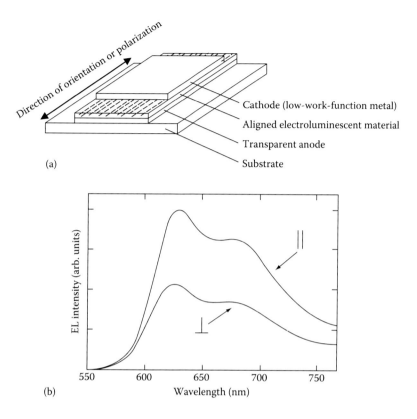

FIGURE 5.1

(a) Simplified schematic representation of the architecture of a single-layer LED emitting polarized light. The electroluminescent polymer is uniaxially oriented in the plane of the device, causing the emitted light to be linearly polarized. (b) Spectral distribution of the polarized EL displayed by the first polarized LED reported by Dyreklev et al. The two curves reflect the EL intensities detected through a linear polarizer oriented parallel ($\parallel$) and perpendicular ($\perp$) to the orientation direction of the light-emitting polymer. (From Dyreklev, P. et al., *Adv. Mater.*, 7, 43, 1995. With permission.)

steps, which are often not compatible with the stringent demands of OLED and PLED technologies (cf. Refs. [22–25], and this chapter and Chapter 6).

In view of the large number of publications on polarized light-emitting materials and polarized LEDs that have appeared to date, and with reference to the excellent reviews by Grell and Bradley [26], Neher [27], and O'Neill and Kelly [28], which summarize many activities, it is clearly outside the scope of this chapter to review in detail the exciting progress that the field has seen. Rather, we have attempted to give a concise overview on the general aspects of emission of linearly polarized light from uniaxially oriented organic matter. The method employed to process the oriented emitting layer usually has a most significant influence on many important device parameters, and thus, the different orientation methodologies were used as a guideline for the organization of this text. We discuss illustrative examples of various devices from the viewpoint of aspects that are particularly relevant for a specific application. We have not made an organizational separation between low-molecular-weight LEDs and PLEDs, because many concepts equally apply to the platform of both materials. As will become evident from the examples discussed herein, the most promising material candidates for linearly polarized LEDs appear to be oligomers that are situated at the border between low-molecular-weight and polymeric species. The material systems and architectures that may be used for LEDs that emit CP light are quite different from those used in linearly polarized LEDs, and thus, a separate section is devoted to the subject of CP emission.

5.2 Representation of Linearly and Circularly Polarized Photoluminescence

A rigorous and complete mathematical treatment of the polarization of light and the interaction of light with oriented matter is outside the scope of this chapter. These subjects have been thoroughly dealt with before and can be found in a number of comprehensive texts [29–32]; the reader is referred to the excellent book by Michl and Thulstrup [3] for a more detailed treatment of optical spectroscopy with polarized light. Here, a conventional, qualitative representation is given to establish the nomenclature and conventions to be used and to facilitate the understanding of the concepts presented.

Using Maxwell's framework of light as an electromagnetic wave, the wavelike behavior of light is represented by an oscillating electric field. The latter is always accompanied by an orthogonal magnetic field of similar frequency. However, as the organic materials discussed here are generally nonmagnetic, the following description of the polarization of light is reduced to the electric field component. For monochromatic light propagating as a plane wave along an optical axis (defined here as the z-axis; cf. Figure 5.2a), the electric field vector $E(r,t)$ as a function of time, t, is given by

$$E(z,t) = E_x^{(0)} \sin(\omega t - kz + \phi_x) + E_y^{(0)} \sin(\omega t - kz + \phi_y) \tag{5.1}$$

where $E_x^{(0)}$, $E_y^{(0)}$, ϕ_x, and ϕ_y describe the amplitude and phase of the wave along the x- and y-axes, respectively; ω is the circular frequency, which is related to the wavelength (λ) and speed of light (c) by $\omega = 2\pi c/\lambda$; and $k = \omega c$ is the propagation vector. The polarization of light is determined by the time course of the electric field vector, as illustrated in Figure 5.2. For monochromatic light propagating along an optical axis (z-axis), $E(r,t)$ lies in the transverse plane (xy-plane). If the propagating electric field vector describes an ellipse (Figure 5.2c), the light is said to be elliptically polarized. One can consider elliptically polarized light as a general form of polarization, and linear and circular polarizations as special cases of this more general form. The polarization is characterized by the orientation and eccentricity of the ellipse and determined by both the magnitudes of $E_x^{(0)}$ and $E_y^{(0)}$ and the relative phases ϕ_x and ϕ_y of these components. The polarization is said to be right-handed when $(\phi_y - \phi_x) <$ 180°, and left-handed when $-180° < (\phi_y - \phi_x) < 0°$. Thus, at a fixed position z, the tip of the electric field vector rotates periodically in the xy-plane, tracing out the ellipse. At a fixed time t, the locus of the tip of the electric field vector follows a helical trajectory in space, lying on the surface of an elliptical cylinder (Figure 5.2c). When the ellipse degenerates into a circle, the wave is said to be CP. This is the case when the phase difference $(\phi_y - \phi_x)$ is $\pm\pi/2$ and $E_x^{(0)} = E_y^{(0)}$ and when the elliptical cylinder depicted in Figure 5.2c becomes circular. When the ellipse degenerates into a straight line, the wave is said to be linearly polarized. This is the case if one of the amplitude components vanishes (the light is then said to be polarized in the direction of the remaining amplitude component) or if the phase difference $(\phi_y - \phi_x)$ is 0 or π.

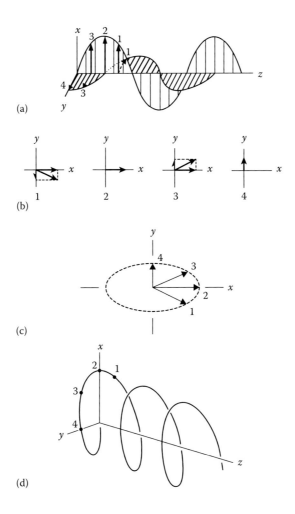

FIGURE 5.2
Wave representation of elliptically polarized light. (a) Orthogonal linear components with unequal amplitudes and 90° relative phase shift; (b) vector sums of the two components shown in (a) at positions labeled 1 through 4; (c) illustration of the fact that the tip of the electric field vector traces an ellipse when viewed along the propagation direction; (d) spatial dependence of the electric field vector represented in (a) through (c) at some point in time. (From Kliger, D.S. et al., *Polarized Light in Optics and Spectroscopy*, San Diego, Academic, 1990. With permission.)

In the process of absorption by a molecule, the electric field vector of linearly polarized light interacts with the electric dipole transition moment of the molecule. The probability of an absorption process is given by the square of the scalar product of the electric dipole transition moment vector and the electric field vector of the light, and is thus proportional to the square of the cosine of the angle between them. This leads to an orientational dependence of the absorption of linearly polarized light and, similarly, of the light emission from PL and EL molecules. It allows one to use spectroscopy with linearly polarized light to investigate the nature of molecular transitions as well as the nature of the orientation of molecular assemblies [3]. Alternatively, as described in this contribution, this orientational dependence can be employed to create optical elements, which absorb or emit highly polarized light. The direction of the electric dipole transition moment vector in the molecular framework is frequently referred to as the absolute polarization direction of the transition or its absolute polarization. However, it is noteworthy to state that the latter not necessarily (but often) coincides with the molecule's main geometric axis. A sample is said to be dichroic or to exhibit linear dichroism if it absorbs or emits light to different degrees depending on its linear polarization. In general, all materials that are at least partially oriented will exhibit linear dichroism in one or another region of the electromagnetic spectrum. The degree of polarization is frequently represented by the optical anisotropy, r_A, and is defined for optical absorption as

$$r_A = \frac{A_p - A_s}{A_p + 2A_s} \tag{5.2}$$

where A_p and A_s are the absorptions observed parallel (p) and perpendicular (s) to the direction of uniaxial alignment, respectively. The degree of emission polarization is defined in analogy to Equation 5.2 as

$$r_E = \frac{E_p - E_s}{E_p + 2E_s} \tag{5.3}$$

where E_p and E_s are the emission intensities observed parallel and perpendicular to the direction of uniaxial alignment, respectively. Ignoring the simplifications that have been made to arrive at Equations 5.2 and 5.3 (cf. Refs. [3,33,34]), many authors employ Equations 5.2 and 5.3 to extract a value for the order parameter S of the system, which is frequently used as a geometric term to describe the orientation of form-anisotropic molecules in a two-dimensional framework ($0 < S < 1$, where a value of $S = 1$ indicates perfect uniaxial orientation). An alternative representation of optical anisotropy that is used by many authors is the linear dichroic ratio or polarization ratio, DR, defined for absorption as

$$DR_A = \frac{A_p}{A_s} \tag{5.4}$$

and for emission as

$$DR_E = \frac{E_p}{E_s} \tag{5.5}$$

Dichroic ratios are very useful to describe the PL emission from a physiological point of view (e.g., brightness of a device as perceived by the human eye), and we have attempted to consistently employ this representation throughout this text. Most authors express DR_E as a ratio of intensities at a given wavelength, usually at a maximum of the emission spectrum, and, unless otherwise stated, we here follow this convention. However, quite frequently, the p- and s-polarized emission spectra are not identical, so that it would be more appropriate to compare the integrated emission spectra in order to allow for an apple-to-apple comparison.

The degree of CP absorption is usually expressed by the difference in absorption of left- and right-handed CP light (circular dichroism, ΔA) or by the so-called anisotropy factor g:

$$\Delta A = A_\text{L} - A_\text{R} \tag{5.6}$$

$$g = \Delta A / A \tag{5.7}$$

where A_L and A_R are absorbance for left and right CP light, respectively, and A is the normal absorbance of the material. Similarly, the degree of circularly polarized photo- and electroluminescence (CPPL and CPEL) is usually expressed by the difference in emission of left- and right-handed CP light (ΔI) or by the so-called dissymmetry factor, g_em:

$$\Delta I = I_\text{L} - I_\text{R} \tag{5.8}$$

$$g_\text{em} = 2(I_\text{L} - I_\text{R})/(I_\text{L} + I_\text{R}) \tag{5.9}$$

where I_R and I_L are the right- and left-handed emission intensities, respectively. Consequently, the maximum value of g_em is ±2, corresponding to purely right- or left-handed CP light. A more detailed and comprehensive discussion elaboration of the theory behind the generation of circularly polarized luminescence (CPL) can be found in the illustrative article of Shi et al. [35] and the literature cited therein.

5.3 Linearly Polarized Photo- and Electroluminescence from Uniaxially Oriented Molecules

As mentioned before, the architectural key feature for highly linearly polarized absorption, PL, and EL is a high degree of uniaxial orientation of form-anisotropic emitting molecules. Light-emitting semiconducting conjugated polymers are preferable because of their high aspect ratios and high chain rigidities. Thus, many of the examples relate to this class of materials. However, high degrees of uniaxial orientation can also be achieved in rigid-rod, low-molecular-weight molecules. Typical structural elements of rodlike, conjugated, luminescent moieties that are frequently employed for the design of low-molecular-weight and polymeric light-emitting materials are shown in Figure 5.3. It should be noted that the solubility of these species is rather limited, and therefore solubilizing side chains (not shown) are usually employed. In the following sections, we discuss different approaches for inducing the order and alignment of these chromophores. These approaches can be subdivided into mechanical alignment (tensile deformation, rubbing, and friction deposition) and self-assembly-type techniques (orientation on orienting substrates, Langmuir–Blodgett films, liquid crystalline [LC] materials).

5.3.1 Tensile Deformation

The tensile deformation of ductile, semicrystalline polymers has long been known as an extremely versatile method for introducing high degrees of uniaxial order in polymeric materials [36]. Orientation of the macromolecules is introduced by solid-state drawing at temperatures close to, but below, the melting temperature of the polymer. While this

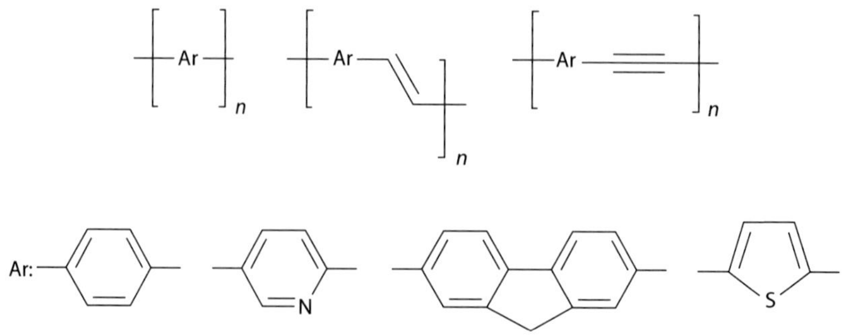

FIGURE 5.3
Typical rodlike conjugated moieties with high aspect ratios, employed in the design of orientable light-emitting chromophores.

method is not applicable to neat low-molecular-weight organic emitters, guest–host systems can easily be produced, in which the form-anisotropic guest molecules adopt the orientation of the polymeric host [37]. Different possibilities exist for incorporating the guest molecules into the polymeric host; the most common procedures are (i) the production of films by casting a solution comprising both components and by evaporating the solvent; (ii) melt mixing of the two components and subsequent melt processing of the resulting blend into a desired shape; (iii) swelling of the polymer with a solution comprising the guest and subsequent evaporation of the solvent; and (iv) diffusion of the guest from the vapor phase into the polymer. The tensile deformation step can always be performed after the incorporation of the guest; however, for techniques (iii) and (iv) that rely on diffusion, it can, in principle, also be done beforehand. Uniaxially oriented guest–host systems comprising dichroic absorbing dyes have been investigated extensively in the past by many research groups, mainly to monitor the orientation process of polymers as a result of mechanical deformation [3,33,34,38]. This concept is also the method of choice in manufacturing dichroic sheet polarizers based on semicrystalline polymers and dichroic dyes [39–42]. Similarly, a number of groups have undertaken to study the properties of such uniaxially oriented polymer systems comprising PL guest molecules. Investigations have, for example, been made on films based on blends of form-anisotropic low-molecular-weight [8,9,43–47], oligomeric [34,48,49], as well as conjugated polymeric PL species [6,7,12,50–52], and different matrix polymers including polystyrene, polyethylene (PE), polyvinylchloride, polyvinylalcohol, or polycarbonate were employed. The extent of polarization, which is usually similar for absorption and emission, observed in these systems depends primarily on the geometric shape of the PL molecules and, in line with the concept of pseudoaffine deformation [53], the draw ratio λ is applied to the films during deformation. The draw ratio is defined as $\lambda = (l - l_0)/l_0$, where l_0 and l are the lengths of a sample before and after deformation, respectively; for high degrees of deformation, λ is often approximated by $\lambda = l/l_0$. Extremely high dichroic ratios of up to 70 were reported in independent studies for gel-processed blends of conjugated polymers such as poly(2-methoxy-5-(2′-ethyl-hexyloxy)–*p*-phenylene vinylene) (MEH–PPV) [50–52] and poly[2,5-dioctyloxy-1,4-diethynyl-phenylene-*alt*-2,5,-bis(2′-ethylhexyloxy)-1,4-phenylene] (EHO-OPPE) [6,7,12] and ultrahigh molecular weight polyethylene (UHMW-PE) (cf. Figures 5.4 and 5.5).

Similarly large anisotropies were later reported for highly emissive blends of alkoxy-substituted bis(phenylethynyl)benzene derivatives and polyolefins such as linear low-density polyethylene (LLDPE) and isotactic polypropylene (*i*-PP) [8,9]. The latter systems

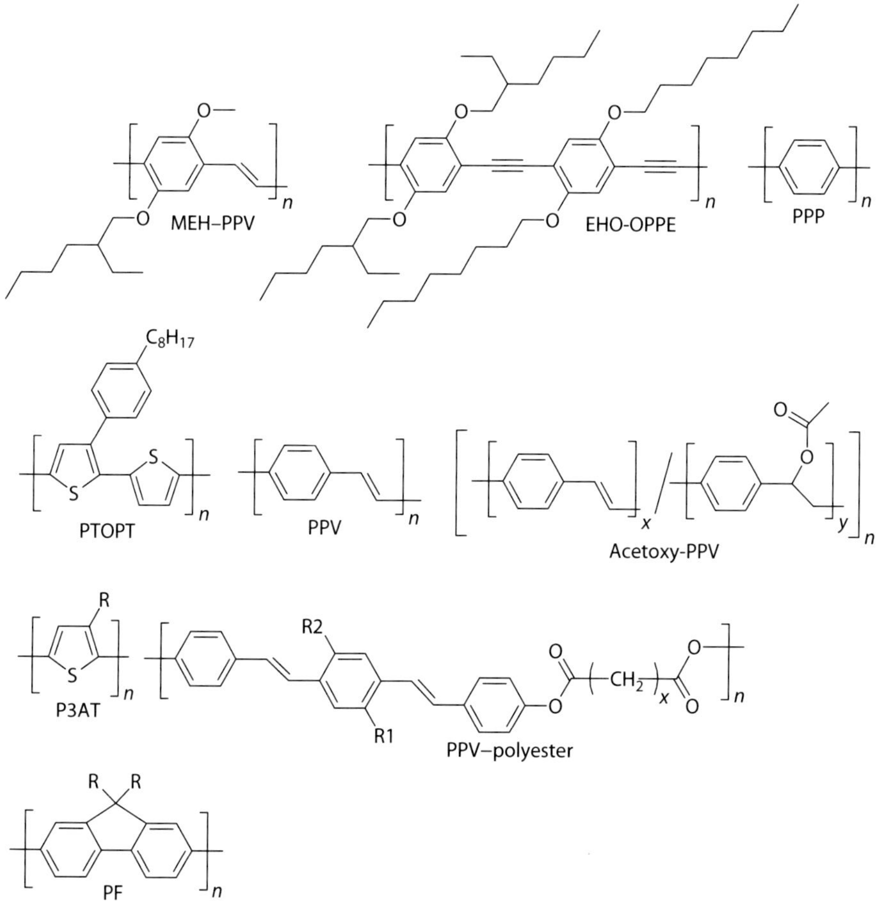

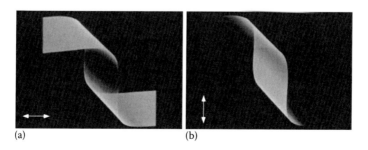

FIGURE 5.4

Chemical structures of photo- and electroluminescent polymers employed for polarized LEDs: MEH–PPV, poly(2-methoxy-5-(2′-ethyl-hexyloxy)–*p*-phenylene vinylene); EHO-OPPE, poly[2,5-dioctyloxy-1,4-diethynyl-phenylene-*alt*-2,5,-bis(2′-ethylhexyloxy)-1,4-phenylene]; PPP, poly(*p*-phenylene); PTOPT, poly(3-(4-octylphenyl)-2,2′-bithiophene); PPV, poly(*p*-phenylene vinylene); P3AT, poly(3-alkylthiophene vinylene); acetoxy-PPV; PPV–polyester; PF, poly(9,9-dialkyl fluorene).

FIGURE 5.5

Polarized PL from a gel-processed, uniaxially drawn film of EHO-OPPE (cf. Figure 5.4) in UHMW-PE. Twisted tapes (drawn to a draw ratio λ = 80) are shown under excitation with UV light (365 nm), and the pictures were taken through a linear polarizer with its polarization axis oriented horizontally (a) and vertically (b). (After Weder, C. et al., *Adv. Mater.*, 9, 1035, 1997.)

reach high levels of anisotropy at very low draw ratios, which is advantageous from a processing point of view.

The guest–host systems discussed above are extremely appealing because unparalleled degrees of optical anisotropy can readily be achieved. However, while many of the emitting materials employed in these blends have, in their neat form, been demonstrated to be useful emitters in actual LEDs [54–57], these blends could hitherto not be successfully employed in a diode configuration. One practical limitation arises from the fact that the thickness of mechanically drawn blend films is at best as low as a few micrometers, thus about one order of magnitude higher than the typical emitting layer in OLEDs [22–24]. However, the main problem is clearly related to the fact that the light-emitting molecules are embedded in an electrically insulating host polymer, which stifles the necessary charge transport through the material.

As demonstrated in the above-mentioned early work by Dyreklev et al. [21], both problems can be circumvented by stretch orienting a film of the neat light-emitting material. This can, for example, be achieved by depositing the light-emitting polymer on top of a deformable substrate, e.g., a ductile polymer film. After alignment of this bilayer structure, the oriented light-emitting layer is separated from the substrate and applied to the LED device. Dyreklev et al.'s study made use of poly(3-(4-octylphenyl)-2,2'-bithiophene) (PTOPT, cf. Figure 5.4) as the light-emitting material, which was deposited and oriented on a ductile PE film. The PE–PTOPT bilayer film was stretched, apparently under ambient conditions, to a draw ratio l/l_0 of 2. The oriented PTOPT was subsequently transferred to the LED device by applying a thermal transfer process, which involved pressing the oriented PTOPT–LLDPE film at elevated temperature to an appropriate substrate, i.e., an indium–tin oxide (ITO)-coated glass. The latter was, however, first covered with an isotropic film of PTOPT to facilitate adequate adhesion of the oriented PTOPT layer on the substrate. A calcium electrode was vacuum deposited on top of the PTOPT, and the resulting LED displayed an emission dichroic ratio, DR_E, of 2.4, with an onset voltage of 2 V and an external efficiency, η_{ext}, of up to 0.1%. Apart from the polarized nature of the emitted light, the device parameters were reported to be similar to nonoriented LEDs; however, the DR_E achieved with this first polarized LED was rather low. The low anisotropy of the device may first of all be related to the presence of an isotropic PTOPT adhesion layer. The latter had a similar thickness as the oriented PTOPT layer and presumably emitted unpolarized light, thereby limiting the anisotropy of the device. The low draw ratio further severely limits the attainable optical dichrism. Unfortunately, deformation to higher draw ratios led to the formation of cracks in the light-emitting materials, which of course are intolerable, as they would lead to shortening of the final device. This problem appears to be inherent for many conjugated polymers, as these materials often display poor mechanical characteristics and high thermal transition temperatures, which translate into low ductility at ambient temperature. However, plasticization of the conjugated polymer or deformation at temperatures close to its glass transition temperature (T_g) would solve this problem. Indeed, an intriguing protocol of tensile deformation of an inherently intractable conjugated polymer—poly(p-phenylene vinylene) (PPV, cf. Figure 5.4)—was reported by Bradley et al. [58]. One way of synthesizing PPV is via the so-called precursor route, where in a first step a nonconjugated sulfonium salt precursor polymer is synthesized. This polyelectrolytic precursor is highly soluble and can be processed into thin films by solution casting or spin coating. The precursor polymer is subsequently converted into the conjugated PPV by thermally induced elimination of a thioalkyl hydrochloride salt. To achieve chain orientation, Bradley et al. applied a constant tensile force on a precursor PPV film during this elimination process, which allowed deforming films up to a draw ratio of 6. Lemmer et al. [59] took this approach a step further by demonstrating that polarized LEDs can be achieved by evaporating gold electrodes onto a stretch-oriented, freestanding PPV film. The resulting EL devices emitted polarized light with a DR_E of up to about 8; however, the thickness of the PPV film prevented a conventional sandwich architecture and an in-plane arrangement of two gold electrodes, which were separated by a distance of 20 µm or more, was used instead. As a result, the onset voltage of these devices was of the order of 500 V.

Thus, in summary, tensile alignment is a very efficient way of achieving high degrees of optical anisotropy. In addition, stretch alignment could be very interesting from a manufacturing standpoint as it potentially allows the use of a roll-to-roll process, where several process steps, e.g., film formation, alignment, and cathode coating, could be done on the same manufacturing line. However, broad exploitation of the approach has hitherto been stifled by the difficulty to produce sufficiently thin layers of the light-emitting material, which exhibit good charge transport characteristics and can adequately be transferred into an LED device.

5.3.2 Rubbing and Friction Deposition

Two alternative procedures to achieve mechanically induced alignment are based on rubbing [60–64] or friction deposition of conjugated light-emitting polymers [65]. It has long been known that rigid-rod polymers can be mechanically aligned by rubbing a film of the polymer with, for example, a cotton or nylon cloth attached to a rotating drum.

An important technological application of this process is in LCDs, where rubbed polyimide (PI) films are used as LC alignment layers [15]. Detailed optical phase retardation and infrared dichroism studies about the effect of the rubbing process on the molecular orientation have been described by van Aerle et al. [66]. The mechanism leading to the orientation of LCs on rubbing-induced alignment layers has been the subject of intensive debate for decades. The speculations ranged from rubbing-induced microgrooves or scratches on the polymer surface [67] to near-surface alignment of polymer chains upon rubbing, which then acts as an alignment layer for LC molecules [68,69]. Today, the latter seems to be the commonly accepted mechanism. Rubbing alignment is also possible in case of conjugated polymers [70,71], and in principle the approach allows to exploit the advantages of mechanical orientation in thin-film architectures suitable for LEDs. Hamaguchi and Yoshino [60] were the first to report polarized EL based on rubbing-aligned conjugated polymer films. Polarized EL was observed in LEDs comprising rubbing-aligned poly(2,5-dinonyloxy-1,4-phenylene vinylene) as the emissive layer and additional layers of PPV and oxadiazol-doped polystyrene for balanced charge injection and transport. An onset voltage of 10 V and DR_E of up to 4.0 were reported for these multilayer devices; interestingly, the authors observed different DR_E values for PL and EL, as well as a current dependence of the DR_E in the LEDs, which suggests that unoriented molecules significantly contribute to the EL emission, particularly at low fields. Much higher degrees of optical anisotropy were subsequently described by Jandke et al. [61]. With the notions that PPV is too rigid to be effectively oriented by rubbing and that PPV precursor polymers are too soft for a rubbing process, these authors have designed and employed a synthetic compromise, featuring conjugated PPV segments and plasticizing, nonconjugated moieties (acetoxy-PPV, Figure 5.4). Rubbed films of this copolymer display excitingly high PL dichroism, characterized by DR_E values of up to 18, corresponding to an order parameter, S, of 0.90. Polarized EL of appreciable brightness (up to 200 cd/m^2) was demonstrated in multilayer devices that comprised ITO, an unoriented 120-nm-thick acetoxy-PPV layer, a 30-nm-thick layer of rubbed acetoxy-PPV, a 30-nm-thick poly(phenylquinoxaline) electron transport layer, and an Al top electrode. Interestingly, the EL of these devices was strongly polarized, as reflected by the DR_E of 12, despite the presence of a comparably thick layer of unoriented acetoxy-PPV. This demonstrates that charge recombination predominantly occurred within the acetoxy-PPV. A potential drawback of the rubbing alignment approach is the mechanical scratches, which were detected by atomic force microscopy. These scratches can lead to leakage currents, which limit the functionality and lifetime of the device [61]. The devices reported by Jandke et al. suggest, however, that this problem can, at least in part, be overcome by an additional transport layer. Another limitation of rubbing or friction deposition for applications in LEDs is the relatively thin layers, in which orientation can efficiently be induced. The available data suggest that by rubbing, not only the surface of the films are aligned but also the depth of the alignment layer can reach 5–60 nm depending on the rubbing density or rubbing pressure. This might be sufficient to efficiently align the emitting layer used in EL devices.

A related approach is the uniaxial orientation of EL polymers through friction deposition. In contrast to rubbing, where the polymer is first deposited on a substrate in isotropic fashion and is oriented in a second (the rubbing) step, friction deposition is a one-step procedure in which the solid polymer, for example in the shape of a block or cylinder, is dragged at elevated temperatures and with a certain pressure over a smooth, rigid (optionally heated) substrate such as glass or metal [72]. This process can result in the deposition of a thin layer of the polymer on the substrate, in which the individual molecules are uniaxially aligned along the drawing direction. With this method, Tanigaki et al. [65] prepared oriented films of poly(*p*-phenylene) (PPP; Figure 5.4), which displayed good optical anisotropy in absorption but suffered from discontinuities. In a recent paper, Nagamatsu et al. [73] described similar attempts to apply this technique to poly(3-alkylthiophene)s (P3AT, Figure 5.4). UV–vis and x-ray diffraction data demonstrated that the polythiophene backbones were indeed uniaxially aligned along the friction-drawing direction. The authors report a DR_A of between 10 and 100; however, it is somewhat unclear whether these values are inflated by the fact that the absorption peaks of the two polarization components display maxima at quite different wavelengths [73]. Although this method may lead to very high degrees of optical anisotropy, discontinuities and mechanical defects may represent a major obstacle for the exploitation of this framework in connection with LEDs.

5.3.3 Orientation of Nonliquid Crystalline Materials on Orienting Substrates

Polarized PL can also be observed from low-molecular-weight or polymeric PL materials that are epitaxially grown on orienting substrates, such as friction-deposited poly(tetrafluoroethylene) [64,72], rubbed PI [74], or photoaligned orientation layers [75,76]. Epitaxial processes have also been used by a number of groups for the preparation of polarized LEDs [77,78]. It should be noted, however, that most conventional orientation layers (*vide supra*) are based on materials that are electrical insulators [79–81]. This represents a problem for their use as an orientation layer in LEDs, as the injection of charges (usually holes) has to also occur through this layer. Approaches to solve this problem include the incorporation of a hole-transporting (HT) compound into the polymer used for the alignment layer (usually by blending; *vide infra*) or the fabrication of alignment layers from inherently (semi)conducting polymers. For example, Era et al. [74] demonstrated polarized EL by epitaxial vapor deposition of *p*-sexiphenyl (PPP, *n* = 6; cf. Figure 5.4) on a rubbed orientation layer of the same material. The dichroic ratio reported for LEDs produced using this architecture (with ITO and Mg:Ag electrodes and an additional oxadiazol derivative–based electron transport layer) was around 5, and the external device efficiency was estimated to be 0.14%. Highly polarized emission from oligo(biphenylthiophenes) (BPnT, Figure 5.6) that were vapor deposited and epitaxially grown on an orientation layer was reported by Yoshida et al. [82]. Building on earlier studies [83] and to solve the problem that traditional orientation layers are nonconducting, the authors made use of a uniaxially aligned semiconducting alignment layer based on rubbed PPP (Figure 5.4). Very high

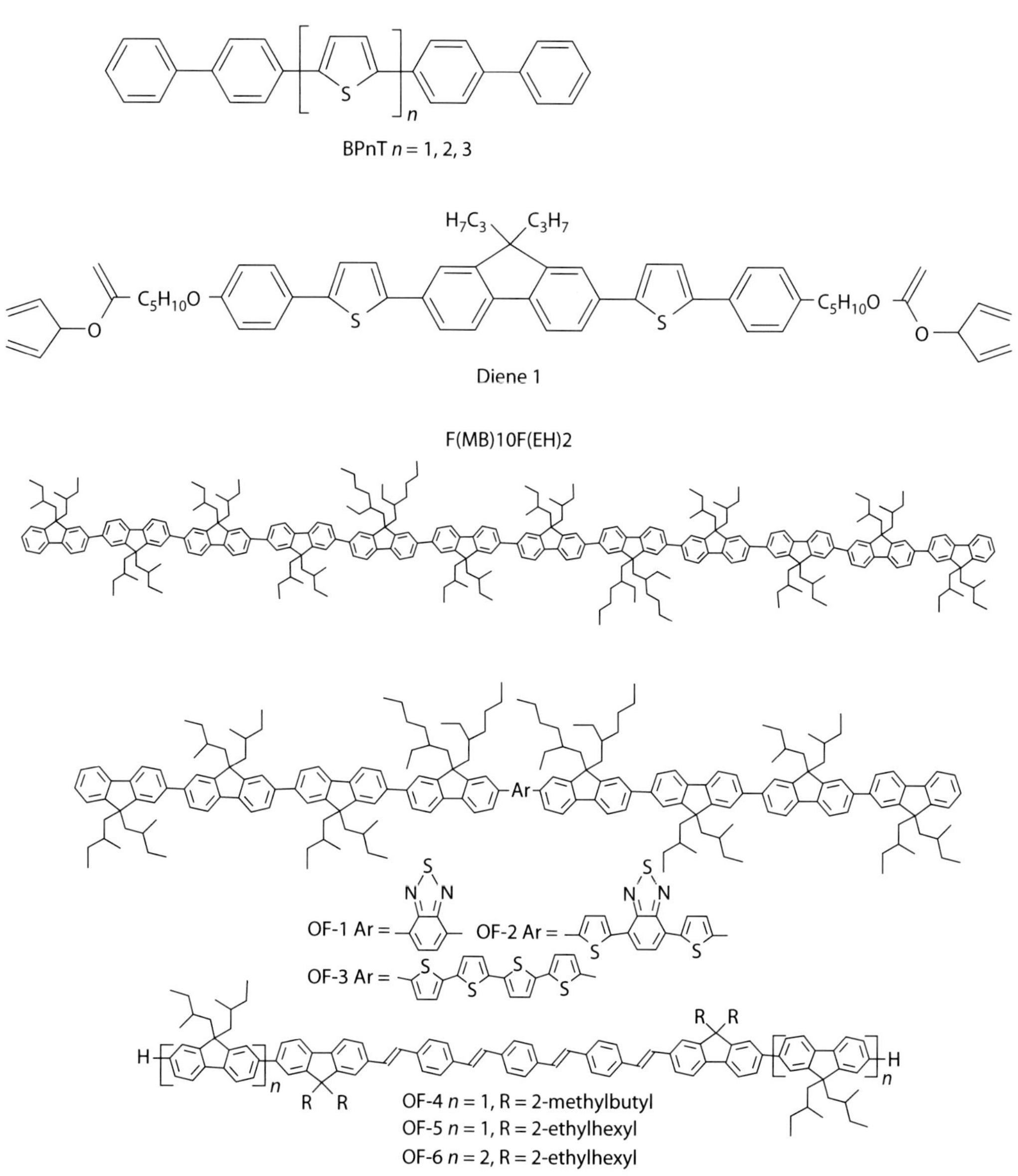

FIGURE 5.6
Chemical structures of low-molecular and oligomeric photo- and electroluminescent materials suitable for polarized LEDs: BPnT, diene 1, F(MB)10F(EH)2, OF-1, OF-2, OF-3, OF-4, OF-5, and OF-6. (From Culligan, S.W. et al., *Adv. Mater.*, 15, 1176, 2003; Geng, Y. et al., *Chem. Mater.*, 15, 4352, 2003. With permission.)

polarization ratios were observed for PL emission ($DR_E = 93.5$); however, it is unclear why the absorption dichroism was substantially lower ($DR_E = 14.7$).

5.3.4 Langmuir–Blodgett Technique

The formation of monomolecular layers at the air–water interface and their deposition onto solid substrates as ordered thin organic films was achieved as far as the early 1930s [84,85] when GE Research Laboratories' Katharine Blodgett and Irving Langmuir discovered a general method that allows the deposition of successive monomolecular layers of amphiphilic molecules on a variety of substrates. Since these initial studies, the method, now known as Langmuir–Blodgett (LB) technique, has become the most versatile tool for the production of ultrathin films with controllable thickness, molecular ordering, and few defects [86]. More recently, deposition by the LB technique has been demonstrated to yield layers or films that exhibit anisotropic optical properties [87–89].

The fundamentals of the main-chain orientation behavior of so-called hairy-rod type polymers—macromolecules with rigid backbone and very often flexible side chains—obtained by the LB technique have been described by Wegner et al. [90,91]. Using the LB technique, a number of side-chain-substituted, rigid-rod conjugated polymers, including PT [88], PPP [87], PPV [92–94], and poly(*p*-phenylene ethynylene) (PPE) [95] derivatives, were processed into multilayer thin films, in which the conjugated chain molecules were predominantly oriented with their backbones parallel to the dipping direction. In an effort to have chemically and thermally more robust devices, these LB-induced long-range orders were stabilized by cross-linking the material in a photochemical reaction [96]. Carefully deposited LB multilayers display anisotropic properties and can be used as emissive layers for LEDs that emit polarized light. However, the degree of optical anisotropy is usually very modest, with DR_E usually between 2 and 5.

5.3.5 Orientation of Liquid Crystalline Light-Emitting Materials

Thus far, the most important and the most successful approach for linearly polarized EL is probably the exploitation of the orientational long-range order of LC materials. Low-molecular-weight molecules, oligomers, and polymeric materials can exhibit intrinsic liquid crystallinity, and the organization of light-emitting LC molecules, typically on orienting substrates such as rubbed PI films, is an attractive way to generate polarized emission of light. Excellent review articles on this subject have appeared in the recent past [26–28]. Early LC systems investigated for polarized light emission include low-molecular-weight [97] oligomeric [98] and polymeric PL liquid crystals, vitrified LCs [99,100], cross-linked LC materials [101], and guest–host systems [102]. The latter have been known since the late 1960s and typically consist of an LC (very often nematic) host matrix into which high-aspect-ratio dye molecules are incorporated. Neat low-molecular-weight liquid crystals used for polarized emission of light encompass a variety of molecules, including benzothiazoles [103] and oxadiazoles [104,105]. The use of emissive layers that remain LC after processing is, however, problematic because liquid flow and intermixing of layers limit the long-term device stability in an LED. Quenching the LC into an ordered glass or fixation of the ordered state through chemical reaction are possibilities that solve this problem. Anisotropic networks formed by the photopolymerization of LC dienes that combine fluorene, thiophene, and phenylene segments (e.g., Diene 1, Figure 5.6) were, for example, investigated by O'Neill and Kelly and others [106,107]. These materials were typically aligned in the nematic LC phase and quenched into a glassy state before cross-linking into a polymer network by exposure to UV light. Alignment was achieved on a photo-alignment layer based on a polymeric coumarin derivative, which had been doped with a rather high fraction (30%) of a hole-conducting triphenylene derivative [106,107]. This elegant approach combines two intriguing concepts. First of all, it adopts the contact-free photoalignment scheme—a protocol originally developed for LCDs [76], in which illu-mination with polarized light generates surface anisotropy in a photoreactive alignment layer—to OLEDs. Second, the problem of limited charge transport through the orientation layer was, at least in part, solved by incorporation of a hole conductor, causing the latter to display almost adequate charge-transport characteristics. LEDs based on this system and ITO and Al electrodes displayed a DR_E of 10 and a maximum brightness of 60 cd/m² at a driving voltage of 11 V. Higher brightness was observed when the doping level of the hole conductor in the orientation layer was increased, but with a reduction in DR_E. Conversely, a polarized LED with a DR_E of 11 was produced from a 20% doped device; however, in this case, the brightness was reduced. While these data suggest that there is some room for improvement of the orientation layer, the dichroic ratios achieved in this work seem to be maximized, and limited by the order parameter S, which can be achieved with low-molecular-weight liquid crystals.

Aspect ratios and achievable order parameters are usually higher in LC polymers, par-ticularly in hairy-rod conjugated polymers, derivatized with solubilizing side chains [108]. These materials are further advantageous because they display usually relatively high nematic-to-glass transition temperatures and good film-forming properties. Thus, most LC polymers employed for polarized LEDs are representatives of this class of materials (*vide infra*). However, a number of other LC polymer architectures have also been investi-gated [108], including polymers with LC side chains and semiflexible or segmented chains in which flexible spacer units are incorporated between nematic or discotic mesogenic moieties (cf. Figure 5.7).

Wendorff and Greiner and others [109–112] were the first to report polarized EL created with LC polymers. The LC polymers employed in their initial studies were PPV derivatives that were segmented with flexible alkyl spacers (PPV–polyester, Figure 5.4). LEDs com-prising these polymers on rubbed PI displayed emission dichroic ratios of up to about 7. However, the macroscopic order parameter was found to strongly depend on the process-ing parameters. The key step for a high degree of orientation is an annealing process, in which the polymer (originally deposited in essentially isotropic fashion by spin coating onto the alignment layer) is brought into a thermotropic nematic state for an extended period. Under these conditions, the material slowly adopts the orientation of the alignment layer, and the orientation can be maintained by finally quenching the polymer to tempera-tures below the T_g. The need for high-temperature annealing, related to the comparably high viscosity of the system, represents a main disadvantage of this general approach (*vide infra*). Another disadvantage of the segmented polymers used in these initial studies are the possible limitations in charge transport caused by the flexible segments, which are not conjugated and interrupt the charge transport along the polymer chain. This prob-lem is, of course, readily solved by employing fully conjugated, preferably thermotropic LC polymers, such as side-chain-substituted PPPs [113], PPVs [114–116], PPEs [117,118], and poly(9,9-dialkyl fluorene) (PF) [119]. Grell et al. [100] have conducted an extensive study of PFs, including poly(9,9-dioctyl fluorene) and poly(9,9-di(2-ethylhexyl) fluorene) (PF, R = *n*-octyl, 2-ethylhexyl; Figure 5.4). The former polymer forms a nematic LC phase when heated to about 170°C. By using a PI alignment layer and employing an annealing or a quenching scheme, the LC polyfluorenes could be processed into stable, oriented emitting layers, which display a DR_E of 6.5. Polarized EL of copolyfluorenes with benzodithiazole

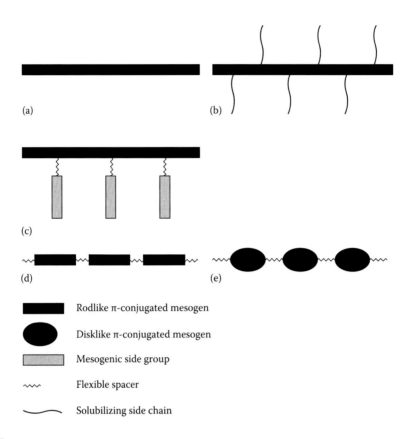

(a)

(b)

(c)

(d)   (e)

▬▬▬ Rodlike π-conjugated mesogen

⬤ Disklike π-conjugated mesogen

▭ Mesogenic side group

∿ Flexible spacer

⌒ Solubilizing side chain

FIGURE 5.7
Schematic representation of typical, (partially) electroluminescent LC polymer architectures. (a) Rodlike structure. (b) Hairy-rod structure. (c) Combined main chain–side chain system. (d) Semiflexible segmented structure. (e) Semiflexible segmented structure with disklike mesogen. (Reprinted from Weder, C. and Smith, P.,
Main-chain liquid-crystalline polymers for optical and electronic devices, in *Encyclopedia of Materials: Science
and Technology*, Buschow, K.H., Cahn, R.W., Flemings, M.C., Ilschner, B., Kramer, E.J., and Mahajan, S., Eds.,
Copyright 2001, with permission from Elsevier.)

(DR$_E$ ~ 5) and dithiophenes (DR$_E$ ~ 5) was also reported; however, the devices could not
quite reach the dichroic ratios of ethylhexyl-substituted polyfluorenes [120]. Poly(9,9-di(2-
ethylhexyl) fluorene) displayed LC behavior with a higher degree of molecular order than
poly(9,9-dioctyl fluorene), presumably owing to its lower effective diameter, which is concomitant with a higher aspect ratio. Poly(9,9-dioctyl fluorene) was shown to orient well
(DR$_E$ = 15) on a rubbed film of PI doped with a starburst-amine hole conductor [121]. This
orientation layer exhibits HT properties that are comparable to those of films of the neat
hole conductors, and as a result, EL devices that combine good polarization (DR$_E$ = 15) with
appreciable brightness (45 cd/m^2) could be realized [121]. Device optimization of the ethylhexyl polyfluorene system increased the polarization ratio in blue emission to DR$_E$ = 21
and the brightness to 100 cd/m^2 [122,123], and in devices based on poly(9,9-di(2-ethylhexyl)
fluorene) end capped with bis(4-methylphenyl)phenylamine, DR$_E$s of up to 26 could be
achieved [124]. Unfortunately, most of the above-referred studies limit the characterization
of LEDs-to-dichroic ratio, emission maximum, and sometimes brightness; however, no
device efficiencies are quoted. Expanding on earlier work by Grell et al. [100], Whitehead
et al. [125] revisited LEDs based on poly(9,9-dioctyl fluorene). Following the work of Jandke

et al. [61] and using a rubbed PPV film as the alignment layer, the dichroic ratio of LEDs comprising this polymer could be improved up to a DR_E of 25 (Figure 5.8). This value is based on a comparison of peak maxima; the dichroism was slightly lower (DR_E = 19) when the integrals of the p- and s-polarized spectra were compared. The devices reached a brightness of up to 327 cd/m² with an efficiency of 0.12 cd/A.

Even higher efficiencies were reported by Miteva et al. [126] who, building on earlier work [124], employed a PF that was end capped with hole-trapping moieties. In contrast to their earlier studies, the number-average molecular weight, M_n, of the polymer was reduced to 12,000 and optimized multilayer LEDs were based on a thin layer of a hole conductor, a PI alignment layer comprising 10% of the hole conductor, the oriented PF, and a Ca/Al top electrode. The integrated DR_E of their best devices was 21, with a brightness of 200 cd/m² (at 19 V) and a luminance efficiency of 0.25 cd/A. The brightness could be further increased to 800 cd/m² (at 19 V) by increasing the concentration of the hole conductor in the PI orientation layer to 20%; however, at the same time, the DR_E was reduced to 15. An exciting further development is the use of these materials in devices that rely on noncontact alignment of a fluorescent LC polymer by a photoinduced alignment of polyfluorenes on photoaddressable alignment layers. Sainova et al. [127] demonstrated that DRs of 10–14 at a luminance of 200 cd/m² and an efficiency of 0.3 cd/A can be achieved if polyfluorenes are deposited on a photoaddressable polymer (PAP), based on a hydroxyethyl methyl methacrylate backbone that was substituted with an azobenzene chromophore and comprising various amounts of a hole-conducting amine. In a subsequent paper, it was shown that device performance can be improved by decreasing the thickness of the alignment layer, and devices with an external efficiency of 0.66 cd/A with a DR_E of about 10 were reported [128]. Of course, one advantage of PAPs is that they allow the formation of patterned or pixelated structures with very high resolution that are able to emit polarized light (Figure 5.9).

Most of the LC polymer systems discussed thus far contain the LC moiety incorporated in the polymer main chain. One of the few examples of polarized emission of light using a side-chain LC system was reported by Chang et al. [129]. A polyacrylate was used as the polymer backbone. The nematic LC side chains consisted of ethyl- and propyl-substituted

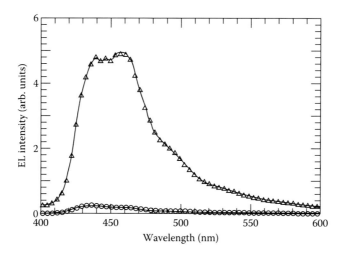

FIGURE 5.8
Electroluminescence from an LED based on ITO-rubbed PPV-aligned poly(9,9-dioctyl fluorene)/Ca recorded through a polarizer oriented parallel (triangles) and perpendicular (circles) to the orientation direction. (From Whitehead, K.S. et al., *Appl. Phys. Lett.*, 76, 2946, 2000. With permission.)

FIGURE 5.9
Polarized fluorescence microscopy images of the fluorescence patterns of a poly(fluorene) deposited on a PAP alignment layer. The patterns were obtained by selective alignment of the PAP, followed by deposition of the light-emitting polymer on top of the PAP layer, and annealing of the multilayer structure. On each picture, the rotation angle of the sample is noted, where 0 means that the analyzer of the polarization microscope is parallel to the molecular orientation after the first photoalignment step. (From Sainova, D. et al., *Adv. Funct. Mater.*, 12, 49, 2002. With permission.)

bis-tolan units. When deposited on a rubbed PEDOT film, polarized EL dichroic ratios were reported to be around 6.

Monodisperse LC oligomers represent an interesting compromise between low-molecular-weight and polymeric LCs, as they seem to combine high aspect ratio, which may translate into high dichroic ratios, with a comparably low viscosity, which allows for rapid orientation. Moreover, the well-defined molecular architecture may allow for efficient purification, which is important for the electronic devices at hand. A systematic structure–property study of monodisperse oligomeric fluorenes with varying lengths and different pendant side chains has been presented by Geng et al. [130]. The T_g and LC behavior were found to be strongly dependent on the structure of these oligofluorenes. Co-oligomeric fluorenes with different branched side chains were shown to exhibit superior stable glassy nematic films when compared with homo-oligomeric or unbranched fluorenes (Figure 5.6, F(MB)10F(EH)2) [131]. Color control was achieved by introducing a variety of band-gap-reducing aromatic groups at the center of these molecules (Figure 5.6, OF-1, OF-2, OF-3) [132]. LEDs were based on a rubbed PEDOT–PSS conductive orientation layer, an oligo(9,9-dialkyl fluorene) emissive layer, a 1,3,5-tri(phenyl-2-benzimidazolyl)benzene electron-transport or hole-blocking layer, a lithium fluoride electron injection layer, and a Mg/Ag cathode layer. The devices exhibit combinations of characteristics that may represent the best overall performances of polarized LEDs reported to date. With F(MB)10F(EH)2 as an emitter (cf. Figure 5.6), the emission is deep blue and is characterized by an integrated DR_E of 25, a luminance yield of up to 1.10 cd/A, and a brightness of up to 900 cd/m^2 [131].

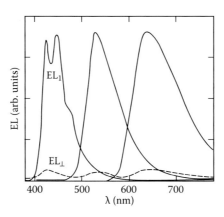

FIGURE 5.10
Polarized EL spectra of LEDs based on a rubbed PEDOT–PSS conductive orientation layer, an oligo(9,9-dialkyl fluorene) emissive layer, a 1,3,5-tri(phenyl-2-benzimidazolyl)benzene electron-transport or hole-blocking layer, a lithium fluoride electron injection layer, and a Mg/Ag cathode layer. The oligomers used as emitters were F(MB)10F(EH)2 (blue emission), OF-1 (green emission), and OF-2 (red emission), and the spectra were recorded through a polarizer oriented parallel (solid lines) and perpendicular (dashed lines) to the orientation direction. (From Geng, Y. et al., *Chem. Mater.,* 15, 4352, 2003. With permission.)

With OF-1 and OF-2, green and red emissions are achieved and the devices display an integrated DR_E of 16 and 14, a luminance yield of up to 5.9 and 0.51 cd/A, and a brightness of up to 1180 and 100 cd/m^2, respectively (Figure 5.10) [132].

5.3.6 Overview of LEDs Emitting Linearly Polarized Light

The key data of the various example devices emitting linearly polarized light discussed in this chapter are compiled in Table 5.1. It can be seen that significant progress has been made, as far as the combination of high DR_E, high luminance, and high quantum yield are concerned. Indeed, devices have become available that cover the entire visible spectral range and combine reasonable luminance and quantum efficiencies. The values of DR_E are also substantial, although for many applications a cleanup polarizer will have to be employed.

5.4 Circularly Polarized Luminescence

Polarized luminescence has attracted widespread attention owing to potential applications in optical information displays, processing, and storage. The state of the art of materials and devices that rely on uniaxially oriented emitting species and generate linearly polarized light is, as discussed in the previous sections, quite advanced, and technological exploitation may be seen in the near future. On the other hand, CP emission of light is more challenging to generate, and comparatively little research effort has been devoted to this subject to date. CPL is typically achieved with helically arranged luminophores [4,133]. This helical arrangement of the chromophores is very difficult to influence or promote by an external force field and, consequently, has to be achieved by self-assembly or self-orientation of the optically active materials. The technological challenges and the limited number of technologically relevant applications where CP light might be of benefit are

TABLE 5.1

Performance of Polarized LEDs Produced by Different Alignment Techniques and with Different EL Materials

Alignment Technique	Device Structure[a]	DR_E[b]	Luminance[c] (cd/m²)	η_{ext}[d] (cd/A)	λ_{max} (nm) Color	Ref.
Conjugated polymer stretched on ductile substrate	ITO/isotropic PTOPT/stretched PTOPT/Ca/Al	2.4	N.A.	0.1	630, 680	[21]
Conjugated polymer stretched as precursor	Stretched PPV/gold–gold; in-plane electrode arrangement	8	N.A.	N.A.	N.A.	[48]
Rubbing of plasticized conjugated copolymer	ITO/isotropic acetoxy-PPV/rubbed acetoxy-PPV/ET/Al	12	200 (Max)	N.A.	511 Green–yellow	[51]
Epitaxial growth of oligomer on rubbed oligomer alignment layer	ITO/rubbed sexiphenyl/vacuum-deposited sexiphenyl/ET/Mg:Ag	5	N.A.	(0.14%)	~430	[62]
LB deposition of conjugated polymer	ITO/Cr/Au/dialkoxy-PPP/Al	3–4	N.A.	(0.05%)	536/524	[71]
Cross-linked LC on noncontact alignment layer	ITO/photoalignment layer:HT/cross-linked LC/Al	10	60 (11 V)	N.A.	~500, 530	[87]
LC polyfluorene on rubbed PPV	ITO/rubbed PPV/LC poly(9,9-dioctyl fluorene)/Ca	25/19 (I)	200 / 327 (Max)	0.12	433, 458 Blue	[103]
LC polyfluorene on rubbed PI:HT	ITO/HT/rubbed PI:HT/LC poly(9,9-dioctyl fluorene)/Ca/Al	21 (I) / 15 (I)	200 (19 V) / 800 (19 V)	0.25 / N.A.	425, 450, 475 Blue	[104]
LC polyfluorene on photoalignment layer	ITO/PEDOT/PAP:HT/LC poly(9,9-dioctyl fluorene)/Ca/Al	14	200	0.3	~425, ~450, ~475	[105]
LC oligofluorene on rubbed PEDOT:HT	ITO/PEDOT:HT/F(MB)10F(EH)2/ET/LiF/Mg:Ag	27–32/25 (I)	214 (6.2 V) / ~1000 (Max)	1.10	424, 448 Blue	[109]
LC oligofluorene on rubbed PEDOT:HT	ITO/PEDOT:HT/OF-1/ET/LiF/Mg:Ag	18/16 (I)	1180 (7.9 V) / ~6000 (Max)	5.91	533 Green	[110]
LC oligofluorene on rubbed PEDOT:HT	ITO/PEDOT:HT/OF-2/ET/LiF/Mg:Ag	14/14 (I)	101 (7.5 V)	0.51	643 Red	[110]

[a] ITO, indium tin oxide; HT, hole-transporting agent; ET, electron-transporting agent. For chemical structures, see Figures 5.4 and 5.6.

[b] Based on ratio of peak maxima unless an (I) indicates that integrated spectra were compared.

[c] At voltage indicated. Values refer to driving conditions under which DR_E and η_{ext} were measured except where (Max) indicates the maximum luminance reported.

[d] External device efficiency in cd/A, except where (%) indicates that the efficiency was expressed by the fraction of photons extracted per charge injected.

probably the main reasons why there has not been as much progress as in the case of linearly polarized light. CP light has been proposed to be of potential interest in several areas. One important potential use are backlights for LCDs, as CP light can be converted very efficiently into linearly polarized light by means of a quarter-wave (λ/4) plate [134,135]. Another example is the use of CP-emitting layers in LEDs. Together with a linear polarizer, the reflection of ambient light at the back metal contact can be efficiently suppressed and the contrast of the device is significantly enhanced [135]. Other applications include optical data storage and photochemical switches [136–139], stereoscopic displays [140], and color image projection systems [141]. As mentioned heretofore, two fundamentally different concepts are employed to design materials that display CPL, namely the incorporation of nonchiral dyes into a chiral matrix and the use of conjugated polymers with chiral side chains. Examples of both platforms are discussed in the following subsections.

5.4.1 Circularly Polarized Luminescence from Achiral Dyes Doped in Chiral Matrices

In particular, most of the early studies on CPL were based on the incorporation of a luminescent achiral chromophore in a chiral nematic or cholesteric liquid crystal. Chiral nematic liquid crystals (CNLCs) are intrinsically birefringent and exhibit a helical supramolecular architecture, which is characterized by the pitch length p (Figure 5.11).

Films of pure CNLCs have a unique transmission behavior as CP light with the same sense of circular polarization as the CNLC is filtered out by reflection, while CP light of the opposite handedness as the CNLC film is transmitted. This selective optical transmission characteristic is referred to as a one-dimensional photonic stopband or a selective reflection band. The stopband is centered at a certain wavelength, λ_c, which is dependent on the pitch length p and the average refractive index $\bar{n}$ of the CNLC:

$$\lambda_c = \bar{n} \cdot p \qquad (5.10)$$

The doping of CNLCs with dyes that emit light with a maximum wavelength λ_E has been of interest for many years, mainly triggered by applications in lasers [142] or twisted

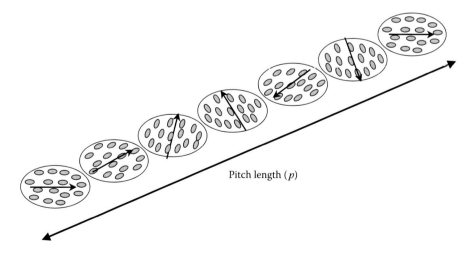

Pitch length (p)

FIGURE 5.11
Supramolecular, helical architecture, and definition of pitch length p of chiral nematic liquid crystals.

nematic LCDs [28]. Several studies [5,10,143] of these systems have shown the importance of the choice of the pitch length p versus λ_E. Commonly, researchers distinguish three different regimes, the Mauguin regime ($\lambda_E \ll \bar{n} \cdot p$), the resonance regime ($\lambda_E \sim \bar{n} \cdot p$), and the regime where $\lambda_E \gg \bar{n} \cdot p$. One of the critical findings for the generation of pure and efficient CP light with CNLC and dye systems is that an overlap of the emission band of the dye (λ_E) and the stopband of the CNLC should be avoided [10,143] because a sharp reversal of the handedness of the circular polarization occurs at the edges of the photonic stopband. This leads to a reduction of the overall CP efficiency. This problem can, for example, be overcome by choosing organolanthanide dyes with narrow emission bandwidths that fall perfectly within the photonic stopband [10]. Modulation of the degree of circular polarization by UV irradiation has been shown by Bobrovski et al. [139] using cholesteric mixtures with photochemically controllable, CP fluorescence. The pitch of the helix could be modified upon irradiation of UV light that causes an *E–Z* photoisomerization of a chiral photochromic dopant. Because of the relative stability of the images created by this technique, these novel photopatternable materials could open new platforms in advanced data storage systems.

In summary, one can say that LC and dye systems can give very pure CP light with dissymmetry factors of up to 1.8; however, this excellent performance is limited to very narrow bandwidths and films of a minimum thickness of 15–35 µm [5]. A more important drawback, as far as LEDs are concerned, is the fact that most CNLC and dye systems display electrical insulating behavior. Conjugated polymers, on the other hand, have the potential to combine adequate charge transport characteristics and suitable optical properties. The combination of an LC and a conjugated polymer has been reported by Katsis et al. [144]. A chiral nematic PPP was reported to display a dissymmetry factor of −1.3 within the spectral region and between 0.3 and 0.9 outside the selective reflection band. It should also be noted here that the film thickness (2 µm) of the chiral nematic PPP film was much less than that of the films of the glass-forming liquid crystals in a previous work (35 µm) [5] of the same group. Although the required film thickness is still about one order of magnitude higher than typically employed in LEDs, this promising result suggests that further work with conjugated polymers may be valuable.

5.4.2 Circularly Polarized Luminescence from Conjugated Polymers

Another approach to CPL is the synthesis of conjugated polymers with intrinsic chirooptical properties. A variety of polymers with CPPL have been synthesized thus far. Most of them are based on well-known conjugated polymers such as poly(thiophene)s [4,133], poly(phenylene vinylene)s [145], poly(thienylene vinylene)s [146], ladder polymers [147,148], PPPs [149], poly(phenylene ethynylene)s, [150], and poly(fluorenes) [151]. All of them have been modified with chiral side chains, which induce the chiro-optical properties.

The highest dissymmetry factor (g_{em}) obtained to date by means of an optically active polymeric emission layer is as high as 0.38 [150]. In this case, the emissive layer was a poly(phenylene ethynylene) derivatized with chiral dimethyloctyl (DMO) side chains. Interestingly, the polymer with the highest *g*-value was not the homopolymer in which all of the repeat units were derivatized with chiral side chains, but a copolymer comprising 50% DMO and 50% racemic 2-ethylhexyl side chains. Although the high dissymmetry factors have not been fully explained, the authors assume that they originate from strand formation during annealing. In other words, the authors believe that the large anisotropy factors are probably caused by helical supramolecular assembly of the PPE chains rather than intramolecular interactions.

Somewhat different conclusions regarding the origin of the chiro-optical properties were drawn by Oda et al. [151,152]. This group also reported a rather high dissymmetry factor (0.28) for LEDs based on a chiral conjugated polymer. The respective emissive layer consisted of a polyfluorene substituted with chiral (*R*)-2-ethylhexyl side chains. The values for CPPL (0.28) and CPEL (0.25) were considerably higher than the value reported for the first CP light-emitting device (CPEL = −0.0013) [145], which was based on a chirally substituted PPV derivative. An interesting aspect of this work is the observation that the further away is the chiral center from the backbone, the weaker is the chiro-optical response. This suggested that intrachain effects are more likely to be responsible for chiro-optical properties rather than interchain exciton couplings [152]. Another finding is the detection of an odd–even effect regarding the sign of the anisotropy factor *g* and the number of carbon atoms along the alkyl side chain between the backbone and the chiral center.

Not many publications about chiral conjugated polymers investigate the detailed origin of the chiro-optical properties. The induction of the overall helical architecture by the chiral side chains can be accomplished in different ways. The detailed helical organization has been investigated in the case of polythiophenes by Langeveld-Voss et al. [133]. It is concluded in this work that the chiro-optical properties stem from "a helical packing of predominantly planar chains" (Figure 5.12c).

Very recently, an exciting approach for controlling the chiral ordering in optically active polythiophenes by a doping process has been reported [153]. It was found that the addition of $Fe(ClO_4)_3$, $NaSO_3CF_3$, or $AgSO_3CF_3$ to chiral polythiophenes had a dramatic effect on the chiral arrangement of the polymer chains. No detailed description of the nature of the helical arrangement of the chains is reported. Nevertheless, this work represents an elegant new tool for controlling the chiral morphology that was previously only possible with time-consuming chemical modification of the conductive polymers.

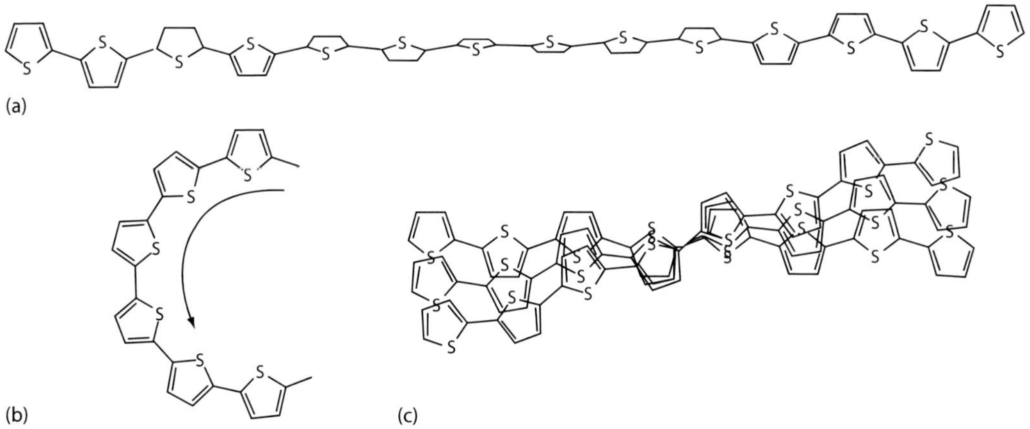

FIGURE 5.12
Three different helical organizations of polythiophenes: (a) helical transoid, (b) helical cisoid, and (c) helical packing of predominantly planar chains. (From Langeveld-Voss, B.M.W. et al., *J. Mol. Struct.*, 21, 285, 2000. With permission.)

5.4.3 Other Systems for Circularly Polarized Luminescence

There has been a report about bridged triarylamine helicenes exhibiting CPL [154]. These molecules preferentially emit and absorb CP light without the help of an LC matrix. Currently, there seems to be ongoing work to further increase the efficiency of these types of CPL materials and to develop the first devices of polarized OLEDs.

The most exciting piece of work for CPL is using a combination of a classical PLED and a CNLC film on top [135]. Simply by using the Al cathode in the back of the device as a "recycling mirror," the researchers produced a very efficient device that created CP light with an outstanding dissymmetry factor of 1.6. This device consists of an emission layer based on an achiral polyfluorene derivative, which, consequently, emits isotropic light. The CP fraction of the latter that has the matching handedness subsequently passes through the CNLC. The fraction with the opposite sense of chirality is reflected at the CNLC. Upon reflection at the Al cathode (recycling mirror), this light changes its handedness and then is able to pass the CNLC. In such a way, each photon should, in principle, be able to leave the system with the same circular polarization, thus creating a very efficient and highly CP emission device. Although the authors did not disclose any data on brightness, efficiency, or turn-on voltage, this photon-recycling device is an important milestone toward highly circularly or linearly polarized EL for LEDs especially because it does not require an emissive layer with chiro-optical properties.

5.5 Conclusions

In just about a decade of research and development, the area of polarized light emission from OLEDs has come a long way. Many exciting proofs-of-concept have been accomplished, and much improvements and breakthroughs have followed. However, to be successful and competitive with existing technologies in the marketplace, new systems must be kept simple and cost-effective. At the same time, they have to be able to clearly outperform current technology. Although this is a very challenging task, it is likely that some of the approaches discussed in this chapter are very promising and appear to be sufficiently mature for technological exploitation in the near future.

Acknowledgment

C.W. acknowledges support from the National Science Foundation under grant no. DMR-0215342.

References

1. E Lommel, Ueber die dichrotische Fluorescenz des Magnesiumplatincyanürs. Experimenteller Beweis der Perpendicularität der Lichtschwingungen zur Polarisationsebene, *Ann. Phys. Chem.*, 8:634–640, 1879.

2. A Jablonski, Über die polarisation der photolumineszenz der doppelbrechenden Kautsky-Phosphore, *Acta Phys. Polon.*, A14:421–434, 1934.
3. J Michl, and EW Thulstrup, *Spectroscopy with Polarized Light*, 1st ed., VCH Publishers, New York, 1986.
4. BMW Langeveld-Voss, R Janssen, MPT Christiaans, SCJ Meskers, HPJM Dekkers, and EW Meijer, Circular dichroism and circular polarization of photoluminescence of highly ordered poly{3,4-di[(*S*)-2-methylbutoxy]thiophene}, *J. Am. Chem. Soc.*, 118:4908–4909, 1996.
5. SH Chen, D Katsis, AW Schmid, JC Mastrangelo, T Tsutsui, and TN Blanton, Circularly polarized light generated by photoexcitation of luminophores in glassy liquid-crystal films, *Nature*, 397:506–508, 1999.
6. C Weder, C Sarwa, C Bastiaansen, and P Smith, Highly polarized luminescence from oriented conjugated polymer/polyethylene blend films, *Adv. Mater.*, 9:1035–1039, 1997.
7. A Montali, C Bastiaansen, P Smith, and C Weder, Polarizing energy transfer in photoluminescent materials for display applications, *Nature*, 392:261–264, 1998.
8. M Eglin, A Montali, ARA Palmans, T Tervoort, P Smith, and C Weder, Ultra-high performance photoluminescent polarizers based on melt-processed polymer blends, *J. Mater. Chem.*, 9:2221–2226, 1999.
9. ARA Palmans, M Eglin, A Montali, C Weder, and P Smith, Tensile orientation behavior of alkoxy-substituted bis(phenylethynyl)benzene derivatives in polyolefin blend films, *Chem. Mater.*, 12:472–480, 2000.
10. M Voigt, M Chambers, and M Grell, Circularly polarized emission from a narrow bandwidth dye doped into a chiral nematic liquid crystal, *Liq. Cryst.*, 29:653–656, 2002.
11. P Smith, and C Weder, Antifalsification paper and other antifalsification items, Swiss Patent Application 1958/98, 1998.
12. C Weder, C Sarwa, A Montali, C Bastiaansen, and P Smith, Incorporation of photoluminescent polarizers into liquid crystal displays, *Science*, 279:835–837, 1998.
13. DS Kliger, JW Lewis, and CE Randall, *Polarized Light in Optics and Spectroscopy*, Academic Press, San Diego, CA, 1990.
14. TJ Nelson, and JR Wullert III, *Electronic Information Display Technologies*, World Scientific Publishing Company, Singapore, 1997.
15. DJ Broer, JAMM van Haaren, P Van de Witte, and C Bastiaansen, New functional polymers for liquid crystal displays review of some recent developments, *Macromol. Symp.*, 154:1–13, 2000.
16. DJ Broer, J Lub, and GN Mol, Wide-band reflective polarizers from cholesteric polymer networks with a pitch gradient, *Nature*, 378:467–469, 1995.
17. DL Wortman, A recent advance in reflective polarizer technology, in *Proceedings of the 1997 International Display Research Conference (Society for Information Display)*, 1997, pp. M-98–M-106.
18. H Jagt, Y Dirix, R Hikmet, and C Bastiaansen, Linear polarizers based on oriented polymer blends, *Adv. Mater.*, 10:934–938, 1998.
19. H Jagt, Y Dirix, R Hikmet, and C Bastiaansen, Linear polarizers based on polymer blends: Oriented blends of poly(ethylene-2,6-naphthalenedicarboxylate) and a poly(styrene/methylmethacrylate) copolymer, *Jpn. J. Appl. Phys.*, 37:4389–4392, 1998.
20. AJ Heeger, and D Braun, Visible light emitting diodes fabricated from soluble semiconducting polymers, U.S. Patent 5,408,109, 1995 (filed February 27, 1991) to The Regents of the University of California.
21. P Dyreklev, M Berggren, O Inganäs, MR Andersson, O Wennerström, and T Hjertberg, Polarized electroluminescence from an oriented substituted polythiophene in a light emitting diode, *Adv. Mater.*, 7:43–45, 1995.
22. A Kraft, AC Grimsdale, and AB Holmes, Electroluminescent conjugated polymers—Seeing polymers in a new light, *Angew. Chem. Int. Ed.*, 37:402–428, 1998.
23. U Mitschke, and P Bäuerle, The electroluminescence of organic materials, *J. Mater. Chem.*, 10:1471–1507, 2000.
24. A Greiner, and C Weder, Light-emitting diodes, in *Encyclopedia of Polymer Science and Technology*, vol. 3, JI Kroschwitz, Ed., Wiley-Interscience, New York, 2003, pp. 87–100.

25. DDC Bradley, Electroluminescent polymers: Materials, physics and device engineering, *Curr. Opin. Solid State Mater. Sci.*, 1:789–797, 1996.
26. M Grell, and DDC Bradley, Polarized luminescence from oriented molecular materials, *Adv. Mater.*, 11:895–905, 1999.
27. D Neher, Polyfluorene homopolymers: Conjugated liquid-crystalline polymers for bright blue emission and polarized electroluminescence, *Macromol. Rapid Commun.*, 22:1365–1385, 2001.
28. M O'Neill, and SM Kelly, Liquid crystals for charge transport, luminescence, and photonics, *Adv. Mater.*, 15:1135–1146, 2003.
29. P Gay, *An Introduction to Crystal Optics*, Longman, London, 1967.
30. D Clarke, and JF Grainger, *Polarized Light and Optical Measurement*, Pergamon Press, Oxford, 1974.
31. W Swindell, *Polarized Light*, Dowden, Hutchinson & Ross, Stroudsburg, PA, 1975.
32. JF Nye, *Physical Properties of Crystals: Their Representation by Tensors and Matrices*, Oxford University Press, New York, 1984.
33. EW Thulstrup, and J Michl, A critical comparison of methods for analysis of linear dichroism of solutes in stretched polymers, *J. Phys. Chem.*, 84:82–93, 1980.
34. T Damerau, and MJ Hennecke, Determination of orientational order parameters of uniaxial films with a commercial 90°C-angle fluorescence spectrometer, *J. Chem. Phys.*, 103:6232–6240, 1995.
35. H Shi, BM Conger, D Katsis, and SH Chen, Circularly polarized fluorescence from chiral nematic liquid crystalline films: Theory and experiment, *Liq. Cryst.*, 24:163–172, 1998.
36. IM Ward, *Structure and Properties of Oriented Polymers*, Applied Science, London, 1975.
37. IM Ward, The measurement of molecular orientation in polymers by spectroscopic techniques, *J. Polym. Sci., Polym. Symp.*, 58:1–21, 1977.
38. EW Thulstrup, and J Michl, Orientation and linear dichroism of symmetrical aromatic molecules imbedded in stretched polyethylene, *J. Am. Chem. Soc.*, 104:5594–5604, 1982.
39. EH Land, Some aspects of the development of sheet polarizers, *J. Opt. Soc. Am.*, 41:957–963, 1951.
40. Y Dirix, TA Tervoort, and C Bastiaansen, Optical properties of oriented polymer/dye polarizers, *Macromolecules*, 28:486–491, 1994.
41. Y Dirix, TA Tervoort, and C Bastiaansen, Optical properties of oriented polymer/dye polarizers. 2. Ultimate properties, *Macromolecules*, 30:2175–2177, 1997.
42. C Bastiaansen, HW Schmidt, T Nishino, and P Smith, Transparence and dichroism of ultradrawn UHMW-PE films in the visible wavelength range, *Polymer*, 34:3951–3954, 1993.
43. JV Brestkin, ES Ediljan, NG Bel'niekevic, G Mann, and SJ Frenkel, Untersuchung der Orientierung bei der Deformation von Polyvinylalkohol mit der Methode der polarisierten Lumineszenz, *Acta Polym.*, 31:646–653, 1980.
44. M van Gurp, G van Ginkel, and YK Levine, On the distribution of dye molecules in stretched poly(vinyl alcohol), *J. Polym. Sci., Part B: Polym. Phys.*, 26:1613–1625, 1988.
45. JJ Dekkers, GP Hoornweg, C Maclean, and NH Velthorst, Fluorescence and phosphorescence polarization of molecules oriented in stretched polymers general description, *Chem. Phys. Lett.*, 19:517–523, 1973.
46. H Springer, R Neuert, D Müller, and G Hinrichsen, Orientation analysis of uniaxially drawn polycarbonate films, doped with fluorescent molecules, by UV-dichroism, fluorescence polarization and birefringence, *Colloid Polym. Sci.*, 262:46–50, 1984.
47. H Springer, J Kussi, HJ Richter, and G Hinrichsen, Dichroism of polyvinylchloride films doped with fluorescent molecules, *Colloid Polym. Sci.*, 259:911–916, 1981.
48. LV Natarajan, FM Stein, and RE Blankenship, Linear dichroism and fluorescence polarization of diphenyl polyenes in stretched polyethylene films, *Chem. Phys. Lett.*, 95:525–528, 1983.
49. M Hennecke, T Damerau, and K Müllen, Fluorescence depolarization in poly(*p*-phenylenevinylene) and related oligomers, *Macromolecules*, 26:3411–3418, 1993.
50. TW Hagler, K Pakbaz, J Moulton, F Wudl, P Smith, and AJ Heeger, Highly ordered conjugated polymers in polyethylene: Orientation by mesoepitaxy, *Polym. Commun.*, 32:339–342, 1991.

51. TW Hagler, K Pakbaz, and AJ Heeger, Enhanced order and electronic delocalization in conjugated polymers oriented by gel processing in polyethylene, *Phys. Rev. B*, 44:8652–8666, 1991.
52. TW Hagler, K Pakbaz, and AJ Heeger, Polarized-electroabsorption spectroscopy of a soluble derivative of poly(*p*-phenylenevinylene) oriented by gel processing in polyethylene: Polarization anisotropy, the off-axis dipole moment, and excited-state delocalization, *Phys. Rev. B*, 49:10968–10975, 1994.
53. PH Hermans, and D Heikens, Orientation in cellulose fibers as derived from measurements of dichroism of dyed fibers, *Recl. Trav. Chim. Pays-Bas*, 71:49–55, 1952.
54. F Wudl, PM Allemand, G Srdanov, Z Ni, and D McBranch, Polymers and an unusual molecular crystal with nonlinear optical properties, in *Materials for Nonlinear Optics: Chemical Perspectives*, SR Marder, JE Sohn, GD Stucky, Eds., ACS, Washington, DC, 1991, pp. 683–692.
55. D Braun, and AJ Heeger, Visible light emission from semiconducting polymer diodes, *Appl. Phys. Lett.*, 58:1982–1984, 1991.
56. A Montali, P Smith, and C Weder, Poly(*p*-phenylene ethynylene)-based light emitting devices, *Synth. Met.*, 97:123–126, 1998.
57. C Schmitz, P Pösch, M Thelakkat, HW Schmidt, A Montali, K Feldman, P Smith, and C Weder, Polymeric light-emitting diodes based on poly(*p*-phenylene ethynylene), poly(triphenyldiamine), and spiroquinoxaline, *Adv. Funct. Mater.*, 11:41–46, 2001.
58. DDC Bradley, RH Friend, H Lindenberger, and S Roth, Infra-red characterization of oriented poly(phenylene vinylene), *Polymer*, 27:1709–1713, 1986.
59. U Lemmer, D Vacar, D Moses, AJ Heeger, T Ohnishi, and T Noguchi, Electroluminescence from poly(phenylene vinylene) in a planar metal–polymer–metal structure, *Appl. Phys. Lett.*, 68:3007–3009, 1996.
60. M Hamaguchi, and K Yoshino, Polarized electroluminescence from rubbing-aligned poly(2,5-dinonyloxy-1,4-phenylenevinylene) films, *Appl. Phys. Lett.*, 67:3381–3383, 1995.
61. M Jandke, P Strohriegl, J Gmeiner, W Brutting, and M Schwoerer, Polarized electroluminescence from rubbing-aligned poly(phenylenevinylene), *Adv. Mater.*, 11:1518–1521, 1999.
62. A Bolognesi, C Botta, and M Martinelli, Oriented poly(3-alkylthiophene) films: Absorption, photoluminescence and electroluminescence behaviour, *Synth. Met.*, 121:1279–1280, 2001.
63. A Bolognesi, C Botta, D Facchinetti, M Jandke, K Kreger, P Strohriegl, A Relini, R Rolandi, and S Blumstengel, Polarized electroluminescence in double-layer light-emitting diodes with perpendicularly oriented polymers, *Adv. Mater.*, 13:1072–1075, 2001.
64. I Moggio, J Le Moigne, E Arias-Marin, D Issautier, A Thierry, D Comoretto, G Dellepiane, and C Cuniberti, Orientation of polydiacetylene and poly(*p*-phenylene ethynylene) films by epitaxy and rubbing, *Macromolecules*, 34:7091–7099, 2001.
65. N Tanigaki, K Yase, and A Kaito, Oriented films of poly(*p*-phenylene) by friction-deposition and oriented growth in polymerization, *Mol. Cryst. Liq. Cryst.*, 267:335–340, 1995.
66. NAJM van Aerle, M Barmentlo, and RWJ Hollering, Effect of rubbing on the molecular orientation within polyimide orienting layers of liquid-crystal displays, *J. Appl. Phys.*, 74:3111–3120, 1993.
67. DW Berreman, Solid surface shape and the alignment of an adjacent nematic liquid crystal, *Phys. Rev. Lett.*, 28:1683–1686, 1972.
68. MB Feller, W Chen, and YR Shen, Investigation of surface-induced alignment of liquid-crystal molecules by optical second-harmonic generation, *Phys. Rev. A: At., Mol., Opt. Phys.*, 43:6778–6792, 1991.
69. MF Toney, TP Russell, JA Logan, H Kikuchi, JM Sands, and SK Kumar, Near-surface alignment of polymers in rubbed films, *Nature*, 374:709–711, 1995.
70. SA Casalnuovo, KC Lim, and AJ Heeger, Reversible optical anisotropy and induced rigidity in polydiacetylene films, *Macromol. Chem., Rapid Commun.*, 5:77–81, 1984.
71. T Kanetake, K Ishikawa, T Koda, Y Tokura, and K Takeda, Highly oriented polydiacetylene films by vacuum deposition, *Appl. Phys. Lett.*, 51:1957–1959, 1987.
72. JC Wittmann, and P Smith, Highly oriented thin films of poly(tetrafluoroethylene) as a substrate for oriented growth of materials, *Nature*, 352:414–418, 1991.

73. S Nagamatsu, W Takashima, K Kaneto, Y Yoshida, N Tanigaki, K Yase, and K Omote, Backbone arrangement in friction-transferred regioregular poly(3-alkylthiophene)s, *Macromolecules*, 36:5252–5257, 2003.
74. M Era, T Tsutsui, and S Saito, Polarized electroluminescence from oriented *p*-sexiphenyl vacuum-deposited film, *Appl. Phys. Lett.*, 67:2436–2438, 1995.
75. K Ichimura, M Momose, K Kudo, and H Akiyama, Surface-assisted photolithography to form anisotropic dye layers as a new horizon of command surfaces, *Langmuir*, 11:2341–2343, 1995.
76. M Schadt, H Seiberle, and A Schuster, Optical patterning of multi-domain liquid-crystal displays with wide viewing angles, *Nature*, 381:212–215, 1996.
77. RE Gill, G Hadziioannou, P Lang, F Garnier, and JC Wittmann, Highly oriented thin films of a substituted oligo(*para*-phenylenevinylene) on friction-transferred PTFE substrates, *Adv. Mater.*, 9:331–338, 1997.
78. Y Yoshida, JP Ni, N Tanigaki, and K Yase, Polarized electroluminescence of oligophenyl thin films prepared on friction transferred poly(*p*-phenylenes), *Mol. Cryst. Liq. Cryst.*, 370:69–72, 2001.
79. D Fenwick, P Smith, and JC Wittmann, Epitaxial and graphoepitaxial growth of materials on highly oriented PTFE substrates, *J. Mater. Sci.*, 31:128–131, 1996.
80. D Fenwick, K Pakbaz, and P Smith, Alignment of fluorescent molecules vapour-deposited on to highly oriented PTFE substrates, *J. Mater. Sci.*, 31:915–920, 1996.
81. K Pichler, RH Friend, PL Burn, and AB Holmes, Chain alignment in poly(*p*-phenylene vinylene) on oriented substrates, *Synth. Met.*, 55:454–459, 1993.
82. Y Yoshida, N Tanigaki, K Yase, and S Hotta, Color-tunable highly polarized emissions from uniaxially aligned thin films of thiophene/phenylene co-oligomers, *Adv. Mater.*, 12:1587–1591, 2000.
83. N Tanigaki, H Kyotani, M Wada, A Kaito, Y Yoshida, EM Han, K Abe, and K Yase, Oriented thin films of conjugated polymers: Polysilanes and polyphenylenes, *Thin Solid Films*, 331:229–238, 1998.
84. KB Blodgett, Monomolecular films of fatty acids on glass, *J. Am. Chem. Soc.*, 56:495–495, 1934.
85. KB Blodgett, Films built by depositing successive monomolecular layers on a solid surface, *J. Am. Chem. Soc.*, 57:1007–1022, 1935.
86. G Roberts, *Langmuir–Blodgett Films*, Plenum Press, New York, 1990.
87. V Cimrova, M Remmers, D Neher, and G Wegner, Polarized light emission from LEDs prepared by the Langmuir–Blodgett technique, *Adv. Mater.*, 8:146–149, 1996.
88. A Bolognesi, G Bajo, J Paloheimo, T Östergård, and H Stubb, Polarized electroluminescence from an oriented poly(3-alkylthiophene) Langmuir–Blodgett structure, *Adv. Mater.*, 9:121–124, 1997.
89. Z Ali-Adib, K Davidson, H Nooshin, and RH Tredgold, Magnetic orientation of phthalocyaninato–polysiloxanes, *Thin Solid Films*, 201:187–195, 1991.
90. S Schwiegk, T Vahlenkamp, G Wegner, and Y Xu, On the origin of main chain orientation of rigid-rod macromolecules during the Langmuir–Blodgett process, *Thin Solid Films*, 210/211:6–8, 1992.
91. S Schwiegk, T Vahlenkamp, Y Xu, and G Wegner, Origin of orientation phenomena observed in layered Langmuir–Blodgett structures of hairy-rod polymers, *Macromolecules*, 25:2513–2525, 1992.
92. M Era, J Koganemaru, T Tsutsui, A Watakabe, and T Kunitake, Spacial distribution of electroluminescence from oriented phenylenevinylene oligomer Langmuir–Blodgett film, *Synth. Met.*, 91:83–85, 1997.
93. A Marletta, D Gonçalvez, ON Oliveira Jr., RM Faria, and FEG Guimarães, Highly oriented Langmuir–Blodgett films of poly(*p*-phenylenevinylene) using a long chain sulfonic counterion, *Macromolecules*, 33:5886–5890, 2000.
94. A Marletta, D Gonçalvez, ON Oliveira Jr., RM Faria, and FEG Guimarães, Circular dichroism and circularly polarized luminescence of highly oriented Langmuir–Blodgett films of poly(*p*-phenylene vinylene), *Synth. Met.*, 119:207–208, 2001.

95. E Arias, T Maillou, I Moggio, D Guillon, J Le Moigne, and B Geffroy, Amphiphilic phenylene–ethynylene polymers and oligomers for polarized electroluminescence, *Synth. Met.*, 127:229–231, 2002.

96. M Remmers, D Neher, and G Wegner, Photo-cross-linkable poly(*p*-phenylene)s. Synthesis, Langmuir–Blodgett multilayer film properties and pattern formation, *Macromol. Chem. Phys.*, 198:2551–2561, 1997.

97. HJ Coles, GA Lester, and H Owen, Fluorescent dye guest–host effects in advanced ferroelectric liquid crystals, *Liq. Cryst.*, 14:1039–1045, 1993.

98. NS Sariciftci, U Lemmer, D Vacar, AJ Heeger, and RAJ Janssen, Polarized photoluminescence of oligothiophenes in nematic liquid crystalline matrices, *Adv. Mater.*, 8:651–653, 1996.

99. BM Conger, JC Mastrangelo, and SH Chen, Fluorescence behavior of low molar mass and polymer liquid crystals in ordered solid films, *Macromolecules*, 30:4049–4055, 1997.

100. M Grell, DDC Bradley, M Inbasekaran, and EP Woo, A glass-forming conjugated main-chain liquid crystal polymer for polarized electroluminescence applications, *Adv. Mater.*, 9:798–802, 1997.

101. AP Davey, RG Howard, and W Blau, Polarised photoluminescence from oriented polymer liquid crystals, *J. Mater. Chem.*, 7:417–420, 1997.

102. GH Heilmeier, and LA Zanoni, Guest–host interactions in nematic liquid crystals: A new electro-optic effect, *Appl. Phys. Lett.*, 13:91–92, 1968.

103. M Funahashi, and JI Hanna, Fast hole transport in a new calamitic liquid crystal of 2-(4′-heptyloxyphenyl)-6-dodecylthiobenzothiazole, *Phys. Rev. Lett.*, 78:2184–2187, 1997.

104. H Tokuhisa, M Era, and T Tsutsui, Polarized electroluminescence from smectic mesophase, *Appl. Phys. Lett.*, 72:2639–2641, 1998.

105. H Mochizuki, T Hasui, T Shiono, T Ikeda, C Adachi, Y Taniguchi, and Y Shirota, Emission behavior of molecularly doped electroluminescent device using liquid-crystalline matrix, *Appl. Phys. Lett.*, 77:1587–1589, 2000.

106. AEA Contoret, SR Farrar, PO Jackson, SM Khan, L May, M O'Neill, JE Nicholls, SM Kelly, and GJ Richards, Polarized electroluminescence from an anisotropic nematic network on a non-contact photoalignment layer, *Adv. Mater.*, 12:971–974, 2000.

107. AEA Contoret, SR Farrar, M O'Neill, JE Nicholls, GJ Richards, SM Kelly, and AW Hall, The photopolymerization and cross-linking of electroluminescent liquid crystals containing methacrylate and diene photopolymerizable end groups for multilayer organic light-emitting diodes, *Chem. Mater.*, 14:1477–1487, 2002.

108. C Weder, and P Smith, Main-chain liquid-crystalline polymers for optical and electronic devices, in *Encyclopedia of Materials: Science and Technology*, KH Buschow, RW Cahn, MC Flemings, B Ilschner, EJ Kramer, and S Mahajan, Eds., Elsevier Science, New York, 2001, pp. 5148–5155.

109. G Lüssem, R Festag, A Greiner, C Schmidt, C Unterlechner, W Heitz, JH Wendorff, M Hopmeier, and J Feldmann, Polarized photoluminescence of liquid-crystalline polymers with isolated arylenevinylene segments in the main-chain, *Adv. Mater.*, 7:923–928, 1995.

110. G Lüssem, F Geffarth, A Greiner, W Heitz, M Hopmeier, JM Oberski, C Unterlechner, and JH Wendorff, Polarized electroluminescence of light emitting liquid crystalline polymers, *Liq. Cryst.*, 21:903–907, 1996.

111. G Lüssem, and JH Wendorff, Liquid crystalline materials for light-emitting diodes, *Polym. Adv. Technol.*, 9:443–460, 1998.

112. JM Oberski, KU Clauswitz, G Lüssem, F Geffarth, JH Wendorff, and A Greiner, Emission of polarized light from liquid crystalline segmented poly(arylenevinylene)s, *Macromol. Symp.*, 154:235–244, 2000.

113. J Oguma, K Akagi, and H Shirakawa, Synthesis and properties of liquid crystalline poly(*p*-phenylene) and poly(*p*-phenylene vinylene) derivatives, *Synth. Met.*, 101:86–87, 1999.

114. H Martelock, A Greiner, and W Heitz, Structural modifications of poly(1,4-phenylenevinylene) to soluble, fusible, liquid-crystalline products, *Macromol. Chem.*, 192:967–979, 1991.

115. Z Bao, Y Chen, R Cai, and L Yu, Conjugated liquid crystalline polymers-soluble and fusible poly(phenylenevinylene) by the Heck coupling reaction, *Macromolecules*, 26:5281–5286, 1993.

116. RE Gill, A Meetsma, and G Hadziioannou, Two novel thermotropic liquid crystalline substituted oligo(*p*-phenylene-vinylene)s: Single crystal x-ray determination of an all-trans oligomeric PPV, *Adv. Mater.*, 8:212–216, 1996.

117. D Steiger, P Smith, and C Weder, Liquid-crystalline, highly luminescent poly(2,5-dialkoxy-*p*-phenylene ethynylene), *Macromol. Rapid. Commun.*, 18:643–649, 1997.

118. L Kloppenburg, D Jones, JB Claridge, HC zur Loye, and UHF Bunz, Poly(*p*-phenylene ethynylene)s are thermotropic liquid crystalline, *Macromolecules*, 32:4460–4463, 1999.

119. M Grell, DDC Bradley, G Ungar, J Hill, and KS Whitehead, Interplay of physical structure and photophysics for a liquid crystalline polyfluorene, *Macromolecules*, 32:5810–5817, 1999.

120. KS Whitehead, M Grell, DDC Bradley, M Inbasekaran, and EP Woo, Polarized emission from liquid crystal polymers, *Synth. Met.*, 111:181–185, 2000.

121. M Grell, W Knoll, D Lupo, A Meisel, T Miteva, D Neher, HG Nothofer, U Scherf, and A Yasuda, Blue polarized electroluminescence from a liquid crystalline polyfluorene, *Adv. Mater.*, 11:671–675, 1999.

122. M Grell, DDC Bradley, and KS Whitehead, Materials and devices for polarized electroluminescence, *J. Kor. Phys. Soc.*, 36:331–336, 2000.

123. T Miteva, A Meisel, M Grell, HG Nothofer, D Lupo, A Yasuda, W Knoll, L Kloppenburg, UHF Bunz, U Scherf, and D Neher, Polarized electroluminescence from highly aligned liquid crystalline polymers, *Synth. Met.*, 111:173–176, 2000.

124. HG Nothofer, A Meisel, T Miteva, D Neher, M Forster, M Oda, G Lieser, D Sainova, A Yasuda, D Lupo, W Knoll, and U Scherf, Liquid crystalline polyfluorenes for blue polarized electroluminescence, *Macromol. Symp.*, 154:139–148, 2000.

125. KS Whitehead, M Grell, DDC Bradley, M Jandke, and P Strohriegl, Highly polarized blue electroluminescence from homogeneously aligned films of poly(9,9-dioctylfluorene), *Appl. Phys. Lett.*, 76:2946–2948, 2000.

126. T Miteva, A Meisel, W Knoll, HG Nothofer, U Scherf, DC Müller, K Meerholz, A Yasuda, and D Neher, Improving the performance of polyfluorene-based organic light-emitting diodes via end-capping, *Adv. Mater.*, 13:565–570, 2001.

127. D Sainova, A Zen, HG Nothofer, U Asawapirom, U Scherf, R Hagen, T Bieringer, S Kostromine, and D Neher, Photoaddressable alignment layers for fluorescent polymers in polarized electroluminescence devices, *Adv. Funct. Mater.*, 12:49–57, 2002.

128. XH Yang, D Neher, S Lucht, HG Nothofer, R Guntner, U Scherf, R Hagen, and S Kostromine, Efficient polarized light-emitting diodes utilizing ultrathin photoaddressable alignment layers, *Appl. Phys. Lett.*, 81:2319–2321, 2002.

129. SW Chang, AK Li, CW Liao, and CS Hsu, Polarized blue emission based on a side chain liquid crystalline polyacrylate containing *bis*-tolane side groups, *Jpn. J. Appl. Phys.*, 41:1374–1378, 2002.

130. YH Geng, SW Culligan, A Trajkovska, JU Wallace, and SH Chen, Monodisperse oligofluorenes forming glassy-nematic films for polarized blue emission, *Chem. Mater.*, 15:542–549, 2003.

131. SW Culligan, Y Geng, SH Chen, K Klubek, KM Vaeth, and CW Tang, Strongly polarized and efficient blue organic light-emitting diodes using monodisperse glassy nematic oligo(fluorene)s, *Adv. Mater.*, 15:1176–1180, 2003.

132. Y Geng, ACA Chen, JJ Ou, SH Chen, K Klubek, KM Vaeth, and CW Tang, Monodisperse glassy-nematic conjugated oligomers with chemically tunable polarized light emission, *Chem. Mater.*, 15:4352–4360, 2003.

133. BMW Langeveld-Voss, RAJ Janssen, and EW Meijer, On the origin of optical activity in polythiophenes, *J. Mol. Struct.*, 21:285–301, 2000.

134. M Schadt, Liquid crystal materials and liquid crystal displays, *Annu. Rev. Mater. Sci.*, 27:305–379, 1997.

135. M Grell, M Oda, KS Whitehead, A Asimakis, D Neher, and DDC Bradley, A compact device for the efficient, electrically driven generation of highly circularly polarized light, *Adv. Mater.*, 13:577–580, 2001.

136. NPM Huck, WF Jager, B de Lange, and BL Feringa, Dynamic control and amplification of molecular chirality by circular polarized light, *Science*, 273:1686–1691, 1996.

137. M Suarez, and GB Schuster, Photoresolution of an axially chiral bicyclo[3.3.0]octan-3-one: Phototriggers for a liquid-crystal-based optical switch, *J. Am. Chem. Soc.*, 117:6732–6738, 1995.

138. C Wang, H Fei, Y Qiu, Y Yang, Z Wei, Y Tian, Y Chen, and Y Zhao, Photoinduced birefringence and reversible optical storage in liquid-crystalline azobenzene side-chain polymers, *Appl. Phys. Lett.*, 74:19–21, 1999.

139. AY Bobrovski, NI Boiko, VP Shibaev, and JH Wendorff, Cholesteric mixtures with photochemically tunable, circularly polarized fluorescence, *Adv. Mater.*, 15:282–287, 2003.

140. DR Hall, Use of stereoscopic systems using chiral liquid crystals, U.S. Patent 5,699,184, 1997.

141. IEJR Heynderickx, and DJ Broer, Illumination system for a color projection device and circular polarizer suitable for use in such an illumination system, and color image projection device comprising such an illumination system and circular polarizer, U.S. Patent 5,626,408, 1997.

142. VI Kopp, B Fan, HKM Vithana, and AZ Genack, Low-threshold lasing at the edge of a photonic stop band in cholesteric liquid crystals, *Opt. Lett.*, 23:1707–1709, 1998.

143. M Voigt, M Chambers, and M Grell, On the circular polarization of fluorescence from dyes dissolved in chiral nematic liquid crystals, *Chem. Phys. Lett.*, 347:173–177, 2001.

144. D Katsis, HP Chen, SH Chen, LJ Rothberg, and T Tsutsui, Polarized photoluminescence from solid films of nematic and chiral-nematic poly(*p*-phenylene)s, *Appl. Phys. Lett.*, 77:2982–2984, 2000.

145. E Peeters, MPT Christiaans, RAJ Janssen, HFM Schoo, HPJM Dekkers, and EW Meijer, Circularly polarized electroluminescence from a polymer light-emitting diode, *J. Am. Chem. Soc.*, 119:9909–9910, 1997.

146. JJLM Cornelissen, E Peeters, RAJ Janssen, and EW Meijer, Chiroptical properties of a chiral-substituted poly(thienylene vinylene), *Acta Polym.*, 49:471–476, 1998.

147. R Fiesel, J Huber, and U Scherf, Synthesis of an optically active poly(*para*-phenylene) ladder polymer, *Angew. Chem. Int. Ed. Engl.*, 35:2111–2113, 1996.

148. R Fiesel, J Huber, U Apel, V Enkelmann, R Hentschke, U Scherf, and K Kabrera, Novel chiral poly(*para*-phenylene) derivatives containing cyclophane-type moieties, *Macromol. Chem. Phys.*, 198:2623–2650, 1997.

149. R Fiesel, and U Scherf, A chiral poly(*para*-phenyleneethynylene) (PPE) derivative, *Macromol. Rapid Commun.*, 19:427–431, 1998.

150. JN Wilson, W Steffen, TG McKenzie, G Lieser, M Oda, D Neher, and UHF Bunz, Chiroptical properties of poly(*p*-phenyleneethynylene) copolymers in thin films: Large *g*-values, *J. Am. Chem. Soc.*, 124:6830–6831, 2002.

151. M Oda, HG Nothofer, G Lieser, U Scherf, SCJ Meskers, and D Neher, Circularly polarized electroluminescence from liquid-crystalline chiral polyfluorenes, *Adv. Mater.*, 12:362–365, 2000.

152. M Oda, HG Nothofer, U Scherf, V Šunjić, D Richter, W Regenstein, and D Neher, Chiroptical properties of chiral substituted polyfluorenes, *Macromolecules*, 35:6792–6798, 2002.

153. ZB Zhang, M Fujiki, M Motonaga, and CE McKenna, Control of chiral ordering in aggregated poly{3-(*S*)-[2-methylbutyl]thiophene} by a doping-dedoping process, *J. Am. Chem. Soc.*, 125:7878–7881, 2003.

154. JE Field, G Muller, JP Riehl, and D Venkataraman, Circularly polarized luminescence from bridged triarylamine helicenes, *J. Am. Chem. Soc.*, 125:11808–11809, 2003.

6

Transparent Electrode for OLEDs

Furong Zhu

CONTENTS

6.1 Transparent Conducting Thin Films

6.1.1 Transparent Conducting Oxides

The thin films of transparent conducting oxides (TCOs) have widespread applications owing to their unique properties of good electric conductivity and high optical transparency in the visible spectrum range. There has been a great deal of activities in the development of TCOs for a variety of applications. In general, properly doped oxide materials, e.g., ZnO, SnO_2, and In_2O_3, are used individually or in separate layers, or as mixtures such as indium–tin oxide (ITO) and indium–zinc oxide (IZO) for making TCO thin films. ITO, aluminum-doped ZnO (AZO), and fluorine-doped SnO_2 (FTO) are commonly used TCO materials for different applications. The distinctive characteristics of these TCOs have been widely used in antistatic coatings, heat mirrors, solar cells [1,2], flat-panel displays [3], sensors [4], and organic light-emitting diodes (OLEDs) [5–7]. The properties of TCO films are often optimized accordingly to meet the requirements in the various applications that involve TCO. A light-scattering effect due to the use of textured TCO substrates helps enhance light absorbance in thin-film amorphous silicon solar cells [8,9]. However, a rough TCO surface is detrimental for OLED applications. The localized high electric fields induced by the rough TCO surface can cause a nonuniform current flow, leading to dark spot formation or a short device operation lifetime.

The conductivity of ZnO, ITO, and SnO_2 can be controlled across an extremely wide range such that they can behave as insulators, semiconductors, or metal-like materials. However, these materials are all n-type electrical conductors in nature. Their applications for optoelectronics are rather restricted. The lack of p-type conducting TCOs prevents the fabrication of a p–n junction composed of transparent oxide semiconductors [2]. The fabrication of highly conducting p-type TCOs is indeed still a challenge.

In comparison with the research in n-type oxide semiconductors, little work has been done on the development of p-type TCOs. The effective p-type doping in TCOs is often compensated owing to their intrinsic oxide structural tolerance to oxygen vacancies and metal interstitials. Recently, significant developments about ZnO, $CuAlO_2$, and Cu_2SrO_2 as true p-type oxide semiconductors have been reported. ZnO exhibits unipolarity or asymmetry in its ability to be doped as n-type or p-type. ZnO is naturally an n-type oxide semiconductor because of a deviation from stoichiometry due to the presence of intrinsic defects such as Zn interstitials and oxygen vacancies. A p-type ZnO has been recently reported, by doping with As or N as a shallow acceptor and codoping with Ga or Zn as donors. However, the origin of the p-type conductivity and the effect of structural defects on n-type to p-type conversion in ZnO films are not completely understood.

The advances in TCO materials development are still in a growth stage. The great potential for p-type TCOs and innovative technologies are predicted to lead to developments beyond anything one can imagine today. These include novel heterostructure applications as part of the rapid emergence of all-oxide electronics. Initial results on ZnO show that a small amount of nitrogen can be incorporated to form a p-type TCO [10]. Theoretical results for the III–V and II–VI materials subsequently indicated that codoping of these materials might allow not only type conversion but also high doping levels [11]. Although the conductivity of the p-type oxide semiconductors is still lower than its n-type counterparts, these new materials offer the potential for a variety of new devices. The new approaches to p-type doping of oxides and integration of these new materials have led to the hope for oxide semiconductor-based p–n junction for novel transparent electronics.

6.1.2 Fundamental Properties of ITO

ITO is one of the most frequently used TCO materials in practical applications. ITO film has attracted much attention because of its unique characteristics, such as good electric conductivity, high optical transparency over the visible wavelength region, excellent adhesion to substrates, stable chemical property, and easy patterning ability. One of the most common uses of ITO coatings has been as transparent electrodes in photovoltaic cells and flat-panel displays, including plasma televisions, liquid crystal displays (LCDs), and OLEDs. In some of these, it is important to ensure as low a resistivity and a high optical transparency as possible. The optical, electrical, structural, and morphological properties of TCO films have direct implications for determining and improving device performance.

6.1.2.1 Preparation of ITO

The deposition techniques that are suitable for the preparation of reproducible thin films of ITO include thermal evaporation deposition [12], magnetron sputtering [13,14], electron beam evaporation [15], spray pyrolysis [16], chemical vapor deposition [17], dip coating [18,19], and pulsed laser deposition methods [20,21]. Among these available techniques for fabricating ITO films, the direct current (dc) or radiofrequency (rf) magnetron sputtering method is most often used to prepare ITO thin films for a wide range of applications. ITO films fabricated using the rf or dc magnetron sputtering method usually require a low oxygen partial pressure in the sputtering gas when both alloy and oxidized targets are used [22,23]. The ITO film quality is determined by a number of factors, such as thickness uniformity, surface morphology, optical transparency, and electrical conductivity. This aside, the deposition technologies, process conditions, and postdeposition treatments also affect the overall optical and electrical properties of ITO.

In addition to the use of reactive oxygen gas during dc/rf magnetron sputtering processes, introducing water vapor or hydrogen gas into the gas mixture during the sputtering processes have also been reported. Harding and Window [24] found that good-quality ITO films can be obtained using an argon–oxygen–hydrogen mixture during deposition. Ishibashi et al. [25] reported that the carrier concentration of the ITO films increased when water vapor or hydrogen gas was used in the dc magnetron sputtering experiments. However, the mechanism for increased carrier concentration in ITO films was not discussed. Baía et al. [26] reported that conductivity of the ITO films sputtered at room temperature increased significantly followed by reannealing the films in vacuum with hydrogen base pressure. The improvement of the film conductivity with regard to the annealing treatment under hydrogen atmosphere is explained as due to the removal of excess oxygen incorporated and the passivation of the defects in the films. These results indicate clearly that the presence of hydrogen species during the preparation or the postdeposition treatment of ITO films can significantly affect the overall optical and electrical properties of ITO films. A better understanding of process conditions on the overall properties of ITO films is of considerable interest.

The following discussion describes the structural, electrical, and optical properties of the ITO films prepared using an oxidized target with In_2O_3 and SnO_2 in a weight proportion of 9:1. The ITO films are deposited on glass substrates using the rf magnetron sputtering method. The base pressure in the system is about 5.0×10^{-8} torr. A sputtering gas mixture of argon–hydrogen is used for the growth of ITO films. The effect of hydrogen partial pressure on the structural and optoelectronic properties of the ITO films is studied over the hydrogen partial pressure range 0–1.6×10^{-5} torr. The ITO films are prepared at a constant substrate temperature of 300°C.

6.1.2.2 Structural Properties of ITO

The structural properties of the ITO films deposited on glass substrates at various hydrogen partial pressures are characterized by x-ray diffraction (XRD) measurements. Figure 6.1 shows a series of XRD patterns of ITO films deposited in a hydrogen partial pressure range of 0–1.6×10^{-5} torr. Figure 6.1 illustrates that ITO films prepared on glass substrates have polycrystalline structures with diffraction peaks corresponding to (221), (222), (400),

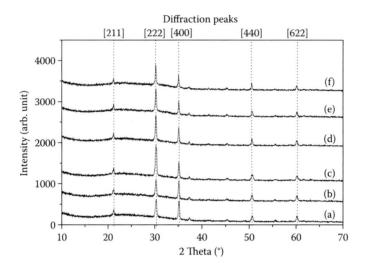

FIGURE 6.1
XRD spectra measured for ITO films grown on glass substrates at different hydrogen pressures of (a) 0 torr, (b) 5.4×10^{-6} torr, (c) 7.9×10^{-6} torr, (d) 1.0×10^{-5} torr, (e) 1.3×10^{-5} torr, and (f) 1.6×10^{-5} torr.

(440), and (622) reflections. In particular, (222) and (400) are the prominent planes for films prepared by this method, indicating that the ITO films have (111) and (100) preferentially orientated textures. The relative intensity of the (222) reflection increases gradually with increasing hydrogen partial pressure.

The crystallinity of the ITO film depends on the fabrication technique and deposition conditions. ITO films prepared by sputtering usually have a preferred orientation in the (100) direction, and those prepared by reactive thermal evaporation have a preferred orientation along the (111) plane [27]. Some studies have shown that the dominant crystal orientations in ITO films can be changed under certain conditions. Thilakan and Kumar [28] reported that the preferred orientation of ITO can be changed from (222) to (400) when the film deposition rate increased. Meng and dos Santos [27] also observed a similar orientation transition from (222) to (400) when the substrate temperature increased from room temperature to about 500°C. From Figure 6.1, it appears that the presence of hydrogen in the sputtering gas mixture preferentially enhances the (222) orientation. This structural change in ITO films may affect the overall optical and electrical properties of the films.

The interplanar distances for the (222), (400), and (440) crystal planes can be obtained by fitting the XRD peaks shown in Figure 6.1. When hydrogen is added to the gas mixture, interplanar separations of ITO films along the orientations of major XRD peaks are generally less than that of the film prepared without hydrogen in a sputtering gas mixture. The deviation of interplanar distances along these crystal directions in the ITO films indicates structural or compressive stress in the film. Adding hydrogen to the sputtering gas mixture has been shown to reduce the structural stress in ITO films prepared by the rf magnetron sputtering method. The interplanar distances obtained from the major XRD peaks change with hydrogen partial pressure and attain minimum values at a hydrogen partial pressure of 7.9×10^{-6} torr. At this hydrogen partial pressure, the minimum value of the plane distance calculated from prominent (222) planes is 2.9369 Å. This value is less than that of 2.9500 Å for In_2SnO_{7-x}, and it is slightly higher than the value of 2.9210 Å for In_2O_3 [29]. The decrease in interplanar distance in ITO films is probably related to a reduced lattice constant of the ITO films. This suggests that a possible stress relaxation occurs in the

ITO films [29], which can be optimized at the hydrogen partial pressure of about 7.9×10^{-6} torr.

It was suggested that the decrease in the lattice constants of the ITO film is attributed to the presence of the oxygen vacancies [30]. Similar results are also reported by Honda et al. [31], indicating that the lattice constants of oxygen-deficient ITO films are smaller than the films without the oxygen deficiency. By fitting XRD peaks, it can be found (Figure 6.1) that an ITO film prepared at hydrogen partial pressure of 7.9×1.0^{-6} torr has a maximum diffraction angle that is related to a minimum lattice constant. This indicates that the presence of hydrogen species in the sputtering gas mixture during the deposition increases the oxygen deficiency in the ITO films. From this analysis, it appears that adding hydrogen in the gas mixture helps reduce the structural stress in films and possibly increases the number of oxygen vacancies in the film. In practical device applications, ITO films are often used in thin-film devices and coated subsequently on semiconductor films or dielectric films; therefore, from a device point of view, ITO films with less stress in a device with a multilayer configuration is more preferable.

6.1.2.3 Electrical Properties of ITO

Using experimentally measured film thicknesses, the corresponding film resistivity can be calculated. The sheet resistance and resistivity of the ITO films as a function of processing condition are plotted in Figure 6.2, which shows that both sheet resistance and resistivity of the films increased considerably when the hydrogen partial pressure is $>1.3 \times 10^{-5}$ torr. The relative minimum values of sheet resistance, 10 Ω/square, and resistivity, 2.7×10^{-4} Ω-cm, can be obtained at the optimal hydrogen partial pressure of 7.9×10^{-6} torr.

Figure 6.3 shows the variation of Hall mobility and the number of charged electron carriers in the ITO films determined by Hall-effect measurements. The solid square symbols represent the carrier concentration N and the circle marks correspond to the Hall mobility, μ, of the films prepared at different ITO depositions. The results shown in Figure 6.3 reveal that the N and μ profiles of the ITO films are different as a function of hydrogen partial pressure. The carrier concentration of the film increases initially with the hydrogen partial pressure, and reaches the maximum value of $1.45 \times 10^{21}/cm^3$ at the optimal hydrogen partial pressure of 7.9×10^{-6} torr. The number of carriers in the ITO film decreases when

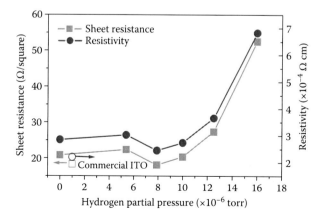

FIGURE 6.2
Sheet resistance and resistivity of ITO films as functions of hydrogen partial pressure.

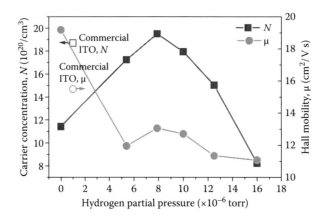

FIGURE 6.3
Carrier concentration and Hall mobility of ITO films as functions of hydrogen partial pressure.

the hydrogen partial pressure is $>7.9 \times 10^{-6}$ torr. Hall-effect measurements reveal that ITO films prepared with a hydrogen–argon mixture have a higher carrier concentration than that of films prepared with only argon gas. In contrast with the variation of N at different hydrogen partial pressures, however, the carrier mobility does not change significantly. Although μ also has a relative maximum value at a hydrogen partial pressure of 7.9×10^{-6} torr, Hall-effect measurements show that ITO films prepared in the presence of hydrogen generally have a lower μ value than that of films fabricated with pure argon gas.

To understand better the mechanism of the carrier concentration variations in films due to the addition of hydrogen into the gas mixture of argon during sputtering processes, secondary ion mass spectroscopy (SIMS) is used to measure the changes of relative oxygen concentrations in films prepared under different conditions. Figure 6.4 shows the typical oxygen depth profiles of an ITO film prepared with only argon gas (curve a), and an ITO film prepared at the hydrogen partial pressure of 7.9×10^{-6} torr (curve b). To compare the relative oxygen contents in different films, the intensities of negatively charged oxygen ions in SIMS are normalized to the corresponding intensities of the indium ones acquired in the same measurements. The depth of the films can be converted using sputtering time

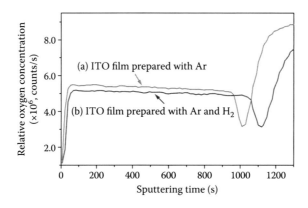

FIGURE 6.4
SIMS depth profiles of (a) ITO film prepared with pure argon gas and (b) ITO film prepared at hydrogen partial pressure of 7.9×10^{-6} torr.

at a sputtering rate of about 0.22 nm/s. From the SIMS results, it reveals that the relative oxygen content in a film prepared with pure argon gas is higher than that of a film prepared using a hydrogen–argon mixture. It indicates that the presence of hydrogen in the sputtering gas mixture of argon makes up for the oxygen lost in films. When hydrogen is added in the sputtering gas mixture, the grow flux during the magnetron sputtering includes a significant amount of energetic hydrogen species with energies over the range 10–250 eV [32]. These active hydrogen species can remove weakly bound oxygen in the depositing films. As a consequence, the addition of hydrogen in the sputtering gas mixture shows a reducing effect on oxide and leads to an increase in the number of oxygen vacancies in the films and hence an increase in the number of charged carriers. Because the electrical conductivity is proportionate to the product of charge carrier concentration and the mobility, the decrease of film resistivity is mainly due to the increase of carrier concentration in ITO films. The above analysis based on SIMS results is consistent with the previous discussions made with XRD measurements.

6.1.2.4 Optical Properties of ITO

The optical properties of ITO films prepared on glass at different hydrogen partial pressures are also characterized. The transmittance of the film is experimentally measured by using a double-beam spectrophotometer over the wavelength range 0.3–1.5 μm. The measured wavelength-dependent transmittance over the visible wavelength range is also used to estimate the optical energy band gap of the films. To study the effect of hydrogen partial pressure on the optical properties of the ITO films, the transmittance values of the ITO films prepared at several oxygen partial pressures in an oxygen–argon mixture over the same wavelength region are presented. Figure 6.5 shows the average transmittance of ITO films of 200 nm thickness measured over a visible light wavelength range of 0.4–0.8 μm, as functions of hydrogen and oxygen partial pressures. The average transmittance of ITO films prepared in oxygen–argon does not change considerably at different oxygen partial pressures. However, the average transmittance of ITO films prepared using hydrogen–argon mixture varies with the hydrogen partial pressure. ITO films with an average transmittance of 89% is obtained at the optimal hydrogen partial pressure of 7.9×10^{-6} torr. The

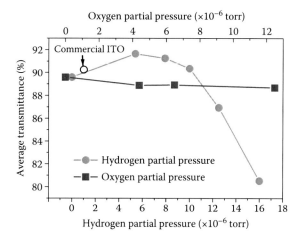

FIGURE 6.5
Average transmittance of ITO films as functions of hydrogen partial pressure, λ, and oxygen partial pressure, ν.

hydrogen partial pressure that produces the most transparent ITO film is almost the same as that which gives the most conducting film, as shown in Figure 6.2. Optical transmittance results in Figure 6.5 together with electrical measurements suggest that 7.9×10^{-6} torr is a suitable hydrogen partial pressure for ITO film preparation under these conditions.

Figure 6.6 shows the optical band gap of ITO films prepared at different hydrogen partial pressures. In Figure 6.6, both direct and indirect optical band gaps of the films are calculated. Their maximum values of 4.21 and 3.35 eV are obtained for films prepared at the hydrogen partial pressure of 7.9×10^{-6} torr. The variation of both direct and indirect optical band gaps shows a similar behavior. The change in the optical band gap shown in Figure 6.6 is mainly due to the absorption edge shift in the transmittance spectrum near the ultraviolet (UV) and visible wavelength regions. It is related to the change in carrier concentration in ITO films prepared at different hydrogen partial pressures. This is also known as the Moss–Burstein shift and can be expressed as

$$E_g - E_{g0} = \frac{(\pi h)^2}{2m_r^*}\left(\frac{3N}{\pi}\right)^{2/3} \tag{6.1}$$

where E_{g0} is the intrinsic optical band gap and m_r^* is the reduced effective mass. Figure 6.7 shows a linear dependence of optical band gap on $N^{2/3}$. From Figure 6.7, the value of the direct intrinsic absorption edge of about 3.75 eV is obtained by extrapolation of N to zero. Different experimental values of 2.98, 3.52, 3.55, and 3.67 eV for intrinsic absorption edges of ITO films have been reported [23,33–35]. The variation of these E_{g0} values is probably due to different deposition conditions used in the film preparations. In these previous experiments, oxidized ITO targets with different weight proportions of In_2O_3 to SnO_2 were employed. However, hydrogen was not introduced in the experiments. The intrinsic absorption edge of ITO films thus obtained is comparable with those in published experimental results. The slightly higher E_{g0} value obtained from the ITO films is probably due to the use of hydrogen–argon mixture during the film preparation. When the hydrogen partial pressure in the gas mixture is higher than the optimal value, the carrier concentration decreases, and the optical band gap shown in Figure 6.6 is also reduced; these observations are consistent with the variation indicated by Equation 6.1.

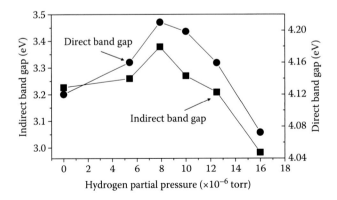

FIGURE 6.6
Calculated direct and indirect optical band gaps of ITO films prepared at different hydrogen partial pressures.

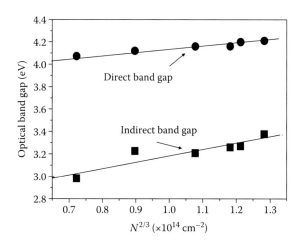

FIGURE 6.7
Optical band gaps as functions of charged carrier concentration in ITO films.

The slope of the straight line for a direct band gap, shown in Figure 6.7, yields $m_r^* = 0.61 m_0$. This is in good agreement with values of between 0.46 and 0.65 m_0 by previous groups [23,34,35]. The optical properties of ITO films, such as optical band gap and effective carrier mass, are affected mainly by the charged carrier density of the films rather than the nature of the dopant itself. The results in Figure 6.3 reveal that the primary effect of using a hydrogen–argon mixture during film deposition seems to increase the number of charge carriers in the ITO film. As a consequence, the fundamental absorption edge shifts toward the shorter wavelength range, and the corresponding increase in the optical band gap is attributed to an increase in carrier concentration.

6.1.2.5 ITO Composition and Surface Electronic Properties

UV photoelectron spectroscopy (UPS) and x-ray photoelectron spectroscopy (XPS) are commonly used to measure the material properties and to understand the technical aspects related to the ITO surface, ITO band structure, electronic structures at the ITO–organic interface, and anode modification in the OLEDs. The XPS peaks of $In3d_{5/2}$ and $Sn3d_{5/2}$ measured for ITO films obtained at different hydrogen partial pressure are given in Figure 6.8a and b, respectively. Typical XPS measurements show that ITO films prepared over the hydrogen partial pressure range 0–1.6×10^{-5} torr have almost identical atomic compositions. A closer examination of the curves in Figure 6.8a and b shows that the binding energies of $In\ 3d_{5/2}$ and $Sn3d_{5/2}$ peaks for films deposited at different hydrogen partial pressures are all at 445.2 and 487.2 eV, respectively. There are no evident shoulders observed at the high binding energy side of In3d peaks, as illustrated in Figure 6.8, which could relate to the formation of In-OH like bonds in the ITO films [36]. The possible reaction between H^+ and weakly and strongly absorbed oxygen may have taken place during film growth. However, the almost identical binding energies of In3d and Sn3d peaks suggest that indium atoms are in the form of In_2O_3.

XPS peaks of In3d and Sn3d obtained from ITO films prepared at different hydrogen partial pressures show typical ITO characteristics. However, the carrier concentration of the ITO films prepared at different hydrogen partial pressures varies considerably over the hydrogen partial pressure range of 0–5.0×10^{-5} torr. The maximum carrier concentration

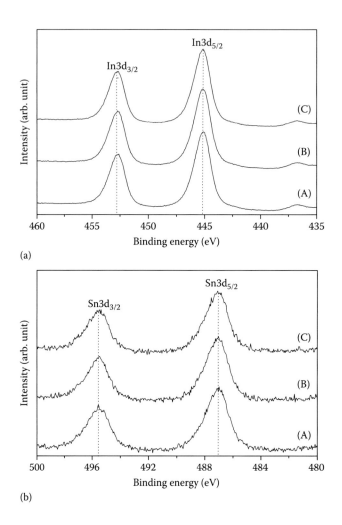

FIGURE 6.8
Comparison of typical (a) In3d and (b) Sn3d XPS peaks measured for ITO films prepared at different hydrogen partial pressures of (A) 0 torr, (B) 7.9×10^{-6} torr, and (C) 1.6×10^{-5} torr.

of 1.45×10^{21} cm^{-3} is obtained at an optimal hydrogen partial pressure of 7.9×10^{-6} torr. The difference in the carrier concentration of the films is probably due to the variations in the number of oxygen vacancies in the films prepared at different hydrogen partial pressures. Meng et al. [37] suggested that the increase in the carrier concentration was not due to tin dopants, for instance the transition of SnO to SnO$_2$, in their annealing experiments. There is no considerable change of chemical bonding energy for In3d and Sn3d peaks observed in the XPS measurements (Figure 6.8). This illustrates that there is no reduction of ITO to form the interstitial metallic atoms. Introduction of hydrogen in the gas mixture during rf magnetron sputtering seems only to vary the oxygen deficiency in ITO films. ITO films prepared are therefore nonstoichometric because of the formation of oxygen vacancies in the films. At the optimal experimental conditions, ITO with a minimum resistivity of 2.7×10^{-4} Ω cm and 89% of transmittance over the visible wavelength region can be achieved.

ITO is an ionically bound semiconducting oxide. Oxygen vacancies are formed relatively easily compared with covalently bound materials. ITO films prepared by rf magnetron sputtering are usually nonstoichometric. The number of the oxygen vacancies is affected by deposition conditions such as sputtering power, substrate temperature, sputtering gas pressure, Sn/In composition in target, and the gases in the mixture. Free electrons provided by tin dopants and ionized oxygen vacancy donors comprise the charge carriers for conduction. Therefore, this material has an n-type character. Banerjee and Das [38] investigated the effect of oxygen partial pressure prepared by electron beam evaporation from a hot-pressed powder of In_2O_3/SnO_2 mixture in weight by 9:1. They found that the increase in film conductivity was due to an enhancement in Hall mobility; however, carrier concentration decreased with the oxygen partial pressure. A similar correlation between oxygen partial pressure and carrier density was also observed by Honda et al. [23]. Experimental results suggest that the improved electrical properties of ITO films made at the optimal oxygen partial pressure are due to increased carrier mobility in the film. The decrease in carrier concentration is attributed to the dissipation of oxygen vacancies when oxygen is used in the gas mixture during the preparation. From the analyses made for the ITO films in the previous sections of this chapter, however, the Hall mobility of the films does not increase considerably at different hydrogen partial pressures. The improvement of the film conductivity can be attributed to the increase of the carrier concentration in comparison with that of ITO made without hydrogen in the gas mixture. The above analyses are consistent with the results obtained from the structural, electrical, optical, and compositional measurements made on the ITO films.

The surface electronic properties and work function of ITO also show strong dependence on the manufacturing process and the method used to clean its surface. In many applications, such as OLEDs and solar cells, the surface electronic properties also play an important role in determining the device performance. In OLEDs, ITO acts as hole-injection electrode that requires a large work function to match the highest occupied molecular orbital (HOMO) of the adjacent organic material for efficient carrier injection. However, in solar cells, ITO serves as an electron collector being more efficient with a low work function.

The electronic properties at the ITO–organic interface are expected to directly affect the characteristic of the devices. As-grown ITO anodes have been found to be less efficient for use as a hole injector than oxygen plasma–treated ITOs. Wu et al. [7] reported that oxygen, hydrogen, and argon plasma treatments on the surface of ITO do not have a significant effect on the bulk properties of the ITO film. Results show that reduction in transmittance, increase in sheet resistance, and substantial changes in ITO surface morphology are due to the formation of indium-rich hillocks and the large nonuniformity of surface compositions. The improvement of OLED performance suggests that the surface electronic and morphological properties of ITO have a great influence on the device performance than that of the bulk properties.

6.2 Anode Modification for Enhancing OLED Performance

6.2.1 ITO Surface Treatment and Modification

Among the many surface treatments of ITO, oxygen plasma treatment is one of the most common techniques [7,39–41]. Irrespective of the complexities of various cleaning processes, which may involve ultrasonic cleaning of the ITO substrate in aqueous and organic solutions, the final and most effective step often invokes the exposure of the precleaned ITO to either UV irradiation or oxygen plasma treatment. It has been reported that oxygen plasma treatment can effectively remove the surface carbon contamination and causes an increase in the work function of ITO [5,42]. This may then lead to a lower energetic barrier at an ITO–hole-transporting layer (HTL) interface and thereby helps enhance hole injection [43,44]. UPS and Kelvin probes are often employed to investigate the changes in the ITO work function due to different surface treatments [45,46]. It shows that an increase in the ITO work function is closely related to the increase in surface oxygen content due to oxygen plasma treatment [5,7,47,48].

ITO is a heavily doped and degenerate n-type oxide semiconductor with both Sn dopants and oxygen vacancies contributing to its conduction. The appropriate ITO surface treatments before the deposition of the organic layers, e.g., oxygen plasma or treatment or UV ozone irradiation, enhance the emission efficiency of OLED and improve its operating lifetime. Although the oxygen plasma treatment helps clean the ITO surface, the removal of hydroxyl functionalities or contaminants on ITO surface does not account fully for the improvement of the OLED performance. Mason et al. [47] reported that oxidative treatments incorporated more oxygen onto the surface, and the work function correlates well with the oxygen addition. The increase in the work function is attributed to the presence of an interfacial dipole resulting from a surface rich in negatively charged oxygen. The enhancement in OLED performance can be attributed to the presence of an interfacial dipole at the ITO–HTL interface favoring the hole injection. Milliron et al. [43] proposed that an increase in the ITO surface dipole layer can be attributed to the oxidation of SnO_x species, which are induced by oxygen plasma. The oxidation process only occurs near the surface region of the ITO and has less effect on the bulk of the ITO.

In parallel to a surface dipole model, a surface band bending of ITO is proposed. Yu et al. [49] reported the O_2 or NH_3 plasma results in a shift of the ITO surface Fermi level, E_F, toward the middle of the band gap, while the E_F remained unchanged in the bulk. This leads to an upward bending in the core levels near the ITO surface region. Thus, the ITO work function increases, leading to a low energy barrier at the ITO–HTL interface. An oxygen plasma–induced electron-transfer process in ITO films was also proposed. Popovich et al. [50] suggested that the oxygen plasma treatment reduces the number of active electron transfer sites at the electrode surface, possibly oxygen vacancies, resulting in slower electron transfer kinetics.

Apart from the existing understanding of the increase in work function of ITO or the presence of an interfacial dipole layer at the ITO–HTL interface due to oxygen plasma, it seems that oxygen plasma also modifies an ITO surface effectively by reducing the oxygen deficiency to produce a low-conductivity region. The improvement in OLED performance also correlates directly with a layer of low conductivity, several nanometers thick. Figure 6.9, panels a and b, shows the current density–voltage (J–V) and luminance–current density (L–J) curves for a set of identical polymer OLEDs made on ITO substrates treated by oxygen plasma at different oxygen flow rates of 0, 40, 60, and 100 sccm. At a given constant current density of 20 mA/cm^2, the luminance and efficiency of identical devices made with oxygen plasma treatments fall within a range of 610–1220 cd/m^2 and 5.4–11.0 cd/A, respectively. These values are 560 cd/m^2 and 5.2 cd/A for the same devices fabricated on nontreated ITO anodes. In this example, the best electroluminescence (EL) performance is found in OLEDs made on an ITO anode treated with plasma using an oxygen flow rate of 60 sccm.

Figure 6.10 shows SIMS depth profiles comparing normalized relative oxygen concentration from the surfaces of a nontreated ITO surface and an ITO surface treated with oxygen

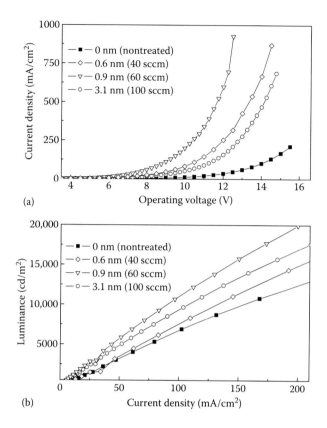

FIGURE 6.9
Current density vs. bias voltage (a) and luminance vs. current density (b) characteristics of identical devices made on ITO anodes treated under different oxygen plasma conditions.

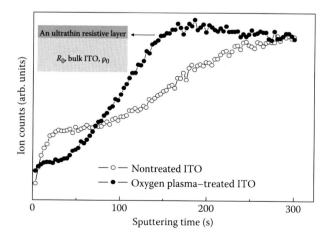

FIGURE 6.10
Comparison of SIMS depth profiles of relative oxygen concentration on the nontreated and oxygen plasma–treated ITO surfaces.

plasma using an oxygen flow rate of 60 sccm. Both nontreated and treated ITO samples are covered with a 5-nm-thick lithium fluoride (LiF) capping layer before being analyzed by a SIMS. This is to prevent any possible contamination on the ITO surfaces when the specimens are exposed to air. The LiF layer is removed by argon ion sputtering during the SIMS measurements. The changes in sheet resistance due to oxygen plasma treatments are measured using an *in situ* four-point probe. The variations in surface content are measured by *ex situ* spectroscopic analyses on the ITO surfaces. It shows clearly that the relative oxygen concentration of the treated ITO surface is higher than that of untreated ITO (middle portion of Figure 6.10). On the basis of the sputter rate used in the measurements, it appears that oxygen plasma treatment can induce an oxygen-rich layer, a few nanometers thick, near the ITO surface. However, the precise thickness of this region is difficult to determine directly by SIMS because of the influence of the interfacial effects that occurred during argon ion sputtering. Using a four-point probe, all treated ITO films are found to have higher sheet resistance than those of nontreated ones. The increase in sheet resistance between the treated and nontreated ITO surfaces are 19, 30, and 85 Ω/square when the ITO is subjected to plasma treatment with different oxygen flows of 40, 60, and 100 sccm, respectively. This implies that the change in sheet resistance corresponds closely with the increase in oxygen content on the treated ITO surface.

ITO is a ternary ionic-bound degenerate oxide semiconductor. Ionized oxygen vacancy donors and tin dopants govern its conductivity. In an ideal situation, free electrons can be generated from either the oxygen vacancies acting as doubly charged donors (providing two electrons each) or from the electrically active tin-ionized donor on an indium site [51,52]. Excess oxidation on an ITO surface by oxygen plasma may cause dissipation of oxygen vacancies. Therefore, oxygen plasma treatment results in a decrease in the electrically active ionized donors in a region near the ITO surface, leading to an overall increase in the sheet resistance as observed from the *in situ* four-point probe measurements.

By comparing the specimens with highly conducting bulk ITO, it can be proposed that treatment of ITO anodes with oxygen plasma induces a low-conductivity layer near the surface. It can be portrayed using a dual-layer model. The cross-sectional view of a nontreated ITO film and an oxygen plasma–induced dual-layer anode are illustrated schematically in Figure 6.11a and b, respectively. A constant film thickness value, d, is assumed for both nontreated and treated ITO, as there is no observable thickness change found in

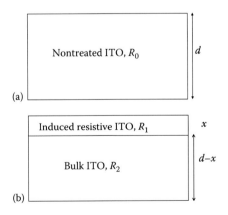

FIGURE 6.11
Schematic diagram of (a) a nontreated ITO anode and (b) an oxygen plasma–induced dual-layer ITO anode consisting of a low conductive layer, x, and bulk ITO layer, $d–x$.

ITO films treated under different conditions. R_0 and R_t are denoted as the sheet resistance measured for nontreated and treated ITO samples. According to the dual-layer parallel resistor model, $R_t = R_1R_2/(R_1 + R_2)$, where R_1 and R_2 are defined as the sheet resistance for a low-conductivity surface layer of thickness, x, and a conductive bulk ITO layer of thickness, $d-x$, as shown in Figure 6.11b.

The sheet resistance of this oxygen-rich ITO layer can be considered to be much larger than that of the bulk ITO, i.e., $R_1 \gg R_2$. Thus, $R_t = R_1R_2/(R_1 + R_2)$ can be simplified as $R_t \approx R_2$. As the bulk ITO (shown in Figure 6.11b) has the same electrical properties of nontreated ITO (shown in Figure 6.11a), it gives $R_0 = (1 - x/d)R_2$ or $R_0 = (1 - x/d)R_t$. For the four-point probe measurements, the ITO films with a layer thickness of approximately $d = 15$ nm are deposited on glass substrates. The nontreated thin ITO film has a sheet resistance of about 400 Ω/square. Substituting for R_0, R_t, and d, the thickness of the oxygen plasma–induced layer, x, is estimated to be 0.6, 0.9, and 3.1 nm, corresponding to increasing flow rates of oxygen. It is interesting to note that with the formation of this thin low-conductivity layer, the EL efficiencies of all OLEDs are greatly enhanced. As a consequence, oxygen plasma–treated ITO behaves somewhat similarly to specimens where there is an ultrathin insulating interlayer serving as an efficient hole-injection anode in OLEDs.

Engineering electrode–organic interfaces can substantially enhance the performance of OLEDs. It is understood that the modification of the anode in OLEDs alters the internal electric field distribution, resulting in changes in both the hole and electron injections. The feasibility of employing surface modification to improve OLED efficiency has also been demonstrated. A variety of stable and densely ordered self-assembled monolayer (SAM) films have been deposited onto ITO surfaces [53,54]. OLEDs incorporating a SAM layer between the ITO electrode surface and HTL have been shown to have good stability and enhanced efficiencies [55,56]. These studies indicate that the surface modification is promising for the improvement of OLED devices. An ITO anode modified with a few-nanometers-thick interlayer for efficient operation of the OLEDs is reported [57–60]. Zhu et al. [57] demonstrated that the presence of an ultrathin insulating interlayer at the ITO–HTL interface favors the efficient operation of the OLEDs.

A method of tailoring the hole–electron current balance in OLEDs by inserting an ultrathin organic tris-(8-hydroxyquinoline) aluminum (Alq_3) interlayer between the anode and HTL has been demonstrated recently [61]. Figure 6.12 (a and b) shows the

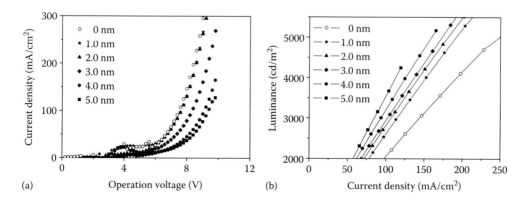

FIGURE 6.12
(a) Current density–voltage and (b) luminance–current density characteristics of OLEDs with a configuration of ITO/Alq_3 interlayer/NPB/Alq_3/Ca/Ag. The thickness of the Alq_3 interlayer was varied over a range of 0–5.0 nm.

J–V and *L–J* characteristics for a set of devices made on a bare ITO (shown as 0 nm in Figure 6.12), and 1.0-, 2.0-, 3.0-, 4.0-, and 5.0-nm-thick Alq$_3$-interlayer-modified ITOs, respectively. As is apparent from the results, there are obvious differences in *J–V* and *L–J* characteristics among devices made with different Alq$_3$ interlayer thicknesses. As the thickness of the Alq$_3$ interlayer increases, the probability of hole transport from ITO to the HTL of *N,N'*-di(naphthalene-1-yl)-*N,N'*-diphenylbenzidine (NPB) decreases, leading to a weaker hole-injection process. Therefore, to achieve the same current density in the OLEDs, the applied voltage can be increased with increasing Alq$_3$ thickness. The results in Figure 6.12a illustrate clearly that the increase in operating voltage is not significant when the Alq$_3$ interlayer thickness is <2.0 nm. However, the required voltage increases substantially when a thicker Alq$_3$ interlayer of 3.0–5.0 nm is inserted between ITO and NPB. For instance, the voltages at a current density of 100 mA/cm² are 8.32, 9.10, and 9.36 V for devices made with 3.0-, 4.0-, and 5.0-nm-thick Alq$_3$ interlayers. This voltage is 7.52 V for the device made with a bare ITO anode. In contrast, the current density to obtain a luminance of 2000 cd/m² for devices with a 0-, 1.0-, 2.0-, 3.0-, 4.0-, and 5.0-nm-thick Alq$_3$ interlayer is 97.4, 83.4, 71.4, 68.2, 60.2, and 56.8 mA/cm², respectively. It is clear that the current density decreases with an increase in interlayer thickness. Table 6.1 is a summary of the EL efficiency and the corresponding voltage at a current density of 100 mA/cm². As the cathode contact for electron injection in all these devices is the same, the results given in Table 6.1 indicate that an organic interlayer plays a role in improving the balance of electron and hole currents and hence enhancing the EL efficiency. The improvement in current balance, which is set by the size of the barrier at the two electrodes, reveals that Alq$_3$ can alter the barrier height at the ITO–HTL interface for carrier injection from ITO to the HTL of NPB.

The enhancement in EL efficiency achieved by inserting an ultrathin interlayer between the ITO and NPB is mainly due to the reduction of hole injection from ITO to NPB in OLEDs. For a simple approximation, EL efficiency (η) can be related directly to a ratio of the recombination current (J_r) to the total current density of OLEDs (J_{tot}). If one denotes the current contributions from holes and electrons in OLEDs as J_h and J_e, respectively, then the sum of hole and electron currents, $J_{tot} = J_h + J_e$, and η can be expressed as

$$\eta \approx \frac{J_r}{J_{tot}} \tag{6.2}$$

TABLE 6.1

Operating Voltage and Corresponding Luminous Efficiency of Identical OLEDs Made on ITO Anode with Different Alq$_3$ Modification Layer Thicknesses, Measured at a Current Density of 100 mA/cm²

ITO Surface Modification Layer Thickness (nm)	Luminous Efficiency (cd/A)	Operation Voltage (V)
0	2.08	7.52
1.0	2.62	7.64
2.0	2.82	7.66
3.0	2.94	8.32
4.0	3.21	9.10
5.0	3.55	9.36

The holes are the majority charge carriers in OLEDs, and the hole current is much larger than the electron current in OLEDs, that is, $J_h \gg J_e$. Thus, the total current $J_{tot} = J_h + J_e$ can be simplified as $J_{tot} \cong J_h$. Therefore, Equation 6.2 can be written as

$$\eta \approx \frac{J_r}{J_h} \tag{6.3}$$

J_r depends on the number of generated electron–hole pairs, and is limited by the minority charge carriers in the device, in this case, the electrons. Under this simplified assumption, J_r can be regarded as a constant if every electron–hole pair decayed by emitting light. When the thickness of the Alq$_3$ interlayer increases, fewer holes are injected into the NPB and thus J_h decreases, leading to an increase in η, as can be seen in Equation 6.3.

The values of the highest occupied molecular orbital (E_{HOMO}) for NPB and Alq$_3$, E_F for ITO, and the vacuum level (E_{VAC}) for each material can be deduced from the UPS measurement and are presented in Figure 6.13a and b. From Figure 6.13a, a barrier of 0.6 eV to hole injection exists at the bare ITO–NPB interface. In comparison, the barrier between the NPB and a 5.0-nm Alq$_3$-modified ITO increases to about 1.08 eV, as shown in Figure 6.13b. The electronic structure shown in Figure 6.13b suggests that holes are less easily transported from the anode to the NPB with the presence of an Alq$_3$ interlayer because the barrier to hole injection is increased. The results in Figure 6.13 explain well the differences in the J–V and L–J characteristics of OLEDs shown in Figure 6.12.

The primary effect of the anode modification on the enhancement in EL efficiency and the increased stability of OLEDs can be attributed to an improved hole–electron current balance. By choosing an interlayer with a suitable thickness of a few nanometers, anode modification enables engineering of the interface electronic properties. The above results indicate that a conventional dual-layer OLEDs of an ITO/NPB/Alq$_3$/cathode have an inherent weakness of instability that can be improved by insertion of an ultrathin interlayer between the ITO and HTL. The improvements are attributed to an improved ITO–HTL interfacial quality and a more balanced hole–electron current that enhances the OLED performance.

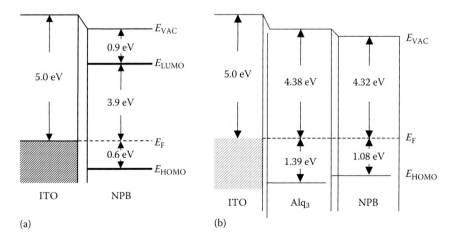

FIGURE 6.13
Interfacial electronic structures of (a) NPB on ITO and (b) NPB on an ultrathin Alq$_3$-modified ITO.

6.2.2 Color Tuning with Graded ITO Thickness

OLED arrays have been used in multicolor and full-color image display devices. An image display includes an array of light-emitting pixels. To achieve full-color OLED arrays, it is conventional to deposit three subpixels that are capable of emitting light in the red (R), green (G), and blue (B) regions of the visible spectrum, containing specific organic emissive materials for each color to form a pixel. Each subpixel is defined by an OLED. The available techniques for depositing different color layers include ink-jet printing, screen printing, spin coating, and thermal evaporation. The organic emissive materials for producing different colors have different life spans. Thus, to ensure proper color mixtures and tones, complicated thin-film transistor arrays are required for the display devices to compensate for the variations in intensity and hue emitted from the subpixels.

The variable or multicolor OLEDs can also be formed using an organic microcavity structure, in which single emissive materials can be used to generate multicolor images, including full-color images. The microcavity OLED architecture comprises a stack of organic layers confined between a top electrode and a bottom electrode. The top and bottom electrodes can be either metallic reflectors or distributed Bragg reflectors. The top and bottom electrodes can be relatively transparent or opaque depending on whether the OLED structure is a top-emitting OLED or a bottom-emitting OLED. By this arrangement, the color can be tuned by varying the thickness of the microcavity length. A multicolor or even full-color pixelated OLED display can also be fabricated using the same concept; for example, a multicolor or full-color pixelated display can be produced by forming an array of OLED structures having microcavities on a substrate. The thickness of the transparent conducting or emissive organic layers in the OLED structures can be varied across the substrate surface so as to achieve color tuning.

Figure 6.14 illustrates an OLED microcavity structure that comprises a stack of organic layers for providing EL, an upper electrode, and a bottom bilayer electrode of metal–transparent conductive layer. The thickness of the transparent conductive layer, e.g., ITO, in the OLED structures can be varied across the substrate surface so as to achieve color tuning. One typical structure of the devices is glass/Ag/ITO (with a graded film thickness)/HTL/EL/semitransparent cathode. If a top-emitting OLED architecture is used, as shown in Figure 6.14, the upper electrode is a semitransparent cathode and the bottom anode can be formed using a metal/TCO bilayer. In this example, the microcavity structure is defined by the upper semitransparent cathode and the metal/TCO bilayer anode formed on the substrate. The shape of the EL spectra of the devices and efficiency enhancement can be achieved by adjusting the thickness of the interposed ITO layer. An array of

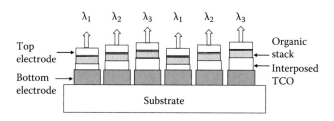

FIGURE 6.14

Cross-sectional view of a multicolor or full-color pixelated display produced by forming an array of OLED structures having microcavities on a substrate. The thickness of the interposed transparent conducting layer in the OLED structures is varied across the substrate surface so as to achieve color tuning, e.g., emitting light with different wavelengths of λ_1, λ_2, and λ_3, respectively.

the microcavity OLEDs illustrated in Figure 6.14 enables achieving multicolor or full-color pixelated display using single organic electroluminescent materials.

Color tuning with graded ITO thickness for achieving multicolor OLED array is developed. To further improve the performance of top-emitting OLED, a bilayer reflective anode of a metal/ITO is used for hole injection in the top-emitting OLEDs. In the bilayer anode, the metal layer serves as a mirror to reflect light to the upper semitransparent cathode. Different anode reflectors of Ag/ITO, Cr/ITO, and Ag:Cr/ITO for top-emitting OLEDs are applied. The results show that the identical top-emitting OLEDs made with different reflective anodes have very similar *J–V* characteristics. This implies that the mirror does not influence the *J–V* characteristic of the top-emitting OLEDs, as the hole-injection properties at anode/ HTL is essentially the same. In the actual device application, however, the top-emitting OLEDs using a Ag:Cr/ITO reflective anode have better device durability in comparison with those made with Ag/ITO and Cr/ITO reflective anodes. An improved performance for top-emitting OLEDs having Ag:Cr/ITO is probably attributed to a preferred combination of the high reflectance of Ag and a good adhesion at the glass–metal interface.

The emission color of the above top-emitting OLEDs can be tuned by varying the thickness of either the organic layers or the TCO. To achieve a desired microcavity length using uniform organic layer arrangement and an easy fabrication route for pixelated OLEDs, the thickness of ITO can be varied by controlling the film deposition time or dry-etching condition. One typical microcavity polymer top-emitting OLED has a configuration of glass/ metal (300 nm)/ITO (50 nm)/HTL/phenyl-substituted poly(*p*-phenylenevinylene) (Ph-PPV) (80 nm)/semitransparent cathode [61], where an HTL of poly(styrene sulfonate)-doped poly(3,4-ethylene dioxythiophene) (PEDOT:PSS) is used. In this example, a layer of Cr:Ag is used as metal reflector. EL peaks measured for the above top-emitting OLEDs exhibit a wide wavelength shift from 547 to 655 nm as the ITO thickness is varied from 20 to 175 nm. Figure 6.15a shows the normalized EL spectra measured for a set of devices with an identical organic layer structure except for a variation in ITO thicknesses. The photographs of the emitting devices, illustrated on the top of the curves in Figure 6.15a, show the EL output

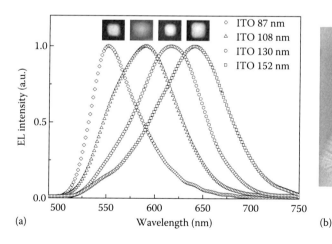

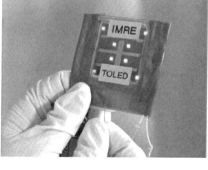

(a) (b)

FIGURE 6.15
(See color insert.) (a) EL spectra of a set of structurally identical OLEDs having a bilayer anode of metal/ITO. Inset pictures are the actual photographs taken for the microcavity OLEDs demonstrating color tuning with graded ITO thickness. (b) Top-emitting OLEDs with microcavity architectures to emit variable color and to enhance light output using a single emissive material.

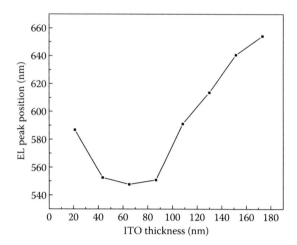

FIGURE 6.16
Measured EL peak position as a function of ITO thickness.

with different colors. The variation in color is attributed to the formation of an optical micro-cavity when embedding an organic stack between a bilayer anode of metal/ITO anode and a semitransparent cathode. Figure 6.15b is a photograph of an array of top-emitting OLEDs emitting variable colors using single emissive material of Ph-PPV.

Figure 6.16 illustrates the correlation between the graded ITO thickness and the EL peak position. The EL peak of the device shows a clear blue shift from 586 to 547 nm as the thickness of the interposed ITO increases from 20 to 65 nm. Likewise, there is a blue shift in EL spectra from 547 to 655 nm when the ITO layer thickness increases from 65 to 175 nm. It demonstrates an easier device fabrication route for multicolor OLED displays using an anode template with a graded ITO thickness.

Electroluminescent devices with optical microcavity structures offer a promising means to achieve the higher-performance organic EL diodes that exhibit very high luminance and can be driven with low dc voltages. The OLED devices with optical microcavity structure offer the possibility to control the spectral properties of emission. By replacing the ITO electrode with highly reflective mirrors, a Fabry–Perot microcavity can be introduced into usual thin-film electroluminescent diodes. In recent years, planar microcavity structures have been used to improve the performance of OLEDs. The emitting layer in organic microcavity devices is embedded between a transparent electrode and a highly reflective distributed Bragg reflector or a quarter wavelength stack, leading to a strong modulation of the emission spectrum and angular dependence [61,62]. In some applications, the microcavity effects are desired to achieve directionality and color saturation.

6.2.3 Non-ITO Anode for OLEDs

ITO has been widely used in flat-panel displays, including LCDs, plasma displays, and OLEDs. The growth in the production of flat-screen televisions has led to a doubling in the demand for ITO materials. The shortage in indium resources worries the fast-growing display industry and creates a need for the development of an efficient solution for low-cost ITO alternatives. AZO is a possible ITO alternative owing to its unique optical and electrical characteristics. AZO thin films also are much cheaper than ITO and have good potential for application in flat-panel displays. The optically transparent and electrically conducting AZO films can be prepared by using the rf/dc magnetron sputtering technique. Figure 6.17 is a typical AFM image taken for an AZO film prepared for OLEDs. A 200-nm-thick AZO film with a root mean square (rms) roughness of ~2 nm, an average transmittance of 83% in the visible wavelength region, and a sheet resistance of ~30 Ω/square can be obtained at a low processing temperature. The feasibility of using other doped ZnO thin films and highly conducting polymers as anodes for OLEDs have also been demonstrated recently [63–65].

The *J–V*, *L–V*, and *E–V* characteristics, measured for the OLEDs made with a commercial ITO anode and an AZO anode, are plotted in Figure 6.18a and b, respectively. The current density measured for an OLED with an AZO anode is lower than that obtained for a device made with an ITO anode at the same operation voltage. The slightly high turn-on voltage observed in the OLED by using an AZO anode is attributed to its lower work function compared with that of the ITO material. This suggests that an anode modification for AZO is required to improve its function as an efficient anode for OLEDs. The electroluminescent efficiency of the OLEDs made with an AZO anode is comparable to that of identical devices made with the commercial ITO anode. Although AZO is not treated specially in this case, the initial results demonstrate its potential OLED applications.

Other possible ITO alternatives that can be made relatively transparent for bottom-emitting OLED or nontransparent for top-emitting OLED are thin films of high-work-function metals and alloys. Metallic materials, including gold, silver, nickel, and their oxides, have been explored to replace ITO for OLEDs. Some high-work-function metals or their oxides, e.g., silver or nickel oxide, may have a work function value comparable to or greater than that of ITO; however, they do not satisfy the requirements of the anode owing to the presence of a large dipole barrier at the metal–organic interface. Such a contact usually induces an increase in the hole-injection barrier and thereby decreases the hole-injection efficiency. It has been demonstrated that a silver layer modified with an ultrathin plasma-polymerized hydrocarbon film (CF_x) can be used as an effective anode to enhance hole injection. The top-emitting OLEDs made with an Ag/CF_x anode shows a maximum EL efficiency of 4.4cd/A, which is greater than that of a conventional bottom-emitting OLED on glass [61,66].

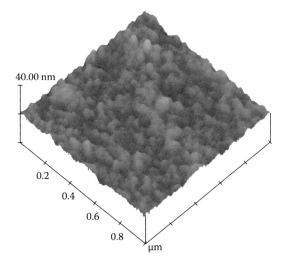

FIGURE 6.17
AFM image of an AZO film prepared for OLEDs.

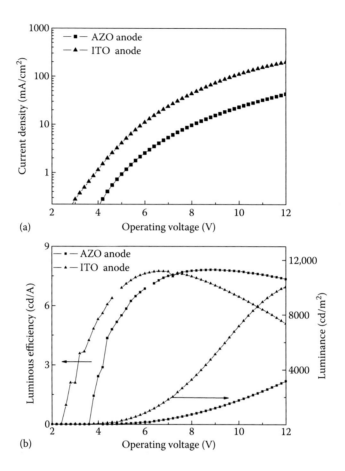

FIGURE 6.18
(a) Current density–voltage, (b) luminance–voltage, and efficiency–voltage characteristics measured for the OLEDs made with an AZO anode and an ITO anode.

6.3 Electrode for Flexible OLEDs

6.3.1 ITO Anode on Flexible Substrates

The demand for more user-friendly displays is propelling efforts to produce head-worn and handheld devices that are flexible, lighter, more cost-effective, and more environmentally benign than those presently available (see Chapter 8 for more details). Flexible thin-film displays enable the production of a wide range of entertainment-related, wireless, wearable computing, and network-enabled devices. The display of the future requires that it should be thin in physical dimension, have small and large formats, flexible, and full color at a low cost. These demands are sorely lacking in today's display products and technologies such as the plasma display and LCD technologies. OLEDs have the potential to replace LCDs as the dominant flat-panel displays. This is because OLEDs have high visibility by self-luminescence, do not require backlighting, and can be fabricated into lightweight, thin, and flexible displays. OLED stands out as a promising technology that can deliver the above challenging requirements.

Next-generation flexible displays are commercially competitive because of their low power consumption, high contrast, light weight, and flexibility. The use of thin flexible substrates in OLEDs will significantly reduce the weight of flat-panel displays and provide the ability to bend or roll a display into any desired shape. Much effort has been focused on fabricating OLEDs on various flexible substrates [67–71]. However, the polymeric flexible substrates, such as polyester and polyethylene terephthalate (PET), are not compatible with high-temperature plasma processes. Usually, a processing temperature of >200°C is required for preparing ITO films with a low electrical resistivity and high optical transparency in the visible wavelength region. ITO films formed at a processing temperature <200°C often have relatively higher resistivity and lower optical transparency than the films prepared at a high substrate temperature. In the application of organic electronics, however, it is often required to coat a layer of TCO on the plastic substrates or the active organic electroluminescent materials that are not compatible with a high processing temperature. The emergence of flexible OLEDs creates a need for the development of low-temperature-processing, high-performance ITO film on plastic or other flexible substrates. Therefore, the development of ITO with high optical transparency and electric conductivity at a low processing temperature is of practical importance.

An argon–hydrogen gas mixture has been employed for the deposition of high-quality ITO on an aluminum-laminated PET substrate (Al-PET) and polymer-reinforced ultrathin flexible glass sheet at a low processing temperature using rf magnetron sputtering. The substrates are not heated during or after the film deposition. The actual substrate temperature, which can be raised due to the plasma process during the film deposition, is <60 ± 5°C. The sputtering power is kept constant at 100 W. The base pressure in the sputtering system is approximately 2.0×10^{-4} Pa. The hydrogen partial pressure is varied from 1.0×10^{-3} to 4.0×10^{-3} Pa to modulate and optimize the properties of the ITO films. The use of a hydrogen–argon gas mixture enables a broader process window for preparation of the ITO films having high optical transparency and high conductivity [13,72]; for example, a 130-nm-thick ITO film with a sheet resistance of ~25 ± 2 Ω/square and an optical transparency of 80% in visible light range can be fabricated at a substrate temperature of 60°C. The transmittance spectra as a function of wavelength over the range of 300–800 nm measured for a bare PET substrate and a 130-nm-thick ITO-coated PET are shown in Figure 6.19. The

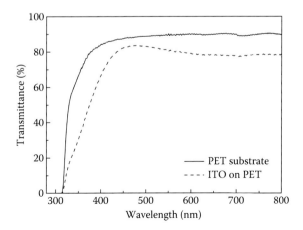

FIGURE 6.19
Wavelength-dependent transmittance of a 130-nm-thick ITO film on an acrylic layer–coated PET substrate.

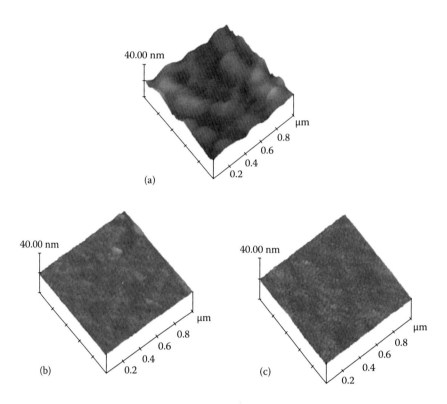

FIGURE 6.20
AFM images of (a) bare PET, (b) PET with an acrylic layer, and (c) a 130-nm-thick ITO film on an acrylic layer–coated PET.

ITO film on polymer-reinforced ultrathin flexible glass sheet also has the similar optical and electric properties.

The typical AFM images generated for bare PET, PET with an acrylic layer, and a 130-nm-thick ITO film on an acrylic layer–coated PET are shown in Figure 6.20 panels a through c, respectively. The surface of bare PET has an rms roughness of ~6.0 ± 0.1 nm. PET with an acrylic layer has a much lower rms roughness of ~0.35 ± 0.1 nm. It shows clearly that the ITO-coated PET foil thus prepared has a very smooth surface with an rms roughness of ~0.37 ± 0.1 nm, which is suitable for OLED fabrication. It reveals that the presence of an acrylic layer improves the adhesion between the anode contact and the substrate when subjected to bending as a function of the number of cycles from flat to a fixed radius of curvature of 12.5 mm. The response of 10,000 cycles of Al-PET/anode to bending shows that there are >5% of the anode layer delaminated from the substrate; however, no anode delamination can be observed for Al-PET/acrylic layer/anode after the same bending test. This is consistent with the adhesion analyses made for the ITO/polymer substrate, which shows an enhancement of the adhesion between the oxide layer and the polymer substrate through an interfacial modification [61].

6.3.2 OLEDs on Polymer-Reinforced Ultrathin Glass Sheets

Ultrathin glass sheet with reinforced polymer coating can be suitable for OLED displays with preformed, curved, or conformed shapes. In comparison to the bare ultrathin glass substrate, the robustness of polymer-reinforced ultrathin flexible glass sheets is improved significantly. The response of 1000 cycles of reinforced ultrathin glass to bending shows that there is >95% of flexible glass sheet passing 30 mm or higher compression; this is equivalent to a minimum radius of curvature of 28 mm or smaller. The reinforcement of the polymer layer helps evenly distribute the stress that is applied to the ultrathin glass. Further observation shows that the reinforcement polymer layer helps repair some of the imperfections along the edges and corners of the glass sheet that may be induced or formed during the substrate cutting or preparation. The cracks originated and propagated from the imperfection from the edges and the corners of the flexible glass sheets are the main reasons that cause substrate breakage. In addition to the top reinforcement coating, the edges of the substrates can also be reinforced using the same technique. This is done by applying the reinforcement polymer at the glass edges and corners after the top reinforcement polymer is coated. The reinforcement polymer layer covers the imperfections so that the cracks are not able to propagate further during the OLED fabrication process.

Figure 6.21 (a through c) shows the J–V, L–V, and E–V characteristics of the OLEDs. The solid triangle and open circle symbols represent the device characteristics measured from a typical OLED made with a low-temperature ITO on polymer-reinforced ultrathin flexible glass sheet and a control device made with a commercial ITO-coated 1.1-mm glass substrate, respectively. A maximum EL efficiency of ~4.1 cd/A at an operating voltage of 4 V is obtained. The experimentally measured OLED characteristics, as shown in Figure 6.21, indicate that the EL performance of the OLEDs made with low-temperature ITO-coated flexible glass sheets is comparable to that of an identical device made with the commercial ITO-coated rigid glass substrate.

6.3.3 Top-Emitting OLEDs on Al-Laminated Plastic Foils

The present OLED technologies employ rigid substrates such as glass; however, flexible device structures are extremely promising for future applications. Substrate materials are essential and a prerequisite for meeting the cost, performance, reliability, and manufacturing goals for flexible displays. Stainless-steel foil [66], ultrathin glass sheet [68], and a variety of plastic films [69,70] have been considered as possible substrate choices for flexible OLED displays. Stainless-steel foil has very good barrier properties but is not suitable for handling multiple folds. Ultrathin glass sheet with reinforced polymer coating [73] can be suitable for OLED displays with preformed, curved, or conformed shapes; however, it has limited flexibility in use. Highly flexible plastic substrates, e.g., PET and polyethylene naphthalate, have also been used for flexible OLEDs. The flexible OLEDs made on ITO-coated PET substrate have been tested under different mechanical stresses, and no significant deterioration in the device performance is observed when they are flexed at various bending radii [71]. However, these flexible OLEDs have very short lifetimes because plastics exhibit low resistance to moisture and oxygen. The development of plastic substrates with an effective barrier against oxygen and moisture has to be achieved before this simple vision of flexible OLEDs can become reality [74].

It is known that most metals possess lower gas permeability than plastics by 6–8 orders of magnitude. An unbreakable and lightweight thin stainless-steel foil substrate has been used for flexible OLEDs [75]. Therefore, a several-micrometers-thick metal layer can serve as a highly effective barrier to minimize the permeation of oxygen and moisture. Hence, the combination of plastic–metal materials is extremely promising for flexible OLED applications. The flexible substrate consisting of a plastic layer laminated onto or coated with a metal layer could be one of the possible solutions for flexible OLEDs. For example, Al-PET foil has very good mechanical flexibility and superior

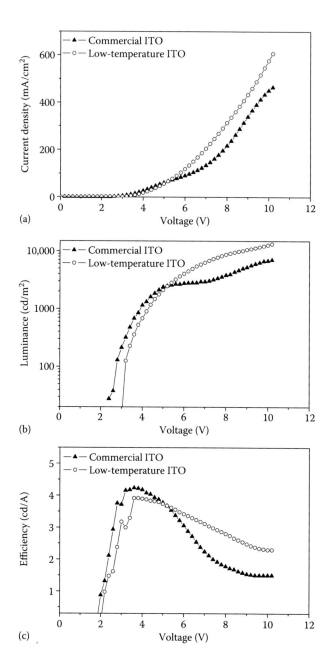

FIGURE 6.21
(a) *J–V*, (b) *L–V*, and (c) *E–V* characteristics measured for OLEDs made with a commercial ITO-coated rigid glass substrate and a low-temperature ITO-coated reinforced ultrathin glass sheet.

barrier properties. This substrate has the potential to meet permeability standards in excess of the most demanding display requirements of ~10^{-6} g/m² day [71]. The robustness of this substrate is also very high. A flexible OLED using an Al-PET substrate may provide a cost-effective approach for mass production, such as roll-to-roll processing, which is a widely used industrial process.

A flexible polymer OLED using top-emitting OLED architecture has been developed recently. The concept of the flexible OLED design is based on an integration of top-emitting OLED on an Al-PET foil. The bilayer anode of Ag/ITO or Ag/CF$_x$ and an upper semitransparent cathode are used in the top-emitting polymer OLEDs based on Ph-PPV. The resulting flexible polymer OLEDs exhibit a luminous efficiency of 4.56 cd/A at an operating voltage of 7.5 V.

The surface of a 0.1-mm-thick Al-PET film is precoated with an UV-curable acrylic layer to improve the smoothness of the Al-PET surface and the adhesion between the anode and the substrate. A 200-nm-thick Ag electrode is deposited on the flexible substrate through a shadow mask with an array of 2 mm × 2 mm openings by thermal evaporation. The silver contact is then covered by a 130-nm-thick ITO film or modified by a 0.3-nm-thick plasma-polymerized CF$_x$ film to form a bilayer anode of Ag/ITO or Ag/CF$_x$ for improving the carrier injection properties in OLEDs. High-performance low-temperature ITO with a sheet resistance of 25 Ω/square and an rms roughness of ~1 nm is deposited on an Ag surface. Cross-sectional views of a control device on a glass and top-emitting OLEDs formed on an Al-PET substrate with bilayer anodes of Ag/ITO and Ag/CF$_x$ are shown in Figure 6.22a through c, respectively. The control device has a similar layer structure of glass/Ag/ITO (130 nm)/PEDOT:PSS (80 nm)/Ph-PPV(80 nm)/semitransparent cathode.

The experimental *J–V* and *L–V* characteristics measured for two top-emitting OLEDs with a configuration of Ag (200 nm)/ITO (130 nm)/PEDOT:PSS (80 nm)/Ph-PPV (80 nm)/

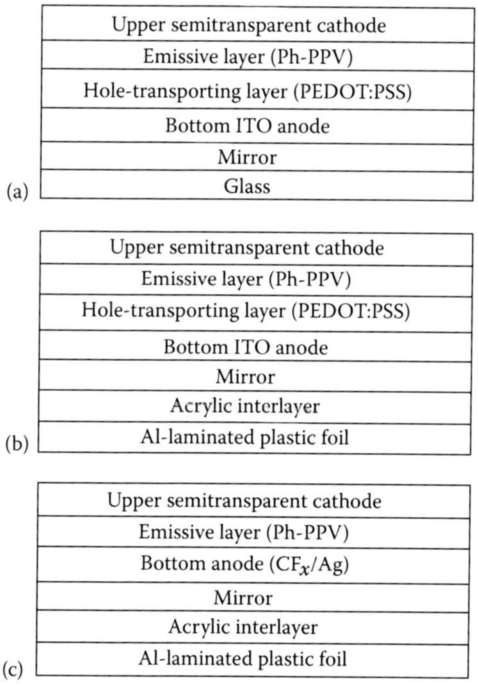

FIGURE 6.22
Schematic diagrams of top-emitting polymer OLED with a configuration of (a) glass/metallic mirror/ITO/PEDOT:PSS/Ph-PPV/semitransparent cathode, (b) Al-PET/acrylic layer/metallic mirror/ITO/PEDOT:PSS/Ph-PPV/semitransparent cathode, and (c) Al-PET/acrylic layer/metallic mirror/anode/Ph-PPV/semitransparent cathode.

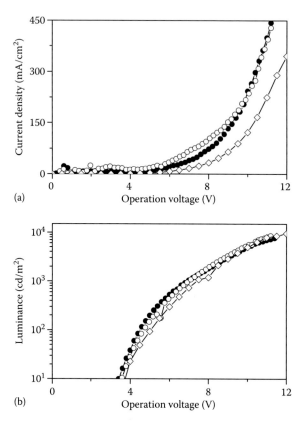

FIGURE 6.23
(a) *J–V* and (b) *L–V* characteristics of three devices with a configuration of glass/Ag (200 nm)/ITO (130 nm)/ PEDOT:PSS (80 nm)/Ph-PPV (80 nm)/semitransparent cathode (closed circles), Al-PET/acrylic layer/Ag (200 nm)/ITO (130 nm)/PEDOT:PSS (80 nm)/Ph-PPV (80 nm)/semitransparent cathode (open circles), and Al-PET/acrylic layer/Ag (200 nm)/CF$_x$ (0.3 nm)/Ph-PPV (110 nm)/semitransparent cathode (open diamonds).

semitransparent cathode on an Al-PET foil (open circles) and on a glass substrate (closed circles) are plotted in Figure 6.23. The top-emitting OLEDs with the same architecture but made on flexible Al-PET foil and a rigid glass have almost identical device performance, which indicates the validity of the bilayer anode for the top-emitting OLEDs. For instance, the current density at 10 V is 236.4 and 243.4 mA/cm^2 for top-emitting polymer OLEDs on an Al-PET foil and a glass substrate, respectively.

It is believed that the escape possibility of light trapped by ITO is low as the output coupling efficiency of the OLEDs is about ~20% due to the critical angle of total reflection within the device [76]. However, this is avoidable by replacing ITO with a metallic anode so that light emitted in the direction toward the anode can be fully reflected back to the upper semitransparent cathode. In contrast to the success in forming a semitransparent cathode on an active layer [66,77], the use of Ag anode in an OLED usually results in a poor device performance owing to the existence of a large barrier for the hole injection. Li et al. [78] demonstrated that the hole injection can be significantly enhanced by inserting a CF$_x$ film between Ag and the organic film, which provides the feasibility to form an ITO-free OLED using a bilayer anode of Ag/CF$_x$. The electrical and optical characteristics of the corresponding device with a configuration of Ag (200 nm)/CF$_x$ (0.3 nm)/Ph-PPV (110 nm)/

semitransparent cathode on an Al-PET substrate (Figure 6.22c) is also plotted in Figure 6.23 (open diamonds). In comparison with the top-emitting OLED made with bilayer anode of Ag/ITO, the top-emitting OLED made with a Ph-PPV layer on CF_x-modified Ag anode requires a slightly higher operating voltage to achieve a similar current density. This is attributed to a thicker Ph-PPV layer of 110 nm used in the device; however, the results indicate that Ag/ITO and Ag/CF_x exhibit a similar hole-injection behavior in OLEDs [79].

The top-emitting OLED with a bilayer anode of Ag/CF_x and an ultrathin Ag layer used in the upper semitransparent cathode forms an optical microcavity, which can tailor the spectral characteristics of the emitters therein by allowing maximum light emission near the resonance wavelengths of an organic microcavity [80,81]. When the mode wavelength of the cavity is fixed at 550 nm, the thickness of the Ph-PPV layer is determined to be about 110 nm [81]. Therefore, a top-emitting OLED with a Ph-PPV layer of 110 nm would provide a higher luminous efficiency.

The results of luminous efficiency as a function of current density measured for a control device with a structure of glass/Ag (200 nm)/ITO (130 nm)/PEDOT:PSS (80 nm)/Ph-PPV (80 nm)/semitransparent cathode (Figure 6.22a) and flexible polymer OLED with a configuration of Al-PET/acrylic layer/Ag (200 nm)/CF_x (0.3 nm)/Ph-PPV (110 nm)/semitransparent cathode (Figure 6.22c) are shown in Figure 6.24a. The control device has a maximum luminance efficiency of 2.70 cd/A, while the top-emitting OLED made with a Ph-PPV layer on a bilayer Ag/CF_x anode has a luminous efficiency of 4.56 cd/A. The enhancement of the efficiency in the top-emitting OLED with a CF_x coating is attributed to two factors: (i) part of the light emitted in a direction toward the anode, which is trapped by ITO in a control device, is almost completely reflected back to the semitransparent cathode by the Ag/CF_x anode leading to an increased luminous efficiency; (ii) the optical microcavity effect redirects the trapped light outside the device. Figure 6.24b is an image of a flexible top-emitting OLED on an Al-PET substrate. The device performance does not deteriorate after repeated bending, suggesting that there is no significant stress-induced change in the characteristics of the OLEDs fabricated on PET foil [71,82]. The results demonstrate the feasibility of fabricating flexible displays using a variety of plastic substrates, including

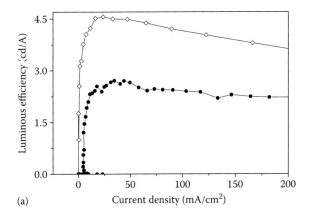

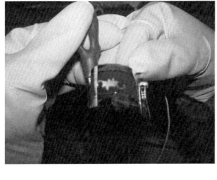

(a)　　Current density (mA/cm^2)　　(b)

FIGURE 6.24
(See color insert.) (a) Luminous efficiency of two top-emitting OLEDs with a configuration of glass/Ag (200 nm)/ITO (130 nm)/PEDOT:PSS (80 nm)/Ph-PPV (80 nm)/semitransparent cathode (closed circles), and Al-PET/ acrylic layer/Ag (200 nm)/CF_x (0.3 nm)/Ph-PPV (110 nm)/semitransparent cathode (open diamonds). (b) Photo showing a flexible top-emitting electroluminescent device on Al-PET substrate.

metal-laminated plastic foils, or a metal film sandwiched between two plastic foils. The flexible device structures enable a display to conform, bend, or roll into any shape and thus make possible other product concepts.

6.4 Optically Destructive Electrode for High-Contrast OLEDs

6.4.1 Black Cathode for High-Contrast OLEDs

There has been a great deal of activities in the development of OLEDs. In a conventional structure of the organic or polymeric OLEDs, the devices have a layer (or layers) of organic material sandwiched between a transparent anode and the cathode. The metallic cathode is typically used and has a high reflection to the ambient light. If the devices are made on transparent substrates, including rigid glass or clear flexible plastic foils, OLEDs with this configuration are usually very reflective. Therefore, the contrast of the devices is very low and the visual image of the OLEDs is poorly legible. In many practical applications, especially in bright ambient condition, the visual contrast is more important than the brightness of the image. Therefore, a sufficient reduction in the reflection of ambient light from the OLEDs is a prerequisite for high-contrast OLED displays.

Much effort has been focused on developing OLEDs with low reflectivity under ambient light. For example, a circular polarizer film can be bonded to the outside of the glass substrate to improve the visual contrast. This is a very simple solution for improving the contrast of OLEDs. In fact, polarizer films have been used to enhance LCD displays to good effect and can be similarly applied for OLEDs. However, the addition of the polarizer film constitutes an additional bonding step to the production of the OLED displays. This aside, polarizer films are subject to humidity and temperature environments; thus, the operating condition of OLED displays is constrained to the limited range of humidity and temperature of the polarizer films. This also results in the inclusion of a material not inherently part of the manufacturing process of the OLEDs. Furthermore, this eventually results in higher costs. This aside, a polarizer also increases the thickness of OLED displays. When using such a circular polarizer in flexible OLED displays, this becomes a genuine concern.

In addition to the straightforward polarizer approach, the feasibility of employing a low-reflectivity cathode to reduce ambient reflection for achieving low-reflectivity OLEDs has been reported. Figure 6.25 is a cross-sectional view of a high-contrast OLED

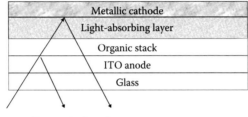

Destructive interference

FIGURE 6.25
Schematic diagram showing an OLED with a black cathode; the presence of an optical absorbing layer causes a destructive interference leading to a low reflection.

with a black cathode. A conducting and light-absorbing contact is used to form an optically destructive interference for reducing the ambient reflection. Hung and Madathil [83] have demonstrated that calcium hexaboride (CaB_6) can be used as ambient light reduction cathode. CaB_6 is highly conductive with a low work function and is substantially black in bulk form. Although the alternative electron injection layer of CaB_6 has an advantage of low reflectivity, the uniform CaB_6 film with stable optical and electrical properties is not very easy to control in the deposition process. A variety of black cathode structures have also been developed to minimize light reflection at the organic–cathode interface. For example, a reflection-less OLED with a multilayer black cathode structure of LiF/Al/ZnO/Al was reported [84]. In this multilayer black cathode, the zinc oxide film was deposited by thermal evaporation. It acts as an optical absorbing layer to reduce the ambient light reflection from the metallic cathode. The use of a high conductive black carbon film in a multilayer cathode system was also demonstrated by Renault et al. [85]. This black cathode consists of a thin electron injector layer of magnesium, an optically absorbing and electrically conductive carbon layer, and a thick aluminum layer. This multilayer black cathode has a similar charge injection property to a conventional cathode of Mg/Al; however, it has a much lower reflectivity. The results show that the reflection decreases from nearly 100% for devices using conventional cathode to ~60% for a multilayer cathode. The black cathodes using conductive light-absorbing layers with mixtures of organic materials and metals have also been reported. It was demonstrated that the presence of an electrically conductive light-absorbing layer at the cathode of an OLED significantly reduced the total reflectance of an OLED [86]. This electrically conductive light-absorbing layer serves to absorb reflected ambient light from the cathode and improves the overall contrast and legibility. A light-absorbing layer is typically made of a mixture of organic material and metal, and is placed between the cathode and the organic layer.

Black Layer™ is another example of a black cathode developed by Luxell Technologies, which uses an interference destructive layer in a low-reflectivity cathode for OLEDs [87,88]. Black Layer consists of a layer of thin absorbing material, a layer of transparent materials, and a thick metal layer. Light reflected from the first and second metal layers is equal in amplitude but differ in phase. Therefore, the destructive interference occurs and significant reduction in reflectance can be achieved if appropriate thickness of each layer is used. There are also other methods using additional light-absorbing layers of a variety of different materials. These methods essentially address the reduction of reflected ambient light by incorporating a low-reflectivity composite cathode [68,89–91].

6.4.2 Gradient Refractive Index Anode for High-Contrast OLEDs

A high-contrast OLED can also be fabricated using a bilayer or multilayer optically destructive anode to reduce the reflectance of the ambient light from the device. The concept is based on using an anode with a gradient or graded refractive index to minimize the ambient light reflection from OLEDs and hence to enhance the visual contrast. The schematic diagram of the device is shown in Figure 6.26. A bilayer anode consisting of a thin film of semitransparent metal oxide, e.g., highly oxygen-deficient ITO, and a normal high-work-function ITO can be used in OLEDs for reducing ambient reflection. The highly oxygen-deficient ITO film is electrically conducting and optically absorbing. The oxygen-deficient ITO layer is inserted between the anode contact (e.g., ITO) and the rigid or flexible transparent substrate to serve as an optically destructive layer for reducing the ambient light reflection from the OLEDs. A typical structure of a high-contrast OLED consists of a rigid/

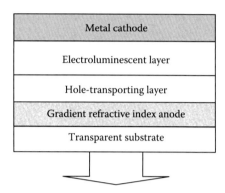

FIGURE 6.26
Schematic diagram of a high-contrast OLED using a gradient refractive index anode.

flexible substrate, an oxygen-deficient conducting oxide layer, an anode, an organic stack of the hole-transporting and emissive layers, and a high-reflectivity metallic cathode, in that order. The semitransparent/transparent conducting metal oxide bilayer anode, both layers may be made of ITO, enables significantly reducing the ambient light reflection from the mirror-like surface of the metallic cathode in OLEDs through light-absorbing and optically destructive interference.

From an optical point of view, an OLED structure can be considered as a multilayer thin-film system composed of absorbing and nonabsorbing materials, as shown in Figure 6.27. Therefore, the optical properties and optimal structure of such a multilayer device can be investigated by applying thin-film optical analysis techniques. On the basis of the theory of optical admittance analysis for analyzing the optical properties of a thin-film system [92], the optical properties of an OLED thin-film system can be simulated to reduce the ambient reflection.

Defining $F(\lambda)$ as the flux of the ambient light incident on the display, the reflectance of the device, R_L, can be calculated as [93]

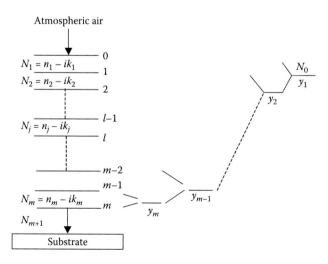

FIGURE 6.27
Schematic representation of an arbitrary multilayer thin-film system.

$$R_{L} = \frac{\int R(\lambda)F(\lambda)d\lambda}{\int F(\lambda)d\lambda} \tag{6.4}$$

where $R(\lambda)$ is the spectral reflectance of the thin-film system. If the wavelength-dependent refractive indices of each layer in the OLED system are known, it becomes possible to optimize the thickness of the composite anode structure through minimizing the reflectance. An optimal structure can thus be designed to sufficiently diminish the reflectance of an OLED, while maintaining the comparable device performance.

The contrast ratio (CR) of an OLED display is very much dependent on the ambient lighting conditions. The actual CR for an OLED display is based on the applications and differs depending on the products, such as car audio, cell phone, etc. Usually for indoor applications, a CR of >10–20:1 is sufficient. According to UDC OLED Technology Roadmap 2001–2010, the targeted CR for OLED displays (300 cd/m^2, under 500 lux) is expected to be 50:1, 100:1, and 200:1 for 2004, 2007, and 2010, respectively. In a pixilated device, CR can be defined as [91]

$$CR = \frac{L_{on} + R_{L} \times L_{ambient}}{L_{off} + R_{L} \times L_{ambient}} \tag{6.5}$$

where L_{on} and L_{off} are the emitted luminance of active ("on") and inactive ("off") pixels, and $L_{ambient}$ is the ambient luminance or the ambient light incident on the display. The corresponding optical parameters, the real part of the refractive index, $n(\lambda)$, and the extinction coefficient, $k(\lambda)$, of each layer, will be measured by variable angle spectroscopic ellipsometry. The luminous reflectance R_{L} and the CR of the OLED displays can then be simulated using Equations 6.4 and 6.5. The optical simulation enables providing in advance the leading design information of an OLED system with an optimal optically destructive anode structure for a desired high CR.

Figure 6.28 shows the wavelength-dependent reflectance measured for a control device (shown as control device in Figure 6.28) and the OLEDs made with the bilayer ITO

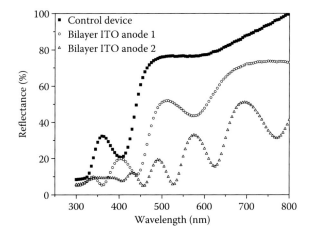

FIGURE 6.28
Ambient light reflection as a function of the wavelength measured for a control device and structurally identical OLEDs made with two different bilayer ITO anodes.

electrodes (anode 1 and anode 2 in Figure 6.28) consisting a 170- and a 400-nm-thick highly oxygen-deficient semitransparent ITO, respectively. The thickness of the upper ITO (anode contact) is kept constant at 130 nm. In this example, the highly oxygen-deficient semitransparent ITO film is used as an absorbing layer. The stack of a semitransparent ITO film and an optimal transparent ITO anode can be deposited subsequently on a glass substrate using rf magnetron sputtering. As is apparent from the results, there are obvious differences in the overall reflectance of OLEDs made with different light-absorbing ITO layer thicknesses. As the thickness of the optically destructive layer increased, the total reflectance of the devices is reduced substantially over the visible wavelength region. The reduction in ambient light reflection from the devices with a bilayer ITO anode structure enables enhancing the visual legibility of OLED displays. OLEDs made with a bilayer ITO anode, consisting of a highly oxygen-deficient ITO and an upper normal high-work function ITO, have similar hole-injection properties as compared to the ones fabricated with a regular single layer high-work function ITO anode. The identical OLEDs made with either a bilayer or a single-layer ITO anode have similar *J–V* characteristics. However, the luminance efficiency of the device with a bilayer anode with 170-nm-thick oxygen-deficient ITO is approximately 60% of that of a control device operated at a same current density. Such a reduction in luminance efficiency can be attributed to the use of an oxygen-deficient ITO layer, as some of the emitted light from the OLED is absorbed by the bilayer anode.

In practical applications, the visual contrast of an emissive display is more important than the brightness of image. As such, improving the visual contrast in OLEDs is another important issue to address with a significant technological implication. The use of a gradient refractive index transparent electrode, e.g., a bilayer ITO anode, is effective for enhancing the visual contrast in OLEDs.

Figure 6.29 shows photographs taken for a control device with a bare ITO anode, the OLEDs made with bilayer optically destructive anodes having 170- and 400-nm-thick highly oxygen-deficient ITOs and a top ITO anode, respectively. High-reflectivity cathodes in a control device are evidently seen in Figure 6.29a; the "black" electrodes shown in Figure 6.29b and c clearly demonstrate the effect of ambient light reduction in the devices. It is obvious from Figure 6.29 that the presence of an optically destructive layer, in this case a highly oxygen-deficient ITO layer, between the anode and the substrate reduces the reflection of the OLEDs. It is demonstrated that a conventional OLED with an inherent weakness of high reflectivity from a mirror-like cathode (Figure 6.29a) can be overcome by employing a bilayer anode with a gradient or graded refractive index. The results indicate that OLEDs with a bilayer anode can provide a substantial enhancement in visual legibility and contrast under an ambient light environment.

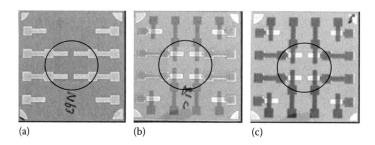

(a)   (b)   (c)

FIGURE 6.29
Photographs taken for OLEDs with different optically destructive ITO layer thicknesses of (a) a bare ITO, (b) 170 nm, and (c) 400 nm under a sunlight environment.

The use of an optically destructive anode shows promise for contrast improvements in OLEDs. The multilayer semitransparent anode is capable of reducing the ambient light reflection from the reflective cathode to enhance the contrast of the devices. A multilayer anode may also be made using a stack of layers with graded refractive indices to minimize the total reflection of the ambient light from the conventional OLEDs. If only one type of material is used, e.g., ITO in the above example, the thickness and the reflective index of the anode can be varied with a desired gradient to form an optically absorbing and electronically conducting anode for high-contrast OLEDs. The ITO with different refractive indices can be easily engineered by varying the hydrogen partial pressure in the argon–hydrogen gas mixture during the film deposition. As the deposition of an ITO with a gradient refractive index is regarded as a part of the anode deposition, there is no additional request for new equipment, materials, and even processing gases. This technique provides a cost-effective approach for achieving high-contrast OLEDs.

The fabrication of an ITO anode with a gradient refractive index for high-contrast OLED can be integrated easily to the existing mass production process for device fabrication. It has stable chemical characteristics and can also be easily patterned using well-developed lithographic techniques for device fabrications. This aside, reproducible ITO films with different optical and electric properties can be prepared by using different techniques, including reactive thermal evaporation deposition, magnetron sputtering, electron beam evaporation, spray pyrolysis, and chemical vapor deposition. The OLEDs with a bilayer optical absorbing/transparent ITO anode shows a sufficient reduction of 10–50% in ambient light reflection compared with a control OLED with a bare ITO. As a bilayer has a normal ITO anode–finished surface, the various surface treatments and the anode modification developed for enhancing OLED performance can be applied to the new anode structure without any change. The multilayer anode provides a feasible and a cost-effective approach for fabrication of visually legible OLED displays.

Although ITO is still one of the most widely used anode materials for OLEDs, other alternatives suited for OLEDs may also be used for making optically destructive anodes for high-contrast OLED displays using this technique. For example, a multilayer optically destructive anode may be fabricated using other oxide materials, including SnO_2, FTO, AZO, IZO, Ga–In–Sn–O, Zn–In–Sn–O, Ga–In–O, and other TCOs suitable for the anode in an OLED. These materials can be used individually or with a combination of different materials.

6.5 Transparent OLEDs

A conventional OLED has a bottom-emitting structure, which includes an opaque metal or metal alloy cathode and a transparent anode on a transparent substrate, enabling light to emit from the bottom of the structure, usually from the glass side. An OLED may also have a top-emission configuration, which is formed on either an opaque or a transparent substrate. Light can be emitted from both the anode and cathode sides when both electrodes are relatively transparent, forming a transparent OLED. Transparent OLED can greatly increase the flexibility of device integration and engineering, making it possible to open up novel product concepts.

Research into the development of high-quality transparent top electrodes, often a cathode, has attracted intensive attention worldwide. The major OLED display companies are actively involved in developing transparent OLED technologies for use in active matrix–driven OLED displays and white OLED (WOLED) lighting. The visible light transparency of a transparent OLED can be >70% when it is not powered. This unique transparency enables it to be integrated into car windshields, architectural windows, eyewear, and other transparent substrates for a wide range of applications such as thin-film organic photosensors, bidirectional light-emitting signage, transparent displays, and lighting. In the following discussion, examples of incorporating a pair of transparent electrodes for application in semitransparent matrix OLED displays and bidirectional semitransparent white WOLED lighting will be reviewed.

6.5.1 Semitransparent Passive Matrix OLED Displays

A prototype semitransparent passive matrix OLED display with a transparent anode and a semitransparent cathode on a glass substrate is developed. It is fabricated using an ITO-coated glass. The ITO anode is 120 nm thick with a sheet resistance of about 30 Ω/square. ITO/glass substrates are cleaned using acetone, isopropanol, and deionized water in an ultrasonicator, subsequently followed by oxygen plasma treatment. An 80-nm-thick PEDOT:PSS is then spin coated on ITO to serve as an HTL. It is cured at 120°C for 15 min in air. An 80-nm-thick Ph-PPV emissive layer is formed on PEDOT:PSS by spin coating. The fabrication of semitransparent OLEDs is then completed by overlaying a multilayer semitransparent cathode of LiF (0.3 nm)/Ca (5 nm)/Ag (15 nm)/Alq$_3$ (52 nm). This is readily prepared by thermal evaporation without incurring radiation damage to the OLED layer structure, particularly the underlying active polymeric Ph-PPV emissive layer [61]. The performance of semitransparent OLEDs is analyzed and compared with a conventional bottom-emission OLED fabricated using the same emissive layer. Both anode and cathode are designed to ensure a perfect hole–electron current balance. This is achieved through a systematic study of interfacial properties at organic–anode and transparent cathode–organic interfaces in semitransparent OLEDs.

The electrode separation process in pixelated OLED displays usually involves a photolithographic step to form insulating bars or structures on prepatterned ITO substrate. A shadow mask is commonly used to pattern the electrode structures for OLED displays. It requires stringent alignment conditions and is challenging when the device fabrication process involves the deposition of different organic and inorganic materials with small dimensions over a large area. Mask warping is a problem for high-resolution pixelated displays. The rapid growth in the development of matrix OLED displays creates a need to explore new electrode patterning technology for new applications. In addition to a conventional photolithography patterning process, a laser ablation technology is developed, which enables to achieve any desired semitransparent cathode layout without damaging the underlying active polymeric or organic layers in an OLED display.

The cathode separation is achieved using a laser (a 355-nm Nd-YAG laser) ablation method. The prototype semitransparent OLED display is fabricated on a substrate of size 50 mm × 50 mm. The display area is 32.25 mm × 11.15 mm and has 100 ITO anode columns and 32 semitransparent cathode rows. The width of the ITO column and cathode row is 270 µm and 300 µm, forming a pixel size of 270 µm × 300 µm. The column and row spacing is 30 µm and 25 µm, respectively. Laser ablation is used to define the cathode structure, in this case the number of cathode lines with a spacing of around 25 µm. The ablation conditions are optimized such that the process enables selective removal of the materials without causing damage to the anode columns and to create the cathode separation without damaging the underlying polymer layers. The laser ablation process can also be programmed to form desired cathode line structures, e.g., aligned perpendicularly to the anode rows. The edge quality of the laser-ablated cathode is dependent on the laser power and scanning velocity.

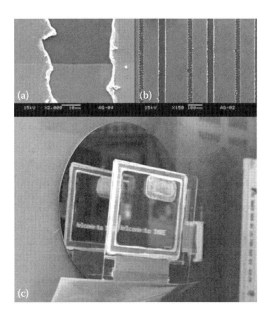

FIGURE 6.30
(See color insert.) Microscopic images of cathode lines: (a) image taken for a laser-ablated cathode edges before the optimization, showing burr and jagged shapes; (b) typical cathode edge profile formed using an optimized ablation process; and (c) photograph of a prototype semitransparent OLED passive matrix display.

Panels a and b of Figure 6.30 are micrographic images of cathode edges formed using a single- and a raster-direction scan, respectively. Some burr and jagged cathode edges are observed, as shown in Figure 6.30a. These edge defects are detrimental to the device, causing electrical shorts between the adjacent cathode lines as well as between the cathode and anode structures, leading to an undesirable current leakage and a cross-talking effect in a passive matrix semitransparent OLED display. With appropriate tuning and the optimization of the laser ablation process conditions, e.g., a power of 17 mW and a repetition rate of 7 kHz with a focused z-distance, a complete separation of cathode rows can be obtained using a laser raster scanning method to ablate the material, as shown in Figure 6.30b. It is demonstrated that the laser ablation process is very reliable, reproducible, and has been used for making semitransparent passive matrix polymer OLED displays.

For the prototype semitransparent OLED display, an AT89C51 microcomputer controller is used, which has 4K bytes of flash-programmable erasable read-only memory and a serial peripheral interface for data transfer. The column source has a serial-in and parallel-out unit for supplying current so as to achieve gray levels for the display. The current source can be continuously tuned from 1 µA to 10 mA. It has high precision and supplies a stable current output. The driver has a high range of operating voltages from 1 to 40 V. In a row scan mode, the controller addresses the display row by row sequentially (at an overall frame rate of about 60 pictures per second) so as to prevent direct coupling of pixels. The current is delivered in pulses during the short row-addressing time to avoid long-term breakdown of the nonselected pixels held in reverse bias. The peak current level can be increased with an increase in the number of rows of the display. Figure 6.30c is a photograph of the prototype passive matrix semitransparent OLED display showing a moving message. The semitransparent OLED technology offers additional flexibility to OLEDs with a potential for a wide variety of information display, industrial, medical, health-care, and other consumer-oriented applications.

6.5.2 Semitransparent WOLED Lighting

The continued advances have resulted in WOLEDs for application in high-resolution displays and diffused planar lighting sources. Compared with other solid-state lighting technologies, WOLED is the only device that can provide a planar diffused light source that is more comfortable to the human eyes, which have evolved to see white light in the solar spectrum. To date, much effort has been focused on developing high-performance phosphorescent WOLEDs for application in lighting. Phosphorescent WOLEDs with a power efficiency of >40 lm/W have been demonstrated on one-sided opaque devices [94–96]. An efficient WOLED with a power efficiency of 55.2 lm/W at 100 cd/m^2, Commission Internationale de l'Eclairage (CIE) coordinates of (0.40, 0.40), and a color rendering index of 81 using a blue iridium carbene complex [97] has been reported. In parallel to the development of new emissive materials, a light outcoupling structure is used to improve the power efficiency of WOLEDs. A phosphorescent WOLED with a threefold increase in power efficiency to ~100 lm/W at 100 cd/m^2 has been demonstrated using a substrate having high refractive index and a periodic outcoupling structure [96].

Semitransparent WOLEDs consisting of an ITO anode and a variety of thin metal cathodes of Al, Ag, Au, Ca/Ag, Yb:Ag, Mg:Ag, and others have been reported [98–107]. These semitransparent WOLEDs usually have a preferential one-sided electroluminescent emission due to asymmetric emission characteristics at the anode–organic and organic–cathode interfaces. A pair of anode and cathode having a symmetrical layer structure of Alq$_3$ (85 nm)/Ag (22 nm)/Alq$_3$ (85 nm)/Ag (22 nm) was proposed for a semitransparent WOLED [108]. Although a bidirectional illumination was obtained, the EL emission has a strong angular dependence due to the microcavity effect. This aside, the overall visible light transparency of the WOLEDs is also limited by the use of 88-nm-thick Ag layers in both electrodes [108].

High-efficiency transparent WOLEDs with efficient and balanced carrier injection at the organic–anode and organic–cathode interfaces are desired, which enables an efficient operation of WOLEDs at a low voltage. Usually, the injection barriers at both contacts are unmatched, which ultimately results in an imbalance in electron and hole currents in the active layers of OLEDs. The conventional WOLED architecture comprises a stack of organic layers confined between a top metal cathode and a bottom ITO anode. The thickness of the top metal cathode is usually in the order of 100–200 nm, and light is emitted from the transparent ITO anode side. When an ultrathin metal (~5–10 nm) cathode is used, the top cathode can be relatively transparent to form a top-emission type WOLED or a semitransparent WOLED. By this arrangement, emitted light can escape from both the ITO anode and the top ultrathin metal cathode sides of the transparent WOLEDs. Realization of a semitransparent phosphorescent WOLED lighting that has high visible light transparency and can emit a pleasant symmetrical bidirectional diffused light still remains an open challenge.

Here, we discuss the recent advances on a semitransparent two-color white semitransparent WOLED with a visible light transparency of >50% and a symmetrical bidirectional EL emission spectra. A two-color white system is produced by the emission of a blue phosphorescent dopant iridium(III)[bis(4,6-difuorophenyl)-pyridinato-N,C^2] picolinate (FIrpic) and an orange phosphorescent dopant bis(2-(9,9-diethyl-9H-fluoren-2-yl)-1-phenyl-1H-benzoimidazol-N,C^3) iridium(acetylacetonate) [Ir(fbi)$_2$acac]; the peak EL emissions of FIrpic and Ir(fbi)$_2$acac dopants at 478 and 562 nm [109] are shown in Figure 6.31. FIrpic and Ir(fbi)$_2$acac dopants were codoped with a host material of 4,4',4"-tri(N-carbazolyl)triphenylamine (TCTA) to form a dual emissive layer (EML) of TCTA:FIrpic:Ir(fbi)$_2$acac (19%, 1%,

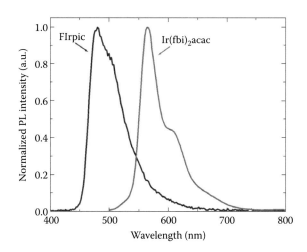

FIGURE 6.31
Peak EL emissions of FIrpic and Ir(fbi)₂acac dopants at 478 and 562 nm.

10 nm)/TCTA:FIrpic (19%, 7.5 nm). An NPB HTL with a hole mobility of 4.0×10^{-4} cm²/V s [110], and an electron-transporting layer (ETL) of 3,5,3″,5″-tetra-3-pyridyl-[1,1′;3′,1″]terphenyl (B3PyPB) with an electron mobility of 1.0×10^{-4} cm²/V s [111] were used. TCTA has a high triplet energy level (E_T = 2.82 eV). B3PyPB has a high HOMO level (6.67 eV) and a high triplet energy level (E_T = 2.77 eV). The use of B3PyPB also helps suppress the Dextor energy transfer from FIrpic (E_T = 2.65 eV) to B3PyPB [111,112]. The layer structure, the schematic energy diagram, and the chemical structures of the functional materials used in the semitransparent phosphorescent WOLEDs are shown in Figure 6.32. The WOLEDs are fabricated on ultrathin bilayer Ag/MoO₃-modified ITO substrates. The bare ITO has a sheet resistance of 15 Ω/square. All functional layers, as shown in Figure 6.32b, were deposited by thermal evaporation in a high vacuum system with a base pressure of $<5.0 \times 10^{-4}$ Pa without breaking the vacuum. All semitransparent WOLEDs are finally encapsulated for current density–voltage–luminance characteristics measurements in air.

The dopant concentrations and the thicknesses of the functional organic layers in the EML stack of the two-color white system were optimized using a control bottom-emission WOLED of the type ITO (80 nm)/MoO₃ (2.5 nm)/NPB/TCTA (5 nm)/EML/B3PyPB/LiF (1.0 nm)/Al (1.5 nm)/Ag (100 nm). It can be considered that a control opaque WOLED does not have a microcavity effect. Therefore, the characteristics of EL emission from the two-color white system can be analyzed using a control opaque WOLED without taking into account the cavity effect. To prevent the possible triplet quenching at the NPB–dopants interface, a 5-nm-thick TCTA interlayer was inserted between the HTL and the EML. It shows that a control WOLED with an EML of TCTA (5 nm)/TCTA:FIrpic:Ir(fbi)₂acac (19%, 1%, 10 nm)/TCTA:FIrpic (19%, 7.5 nm) and a pair of 40-nm-thick NPB (HTL) and B3PyPB (ETL) results in a high power efficiency of 37 lm/W (measured at 100 cd/m²) with CIE color coordinates of (0.34, 0.42). The effect of organic layer thickness variation on the CIE color coordinates of anode and cathode emissions in semitransparent WOLED is presented in Figure 6.33. The parameters of the EML of TCTA (5 nm)/TCTA:FIrpic: Ir(fbi)₂acac (19%, 1%, 10 nm)/TCTA:FIrpic (19%, 7.5 nm) obtained by experimental optimization are used in device modeling. A pair of 40-nm-thick NPB HTL and B3PyPB ETL can be selected by considering an optimal combination of high power efficiency and preferable CIE color

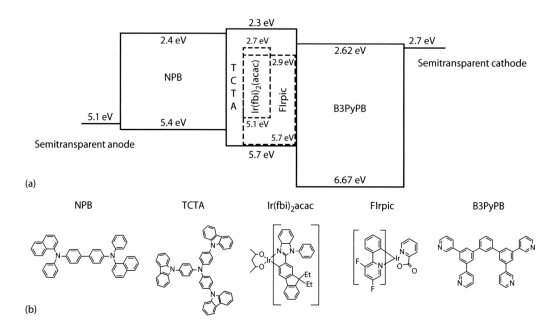

(a)

(b)

FIGURE 6.32
(a) Schematic energy diagram and (b) chemical structures of functional materials used in phosphorescent semi-transparent WOLEDs.

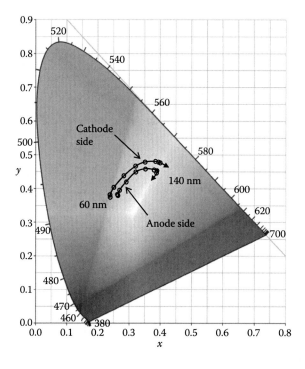

FIGURE 6.33
(See color insert.) Effect of organic layer thickness variation on CIE color coordinates of anode and cathode emissions in semitransparent WOLED. In this design, a pair of 40-nm-thick NPB HTL and B3PyPB ETL is chosen by considering an optimal combination of high power efficiency and preferable CIE color coordinates.

coordinates [113]. A combination of LiF (1.0 nm)/Al (1.5 nm) cathode contact is used to facilitate the efficient electron injection [114]. A compound semitransparent cathode of LiF (1.0 nm)/Al (1.5 nm)/Ag (15 nm)/NPB (50 nm) was optimized by the optical simulation. The use of a 15-nm-thick Ag in the cathode stack helps improve the lateral conductivity of the electrode. The layers of LiF (1.0 nm)/Al (1.5 nm)/Ag (15 nm)/NPB (50 nm) can be fabricated easily by sequential thermal evaporation in the same vacuum system, serving as an efficient semitransparent cathode for application in semitransparent WOLEDs.

Symmetrical and bidirectional emission characteristics, for example, the EL spectra and the power efficiency, can then be realized by optimizing the optical outcoupling characteristics at the anode and cathode sides of the semitransparent WOLEDs. To illustrate the point, a set of semitransparent WOLEDs fabricated with the same cathode of LiF (1.0 nm)/ Al (1.5 nm)/Ag (15 nm)/NPB (50 nm) but different anode structures of (a) ITO/MoO$_3$, (b) ITO/Ag (7.5 nm)/MoO$_3$, (c) ITO/Ag (10.0 nm)/MoO$_3$, and (d) ITO/Ag (12.5 nm)/MoO$_3$ was fabricated. A semitransparent WOLED made with an anode of ITO/MoO$_3$ served as a control device for comparison studies. The corresponding EL characteristics of different semitransparent WOLEDs, measured at a current density of 10 mA/cm^2, are plotted in Figure 6.34. It can be seen clearly that the control semitransparent WOLED has an obvious preferential emission due to the asymmetrical optical properties at the anode and cathode, with EL intensity from the anode side almost two times stronger than that of the cathode side. The EL intensities and the CIE coordinates of the bidirectional illumination in a semitransparent WOLED can be varied by adjusting the optical outcoupling characteristics from both sides. The experimental results agree with the theoretical simulation

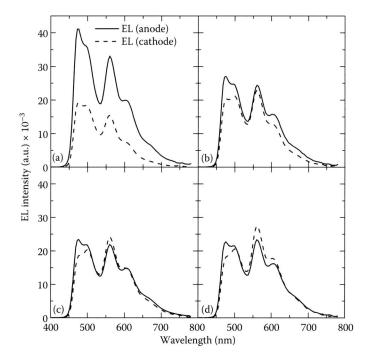

FIGURE 6.34
EL spectra from the anode and the cathode sides of the semitransparent WOLEDs, made with different anode structures of (a) ITO/MoO$_3$, (b) ITO/Ag (7.5 nm)/MoO$_3$, (c) ITO/Ag (10.0 nm)/MoO$_3$, and (d) ITO/Ag (12.5 nm)/ MoO$_3$, measured at a current density of 10 mA/cm^2.

in showing that a pair of an ITO (80 nm)/Ag (10 nm)/MoO$_3$ (2.5 nm) anode and an LiF (1.0 nm)/Al (1.5 nm)/Ag (15 nm)/NPB (50 nm) cathode is a suitable transparent electrode choice for symmetrical and bidirectional WOLEDs. This device design yields comparable CIE color coordinates and an almost balanced luminance brightness from the anode and cathode, measured at the current density of 10 mA/cm^2.

Figure 6.35 presents the power efficiency as a function of the luminance for a set of structurally identical semitransparent WOLEDs, fabricated with different anodes of (a) ITO/MoO$_3$, (b) ITO/Ag (7.5 nm)/MoO$_3$, (c) ITO/Ag (10 nm)/MoO$_3$, and (d) ITO/Ag (12.5 nm)/MoO$_3$. It can be seen that there is an obvious deviation in the power efficiency of the EL emission measured from the anode and cathode sides. For example, at a luminance of 100 cd/m^2, a power efficiency of ~17.3 lm/W was measured from the anode side. However it yielded only ~7.1 lm/W from the cathode side measured at the same brightness. A more than two times difference in the power efficiency of the EL emission is mainly due to the optically asymmetrical transparent electrodes used. A more balanced power efficiency of the bidirectional emission can be realized by optimizing the optical outcoupling characteristics at the anode and cathode sides of the semitransparent WOLEDs. A semitransparent WOLED fabricated with a pair of an ITO (80 nm)/Ag (10 nm)/MoO$_3$ (2.5 nm) anode and an LiF

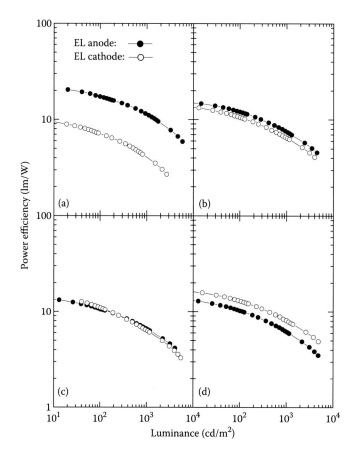

FIGURE 6.35
Comparison of power efficiency as a function of the luminance for a set of structurally identical semitransparent WOLEDs, fabricated with different anodes of (a) ITO/MoO$_3$, (b) ITO/Ag (7.5 nm)/MoO$_3$, (c) ITO/Ag (10 nm)/ MoO$_3$, and (d) ITO/Ag (12.5 nm)/MoO$_3$.

(1.0 nm)/Al (1.5 nm)/Ag (15 nm)/NPB (50 nm) cathode possesses an almost identical power efficiency of the EL emission from both sides, measured at a luminance range from 10^0 to 10^4 cd/m^2. The semitransparent WOLEDs with perfect symmetrical and bidirectional illumination characteristics thus developed offer new features and design freedoms for application in planar diffused lighting. A summary of the power efficiency and CIE color coordinates of a set of semitransparent WOLEDs that were measured at 100 and 1000 cd/m^2 is given in Table 6.2.

The EL emission of a semitransparent WOLED, made with a pair of physically symmetrical anode and cathode of Alq$_3$/Ag (22 nm)/Alq$_3$/Ag (22 nm) [108], has a strong angular dependence due to the microcavity effect. It is known that the resonant wavelength of a cavity mode can be described by the Fabry–Perot cavity condition [115,116]:

$$\sum_i \frac{4\pi \cdot d_i \cdot n_i(\lambda)}{\lambda} - \varphi_{\text{cathode}}(0,\lambda) - \varphi_{\text{anode}}(0,\lambda) = 2m\pi \tag{6.6}$$

where λ is the emission wavelength, $\varphi_{\text{cathode}}(0, \lambda)$ and $\varphi_{\text{anode}}(0, \lambda)$ are the wavelength-dependent phase changes in the reflection at the organic–anode and organic–cathode interfaces at normal incidence, m is the mode number ($m = 0$ was used in our simulation), and $n_i(\lambda)$ and d_i denote the refractive index and thickness of organic layers in the cavity structure. In addition to the angular-dependent EL emission, an unbalanced illumination from the anode and cathode sides of a semitransparent WOLED is often observed. This is attributed to the asymmetric light outcoupling properties at the organic–anode and organic–cathode interfaces in the semitransparent WOLEDs.

Compared with the high refractive index of Ag, ITO and NPB have relatively lower refractive indices of 1.7–2.1 in the visible wavelength range. Therefore, the use of an ITO (80 nm)/Ag (10 nm)/MoO$_3$ (2.5 nm) anode and an LiF (1.0 nm)/Al (1.5 nm)/Ag (15 nm)/NPB (50 nm) cathode creates a pair of electric and optical symmetrical transparent electrodes for semitransparent WOLEDs. In addition to the symmetrical EL characteristics discussed earlier, it is found that EL emission from both sides of the semitransparent WOLEDs thus developed does not have an obvious angular dependence. Figure 6.36 plots the normalized EL spectra, measured from both sides of a semitransparent WOLED with an ITO (80 nm)/

TABLE 6.2

Summary of Power Efficiency and CIE Color Coordinates of a Set of Semitransparent WOLEDs Fabricated with Different Anodes of (a) ITO/MoO$_3$, (b) ITO/Ag (7.5 nm)/MoO$_3$, (c) ITO/Ag (10 nm)/MoO$_3$, and (d) ITO/Ag (12.5 nm)/MoO$_3$

Semitransparent WOLEDs		Power Efficiency (lm/W) Measured at 100 cd/m^2/1000 cd/m^2	CIE Coordinates Measured at 100 cd/m^2/1000 cd/m^2
(a)	Anode	17.3/11.3	(0.34, 0.42)/(0.33, 0.42)
	Cathode	7.1/4.0	(0.31, 0.43)/(0.30, 0.43)
(b)	Anode	12.0/7.5	(0.35, 0.43)/(0.34, 0.43)
	Cathode	10.6/6.5	(0.35, 0.45)/(0.34, 0.45)
(c)	Anode	10.6/6.6	(0.36, 0.43)/(0.35, 0.43)
	Cathode	11.1/6.4	(0.38, 0.46)/(0.37, 0.46)
(d)	Anode	10.3/6.2	(0.36, 0.43)/(0.35, 0.43)
	Cathode	13.0/8.1	(0.38, 0.46)/(0.37, 0.46)

Note: EL characteristics measured at 100 and 1000 cd/m^2 are also listed for comparison.

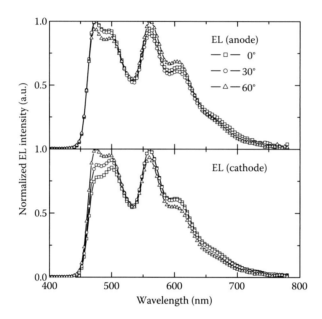

FIGURE 6.36
Normalized EL spectra measured for a semitransparent WOLED fabricated with an ITO (80 nm)/Ag (10 nm)/ MoO₃ (2.5 nm) anode and an LiF (1.0 nm)/Al (1.5 nm)/Ag (15 nm)/NPB (50 nm) cathode at different viewing angles of 0°, 30°, and 60°.

Ag (10 nm)/MoO₃ (2.5 nm) anode and an LiF (1.0 nm)/Al (1.5 nm)/Ag (15 nm)/NPB (50 nm) cathode. It can be seen clearly that there is no obvious shift in the EL emission measured at different viewing angles of 0°, 30°, and 60°. This suggests that the semitransparent WOLED has a weak microcavity effect and is beneficial for application in lighting.

In Figure 6.36, it shows that the EL spectra measured from the anode and cathode sides have a slight change in the relative intensity of blue and orange contributions in the two-color system at different angles. This could be attributed to the weak microcavity effect. Although a pair of an ITO (80 nm)/Ag (10 nm)/MoO₃ (2.5 nm) anode and an LiF (1.0 nm)/ Al (1.5 nm)/Ag (15 nm)/NPB (50 nm) cathode is the most suitable for semitransparent WOLED, a weak microcavity effect can be expected. Following Equation 6.6, the resonant wavelength of the semitransparent WOLED is around 550 nm, which does not overlap with the EL emission of 478 nm (from blue phosphorescent FIrpic dopant) and 562 nm (from orange phosphorescent Ir(fbi)₂acac dopant). This explains well that the EL spectra from both sides of a weak microcavity semitransparent WOLED are less sensitive to viewing angles, as shown in Figure 6.36.

The visible light transparency of a semitransparent WOLED of glass/ITO (80 nm)/Ag (10 nm)/MoO₃ (2.5 nm)/NPB (40 nm)/EML/B3PyPB (40 nm)/LiF (1.0 nm)/Al (1.5 nm)/Ag (15 nm)/NPB (50 nm) was measured. The measured (solid symbols) and simulated (open symbols) visible light transparency of the same device are plotted in Figure 6.37a. The experimental results agree well with the simulation in showing that the semitransparent WOLED has an average transmission of >50% in visible light wavelength range. Figure 6.37 panels b and c are the pictures taken for the semitransparent WOLED emitting light and illustrating the semitransparent feature without power. The external quantum efficiency (EQE) is also determined from the EL measurement. It is found that the EQE (sum of the EL emission from both sides) measured for a set of semitransparent WOLEDs, as shown in Figure 6.34a

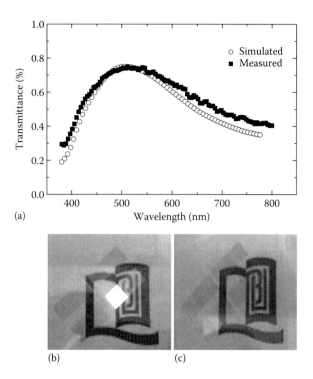

FIGURE 6.37

(a) Comparison of measured (solid symbols) and simulated (open symbols) visible light transparency of a semi-transparent WOLED of glass/ITO (80 nm)/Ag (10 nm)/MoO$_3$ (2.5 nm)/NPB (40 nm)/EML/B3PyPB (40 nm)/ LiF (1.0 nm)/Al (1.5 nm)/Ag (15 nm)/NPB (50 nm). Pictures taken for the device emitting light (b) and illustrating the semitransparent feature without power (c).

through d, is about 10%. This is comparable to the EQE measured for a control bottom-emission WOLED. This suggests that a pair of transparent anode and cathode used is both optically and electrically favorable for an efficient operation of the semitransparent WOLEDs.

6.6 Conclusions

OLED technology has significant advantages in large-area electronics, allowing for an astonishing amount of electronic complexity to be integrated onto lightweight, flexible substrates for a broad range of applications. OLEDs are self-emissive, free of UV and IR emissions, and have a high color-rendering-index, continuous spectrum with a potential of more cost-effective fabrication processes than those presently available for displays and lighting. The design and fabrication flexibility provided by the organic semiconductors and processes also have significant cost benefit, making OLED technology possible for new applications in automation brightness control, information displays, monitoring systems, portable electronics, sensors, and lighting. The new transparent WOLED lighting can be almost invisible by day and can emit a pleasant diffused light at night, allowing the surface light source to shine in both directions, an exciting new lighting technology that could bring novel device concepts and a new edge of area lighting.

References

1. K.L. Chopra, S. Major, and D.K. Pandya. Transparent conductors—A status review. *Thin Solid Films* 102: 1–46, 1983.
2. I. Hamburg, and C.G. Granvist. Evaporated Sn-doped In$_2$O$_3$ films: Basic optical properties and applications to energy-efficient windows. *J Appl Phys* 60: R123–R159, 1986.
3. B.H. Lee, I.G. Kim, S.W. Cho, and S.H. Lee. Effect of process parameters on the characteristics of indium tin oxide thin film for flat panel display application. *Thin Solid Films* 302: 25–30, 1997.
4. J.Q. Hu, F.R. Zhu, J. Zhang, and H. Gong. A room temperature indium tin oxide/quartz crystal microbalance gas sensor for nitric oxide. *Sens Actuators B* 93: 175–180, 2003.
5. J.S. Kim, M. Granström, R.H. Friend, N. Johansson, W.R. Salaneck, R. Daik, W.J. Feast, and F. Cacialli. Indium–tin oxide treatments for single- and double-layer polymeric light-emitting diodes: The relation between the anode physical, chemical, and morphological properties and the device performance. *J Appl Phys* 84: 6859–6870, 1998.
6. I.D. Parker. Carrier tunneling and device characteristics in polymer light-emitting diodes. *J Appl Phys* 75: 1656–1666, 1994.
7. C.C. Wu, C.I. Wu, J.C. Sturm, and A. Kahn. Surface modification of indium tin oxide by plasma treatment: An effective method to improve the efficiency, brightness, and reliability of organic light emitting devices. *Appl Phys Lett* 70: 1348–1350, 1997.
8. B. Schröder. Thin-film technology based on hydrogenated amorphous-silicon. *Mater Sci Eng A* 139: 319–333, 1991.
9. F.R. Zhu, T. Fuyuki, H. Matsunami, and J. Singh. Assessment of combined TCO/metal rare contact for thin film amorphous silicon solar cells. *Sol Energy Mater Solar Cells* 39: 1–9, 1995.
10. Y.F. Yan, S.B. Zhang, and S.T. Pantelides. Control of doping by impurity chemical potentials: Predictions for p-type ZnO. *Phys Rev Lett* 86: 5723–5726, 2001.
11. G. Xiong, J. Wilkinson, B. Mischuck, S. Tüzemen, K.B. Ucer, and R.T. Williams. Control of p- and n-type conductivity in sputter deposition of undoped ZnO. *Appl Phys Lett* 80: 1195–1197, 2002.
12. A. Salehi. The effects of deposition rate and substrate temperature of ITO thin films on electrical and optical properties. *Thin Solid Films* 324: 214–218, 1998.
13. K.R. Zhang, F.R. Zhu, C.H.A. Huan, and A.T.S. Wee. Effect of hydrogen partial pressure on optoelectronic properties of indium tin oxide thin films deposited by radio frequency magnetron sputtering method. *J Appl Phys* 86: 974–980, 1999.
14. K.R. Zhang, F.R. Zhu, C.H.A. Huan, A.T.S. Wee, and T. Osipowicz. Indium-doped zinc oxide films prepared by simultaneous rf and dc magnetron sputtering. *Surf Interface Anal* 28: 271–274, 1999.
15. J.K. Sheu, Y.K. Su, G.C. Chi, M.J. Jou, and C.M. Chang. Effects of thermal annealing on the indium tin oxide Schottky contacts of n-GaN. *Appl Phys Lett* 72: 3317–3319, 1999.
16. S. Major, and K.L. Chopra. Indium-doped zinc-oxide films as transparent electrodes for solar-cells. *Sol Energy Mater* 17: 319–327, 1988.
17. J. Hu, and R.G. Gordon. Atmospheric pressure chemical vapor deposition of gallium doped zinc oxide thin films from diethyl zinc, water, and triethyl gallium. *J Appl Phys* 72: 5381–5392, 1992.
18. Y. Takahashi, S. Okada, R.B.H. Tahar, K. Nakano, T. Ban, and Y. Ohya. Dip-coating of ITO films. *J Non-Cryst Solids* 218: 129–134, 1997.
19. K. Nishio, T. Sei, and T. Tsuchiya. Preparation and electrical properties of ITO thin films by dip-coating process. *J Mater Sci* 31: 1761–1766, 1996.
20. H.S. Kwok, X.W. Sun, and D.H. Kim. Pulsed laser deposited crystalline ultrathin indium tin oxide films and their conduction mechanisms. *Thin Solid Films* 335: 299–302, 1998.
21. H. Kim, A. Piqué, J.S. Horwitz, H. Mattoussi, H. Murata, Z.H. Kafafi, and D.B. Chrisey. Indium tin oxide thin films for organic light-emitting devices. *Appl Phys Lett* 74: 3444–3446, 1999.

22. C. May, and J. Strümpfel. ITO coating by reactive magnetron sputtering—Comparison of properties from DC and MF processing. *Thin Solid Films* 351: 48–52, 1999.

23. S. Honda, M. Watamori, and K. Oura. The effects of oxygen content on electrical and optical properties of indium tin oxide films fabricated by reactive sputtering. *Thin Solid Films* 281–282: 206–208, 1996.

24. G.L. Harding, and B. Window. DC magnetron reactively sputtered indium–tin-oxide films produced using argon–oxygen–hydrogen mixtures. *Solar Energy Mater* 20: 367–379, 1990.

25. S. Ishibashi, Y. Higuchi, Y. Ota, and K. Nakamura. Preparation and post-annealing effects on the optical properties of indium tin oxide thin films. *J Vac Sci Technol A* 8: 1399–1402, 1990.

26. I. Baía, M. Quintela, L. Mendes, P. Nunes, and R. Martins. Performances exhibited by large area ITO layers produced by rf magnetron sputtering. *Thin Solid Films* 337: 171–175, 1999.

27. L.J. Meng, and M.P. dos Santos. Properties of indium tin oxide films prepared by rf reactive magnetron sputtering at different substrate temperature. *Thin Solid Films* 322: 56–62, 1998.

28. P. Thilakan, and J. Kumar. Studies on the preferred orientation changes and its influenced properties on ITO thin films. *Vacuum* 48: 463–466, 1997.

29. L.J. Meng, and M.P. dos Santos. Properties of indium tin oxide (ITO) films prepared by r.f. reactive magnetron sputtering at different pressures. *Thin Solid Films* 303: 151–155, 1997.

30. M. Buchanan, J.B. Webb, and D.F. Williams. Preparation of conducting and transparent thin films of tin-doped indium oxide by magnetron sputtering. *Appl Phys Lett* 37: 213–215, 1980.

31. S. Honda, A. Tsujimoto, M. Watamori, and K. Oura. Depth profiling of oxygen concentration of indium tin oxide films fabricated by reactive sputtering. *Jpn J Appl Phys* 33: L1257–L1260, 1994.

32. M. Katiyar, Y.H. Yang, and J.R. Ableson. Hydrogen–surface reactions during the growth of hydrogenated amorphous silicon by reactive magnetron sputtering: A real time kinetic study by in situ infrared absorption. *J Appl Phys* 77: 6247–6256, 1995.

33. W.G. Haines, and H.R. Bube. Effects of heat treatment on the optical and electrical properties of indium–tin oxide films. *J Appl Phys* 49: 304–307, 1978.

34. S. Ray, R. Banerjee, N. Basu, A.K. Batabyal, and A.K. Barua. Properties of tin doped indium tin oxide thin films prepared by magnetron sputtering. *J Appl Phys* 54: 3497–3501, 1983.

35. Y. Ohhata, F. Shinoki, and S. Yoshida. Optical properties of r.f. reactive sputtered tin-doped In_2O_3 films. *Thin Solid Films* 59: 255–261, 1979.

36. Y. Shigesato, Y. Hayashi, A. Masui, and T. Haranou. The structural changes of indium–tin oxide and a-WO_3 films by introducing water to the deposition processes. *Jpn J Appl Phys* 30: 814–819, 1991.

37. L.J. Meng, A. Macarico, and R. Martins. Study of annealed indium tin oxide-films prepared by rf reactive magnetron sputtering. *Vacuum* 46: 673–680, 1995.

38. R. Banerjee, and D. Das. Properties of tin oxide films prepared by reactive electron beam evaporation. *Thin Solid Films* 149: 291–301, 1987.

39. S.A. Van Slyke, C.H. Chen, and C.W. Tang. Organic electroluminescent devices with improved stability. *Appl Phys Lett* 69: 2160–2162, 1996.

40. J.S. Kim, R.H. Friend, and F. Cacialli. Improved operational stability of polyfluorene-based organic light-emitting diodes with plasma-treated indium–tin–oxide anodes. *Appl Phys Lett* 74: 3084–3086, 1999.

41. Y. Kurosaka, N. Tada, Y. Ohmori, and K. Yoshino. Improvement of electrode/organic layer interfaces by the insertion of monolayer-like aluminum oxide film. *Jpn J Appl Phys* 37: L872–L875, 1998.

42. K. Sugiyama, H. Ishii, Y. Ouchi, and K. Seki. Dependence of indium–tin–oxide work function on surface cleaning method as studied by ultraviolet and x-ray photoemission spectroscopies. *J Appl Phys* 87: 295–298, 2000.

43. D.J. Milliron, I.G. Hill, C. Shen, A. Kahn, and J. Schwartz. Surface oxidation activates indium tin oxide for hole injection. *J Appl Phys* 87: 572–576, 2000.

44. F. Steuber, J. Staudigel, M. Stössel, J. Simmerer, and A. Winnacker. Reduced operating voltage of organic electroluminescent devices by plasma treatment of the indium tin oxide anode. *Appl Phys Lett* 74: 3558–5580, 1999.

45. Y. Park, V. Choong, Y. Gao, B.R. Hsieh, and C.W. Tang. Work function of indium tin oxide transparent conductor measured by photoelectron spectroscopy. *Appl Phys Lett* 68: 2699–2701, 1996.
46. I.H. Campbell, J.D. Kress, R.L. Martin, D.L. Smith, N.N. Barashkov, and J.P. Ferraris. Controlling charge injection in organic electronic devices using self-assembled monolayers. *Appl Phys Lett* 71: 3528–3530, 1997.
47. M.G. Mason, L.S. Hung, C.W. Tang, S.T. Lee, K.W. Wong, and M. Wang. Characterization of treated indium–tin–oxide surfaces used in electroluminescent devices. *J Appl Phys* 86: 1688–1692, 1999.
48. J.A. Chaney, and P.E. Pehrsson. Work function changes and surface chemistry of oxygen, hydrogen, and carbon on indium tin oxide. *Appl Surf Sci* 180: 214–226, 2001.
49. H.Y. Yu, X.D. Feng, D. Grozea, Z.H. Lu, R.N.S. Sodhi, A.-M. Hor, and H. Aziz. Surface electronic structure of plasma-treated indium tin oxides. *Appl Phys Lett* 78: 2595–2597, 2001.
50. N.D. Popovich, S.S. Wong, S. Ufer, V. Sakhrani, and D. Paine. Electron-transfer kinetics at ITO films—Influence of oxygen plasma. *J Eelectrochem Soc* 150: H255–H259, 2003.
51. F.R. Zhu, C.H.A. Huan, K.R. Zhang, and A.T.S. Wee. Investigation of annealing effects on indium tin oxide thin films by electron energy loss spectroscopy. *Thin Solid Films* 359: 244–250, 2000.
52. R.B.H. Tahar, T. Ban, Y. Ohya, and Y. Takahashi. Tin doped indium oxide thin films: Electrical properties. *J Appl Phys* 83: 2631–2645, 1998.
53. P.K.H. Ho, M. Granstrom, R.H. Friend, and N.C. Greenham. Ultrathin self-assembled layers at the ITO interface to control charge injection and electroluminescence efficiency in polymer light-emitting diodes. *Adv Mater* 10: 769–774, 1998.
54. C. Yan, M. Zharnikov, A. Golzhauser, and M. Grunze. Preparation and characterization of self-assembled monolayers on indium tin oxide. *Langmuir* 16: 6208–6215, 2000.
55. S.F.J. Appleyard, and M.R. Willis. Electroluminescence: Enhanced injection using ITO electrodes coated with a self assembled monolayer. *Opt Mater* 9: 120–124, 1998.
56. J.E. Malinsky, G.E. Jabbour, S.E. Shaheen, J.D. Anderson, A.G. Richter, T.J. Marks, N.R. Armstrong, B.K. Pulak Dutta, and N. Peyghambarian. Self-assembly processes for organic LED electrode passivation and charge injection balance. *Adv Mater* 11: 227–231, 1999.
57. F.R. Zhu, B.L. Low, K.R. Zhang, and S.J. Chua. Lithium-fluoride-modified indium tin oxide anode for enhanced carrier injection in phenyl-substituted polymer electroluminescent devices. *Appl Phys Lett* 79: 1205–1207, 2001.
58. H. Tang, F. Li, and J. Shinar. Bright high efficiency blue organic light-emitting diodes with Al_2O_3/Al cathodes. *Appl Phys Lett* 71: 2560–2562, 1997.
59. Z.B. Deng, X.M. Ding, S.T. Lee, and W.A. Gambling. Enhanced brightness and efficiency in organic electroluminescent devices using SiO_2 buffer layers. *Appl Phys Lett* 74: 2227–2229, 1999.
60. X.M. Ding, L.M. Hung, L.F. Cheng, Z.B. Deng, X.Y. Hou, C.S. Lee, and S.T. Lee. Modification of the hole injection barrier in organic light-emitting devices studied by ultraviolet photoelectron spectroscopy. *Appl Phys Lett* 76: 2704–2706, 2000.
61. Y.Q. Li, L.W. Tan, X.T. Hao, K.S. Ong, F.R. Zhu, and L.S. Hung. Flexible top-emitting electroluminescent devices on polyethylene terephthalate substrates. *Appl Phys Lett* 86: 153508-153508-3, 2005.
62. R.H. Jordan, A. Dodabalapur, and R.E. Slusher. Efficiency enhancement of microcavity organic light emitting diodes. *Appl Phys Lett* 69: 1997–1999, 1996.
63. H. Kim, J.S. Horwitz, W.H. Kim, S.B. Qadri, and Z.H. Kafafi. Anode material based on Zr-doped ZnO thin films for organic light-emitting diodes. *Appl Phys Lett* 83: 3809–3811, 2003.
64. X. Jiang, F.L. Wong, M.K. Fung, and S.T. Lee. Aluminum-doped zinc oxide films as transparent conductive electrode for organic light-emitting devices. *Appl Phys Lett* 83: 1875–1877, 2003.
65. W.H. Kim, A.J. Mäkinen, N. Nikolov, R. Shashidhar, H. Kim, and Z.H. Kafafi. Molecular organic light-emitting diodes using highly conducting polymers as anodes. *Appl Phys Lett* 80: 3844–3846, 2002.
66. Z.Y. Xie, L.S. Hung, and F.R. Zhu. A flexible top-emitting organic light–emitting diode on steel foil. *Chem Phys Lett* 38: 691–696, 2003.

67. C. Fou, O. Onitsuka, M. Ferreira, M.F. Rubner, and B.R. Hsieh. Fabrication and properties of light-emitting diodes based on self-assembled multilayers of poly(phenylene vinylene). *J Appl Phys* 79: 7501–7509, 1996.

68. A.N. Krasnov. High-contrast organic light-emitting diodes on flexible substrates. *Appl Phys Lett* 80: 3853–3855, 2002.

69. G. Gustafsson, G.M. Treacy, Y. Cao, F. Klavetter, N. Colaneri, and A.J. Heeger. The "plastic" led: A flexible light-emitting device using a polyaniline transparent electrode. *Synth Met* 57: 4123–4127, 1993.

70. G. Gu, P.E. Burrows, S. Venkatesh, and S.R. Forrest. Vacuum-deposited, nonpolymeric flexible organic light-emitting devices. *Opt Lett* 22: 172–174, 1997.

71. R. Paetzold, K. Heuster, D. Henseler, S. Roeger, G. Weittmann, and A. Winnacker. Performance of flexible polymeric light-emitting diodes under bending conditions. *Appl Phys Lett* 82: 3342–3344, 2003.

72. K.R. Zhang, F.R. Zhu, C.H.A. Huan, and A.T.S. Wee. Indium tin oxide films prepared by radio frequency magnetron sputtering method at a low processing temperature. *Thin Solid Films* 376: 255–263, 2000.

73. A. Plichta, A. Weber, and A. Habeck. Ultra thin flexible glass substrates. *Mater Res Soc Symp Proc* 769: H9.1.1–H9.1.9, 2003.

74. A.B. Chwang, M.R. Rothman, S.Y. Mao, R.H. Hewitt, M.S. Weaver, J.A. Silvermail, K. Rajan, M. Hack, J.J. Brown, X. Chu, L. Moro, T. Krajewski, and N. Rutherford. Thin film encapsulated flexible organic electroluminescent displays. *Appl Phys Lett* 83: 413–415, 2003.

75. C.C. Wu, S.D. Theiss, G. Gu. M.H. Lu, J.C. Sturm, S. Wagner, and S.R. Forrest. Integration of organic LEDs and amorphous Si TFTs onto flexible and lightweight metal foil substrates. *IEEE Electron Device Lett* 18: 609–612, 1997.

76. C.F. Madigan, M.H. Lu, and J.C. Sturm. Improvement of output coupling efficiency of organic light-emitting diodes by backside substrate modification. *Appl Phys Lett* 76: 1650–1652, 2000.

77. L.S. Hung, C.W. Tang, and M.G. Mason. Enhanced electron injection in organic electroluminescence devices using an Al/LiF electrode. *Appl Phys Lett* 70: 152–154, 1997.

78. Y.Q. Li, J.X. Tang, Z.Y. Xie, L.S. Hung, and S.S. Lau. An efficient organic light-emitting diode with silver electrodes. *Chem Phys Lett* 386: 128–131, 2004.

79. J.X. Tang, Y.Q. Li, L.S. Hung, and C.S. Lee. Photoemission study of hole-injection enhancement in organic electroluminescent devices with Au/CF_x anode. *Appl Phys Lett* 84: 73–75, 2004.

80. H. Riel, S. Karg, T. Beielein, B. Ruhstaller, and W. Rieß. Phosphorescent top-emitting organic light-emitting devices with improved light outcoupling. *Appl Phys Lett* 82: 466–469, 2003.

81. H. Becker, S.E. Burns, N. Tessler, and R.H. Friend. Role of optical properties of metallic mirrors in microcavity structures. *J Appl Phys* 81: 2825–2829, 1997.

82. J. Lewis, S. Grego, B. Chalamala, E. Vrik, and D. Temple. Highly flexible transparent electrodes for organic light-emitting diode-based displays. *Appl Phys Lett* 85: 3450–3452, 2004.

83. L.S. Hung, and J.K. Madathil. Reduction of ambient-light-reflection in organic light-emitting devices. U.S. Patent: 6,429,451 B1, 2002.

84. L.S. Hung, and J. Madathil. Reduction of ambient light reflection in organic light-emitting diodes. *Adv Mater* 13: 1787–1790, 2001.

85. O. Renault, O.V. Salata, M. Etchells, P.J. Dobson, and V. Christou. A low reflectivity multilayer cathode for organic light-emitting diodes. *Thin Solid Films* 379: 195–198, 2000.

86. H. Aziz, Y.F. Liew, H.M. Grandin, and Z.D. Popovic. Reduced reflectance cathode for organic light-emitting devices using metalorganic mixtures. *Appl Phys Lett* 83: 186–188, 2003.

87. A.N. Krasnov. ELDs rise on organic wings. *Inf Display* 18: 18–21, 2002.

88. P.G. Hofstra, and A. Krasnov. Organic electroluminescent devices. U.S. Patent: 6,411,019 B1, 2002.

89. Z.Y. Xie, and L.S. Hung. High-contrast organic light-emitting diodes. *Appl Phys Lett* 84: 1207–1209, 2004.

90. X.D. Feng, R. Khangura, and Z.H. Lu. Metal–organic–metal cathode for high-contrast organic light-emitting diodes. *Appl Phys Lett* 85: 497–499, 2004.

91. J. Heikenfeld, and A.J. Steckl. Contrast enhancement in black dielectric electroluminescent devices. *IEEE Trans Electron Devices* 49: 1348–1352, 2002.
92. F.R. Zhu, and J. Singh. On the optical design of thin film amorphous silicon solar cells. *Solar Energy Mater Solar Cells* 31: 119–131, 1993.
93. F.R. Zhu, P. Jennings, J. Cornish, G. Hefter, and K. Luczak. Optimal and optical design of thin-film photovoltaic devices. *Solar Energy Mater Solar Cells* 49: 163–169, 1997.
94. Y. Sun, and S.R. Forrest. High-efficiency white organic light emitting devices with three separate phosphorescent emission layers. *Appl Phys Lett* 91: 263503-1–263503-3, 2007.
95. S.J. Su, E. Gonmori, H. Sasabe, and J. Kido. Highly efficient organic blue- and white-light-emitting devices having a carrier- and exciton-confining structure for reduced efficiency roll-off. *Adv Mater* 20: 4189–4194, 2008.
96. S. Reineke, F. Lindner, G. Schwartz, N. Seidler, K. Walzer, B. Lüssem, and K. Leo. White organic light-emitting diodes with fluorescent tube efficiency. *Nature* 459: 234–238, 2009.
97. H. Sasabe, J. Takamatsu, T. Motoyama, S. Watanabe, G. Wagenblast, N. Langer, O. Molt, E. Fuchs, C. Lennartz, and J. Kido. High-efficiency blue and white organic light-emitting devices incorporating a blue iridium carbene complex. *Adv Mater* 22: 5003–5007, 2010.
98. G. Gu, V. Bulovic, P.E. Burrows, S.R. Forrest, and M.E. Thompson. Transparent organic light emitting devices. *Appl Phys Lett* 68: 2606–2608, 1996.
99. A. Yamamori, S. Hayashi, T. Koyama, and Y. Taniguchi. Transparent organic light-emitting diodes using metal acethylacetonate complexes as an electron injective buffer layer. *Appl Phys Lett* 78: 3343–3345, 2001.
100. C.H. Chung, Y.W. Ko, Y.H. Kim, C.Y. Sohn, H.Y. Chu, and J.H. Lee. Improvement in performance of transparent organic light-emitting diodes with increasing sputtering power in the deposition of indium tin oxide cathode. *Appl Phys Lett* 86: 09350-1–09350-3, 2005.
101. S.F. Hsu, C.C. Lee, S.W. Hwang, and C.H. Chen. Highly efficient top-emitting white organic electroluminescent devices. *Appl Phys Lett* 86: 253508-1–253508-3, 2005.
102. C.J. Lee, R.B. Pode, J.I. Han, and D.G. Moon. Ca/Ag bilayer cathode for transparent white organic light-emitting devices. *Appl Surf Sci* 253: 4249–4253, 2007.
103. K.S. Yook, S.O. Jeon, C.W. Joo, and J.Y. Lee. Transparent organic light emitting diodes using a multilayer oxide as a low resistance transparent cathode. *Appl Phys Lett* 93: 013301-1–013301-3, 2008.
104. T.Y. Zhang, L.T. Zhang, W.Y. Ji, and W.F. Xie. Transparent white organic light-emitting devices with a LiF/Yb:Ag cathode. *Opt Lett* 34: 1174–1176, 2009.
105. J. Lee, S. Hofmann, M. Furno, M. Thomschke, Y.H. Kim, B. Lüssem, and K. Leo. Systematic investigation of transparent organic light-emitting diodes depending on top metal electrode thickness. *Org Elect* 12: 1383–1388, 2011.
106. H. Cho, J.M. Choi, and S.H. Yoo. Highly transparent organic light-emitting diodes with a metallic top electrode: The dual role of a Cs_2CO_3 layer. *Opt Exp* 19: 1113–1121, 2011.
107. T.H. Gil, C. May, S. Scholz, S. Franke, M. Toerker, H. Lakner, K. Leo, and S. Keller. Origin of damages in OLED from Al top electrode deposition by DC magnetron sputtering. *Org Elect* 11: 322–331, 2010.
108. W.Y. Ji, L.T. Zhang, W.F. Xie, H.H. Zhang, G.Q. Lui, and J.B. Yao. Semitransparent white organic light-emitting devices with symmetrical electrode structure. *Org Elect* 12: 2192–2197, 2011.
109. Q. Wang, J.Q. Ding, D.G. Ma, Y.X. Cheng, and L.X. Wang. Highly efficient single-emitting-layer white organic light-emitting diodes with reduced efficiency roll-off. *Appl Phys Lett* 94: 103503-1–103503-3, 2009.
110. K.L. Tong, S.W. Tsang, K.K. Tsung, S.C. Tse, and S.K. So. Hole transport in molecularly doped naphthyl diamine. *J Appl Phys* 102: 093705-1–093705-4, 2007.
111. H. Sasabe, E. Gonmori, T. Chiba, Y.J. Li, D. Tanaka, S.J. Su, T. Takeda, Y.J. Pu, K. Nakayama, and J. Kido. Wide-energy-gap electron-transport materials containing 3,5-dipyridylphenyl moieties for an ultra high efficiency blue organic light-emitting device. *Chem Mater* 20: 5951–5953, 2008.
112. H. Sasabe, and J. Kido. Multifunctional materials in high-performance OLEDs: Challenges for solid-state lighting. *Chem Mater* 23: 621–630, 2011.

113. W.H. Choi, H.L. Tam, F.R. Zhu, D.G. Ma, H. Sasabe, and J. Kido. High performance semitransparent phosphorescent white organic light emitting diodes with bi-directional and symmetrical illumination. *Appl Phys Lett* 102: 153308-1–153308-4, 2013.

114. Y.H. Chen, J.S. Chen, D.G. Ma, D.H. Yan, L.X. Wang, and F.R. Zhu. High power efficiency tandem organic light-emitting diodes based on bulk heterojunction organic bipolar charge generation layer. *Appl Phys Lett* 98: 243309-1–243309-3, 2011.

115. W.Y. Ji, L.T. Zhang, T.Y. Zhang, G.Q. Liu, W.F. Xie, S.Y. Liu, H.Z. Zhang, L.Y. Zhang, and B. Li. Top-emitting white organic light-emitting devices with a one-dimensional metallic–dielectric photonic crystal anode. *Opt Lett* 34: 2703–2705, 2009.

116. D.P. Puzzo, M.G. Helander, P.G.O'Brien, Z.B. Wang, N. Soheilnia, N. Kherani, Z.H. Lu, and G.A. Ozin. Organic light-emitting diode microcavities from transparent conducting metal oxide photonic crystals. *Nano Lett* 11: 1457–1462, 2011.

7

Vapor-Deposited Organic Light-Emitting Devices

Michael S. Weaver

CONTENTS

7.1 Vapor-Deposited Organic Light-Emitting Devices

The first observations of electroluminescence from organic materials were made in the 1950s [1]. Interest in this phenomenon was fueled by the work of Pope et al. [2], who observed electroluminescence from single crystals of anthracene. A voltage was applied between silver paste electrodes that were placed on the opposite sides of an anthracene crystal. Bright blue emission was observed. However, these devices were impractical for commercial applications because of the high voltages required for their operation and the need for exceptionally pure crystals. Owing to innovations in the fields of vacuum and thin-film coating technologies in 1982, Vincett et al. [3] fabricated light-emitting devices based on evaporated thin films of anthracene. These were an order of magnitude thinner than the single crystals used by Pope et al. By using very thin vapor-deposited films, high fields are generated across the devices at much lower voltages, thereby substantially improving the device efficiency.

By taking advantage of modern thin-film organic vapor deposition techniques, the field of organic electroluminescence or organic light-emitting devices (OLEDs) gained new impetus in the 1980s. In 1987, Tang and VanSlyke [4,5] reported a key breakthrough in terms of improved device performance. They separated the functions of charge transport and emission in a device by introducing monopolar charge transport layers. This device, along with the chemical structures of the materials used, is shown in Figure 7.1. OLED

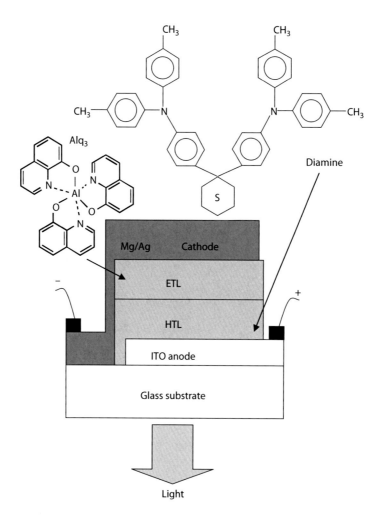

FIGURE 7.1

Two-layer vapor-deposited OLED first demonstrated by Tang and VanSlyke. Diamine acts as the hole-transporting layer; Alq$_3$ acts as the electron-transporting and emitting layer. External quantum efficiency, ~1%. (From Tang, C.W. and VanSlyke, S.A., *Appl. Phys. Lett.*, 51, 913, 1987.)

architectures are now far more complicated compared with the early devices. The idea of using multilayer [6] structures to separate the functions of charge injection, charge transport, and emission can be extended to three layers or more by using an emitter layer sandwiched between a hole-injection layer (HIL) or hole-transport layer (HTL) and an electron-transport layer (ETL).

Another important early advance made by Tang et al. [7] is the use of fluorescent doping, i.e., the addition of a small percentage of an emissive fluorescent material into a host matrix. This can be used to alter the color of emission, in addition to improving the efficiency and the lifetime of devices. The technique of simultaneously vapor depositing a host and emissive dopant material is now widely used in the field of OLEDs.

During the last 50 years, remarkable improvements in the performance of vapor-deposited OLEDs have been made. Operating voltages have been decreased from a few kilovolts to a few volts; at the same time, efficiencies have exceeded 100 lm/W. These improvements in device performance have made commercial displays based on vapor-deposited OLEDs viable. This technology is now competing with liquid crystal displays (LCDs) in an expanding flat-panel display marketplace.

Early displays [8] based on vapor-deposited OLEDs were simple alphanumeric devices. More recently, there have been rapid increases in the complexity of these devices. In 1996, Pioneer Corporation demonstrated a monochrome 64 × 256 pixel OLED display [9] that was subsequently developed into a product and was incorporated into automobile stereos (see Figure 7.2). Today, full-color, high-resolution vapor-deposited OLED displays as large as 55″–56″ have been developed [10–13].

In addition to display applications, researchers are also exploring the use of vapor-deposited organic materials in devices such as photovoltaics [14,15], organic lasers [16], and organic thin-film transistors (TFTs) [17,18].

The aim of this chapter is to give the reader a broad overview of the field of vapor-deposited small-molecule OLEDs. It is beyond the scope of this chapter to cover every aspect of these devices. However, key references are given throughout the text for those readers who are interested in delving more deeply into this topic. Section 7.2 describes the key elements of a typical OLED. Alternative device architectures are also briefly described. Section 7.3 describes the typical fabrication methods and materials used in the construction of vapor-deposited OLEDs. Section 7.4 describes the physics of an OLED in addition to the improvement of the performance over time made through advances in device architectures and materials. Section 7.5 discusses OLED displays, and Section 7.6 looks at the future exciting possibilities for the field of vapor-deposited organic devices.

FIGURE 7.2
World's first commercially available OLED display (1999), manufactured by Pioneer Corporation and incorporated into automotive stereos.

7.2 Vapor-Deposited Organic Light-Emitting Device Architectures

Figure 7.1 shows one of the simplest possible OLED architectures. It consists of two organic layers sandwiched between a transparent anode and a metal cathode. When a voltage is applied across the device, electrons and holes are injected into the organic layers at the electrodes and move through the device under the influence of a high applied electrical field (of the order of 10^6 V/cm). These charges can then combine via a coulombic interaction, forming excited molecular species, some of which may then emit light as they decay to their ground state. This process is described in more detail in Section 7.4.

The following sections will first describe the major components of a typical bottom-emitting OLED: transparent anode, organic layers, and metal cathode. Alternative device architectures are also briefly described.

7.2.1 Anode

The anode in OLED devices is typically indium–tin oxide (ITO), because it is a stable, transparent, and highly conductive material (for more details, see Chapter 6). It is also easily patterned using standard lithographic techniques to define different emitting regions or pixels on a substrate. A thin inorganic barrier layer (e.g., SiO_2) is often used between the polished glass substrate and the ITO film to prevent the migration of ions from the glass into the subsequently deposited device. Various groups have looked at different types of ITO (e-beam and sputtered) and the effects of annealing in O_2 and other surface treatments [19–22]. In the case of transparent conductive metal oxides (e.g., ITO), the stoichiometry and thickness of the oxide is controlled to realize a specific transparency, outcoupling constant, and conductivity of the film. Typically in OLED displays, the film is 50- to 200-nm thick. Additionally, as OLEDs are field-dependent devices, and the organic films are of the same order of magnitude in thickness as the anode, the surface roughness of the anode must be low. This is to prevent shorting of the OLEDs and nonuniform light emission. Typically, a surface roughness of <2 nm root mean square (rms) is required. In the case of flexible substrates, achieving this uniformity can be problematic. For example, commercially available heat-stabilized polyethylene terephthalate (PET) substrates have surface asperities up to 150 nm [23]. Here, additional planarizing layers must be employed before deposition of the anode. See Section 7.6 for more discussion on flexible substrates and devices.

Semitransparent layers of various polymers [24], metals [25], and metal oxides [26] have been used as alternatives to ITO. However, ITO is the present industry standard owing to its favorable properties and its widespread use in the more mature and widespread LCD industry.

7.2.2 Organic Materials

A wide range of small-molecule organic materials have been used in vapor-deposited OLEDs (see Chapter 3). Some of the requisite properties of the materials used in vapor deposition are as follows:

- The organic materials must evaporate without decomposing during the fabrication process. The typical deposition temperature range is between 150°C and 450°C. Factors that contribute to the ultimate temperature used in addition to the physical properties of the material include the vacuum pressure, source-to-substrate geometry, and required deposition rate.

- The organic material, once deposited, must form high-quality, defect-free films with precisely controlled thicknesses, typically in the region of 5–200 nm.
- The films must be stable for long periods. Some materials, particularly those with a low glass transition temperature (T_g), may crystallize over time [27,28]. Crystallization may be accelerated when the temperature of the thin film is raised during device operation [29–31]. Therefore, a high T_g is often desirable for the long-term durability of the OLED, e.g., $T_g > 110°C$.

The morphology of the organic films can be assessed using optical microscopy (in particular, techniques such as Nomarski microscopy, atomic force microscopy, and surface profiling techniques). It should also be noted that the purity of the organic materials used is of crucial importance for efficient charge transport and emission in addition to the lifetime of the OLED.

As described earlier, the multiple organic layers in OLED devices must collectively fulfill three main functions: charge injection, charge transport through the device, and light emission. Although these functions may be separated and materials optimized for each property independently, choosing an appropriate combination of emitting and charge transporting films to avoid exciplex formation [32], for example, is preferred for building efficient devices [33].

7.2.3 Cathode

The cathode is typically a low-work-function metal or metal alloy that facilitates the injection of electrons into the organic material adjacent to it. A low-work-function metal is necessary to minimize the barrier to electron injection into the adjacent organic material that typically has lowest unoccupied molecular orbital (LUMO) levels of ~3 eV. Low-work-function metals that have been used include In, Mg, Ca, and Ba. However, all of these are reactive under ambient conditions and require careful encapsulation after completion of the device. Tang and VanSlyke [4] found that, while Mg is a difficult material to deposit reproducibly onto many organic materials, evaporating a small amount of Ag from a second source during the Mg evaporation (coevaporation is typically in a 10:1 ratio) resulted in more reproducible results and much improved film formation. Murayama et al. [34] found that aluminum cathodes with a codeposited 0.1% Li concentration resulted in increased device efficiencies and reduced drive voltages. Recent work has increased our understanding of these systems, particularly with respect to the way lithium diffuses. The result has been that the use of multilayer Al/Li/Al cathodes [35] and the use of lithium-doped organic electron injecting layers [36] each provide for good injection characteristics. Researchers have also reported that the presence of a thin insulating layer such as LiF [37] or the lithium quinolate complex 8-hydroxyquinolinolatolithium (Liq) [38] between the cathode and the organic layers leads to improved device performance. LiF/Al, Liq/Al, or Li_2O/Al [39] cathodes are now widely used in the vapor-deposited OLED community.

7.2.4 Alternative Device Architectures

Described above are examples of typical materials used in traditional bottom-emitting OLEDs, i.e., in a device architecture where light exits through a transparent anode that is in intimate contact with a transparent substrate. Alternative device architectures are also possible, for example, top-emitting OLEDs (TOLEDs). Here, the cathode is transparent, thereby allowing light to exit though the top of the device. Cathodes in such a device architecture can be formed by using a thin metal contact [40], a thin metal (e.g., ~10 nm of Mg/Ag) in conjunction with a conductive metal oxide (e.g., ITO) [41], or a metal-free electrode [42]. In a top-emitting configuration, the anode is usually a high-reflectance, high-work-function material or composite of materials (e.g., Pt, Ag/ITO, or Al/Ni). Transparent OLEDs can also be made if the anode itself is also transparent [43]. For further discussion of top-emitting devices, see Section 7.5.

Other device architectures include inverted OLEDs. Here, the cathode is in intimate contact with the substrate. The organic layers are then deposited onto the cathode in reverse order, i.e., starting with the electron-transport material and ending with the HIL. The device is completed with an anode contact. In this case, as above, one of the electrodes is transparent, and light exits from the device through that contact. For example, Bulovic et al. [44] fabricated a device in which Mg/Ag was the bottom contact and ITO was the top electrode. The advantage of this type of architecture is that it allows for easier integration with n-type TFTs (see Section 7.5 for a discussion of active matrix–driven OLED displays).

7.3 Device Fabrication

Figure 7.3 shows a simple schematic example of the basic steps required to fabricate a bottom-emitting, vapor-deposited OLED test pixel similar to the device shown in Figure 7.1.

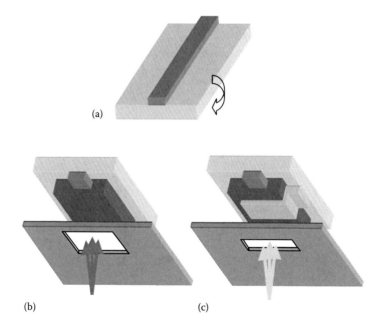

(a)

(b)   (c)

FIGURE 7.3
Schematic representation of the basic steps required in fabricating a vapor-deposited OLED test pixel: (a) anode patterning via lithography, and deposition of the (b) organic and (c) metal cathode layers through shadow masks.

7.3.1 Anode Preparation

All substrate preparations before the deposition of the organic materials are carried out in a clean room environment to minimize particulates on the anode surface. OLEDs are typically of the order of 80–300-nm thick, and therefore any particles on the anode surface may lead to breaks in the continuity of the film, resulting in electrical shorts or visible defects. In the case of simple test devices, the clean ITO substrate is first patterned using standard photolithographic techniques to form transparent anodes as shown in Figure 7.3a. For a display, the substrate topography is more complicated and may include transistors or bus lines, etc. This situation is discussed in detail in Section 7.5. Even in the case of these more complicated structures, as, for example, in a 210 dots per inch (dpi) full-color display where the anode pitch is 40 μm, patterned anodes are easily realized using conventional lithographic techniques. Once the anode is patterned, the substrate is then transferred to a vacuum deposition chamber. Before the growth of the organic layers, the ITO surface is usually subjected to some additional pretreatment step, e.g., oxygen plasma [45] or ultraviolet (UV) ozone exposure [7]. This pretreatment, before device fabrication, leads to an enhanced device performance that is believed to be due, in part, to a resultant increase in the work function of the ITO surface [20].

The typical thickness of the ITO layer is 80–300 nm, which is of the same order of thickness as the total organic layer. Therefore, devices risk the possibility of nonuniform emission or shorts developing due to the presence of a thinner organic film and higher electrical fields at the sides of the patterned ITO electrodes (substrate to ITO step).

This is especially true when fabricating displays, where high yields are necessary. To counter this problem, some manufacturers add additional complexity to the pixel by adding a grid material. Here, again using lithography, a nonconductive material is deposited over the pixel and patterned so as to leave an opening to the ITO that will eventually define the active area of the pixel. This method is used to limit the possibility of shorting at the edges of the ITO-to-substrate step. Photoresist materials or inorganic oxides, e.g., SiO_2, are often used as grid material. The fabrication conditions are chosen so as to give the grid edge a tapered profile so that when the organic and metal films are deposited into the opening, shadowing effects do not occur at the edges of the film, leaving nonuniform thinner films in these regions that may be prone to electrical shorts.

7.3.2 Deposition of Organic Layers

The organic layers are deposited sequentially onto the anode through a shadow mask (see Figure 7.3b) by thermal evaporation [46] from resistively heated source boats in a high-vacuum environment. In research tools, a given evaporation chamber will often contain several different materials. In manufacturing tools, multiple evaporation chambers are often used with a limit of only one or two types of material per chamber to minimize the risk of cross contamination between evaporation sources. The shadow mask defines the area on the substrate over which the organic layers are deposited. If the final device is a full-color display, at least three shadow masks are typically needed to independently define the red, green, and blue pixels. For high-resolution displays, high accuracy is required in positioning the shadow mask and depositing the three colors. For example, a 210-dpi display has a subpixel pitch of 40 μm. Therefore, to prevent an overlap of one organic layer color over to the next pixel and to maximize the fill factor or active area of the display, a mask alignment accuracy of 5 μm is stated as a requirement by most manufacturers. This alignment process is typically achieved through the aid of (i) a moving stage, (ii) cameras to locate the correct relative position of mask to substrate, and (iii) an electromagnet to fix the mask in position. Masking accuracy, in addition to reduced feathering, i.e., poor definition of the edge of the organic film, is improved by using contact alignment. The alignment procedure usually takes place in a chamber that is remote from the organic deposition chambers. Often additional masks are used to define the deposition area for the organic layers common to all the three colors, e.g., the charge injection and transport layers. In this case and when monochrome devices are fabricated, the alignment tolerances are not as stringent.

The organic deposition sources are made of a variety of materials, including ceramics (e.g., boron nitride, aluminum oxide, and quartz) or metallic boats (e.g., tantalum or molybdenum). Deposition is carried out in high vacuum at a base pressure of around 10^{-7} to 10^{-8} torr. The vacuum conditions under which OLEDs are fabricated are extremely important [47–49], and evaporation rates, monitored using quartz oscillators, are typically in the range 0.01–0.5 nm/s in research and development tools. In manufacturing, higher rates or multiple sources may be used to reduce the tact time.

Depending on the organic material in question, evaporation takes place from either the liquid or the solid state. In general, this occurs from the liquid state if, at the melting point, the material does not reach a vapor pressure $>10^{-3}$ torr. For example, 4,4′-*bis*[*N*-(1-napthyl)-*N*-phenyl-amino] biphenyl (NPB), a hole-transport material used in OLEDs, melts before evaporating under typical growth rates. Most of the other organic materials achieve a higher vapor pressure well before their melting point, and therefore evaporation is achieved via sublimation of the material.

The deposition geometry, i.e., the relative position of the substrate and sources, is of paramount importance when fabricating OLEDs (see Figure 7.4). For a substrate on a plane at

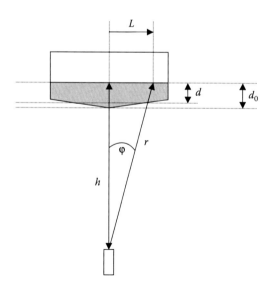

FIGURE 7.4
Schematic diagram of the deposition geometry for a substrate on a plane at right angles to a point evaporation source. d is the deposited film thickness; d_0 is the thickest point of the film and h is the substrate-to-source distance at this point (i.e., the shortest distance between the source and substrate); r is the source-to-substrate distance; φ is the angle subtended by the line joining source and substrate measurement points and the normal from the substrate plane to source position; and L is the distance between the source and the point on the substrate at which d is being measured.

right angles to a point evaporation source, the deposited film thickness d can be expressed by the following relationship:

$$d = \frac{M_e \cos \varphi}{4\pi\rho r^2} \qquad (7.1)$$

where M_e is the total evaporated mass, r is the source-to-substrate distance, ρ is the deposited film density, and φ is the angle subtended by the line joining source and substrate measurement points and the normal from the substrate plane-to-source position.

OLEDs typically operate under applied fields of 10^6 V/cm and are of the order of 100 nm in thickness. The current through the device and the light emitted from the device, above the threshold voltage, increase with a power law dependence on the applied field. Therefore, any significant variations in film thickness across a display will lead to visible nonuniformities in emission. In commercial evaporators, to reduce production costs, manufacturers maximize the number of displays per substrate and the size of the substrate. They typically require a uniformity tolerance of ~3% to 5% for the organic films across a substrate. A complete discussion of the relationship between film uniformity and deposition geometry is beyond the scope of this chapter. For an excellent discussion of this and the various methods used to maximize film uniformities, see Milton Ohring's discussion [50]. In brief, the relationship between the film thickness d (see Figure 7.4) at a given point on a substrate located parallel to a surface source can be expressed as

$$\frac{d}{d_0} = \frac{1}{[1+(L/h)^2]^{2*}} \qquad (7.2)$$

where d_0 is the thickest point of the film and h is the substrate-to-source distance at this point (i.e., the shortest distance between the source and the substrate). L is the distance between the source and the point on the substrate at which d is measured. As is apparent from this relationship, as the substrate size increases, to minimize the difference between d and d_0, i.e., to obtain a uniform film within acceptable tolerances, the source-to-substrate distance (h) must be increased. Using this simple relationship to achieve a uniformity tolerance of 5% across a 300-mm substrate, a throw distance from a point source of 80 cm is needed. As a result of the larger throw distances used in manufacturing, consequently to achieve reasonable deposition rates, the deposition source temperatures are often significantly higher perhaps by as much as 50°C. This places additional stability requirements on the materials that are used in manufacturing. Typically, a manufacturer expects to continuously deposit from a source for 24–168 h at this elevated temperature. Therefore, the materials used must be able to withstand this environment without degradation that may reduce the quality of the deposited material or impact the ability to adequately access stable deposition conditions; that is, the material's purity must be high and its decomposition temperature significantly above the evaporation temperature.

Other manufacturers have developed linear sources [51]. Here, the deposition source is a long linear evaporation source that stretches beyond the width of the substrate. Depending on the design, either the source or the substrate is translated relative to each other in one direction. The advantage of this geometry is that the substrates no longer need to be at such a great distance from the source to achieve the requisite film uniformity. In addition, substrate rotation is no longer necessary. Materials use and deposition rates are expected

---

* For an ideal point source, the factorial is 3/2.

to be higher using this technique. In the past, organic thin-film nonuniformities of <5% over 300 × 400 mm substrates at a throw distance of 100 mm have been demonstrated [52]. Today, manufacturers are planning to deposit OLEDs on substrates as large as Gen 8 (2880 × 3130 mm) [10].

7.3.3 Alternative Organic Deposition Techniques

Various other techniques can also be used to deposit small-molecule materials as thin films. Organic molecular beam deposition [52], spin coating [53–55], doctor blade, ink-jet printing [56,57], thermal transfer [58], Langmuir–Blodgett films [59,60], chemical vapor deposition [61], and self-assembled films [62] have also been used to deposit thin organic layers. Organic vapor phase deposition (OVPD), in particular, looks to be a very exciting new technique for depositing small-molecule materials [63]. Vapor-phase epitaxy is used extensively in the inorganic semiconductor industry, and the analogous technique OVPD is now being developed for organic materials.

To deposit materials using OVPD, the organic material is vaporized and then an inert carrier gas such as nitrogen is used to dilute and carry the material to the cooled substrate. The walls of the deposition chamber are kept hot, thus preventing the organic material from condensing on anything but the substrate surface. Two regimes of growth can occur. The first is mass transport, where the growth rate is determined by the arrival rate of the source materials at the substrate. The second is the kinetic regime, in which, due to viscous flow effects, a boundary layer forms at the substrate owing to the slower moving gas. The material must diffuse through this region, and thus the growth rate in this regime is dependent on the concentration of the material and the depth and shape of the boundary layer.

OVPD has a potentially significant advantage over vapor deposition in manufacturing OLEDs, as OVPD is far less wasteful in terms of materials usage. In addition to the obvious advantage that OVPD has in terms of manufacturing compatibility, it also has the potential to be a more accurate deposition tool. This is because the deposition rate is controlled by the mass flow controllers rather than by the temperature of a thermal source. When thermally evaporating organic materials, small changes in temperature have significant effects on the deposition rate owing to the rapid changes in vapor pressure at the material's sublimation point. OVPD avoids this problem by controlling the flow of the carrier gas.

OVPD can be taken one step further in the form of organic vapor jet deposition (OVJP) [64]. With OVJP, the transport of the organic materials is the same as for OVPD. However, at the point of deposition onto the target substrate, the vapor is passed through one or more microscopic nozzles to form a highly collimated gas jet. The organic materials contained in the vapor jet can physisorb onto the cooled substrate. Translating the jet(s) relative to the substrate allows the deposited material to be patterned without the use of shadow masks, etc. Feature sizes of 16 μm have been demonstrated using this approach [65].

7.3.4 Deposition of Cathode

Metals and metal oxides used in the cathode are typically deposited at higher deposition rates than the organic layers. Layer thicknesses vary depending on the device architecture, e.g., 10 nm to 1 μm. Aluminum is often the first choice as a cathode material (see Section 7.2 for other cathode materials). Aluminum evaporates from a liquid phase. In many research tools, the primary method of deposition is via thermal evaporation from resistively heated sources. For aluminum deposition, boron nitride crucibles or tungsten wires are often used as sources. An e-beam evaporation source can also be used. This has the advantage of having a reduced risk of source contamination due to the amount of material in contact with the source boat in the liquid phase, in addition to the reduced contaminant level, e.g., oxygen content, of the deposited film as a consequence of the higher evaporation rates that can be realized. However, care must be taken to avoid the possibility of secondary electrons causing damage to the organic layers. Sputtering deposition techniques can also be used to deposit the cathode, depending on the material [66]. However, care has to be taken when employing this technique to avoid damage from radiation, charging, and heating.

The metal cathode is deposited onto the organic layers through a shadow mask (see Figure 7.3c). For active-matrix OLED (AMOLED) displays, a single unbroken cathode is often used over the entire display area.

7.3.5 Encapsulation of Organic Light-Emitting Device

The final step in the OLED fabrication process is encapsulation. This step is necessary to ensure a long device lifetime. OLEDs [4,67,68] built on glass substrates have been shown to have lifetimes (generally defined as the time taken to decay to half the initial luminance at constant current) of tens of thousands of hours [69,70]. There have been many proposed mechanisms for the decay in luminance; however, most theories agree that one of the dominant degradation mechanisms in unencapsulated OLEDs, which have far shorter lifetimes than encapsulated devices, is exposure of the organic–cathode interface to atmospheric oxygen and water. This leads to oxidation and delamination of the metal cathode [71,72] as well as potential chemical reactions within the organic layers. As most of the OLED works to date have been focused on the development and manufacture of glass-based displays, the degradation problem is ameliorated by sealing the display in an inert atmosphere, e.g., a nitrogen or argon glove box (<1 ppm water and oxygen), using a glass or metal lid attached by a bead of UV-cured epoxy resin [73]. A desiccant such as CaO or BaO is often included in the package to react with any residual water incorporated in the package or diffusing through the epoxy seal. In addition to encapsulation techniques using a lid, thin-film encapsulation techniques are also possible. For a more detailed description of these, see Section 7.6 or, for example, Lewis and Weaver [74].

7.4 Device Operation

Figure 7.5 shows a schematic example of the electroluminescent process in a typical two-layer OLED device architecture. When a voltage is applied to the device, five key processes must take place for light emission to occur from the device.

1. *Charge injection*: Holes must be injected from the anode into the HTL while electrons are injected from the cathode into the ETL.
2. *Charge transport*: The holes and electrons must move through the device under the influence of the applied electrical field. The mobility of holes in typical hole-transport organic materials is approximately 10^{-3} cm^2/(V s) [75]. For electrons, the mobility is usually one or more orders of magnitude lower [76].
3. *Exciton formation*: The holes and electrons must combine in the emitter region of the device via a coulombic interaction to form excitons [77] (neutral excited species); other excited states are also possible, such as excimer [78] or exciplex excited states [32,79].

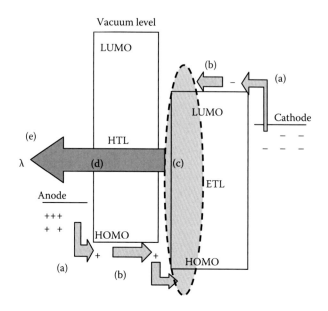

FIGURE 7.5
Schematic diagram of the light emission process in a typical two-layer OLED device architecture. For light emission to exit from the device, five key processes must take place: (a) charge injection of holes and electrons at the anode and cathode, (b) charge transport through the device, (c) exciton formation, (d) exciton decay, and (e) light emission.

4. *Exciton decay*: When an exciton decays radiatively, a photon is emitted. When the excitons form in fluorescent materials, radiative decay is limited to singlet excitons and emission occurs close to the recombination region [7] of the OLED owing to the relatively short lifetime of the excited state (of the order of 10 ns). For phosphorescent materials, emission can occur from triplet excitons. Because of the longer excited state lifetime (of the order of hundreds of nanoseconds), triplet excitons can diffuse further before decaying.

5. *Light emission*: Light is observed from photons that exit the OLED structure. Typically, many photons are lost owing to processes such as total internal reflection and self-absorption of the internal layers [80]. In typical bottom-emitting device architectures, only 20%–30% of the photons created exit the device through the front of the substrate.

Figure 7.6 shows typical current density–voltage–luminance (*J–V–L*) and emission characteristics of an OLED device. OLEDs have a similar electrical characteristic to that of a rectifying diode. In forward bias, the device starts with a small current at low voltages. In this region, charge carriers are injected into the device; however, little exciton formation, hence light emission, occurs. As the voltage is increased, the current quickly increases, obeying a power law dependence on the voltage. Here, many charges are injected into the device, moving through the charge-transport layers and then forming excited species that radiatively decay to produce light. The electroluminescence spectra from OLEDs are generally broad with full width half maxima usually >50 nm. The light emitted from the OLED is directly proportional to the current passing through it. Under reverse bias, there is a small leakage current and, in most cases, no light emission. Many groups have attempted to model the electrical and light-emitting behavior of OLEDs [81].

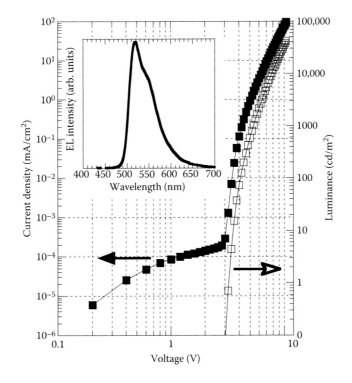

FIGURE 7.6
Typical current density (filled squares)–voltage–luminance (open squares) (*J–V–L*) and emission characteristics (inset) of an OLED device. This *J–V–L* data is from the device discussed later in Figure 7.8.

In general, a two-layer device structure is more efficient than single-layer architectures. There are two key reasons for this. First, each layer can be separately optimized for the injection and transport of one carrier type. Second, exciton formation and radiative decay take place close to the HTL–ETL interface away from the quenching sites at the organic–metal contacts.

To facilitate good charge transport in an OLED, the organic materials must satisfy three key requirements: they must have a high mobility for either electrons or holes, a good injection efficiency from the contact electrode, and suitable band offsets with other organic layers within the device. These processes are discussed in detail by, for example, Kalinowski [82] and Greenham and Friend [83].

For hole transport, many of the first materials used in OLEDs were originally developed for use in xerography. Various models have been proposed to describe charge transport in such materials [84,85]. One of the most common classes of material used is the arylamines, one common example of which is NPB (see Figure 7.7). This family of materials exhibits high hole mobilities [86] and a reasonable energy level (ionization potential) that is lined up with the work function of ITO (>4.7 eV), thus giving a relatively small barrier to hole injection. NPB, for example, has a hole mobility of 10^{-3} to 10^{-4} cm^2/(V s) and an ionization potential that defines the highest occupied molecular orbital (HOMO) level of 5.4 eV [75]. Often in small-molecule OLEDs, an additional HIL is added between the hole-transport material and the anode. This is designed to facilitate a smaller energy barrier to hole injection and to counter issues at this interface such as recrystallization and delamination of the organic layer at the anode, which can occur when thin films are deposited by vapor

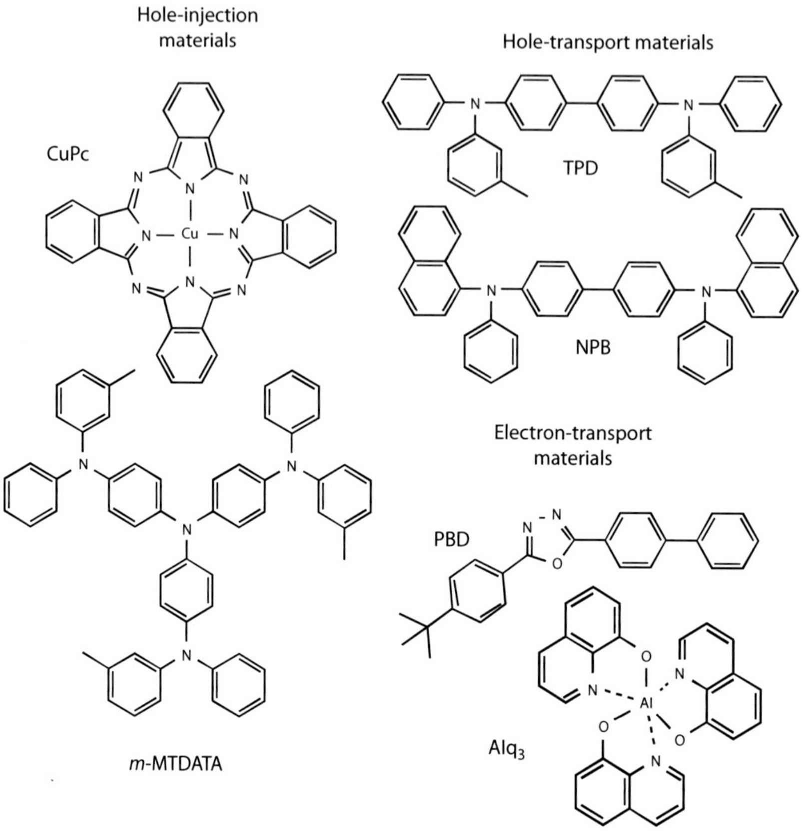

FIGURE 7.7
Examples of hole-injecting, hole-transporting, and electron-transporting materials.

deposition onto the anode surface. For example, copper phthalocyanine (CuPc), shown in Figure 7.7, has been frequently used on ITO anodes to increase the adhesion of subsequent organic layers and to increase device lifetime [87,88]. Thin layers of carbon [89] between the ITO and the HTL have also been shown to decrease the operating voltage and improve the device characteristics. In addition, thin fluorocarbon films on ITO have been used to enhance the stability of the anode interface [90].

The criteria for good electron-transport materials are that they should transport electrons, block holes, and have a small barrier to electron injection from the metal cathode. A commonly used ETL in vacuum-deposited OLEDs is tris-(8-hydroxyquinoline) aluminum (Alq_3), as shown in Figure 7.7. Alq_3, for example, has a LUMO energy level of 3 eV [75] and an electron mobility of ~5×10^{-5} cm²/(V s) [76].

7.4.1 Improving Device Efficiency

For a Lambertian emitting OLED source, where V is the operating voltage and η_{le} is the luminance efficacy (in cd/A), the power efficacy (η) is given by

$$\eta = \eta_{le}\,\pi/V \tag{7.3}$$

$$\eta_{le} = k\eta_{Int} \cdot \eta_{Out} \qquad (7.4)$$

where η_{Int} is the internal quantum efficiency (% excitons to photons), η_{Out} is the outcoupling efficiency (a measure of how many generated photons are emitted from the device), and k is a constant depending on the photopic response of the human eye. Hence

$$\eta = k\eta_{Int} \cdot \eta_{Out}\, \pi/V \qquad (7.5)$$

As a result, power efficacy is a function of the internal quantum efficiency, η_{Int}, the light extraction, η_{Out}, and the voltage, V. Thus, to improve device performance, advances in these three key areas are required. Examples of strategies used to maximize power efficacy are described below.

7.4.2 Improving Internal Quantum Efficiency

A wide range of dyes, many of which were originally developed as laser dyes, have been used as emitter materials in OLEDs. One criterion for emitter materials is a high photoluminescence (PL) efficiency with a specific desired color of emission depending on the application. However, high PL quantum efficiencies in dilute solutions do not always translate to high electroluminescence quantum efficiencies when incorporated into devices. This may be due to quenching via interactions with different molecules [91], oxygen, or the electrodes [92], or due to concentration quenching in the solid state (generally through the formation of aggregate states). One solution to this latter problem is to add substituent groups to the molecule so as to prevent aggregation by increasing the steric hindrance and, hence, limit the formation of aggregate states [93,94]. Unfortunately, this can lead to poor charge transport through the material, which in turn would lead to an increased device operating voltage and a poor power efficiency.

An elegant solution to this problem is to dope the emissive dye into an organic matrix or host. This effectively dilutes the concentration of the emissive dye (the dopant) and thus prevents aggregation. As long as the dopant is red shifted compared with the host, i.e., there is adequate overlap of the host emission and dopant absorption spectra to facilitate the Förster transfer of the excitons (in fluorescent materials), excitons formed in the host material will tend to migrate to the dopant before relaxation. This results in emission that is predominantly from the dopant. Tang et al. [7] found that adding a small percentage of 4-(dicyanomethylene)-2-methyl-6-(4-dimethylaminostyryl)-4H-pyran (DCM1) (see Figure 7.8) to an Alq$_3$ layer shifted the emission from green to orange–red. Doped devices are therefore fabricated by vapor depositing the host and dopant material at the same time. The ratio of host to dopant is determined by controlling the relative rate of evaporation of the two materials to form a doped layer within the device. The key benefit of this approach, which is now widely used within the industry [95], is that it allows the processes of charge transport, exciton formation, and emission to be optimized separately within the host and dopant materials. This also allows emitters to be used that would not otherwise form good films.

7.4.3 Improved Efficiency through Doping

The doping of the emissive layer in an OLED has been used extensively as a way of improving efficiency and lifetime, in addition to being used to modify the emission color. Tang et al. [7] first introduced fluorescent dyes, 3-(2-benzothiazolyl)-7-diethylaminocoumarin

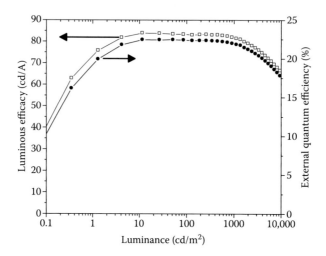

FIGURE 7.8
Luminous efficacy (open squares) and external quantum efficiency (filled circles) versus luminance for a PHOLED incorporating a high-efficiency green phosphorescent emitter.

(coumarin 540 or coumarin 6) and DCMs, as dopants in Alq_3 to improve the efficiency and color purity of devices (see Figure 7.8). Since then, a wide range of fluorescent dopants have been used in OLEDs [96,97]. The ground state of most materials has a singlet spin state $S = 0$. Emission of a photon in fluorescent materials conserves spin; therefore, only singlet $S = 0$ excited states typically emit light. Decay from the triplet $S = 1$ excited states is typically a nonradiative process for most organic materials, and thus these triplet excitons are lost from the perspective of light emission. The maximum possible internal quantum efficiency that can be obtained in an OLED using fluorescent materials is limited by the ratio of these excited states $S = 0$ and $S = 1$ or the so-called exciton singlet-to-triplet ratio, which is approximately 1:3 [98]. This limits fluorescent OLEDs to a maximum internal quantum efficiency of approximately 25%. Recently, phosphorescent OLEDs (PHOLEDs) [99], which incorporate heavy metal organometallic compounds as dopants, have surpassed this efficiency [100].

In a PHOLED system, all the singlet excited state (S_1) excitons may be converted into the triplet excited state (T_1) through intersystem crossing (S_1 to T_1) via the presence of a heavy metal atom. In these devices, the triplet states can decay to the singlet ground state (T_1 to S_0) and emit radiatively, enabling record high conversion efficiencies. The first generation of PHOLEDs contained platinum 2,3,7,8,12,13,17,18-octaethyl-12H,23H-porphyrin (PtOEP) as the phosphor [99] (see Figure 7.9). Porphine complexes possess long-lived triplet states that have been used in oxygen detection. This lifetime is reduced by addition of a platinum atom to the porphine ring due to the increased spin–orbit coupling. The result is an enhancement in efficiency due to the combined effect of forming 75% of the excitons directly as triplets in addition to the 25% that intersystem cross from the singlet excited state to the triplet excited state. The first published PtOEP devices had an external quantum efficiency of 4%. This was subsequently increased to 6% with the introduction of a blocking layer within the device structure [101]. Although at the time, this was an impressive device efficiency, it could only be realized at low drive currents and luminance levels. At high luminance levels due to the long-lived nature of the triplet excitons in PtOEP (>10 μs), the available dopant sites in the device at low dopant concentrations become saturated, which results in a roll-off in efficiency with increased current.

Later generations of PHOLEDs have improved considerably on the early promise of PtOEP. In organometallic complexes, the presence of a heavy metal atom, in addition to allowing intersystem crossing from the singlet S_1 to the triplet T_1 excited state on the organic ligands, can also participate in the transfer of an electron to an organic ligand. This is known as metal–ligand charge transfer (MLCT). The resultant excitons have a larger overlap with the metal atom than is the case with ligand excitons. The spin–orbit coupling is therefore enhanced, resulting in a mixing between the MLCT triplet and the singlet. To ensure high efficiency, particularly at high luminance levels, it is essential to minimize the triplet excited state lifetime. To achieve this, the MLCT triplet energy should be lower than that of the ligand. The $5d^6$ complexes that use Ir^{3+} provide for this possibility. PHOLEDs incorporating phosphorescent organometallic iridium compounds have exhibited green electroluminescence with high external quantum efficiencies [102,103]. Figure 7.8 shows an example of a high-efficiency green PHOLED. This emitter has an external quantum efficiency of 22%. Allowing for effects such as outcoupling, the internal quantum efficiency of such devices has been estimated to be close to 100%. Figure 7.9 contains examples of Ir^{3+} organometallic complexes that exhibit light emission spanning the visible spectrum from blue [104] through green [100] to red [105].

PHOLEDs also show excellent stability under display drive conditions. First-generation phosphorescent dopants such as PtOEP, *fac* tris(2-phenylpyridine)iridium (Ir(ppy)$_3$), and iridium(III)*bis*(2-phenylquinolyl-*N*,*C*^{2})acetylacetonate (PQ$_2$Ir(acac)) have demonstrated lifetimes of several thousands of hours [70,105]. Recent PHOLEDs have shown lifetimes in excess of 400,000 h at display brightness [106].

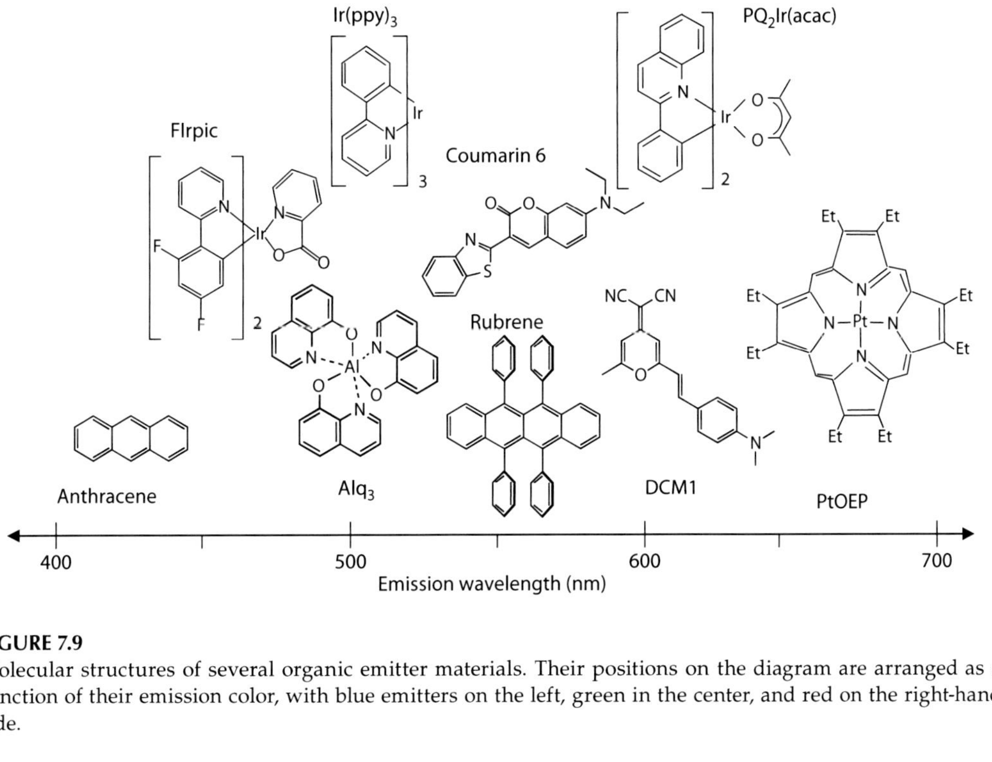

FIGURE 7.9
Molecular structures of several organic emitter materials. Their positions on the diagram are arranged as a function of their emission color, with blue emitters on the left, green in the center, and red on the right-hand side.

7.4.4 Improving Power Efficacy

In typical OLEDs, the applied voltage V is usually 2–5 V, when illuminated at 1000 cd/m². The voltage drop across the emission layer itself is usually 2–3 V, depending on the emission wavelength. The remaining voltage is dropped predominantly across the ETL, across the HTL, and at the heterojunction interfaces. Current transport in low-mobility organic films is space charge limited [107], and high electric fields are required to inject the necessary charge to generate the desired photon flux. Band misalignments at the heterojunction interfaces also result in voltage loss. However, the drive voltage can be significantly reduced by conductivity doping of the transport layers [108]. More recently, conductivity doping was demonstrated using green Ir(ppy)$_3$-doped PHOLEDs [109]. It was observed that the drive voltage necessary to produce 100 cd/m² was 2.65 V, i.e., only slightly higher than the voltage of the emitted photon V_λ. This device used p-type (tetrafluoro-tetracyanoquinodimethane, F$_4$-TCNQ) and n-type (Li) doping of the HTL and ETL, respectively.

7.4.5 Outcoupling

Once photons are created within conventional bottom-emitting OLEDs, the external quantum efficiency, i.e., the number of photons emitted from the viewing side (or glass substrate) per injected charge, is limited by a number of loss mechanisms. Photons can be lost through self-absorption of the organic layers, waveguiding within the device, surface plasmon modes, and absorption of the photons in the cathode. To a first approximation, the outcoupling efficiency, i.e., the fraction of light emitted by the device, χ, is given by

$$\chi = 1 - (1 - (1/n_i^2))^{1/2} \tag{7.6}$$

where n_i is the refractive index of the emissive layer [110]. For typical materials used in an OLED with $n_i = 1.7$, Equation 7.6 produces an estimated outcoupling efficiency of 19%. Most of the remaining light is waveguided in the substrate and the organic layers.

Hence, we find that today, the most significant limitation to the efficiency of OLEDs is the internal reflection of about 80% of the emitting light in the glass substrate. In this case, without light extraction enhancement outcoupling, $\eta_{ext} \sim 20\%$ presents a fundamental limit for devices with 100% internal efficiency.

However, Equation 7.6 is a very approximate and inaccurate, but often quoted, estimate. The relationship given above is only a first-order approximation, and the actual amount of light emitted is color, material, and thickness dependent. A more sophisticated analysis is needed to account for the coupling of excited states to the device cavity modes [111]. Methods employed to overcome efficiency limitations due to light trapping have primarily concentrated on expanding the escape cone of the substrate and suppressing the waveguide modes. These methods include introducing rough or textured surfaces [112], mesa structures [113], and lenses [114], and the use of reflecting surfaces or distributed Bragg reflectors [115,116]. Consequently, many of the methods used to improve LED outcoupling [117] have also been applied to OLEDs. For polymer LEDs (PLEDs) [118], it was shown that a corrugated substrate increased the light output by a factor close to 2 by Bragg scattering in the forward direction. A similar improvement was achieved by placing a single millimeter-sized hemispherical lens [119,120] on the substrate aligned with the OLED on its opposite surface. Also, shaping of the device into a mesa structure showed an increase of η_{ext} by a factor of 2 [121]. The incorporation of monolayers of silica spheres with diameters of 550 nm as a scattering medium in a device, or the positioning of these monolayers on the substrate, also showed enhanced light output [122]. Yamasaki et al. [123] showed that the external quantum efficiency can be doubled by incorporating a thin layer of a very-low-refractive-index silica aerogel ($n_i \sim 1.03$) in the device.

Another method of enhancing the outcoupling efficiency has been the use of an ordered array of microlenses [124]. In a conventional planar structure, the light generated in the OLED is either emitted externally or waveguided in the substrate or the ITO–organic layer, and lost (Figure 7.10). Substrate patterning destroys the substrate waveguide, redirecting the waveguiding modes externally, thus increasing the outcoupling efficiency. The lenses are produced using a simple fabrication process, and require no alignment with the OLEDs. Furthermore, the emission spectrum of the lensed OLEDs exhibits no angular dependence. In particular, the light output for high angles of observation with respect to the surface normal is considerably increased. Experimentally, the external quantum efficiency of an electrophosphorescent device is found to increase from 9.5% using a flat glass substrate to 14.5% using a substrate with a micromolded lens array.

Although a significant increase of η_{ext} was observed for the reported methods above, they are often accompanied by changes in the radiation pattern. This can lead to undesirable angular-dependent emission spectra or, in the case of displays, a blurring of the display as light emitted from neighboring pixels exits the display distant to the originating pixel. For a lighting application, this is not a major concern except from the perspective of added cost. However, for displays, the only significant manufacturable increase in the outcoupling scheme to date has been through the use of microcavity architectures in mobile applications where angular dependence issues are less problematic.

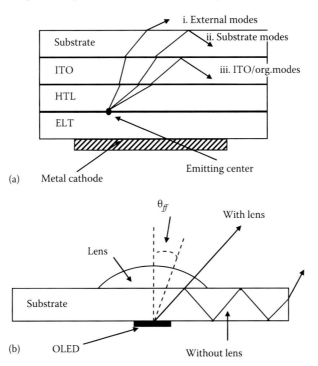

FIGURE 7.10
Outcoupling in an OLED. (a) Three radiative modes in an OLED: (i) external modes, (ii) substrate modes, and (iii) ITO–organic modes. (b) Attaching a lens to the backside of an OLED converts some of the light from substrate to external modes.

7.4.6 Lifetime

One issue that limited the early adoption of OLEDs in commercial products was device stability both during storage and in operation. Suggested causes of degradation include indium migration from the ITO anode [125], morphological instability of the organic materials [126], fixed charge accumulation within the device [127], damage to the electrodes, and the formation of nonemissive dark spots [73,128,129]. Water and oxygen are known to cause problems in OLEDs. Therefore, a great deal of effort has been directed toward the encapsulation of devices. Encapsulation is typically carried out under a nitrogen atmosphere inside a glove box.

In addition to extrinsic environmental causes of degradation in OLEDs, some groups have explored the limitations of the individual device materials to transport charge and emit light. For example, Aziz et al. [130] have proposed that in simple Alq_3 devices, hole transport through the Alq_3 layer is the dominant cause of device degradation due to the instability of the Alq_3^+ cationic species. Other causes of device degradation include the thermal stability of materials [131]. However, this issue applies mainly to the high-temperature operation of OLEDs. Useful overviews of the factors affecting device reliability are given by Forrest et al. [132], Popovic and Aziz [133], and So and Kondakov [134].

A minimum target set by many manufacturers for test pixel architectures before adoption in a commercial display is a lifetime of 10,000 h at display brightness. However, the lifetime of a similar pixel obtained in a display is often less than this value owing to additional complications such as pixel yield and added heat load [31] to the pixel from the display. Therefore, the realization of a higher performance specification in a pixel is often required before a new material or device architecture is transferred to a commercial display. Great strides in lifetime have been made within the OLED community in recent years, with several manufacturers demonstrating lifetimes in excess of 20,000 h [106], where lifetime is defined as a loss of 5% of the initial luminance at display brightness.

7.5 Vapor-Deposited Organic Light-Emitting Device Displays

One of the most obvious markets for thin-film vapor-deposited organic materials is in flat-panel displays [135], a market currently dominated by LCDs. During the last two decades, a great improvement in the lifetime and efficiency of OLEDs has been achieved. Today, OLED displays can be found in many applications, such as automobile stereos, mobile phones, digital cameras, and TVs. However, to exploit the advantages of the technology fully, it is necessary to pattern the OLEDs to form monochrome, or more preferentially, full-color displays. This section will consider the difficulties involved in addressing such displays (either passively or actively) and the variety of patterning methods that can be used to produce full-color displays.

7.5.1 Passive Addressing Schemes

Displays based on OLEDs may be addressed either passively or actively [136], and the drive requirements are different in each case. In passive-matrix addressing, the display is addressed one line at a time; thus, if a display has 480 lines, then a pixel can only be emitting for 1/480th of the time. This has important implications for the materials and device structure chosen because it means that to achieve an average luminance of 200 cd/m^2, the instantaneous luminance on the pixel must be (200×480) 96,000 cd/m^2. This leads to high current densities in the metal tracks, requiring expensive driver integrated circuits (ICs) capable of handling high currents and generating significant heating problems. It is therefore necessary to design materials and device structures capable of operating under pulsed operation. By successfully operating their displays at a duty ratio of 0.002, researchers at Idemitsu Kosan were able to conclude that it is possible to operate passively addressed panels with <500 lines at video rate [137].

Previously, there had been a perception in the OLED research field that PHOLEDs, although highly desirable for use in active-matrix applications because of their high efficiency, were unsuitable for passive-matrix applications. The first generation of PHOLEDs contained PtOEP as the phosphorescent dopant. However, these devices had a spectral dependence on applied current. This was a result of the long radiative lifetime of PtOEP (>10 μs). As the applied current was increased, dopant sites within the device became saturated, resulting in the inability of excitons to transfer from the host material. Some of these excitons then decayed, emitting light characteristic of the host material. The devices also exhibit a steep roll-off in efficiency that has been mainly interpreted as the result of triplet–triplet annihilation. However, the latest generations of PHOLEDs have efficiency roll-offs at high drives that are comparable to or better than the conventional fluorescent OLED or PLED devices [138], and they show higher efficiency at the high luminance values needed in a passively addressed display. The main advantages of passive addressing are that it is relatively inexpensive and that, in principle, it is possible to drive small- to medium-sized displays. However, large-area, high-resolution displays (>5″) are problematic because of the high current densities and hence increased power consumption (I^2R) described above. For a more detailed discussion of these issues, see, for example, Gu and Forrest [139].

For passive-matrix OLED (PMOLED) displays, the individual pixels have to be addressed via addressable row and column electrodes, hence the necessity to pattern the metal cathode into rows. The thickness of these rows is dependent on the display resolution. If we again consider a 150-dpi display, a cathode row pitch of 169 μm is required. It is possible on a small scale to pattern these features using a string cathode mask, i.e., where very thin parallel metal strings are used to break the deposited film into patterned cathode rows. However, for high-resolution displays, or indeed any display of appreciable size, this technique is problematic because of the low fill factor that can be realized. This is due to the width of the strings that is necessitated by the need for mechanical integrity and the issue of a poorly defined deposition footprint due to feathering at the edges of the deposited metal film.

An alternative method to define the cathode rows is by an integrated shadow mask (ISM), as shown in Figure 7.11. This shows a cross section of one individual element of the ISM. This mushroom feature is patterned on the substrate orthogonal to the anode columns. These features are placed at intervals equal to the required display pitch. After deposition of the organic layers, the subsequently deposited metal is then broken into regularly patterned electrically isolated rows. The ISM is usually constructed from photoresist material that can be patterned with high precision using lithography, thus enabling a higher fill factor and consequently higher-resolution PMOLED displays to be realized. Other proposed schemes for patterning the metal cathode on OLED displays include stamping [140], laser ablation [141], and lithography [142].

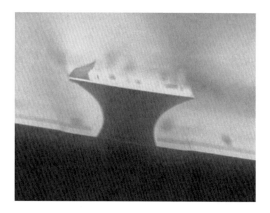

FIGURE 7.11
Scanning electron micrograph image of an ISM cross section.

7.5.2 Active Addressing Schemes

Many LCDs are based on active-matrix addressing, in which an active device circuit containing one or more TFTs is connected to each pixel. The TFT circuit at each pixel effectively acts as an individual electrical switch that provides the means to store display information on a storage capacitor for the entire frame time, such that the pixel can remain emitting during this entire time rather than for a small fraction of time, as is the case in passive addressing.

A remaining challenge in manufacturing AMOLED displays is the requirement for a backplane that provides a constant uniform drive current. Presently, the pixel and driver circuitry must compensate for the initial nonuniformities of the low-temperature polysilicon TFTs or the threshold voltage shift of the amorphous silicon (α-Si) TFTs [143,144]. Some pixel designs incorporate more than two transistors along with a conventional bottom-emitting OLED [143,145]. This can significantly diminish the aperture ratio, forcing the OLED to operate at a high luminance level, reducing lifetime. One solution is to build TOLEDs over a planarized backplane [41,146]. Figure 7.12 shows a 24″ OLED display by Sony that uses a TOLED device architecture.

Previously, it was a commonly held belief that TOLEDs are less efficient than their bottom-emitting counterparts, such that any gain in aperture ratio must be large to offset the efficiency loss. However, this assumption has been disproved. Lu et al. [147] compared

FIGURE 7.12
24″ AMOLED display fabricated by Sony using top-emission pixel architectures.

the characteristics of equivalent green (Ir(ppy)$_3$) top-emitting PHOLEDs, bottom-emitting PHOLEDs, and transparent PHOLEDs. They observed that the TOLED had a 15% higher luminous efficacy than the bottom-emitting PHOLED in an equivalent device architecture.

Also, it has been observed that top-emission OLEDs can be fabricated with lifetimes equivalent to conventional bottom-emitting devices [148]. In a display where top-emitting architectures have a larger fill factor, less current density is required per pixel to produce the same amount of light as compared with an equivalent bottom-emitting design. Therefore, top-emitting displays may enable longer lifetimes in addition to any potential power savings. Figure 7.13 shows an example of a transparent 2.2″ display shown at the 2003 Society for Information Display Conference by Samsung.

OLEDs appear ideally suited to active-matrix addressing as they are a low-voltage technology. Active addressing offers several advantages over passive addressing. Active addressing (i) eliminates the problem of cross talk due to reverse bias leakage currents, (ii) extends display lifetimes, and (iii) improves efficiency owing to the lower operating voltages and currents that are necessary. In addition, it reduces power losses and heating problems due to resistive heating in the ITO tracks (I^2R losses). There has, therefore, been considerable interest in combining vapor-deposited small-molecule materials with poly-Si TFTs [143,149,150], and in 1995 TDK announced the first full-color (240 × 320 pixel) actively addressed OLED display [151]. Polycrystalline silicon (p-Si) TFTs, until recently, were preferred, as it was widely believed that conventional α-Si TFTs could not accommodate the high currents required. However, in 2002, AU Optronics successfully demonstrated a 4″ full-color AMOLED display based on an α-Si TFT backplane and incorporating a high-efficiency red phosphorescent material. The use of the red phosphorescent subpixel reduced display power consumption by 42%, compared with a comparable display based on only fluorescent materials [152]. This display illustrated the benefits of PHOLEDs, and the possibility of using α-Si TFT backplanes in full-color AMOLEDs. Tsujimura et al. [153] demonstrated a 20″ AMOLED using an α-Si TFT backplane. If the stability issues of α-Si TFTs can be addressed, then the potential cost savings to AMOLED displays are significant due to the maturity of the α-Si backplane industry.

A future step in the development of actively addressed OLED displays may be an all-organic TV. At present, poly-Si TFTs are relatively expensive (particularly for larger-area displays) and therefore detract from cheap production costs, which is one of the widely perceived main advantages of OLED technology. However, if organic materials can be

FIGURE 7.13
2.2″ Full-color transparent AMOLED display fabricated by Samsung SDI incorporating phosphorescent dopants.

used to make the TFT, as well as the OLED, using similar fabrication processes, then the cost of production can be reduced. Field-effect transistors have been fabricated using organic materials [154], and in 1998 Sirringhaus et al. [155] constructed the first all-plastic polymer TFT and PLED. In the same year, a vapor-deposited OLED was fabricated for the first time in combination with an organic thin-film field-effect transistor [156]. These types of ICs open up the possibility of producing low-cost, flexible plastic displays [157].

7.5.3 Full-Color Displays

Various ways of making full-color displays have been proposed. These are summarized in Figure 7.14. Perhaps the most obvious method is simply to fabricate red, green, and blue subpixels side by side on the same substrate (Figure 7.14a). Many companies have adopted this approach; for example, Pioneer demonstrated a full-color QVGA (320 × 240 pixels) display at the Japan Electronics Show in 1998. Figure 7.15 is an example of a full-color display patterned using a side-by-side approach.

In conventional LCDs, emission from a white backlight is filtered using absorption filters to produce red, green, and blue emission. The same technique can also be used with

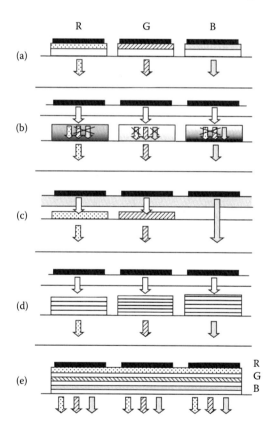

FIGURE 7.14
Various methods of pixel design to achieve full-color displays: (a) red, green, and blue subpixels side by side on the same substrate; (b) white OLED backlight is filtered using absorption filters to produce red, green, and blue emission; (c) blue OLEDs and down-conversion PL red or green color-change filters; (d) a relatively broad emitter (e.g., white) in combination with microcavities to selectively emit red, green, and blue light; and (e) red, green, and blue pixels stacked on top of each other.

FIGURE 7.15
Commercially produced Fujitsu cell phone incorporating a Pioneer-manufactured OLED subdisplay that contains both phosphorescent (red) and fluorescent (green and blue) pixels.

OLEDs (Figure 7.14b). This approach has one advantage over the side-by-side approach (Figure 7.14a) in that for high-resolution displays precision shadow masking is unnecessary as the pixels (color filters) are patterned via lithography. However, for each pixel color, approximately two-thirds of the light is wasted due to the need to absorb the unwanted elements of the white emission to render a saturated color from each pixel element. This waste in energy can be partly alleviated by the use of a four-pixel configuration: red, green, blue, and white (RGBW). Here, the red, green, and blue colors are rendered in the same way by passing white light through color filters. A fourth unfiltered white subpixel is added. The power saving of this configuration depends on the display image content but can be, on average, 50% more efficient than without the white subpixel [158].

White emission is generally produced by combining the emission from two or more dyes [159]. For example, TDK combined the emission from two separate dopants, coumarin 6 and rubrene (see Figure 7.9), to approximate white emission. Conventional color filters were then used to produce red, yellow, and green emission [160]. The white emitter or the color filter approach was used in a commercial AMOLED display by Sanyo–Eastman Kodak. Figure 7.16 shows this display. Recently, LGD has demonstrated 55″ full-HD AMOLED TVs using a white-plus-color-filter approach [10].

Another technique for fabricating full-color displays is based on the fact that blue light can be converted into green or red light by using dyes that absorb blue light and then emit green or red light via PL. Researchers at Idemitsu Kosan [161] have demonstrated color displays based on blue emitters [162] and color-change filters (Figure 7.14c). They have developed a range of distyrylarylene derivatives that give high efficacies (up to 6 lm/W) with lifetimes of >20,000 h. More recently, they demonstrated a (0.15, 0.08) blue with an efficacy of 5.5 cd/A at 10 mA/cm^2 and a lifetime of 11,000 h to 50% of an initial display luminance of 500 cd/m^2 [163]. In 1998, they showed a 10″ VGA (640 × 480 pixels) color display using the blue OLED plus color-change filter approach [164].

When a relatively broad emitter is confined in a microcavity, the dimensions of the microcavity influence the emission spectrum and peak wavelength (Figure 7.14d) [165]. This idea has been developed by researchers at several companies to produce red, green,

FIGURE 7.16
Commercially produced digital camera containing a full-color AMOLED display from Sanyo–Eastman Kodak.

and blue emission [166,167]. The main advantage of the technique is that red, green, and blue emission can be obtained from one broad organic emitter. However, one disadvantage of this approach is that the emission from such devices is strongly directional, leading to a narrow viewing angle; that is, the emission intensity is usually at a peak orthogonal to the display surface and decreases with a distortion of the color (blue spectral shift) as the viewing angle is increased.

An elegant way of achieving full-color displays is to stack the red, green, and blue pixels on top of each other [168] (Figure 7.14e). As there are no subpixels, this has the advantage of increasing the resolution of the display by a factor of 3. However, it requires semitransparent electrodes that are also compatible with high current densities. A similar approach that would avoid this problem might be to develop structures in which the emission color can be varied simply by varying the applied voltage [169,170].

Another intriguing possibility for patterning the pixels in an OLED display without the use of high-precision shadow masking is the technique of laser-induced thermal imaging [58,171,172]. This approach to patterning makes use of a donor film coated with a light-to-heat conversion layer, which in turn is coated with the organic layer (or multiple layers) that are the intended components of the OLED. The organic layer is placed in contact with the OLED substrate, and the donor film is scanned with the intended deposition pattern using a laser. The donor film is typically a transparent flexible material, e.g., a polyester, and the conversion film is chosen to absorb the laser light in the infrared region of the spectrum and convert this light to heat. Typical coatings include carbon. As the film is scanned, the light is converted to heat and the OLED material is transferred from the conversion film to the substrate. To achieve a high-quality OLED, the deposited materials must be chosen to have the requisite mechanical properties, e.g., weak cohesion and the correct balance of donor film or deposited film adhesive forces. The use of this technique is still in the early stages of development for OLEDs but holds great promise.

Patterning small-molecule organic displays via ink-jet printing is also explored within the OLED industry. Ink-jet printing has been researched extensively within the PLED community as a means of patterning full-color PLED displays culminating in a demonstration of a 56″ display in 2013 by Panasonic [12]. The application of this technique to small-molecule OLED displays is under investigation by a number of groups.

Each of the techniques described above has unique strengths and weaknesses, and the optimum device structure for commercial full-color displays is also heavily influenced by the ease with which it can be mass produced. Currently, full-color OLED displays are being manufactured commercially by using two of the above-described techniques, i.e., (i) side-by-side pixels deposited by high-precision shadow masking and (ii) using white OLEDs and color absorption filters.

7.6 Future-Generation Vapor-Deposited Organic Devices

7.6.1 Flexible Displays

OLED displays can also be fabricated on flexible substrates [173–175] such as metal foils or plastic (see Chapter 8). This enables entirely new display features such as conformability, ruggedness, flexibility, and reduced weight.

To build an OLED display on a flexible substrate, a number of important issues have to be investigated in addition to those encountered when processing rigid glass or silicon substrates. Issues such as chemical stability, temperature limits, and mechanical stability all have to be addressed. Nevertheless, Figure 7.17 shows a 128 × 64 (60-dpi) monochrome passive-matrix vapor-deposited OLED display on a flexible substrate. It was fabricated by Universal Display Corporation and was exhibited in May 2000. It is believed to be the first report of such a display built on a plastic substrate. The plastic is 175-μm thick and the pixels are 500 × 400 μm in size. Figure 7.18 shows a close-up view of a typical pixel. In this particular architecture, a metal bus line has also been added to lower the column resistivity of the display. This significantly reduces the power losses incurred by the ITO electrodes and should enable scale up to larger-area, passive-matrix displays.

To realize the enormous potential of this flexible display technology, however, a number of important issues still require solutions. In particular, fabrication technologies, suitable for plastic substrates and extended device operating lifetimes, must be demonstrated. The problem incurred when using a flexible plastic substrate material is that polymers are very poor barriers to the diffusion of water and oxygen due to their low density. To realize the long lifetimes needed for displays, the OLED package must have an estimated permeability of $<10^{-6}$ g/(m^2 day) at 25°C [23]. Typical plastic substrate materials have permeabilities of 0.1–1.0 g/(m^2 day) at 25°C, and are unsuitable for a commercial OLED product. One approach to overcome this issue is to use thin-film coatings of dense dielectric materials

FIGURE 7.17
128 × 64 Monochrome PMOLED display on a 175-μm-thick flexible PET substrate. Image is displayed at a luminance of 200 cd/m^2.

FIGURE 7.18
Magnified view of a flexible PMOLED on a PET substrate. Pixel pitch is 317 µm, and active pixel area is defined by each square.

to inhibit diffusive processes. Here, a barrier film is deposited directly onto the completed OLED, e.g., a thick Si_xN_y film [176]. However, such a film has to be almost defect free in terms of pinholes and grain boundaries within the inorganic layer.

One solution being explored is the use of a multilayer barrier coating [23,177] on the plastic substrate and the OLED. This consists of a hybrid organic–inorganic multilayer barrier coating [23,178]. The composite barrier consists of alternating layers of polyacrylate films and an inorganic oxide. Acrylic monomer is deposited by flash evaporation in vacuum onto the OLED surface [178]. The condensed monomer is cured using UV light to form a nonconformal highly cross-linked polyacrylate film that acts to planarize the substrate (or OLED) surface. The surface roughness of the coated sample is <1 nm rms [23]. A 10–30-nm-thick Al_2O_3 film is subsequently deposited onto the polymer layer as a barrier to the diffusion of water and oxygen. By repeating the alternating process to deposit multiple layers, the polymer layers decouple any defects in the oxide layers, thereby preventing propagation of defects through the multilayer structure. Both the optical and barrier properties of the composite layer can be tailored by varying the total number and thickness of the polymer and inorganic layers in the thin-film coating, yielding an engineered barrier [179]. Figure 7.19 shows a scanning electron micrograph of a fracture cross section of a generic multilayer barrier structure. In this particular configuration, 10 layers are used.

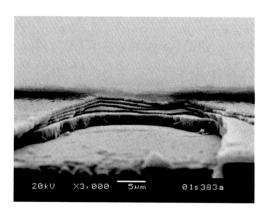

FIGURE 7.19
Scanning electron micrograph image of the cross section of a multilayer PET–Al_2O_3 barrier film.

For a more thorough review of thin-film encapsulation of OLEDs, see, for example, Lewis and Weaver [74].

OLEDs grown and encapsulated using these techniques are beginning to show significant promise. Chwang et al. [180] demonstrated the effects of flexing a 64 × 64 (180-dpi) passive-matrix flexible OLED (FOLED) display fabricated on a PET substrate with thin-film encapsulation. In addition, lifetimes of thin-film-encapsulated OLED test pixels on flexible substrates have now been demonstrated to be thousands of hours [176,181,182].

7.6.2 Lighting

In 2010, in the United States alone, lighting was estimated to consume 7.5 quads of primary energy (approximately 700 TWh), or about 22% of the total electricity generated. Linear fluorescent lighting is the leading energy consumer at 294 TWh, followed by high-intensity discharge lighting at 183 TWh and incandescent lighting with about 156 TWh. The efficacy for incandescent lights is in the range 10–20 lm/W, and for fluorescent lamps in the range 30–90 lm/W. Within the past 30–50 years, little progress has been made in the energy efficiency of conventional sources of light—incandescent, fluorescent, and halogen. At present, approximately 70% of the energy used by these sources is wasted as heat; thus, significant savings can be made by the introduction of new, higher-performance illumination sources. The introduction of solid-state lighting technologies such as LEDs and OLEDs could significantly reduce the energy usage in lighting applications, in addition to opening up new lighting possibilities, large-area light walls, or architectural lighting features. However, to achieve the price (such as $10 per 1000 lm) and performance (>60 lm/W) required to enable the wholesale OLEDs to compete in general illumination applications, improvements in device performance are needed, particularly in terms of lifetime under high drive conditions.

In the vapor-deposited OLED community, a number of approaches have been employed to produce white light emission. White OLEDs have been demonstrated based on multilayer structures, e.g., stacked backlights [168,183], multidoping of single-layer structures [159], phosphorescent monomer–excimer emission layers [184], and on doping of phosphorescent materials into separate bands within the emission zone [185]. White OLED efficacies exceeding 110 lm/W have since been demonstrated [186,187].

All of these approaches have varying degrees of merit but each deals with the emission of white light conceptually at a pixel level. For lighting, the emission can be generated using a white-producing system as opposed to using individual pixels. For example, white light for illumination purposes can be generated in a similar way to many displays, i.e., using two or three individual color elements that, when combined, produce white emission, e.g., yellow and blue or red, green, and blue striped arrays [188–190]. The advantage of this approach is that the individual elements can be separately driven and optimized. This has the potential benefits of improved efficiency and lifetime, and allows for the ability to compensate for differential aging of the white emission as a function of time.

Although the power efficacy and lifetime requirements are difficult targets, OLED device lighting architectures have fewer constraints in certain areas than when used in displays. For example, lighting applications allow for much greater flexibility in improving outcoupling from an OLED because factors such as blurring between pixels are no longer an issue.

Today, white OLEDs have demonstrated 128 lm/W at 1000 cd/m^2 [187] and 136 lm/W at 100 cd/m^2 [191]. In theory, if the following metrics could be realized by an OLED: (i) luminous efficacy of radiation of 350 lm/W for a broad white OLED spectrum; (ii) 95%

internal quantum efficiency; (iii) outcoupling of 60%, and an electrical efficacy of 70%, this would allow OLEDs to realize a power efficacy of 180 lm/W. This would be an extremely competitive efficacy for lighting applications. However, the challenge for OLEDs will be to realize large lighting elements with a similar efficacy at high brightness with long lifetimes and manufactured at a low cost. For a discussion of white OLEDs, see D'Andrade and Forrest [192].

7.6.3 Displays and Beyond

The field of vapor-deposited OLEDs has grown markedly during the last two decades to the point where virtually every large display manufacturer has a research program or is in the process of commercializing its OLED displays. Issues, of course, remain to be resolved. For television applications, lifetimes need to be improved further. As a disruptive flat-panel display technology, OLEDs on glass-based substrates need to be cost-competitive with LCDs. One issue facing vapor-deposited OLED display manufacturers is employing very-large-area shadow masks that are necessary as the mother glass size is increased to reduced display unit costs. Issues such as the shadow mask integrity, alignment, and thermal expansion all need to be addressed. However, great progress is rapidly being made in solving these problems.

Now that organic materials have been developed that have the requisite qualities for commercialization and that the manufacturing infrastructure is being put into place, what is next for vapor-deposited organic materials? Already a number of research groups worldwide are attempting to use similar materials for electrically pumped lasers. Optical pumping has already been demonstrated [16]. The realization of this goal is a formidable challenge as the estimated threshold current densities required, for amorphous organic materials, are of the order of <1000 A/cm^2 [193]. However, research is in progress.

In photovoltaics, organic solar cells with conversion efficiencies of 4%–12% have been demonstrated [15,194–197]. The challenge here is not only to improve efficiencies and lifetimes but also to compete on cost with silicon solar cells. Another intriguing possibility is the idea of fabricating organic computers. Organic TFTs are under development that at present have mobilities of the order of 1 –20 cm^2/(V s) [198,199]. However, improvements are still necessary for the commercialization of organic TFTs; nevertheless, it could be argued that the demands on the organic materials incorporated into these devices are less severe than those placed on their OLED counterparts.

The field of vapor-deposited OLEDs has seen amazing progress during the last three decades, from Tang's first efficient device, to OLED smartphone displays in everyday use, and to the demonstration of 55″–56″ three-dimensional OLED TVs. However, the possibilities for the next 30 years are even more exciting.

References

1. A. Bernanose. "Electroluminescence of organic compounds." *Br. J. Appl. Phys.*, 6(Suppl. 4):S54–S55, 1955.
2. M. Pope, H.P. Kallmann, and P.J. Magnante. "Electroluminescence in organic crystals." *J. Chem. Phys.*, 38:2042–2043, 1963.

3. P.S. Vincett, W.A. Barlow, R.H. Hann, and G.G. Roberts. "Electrical conduction and low voltage blue electroluminescence in vacuum-deposited organic films." *Thin Solid Films*, 94:171–183, 1982.
4. C.W. Tang, and S.A. VanSlyke. "Organic electroluminescent diodes." *Appl. Phys. Lett.*, 51:913–915, 1987.
5. S.A. VanSlyke, and C.W. Tang. "Organic electroluminescent devices having improved power conversion efficiencies." U.S. Patent 4,539,507, September 3, 1985.
6. C. Adachi, S. Tokito, T. Tsutsui, and S. Saito. "Electroluminescence in organic films with three-layer structure." *Jpn. J. Appl. Phys.*, 27(Part 2):L269–L271, 1988.
7. C.W. Tang, S.A. VanSlyke, and C.H. Chen. "Electroluminescence of doped organic thin films." *J. Appl. Phys.*, 65(9):3610–3616, 1989.
8. H. Tokailin, M. Matsuura, H. Higashi, C. Hosokawa, and T. Kusumoto. "Characteristics of blue organic electroluminescent devices with distyryl arylene derivatives." In *Proceedings of the International Society for Optical Engineering*, Vol. 1910, San Jose, CA, 1993, pp. 38–47.
9. T. Wakimoto, R. Murayama, K. Nagayama, Y. Okuda, H. Nakada, and T. Tohma. "Organic LED dot-matrix display." In *Proceedings of the Society for Information Displays*, Technical Digest 27, San Diego, CA, 1996, pp. 849–852.
10. C.W. Han, K.M. Kim, S.J. Bae, H.S. Choi, J.M. Lee, T.S. Kim, Y.H. Tak, S.Y. Cha, and B.C. Ahn. "55-inch FHD OLED TV employing new tandem WOLEDs." In *Proceedings of the Society for Information Displays*, Technical Digest, Boston, 2012, pp. 279–281.
11. Samsung OLED TV. Available at: www.oled-info.com/samsung-real-oled-tv.
12. Panasonic 56″ 4K printed OLED TV prototype. Available at: www.oled-info.com/panasonic -shows-56-4k-printed-oled-tv-prototype.
13. Sony OLED TV. Available at: www.oled-info.com/sony-oled.
14. C.W. Tang. "Two-layer organic photovoltaic cell." *Appl. Phys. Lett.*, 48:183–185, 1986.
15. Heliatek organic solar cell with a cell efficiency of 12%. Available at: www.heliatek.com/wp -content/uploads/2013/01/130116_PR_Heliatek_achieves_record_cell_effiency_for_OPV.pdf.
16. V.G. Kozlov, V. Bulovic, P.E. Burrows, and S.R. Forrest. "Laser action in organic semiconductor waveguide and double-heterostructure devices." *Nature*, 389:362–364, 1997.
17. K. Kudo, M. Yamashina, and T. Morizumi. "Field effect measurement of organic dye films." *Jpn. J. Appl. Phys.*, 23:130, 1984.
18. H. Klauk. "Organic thin-film transistors." *Chem. Soc. Rev.*, 39:2643–2666, 2010.
19. K. Furukawa, Y. Terasaka, H. Ueda, and M. Matsumura. "Effect of a plasma treatment of ITO on the performance of organic electroluminescent devices." *Synth. Met.*, 91:99–101, 1997.
20. M.G. Mason, L.S. Hung, C.W. Tang, S.T. Lee, K.W. Wong, and M. Wang. "Characterization of treated indium–tin-oxide surfaces used in electroluminescent devices." *J. Appl. Phys.*, 86:1688–1692, 1999.
21. Z.B. Deng, X.M. Ding, S.T. Lee, and W.A. Gambling. "Enhanced brightness and efficiency in organic electroluminescent devices using SiO_2 buffer layers." *Appl. Phys. Lett.*, 74:2227–2229, 1999.
22. I.-M. Chan, T.-Y. Hsu, and F.C. Hong. "Enhanced hole injections in organic light emitting devices by depositing nickel oxide on indium tin oxide anode." *Appl. Phys. Lett.*, 81:1899–1901, 2002.
23. P.E. Burrows, G.L. Graff, M.E. Gross, P.M. Martin, M. Hall, E. Mast, C. Bonham, W. Bennet, L. Michalski, M.S. Weaver, J.J. Brown, D. Fogarty, and L.S. Sapochak. "Gas permeation and lifetime tests on polymer-based barrier coatings." In *Proceedings of the International Society for Optical Engineering*, Vol. 4105, San Diego, CA, 2000, pp. 75–83.
24. Y. Yang, and A.J. Heeger. "Polyaniline as a transparent electrode for polymer light-emitting diodes: Lower operating voltage and higher efficiency." *Appl. Phys. Lett.*, 64:1245–1247, 1994.
25. H. Riel, S. Karg, T. Beierlein, B. Ruhstaller, and W. Riess. "Phosphorescent top-emitting organic light-emitting devices with improved light outcoupling." *Appl. Phys. Lett.*, 82:466–468, 2002.
26. S. Tokito, K. Nada, and Y. Taga. "Metal oxides as a hole-injecting layer for an organic electroluminescent device." *J. Phys. D: Appl. Phys.*, 29:2750–2753, 1996.

27. S.W. Yin, Z. Shuai, and Y. Wang. "A quantitative structure–property relationship study of the glass transition temperature of OLED materials." *J. Chem. Inf. Comput. Sci.*, 43:970–977, 2003.

28. D.F. O'Brien, P.E. Burrows, S.R. Forrest, B.E. Koene, D.E. Loy, and M.E. Thompson. "Hole transporting materials with high glass transition temperatures for use in organic light-emitting devices." *Adv. Mater.*, 10:1108–1112, 1998.

29. X. Zhou, J. He, L.S. Liao, M. Lu, X.M. Ding, X.Y. Hou, M. Zhang, X.Q. He, and S.T. Lee. "Real-time observation of temperature rise and thermal breakdown processes in organic LEDs using an IR imaging and analysis system." *Adv. Mater.*, 12:265–269, 2000.

30. M. Ishii, and Y. Taga. "Influence of temperature and drive current on degradation mechanisms in organic light-emitting diodes." *Appl. Phys. Lett.*, 80:3430–3432, 2002.

31. J.C. Sturm, W. Wilson, and M. Iodice. "Thermal effects and scaling in organic light emitting flat panel displays." *IEEE J. Sel. Top. Quant. Electron.*, 4(1):75–82, 1998.

32. K. Itano, H. Ogawa, and Y. Shirota. "Exciplex formation at the organic solid-state interface: Yellow emission in organic light-emitting diodes using green-fluorescent tris(8-quinolinolato) aluminum and hole-transporting molecular materials with low ionization potentials." *Appl. Phys. Lett.*, 72:636–638, 1998.

33. C. Adachi, T. Tsutsui, and S. Saito. "Blue light-emitting organic electroluminescent devices." *Appl. Phys. Lett.*, 56:799–801, 1989.

34. R. Murayama, S. Kawami, T. Wakimoto, H. Sato, H. Nakada, T. Namiki, K. Imai, and M. Nomura. "Organic EL devices doped with a quinacridone derivative showing higher brightness and luminescent efficiency." *Jpn. Soc. Appl. Phys.*, *Extended Abstracts (54th Autumn Meeting, 1993)*, 3:1127, 1993.

35. E.I. Haskal, A. Curioni, P.F. Seidler, and W. Andreoni. "Lithium–aluminum contacts for organic light-emitting devices." *Appl. Phys. Lett.*, 71:1151–1153, 1997.

36. J. Kido, and T. Matsumoto. "Bright organic electroluminescent devices having a metal-doped electron-injecting layer." *Appl. Phys. Lett.*, 73:2866–2868, 1998.

37. L.S. Hung, C.W. Tang, and M.G. Mason. "Enhanced electron injection in organic electroluminescent devices using an Al/LiF electrode." *Appl. Phys. Lett.*, 70:152–155, 1997.

38. C. Schmitz, H. Schmidt, and M. Thelakkat. "Lithium–quinolate complexes as emitter and interface materials in organic light-emitting diodes." *Chem. Mater.*, 12:3012–3019, 2000.

39. T. Wakimoto. "Organic electroluminescent device." U.S. Patent 5,739,635, April 14, 1998.

40. L.S. Hung, C.W. Tang, M.G. Mason, P. Raychaudhuri, and J. Madathil. "Application of an ultra-thin LiF/Al bilayer in organic surface-emitting diodes." *Appl. Phys. Lett.*, 78:544–546, 2001.

41. V. Bulovic, G. Gu, P.E. Burrows, S.R. Forrest, and M.E. Thompson. "Transparent light-emitting devices." *Nature*, 380:29, 1996.

42. G. Parthasarathy, P.E. Burrows, V. Khalfin, V.G. Kozlov, and S.R. Forrest. "A metal-free cathode for organic semiconductor devices." *Appl. Phys. Lett.*, 72:2138–2140, 1998.

43. G. Gu, V. Bulovic, P.E. Burrows, and S.R. Forrest. "Transparent organic light emitting devices." *Appl. Phys. Lett.*, 68:2606–2608, 1996.

44. V. Bulovic, P. Tian, P.E. Burrows, M.R. Gokhale, S.R. Forrest, and M.E. Thompson. "A surface-emitting vacuum-deposited organic light emitting device." *Appl. Phys. Lett.*, 70:2954–2956, 1997.

45. C.C. Wu, C.I. Wu, J.C. Sturm, and A. Kahn. "Surface modification of indium tin oxide by plasma treatment: An effective method to improve the efficiency, brightness, and reliability of organic light emitting devices." *Appl. Phys. Lett.*, 70:1348–1350, 1997.

46. E.B. Graper. In *Handbook of Thin Film Process Technology*, 1st ed., *Resistance evaporation*, D.A. Glocker, and S.I. Shah, Eds., IOP Publishing, Bristol, 1995, pp. A1.0:1–A1.1:7.

47. A. Böhler, S. Dirr, H.-H. Johannes, D. Ammermann, and W. Kowalsky. "Influence of the process vacuum on the device performance of organic light-emitting diodes." *Synth. Met.*, 91:95–97, 1997.

48. H. Yamamoto, J. Brooks, M.S. Weaver, J.J. Brown, T. Murakami, and H. Murata. "Improved initial drop in operational lifetime of blue phosphorescent organic light emitting device fabricated under ultra high vacuum condition." *Appl. Phys. Lett.*, 99:033301, 2011.

49. H. Yamamoto, C. Adachi, M.S. Weaver, and J.J. Brown. "Identification of device degradation positions in multi-layered phosphorescent organic light emitting device using water probes." *Appl. Phys. Lett.*, 100:183306, 2012.

50. M. Ohring. *The Materials Science of Thin Films*, 1st ed., Academic Press, London, 1992, pp. 79–145.

51. S.A. VanSlyke, A. Pignato, D. Freeman, N. Redden, D. Waters, H. Kikuchi, T. Negishi, H. Kanno, Y. Nishio, and M. Nakai. "Linear source deposition of organic layers for full-color OLED." In *Proceedings of the Society for Information Display*, Digest of Technical Papers, Vol. 33(Suppl. 2), Boston, 2002, pp. 886–889.

52. S.R. Forrest. "Ultrathin organic films grown by organic molecular beam deposition and related techniques." *Chem. Rev.*, 97:1793–1896, 1997.

53. C.-L. Lee, K.B. Lee, and J.-J. Kim. "Polymer phosphorescent light-emitting devices doped with tris(2-phenylpyridine) iridium as a triplet emitter." *Appl. Phys. Lett.*, 77:2280–2282, 2000.

54. D.F. O'Brien, C. Giebler, R.B. Fletcher, A.J. Cadby, L.C. Palilis, D.G. Lidzey, P.A. Lane, D.D.C. Bradley, and W. Blau. "Electrophosphorescence from a doped polymer light emitting diode." *Synth. Met.*, 116:379–383, 2001.

55. J.P.J. Markham, S.-C. Lo, S.W. Magennis, P.L. Burn, and I.D.W. Samuel. "High-efficiency green phosphorescence from spin-coated single-layer dendrimer light-emitting diodes." *Appl. Phys. Lett.*, 80:2645–2647, 2002.

56. T.R. Hebner, C.C. Wu, D. Marcy, M.H. Lu, and J.C. Sturm. "Ink-jet printing of doped polymers for organic light emitting devices." *Appl. Phys. Lett.*, 72:519–521, 1998.

57. T.R. Hebner, and J.C. Stur. "Local tuning of organic light-emitting diode by dye droplet application." *Appl. Phys. Lett.*, 73:1775–1777, 1998.

58. T.A. Isberg, C.A. Jalbert, J.S. Staral, W.A. Tolbert, and M.B. Wolk. "Process for preparing high resolution emissive arrays and corresponding articles." U.S. Patent 5,998,085, December 7, 1999.

59. G.G. Roberts, M. McGinnity, W.A. Barlow, and P.S. Vincett. "Electroluminescence, photoluminescence and electroabsorption of a lightly substituted anthracene Langmuir film." *Solid State Commun.*, 32:683–686, 1979.

60. M. Era, C. Adachi, T. Tsutsui, and S. Saito. "Organic electroluminescent device with cyanine dye Langmuir–Blodgett film as an emitter." *Thin Solid Films*, 210/211:468–470, 1992.

61. M.S. Weaver, and D.D.C. Bradley. "Organic electroluminescence devices fabricated with chemical vapour deposited films." *Synth. Met.*, 83:61–66, 1996.

62. F. Papadimitrakopoulos, D.L. Thomsen, and K.A. Higginson. "Quinoline-based polymeric metal chelate light-emitting diodes." In *Proceedings of the International Society for Optical Engineering*, Vol. 3148, San Diego, CA, 1997, pp. 170–177.

63. M.A. Baldo, M. Deutsch, P.E. Burrows, H. Gossenberger, M. Gerstenberg, V. Ban, and S.R. Forrest. "Organic vapor phase deposition." *Adv. Mater.*, 10:1505–1514, 1998.

64. M. Shtein, P. Peumans, J.B. Benziger, and S.R. Forrest. "Direct, mask- and solvent-free printing of molecular organic semiconductors." *Adv. Mat.*, 16:1615–1620, 2004.

65. G.J. McGraw, D.L. Peters, and S.R. Forrest. "Organic vapor jet printing at micrometer resolution using microfluidic nozzle arrays." *Appl. Phys. Lett.*, 98:013302-1-3, 2011.

66. L.S. Hung, L.S. Liao, C.S. Lee, and S.T. Lee. "Sputter deposition of cathodes in organic light emitting diodes." *J. Appl. Phys.*, 86:4607–4612, 1999.

67. C. Adachi, M.A. Baldo, S.R. Forrest, and M.E. Thompson. "High-efficiency organic electrophosphorescent devices with tris(2-phenylpyridine)iridium doped into electron-transporting materials." *Appl. Phys. Lett.*, 77:904–906, 2000.

68. J.H. Burroughs, D.D.C. Bradley, A.R. Brown, R.N. Marks, K. MacKay, R.H. Friend, P.L. Burn, and A.B. Holmes. "Light-emitting diodes based on conjugated polymers." *Nature*, 347:539–541, 1990.

69. J. Shi, and C.W. Tang. "Doped organic electroluminescent devices with improved stability." *Appl. Phys. Lett.*, 70:1665–1667, 1997.

70. P.E. Burrows, S.R. Forrest, T.X. Zhou, and L. Michalski. "Operating lifetime of phosphorescent organic light emitting devices." *Appl. Phys. Lett.*, 76:2493–2495, 2000.

71. Y.F. Liew, H. Aziz, N.-X. Hu, H.S.O. Chan, G. Xu, and Z. Popovic. "Investigation of the sites of dark spots in organic light-emitting devices." *Appl. Phys. Lett.*, 77:2650–2652, 2000.

72. D. Kolosov, D.S. English, V. Bulovic, P.F. Barbara, S.R. Forrest, and M.E. Thompson. "Direct observation of structural changes in organic light emitting devices during degradation." *J. Appl. Phys.*, 90:3242–3247, 2001.

73. P.E. Burrows, V. Bulovic, S.R. Forrest, L.S. Sapochak, D.M. McCarty, and M.E. Thompson. "Reliability and degradation of organic light emitting devices." *Appl. Phys. Lett.*, 65:2922–2924, 1994.

74. J.S. Lewis, and M.S. Weaver. "Thin film permeation barrier technology for flexible organic light emitting devices." *IEEE J. Sel. Top. Quant. Electron.*, 10:47–57, 2004.

75. W. Brutting, S. Berleb, and A.G. Muckl. "Device physics of organic light-emitting diodes based on molecular materials." *Org. Electron.*, 2:1–36, 2001.

76. C. Hosokawa, H. Tokailin, H. Higashi, and T. Kusumoto. "Transient behaviour of thin film electroluminescence." *Appl. Phys. Lett.*, 60:1220–1222, 1992.

77. M. Pope, and C.E. Swenberg. *Electronic Processes in Organic Crystals*, 1st ed., Clarendon Press, Oxford, 1982.

78. J. Kalinowski, G. Giro, M. Cocchi, V. Fattori, and P. DiMarco. "Unusual disparity in electroluminescence and photoluminescence spectra of vacuum-evaporated films of 1,1-*bis* ((di-4-tolylamino) phenyl) cyclohexane." *Appl. Phys. Lett.*, 76:2352–2354, 2000.

79. L.C. Palilis, A.J. Makinen, M. Uchida, and Z.H. Kafafi. "Highly efficient molecular organic light-emitting diodes based on exciplex emission." *Appl. Phys. Lett.*, 82:2209–2211, 2003.

80. M.H. Lu, and J.C. Sturm. "Optimization of external coupling and light emission in organic light-emitting devices: Modelling and experiment." *J. Appl. Phys.*, 91:595–604, 2002.

81. D.D.C. Bradley. "Electroluminescent polymers: Materials, physics and device engineering." *Curr. Opin. Solid State Mater. Sci.*, 1:789–797, 1996.

82. J. Kalinowski. "Electronic processes in organic electroluminescence." In *Organic Electroluminescent Materials and Devices*, S. Miyata, and H.S. Nalwa, Eds., Gordon & Breach, Amsterdam, 1997, pp. 1–72.

83. N.C. Greenham, and R.H. Friend. "Semiconductor device physics of conjugated polymers." *Solid State Phys.*, 49:2–149, 1995.

84. P.M. Borsenberger, and D.S. Weiss. *Organic Photoreceptors for Imaging Systems*, Marcel Dekker, New York, 1993.

85. P.E. Burrows, Z. Shen, V. Bulovic, D.M. McCarty, S.R. Forrest, J.A. Cronin, and M.E. Thompson. "Relation between electroluminescence and current transport in organic heterojunction light-emitting devices." *J. Appl. Phys.*, 79:7991–8006, 1996.

86. M. Stolka, J.F. Yanus, and D.M. Pai. "Hole transport in solid solutions of a diamine in polycarbonate." *J. Phys. Chem.*, 88:4707–4714, 1984.

87. C.W. Tang. "Organic electroluminescent cell." U.S. Patent 4,356,429, October 26, 1982.

88. S.A. VanSlyke, C.H. Chen, and C.W. Tang. "Organic electroluminescent devices with improved stability." *Appl. Phys. Lett.*, 69:2160–2162, 1996.

89. A. Gyoutoku, S. Hara, T. Komatsu, M. Shirinashihama, H. Iwanaga, and K. Sakanoue. "An organic electroluminescent dot-matrix display using carbon underlayer." *Synth. Met.*, 91:73–75, 1997.

90. J.X. Tang, Y.Q. Li, L.R. Zheng, and L.S. Hung. "Anode/organic interface modification by plasma fluorocarbon films." *J. Appl. Phys.*, 95:4397–4403, 2004.

91. Y. Kawamura, J. Brooks, J.J. Brown, H. Sasabe, and C. Adachi. "Intermolecular interaction and a concentration-quenching mechanism of phosphorescent Ir(III) complexes in a solid film." *Phys. Rev. Lett.*, 96:017404-1–017404-4, 2006.

92. V.E. Choong, Y. Park, N. Shivaparan, C.W. Tang, and Y. Gao. "Deposition-induced photoluminescence quenching of tris-(8-hydroxyquinoline) aluminum." *Appl. Phys. Lett.*, 71:1005–1007, 1997.

93. V. Adamovich, J. Brooks, A. Tamayo, A.M. Alexander, P.I. Djurovich, B.W. D'Andrade, C. Adachi, S.R. Forrest, and M.E. Thompson. "High efficiency single dopant white electrophosphorescent light emitting diodes." *New J. Chem.*, 26:1171–1178, 2002.

94. Y. Wang, N. Herron, V.V. Grushin, D. LeCloux, and V. Petrov. "Highly efficient electroluminescent materials based on fluorinated organometallic iridium compounds." *Appl. Phys. Lett.*, 79:449–451, 2001.

95. Y. Sato, T. Ogata, S. Ichinosawa, and Y. Murata. "Characteristics of organic electroluminescent devices with new dopants." *Synth. Met.*, 91:103–107, 1997.

96. T. Sano, H. Fijii, Y. Nishio, Y. Hamada, H. Takahashi, and K. Shibata. "Organic electroluminescent devices doped with condensed polycyclic aromatic compounds." *Synth. Met.*, 91:27–30, 1997.

97. Y. Hamada, T. Sano, K. Shibata, and K. Kuroki. "Influence of the emission site on the running durability of organic electroluminescent devices." *Jpn. J. Appl. Phys.*, 34:L824–L826, 1995.

98. M.A. Baldo, D.F. O'Brien, M.E. Thompson, and S.F. Forrest. "The excitonic singlet–triplet ratio in a semiconducting organic thin film." *Phys. Rev. B*, 60:14422–14428, 1999.

99. M.A. Baldo, D.F. O'Brien, Y. You, A. Shoustikov, S. Sibley, M.E. Thompson, and S.R. Forrest. "Highly efficient phosphorescent emission from organic electroluminescent devices." *Nature*, 395:151–154, 1998.

100. M.A. Baldo, S. Lamansky, P.E. Burrows, M.E. Thompson, and S.R. Forrest. "Very high-efficiency green organic light-emitting devices based on electrophosphorescence." *Appl. Phys. Lett.*, 75:4–6, 1999.

101. D.F. O'Brien, M.A. Baldo, M.E. Thompson, and S.R. Forrest. "Improved energy transfer in electrophosphorescent devices." *Appl. Phys. Lett.*, 74:442–444, 1999.

102. C. Adachi, M.A. Baldo, M.E. Thompson, and S.R. Forrest. "Nearly 100% internal phosphorescence efficiency in an organic light emitting device." *J. Appl. Phys.*, 90:5048–5051, 2001.

103. M. Ikai, S. Tokito, Y. Sakamoto, T. Suzuki, and Y. Taga. "Highly efficient phosphorescence from organic light-emitting devices with an exciton-block layer." *Appl. Phys. Lett.*, 79:156–158, 2001.

104. C. Adachi, R.C. Kwong, P. Djurovich, V. Adamovich, M.A. Baldo, M.E. Thompson, and S.R. Forrest. "Endothermic energy transfer: A mechanism for generating high-energy phosphorescent emission in organic materials." *Appl. Phys. Lett.*, 79:2082–2084, 2001.

105. R.C. Kwong, M.R. Nugent, L. Michalski, T. Ngo, K. Rajan, Y.-J. Tung, M.S. Weaver, T.X. Zhou, M. Hack, M.E. Thompson, S.R. Forrest, and J.J. Brown. "High operational stability of electrophosphorescent devices." *Appl. Phys. Lett.*, 81:162–164, 2002.

106. PHOLEDs: Features and performance. Available at: www.udcoled.com/default.asp?content ID=604.

107. R.N. Marks. "The optical and electronic response of poly(p-phenylene vinylene) thin film devices." PhD thesis, University of Cambridge, Cambridge, UK, 1993.

108. J. Blochwitz, M. Pfeiffer, M. Hofman, and K. Leo. "Non-polymeric OLEDs with a doped amorphous hole transporting layer and operating voltages down to 3.2 V to achieve 100 cd/m^2." *Synth. Met.*, 127:169–173, 2002.

109. M. Pfeiffer, S.R. Forrest, K. Leo, and M.E. Thompson. "Electrophosphorescent p–i–n organic light-emitting devices for very-high-efficiency flat-panel displays." *Adv. Mater.*, 14:1633–1636, 2002.

110. T. Tsutsui, M. Yahiro, H. Yokogawa, K. Kawano, and M. Yokoyama. "Doubling coupling-out efficiency in organic light-emitting devices using a thin silica aerogel layer." *Adv. Mater.*, 13:1149–1152, 2001.

111. W. Brutting, J. Frischeisen, T. Schmidt, B.J. Scholz, and C. Mayr. "Device efficiency of organic light-emitting diodes: Progress by improved outcoupling." *Phys. Status Solidi A*, 210(1):44–65, 2013.

112. V. Bulovic, V.B. Khalfin, G. Gu, P.E. Burrows, D.Z. Garbuzov, and S.R. Forrest. "Weak microcavity effects in organic light-emitting devices." *Phys. Rev. B*, 58:3730–3740, 1998.

113. S.R. Forrest, P.E. Burrows, and D.Z. Garbuzov. "Displays having mesa pixel configuration." U.S. Patent 6,091,195, July 18, 2000.

114. R.J. Nelson, and R.G. Sobers. "Minority-carrier lifetime and internal quantum efficiency of surface-free GaAs." *J. Appl. Phys.*, 49:6103–6108, 1978.

115. I. Schnitzer, E. Yablonovitch, C. Caneau, T.J. Gmitter, and A. Scherer. "30% external quantum efficiency from surface textured, thin-film light-emitting diodes." *Appl. Phys. Lett.*, 63:2174–2176, 1993.

116. R.H. Haitz. "Light-emitting diode with diagonal faces." U.S. Patent 5,087,949, February 11, 1992.

117. T.H. Gessmann, and E.F. Schubert. "High efficiency AlGaInP light-emitting diodes for solid state lighting applications." *J. Appl. Phys.*, 95:2203–2216, 2004.

118. F.A. Kish Jr., and S.A. Stockman. "Transparent substrate light emitting diodes with directed light output." U.S. Patent 6,015,719, January 18, 2000.

119. S.H. Fan, P.R. Villeneuve, J.D. Joannopoulos, and E.F. Schubert. "High extraction efficiency of spontaneous emission from slabs of photonic crystals." *Phys. Rev. Lett.*, 78:3294–3297, 1997.

120. M. Boroditsky, R. Vrijen, T.F. Krauss, R. Coccioli, R. Bhat, and E. Yablonovitch. "Spontaneous emission extraction and Purcell enhancement from thin-film 2-D photonic crystals." *J. Lightwave Technol.*, 17:2096–2112, 1999.

121. B.J. Matterson, J.M. Lupton, A.F. Safonov, M.G. Salt, W.L. Barnes, and I.D.W. Samuel. "Increased efficiency and controlled light output from a microstructured light-emitting diode." *Adv. Mater.*, 13:123–127, 2001.

122. C.F. Madigan, M.H. Lu, and J.C. Sturm. "Improvement of output coupling efficiency of organic light-emitting diodes by backside substrate modification." *Appl. Phys. Lett.*, 76:1650–1652, 2000.

123. T. Yamasaki, K. Sumioka, and T. Tsutsui. "Organic light-emitting device with an ordered monolayer of silica microspheres as a scattering medium." *Appl. Phys. Lett.*, 76:1243–1245, 2000.

124. S. Möller, and S.R. Forrest. "Improved light out-coupling in organic light emitting diodes employing ordered microlens arrays." *J. Appl. Phys.*, 91:3324–3327, 2001.

125. S.T. Lee, Z.Q. Gao, and L.S. Hung. "Metal diffusion from electrodes in organic light-emitting diodes." *Appl. Phys. Lett.*, 75:1404–1406, 1999.

126. K.A. Higginson, X.-M. Zhang, and F. Papadimitrakopoulos. "Thermal and morphological effects on the hydrolytic stability of aluminum tris(8-hydroxyquinoline) (Alq_3)." *Chem. Mater.*, 10:1017–1020, 1998.

127. D.Y. Kondakov, J.R. Sandifer, C.W. Tang, and R.H. Young. "Nonradiative recombination centers and electrical aging of organic light-emitting diodes: Direct connection between accumulation of trapped charge and luminance loss." *J. Appl. Phys.*, 93:1108–1119, 2003.

128. H. Aziz, Z. Popovic, S. Xie, A.-M. Hor, N.-X. Hu, C. Tripp, and G. Xu. "Humidity-induced crystallization of tris(8-hydroxyquinoline) aluminum layers in organic light-emitting devices." *Appl. Phys. Lett.*, 72:756–758, 1998.

129. B.H. Cumpston, and K.F. Jensen. "Electromigration of aluminum cathodes in polymer-based electroluminescent devices." *Appl. Phys. Lett.*, 69:3941–3943, 1996.

130. H. Aziz, Z.D. Popovic, N.-X. Hu, A.-M. Hor, and G. Xu. "Degradation mechanism of small molecule based organic light emitting diodes." *Science*, 283:1900–1902, 1999.

131. C. Adachi, K. Nagai, and N. Tamoto. "Molecular design of hole transport materials for obtaining high durability in organic electroluminescent diodes." *Appl. Phys. Lett.*, 66:2679–2681, 1995.

132. S.R. Forrest, P.E. Burrows, and M.E. Thompson. "Growth and characterization of electroluminescent display devices using vacuum-deposited organic materials." In *Organic Electroluminescent Materials and Devices*, S. Miyata, and H.S. Nalwa, Eds., Gordon & Breach, Amsterdam, 1997, pp. 447–453.

133. Z.D. Popovic, and H. Aziz. "Reliability and degradation of small molecule-based organic light-emitting devices (OLEDs)." *IEEE J. Sel. Top. Quant. Electron.*, 8:362–371, 2002.

134. F. So, and D. Kondakov. "Degradation mechanisms in small-molecule and polymer organic light-emitting diodes." *Adv. Mat.*, 22:3762–3777, 2010.

135. J. Kido. "Organic displays." *Phys. World*, 12(Suppl. 3):27–30, 1999.

136. Y.A. Ono. *Electroluminescent Displays*, World Scientific, Singapore, 1995, pp. 98–117.

137. C. Hosokawa, M. Eida, M. Matsuura, K. Fukuoka, H. Nakamura, and T. Kusumoto. "Organic multi-color electroluminescence display with fine pixels." *Synth. Met.*, 91:3–7, 1997.

138. R.C. Kwong, M.R. Nugent, L. Michalski, T. Ngo, K. Rajan, Y.J. Tung, A.B. Chwang, M.S. Weaver, T.X. Zhou, M. Hack, and J.J. Brown. "Display properties of high-efficiency electrophosphorescent diodes." In *Proceedings of the Society for Information Display*, Digest of Technical Papers, Vol. 33(Suppl. 2), Boston, 2002, pp. 1365–1367.

139. G. Gu, and S.R. Forrest. "Design of flat-panel displays based on organic light-emitting devices." *IEEE J. Sel. Top. Quant. Electron.*, 4:83–99, 1998.

140. C. Kim, P.E. Burrows, and S.R. Forrest. "Micropatterning of organic electronic devices by cold-welding." *Science*, 288:831–833, 2000.

141. S. Noach, E.Z. Faraggi, G. Cohen, Y. Avny, R. Neumann, D. Davidov, and A. Lewis. "Microfabrication of an electroluminescent polymer light emitting diode pixel array." *Appl. Phys. Lett.*, 69:3650–3652, 1996.

142. D.G. Lidzey, M.S. Weaver, M.A. Pate, T.A. Fisher, D.D.C. Bradley, and M.S. Skolnick. "Photoprocessed and micropatterned conjugated polymer LEDs." *Synth. Met.*, 82:141–148, 1996.

143. R.M.A. Dawson, Z. Shen, D.A. Furst, S. Connor, J. Hsu, M.G. Kane, R.G. Stewart, A. Ipri, C.N. King, P.J. Green, R.T. Flegal, S. Pearson, W.A. Barrow, E. Dickey, K. Ping, C.W. Tang, S. VanSlyke, F. Chen, and J. Shi. "Design of an improved pixel for a polysilicon active-matrix organic LED display." In *Proceedings of the Society for Information Display*, Digest of Technical Papers, Vol. 29, Anaheim, CA, 1998, pp. 11–14.

144. M.J. Powell, C. van Berkel, and J.R. Hughes. "Time and temperature dependence of instability mechanisms in amorphous silicon thin-film transistors." *Appl. Phys. Lett.*, 54:1323–1325, 1989.

145. Y. He, R. Hattori, and J. Kanicki. "Four-thin film transistor pixel electrode circuits for active-matrix organic light-emitting displays." *Jpn. J. Appl. Phys.*, 40(Part 1):1199–1208, 2001.

146. P. Burrows, G. Gu, S.R. Forrest, P.E. Vicenzi, and T.X. Zhou. "Semitransparent cathodes for organic light emitting devices." *J. Appl. Phys.*, 87:3080–3085, 2000.

147. M.H. Lu, M.S. Weaver, T.X. Zhou, M. Rothman, R.C. Kwong, M. Hack, and J.J. Brown. "High efficiency top-emitting organic light-emitting devices." *Appl. Phys. Lett.*, 81:3921–3923, 2002.

148. M.H. Lu, M.S. Weaver, T.X. Zhou, M. Rothman, R.C. Kwong, M. Hack, and J.J. Brown. "Highly efficient top-emitting electrophosphorescent organic light-emitting devices." In *Proceedings of the 2nd International Meeting on Information Displays*, Daegu, 2002, pp. 90–93.

149. C.W. Tang, and B.C. Hseih. "Method of fabricating a TFT-EL pixel." U.S. Patent 5,550,066, August 27, 1996.

150. L.K. Lam, P. Fleming, and D.G. Ast. "Low temperature polysilicon TFT active matrix OLED display on Corning 1737 glass." *Proceedings of Asia Display '98*, 225–228, 1998.

151. *Nikkei Sangyo Shinbun*, October 27, 1995 (in Japanese).

152. J.J. Lih, C.F. Sung, M.S. Weaver, M. Hack, and J.J. Brown. "A phosphorescent active-matrix OLED display driven by amorphous silicon backplane." In *Proceedings of the Society for Information Displays International Symposium*, Digest of Technical Papers, Vol. 34(Book 1), Baltimore, 2003, pp. 14–17.

153. T. Tsujimura, Y. Kobayashi, K. Murayama, A. Tanaka, M. Morooka, E. Fukumoto, H. Fujimoto, J. Sekine, K. Kanoh, K. Takeda, K. Miwa, M. Asano, N. Ikeda, S. Kohara, S. Ono, C.-T. Chung, R.-M. Chen, J.-W. Chung, C.-W. Huang, H.-R. Guo, C.-C. Yang, C.-C. Hsu, H.-J. Huang, W. Riess, H. Riel, S. Karg, T. Beierlein, D. Gundlach, S. Alvarado, C. Rost, P. Mueller, F. Libsch, M. Maestro, R. Polastre, A. Lien, J. Sanford, and R. Kaufman. "A 20-inch OLED display driven by super-amorphous-silicon technology." In *Proceedings of the Society for Information Displays International Symposium*, Digest of Technical Papers, Vol. 34(Book 1), Baltimore, 2003, pp. 6–9.

154. J.H. Burroughes, C.A. Jones, and R.H. Friend. "New semiconductor device physics in polymer diodes and transistors." *Nature*, 335:137–141, 1988.

155. H. Sirringhaus, N. Tessler, and R.H. Friend. "Integrated optoelectronic devices based on conjugated polymers." *Science*, 280:1741–1744, 1998.

156. A. Dodabalapur, Z. Bao, A. Makhija, J.G. Laquindanum, V.R. Raju, Y. Feng, H.E. Katz, and J. Rogers. "Organic smart pixels." *Appl. Phys. Lett.*, 73:142–144, 1998.

157. K. Nomoto, M. Noda, N. Kobayashi, M. Katsuhara, A. Yumoto, S.-I. Ushikura, R.-I. Yasada, N. Hirai, G. Yukawa, and I. Yagi. "Rollable OLED display driven by organic TFTs." In *Proceedings of the Society for Information Display*, Digest of Technical Papers, Vol. 42(1), Boston, 2012, pp. 488–491.

158. J.P. Spindler, T.K. Hatwar, M.E. Miller, A.D. Arnold, M.J. Murdoch, P.J. Kane, and S.A. Van Slyke. "Lifetime- and power-enhanced RGBW displays based on white OLEDs." In *Proceedings of the 2005 Society for Information Displays International Symposium*, Digest of Technical Papers, Vol. 36, Boston, 2005, pp. 36–39.

159. J. Kido, K. Hongawa, K. Okuyama, and K. Nagai. "White light-emitting organic electroluminescent devices using the poly(N-vinylcarbazole) emitter layer doped with three fluorescent dyes." *Appl. Phys. Lett.*, 64:815–817, 1994.

160. M. Arai, K. Nakaya, O. Onitsuka, T. Inoue, M. Codama, M. Tanaka, and H. Tanabe. "Passive matrix display of organic LEDs." *Synth. Met.*, 91:21–25, 1997.

161. C. Hosokawa, M. Eida, M. Matsuura, K. Fukuoka, H. Nakamura, and T. Kusumoto. "Organic multicolor EL display with fine pixels." In *Proceedings of the Society for Information Displays International Symposium*, Digest of Technical Papers, Vol. 28, Boston, 1997, pp. 1073–1076.

162. C. Hosokawa, K. Fukuoka, H. Kawamura, T. Sakai, M. Kubota, M. Funahashi, F. Moriwaki, and H. Ikeda. "Improvement of lifetime in organic electroluminescence." In *Proceedings of the Society for Information Displays International Symposium*, Digest of Technical Papers, Vol. 35(Book 2), Seattle, WA, 2004, pp. 780–783.

163. T. Kato. "New deep blue fluorescent materials and their applications to high performance OLEDs." In *FPD International 2011 Forum*, Yokohama, Japan, Papers B-03, 2011.

164. C. Hosokawa, M. Matsuura, M. Eida, K. Fukuoka, H. Tokailin, and T. Kusumoto. "Full color organic EL display." In *Proceedings of the 1998 Society for Information Displays International Symposium*, Digest of Technical Papers, Vol. 29, Anaheim, CA, 1998, pp. 7–10.

165. S. Tokito, Y. Taga, and T. Tsutsui. "Strongly modified emission from organic electroluminescent device with a microcavity." *Synth. Met.*, 91:49–52, 1997.

166. T. Nakayama, Y. Itoh, and A. Kakuta. "Organic photo- and electroluminescent devices with double mirrors." *Appl. Phys. Lett.*, 63:594–595, 1993.

167. A. Dodabalapur, L.J. Rothberg, T.M. Miller, and E.W. Kwock. "Microcavity effects in organic semiconductors." *Appl. Phys. Lett.*, 64:2486–2488, 1994.

168. P.E. Burrows, S.R. Forrest, S.P. Sibley, and M.E. Thompson. "Color-tunable organic light-emitting devices." *Appl. Phys. Lett.*, 69:2959–2961, 1996.

169. M. Uchida, Y. Ohmori, T. Noguchi, T. Ohnishi, and K. Yoshino. "Color-variable light-emitting diode utilizing conducting polymer containing fluorescent dye." *Jpn. J. Appl. Phys.*, 32:L921–L924, 1993.

170. T. Mori, K. Obata, K. Imaizumi, and T. Mizutani. "Preparation and properties of an organic light emitting diode with two emission colors dependent on the voltage polarity." *Appl. Phys. Lett.*, 69:3309–3311, 1996.

171. S.T. Lee, J.Y. Lee, M.H. Kim, M.C. Suh, T.M. Kang, Y.J. Choi, J.Y. Park, J.H. Kwon, and H.K. Chung. "A new patterning method for full-color polymer light-emitting devices: Laser induced thermal imaging (LITI)." In *Proceedings of the Society for Information Display*, Digest of Technical Papers, Vol. 33(Suppl. 2), Boston, 2002, pp. 784–787.

172. S.T. Lee, B.D. Chin, M.H. Kim, T.M. Kang, M.W. Song, J.H. Lee, H.D. Kim, H.K. Chung, M.B. Wolk, E. Bellman, J.P. Baetzold, S. Lamansky, V. Savvateev, T.R. Hoffend, J.S. Staral, R.R. Roberts, and Y. Li. "A novel patterning method for full-color organic light-emitting devices: Laser induced thermal imaging (LITI)." In *Proceedings of the Society for Information Display*, Digest of Technical Papers, Vol. 35(Suppl. 2), Seattle, WA, 2004, pp. 1008–1011.

173. G. Gu, P.E. Burrows, S. Venkatesh, and S.R. Forrest. "Vacuum-deposited, non-polymeric flexible organic light-emitting devices." *Opt. Lett.*, 22:172–174, 1997.

174. G. Gu, P.E. Burrows, and S.R. Forrest. "Vacuum-deposited, non-polymeric flexible organic light emitting devices." U.S. Patent 5,844,363, December 1, 1998.

175. M.S. Weaver, W.D. Bennet, C. Bonham, P.E. Burrows, G.L. Graff, M.E. Gross, M. Hall, R.H. Hewitt, S.Y. Mao, E. Mast, P.M. Martin, L.A. Michalski, T. Ngo, K. Rajan, M.A. Rothman, and J.A. Silvernail. "Flexible organic LEDs." In *Proceedings of the Society for Information Display, Electronic Information Displays Conference*, London, 2000, pp. 191–222.

176. A. Yoshida, S. Fujimura, T. Miyake, T. Yoshizawa, H. Ochi, A. Sugimoto, H. Kubota, T. Miyadera, S. Ishizuka, M. Tsuchida, and H. Nakada. "3-Inch full-color OLED display using a plastic substrate." In *Proceedings of the 2003 Society for Information Displays International Symposium*, Digest of Technical Papers, Vol. 34(Book 2), Baltimore, 2003, pp. 856–859.

177. P.M. Martin. "Barrier coatings from potato chips to display chips." *Vac. Technol. Coating*, 10:20–24, 2000.

178. J.D. Affinito, M.E. Gross, P.A. Mournier, M.K. Shi, and G.L. Graff. "Ultrahigh rate, wide area, plasma polymerized films from high molecular weight/low vapour pressure liquid or solid monomer precursors." *J. Vac. Sci. Technol. A*, 17:1974–1981, 1999.

179. L.L. Moro, T.A. Krajewski, N.M. Rutherford, O. Phillips, R.J. Visser, M.E. Gross, W.D. Bennet, and G.L. Graff. Process and design of a multilayer thin film encapsulation of passive matrix OLED displays, Proceedings of SPIE, Conference Volume 5214, Organic Light-Emitting Materials and Device VII, 83 (Feb 16, 2004).

180. A.B. Chwang, M.A. Rothman, S.Y. Mao, R.H. Hewitt, M.S. Weaver, J.A. Silvernail, K. Rajan, M. Hack, J.J. Brown, X. Chu, L. Moro, T. Krajewski, and N. Rutherford. "Thin film encapsulated flexible organic electroluminescent displays." In *Proceedings of the Society for Information Display*, Digest of Technical Papers, Vol. 34(2), Baltimore, 2003, pp. 868–871.

181. A.C. Chwang, M.A. Rothman, S.Y. Mao, R.H. Hewitt, M.S. Weaver, J.A. Silvernail, K. Rajan, M. Hack, J.J. Brown, X. Chu, M. Lorenza, T. Krajewski, and N. Rutherford. "Thin film encapsulated flexible OLED displays." *Appl. Phys. Lett.*, 83:413–415, 2003.

182. P. Mandlik, H. Pang, P.A. Levermore, J.A. Silvernail, K. Rajan, E. Krall, R. Ma, and J.J. Brown. "Flexible phosphorescent OLED lighting devices encapsulated with a novel thin film barrier." In *Society of Vacuum Coaters Technical Conference Proceedings*, Santa Clara, CA, 2012.

183. Z. Shen, P.E. Burrows, V. Bulovic, S.R. Forrest, and M.E. Thompson. "Three-color, tunable, organic light-emitting devices." *Science*, 276:2009–2011, 1997.

184. B.W. D'Andrade, J. Brooks, V. Adamovich, M.E. Thompson, and S.R. Forrest. "White emission using triplet excimers in electrophosphorescent organic light-emitting devices." *Adv. Mater.*, 14:1032–1036, 2002.

185. B.W. D'Andrade, M.E. Thompson, and S.R. Forrest. "Controlling exciton diffusion in multi-layer white phosphorescent organic light emitting devices." *Adv. Mater.*, 14:147–151, 2002.

186. M.S. Weaver, P.A. Levermore, V.I. Adamovich, K. Rajan, C. Lin, G.S. Kottas, S. Xia, R. Ma, R.C. Kwong, M. Hack, and J.J. Brown. "White phosphorescent OLED lighting: A green technology." In *Proceedings of the IMID Conference*, 30.1, Seoul, 2010, pp. 208–209.

187. T. Komoda, N. Ide, V. Kittichungchit, K. Yamae, H. Tsuji, and Y. Matsuhisa. "High-performance white OLEDs with high color rendering index for next-generation solid-state lighting." *J. SID*, 19, pp. 838, 2011.

188. A.R. Duggal. "Color tunable organic electroluminescent light source." U.S. Patent 6,661,029, December 9, 2003.

189. H. Antoniadis, and K. Pichler. "Organic light emitting diode light source." U.S. Patent 6,680,578, January 20, 2004.

190. M. Hack, M.H. Lu, and M.S. Weaver. "Organic light emitting devices for illumination." U.S. Patent Application 20,040,032,205, February 19, 2004.

191. P.A. Levermore, V.I. Adamovich, K. Rajan, W. Yeager, C. Lin, S. Xia, G.S. Kottas, M.S. Weaver. R. Kwong, R. Ma, M. Hack, and J.J. Brown. "Highly efficient phosphorescent OLED lighting panels for solid state lighting." In *Proceedings of Society for Information Display*, Digest of Technical Papers, Vol. 41(2), 2010, pp. 786–789.

192. B. D'Andrade, and S.R. Forrest. "White organic light-emitting devices for solid state lighting." *Adv. Mater.*, 16:1585–1595, 2004.

193. M.A. Baldo, R.J. Holmes, and S.R. Forrest. "Prospects for electrically pumped organic lasers." *Phys. Rev. B*, 66:035320.1–035321.16, 2002.
194. J. Xue, S. Uchida, B.P. Rand, and S.R. Forrest. "4.2% efficient organic photovoltaic cells with low series resistances." *Appl. Phys. Lett.*, 84:3013–3015, 2004.
195. M.J. Currie, J.K. Mapel, T.D. Heidel, S. Goffri, and M.A. Baldo. "High-efficiency organic solar concentrators for photovoltaics." *Science*, 321(5886):226–228, 2008.
196. Y. Liang, Z. Xu, J. Xia, S.-T. Tsai, Y. Wu, G. Li, C. Ray, and L. Yu. "For the bright future—Bulk heterojunction polymer solar cells with power conversion efficiency of 7.4%." *Adv. Mater.*, 22(20):E135–E138, 2010.
197. X. Zhan, and D. Zhu. "Conjugated polymers for high-efficiency organic photovoltaics." *Polym. Chem.*, 1:409–419, 2010.
198. D.J. Gundlach, H. Klauk, C.D. Sheraw, C.C. Kuo, J.R. Huang, and T.N. Jackson. "High-mobility, low voltage organic thin film transistors." In *Proceedings of the International Electron Devices Meeting Technical Digest*, December 1999, pp. 111–114.
199. M. Yamagishi, Y. Tominari, R. Hirahara, Y. Nakazawa, T. Nishikawa, T. Kawase, T. Shimoda, and S. Ogawa. "Very high-mobility organic single-crystal transistors with in-crystal conduction channels." *Appl. Phys. Lett.*, 90(10):102120–102120-3, 2007.

8

Material Challenges for Flexible OLED Displays

Peter F. Carcia

CONTENTS

8.1 Introduction

In a product concept video by one of the leading display manufacturers, a transparent, folded sheet of plastic, small enough to store in the billfold of a wallet, unfolds to the size of a smartphone and then to a tablet, suitable for viewing a newspaper. This is one futuristic vision of a flexible organic light-emitting device (OLED) display. Another flexible product concept is that of an OLED display with a rollable screen that can be retracted into a compact cylinder when not in use. However, even a flat OLED display built on a compliant, flexible plastic substrate is also expected to have the advantages of greater durability and lighter weight than one on rigid glass.

For all of these formats—foldable, rollable, or flat—there are a number of considerations beyond the obvious mechanical issues of bending and more extreme folding of display materials [1]. The choice of substrate material [2] and its compatibility with the OLED display fabrication process is one. If the choice of flexible substrate is a polymer, then a barrier packaging scheme [3] that excludes gas permeation is essential for acceptable OLED device lifetime—typically years—since polymers are highly permeable to moisture and air, which react with and degrade OLED materials. (Glass is impermeable to gases and therefore acts as its own barrier.) Furthermore, in an OLED display with active-matrix (AM) addressing, each color pixel is driven by an individual electronic circuit, and technologies based on silicon, organic, or oxide semiconductors are currently being investigated for driving an OLED [4]. What are the advantages and suitability of each technology for flexible OLED displays? These are the topics that this chapter will discuss: (i) substrate materials, (ii) barrier material approaches for packaging OLEDs, and (iii) thin-film transistor technologies to drive flexible AMOLEDs. Obviously, it is not possible to give much more than an overview of each of these topics in this short chapter. For more in-depth discussion, each section in this chapter provides references to recent review articles on that particular topic.

8.2 Substrates

8.2.1 Overview

In general, the base substrate for a flexible OLED display could be a polymer, a metal foil, e.g., stainless steel, or "flexible" glass, a product that Corning is developing [5]. However, for a foldable display, a polymer would be the best choice, since thin (0.1 mm) flexible glass is currently limited to a bending radius of ~5 cm before incurring significant bend stress [5], and metal foils are heavy, not transparent, which restricts the display structure to top emission only, and typically have rough surfaces, unsuitable for growing thin films without planarization. Compared with a stiffer metal substrate, use of a compliant polymer can also substantially reduce the strain in the thin-film layers of an OLED display and is more compatible with encapsulation schemes to neutralize strain [2]. We will therefore discuss the properties and limitations of polymer candidate substrates in this section. In that regard, we will draw extensively on the authoritative published work of our colleague, Dr. Bill MacDonald of DuPont–Teijin Films [2,6,7].

Fabricating an AMOLED display requires a base substrate that is optically clear, stable at high device processing temperature, dimensionally stable with changes in temperature, chemically resistant to solvents, has an ultrasmooth surface, and is hermetic to atmospheric gases. An optically clear substrate offers the possibility of both bottom- and top-emission OLEDs. In general, thin films processed at higher temperature are denser, more conductive, and exhibit higher performance. For example, thin-film transistors (TFTs) that drive OLED pixels typically show higher performance when they are processed at elevated temperatures (>300°C). For polymer substrates, the process window is at least up to the glass transition temperature (T_g); thus, higher T_gs are favored. To achieve proper registration between the device layers, dimensional changes in the substrate should be small and predictable. Furthermore, device processing commonly involves use of chemical solvents that the substrate must be able to tolerate. A smooth substrate without particulates, defects, and scratches is also critical for a reliable, high-quality display. Moreover, because the constituents of an OLED display are especially sensitive to atmospheric gases, particularly water vapor, the substrate should be impermeable to these gases. A glass substrate meets all of these process requirements for OLED fabrication; however, glass is not flexible. While no single polymer base film matches all the properties of glass, the challenge is to modify the polymer, e.g., adding layers, to approach the properties of glass.

8.2.2 Leading Polymer Candidates

Figure 8.1 summarizes the leading candidate polymers under consideration for application as a substrate in electronics [7]. They range categorically from the semicrystalline polyesters, as typified by polyethylene terephthalate (PET) and polyethylene naphthalate (PEN); to amorphous thermoplastic polymers, such as polycarbonate (PC) and polyethersulfone (PES); to high-temperature, solvent-cast, amorphous polymers, of which polyimides (PIs), aromatic fluorene containing polyarylates (PAR), and polynorborene (also known as polycyclic olefins, PCO) are examples. From this graph, the trend is increasing T_g, as we move from the semicrystalline polyesters to thermoplastic polymers, to the higher T_g solvent-cast polymers. Although it may seem that the polyesters with lower T_g are less attractive candidates, annealing PET and PEN above their T_g produces heat-stabilized (HS) products (HS-PET and HS-PEN) [2] with good dimensional stability and higher upper-use temperature—150°C for PET and 200°C for PEN, comparable to the thermoplastics. (The melting temperature for PET and PEN is ~250°C.)

Table 8.1 gives a comparison of specific properties—optical transmission (%T), coefficient of thermal expansion (CTE), water absorption (%), and mechanical moduli—for

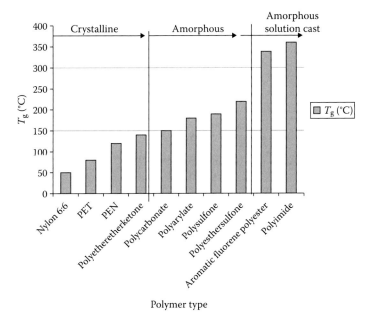

FIGURE 8.1
Glass transition temperature (T_g) of candidate polymers for flexible electronics. (Reprinted with the permission from B.A. MacDonald, K. Rollins, R. Eveson, K. Rakos, B.A. Rustin, M. Handa, *Mat. Res. Soc. Symp. Proc.*, 769: H9.3.1–H9.3.8, 2003. Copyright 2011 Cambridge University Press.)

TABLE 8.1

Property Summary for Representative Polymer Candidates for Flexible Electronics

Base Polymer	HS-PET	HS-PEN	PC	PES	PAR	PCO	PI
CTE ($-55°C$ to $85°C$) (ppm/$°C$)	15	13	60–70	54	53	74	30–60
% Transmission (400–700 nm)	>85	>85	>90	90	90	91.6	Yellow
Water absorption (%)	0.14	0.14	0.4	1.4	0.4	0.03	1.8
Young's modulus (GPa)	5.3	6.1	1.7	2.2	2.9	1.9	2.5
Tensile strength (MPa)	225	275	NA	83	100	50	231

Source: W.A. MacDonald, *J. Mater. Chem.*, 14:4–10, 2004. Reproduced by permission of The Royal Society of Chemistry.

representative polymers [6]. CTE and water absorption both are relevant to dimensional changes in the base polymer film during high-temperature processing.

8.2.2.1 Optical Properties

With the exception of PI, which is yellow, all of the polymers have adequate (>85%) optical transmission. Birefringence, which does restrict use of the semicrystalline polyester polymers in polarization-sensitive displays, e.g., liquid crystal display (LCD), is not an issue for OLED displays.

8.2.2.2 Dimensional Stability

The CTE for HS polyesters is <20 ppm/$°C$, whereas it is >50 ppm/$°C$ for the other amorphous polymers. As a comparison, the CTE for inorganic materials found in an OLED display are ~10 ppm/$°C$ for ITO, 22 ppm/$°C$ for Al, and 7 ppm/$°C$ for Al_2O_3. The thermal strain that develops when one of these materials is deposited on a polymer is approximately equal to the product of the difference in their CTEs and the temperature difference associated with the particular processing step.

It is noteworthy that before heat stabilization, the dimensional stability of the polyesters below T_g is dominated by shrinkage due to irreversible strain relaxation in the crystalline regions of the polymer. After thermal annealing, this shrinkage is reduced to <0.1% in HS-PEN. Furthermore, shrinkage of only 25 ppm at 150°C, acceptable for some small displays, is achieved when HS-PEN (DuPont–Teijin Films Q65) is further annealed for an extended time at 200°C [2]. In contrast, the amorphous polymers exhibit very low shrinkage.

Water absorption, which can also cause undesirable dimensional changes during device processing, varies among the polymers, with low absorption for PCO and relatively high absorption for PI. In polyesters, moisture absorption can take up to 12 h to reach equilibrium values, and for every 100 ppm they absorb, the polymer expands equally in all directions by 45 ppm. Coating both sides of the polymer base substrate with a thin-film inorganic, gas diffusion barrier can retard water absorption, minimizing absorption-related dimensional changes. For comparison with glass, the CTE for an LCD display grade is <4 ppm, and dimensional change due to water absorption is negligible.

To put the magnitude of these dimensional changes due to CTE mismatch, shrinkage, and water absorption in perspective, a 50 ppm change during the fabrication of an OLED display on a polymer web width of 1 m can cause a 50 μm misregistration in device layers. Moreover, in a high-definition display, in which a pixel transistor size is ~100 μm, a dimensional shift of 50 μm is not tolerable.

8.2.2.3 Mechanical Moduli

Mechanically, because of their crystalline character, the polyesters are stiffer than the other amorphous polymers. For the same thickness, therefore, the polyesters may have an advantage for batch processing, which is the current commercial process for producing displays on large sheets of glass. Of course, it is also possible to temporarily attach (glue) a polymer base substrate to rigid glass, to constrain dimensional changes during batch processing and then remove the flexible substrate, when the display fabrication is complete. However, this adds cost, and one attraction of flexible polymer base substrates, with the evolution of solution deposition and graphic-art printing methods applied to device fabrication, is low-cost, high-volume roll-to-roll (R2R) manufacturing. In that case, stiffer polymers, such as the semicrystalline polyesters, will stretch less because of winding tension.

8.2.2.4 Solvent Resistance

Common solvents used during device fabrication include ketones, alcohols, esters, hydrocarbons, acids, and alkali. In general, amorphous polymers have poor solvent resistance compared with the semicrystalline polymers. This necessitates coating the amorphous polymers with a hard coat, which imparts both chemical and scratch or mechanical resistance. Amorphous polymers with a hard coat and crystalline polymers with or without a coating are then both tolerant to the solvent chemicals used in device fabrication.

8.2.2.5 Surface Quality

The surface quality of commercial polymer base films can vary significantly, from very good to poor, even for the same film type, and often does not meet the requirements for OLED fabrication. Surface quality is directly related to cleanliness conditions during the film manufacture and postproduction cleaning steps. A polymer base film is not usually manufactured in a cleanroom or to the standards demanded for electronic and display device fabrication. Furthermore, the low volume of polymer base film used in electronic applications does not encourage making costly upgrades in manufacturing. Ignoring cost considerations, improvements can likely be made to film surface quality by adopting cleanroom-like manufacturing standards and practices, adapted to polymer film production. A more immediate solution to this problem, however, is to coat the film surface in a cleanroom with a planarizing layer. The idea is that the liquid-like coating covers defects and also functions as a scratch-resistant hard coat, after curing. Of course, the better the initial surface quality of the polymer film, the thinner the planarizing layer thickness that is needed to overcoat scratches and defects—100s versus 1000s of nanometers is more desirable to avoid delamination issues during processing due to thermomechanical differences between the polymer base film and the planarizing layer.

8.2.2.6 Summary Comparison

Table 8.2 presents a summary comparison of the three categories of bare polymer substrate candidates for flexible electronics [2]. Note that only the HS version of the polyesters is considered, and some of the noted deficiencies of other polymers can be remedied or improved by adding a coating. If a high processing temperature is the key property, then PI may be the best candidate. A hardened PI substrate to accommodate a polysilicon backplane is what Samsung has apparently chosen for its initial flexible display technology. For

TABLE 8.2

Comparison of Relative Property Merits of Polymer Candidates for Flexible Electronics

Property	Heat-Stabilized PET Teonex© Q65	Heat-Stabilized PEN Melinex© ST504/506	PC	PES	Polyarylate	Polyimide
CTE (−55°C to 85°C) (ppm/°C)	√√	√√	√	√	√	√√
% Transmission (400–700 nm)	√√	√√	√√√	√√	√√	×
Water absorption (%)	√√	√√	√	×	√	×
Young's modulus (GPa)	√√	√√	√	√	√	√
Tensile strength (MPa)	√√	√√	√	√	√	√
Solvent resistance	√√	√√	×	×	×	√√
Upper operating temperature	√√	√	√	√√	√√√	√√√

Source: W.A. MacDonald, M.K. Looney, D. Mackerron, R. Eveson, R. Adam, K. Hashimoto, and K. Rakos: Latest advances in substrates for flexible electronics. *Journal of the SID*. 2007. 15. W1075–W1083. Copyright Wiley-VCH Verlag GmbH & Co. kGaA. Reproduced with permission.

Note: √√√, Excellent; √√, good; √, moderate; ×, poor.

a backplane technology, e.g., organic or oxide semiconductors, that can be processed at a lower temperature, HS-PEN may be more attractive. The table does not reflect important cost considerations. For a more cost-driven application, such as flexible OLED lighting, a PET substrate could be attractive because of its low cost.

8.3 Barrier Coatings

8.3.1 OLED Barrier Requirements

Very early in the development of OLED displays, researchers recognized that these devices would need a robust packaging strategy to protect them from rapid degradation by reaction with atmospheric gases [8]. Specifically, organic layers are air sensitive, and efficient devices need a low-work-function metal cathode, e.g., Ca, Ba, Li, or Cs, that all react with air or moisture. Early experiments and models predicted that the moisture permeation rates for the substrate had to be less than ~10^{-5} g-H_2O/m^2-day for adequate OLED lifetimes, up to 10,000 h [8]. Although polymers are commonly used in food packaging to extend product lifetime [9], their permeation rates are ~0.1–10 g/m^2-day, much too high to protect OLEDs. Free volume, which is a characteristic of all polymeric materials, limits their barrier properties. To meet the lifetime requirements of OLEDs, a novel barrier technology is needed for flexible polymer substrates. Until then, only glass, which is impermeable, is a suitable substrate for OLED displays. (Flexible metal foil could be used as the substrate since thick foils are impermeable; however, the top layer would need to be transparent—a nonmetal—and have barrier properties.)

In principle, a defect-free, thin inorganic coating on a polymer film should be impermeable. However, transparent inorganic thin films, such as sputtered or evaporated SiO_x, SiN_x, and Al_2O_3, coated on polymers to extend lifetime for food packaging, have defects [10] and consequently only reduce permeation by a factor of ~10–50× compared with the bare polymer film. This is not an adequate barrier for protecting OLEDs. One source of defects in vacuum-deposited inorganic thin films is due to substrate imperfections [11], such as particles and scratches, which are mirrored in the film. Also, vapor-deposited films have a microstructure with boundaries and gaps that evolve from the nucleation process and self-shadowing during growth [12]. All of these film imperfections lead to facile pathways for gas permeation, and consequently only a modest reduction in gas permeation for these coated polymers. Note that the same issues apply whether the barrier film is being applied to the polymer substrate or directly to the OLED device.

8.3.2 Organic–Inorganic Multilayer Barrier Films

8.3.2.1 Multilayers

To improve on the shortcomings of a single-layer barrier film for OLEDs, the group at Battelle Pacific Northwest National Laboratories (PNNL) demonstrated a multilayer barrier technology that gave promising results for encapsulating OLEDs [8,13]. This technology was subsequently spun-off into a private company, Vitex, and currently commercial quantities of this multilayer barrier are offered by 3M Corp., which was also an early adopter of the technology. This multilayer technology is still today a promising candidate for OLED encapsulation, although the lengthy time to deposit multiple layers presents a challenge for the cost-effective manufacturing of flexible OLEDs. In the following sections, we will describe more details about this multilayer technology, some of it variants, and we will also describe a relatively new single-layer encapsulation technology based on atomic layer deposition (ALD) [14].

An electron image of a multilayer barrier cross-section is shown in Figure 8.2. The layers consist of alternating polymer and inorganic oxide layers. The first layer, which can be up to a few microns thick, is a liquid-like polymer (e.g., an acrylate), which after curing smooths or planarizes the substrate. That is, this planarizing layer covers over substrate imperfections and irregularities, where the substrate could be either the polymer substrate or the OLED device itself. After the first polymer smoothing layer is cured, an inorganic oxide layer, typically <50 nm thick, is deposited by physical (e.g., sputtering or evaporation) or chemical vapor deposition (CVD). Battelle Labs has used sputtered Al_2O_3. Sputtering on a smooth polymer layer is intended to promote growth of an inorganic barrier layer with

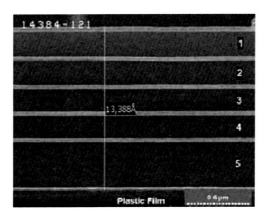

FIGURE 8.2
Electron microscope image of a polymer multilayer barrier film consisting of 5 dyads of polymer–inorganic thin film layers. (Reprinted from *Mater. Today*, 9, J. Lewis, Material challenge for flexible organic devices, 38–45, Copyright 2006, with permission from Elsevier.)

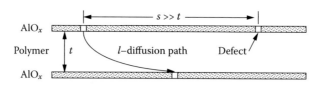

FIGURE 8.3
Illustration of increased gas permeation diffusion path in a bilayer laminated barrier structure. (Reprinted with permission from G.L. Graff, R.E. Williford, P.E. Burrows, *J. Appl. Phys.*, 96: 1840–1849, 2004. Copyright 2004, American Institute of Physics.).

fewer defects. The pair of layers, called a dyad, is repeated multiple times to enhance barrier properties. Successive polymer layers are generally thinner than the initial smoothing layer, approximately several hundred nanometers, and their purpose is to decouple defects that could occur in one of the inorganic barrier layers. This creates very long pathways for permeating gas molecules to diffuse between defects in adjacent inorganic layers. This rationale for increasing the path for diffusing species is illustrated schematically in Figure 8.3. Clearly, the need to deposit several dyads to encapsulate an OLED could be a time-consuming process, and this is one of the reasons for its high cost and a hurdle to its adoption in OLED manufacturing, where the barrier is directly deposited on the OLED device.

Close to the Battelle approach is a polymer multilayer barrier reported by the Philips group [15]. In this implementation, the inorganic barrier and polymer layers are also separate layers, as in the Battelle–Vitex structure; however, the barrier layer is SiN_x produced by CVD.

From the information provided on their website, the barrier technology of Mitsubishi Plastics [16] also consists of multilayers. In their approach, the barrier is a vapor-deposited SiO_x on a thin polymer base film. A low water vapor transmission rate (WVTR) is achieved by laminating several of these films together. A WVTR of $<10^{-4}$ g/m²-day is claimed for the "X-barrier" version.

8.3.2.2 Modulated and Composite Barrier Structure

Another implementation of this polymer multilayer barrier approach involves a continuous, R2R, modulated CVD process, used by a group at GE [17,18]. When the barrier structure calls for an inorganic layer, the chemical precursor concentrations in the CVD process are adjusted to produce a layer of SiN_x. Shifting to a polymer interlayer is accomplished by gradually changing the precursor chemistry during deposition, so that the deposited layer has a more "polymer" or "organic" character. In this way, a polymer–inorganic multilayer structure is created by modulating chemical precursor chemistry continuously during deposition, and the multilayer structure is also continuously modulated. Good results for encapsulating OLEDs are reported by this group.

Related to the modulated CVD approach of GE is the barrier developed in a collaboration of Universal Display Corp (UDC) and Prof. Wagner's group at Princeton University [19]. In this method, the barrier is again produced by plasma-enhanced CVD (PECVD), but instead of modulating precursor chemistry, it is fixed at a composition that gives an "optimum" ratio of inorganic to organic character, or a two-phase film composition. The hybrid film is reported to be a mixture of SiO_2 and silicone, and the tested barrier is relatively thick (~5 μm). One can imagine a film microstructure with a continuous inorganic phase, e.g., SiO_2, traversing the film thickness in a tortuous path through an organic silicone matrix. Unfortunately, few technical details have been published about this hybrid barrier material. However, the simplicity of a fixed process chemistry, without need for a separate curing step and disparate depositions with an organic precursor and by sputtering, has attracted interest in this hybrid material barrier for protecting flexible OLED displays.

8.3.2.3 Lag-Time/Desiccant Effects

While the initial explanation for low permeation in polymer–inorganic multilayers was an increase in the overall diffusion path length for molecular gases, subsequent detailed analysis by Graff et al. [20] (PNNL) showed that an extended lag time, or a desiccant-like effect, is responsible, not a reduction in equilibrium diffusion. The gas permeant is actually accumulating, albeit slowed down by the interleaving barrier layers, in the organic phase or at the interfaces between the inorganic and organic phases [21]. Only after the permeant storage capacity is exceeded, corresponding to a lag time, do equilibrium permeation rates prevail. Graff's modeling showed that increasing the number of dyads only marginally reduces equilibrium permeation rates, although the lag times could increase significantly. With good inorganic barrier layers, i.e., few defects, lag times could exceed 1 year. Until the lag time is exceeded, the "apparent" low permeation rate of molecular gases through a multilayer barrier meets the needs for OLED device operation. It seems likely that this same desiccant-like effect, elucidated by Graff et al., would also dominate the barrier performance of other organic–inorganic multilayer barriers and the UDC–Princeton mixed-phase film.

One of the key drawbacks of these two-phase, organic–inorganic, barrier approaches is that films must have numerous dyads or they must be thick to have long lag times. This is unattractive for manufacturing flexible OLED displays. Furthermore, what could easily be overlooked in Graff's analysis is that increasing the number of dyads alone, or making barrier films thick, is not a sufficient condition to achieve long lag times. The inorganic layers must have a low defect density, i.e., they must be good barriers, for a long lag time [20]. With regard to evaluation of the barrier film lifetime, the extrapolation of accelerated lifetime testing of multilayer barrier films with a long lag time should be treated with caution. The results of testing a barrier in damp heat conditions (high temperature and humidity, e.g., 85°C/85% RH) for ~1000–2000 h (typical) may not accurately reflect an extrapolated lifetime for these films, particularly if failure occurs abruptly, as would be the case for a transition from a lag-time regime to equilibrium diffusion.

8.3.3 Single-Layer Barrier Films

Because single-layer barrier films, prepared by physical vapor deposition on a polymer base film, only modestly reduce permeation by 10–50×, much better single-layer barrier properties are needed for OLEDs even in multilayer structures. The Si integrated circuit (IC) industry recently faced a similar, related challenge [22]. Much of the unique, high reliability of Si-based electronics is attributable to a defect-free, pinhole-free SiO_2 insulating layer, which is thermally grown at ~1100°C directly on Si. This perfect oxide layer performs as the gate dielectric in Si field-effect transistors. However, the relentless drive to enhance performance (Moore's law) has reduced the gate oxide thickness (<5 nm) to its limit for acceptably low leakage current. This fueled the search for other oxides with a higher dielectric constant and, more important, for a process to deposit thin films without defects or pinholes. ALD was the process adopted by the IC industry for depositing an HfO_2 gate dielectric to replace thermally grown SiO_2. ALD is a nondirectional process with a growth mechanism that achieves a featureless, conformal microstructure [14]. Films grow layer by layer (Figure 8.4) and avoid columnar growth with a granular microstructure, common to sputtered and evaporated films.

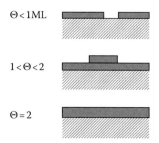

FIGURE 8.4
Thin-film layer-by-layer growth mode versus surface coverage (Θ).

8.3.3.1 Barrier Films Grown by ALD

Specifically, ALD deposits a film by sequential, self-limiting surface reactions from gas-phase precursors, or reactants. As an example, Al_2O_3 can be grown in a reactor by thermal ALD, using trimethylaluminum (TMA) and water. The process steps, written as two half-reactions, where the asterisk designates a surface species,

$$AlOH^* + Al(CH_3)_3 \rightarrow AlOAl(CH_3)_2^* + CH_4$$

$$AlCH_3^* + H_2O \rightarrow AlOH^* + CH_4$$

are as follows: (i) dose the reactor with TMA, which ideally deposits only a monolayer of TMA on the heated substrate surface; (ii) allow adequate purging with an inert gas to remove excess TMA from the reactor atmosphere; (iii) dose with H_2O, which reacts with adsorbed TMA to form a near monolayer of aluminum oxide on the substrate with the release of a gaseous CH_4 by-product; and (iv) purge the reactor with inert gas to remove excess H_2O. Steps 1–4 are repeated as many times as needed to grow the desired thickness of Al_2O_3. The steps in the ALD process are illustrated schematically in Figure 8.5.

What is unique about ALD is that dense, layer-by-layer growth can occur at low temperature, on a noncrystalline substrate, and the film's structure can be "glassy" or amorphous with a featureless microstructure, ideal for a gas permeation barrier or a gate dielectric in a field-effect transistor device. Other advantages of ALD include an inherently low defect density, precise thickness control, and near-atmospheric pressure operation,

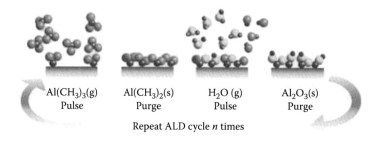

FIGURE 8.5
Steps in a single cycle of ALD process to produce Al_2O_3 from trimethyl aluminum and water. (Reprinted from http://www.Beneq.com with permission from Beneq Oy.)

only requiring a mechanical pump. Often ALD coatings are described as pinhole free, which is deduced from their superior electrical properties, and is the key reason why thin gate dielectrics grown by ALD have replaced thermally grown SiO_2 on Si in the highest-performance TFTs for ICs. Because we recognized the commonality for thin-film integrity or perfection needed in both ICs and ultralow gas permeation, our group was among the first to demonstrate ultrabarrier performance (WVTR $<10^{-5}$ g/m^2-day) in single-layer Al_2O_3 grown by ALD [23,24].

At the 2003 Conference on Atomic Layer Deposition in San Jose, California, McLean et al. [25] presented the first experimental evidence for ultralow gas permeation in a barrier film grown by ALD on polyester and PI substrates. This work was done in collaboration with ASML, at the time a manufacturer of ALD tools for the semiconductor industry. The barrier films were HfO_2 (10 or 25 nm thick) grown at 120°C and 150°C on either PEN or Kapton® PI and evaluated as oxygen barriers. Using a commercial instrument (MOCON OxTran, Minneapolis, MN), we found that the oxygen transport rate through our barrier films was below the limit of the instrument, <0.005 mL-O_2/m^2-day.

In a subsequent collaboration with Prof. Steven George and students at University of Colorado, Carcia et al. [24] demonstrated that Al_2O_3 barrier films grown by ALD at 125°C on PEN reduced WVTRs to ~10^{-5} g/m^2-day. These results, illustrated in Figure 8.6, were obtained after 3000 h combined testing at 38°C/85% RH and 60°C/85% RH, using an optical Ca-test method, which was developed in our laboratory. Several of these test samples were remeasured after storing them in the laboratory at ambient conditions for several years. Figure 8.7 shows the results for a typical sample; this one was stored for 3 years [26]. There is no visible degradation to the Ca metal squares, which the barrier protects, and the calculated WVTR from the increase in optical transmission through the thin (50 nm) Ca layer is estimated to be ~10^{-5} g/m^2-day.

Since these initial demonstrations, we have continued to refine our understanding of permeation in thin films grown by ALD on polymers. The barrier structure we propose for flexible OLEDs consists of two thin polymer sheets coated on one or both sides with a transparent ALD barrier layer laminated together with an adhesive. Locating the two primary barrier layers to the inside of the structure enhances its mechanical robustness and scratch resistance. Using two redundant barrier layers should also improve reliability. This structure, which is illustrated in Figure 8.8, is proposed as the substrate for fabrication of

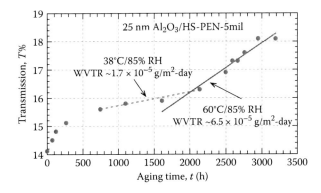

FIGURE 8.6

Representative change in optical transmission in Ca-test structure protected with a PEN lid coated with 25-nm-thick ALD Al_2O_3. (Reprinted with permission from P.F. Carcia, R.S. McLean, M.H. Reilly, M.D. Groner, S.M. George, *Appl. Phys. Lett.*, 89: 031915, 2006. Copyright 2006, American Institute of Physics.)

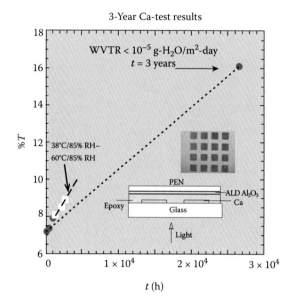

FIGURE 8.7
Cumulative change in optical transmission, corresponding to WVTR ~10^{-5} g/m^2-day during 3 years, for a Ca-test structure protected by a PEN lid coated on one side with 25-nm ALD Al$_2$O$_3$. (Modified from P.F. Carcia, Thin-film diffusion barriers for electronic application, in *Comprehensive Materials Processing*, vol. 4 (D. Cameron, ed., Elsevier, London), 463–498, 2014.)

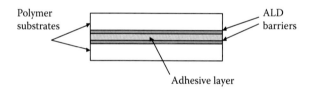

FIGURE 8.8
Proposed structure for ALD barrier on polymer.

an OLED display and a second identical structure can be adhered to the completed display to protect it from moisture infiltration from the top. One advantage of this approach is that the barrier structure can be manufactured offline. Alternatively, a top ALD barrier could be directly deposited on the OLED display; however, then, the time for applying the barrier coating inline becomes a limitation since ALD deposition can be relatively slow.

Recently, we demonstrated the effectiveness of this barrier structure as a flexible lid to protect a sensitive blue-emitting OLED, fabricated on a glass substrate [27]. In collaboration with DuPont Displays, we tested the barrier performance of a flexible lid for protecting blue-emitting OLED devices. The devices were bottom emission, i.e., light exits the transparent conducting anode electrode on glass. The opaque top cathode electrode consisted of a thin evaporated layer of an alkali metal salt, overcoated with thick Al. The completed devices had either a glass lid with an edge-bead, ultraviolet-curable epoxy, or a PET/barrier plastic lid with either an edge-bead or full-coverage epoxy. The barrier structure consisted of two laminated sheets of PET, each coated by ALD with 25-nm-thick Al$_2$O$_3$.

The comparative lifetimes of these OLED devices were evaluated in ambient conditions by tracking the change in light output with the time for operation at high voltage and high current, which accelerates device degradation. Specifically, $V = 7.3$ V and $I = 8.7$ mA (165 mA/cm^2), corresponding to a light output of ~8200 cd/cm^2.

For these test conditions, we found that blue emission for OLED devices with a glass lid decreased to 50% of its initial value in ~500 h, whereas emission from devices protected with an ALD Al$_2$O$_3$/PET flexible barrier lid was reduced to 50% at 440 and 470 h, corresponding to between 88% and 94% of the glass lid lifetime, a very encouraging result.

While batch coating Si wafers by ALD is a mature manufacturing process in the IC industry, the extrapolation to a continuous R2R process on polymer base film is still currently in development [28] and a challenge that must be met to realize a cost-competitive, superior single layer ALD barrier product.

8.4 AMOLED TFT Backplanes

8.4.1 Simple TFT Pixel Circuits

In an AMLCD or AMOLED display (see Chapter 9), each pixel is individually addressed with its own electronic circuit [4]. Figure 8.9 shows simple transistor circuits for addressing an LCD and an OLED display pixel. The AMLCD pixel uses a voltage drive, whereas the AMOLED requires a stable current drive. For LCDs, the single transistor acts as a switch, which is turned "on" when that row of pixels is selected or scanned. The liquid crystal (C_{LC}) is then charged to the voltage supplied from the data or column line, and the storage capacitor (C_S) holds that voltage until the next scan cycle. To drive an OLED requires at least two transistors. When T_2 is switched "on" or selected, a voltage is applied from the data or column line to the gate electrode of the drive transistor T_1, which allows a programmed current to flow continuously through the OLED. The storage capacitor (C_S) maintains the gate voltage until the next switching cycle. For both pixel types, the select or switching transistor is only "on" for a short time, and thus experiences low electrical stress compared with the OLED drive transistor, which must provide a specific current for as long as the pixel is "on." Because the OLED light emission sensitively depends on the current, which can vary with changing transistor parameters, such as the gate threshold voltage, compensating circuits with up to five TFTs and two capacitors are being considered for controlling the pixel current.

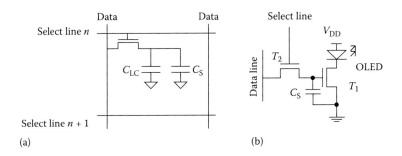

FIGURE 8.9
(a) Single-transistor circuit for switching pixel in LCD display. (b) Two-transistor circuit for switching pixel in an OLED display.

8.4.2 Si-Based TFTs

For LCD displays, amorphous silicon (a-Si:H) transistors, even with their low mobility (<1 cm^2/V-s), provide adequate performance [4,29]. (Recently, there has been some discussion of replacing a-Si:H with higher-mobility TFTs to meet the future needs of higher LCD display refresh rates.) a-Si:H backplanes are attractive because they can be fabricated uniformly over large areas at a low cost. However, the low mobility of a-Si:H transistors is only marginally acceptable for driving OLEDs [30,31]. Typical current densities for an OLED are 10–100 mA/cm^2, and for a pixel size of 100 μm × 100 μm, this corresponds to a pixel current of 1–10 μA. Using the equation for a transistor operating in saturation mode, one can show that the mobility requirement to deliver 5 μA/pixel leads to the condition

$$\mu^* \, (W/L) = 12.5 \text{ cm}^2/\text{V-s} \tag{8.1}$$

where W and L are the transistor dimensions and μ is mobility. For subpixel-size transistors and a large light emission aperture, μ should be >1 cm^2/V-s, according to Equation 8.1. However, more critically, when a-Si:H TFTs experience a sustained voltage stress, a permanent shift occurs in the threshold voltage, causing an unacceptable change in OLED light emission. In that regard, low-temperature polycrystalline silicon (LTPS) transistors [31] are more attractive for OLEDs as a stable alternative to a-Si:H.

LTPS is made by annealing and crystallizing a-Si:H, commonly using a scanning excimer laser [32]. (The a-Si:H films are deposited by a vacuum process—PECVD.) Composed of polycrystalline Si grains, LTPS transistors have a much higher mobility (30–300 cm^2/V-s), are voltage-bias stable, and therefore are more attractive than a-Si:H for driving OLEDs. However, because of added process complexity, LTPS transistors are more expensive to manufacture than a-Si:H, and nonuniformity in transistor properties due to variation in grain size is an issue, especially over larger areas. Nonuniform mobility over a display can also cause an unacceptable variation in OLED light intensity. The higher mobility of LTPS, however, does allow integration of transistors on the same substrate for display driver/logic and pixel addressing, which is advantageous for smaller OLED displays, e.g., cell phones and digital cameras, where there is limited space to accommodate a separate driver circuit board. The mobility of a-Si:H TFTs is too low for OLED drive/logic circuits.

For flexible OLEDs, however, the process temperature for producing a-Si:H (T > 300°C) and LTPS (annealing T > 600°C) backplanes is too high for most polymer substrates. Although there have been demonstrations of OLED displays, which were fabricated at a lower process temperature with a-Si:H on PEN [33] and PI [34] substrates, the long-term voltage-bias stability remains an issue. Alternatively, an OLED display can first be fabricated with a-Si:H or LTPS on a high-temperature substrate, e.g., glass, and then transferred to a flexible polymer substrate. This process technology exists but adds to the cost and is unlikely to be practical for large displays. Taking account of the broader objective for flexible OLED displays of low-cost R2R manufacturing, which replaces vacuum deposition with solution deposition and lithography with printing, it seems unlikely that either a-Si:H or LTPS is a suitable transistor technology. Then, what are the more suitable transistor technologies for flexible OLEDs?

8.4.3 Oxide TFTs

One class of materials that has recently emerged as a transistor candidate with high mobility and low process temperature, compatible with most flexible polymer substrates, is oxides [35–38]. Chapter 9 is exclusively devoted to the oxide TFTs for OLED applications. Specifically, these are binary, ternary, and quaternary combinations, ZnO, In_2O_3, SnO_2, Ga_2O_3, and CdO, which comprise the transparent conducting oxide (TCO) materials [39], ubiquitous in displays, e.g., $0.9\text{-}In_2O_3\text{-}0.1\text{-}SnO_2$ or ITO, and solar cells, e.g., Al-ZnO (AZO). TCOs find uses whenever an electrical conductor is needed to collect charge carriers (solar cells) or apply an excitation voltage (OLED) and simultaneously be optically transparent for light collection (solar) or emission (OLED).

While there was work by two Japanese groups on oxide TFTs—ZnO [40] and $InGaO_3$–$(ZnO)_5$ [41]—both groups were focused on producing single crystalline thin films by high-temperature, nonstandard deposition techniques [42]. The theme of their research was "transparent electronics." However, in 2003, Hoffman et al. [43] and Carcia et al. [44] reported that high-performance ZnO transistors could be fabricated by sputtering, a common deposition technique used in electronics manufacturing. Carcia et al. [45] further reported room-temperature fabrication of oxide transistors on a flexible PI substrate. Subsequently, the Hosono group reported room-temperature growth of oxide TFTs [46] by pulse laser deposition on PET, and then a number of universities and display companies joined in the development of oxide transistors, composed principally of binary, ternary, and quaternary mixtures of ZnO, In_2O_3, SnO_2, and Ga_2O_3. Many of those studies have focused on the $InZnGaO_4$ (IGZO) composition [46] reported by the Hosono group, although it is debatable whether this particular composition, chosen because of its connection to the earlier single crystal work, is optimum. One consensus of oxide TFT research is that high mobility (1–100 cm²/V-s) can be achieved even for room-temperature deposition on flexible substrates.

As an example, Figure 8.10 panels a and b are the transfer and transistor output curves for sputtered ZnO TFTs grown in our laboratory on a flexible PEN substrate. Figure 8.11a shows the structure of this TFT, which had a 42.5-nm-thick HfO_2 gate dielectric grown by ALD at 150°C; the metal layers and the ZnO semiconductor were sputtered at room temperature. Figure 8.11b is a photo of an array of these transistors on transparent, 125-μm-thick, flexible PEN. The mobility of this TFT is ~21 cm²/V-s, with on/off ratio >10⁶, high output current ~0.5 mA, and leakage current <0.1 nA. The device also exhibits negligible hysteresis. Often overlooked is the critical role in device performance of the gate dielectric, which in this TFT is high-quality HfO_2 grown by ALD.

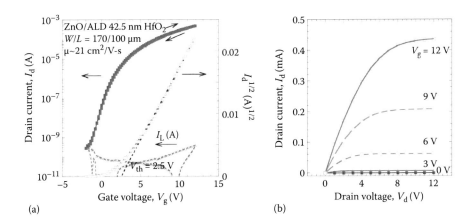

FIGURE 8.10
(a) Transfer and (b) current characteristics for a ZnO transistor grown at room temperature by rf sputtering on a flexible PEN substrate. Gate dielectric is HfO_2 layer grown by ALD at 150°C.

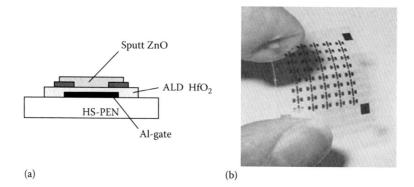

FIGURE 8.11
(a) Structure of ZnO transistor in Figure 8.10. (b) Photograph of array of ZnO transistors in (a) on flexible PEN substrate.

A surprising result for oxide TFTs is that even noncrystalline or "amorphous" oxide semiconductors can have relatively high electron mobility and perform well as a transistor channel [46]. The amorphous or "glassy" structure is a consequence of mixing multiple cations whose oxides have different crystallographic structures, the high quenching rate of sputtering, and the low substrate temperature. Also, because of the nondirectional, ionic contribution to bonding in these oxides—the outer valence electrons exist in symmetrical s-orbitals—structural order, or crystallinity, is not needed for high mobility. However, in that regard, it should be noted that sputtered oxide thin films will still have columnar microstructure [12], i.e., they will contain boundaries that can trap charge, even though the atomic arrangement is disordered or "amorphous." Growth boundaries occur in sputtered films because of low adatom mobility during nucleation and growth at low temperature and self-shadowing during sputtering. Furthermore, uniformity in deposition over large areas and time will be challenging for sputtering from a target with multiple chemical elements. Furthermore, practical issues such as target cracking, which ceramic oxide targets are prone to do, can also affect the manufacturing yields.

Unfortunately, oxide transistors are not immune to instability [47] effects due to illumination and voltage bias, which shift the threshold voltage—undesirable for OLED displays. One source of instability is attributed to the creation of oxygen vacancies [4], which can trap and release electrons and also modify transport properties. The capacity of TCO materials to accommodate large concentrations of oxide vacancies without a change in their crystallographic structure has long been known. Improvements in stability can be made by passivating the oxide semiconductor with a "barrier" layer [37]; oxide growth at a higher temperature (>300°C); the addition of other elements, such as Hf or Zr [31], which bond strongly to oxygen; or using double-oxide semiconductor layers [48]. Of course, some of these measures will add process steps, and a too high temperature growth is incompatible with low-temperature flexible plastic substrates.

To further improve the reliability of oxide TFTs, some groups have reverted to crystallizing the IGZO thin films [49] by processing up to 500°C. Films that form have a high degree of crystalline order along the c-axis, normal to the film plane, but random orientation in the plane. The c-axis aligned crystalline (CAAC) IGZO films are found to have lower hydrogen defect content and a featureless microstructure, i.e., no grain boundaries. Transistors made with CAAC IGZO are reported to have superior reliability and stability, and lower off-currents than low-temperature IGZO devices. Unfortunately, these improvements are

FIGURE 8.12
(See color insert.) Flexible OLED display made by South China University of Technology. (Courtesy of Dr. Lei Wang.)

unlikely to be relevant for OLED displays on a flexible polymer substrate because of the high processing temperature (Figure 8.12).

8.4.4 Organic TFTs

Organic TFTs [50], which today have been studied for more than 25 years, are also an attractive backplane candidate for flexible AMOLEDs. This is particularly so because they can be fabricated entirely with bendable organic materials and organic semiconductors based on conjugated organic molecules and polymers. Prototypical of a polymer candidate material is polythiophene, while vapor-deposited pentacene is a much-studied small-molecule candidate. The polymer materials can be easily chemically modified to make them soluble in a variety of organic solvents, and thin films can be then prepared by spin coating, dip coating, screen printing, and ink-jet printing. Progress has also been made to solubilize small-molecule organic semiconductors and OLED emitters. The materials themselves are inexpensive, and because they can be deposited from solution, they can be printed by low-cost, continuous R2R processing onto flexible polymer substrates. Although initial mobilities were low ($\ll 1$ cm^2/V-s) and the stability poor, there have been improvements to both. Polycrystalline films of pentacene have been vapor deposited with mobility in the range 1–3 cm^2/V-s. Because larger grain size can be achieved with small-molecule organic materials, their mobility is generally higher than for polymer materials. Moreover, although n-channel devices are possible with organic semiconductors, their mobility is much lower than for p-channel devices.

Laboratory-scale OLED displays on a flexible PEN substrate have been reported [51], and in 2010 Sony announced the demonstration of a rollable (4-mm-diameter), 4.1″ OLED display, driven by organic TFTs with p-type mobility of 0.4 cm^2/V-s, resolution of 121 ppi (pixels per inch), and $432 \times 240 \times$ RGB pixels [52]. The thickness of the panel was only 80 μm, and the size of a pixel was 210 μm $\times$ 210 μm.

8.4.5 Current Trends in AMOLED Backplane Technology

We can gain insight into the current importance and acceptance of a particular backplane technology from surveying the titles of presentations at the recent (2014) Display Week Symposium held in San Diego, California, and sponsored by the Society for Information Display. Leading display manufacturers presenting at the symposium included AU Optronics, Samsung, LG, Sony, Shenzhen China Star, Panasonic, Nokia, BOE Technology, and Plastic Logic.

At that symposium, the number of presentations on AMOLEDs appeared to be weighted in favor of oxide transistor backplane technology. Many of those presentations focused on fabricating large (55″) OLED TVs on rigid (glass) substrates. Samsung appears to be alone in attempts to scale up LTPS backplanes for large displays, using advanced excimer laser annealing. For smaller AMOLED displays, the integration of pixel switching and logic functions on a single substrate favors an LTPS backplane. Moreover, OLED displays for cell phones and tablets are currently manufactured with an LTPS backplane on rigid substrates.

In the case of flexible or foldable AMOLED displays, the subject of this chapter, this activity was mainly focused on smaller displays, and both oxide and organic semiconductor backplanes were prominent, although LTPS backplanes fabricated on glass and then release-laminated to a flexible substrate were also discussed. The focus on smaller flexible display was probably more a statement of the early stage of flexible technology rather than product or substrate preference size. In a couple of instances, solution processing was used for fabricating the flexible oxide and organic TFT backplanes, and LG and Konica Minolta both had presentations on R2R processing related to fabricating OLEDs on a flexible plastic substrate. The LG work was on producing oxide TFTs, while Konica–Minolta emphasized OLED fabrication on a barrier-coated flexible substrate. In spite of a number of previously published papers in the literature on a-Si:H backplanes to drive an OLED, the symposium was relatively silent on their application in AMOLEDs, suggesting that the low mobility and poor stability of a-Si:H TFTs make them less favorable as a candidate technology for driving OLEDs.

Although there were a few presentations on barrier technology, no preferred barrier technology is apparent and each manufacturer seems to be pursuing in-house approaches without consensus.

8.5 Summary Remarks

In this overview of material challenges for flexible OLED displays, we focused on the choice of substrate material, a packaging strategy to ensure adequate OLED device lifetime, and a transistor technology to act as a backplane to energize or provide current drive for the OLEDs. We made the choice that polymers were the most suitable substrate for a flexible OLED display, and we reviewed the properties of leading polymer candidates. While flexibility and optical transparency of polymers make them attractive for flexible displays, the large dimensional changes that occur with temperature, humidity, and stress in polymers present unique processing challenges. Also, because polymers are permeable to air and moisture, which will degrade OLEDs, polymers will need to be modified to have ultralow permeation, if devices are to last several years—a reasonable expectation today.

We reviewed the pros and cons of polymer–multilayer barrier coatings, the most mature and developed barrier technology currently available. We also introduced an emerging single-layer barrier technology based on ALD. Finally, we pointed out the need to replace an AM backplane based on amorphous Si:H TFTs, which currently drive LCD displays. Either oxide or polycrystalline Si transistors, with higher mobility than a-Si:H, are a better fit for OLEDs, although research and development on flexible organic TFTs to drive OLEDs continues. Because oxides and organic TFTs can be processed at low temperature, they are more compatible with polymer substrates and flexible OLED displays.

While this overview strives to give an objective snapshot of current challenges and the technology solutions under consideration, it should be stated that a flexible OLED display made by low-cost R2R manufacturing has been a product vision for more than 20 years, since the discovery of polymer OLEDs [53]. There have been many technological developments in that interim, and more are likely needed, before we can fold our phone/tablet and store it away in our wallet.

References

1. Z. Suo, E.Y. Ma, H. Gleskova, S. Wagner, Mechanics of rollable and foldable film-on-foil electronics, *Appl. Phys. Lett.*, 74: 1177–1179, 1999.
2. W.A. MacDonald, M.K. Looney, D. MacKerron, R. Eveson, R. Adam, K. Hashimoto, K. Rakos, Latest advances in substrates for flexible electronics, *J. SID*, 15: W1075–W1083, 2007.
3. J. Lewis, Material challenge for flexible organic devices, *Mater. Today*, 9: 38–45, 2006.
4. R.A. Street, Thin-film transistors, *Adv. Mater.*, 21: 2007–2022, 2009.
5. Corning Willow Glass fact sheet, available at: http://www.corning.com/displaytechnologies.
6. W.A. MacDonald, Engineered films for display technologies, *J. Mater. Chem.*, 14: 4–10, 2004.
7. B.A. MacDonald, K. Rollins, R. Eveson, K. Rakos, B.A. Rustin, M. Handa, New developments in polyester film for flexible electronics, *Mater. Res. Soc. Symp. Proc.*, 769: H9.3.1–H9.3.8, 2003.
8. P.E. Burrows, G.L. Graff, M.E. Gross, P.M. Martin, M. Hall, E. Mast, C. Bohnam, W. Bennett, L. Michalski, M. Weaver, J.J. Brown, D. Fogarty, L.S. Sapochak, Gas permeation and lifetime tests on polymer-based barrier coatings, in *Proceedings of SPIE*, 4105 (Organic Light-Emitting Devices IV, Z.H. Kafafi, ed.), 75–81, 2001.
9. J. Lange, Y. Wyser, Recent innovations in barrier technologies for plastic packaging—A review, *Packag. Technol. Sci.*, 16: 149–158, 2003.
10. G. Rossi, G. Nulman, Effect of local flaws in polymeric permeation reducing barriers, *J. Appl. Phys.*, 74: 5471–5475, 1993.
11. C.A. Bishop, *Roll-to-Roll Vacuum Deposition of Barrier Coatings* (Scrivener, Salem, MA), 2010.
12. M. Ohring, *The Materials Science of Thin Films* (Academic Press, San Diego, CA), 197, 1992.
13. J.D. Affinito, M.E. Gross, C.A. Coronado, G.L. Graff, I.N. Greenwell, P.M. Martin, A new method for fabricating transparent barrier layers, *Thin Solid Films*, 290–291: 63–67, 1996.
14. S.M. George, Atomic layer deposition: An overview, *Chem. Rev.*, 110: 111–131, 2010.
15. F.J.H. van Assche, E.W.A. Young, J.J. Michels, G.H. Rietjens, P. van de Weijer, P.C.P. Bouten, A.M.B. van Mol, Thin-film barrier on foil for roll-to-roll OLEDs, abstract in AVS International Symposium, Nov. 8–13, 2009, San Jose, CA.
16. SiO$_x$ vacuum coated high gas barrier film, available at: https://www.mpi.co.jp/english/products/packaging_materials/pm004.html, accessed July 13, 2014.
17. M. Yan, T.W. Kim, A.G. Erlat, M. Pellow, D.F. Foust, J. Liu, M. Schaepkens, C.M. Heller, P.A. McConnelee, T.P. Feist, A.R. Duggal, A transparent, high barrier, and high heat substrate for organic electronics, *Proc. IEEE*, 93: 1468–1477, 2005.

18. A.G. Erlat, M. Yan, A.R. Duggal, Substrates and thin-film barrier technology for flexible electronics, in *Flexible Electronics: Materials and Applications* (W.S. Wong, A. Salleo, eds., Springer Science) Springer Science+Business Media, New York, 413–449, 2009.

19. P. Mandlik, J. Gartside, L. Han, I.-C. Cheng, S. Wagner, J.A. Silvernail, R.-Q. Ma, M. Hack, J.J. Brown, A single-layer permeation barrier for organic light-emitting displays, *Appl. Phys. Lett.*, 92: 103309-1-3, 2008.

20. G.L. Graff, R.E. Williford, P.E. Burrows, Mechanisms of vapor permeation through multilayer barrier films: Lag time versus equilibrium permeation, *J. Appl. Phys.*, 96: 1840–1849, 2004.

21. B.D. Vogt, H.-J. Lee, V.M. Prabhu, D.M. DeLongchamp, E.K. Lin, W. Wu, X-ray and neutron reflectivity measurements of moisture transport through model multilayered barrier films for flexible displays, *J. Appl. Phys.*, 97: 114509, 2005.

22. R.M. Wallace, G. Wilk, Alternative gate dielectrics for microelectronics, *MRS Bull.*, 27: 186, 2002.

23. M.D. Groner, S.M. George, R.S. McLean, P.F. Carcia, Gas diffusion barriers on polymers using Al_2O_3 atomic layer deposition, *Appl. Phys. Lett.*, 88: 051907, 2006.

24. P.F. Carcia, R.S. McLean, M.H. Reilly, M.D. Groner, S.M. George, Ca test of Al_2O_3 gas diffusion barriers grown by atomic layer deposition on polymers, *Appl. Phys. Lett.*, 89: 031915, 2006.

25. R.S. McLean, P.F. Carcia, S.G. Park, Y. Sensaki, Gas permeation barrier films grown by ALD on polyester and polyimide substrates, in *Proceedings of AVS 3rd International Conference on Atomic Layer Deposition 2003* (San Jose, CA), August 4–6, 2003.

26. P.F. Carcia, R.S. McLean, B.B. Sauer, M.H. Reilly, Atomic layer deposition ultra-barriers for electronic applications—Strategies and implementation, *J. Nanosci. Nanotechnol.*, 11: 7994–7998, 2011.

27. P.F. Carcia, Thin-film diffusion barriers for electronic applications, in *Comprehensive Materials Processing*, vol. 4 (D. Cameron, ed., Elsevier, London), 463–498, 2014.

28. P. Poodt, D.C. Cameron, E. Dickey, S.M. George, V. Kuznetsov, G.N. Parson, F. Roozeboom, G. Sundaram, A. Vermeer, Spatial atomic layer deposition: A route towards further industrialization of atomic layer deposition, *J. Vac. Sci. Technol. A*, 30: 01802-1, 2012.

29. J. Jang, Preparation and properties of hydrogenated amorphous silicon thin-film transistors, in *Thin-Film Transistors* (C.R. Kagan, P. Andry, eds., Marcel Dekker, New York), 35–69, 2003.

30. T. Tsujimura, Amorphous/microcrystalline silicon thin film transistor characteristics for large size OLED television driving, *Jap. J. Appl. Phys.*, 43: 5122–5128, 2004.

31. H.J. Kim, B.D. Chin, A review of recent advances in backplanes and color patterning technologies for AMOLED display, *IEEE Photon. Soc. Newsl.*, April 2013: 4–12, 2013.

32. A.T. Voutsas, M.K. Hatalis, Technology of polysilicon thin-film transistors, in *Thin-Film Transistors* (C.R. Kagan, P. Andry, eds., Marcel Dekker, New York), 139–207, 2003.

33. K.R. Sarma, J. Roush, J. Schmidt, C. Chanley, S. Dodd, Flexible active matrix organic light emitting diode (AMOLED) displays, in *Proc. of ASID '06* (New Delhi), 337–342, October 8–12, 2006.

34. K. Long, A.Z. Kattamis, I.-C. Cheng, H. Gleskova, S. Wagner, J.C. Sturm, M. Stevenson, G. Yu, M. O'Regan, Active-matrix amorphous-silicon TFT arrays at 180°C on clear plastic and glass substrates, *IEEE Trans. Electron Devices*, 53: 1789–1793, 2006.

35. J.-S. Park, Oxide TFTs for AMOLED TVs, *Inf. Display*, 29: 16–19, 2013.

36. J.K. Jeong, H.-J. Chung, Y.-G. Mo, H.D. Kim, A new era of oxide thin-film transistors for large-sized AMOLED displays, *Inf. Display*, 24: 20–23, 2008.

37. E. Fortunato, P. Barquinha, R. Martins, Oxide semiconductor thin-film transistors: A review of recent advances, *Adv. Mater.*, 24: 2945–2986, 2012.

38. J.S. Park, W.-J. Maeng, H.-S. Kim, J.-S. Park, Review of recent developments in amorphous oxide semiconductor thin-film transistor devices, *Thin Solid Films*, 520: 1679–1693, 2012.

39. See articles on transparent conducting oxides, *MRS Bull.*, 25: 15–102, 2000.

40. A. Ohtomo, M. Kawasaki, Novel semiconductor technologies of ZnO films towards ultraviolet LEDs and invisible FETs, *IEICE Trans. Electron.*, E83-C: 1614–1617, 2000.

41. K. Nomura, H. Ohta, K. Ueda, T. Kamiya, M. Hirano, H. Hosono, Thin-film transistor fabricated in single-crystalline transparent oxide semiconductor, *Science*, 300: 1269–1272, 2003.

42. G.T. Huang, Computers that do Windows, *MIT Technol. Rev.*, September 30, 2003; Research news: Fast and invisible, *Mater. Today*, 6 (7–8): 6, 2003.

43. R.L. Hoffman, B.J. Norris, J.F. Wager, ZnO-based transparent thin-film transistors, *Appl. Phys. Lett.*, 82: 733–735, 2003.

44. P.F. Carcia, R.S. McLean, M.H. Reilly, G. Nunes, Transparent ZnO thin-film transistor fabricated by rf magnetron sputtering, *Appl. Phys. Lett.*, 82: 1117–1119, 2003.

45. P.F. Carcia, R.S. McLean, M.H. Reilly, I. Malajovich, K.G. Sharp, S. Agrawal, G. Nunes, ZnO thin film transistors for flexible electronics, *Mater. Res. Soc. Symp. Proc.*, 769: H7.2.1–H7.2.6, 2003.

46. K. Nomura, H. Ohta, A. Takagi, T. Kamiya, M. Hirano, H. Hosono, Room-temperature fabrication of transparent flexible thin-film transistors using amorphous oxide semiconductors, *Nature*, 432: 488–492, 2004.

47. T.-C. Fung, K. Abe, H. Kumomi, J. Kanicki, Electrical instability of RF sputter amorphous In-Ga–Zn–O thin-film transistors, *J. Display Technol.*, 5: 452–461, 2009.

48. H.Y. Jung, Y. Kang, A.Y. Hwang, C.K. Lee, S. Han, D.-H. Kim, J.-K. Bae, W.-S. Shin, J.K. Jeong, Origin of the improved mobility and photo-bias stability in a double-channel metal oxide transistor, *Sci. Rep.*, 4 (1–8): 3765, 2014.

49. S. Yamazaki, H. Suzawa, K. Inoue, K. Kato, T. Hirohashi, K. Okazaki, N. Kimizuka, Properties of crystalline In–Ga–Zn-oxide semiconductor and its transistor characteristics, *Jpn. J. Appl. Phys.*, 53: 04ED18, 2014.

50. H. Klauk, Organic thin-film transistors, *Chem. Soc. Rev.*, 39: 2643–2666, 2010.

51. L. Zhou, A. Wanga, S.-C. Wu, J. Sun, S. Park, T.N. Jackson, All organic active matrix flexible display, *Appl. Phys. Lett.*, 88: 083502 (3 pages), 2006.

52. Sony develops a "rollable" OTFT-driven OLED display that can wrap around a pencil, available at: http://www.sony.net/SonyInfo/News/Press/201005/10-070E/, accessed July 15, 2014.

53. J.H. Burroughes, D.D.C. Bradley, A.R. Brown, R.N. Marks, K. Mackay, R.H. Friend, P.L. Burns, A.B. Holmes, Light-emitting diodes based on conjugated polymers, *Nature*, 347: 539–541, 1990.

9

Oxide Thin-Film Transistors for Active Matrix OLEDs

Linfeng Lan, Weijing Wu, and Lei Wang

CONTENTS

9.1 Introduction

Owing to the need for thin and lightweight panels with vivid images, wide viewing angles, and high-contrast-ratio representation, as described in Chapter 1, active matrix organic light emitting diode (AMOLED) displays for TV and smartphone applications are increasingly important in our daily life [1]. The thin-film transistor (TFT) technology with good electrical performance is one of the keys in realizing AMOLED panels. For the AMOLED displays, TFTs must have field-effect mobilities higher than 5 $cm^2 V^{-1} s^{-1}$ because OLEDs are current-driven devices, which require relatively high currents. In this regard, conventional amorphous-silicon TFTs, which are widely adopted in the backplane of liquid crystal displays (LCDs), cannot meet the requirements of the AMOLED displays because of their low mobility; low-temperature polysilicon (LTPS) TFTs have higher mobilities (>50 cm^2 $V^{-1} s^{-1}$) and superior operational stability but also suffer from poor uniformity over large areas and high process temperatures (400°C–500°C) [2,3].

Recently, oxide TFTs have gained considerable attention as an attractive alternative to LTPS TFTs because they can provide much better uniformity over large areas [4,5]. Other important advantages are that they can be deposited using conventional semiconductor process methods, such as sputtering at room temperature, and have high mobility even in an amorphous state. Because of these remarkable characteristics, oxide TFTs are excellent candidates for large-area, ultradefinition, and fast-frame-rate LCD and AMOLED panels. In recent years, by properly implementing oxide TFT arrays, several companies, such as Samsung [6], AUO [7], Sony [8], and New Vision [9], have demonstrated working AMOLED prototypes. In 2014, LG successfully developed 55″ OLED TV sets with oxide TFTs that were sold worldwide [10]. In this chapter, we describe oxide TFT AMOLED technology, including oxide semiconductor materials used in TFTs, TFT structures, pixel circuit designs, and TFT fabrication processes.

9.2 Development in Oxide Semiconductors

9.2.1 Binary Oxide Materials

Binary oxide semiconductors, including ZnO, In_2O_3, SnO_2, and Ga_2O_3, are some of the simplest oxide semiconductors and base materials for multicomponent oxide semiconductors. These binary oxides have a wide band gap that allows good transmission of visible light. The metal ions in these oxides are not as electropositive as those in the alkaline-earth metals. They are predominantly ionic bonded, except for ZnO.

ZnO is the most commonly used binary oxide semiconductor. ZnO typically has the hexagonal wurtzite structure in which each Zn or O atom is surrounded by four neighbors of the other type, as shown in Figure 9.1a. The lattice constants of ZnO are a = 3.2475–3.2501 Å, c = 5.2042–5.2075 Å, and c/a = 1.5930–1.6035 [11]. The measured direct band gap energy (E_g) of ZnO is 3.44 eV. However, the calculated E_g is only 0.23–1.15 eV using the uncorrected local-density approximation (LDA) [12]. Using newly developed screened exchange (sX) and hybrid density functional methods, the calculated E_g is 3.41 eV [13], as shown in Figure 9.1b, much closer to the measured experimental value.

Despite many years of experimental investigations and theoretical calculations, some of the basic properties of ZnO, especially the origins of the n-type conductivity in undoped ZnO films, are still unclear. A few explanations for its intrinsic n-type conductivity have been offered thus far, including bulk intrinsic defects, such as oxygen vacancies (V_O) [13–15], unintentional impurities like hydrogen [16], and defect complexes [17–19]; however, experimental confirmation is largely lacking.

ZnO-based TFTs have been intensively studied because of their high-mobility, low-temperature process, transparency in visible light, low manufacturing cost, etc. The first

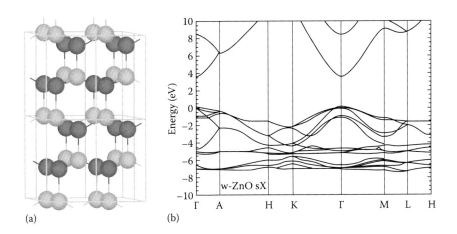

FIGURE 9.1
(a) Hexagonal wurtzite structure of ZnO; (b) band structure of ZnO in the wurtzite structure evaluated using sX functional. (From Clark, S.J. et al., *Phys. Rev. B*, 81, 115311, 2010.)

report on TFTs using single-crystal ZnO was published as early as 1968 [20]. However, this kind of TFT did not draw much attention until 2003 when Hoffman et al. [21] demonstrated sputtered ZnO-based transparent TFTs with mobilities as high as 2.5 cm² V⁻¹ s⁻¹ and on/off current ratios as high as 10⁷. In 2005, Fortunato et al. [22] developed fully transparent ZnO TFTs processed at room temperature, exhibiting high mobility near 20 cm² V⁻¹ s⁻¹. However, such devices required relatively high gate turn-on voltages, as they are operated well in the enhancement mode. Lim et al. [23,24] were able to control threshold voltage (V_{th}) by nitrogen (N) doping during ZnO deposition using NH_4OH injection by atomic layer deposition (ALD). This N-doping was shown to enhance mobility as well as the electrical stability. The highest reported Hall mobility of ZnO thin films is 440 cm² V⁻¹ s⁻¹ [25], much higher than the highest reported field-effect mobility of ZnO TFTs, probably due to the difficulty in controlling defects at the gate insulator/ZnO interface.

In_2O_3 is another famous binary oxide semiconductor with high mobility and high carrier density. In_2O_3 has the bixbyite structure, in which the oxygen form a close-packed lattice and the In ions lie at sixfold and fourfold interstices. The In sites have sixfold oxygen coordination. The overall symmetry is cubic but the unit cell is large, containing 40 atoms. Its calculated E_g is about 3.7 eV, a little wider than that of ZnO.

Similar to ZnO, first-principle calculations [26,27] suggest that hydrogen in In_2O_3 may act as an unintentional dopant. However, these calculated defect levels are affected by the choice of the functionals employed. For example, some calculations [26] based on LDA or LDA plus Hubbard (LDA + U) [28] for strongly correlated materials in the d-orbital (e.g., In_2O_3) show [26] that hydrogen is the only cause of n-type conductivity in In_2O_3, while others using hybrid functionals reported [28] that oxygen vacancies are responsible for the n-type doping in In_2O_3. One recent report [29] using the LDA + U method also suggests that both hydrogen and oxygen vacancies are possible n-type dopants in In_2O_3. However, both hydrogen and oxygen vacancy models cannot account for the fact that thin-film conductivities can be 4–5 orders of magnitude higher, reaching transparent conducting oxide–like conductivities >1000 S·cm in pure In_2O_3 thin films. Most recently, Lany et al. [30] demonstrated that surface donors rather than bulk defects dominate the conductivity of In_2O_3 thin films.

Although the calculated effective electron mass (m^*/m_e, lower m^*/m_e means higher mobility because the carrier is easier to transport) of In_2O_3 is 0.35 [31], higher than that of ZnO (0.28), the reported field-effect mobility of In_2O_3 TFTs is much higher than that of ZnO TFTs, which may be attributed to the less sensitivity to the structure of In_2O_3 for its large overlap of the s-orbitals. It is not difficult to attain field-effect mobility >30 cm^2 V^{-1} s^{-1} for In_2O_3 TFTs, as reported [32,33]. In fact, as early as in 2006, Wang et al. [34] reported a field-effect mobility >120 cm^2 V^{-1} s^{-1} for In_2O_3 TFTs by combining In_2O_3 thin films with nanoscale organic dielectrics, very promising for low-cost and high-performance TFT technology. In 2007, their group demonstrated flexible In_2O_3 TFTs with a mobility as high as 160 cm^2 V^{-1} s^{-1} [35]. Recently, Nayak et al. [36] demonstrated a solution-processed In_2O_3 TFT with a mobility of as high as 127 cm^2 V^{-1} s^{-1} and an on/off current ratio of 10^6 using chemically derived aluminum oxide dielectric, as shown in Figure 9.2 [36].

Although In_2O_3 has excellent electron mobility, it has some drawbacks as applied to AMOLED displays. First, room temperature–grown In_2O_3 films are still polycrystalline. Second, the carrier density of In_2O_3 films is too large to control, making the TFT devices hard to turn off. Third, the electrical and optical stability still need to be improved. Lastly, indium is a toxic and rare element in the earth's crust, which could cause severe supply issues in large-scale application.

SnO_2 is more attractive compared with In_2O_3 in terms of cost and environmental concerns. SnO_2 has a rutile structure, in which each tin atom is surrounded by six oxygens in an octahedral array and each oxygen atom is surrounded by three tins in a planar array. Its direct band gap is calculated to be 3.6 eV.

SnO_2 is the earliest oxide semiconductor used in the channel layer for TFTs. The first SnO_2 TFT dates back to 1964, when Klasens and Koelmans [37] proposed a TFT composed of an evaporated SnO_2 semiconductor on glass, with aluminum source-drain and gate electrodes, and an anodized Al_2O_3 gate dielectric. Although SnO_2 TFT has longer history compared with ZnO and In_2O_3 TFTs, reports on SnO_2 TFTs are much fewer than those about ZnO and In_2O_3 TFTs. In 2004, Presley et al. [38] demonstrated transparent TFTs with a radiofrequency magnetron sputtered SnO_2 channel layer, achieving mobilities of 2.0 cm^2 V^{-1} s^{-1}. In 2007, Lee et al. [39] fabricated a depletion-mode TFT using an ink-jet porous SnO_2 channel layer with a field-effect mobility of 3.62 cm^2 V^{-1} s^{-1}. In 2009, Cheong et al. [40] reported a top-gate (TG) SnO_2 transparent TFT with a mobility of 17.4 cm^2 V^{-1} s^{-1}. In

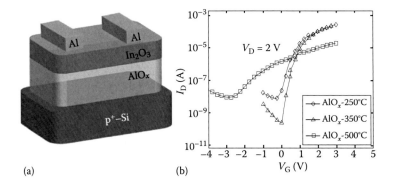

(a) (b)

FIGURE 9.2
(See color insert.) (a) Schematic of the TFT structure; (b) transfer characteristic curves of the In_2O_3 TFTs fabricated using AlO_x films prepared at different temperatures as the gate dielectrics. (From Nayak, P.K. et al., *Appl. Phys. Lett.*, 103, 033518, 2013.)

2013, Jang et al. [41] demonstrated a TFT from solution-processed SnO_2 gel-like precursors, exhibiting mobilities of as high as 103 $cm^2 V^{-1} s^{-1}$. These results illustrate good performance of SnO_2 TFTs.

Although ZnO, In_2O_3, and SnO_2 all have relatively wide band gaps (3.3–3.8 eV), they would not be wide enough for under UV illumination. In these cases, oxide semiconductors with band gaps wider than 4 eV are needed. Ga_2O_3 is a good candidate owing to its wider band gap of 4.52–4.9 eV [42,43]. Ga_2O_3 has more than one crystalline phase. For example, the β-Ga_2O_3 structure has a band gap of 4.8–4.9 eV, the second widest after diamond among semiconductors. The crystal structure of β-Ga_2O_3 belongs to the monoclinic system, with lattice parameters of $a = 1.223$ nm, $b = 0.304$ nm, $c = 0.580$ nm, and $\beta = 103.7°$, and the space group of C2/m [44]. The Ga sites are fourfold or sixfold coordinated.

The first Ga_2O_3 transistor was presented by Matsuzaki et al. [45]; however, its mobility is only 5×10^{-2} $cm^2 V^{-1} s^{-1}$. In 2012, Higashiwaki et al. [46] reported a demonstration of single-crystal Ga_2O_3 metal–semiconductor field-effect transistor with a high breakdown voltage of >250 V and an estimated mobility of ~100 $cm^2 V^{-1} s^{-1}$. However, the mobility of Ga_2O_3 is strongly dependent on the crystalline structure of Ga_2O_3 films because the Ga^{3+} ion is smaller than the In^{3+} ion, resulting in no orbital overlap between the neighboring Ga^{3+} ions.

Other binary oxide semiconductors including Cu_2O, SnO, and CdO. Cu_2O and SnO are typically p-type semiconductors. The highest field-effect hole mobilities reported for Cu_2O and SnO are 4.3 and 4.8 $cm^2 V^{-1} s^{-1}$, respectively, which are close to those of some n-type oxide semiconductors. However, the reliability and stability of Cu_2O and SnO TFTs are poor, and need to be improved before their application in display back planes. CdO has a low effective electron mass; however, its band gap is relatively narrow (~2.1 eV). Furthermore, it is toxic and expensive. Therefore, CdO is not a good choice in AMOLED applications.

Although significant progress has been made in the development of binary oxide TFTs to the point that their mobility can even surpass those of LTPS TFTs, they still have some serious drawbacks. Crystalline structures are indispensable for high-mobility binary films; however, they also result in poor device uniformity. Furthermore, the electrical stability of the binary oxide TFTs is inferior, holding back their adoption in commercial AMOLED displays.

9.2.2 Multicomponent Oxide Materials

It is possible to design an amorphous material by mixing two or more crystalline oxides. For example, In_2O_3 and ZnO have bixbyite and wurtzite structures, respectively, and also have different coordination numbers to the oxygen. This allows In–Zn oxide (IZO) to have an amorphous phase. Consequently, IZO can be used as the transparent conducting oxide in both electrodes and the semiconductor layer in TFTs. By varying the composition and deposition conditions, IZO has exhibited a wide range of resistivities, ranging from 10^{-4} to 10^8 Ω cm [47]. Generally, the mobility and carrier density increase as the In-to-Zn ratio increases. However, IZO films with In/(In + Zn) atomic ratios >0.80 will become conductive and polycrystalline. However, there are some exceptions. Early in 2007, Fortunato et al. [48] demonstrated normally off bottom-gate IZO TFTs using a ceramic oxide target with In_2O_3–ZnO (9:1), achieving their highest mobility, 107.2 $cm^2 V^{-1} s^{-1}$, with an amorphous structure. Most recently, Park et al. [49] demonstrated an IZO TFT with a 9:1 In:Zn atomic ratio, achieving a mobility of 157 $cm^2 V^{-1} s^{-1}$ also with amorphous structure by adopting a self-aligned coplanar TG structure and modifying the surface of IZO material using N_2O plasma, but the TFT showed slightly normally on characteristics. Owing to these advantages, IZO has been one of the most extensively investigated semiconductor oxides for display applications.

Although IZO TFTs have excellent performance in laboratory tests, they are seldom adopted for TFT backplanes in mass production. One possible reason is that the reproducibility of IZO TFTs is rather poor because the performance of IZO TFTs is very sensitive to the fabrication process, oxygen vacancies, and the atomic ratios of In to Zn.

In 2004, Nomura et al. [50] introduced a new class of amorphous oxide semiconductors, IGZO, by doping Ga into IZO, demonstrating new class of high-performance transistors ($\mu \approx 8.3$ cm^2 V^{-1} s^{-1}).

Their approach showed that the carrier concentration of amorphous IGZO (a-IGZO) can be lowered to $<10^{17}$ cm^{-3}, while keeping its electron mobility high. The origin of high mobility in the amorphous state was attributed to the electronic orbital structure in which direct overlap between neighboring metal s-orbitals can occur, as shown in Figure 9.3. This forms a conducting path for free electrons, which is not significantly affected even in a distorted amorphous structure. The conduction band in IGZO is mainly formed by the overlap of In 5s orbitals, and exhibits isotropic properties (spherical symmetry). This makes IGZO insensitive to structural deformations, maintaining its high mobility in the amorphous state. In contrast, Si undergoes a significant reduction in mobility from 1000 cm^2 V^{-1} s^{-1} (in a single-crystal state) to <1 cm^2 V^{-1} s^{-1} (in an amorphous state). The decrease in carrier concentration has been attributed to the high ionic potential of Ga^{3+} ions, which allows them to tightly bind oxygen ions and suppress the formation of oxygen vacancies. The development of IGZO materials is a breakthrough in TFT technologies for displays, especially AMOLEDs.

To further enhance device performance, optimization of the cation composition in IGZO TFTs has been actively studied. The performance dependence on the In, Ga, and Zn compositions was investigated in detail, as shown in Figure 9.4 [51]. Higher In/(Ga + Zn) ratio

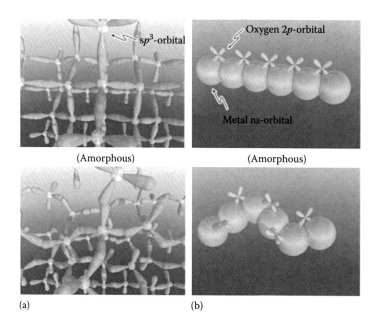

(a)   (b)

FIGURE 9.3
Schematic orbital drawings for the carrier transport paths in (a) covalent semiconductors (silicon crystalline) and (b) posttransition metal oxide semiconductors (crystalline). (From Nomura, K. et al., *Nature*, 432, 488, 2004.)

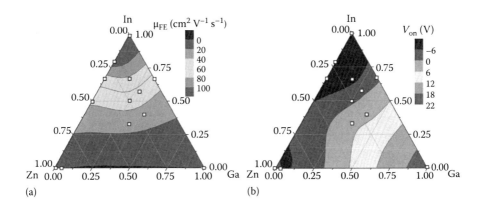

FIGURE 9.4
(See color insert.) (a) μ_{FE} and (b) V_{on} obtained for TFTs with different oxide semiconductor compositions, in the indium–gallium–zinc oxide system. (From Jang, J. et al., *Adv. Mater.*, 25, 1042, 2013.)

yields higher electron mobility but will make the device more difficult to turn off (more negative V_{on}). Owing to the stronger gallium–oxygen bond, IGZO carrier density decreases as its gallium content increases, resulting in a considerable positive V_{on} shift.

Alternative oxide semiconductor compounds have also been investigated by substituting Ga with other low-electronegative elements such as Al [52,53], Hf [54,55], Zr [56], Ta [57,58], La [59], etc. These elements were shown to act as strong oxygen binders and carrier suppressors, leading to improvements in the bias stability of their resulting TFTs.

Besides In-based oxide semiconductors, there is a particular need to develop In-free oxide semiconductors because In is a rare and expensive element. Zn–Sn oxide (ZTO) [60,61] is among the most famous where Sn^{4+} is one of the heaviest posttransition cations that satisfies $(n - 1)d^{10}ns^0$ ($n \geq 5$). The electron mobility for ZTO TFTs can be >10 cm² V⁻¹ s⁻¹; however, the processing temperature of ZTO is also higher than those for IZO and IGZO. There have also been several attempts to dope ZTO with other elements, such as Al and Zr [62,63]. However, their performance is lower than that of IGZO TFTs.

9.3 Structure and Characteristics of Oxide TFTs

9.3.1 Device Structure of Oxide TFTs

Oxide TFTs are three-terminal field-effect devices, whose working principle relies on the modulation of the current flowing in a semiconductor placed between two electrodes (source and drain). A dielectric layer (gate insulator) is inserted between the semiconductor and a transversal electrode (gate), being the current modulation achieved by the capacitive injection of carriers close to the dielectric/semiconductor interface, known as field effect [22]. Figure 9.5 shows the three classes of oxide TFTs: (i) back-channel etch (BCE), (ii) etch stopper layer (ESL), and (iii) TG structures. Each class can be classified more precisely into top and bottom contacts in which the source and drain electrodes at the top or bottom of the oxide semiconductor layer.

The BCE structure is almost the same as the conventional a-Si TFT, so it is possible to adapt the conventional a-Si TFT production lines to the oxide TFT fabrication in theory.

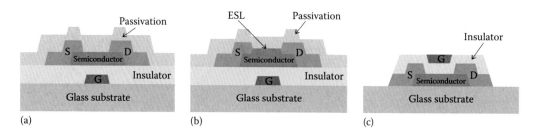

FIGURE 9.5
(See color insert.) Typical device structures used to fabricate amorphous oxide semiconductor TFTs with (a) back-channel etch, (b) etch stopper layer, and (c) top-gate structures.

However, it is a challenge to pattern the source and drain (S/D) electrodes directly on the oxide semiconductor layer via wet etch as done in a-Si TFT fabrication, because of their weak bonding in most of the wet etchants, even in weak acids such as acetic acid. Thus, the S/D electrodes should be etched via dry etch; however, the plasma bombardment and the residual ions such as Cl^- or F^- used during dry etch would cause performance degradation. The most commonly used BCE method was the lift-off process in which the S/D electrodes were patterned by lift-off. However, this method is difficult to control and will inevitably leave some residues on the panel. Therefore, it is a challenge to apply the lift-off method to display fabrication, especially for large-size, high-resolution displays. Recently, Xu et al. [64] demonstrated oxide TFTs with BCE structure using the wet-etch method to pattern Mo/Al/Mo S/D electrodes directly on the IZO channel layer. The wet etch of the bottom Mo layer was carried out in an H_2O_2 solution, which has high selectivity to IZO. A good etching profile with little residue left on the channel was obtained. However, the edges of the Mo lines will be oxidized by lateral penetrated H_2O_2 etchant from the edges of the overlying photoresist. By adding alkali to the H_2O_2 etchant, the etch rate of Mo could be accelerated, and the oxidizing effect would decrease [65].

Although the BCE structure is regarded to be one of the preferred configurations for cost-competitive TFTs, it has not been adapted to mass production owing to its uncontrollability and poor reproducibility. The alternative method is to add an ESL, which is formed and patterned before S/D electrode deposition to protect the channel during etching of the S/D electrodes. The ESL structure is free from back-channel damage; however, it requires an extra deposition process and patterning mask. Unlike silicon or organic TFTs, the back channel of oxide TFTs is extremely sensitive to the air or the protection layer. Therefore, ESL preparation is a key factor for attaining highly stable oxide TFTs. Conventional SiN_x prepared by PECVD is not suitable for ESL because of its high hydrogen content. Instead, PECVD-deposited SiO_2 is the most commonly used ESL for oxide TFTs; however, the plasma damage and the hydrogen doping will also severely degrade the device performance. Therefore, the SiO_2 ESL should be deposited at low-temperature (<250°C) to reduce hydrogen content. Currently, the ESL structure has been employed for the mass production of oxide TFT backplanes.

The TG structure, the third class of TFT, was studied in the early stage of oxide TFTs where the IGZO film was epitaxially grown onto a single-crystal yttria-stabilized zirconia substrate [66]. The TG structure was the only choice for epitaxial semiconductors because the epitaxial semiconductor film cannot be grown on ordinary gate insulators, only on specified substrates. In oxide TFTs, the TG structure is intensely studied because of its unique advantages. For example, the upper-gate insulator and electrodes may act as a passivation that protects the channel layer from external damage. Furthermore, for bottom-emitting OLED applications, because light from the backlight unit will be blocked by the gate electrode from reaching the semiconductor, the stability of oxide semiconductor for light illumination should be strengthened. Another advantage is that there is no overlap between the gate and the S/D electrodes (self-aligned method), and consequently, it does not create a signal delay from the parasitic capacitance.

There is another special double-gate structure besides the conventional structures mentioned above. In such structure, an additional gate can effectively control a large portion in the vertical direction of the semiconductor bulk. Lim et al. [67] obtained field-effect mobility values about twice as high as single-gate TFTs, and subthreshold swing (SS) values lower by approximately half that of a single-gate structure. Moreover, double-gate TFTs are reported to exhibit excellent device stability [68]. The main drawback of the double-gate structure is the increase of the fabrication cost.

9.3.2 Fabrication Process of Oxide TFTs

As shown in Figure 9.5, the fabrication process of oxide TFTs includes the fabrication of gate, gate insulator, oxide semiconductor layer, ESL, S/D electrodes, etc. The fabrication process of these layers was similar to that of a-Si TFTs except the semiconductor layer. Unlike the a-Si film, which is fabricated by PECVD, the oxide semiconductor film is fabricated using physical vapor deposition such as sputtering, pulse laser deposition, or ALD. Because the composition of oxide semiconductors is more complicated than Si, it is much more difficult to control the film deposition conditions. For example, IGZO, which is the most widely investigated oxide semiconductor, is composed of four components, In, Ga, Zn, and O. Deviation of the composition would cause performance change.

The oxygen concentration is one of the key parameters that influences TFT characteristics because oxygen vacancies are the major source of the traps. By optimizing oxygen content, very high mobility (46 cm^2 V^{-1} s^{-1}) can be obtained [69]. In addition, undesired subthreshold photocurrent in the presence of visible light can be significantly reduced by properly optimizing the sputtering power and O$_2$/Ar gas flow ratio during deposition [70]. Some groups have used N$_2$O plasma treatment to reduce oxygen vacancies and control the free carriers [49,71]. Recently, Raja et al. [72] observed that nitrogen doping (adding N$_2$ gas during sputtering) enhances device stability by well controlling oxygen vacancy and trap sites in channel and channel/dielectric interface.

Annealing is another important process in the oxide TFT fabrication. For example, IGZO TFTs have a high mobility of >10 cm^2 V^{-1} s^{-1}, even when fabricated at room temperature; however, the uniformity and stability are poor for unannealed TFTs [73,74]. Therefore, most IGZO TFTs in prototype displays have been annealed at a temperature >300°C (lower annealing temperatures have been applied recently). Using high-temperature annealing is necessary to oxidize a-IGZO and reduce the concentration of native donor defects in as-deposited a-IGZO, even in a pure O$_2$ atmosphere; electrical conductivity increases with annealing temperature up to 300°C and then starts decreasing [73]. This result indicates that O$_2$ molecules do not have sufficient oxidizing power to passivate the defects in a-IGZO below 300°C. It has also been found that annealing is more effective in wet oxygen than in dry oxygen, which was attributed to the fact that wet annealing suppresses the desorption of H$_2$O, Zn, and O species significantly up to 300°C. Thermal annealing removes weak chemical bonds, particularly Zn–O-related bonds, and forms stable IGZO. Constant-current stress tests revealed that annealed IGZO TFTs are much more stable than unannealed ones.

9.3.3 Characteristics of Oxide TFTs

The electrical characteristics of oxide TFTs that determine device performance are evaluated in terms of several parameters such as mobility (μ), on/off current ratio (I_{on}/I_{off}), threshold voltage (V_{th}), turn-on voltage (V_{on}), and SS. These parameters are, in general, deduced from the output characteristics, where the source-to-drain current (I_{DS}) is plotted against the source-to-drain voltage (V_{DS}) for various gate voltages (V_G), and from the transfer characteristics, where I_{DS} is plotted against V_G for various V_{DS}.

Figure 9.6 shows the TFT output characteristic where the source-to-drain current (I_{DS}) is plotted against the source-to-drain voltage (V_{DS}) for various gate voltages (V_G). When $V_{DS} <$ ($V_G - V_T$), I_{DS} increases with V_{DS}, regarded as a linear regime. When $V_{DS} > (V_G - V_T)$, I_{DS} pins at a constant value because of the channel pinch-off effect (the channel near the drain electrode is turned off because the gate-to-drain voltage, V_{GD}, is lower than V_T) near the drain electrode, regarded as a saturated regime. The output current at different V_G can be seen clearly from the output curves. The total resistance between source and drain electrodes can be calculated by V_{DS}/I_{DS}, and the source and drain contact resistance can be extracted by measuring total resistances with various channel lengths. High contact resistance will cause a current crowding effect (I_{DS} will increase significantly at $V_{DS} > 0$ rather than at $V_{DS} = 0$) in the low V_{DS} regime of the output characteristic.

Figure 9.7 shows the TFT transfer characteristic where I_{DS} is plotted against V_G for a constant V_{DS}. When V_{DS} fixes at lower voltages (usually 0.1–1 V), the device operates at the linear regime, as shown in Figure 9.7a, and the relationship between I_{DS} and V_G can be expressed by

$$I_{DS} = \frac{W}{L}\mu_{lin}C_i\left(V_G - V_{th} - \frac{V_{DS}}{2}\right)V_{DS} \tag{9.1}$$

where μ_{lin} denotes the linear mobility in the channel, W is the channel width, L is the channel length, V_{th} is the threshold voltage, and C_i is the capacitance of the gate insulator per unit area. I_{DS} is in linear relationship with V_G, as shown in the plot of I_{DS} vs. V_G in linear scale in the right side of Figure 9.7a. Thus, μ_{lin} can be deduced from the slope of the I_{DS} vs. V_G curve.

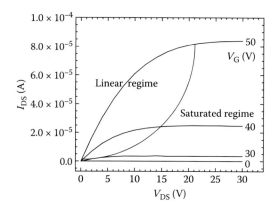

FIGURE 9.6
Output characteristic of an oxide TFT.

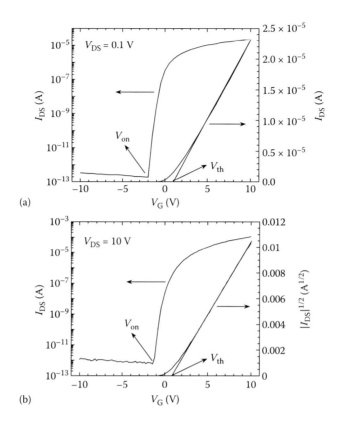

FIGURE 9.7

Transfer characteristics of an oxide TFT for (a) $V_{DS} = 0.1$ V and (b) $V_{DS} = 10$ V.

When V_{DS} fixes at higher voltages ($V_{DS} \gg (V_G - V_{th})$), the device operates at the saturation regime, as shown in Figure 9.7b, and the relationship between I_{DS} and V_G can be expressed by

$$I_{DS} = \frac{W\mu_{sat}C_i}{2L}(V_G - V_{th})^2 \tag{9.2}$$

where μ_{sat} denotes the saturated mobility in the channel. It can be deduced from Equation 9.2 that the square root of I_{DS} ($I_{DS}^{1/2}$) is in linear relationship with V_G (for $V_{DS} \gg (V_G - V_{th})$), as shown in the plot of $I_{DS}^{1/2}$ vs. V_G in the right side of Figure 9.7b. Thus, μ_{sat} can be deduced from the slope of the $I_{DS}^{1/2}$ vs. V_G curve.

The on/off current ratio (I_{on}/I_{off}) is simply defined as the ratio of the maximum to the minimum values of I_{DS}. The minimum I_{DS} is generally given by the noise level of the measurement equipment and by the gate leakage current (I_G), while the maximum I_{DS} depends on the semiconductor material itself and on the effectiveness of capacitive injection by the field effect. An I_{on}/I_{off} of >10^6 is required for AMOLED displays.

The threshold voltage (V_{th}) can be determined using different methods, such as the gate voltage axis intercept of the linear extrapolation of the I_{DS} vs. V_G plot (for low V_{DS}) or of the $I_{DS}^{1/2}$ vs. V_G plot (for high V_{DS}), or the gate voltage corresponding to a certain predefined practical constant drain current. However, large ambiguity can arise on the determination of V_{th}. Therefore, researchers prefer to use another parameter—the turn-on voltage (V_{on}), defined by the voltage where I_{DS} begins to rise in the transfer curve (in logarithmic scale)—to give a more precise definition of the on/off state of the TFTs. V_{on} represents the V_G necessary to fully turn off the transistor, and can be read directly from the transfer curves (in logarithmic scale) as shown in Figure 9.7.

Another important TFT parameter is the SS, which reflects the amount of V_G change required to increase I_{DS} by a decade in the subthreshold region. From the transfer characteristics, SS can be extracted using the equation

$$SS = \frac{dV_G}{d(\log I_{DS})} \tag{9.3}$$

The SS value provides important information about the operation speeds and operation voltages of an oxide TFT. Typically, $SS \ll 1$, around 0.10–0.30 V dec^{-1}, and small values result in higher speeds and lower operation voltage (lower power consumption). The SS value also provides information about the trap density in the bulk of the oxide semiconductor layer or at the insulator/semiconductor interface, affecting the device stability. The total trap density N_t near the semiconductor/gate dielectric interface can be expressed by

$$N_t = \left[\frac{SS \log(e)}{kT/q} - 1 \right] \frac{C_i}{q} \tag{9.4}$$

where k is the Boltzmann constant and q is the charge of an electron.

9.3.4 Long-Term Device Stability of Oxide TFTs

The long-term stability and reliability of TFTs are two of the most important concerns for their applications in AMOLED displays. Therefore, a highly stable device should be verified under variable conditions, such as gate bias stress with and without light illumination and temperature stress.

Positive bias stress (PBS) has usually been tested for the AMOLED applications, because the TFT, especially the driving TFT, must supply a stable current for the entire operating time of the OLED. Accordingly, instability measurement is generally carried out under a $V_G = V_{DS} > 0$ condition to supply a constant current. The oxide TFTs under PBS exhibits significant V_{th} shift (ΔV_{th}), as shown in Figure 9.8. Many efforts have been made on the understanding of mechanism for the instability of oxide TFTs under different PBS test conditions. Earlier, researches focused on the device structure and fabrication process including the trapping of charges into the gate dielectric or at the dielectric/channel interface [75,76], the adsorption–desorption effect on the back channel [77], and the inducing defects during passivation fabrication [78]. Cross and DeSouza [75] investigated the stability of TFTs incorporating sputtered ZnO as the channel layer under gate bias stress. They have found that positive stress results in a positive shift of the transfer characteristics, while negative stress results in a negative shift. Low bias stress has no effect on the subthreshold characteristics. This instability is believed to be a consequence of charge trapping at/near the channel/insulator interface. Higher biases and longer stress times cause degradation of the subthreshold slope, which is thought to arise as a consequence of defect state

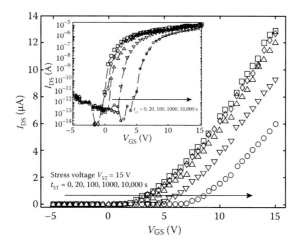

FIGURE 9.8
Linear transfer I_{DS}–V_{GS} curve of a-IGZO TFTs as a function of stress time t_{ST}. The inset shows the bias stress–induced shift of log(I_{DS}–V_{GS}) curve. Sweep was done at $V_{DS} = 0.5$ V in both curves. (From Lee, J.-M. et al., *Appl. Phys. Lett.*, 93, 093504, 2008.)

creation within the ZnO channel material. Similar results were reported by Lee et al. [76], as shown in Figure 9.8. This positive V_{th} shift upon application of a positive V_G stress may be interpreted to occur by either negative charge trapping at the channel/gate insulator interface or negative charge injection into the gate dielectric bulk. Moreover, the trapping of the negative charge will cause the reduction of the electron carrier density, requiring higher gate voltage to turn on the device. The charge trapping and subsequent redistribution of the trap charge in a deeper state can be described by the following stretched-exponential equation:

$$\Delta V_{th} = \Delta V_{th0}\left\{1 - \exp\left[-\left(\frac{t}{\tau}\right)^{\beta}\right]\right\} \tag{9.5}$$

where ΔV_{th0} is the ΔV_{th} at infinite time, t is the stress time, τ is the relaxation time constant, and β is the stretched-exponential exponent.

With the continuous improvement of the device fabrication methods, the influence of the device on the instabilities has constantly decreased, and researchers are now focusing on the native defects in the oxide semiconductors such as oxygen vacancies or surface states [30,79,80], defect creations [81,82], joule-heating effects [83,84], and others. Recent studies show that the doubly charged oxygen vacancies (V_O^{2+}) are the most possible electron traps causing electrical instability under PBS. However, the origins of the native defects remain unresolved, at least from the experimental aspect, and oxide TFTs with better stability is still the main research and development goal in this field.

In general, oxide semiconductors have n-type characteristics; thus, applying negative voltage at the gate is required for turning off the device. In particular, the total duration time of the negative gate bias applied on the switching transistor is much longer than that of the positive gate bias, because the switching TFTs are only turned on once while keeping the "off" state at the rest stage of every scanning period. As most of the oxide semiconductors are n-type, almost no carriers will be induced during negative-gate bias stress

(NBS). As a result, the instability under NBS (without light illumination) is usually better than that under PBS [76].

However, because the TFTs are exposed to visible light originating from the underlying backlight unit in a self-emitting radiation during the operation of AMOLED, the susceptibility of oxide TFTs with respect to illumination should also be minimized. Theoretically, the oxide TFTs should be stable under visible light illumination because of their wide band gap. In reality, an apparent increase of off-current is found under visible light illumination. As a result, the pixels do not turn off when they should, and hence blurred images will be shown on the display. In addition, the switching TFTs are, most of the time, experiencing a negative gate bias, maintaining the "off" state. Therefore, the device stability under negative bias illumination stress (NBIS) should be considered in AMOLED applications.

Shin et al. [85] reported for the first time the bias stability characteristics of ZnO TFTs under visible light illumination. The transfer curves exhibited almost no change under positive V_G stress with illumination, while large negative shifts in V_{th} were observed under negative V_G stress. The most plausible degradation mechanism of the ZnO TFTs under NBIS conditions was suspected due to photogenerated holes being trapped at the gate insulator and/or insulator/channel interface.

An alternative instability mechanism under negative bias illumination has recently been proposed by Ghaffarzadeh et al. [86,87]. Passivated Hf-doped IZO (HIZO) TFTs were observed to undergo negative V_{th} shifts under NBIS conditions due to persistent photoconductivity, which leads to a long recovery time of the negatively shifted V_{th}. The origin of the persistent photoconductivity was attributed to the ionization of oxygen vacancy (V_O) sites, based on temperature and wavelength-dependent photoluminescence emission from the HIZO semiconductor. It was then suggested that photogenerated hole carriers may be physically localized at V_O sites, of which the energy level is at approximately ~2.3 eV above the valence band maximum of HIZO, as shown in Figure 9.9. As a result, the V_O sites

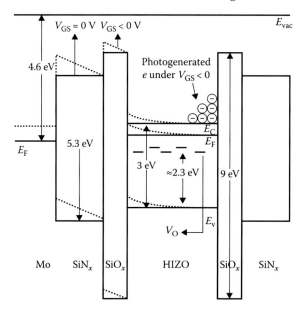

FIGURE 9.9
Approximate energy band diagrams in a TFT stack. Ionized V_O sites are proposed as origin of persistent photoconductivity. Negative gate bias bends the energy bands upward and separates the photogenerated electron–hole pairs. (From Ghaffarzadeh, K. et al., *Appl. Phys. Lett.*, 97, 113504, 2010.)

become singly $\left(V_O^+\right)$ or doubly ionized $\left(V_O^{2+}\right)$ under illumination, and these in turn contribute free electrons, which are responsible for the persistent photoconductivity. Therefore, it may be anticipated that the NBIS as well as PBS instability of oxide TFTs can be improved by decreasing the concentration of V_O sites within the channel and at interfaces.

Usually, the device stability of oxide TFTs is tested at room temperature. However, because the device will be heated during operation, the temperature stress should be considered in oxide TFT stability studies. Moderate temperature (60°C–100°C) is often added during bias stress test, resulting in more serious V_{th} shift owing to the higher activity of the defect states at higher temperature.

9.4 Pixel Circuits and Gate Driver Integrated by TFTs

9.4.1 Conventional 2T1C Pixel Circuit

9.4.1.1 Pixel Circuit Operation

To maintain a continuous current during the whole frame, a pixel needs at least two TFTs with an appropriate storage capacitor. Figure 9.10 shows the simplest pixel circuit with two n-type TFTs. T1 is a switching TFT and T2 is a driving TFT. T1 operates in the linear region, while T2 operates in the saturation region. When the V_{gate} line is selected (programming state), T1 is on and data signal is written via T1 to the gate of T2. When V_{gate} is low (driving state), T1 is off and the written voltage V_{data} is therefore stored in C_s for a whole frame period. The emitting current of OLED is determined by the following equation:

$$I_{DS2} = \frac{1}{2}\mu_n C_{ox}\left(\frac{W}{L}\right)_2 (V_{GS2} - V_{th2})^2 \tag{9.6}$$

where W, L, μ_n, C_{ox}, and V_{th2} are the channel width, channel length, field-effect mobility, gate oxide capacitance per unit area, and threshold voltage of T2, respectively. As shown in Equation 9.6, the OLED current in the emission period depends on the threshold voltage

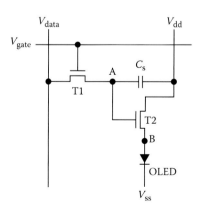

FIGURE 9.10
Pixel circuit with two TFTs for AMOLED.

of the driving TFT (T2). If the threshold voltage of the driving TFT shifts, the emission current of OLED will decline, which results in poor display quality. Therefore, a pixel circuit with compensation function is needed to ensure the uniformity of the display panel, which will be discussed later in this chapter.

9.4.1.2 Parameter Selection

Oxide TFT and OLED device performance parameters are listed in Tables 9.1 and 9.2. Assuming the designed AMOLED display array is 480 × 320 (red, green, blue—RGB), the subpixel size is 70 μm × 210 μm. The brightness of the display panel is 400 cd m^{-2}; thus, $L_{subpixel}$ = 400 cd m^{-2}. From Equation 9.6, we can calculate the required drive current of each pixel:

$$I_{OLED} = \frac{L_{subpixel} \cdot A_{subpixel}}{\eta_{LE}} \tag{9.7}$$

where $A_{subpixel}$ is the area of subpixel and η_{LE} is the OLED luminous efficiency. The required drive currents of three RGB pixels are shown in Table 9.3.

When the 2T1C pixel is working, T2 works in the saturation region; then, the power supply of the panel, V_{dd}, should meet

$$V_{dd} \geq V_{G2} - V_{th2} \tag{9.8}$$

From Equation 9.6, we can obtain the aspect ratio of the driving TFT in the RGB subpixel:

$$\left(\frac{W}{L}\right)_2 = \frac{2I_{DS2}}{\mu_n C_{ox}(V_{GS2} - V_{th2})^2} \tag{9.9}$$

TABLE 9.1

Performance of Oxide TFT Device

TFT Parameters	μ (cm^2 V^{-1} s^{-1})	C_{ox} (nF cm^{-2})	V_{th} (V)	I_{off} (fA) (W/L = 10 μm/10 μm)
Value	10	20	0	100

TABLE 9.2

Performance of RGB OLED (1000 cd m^{-2})

OLED Parameters	R	G	B
V_{OLED} (V)	3.9	3.7	5.2
J_{OLED} (A/m^2)	130	60	165
η_{LE} (cd A^{-1})	7.7	17.1	6.1

TABLE 9.3

Required Drive Currents of Three RGB Pixels

Pixel	R	G	B
I_{OLED} (μA)	0.76	0.35	0.97

TABLE 9.4

Aspect Ratio of Driving TFT in RGB Subpixel

Pixel	R	G	B
W/L	0.45	0.18	1.22

The required W/L of three RGB pixels is shown in Table 9.4. The data signal voltage range is ΔV_g (8 V or so in our design), and the grayscale $N_g = 256$. The T2 gate voltage change must meet the following equation in one frame:

$$I_{off} \cdot \frac{1}{f} \leq C_s \cdot \frac{\Delta V_g}{N_g} \tag{9.10}$$

where f is the frame frequency, I_{off} is the off current, and C_s is the storage capacitor in Figure 9.10. Then, we can obtain

$$C_s \geq I_{off} \cdot \frac{1}{f} \frac{N_g}{\Delta V_g} \tag{9.11}$$

After calculation, the minimum value of the storage capacitor is 13.3 fF. Taking into account the effect of the capacitance coupling, we take 300 fF for the design, therefore obtaining the area of the storage capacitor, $A_s = 1500\ \mu m^2$.

When the capacitor is charging, T1 is working in the linear region and the charging current flowing through T1 is

$$I_{DS1} = \mu_n C_{ox} \left(\frac{W}{L} \right)_1 (V_{GS1} - V_{th1}) V_{DS1} \tag{9.12}$$

In Equation 9.12, μ_n, C_{ox}, and $(W/L)_1$ are the effective electron mobility, gate capacitance per unit area, and the aspect ratio of T1, respectively. V_{GS1}, V_{th1}, and V_{DS1} are the gate-source voltage, threshold voltage, and drain-source voltage of T1, respectively. To charge the storage capacitance within the addressing time, the on current of T1 must satisfy

$$5 \left(\frac{\partial I_{DS}}{\partial V_{DS1}} \right)^{-1} C_s \leq \frac{1}{M} \times \frac{1}{f} \tag{9.13}$$

where M is the number of rows in the display. From Equations 9.12 and 9.13, we can obtain

$$\left(\frac{W}{L} \right)_1 \geq \frac{5MfC_s}{\mu_n C_{ox}(V_{GS1} - V_{th1})} \tag{9.14}$$

According to Equation 9.14, we calculate $(W/L)_1 \geq 0.96$. As long as the aspect ratio of T1 is greater than 0.96, the display requirement can be met; here, we take $(W/L)_1 = 10\ \mu m/10\ \mu m = 1$. However, if the W/L of T1 is too large, the leakage current would increase when T1 is off, which would affect the normal grayscale display. If T1 is too small, the capacitor may not be fully charged or discharged during the programming time, which also affects the normal display.

9.4.1.3 Nonuniformity of 2T1C Pixel

For the 2T1C pixel circuit, long bias causes the threshold voltage of the oxide TFT drift so that the light emission current will decline. On the other hand, the aging of the OLED device will occur with long bias; then, luminous efficiency decreases, resulting in a reduction in the luminance of an AMOLED panel.

Figure 9.11 is the curve of the threshold voltage shift of oxide TFT with long forward bias (+15 V). It can be seen from the figure, after 11 h of bias stress, that the TFT threshold voltage drifts from 0.5 to 2.53 V. Here, the TFT in our experiment has an etch stopper structure with five lithographic processes. For the 2T1C pixel circuit, when OLED is in the emission period, the driving TFT is working in the saturation region; the OLED light-emitting current is given by Equation 9.15:

$$I_{OLED} = \frac{1}{2}\mu_n C_{ox} \left(\frac{W}{L}\right)_2 (V_{GS2} - V_{th2})^2 = \frac{1}{2}\mu_n C_{ox} \left(\frac{W}{L}\right)_2 (V_{data} - V_B - V_{th2})^2 \qquad (9.15)$$

where V_{data} is the data signal provided by the source IC, V_B is the OLED anode voltage, and V_{th2} is the threshold voltage of T2. When the threshold voltage of T2 increases, the OLED light emission current will decline.

Figure 9.12 is the aging test curve of an OLED device with an area of 3 mm × 3 mm. In the experiment, OLED is emitting under a constant current of 2 mA. It can be seen from the figure that, after 180 h of aging test, the OLED anode voltage shifts from 6.6 to 7.1 V.

Figures 9.13 and 9.14 show the effect of OLED degradation for a 2T1C pixel circuit. As seen from Figure 9.13, the OLED current will decrease with the same data voltage (V_A) due to the OLED degradation. As shown in Figure 9.14, the luminance of OLED will decrease with the same OLED current due to the OLED degradation.

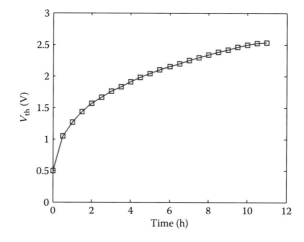

FIGURE 9.11
Threshold voltage shift of metal oxide TFT with forward bias (+15 V).

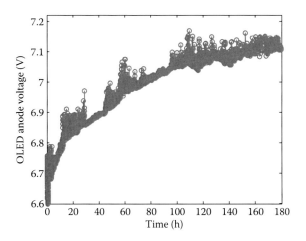

FIGURE 9.12
Aging test curve of an OLED device.

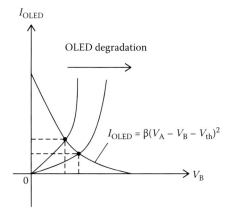

FIGURE 9.13
I_{OLED} vs. V_B with OLED degradation.

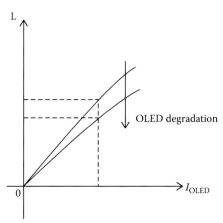

FIGURE 9.14
Luminance vs. I_{OLED} with OLED degradation.

9.4.1.4 Full-Color AMOLED Panel Based on 2T1C Pixel

On the basis of our oxide TFT and OLED devices, we designed a 4.8″ full-color AMOLED display panel (Figure 9.15). The resolution is 480 × 320 (RGB). The OLED device is bottom emitted with an aspect ratio of 36.6%.

The oxide TFT process is applied on a 200 mm × 200 mm glass substrate, the TFT has an etch stopper structure, and full-color is achieved by patterning small-RGB-molecule OLED with a fine metal mask.

In the design process of full-color AMOLED panels, white balance should be considered carefully. R, G, B color coordinates are determined by the light-emitting material. Here, the color parameters are R (Y_r, x_r, y_r), G (Y_g, x_g, y_g), and B (Y_b, x_b, y_b). Using the chromaticity coordinate system and the color additive, we can calculate the synthetic light color coordinates W (Y_w, x_w, y_w).

For RGB arranged subpixels, the relationship between the luminance and chrominance color of synthesized light and RGB are

$$x_w = \left(x_r \times \frac{Y_r}{y_r} + x_g \times \frac{Y_g}{y_g} + x_b \times \frac{Y_b}{y_b} \right) \bigg/ \left(\frac{Y_r}{y_r} + \frac{Y_g}{y_g} + \frac{Y_b}{y_b} \right) \tag{9.16}$$

$$y_w = (Y_r + Y_g + Y_b) \bigg/ \left(\frac{Y_r}{y_r} + \frac{Y_g}{y_g} + \frac{Y_b}{y_b} \right) \tag{9.17}$$

$$Y_w = (Y_r + Y_g + Y_b)/3 \tag{9.18}$$

From Equations 9.16 through 9.18, we obtain

$$\begin{bmatrix} (x_w - x_r)/y_r & (x_w - x_g)/y_g & (x_w - x_b)/y_b \\ y_w/y_r - 1 & y_w/y_g - 1 & y_w/y_b - 1 \\ 1/3 & 1/3 & 1/3 \end{bmatrix} \cdot \begin{bmatrix} Y_r \\ Y_g \\ Y_b \end{bmatrix} = \begin{bmatrix} 0 \\ 0 \\ Y_w \end{bmatrix} \tag{9.19}$$

FIGURE 9.15
(See color insert.) Photograph of 4.8″ full-color AMOLED display.

Solving Equation 9.19, we can establish the luminance relationship between RGB and white color:

$$Y_r = \alpha^* Y_w \tag{9.20}$$

$$Y_g = \beta^* Y_w \tag{9.21}$$

$$Y_b = \gamma^* Y_w \tag{9.22}$$

Here α, β, and γ are

$$
\begin{cases}
\alpha = \dfrac{[(x_w - x_b)(y_w - y_g) - (x_w - x_g)(y_w - y_b)] \times 3y_r}{y_w(y_r(x_g - x_b) + y_g(x_b - x_r) + y_b(x_r - x_g))} \\[3mm]
\beta = \dfrac{[(x_w - x_r)(y_w - y_b) - (x_w - x_b)(y_w - y_r)] \times 3y_g}{y_w(y_r(x_g - x_b) + y_g(x_b - x_r) + y_b(x_r - x_g))} \\[3mm]
\gamma = \dfrac{[(x_w - x_g)(y_w - y_r) - (x_w - x_r)(y_w - y_g)] \times 3y_b}{y_w(y_r(x_g - x_b) + y_g(x_b - x_r) + y_b(x_r - x_g))}
\end{cases}
\tag{9.23}
$$

In our design, the chrominance coordinates of RGB-emitting material are R (0.66, 0.34), G (0.28, 0.65), and B (0.15, 0.23). To obtain a peak white luminance of 400 cd m^{-2} with chrominance coordinates of (0.33, 0.33), the RGB subpixel luminances are Y_r = 387.2 cd m^{-2}, Y_g = 368.9 cd m^{-2}, and Y_b = 443.9 cd m^{-2}.

However, long-term instability of oxide backpanels would cause OLED luminance degradation. The blue subpixel degrades faster than other subpixels, which causes the color of the display panel to shift to yellow. For the commercial use of AMOLED, compensation pixels are usually needed, which will be discussed in the next section.

9.4.2 Compensation Pixel Circuits

9.4.2.1 Basic Principle of Compensation Pixel Circuit for AMOLED Display

The driving schemes of pixel circuits can be classified into current programming (the input signal from the source driver is current) [88–90] and voltage programming methods (the input signal, voltage) [91–95]. The current programming methods can compensate for both the mobility variations and the threshold voltage variations of TFTs. However, the voltage programming methods are more attractive to AMOLED display owing to their fast programming time [96]. The conventional 2T1C pixel circuit is the simplest one, with various advantages such as high speed, high aperture ratio, and compatibility with TFT LCD displays. However, it may not be suitable for AMOLED displays with high quality because the OLED current is nonuniform owing to the variations in the threshold voltage of the driving TFT. In this section, we focus on the voltage programming method with the compensation pixel circuits.

Normally, the driving scheme of compensation pixel circuits can be divided into four stages: initialization, threshold voltage detection, data input, and OLED emission. Therefore, the threshold voltage detection is the most important one. There are two different methods for detecting the threshold voltage of the driving TFT, as shown in Figures 9.16 and 9.17. As seen from Figure 9.16a for n-type TFTs, the voltage at node A (V_A) stored in the

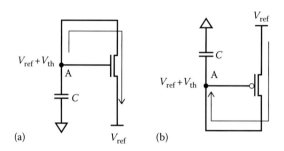

FIGURE 9.16
Threshold voltage detection method: (a) n-type TFTs; (b) p-type TFTs.

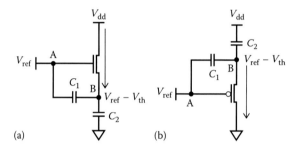

FIGURE 9.17
Another threshold voltage detection method: (a) n-type TFTs; (b) p-type TFTs.

storage capacitor discharges through the driving TFT until it is turned off, i.e., $V_A = V_{ref} + V_{th}$. As a result, the gate-source voltage of the driving TFT is V_{th}. V_A will be V_{th} if V_{ref} is set to be ground. The case for p-type TFTs is similar as shown in Figure 9.16b. For the other method shown in Figure 9.17a for n-type TFTs, the voltage at node B (V_B) will be charged until the driving TFT is turned off, i.e., $V_B = V_{ref} - V_{th}$. Obviously, the threshold voltage of

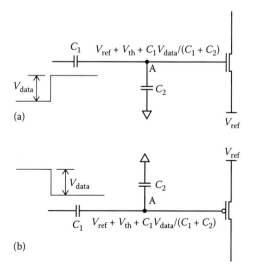

FIGURE 9.18
Data input method: (a) n-type TFTs; (b) p-type TFTs.

the driving TFT is stored at C_1. The case for p-type TFTs is similar, as shown in Figure 9.17b. For the condition of Figure 9.16, there are two methods to input data voltage. For the first method, V_{ref} is set to be V_{data} in the threshold voltage detection stage. As a result, V_A becomes $V_{data} + V_{th}$ at the end of this stage. Obviously, the threshold voltage detection and the data input are in the same stage for this method. For the second method, as shown in Figure 9.18, V_{data} is input by the capacitor coupling effect in the data input stage, which is after the threshold voltage detection stage, and then $V_A = V_{ref} + V_{th} + C_1V_{data}/(C_1 + C_2)$. For the method described in Figure 9.17, the data voltage can be input by a capacitor coupling effect. More in details, V_A becomes $V_{ref} + V_{data}$ from V_{ref} in the data input stage, and then V_B will become $V_{ref} - V_{th} + C_1V_{data}/(C_1 + C_2)$, and then $V_{AB} = V_{th} + C_2V_{data}/(C_1 + C_2)$.

9.4.2.2 Compensation Pixel Circuit with p-Type TFTs

Figure 9.19 is the first reported compensation pixel circuit developed by Dawson et al. [91], to our knowledge. The principles of this circuit are explained as below. MN1 is turned on when the Select signal goes to low level in period 1. MN3 is turned on when the AZ signal goes to low level in period 2. Obviously, period 1 and period 2 compose the initialization stage. MN4 is turned off and then threshold voltage is detected in period 3. As a result, V_A becomes $V_{DD} + V_{th}$, and V_{th} is negative for p-type TFT. MN3 is turned off in period 4 since the AZB signal goes to a high level. Similarly, period 3 and period 4 compose the threshold voltage detection stage. In period 5, the data line jumps and the changed value is ΔV_{data}, and then V_A becomes $V_{dd} + V_{th} + C_1\Delta V_{data}/(C_1 + C_2)$ by the capacitor coupling effect, where ΔV_{data} is negative. In period 6, MN1 is turned off since the Select signal goes to a high level. Period 5 and period 6 compose the data input stage. The data line recovers to be V_{ref} in period 7. In period 8, the OLED emits light by turning on MN4 with AZB signal going to low level. Period 7 and period 8 compose the OLED emission stage. Note that there is only one signal jump in each period in order to avoid the signal competition. The corresponding OLED current is equal to the drain current of T5, which is working in the saturation region; as a result

$$
\begin{aligned}
I_{OLED} &= \beta(V_{GS} - V_{th})^2 \\
&= \beta(V_{C2} - V_{th})^2 \\
&= \beta\left(\frac{C_1}{C_1 + C_2}\Delta V_{data}\right)^2
\end{aligned}
\tag{9.24}
$$

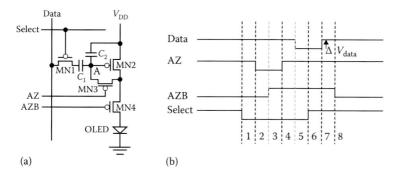

(a)   (b)

FIGURE 9.19
(a) Pixel circuit; (b) timing diagram.

with $\beta = 0.5\mu C_{ox}W/L$, where W, L, μ, and C_{ox} are the channel width, channel length, field-effect mobility, and gate oxide capacitance per unit area, respectively. It is seen that the threshold voltage of the driving TFT (MN2) has been compensated.

Figure 9.20 shows another compensation pixel circuit [92]. The driving scheme of this circuit can be divided into three stages. (i) Initialization period: the voltage at node A (V_A) is initialized as V_{int}. (ii) Threshold voltage detection and data input period: V_A is charged up to $V_{data} + V_{th}$ from V_{int} through T2, T1, and T3. (iii) Emission period: similarly, the OLED current is expressed as

$$I_{OLED} = \beta(V_{data} - ELVDD)^2 \tag{9.25}$$

Obviously, for such driving method, data at the next row will not be renewed until the V_{th} of the driving TFT at the previous row has been detected. As a result, less time is allocated to V_{th} compensation as the number of rows increases or frame rate increases. Therefore, this pixel circuit may not be suitable for the high-resolution AMOLED displays, especially three-dimensional AMOLED displays with a higher frame rate.

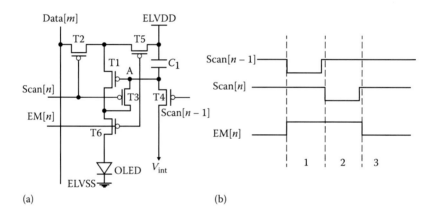

FIGURE 9.20
(a) Pixel circuit; (b) timing diagram.

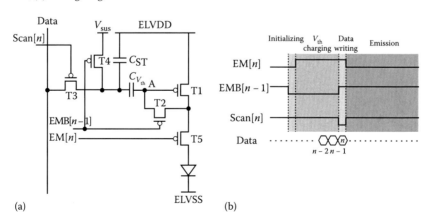

FIGURE 9.21
(a) Pixel circuit; (b) timing diagram.

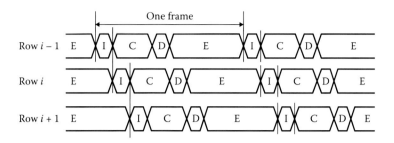

FIGURE 9.22
Pipeline technology for compensation pixel circuit. I: initialization, C: compensation of threshold voltage, D: data input, and E: emission of OLED.

The pixel circuit shown in Figure 9.21 is similar to that shown in Figure 9.19, except that there is T4 connected to V_{sus} at node A [92]. The advantage for this circuit is that the threshold voltage detection period is independent of the data signal line. As a result, the pipeline technology can be used to improve the speed of the compensation pixel circuits [93], which is shown in Figure 9.22. The OLED current in the emission period is expressed as

$$I_{OLED} = \beta(V_{data} - V_{sus})^2 \qquad (9.26)$$

9.4.2.3 Compensation Pixel Circuit with n-Type TFTs

A new threshold voltage compensation pixel circuit employing crystalline IGZO TFTs is shown in Figure 9.23 [94]. In period I, the signals are reset and the drain-source voltage of the driving TFT (DrTr) becomes higher than V_{th}. In period II, the threshold voltage is detected and data are input simultaneously. Thus, the voltage at node B is $V_0 - V_{th}$, and the voltage across the storage capacitor is $V_{data} - V_0 + V_{th}$. The pixel emits light in period III in which the storage capacitor is connected to the gate and the source of DrTr. Therefore, the gate-source voltage of DrTr is represented by $V_{data} - V_0 + V_{th}$, which shows that the

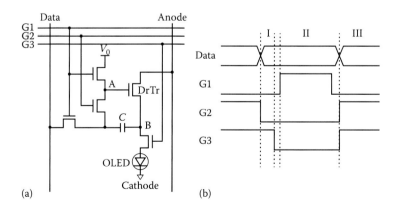

FIGURE 9.23
(a) Pixel circuit; (b) timing diagram.

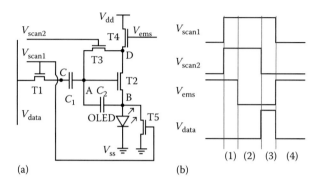

FIGURE 9.24
(a) Pixel circuit; (b) timing diagram.

threshold voltage has been compensated. Moreover, the corresponding OLED current is expressed as

$$I_{OLED} = \beta(V_{data} - V_0)^2 \tag{9.27}$$

Figure 9.24 shows another compensation pixel circuit [95]. In period 1, the previous stored voltage at node A (V_A) would be charged up to a specific value through T4 and T3. In period 2, V_A is discharged through T3, T2, and T5 until T2 is turned off. As a result, the V_{th} of T2 is stored at C_2. In period 3, data is input by the capacitor coupling effect between C_1 and C_2. The corresponding OLED current in the emission period (i.e., period 4) is similarly expressed as

$$I_{OLED} = \beta \left(\frac{C_1}{C_1 + C_2} V_{data} \right)^2 \tag{9.28}$$

9.4.3 Gate Driver in Panel

Gate driver integrated in panel has attracted great attention and is a new trend for flat-panel technology. Compared with gate driver IC bonding on panel, gate driver integrated by TFTs in panel (GIP) technology has many advantages, such as reduction of the manufacturing cost, achievement of narrow bezel, and improvement of the panel's reliability [97,98].

Oxide TFT technology should be more suitable for flat-panel displays (FPD) because oxide TFT has a higher current driving capability than that of the hydrogenated amorphous silicon and superior scalability with a low production cost in comparison with LTPS TFT. In fact, gate driver integrated by oxide TFTs is being developed for FPD [99,100].

However, GIP with oxide TFTs still faces some challenges. The oxide TFTs are inherently depletion-mode devices, which cannot ensure that the TFT device is turned off completely when the gate-source voltage (V_{GS}) equals zero. Hence, the leakage current in the off state must be considered or it will affect the GIP circuit's normal operation. Besides, the leakage current may cause additional power consumption, which also needs to be considered in mobile applications.

9.4.3.1 Typical GIP Circuits Employing a-Si TFTs and LTPS TFTs

The main function of the GIP circuit is to turn on the pixels row by row. Then, the data is input into the pixels of every row during the turning time. A typical GIP circuit schematic and its timing diagram are shown in Figure 9.25. This typical GIP circuit employing a-Si or the LTPS TFT technology has some advantages, such as simple structure, fewer devices, and small number of driving signals [100].

As Figure 9.25 shows, this GIP circuit consists of three parts: input, storage, and output. The input part includes T6, and mainly transfers the previous output stage's signal. The storage part includes T3, T4, and T5, and mainly stores the invert signals Q and Qb. The output part includes T1 and T2, and mainly generates output signals.

As the timing diagram shows in Figure 9.25, the operation of the typical GIP circuit also includes three periods: set, output generation, and reset.

1. Set period: The driving signals CLK and CLKB are set high and low, respectively. At the same time, the high voltage of the previous output stage is input through T6. Thus, node Q becomes high and the inverter signal Qb becomes low, because T3, T4, T5, and T6 constitute the SR latch.

2. Output generation period: CLK and OUT (n − 1) are set low. Thus, node Q and node Qb become high and low, respectively. In the meantime, CLKB becomes high and it is input through T1. Therefore, the output signal generates to drive the panel during this period. It is noted that the voltage of node Q becomes higher than that in the set period owing to the bootstrap effect of the storage capacitor C_1.

3. Reset period: The driving signals CLK and CLKB are set high and low, respectively. Meanwhile, OUT (n − 1) keeps low. Therefore, node Q discharges to low through T6 and Qb charges to high. Then, T2 is turned on, which pulls down the output signal OUT (n) to the low level VGL.

The GIP circuit mentioned above is composed of n-type TFTs. P-type TFTs can also compose the GIP circuit. P-type TFTs are mainly fabricated by polycrystalline silicon technology. The stability of p-type TFTs is better than that of n-type TFTs in LTPS technology. Figure 9.26 shows the schematic of GIP circuit employing all p-type TFTs and the timing diagram [101].

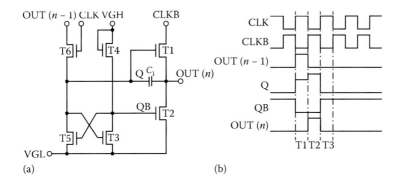

FIGURE 9.25
(a) Typical GIP circuit schematic and (b) its timing diagram.

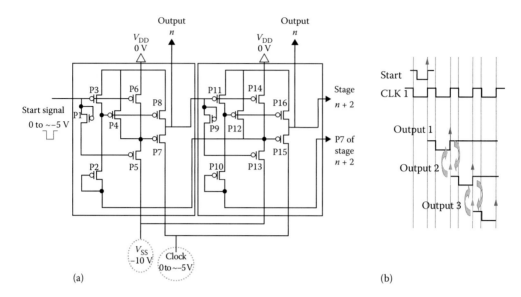

FIGURE 9.26
(a) Schematic of p-type GIP circuit; (b) the timing diagram.

Each stage of this GIP circuit includes eight p-type TFTs. Two TFTs, P7 and P8, are used for output buffers. Four TFTs, P1, P3, P5, and P6, form an inverter, and the TFTs P2 and P4 operate as the reset. In addition, one control clock signal and two power lines, V_{DD} and V_{SS}, are needed. As seen from Figure 9.26, the driving mechanism is very simple, which is similar to the n-type GIP circuit mentioned above. The difference is that the reset part is controlled by the next stage shift register.

9.4.3.2 GIP Circuits with Oxide TFTs

There are some problems if the GIP circuit shown in Figure 9.25 is realized by oxide TFTs. As seen in Figure 9.27, there are mainly three leakage current paths, which are denoted by the three lines with arrows. The voltage between the gate and source of T2, T5, and T6 is zero during the working period so that T2, T5, and T6 are not turned off completely, which may cause the leakage current paths. The larger the TFT size is, the higher leakage current will be [100].

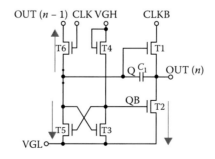

FIGURE 9.27
Typical GIP circuit using oxide TFT.

As the lines with arrows show, when node Q is high, the leakage current from T6 and T5 could decrease the voltage of node Q so T1 is not conductive enough to transfer the signal CLKB to the OUT (*n*) node. Besides, the leakage current from T2 also distorts the output signal. The leakage current paths may lower the driving ability of the OUT (*n*) signal. Furthermore, the leakage current may cause serious power consumption. Therefore, some additional considerations should be taken for oxide TFTs used in GIP circuits.

Figure 9.28 shows a new gate driver circuit developed by Kim et al. [102]. The circuit includes four parts: charging TFTs (including T1a, T1b), discharging TFT (T2), an inverter (T4, T5), and the output section (part A and part B). The operation of this circuit includes three periods and every period includes two stages.

1. Set period: First, the high voltage of part A of the output section at the previous *n* − 2 stage is input to the gates of the charging TFTs (T1a, T1b), and the high voltage of part B of the output section at the previous *n* − 2 stage is also input to the drain of T1a at the same time. Other signals including CLK (*X*), CLK (*X* − 1), Vg (*n* + 2), and Vc (*n* + 2) keep low so node Q is charging to high voltage. Second, Vg (*n* − 2) and Vc (*n* − 2) remain high voltage and CLK (*X* − 1) and Vc (*n* − 1) are set high. Thus, Vc (*n* − 1) is supplied to node Q through T3 to keep node Q at the high level.

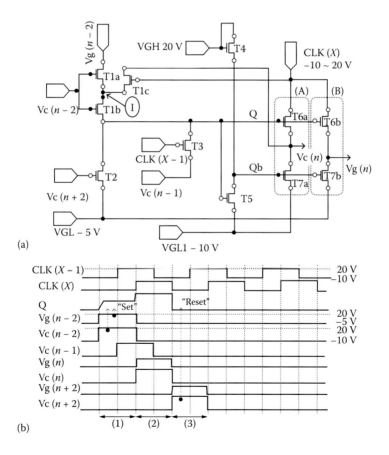

(a)

(b)

FIGURE 9.28
(a) Schematic of B. Kim's circuit; (b) timing diagram of B. Kim's circuit.

2. Output generation period: First, CLK (X − 1) and Vc (n − 1) remain high. CLK (x) turns high and it is input through T6a and T6b, so the output section of the present stage can generate the pulse to drive the panel. At the same time, node Q will increase because of the parasitic capacitance between the gate and drain of T6a and T6b. Second, CLK (X − 1) and Vc (n − 1) turn low. CLK (X) remains high to keep the output section at a high voltage level. In this period, Vc (n) is supplied to node I through T1c when CLK (X) keeps high so that the V_{GS} of T1b is negative, which turns off T1b completely. It is noted that Vg (n − 2), Vc (n − 2), and Vc (n + 2) are kept to VGL, VGL1, and VGL1, respectively, and VGL1 is lower than VGL. Therefore, the V_{GS} of T1a, T1b, and T2 becomes negative during this period and they are turned off completely. Therefore, node Q keeps a high level without distortion. Moreover, the V_{GS} of T7b is also negative, so the leakage from T7b is negligible.

3. Reset period: The signal Vc (n + 2) turns high to turn on T2, so node Q discharges to VGL. Node Q turns off T5 so Qb becomes high to turn on T7a and T7b. Then, Vc (n) and Vg (n) discharge to VGL and VGL1, respectively.

In short, the oxide TFT is a kind of depletion-mode device. Employing two low voltage–level signals to turn off the depletion-mode TFT is an effective way to operate an oxide TFT GIP circuit. Kim et al.'s other circuit is a building block for oxide GIP. Figure 9.29 shows this circuit's schematic and the timing diagram. Compared with previous ones, the circuit has low power consumption. This circuit mainly employs the series-connected two-transistor (STT) structure for charging and discharging TFTs. As shown in Figure 9.29a, TFT-a and TFT-b are connected in series and the high voltage from output section is input back to node S through TFT-c during the generation period. Thus, in the STT structure,

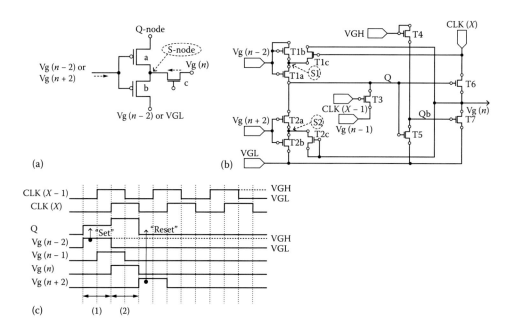

FIGURE 9.29
(a) STT structure; (b) low power consumption circuit; and (c) circuit's timing diagram.

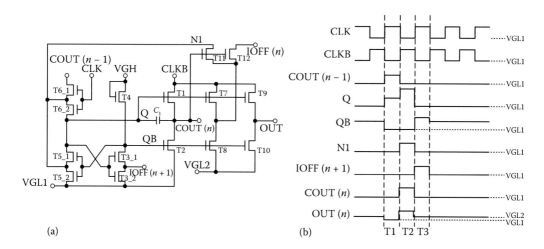

(a)   (b)

FIGURE 9.30
(a) Circuit schematic and (b) its timing diagram.

a negative voltage can be applied to the V_{GS} of depletion-mode oxide TFTs, which can minimize leakage current during the output generation period [103,104].

Using this method, oxide TFT GIP circuits can generate high-voltage pulse for driving the pixel panel successfully without distortion. In addition, using the STT structure can also reduce the power consumption.

A new and practical oxide TFT GIP circuit created by Wu et al. [100] is shown in Figure 9.30. Compared with previous two GIP circuits using oxide TFTs, this circuit employs the STT structure and two low-voltage level power signals, which can effectively reduce the leakage current occurring in the set, output generation, and reset periods. Moreover, the timing diagram is very simple. In Figure 9.30, it is shown that the circuit includes three parts: output section, SR latch, and feedback section. T3–T6 constitute the SR latch, which generates the inverter signals Q and Qb. The output section is divided into three stages. The first stage, made of T1 and T2, provides the carry signal COUT (n) connecting to the input of the next stage shift register. The second stage, consisting of T7 and T8, exports the feedback signals to the STT structure. The third stage, including T9 and T10, generates the actual output pulse to drive the panel.

In summary, the method employing the STT structure and two low-voltage-level power lines can greatly reduce different kinds of leakage current and generate a stable pulse to drive the panel, which should possibly be used in commercial oxide TFT panels.

Another new GIP circuit uses a direct current (DC)-type output stage and one-clock signal, which can effectively eliminate the clock skew problem and the capacitive load on clock signals. Figure 9.31 shows the circuit's schematic and its timing diagram. To reduce the capacitive load of the clock signal, the DC power line V_{DD} is connected to the drain node of pull-up TFTs (N19 and N21) in the output stage. The clock signal is connected to two small TFTs (N3 and N9), of which channel width is very small so that the capacitive load of the clock signal can be largely reduced. By these means, the GIP circuit can achieve low power consumption [97].

In summary, a gate driver circuit integrated on panel is an important technology in FPD. The GIP circuit with a simple structure, high speed, and low power consumption is a developing trend for AMOLED displays.

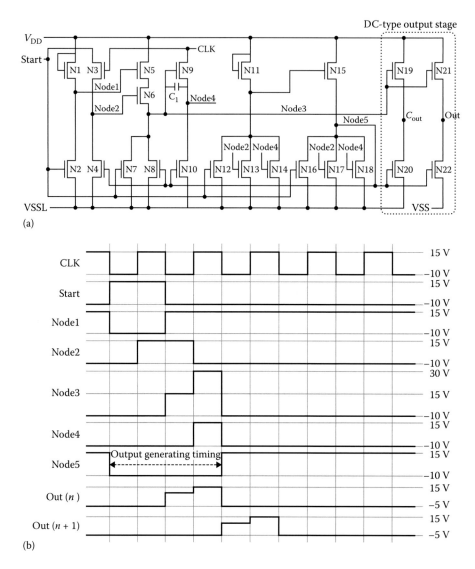

FIGURE 9.31
(a) One clock signal's GIP circuit; (b) timing diagram.

9.5 Conclusion

Multicomponent oxide materials have been intensively developed because of their amorphous nature (uniformity over large area), which is critical in mass production. We discussed the oxide TFT structures, process, and especially device stability. The long-term stability and reliability of oxide TFTs are the most important challenges for their applications in AMOLED displays. We fabricated a stable oxide TFT panel based on an ESL structure, which was used to successfully drive a 4.8″ full-color AMOLED display with resolution of 480 × RGB × 320. Finally, the pixel circuit of the pixel design was discussed. Oxide TFT is a very promising technology for AMOLED applications.

Acknowledgments

This work was supported by the National Natural Science Foundation of China (grant 61306099), Fundamental Research Funds for the Central Universities (grant 2012ZM0003), and the Foundation of Outstanding Doctoral Dissertation in Guangdong Province (grant SYBZZXM201113). The authors would like to thank Profs. Y. Cao, J.B. Peng, and H.L. Ning in South China University of Technology for their technical assistance during the project, and Letitia Li (MIT) and Tony Liang (DuPont) for the editing. The authors also would like to thank Dr. M. Xu, Dr. J.H. Zou, Dr. H. Tao, D.X. Luo, M. Li, H. Xu, and M.J. Zhao in South China University of Technology and New Vision company, without whose hard work the results presented in this chapter could not have been achieved.

References

1. T. Arai, and T. Sasaoka, Emergent oxide TFT technologies for next-generation AM-OLED displays, *SID Symp. Dig.*, 42:710, 2011.
2. T. Kamiya, Z. A. K. Durrani, H. Ahmed, T. Sameshima, Y. Furuta, H. Mizuta, and N. Lloyd, Reduction of grain-boundary potential barrier height in polycrystalline silicon with hot-vapor annealing probed using point-contact devices, *J. Vac. Sci. Technol. B*, 21:1000, 2003.
3. C. Yang, K. Hong, J. Jang, D. S. Chung, T. K. An, W.-S. Choi, and C. E. Park, Solution-processed flexible ZnO transparent thin-film transistors with a polymer gate dielectric fabricated by microwave heating, *Nanotechnology*, 20:465201, 2009.
4. M. Stewart, R. S. Howell, L. Pires, and M. K. Hatalis, Polysilicon TFT technology for active matrix OLED displays, *IEEE Trans. Electron Devices*, 48(5):845–851, 2001.
5. J. H. Lee, W. J. Nam, B. K. Kim, H. S. Choi, Y. M. Ha, and M. K. Han, A new poly-Si TFT current-mirror pixel for active matrix organic light emitting diode, *IEEE Electron Device Lett.*, 27:830, 2006.
6. Y. G. Mo, M. Kim, C. K. Kang, J. H. Jeong, Y. S. Park, C. G. Choi, H. D. Kim, and S. S. Kim, Amorphous oxide TFT backplane for large size AMOLED TVs, *SID 2010 Dig.*, 1037–1040, 2010.
7. H. H. Lu, H. C. Ting, T. H. Shih, C. Y. Chen, C. S. Chuang, and Y. Lin, 32-Inch LCD panel using amorphous indium–gallium–zinc–oxide TFTs, *SID 2010 Dig.*, 1136–1138, 2010.
8. N. Morosawa, M. Nishiyama, Y. Ohshima, A. Sato, Y. Terai, K. Tokunaga, J. Iwasaki, K. Akamatsu, Y. Kanitani, S. Tanaka, T. Arai, and K. Nomoto, High mobility self-aligned top-gate oxide TFT for high-resolution AMOLED, *SID 2013 Dig.* 10.1, 2013.
9. H. Xu, D. X. Luo, M. Li, M. Xu, J. H. Zou, H. Tao, L. F. Lan, L. Wang, J. B. Peng, and Y. Cao, *J. Mater. Chem. C*, 2:1255, 2014.
10. W. J. Nam, J. S. Shim, H. J. Shin, J. M. Kim, W. S. Ha, K. H. Park, H. G. Kim, B. S. Kim, C. H. Oh, B. C. Ahn, B. C. Kim, and S. Y. Cha, 55-Inch OLED TV using InGaZnO TFTs with WRGB pixel design, *SID 2013 Dig.* 21.2, 2013.
11. Ü. Özgür, Y. Alivov, C. Liu, A. Teke, M. A. Reshchikov, S. Doğan, V. Avrutin, S.-J. Cho, and H. Morkoç, A comprehensive review of ZnO materials and devices, *J. Appl. Phys.*, 98:041301, 2005.
12. M. Usuda, N. Hamada, T. Kotani, and M. van Schilfgaarde, All-electron GW calculation based on the LAPW method: Application to wurtzite ZnO, *Phys. Rev. B*, 66:125101, 2002.
13. S. J. Clark, J. Robertson, S. Lany, and A. Zunger, Intrinsic defects in ZnO calculated by screened exchange and hybrid density functional, *Phys. Rev. B*, 81:115311, 2010.
14. S. Lany, and A. Zunger, Anion vacancies as a source of persistent photoconductivity in II–VI and chalcopyrite semiconductors, *Phys. Rev. B*, 72:035215, 2005.

15. S. Lany, and A. Zunger, Dopability, intrinsic conductivity, and nonstoichiometry of transparent conducting oxides, *Phys. Rev. Lett.*, 98:045501, 2007.
16. C. G. Van de Walle, Hydrogen as a cause of doping in zinc oxide, *Phys. Rev. Lett.*, 85:1012, 2000.
17. F. Oba, M. Choi, A. Togo, and I. Tanaka, Point defects in ZnO: An approach from first principles, *Sci. Technol. Adv. Mater.*, 12:034302, 2011.
18. D. C. Look, G. C. Farlow, P. Reunchan, S. Limpijumnong, S. B. Zhang, and K. Nordlund, Evidence for native-defect donors in n-type ZnO, *Phys. Rev. Lett.*, 95:225502, 2005.
19. Y.-S. Kim, and C. H. Park, Rich variety of defects in ZnO via an attractive interaction between O vacancies and Zn interstitials: Origin of n-type doping, *Phys. Rev. Lett.*, 102:086403, 2009.
20. G. F. Boesen, and J. E. Jacobs, ZnO field-effect transistor, *Proc. IEEE*, 56(11):2094, 1968.
21. R. L. Hoffman, B. J. Norris, and J. F. Wager, ZnO-based transparent thin-film transistors, *Appl. Phys. Lett.*, 82(5):733, 2003.
22. E. Fortunato, P. Barquinha, A. Pimentel, A. Goncalves, A. Marques, L. Pereira, and R. Martins, Fully transparent ZnO thin-film transistor produced at room temperature, *Adv. Mater.*, 17(5):590, 2005.
23. S. J. Lim, S. J. Kwon, H. Kim, and J. S. Park, Anomalous light-induced enhancement of photo-luminescence from Si nanocrystals fabricated by thermal oxidation of amorphous Si, *Appl. Phys. Lett.*, 91(3):033111, 2007.
24. S. J. Lim, J.-M. Kim, D. Kim, S. Kwon, J.-S. Park, and H. Kim, Atomic layer deposition ZnO:N thin film transistor: The effects of N concentration on the device properties, *J. Electrochem. Soc.*, 157(2):H214, 2010.
25. A. Tsukazaki, A. Ohtomo, and M. Kawasaki, High-mobility electronic transport in ZnO thin films, *Appl. Phys. Lett.*, 88:152106, 2006.
26. S. Limpijumnong, P. Reunchan, A. Janotti, and C. G. Van de Walle, Hydrogen doping in indium oxide: An ab initio study, *Phys. Rev. B*, 80:193202, 2009.
27. P. Reunchan, X. Zhou, S. Limpijumnong, A. Janotti, and C. G. Van de Walle, Vacancy defects in indium oxide: An ab-initio study, *Curr. Appl. Phys.*, 11:S296, 2011.
28. K. A. Johnson, and N. W. Ashcroft, Corrections to density-functional theory band gaps, *Phys. Rev. B*, 58:15548, 1998.
29. P. Agoston, K. Albe, R. M. Nieminen, and M. J. Puska, Intrinsic n-type behavior in transparent conducting oxides: A comparative hybrid-functional study of In_2O_3, SnO_2, and ZnO, *Phys. Rev. Lett.*, 103:245501, 2009.
30. S. Lany, A. Zakutayev, T. O. Mason, J. F. Wager, K. R. Poeppelmeier, J. D. Perkins, J. J. Berry, D. S. Ginley, and A. Zunger, Surface origin of high conductivities in undoped In_2O_3 thin films, *Phys. Rev. Lett.*, 108:016802, 2012.
31. K. Ellmer, Resistivity of polycrystalline zinc oxide films: Current status and physical limit, *J. Phys. D Appl. Phys.*, 34:3097–3108, 2001.
32. H. Zhang, H. Cao, A. Chen, L. Liang, Z. Liu, and Q. Wan, Enhancement of electrical performance in In_2O_3 thin-film transistors by improving the densification and surface morphology of channel layers, *Solid-State Electron.*, 54:479–483, 2010.
33. H. Z. Zhang, L. Y. Liang, A. H. Chen, Z. M. Liu, Z. Yu, H. T. Cao, and Q. Wan, High-performance transparent thin-film transistor based on Y_2O_3/In_2O_3 with low interface traps, *Appl. Phys. Lett.*, 97:122108, 2010.
34. L. Wang, M. Yoon, G. Lu, Y. Yang, A. Facchetti, and T. Marks, High-performance transparent inorganic–organic hybrid thin-film n-type transistors, *Nat. Mater.*, 5:893–900, 2006.
35. L. Wang, M.-H. Yoon, A. Facchetti, and T. J. Marks, Flexible inorganic/organic hybrid thin-film transistors using all-transparent component materials, *Adv. Mater.*, 19:3252–3256, 2007.
36. P. K. Nayak, M. N. Hedhili, D. Cha, and H. N. Alshareef, High performance In_2O_3 thin film transistors using chemically derived aluminum oxide dielectric, *Appl. Phys. Lett.*, 103:033518, 2013.
37. H. A. Klasens, and H. Koelmans, A tin oxide field-effect transistor, *Solid-State Electron.*, 7:701, 1964.
38. R. Presley, C. Munsee, C. Park, D. Hong, J. Wager, and D. Keszler, Tin oxide transparent thin-film transistors, *J. Phys. D Appl. Phys.*, 37:2810–2813, 2004.

39. D.-H. Lee, Y.-J. Chang, W. Stickle, and C.-H. Chang, Functional porous tin oxide thin films fabricated by inkjet printing process, *Electrochem. Solid-State Lett.*, 10(11):K51–K54, 2007.

40. W. Cheong, S. Yoon, C. Hwang, and H. Chu, High-mobility transparent SnO_2 and $ZnO–SnO_2$ thin-film transistors with SiO_2/Al_2O_3 gate insulators, *Jpn. J. Appl. Phys.*, 48:04C090, 2009.

41. J. Jang, R. Kitsomboonloha, S. Swisher, E. Park, H. Kang, and V. Subramanian, Transparent high-performance thin film transistors from solution-processed SnO_2/ZrO_2 gel-like precursors, *Adv. Mater.*, 25:1042–1047, 2013.

42. N. Ueda, H. Hosono, R. Waseda, and H. Kawazoe, Anisotropy of electrical and optical properties in beta-Ga_2O_3 single crystals, *Appl. Phys. Lett.*, 71:933, 1997.

43. M. Orita, H. Ohta, M. Hirano, and H. Hosono, Deep-ultraviolet transparent conductive beta-Ga_2O_3 thin films, *Appl. Phys. Lett.*, 77:4166, 2000.

44. S. Geller, Crystal structure of β-Ga_2O_3, *J. Chem. Phys.*, 33:676, 1960.

45. K. Matsuzaki, H. Yanagi, T. Kamiya, H. Hiramatsu, K. Nomura, M. Hirano, and H. Hosono, Field-induced current modulation in epitaxial film of deep-ultraviolet transparent oxide semiconductor Ga_2O_3, *Appl. Phys. Lett.*, 88:092106, 2006.

46. M. Higashiwaki, K. Sasaki, A. Kuramata, T. Masui, and S. Yamakoshi, Gallium oxide (Ga_2O_3) metal-semiconductor field-effect transistors on single-crystal β-Ga_2O_3 (010) substrates, *Appl. Phys. Lett.*, 100:013504, 2012.

47. J.-Y. Kwon, D.-J. Lee, and K.-B. Kim, Transparent amorphous oxide semiconductor thin film transistor, *Electron. Mater. Lett.*, 7(1):1–11, 2011.

48. E. Fortunato, P. Barquinha, A. Pimentel, L. Pereira, G. Gonçalves, and R. Martins, Amorphous IZO TTFTs with saturation mobilities exceeding 100 cm²/Vs, *Phys. Stat. Sol. (Rapid Res. Lett.)*, 1(1):R34–R36, 2007.

49. J. C. Park, H.-N. Lee, and S. Im, Self-aligned top-gate amorphous indium zinc oxide thin-film transistors exceeding low-temperature poly-Si transistor performance, *ACS Appl. Mater. Interfaces*, 5:6990–6995, 2013.

50. K. Nomura, H. Ohta, A. Takagi, T. Kamiya, M. Hirano, and H. Hosono, Room-temperature fabrication of transparent flexible thin-film transistors using amorphous oxide semiconductors, *Nature*, 432:488, 2004.

51. E. Fortunato, P. Barquinha, and R. Martins, Oxide semiconductor thin-film transistors: A review of recent advances, *Adv. Mater.*, 24:2945–2986, 2012.

52. Y. Hwang, J. Jeon, S. Seo, and B. Bae, Solution-processed, high performance aluminum indium oxide thin-film transistors fabricated at low temperature, *Electrochem. Solid-State Lett.*, 12(9):H336–H339, 2009.

53. Y. Hwang, J. Jeon, and B. Bae, Post-humid annealing of low-temperature solution-processed indium based metal oxide TFTs, *Electrochem. Solid-State Lett.*, 14(7):H303–H305, 2011.

54. C. Kim, S. Kim, J. Lee, J. Park, S. Kim, J. Park, E. Lee, J. Lee, Y. Park, J. Kim, S. Shin, and U.-I. Chung, Amorphous hafnium–indium–zinc oxide semiconductor thin film transistors, *Appl. Phys. Lett.*, 95:252103, 2009.

55. E. Chong, K. Jo, and S. Lee, High stability of amorphous hafnium–indium–zinc–oxide thin film transistor, *Appl. Phys. Lett.*, 96:152102, 2010.

56. P. Jin-Seong, K. KwangSuk, P. Yong-Gil, M. Yeon-Gon, K. Hye Dong, and J. Jae Kyeong, Novel ZrInZnO thin-film transistor with excellent stability, *Adv. Mater.*, 21(3):329, 2009.

57. H. Park, B. Kim, J. Park, and K. Chung, Device performance and bias instability of Ta doped InZnO thin film transistor as a function of process pressure, *Appl. Phys. Lett.*, 102:102102, 2013.

58. L. Lan, N. Xiong, P. Xiao, M. Li, H. Xu, R. Yao, S. Wen, and J. Peng, Enhancement of bias and illumination stability in thin-film transistors by doping InZnO with wide-band-gap Ta_2O_5, *Appl. Phys. Lett.*, 102:242102, 2013.

59. J. Park, S. Kim, C. Kim, and H. Lee, Low-temperature fabrication and characteristics of lanthanum indium zinc oxide thin-film transistors, *IEEE Electron Device Lett.*, 33(5):685, 2012.

60. H. Q. Chiang, J. F. Wager, R. L. Hoffman, J. Jeong, and D. A. Keszler, High mobility transparent thin-film transistors with amorphous zinc tin oxide channel layer, *Appl. Phys. Lett.*, 86:013503, 2005.

61. W. B. Jackson, R. L. Hoffman, and G. S. Herman, High-performance flexible zinc tin oxide field-effect transistors, *Appl. Phys. Lett.*, 87:193503, 2005.
62. D.-H. Cho, S. Yang, C. Byun, J. Shin, M. K. Ryu, S.-H. K. Park, C.-S. Hwang, S. M. Chung, W.-S. Cheong, S. M. Yoon, and H.-Y. Chu, Transparent Al-Zn-Sn-O thin film transistors prepared at low temperature, *Appl. Phys. Lett.*, 93:142111, 2008.
63. Y. S. Rim, D. L. Kim, W. H. Jeong, and H. J. Kim, Effect of Zr addition on ZnSnO thin-film transistors using a solution process, *Appl. Phys. Lett.*, 97:233502, 2010.
64. H. Xu, L. Lan, M. Xu, J. Zou, L. Wang, D. Wang, and J. Peng, High performance indium–zinc–oxide thin-film transistors fabricated with a back-channel-etch-technique, *Appl. Phys. Lett.*, 99:253501, 2011.
65. M. Zhao, L. Lan, H. Xu, M. Xu, M. Li, D. Luo, L. Wang, S. Wen, and J. Peng, Wet-etch method for patterning metal electrodes directly on amorphous oxide semiconductor films, *ECS Solid State Lett.*, 1(5):P82–P84, 2012.
66. K. Nomura, H. Ohta, K. Ueda, T. Kamiya, M. Hirano, and H. Hosono, Thin-film transistor fabricated in single-crystalline transparent oxide semiconductor, *Science*, 300:1269–1272, 2003.
67. H. Lim, H. Yin, J. S. Park, I. Song, C. Kim, J. Park, S. Kim, S. W. Kim, C. B. Lee, Y. C. Kim, Y. S. Park, and D. Kang, *Appl. Phys. Lett.*, 93(6):063505, 2008.
68. P. Joon Seok, S. Kyoung Seok, K. Tae Sang, J. Ji Sim, L. Kwang-Hee, M. Wan-Joo, K. Hyun-Suk, K. Eok Su, P. Kyung-Bae, S. Jong-Baek, K. Jang-Yeon, R. Myung Kwan, and L. Sangyun, High performance and stability of double-gate Hf–In–Zn–O thin-film transistors under illumination, *IEEE Electron Device Lett.*, 31(9):960, 2010.
69. P. Barquinha, L. Pereira, G. Goncalves, R. Martins, and E. Fortunato, The effect of deposition conditions and annealing on the performance of high-mobility GIZO TFTs, *Electrochem. Solid-State Lett.*, 11(9):H248, 2008.
70. H.-S. Kim, K.-B. Park, K. S. Son, J. S. Park, W.-J. Maeng, T. S. Kim, K.-H. Lee, E. S. Kim, J. Lee, J. Suh, J.-B. Seon, M. K. Ryu, S. Y. Lee, K. Lee, and S. Im, The influence of sputtering power and O_2/Ar flow ratio on the performance and stability of Hf–In–Zn–O thin film transistors under illumination, *Appl. Phys. Lett.*, 97:102103, 2010.
71. J. Park, S. Kim, C. Kim, S. Kim, I. Song, H. Yin, K. Kim, S. Lee, K. Hong, J. Lee, J. Jung, E. Lee, K. Kwon, and Y. Park, High-performance amorphous gallium indium zinc oxide thin-film transistors through N_2O plasma passivation, *Appl. Phys. Lett.*, 93:053505, 2008.
72. J. Raja, K. Jang, N. Balaji, W. Choi, T. Trinh, and J. Yi, Negative gate-bias temperature stability of N-doped InGaZnO active-layer thin-film transistors, *Appl. Phys. Lett.*, 102:083505, 2013.
73. K. Nomura, T. Kamiya, H. Ohta, M. Hirano, and H. Hosono, Defect passivation and homogenization of amorphous oxide thin-film transistor by wet O_2 annealing, *Appl. Phys. Lett.*, 93:192107, 2008.
74. K. Nomura, T. Kamiya, M. Hirano, and H. Hosono, Origins of threshold voltage shifts in room-temperature deposited and annealed a-In–Ga–Zn–O thin-film transistors, *Appl. Phys. Lett.*, 95:013502, 2009.
75. R. B. M. Cross, and M. M. DeSouza, Investigating the stability of zinc oxide thin film transistors, *Appl. Phys. Lett.*, 89:263513, 2006.
76. J.-M. Lee, I.-T. Cho, J.-H. Lee, and H.-I. Kwon, Bias-stress-induced stretched-exponential time dependence of threshold voltage shift in InGaZnO thin film transistors, *Appl. Phys. Lett.*, 93:093504, 2008.
77. J. S. Park, J. K. Jeong, H. J. Chung, Y. G. Mo, and H. D. Kim, Electronic transport properties of amorphous indium-gallium-zinc oxide semiconductor upon exposure to water, *Appl. Phys. Lett.*, 92:072104, 2008.
78. M. Li, L. Lan, M. Xu, H. Xu, D. Luo, N. Xiong, and J. Peng, Impact of deposition temperature of the silicon oxide passivation on the performance of indium zinc oxide thin-film transistors, *Jpn. J. Appl. Phys.*, 51:076501, 2012.
79. S. Lee, and D. C. Paine, Identification of the native defect doping mechanism in amorphous indium zinc oxide thin films studied using ultra high pressure oxidation, *Appl. Phys. Lett.*, 102:052101, 2013.

80. A. U. Adler, T. C. Yeh, D. B. Buchholz, R. P. H. Chang, and T. O. Mason, Quasi-reversible point defect relaxation in amorphous In-Ga-Zn-O thin films by in situ electrical measurements, *Appl. Phys. Lett.*, 102:122103, 2013.

81. T. Kamiya, K. Nomura, and H. Hosono, Present status of amorphous In–Ga–Zn–O thin-film transistors, *Sci. Technol. Adv. Mater.*, 11:044305, 2010.

82. T. Fung, C. Chuang, C. Chen, K. Abe, H. Kumomi, and J. Kanicki, 2-D numerical simulation of high performance amorphous In-Ga-Zn-O TFTs for flat panel displays, in *AMFPD' 08 Dig. Tech. Paper*, 251, 2008.

83. M. Fuji, Y. Uraoka, T. Fuyuki, J. Jung, and J. Kwon, Experimental and theoretical analysis of degradation in Ga_2O_3–In_2O_3–ZnO thin-film transistors, *Jpn. J. Appl. Phys.*, 48:04C091, 2009.

84. S. Urakawa, S. Tomai, Y. Ueoka, H. Yamazaki, M. Kasami, K. Yano, D. Wang, M. Furuta, M. Horita, Y. Ishikawa, and Y. Uraoka, Thermal analysis of amorphous oxide thin-film transistor degraded by combination of joule heating and hot carrier effect, *Appl. Phys. Lett.*, 102:053506, 2013.

85. J.-H. Shin, J.-S. Lee, C.-S. Hwang, S.-H. K. Park, W.-S. Cheong, M. Ryu, C.-W. Byun, J.-I. Lee, and H. Y. Chu, Light effects on the bias stability of transparent ZnO thin film transistors, *ETRI J.*, 31:62, 2009.

86. K. Ghaffarzadeh, A. Nathan, J. Robertson, S. Kim, S. Jeon, C. Kim, U.-I. Chung, and J.-H. Lee, Persistent photoconductivity in Hf–In–Zn–O thin film transistors, *Appl. Phys. Lett.*, 97:143510, 2010.

87. K. Ghaffarzadeh, A. Nathan, J. Robertson, S. Kim, S. Jeon, C. Kim, U.-I. Chung, and J.-H. Lee, Instability in threshold voltage and subthreshold behavior in Hf–In–Zn–O thin film transistors induced by bias- and light-stress, *Appl. Phys. Lett.*, 97:113504, 2010.

88. Y. He, R. Hattori, and J. Kanicki, Improved a-Si:H TFT pixel electrode circuits for active-matrix organic light emitting displays, *IEEE Trans. Electron Devices*, 48(7):1322–1325, 2001.

89. J. H. Lee, W. J. Nam, S. H. Jung, and M. K. Han, A new current scaling pixel circuit for AMOLED, *IEEE Electron Device Lett.*, 25(5):280–282, 2004.

90. Y. C. Lin, H. P. D. Shieh, and J. Kanicki, A novel current-scaling a-Si:H TFTs pixel electrode circuit for AMOLEDs, *IEEE Trans. Electron Devices*, 52(6):1123–1131, 2005.

91. R. M. A. Dawson, Z. Shen, D. A. Furest, S. Connor, J. Hsu, M. G. Kane, R. G. Stewart, A. Ipri, C. N. King, P. J. Green, R. T. Flegal, S. Pearson, C. W. Tang, S. Van Slyke, F. Chen, J. Shi, M. H. Lu, and J. C. Sturm, The impact of the transient response of organic light emitting diodes on the design of active matrix OLED displays, in *IEDM Tech. Dig.*, 875–878, 1998.

92. D. W. Park, C. K. Kang, Y. S. Park, B. Y. Chung, K. H. Chung, K. Kim, B. H. Kim, and S. S. Kim, High-speed pixel circuits for large-sized 3-D AMOLED displays, *J. Soc. Inf. Display*, 19(4):329–334, 2011.

93. G. R. Chaji, and A. Nathan, Parallel addressing scheme for voltage programmed active-matrix OLED displays, *IEEE Trans. Electron Devices*, 54(5):1095–1100, 2007.

94. T. Tanabe, S. Amano, H. Miyake, A. Suzuki, R. Komatsu, J. Koyama, S. Yamazaki, K. Okazaki, M. Katayama, H. Matsukizono, Y. Kanzaki, and T. Matsuo, New threshold voltage compensation pixel circuits in 13.5″ quad full high definition OLED display of crystalline In–Ga–Zn–oxide FETs, *SID 2012 Dig.*, 43(1):88–91, 2012.

95. W. J. Wu, L. Zhou, R. H. Yao, and J. B. Peng, A new voltage-programmed pixel circuit for enhancing the uniformity of AMOLED displays, *IEEE Electron Device Lett.*, 32(7):931–933, 2011.

96. A. Nathan, G. R. Chaji, and S. J. Ashtiani, Driving schemes for a-Si and LTPS AMOLED displays, *J. Display Technol.*, 1(2):267–277, 2005.

97. S.-J. Yoo, S.-J. Hong, J.-S. Kang, H.-J. In, and O.-K. Kwon, A low-power single-clock-driven scan driver using depletion-mode a-IGZO TFTs, *IEEE Electron Device Lett.*, 33(3):402–404, 2012.

98. C. W. Liao, C. D. He, T. Chen, D. Dai, S. Chung, T. S. Jen, and S. D. Zhang, Implementation of an a-Si: H TFT gate driver using a five-transistor integrated approach, *IEEE Trans. Electron Devices*, 38(6):1–7, 2012.

99. C. W. Liao, C. D. He, T. Chen, D. Dai, S. Chung, T. S. Jen, and S. D. Zhang, Design of integrated amorphous-silicon thin-film transistor gate driver, *IEEE J. Display Technol.*, 9(1):7–16, 2013.

100. Z. Y. Wu, L. Y. Duan, G. C. Yuan, C. S. Jiang, Y. Z. Li, and L. C. Yan, An integrated gate driver circuit employing depletion-mode IGZO TFTs, *SID 2012 Dig.*, 43(1):5–7, 2012.
101. S. H. Jung, H. S. Shin, J. H. Lee, and M. K. Han, An AMOLED pixel for the VT compensation of TFT and a p-type LTPS shift by employing 1 phase clock signal, *SID 2005 Dig.*, 36(1):300–303, 2005.
102. B. Kim, C.-I. Ryoo, S.-J. Kim, J.-U. Bae, H.-S. Seo, C.-D. Kim, and M.-K. Han, New depletion-mode IGZO TFT shift register, *IEEE Electron Device Lett.*, 32(2):158–160, 2011.
103. B. Kim, S. C. Choi, S.-H. Kuk, Y. H. Jang, K.-S. Park, C.-D. Kim, and M.-K. Han, A novel level shifter employing IGZO TFT, *IEEE Electron Device Lett.*, 32(2):167–169, 2011.
104. B. Kim, S. C. Choi, J. S. Lee, S. J. Kim, Y. H. Jang, S. Y. Young, C. D. Kim, and M. K. Han, A depletion-mode In–Ga–Zn–O thin-film transistor shift register embedded with a full swing level shifter, *IEEE Trans. Electron Devices*, 58(9):3012–3017, 2011.

10

Microstructural Characterization and Performance Measurements

Zhigang Rick Li and Jeff Meth

CONTENTS

From the early stage of organic light-emitting diodes (OLEDs) development to today's large commercial-scale manufacturing of OLEDs (e.g., widely used in smartphones), many advanced microstructural characterization techniques and sophisticated performance measurement devices have been broadly used [1–14]. In this chapter, we will briefly describe these techniques and devices, and review how scientists and engineers use them to improve the performance of OLEDs.

10.1 Microstructural Characterization Techniques

In the last century, many microstructural characterization techniques have been developed, such as electron microscopy, atomic tunneling microscopy, photoelectron spectroscopy, and Raman spectroscopy. The structure of OLED-based displays is such that many pixels are arranged orderly in the x–y plane. The size and number of pixels determine the resolution and size of the display. Along the z-axis, several layers are stacked on each other. These layers are made up of very different materials, varying from soft, organic to hard, inorganic materials. Many of the layers are ultrathin, in the range of nanometers. The microstructure of the layers and the interfaces between them strongly influence the performance of the device [15]. Also, the presence of included defects, such as submicron-sized particles, critically degrades the diode's lifetime. Advanced microstructural characterization techniques have been providing the required structural information about the OLED in recent years.

10.1.1 Transmission Electron Microscopy

Transmission electron microscopy (TEM) is a powerful and mature microstructural characterization technique. The principles and applications of TEM have been described in many books [16–19]. The image formation in TEM is similar to that in optical microscopy; however, the resolution of TEM is far superior to that of an optical microscope because of the enormous differences in the wavelengths of the sources used in these two microscopes. Today, most conventional TEMs can be routinely operated at a resolution better than 0.2 nm, which provides the needed microstructural information about ultrathin layers and their interfaces in OLEDs. Electron beams can be focused to nanometer size, so nanochemical analysis of materials can be performed. These unique abilities to provide structural and chemical information down to atomic–nanometer dimensions make it an indispensable technique in OLED development. However, TEM specimens need to be very thin to make them transparent to electrons. This is one of the most formidable obstacles in using TEM in this field. Current versions of OLEDs are composed of hard glass substrates, soft organic materials, and metal layers. Conventional TEM sample preparation techniques are no longer suitable for these samples [20–22]. These difficulties have been overcome by using the advanced dual beam (DB) microscopy technique, which will be discussed later. In recent years, aberration-corrected scanning transmission electron microscopy (ACSTEM) has been installed in many institutions [23]. The ACSTEM's resolution reaches impressive values of better than 0.06 nm. The electron beam can be focused with beam size of 0.1–0.2 nm when using electron energy loss spectroscopy (EELS) and energy dispersive spectroscopy (EDS).

TEM has been used in OLED research and development. Indium–tin oxide (ITO) is commonly used as a transparent conducting electrode in displays (see Chapter 6). To achieve required properties, in particular low resistivity, ITO films must be either directly deposited on a heated substrate, or postannealed at high temperature, which exclude most inexpensive plastic substrates from use. Carcia et al. [24] examined the microstructure of the ITO films grown by radiofrequency (RF)–magnetron sputtering on polyester at room temperature. Figure 10.1 shows a high-resolution TEM image of an ITO thin film on polyester substrate in cross-sectional view. It is found that crystalline features formed only near the air interface and are not uniform throughout the film thickness. Phosphorescent dye-doped films were studied in detail by Noh et al. [25]. The TEM results indicate that the formation of aggregates may affect the performance of the devices. Park et al. [26] reported that a dramatic increase in photostability of a blue light-emitting polymer was achieved by the addition of gold nanoparticles. The TEM was used to characterize the microstructure of polymers doped with gold nanoparticles. They found that the sizes of gold nanoparticles in polymer films are in the range of 5–10 nm only. In the development of highly efficient gas diffusion barriers for OLEDs, Meyer et al. [27] showed cross-section TEM images of nanolaminate structures composed of alternating Al_2O_3 and ZrO_2 sublayers

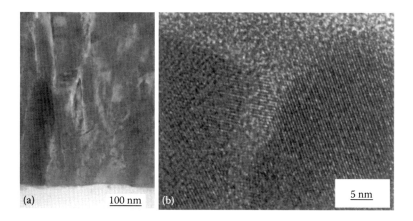

FIGURE 10.1
(a) Lower-magnification TEM image of ITO thin film on polyester substrate in cross-sectional view. (b) Atomic-resolution image of ITO film near the surface showing the crystalline feature.

grown by atomic layer deposition. TEM results indicate that the individual sublayers are well defined and the roughness increases toward to the top of barrier.

10.1.2 Scanning Electron Microscopy

The scanning electron microscope (SEM) is another important and widely used electron beam instrument for the microstructural characterization of materials [28]. A modern SEM is very easy to use, and its images are also usually easy to interpret. The SEM has significant advantages over optical microscopy, with nanometer resolution, three-dimensional appearance due to the large depth of the view, and elemental composition analysis ability because of backscattering imaging and EDS. The SEM, like TEM, also uses a high-energy electron beam focused on the sample. In the case of SEM, however, it provides microstructural information about the surface or the materials near the surface. An important advantage of SEM over TEM is that it can accommodate a relatively large OLED sample with modest sample preparation. Since 1979, environmental SEM has been developed and can be routinely used to observe wet samples [29]. Typically, a cross section for SEM is prepared using mechanical polishing methods or microtome. These methods can produce artifacts into the materials studied, for example, deformation of the specimens around the voids. Recently, a new cross-section polisher (CP) has been developed and is being used more and more for many applications [30]. A new CP simplifies the preparation of the samples and makes it possible to prepare significantly better representative cross sections with fewer artifacts and distortion. It can produce many large cross-section areas compared with DB microscopy (see Section 10.1.3) in reasonable sample preparation time. It is clear that this new technique will be very useful in the microstructural investigation of OLED devices.

Using modern SEM with nanometer resolution, one can detect foreign particles on the substrate. Existence of these particles on ITO films would severely affect the performance of the OLED. The SEM image of such a particle with size >30 µm is shown in Figure 10.2, compared with the distance between cathode and anode of usually <200 nm. This kind of defects can generate visible dark spots or nonemissive areas. Do et al. [31] used SEM and other microstructural characterization techniques to study degradation processes of Al electrodes in the unprotected OLEDs. They concluded that one of the most crucial factors

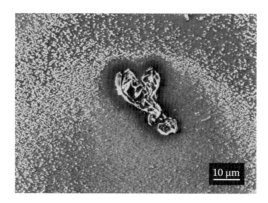

FIGURE 10.2
SEM image of small particle on ITO anode layer.

of the degradation process in those devices was deformation of metal and organic layers due to heat, gas evolution, and oxidation caused by applied voltage. Examples of utilization of SEM in the development of OLED were also reported by several other groups [32–34], and we do see SEM has been widely used in the commercial-scale production of OLEDs.

10.1.3 Dual Beam Microscopy

A relatively new kind of microscopy, DB microscopy, has been used in material characterization and development [35]. The unique combination of scanning electron beam and focused ion beam (DB) technologies working jointly on a single platform can greatly expand the application range to include in situ modification (etch and deposition) of materials, which other electron microscopy techniques lack. The DB microscopy technique overcomes the limits of top-surface imaging and provides a three-dimensional, subsurface view of materials. The site-specific analysis ability of DB microscopy makes it a very useful technique in OLED development. Often, fabricated OLEDs contain many defects such as microscopic particles on the surface. By using DB microscopy, these defects can be effectively located and then analyzed in great detail. Another important use of DB microscopy is to prepare ultrathin samples for TEM, as mentioned before, so one can maximally explore the capability of ultrahigh-resolution TEM [36].

The OLED is composed of hard and soft layers so that conventional cross-sectional TEM sample preparation techniques cannot be applied. Figure 10.3 is a first DB microscopy-prepared TEM image of an OLED in cross-sectional view [37]. The glass substrate, ITO, organic layers, and Al cathode are indicated in the image. The microstructure and interfaces of all these layers can be well studied now. The nanometer-sized spots in organic layers are indium-rich particles. The combination of DB microscopy and TEM will continuously advance OLED research and development in the future.

10.1.4 Scanning Probe Microscopy

Scanning probe microscopy (SPM) is another relatively new class of microstructural characterization technique that probes materials on micrometer to subnanometer scale [38]. The SPM includes atomic force microscopy (AFM), scanning tunneling microscopy (STM), and tens of other related imaging techniques. Each SPM uses a sharp probe to scan the surface of the sample, point-by-point and line-by-line, to form an image of the surface. The simplest

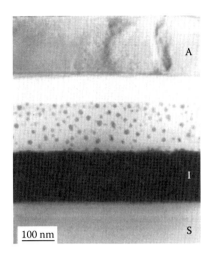

100 nm

FIGURE 10.3
DB microscopy–prepared TEM image of an OLED in cross-sectional view. Glass substrate (S), ITO (I) anode, organic layers, and Al (A) cathode are clearly seen in the image.

map is of three-dimensional topography. Other maps distinguish regions that are physically or chemically different from one another, revealing information about electrical, mechanical, magnetic, optical, and many other properties of the materials. Unlike electron microscopy, which requires vacuum and, often, some sample preparation, most of the SPM works in air and even in a liquid, with minimal or no sample preparation. The SPM measures surfaces in all the three dimensions: x, y, and z. Lateral topographic resolution for most SPM scanning techniques is typically 2–10 nm, and vertical resolution is typically better than 0.1 nm, far superior to the SEM. The SPM is the most powerful tool for surface metrology in our time. However, there are several limitations for SPM. For example, the sample must not have local variations in surface height of tens of micrometers. In 2012, Bruker announced the release of the photoconductive atomic force microscopy (pcAFM) module. The new accessory enables sample illumination while performing nanoscale electrical characterization [39].

There are many publications reporting how SPM was used to characterize the surface of organic materials in OLEDs [40–42]. Halls et al. [40] used AFM to study the surface morphology of polyfluorene composites. In 1996, Cumpston and Jensen [41] investigated the failure mechanism in OLEDs. The circular, nonemissive dark spot defects were imaged and analyzed, and the formation of such defects was explained based on AFM results. The nature of the interface between the light-emitting layer and the metal electrode is of importance in determining device performance. Ahn and Lee [42] studied the effects of annealing of polythiophene derivative for polymer light-emitting diodes (PLEDs). The AFM images clearly indicate the differences in roughness of the films before and after annealing at four different temperatures.

10.1.5 Laser Scanning Confocal Microscopy

Laser scanning confocal microscopy (LSCM) is a valuable tool for characterizing biological and other materials [43,44]. In conventional fluorescence microscopy, much of the structural detail that could otherwise be resolved is obscured due to the contribution of fluorescence from regions outside the plane of focus. By using a pinhole restricting the collection of light from out-of-focus regions, confocal microscopy provides both significantly higher lateral- and vertical-resolution images. The plane of focus (Z-plane) is selected by a computer-controlled fine-stepping motor that moves the microscope stage up and down. Typical focus motors can adjust the focal plane in as little as 0.1-μm increment. A three-dimensional reconstruction of a specimen can be generated by stacking the two-dimensional optical sections collected in series. The LSCM is a nondestructive microstructure characterization technique, which can directly generate, with and without applying the voltage, both electroluminescent and fluorescent images of the organic films in encapsulated OLEDs. More applications of LSCM in the OLEDs may be found in the near future.

10.1.6 Electron Spectroscopy for Chemical Analysis

Electron spectroscopy for chemical analysis (ESCA), also known as x-ray photoelectron spectroscopy (XPS) was developed in the 1960s [45]. This technique is sensitive to 0.05–0.5 atomic percentage or a fraction of a monolayer with sampling depth on the order of 10 nm. In ESCA, the kinetic energies of photoelectrons excited by x-rays are measured using an energy analyzer. The most commonly used x-ray sources are Al Kα (1486.6 eV, with 0.85-eV line width) and Mg Kα (1253.6 eV, with 0.7-eV line width) x-rays. By measuring the kinetic energies, one can calculate the electron binding energies, which are characteristic of the atoms from which the electrons originate and the chemical state or oxidation state of the atom. An ESCA spectrum is a plot of the number of emitted electrons versus binding energy. The peak areas can be used to determine the surface elemental composition. Hence, ESCA can provide both chemical binding information and elemental concentration. The ESCA detects all elements in the periodic table except hydrogen and helium. Different from SPM, ESCA is carried out in ultrahigh vacuum conditions. Recently, C_{60} ions have been used to sputter materials away to investigate the composition in depth [46]. Experimental results indicate less damage to the structure when using C_{60}^{+} sputtering compared with conventional Ar^{+} sputtering. Lin et al. pointed out that the new technique provides a direct method for analyzing the distribution of materials in OLED with high depth resolution.

ESCA has been very heavily used in characterization of interfaces existing in the OLED [47–53]. By the oxygen plasma treatment of ITO anode, its work function can change, which affects the performance of the OLED significantly [11]. Milliron et al. [50] found that the oxygen plasma treatment of ITO generates a new type of oxygen species using the ESCA technique, which could be responsible for the change in the work function. Addition of poly(3,4-ethylene dioxythiophene)–poly(styrene sulfonate), referred to as PEDOT–PSS, on ITO can significantly improve the lifetime of PLEDs. However, the indium contamination of the polymers has become a new issue. Wong et al. [51] used ESCA to measure indium incorporation in the emissive polymer layer. Schlatmann et al. [53] and Jong et al. [54] reported that PLEDs, before operation, contain high concentrations of metal impurities of indium and aluminum, indicated by ESCA depth profiling data and Rutherford backscattering spectroscopy results. Degradation-induced changes in the structural and optical properties of the polyfluorene-based blue PLEDs are examined using ESCA and other techniques by Bliznyuk et al. [52]. Two primary mechanisms of degradation, photooxidation of the polymer matrix and aggregate formation, are identified. These findings are helping researchers and engineers to further improve the lifetimes of the OLEDs.

10.1.7 Raman Spectroscopy

Raman spectroscopy comprises the family of spectral measurements made on molecular media based on inelastic scattering of monochromatic radiation [55,56]. During this process, energy is exchanged between the photon and the molecule such that the scattered photon is of higher or lower energy than the incident photon. The difference in energy is associated with a change in the population of the rotational or the vibrational energy levels of the molecule. In 1930, Raman received the Nobel Prize in Physics for the discovery that bears his name. Raman spectroscopy can be used for routine qualitative and quantitative measurements of both inorganic and organic materials, and it is successfully employed to solve complex analytical problems such as chemical identification, defect analysis, determination of crystalline structure, investigation of phase separation, and analysis of charge transfer and stress. It requires little or no sample preparation, and can be directly applied for all samples types, including powders, fibers, liquids, gases, and sample in containers. New confocal Raman spectroscopy has been introduced in the marketplace by coupling a Raman spectrometer to a confocal microscope. The confocal Raman system allows one to collect the spectrum at a particular position of a sample. It also enables one to analyze small samples—down to the micron scale. With an automated sample stage, a series of spectra can be automatically acquired to generate "chemical" images of a sample based on material composition and structure. The Raman imaging technique has widespread applications in the study of heterogeneous samples, laminated materials, domain structures, and nanomaterials [57].

Several excellent examples [58–61] using Raman spectroscopy in the field are illustrated. Kim et al. [58] studied the nature of nonemissive black spots in the PLEDs. In their devices, PEDOT–PSS as a hole-transporting layer was spin-coated on the top of the treated ITO, and Ca was used as a cathode with an Al capping layer. Micro-Raman as a nondestructive technique delivers valuable chemical and structural information with submicron resolution from the surface films as well as from the buried layers. Kim et al. demonstrated how micro-Raman spectroscopy could be used to yield new insight into organic device degradation through spectroscopic cross-mapping of the active layer in the working devices. Raman spectra were taken spot-by-spot across the black spots. On the basis of the micro-Raman results, they concluded that the nonemissive disks are characterized by a localized electrochemical reaction with reduction of the normal doped PEDOT–PSS to the dedoped material and oxidation of the active metal.

10.1.8 Secondary Ion Mass Spectrometry

Today, secondary ion mass spectrometry (SIMS) is frequently used for analysis of trace elements in solid materials, especially in semiconductors and in thin films [62,63]. Primary beam species useful in SIMS include Cs^+, O_2^+, O, Ar^+, and Ga^+ at energies from 1 to 30 keV. The bombarding primary ion beam produces monatomic and polyatomic particles of sample material and resputtered primary ions, along with electrons and photons. The secondary particles carry negative, positive, and neutral charges, and they have kinetic energies that range from zero to several hundred electron volts. During SIMS analysis, the sample surface is slowly sputtered away. When the sputtering rate is extremely slow, the entire analysis can be performed while removing less than a tenth of an atomic layer. This extremely slow sputtering mode is called static SIMS. Monitoring the secondary ion count rate of selected elements as a function of time leads to depth profiles, or dynamic SIMS. Ion images show secondary ion intensities as a function of location on sample surfaces. Ion images can be acquired in two operating modes, called ion microscope or stigmatic imaging, and ion microbeam imaging or raster scanning. Ion microscopy requires a combination of an ion microscope and a mass spectrometer capable of transmitting a mass-selected ion beam from the sample to the detector without loss of lateral position information. Image detectors indicate the position of the arriving ions.

Chua et al. [64] studied the lifetime issue in OLED by SIMS. In their devices, a thin, 3-nm-thick parylene layer is deposited by chemical vapor deposition at room temperature on the ITO-coated glass substrate to form a bilayer anode of an OLED. The SIMS depth profiling was taken for C, In, O, Ag, and Ca (cathode materials) in the normal and the modified device structures at the bright and the dark nonemissive areas in electrically stressed devices. The device was taken out from the evaporator after cathode deposition and electrically stressed at a constant voltage. The position of dark spots formed in the active area was monitored. The sample was then loaded into the SIMS chamber for analysis. Chua et al. found that the parylene film not only reduces the occurrence of the dark spots, acting as a barrier for oxygen diffusion from either the ITO or the atmosphere and stabilizing the migration of the electrodes during electrical stress, but also improves the injection of the holes from the anode.

10.2 Performance Measurement Techniques

Characterizing the performance of the OLED devices requires an understanding of how the device functions and how the performance is measured. First, we discuss the subjective visual response in relation to the objective emission of light. Then, we describe basic measurements and efficiency calculations. Next, we describe energy levels in OLED devices. Finally, we discuss the lifetime measurements.

10.2.1 Human Vision, Light, and Color

One primary distinction that needs to be made is the difference between physical units and human-eye units [65,66]. Physical units are watts, meters, and steradians (solid angle). These units exist independently of the human eyeball. The human eye is sensitive to a small range of wavelengths (400–700 nm), and it is not uniformly responsive across the spectrum. This response is known as the photopic response. It is a fixed function of wavelength, and has been standardized over the years for the average human. Not all people have the same photopic response. As people have rods and cones, there are really two human response curves. The so-called black and white vision, which we use under low lighting conditions, is the scotopic response curve provided by the rods. Under normal lighting conditions, we see colors owing to the cones in the eye. The response curve of the cones is the photopic response, and is applicable to OLEDs. If we take a physical unit, then convolve it with the photopic response, we get the corresponding human-eye unit. Table 10.1 clarifies the analogy.

In human-eye units, a watt becomes a lumen. At 555 nm, the peak sensitivity of the photopic response, there are 683 lm/W. A watt is a joule per second, and a photon has

TABLE 10.1

Summary of Photometric Quantities and Associated Units

Quantity	Physical Name	Physical Units	Human-Eye Name	Human-Eye Units
Flux	Radiant flux	W	Luminous flux	lm
Flux density	Irradiance	W/m^2	Illuminance	lm/m^2
Flux density per solid angle	Radiance	$W/(m^2 \, sr)$	Luminance	$lm/(m^2 \, sr) \, (cd/m^2) \, (nit)$
Flux per solid angle	Radiant intensity	W/sr	Luminous intensity	$lm/(sr \, cd)$

energy in units of joules. A watt tells us how many photons per second are coming out of the OLED. A lumen tells us how many of those photons we actually see, accounting for the wavelength sensitivity of the eye.

The candela (cd) is equal to a lumen per steradian (sr). A steradian is the measure of a solid angle. A sphere has 4π sr on its surface area. This stems from the fact that the surface area of a sphere is SA = $4\pi r^2$. The SA divided by r^2 gives the total solid angle of the sphere. Similarly, to calculate the solid angle of a piece of a sphere's surface, one takes the area of interest, and divides by the radius squared. This gives the solid angle in steradians. A steradian is dimensionless because it is the ratio of two areas.

It is easy to understand watts, the number of photons per second, and lumens, the weighted number of photons per second that we can see. Radiance is more confusing. Imagine that there are a certain number of photons generated by a point source per second. These photons then travel out in all directions (in the isotropic case). The number of photons per second per solid angle is the radiant intensity. If you hold a piece of paper a distance away from the light source (thus fixing the distance and the area), that paper is irradiated with irradiance. If you measure the light coming off the paper, you are measuring the radiance of that piece of paper. Irradiance refers to what is impinging on a surface. Radiance refers to what is emitting from a surface.

Radiant intensity can be described as the amount of power (watt) heading in your direction, i.e., per steradian, from a light source. The total amount of power emitted by the source is the radiant flux (watt). If you integrate the radiant intensity over all solid angles, you get the total radiant flux. If it is weighted by the photopic response, then it is the luminous intensity and the luminous flux.

The candela is the amount of visible light coming in your direction from the OLED. The OLED pixels are small and flat, and have a fixed area. When you look at it, there is a certain amount of light coming out of the OLED in your direction. When this amount of light, in cd, is divided by the area of the OLED, you get the luminance of the OLED in cd/m^2.

Color is a subjective perception of optical stimuli [67]. There has been much work aimed at objectifying our visual experience [68]. For describing information on a visual screen, color is projected onto a two-dimensional plane known as the chromaticity diagram, standardized by the Commission International de l'Eclairage (CIE). This objectification enables the industry to develop standards. To produce a full-color display, it is generally accepted that three colors are needed. For an emissive display, red, blue, and green are the primary colors. For a reflective display, cyan, magenta, and yellow are the primary colors. The three colors form a triangle in color space. Any color inside this triangle can be reproduced by the display. Pushing the vertices of this triangle to the edge of the color space increases the fill factor, thus increasing the gamut of colors that can be reproduced. One advantage of the OLED technology is that the emitted colors are quite pure, which corresponds to a large fill factor. Generally, with only one green emitter, it is difficult to capture the full range of greens in a natural scene on a display, or to capture subtle distinctions in skin tone. There are certain aspects to color that are not fully understood, as demonstrated by Land's famous set of experiments. The colors generated from the OLED depend on the energy differences between the excited and ground states of emissive molecules and can be largely tailored or tuned by modifying the chemical structure [3,5,10,14,15,69,70].

10.2.2 Basic OLED Measurement and Efficiency

The primary OLED measurement is the current density–voltage–luminance (*J–V–L*) family of curves with an example illustrated in Figure 10.4. One applies a fixed voltage to a

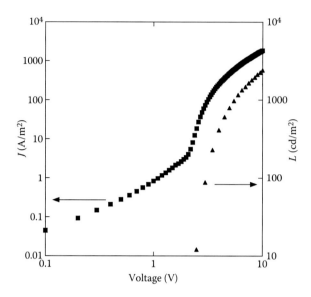

FIGURE 10.4
J–V–L family of curves of OLED.

device, and then measures the resulting current and light output. The current is converted to current density to account for the lateral area of the OLED. The light output is measured by a commercial luminance meter [66]. In a typical curve, there is a subthreshold region, a turn-on region, and a saturated region. In the subthreshold region, the current is primarily ohmic through the device, and no light is emitted. In the turn-on region, the space-charge density is sufficiently large enough across the entire device to generate current and light output. However, many of these charges are occupying trap states. In this region, the current density is proportional to some high power of the voltage (8–12), and the luminance increases dramatically with voltage. The transition between subthreshold and turn-on occurs at the turn-on voltage. After the traps are filled, one reaches the saturation regime. A steady state of space-charge current exists in the device.

The optical and electronic properties of organic materials are described in several review papers [4,7,14,71,72]. The absorption and photoluminescent (PL) emission generated by ultraviolet light reflect intrinsic properties of organic materials. The absorption and PL spectra of many important organic materials for OLEDs were measured and reported [10,14,73,74]. The differences between photoluminescence and electroluminescence (EL) have been pointed out by several authors [4,5,10,14]. The EL spectrum (see Figure 10.5), the emitted light from OLED, as a function of wavelength, can be converted into CIE coordinates.

Important electrical information about OLEDs, such as charge transport, charge injection, and carrier mobility, can be obtained from bias-dependent impedance spectroscopy, which in turn provides insight into the operating mechanisms of the OLED [14,15,73, 75–78]. Campbell et al. [75] reported electrical measurements of a PLED with a 50-nm-thick emissive layer. Marai et al. [76] studied the electrical measurement of capacitance–voltage and impedance–frequency of ITO/1,4-*bis*-(9-anthrylvinyl)-benzene/Al OLED under different bias voltage conditions. They found that the current is space-charge limited with traps, and the conductivity exhibits a power-law frequency dependence.

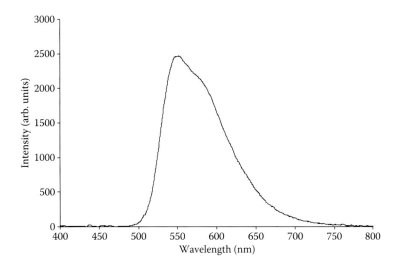

FIGURE 10.5
Typical electroluminescence spectrum from a PLED, comprising an ITO anode, a Baytron-P (Bayer) buffer layer, a PDY-131 (Covion) emitting layer, and an ytterbium cathode. Peak at 550 nm is associated with the emission from the polymeric excited state to the ground state, without exciting any vibrations. Shoulder at 580 nm is associated with relaxation to the vibrationally excited ground state. Higher overtones in the Franck–Condon progression, along with trapped states, account for the tail into the red spectrum of an OLED.

The impedance of PLEDs can be viewed in many different formats: G–B, R–X, or as a function of frequency (see Figure 10.6). The diode in this example was ITO/PEDOT–PSS/alkoxy-PPV/Ca/Al, measured from 10^2 to 10^7 Hz, with forward bias from 0 to 5 V. The R–X plot shows the most variation in the subthreshold region, while the G–B plot shows the most variation above the threshold. One sees from the G–B plot that the high-frequency response of the diode is independent of bias (>1 MHz). To fit the data, one models each material phase or interface as a parallel R–C combination (or perhaps using a constant phase element). These combinations are then added in series, and an overall series resistance and series inductance are derived. For the data in Figure 10.6, three R–C elements are used. One R–C element is associated with the Schottky barrier. Another is associated with the high frequency bias-independent arc, which we believe is associated with the capacitance of the alkoxy-PPV. The thinness of the film produces a relatively large capacitance, which, along with its large resistance, leads to a high frequency relaxation. The third is associated with the alternating-current conductivity of the PEDOT–PSS. This third R–C can be modeled more accurately as a lossy, amorphous semiconductor using the pair approximation, which is equivalent to the well-known Z_{ARC} model in the impedance literature; however, the increased accuracy comes at the expense of an added parameter, which is eschewed to simplify the analysis. Changes induced by aging can be followed by changes in these elements.

There are three principal efficiency measurements in OLED: external quantum efficiency, luminous efficiency, and power efficiency [13].

Although one would prefer to know the internal quantum efficiency, it is only possible to measure the external quantum efficiency. Much of the light generated by an OLED is wave-guided out from the edges of the device. Optical engineering applied to outcoupling is one of the primary avenues for improving OLED performance.

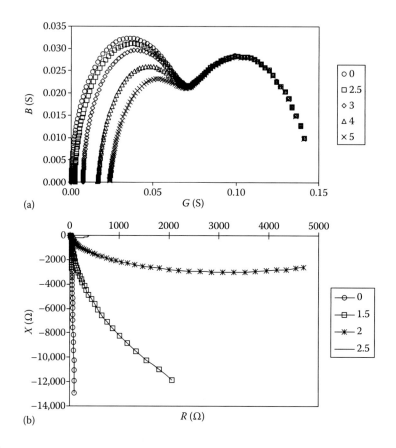

FIGURE 10.6
Impedance of an ITO/PEDOT–PSS/alkoxy-PPV/Ca/Al LED. The bias voltage is indicated in the legend: (a) admittance of the diode, showing sensitivity above threshold voltage; (b) impedance of the diode, showing sensitivity below threshold voltage. Threshold voltage of this device: 2.5 V.

The OLED emits a certain number of photons per second. The energy of one photon is $E = h\nu$. The energy of N photons is $E_N = Nh\nu$, with units of joules. Therefore, by knowing how many optical watts are generated by the OLED, one can calculate the number of photons per second, as long as the frequency (or wavelength) is known. The number of photons per second equals the light output in watts divided by $h\nu$: $N_\varphi = W/h\nu$.

The current determines the number of electrons that pass through the device per second. As 1 A = 1 C/s, the number of electrons per second equals the current divided by e: $N_e = I/e$. Knowing the device area allows one to use current density instead.

The external quantum efficiency is given by $\eta = N_\varphi/N_e = We/(h\nu*I)$.

The luminous efficiency is the simplest to calculate. It is the ratio of the luminance (cd/m²) to the current density (A/m²), and has units of cd/A. Finally, the power efficiency is the ratio of the light output in lumens divided by the electrical input in watts. For reference, a typical incandescent light bulb is ~15 lm/W, while a fluorescent light bulb is ~60 lm/W. With full output coupling, inorganic LEDs can reach 250 lm/W. Complications here arise from two areas. First, the OLED is not necessarily a sharp line emitter. It is necessary to weight the output by the emission spectrum to get the number of photons. Second, the emission is not isotropic. A model needs to be constructed for the OLED emission geometry to relate the external quantum efficiency to the internal quantum efficiency. That

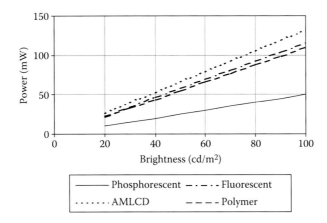

FIGURE 10.7
Power consumption simulation for a 2.2″ full-color OLED display using Universal Display's phosphorescent OLEDs, small-molecule fluorescent devices, and polymer OLEDs along with a comparison of the power consumed by an active-matrix liquid crystal display backlight. R/G/B = 3:6:1, 50% polarizer efficiency, and 30% of pixels lit. (From Mahon, J.K., *Adv. Imaging*, June, 28, 2003. With permission.)

model will take into account the fact that light only goes out from one side of the OLED, because there is a relatively thick metal electrode on the back surface that reflects the light. The model will also account for the fact that the ITO-coated glass has a finite thickness, and that light emitted at a wide angle will be totally internally reflected at the glass–air interface, and will exit the device from the edge of the glass. This model will be geometry dependent, and not device dependent, so it will be a fixed factor for all the devices [79]. Creating a tuned optical cavity also improves outcoupling efficiency.

Efficiency measurement results of OLEDs can be found in many published papers [3,4,7,8,11,12,14,15,80–83]. Mahon [83] assessed the efficiencies of 2.2″ active-matrix OLEDs made by Samsung with a polycrystalline–Si backplane and Universal Display's phosphorescent red and green materials system. Figure 10.7 shows a power consumption simulation for a 2.2″ full-color OLED display using UDC's phosphorescent OLEDs, small-molecule fluorescent devices, and polymer OLEDs along with a comparison of the power consumed by an active-matrix liquid crystal display (AMLCD) backlight. The use of phosphorescent OLED technology leads to a 100 cd/m^2 display consuming only 50 mW, compared with 110–130 mW for other OLED technologies and the AMLCD backlight.

10.2.3 Getting on the Same Energy Scale

In OLEDs, the positive and the negative charges (holes and electrons) are injected from the anode and the cathode, respectively. These charges then move through the hole and electron transport layers, and meet in the emissive layer, creating a bound exciton. Radiative recombination of these excitons results in light emission. The electronic structure of the electrodes, transport layers, and emissive layers must be carefully considered when designing for high performance [4,6–8,14,15,73]. The relative positions of the highest occupied molecular orbitals (HOMOs) and the lowest unoccupied molecular orbitals (LUMOs) of these organic layers ultimately govern the efficiencies, turn-on voltages, and emission spectrums, etc. [14,15,84–98]. In small-molecule OLEDs, where these materials are glasses (as opposed to crystalline), and the emitting layers consist of guests doped into host matrices, the inhomogeneous broadening of the energy levels must also be considered.

It is important to clarify how the HOMO and the LUMO energies are related to quantities such as the work function and the vacuum energy. An energy-level diagram relating photoelectron spectroscopy (PES) and cyclic voltammetry (CV) measurements is shown in Figure 10.8. We choose the zero point of energy to correspond to the vacuum energy. This is the energy of an electron, in vacuum, not near any other particles, and having no kinetic energy. This is the zero energy vacuum state. It is important to be aware of the surface dipole term to avoid confusion.

When working with metal electrodes, the energy of the electrons in the metal is lower than the vacuum level by the work function of the metal, which tends to be 3–5 eV. Work functions of some materials relevant to LED devices are collected in Table 10.2 [11]. The work function can vary depending on the crystal facet from which emission is measured (or if the metal is amorphous), and sample preparation details. The photoelectric (PE) effect is exploited in XPS (ESCA) or UPS to measure the work function. It is very critical to realize that, in these experiments, what is measured is the energy required to remove an electron to a point just outside the surface of the solid, not to infinity. At this range, the dipolar forces at the surface are still active, and one can learn about surface dipoles in the material.

The ionization potential of a molecule is the energy difference between the ground state of the molecule (HOMO) and the vacuum level. It is measured using UPS or XPS. The electron affinity of the molecule is the energy from the vacuum level to the LUMO. It is measured using inverse photoelectron spectroscopy [15]. The values obtained in the gas

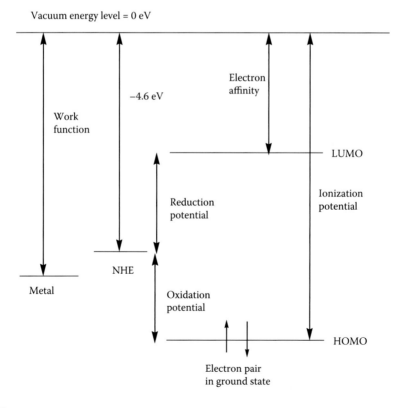

FIGURE 10.8
Energy-level diagram relating PES and CV measurements.

TABLE 10.2

Work Functions of Some Materials Relevant to LED Devices

Material	Work Function, Φ (eV)
Calcium	2.9
Magnesium	3.7
Aluminum	4.3
Silver	4.3
ITO	4.5–5.0
Polyaniline	4.4–5.0
Gold	4.8–5.4

Note: Work function can vary depending on the crystal facet from which emission is measured, and on sample preparation details.

phase are different from those obtained in the solid state, and shifts due to amorphous versus crystalline regions can be noticed.

CV is the electrochemical method for determining energy levels of molecules. In CV, the reduction potential is the voltage necessary to add an electron into the LUMO. The oxidation potential is the energy necessary to take an electron out of the HOMO. The sum of the absolute values of the oxidation potential and the reduction potential corresponds (not exact, but close) to the optical transition energy. The potentials measured by CV are measured relative to the standard calomel electrode (SCE) or the normal hydrogen electrode (NHE), or some other reference electrode. The NHE is at −4.6 eV compared with the zero vacuum level [97]. This very important link helps connect chemistry and physics. CV measurements are usually performed in solution, so there can be an additional shift of the energy levels upon going from solution to the solid state.

10.2.4 Lifetime Measurements

One of the issues related to OLEDs is the long-term stability, also known as device lifetime. A normal cathode ray tube can last for about 100,000 h, or about 12 years of continuous operation. The backlights for LCD displays were made of fluorescent lamps, with lifetimes of 50,000 h. The current LED backlights for LCD displays have lifetimes of 50,000 h. These lifetimes ensure that device issues do not limit the tenure of a set in one's home. Rather, it has been technology updates (HDTV, internet connectivity) that have made older sets obsolete. Thus, OLEDs need to have that same characteristic to be accepted by the mass market. In the early stages of OLED development, the short lifetime of OLEDs of only tens of hours made them impractical. Burrows et al. [99] introduced a simple encapsulation technique to protect OLEDs from the deleterious effects of the atmosphere, and demonstrated a greatly enhanced lifetime. This technique was widely adopted. Many studies of degradation mechanisms have been studied and proposed [4–7,98–115]. Xu [100] pointed out that the typical OLED degrades in two separate and distinct ways. The first is the intrinsic degradation, a gradual decrease in luminance of the display phosphors that occurs during the operation. The other is the aging problem known as dark spot degradation, which is characterized by circular, nonemissive areas that gradually cover a pixel. The dark spot problem has been associated with poor encapsulation, and that has been solved over the years as glass sealing systems and getters have been employed. One of the major problems today is the fact that the blue light-emitting materials employed in OLEDs age much more rapidly than red and green ones, which makes the overall lifetime of the displays too short to be widely used in our daily life. AMOLEDs for smartphones have found widespread acceptance; however, OLED TVs are working their way into the marketplace.

The basic OLED performance such as brightness decay, i.e., lifetime to 50% (LT 50%) or 70% of initial brightness, can be measured with a direct current or pulsed-constant light–current–voltage test [101]. This test uses forward bias constant sourcing with high-speed voltage and light measurement during a single pulse. Pulsed testing of long-term characteristics requires continuous pulse trains with sample measurements at regular intervals of minutes or hours. Modular *J–V–L* systems can include a calibrated luminescence meter for calibrated light measurements; however, a photometer alone will work in most applications because percentage change from initial brightness is sufficient. The relationship between voltage increase and luminance loss over time is also of interest in calculating end-of-life efficiencies.

Today, many red and green OLEDs have very long lifetimes (see one example in Table 10.3). UDC's solution-processable material systems (red, green, and light blue) have progressed significantly since the first edition of this book, as shown in Figure 10.9. These

TABLE 10.3

Performance of Several OLEDs Made by UDC (at 1000 cd/m^2)

	1931 CIE Color Coordinator	Luminous Efficiency (cd/A)	Operating Lifetime (h, LT 50%)
Deep red	(0.69, 0.31)	17	250,000
Red	(0.64, 0.36)	30	900,000
Green–yellow	(0.46, 0.53)	72	1,400,000
Green	(0.34, 0.62)	78	400,000
Light blue	(0.18, 0.42)	47	20,000

Source: http://www.oled-info.com/universal-display-pholed-performance-chart-sid-2011.

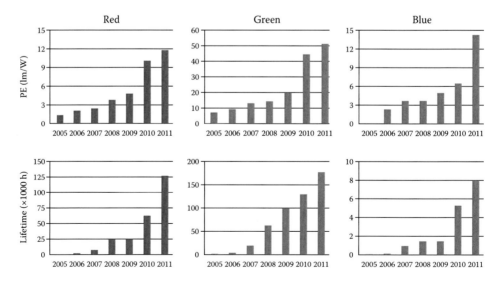

FIGURE 10.9
Power efficiency and lifetime improvement of UDC's solution-processable material system since 2005. (From http://www.oled-info.com/universal-display-pholed-performance-chart-sid-2011.)

TABLE 10.4

Sumitomo/CDT PLED Performance Data (2012)

	CIE Color Coordinator	Luminous Efficiency (cd/A)	Operating Lifetime (h, LT 50%)
Red	(0.65, 0.45)	21.8	350,000
Green	(0.32, 0.63)	56	296,000
Blue	(0.14, 0.14)	9.4	20,000

Source: http://www.cdtltd.co.uk.

advances have included luminous efficiency increases and operating voltage reductions, affecting the power efficiency. Instead of testing the OLED at room temperature, an accelerated experiment may be useful. The operational lifetime of PLED was studied at several temperatures (25°C–85°C) by Parker et al. [110]. They found the changes in luminance were significantly accelerated at higher temperatures. From the lifetime data at several temperatures, the OLED lifetime at room temperature can be readily estimated. Accelerated tested has become the norm in the industry over the last decade. Table 10.4 shows Sumimoto/CDT's PLED performance data (2012).

Otsuka Electronics Company produces an OLED panel lifetime evaluation system [102]. Their system can measure the historical change of chromaticity, luminance, current, and voltage of the OLED panel emitted for a long period under the preset condition of constant voltage, constant current, or designated luminance value (initial value) in the constant temperature environment. Furthermore, a maximum of 100 samples can be measured individually at the same time under different measurement conditions.

The intrinsic loss is still an area for fundamental research. The difficulty lies in identifying the molecular mechanisms responsible. Exciton–polaron annihilation is one theory proposed in the literature [116,117] by the Forrest group. For well-designed diodes, the processes responsible for degradation occur at extremely low quantum efficiencies, which makes understanding them in detail quite difficult. The very detailed and thorough work by Kondakov and coworkers [118–124] has added much knowledge to our understanding of the decay processes within OLEDs during operation, as electrical current causes chemical changes in host and transport materials. Work has also been performed along these lines by the Leo group [125–129]. They use matrix-assisted laser desorption/ionization time-of-flight mass spectroscopy (MALDI–TOF–MS) to examine chemical degradation products in OLEDs. The mechanisms discovered by all this work demonstrate how difficult it will be to solve this problem.

In conclusion, we have presented many microstructural characterization and performance measurement techniques, together with examples of how to use these techniques to improve the performance of OLEDs. Seeing the enormous progress recently, we expect more and more OLED-related products will be used in our daily life.

Acknowledgments

We would like to thank L. Zhang, R. McLean, L. Liang, D. Howell, T. Liang (DuPont), X. Dong (Eli Lilly), and J. Wang (DuPont China) for their valuable discussion.

References

1. C.W. Tang, and S.A. VanSlyke, Organic electroluminescent diodes, *Appl. Phys. Lett.*, 52:913–915, 1987.
2. J.H. Burroughes, D.D. Bradley, A.R. Brown, R.N. Marks, K. Mackay, R.H. Friend, P.L. Burns, and A.B. Holmes, Light-emitting diodes based on conjugated polymers, *Nature*, 347:539–541, 1990.
3. Y. Yang, Polymer electroluminescent devices, *MRS Bull.*, 22:31–38, 1997.
4. A. Dodabalapur, Organic light emitting diodes, *Solid State Commun.*, 102:259–267, 1997.
5. M.T. Bernius, M. Inbasekaran, J. O'Brien, and W. Wu, Progress with light-emitting polymers, *Adv. Mater.*, 12:1737–1750, 2000.
6. L.R. Sheats, H. Antoniadis, M. Hueschen, W. Leonard, J. Miller, R. Moon, D. Roitman, and A. Stocking, Organic electroluminescent devices, *Science*, 273:884–888, 1996.
7. R.H. Friend, R.W. Gymer, A.B. Holmes, J.H. Burroughes, R.N. Marks, C. Taliani, D.D. Bradley, D.A. Dos Santos, J.L. Bredas, M. Logdlund, and W.R. Salaneck, Electroluminescence in conjugated polymers, *Nature*, 397:121–128, 1999.
8. D. Neher, Polyfluorene homopolymers: Conjugated liquid-crystalline polymers for bright blue emission and polarized electroluminescence, *Macromol. Rapid Commun.*, 22:1365–1385, 2001.
9. G. Gustafsson, Y. Cao, G.M. Treacy, F. Klavetter, N. Colaneri, and A.J. Heeger, Flexible light-emitting diodes made from soluble conducting polymers, *Nature*, 357:477–479, 1992.
10. A. Kraft, A.C. Grimsdale, and A.B. Holmes, Electroluminescent conjugated polymers—Seeing polymers in a new light, *Angew. Chem., Int. Ed. Engl.*, 37:402–428, 1998.
11. J.S. Kim, M. Granstrom, R.H. Friend, N. Johansson, W.R. Salaneck, R. Daik, W.J. Feast, and F. Cacialli, Indium–tin oxide treatments for single- and double-layer polymeric light-emitting diodes: The relation between the anode physical, chemical and morphological properties and the device performance, *J. Appl. Phys.*, 84:6859–6870, 1998.
12. H. Becker, H. Spreitzer, W. Kreuder, E. Kluge, H. Vestweber, H. Schenk, and K. Treacher, Advances in polymers for PLEDs: From a polymerization mechanism to industrial manufacturing, *Synth. Met.*, 122:105–110, 2001.
13. S.R. Forrest, D.D.C. Bradley, and M.E. Thompson, Measuring the efficiency of organic light-emitting devices, *Adv. Mater.*, 15:1043–1048, 2003.
14. S. Miyata, and H.S. Nalwa, Eds., *Organic Electroluminescent Materials and Devices*, Gordon and Breach, Amsterdam, 1997.
15. W.R. Salanek, K. Seki, A. Kahn, and J. Pireaux, Eds., *Conjugated Polymer and Molecular Interfaces*, Marcel Dekker, New York, 2002.
16. J.W. Cowley, *Diffraction Physics*, 2nd ed., North-Holland, New York, 1981.
17. D.B. Williams, and C.B. Carter, *Transmission Electron Microscopy*, Plenum Press, New York, 1996.
18. P. Hirsch, A. Howie, R.B. Nicholson, D.W. Pashley, and M.J. Whelan, *Electron Microscopy of Thin Crystals*, 2nd ed., Kreiger Huntington, New York, 1977.
19. Z.G. Li, Ed., *Industrial Applications of Electron Microscopy*, Marcel Dekker, New York, 2002.
20. J. Bravman, R. Anderson, and M. McDonald, Eds., Specimen preparation for transmission electron microscopy of materials (I), *MRS Symp. Proc.*, Vol. 115, Pittsburgh, PA, 1988.
21. J. Bravman, and R. Sinclair, The preparation of cross-section specimens for transmission electron microscopy, *J. Electron. Microsc. Tech.*, 1:53–61, 1984.
22. Z.G. Li, P. Carcia, and P. Donohue, Microstructure of LaB_6-based thick film resistors, *J. Mater. Res.*, 7:2225–2229, 1992.
23. R. Erni, *Aberration-Corrected Imaging in Transmission Electron Microscopy*, Imperial College Press, London, 2010.
24. P.F. Carcia, R.S. Mclean, M.H. Reilly, Z.G. Li, L.J. Pillione, and R.F. Messier, Low-stress indium–tin-oxide thin films rf magnetron sputtered on polyester substrates, *Appl. Phys. Lett.*, 81:1800–1802, 2002.

25. Y.Y. Noh, C.L. Lee, and J.J. Kim, Energy transfer and performance in phosphorescent dye doped polymer light emitting diodes, *J. Chem. Phys.*, 118:2853–2864, 2003.

26. J.H. Park, Y.T. Lim, O.O. Park, and Y.C. Kim, Enhancement of photostabilities in blue-light-emitting polymers doped with gold nanoparticles, *Macromol. Rapid Commun.*, 24:331–334, 2003.

27. J. Meyer, D. Schneidenbach, T. Winkler, S. Hamwi, T. Weimann, P. Hinze, S. Ammeemann, H. Joannes, T. Riedl, and W. Kowalsky, Reliable thin film encapsulation for organic light emitting diodes grown by low temperature atomic layer deposition, *Appl. Phys. Lett.*, 94:233305–233308, 2009.

28. J.I. Goldstein, D.E. Newbury, P. Echlin, D.C. Joy, A.D. Roming, C.E. Lyman, C. Fiori, and E. Lifshin, *Scanning Electron Microscopy and X-Ray Microanalysis*, 2nd ed., Plenum Press, New York, 1992.

29. J.S. Shah, and A. Beckett, A preliminary evaluation of moist environment ambient temperature scanning electron microscopy (MEATSEM), *Micron*, 10:13–23, 1979.

30. N. Erdman, C. Nielsen, and V.E. Robertson, Shedding new light on cathodoluminescence—A low voltage perspective, *Microsc. Microanal.*, 18:1–7, 2012.

31. L.M. Do, E.H. Han, Y. Niidome, and M. Fujihira, Observation of degradation processes of Al electrodes in organic electroluminescence devices by electroluminescence microscopy, atomic force microscopy, scanning electron microscopy, and Auger electron spectroscopy, *J. Appl. Phys.*, 76:5118–5121, 1994.

32. M.S. Weaver, L.A. Michalski, K. Rajan, M.A. Rothman, J.A. Silvernail, and J.J. Brown, Organic light emitting devices with extended operating lifetimes on plastic substrates, *Appl. Phys. Lett.*, 81:2929–2931, 2002.

33. T. Gil, C. May, S. Scholz, S. Franke, M. Toerker, H. Lakner, K. Leo, and S. Keller, Origin of damages in OLED from Al top electrode deposition by DC magnetron sputtering, *Org. Electron.*, 11:322–331, 2010.

34. M. Nagai, Impact of particulate contaminants on the current leakage defect in OLED devices, *J. Electrochem. Soc.*, 154:J387–J392, 2007.

35. S. Rubanov, and P.R. Munroe, Applications of focused ion beam using FEI DualBeam DB235, *Microsc. Microanal.*, 9(Suppl. 2):884–885, 2003.

36. L.A. Giannuzzi, J.L. Brown, S.R. Brown, R.B. Irwin, and F.A. Stevie, Focused ion beam milling and micromanipulation lift-out for site specific cross-sectional TEM specimen preparation, *Mater. Res. Soc. Symp. Proc.*, 480:19–27, 1997.

37. Z.G. Li, Applications of microscopy in organic light-emitting diode research and development, *Proc. Microsc. Microanal.*, 1352–1353, 2004.

38. B.D. Ratner, and V.V. Tsukruk, Eds., *Scanning Probe Microscopy*, Oxford University Press, New York, 1998.

39. Available at http://www.Bruker.com (news release, August 14, 2012).

40. J.M. Halls, A.C. Arias, J.D. MacKenzie, W.S. Wu, M. Inbasekaran, E.P. Woo, and R.F. Friend, Photodiodes based on polyfluorene composites: Influence of morphology, *Adv. Mater.*, 12:498–502, 2000.

41. B.H. Cumpston, and K.F. Jensen, Electromigration of aluminum cathodes in polymer-based electroluminescent devices, *Appl. Phys. Lett.*, 69:3941–3943, 1996.

42. T. Ahn, and H. Lee, Effects of annealing of polythiophene derivative for polymer light emitting diodes, *Appl. Phys. Lett.*, 80:392–394, 2002.

43. C.J.R. Sheppard, and D.M. Shotton, *Confocal Laser Scanning Microscopy*, Springer-Verlag, New York, 1997.

44. J. Pawley, Ed., *The Handbook of Biological Confocal Microscopy*, IMR Press, Madison, WI, 1989.

45. D.B. Briggs, and M.P. Seah, *Auger and X-Ray Photoelectron Spectroscopy*, Wiley & Sons, New York, 1990.

46. W. Lin, W. Wang, Y. Lin, B. Yu, Y. Chen, M. Hsu, J. Jou, and J. Shyue, Migration of small molecules during the degradation of organic light-emitting diodes, *Org. Electron.*, 10:581–586, 2009.

47. T. Kugler, M. Logdlun, and W.R. Saneck, Photoelectron spectroscopy and quantum chemical modeling applied to polymer surfaces and interfaces in light-emitting devices, *Acc. Chem. Res.*, 32:225–234, 1999.

48. P.K.H. Ho, J.S. Kim, J.H. Burroughes, H. Becker, S.F.Y. Li, T.M. Brown, F. Cacialli, and R.H. Friend, Molecular-scale interface engineering for polymer light-emitting diodes, *Nature*, 404:481–484, 2000.

49. G. Wang, and X. Tao, Modification of conductive polymer for polymeric anodes of flexible organic light-emitting diodes, *Nanoscale Res. Lett.*, 4:613–617, 2009.

50. D.J. Milliron, I.G. Hill, C. Shen, and A. Kahn, Surface oxidation activates indium tin oxide for hole injection, *J. Appl. Phys.*, 87:572–576, 2000.

51. K.W. Wong, H.L. Yip, Y. Luo, K.Y. Wong, W.M. Lau, K.H. Low, H.F. Chow, Z.Q. Gao, W.L. Yeung, and C.C. Chang, Blocking reactions between indium–tin oxide and poly (3,4-ethylene dioxythiophene): Poly(styrene sulphonate) with a self-assembly monolayer, *Appl. Phys. Lett.*, 80:2788–2790, 2002.

52. V.N. Bliznyuk, S.A. Carter, J.C. Scott, G. Klärner, R.D. Miller, and D.C. Miller, Electrical and photoinduced degradation of polyfluorene based films and light-emitting devices, *Macromolecules*, 32:361–369, 1999.

53. A.R. Schlatmann, D.W. Floet, A. Hilberer, F. Garten, P.J.M. Smulders, T.M. Klapwijk, and G. Hadziioannou, Indium contamination from the indium–tin–oxide electrode in polymer light-emitting diodes, *Appl. Phys. Lett.*, 69:1764–1766, 1996.

54. M.P. de Jong, L.J. van Ijzendoorm, and M.J.A. de Voigt, Stability of the interface between indium–tin–oxide and poly(3,4-ethylenedioxythiophene)/poly(styrenesulfonate) in polymer light-emitting diodes, *Appl. Phys. Lett.*, 77:2255–2257, 2000.

55. C.V. Raman, and K.S. Krishnan, The optical analog of the Compton effect, *Nature*, 121:711, 1928.

56. C.V. Raman, and K.S. Krishnan, The production of new radiations by light scattering, *Proc. R. Soc. Lond.*, 122:23–34, 1929.

57. P. Larkin, *Infrared and Raman Spectroscopy: Principles and Spectral Interpretation*, Elsevier Science Ltd., Amsterdam, 2011.

58. J.S. Kim, P.K.H. Ho, C.E. Murphy, N. Baynes, and R.H. Friend, Nature of non-emissive black spots in polymer light-emitting diodes by in-situ micro-Raman spectroscopy, *Adv. Mater.*, 14:206–209, 2001.

59. M. Arif, S. Guha, A. Tsami, and U. Scherf, Probing electronic excitations in organic lighting diodes via Raman scattering, *Appl. Phys. Lett.*, 90:252105–252107, 2007.

60. M. Ariu, D.G. Lidzey, and D.D.C. Bradley, Influence of film morphology on the vibrational spectra of dioctyl substituted polyfluorene (PFO), *Synth. Met.*, 111–112:607–610, 2000.

61. S. Garreau, G. Louarn, J.P. Buisson, G. Froyer, and S. Lefrant, In-situ spectroelectrochemical Raman studies of poly(3,4-ethylenedioxythiophene), *Macromolecules*, 32:6807–6812, 1999.

62. J.C. Vickerman, and D. Briggs, *TOF–SIMS: Surface Analysis by Mass Spectrometry*, IM Publications and Surface Spectra Limited, London, 2001.

63. H. Itoh, Depth profiling analysis of organic materials by using ToF–SIMS and gradient shaving preparation, *J. Surf. Anal.*, 15:235–238, 2009.

64. S.J. Chua, L. Ke, R.S. Kumar, and K. Zhang, Stabilization of electrode migration in polymer electroluminescent devices, *Appl. Phys. Lett.*, 81:1119–1121, 2002.

65. A. Ryer, *Light Measurement Handbook*, International Light, Newburyport, MA, 1997.

66. P.A. Keller, *Electronic Display Measurement*, Wiley & Sons, New York, 1997.

67. S. Sherr, *Electronic Display*, 2nd ed., Wiley & Sons, New York, 1993, pp. 1–78.

68. R.S. Berns, *Billmeyer and Saltzman's Principles of Color Technology*, 3rd ed., Wiley & Sons, New York, 2000.

69. I.S. Millard, High-efficiency polyfluorene polymers suitable for RGB applications, *Synth. Met.*, 111–112:119–123, 2000.

70. Y. Kawamura, S. Yanagida, and S.R. Forrest, Energy transfer in polymer electrophosphorescent light emitting devices with single and multiple doped luminescent layers, *J. Appl. Phys.*, 92:87–93, 2002.

71. G. Hadiioannou, and P.F. van Hutten, Eds., *Semiconducting Polymers*, Wiley & Sons, New York, 2000.

72. M. Pope, and C.E. Swenberg, *Electronic Processes in Organic Crystals and Polymers*, 2nd ed., Oxford University Press, New York, 1999.

73. N.C. Greenham, and R.H. Friend, Semiconductor device physics of conjugated polymers, *Solid State Phys.*, 49:32–62, 1995.

74. P.L. Burn, A.B. Holmes, A. Kraft, D.D.C. Bradley, A.R. Brown, R.H. Friend, and R.W. Gymet, Chemical tuning of electroluminescent copolymers to improve emission efficiencies and allow patterning, *Nature*, 356:47–49, 1992.

75. I.H. Campbell, D.L. Smith, and J.P. Ferraris, Electrical impedance measurements of polymer light emitting diodes, *Appl. Phys. Lett.*, 66:3030–3032, 1995.

76. F. Marai, S. Romdhane, L. Hassine, M. Majdoub, and H. Bouchriha, Static and dynamic electrical investigations on AVB polymer light emitting diodes, *Synth. Met.*, 132:117–122, 2003.

77. H.C.F. Martens, and H.B. Brom, Frequency-dependent electrical response of holes in poly(p-phenylene vinylene), *Phys. Rev. B*, 60:R8489–R8492, 1999.

78. H.C.F. Martens, J.N. Huiberts, and H.B. Brom, Simultaneous measurement of electron and hole mobilities in polymer light-emitting diodes, *Appl. Phys. Lett.*, 77:1852–1853, 2000.

79. N.C. Greenham, R.H. Friend, and D.D.C. Bradley, Angular dependence of the emission from a conjugated polymer light-emitting diode—Implications for efficiency calculations, *Adv. Mater.*, 6:491–494, 1994.

80. A.W. Grice, D.D.C. Bradley, M.T. Bernius, M. Inbasekaran, W.W. Wu, and E.P. Woo, High brightness and efficiency blue light-emitting polymer diodes, *Appl. Phys. Lett.*, 73:629–631, 1998.

81. Y.H. Niu, Q. Hou, and Y. Cao, High efficiency polymer light-emitting diodes with stable saturated red emission based on blends of dioctylfluorene–benzothiadiazole–dithienylbenzothiadiazole terpolymers and poly[2-methoxy,5-(2-ethylhexoxy)-1,4-phenylene vinylene], *Appl. Phys. Lett.*, 82:2163–2165, 2003.

82. W. Zhu, Y. Mo, M. Yuan, W. Yang, and Y. Cao, Highly efficient electrophosphorescent devices based on conjugated polymers doped with iridium complexes, *Appl. Phys. Lett.*, 80:2045–2047, 2002.

83. J.K. Mahon, The OLED technology platform: Strengthening the promise, *Adv. Imaging*, June:28–32, 2003.

84. A.J. Campbell, D.D.C. Bradley, and H. Antoniadis, Quantifying the efficiency of electrodes for positive carrier injection into poly(9,9-dioctylfluorene) and representative copolymers, *J. Appl. Phys.*, 89:3343–3351, 2001.

85. M.A. Baldo, A. Lamansky, P.E. Burrows, M.E. Thompson, and S.R. Forrest, Very high-efficiency green organic light-emitting devices based on electrophosphorescence, *Appl. Phys. Lett.*, 75:4–6, 1999.

86. F.C. Chen, G. He, and Y. Yang, Triple exciton confinement in electrophosphorescent polymer light-emitting diodes, *Appl. Phys. Lett.*, 82:1006–1008, 2003.

87. H. Kobayashi, S. Kanbe, S. Seki, H. Kigchi, M. Kimura, I. Yudasaka, S. Miyashita, T. Shimoda, C.R. Towns, J.H. Burroughes, and R.H. Friend, A novel RGB multicolor light-emitting polymer display, *Synth. Met.*, 111–112:125–128, 2000.

88. Q. Huang, J. Cui, H. Yan, J.G.C. Veinot, and T.J. Marks, Small molecule organic light-emitting diodes can exhibit high performance without conventional hole transport layers, *Appl. Phys. Lett.*, 81:3528–3530, 2002.

89. C. Adachi, M.A. Baldo, M.E. Thompson, and S.R. Forrest, Nearly 100% internal phosphorescent efficiency in an organic light-emitting device, *J. Appl. Phys.*, 90:5048–5050, 2001.

90. R.B. Fletcher, D.G. Lidzey, D.D.C. Bradley, S. Walker, M. Inbasekaran, and E.P. Woo, High brightness conjugated polymer LEDs, *Synth. Met.*, 111–112:151–153, 2000.

91. E. Bellmann, S.E. Shaheen, R.H. Grubbs, S.R. Marder, B. Kippelen, and N. Peyghambarian, Organic two-layer light-emitting diodes based on high-Tg hole-transporting polymers with different redox potentials, *Chem. Mater.*, 11:399–407, 1999.

92. S.A. Carter, J.C. Scott, and P.J. Brock, Enhanced luminance in polymer composite light emitting devices, *Appl. Phys. Lett.*, 71:1145–1147, 1997.
93. K.J. Reynolds, G.L. Frey, and R.H. Friend, Solution-processed niobium diselenide as conductor and anode for polymer light-emitting diodes, *Appl. Phys. Lett.*, 82:1123–1125, 2003.
94. D. Sainova, T. Miteva, H.G. Nothofer, and U. Scherf, Control of color and efficiency of composite light emitting diodes based on polyfluorenes blended with hole-transporting molecules, *Appl. Phys. Lett.*, 76:1810–1812, 2000.
95. T.W. Lee, and O.O. Park, The effect of different heat treatments on the luminescence efficiency of polymer light emitting diodes, *Adv. Mater.*, 12:801–804, 2000.
96. M.M. Alam, and S. Jenekhe, Polybenzobisazoles are efficient electron transport materials for improving the performance and stability of polymer light emitting diodes, *Chem. Mater.*, 14:4775–4780, 2002.
97. A.J. Bard, and L.R. Faulkner, *Electrochemical Methods*, Wiley & Sons, New York, 1980.
98. L.S. Hung, and C.H. Chen, Recent progress of molecular organic electroluminescent materials and devices, *Mater. Sci. Eng., R*, 39:143–222, 2002.
99. P.E. Burrows, V. Bulovic, S.R. Forrest, L.S. Sapochak, D.M. McCarty, and M.E. Thompson, Reliability and degradation of organic light-emitting devices, *Appl. Phys. Lett.*, 65:2922–2924, 1994.
100. G. Xu, Fighting OLED degradation, *Inform. Display*, 19:18–21, 2003.
101. C. Cimico, Electro-optical characterization of organic LEDs, *Photonics Spectra*, July:66–68, 2003.
102. Available at http://www.photal.co.jp/.
103. H. Aziz, Z. Popvic, S. Xie, A.M. Hor, N.X. Hu, and C. Tripp, Humidity-induced crystallization of *tris*(8-hydroxyquinoline) aluminum layers in organic light-emitting devices, *Appl. Phys. Lett.*, 72:756–758, 1998.
104. J. McElvain, H. Antoniadis, M.R. Hueschen, J.N. Miller, D.M. Roitman, J.R. Sheats, and R.L. Moon, Formation and growth of black spots in organic light-emitting diodes, *J. Appl. Phys.*, 10:6002–6007, 1996.
105. J.C. Scott, J.H. Kaufman, P.J. Brock, R. DiPietro, J. Salem, and J.A. Goitia, Degradation and failure of MEH–PPV light-emitting diodes, *J. Appl. Phys.*, 79:2745–2752, 1996.
106. M. Fujihira, L.M. Do, A. Koike, and E.M. Han, Growth of dark spots by interdiffusion across organic layers in organic electroluminescent devices, *Appl. Phys. Lett.*, 68:1787–1789, 1996.
107. J.S. Kim, R.H. Friend, and F. Cacialli, Improved operational stability of polyfluorene-based organic light-emitting diodes with plasma-treated indium–tin–oxide anodes, *Appl. Phys. Lett.*, 74:3084–3086, 1999.
108. N.C. Greenham, S.C. Moratti, D.D.C. Bradley, R.H. Friend, and A.B. Holmes, Efficient light-emitting diodes based on polymers with high electron affinities, *Nature*, 365:628–630, 1993.
109. P.E. Burrows, S.R. Forrest, T.X. Zhou, and L. Michalski, Operating lifetime of phosphorescent organic light-emitting devices, *Appl. Phys. Lett.*, 76:2493–2495, 2000.
110. I.D. Parker, Y. Cao, and Y. Yang, Lifetime and degradation effects in polymer light-emitting diodes, *J. Appl. Phys.*, 85:2441–2447, 1999.
111. S. Karg, J.C. Scott, S.R. Salem, and M. Angelopoulos, Increased brightness and lifetime of polymer light-emitting diodes with polyaniline anodes, *Synth. Met.*, 80:111–117, 1996.
112. S.A. Carter, M. Angelopoulos, S. Karg, J. Brock, and J.C. Scott, Polymeric anodes for improved polymer light-emitting diode performance, *Appl. Phys. Lett.*, 70:2067–2069, 1997.
113. S.H. Kim, Y. Chu, T. Zyung, L.M. Do, and D.H. Hwang, The growth mechanism of black spots in polymer EL device, *Synth. Met.*, 111–112:253–256, 2000.
114. S.F. Lim, L. Ke, W. Wang, and S.J. Chua, Correlation between dark spot growth and pinhole size in organic light-emitting diodes, *Appl. Phys. Lett.*, 78:2116–2118, 2001.
115. B.H. Cumpston, I.D. Parker, and K.F. Jensen, In situ characterization of the oxidative degradation of a polymeric light emitting devices, *J. Appl. Phys.*, 81:3716–3720, 1997.
116. N.C. Giebink, B.W. D'Andrade, M.S. Weaver, P.B. Mackenzie, J.J. Brown, M.E. Thompson, and S.R. Forrest, Intrinsic luminance loss in phosphorescent small-molecule organic light emitting devices due to bimolecular annihilation reactions, *J. Appl. Phys.*, 103:044509, 2008.

117. N.C. Giebink, B.W. D'Andrade, M.S. Weaver, J.J. Brown, and S.R. Forrest, Direct evidence for degradation of polaron excited states in organic light emitting diodes, *J. Appl. Phys.*, 105:124514, 2009.

118. D.Y. Kondakov, J.R. Sandifer, C.W. Tang, and R.H. Young, Nonradiative recombination centers and electrical aging of organic light-emitting diodes: Direct connection between accumulation of trapped charge and luminance loss, *J. Appl. Phys.*, 93:1108–1119, 2003.

119. D.Y. Kondakov, W.C. Lenhart, and W.F. Nichols, Operational degradation of organic light-emitting diodes: Mechanism and identification of chemical products, *J. Appl. Phys.*, 101:024512, 2007.

120. D.Y. Kondakov, Characterization of triplet–triplet annihilation in organic light-emitting diodes based on anthracene derivatives, *J. Appl. Phys.*, 102:114504, 2007.

121. D.Y. Kondakov, Role of chemical reactions of arylamine hole transport materials in operational degradation of organic light-emitting diodes, *J. Appl. Phys.*, 104:084520, 2008.

122. D.Y. Kondakov, C.T. Brown, T.D. Pawlik, and V.V. Jarikov, Chemical reactivity of aromatic hydrocarbons and operational degradation of organic light-emitting diodes, *J. Appl. Phys.*, 107:024507, 2010.

123. D.Y. Kondakov, and R.H. Young, Variable sensitivity of organic light-emitting diodes to operation-induced chemical degradation: Nature of the antagonistic relationship between lifetime and efficiency, *J. Appl. Phys.*, 108:074513, 2010.

124. F. So, and D. Kondakov, Degradation mechanisms in small-molecule and polymer organic light-emitting diodes, *Adv. Mater.*, 22:3762–3777, 2010.

125. S. Scholz, C. Corten, K. Walzer, D. Kuckling, and K. Leo, Photochemical reactions in organic semiconductor thin films, *Org. Electron.*, 8:709–717, 2007.

126. S. Scholz, K. Walzer, and K. Leo, Analysis of complete organic semiconductor devices by laser desorption/ionization time-of-flight mass spectrometry, *Adv. Funct. Mater.*, 18:2541–2547, 2008.

127. S. Scholz, B. Lussem, and K. Leo, Chemical changes on the green emitter tris(8-hydroxy-quinolinato)aluminum during device aging of *p-i-n*-structured organic light emitting diodes, *Appl. Phys. Lett.*, 95:183309, 2009.

128. R. Seifert, S. Scholz, B. Lussem, and K. Leo, Comparison of ultraviolet- and charge-induced degradation phenomena in blue fluorescent organic light emitting diodes, *Appl. Phys. Lett.*, 97:013308, 2010.

129. I.R. de Moraes, S. Scholz, B. Lussem, and K. Leo, Analysis of chemical degradation mechanism within sky blue phosphorescent organic light emitting diodes by laser-desorption/ionization time-of-flight mass spectrometry, *Org. Electron.*, 12:341–347, 2011.

Index

Page numbers followed by f, t, C and S indicate figures, tables, Chart and Scheme, respectively.

Printed and bound by CPI Group (UK) Ltd, Croydon, CR0 4YY

28/10/2024

01779678-0002